SOLUTIONS MANUAL TO ACCOMPANY

MECHANICS OF MATERIALS

SOLUTIONS MANUAL TO ACCOMPANY

MECHANICS OF MATERIALS

Madhukar Vable
Michigan Technological University

New York Oxford
OXFORD UNIVERSITY PRESS
2002

Oxford University Press

Oxford New York
Auckland Bangkok Buenos Aires Cape Town Chennai
Dar es Salaam Delhi Hong Kong Istanbul Karachi Kolkata
Kuala Lumpur Madrid Melbourne Mexico City Mumbai Nairobi
São Paulo Shanghai Singapore Taipei Tokyo Toronto

and an associated company in Berlin

Published by Oxford University Press, Inc.
198 Madison Avenue, New York, New York, 10016
http://www.oup-usa.org

Oxford is a registered trademark of Oxford University Press

ISBN: 0-19-515844-X

Printing number: 9 8 7 6 5 4 3 2 1

Printed in the United States of America
on acid-free paper

Contents

Note to the Instructor

Brief descriptions of the major elements in this manual are given below.

(i) Sample syllabi.

A sample syllabus for 15 week three credit course and a 10 week four credit course are presented. The syllabi contains list of suggested topics, class problems, and home assignment problems. Both syllabi are also available on spreadsheets and in text form for your convenience in modifying as you see appropriate. These files are in the directory called Spread on the computer disc.

(ii) Solution to problems in the text.

The problem statement is presented first. It is followed by a line with values of the variables that are given and variables with a question mark identifies the variables that are to be determined in the problem. All answers are boxed.

I have tried to avoid shortcuts in presenting the solutions and have given more details than are generally found in a solutions manual. This is done in recognition of the growth of learning centers where students go for help from other senior students (coaches). The solution manuals are often kept in the learning centers for use by the coaches and at times seen by the students also.

(iii) Solution to the second static review exam.

Solution to the first static review exam is included in the textbook for student use for self diagnosis. Solution to the second static review exam is at the end of this manual for your use in the way you see appropriate.

(iv) Solution to computer problems on a spreadsheet.

The computer disc contains powerpoint slides of all the art work in the book and spreadsheets solutions to computer problems in the directory called Spread. The results of the spreadsheet are presented as a table in the respective problem solution. Hopefully the solutions on the spreadsheet will be useful to you. For example, computer problem 2.70 is done on the spreadsheet prob2.70.wks. The spreadsheet used is Microsoft Works.

Solutions to all problems have been double checked by me and others. However, given the complexity of writing a new textbook, the possibility of errors in the first print must be admitted. Errata will be posted on my webpage (www.me.mtu.edu/~mavable) and Oxford University Press webpage (www.oup-usa.org) where you can use their search engine for my name. If you have corrections, comments, or suggestions, then please send me an e-mail at: mavable@mtu.edu.

Thank you for your consideration.

Madhu Vable

Sample Syllabus (Semester)

The table below describes a sample syllabus for 15 weeks, three credit course with two term exams and a final. This syllabus is available on a spreadsheet Semester.wks and in text form in Semester.txt. These two files are in the directory called Spread on the computer disc. Problems with * are computer problems. Problems with + are stretch yourself problems.

Class	Section	Topic	Class Problems	Home Problems
1		Introduction		Static review exam 2
2	1.0-1.1	Stress on a surface	1.5, 1.24, 1.31	1.6, 1.8, 1.12, 1.19, 1.25
3	1.1	Stress on a surface	1.42,1.43	1.28,1.29, 1.36, 1.44, 1.45
4	1.1	Internal distributed force systems	1.57,1.69	1.53,1.67, 1.71
5	1.2	Stress at a point	1.78, 1.87	1.77, 1.83, 1.88, 1.94
6	2.0-2.4	Small strain	2.5, 2.28, 2.40	2.4, 2.6, 2.13, 2.25, 2.31, 2.36, 2.42
7	2.5-2.6	Strain at a point	2.54, 2.60, 2.70*	2.52, 2.55, 2.58, 2.61, 2.64, 2.69, 2.71*
8	30.-3.1	Tension test	3.23, 3.27	3.4, 3.7, 3.17, 3.21, 3.24, 3.26
9	3.2	Logic in mechanics of materials	3.42, 3.49, 3.58*	3.40, 3.43, 3.44, 3.47, 3.52+, 3.60*, 3.65
10	3.3-3.6	Generalized Hooke's law/Factor of safety	3.78, 3.106	3.62, 3.65, 3.72, 3.82, 3.83, 3.95+, 3.103, 3.107
11	4.1	Axial members	4.8, 4. 12, 4.19	4.2, 4.7, 4.11, 4.17, 4.20, 4.21
12	4.1	Axial members	4.28, 4.32, 4.41*	4.30, 4.31, 4.34+, 4.39*
13	4.2	Composite bars	4.47, 4.49	4.43, 4.48, 4.51+
14	4.3	Structural analysis	4.63, 4.68	4.58, 4.62, 4.73, 4.76
15	4.3	Structural analysis	4.78, 4.81	4.79, 4.82
16	4.4-4.5	Initial Strain/ Temperature effects	4.88, 4.94	4.87, 4.90, 4.92, 4.96
17	4.6--4.7	Thin-walled pressure vessels	4.105, 4.107	4.104, 4.110
18		Exam 1		
19	5.0-5.1	Prelude to theory of torsion of shafts	5.4, 5.11	5.1, 5.2, 5.9, 5.12
20	5.2	Torsion of shafts	5.15, 5.16, 5.21, 5.25, 5.32	5.14, 5.17,5.18, 5.20, 5.24, 5.29, 5.31, 5.42+, 5.47*
21	5.3	Statically indeterminate shafts	5.55, 5.61	5.53, 5.56, 5.63, 5.65
22	5.4	Composite shafts	5.73, 5.77	5.69, 5.70, 5.72, 5.74
23	6.0-6.1	Prelude to theory of bending of beams	6.5, 6.8	6.1, 6.6, 6.7, 6.9, 6.13
24	6.2	Bending of beams	6.15, 6.20, 6.26	6.14, 6.21, 6.23, 6.27
25	6.2	Bending of beams	6.30, 6.34, 6.41, 6.46	6.29, 6.32, 6.39, 6.44, 6.47
26	6.3-6.4	Shear and moment by equilibrium	6.51, 6.58	6.50, 6.52, 6.54, 6.59
27	6.4-6.5	Shear and moment diagrams	6.70, 6.77	6.63, 6.64, 6.68, 6.76, 6.79, 6.83
28	6.6	Shear stresses in beams	6.88, 6.90, 6.97	6.89, 6.94, 6.96
29	6.6	Shear stresses in beams	6.104, 6.115*	6.101, 6.102, 6.106, 6.112+, 6.116*
30	7.0-7.1	Beam deflection	7.3, 7.4, 7.7	7.2, 7.5, 7.8
31	7.0-7.1	Beam deflection	7.10, 7.14	7.9, 7.12, 7.13
32		Exam 2		
33	8.0-8.1	Stress transformation: Wedge method	8.2, 8.6, 8.16, 8.20	8.1, 8.4, 8.10, 8.15, 8.19, 8.26
34	8.2	Method of equations/ Mohr's circle	8.35, 8.38	8.29, 8.36, 8.37
35	8.3	Method of equations/ Mohr's circle	8.44, 8.54, 8.56	8.34, 8.42, 8.49, 8.52, 8.55, 8.60, 8.65+
36	9.0-9.2	Strain transformation: Line Method	9.5, 9.17, 9.20	9.2, 9.6, 9.7, 9.8, 9.19, 9.21
37	9.3-9.4	Method of equations/ Mohr's circle	9.35, 9.54	9.23, 9.34, 9.39, 9,50
38	9.5	Strain gages	9.60, 9.73	9.57, 9.61, 9.65, 9.67, 9.69, 9.72, 9.83*
39	10.0-10.1	Combined loading	10.7, 10.10, 10.25	10.1, 10.4, 10.12
40	10.1-10.2	Analysis and design of structures	10.34, 10.41	10.36, 10.40, 10.37
41	10.2	Analysis and design of structures	10.45	10.43, 10.44
42	10.2	Analysis and design of structures	10.49	10.51
43	11.0-11.2	Euler buckling	11.5, 11.13	11.2, 11.6, 11.9, 11.12
44	11.2	Euler buckling	11.16, 11.26	11.15, 11.19, 11.22
45		Review		
		Final		

Sample Syllabus (Quarter)

The table below describes a sample syllabus for 10 weeks, four credit course with two term exams and a final. This syllabus is available on a spreadsheet Quarter.wks and in text form in Quarter.txt. These two files are in the directory called Spread on the computer disc. Problems with * are computer problems. Problems with + are stretch yourself problems.

Class	Section	Topic	Class Problems	Home Problems
1		Introduction		Static review exam 2
2	1.0-1.1	Stress on a surface	1.5, 1.24, 1.31	1.6, 1.8, 1.12, 1.19, 1.25
3	1.1	Stress on a surface	1.42,1.43	1.28,1.29, 1.36, 1.44, 1.45
4	1.1	Internal distributed force systems	1.57,1.69	1.53,1.67, 1.71
5	1.2	Stress at a point	1.78, 1.87	1.77, 1.83, 1.88, 1.94
6	2.0-2.4	Small strain	2.5, 2.28, 2.40	2.4, 2.6, 2.13, 2.25, 2.31, 2.36, 2.42
7	2.5-2.6	Strain at a point	2.54, 2.60, 2.70*	2.52, 2.55, 2.58, 2.61, 2.64, 2.69, 2.71*
8	30.-3.1	Tension test	3.23, 3.27	3.4, 3.7, 3.17, 3.21, 3.24, 3.26
9	3.2	Logic in mechanics of materials	3.42, 3.49, 3.58*	3.40, 3.43, 3.44, 3.47, 3.52+, 3.60*, 3.65
10	3.3-3.6	Generalized Hooke's law/Factor of safety	3.78, 3.106	3.62, 3.65, 3.72, 3.82, 3.83, 3.95+, 3.103, 3.107
11	4.1	Axial members	4.8, 4. 12, 4.19	4.2, 4.7, 4.11, 4.17, 4.20, 4.21
12	4.1	Axial members	4.28, 4.32, 4.41*	4.30, 4.31, 4.34+, 4.39*
13	4.3	Structural analysis	4.63, 4.68	4.58, 4.62, 4.73, 4.76
14	4.6--4.7	Thin-walled pressure vessels	4.105, 4.107	4.104, 4.110
15		Exam 1		
16	5.0-5.1	Prelude to theory of torsion of shafts	5.4, 5.11	5.1, 5.2, 5.9, 5.12
17	5.2	Torsion of shafts	5.15, 5.16, 5.21, 5.25, 5.32	5.14,5.17, 5.18, 5.20, 5.24, 5.29, 5.31, 5.42+, 5.47*
18	5.3	Statically indeterminate shafts	5.55, 5.61	5.53, 5.56, 5.63, 5.65
19	6.0-6.1	Prelude to theory of bending of beams	6.5, 6.8	6.1, 6.6, 6.7, 6.9, 6.13
20	6.2	Bending of beams	6.15, 6.20, 6.26	6.14, 6.21, 6.23, 6.27
21	6.2	Bending of beams	6.30, 6.34, 6.41, 6.46	6.29, 6.32, 6.39, 6.44, 6.47
22	6.3-6.4	Shear and moment by equilibrium	6.51, 6.58	6.50, 6.52, 6.54, 6.59
23	6.4-6.5	Shear and moment diagrams	6.70, 6.77	6.63, 6.64, 6.68, 6.76, 6.79, 6.83
24	6.6	Shear stresses in beams	6.88, 6.90, 6.97	6.89, 6.94, 6.96
25	6.6	Shear stresses in beams	6.104, 6.115*	6.101, 6.102, 6.106, 6.112+, 6.116*
26	7.0-7.1	Beam deflection	7.3, 7.4, 7.7	7.2, 7.5, 7.8
27	7.0-7.1	Beam deflection	7.10, 7.14	7.9, 7.12, 7.13
28		Exam 2		
29	8.0-8.1	Stress transformation: Wedge method	8.2, 8.6, 8.16, 8.20	8.1, 8.4, 8.10, 8.15, 8.19, 8.26
30	8.2	Method of equations/ Mohr's circle	8.35, 8.38	8.29, 8.36, 8.37
31	8.3	Method of equations/ Mohr's circle	8.44, 8.54, 8.56	8.34, 8.42, 8.49, 8.52, 8.55, 8.60, 8.65+
32	9.0-9.2	Strain transformation: Line Method	9.5, 9.17, 9.20	9.2, 9.6, 9.7, 9.8, 9.19, 9.21
33	9.3-9.4	Method of equations/ Mohr's circle	9.35, 9.54	9.23, 9.34, 9.39, 9,50
34	9.5	Strain gages	9.60, 9.73	9.57, 9.61, 9.65, 9.67, 9.69, 9.72, 9.83*
35	10.0-10.1	Combined loading	10.7, 10.10, 10.25	10.1, 10.4, 10.12
36	10.1-10.2	Analysis and design of structures	10.34, 10.41	10.36, 10.40, 10.37
37	10.2	Analysis and design of structures	10.45	10.43, 10.44
38	11.0-11.2	Euler buckling	11.5, 11.13	11.2, 11.6, 11.9, 11.12
39	11.2	Euler buckling	11.16, 11.26	11.15, 11.19, 11.22
40		Review		

Problem Set 1.1

1.1 In a tug of war, each man shown in Fig. P1.1 exerts a force of 200 lbs. If the diameter of the rope is 1/2 inch, determine the axial stress in the rope.

Fig. P1.1

Solution P=200 lbs d=1/2 in $\sigma = ?$

The free body diagram of the rope after an imaginary cut can be drawn as shown below.

By equilibrium of forces we have: $N = 200\text{lbs}$

The area of cross-section of the rope is: $A = \frac{\pi}{4}\left(\frac{1}{2}\right)^2 = 0.1963\text{in}^2$

The average axial stress in the rope is: $\sigma = \frac{N}{A} = \frac{200}{0.1963} = 1018.6\text{psi}$ or $\boxed{\sigma = 1019 \text{ psi(T)}}$

1.2 A weight is being raised using a cable and a pulley as shown. If the weight W = 200 lbs, determine the axial stress assuming (a) cable diameter is 1/8 inch. (b) cable diameter is 1/4 inch.

Fig. P1.2

Solution W=200 lbs (a) d=1/8 in $\sigma = ?$ (b) d=1/4 in $\sigma = ?$

The free body diagram of the weight is shown below.

By equilibrium of forces we have: $N = W$

(a) The area of cross-section is: $A = \frac{\pi}{4}\left(\frac{1}{8}\right)^2 = 0.01227\text{in}^2$

The average axial stress in the rope is: $\sigma = \frac{N}{A} = \frac{200}{0.01227} = 16297\text{psi}$ or $\boxed{\sigma = 16.3\text{ksi(T)}}$

(b)The area of cross-section is: $A = \frac{\pi}{4}\left(\frac{1}{4}\right)^2 = 0.04909\text{in}^2$

The average axial stress in the rope is: $\sigma = \frac{N}{A} = \frac{200}{0.04909} = 4070\text{psi}$ or $\boxed{\sigma = 4.1\text{ksi(T)}}$

1.3 The cable shown in Fig. P1.2 has a diameter of 1/5 inch. If the maximum stress in the cable must be limited to 4 ksi (T), what is the maximum weight that can be lifted?

Solution d=1/5 in $\sigma_{max} \le 4\text{ksi(T)}$ $W_{max} = ?$

From Eq. 1 in problem 1.2 we have: $N = W$

The area of cross-section of the rope is: $A = \frac{\pi}{4}\left(\frac{1}{5}\right)^2 = 31.41(10^{-3})\text{in}^2$

The average axial stress is: $\sigma = \frac{N}{A} = \frac{W}{31.41(10^{-3})} \le 4(10^3)$ or $W \le 125.66\text{lb}$ or $\boxed{W_{max} = 125.6\text{lb}}$

1.4 The weight shown in Fig. P1.2 is W = 250 lbs. If the maximum stress in the cable must be limited to 5 ksi (T), determine the minimum diameter of the cable to the nearest 1/16 inch.

Solution $\sigma_{max} \le 5\text{ksi(T)}$ W = 250 lbs. d_{min} = ? to nearest 1/16th of an inch

---

From Eq. 1 in problem 1.2 we have: $N = W = 250\text{lbs}$

The average axial stress is: $\sigma = \frac{N}{A} = \frac{W}{(\pi d^2/4)} \le 5(10^3)$ or $d \ge 0.252\text{in}$ or $\boxed{d_{min} = \frac{5}{16}\text{inch}}$

1.5 A 6 kg light shown in Fig. P1.5 is hanging from the ceiling by wires of diameter of 0.75 mm. Determine the tensile stress in the wires AB and BC.

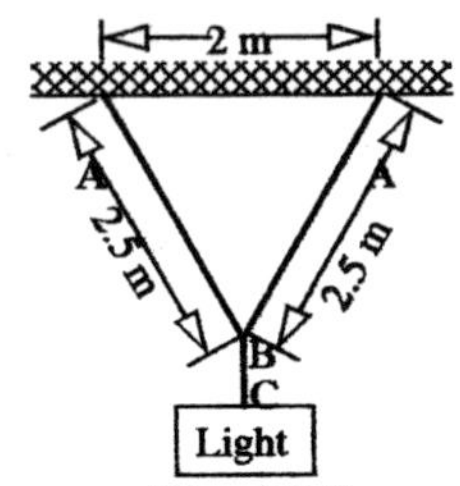

Fig. P1.5

Solution m = 6 kg. d = 0.75 mm σ_{AB} = ? σ_{BC} = ?

The weight is: $W = mg = 58.86\text{N}$

The following free body diagrams can be drawn.

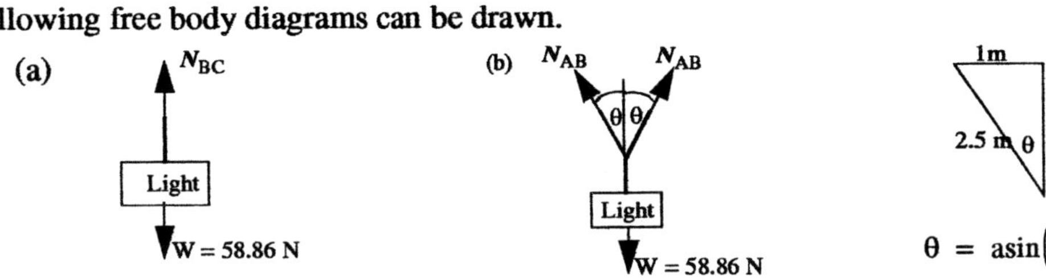

$\theta = \text{asin}\left(\frac{1}{2.5}\right) = 23.6^\circ$

By force equilibrium in the y-direction in Figs. (a) and (b) we have the following.

$$N_{BC} = W \quad (1)$$

$$2N_{AB}\cos\theta = W \quad \text{or} \quad N_{AB} = \frac{W}{2\cos 23.6} = 0.5456W \quad (2)$$

The area of cross-section is: $A = \frac{\pi}{4}(0.00075)^2 = 0.4418(10^{-6})\text{m}^2$

The average axial stress in BC is: $\sigma_{BC} = \frac{N_{BC}}{A} = \frac{58.86}{0.4418(10^{-6})} = 133.2(10^6)$ or $\boxed{\sigma_{BC} = 133.2\text{MPa(T)}}$

The average axial stress in AB is: $\sigma_{AB} = \frac{N_{AB}}{A} = \frac{(0.5456)(58.86)}{0.4418(10^{-6})} = 72.7(10^6)$ or $\boxed{\sigma_{AB} = 72.7\text{MPa(T)}}$

1.6 A 8 kg light shown in Fig. P1.5 is hanging from the ceiling by wires. If the tensile stress in the wires cannot exceed 50 MPa, determine the minimum diameter of the wire to the nearest tenth of a millimeter.

Solution m = 8 kg. $\sigma_{max} \le 50\text{MPa}$ d_{min} = ? nearest tenth of a millimeter

---

The weight is: $W = mg = 78.48\text{N}$

From Eqs. (1) and (2) in problem 1.5 we have: $N_{BC} = W$ $N_{AB} = 0.5456W$

We see that the internal force in BC will be more. As the area of cross-section is the same, the maximum axial stress will be in BC.

The average axial stress in the rope is: $\sigma_{max} = \frac{N_{BC}}{A} = \frac{78.48}{(\pi d^2/4)} \le 50(10^6)$ or $d \ge 1.41(10^{-3})$m or

Rounding to the nearest tenth of a millimeter that satisfy the inequality, we obtain: $\boxed{d_{min} = 1.5\ \text{mm}}$

1.7 Wires of diameter 0.5 mm are to be used for hanging lights such as shown in Fig. P1.5. If the tensile stress in the wires cannot exceed 80 MPa, determine the maximum mass of the light that can be hung using these wires.

Solution $d = 0.5$ mm $\sigma_{max} \le 80$MPa $m_{max} = ?$

From Eqs. (1) and (2) in problem 1.5 we have: $N_{BC} = W = mg$ $N_{AB} = 0.5456W$

We see that the internal force in BC will be more. As the area of cross-section is the same, the maximum axial stress will be in BC.

The area of cross-section is: $A = \frac{\pi}{4}(0.0005)^2 = 0.1963(10^{-6})\text{m}^2$

The average axial stress is: $\sigma_{max} = \frac{N_{BC}}{A} = \frac{m(9.81)}{0.1963(10^{-6})} \le 80(10^6)$ or $m \le 1.601$kg or $\boxed{m_{max} = 1.6\text{kg}}$

1.8 A 3-lb picture is hung using a wire of diameter of 1/8 inch as shown in Fig. P1.8. What is the average normal stress in the wire.

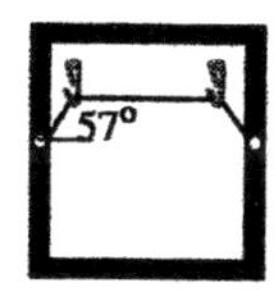

Fig. P1.8

Solution W= 3-lbs d = 1/8 inch $\sigma = ?$

The free body diagram of the picture frame is shown in Fig. (a). By equilibrium of forces in the y-direction we have the following equation.

$$2N\sin 57 = W \quad \text{or} \quad N = 0.5962W$$ 1

(a)

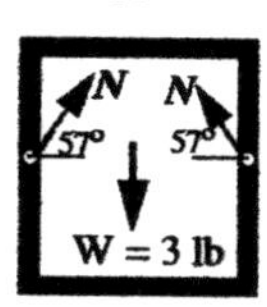

The area of cross-section is: $A = \frac{\pi}{4}\left(\frac{1}{8}\right)^2 = 12.27(10^{-3})\text{in}^2$

The average axial stress in the wire is: $\sigma = \frac{N}{A} = \frac{(0.5962)(3)}{12.27(10^{-3})} = 145.7\text{psi}$ or $\boxed{\sigma = 145.7\text{psi(T)}}$

1.9 A 5-lb picture is hung using a wire as shown in Fig. P1.8. If the tensile stress in the wires cannot exceed 800 psi, determine the minimum diameter of the wire to the nearest 1/16th inch.

Solution W= 5-lbs $\sigma_{max} \le 800$psi $d_{min} =$ nearest 1/16th inch.

From Eq. (1) in problem 1.8 we have: $N = 0.5962W$.

The average axial stress in the wire is: $\sigma = \frac{N}{A} = \frac{(0.5962)(5)}{(\pi d^2/4)} \le 800$ or $d \ge 0.0689$ in or

To the nearest (1/16th) inch that satisfy the inequality, we obtain: $d_{min} = \frac{1}{8}$ inch

1.10 Wires of diameter 16 mils are used for hanging picture such as shown in Fig. P1.8. If the tensile stress in the wire cannot exceed 750 psi, determine the maximum weight of the picture that can be hung using these wires. One mil equals 0.001 inch.

Solution d= 0.016 inch $\sigma_{max} \le 750$psi W_{max} =?

From Eq. (1) in problem 1.8 we have: $N = 0.5962W$.

The area of cross-section is: $A = \frac{\pi}{4}(0.016)^2 = 0.201(10^{-3})\text{in}^2$

The average axial stress is: $\sigma_{max} = \frac{N}{A} = \frac{0.5962W}{0.201(10^{-3})} \le 750$ or $W \le 0.2529\text{lb}$ or $W_{max} = 0.25\text{lbs}$

1.11 For swim meets, diving boards are raised to lean against the left wall using a cable and pulley as shown in Fig. P1.11. Determine the axial stress in the cable in terms of the length L of the diving board, γ the specific weight per unit length of the diving board, diameter d of the cable and the angles θ and α shown in Fig. P1.11.

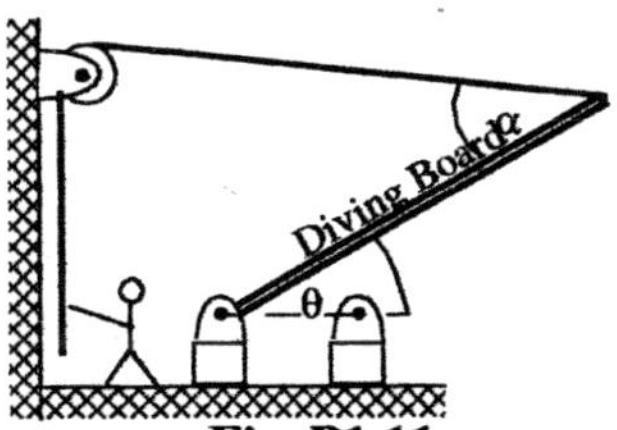

Fig. P1.11

Solution $\sigma = f(L,\gamma,d,\theta,\alpha)$=?

The free body diagram of the diving board is shown below.

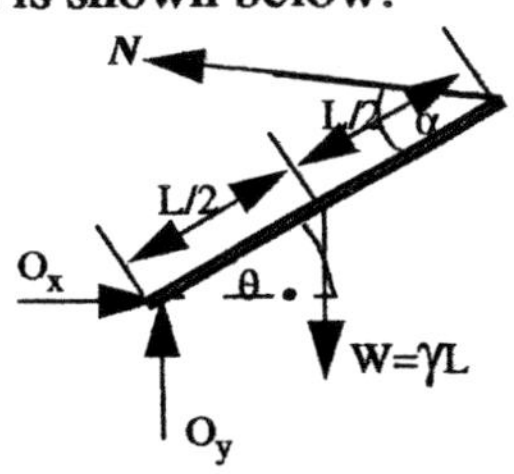

By equilibrium of moment about point O we obtain the following.

$$(\gamma L)\left(\frac{L}{2}\cos\theta\right) - (N\sin\alpha)L = 0 \quad \text{or} \quad N = \frac{\gamma L\cos\theta}{2\sin\alpha}$$

The average axial stress is: $\sigma = \frac{N}{A} = \frac{(\gamma L\cos\theta/(2\sin\alpha))}{(\pi d^2/4)}$ or $\sigma = \frac{N}{A} = \frac{2\gamma L\cos\theta}{\pi d^2 \sin\alpha}$

1.12 A hollow circular column supporting a building is attached to a metal plate and bolted into concrete foundation as shown in Fig. P1.12. The column outside diameter is 4 inches and an inside diameter is 3.5 inches.The metal plate dimensions are 8 in x8 in x 0.5 in. The load P is estimated at 150 kips. Determine (a) the compressive stress in the column. (b) average bearing stress between the metal plate and the concrete.

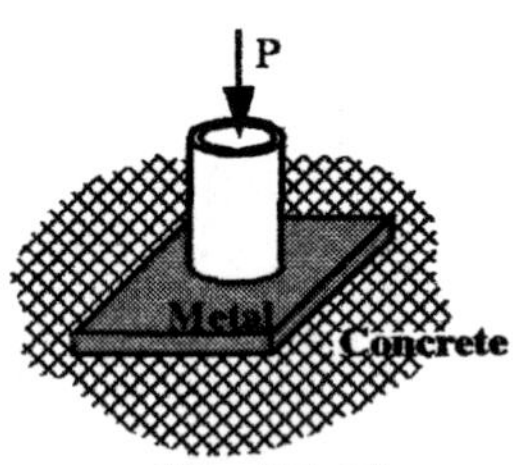

Fig. P1.12

Solution d_o = 4 inch d_i = 3.5 inch plate: 8 x8 x 0.5 P = 150 kips σ_{col} = ? σ_b= ?

The following free body diagrams can be drawn.

(a)

(b)

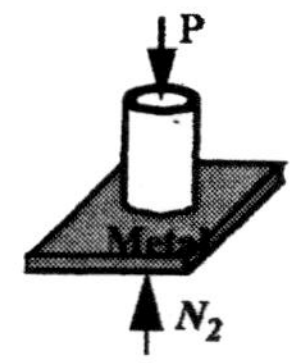

By equilibrium of forces in Figs. (a) and (b) we obtain the following equations.

$$N_1 = P \quad \text{and} \quad N_2 = P \qquad (1)$$

The area of cross-section of the column is: $A_{col} = \frac{\pi}{4}(4^2 - 3.5^2) = 2.945 \ \text{in}^2$

The average normal stress in the column is: $\sigma_{col} = \frac{N_1}{A_{col}} = \frac{(150)}{2.945} = 50.93\text{ksi}$ or $\boxed{\sigma_{col} = 50.9\text{ksi(C)}}$

The area of cross-section of the plate is: $A_{plate} = (8)(8) = 64 \ \text{in}^2$

The average bearing stress is: $\sigma_b = \frac{N_2}{A_{plate}} = \frac{(150)}{64} = 2.344\text{ksi}$ or $\boxed{\sigma_b = 2.3\text{ksi(C)}}$

1.13 A hollow circular column supporting a building is attached to a metal plate and bolted into concrete foundation as shown in Fig. P1.12. The column outside diameter is 4 inches and an inside diameter is 3.5 inches.The metal plate dimensions are 10 in x 10 in x 0.75 inch. If the allowable average compressive stress in the column is 30 ksi and the allowable average bearing stress in concrete is 2 ksi, determine the maximum load P that can be applied to the column.

Solution $d_o = 4$ inch $\quad d_i = 3.5$ inch $\quad$ plate: 10 in x 10 in x 0.75

$\sigma_{col} \le 30\text{ksi} \quad \sigma_b \le 2\text{ksi} \quad P = ?$

From Eq. 1 in problem 1.12 we have: $N_1 = P$ and $N_2 = P$

The area of cross-section of the column is: $A_{col} = \frac{\pi}{4}(4^2 - 3.5^2) = 2.945 \ \text{in}^2$

The average normal stress in the column is: $\sigma_{col} = \frac{N_1}{A_{col}} = \frac{P}{2.945} \le 30$ or $\quad P \le 88.35\text{kips}$

The area of cross-section of the plate is: $A_{plate} = (10)(10) = 100 \ \text{in}^2$

The average bearing stress is: $\sigma_b = \frac{N_2}{A_{plate}} = \frac{P}{100} \le 2$ or $\quad P \le 200\text{kips}$

The maximum value of P that satisfies the two inequalities above is: $\boxed{P = 88.3\text{kips}}$

1.14 A hollow square column supporting a building is attached to a metal plate and bolted into concrete foundation as shown in Fig. P1.14. The column outside dimension is 120 mm x 120 mm and has a thickness of 10 mm. The load P is estimated at 600 kN. The metal plate dimensions are 250 mm x 250 mm x 15 mm. Determine (a) the compressive stress in the column. (b) the average bearing stress between the metal plate and the concrete.

Fig. P1.14

Solution $P = 600$ kN plate: 250 mm x 250 mm x 15 mm $\sigma_{col} = ?$ $\sigma_b = ?$

Fig(a) below shows the geometry of the cross-section. Figs. (b) and (c) show the free body diagrams.

(a)

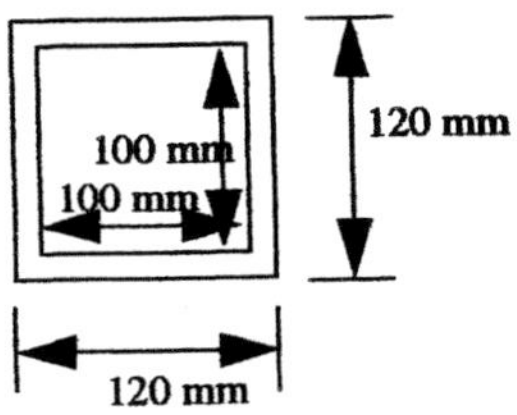

(b)

(c)

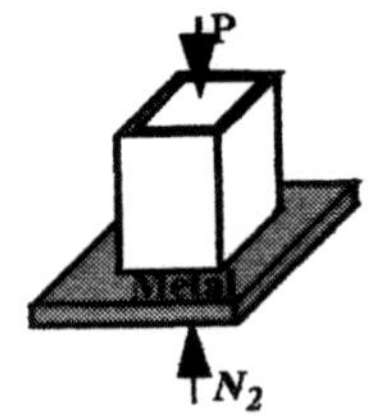

By equilibrium of forces in Figs. (a) and (b) we obtain the following equations.

$$N_1 = P \quad \text{and} \quad N_2 = P$$ 1

The area of cross-section of the column is: $A_{col} = (0.12)(0.12) - (0.1)(0.1) = 4.4(10^{-3})\ m^2$

The average normal stress in the column is: $\sigma_{col} = \dfrac{N_1}{A_{col}} = \dfrac{(600)(10^3)}{4.4(10^{-3})} = 136.4(10^6)\ N/m^2$ or

$$\boxed{\sigma_{col} = 136.4\text{MPa(C)}}$$

The area of cross-section of the plate is: $A_{plate} = (0.25)(0.25) = 62.5(10^{-3})\ m^2$

The average bearing stress is: $\sigma_b = \dfrac{N_2}{A_{plate}} = \dfrac{(600)(10^3)}{62.5(10^{-3})} = 9.6(10^6)\ N/m^2$ or $\boxed{\sigma_b = 9.6\text{MPa(C)}}$

1.15 A column with cross-section shown in Fig. P1.15 supports a building. The column is attached to a metal plate and bolted into concrete foundation. The load P is estimated at 750 kN. The metal plate dimensions are 300 mm x 300 mm x 20 mm. Determine (a) the compressive stress in the column. (b) the average bearing stress between the metal plate and the concrete.

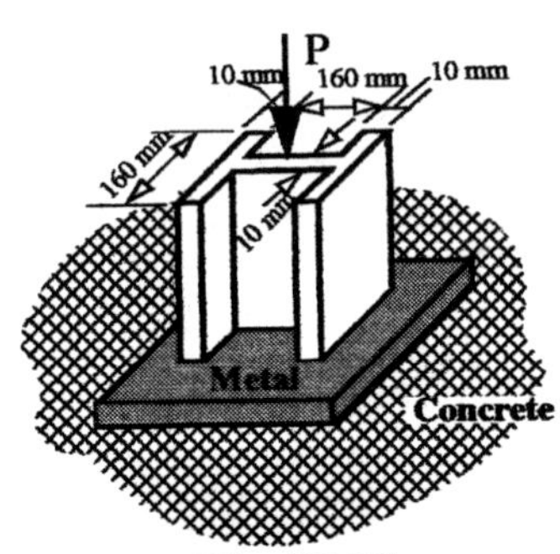

Fig. P1.15

Solution $P = 750$ kN plate: 300 x 300 x 20 mm (a) $\sigma_{col} = ?$ (b) $\sigma_p = ?$

The area of cross-section is: $A_{col} = (160)\,(10) + (160)\,(10) + (160)\,(10) = 4800\ mm^2 = 4.80\ (10^{-3})\ m^2$

(a)

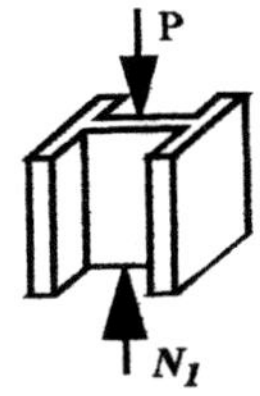

(b)

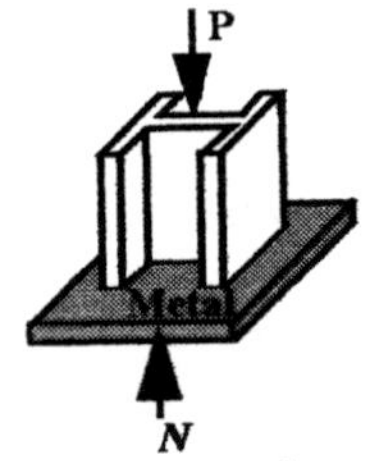

Fig. (a) shows the free body diagram of the column after an imaginary cut is made through it. By equilibrium of forces we have: $N_1 = P = 750\ (10^{-3})$ N

The average axial stress is: $\sigma_{col} = \dfrac{N_1}{A_{col}} = \dfrac{750(10^3)}{4.8(10^{-3})} = 156.25(10^6) N/m^2$ or $\boxed{\sigma_{col} = 156\ \text{MPa(C)}}$

The area of cross-section of the metal plate is: $A_{plate} = (300)\,(300) = 90\,(10^{-3})\ m^2$

Fig. (b) shows the free body diagram after an imaginary cut is made between the plate and the ground. By force equilibrium we have: $N_2 = P = 750\,(10^{3})\ N.$

The average bearing stress is: $\sigma_b = \dfrac{N_2}{A_{plate}} = \dfrac{750(10^3)}{90(10^{-3})} = 8.33(10^6)N/m^2$ $\boxed{\sigma_b = 8.33\ MPa(C)}$

1.16 A 70 kg. person is standing on a bathroom scale that has a dimensions 150 mm x 100 mm x 40 mm. Determine the bearing stress between the scales and the floor.

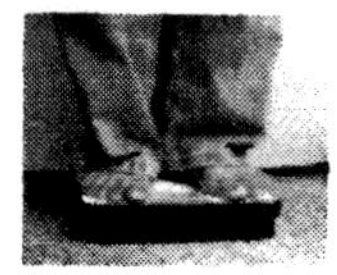

Fig. P1.16

Solution $m = 70$ kg Scale: 150 x 100 $\sigma_b = ?$

The weight of the person is: $W = mg = (70)\,(9.81) = 686.7$ N

By making an imaginary cut between the scales and the floor we obtain the following free body diagram.

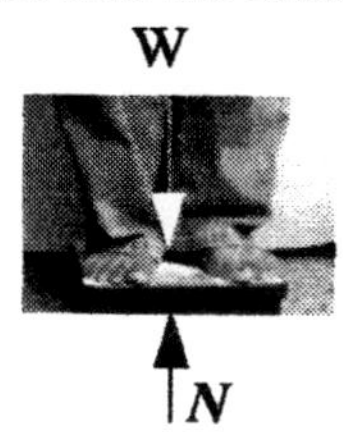

By equilibrium of forces we obtain: $N = W = 686.7$N

The area of cross-section is: $A = (150)\,(100) = 15\,(10^3)\ mm^2 = 15.0\,(10^{-3})\ m^2$

The average bearing stress is: $\sigma_b = \dfrac{N}{A} = \dfrac{686.7}{15(10^{-3})} = 45.78(10^{-3})\ N/m^2$ $\boxed{\sigma_b = 45.8\ kPa\ (C)}$

1.17 A 30 ft tall brick-chimney has an outside diameter of 3 ft. and a wall thickness of 4 inch. If the specific weight of the bricks is 80 lbs/ft^3, determine the average bearing stress at the base of the chimney.

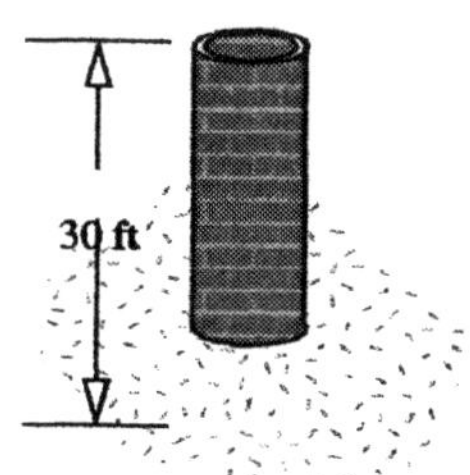

Fig. P1.17

Solution $L = 30$ ft. $d = 3$ ft $t = 4$ in $= 1/3$ ft $\gamma = 80$ lbs/ft^3 $\sigma_b = ?$

The volume of the brick material is: $V = \left[\frac{\pi}{4}(d_o^2 - d_i^2)\right]L = \frac{\pi}{4}(3^2 - 2.33^2)(30) = 83.77\ ft^3$

The weight is: $W = \gamma\,V = (80)\,(83.77) = 6702$ lbs.

By making an imaginary cut at the base of the chimney we obtain the following free body diagram.

By force equilibrium we have: $N = W = 6702$ lbs

The area of cross-section is: $A = = \frac{\pi}{4}(3^2 - 2.33^2) = 2.792\ ft^2 = 402.1\ in^2$

The average bearing stress is: $\sigma_b = \frac{N}{A} = \frac{6702}{402.1} = 16.7$ psi

$\boxed{\sigma_b = 16.7 \text{ psi (C)}}$

1.18 Determine the average bearing stress at the bottom of a block shown in Fig. P1.18 in terms of the specific weight γ, and the length dimensions a and h.

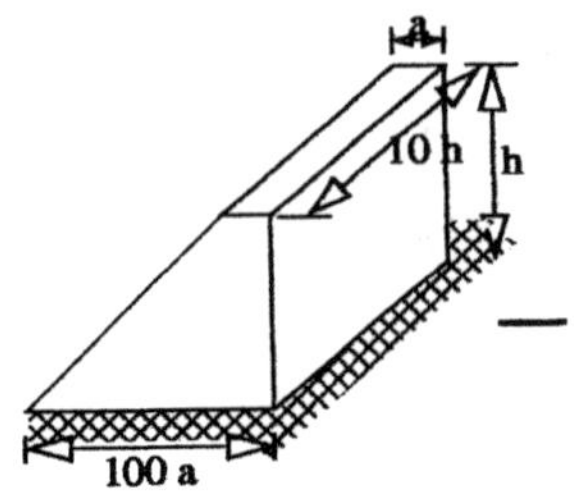

Fig. P1.18

Solution $\sigma_b = f(\gamma, a, h) = ?$

The volume of material is: $V = \left[\frac{1}{2}(100a + a)h\right](10h) = 505ah^2$

The weight of the material is: $W = \gamma V = 505\gamma ah^2$

By making an imaginary cut at the base of the block, we obtain the following free body diagram.

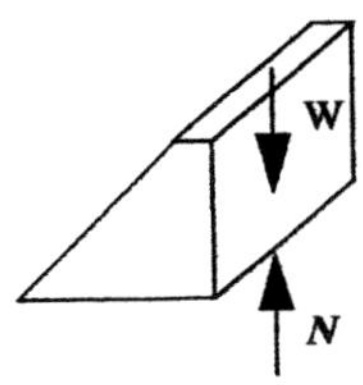

By force equilibrium we have: $N = W = 505\gamma ah^2$

The area of cross-section is: $A = (100\ a)\ (10\ h) = 1000\ ah$

The average bearing stress is: $\sigma_b = \frac{N}{A} = \frac{505\gamma ah^2}{1000ah} = 0.505\gamma h$

$\boxed{\sigma = 0.505\gamma h \text{ (C)}}$

1.19 The Washington monument is an Obelisk with a hollow rectangular cross-section that tapers along its length. The approximation of the monument geometry is shown on the right in Fig. P1.19. The thickness at the base is 4.5 m. Assume that the thickness varies proportionately with the outside taper. The monument is constructed from marble and granite.Using a specific weight for these material as 28 kN/m^3, determine the average bearing stress at the base of the monument.

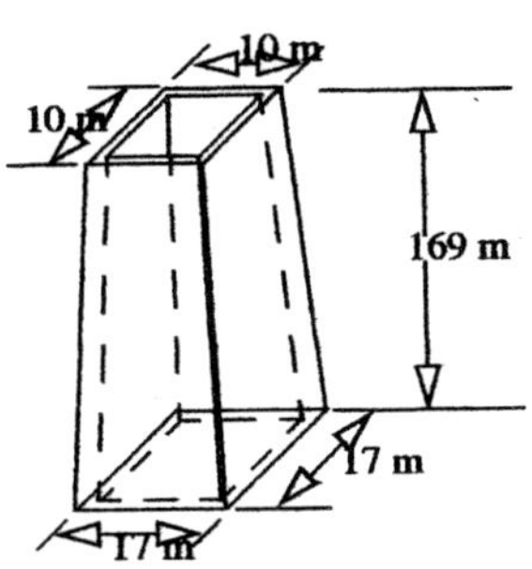

Fig. P1.19

Solution $t_0 = 4.5$ m $t_L = 2.5$ m $\gamma = 28$ kN/m2 $\sigma_b = ?$

The variation of the taper can be found using the geometry shown in Fig. (a).The outer dimension h_o and inner dimension h_i are linear functions of x and can be written as:

$$\frac{h_o}{2} = m_1x + C_1 \quad \text{and} \quad \frac{h_i}{2} = m_2x + C_2 \qquad 1$$

(a)

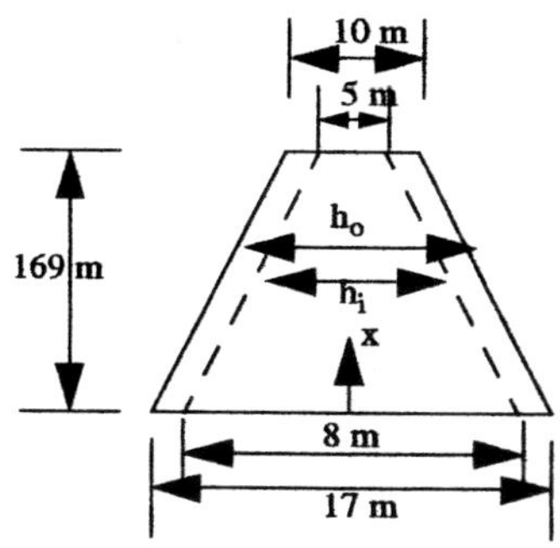

Noting that @ $x = 0$, $h_o = 17$ and $h_i = 8$, we obtain: $C_1 = 8.5$ and $C_2 = 4$

Noting that @ $x = 169$, $h_o = 10$ and $h_i = 5$ we obtain the following equations:

$$5 = m_1(169) + 8.5 \text{ and } 2.5 = m_2(169) + 4$$

which yields: $m_1 = -3.5/169$ and $m_2 = -1.5/169$. We obtain the following equations for the outer and inner dimension:

$$h_o = (17 - 7x/169) \qquad \text{and} \qquad h_i = (8 - 3x/169) \qquad \mathbf{2}$$

The volume of the monument material is: $V = \int_0^{169} (h_o^2 - h_i^2)dx = \int_0^{169} \left[\left(17 - \frac{7x}{169}\right)^2 - \left(8 - \frac{3x}{169}\right)^2\right]dx$ or

$$V = \frac{1}{3}\left(17 - \frac{7x}{169}\right)^3\left(-\frac{169}{7}\right)\Bigg|_0^{169} - \frac{1}{3}\left(8 - \frac{3x}{169}\right)^3\left(\frac{-169}{3}\right)\Bigg|_0^{169} = 24223.3 \text{ m}^3$$

The weight of the monument is: $W = \gamma V = 28\,(10^3)\,(24223.3) = 678253\,(10^3)$ N.

The force at the base of the monument is equal to the weight of the monument i.e., $N = W = 678253\,(10^3)$N

The area of cross-section of the base is: $A = 17^2 - 8^2 = 225 \text{ m}^2$

The bearing stress is: $\sigma_b = \frac{N}{A} = \frac{678253\,(10^3)}{225} = 3014\,(10^3)\ \text{N/m}^2$ or $\boxed{\sigma_b = 3 \text{ MPa (C)}}$

1.20 The post shown has a rectangular cross-section 2 in x 4 in. The length L of the post buried in ground is 12 inches and the shear strength of the soil is 2 psi. Determine the force P needed to pull the post out of the ground.

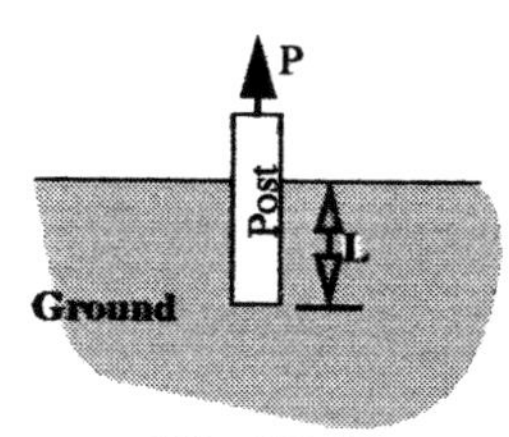

Fig. P1.20

Solution Post 2 in x 4 in $\quad L = 12$ in $\quad \tau = 2$ psi $\quad P_{min} = ?$

The following free body diagram can be drawn by making an imaginary cut around the surface of the post.

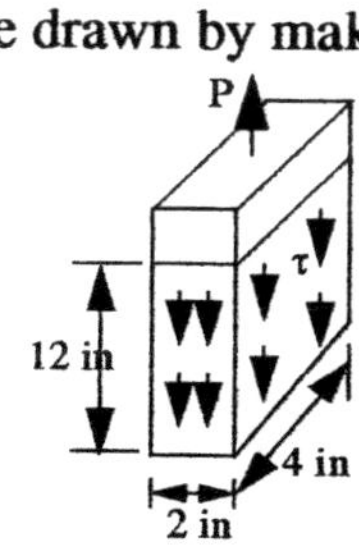

By equilibrium of forces the applied force should equal the average shear stress multiplied by the surface area as shown below.

$P = (\tau)[12(2+4+2+4)] = 144\tau$ or $\tau = \frac{P}{144} \le 2$ psi or $P \le 288$ or $\boxed{P_{min} = 288 \text{ lbs}}$

1.21 The post shown in has a circular cross-section of diameter 100 mm. The length L of the post buried in ground is 400 mm. It took a force of 1250 N to pull the post out of ground. What was the average shear strength of the soil?

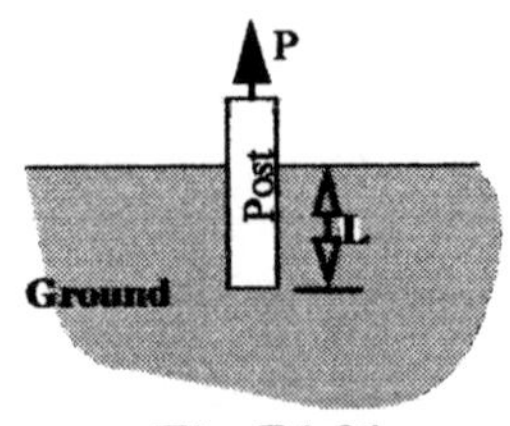

Fig. P1.21

Solution d = 100 mm = 0.1 m   L = 400 mm = 0.4 m   P = 1250 N   τ = ?

The following free body diagram can be drawn by making an imaginary cut around the surface of the post.

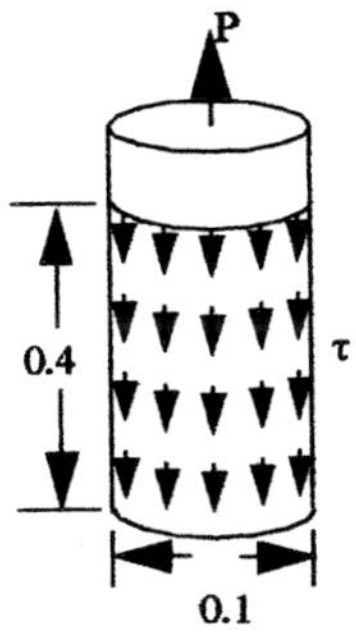

By equilibrium of forces the applied force should equal the average shear stress multiplied by the surface area as shown below.

$P = \tau(0.4)[\pi(0.1)] = 0.04\pi\tau$ or $\tau = \frac{1250}{0.04\pi} = 9947 \text{ N/m}^2$ or $\boxed{\tau = 9947 \text{ Pa}}$

1.22 The cross-section of the post shown in is an equilateral triangle with each side of dimension 'a'. If the shear strength of the soil is τ, determine the force P needed to pull the post out of the ground in terms of τ, L, and a.

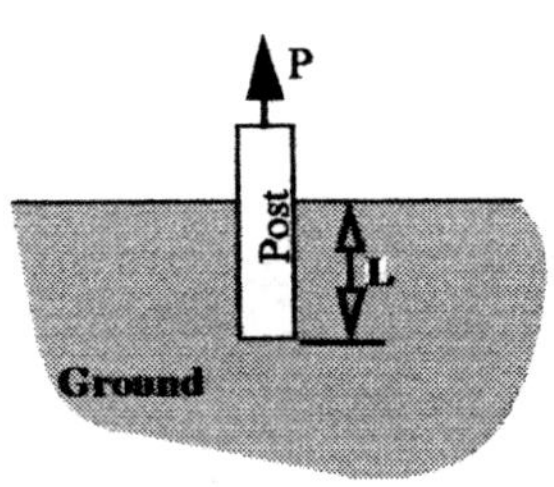

Solution $P = f(\tau, h, a) = ?$

The following free body diagram can be drawn by making an imaginary cut around the surface of the post.

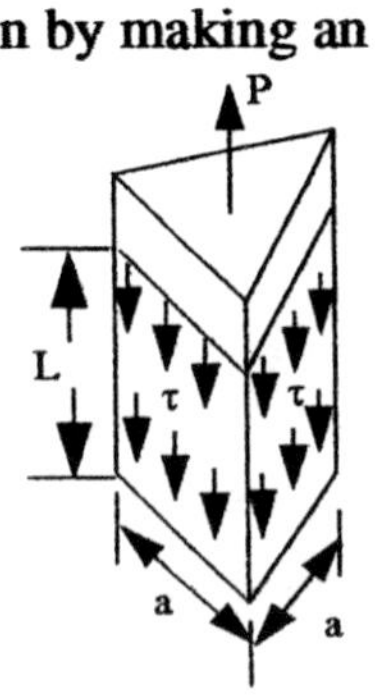

By equilibrium of forces the applied force should equal the average shear stress multiplied by the surface area as shown below.

$P = \tau(3a)(L)$ or $\boxed{P = 3aL\tau}$

1.23 A force P = 10 lbs is applied to the handle of a hammer in an effort of pulling the nail out of the wood as shown in Fig. P1.23. The nail has a diameter of 1/8 inch. Determine the shear stress acting on the nail.

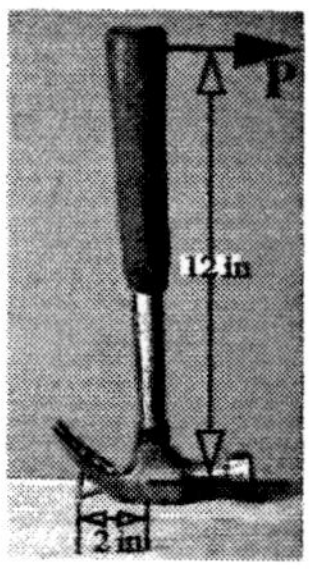

Fig. P1.23

Solution P = 10 lbs d = 118 in L = 2 in $\tau = ?$

The following free body diagram can be created.

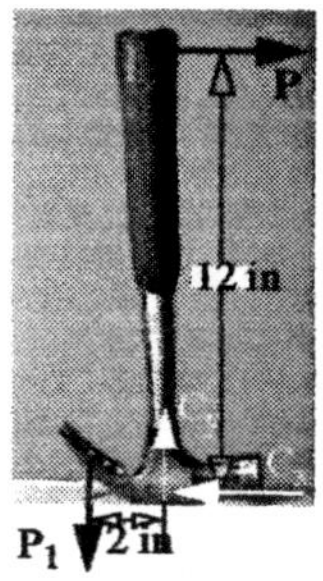

By moment equilibrium about point C, we obtain: $P_1(2) = P(12)$ or $P_1 = 6P = 60$ lbs

The surface area of the nail in the wood is: $A = L\pi D = (2)\pi(1/8) = 0.7854\ in^2$

The shear force acting on the nail is $V = P_1$. The shear stress acting on the nail can be found as:

$\tau = \frac{V}{A} = \frac{60}{0.7854} = 76.39$ psi or $\boxed{\tau = 76.4\ psi}$

1.24 The device shown in Fig. P1.24 is used for determining the shear strength of the wood. The dimensions of the wood block are 6 in x 8 in x 1.5 in. If the force required to break the wood block is 12 kips, determine the average shear strength of the wood.

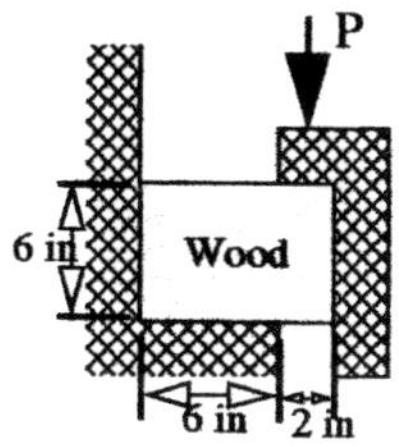

Fig. P1.24

Solution P = 12 kips Wood: 6 in x 8 in x 1.5 in. $\tau = ?$

The free body diagram of the wood that would shear off is shown below.

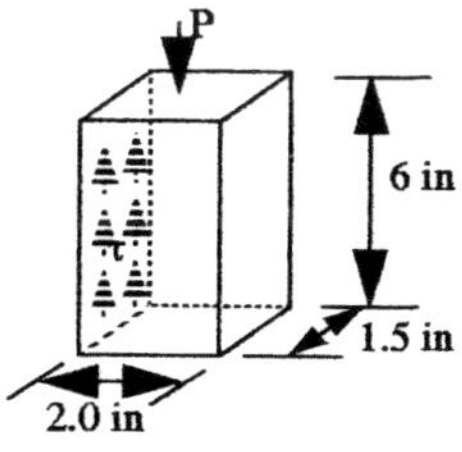

By equilibrium of forces we have the following equation.

$P = \tau(1.5)(6)$ or $\tau = P/9 = 12/9$ or $\boxed{\tau = 1.33\text{ksi}}$

1.25 The device shown in Fig. P1.24 is used for determining the shear strength of the wood. The dimensions of the wood block are 6 in x 8 in x 2 in. Estimate the force P that should be applied to break the block if the shear strength of the wood is 1.1 ksi.

Solution Wood: 6 in x 8 in x 2.0 in. $\tau \le 1.1\text{ksi}$ $P = ?$

The free body diagram of the wood that would shear off is shown below.

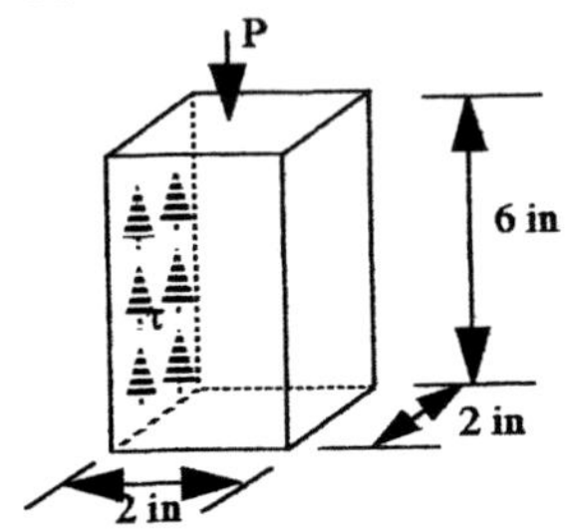

By equilibrium of forces we have the following equation.

$P = \tau(2)(6)$ or $\tau = P/12 \le 1.1$ or $P \le 13.2$ kips or $\boxed{P_{max} = 13.2 \text{ kips}}$

1.26 The schematic of a punch and die arrangement shown is used to punch out thin plate objects of different shapes. The cross-section of the punch and the die shown is a circle of diameter 1 inch. A force of P = 6 kips is applied to the punch. If the plate thickness t = 1/8 inch, what would be the average shear stress in the plate along the path of the punch.

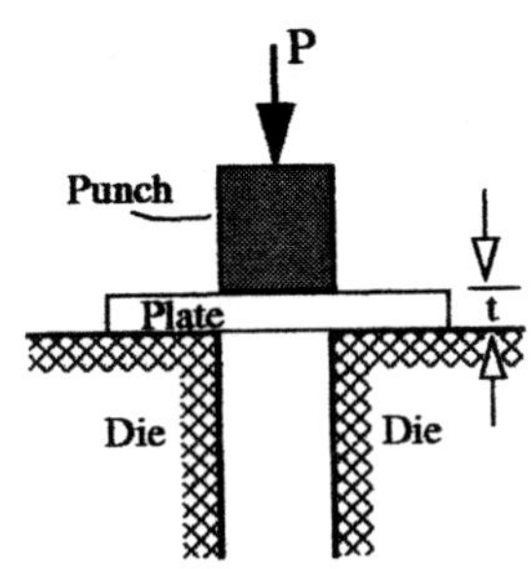

Fig. P1.26

Solution $P = 6$ kips $t = 1/8$ inch $d = 1$ inch $\tau = ?$

The free body diagram of the punch as it will go through the plate is shown below.

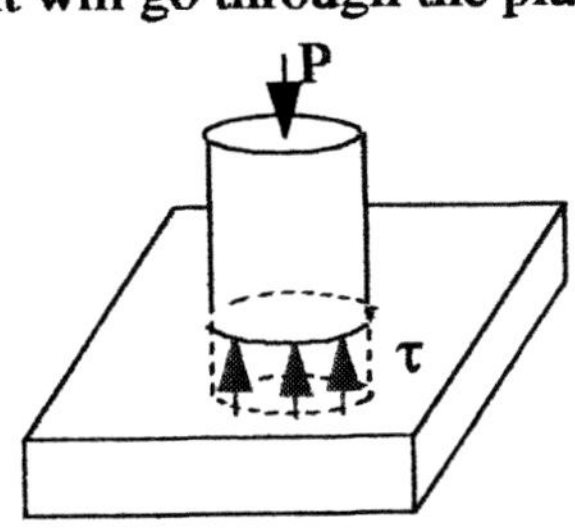

By equilibrium of forces we have the following equation.

$P = \tau(\pi d)(t)$ or $\tau = \dfrac{P}{\pi d t} = \dfrac{6}{\pi(1)(1/8)} = 15.28\text{ksi}$ or $\boxed{\tau = 15.3\text{ksi}}$

1.27 The schematic of a punch and die arrangement shown in Fig. P1.26 is used to punch out thin plate objects of different shapes. The cross-section of the punch and the die is a square of dimension 10 mm x 10 mm. The plate shown has a thickness of t = 3 mm and shear strength of 200 MPa. What is the average force P needed to drive the punch through the plate.

Solution $t = 3$ mm plate: 10 mm x 10 mm $\tau \le 200MPa$ $P = ?$

The free body diagram of the punch as it will go through the plate is shown below. By equilibrium of forces we have the following equation.

$P = \tau(4a)(t) = (200)(10^6)(4)(0.01)(0.003) = 24(10^3)N$ or $\boxed{P = 24kN}$

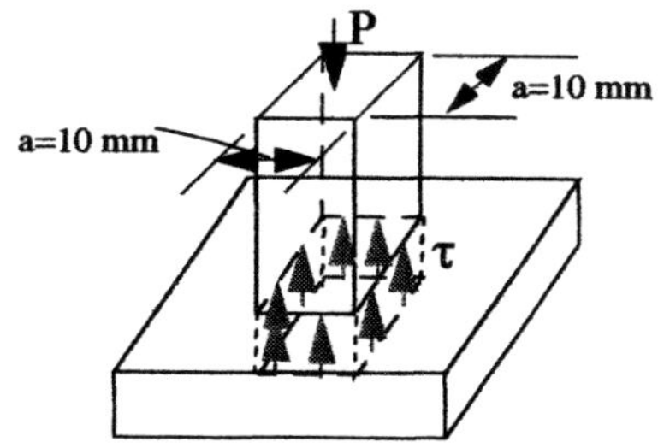

1.28 A schematic of a punch and die for punching washers is shown in Fig. P1.28. Determine the force P needed to punch out washers, in terms of the plate thickness t, plate shear strength τ, and the inner and outer diameter of the washers d_i and d_o.

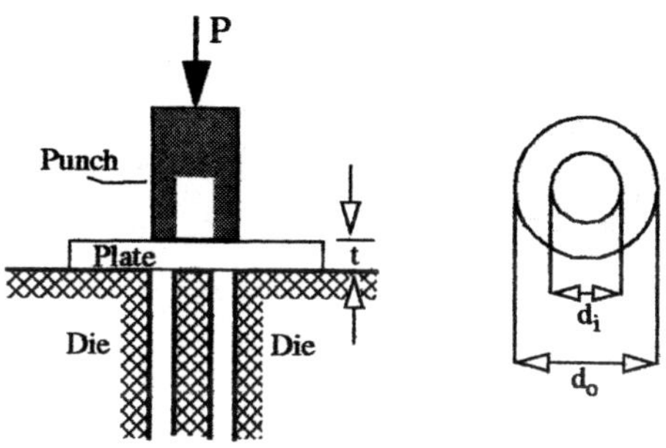

Fig. P1.28

Solution $P = f(\tau, t, d_i, d_o) = ?$

The surface area on which the shear stress will is the inner and outer cylindrical surface area punched through the thickness of the plate i.e., $A = \pi(d_o + d_i)t$. By equilibrium of forces we have: $P = \tau A$ or

$\boxed{P = \tau\pi(d_o + d_i)t}$

1.29 Two cast iron pipes are adhesively bonded together over a length of 200 mm. The outer diameters of the two pipes are 50 mm and 70 mm and wall thickness of each pipe is 10 mm. The two pipes separated while transmitting a force of 100 kN. What was the shear stress in the adhesive just before the two pipes separated.

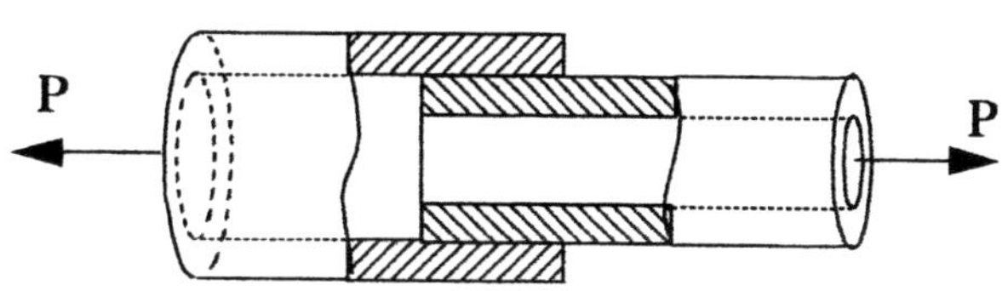

Fig. P1.29

Solution $(d_o)_{small} = 50$ mm $(d_i)_{small} = 30$ mm $(d_o)_{big} = 70$ mm $(d_i)_{big} = 50$ mm

$P = 100$ kN $L = 200$ mm $\tau = ?$

The free body diagram of the smaller pipe is shown below.

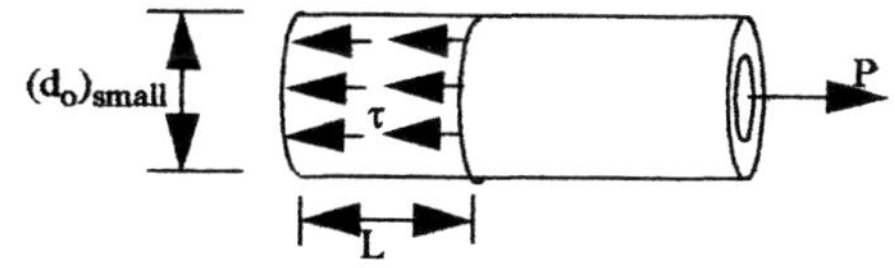

By equilibrium of forces we have:

$$P = \tau[\pi(d_o)_{small}]L \quad \text{or} \quad \tau = \frac{P}{[\pi(d_o)_{small}]L} = \frac{100(10^3)}{\pi(0.05)(0.2)} = 3183(10^3)\text{N/m}^3 \quad \text{or} \quad \boxed{\tau = 3.18\text{MPa}}$$

1.30 Two cast iron pipes are adhesively bonded together over a length of 200 mm. The outer diameters of the two pipes are 50 mm and 70 mm and wall thickness of each pipe is 10 mm. The two pipes separated while transmitting a torque of 2 kN-m. What was the shear stress in the adhesive just before the two pipes separated?

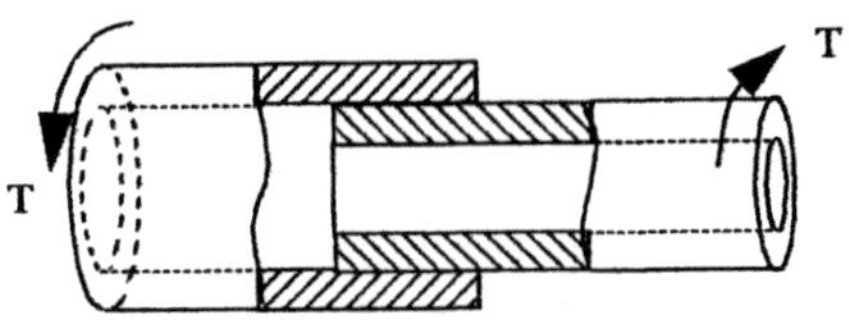

Fig. P1.30

Solution $(d_o)_{small} = 50$ mm $(d_i)_{small} = 30$ mm $(d_o)_{big} = 70$ mm $(d_i)_{big} = 50$ mm

$T = 2$ kN-m $L = 200$ mm $\tau = ?$

The free body diagram of the smaller pipe and the differential circular surface area over is shown below.

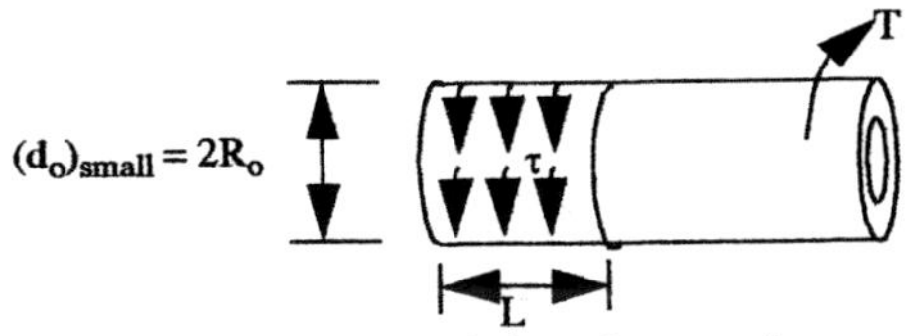

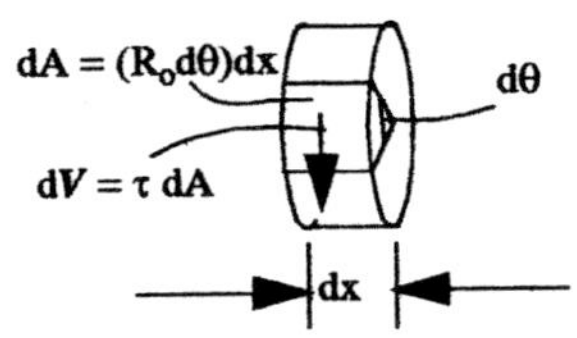

By moment equilibrium about the axis of the pipe we have the following.

$$T = \int_A R_o\, dV = \int_A R_o \tau dA = \int_0^L \int_0^{2\pi} R_o \tau (R_o d\theta)dx = \tau R_o^2 \int_0^L \left[\int_0^{2\pi} d\theta\right] dx = \tau R_o^2 [2\pi](L) \quad \text{or}$$

$$= \frac{T}{2\pi R_o^2 L} = \frac{2(10^3)}{2\pi(0.025)^2(0.2)} = 2.546(10^6)\text{N/m}^2 \text{ or } \quad \boxed{\tau = 2.55\text{MPa}}$$

1.31 Two cast iron pipes are held together by a bolt as shown. The outer diameters of the two pipes are 50 mm and 70 mm and wall thickness of each pipe is 10 mm.The diameter of the bolt is 15 mm. The bolt broke while transmitting a torque of 2 kN-m. On what surface(s) did the bolt break? What was the shear stress in the bolt on the surface where it broke?

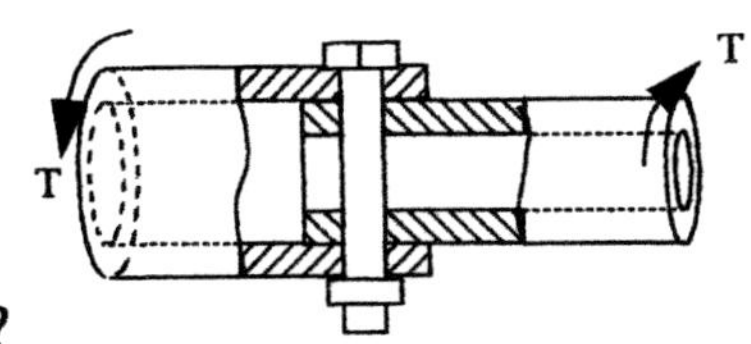

Fig. P1.31

Solution $T = 2$ kN-m $(d_o)_{small} = 50$ mm $d_{bolt} = 15$ mm $\tau = ?$

The free body diagram with the bolt broken is shown below.

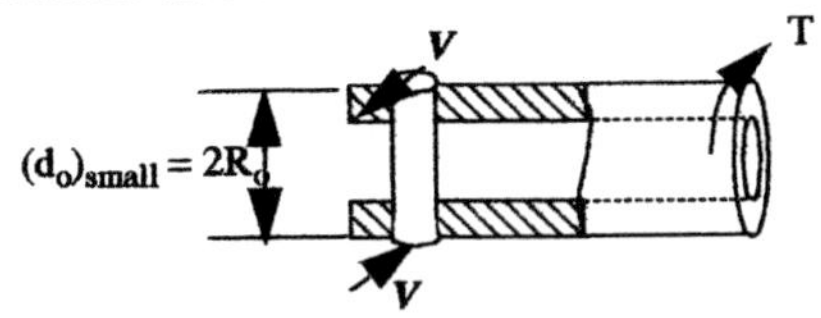

By equilibrium of moment we have: $T = (2R_o)[V] = (2R_o)[(\tau)(\pi d_{bolt}^2/4)]$ or

$$\tau = \frac{2T}{(\pi d_{bolt}^2)R_o} = \frac{(2)(2)(10^3)}{\pi(0.015)^2(0.025)} = 226.3(10^6)\text{N/m}^2 \quad \boxed{\tau = 226.3\text{MPa}}$$

1.32 The forces acting on a pin in a truss joint are shown. The diameter of the pin is 20 mm. Determine the maximum transverse shear stress in the pin.

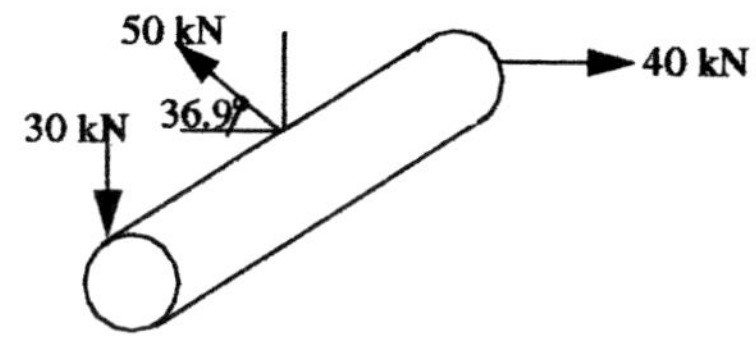

Fig. P1.32

Solution $d = 20$ mm $\quad \tau_{max} = ?$

The area of cross-section of the pin is: $A = \pi(0.02)^2/4 = 314.6(10^{-6})\text{m}^2$
The following free body diagrams can be drawn.

From the above free body diagram we see that the maximum shear force is: $V_2 = 40$kN. The maximum shear stress is: $\tau_{max} = \dfrac{V_2}{A} = \dfrac{40(10^3)}{314.6(10^{-6})} = 127.3(10^6)\text{N/m}^2$ or $\boxed{\tau_{max} = 127.3\text{MPa}}$

1.33 A weight W = 200lbs is being raised using a cable and a pulley as shown. The cable diameter is 1/4 inch and the pin in the pulley has diameter of 3/8 inch. Determine the axial stress in the cable and shear stress in the pin assuming the pin is in double shear.

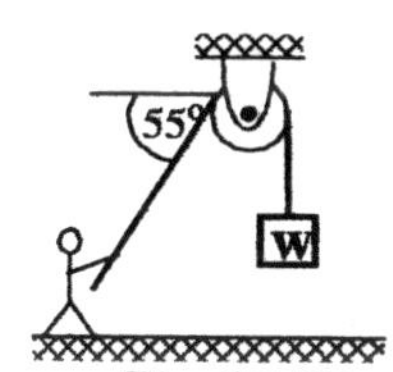

Fig. P1.33

Solution $W = 200$lbs $\quad d_c = 1/4$ in $\quad d_p = 3/8$ in $\quad \sigma_C = ?$ $\quad \tau_p = ?$

The following free body diagrams can be drawn.

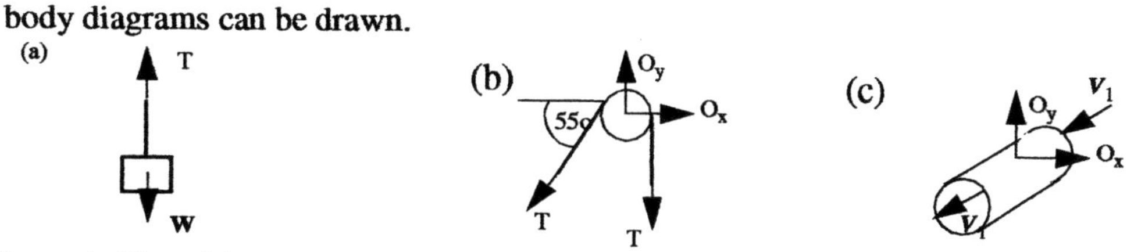

By equilibrium of forces in Figs. (a) and (b) we have the following.

$$T = W \tag{1}$$

$$O_x = T\cos 55 = 0.5736W \tag{2}$$

$$O_y - T - T\sin 55 = 0 \quad \text{or} \quad O_y = T(1 + \sin 55) = 1.8192W \tag{3}$$

The area of cross-section of the cable is: $A_C = \pi(1/4)^2/4 = 49.08(10^{-3})\text{in}^2$

The average axial stress in the cable is: $\sigma_C = \dfrac{T}{A_C} = \dfrac{200}{49.08(10^{-3})} = 4074.4$ psi or $\boxed{\sigma_C = 4074.\text{ psi(T)}}$

From equilibrium of forces in Fig. (c) we have.

$$2V = \sqrt{O_x^2 + O_y^2} \tag{4}$$

Substituting W = 200 lbs in Eqs. (2) and (3) we obtain: $O_x = 114.7$ lbs and $O_y = 363{,}8$ lbs. Substituting these values into Eq. (4) we obtain $V = 190.7$ lbs.

The area of cross-section of the pin is: $A_p = \pi(3/8)^2/4 = 110.4(10^{-3})\text{in}^2$

The average shear stress in the pin is: $\tau = \dfrac{V}{A_P} = \dfrac{190.7}{110.4(10^{-3})} = 1727.4\text{psi}$ $\boxed{\tau = 1727\text{psi}}$

1.34 The cable in Fig. P1.33 has a diameter of 1/5 inch and the pin in the pulley has a diameter of 3/8 inch. If the maximum normal stress in the cable must be limited to 4 ksi (T) and the maximum shear stress in the pin is to be limited to 2 ksi, determine the maximum weight that can be lifted. The pin is in double shear.

Solution $d_c = 1/5$ in $d_p = 3/8$ in $\sigma_C \le 4\text{ksi}$ $\tau_p \le 2\text{ksi}$ $\tau_p = ?$ $W_{max} = ?$

From Eq. (1) in problem 1.33 we have: $T = W$

The area of cross-section of the cable is: $A_C = \pi(1/5)^2/4 = 31.42(10^{-3})\text{in}^2$

The average axial stress in the cable is: $\sigma_C = \dfrac{T}{A_C} = \dfrac{W}{31.42(10^{-3})} \le 4(10^3)\text{psi}$ or $W \le 125.66\text{lbs}$

From Eqs. (2), (3), and (4) in problem 1.33 we have the following.

$O_x = 0.5736W$ $O_y = 1.8192W$ $2V = \sqrt{O_x^2 + O_y^2}$ or

$2V = W\sqrt{(0.5736)^2 + (1.8192)^2}$ or $V = 0.9537W$

The area of cross-section of the pin is: $A_p = \pi(3/8)^2/4 = 110.4(10^{-3})\text{in}^2$

The average shear stress in the pin is: $\tau = \dfrac{V}{A_P} = \dfrac{0.9537W}{110.4(10^{-3})} \le 2(10^3)\text{psi}$ or $W \le 231.5\text{lb}$

The maximum value of W that satisfies both inequalities above is: $\boxed{W_{max} = 125.6\text{lb}}$

1.35 The manufacturer of the plastic carrier for dry-wall panels prescribes a maximum load P of 200 lbs. If the cross-sectional areas at section AA and BB are 1.3 in^2 and 0.3 in^2 respectively, determine the average shear stress at section AA and average normal stress at section BB at the maximum load P.

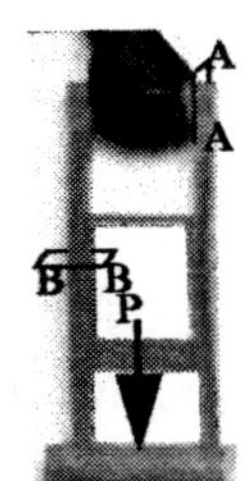

Fig. P1.35

Solution P = 200 lbs. $A_{AA} = 1.3\text{ in}^2$ $A_{AA} = 0.3\text{ in}^2$ $\tau_{AA} = ?$ $\sigma_{BB} = ?$

The free body diagrams of the plastic carrier after making the imaginary cuts though AA and BB are shown below. By force equilibrium we obtain: $V_A = P/2 = 100\text{lbs}$ and $N_A = P/2 = 100\text{lbs}$

The average shear stress at section AA is: $\tau_{AA} = \dfrac{V_A}{A_{AA}} = \dfrac{100}{1.3} = 76.92\text{psi}$ or $\boxed{\tau_{AA} = 76.9\text{psi}}$

(a)

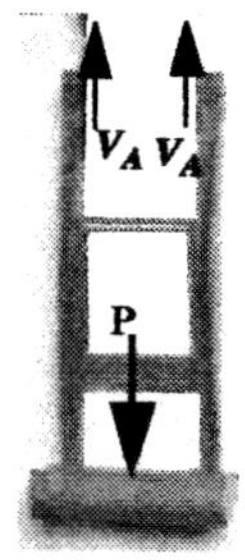

(b)

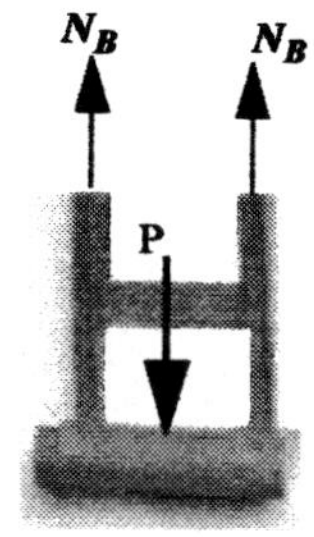

The average normal stress at section BB is: $\sigma_{BB} = \dfrac{N_B}{A_{BB}} = \dfrac{100}{0.3} = 333.33\text{psi}$ or $\boxed{\sigma_{BB} = 333.3 \text{ psi (T)}}$

1.36 A bolt passing through a piece of wood is shown in Fig. P1.36. Determine (a) the axial stress in the bolt. (b) the average shear stress in the bolt-head. (c) the average bearing stress between the bolt head and the wood. (d) the average shear stress in the wood.

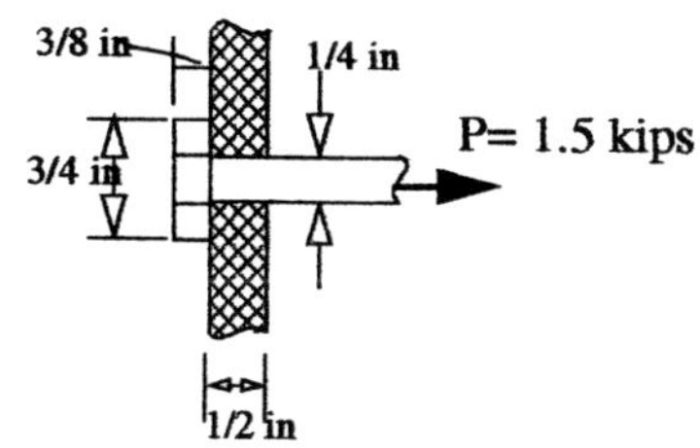

Fig. P1.36

Solution $\sigma_{bolt} = ?$ $\tau_{bolt} = ?$ $\sigma_{bear} = ?$ $\tau_{wood} = ?$

The following free body diagrams can be drawn.

(a) (b) (c) (d)

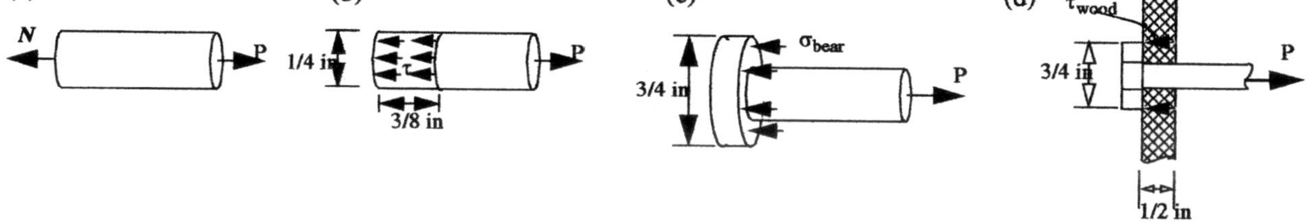

The area of cross-section of the bolt is: $A_{bolt} = \pi(3/4)^2/4 = 49.09(10^{-3})\text{in}^2$

By equilibrium of forces in Fig. (a) we obtain: $N = P = 1.5$ kips

The average normal stress at section BB is:

$$\sigma_{bolt} = \frac{N}{A_{bolt}} = \frac{1.5}{49.09(10^{-3})} = 30.556\text{ksi} \text{ or } \boxed{\sigma_{bolt} = 30.56 \text{ ksi (T)}}$$

By equilibrium of forces in Fig. (b) we obtain the following.

$$\tau[\pi(1/4)](3/8) = P = 1.5 \quad \text{or} \quad \tau = \frac{1.5}{(3\pi/32)} = 5.093\text{ksi} \text{ or } \boxed{\tau = 5.09 \text{ ksi}}$$

As per Fig(c), the area of cross-section over which the bearing stress acts is as given below.

$$A_{bear} = \pi[(3/4)^2 - (1/4)^2]/4 = 0.3927\text{in}^2$$

By equilibrium of forces in Fig. (c) we obtain the following.

$$\sigma_{bear}A_{bear} = P = 1.5 \quad \text{or} \quad \sigma_{bear} = \frac{1.5}{0.3927} = 3.8197\text{ksi} \text{ or } \boxed{\sigma_{bear} = 3.82 \text{ ksi (C)}}$$

By equilibrium of forces in Fig. (d) we obtain the following.

$$\tau_{wood}[\pi(3/4)](1/2) = P = 1.5 \quad \text{or} \quad \tau_{wood} = \frac{1.5}{(3\pi/8)} = 1.273\text{ksi} \text{ or } \boxed{\tau_{wood} = 1.27\text{ksi}}$$

1.37 The axial force P = 12 kips acts on a rectangular member shown in Fig. P1.37. Determine the average normal and shear stress on the inclined plane AA.

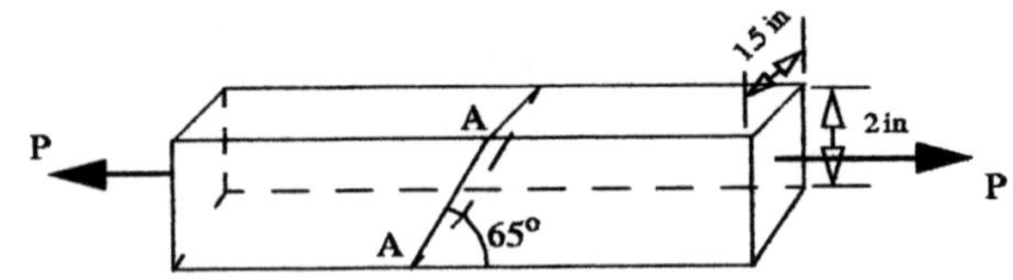

Fig. P1.37

Solution σ_{AA}= ? τ_{AA}= ?

The following free body diagram can be drawn after making an imaginary cut through section AA.

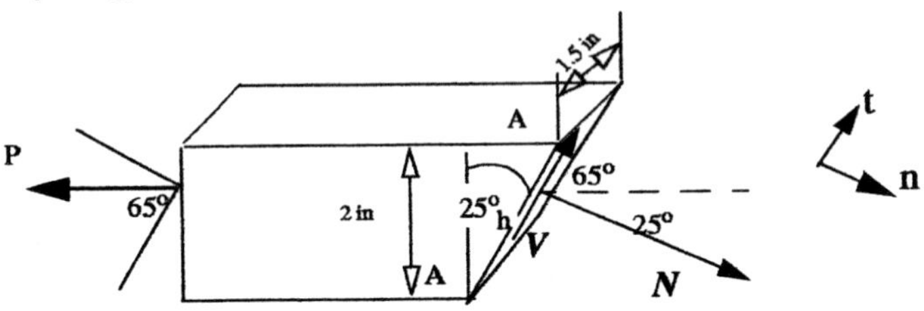

By equilibrium of forces in the n-direction we obtain: $N = P\sin 65 = 12\sin 65 = 10.876\text{kips}$.
By equilibrium of forces in the t-direction we obtain: $V = P\cos 65 = 12\cos 65 = 5.0714\text{kips}$.

The area of cross-section of plane AA is: $A_{AA} = (h)(1.5) = \left(\frac{2}{\cos 25}\right)(1.5) = 3.3107\text{in}^2$.

The average normal stress on plane AA is: $\sigma_{AA} = \frac{N}{A_{AA}} = \frac{10.876}{3.3107} = 3.286\text{ksi}$ or $\boxed{\sigma_{AA} = 3.286\text{ksi(T)}}$

The average shear stress on plane AA is: $\tau_{AA} = \frac{V}{A_{AA}} = \frac{5.0714}{3.3107} = 1.532\text{ksi}$ or $\boxed{\tau_{AA} = 1.53\text{ksi}}$

1.38 Two rectangular bars of thickness 10 mm are loaded as shown. If the normal stress on plane AA is 150 MPa (C), determine the force F_1 and the normal and shear stress on plane BB.

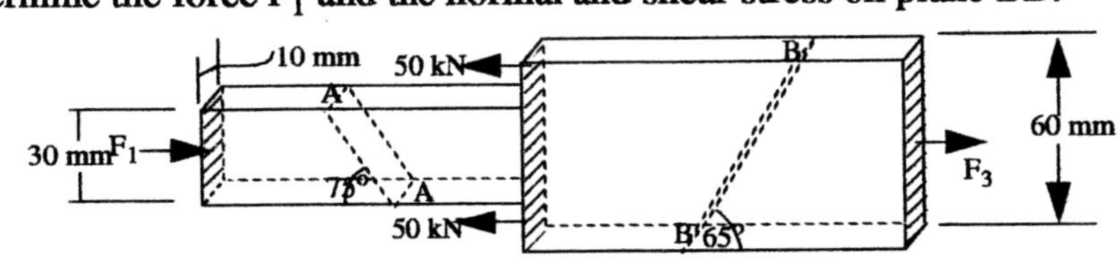

Fig. P1.38

Solution σ_{AA}= 150 MPa (C) F_1 = ? σ_{BB}= 1? τ_{BB}= ?

Figures below show the free body diagrams after the imaginary cuts are made through AA and BB.
By equilibrium of forces in the n-direction in Fig. (a) we obtain: $N_{AA} = F_1\sin 75 = 0.9659F_1$.

The area of cross-section of plane AA is: $A_{AA} = (h)(0.01) = \left(\frac{0.03}{\sin 75}\right)(0.01) = 0.3106(10^{-3})\text{m}^2$.

The average normal stress on plane AA is: $\sigma_{AA} = \frac{N_{AA}}{A_{AA}} = \frac{0.9659F_1}{0.3106(10^{-3})} = 150(10^6)$ or $F_1 = 48,230\text{N}$ or

$\boxed{F_1 = 48.2\text{kN}}$

By equilibrium of forces in Fig. P1.38 we obtain $F_1 + F_3 - 100 = 0$ or $F_3 = 51.77\ \text{kN}$

By equilibrium of forces in the n-direction in Fig. (b) we obtain: $N_{BB} = F_3\sin 65 = 51.77\sin 65 = 46.95\text{kN}$

By equilibrium of forces in the t-direction in Fig. (b) we obtain: $V_{BB} = F_3\cos 65 = 51.77\cos 65 = 21.89\text{kN}$

The area of cross-section of plane BB is: $A_{BB} = (h_B)(1.5) = \left(\frac{0.06}{\cos 65}\right)(0.01) = 0.662(10^{-3})\text{m}^2$.

The average normal stress on plane BB is: $\sigma_{BB} = \frac{N_{BB}}{A_{BB}} = \frac{(46.95)(10^3)}{0.662(10^{-3})} = 70.91(10^6)\text{N/m}^2$ or

$$\boxed{\sigma_{BB} = 70.9\text{MPa(T)}}$$

The average shear stress on plane AA is: $\tau_{BB} = \frac{V_{BB}}{A_{BB}} = \frac{(21.89)(10^3)}{0.662(10^{-3})} = 33.065(10^6)\text{N/m}^2$ or

$$\boxed{\tau_{BB} = 33.1\text{MPa}}$$

1.39 (a) In terms of P, a, b and θ, determine the average normal and shear stress on the inclined plane AA shown in Fig. P1.39. (b) Plot the normal and shear stress as a function of θ and determine the maximum value of the normal and shear stress. (c) At what angles of the incline plane the maximum normal and maximum shear stress occur.

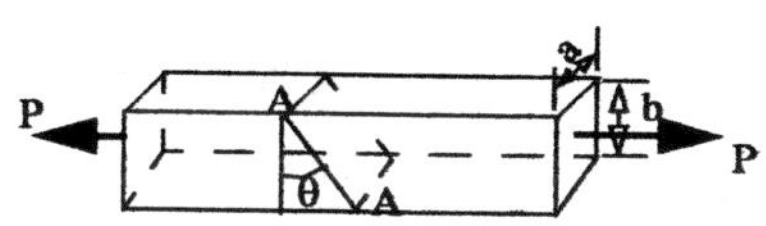

Fig. P1.39

Solution $\sigma = f_1(P, a, b, \theta)$ $\tau = f_2(P, a, b, q)$ $\theta_\sigma = ?$ when σ is max. $\theta_\tau = ?$ when τ is max.

By making an imaginary cut through AA we obtain the following free body diagram.

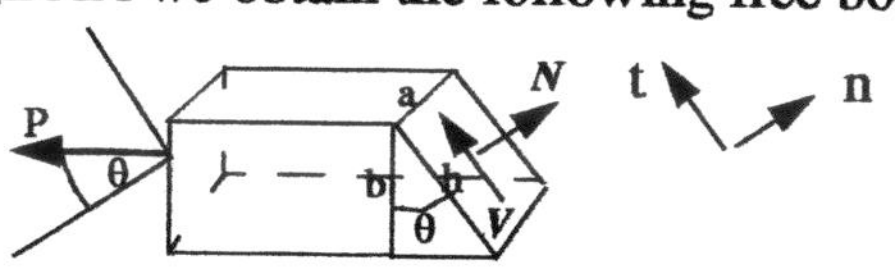

By equilibrium of forces in the n-direction we obtain: $N - P\cos\theta = 0$ or $N = P\cos\theta$

By equilibrium of forces in the t-direction we obtain: $P\sin\theta + V = 0$ or $V = -P\sin\theta$

The area of the incline can be found as: $A = ah = a(b/\cos\theta)$

The average normal stress on the incline plane is: $\sigma = \frac{N}{A} = \frac{P\cos\theta}{a(b/\cos\theta)}$ $\boxed{\sigma = \frac{P}{ab}\cos^2\theta}$

The average shear stress on the incline plane is: $\tau = \frac{V}{A} = \frac{-P\sin\theta}{a(b/\cos\theta)}$ $\boxed{\tau = \frac{-P}{ab}\sin\theta\cos\theta}$

The minus sign in the shear stress expression indicates that is acts down the incline. The normal stress and shear stress are plotted below. From the plot it is seen that normal stress is maximum at θ = 0 and shear stress magnitude is maximum at θ = 45°. Thus the answers are: $\boxed{\theta_\sigma = 0^\circ \quad \theta_\tau = 45^\circ}$

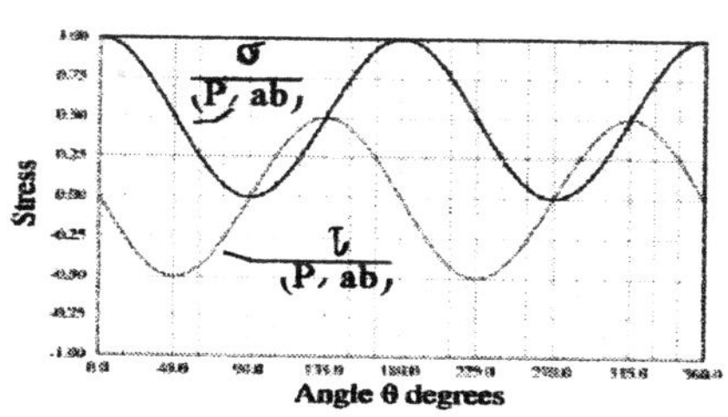

1.40 An axial load is applied to a 1 inch diameter circular rod. The shear stress on section AA was found to be 20 ksi. The section AA is at 45° to the axis of the rod. Determine the applied force P and the average normal stress acting on section AA.

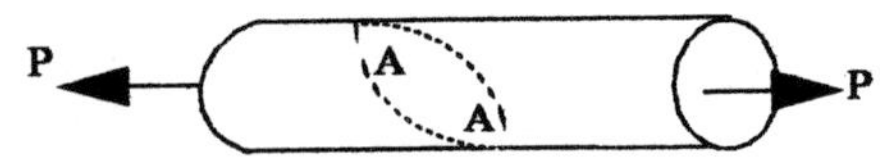

Fig. P1.40

Solution $d = 1$ $\tau_A = 20\text{ksi}$ $P = ?$ $\sigma_A = ?$

By making an imaginary cut through AA we obtain the following free body diagram.

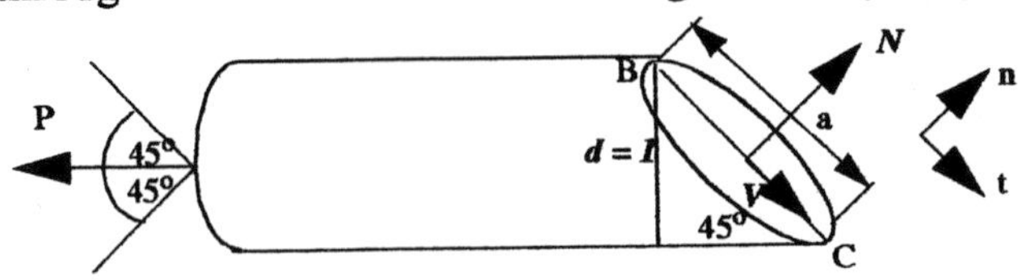

The length BC can be found as: $a = d/(\sin 45) = 1.414$. The area of the elliptic inclined section can be found as: $A = \frac{\pi}{4}ad = \frac{\pi}{4}(1.414) = 1.1107\text{in}^2$

The shear force can be found as: $V = \tau_A A = (20)(1.1107) = 22.214\text{kips}$

By equilibrium of forces in the t-direction we obtain: $P\sin 45 - V = 0$ or $P = \frac{V}{\sin 45} = \frac{22.214}{\sin 45}$

$$\boxed{P = 31.4 \text{ kips}}$$

By equilibrium of forces in the n-direction we obtain: or $N = P\cos 45 = 31.4\cos 45 = 22.214\text{kips}$

The average normal stress on the incline plane is: $\sigma = \frac{N}{A} = \frac{22.214}{1.1107} = 20$ $\boxed{\sigma = 20\text{ksi(T)}}$

1.41 A simplified model of a child's arm lifting weight is shown in Fig. P1.41. The area of cross-section of the biceps muscle is estimated as 2 in^2. Determine the average normal stress in the muscle and the average shear force at the elbow joint A.

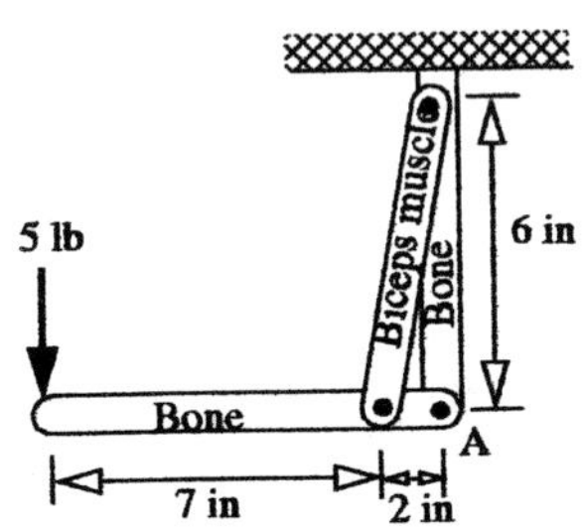

Fig. P1.41

Solution $A = 2\text{in.}^2$ $\sigma = ?$ $V = ?$

The free body diagram of the fore arm is shown below.

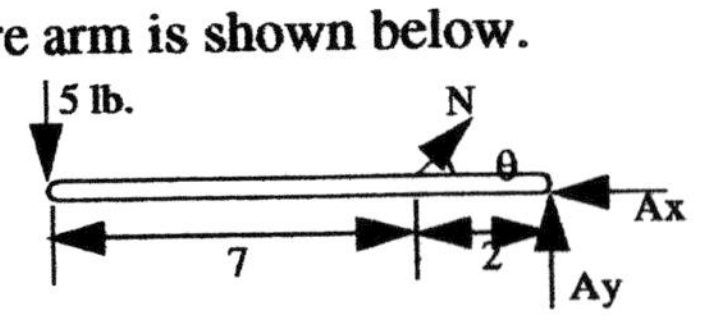

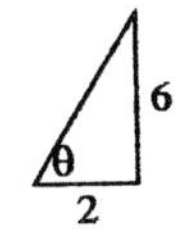

The angle θ can be found as $\theta = \text{atan}(6/2) = 71.56^\circ$

By equilibrium of moment about point A we obtain:

$(N\sin\theta)(2) - (5)(9) = 0$ or $N = \frac{45}{2\sin 71.56} = 23.717\text{lbs}$

The average normal stress in the muscle is: $\sigma = \frac{N}{A} = \frac{23.717}{2} = 11.858$ or $\boxed{\sigma = 11.9\text{psi(T)}}$

By equilibrium of forces in the x-direction we obtain: $A_x = N\cos\theta = 23.717\cos 71.56 = 7.5\text{lbs}$

By equilibrium of forces in the y-direction we obtain: $A_y = 5 - N\sin\theta = 5 - 23.717\sin 71.56 = -17.5\text{lbs}$

$V = \sqrt{A_x^2 + A_y^2} = 19.04$ $\boxed{V = 19\text{lbs}}$

1.42 Figure shows a truss and the sequence of assembly of members at pin H, G, and C. All members of the truss have a cross-sectional area of 250 mm^2 and all pins have a diameter of 10 mm. Determine the axial stresses in members HA, HB, HG and HC of the truss shown.

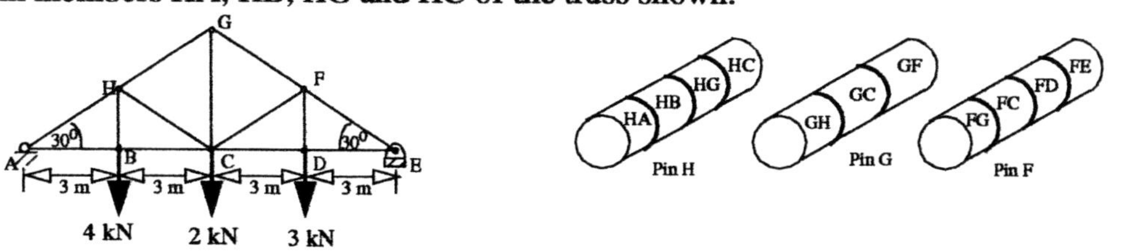

Fig. P1.42

Solution $A_m = 250\text{ mm}^2$ $d_p = 15\text{ mm}$ $\sigma_{HA} = ?$ $\sigma_{HB} = ?$ $\sigma_{HG} = ?$ $\sigma_{HC} = ?$

The free body diagram of the entire truss can be drawn as shown below.By equilibrium of moment about point E, we obtain: Ay(12) - 4(9) - 2(6) - 3(3) = 0 or

$$A_y = 4.75\text{ kN}$$ 1

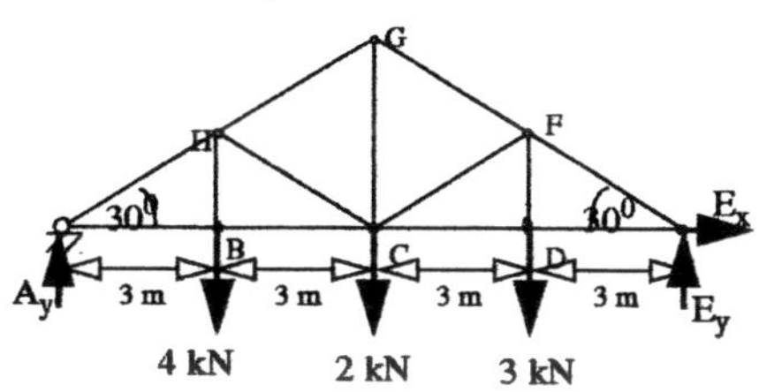

By equilibrium of forces in the y-direction, we obtain: $A_y + E_y - 4 - 2 - 3 = 0$ or

$$E_y = 4.25\text{ kN}$$ 2

By equilbrium of forces in the x-direction, we obtain:

$$E_x = 0$$ 3

Making an imaginary cut through members HG, HC, and BC we obtain the free body diagram in Fig. (a). By equilibrium of moment about point C, we obtain: $[(N_{HG}\sin 30) + A_y](6) - 4(3) = 0$ or

$N_{HG} = (12.0 - 6(4.75)) / 6\sin 30 = -5.5\text{ kN}$

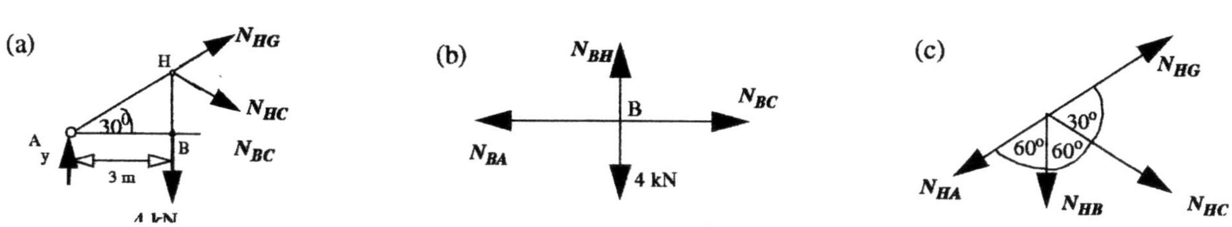

By equilibrium of moment about point A, we obtain: $N_{HC}\sin 30(6) + 4(3) = 0$ or $N_{HC} = -4\text{ kN}$

Fig. (b) is the free body diagram of joint B. By equilibrium of forces in the y-direction we obtain:

$$N_{BH} = N_{HA} - 4\text{ kN}$$

Fig. (c) represents the free body diagram of joint H. By equilibrium of forces in the x-direction we obtain:

N_{HG} cos30 + N_{HC} cos30 - N_{HA} sin60 = 0or (-5.5 - 4) cos 30 = N_{HA} sin60 or N_{HA} = -9.5 kN

The axial stresses in each member can be found as shown below.

$\sigma_{HA} = N_{HA}/A = (-9.5)(10^3)/(250)(10^{-6}) = -38(10^6)\ N/m^2$ **σ_{HA} = 38 MPa (C)**

$\sigma_{HB} = N_{HB}/A = (4)(10^3)/(250)(10^{-6}) = 16(10^6)\ N/m^2$ **σ_{HB} = 16 MPa (T)**

$\sigma_{HG} = N_{HG}/A = (-5.5)(10^3)/(250)(10^{-6}) = -22(10^6)\ N/m^2$ **σ_{HG} = 22 MPa (C)**

$\sigma_{HC} = N_{HA}/A = (-4)(10^3)/(250)(10^{-6}) = -16(10^6)\ N/m^2$ **σ_{HC} = 16 MPa (C)**

1.43 Determine the maximum shear stress in pin H in the truss shown in Fig. P1.42

Solution $d_p = 15mm^2$ $(\tau_H)_{max} = ?$

From Eqs (4), (5), (6) and (7) in problem 1-42 we know the forces acting on pin H.
The free body diagram of pin H is shown in Fig. (a). By making imaginary cuts at different sections of the pin the free body diagrams shown in Figs. (b), (c), and (d) are obtained.

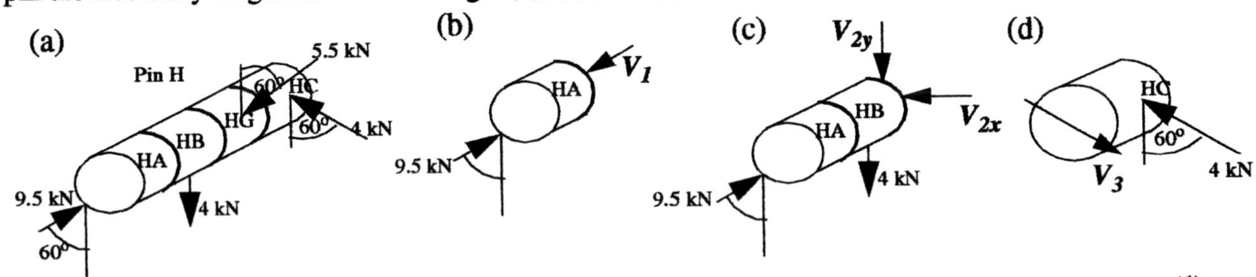

From Fig. (b) $V_1 = 9.5$ kN (1)

From Fig. (c) $V_{2x} - 9.5\sin60 = 0$ $V_{2x} = 8.22$ kN

$V_{2y} + 4 - 9.5\cos60 = 0$ $V_{2y} = 0.75$ kN

$$V_2 = \sqrt{V_{2x}^2 + V_{2y}^2} = 8.25\text{kN} \quad (2)$$

From Fig (d) $V_3 = 4$ kN (3)

From equations (1), (2) and (3), we see that the maximum shear force is 9.5kN, thus the maximum sheer stress will be between members HA and HB.

$$(\tau_H)_{max} = \frac{V_1}{A_{pin}} = \frac{9.5(10^3)}{\frac{\pi}{4}(0.015)^2} = 53.76(10^6)$$

$(\tau_H)_{max}$ = 53.76 MPa

1.44 Fig. P1.42 shows a truss and the sequence of assembly of members at pin H, G, and C. All members of the truss have a cross-sectional area of 250 mm^2 and all pins have a diameter of 10 mm. Determine the axial stresses in members FG, FC, FD, and FE of the truss shown in Fig. P1.42.

Fig. P1.44

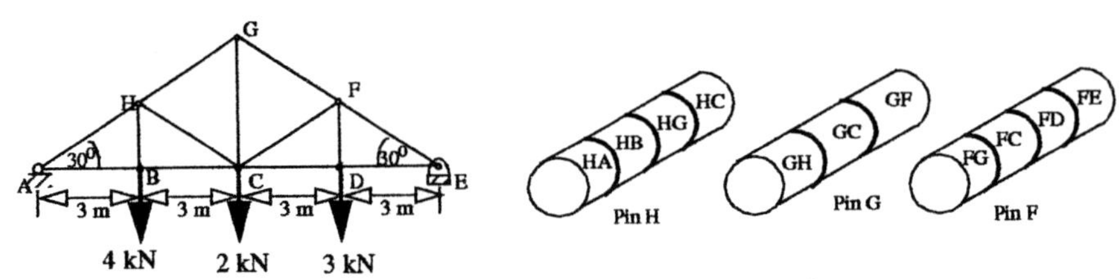

Solution $A_m = 250\ mm^2$ $d_p = 15$ mm $\sigma_{FG} = ?$ $\sigma_{FC} = ?$ $\sigma_{FD} = ?$ $\sigma_{FE} = ?$

From Eqs.(2) and (3) in problem 1.42 we have $E_y = 4.25$ kN and $E_x = 0$. We can make an imaginary cut through members FG, FC, and DC to obtain the free body diagram shown in Fig. (a).

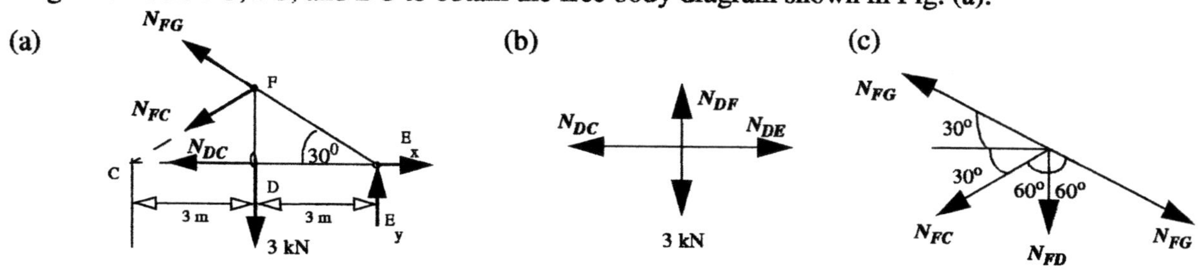

By equilibrium of moment about point C in Fig. (a), we obtain: N_{FG} sin30)(6) + (4.25)(6) - 3(3) = 0 or

$$N_{FG} = -5.5 \text{ kN}$$ 1

By equilibrium of moment about point E, we obtain: N_{FC} sin30)(6) + 3(3) = 0

$$N_{FC} = -3 \text{ kN}$$ 2

Fig. (b) is the free body diagram of joint D. By equilibrium of forces in the y-direction we obtain:

$$N_{DF} = 3 \text{ kN}$$ 3

Fig. (c) represents the free body diagram of joint F. By equilibrium of forces in the x-direction we obtain: $N_{FE}\sin 60 - N_{FG}\cos 30 - N_{FC}\cos 30 = 0$ or

$$N_{FE} = -8.5 \text{ kN}$$ 4

The axial stresses in each member can be found as shown below.

$\sigma_{FG} = N_{FG} / A = -5.5(10^3) / 250(10^{-6}) = -22(10^6)\ N/m^2$ — $\boxed{\sigma_{FG} = 22 \text{ MPa (C)}}$

$\sigma_{FC} = N_{FC} / A = -3(10^3) / 250(10^{-6}) = -12(10^6)\ N/m^2$ — $\boxed{\sigma_{FC} = 12 \text{ MPa (C)}}$

$\sigma_{FD} = N_{FD} / A = 3(10^3) / 250(10^{-6}) = 12(10^6)\ N/m^2$ — $\boxed{\sigma_{FD} = 12 \text{ MPa (T)}}$

$\sigma_{FE} = N_{FE} / A = -8.5(10^3) / 250(10^{-6}) = -34(10^6)\ N/m^2$ — $\boxed{\sigma_{FE} = 34 \text{ MPa (C)}}$

1.45 Determine the maximum shear stress in pin F in the truss shown in Fig. P1.42.

Solution $d_p = 15mm^2$ $(\tau_F)_{max} = ?$

From Eqs (1), (2), (3) and (4) in problem 1-44 we know the forces acting on pin H.
The free body diagram of pin F is shown in Fig. (a). By making imaginary cuts at different sections of the pin the free body diagrams shown in Figs. (b), (c), and (d) are obtained.

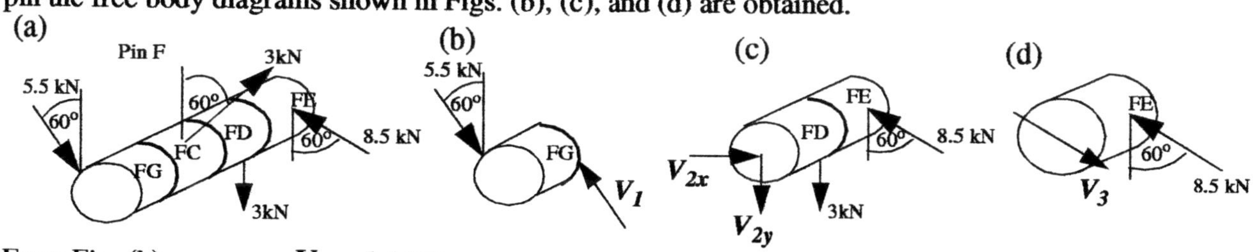

From Fig. (b) $V_1 = 5.5$ kN (1)

From Fig. (c) $V_{2x} - 8.5 \sin60 = 0$ $V_{2x} = 7.36$ kN

$-V_{2y} - 3 + 8.5 \cos60 = 0$ $V_{2y} = 1.25$ kN

$$V_2 = \sqrt{V_{2x}^2 + V_{2y}^2} = 8.50\text{kN} \qquad \textbf{(2)}$$

From Fig (d) $V_3 = 8.5$ kN **(3)**

From equations (1), (2) and (3), we see that the maximum shear force is 8.5kN, thus the maximum sheer stress will be between members FD and FE.

$$(\tau_F)_{max} = \frac{V_3}{A_{pin}} = \frac{8.5(10^3)}{\frac{\pi}{4}(0.015)^2} = 48.1(10^6)$$

$\boxed{(\tau_F)_{max} = 48.1 \text{ MPa}}$

1.46 Fig. P1.42 shows a truss and the sequence of assembly of members at pin H, G, and C. All members of the truss have a cross-sectional area of 250 mm² and all pins have a diameter of 10 mm. Determine the axial stresses in members GH, GC, and GF of the truss shown Fig. P1.42.

Solution $A_m = 250\text{mm}^2$ $\sigma_{GH} = ?$ $\sigma_{GC} = ?$ $\sigma_{GF} = ?$

From Eq. (3) problem 1-42

$$N_{HG} = -5.5 \text{ kN} = N_{GH} \qquad 1$$

The free body diagram of the joint G can be drawn as shown below.

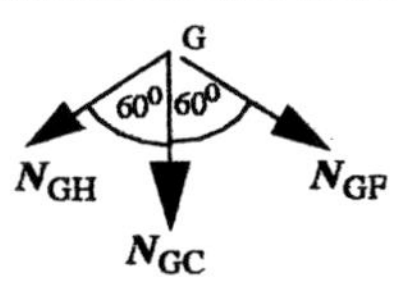

By equilibrium of forces in the x-direction we obtain: $-N_{GH} \sin60 + N_{GF} \sin60 = 0$ or

$$N_{GF} = N_{GH} = -5.5 \text{ kN} \qquad 2$$

By equilibrium of forces in the y-direction we obtain:: $-N_{GH} \cos60 - N_{GF} \cos60 - N_{GC} = 0$ or

$$N_{GC} = 5.5 \text{ kN} \qquad 3$$

The axial stresses in each member can be found as shown below.

$\sigma_{GH} = -5.5(10^{-3}) / 250(10^{-6}) = -22(10^6) \text{ N/m}^2$ $\boxed{\sigma_{GH} = 22 \text{ MPa (C)}}$

$\sigma_{GC} = 5.5(10^{-3}) / 250(10^{-6}) = 22(10^6) \text{ N/m}^2$ $\boxed{\sigma_{GC} = 22 \text{ MPa (T)}}$

$\sigma_{GF} = -5.5(10^{-3}) / 250(10^{-6}) = -22(10^6) \text{ N/m}^2$ $\boxed{\sigma_{GF} = 22 \text{ MPa (C)}}$

1.47 Determine the maximum shear stress in pin G in the truss shown in Fig. P1.42.

Solution $d_p = 15\text{mm}$ $(\tau_G)_{max} = ?$

From equations (1), (2) and (3) in problem 1-46, we know the forces acting on pins. The free body diagram of pin G is shown in Fig. (a). By making imaginary cuts at different sections of the pin the free body diagrams shown in Figs. (b) and (c) are obtained.

By force equilibrium in Figs. (b) and (c) we obtain: $V_1 = 22$ kN $V_2 = 22$ kN

Magnitude of sheer stress is constant though the direction changes. The maximum shear stress is

$$(\tau_G)_{max} = \frac{(5.5)(10^3)}{\frac{\pi}{4}(0.015)^2} = 31.1(10^6) = 31.1\text{MPa}$$

$\boxed{(\tau_G)_{max} = 31.1 \text{ MPa}}$

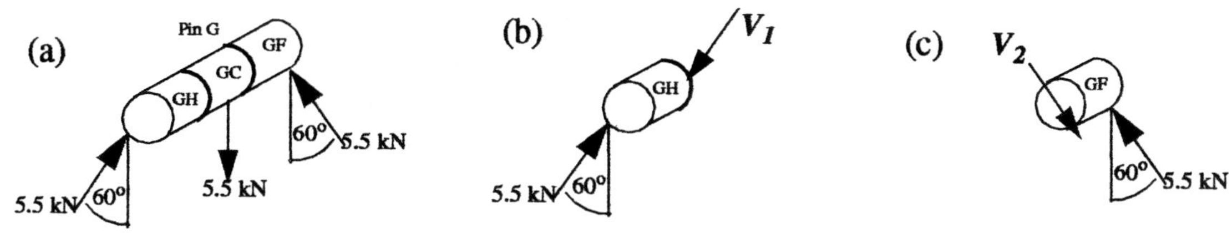

1.48 The pin at C is has a diameter of 1/16 inch and is in double shear. The area of cross-sections for members AB and BC are 2 in^2 and 2.5 in^2. Determine the axial stress in member AB and the shear stress in pin C.

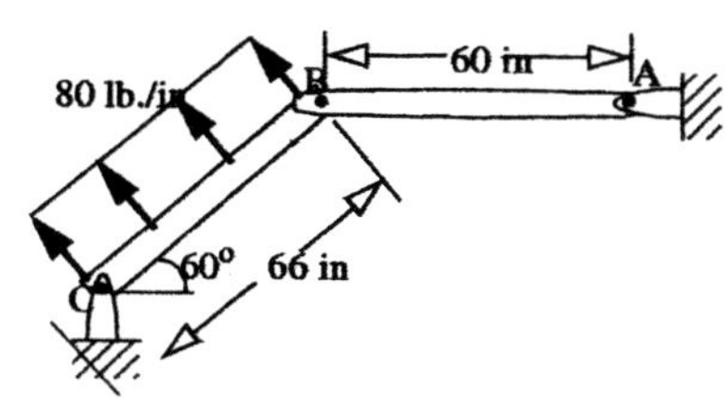

Fig. P1.48

Solution $d_c = 1/2$ in. $A_{AB} = 2$ in^2 $A_{BC} = 2.5$ in^2 $\sigma_{AB} = ?$ $\tau_c = ?$

The free body diagram of member BC can be drawn with the distributed force replaced by an equivalent force as shown in Fig (a).

(a)

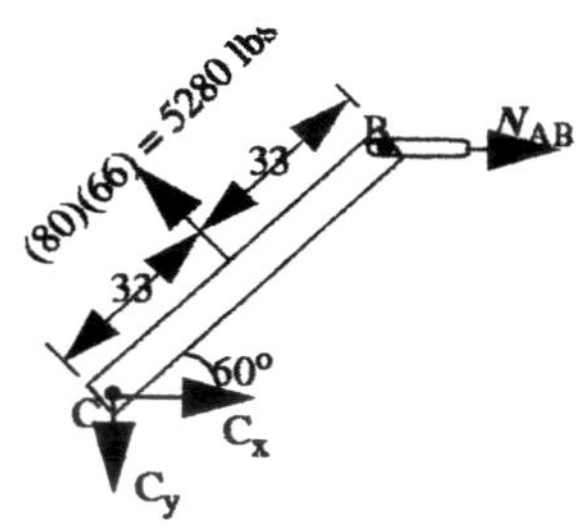

(b)

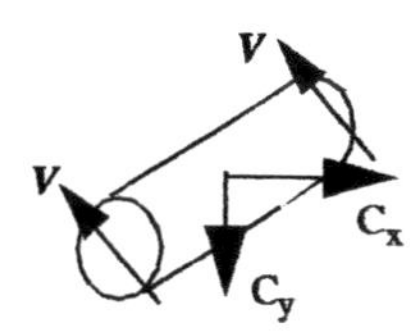

By equilibrium of moment about point C we obtain: $(5280)(33) - N_{AB}(66\sin60) = 0$ or

$N_{AB} = 3048.4\text{lbs}$ 1

By equilibrium of forces in the x-direction we obtain: $C_x + N_{AB} - 5280\cos30 = 0$ or

$C_x = 1524.2$ lbs. 2

By equilibrium of forces in the y-direction we obtain $-C_y + 5280\sin30 = 0$

$C_y = 2640$ lbs. 3

The axial stress is: $\sigma_{AB} = \frac{N_{AB}}{A_{AB}} = \frac{3048.4}{2} = 1524.2\text{psi}$ or $\boxed{\sigma_{AB} = 1524 \text{ psi(T)}}$

The free body diagram in Fig. (b) shows pin C in double shear. By equilibrium of force we obtain:

$$2V = \sqrt{C_x^2 + C_y^2} = 1524.2$$

The shear stress in pin C is: $\tau_C = \frac{V}{A_C} = \frac{1524.2}{(\pi(0.5)^2/4)} = 7763\text{psi}$ $\boxed{\tau_C = 7763\text{psi}}$

1.49 All pins shown are in single shear and have a diameter of 10 mm. All members have square cross-sections. Determine the maximum shear stress in the pins and the axial stress in member BD.

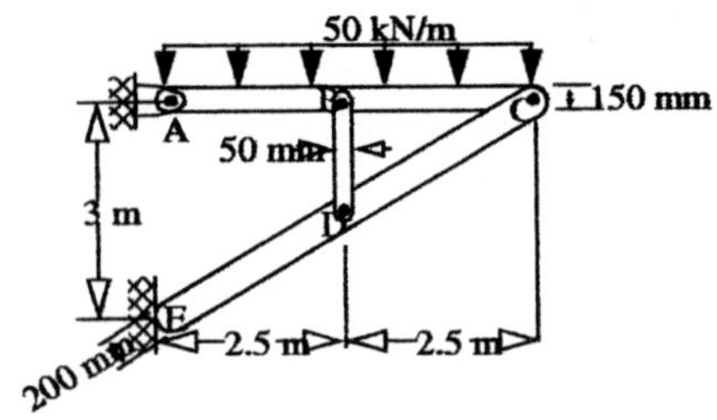

Fig. P1.49

Solution $d = 40\text{mm}$ $\sigma_{BD} = ?$ $\tau_{max} = ?$

The free body diagram of the entire structure can be drawn as shown in Fig. (a). By equilibrium of moment about point A we obtain: $E_x(3) - 250(2.5) = 0$ or $E_x = 208.33$ kN

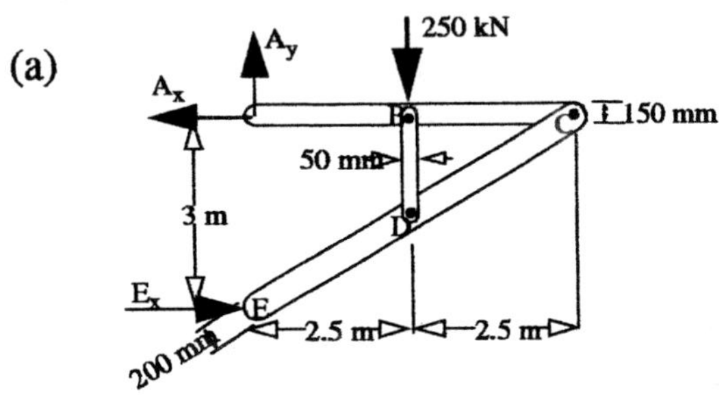

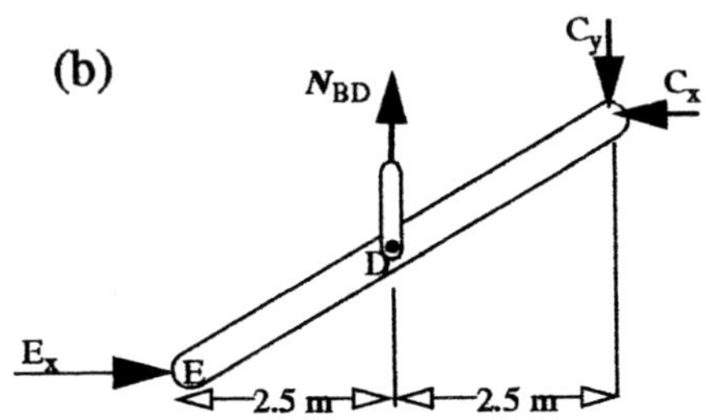

By equilibrium of force in the x-direction we obtain: $A_x = E_x = 208.33$ kN or $A_x = 208.33$ kN
By equilibrium of force in the y-direction we obtain: $A_y = 250$ kN
The free body diagram of member CE can be drawn as shown in Fig. (b)
By equilibrium of moment about point C we obtain: $N_{BD}(2.5) - 208.33(3) = 0$ or $N_{BD} = 250$ kN

The axial stress in member BD is: $\sigma_{BD} = \dfrac{250(10^3)}{(0.05)(0.05)} = 100(10^6)\ \text{N/m}^2$ or $\boxed{\sigma_{BD} = 100\text{MPa (T)}}$

By equilibrium of force in the x-direction in Fig. (b) we obtain: $C_x = E_x = 208.33$ kN or $C_x = 208.33$ kN
By equilibrium of force in the x-direction in Fig. (b) we obtain: $C_y = N_{BD} = 250$ kN or $C_y = 250$ kN

The resultant force at A: $F_A = \sqrt{A_x^2 + A_y^2} = 325\text{kN}$

The resultant force at $F_C = \sqrt{C_x^2 + C_y^2} = 325\text{kN}$

The force on the pin B and D will equal N_{BD}. Thus, pins A & C will have maximum shear.

$$\tau_{max} = \frac{325(10^3)}{(0.05)(0.05)} = 258.6(10^6)\ \text{N/m}^2$$

$\boxed{\tau_{max} = 259\text{MPa}}$

1.50 A C cross-section has a uniform normal stress distribution of 10 ksi (T). Determine the equivalent internal normal force (N) and the location (y_N) from the bottom at which it should be placed so as to require no internal moment.

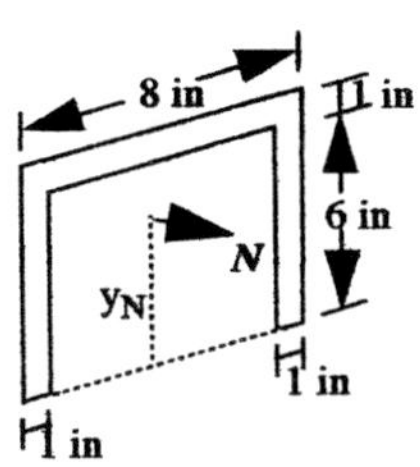

Fig. P1.50

Solution $\sigma = 10$ ksi (T) $N = ?$ $y_N = ?$

The stress in each part of the cross-section can be replaced by an equivalent force as shown below.

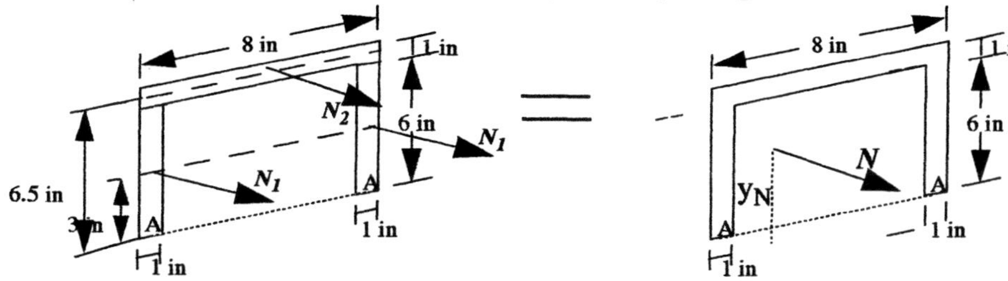

The equivalent forces to the stress are: $N_1 = (10)(6)(1) = 60$ kips and $N_2 = (10)(8)(1) = 80$ kips

By force equivalency in the two figures we obtain: $N = 2N_1 + N_2 = 200$ kips or $\boxed{N = 200 \text{ kips}}$

By moment equivalency about line AA we obtain: $2N_1(3) + N_2(6.5) = N(y_N)$

or $(200)y_N = 2(60)(3) + (80)(6.5)$ or $\boxed{y_N = 4.4 \text{ in}}$

1.51 A steel strip is attached to a wooden bar as shown. The axial stress in steel and wood were found to be at uniform values of 15 ksi (T) and 1 ksi (T) respectively. Determine the equivalent axial force N and its location y_N from the bottom so as to require no internal moment.

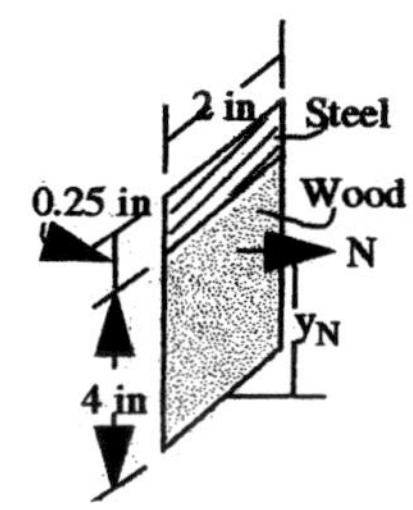

Fig. P1.51

Solution $\sigma_s = 15$ ksi (T) $\sigma_w = 1$ ksi (T) $N = ?$ $y_N = ?$

The stress in each part of the cross-section can be replaced by an equivalent force as shown below

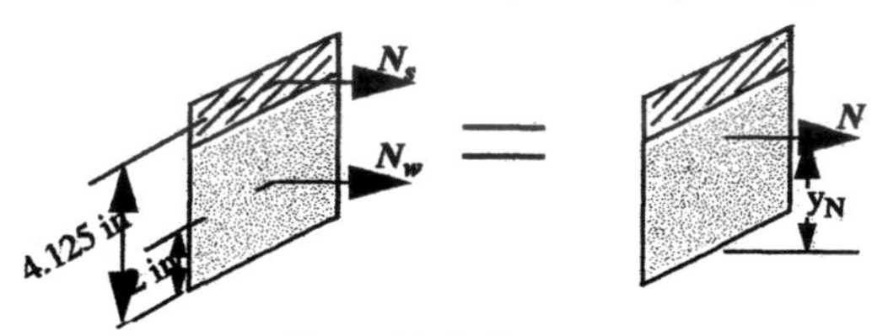

The equivalent forces to the stresses are: $N_s = (15)(2)(0.25) = 7.5$ kips and $N_w = (1)(2)(4) = 8$ kips

By force equivalency in the two figures we obtain: $N = N_s + N_w$ $\boxed{N = 15.5 \text{ kips}}$

By moment equivalency about line AA we obtain: $N_s(4.125) + 2\,(N_w) = Ny_N$ $\boxed{y_N = 3.03 \text{ in}}$

1.52 Determine the value of F_2 and F_3 assuming section AA is of problem 1.50.

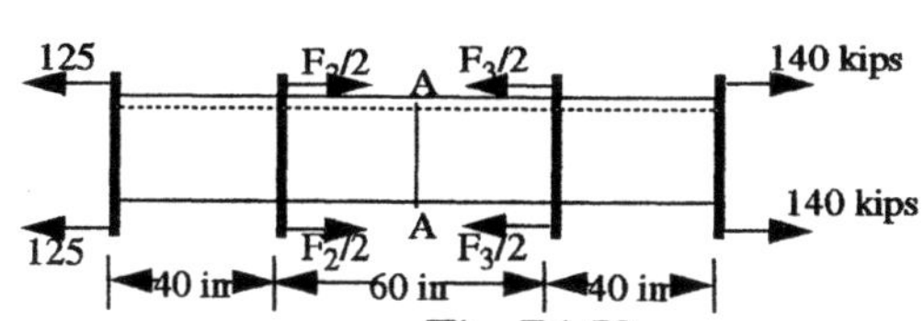

Fig. P1.52

Solution $F_2 = ?$ $F_3 = ?$

By making an imaginary cut through section AA, we obtain the free body diagrams below.By force equilibrium in Fig (a):$-250 + F_2 + N = 0$ or

$$F_2 = 250 - N$$

1

By force equilibrium in (b): $280 - F_3 - N = 0$ or

$$F_3 = 280 - N \qquad 2$$

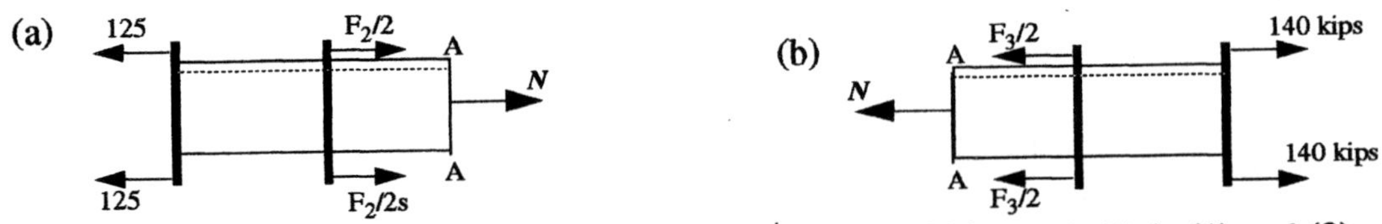

Substituting the value of internal force problem 1-50 i.e., $N = 200$ kips in Eq's (1) and (2) we obtain:

$\boxed{F_2 = 50 \text{ kips}}$ $\boxed{F_3 = 80 \text{ kips}}$

1.53 Determine the value of F_2 and F_3 assuming section AA is of problem 1.51

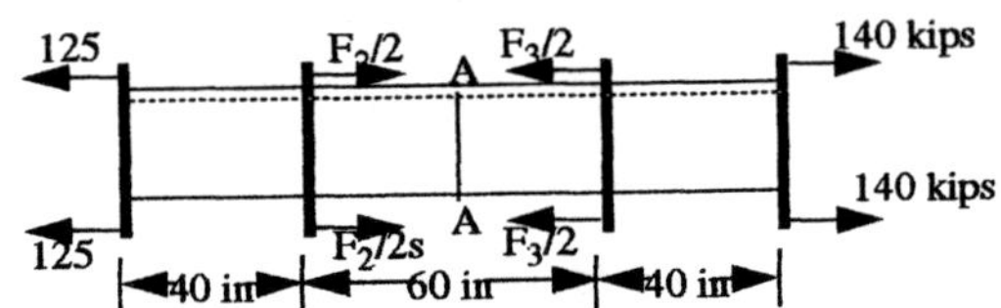

Fig. P1.53

Solution $F_2 = ?$ $F_3 = ?$

Substituting the value of internal force problem 1-51 i.e., $N = 15.5$ kips in Eq's (1) and (2) of problem 1-52 we obtain:

$\boxed{F_2 = 234.5 \text{ kips}}$ $\boxed{F_3 = 264.5 \text{kips}}$

1.54 Determine the axial forces F_1 and F_4 in the composite bar shown. The normal stresses in steel and wood are 180 MPa (C) and 12 MPa (C) respectively at section AA.

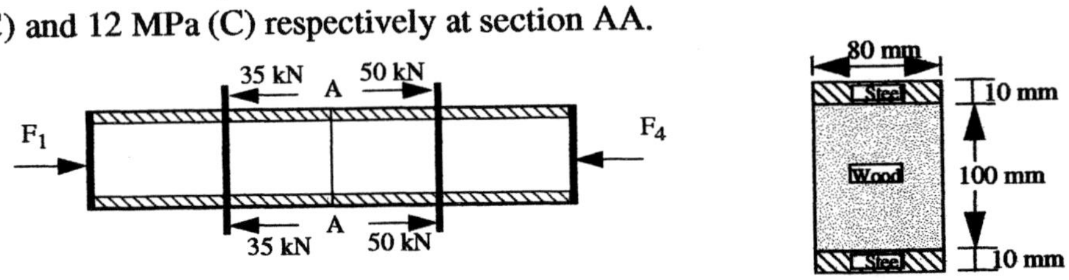

Fig. P1.54

Solution $\sigma_s = 180$ MPa (C) $\sigma_w = 12$ MPa (C) $F_1 = ?$ $F_4 = ?$

The stress distribution can be replaced by equivalent forces as shown below in Fig (a).

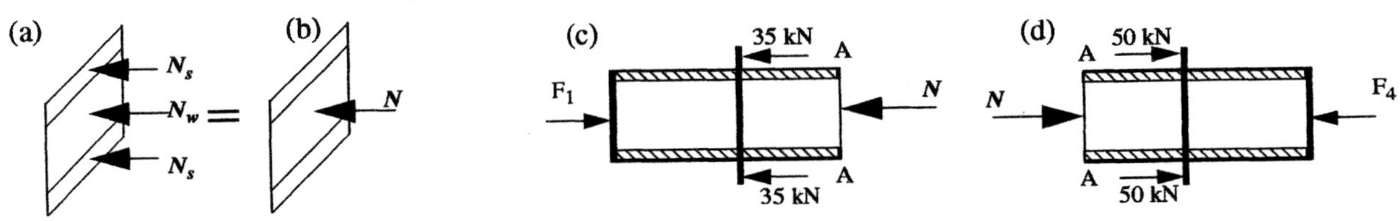

The equivalent forces to the stresses are:

$N_s = (180)(10^6)(.08)(.01) = 144(10^3)$ N = 144 kN and $N_w = (12)(10^6)(.08)(.01) = 96(10^3)$ N= 96 kN

By force equivalency of Figs. (a) and (b) we obtain: $N = 2N_s + N_w = 384$ kN

By making an imaginary cut through section AA we obtain the free body diagrams shown in Figs. (c) and (d). By force equilibrium in Fig (c) we obtain: $F_1 - 70 - N = 0$ $\boxed{F_1 = 454 \text{ kN}}$

By force equilibrium in Fig (d) we obtain: $-F_4 + 100 + N = 0$ $\boxed{F_4 = 484 \text{ kN}}$

1.55 A cantilever beam with a square cross-section is loaded with linearly varying distributed load with maximum intensity of *w* N/m as shown on the left in Fig. P1.55. The section AA has an average shear stress τ and normal stress distribution as shown. Determine (a) the internal moment about the z-axis that is equivalent to the normal stress distribution. (b) the intensity w and the average shear stress τ on section AA.

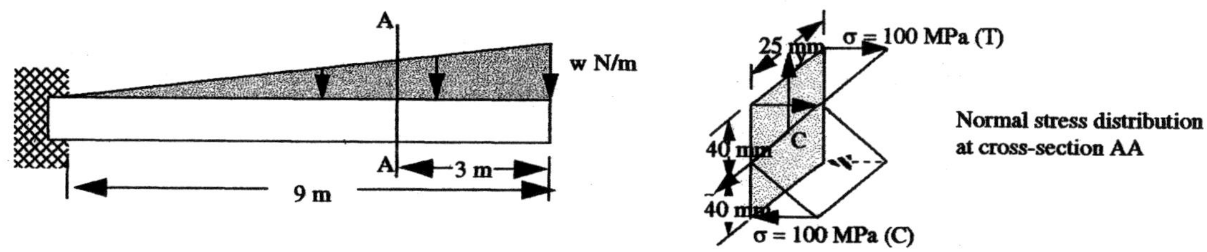

Fig. P1.55

Solution $M_z = ?$ $\quad w = ?$ $\quad \tau = ?$

The stress distribution can be replaced by equivalent forces as shown below.

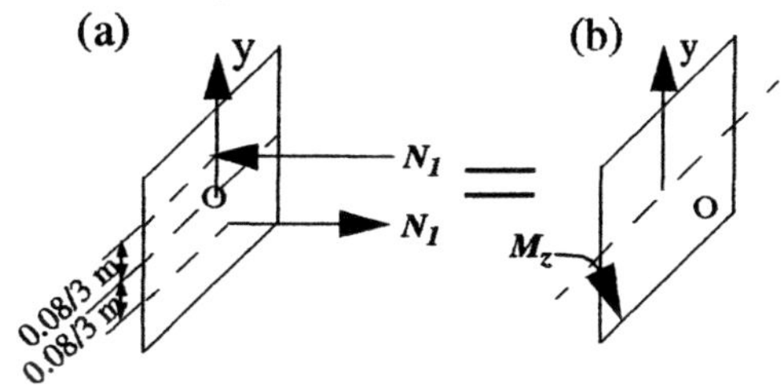

The equivalent forces to the stresses are: $N_1 = 1/2(100)(10^6)(.04)(.025) = 50(10^3)$ N

By moment equivalency about point O we obtain: $M_z = N_1(.08/3) + N_1\ (.08/3) = 2667$ N-m.

The free body diagram of the beam is shown in Fig. (c). By force and moment equilibrium we obtain: $R_w = 9w/2$ and $M_w = 27w$.

By making a imaginary cut through section at A, we obtain the free body diagram in Fig. (d) and replace it by an equivalent diagram in Fig. (e). By moment equilibrium about pint A we obtain:

$M_w - R_w(6) + (2w)(2) - M_z = 0$ or $27w - 27w + 4w - 2667 = 0$ or $\boxed{w = 666.67 \text{ N/m}}$

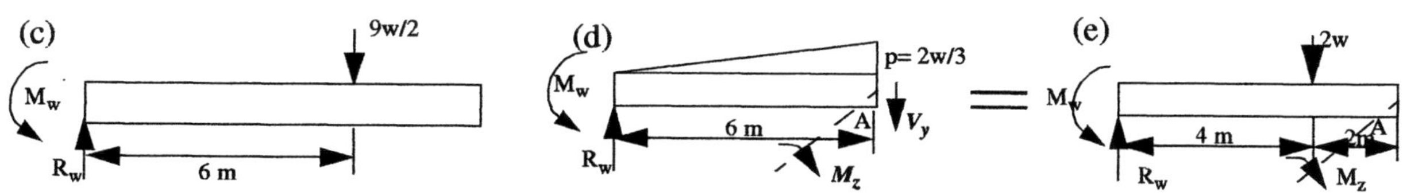

By force equilibrium we obtain: $R_w - 2w - V = 0$ or $V = 9w/2 - 2w = 5w/2 = 5/2(666.67) = 1666.67$N

The average shear stress is: $\tau = V/A = 1666.67/[(.08)(.025)] = 833.3(10^3)$ N/m^2 $\boxed{\tau = 833 \text{ kPa}}$

1.56 A simply supported beam with an I-cross-section has a moment couple M applied at one end. The section AA has an average shear stress τ and normal stress distribution as given. Determine (a) the internal moment about the z-axis that is equivalent to the normal stress distribution. (b) the external moment M, the external force P, and the average shear stress τ on section AA..

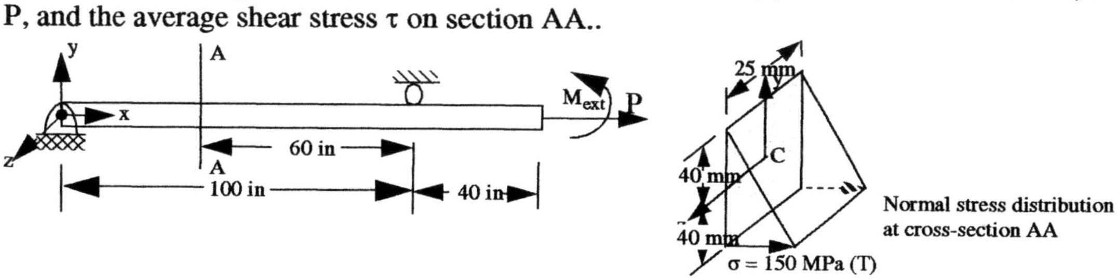

Fig. P1.56

Solution $M_z = ?$ $M_{ext} = ?$ P= ? $\tau = ?$

The stress distribution can be replaced by equivalent forces as shown below.

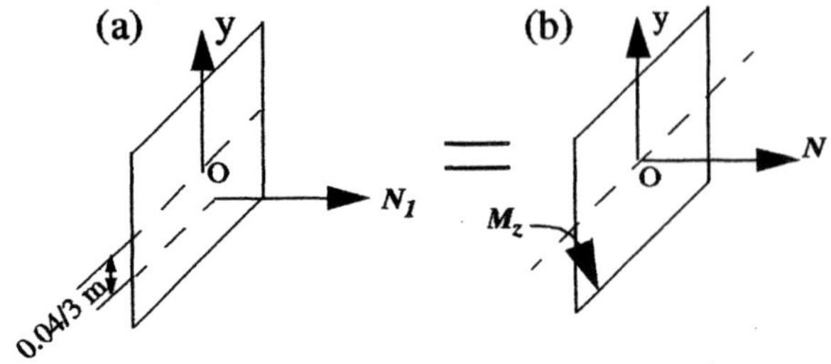

The equivalent force to the stress distribution is: $N_1 = 1/2\,(150)(10^6)(.08)(.025) = 150(10^3)$ N

By force equivalency of Figs. (a) and (b) we obtain: $N = N_1 = 150(10^3)$ N

By moment equivalency about point O we obtain $M_z = N_1(.04/3) = 2(10^3)$ N-m$M_z = 2$ kN-m

Fig.(c) is the free body diagram of the entire beam. By moment equilibrium about point C we obtain the following. $B_y(1) = M_{ext}$ $B_y = M_{ext}$

By force equilibrium in the x-direction we obtain: $B_x = P$

By force equilibrium in the y-direction we obtain: $C_y = B_y = M_{ext}$

(c)

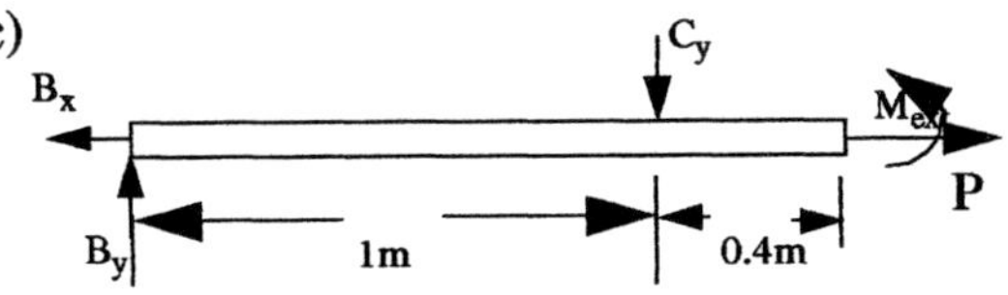

(d)

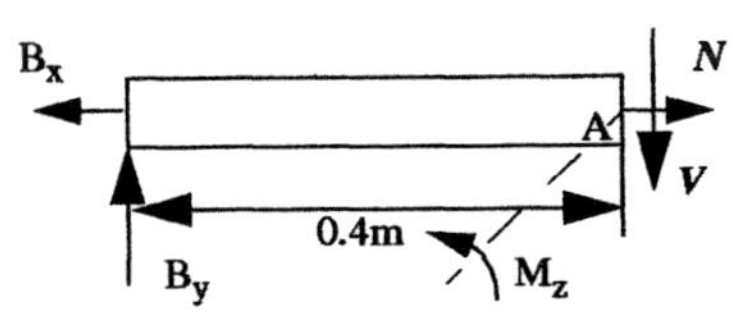

Fig. (d) is the free body diagram after making the imaginary cut through section AA.

By force equilibrium in the x-direction we obtain: $P = N = 150(10^3)$ **P = 150 kN**

By moment equilibrium about point A we obtain$B_y(0.4) - M_z = 0$. Noting that $B_y = M_{ext}$ we obtain:

$M_{ext}(0.4) = 2(10^3)$ or **M_{ext} = 5 kN-m**

By force equilibrium in the y-direction we obtain: $V = B_y = M_{ext} = 5(10^3)$

The area of cross-section is: $A = (0.08)\,(0.025) = 2(10^{-3})$ m^2.

The average shear stress is: $\tau = V/A = 5(10^3)/2(10^{-3}) = 2.5(10^6)$ N/m^2 **τ = 2.5 MPa**

1.57 The cross-section AA on the structure shown has an average shear stress τ and a normal stress as shown. (a) Replace the normal stress distribution by an equivalent internal normal force N and an internal moment M_z acting at the centroid C of the T-cross-section.(b) Determine the value of the average shear stress τ, the intensity of the external distributed force w, and the force P.

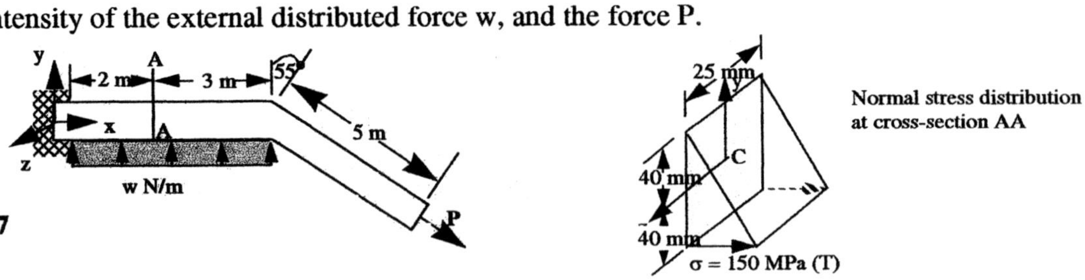

Fig. P1.57

Solution $N = ?$ $M_z = ?$ P= ? w = ? $\tau = ?$

From Fig. (b) in problem 1-56, we know the equivalent internal normal force and internal bending moment are: $N = 150(10^3)$ N and $M_z = 2(10^3)$ N-m. Fig. (a) below shows the direction in which these internal quantities act as per the coordinate system used in problem 1-56. Fig. (b) below is the free body diagram after making an imaginary cut through section AA. The internal quantities shown are opposite to those shown in Fig. (a).

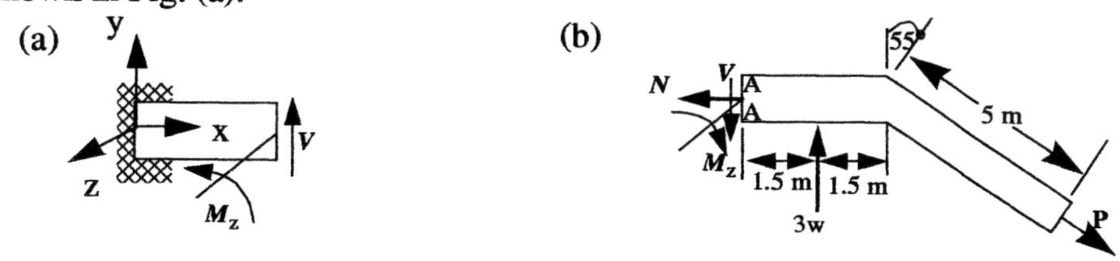

By force equilibrium in the x-direction in Fig. (b) we obtain: $P\sin35 - N = 0$ or $P = 150(10^3) / \sin35$ or

P = 262 kN

By moment equilibrium about point A we obtain $P\cos35(3) - 3w(1.5) + M_z = 0$ or

$4.5w = (251.5)(10^3)(\cos35)(3) + 2(10^3)$ or **w = 143 kN/m**

By force equilibrium in the y-direction we obtain: $V + 3w - P\cos35 = 0$ or

$V = 3(143)(10^3) - (261.5)(10^3)\cos35 = 215.56(10^3)$ Newton.

The area of cross-section is: $A=(0.08)(0.025) = 2(10^{-3})$ m².

The average shear stress is: $\tau = V/A = 215.56(10^3) / 0.002 = 107.78(10^6)$ N/m² **τ = 107.8 MPa**

1.58 A simply supported beam with a C-cross-section has a moment couple M applied at one end. The section AA has an average shear stress τ and normal stress distribution as given. (a) Replace the normal stress distribution by an equivalent internal moment M_z acting at the centroid C. (b) Determine the external moment M and the average shear stress τ on section AA.

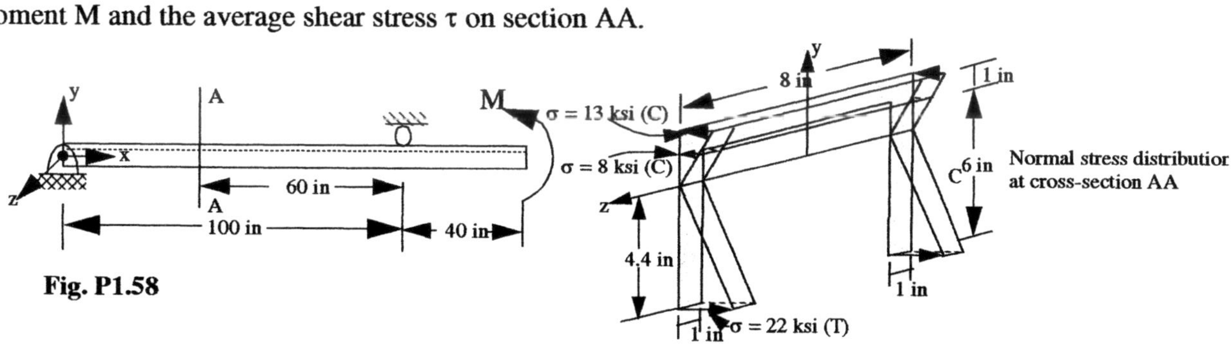

Fig. P1.58

Solution $M_z = ?$ $M_{ext} = ?$ $\tau = ?$

Figure below shows the stress distribution replaced by equivalent forces and moments. To find the internal forces, we can add material between centroid and the bottom of flange and calculate equivalent force N_2 fora triangular distribution. N_3 corresponds to the subtraction of a triangular distribution that was added for calculation of N_2.

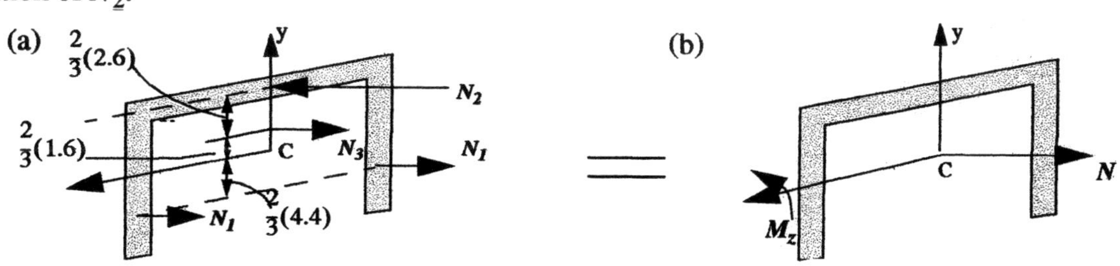

The internal forces equivalent to the stress distribution can be calculated as shown below.

$N_1 = 1/2\ (22)(4.4) = 48.4$ kips $\quad N_2 = 1/2\ (13)(2.6)(8) = 135.2$ kips $\quad N_3 = 1/2\ (8)(1.6)(6) = 38.4$ kips

By force equivalency of Figs. (a) and (b) we obtain: $\quad N = 2N_1 + N_3 - N_2 = 0$

By moment equivalency about point C we obtain $M_z = 2(N_1)(2/3)(4.4) + N_2(2/3)(2.6) - N_3(2/3)(1.6)$ or

$\boxed{M_z = 477.3 \text{ in-kips}}$

Fig.(c) is the free body diagram of the entire beam. By moment equilibrium about point C we obtain the following. $\quad B_y\,(100) = M_{ext} \qquad B_y = M_{ext}/100$

By force equilibrium in the x-direction we obtain: $\quad B_x = 0$

By force equilibrium in the y-direction we obtain: $\quad C_y = B_y = M_{ext}/100$

(c)

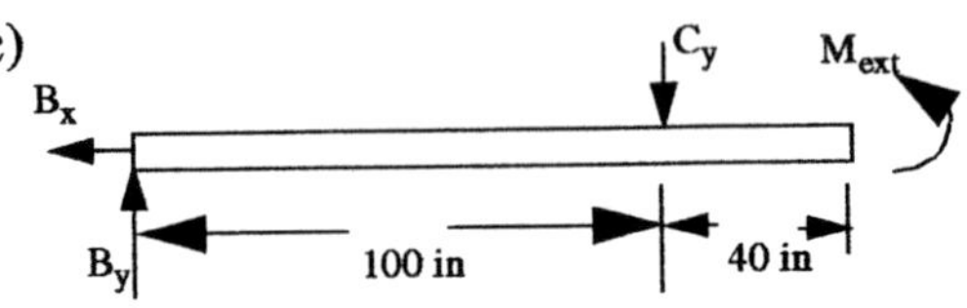

(d)

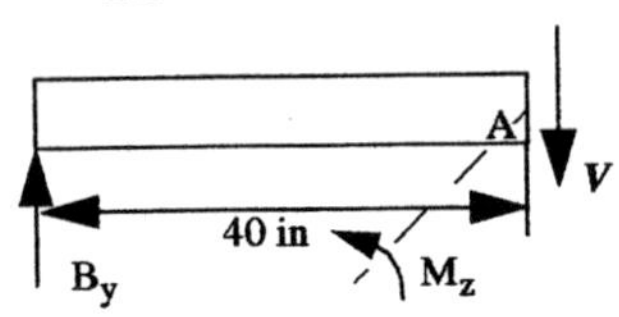

Fig. (d) is the free body diagram after making the imaginary cut through section AA.

By moment equilibrium about point A we obtain $B_y(40) - M_z = 0$. Noting that $B_y = M_{ext}/100$ we obtain:

$477.3 - (M_{ext}/100)(40) = 0$ or $\quad \boxed{M_{ext} = 1193 \text{ in-kips}}$

By force equilibrium in the y-direction we obtain: $V = B_y = M_{ext}/100 = 11.933$ kips)

The area of cross-section is: $A = 2(6)(1) + (8)(1) = 20 \text{ in}^2$.

The average shear stress is: $\tau = V/A = 11.93/20 = 0.59667$ ksi $\quad \boxed{\tau = 597 \text{ psi}}$

1.59 The cross-section AA on the structure shown has an average shear stress τ and a normal stress as shown. (a) Replace the normal stress distribution by an equivalent internal normal force N and an internal moment M_z acting at the centroid C of the T-cross-section.(b) Determine the value of the average shear stress τ, the intensity of the external distributed force w, and the force P.

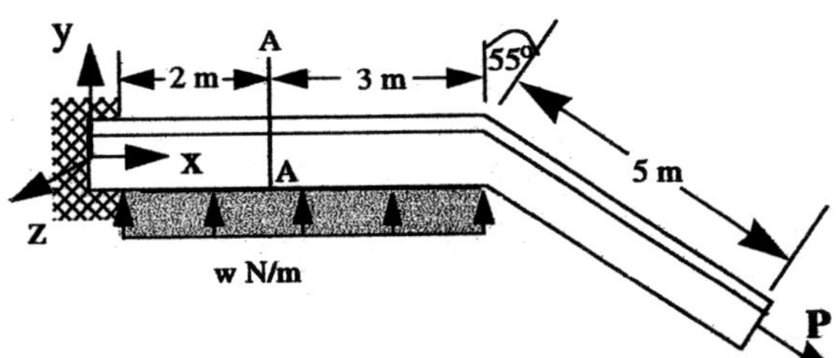

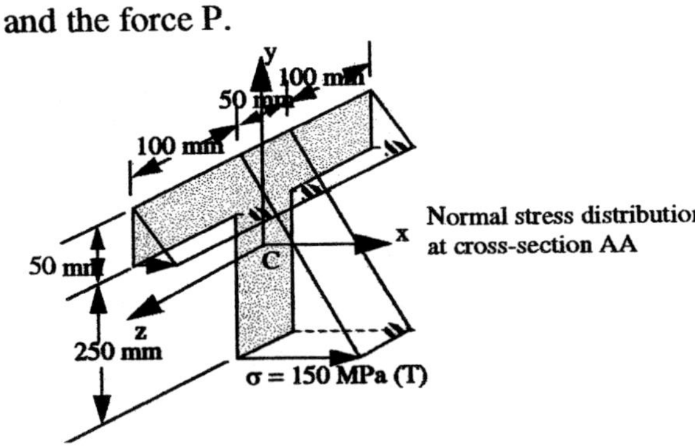

Fig. P1.59

Solution $\quad N = ? \quad w = ? \quad M_z = ? \quad P = ? \quad \tau = ?$

Figure below shows the stress distribution replaced by equivalent forces and moments.

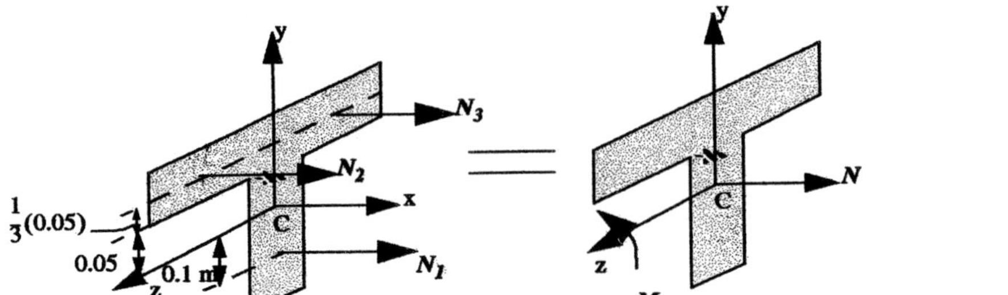

The internal forces equivalent to the stress distribution can be calculated as shown below.

$N_1 = 1/2(150)(10^6)(0.3)(0.05) = 1125(10^3)$ N $\quad N_2 = 1/2(25)(10^6)(0.1)(0.05) = 62.5(10^3)$ N

By force equivalency of Figs. (a) and (b) we obtain: $N = N_1 + 2N_2 = 1250(10^3)$ Newton $\boxed{N = 1250 \text{ kN}}$

By moment equivalency about point C we obtain: $M_z = N_1(0.01) - 2N_2(0.05 + 0.05/3) = 104.17(10^3)$

$\boxed{M_z = 104.2 \text{ kN-m}}$

Fig. (b) in problem 1-57 represents the free body diagram after making an imaginary cut through section AA. By equilibrium of forces in the x-direction we obtain: Psin35 = N = 1250 kN $\boxed{P = 2179.3 \text{ kN}}$

By moment equilibrium about point A we obtain(Pcos35)(3) - 3w(1.5) + $M_z = 0$

or 4.5w = 2179.3 cos35 (3) + 104.2 $\boxed{w = 1213.3 \text{ kN/m}}$

By equilibrium of forces in the x-direction we obtain: V = 3w - Pcos35 =3(1213.3)-2179.3 cos35=1854.7kN

The area of cross-section is: A=(0.25)(0.05) + (0.25)(0.05) = 0.025 m^2.

The average shear stress is: $\tau = V/A = (1854.7(10^3))/(.025) = 74.188(10^6)$ N/m^2 $\boxed{\tau = 74.2 \text{ MPa}}$

1.60 The bottom screw in a hook supports 60% of the load P, while the remaining 40% of P is carried by the top screw. The shear strength of the screws is 50 MPa. Develop a table for maximum load P that the hook can support for screw diameters that vary from 1 mm to 5 mm in steps of 1 mm.

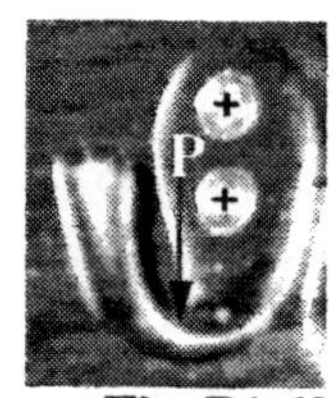

Fig. P1.60

Solution $\tau_{max} \le 50\text{MPa}$ $\qquad$ P vs. d = ?

The shear force on the bottom screw is $V_{bottom} = 0.6P$

$$\tau_{max} = \frac{V_{bottom}}{A} = \frac{0.6P}{(\pi d^2/4)} \le 50(10^6) \quad \text{or} \quad \frac{P}{d^2} \le 65.45(10^6)$$

The table below list the values of d and P that satisfies the above inequality.

d (mm)	P (Newton)
1	65
2	261
3	589
4	1047
5	1636

1.61 A tire swing is suspended using three chains. Each chain makes an angle of 12° with the vertical. The chain is made from links as shown. For purpose of design, assume that more than one person may use the swing and hence the swing is to be designed to carry a weight of 500-lb. If the maximum average normal stress in the links is not to exceed 10 ksi, determine to the nearest 1/16 inch the diameter of the wire that should be used for constructing the links.

Fig. P1.61

Solution $\sigma_{max} \le 10\text{ksi}$ W = 500-lb d = ? to nearest 1/16th of an inch

---

The free body diagram of the tire and the chain link is shown below.
By force equilibrium in the y-direction we obtain: $3T\cos 12 = 500$ or T = 170.4bs
By force equilibrium on the link we obtain: $2N = T$ or $N = 85.195$ lbs.

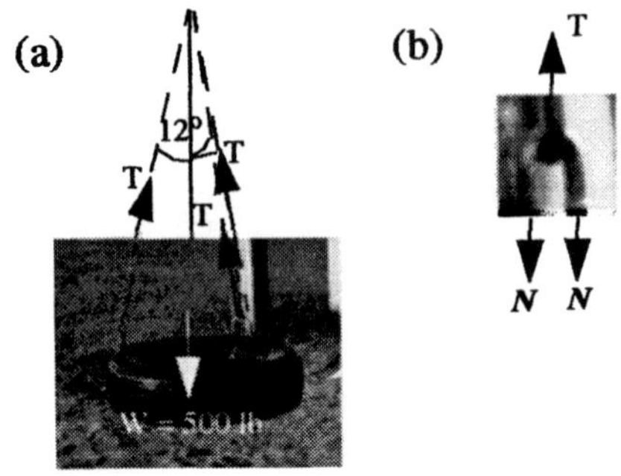

The average axial stress in the link is: $\sigma = \dfrac{N}{A} = \dfrac{85.195}{(\pi d^2/4)} \le 10(10^3)$ or $d \ge 0.1041\text{inch}$

The minimum d to the nearest 1/16th of an inch is: $\boxed{d_{min} = \frac{1}{8}\text{inch}}$

1.62 Two cast iron pipes are held together by a bolt as shown. The outer diameters of the two pipes are 50 mm and 70 mm and the wall thickness of each pipe is 10 mm. The diameter of the bolt is 15 mm. What is the maximum force P this assembly can transmit if the maximum permissible stresses in the bolt and cast iron are 200 MPa in shear and 150 MPa in tension, respectively.

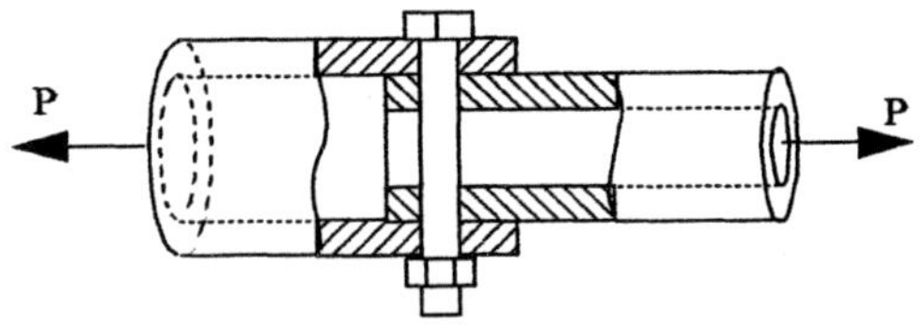

Fig. P1.62

Solution $\sigma_{cast} \le 150\text{MPa}$ $\tau_{bolt} \le 200\text{MPa}$ $d_{bolt} = 15\text{mm}$ $P_{max} = ?$

The smaller pipe will have higher normal stress. The free body diagrams of the smaller pipe and the pipe with bolts cut by imaginary sections are shown below.

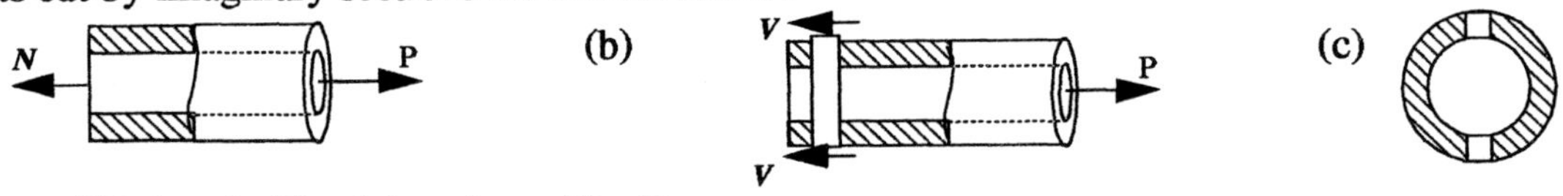

By force equilibrium in Fig. (a) we have $N = P$
The area of cross-section of the cast iron pipe is minimum at the bolt hole as shown in Fig. (c).The cross-sectional area is: $A_{cast} = \pi(0.05^2 - 0.03^2)/4 - 2(0.01)(0.015) = 0.9566(10^{-3})\text{m}^2$
The average normal stress in the cast iron pipe is as shown below.

$$\sigma_{cast} = \frac{N}{A_{cast}} = \frac{P}{0.9566(10^{-3})} \le 150(10^6) \quad \text{or} \quad P \le 143.49(10^3)\text{Newtons} \qquad (1)$$

By force equilibrium in Fig. (b) we have $2V = P$ or $V = P/2$
The area of cross-section of the bolt is: $A = \pi(0.015)^2/4 = 176.7(10^{-6})\text{m}^2$

$$\tau_{bolt} = \frac{V}{A_{bolt}} = \frac{(P/2)}{(176.7)(10^{-6})} \le 200(10^6) \quad \text{or} \quad P \le 70.68(10^3)\text{Newtons} \qquad (2)$$

The maximum value of P that satisfies inequalities in Eqs. (1) and (2) is: $\boxed{P_{max} = 70.6\text{kN}}$

1.63 A normal stress of 20 ksi is to be transferred from one plate to another by riveting a plate on top. The shear strength of the 1/2 inch rivets used is 40 ksi. Assuming all rivets carry equal shear stress, determine the minimum even number of rivets that must be used.

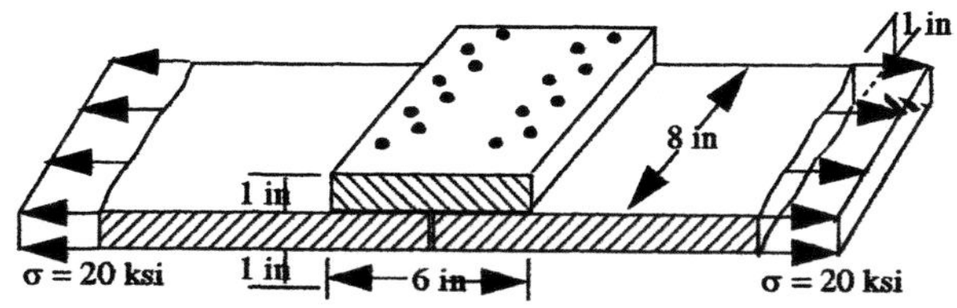

Fig. P1.63

Solution $\sigma = 20$ ksi $d_{rivet} = 1/2$ $\tau_{rivet} \le 40\text{ksi}$ 2n= number of rivets =?

We can draw the following free body diagram.

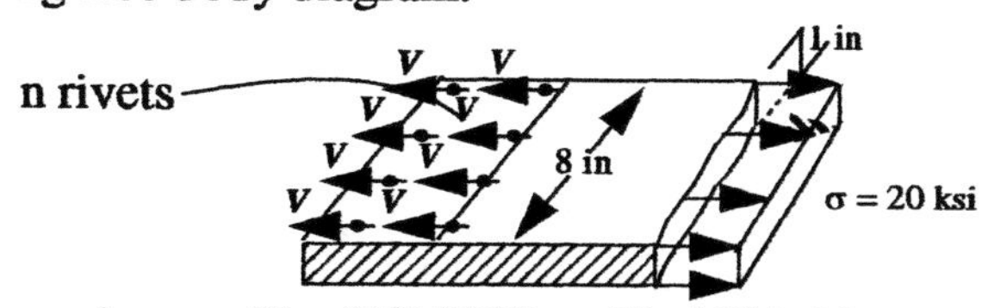

By equilibrium of forces we have: $nV = (20)(8)(1)$ or $V = 160/n$ kips

The area of cross-section of each rivet is: $A_{rivet} = (\pi/4)(1/2)^2 = 0.1963\ \text{in}^2$

The average stress in each rivet can be written as:

$$\tau_{rivet} = \frac{V}{A_{rivet}} = \frac{160}{n(0.1963)} \le 40 \quad \text{or} \quad n \ge \frac{160}{(40)(0.1963)} \quad \text{or} \quad n \ge 20.37 \text{ or } n = 21 \text{ rivets}$$

$\boxed{2n = 42 \text{ rivets}}$

1.64 Two possible joining configurations are to be evaluated. The forces on the joint A in a truss were calculated and are as shown. The diameter of the pin is 20 mm. Determine which joint assembly is better by calculating the maximum shear stress in the pin for each case

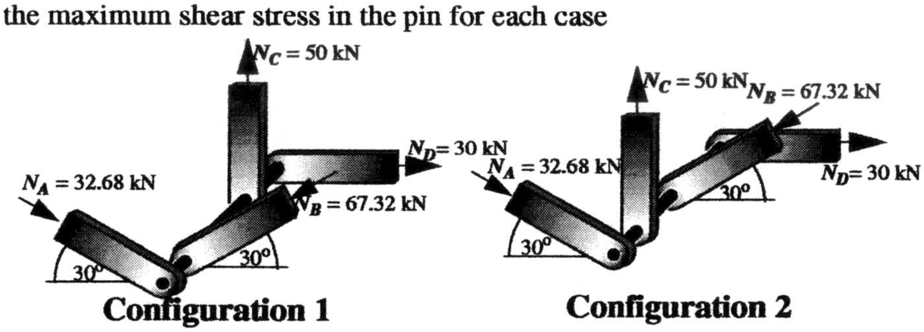

Solution $d_p = 20\text{mm}$ configuration = ?

Configuration 1

Fig. (a) below shows the free body diagram of the entire pin. Figs. (b), (c), and (d) show the free body diagrams after making the imaginary cut at different sections of the pin.

From Fig. (b) $V_1 = 32.68$ kN

From Fig. (c) $V_{2x} = 30$ kN $V_{2y} = 50$ kN. The resultant shear force is: $V_2 = \sqrt{30^2 + 50^2} = 58.31\text{kN}$

From Fig. (d) $V_3 = 30$ kN.

The maximum shear force is V_2. Thus, the maximum shear stress is:

$$(\tau_1)_{max} = \frac{V_2}{A} = \frac{58.31(10^3)}{\pi(0.02^2)/4} = 185.6(10^6)\ \text{N/m}^2 \tag{1}$$

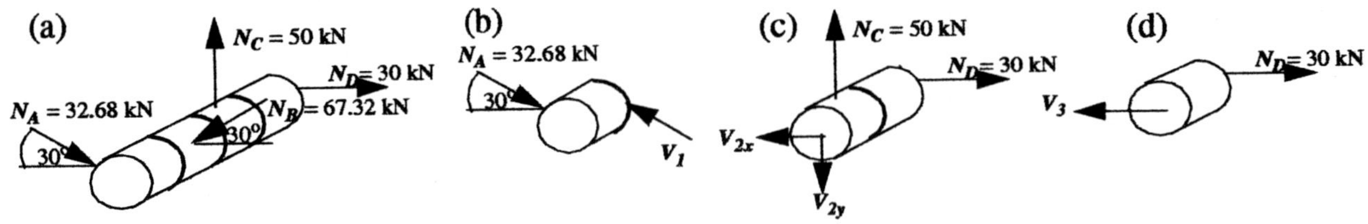

Configuration 2

Fig. (e) shows the free body diagram of the entire pin. Figs. (f), (g), and (h) show the free body diagrams after making the imaginary cut at different sections of the pin.

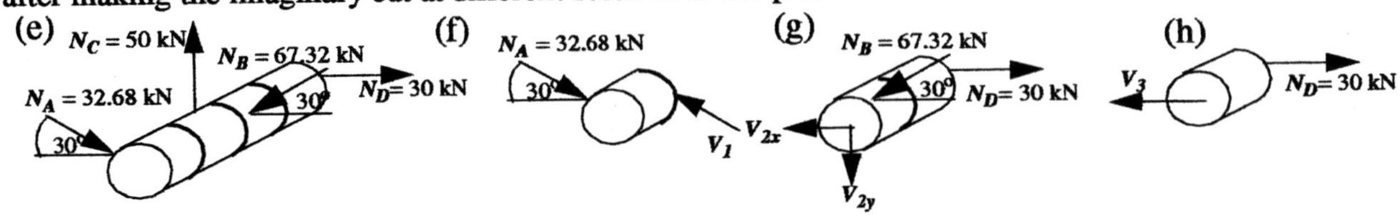

From Fig (f) $\quad V_1 = 32.68$ kN

From Fig (g) $\quad V_{2x} = 28.3$ kN $\quad V_{2y} = 33.66$ kN

The resultant shear force is: $V_2 = \sqrt{28.3^2 + 33.66^2} = 43.97\text{kN}$

From Fig (h) $\quad V_3 = 30$ kN

The maximum shear force is V_2. Thus, the maximum shear stress is:

$$(\tau_2)_{max} = \frac{V_2}{A} = \frac{43.97(10^3)}{\pi(0.02^2)/4} = 139.98(10^6)\ \text{N/m}^2 \qquad 2$$

Comparing Eqs. (1) and (2) we see that configuration 2 has lower shear stress. **Configuration 2 is better.**

1.65 Truss analysis showed the following forces at joint A. Determine the sequence in which the three members at joint A should be assembled, so that the shear stress in the pin is minimum.

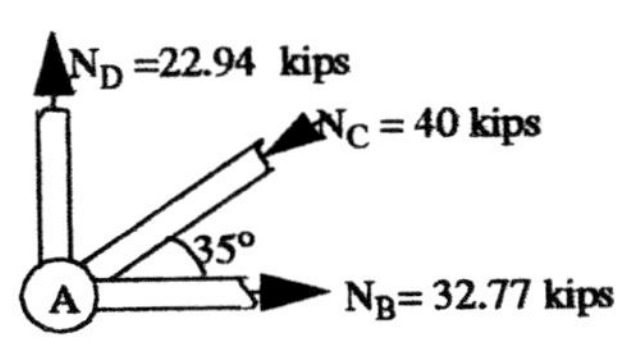

Fig. P1.65

Solution $\qquad$ Sequence = ?

There are 6 possible sequences. Three of the sequences are shown below. The remaining three are the mirror image of the three shown and hence will result in the same shear stress.

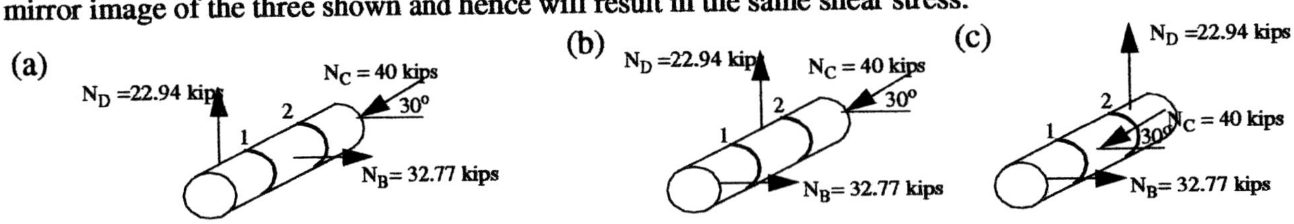

The shear force at the imaginary cuts 1 and 2 will equal to the force at the nearest end. Thus, if smaller forces are on the outside, then the shear stress will be smaller. Configuration shown in Fig.(c) will yield the smallest shear stress.

Member should be assembled as in Fig.(c).

1.66 An 8 in x 8 in reinforced concrete bar needs to be designed to carry a compressive axial force of

235 kips. The reinforcement is done using 1/2 inch round cast iron bars. Assuming normal stress in concrete to be a uniform maximum value of 3 ksi and in iron bars to be a uniform value of 20 ksi, determine the minimum number of iron bars that are needed.

Solution $N \geq 235\text{kips}$ $d_s = 1/2$ in. bar = 8in x 8in

$\sigma_{conc} = 3$ ksi (C) $\sigma_s = 20$ ksi (C) n_{min} = number of steel bars = ?

The cross-section of the reinforced concrete bar is shown.

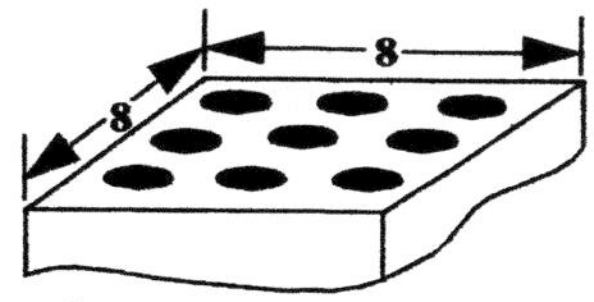

The material area of steel is: $A_s = n\,(\pi/4)(1/2)^2 = 0.19635n\ in^2$

The material area of concrete is: $A_{conc} = (8)(8) - 0.19635n$

The axial stress in steel is: $\sigma_s = N_s/A_s = 20$ ksi or $N_s = 3.927n$ kips

The axial stress in concrete is: $\sigma_{conc} = N_{conc}/A_{conc} = 3$ ksi or $N_{conc} = (192 - 0.58905n)$ kips.

The total axial force that the bar can carry is:

$N = N_s + N_{conc} = 192 + 3.3379n \geq 235$ or $n \geq 12.88$ or $n_{min} = 13$

1.67 A wooden axial member has a cross-section of 2 in x 4 in. The member was glued along the line AA as shown in Fig. P1.67. Determine the maximum force P that can be applied to the repaired axial member if the maximum normal stress in the glue cannot exceed 800 psi and the maximum shear stress in the glue cannot exceed 350 psi.

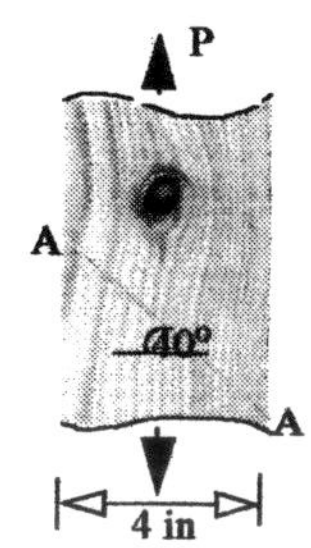

Solution cross-section = 2in x 4in $\sigma \leq 800$ psi $\tau \leq 350$ psi $P_{max} = ?$

The free body diagram of a lower part of the wedge is shown below.

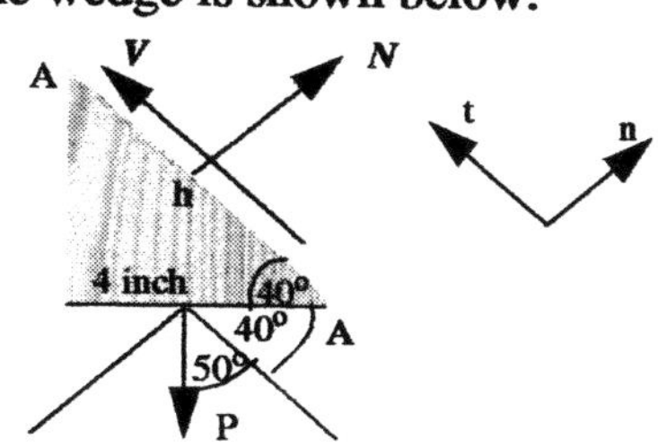

By equilibrium of forces in the n-direction we have: $N = P\sin 50^\circ = 0.7660P$

By equilibrium of forces in the t-direction we have: $V = P\cos 50^\circ = 0.6428P$

The area of the incline plane is: $A = 2h = 2(4/\cos 40) = 10.44\ in^2$

The normal stress on the incline is: $\sigma = \frac{N}{A} = \frac{0.766P}{10.44} \leq 800$ or $P \leq 10902\text{lbs}$

The shear stress on the incline is: $\tau = \frac{V}{A} = \frac{0.6428P}{10.44} \leq 350$ or $P \leq 5684\text{lbs}$

The maximum value of P that satisfies the above two inequalities is: $\boxed{P_{max} = 5684\ \text{lb}}$

1.68 An adhesively bonded joint in wood is fabricated as shown in Fig. P1.68. The length of bonded region L = 5 inches. Determine the maximum force P the joint can support if the shear strength of adhesive is 300 psi and the wood strength is 6 ksi in tension.

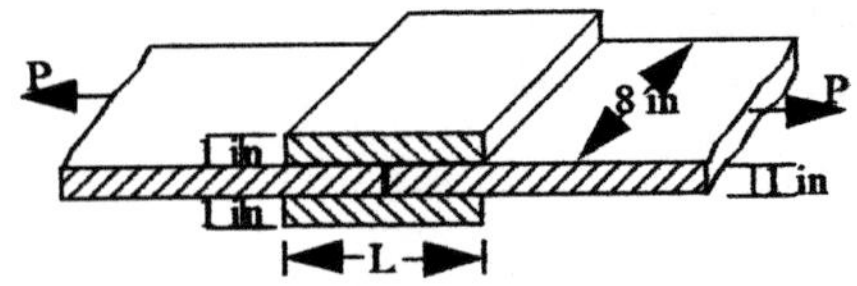

Fig. P1.68

Solution $L = 5$ in. $\sigma_W \le 6000\text{psi}$ $\tau_{glue} \le 300\text{psi}$ $P_{max} = ?$

Fig. (a) shows the free body diagram after making an imaginary cut though the wood.

(a)

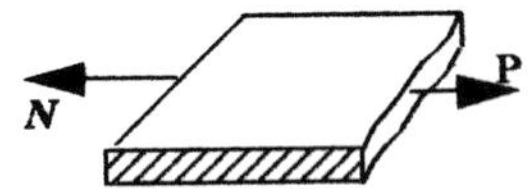

(b)

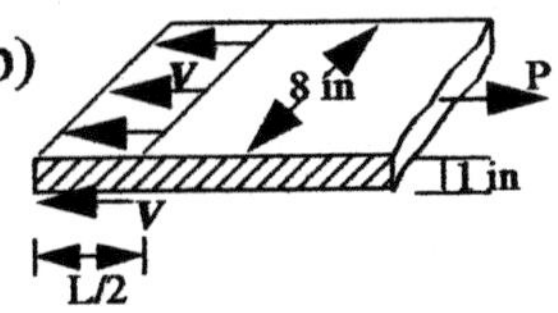

By equilibrium of forces we have: $N = P$

The average normal stress in wood is: $\sigma_W = \dfrac{N}{A_W} = \dfrac{P}{(8)(1)} \le 6000$ or $P \le 48000\ \text{lbs}$

Fig. (b) shows the free body diagram after imaginary cuts are made through the glue on the top and bottom surface of the wood. By equilibrium of forces we have: $2V = P$

The average shear stress in the glue is: $\tau_{glue} = \dfrac{V}{A_{glue}} = \dfrac{P}{(2)(8)(5/2)} \le 300$ or $P \le 12000\ \text{lbs}$

The maximum value of P that satisfies equations the two inequalities above is: $P_{max} = 12\ \text{kips}$

1.69 The joint in Fig. P1.68 is to support a force P = 25 kips, what should be the length L of the bonded region if the adhesive strength in shear is 300 psi.

Solution $P = 25$ kips $\tau_{glue} \le 300\text{psi}$ $L = ?$

From Fig. (b) in problem 1-68: $2V = P = 25(10^3)$ or $V = 12.5(10^3)$ lbs.

The average shear stress is: $\tau_{glue} = \dfrac{V}{A_{glue}} = \dfrac{12.5(10^3)}{(2)(8)(L/2)} \le 300$ or $L \ge 5.208$ inch or $\boxed{L = 5.2\ \text{in}}$

1.70 It is proposed to use 1/2 inch bolts in a 10 inch diameter coupling for transferring a torque of 100 in-kips from a 4 inch diameter shaft onto another. The maximum average shear stress in the bolts is to be limited to 20 ksi. How many bolts and at what radius should the bolts be placed on the coupling?

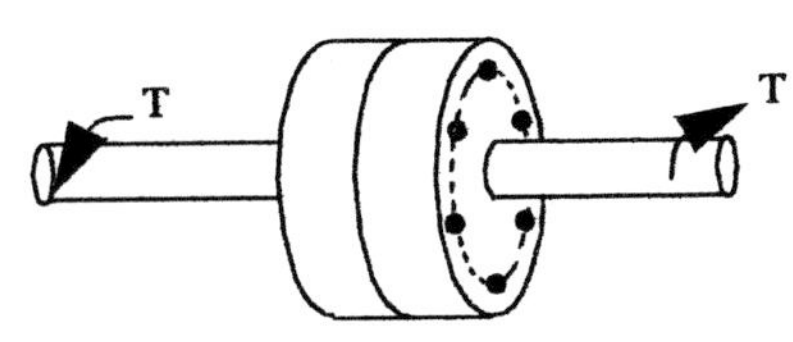

Fig. P1.70

Solution $d_{bolt} = 1/2$ inch $T = 100$ in-kips $\tau_{bolt} \le 20\text{ksi}$ n = number of bolts = ? r = ?

The free body diagram of the coupling can be drawn after making an imaginary cut through the bolts as shown below. By equilibrium of moment about the shaft axis we obtain:

$T = nVr = 100(10^3)$ or $V = 100(10^3)/(nr)$

The shear stress in the bolt can be written as given below.

$\tau_{bolt} = \dfrac{V}{A} = \dfrac{100(10^3)}{(nr)(\pi d^2/4)} \le 20(10^3)$ or $nr \ge \dfrac{5}{(\pi 0.5^2/4)}$ or $nr \ge 25.465$ 1

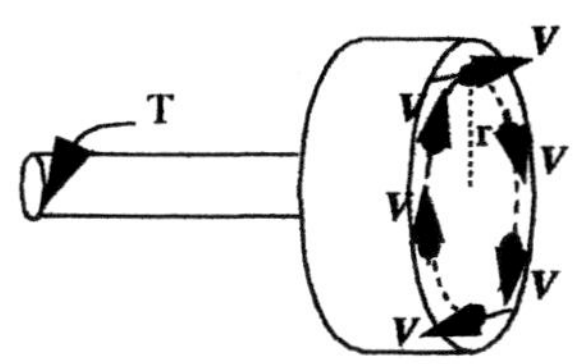

1

As per Eq. 1 there are several for n and r.
The minimum value of r is r_{min} = 2+0.25=2.25 corresponding to n = 11.3 i.e., n = 12.
The maximum value of r is r_{max} = 5-0.25=4.75 corresponding to n = 5.36 i.e., n = 6.
the table below gives all possible answers between n = 6 to n = 12.

n (number of bolts)	6	7	8	9	10	11
r in inches	4.25	3.64	3.18	2.83	2.55	2.31

For minimizing machining and assembly cost 6 bolts should be used.

1.71 A human hand can comfortably apply a torsional moment of 15 in-lbs.
(a)What should be the breaking shear strength of a seal between a lid and the bottle assuming the lid has a diameter of 1.5 inches and a height of 1/2 inch?
(b) If the same sealing strength as in part a is used on a lid that is 1 inch in diameter and 1/2 inch in height, what would be the torque needed to open the bottle?

Solution T = 15 in-lb
(a) d = 1.5 inch h = 1/2 inch τ = ?
(b) d = 1 inch h = 1/2 inch τ = ?

The free body diagram of the bottle with lid removed is shown below.

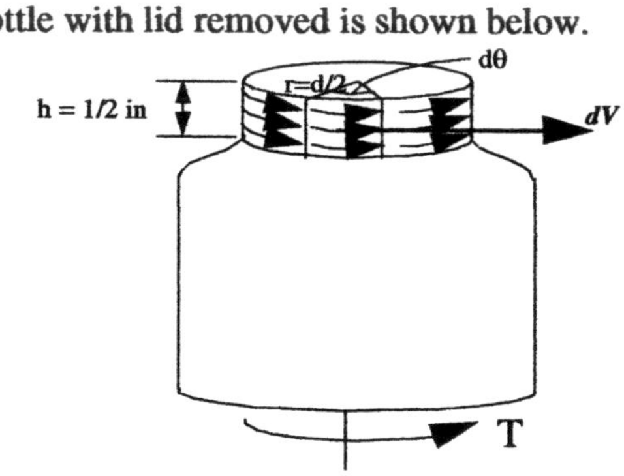

By moment equilibrium about the axis of the bottle we obtain:

$$T = \int_A \left(\frac{d}{2}\right)(dV) = \int_0^{2\pi} \left(\frac{d}{2}\right)\left[\tau h\left(\frac{d}{2}\right)(d\theta)\right] = \left(\frac{d^2}{4}\right)h\tau \int_0^{2\pi} d\theta = \frac{\tau(\pi d^2 h)}{2} \qquad 1$$

(a) Substituting d= 1.5 inch, h= 0.5 inch, T= 15 in-lbs, into Eq. (1) we obtain the following.

$$15 = \frac{\tau\pi(1.5)^2(0.5)}{2} \quad \text{or} \quad \tau = 8.49\text{psi or} \qquad \boxed{\tau = 8.5 \text{ psi}}$$

$15 = \tau\,[\pi(1.5)^2(.5)/2]$ τ = 8.49 psi

(b) Substituting d= 1.0 inch, h= 0.5 inch, τ= 8.49 psi, into Eq. (1) we obtain the following

$$T = \frac{\tau\pi(1.0)^2(0.5)}{2} \quad \text{or} \quad T = 6.67 \text{ in-lbs or} \quad \boxed{T = 6.7 \text{ in-lb}}$$

1.72 The normal stress in the members of the truss shown in Fig. P1.72 is to be limited to 160 MPa in tension or compression. All members have a circular cross-section. The shear stress in the pins is to be limited to 110 MPa. Determine the minimum diameter to the nearest millimeter of members ED, EG, and EF of the truss shown in Fig. P1.72.

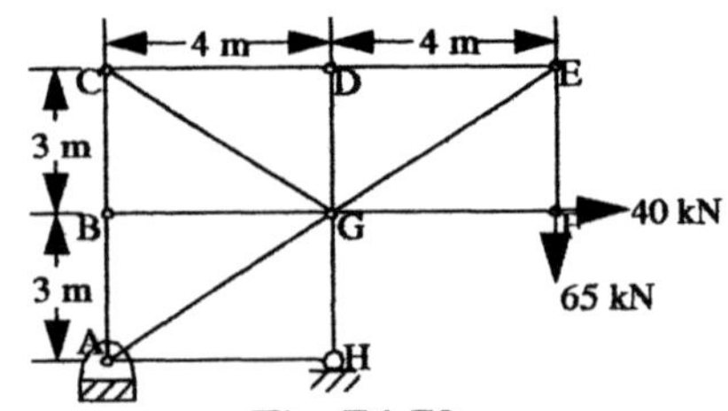

Fig. P1.72

Solution $\sigma \leq 160\text{MPa}$ $\quad d_{EP} = ?$ $\quad d_{EG} = ?$ $\quad d_{EF} = ?$

Figs. (a) and (b) shows the free body diagrams of joints F and E.
By equilibrium of force in Fig. (a) in the y-direction we obtain: $N_{EF} = 65\text{kN}$

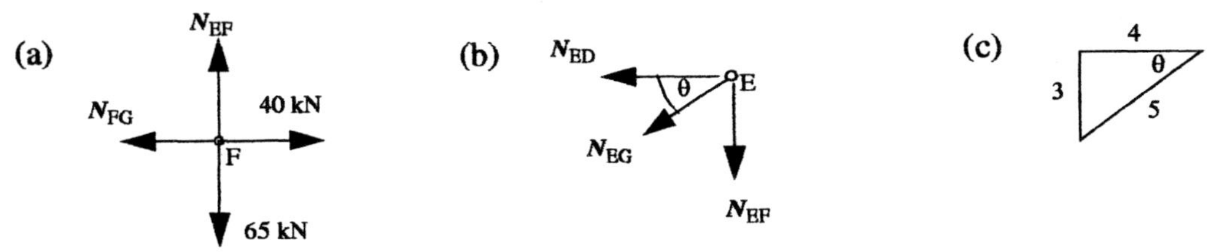

By equilibrium of force in Fig. (b) in the y-direction we obtain: $N_{EG}\sin\theta + N_{EF} = 0$ or $N_{EG} = -108.3$ kN
By equilibrium of force in Fig. (b) in the y-direction we obtain: $N_{ED} + N_{EG}\cos\theta = 0$ or $N_{ED} = 86.67\text{kN}$

$$|\sigma_{ED}| = \left|\frac{N_{ED}}{(\pi d_{ED}^2)/4}\right| = \frac{86.67(10^3))}{(\pi d_{ED}^2)/4} \leq 160(10^6)) \quad \text{or} \quad d_{ED} \geq 26.26(10^{-3})\text{m} \quad \text{or} \quad \boxed{d_{ED} = 27 \text{ mm}}$$

$$|\sigma_{EG}| = \left|\frac{N_{EG}}{(\pi d_{EG}^2)/4}\right| = \frac{108.3(10^3))}{(\pi d_{EG}^2)/4} \leq 160(10^6)) \quad \text{or} \quad d_{EG} \geq 29.36(10^{-3})\text{m} \quad \text{or} \quad \boxed{d_{EG} = 30 \text{ mm}}$$

$$|\sigma_{EF}| = \left|\frac{N_{EF}}{(\pi d_{EF}^2)/4}\right| = \frac{65(10^3))}{(\pi d_{EF}^2)/4} \leq 160(10^6)) \quad \text{or} \quad d_{EF} \geq 22.7(10^{-3})\text{m} \quad \text{or} \quad \boxed{d_{EF} = 23 \text{ mm}}$$

1.73 The normal stress in the members of the truss shown in Fig. P1.72 is to be limited to 160 MPa in tension or compression. All members have a circular cross-section. The shear stress in the pins is to be limited to 110 MPa. Determine the minimum diameter of the pin E to the nearest millimeter and the sequence of assembly of members ED, EG, and EF.

Solution $\tau_E \leq 250\text{MPa}$ $\quad d_E = ?$

There are 6 possible assembly sequences. Three of the sequences are shown below. The remaining three are mirror images and will result in the same shear stress. All the forces appearing on the pins are known from problem 1-72.
The shear force at the imaginary cuts 1 and 2 will equal to the force at the nearest end. Thus, if smaller forces are on the outside, then the shear stress will be smaller. Assembly in Fig. (b) will yield the smallest shear stress. $\boxed{\text{The sequence of assembly shown in Fig. (b) should be used.}}$

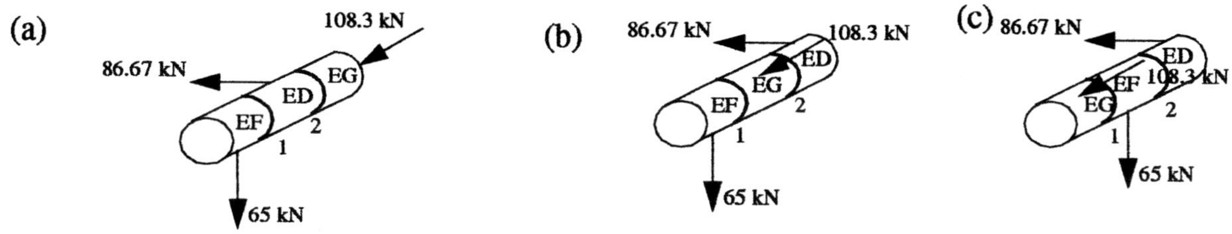

In Fig. (b) the max shear force is at cut 2 and its value is: $V = 86.67$ kN

$$\tau_E = \left|\frac{V}{(\pi d_E^2)/4}\right| = \frac{86.67(10^3)}{(\pi d_E^2)/4} \le 250(10^6) \quad \text{or} \quad d_E \ge 21.01(10^{-3})\text{m} \quad \text{or} \quad \boxed{d_E = 22 \text{ mm}}$$

1.74 The normal stress in the members of the truss shown in Fig. P1.72 is to be limited to 160 MPa in tension or compression. All members have a circular cross-section. The shear stress in the pins is to be limited to 110 MPa. Determine the minimum diameter to the nearest millimeter of members CG, CD, and CB of the truss shown in Fig. P1.72.

Solution $\sigma \le 160\text{MPa}$ $d_{CG} = ?$ $d_{CD} = ?$ $d_{CB} = ?$

Figs. (a) and (b) shows the free body diagrams of joints D and C.

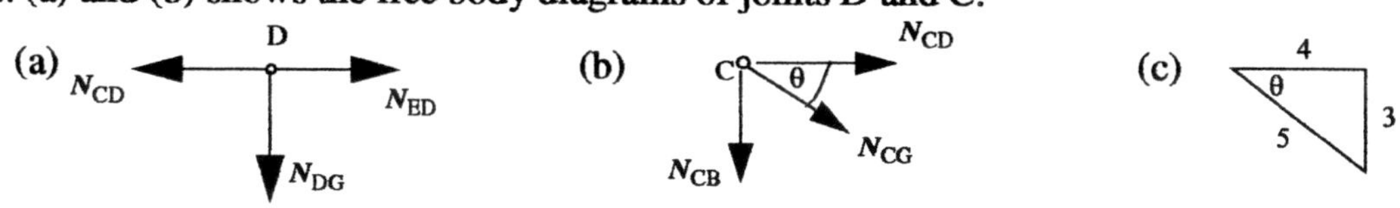

From problem 1-72 we know $N_{ED} = 86.67$ kN

By equilibrium of force in Fig. (a) in the x-direction we obtain: $N_{CD} = N_{ED} = 86.67$ kN

By equilibrium of force in Fig. (b) in the x-direction we obtain: $N_{CG}\cos\theta + N_{CD} = 0$ or $N_{CG} = -108.3$ kN

By equilibrium of force in Fig. (b) in the y-direction we obtain: $N_{CB} + N_{CG}\sin\theta = 0$ or $N_{CB} = 65.0$kN

$$|\sigma_{CG}| = \left|\frac{N_{CG}}{(\pi d_{CG}^2)/4}\right| = \frac{108.3(10^3)}{(\pi d_{CG}^2)/4} \le 160(10^6) \quad \text{or} \quad d_{CG} \ge 29.36(10^{-3})\text{m} \quad \text{or} \quad \boxed{d_{CG} = 30 \text{ mm}}$$

$$|\sigma_{CD}| = \left|\frac{N_{CD}}{(\pi d_{CD}^2)/4}\right| = \frac{86.67(10^3)}{(\pi d_{CD}^2)/4} \le 160(10^6) \quad \text{or} \quad d_{CD} \ge 26.26(10^{-3})\text{m} \quad \text{or} \quad \boxed{d_{CD} = 27 \text{ mm}}$$

$$|\sigma_{CB}| = \left|\frac{N_{CB}}{(\pi d_{CB}^2)/4}\right| = \frac{65(10^3)}{(\pi d_{CB}^2)/4} \le 160(10^6) \quad \text{or} \quad d_{CB} \ge 22.7(10^{-3})\text{m} \quad \text{or} \quad \boxed{d_{CB} = 23 \text{ mm}}$$

1.75 The normal stress in the members of the truss shown in Fig. P1.72 is to be limited to 160 MPa in tension or compression. All members have a circular cross-section. The shear stress in the pins is to be limited to 110 MPa. Determine the minimum diameter of the pin C to the nearest millimeter and the sequence of assembly of members CG, CD, and CB.

Solution $\tau_E \le 250\text{MPa}$ $d_C = ?$

Based on the argument in 1-73, the configuration that will produce the smallest maximum shear stress will be the one with members carrying the smallest forces on the outside edge of the pin. From problem 1-74 we have: $N_{CD} = 86.67$ kN (T) $N_{CG} = 108.3$ kN (C $N_{CB} = 65.0$ kN(T). Free body diagram of pin C is

shown below.

(a)

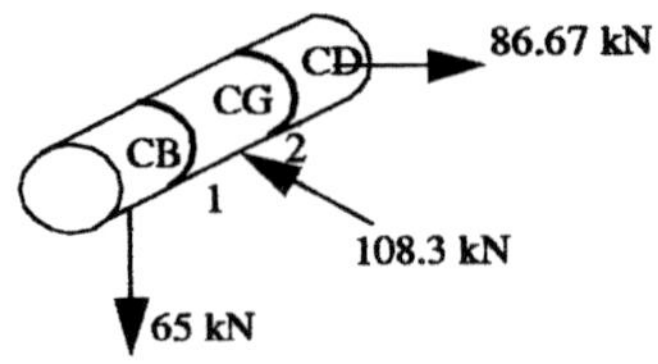

The sequence of assembly shown in Fig. (a) should be used.

In Fig. (a) the max shear force is at cut 2 and its value is: $V = 86.67$ kN

$$\tau_C = \left|\frac{V}{(\pi d_C^2)/4}\right| = \frac{86.67(10^3)}{(\pi d_C^2)/4} \le 250(10^6) \quad \text{or} \quad d_C \ge 21.01(10^{-3})\text{m} \quad \text{or} \quad \boxed{d_C = 22 \text{ mm}}$$

1.76 Truss analysis showed the following forces at joint A. Determine the sequence in which the four members at joint A should be assembled to minimize the shear stress in the pin.

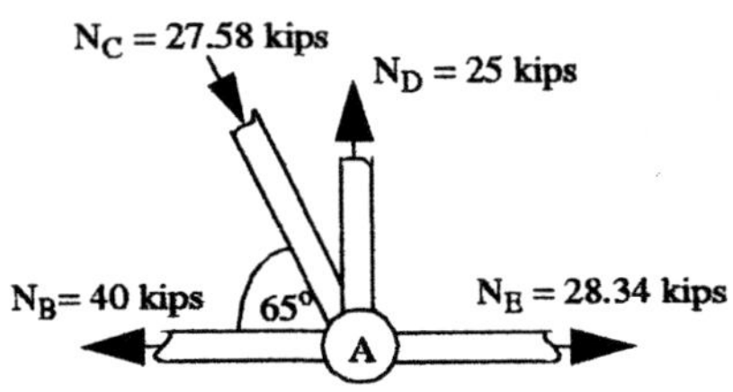

Fig. P1.76

Solution assembly sequence = ?

There are factorial four i.e., twenty-four possible ways of assembling the four member. Twelve of the twenty-four sequence are mirror images of the other twelve. To consider the twelve assembly sequence, we consider the figure below, where a member may be in any of the four location shown.

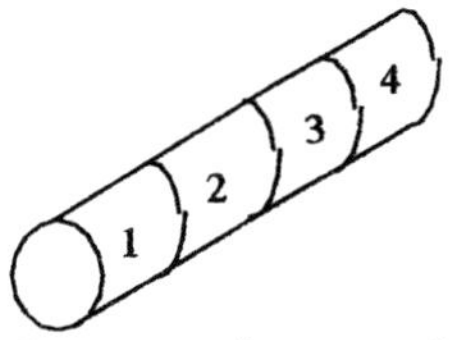

The table below describe the location of the four members and the maximum shear stress location and value of maximum shear stress for each sequence.

Sequence Number	Member Location 1	Member Location 2	Member Location 3	Member Location 4	Location of Maximum shear force between members	Maximum Shear Force (kips)
1	AB	AC	AD	AE	1 and 2	40
2	AB	AC	AE	AD	1 and 2	40
3	AC	AB	AD	AE	2 and 3	37.8
4	AC	AB	AE	AD	2 and 3	37.8
5	AB	AD	AE	AE	2 and 3	47.17
6	AB	AD	AE	AC	2 and 3	47.17
7	AD	AB	AC	AE	2 and 3	47.17
8	AD	AB	AE	AC	2 and 3	47.17
9	AB	AE	AC	AD	1 and 2	40
10	AB	AE	AD	AC	1 and 2	40
11	AE	AB	AC	AD	1 and 2	28.34
12	AE	AB	AD	AC	1 and 2	28.34

The maximum shear stress is smallest for sequence 11 and 12, thus we have the following conclusion.

The members should be assembled in the following sequence or the mirror image of the given sequence
AE, AB, AD, AC or AE, AB, AD, AC

Problem Set 1.2

1.77 Show the non-zero stress components on the A,B, and C faces of the cube.

$$\begin{bmatrix} \sigma_{xx} = 100\text{MPa(T)} & \tau_{xy} = 200\text{MPa} & \tau_{xz} = -125\text{MPa} \\ \tau_{yx} = 200\text{MPa} & \sigma_{yy} = 175\text{MPa(C)} & \tau_{yz} = 225\text{MPa} \\ \tau_{zx} = -125\text{MPa} & \tau_{zy} = 225\text{MPa} & \sigma_{zz} = 150\text{MPa(C)} \end{bmatrix}$$

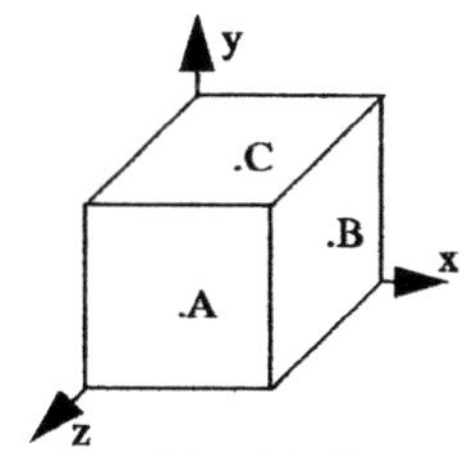

Fig. P1.77

Solution

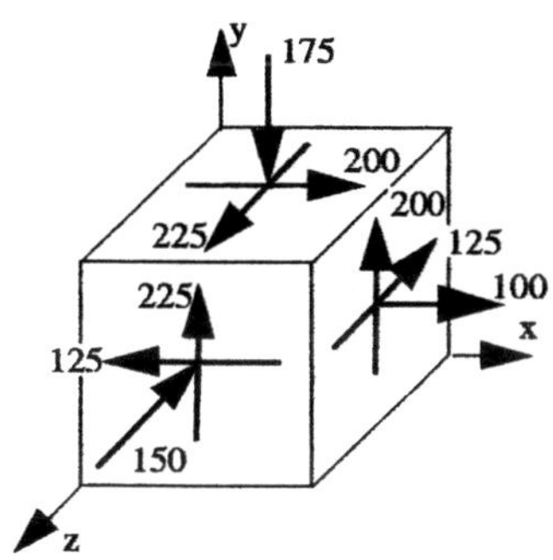

1.78 Show the non-zero stress components on the A,B, and C faces of the cube.

$$\begin{bmatrix} \sigma_{xx} = 90\text{MPa(C)} & \tau_{xy} = -200\text{MPa} & \tau_{xz} = 0 \\ \tau_{yx} = -200\text{MPa} & \sigma_{yy} = 175\text{MPa(C)} & \tau_{yz} = 225\text{MPa} \\ \tau_{zx} = 0 & \tau_{zy} = 225\text{MPa} & \sigma_{zz} = 150\text{MPa(C)} \end{bmatrix}$$

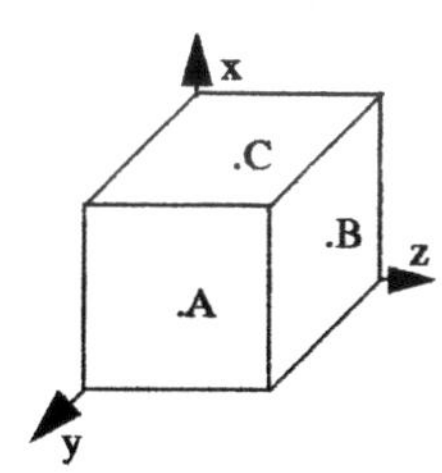

Fig. P1.78

Solution

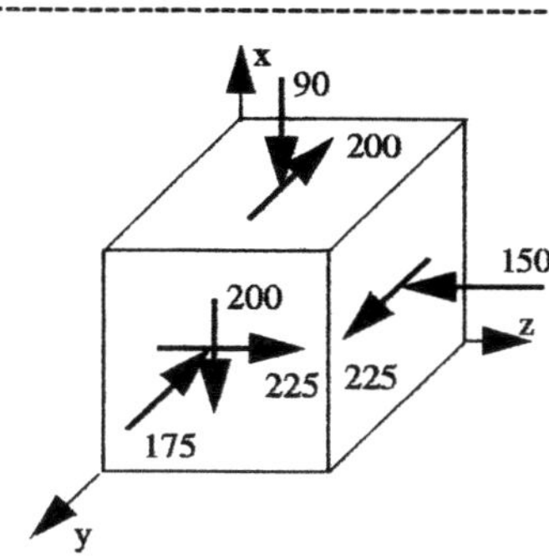

1.79 Show the non-zero stress components on the A,B, and C faces of the cube.

$$\begin{bmatrix} \sigma_{xx} = 0 & \tau_{xy} = -15\text{ksi} & \tau_{xz} = 0 \\ \tau_{yx} = -15\text{ksi} & \sigma_{yy} = 10\text{ksi(C)} & \tau_{yz} = 25\text{ksi} \\ \tau_{zx} = 0 & \tau_{zy} = 25\text{ksi} & \sigma_{zz} = 20\text{ksi(T)} \end{bmatrix}$$

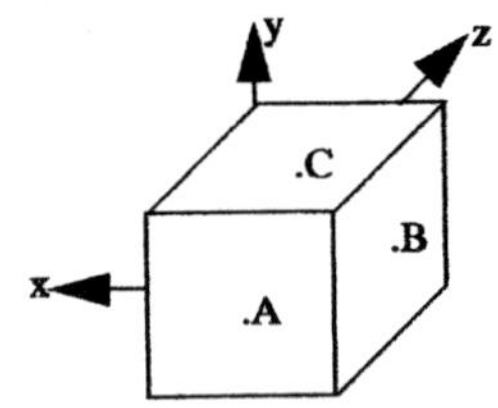

Fig. P1.79

Solution

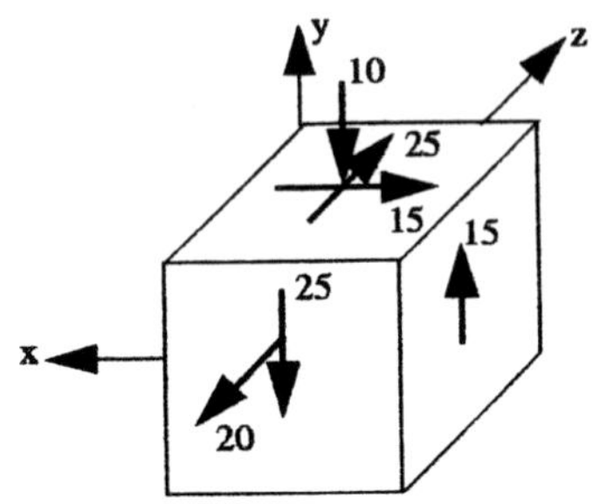

1.80 Show the non-zero stress components on the A,B, and C faces of the cube.

$$\begin{bmatrix} \sigma_{xx} = 0 & \tau_{xy} = -15\text{ksi} & \tau_{xz} = 0 \\ \tau_{yx} = -15\text{ksi} & \sigma_{yy} = 10\text{ksi(C)} & \tau_{yz} = 25\text{ksi} \\ \tau_{zx} = 0 & \tau_{zy} = 25\text{ksi} & \sigma_{zz} = 20\text{ksi(T)} \end{bmatrix}$$

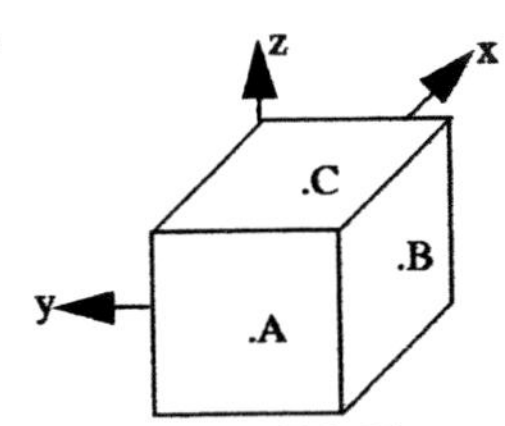

Fig. P1.80

Solution

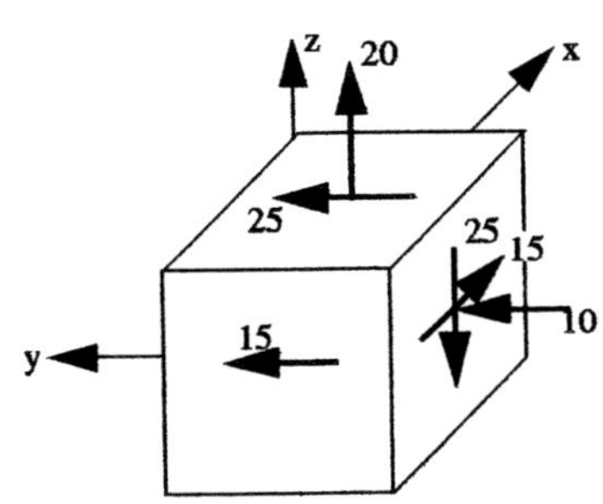

1.81 Show the non-zero stress components on the A,B, and C faces of the cube.

$$\begin{bmatrix} \sigma_{xx} = 70\text{MPa(T)} & \tau_{xy} = -40\text{MPa} & \tau_{xz} = 0 \\ \tau_{yx} = -40\text{MPa} & \sigma_{yy} = 85\text{MPa(C)} & \tau_{yz} = 0 \\ \tau_{zx} = 0 & \tau_{zy} = 0 & \sigma_{zz} = 0 \end{bmatrix}$$

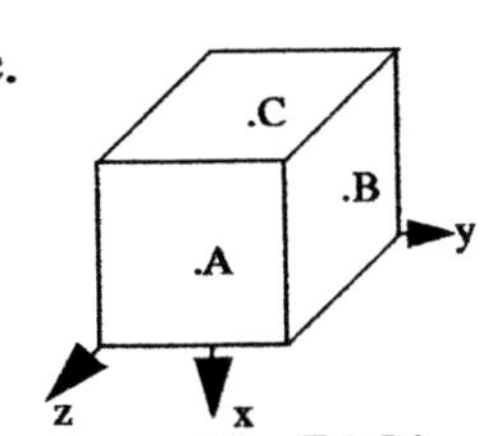

Fig. P1.81

Solution

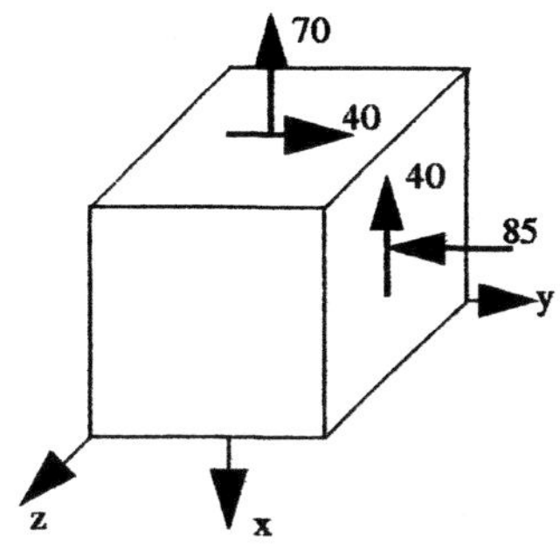

1.82 Show the non-zero stress components on the A,B, and C faces of the cube.

$$\begin{bmatrix} \sigma_{xx} = 70\text{MPa(T)} & \tau_{xy} = -40\text{MPa} & \tau_{xz} = 0 \\ \tau_{yx} = -40\text{MPa} & \sigma_{yy} = 85\text{MPa(C)} & \tau_{yz} = 0 \\ \tau_{zx} = 0 & \tau_{zy} = 0 & \sigma_{zz} = 0 \end{bmatrix}$$

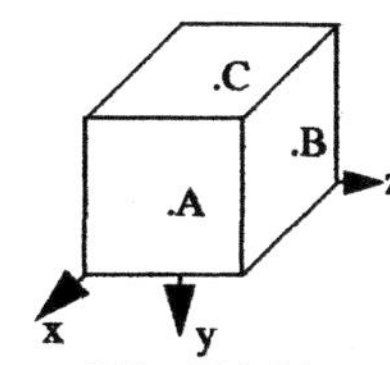

Fig. P1.82

Solution

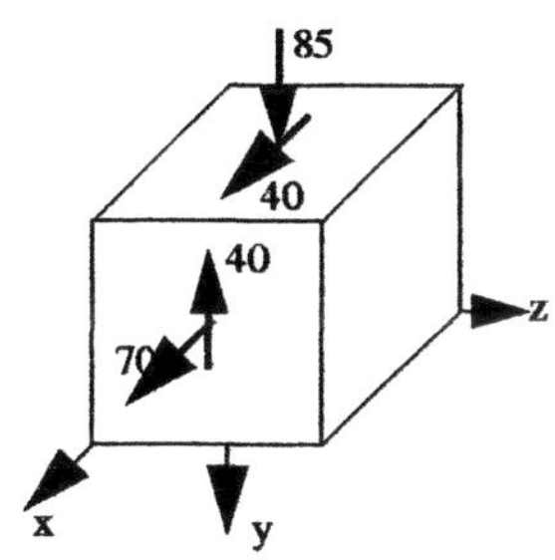

1.83 Show the stress components of a point in plane stress on the square

$$\begin{bmatrix} \sigma_{xx} = 100\text{MPa(T)} & \tau_{xy} = -75\text{MPa} \\ \tau_{yx} = -75\text{MPa} & \sigma_{yy} = 85\text{MPa(T)} \end{bmatrix}$$

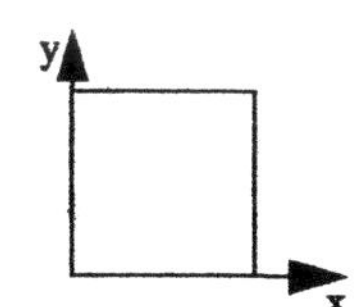

Fig. P1.83

Solution

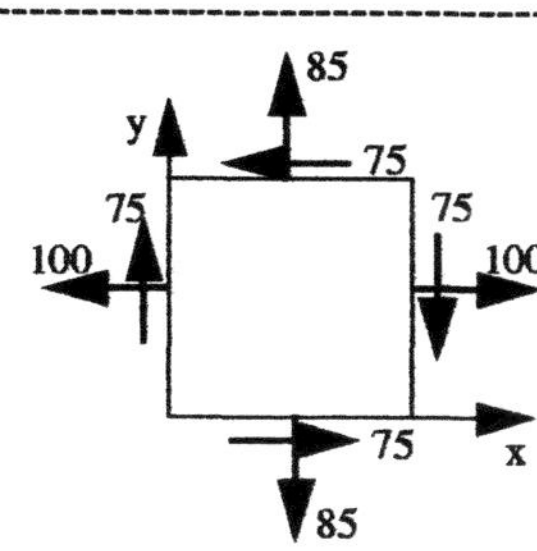

1.84 Show the stress components of a point in plane stress on the square

$$\begin{bmatrix} \sigma_{xx} = 100\text{MPa(T)} & \tau_{xy} = -75\text{MPa} \\ \tau_{yx} = -75\text{MPa} & \sigma_{yy} = 85\text{MPa(T)} \end{bmatrix}$$

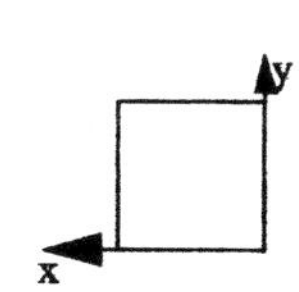

Fig. P1.84

Solution

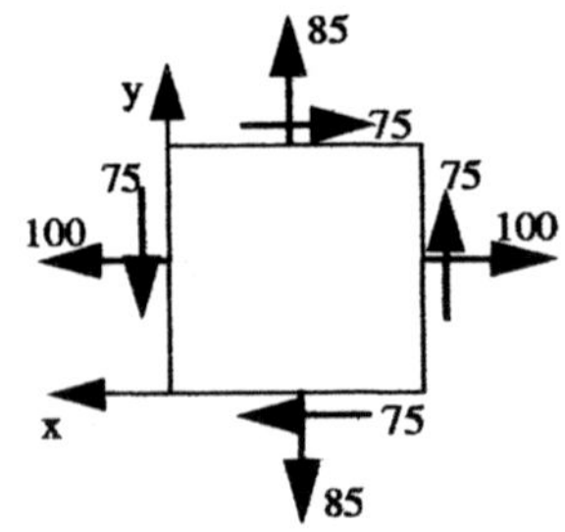

1.85 Show the stress components of a point in plane stress on.the square

$$\begin{bmatrix} \sigma_{xx} = 27\text{ksi(C)} & \tau_{xy} = 18\text{ksi} \\ \tau_{yx} = 18\text{ksi} & \sigma_{yy} = 85\text{ksi(T)} \end{bmatrix}$$

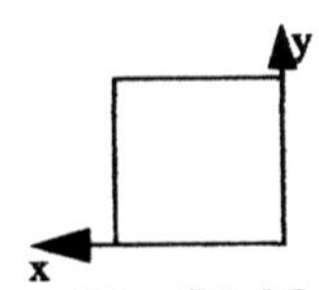

Fig. P1.85

Solution

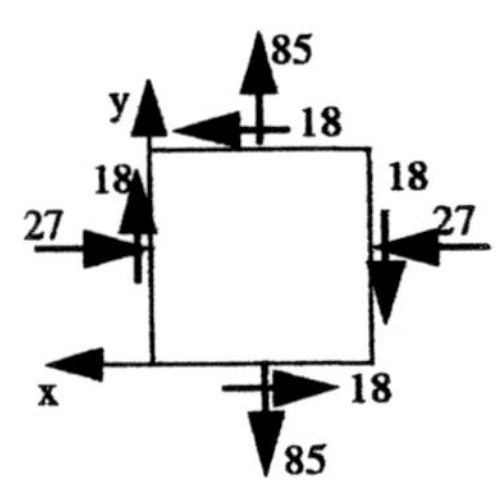

1.86 Show the stress components of a point in plane stress on.the square

$$\begin{bmatrix} \sigma_{xx} = 27\text{ksi(C)} & \tau_{xy} = 18\text{ksi} \\ \tau_{yx} = 18\text{ksi} & \sigma_{yy} = 85\text{ksi(T)} \end{bmatrix}$$

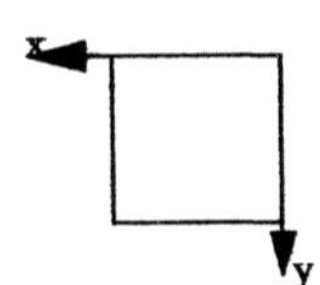

Solution

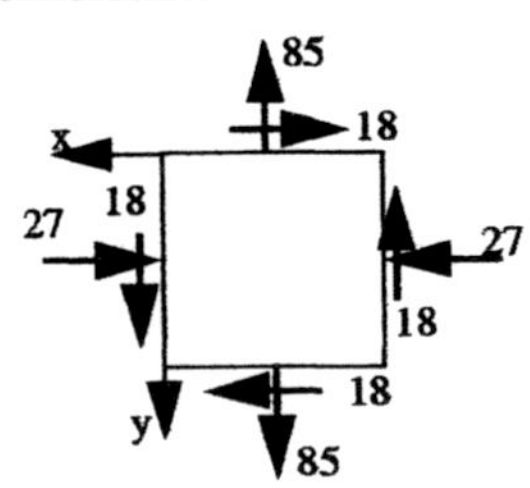

1.87 Show the non-zero stress components in the r, θ, and x cylindrical coordinate system on the A,B, and C faces of the stress element shown.

$$\begin{bmatrix} \sigma_{rr} = 145\text{MPa(C)} & \tau_{r\theta} = 100\text{MPa} & \tau_{rx} = -125\text{MPa} \\ \tau_{\theta r} = 100\text{MPa} & \sigma_{\theta\theta} = 160\text{MPa(T)} & \tau_{\theta x} = 165\text{MPa} \\ \tau_{xr} = -125\text{MPa} & \tau_{x\theta} = 165\text{MPa} & \sigma_{xx} = 150\text{MPa(T)} \end{bmatrix}$$

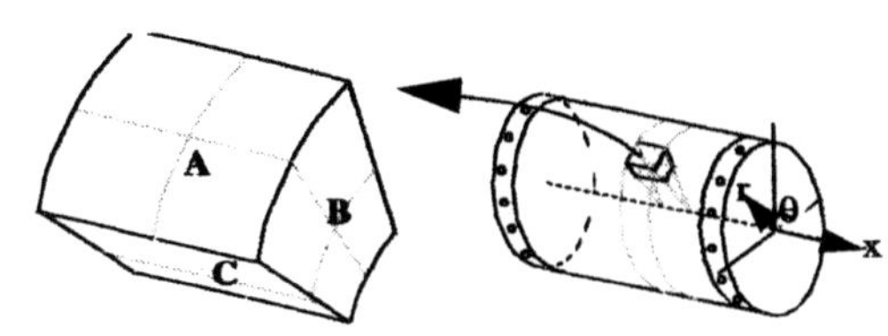

Solution

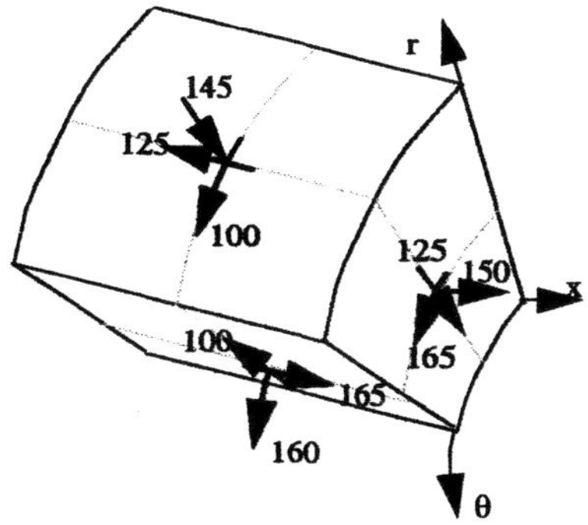

1.88 Show the non-zero stress components in the r, θ, and x cylindrical coordinate system on the A,B, and C faces of the stress element shown

$$\begin{bmatrix} \sigma_{rr} = 10\text{ksi(C)} & \tau_{r\theta} = 22\text{ksi} & \tau_{rx} = 32\text{ksi} \\ \tau_{\theta r} = 22\text{ksi} & \sigma_{\theta\theta} = 0 & \tau_{\theta x} = 25\text{ksi} \\ \tau_{xr} = 32\text{ksi} & \tau_{x\theta} = 25\text{ksi} & \sigma_{xx} = 20\text{ksi(T)} \end{bmatrix}$$

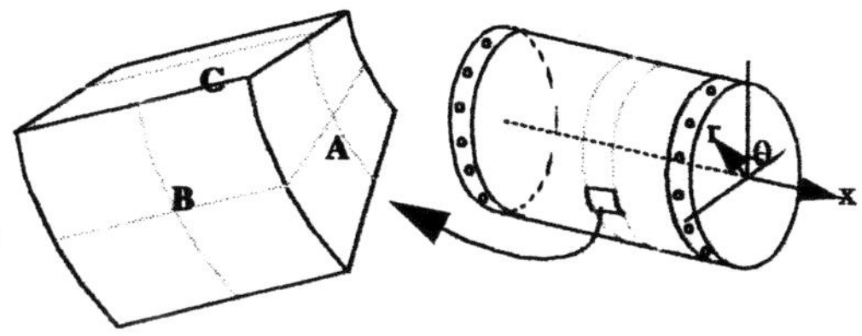

Fig.P1.88

Solution

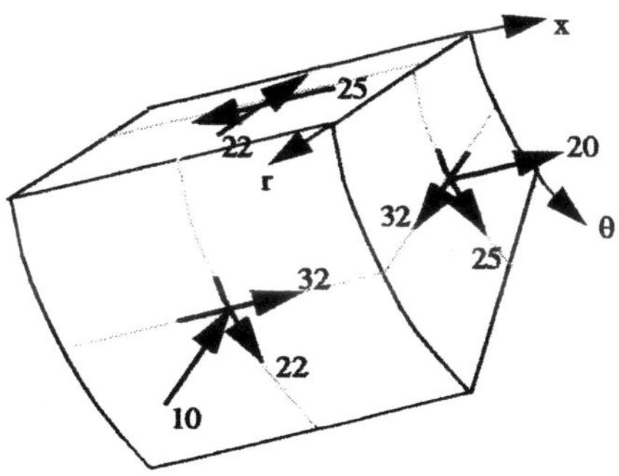

1.89 Show the stress components of a point in plane stress on the stress element in polar coordinates

$$\begin{bmatrix} \sigma_{rr} = 125\text{MPa(T)} & \tau_{r\theta} = -65\text{MPa} \\ \tau_{\theta r} = -65\text{MPa} & \sigma_{\theta\theta} = 90\text{MPa(C)} \end{bmatrix}$$

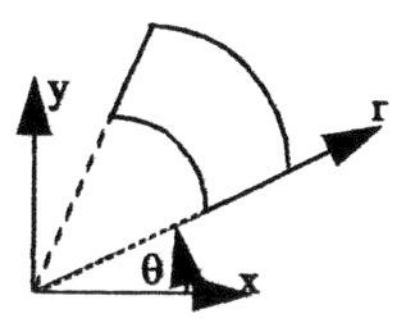

Fig. P1.89

Solution

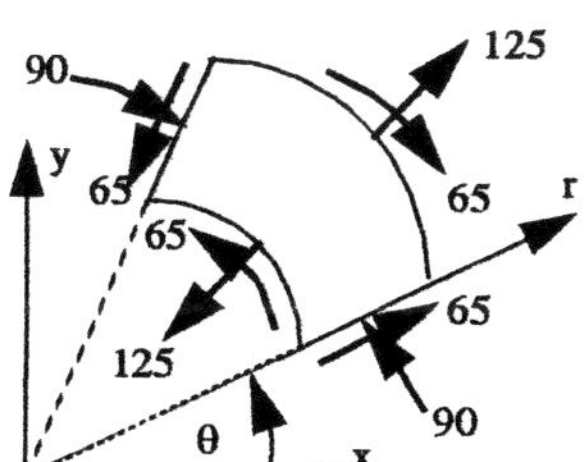

1.90 Show the stress components of a point in plane stress on the stress element in polar coordinates

$$\begin{bmatrix} \sigma_{rr} = 125\text{MPa(T)} & \tau_{r\theta} = -65\text{MPa} \\ \tau_{\theta r} = -65\text{MPa} & \sigma_{\theta\theta} = 90\text{MPa(C)} \end{bmatrix}$$

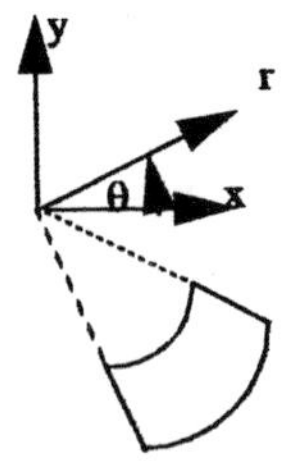

Fig. P1.90

Solution

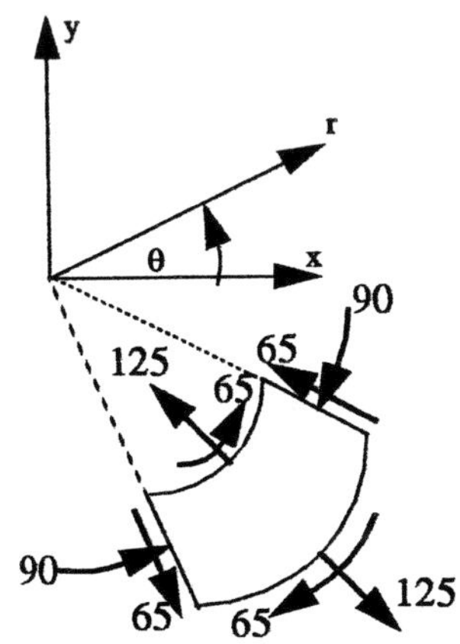

1.91 Show the stress components of a point in plane stress on the stress element in polar coordinates

$$\begin{bmatrix} \sigma_{rr} = 25\text{ksi(C)} & \tau_{r\theta} = 12\text{ksi} \\ \tau_{\theta r} = 12\text{ksi} & \sigma_{\theta\theta} = 18\text{ksi(T)} \end{bmatrix}$$

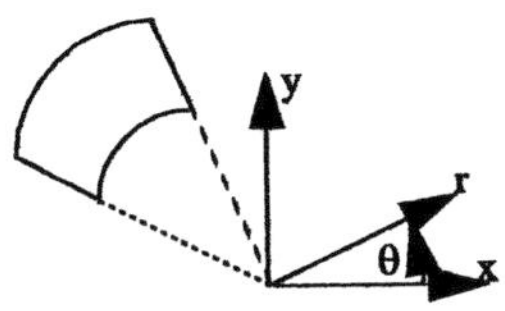

Fig. P1.91

Solution

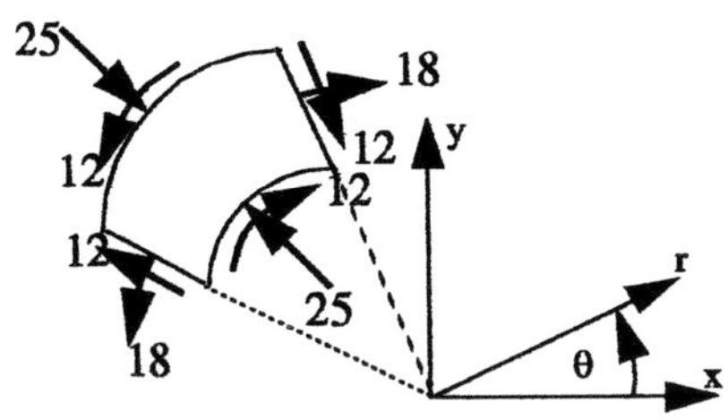

1.92 Show the stress components of a point in plane stress on the stress element in polar coordinates

$$\begin{bmatrix} \sigma_{rr} = 25\text{ksi(C)} & \tau_{r\theta} = 12\text{ksi} \\ \tau_{\theta r} = 12\text{ksi} & \sigma_{\theta\theta} = 18\text{ksi(T)} \end{bmatrix}$$

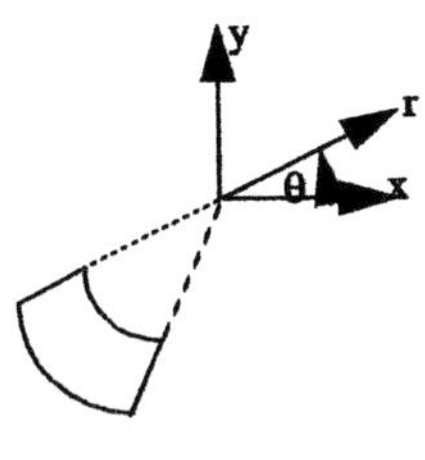

Solution

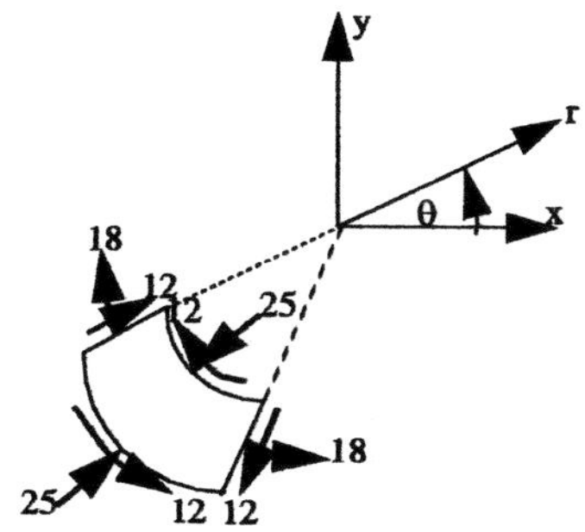

1.93 Show the non-zero stress components in the r, θ, and φ spherical coordinate system on the A,B, and C faces of the stress element shown.

$$\begin{bmatrix} \sigma_{rr} = 135\text{MPa(C)} & \tau_{r\theta} = 100\text{MPa} & \tau_{r\phi} = -125\text{MPa} \\ \tau_{\theta r} = 100\text{MPa} & \sigma_{\theta\theta} = 160\text{MPa(C)} & \tau_{\theta\phi} = 175\text{MPa} \\ \tau_{\phi r} = -125\text{MPa} & \tau_{\phi\theta} = 175\text{MPa} & \sigma_{\phi\phi} = 150\text{MPa(T)} \end{bmatrix}$$

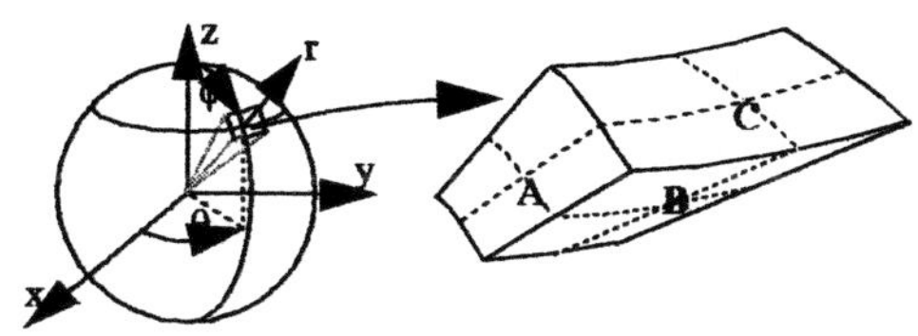

Solution

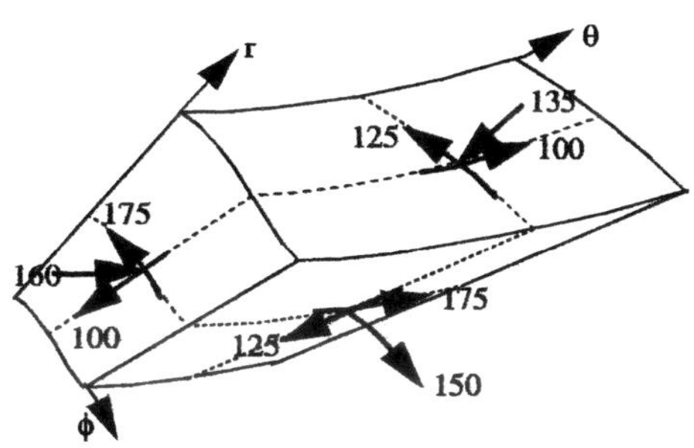

1.94 Show the non-zero stress components in the r, θ, and φ spherical coordinate system on the A,B, and C faces of the stress element shown.

$$\begin{bmatrix} \sigma_{rr} = 0 & \tau_{r\theta} = -18\text{ksi} & \tau_{r\phi} = 0\text{ksi} \\ \tau_{\theta r} = -18\text{ksi} & \sigma_{\theta\theta} = 10\text{ksi(C)} & \tau_{\theta\phi} = 25\text{ksi} \\ \tau_{\phi r} = 0 & \tau_{\phi\theta} = 25\text{ksi} & \sigma_{\phi\phi} = 20\text{ksi(T)} \end{bmatrix}$$

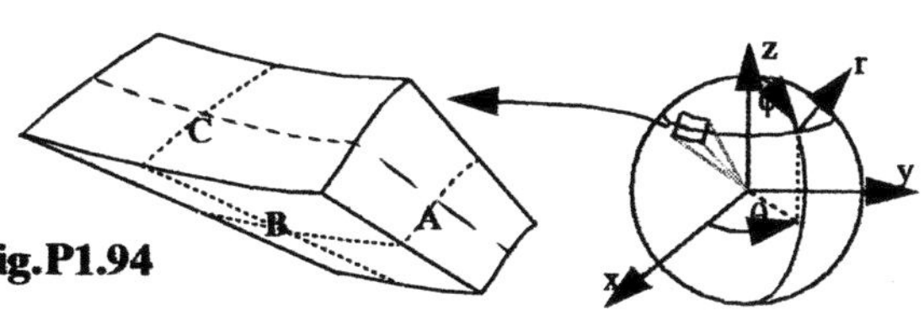

Fig.P1.94

Solution

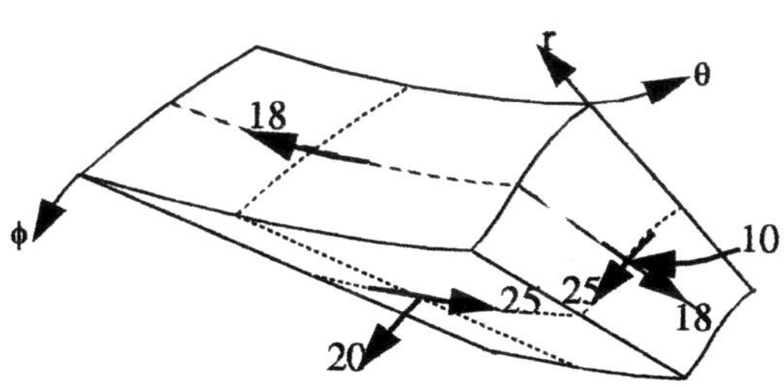

1.95 Show that the normal stress σ_{xx} on a surface can be replaced by equivalent internal normal force N and internal bending moments M_y and M_z as given below and shown in Figure 1.95.

$$N = \int_A \sigma_{xx} dA \qquad \textbf{1.5a}$$

$$M_y = -\int_A z\sigma_{xx} dA \qquad \textbf{1.5b}$$

$$M_z = -\int_A y\sigma_{xx} dA \qquad \textbf{1.5c}$$

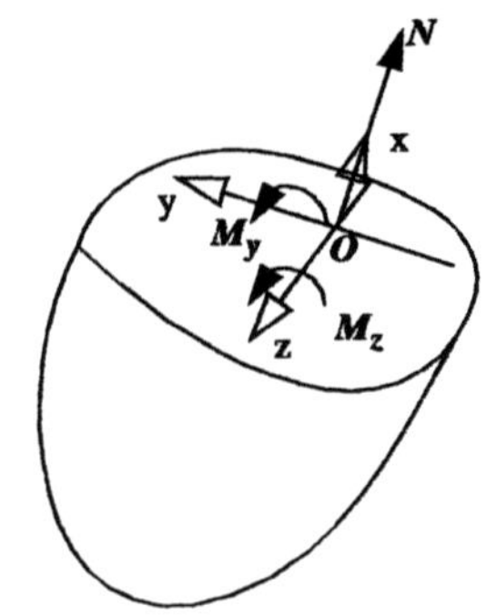

Fig. P1.95

Solution

Consider a differential area dA on which the normal force is $dN = \sigma_{xx}$ dA as shown below.

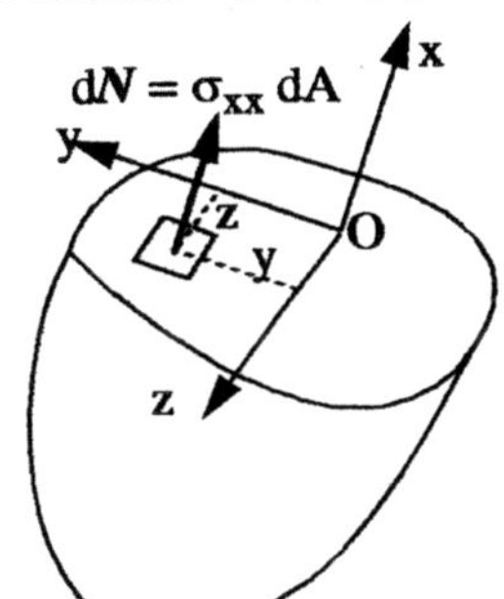

For static equivalency, the force in the x direction and the moments about the y and z axis for the above figure and figure Fig. P1.95 must be same. Thus, we obtain:

$$N = \int_A dN = \int_A \sigma_{xx} dA \qquad 1$$

$$M_y = -\int_A z dN = -\int_A z\sigma_{xx} dA \qquad 2$$

$$M_z = -\int_A y dN = -\int_A y\sigma_{xx} dA \qquad 3$$

Eq's (1),(2), and (3) are the same as Eq's 1.5a,1.5b, and 1.5c.

1.96 The normal stress on a cross-section is given by $\sigma_{xx} = a + b\,y$, where, y is measured from the centroid of the cross-section. If A is the area of cross-section, I_{zz} is the area moment of inertia about the z axis, N and M_z are the internal axial force and internal bending moment given by Equations 1.5a. and 1.5c, show the following:

$$\sigma_{xx} = \frac{N}{A} - \left(\frac{M_z}{I_{zz}}\right) y \qquad \textbf{1.6}$$

We will see the above equation in combined axial and symmetric bending problem later in the book.

Solution

Substituting $\sigma_{xx} = a + b\,y$ into Eq's 1.5a and 1.5c, we obtain the following:

$$N = \int_A (a+by)dA = a\int_A dA + b\int_A y dA \text{ and } M_z = -\int_A y(a+by)dA = -a\int_A y dA - b\int_A y^2 dA$$

The first moment of area $\int_A y dA$ is zero because the origin is at the centroid. Noting that $A = \int_A dA$ and $I_{zz} = \int_A y^2\, dA$ we obtain:

$$N = aA \quad \text{or} \quad a = N/A \quad \text{and} \quad M_z = -bI_{zz} \quad \text{or} \quad b = -(M_z/I_{zz})$$

Substituting the values of a and b in $\sigma_{xx} = a + b\,y$ we obtain Eq. 1.6.

1.97 The normal stress on a cross-section is given by $\sigma_{xx} = a + b\,y + c\,z$, where y and z are measured from the centroid of the cross-section. If A is the area of cross-section, I_{yy}, I_{zz}, and I_{yz} are the area moment of inertia, N, M_y and M_z are the internal axial force and internal bending moments given by Equations 1.5a., 1.5b, and 1.5c, show the following:

$$\sigma_{xx} = \frac{N}{A} - \left(\frac{M_z I_{yy} - M_y I_{yz}}{I_{yy}I_{zz} - I_{yz}^2}\right) y - \left(\frac{M_y I_{zz} - M_z I_{yz}}{I_{yy}I_{zz} - I_{yz}^2}\right) z \qquad \mathbf{1.7}$$

The above equation is used in unsymmetrical bending of beams. Note that if either y or z is an axis of symmetric then $I_{yz} = 0$. In such a case the above equation simplifies considerably.

Solution

Substituting $\sigma_{xx} = a + b\,y + c\,z$, into Eq's 1.5a, 1.5b, and 1.5c, we obtain the following:

$$N = \int_A (a+by+cz)dA = a\int_A dA + b\int_A y dA + c\int_A z dA \qquad \mathbf{1}$$

$$M_y = -\int_A z(a+by+cz)dA = -a\int_A z dA - b\int_A yz dA - c\int_A z^2 dA \qquad \mathbf{2}$$

$$M_z = -\int_A y(a+by+cz)dA = -a\int_A y dA - b\int_A y^2 dA - c\int_A yz dA \qquad \mathbf{3}$$

The first moment of area $\int_A y dA$ and $\int_A z dA$ are zero because the origin is at the centroid. Noting that $A = \int_A dA$, and $I_{zz} = \int_A y^2\, dA$, $I_{yy} = \int_A z^2\, dA$, $I_{yz} = \int_A yz\, dA$ we obtain:

$$N = aA \quad \text{or} \quad a = \frac{N}{A} \qquad \mathbf{4}$$

$$M_y = -bI_{yz} - cI_{yy} \qquad \mathbf{5}$$

$$M_z = -bI_{zz} - cI_{yz} \qquad \mathbf{6}$$

Solving Eq's (5) and (6) for the constants b and c, we obtain:

$$b = -\left(\frac{M_z I_{yy} - M_y I_{yz}}{I_{yy}I_{zz} - I_{yz}^2}\right) \qquad c = -\left(\frac{M_y I_{zz} - M_z I_{yz}}{I_{yy}I_{zz} - I_{yz}^2}\right) \qquad \mathbf{7}$$

Substituting the values of a, b, and c in $\sigma_{xx} = a + by + cz$, we obtain Eq. 1.7.

1.98 An infinitesimal element in plane stress is shown. F_x and F_y are the body forces acting at the point and have the dimensions of force per unit volume. By converting stresses into forces and writing equilibrium equations show

$$\frac{\partial \sigma_{xx}}{\partial x} + \frac{\partial \tau_{yx}}{\partial y} + F_x = 0 \qquad \textbf{1.8a}$$

$$\frac{\partial \tau_{xy}}{\partial x} + \frac{\partial \sigma_{yy}}{\partial y} + F_y = 0 \qquad \textbf{1.8b}$$

$$\tau_{xy} = \tau_{yx} \qquad \textbf{1.8c}$$

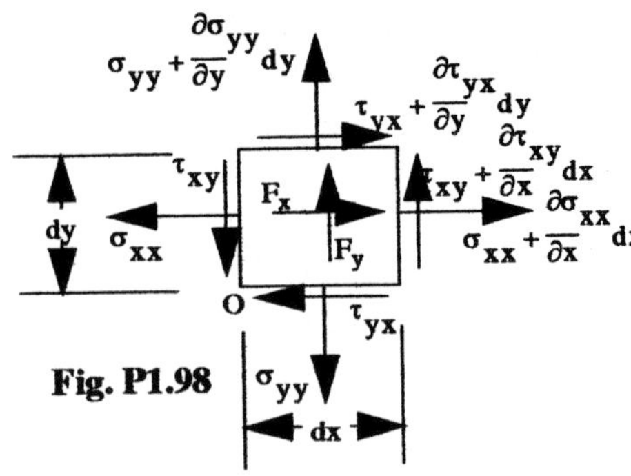

Fig. P1.98

Solution

Converting stresses into forces by multiplying by the areas of the plane on which they act and multiplying body force by the differential volume we obtain the following free body diagram below.
By force Equilibrium in the x-direction we obtain:

$$\left(\sigma_{xx} + \frac{\partial \sigma_{xx}}{\partial x}dx\right)(dydz) - (\sigma_{xx})(dydz) + \left(\tau_{yx} + \frac{\partial \tau_{yx}}{\partial y}dy\right)(dxdz) - (\tau_{yx})(dxdz) + (F_x)(dxdydz) = 0 \text{ or}$$

$$\frac{\partial \sigma_{xx}}{\partial x}(dxdydz) + \frac{\partial \tau_{yx}}{\partial y}(dxdydz) + (F_x)(dxdydz) = 0 \qquad 1$$

Noting that (dxdydz) is a common factor and can be removed, we obtain Eq. 1.8a.

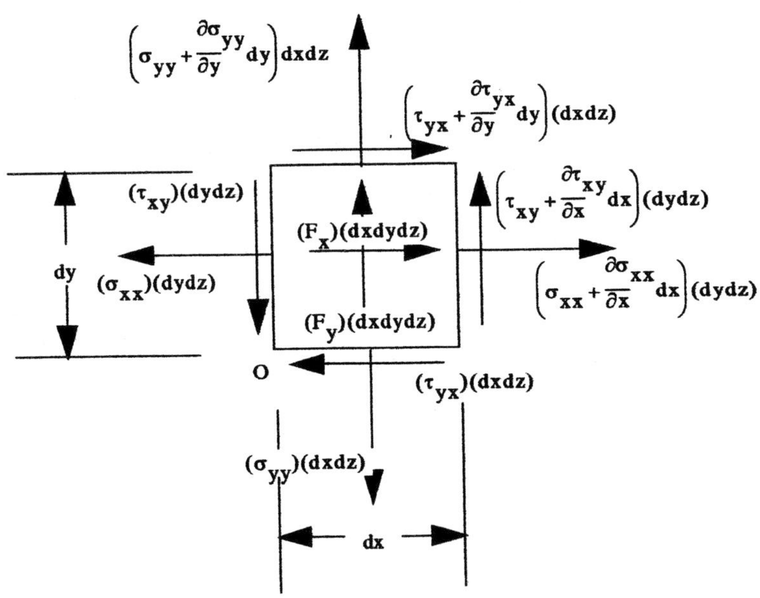

By force equilibrium in the y-direction we obtain:

$$\left(\sigma_{yy} + \frac{\partial \sigma_{yy}}{\partial y}dy\right)dxdz - (\sigma_{yy})(dxdz) + \left(\tau_{xy} + \frac{\partial \tau_{xy}}{\partial x}dx\right)(dydz) - (\tau_{xy})(dydz) + (F_y)(dxdydz) = 0$$

$$\frac{\partial \sigma_{yy}}{\partial y}(dxdydz) + \frac{\partial \tau_{xy}}{\partial x}(dxdydz) + (F_y)(dxdydz) = 0 \qquad 2$$

We now consider moment equilibrium about point about the center of the differential element. The body forces do not produce any moment as these pass through the center. The forces from the normal stresses do not produce any moment as these also pass through the center of the differential element. The moment from the variation of shear stresses can be written as

$$\left(\tau_{xy}+\frac{\partial\tau_{xy}}{\partial x}dx\right)(dydz)\left(\frac{dx}{2}\right)+(\tau_{xy})(dydz)\left(\frac{dx}{2}\right)-\left(\tau_{yx}+\frac{\partial\tau_{yx}}{\partial y}dy\right)(dxdz)\left(\frac{dy}{2}\right)-(\tau_{yx})(dxdz)\left(\frac{dy}{2}\right)=0$$

Neglecting the terms containing the product of four differentials as these term tend to zero faster in the limit than the terms with product of three differentials. We obtain: $(\tau_{xy})(dydz)(dx)-(\tau_{yx})(dxdz)(dy) = 0$. Removing the common term (dxdydz) we obtain Eq. 1.8a.

Problem Set 2.1

2.1 A 80 cm stretch cord is used to tie the rear of the canoe to the car hook as shown. In the stretched position the cord forms the side AB of the triangle shown. Determine the average normal strain in the stretch cord.

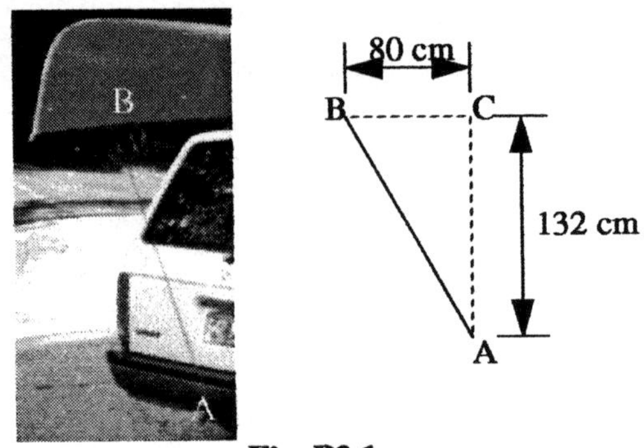

Fig. P2.1

Solution L_o=80cm ε=?

The final length of the cord is: $L_f = AB = \sqrt{AC^2 + BC^2} = \sqrt{80^2 + 132^2} = 154.35\text{cm}$

The average normal strain is: $\varepsilon = \dfrac{L_f - L_o}{L_o} = \dfrac{154.35 - 80}{80} = 0.9294$ or $\boxed{\varepsilon = 0.9294\ \text{cm/cm}}$

2.2 The diameter of a spherical balloon changes from 250 mm to 252 mm. Determine the change in the average circumferential normal strain.

Fig. P2.2

Solution d_o=250mm d_f=252mm ε=?

The average normal strain is: $\varepsilon = \dfrac{\pi d_f - \pi d_o}{\pi d_o} = \dfrac{2}{250} = 0.008$ or $\boxed{\varepsilon = 0.008\dfrac{\text{mm}}{\text{mm}}}$

2.3 Two rubber bands are used for packing an air mattress for camping. The undeformed length of the rubber band is seven inches. Determine the average normal strain in the rubber bands if the diameter of the mattress is 4.1 inch at the section where the rubber bands are on the mattress.

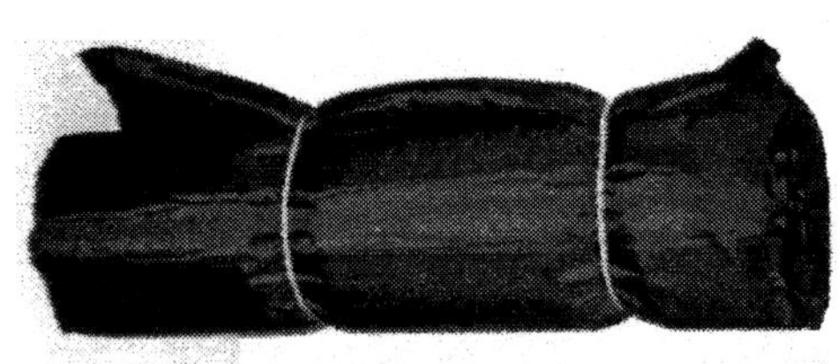

Fig. P2.3

Solution L_o=7in d=4.1in ε=?

The final length is: $L_f = \pi d = 4.1\pi = 12.88$

The average normal strain is: $\varepsilon = \dfrac{L_f - L_o}{L_o} = \dfrac{12.88 - 7}{7} = 0.8401$ $\boxed{\varepsilon = 0.8401\ \text{in/in}}$

2.4 A canoe on top of a car is tied down using rubber stretch cords as shown. The undeformed length of the stretch cord is 40 inches. Determine the average normal strain in the stretch cord assuming that the path of the stretch cord over the canoe can be approximated as shown below.The average normal strain can be found as shown below.

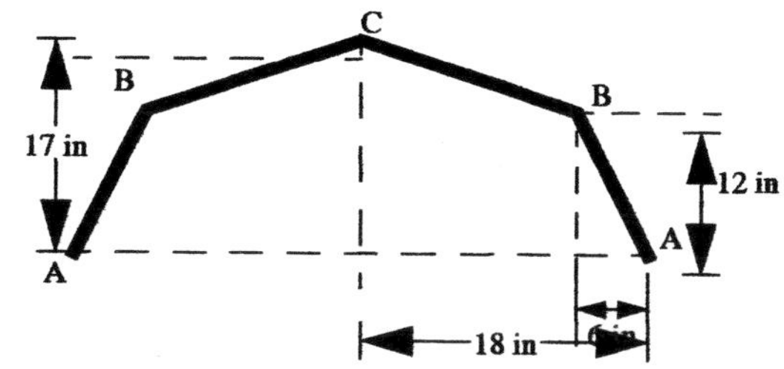

Solution L_o=40 in ε=?

The figure below can be used for calculating the lengths of segments AB and BC as

$$AB = \sqrt{6^2 + 12^2} = 13.416 \text{ and } BC = \sqrt{5^2 + 12^2} = 13$$

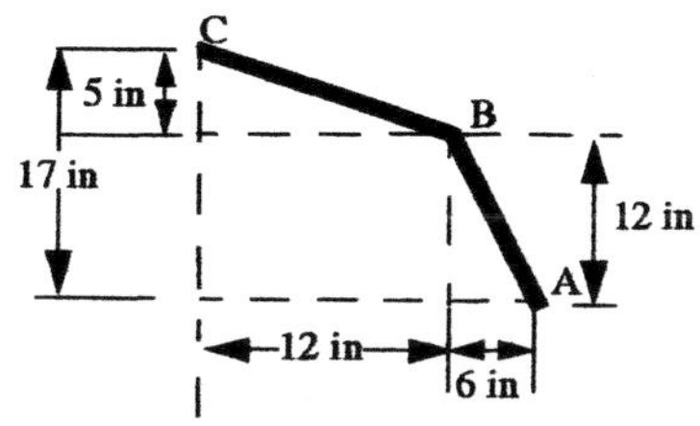

The final length of the cord is: $L_f = 2(AB + BC) = 52.83$

The average normal strain is: $\varepsilon = \dfrac{L_f - L_o}{L_o} = \dfrac{52.83 - 40}{40} = 0.3208$ or $\boxed{\varepsilon = 0.321 \text{ in/in}}$

2.5 Due to the application of the forces, the displacement of the rigid plates in the x direction were observed as given below. Determine the axial strains in rods in sections AB, BC, and CD.

$u_B = -1.8$ mm $u_C = 0.7$ mm $u_D = 3.7$ mm

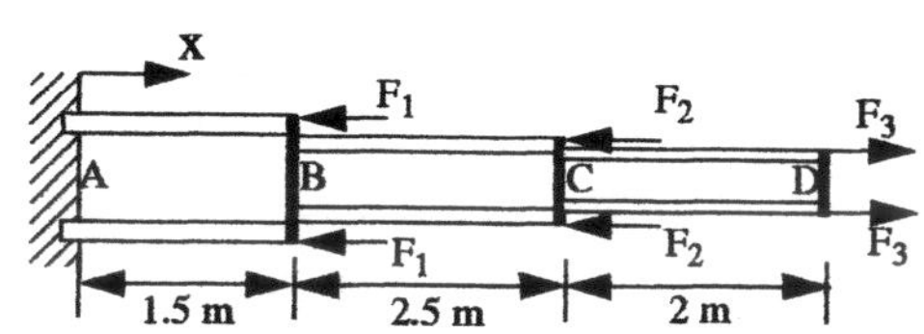

Solution ε_{AB}=? ε_{BC}=? ε_{CD}=?

The average normal strains in the rods can be found as shown below.

$$\varepsilon_{AB} = \frac{u_B - u_A}{(x_B - x_A)} = \frac{(-1.8 - 0)}{(1.5)(10^3)} = -1.2(10^{-3}) \quad \text{or} \quad \boxed{\varepsilon_{AB} = -1200\mu\frac{mm}{mm}}$$

$$\varepsilon_{BC} = \frac{u_C - u_B}{(x_C - x_B)} = \frac{0.7 - (-1.8)}{2.5(10^3)} = 1(10^{-3}) \quad \text{or} \quad \boxed{\varepsilon_{BC} = 1000\mu\frac{mm}{mm}}$$

$$\varepsilon_{CD} = \frac{u_D - u_C}{(x_D - x_C)} = \frac{3.7 - (0.7)}{2(10^3)} = 1.5(10^{-3}) \quad \text{or} \quad \boxed{\varepsilon_{CD} = 1500\mu\frac{mm}{mm}}$$

2.6 Due to the application of the forces, the average normal strains in the bars were found to be as given below. Determine the movement of point at D with respect to the left wall.

$\varepsilon_{AB} = -800\mu$ $\varepsilon_{BC} = 600\mu$ $\varepsilon_{CD} = 1100\mu$

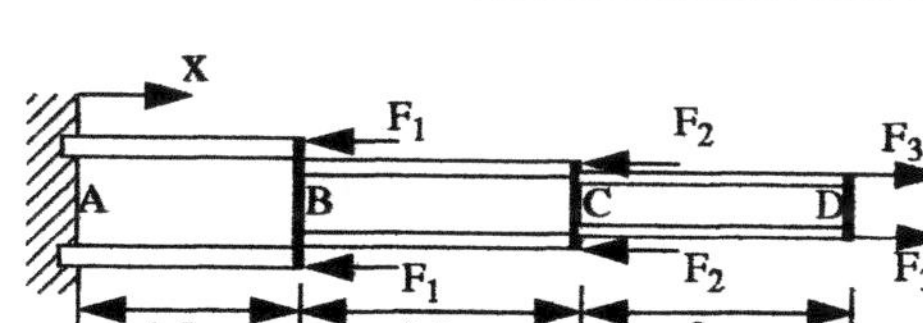

Fig. P2.6

Solution $u_D - u_A = ?$

The relative displacements of the ends of each rod can be found as shown below.

$$u_B - u_A = \varepsilon_{AB}(x_B - x_A) = (-800)(10^{-6})(1.5)(10^3) = -1.2\text{mm}$$ 1

$$u_C - u_B = \varepsilon_{BC}(x_C - x_B) = (600)(10^{-6})(2.5)(10^3) = 1.5\text{mm}$$ 2

$$u_D - u_C = \varepsilon_{CD}(x_D - x_C) = (1100)(10^{-6})(2)(10^3) = 2.2\text{mm}$$ 3

Adding (1),(2) and (3): $u_D - u_A = 2.2 + 1.5 - 1.2 = 2.5\text{mm}$ or $\boxed{u_D - u_A = 2.5\text{mm}}$

2.7 Due to the application of the forces, the rigid plate in Fig. P2.8 is observed to move 0.0236 inch to the right. Determine the average normal strain in bars A and B.

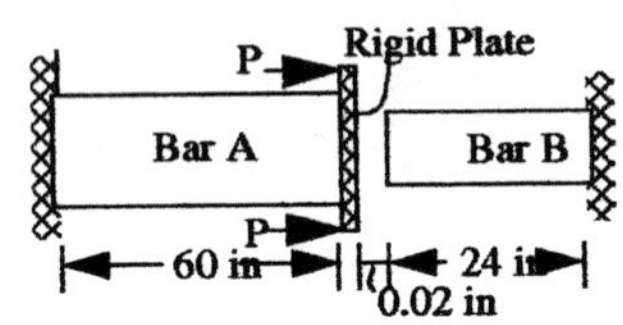

Solution $\varepsilon_A = ?$ $\varepsilon_B = ?$

The exaggerated deformed geometry can be drawn as shown below.

(a)

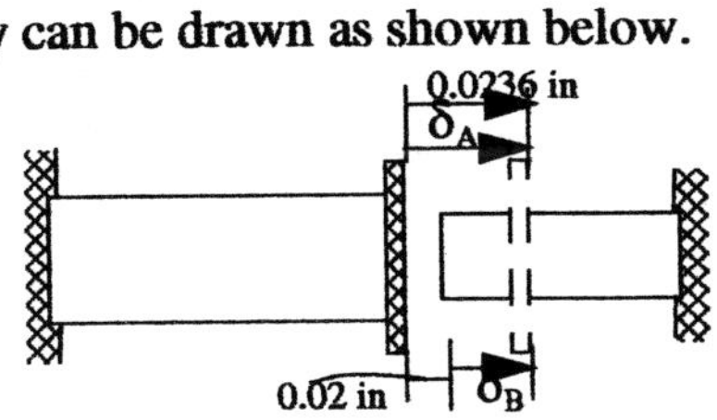

From the above figure we obtain the following. $\delta_A = 0.0236$

$\delta_B = \delta_A - 0.02 = 0.0036(\text{Contraction})$

The average normal strain can be found as shown below.

$$\varepsilon_A = \frac{\delta_A}{L_A} = \frac{0.0236}{60} = 393.3(10^{-6})$$ $\boxed{\varepsilon_A = 393.3\mu\frac{\text{in}}{\text{in}}}$

$$\varepsilon_B = \frac{\delta_B}{L_B} = \frac{-0.0036}{24} = -1500(10^{-6})$$ $\boxed{\varepsilon_B = -1500\mu\frac{\text{in}}{\text{in}}}$

2.8 Due to the application of the forces, the average normal strain in bar A in Fig. P2.8 was found to be 2000μ in/in. Determine the normal strain in bar B.

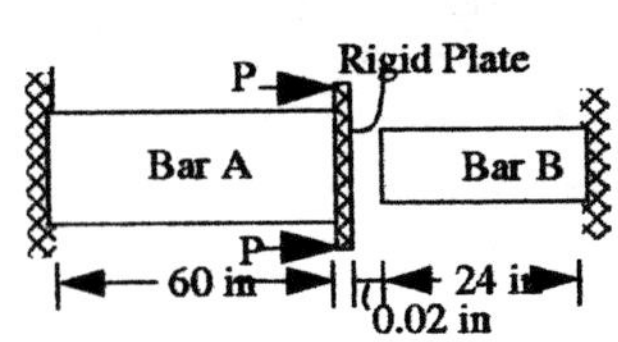

Fig. P2.8

Solution $\varepsilon_A = 2000\mu$ $\varepsilon_B = ?$

The deformation of bar a is: $\delta_A = \varepsilon_A L_A = (2000)(10^{-6})(60) = 120(10^{-3}) = 0.120\text{in}$

From Fig(a) in problem 2.7: $\delta_B = \delta_A - 0.02 = 0.10\text{in}(\text{Contraction})$

$$\varepsilon_B = \frac{\delta_B}{L_B} = \frac{-0.1}{24} = -4167(10^{-6})$$ $\boxed{\varepsilon_B = -4167\ \mu\frac{\text{in}}{\text{in}}}$

2.9 Due to the application of force P, point B in figure shown was observed to move upwards by 0.06 inch. If length of bar A is 24 inches determine the average normal strain in bar A.

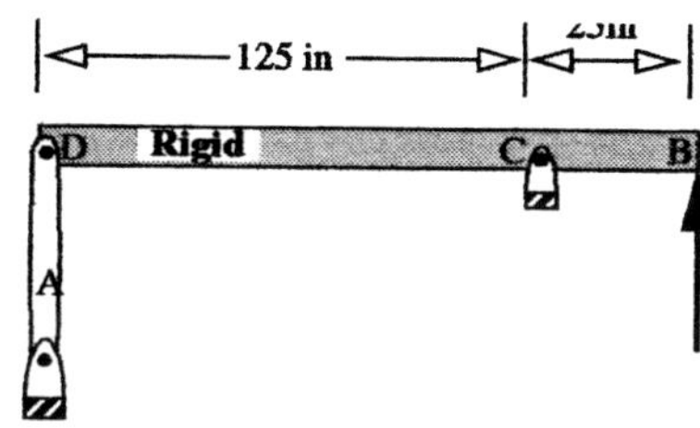

Solution $\delta_B = 0.06$ inch $\qquad L_A = 24$ inches $\qquad \varepsilon_A = ?$

We draw the exaggerated deformed shape as shown below.

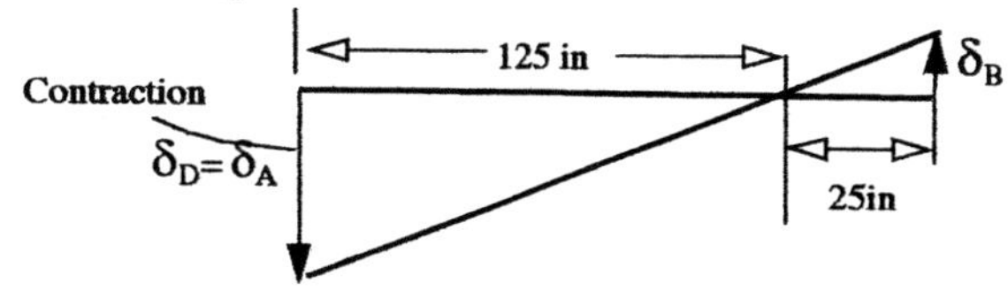

Using similar triangle, the deformation equations can be written as given below.

$$\frac{\delta_A}{125} = \frac{\delta_B}{25} \qquad \text{or} \qquad \delta_A = 5\delta_B = 5(0.06) = 0.3 \text{ contraction} \qquad (1)$$

The average normal strain can be found as shown below.

$$\varepsilon_A = \frac{\delta_A}{L_A} = -\left(\frac{0.3}{24}\right) = -0.0125$$

$$\boxed{\varepsilon_A = -0.0125 \text{ in/in}}$$

2.10 Due to the application of force P, the average normal strain in bar A in Fig. P2.10 was found to be - 6000 μ. If length of bar A is 36 inches determine the movement of point B.

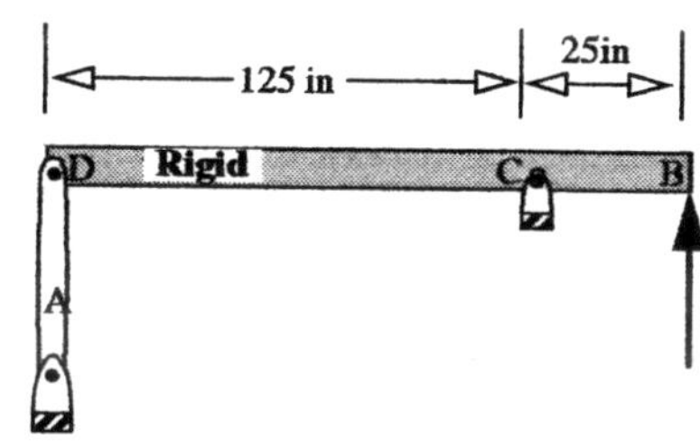

Fig. P2.10

Solution $\varepsilon_A = -6000\mu$ $\qquad L_A = 36$ inches $\qquad \delta_B = ?$

The deformation of A can be found as: $\delta_A = \varepsilon_A L_A = (6000)(10^{-6})(36) = 0.216\text{inch contraction}$

From Eq.(1) in problem 2.9 we obtain:

$$\delta_B = \left(\frac{25}{125}\right)\delta_A = \frac{0.216}{5} = 0.0432\text{inch or}$$

$$\boxed{\delta_B = 0.0432\text{inch upwards}}$$

2.11 Due to the application of force P, point B in figure shown was observed to move upwards by 0.06 inch. If length of bar A is 24 inches, determine the average normal strain in bar A.

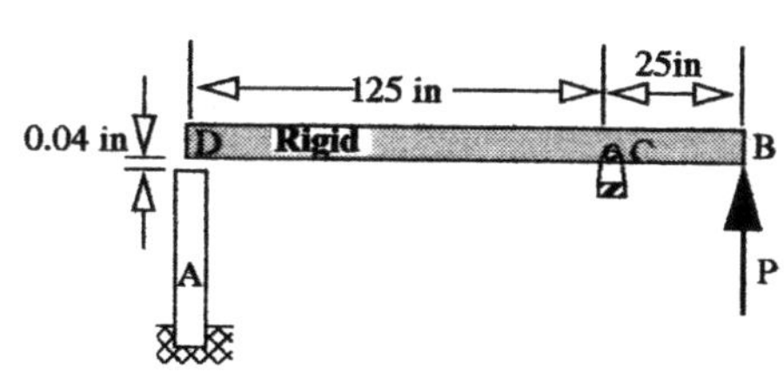

Solution $\delta_B = 0.06$ inch $\qquad L_A = 24$ inches $\qquad \varepsilon_A = ?$

Consider point D on the rigid bar. We draw the exaggerated deformed shape as shown below.

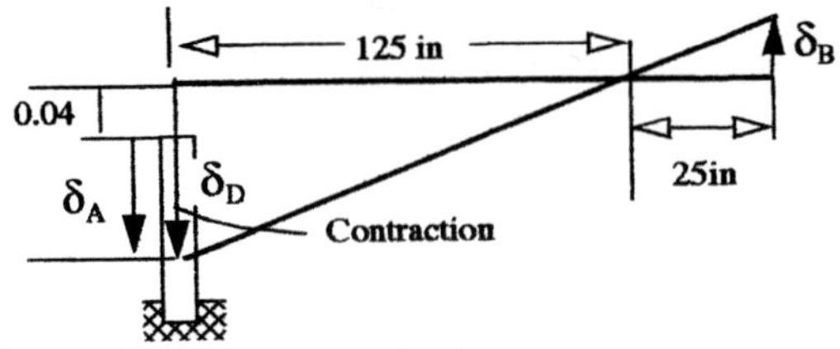

The deformation equations can be written as shown below.

$$\delta_D = \delta_A + 0.04 \qquad 1$$

Using similar triangle, the deformation equations can be written as given below.

$$\frac{\delta_D}{125} = \frac{\delta_B}{25} \quad \text{or} \quad \frac{\delta_A + 0.04}{125} = \frac{\delta_B}{25} \quad \text{or} \qquad 2$$

$\delta_A = 5\delta_B - 0.04 = 5(0.06) - 0.04 = \quad = 0.26\text{inch contraction}$

The average normal strain can be found as shown below.

$$\varepsilon_A = \frac{\delta_A}{L_A} = -\left(\frac{0.26}{24}\right) = -0.0108 \qquad \boxed{\varepsilon_A = -0.0108 \text{ in/in}}$$

2.12 Due to the application of force P, the average normal strain in bar A in Fig. P2.12 was found to be - 6000 μ. If the length of bar A is 36 inches, determine the movement of point B.

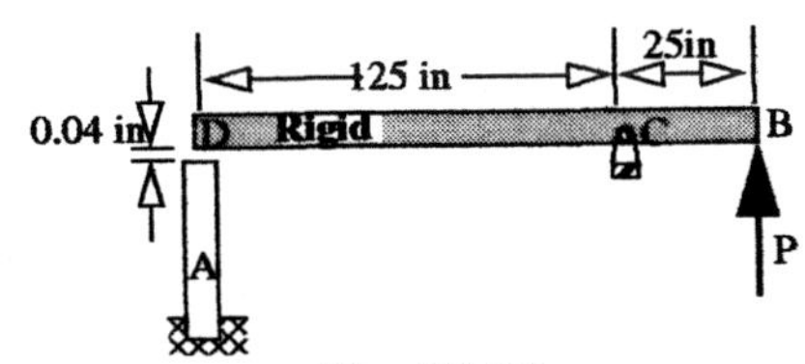

Fig. P2.12

Solution $\varepsilon_A = -6000\mu$ $\quad L_A = 36$ inches $\quad \delta_B = ?$

The deformation of A can be found as: $\delta_A = \varepsilon_A L_A = (6000)(10^{-6})(36) = 0.216\text{inch contraction}$

From Eq.(2) in problem 2.11 we obtain:

$$\delta_B = \frac{\delta_A + 0.04}{5} = \frac{0.216 + 0.04}{5} = 0.0512\text{inch or} \qquad \boxed{\delta_B = 0.0512\text{inch upwards}}$$

2.13 Due to the application of force P, point B in figure shown was observed to move upwards by 0.06 inch. If lengths of bars A and F are 24 inches, determine the average normal strain in bar A and F.

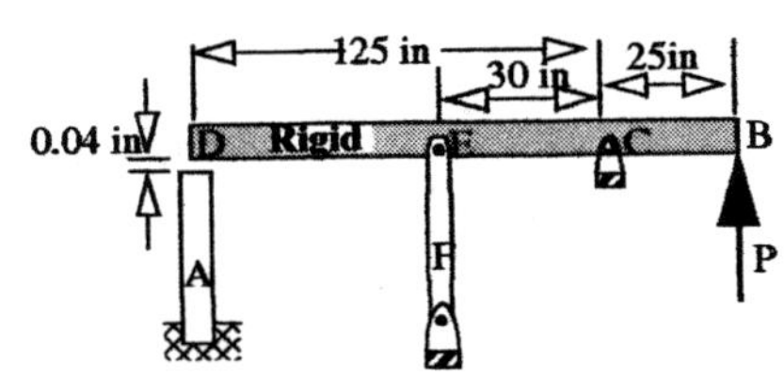

Solution $\delta_B = 0.06$ $\quad L_A = 24$ inches $\quad L_F = 24$ inches $\quad \varepsilon_A = ?$ $\quad \varepsilon_F = ?$

Consider point D on the rigid bar. We draw the exaggerated deformed shape as shown below.

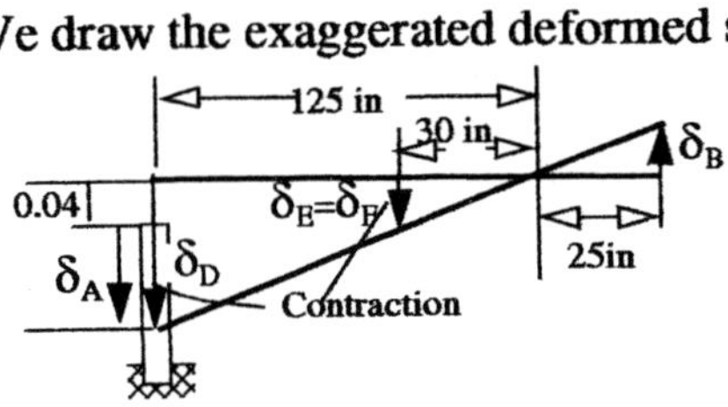

The deformation equations can be written as shown below.

$$\delta_D = \delta_A + 0.04 \qquad 1$$

Using similar triangle, the deformation equations can be written as given below.

$$\frac{\delta_D}{125} = \frac{\delta_E}{30} = \frac{\delta_B}{25} \quad \text{or} \quad \frac{\delta_A + 0.04}{125} = \frac{\delta_F}{30} = \frac{\delta_B}{25} \quad \text{or} \qquad 2$$

$\delta_A = 5\delta_B - 0.04 = 5(0.06) - 0.04 = \;= 0.26\text{inch contraction}$ and

$\delta_F = \frac{30}{25}\delta_B = \frac{6}{5}(0.06) = 0.072\text{inch contraction}$

The average normal strain can be found as shown below.

$\varepsilon_A = \frac{\delta_A}{L_A} = -\left(\frac{0.26}{24}\right) = -0.0108$ $\boxed{\varepsilon_A = -0.0108 \text{ in/in}}$

$\varepsilon_F = \frac{\delta_F}{L_F} = -\left(\frac{0.072}{24}\right) = -0.003$ $\boxed{\varepsilon_F = -0.003 \text{ in/in}}$

2.14 Due to the application of force P, the average normal strain in bar A was found to be - 6000 μ. If lengths of bars A and F are 36 inches determine the movement of point B and the average normal strain in bar F.

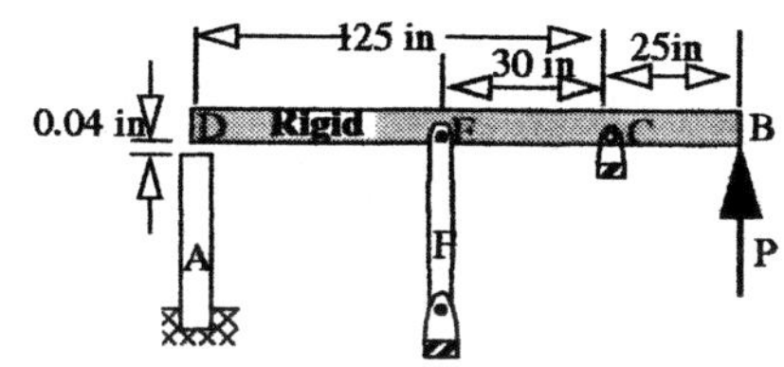

Solution $L_A = 36$ inches $L_F = 36$ inches $\varepsilon_A = -6000\ \mu$

$\delta_B = ?$ $\varepsilon_F = ?$

The deformation of A can be found as: $\delta_A = \varepsilon_A L_A = (6000)(10^{-6})(36) = 0.216\text{inch contraction}$

From Eq. (1) in problem 2.13

$\delta_B = \frac{\delta_A + 0.04}{5} = \frac{0.216 + 0.04}{5} = 0.0512\text{inch}$ $\boxed{\delta_B = 0.0512\text{inch upward}}$

$\delta_F = \frac{30}{25}\delta_B = \frac{6}{5}(0.0512) = 0.06144\text{inch contraction}$

The average normal strain can be found as shown below.

$\varepsilon_F = \frac{\delta_F}{L_F} = -\left(\frac{0.06144}{36}\right) = -1.707(10^{-3})$ $\boxed{\varepsilon_F = -1707\mu\text{in/in}}$

2.15 Due to the application of force P, point B in figure shown was observed to move left by 0.75 mm. If length of bar A is 1.2 m, determine the average normal strain in bar A.

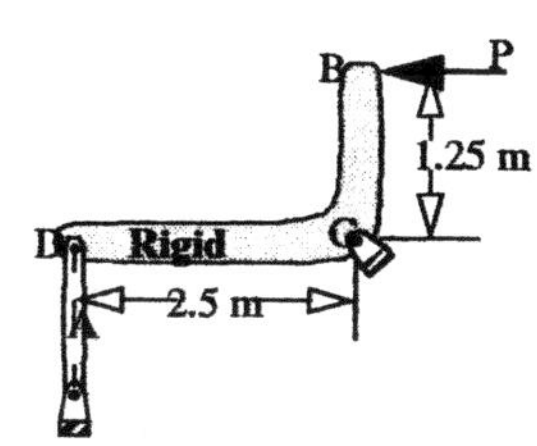

Solution $\delta_B = 0.75$ mm $L_A = 1.2$ m $\varepsilon_A = ?$

We draw the exaggerated deformed shape as shown below.

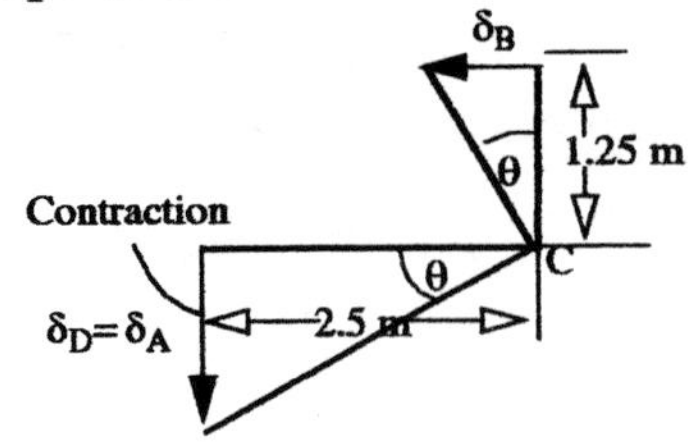

Using similar triangle, the deformation equations can be written as given below.

$$\frac{\delta_A}{2.5} = \frac{\delta_B}{1.25} \qquad \text{or}$$ 1

$$\delta_A = 2\delta_B = 2(0.75) = 1.5\text{mm} = 1.5(10^{-3})\text{m} \quad \text{contraction}$$

The average normal strain can be found as shown below.

$$\varepsilon_A = \frac{\delta_A}{L_A} = -\left(\frac{1.5(10^{-3})}{1.2}\right) = -1.2(10^{-3})$$

$$\boxed{\varepsilon_A = -1250\mu \text{ mm/mm}}$$

2.16 Due to the application of force P, the average normal strain in bar A in Fig. P2.16 was found to be - 2000 μ. If lengths of bar A is 2 m determine the movement of point B.

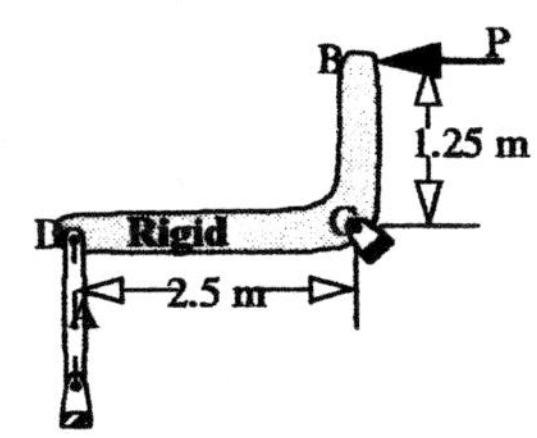

Fig. P2.16

Solution $\varepsilon_A = -2000\mu$ $L_A = 2$ m $\delta_B = ?$

The deformation of A can be found as: $\delta_A = \varepsilon_A L_A = (2000)(10^{-6})(2) = 4\text{mm}$ contraction

From Eq.(1) in problem 2.15 we obtain:

$$\delta_B = \left(\frac{1.25}{2.5}\right)\delta_A = \frac{4}{2} = 2\text{mm} \quad \text{or}$$

$$\boxed{\delta_B = 2\text{mm to the left}}$$

2.17 Due to the application of force P, point B in the figure shown was observed to move left by 0.75 mm. If length of bar A is 1.2 m, determine the average normal strain in bar A.

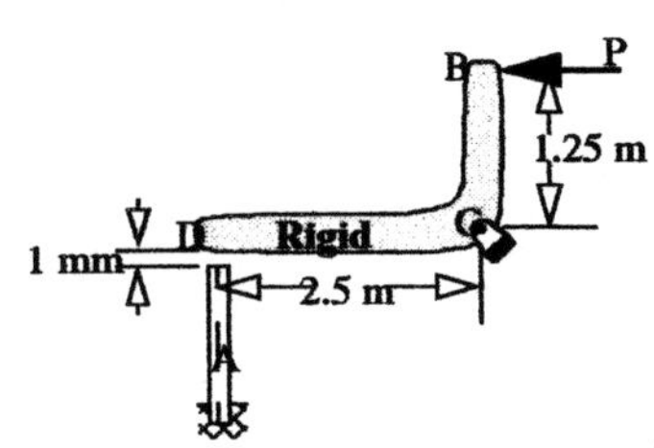

Solution $\delta_B = 0.75$ mm $L_A = 1.2$ m $\varepsilon_A = ?$

Consider point D on the rigid bar. We draw the exaggerated deformed shape as shown below.
We can write the deformation equations as shown below.

$$\delta_D = \delta_A + 1$$ 1

The deformation equations can be written as given below.

$$\frac{\delta_D}{2.5} = \frac{\delta_B}{1.25} \qquad \frac{\delta_A + 1}{2.5} = \frac{\delta_B}{1.25}$$ 2

or $\delta_A = 2\delta_B - 1 = 2(0.75) - 1 = 0.5\text{mm} = 0.5(10^{-3})\text{m contraction}$

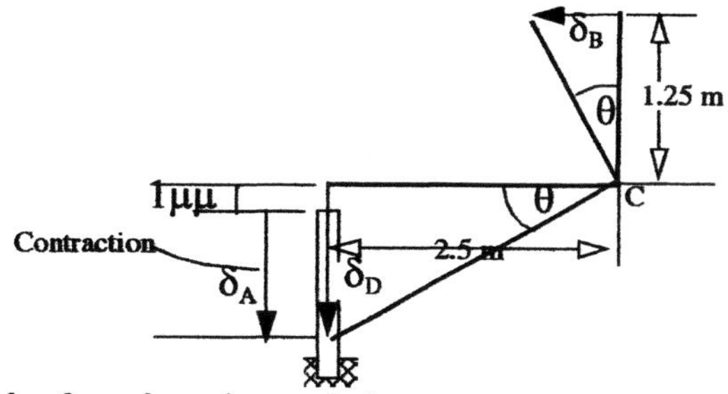

The average normal strain can be found as shown below.

$$\varepsilon_A = \frac{\delta_A}{L_A} = -\left(\frac{0.5(10^{-3})}{1.2}\right) = -0.4167(10^{-3})$$

$$\boxed{\varepsilon_A = -416.7\mu \text{ mm/mm}}$$

2.18 Due to the application of force P, the average normal strain in bar A was found to be - 2000 μ. If length of bar A is 2 m, determine the movement of point B.

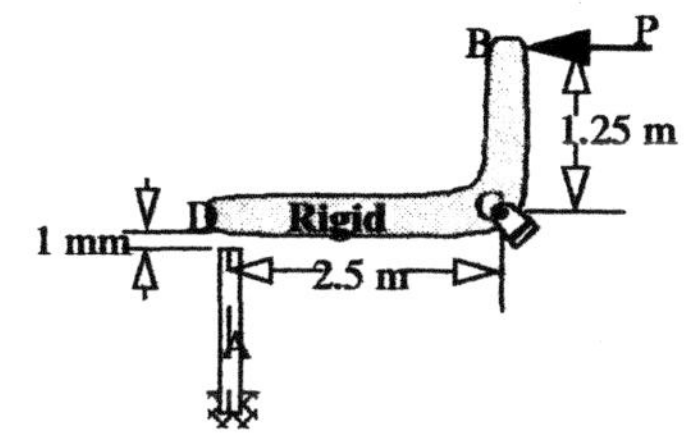

Solution $\varepsilon_A = -2000\mu$ $L_A = 2$ m $\delta_B = ?$

The deformation of A can be found as: $\delta_A = \varepsilon_A L_A = (2000)(10^{-6})(2) = 4\text{mm contraction}$

From Eq.(2) in problem 2.17 we obtain:

$$\delta_B = \left(\frac{1.25}{2.5}\right)(\delta_A + 1) = \frac{5}{2} = 2.5\text{mm or}$$

$$\boxed{\delta_B = 2.5\text{mm to the left}}$$

2.19 Due to the application of force P, point B in the figure shown was observed to move left by 0.75 mm. If lengths of bar A and F are1.2 m, determine the average normal strain in bars A and F.

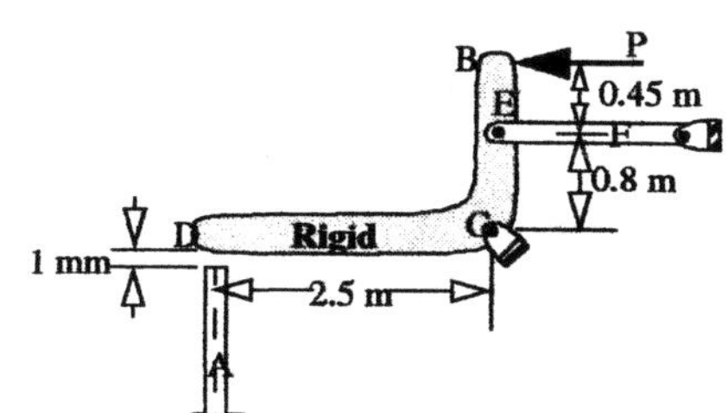

Solution $\delta_B = 0.75$ mm $L_A = 1.2$ m $L_F = 1.2$ m $\varepsilon_A = ?$ $\varepsilon_F = ?$

Consider point D on the rigid bar. We draw the exaggerated deformed shape as shown below.

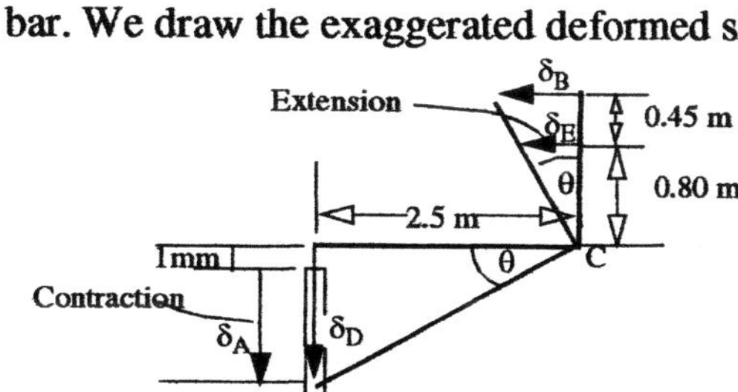

We can write the deformation equations as shown below.

$$\delta_D = \delta_A + 1$$ 1

$$\frac{\delta_D}{2.5} = \frac{\delta_F}{0.8} = \frac{\delta_B}{1.25} \quad \text{or} \quad \frac{\delta_A + 1}{2.5} = \frac{\delta_F}{0.8} = \frac{\delta_B}{1.25} \quad \text{or}$$ 2

$\delta_A = 2\delta_B - 1 = 2(0.75) - 1 = 0.5\text{mm} = 0.5(10^{-3})\text{m}$ contraction and

$\delta_F = 0.8\delta_B/1.25 = 0.64 \text{ mm} = 0.64(10^{-3})\text{m}$ extension

The average normal strains can be found as shown below.

$$\varepsilon_A = \frac{\delta_A}{L_A} = -\left(\frac{0.5(10^{-3})}{1.2}\right) = -0.4167(10^{-3}) \quad \text{or} \quad \boxed{\varepsilon_A = -416.7\mu \text{ mm/mm}}$$

$$\varepsilon_F = \frac{\delta_F}{L_F} = \left(\frac{0.64(10^{-3})}{1.2}\right) = 0.5333(10^{-3}) \quad \text{or} \quad \boxed{\varepsilon_F = 533.3\mu \text{ mm/mm}}$$

2.20 Due to the application of force P, the average normal strain in bar A was found to be - 2000 μ. Bars A and F are 2 m long. Determine the movement of point B and average normal strain in bar F.

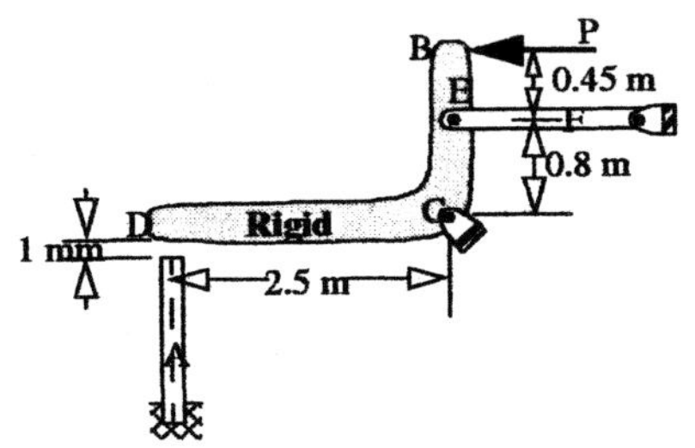

Solution $\varepsilon_A = -2000\ \mu$ $\quad L_A = 2$ m $\quad L_F = 2$ m $\quad \delta_B = ?$ $\quad \varepsilon_F = ?$

The deformation of A can be found as: $\delta_A = \varepsilon_A L_A = (2000)(10^{-6})(2) = 4\text{mm}$ contraction

From Eq.(2) in problem 2.19 we obtain:

$$\delta_B = \left(\frac{1.25}{2.5}\right)(\delta_A + 1) = \frac{5}{2} = 2.5\text{mm or} \quad \boxed{\delta_B = 2.5\text{mm to the left}}$$

From Eq.(2) in problem 2.19 we obtain: $\delta_F = \delta_F = 0.8\delta_B/1.25 = 1.6 \text{ mm} = 1.6(10^{-3})\text{m}$ extension

The average normal strains can be found as shown below.

$$\varepsilon_F = \frac{\delta_F}{L_F} = \left(\frac{1.6(10^{-3})}{2.0}\right) = 0.8(10^{-3}) \quad \text{or} \quad \boxed{\varepsilon_F = 800\mu \text{ mm/mm}}$$

2.21 A rectangular plastic plate deforms into the shaded shape. Determine the average shear strain at point A

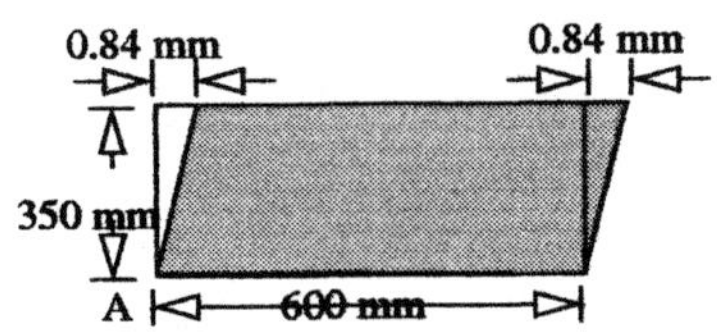

Solution $\gamma_A = ?$

The shear strain represents the change of angle from the right angle at A as shown in the figure below.

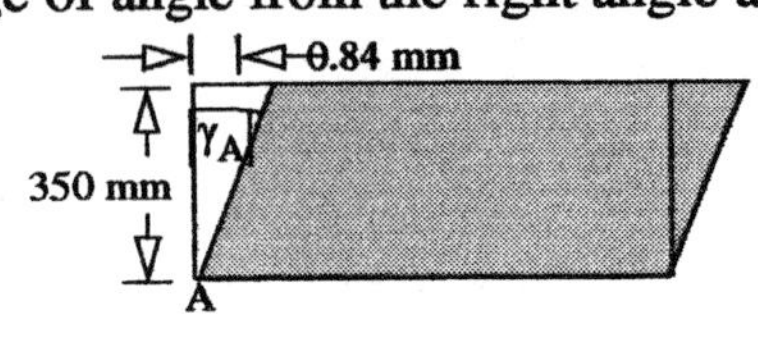

From geometry, we obtain the magnitude of γ_A as shown below.

$$\tan|\gamma_A| = \frac{0.840}{350} \qquad \text{or} \qquad |\gamma_A| = 0.0024 \text{ rads} \qquad \mathbf{1}$$

Shear strain is positive as angle decreases.

$$\boxed{\gamma_A = 2400\mu \text{ rads}}$$

2.22 A rectangular plastic plate deforms into the shaded shape. Determine the average shear strain at point A.

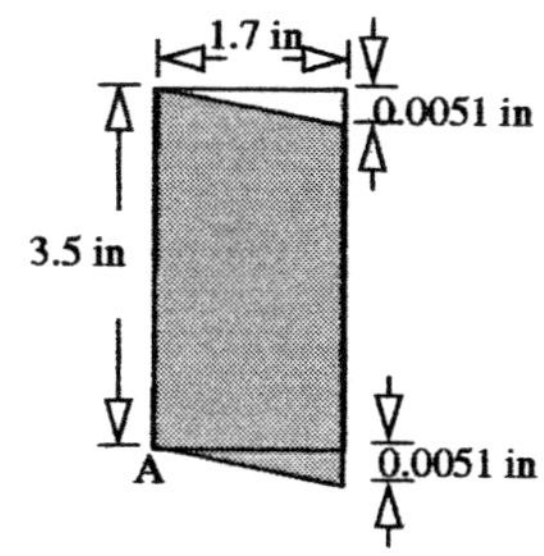

Solution γ_A=?

The shear strain represents the change of angle from the right angle at A as shown in the figure below.

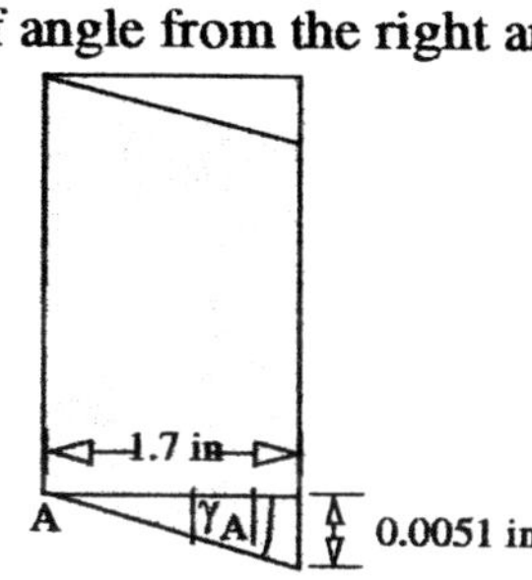

From geometry, we obtain the magnitude of γ_A as shown below.

$$\tan|\gamma_A| = \frac{0.0051}{1.7} \qquad \text{or} \qquad |\gamma_A| = 0.003 \text{ rads} \qquad \mathbf{1}$$

Shear strain is negative as angle increases.

$$\boxed{\gamma_A = -3000\mu \text{ rad}}$$

2.23 A rectangular plastic plate deforms into the shaded shape. Determine the average shear strain at point A.

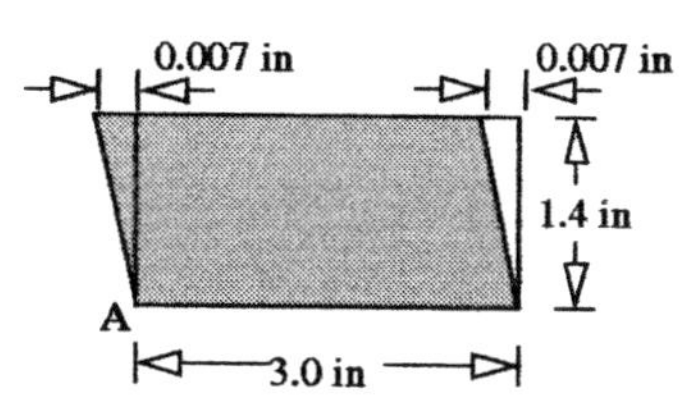

Solution γ_A=?

The shear strain represents the change of angle from the right angle at A as shown in the figure below.

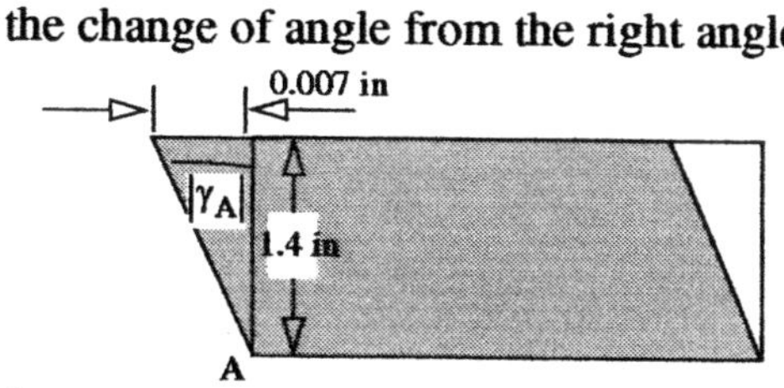

From geometry, we obtain the magnitude of γ_A as shown below.

$$\tan|\gamma_A| = \frac{0.007}{1.4} \qquad \text{or} \qquad |\gamma_A| = 0.005 \text{ rads} \qquad \mathbf{1}$$

Shear strain is negative as angle increases.

$\boxed{\gamma_A = -5000\mu \text{ rad}}$

2.24 A rectangular plastic plate deforms into the shaded shape. Determine the average shear strain at point A.

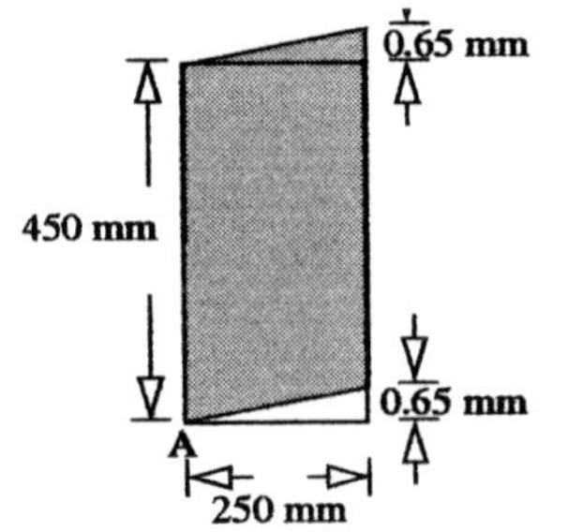

Solution γ_A=?

The shear strain represents the change of angle from the original right angle at A as shown in the figure below.

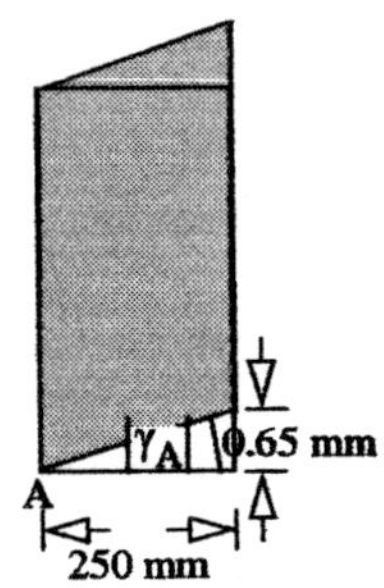

From geometry, we obtain the magnitude of γ_A as shown below.

$$\tan|\gamma_A| = \frac{0.65}{250} \qquad \text{or} \qquad |\gamma_A| = 0.00026 \text{ rads}$$ 1

Shear strain is positive as angle decreases.

$\boxed{\gamma_A = 260\mu \text{ rad}}$

2.25 A rectangular plastic plate deforms into the shaded shape. Determine the average shear strain at point A.

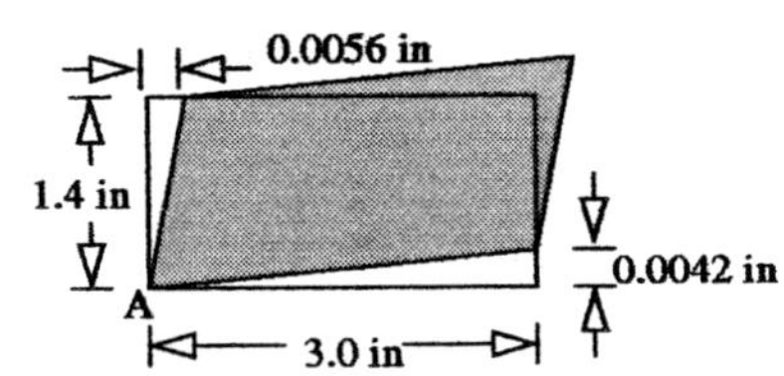

Solution γ_A=?

The angles ϕ_1 and ϕ_2 are measures of the change of angle as shown in the figure below.

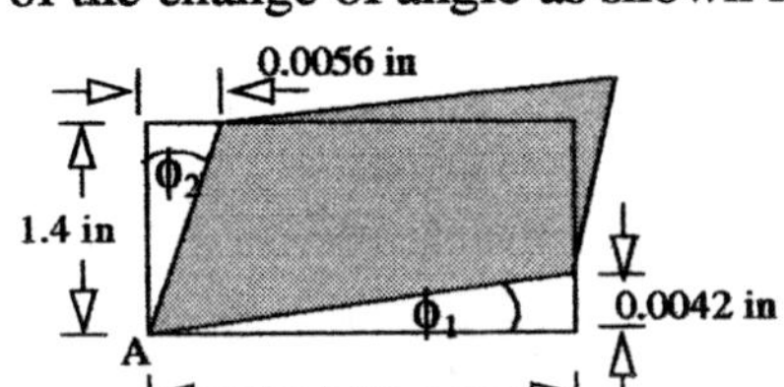

From geometry, we obtain the angles ϕ_1 and ϕ_2 as shown below.

$$\tan\phi_1 = \frac{0.0042}{3.0} \qquad \text{or} \qquad \phi_1 = 0.0014 \text{ rads}$$ 1

$$\tan\phi_2 = \frac{0.0056}{1.4} \quad \text{or} \quad \phi_2 = 0.004 \text{ rads}$$ 2

The decrease in the angle from the original right angle at A represents positive shear strain, i.e.,

$\gamma_A = \phi_1 + \phi_2 = 0.0054$ rads or $\boxed{\gamma_A = 5400\mu \text{ rad}}$

2.26 A rectangular plastic plate deforms into the shaded shape. Determine the average shear strain at point A

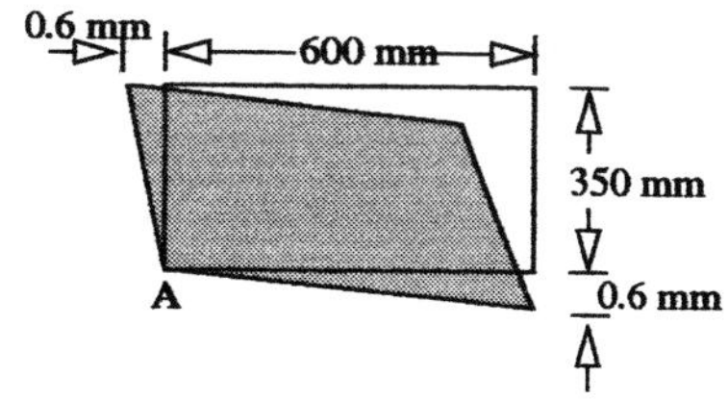

Solution γ_A=?

The angles ϕ_1 and ϕ_2 are measures of the change of angle as shown in the figure below.

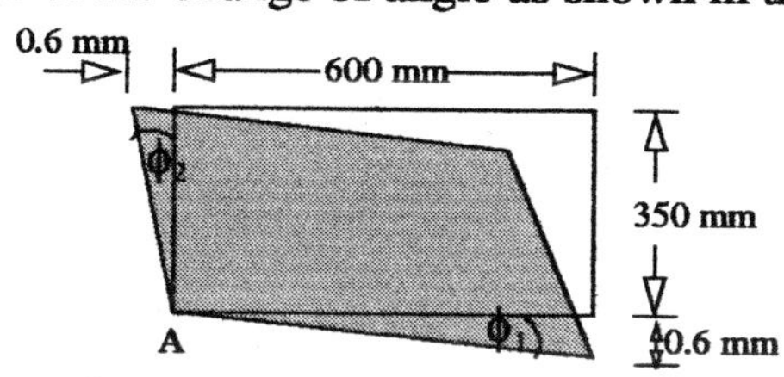

From geometry we obtain the angles ϕ_1 and ϕ_2 as shown below.

$$\tan\phi_1 = \frac{0.6}{600} \quad \text{or} \quad \phi_1 = 1.0(10^{-3}) \text{ rads}$$ 1

$$\tan\phi_2 = \frac{0.6}{350} \quad \text{or} \quad \phi_2 = 1.714(10^{-3}) \text{ rads}$$ 2

The increase in angle from the original right angle at A represents negative shear strain, i.e.,

$\gamma_A = -(\phi_1 + \phi_2) = -2.714(10^{-3})$ rads or $\boxed{\gamma_A = -2714\mu \text{ rad}}$

2.27 A thin triangular plate ABC forms a right angle at point A. During deformation, point A moves vertically down by δ_A. Determine the average shear strain at point A.

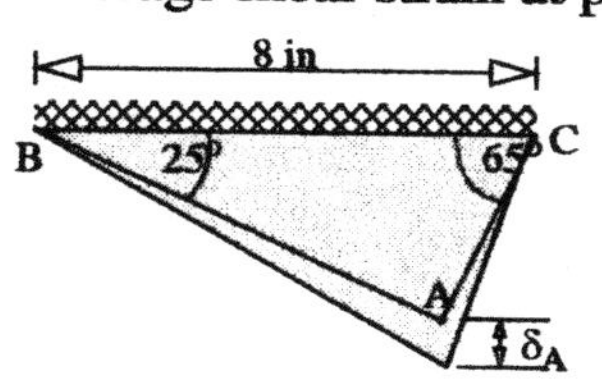

$\delta_A = 0.005$ in

Solution δ_A=0.005in γ_A=?

The exaggerated deformed geometry can be drawn as shown below.

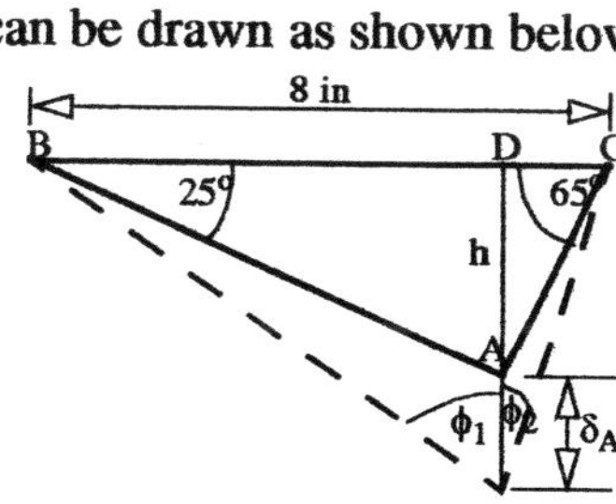

From triangle ABC $\quad AB = 8\cos 25 = 7.2505$
From triangle ABD

$$BD = AB\cos 25 = 6.5712$$ 1

$$h = AB\sin 25 = 3.0642$$ 2

$$BC = BD + DC \quad \text{or} \quad DC = 1.4288$$ 3

$$\tan\phi_1 = \frac{BD}{h+\delta_A} = \frac{6.5712}{(3.0642+0.005)} \quad \text{or} \quad \phi_1 = 1.1338\text{rads}$$

$$\tan\phi_2 = \frac{DC}{h+\delta_A} = \frac{1.4288}{(3.0642+0.005)} \quad \text{or} \quad \phi_2 = 0.4357\text{rads}$$

The decrease in the angle from the original right angle at A represents positive shear strain, i.e.,

$$\gamma_A = \frac{\pi}{2} - (\phi_1 + \phi_2) = 1.2963(10^{-3})\text{rads} \qquad \boxed{\gamma_A = 1296\mu\text{rad}}$$

2.28 A thin triangular plate ABC forms a right angle at point A. During deformation, point A moves vertically down by δ_A. Determine the average shear strain at point A.

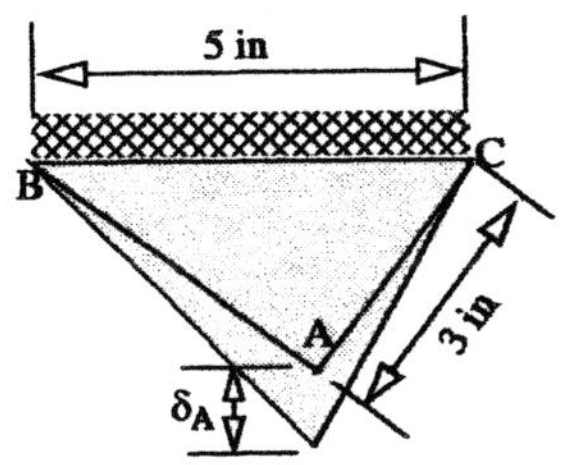

$\delta_A = 0.008$ in

Solution $\quad \delta_A = 0.008 \qquad \gamma_A = ?$

The exaggerated deformed geometry can be drawn as shown below.

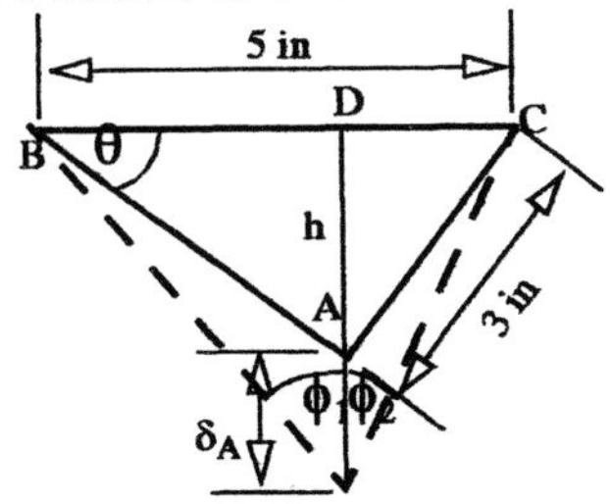

From triangle ABC $\quad AB = \sqrt{5^2 - 3^2} = 4 \qquad \cos\theta = 0.8 \qquad \sin\theta = 0.6$
From triangle ABD

$$BD = AB\cos\theta = 3.2$$ 1

$$h = AB\sin\theta = 2.4$$ 2

$$BC = BD + DC \quad \text{or} \quad DC = 1.8$$ 3

$$\tan\phi_1 = \frac{BD}{h+\delta_A} = \frac{3.2}{(2.4+0.008)} \quad \text{or} \quad \phi_1 = 0.9257\text{rads}$$

$$\tan\phi_2 = \frac{DC}{h+\delta_A} = \frac{1.8}{(2.4+0.008)} \quad \text{or} \quad \phi_2 = 0.6419\text{rads}$$

The decrease in the angle from the original right angle at A represents positive shear strain, i.e.,

$$\gamma_A = \frac{\pi}{2} - (\phi_1 + \phi_2) = 3.1947(10^{-3})\text{rads} \qquad \boxed{\gamma_A = 3195\mu\text{rads}}$$

2.29 A thin triangular plate ABC forms a right angle at point A. During deformation, point A moves vertically down by δ_A that is given in each problem. Determine the average shear strain at point A.

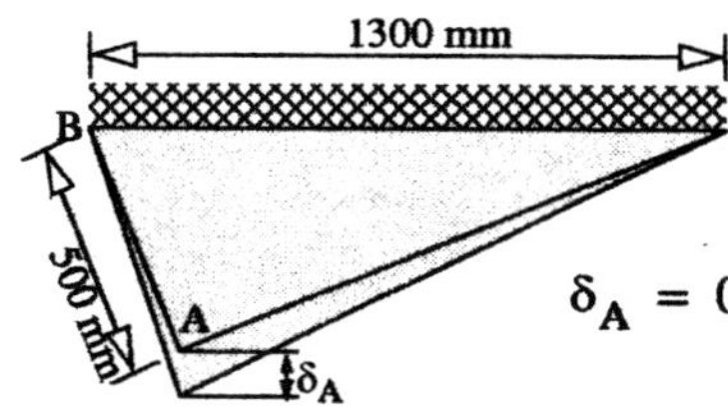

$\delta_A = 0.75$ mm

Solution δ_A=0.75mm γ_A= ?

The exaggerated deformed geometry can be drawn as shown below.

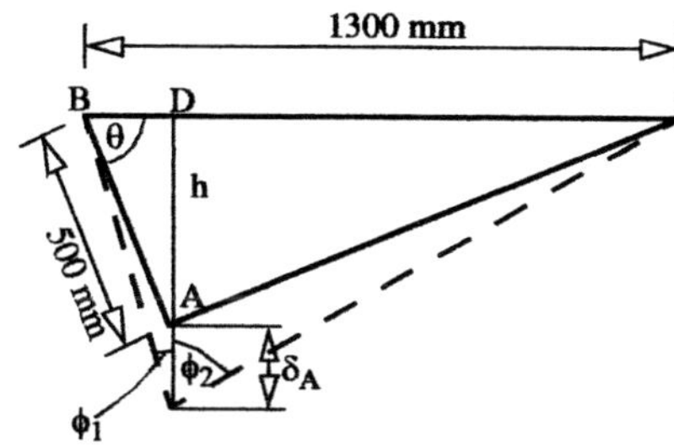

From triangle ABC $AC = \sqrt{1300^2 - 500^2} = 1200$ $\cos\theta = \frac{5}{13}$ $\sin\theta = \frac{12}{13}$

From triangle ABD

$BD = AB\cos\theta = 192.308$ **1**

$h = AB\sin\theta = 461.538$ **2**

$BC = BD + DC$ or $DC = 1107.69$ **3**

$$\tan\phi_1 = \frac{BD}{h+\delta_A} = \frac{192.308}{(461.538+0.75)} \quad \text{or} \quad \phi_1 = 0.3942\text{rads}$$

$$\tan\phi_2 = \frac{DC}{h+\delta_A} = \frac{1107.69}{(461.538+0.75)} \quad \text{or} \quad \phi_2 = 1.1754\text{rads}$$

The decrease in the angle from the original right angle at A represents positive shear strain, i.e.,

$$\gamma_A = \frac{\pi}{2} - (\phi_1 + \phi_2) = 1.1529(10^{-3})\text{rads}$$

$\boxed{\gamma_A = 1153\mu\text{rads}}$

2.30 A thin triangular plate ABC forms a right angle at point A. During deformation, point A moves horizontally by δ_A that is given in each problem. Determine the average shear strain at point A.

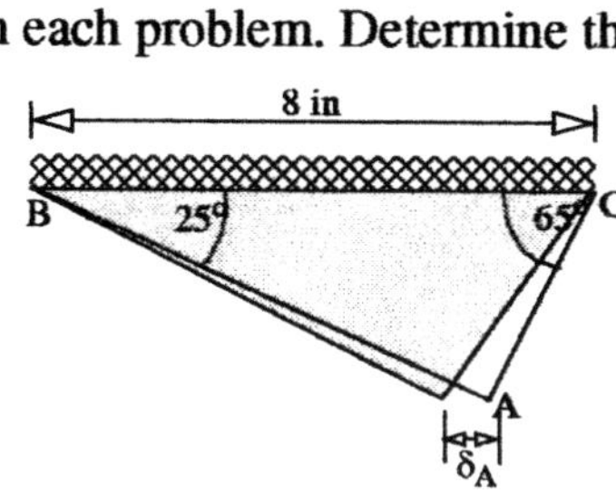

$\delta_A = 0.005$ in

Solution δ_A=0.005 γ_A= ?

The exaggerated deformed geometry can be drawn as shown below.

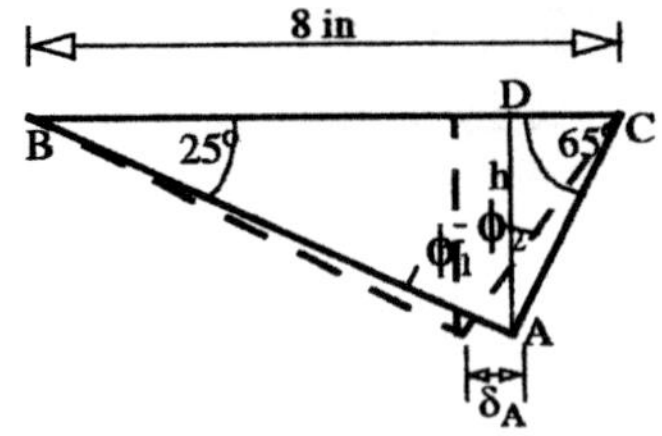

From Equations (1),(2) and (3) of problem 2.27: BD = 6.5712 h = 3.0642 DC = 1.4288

$$\tan\phi_1 = \frac{BD-\delta_A}{h} = \frac{6.5712-0.005}{3.0642} \quad \text{or} \quad \phi_1 = 1.13417\text{rads}$$

$$\tan\phi_2 = \frac{DC+\delta_A}{h} = \frac{1.4288+0.005}{3.0642} \quad \text{or} \quad \phi_2 = 0.43766\text{rads}$$

The decrease in the angle from the original right angle at A represents positive shear strain, i.e.,

$$\gamma_A = \frac{\pi}{2}-(\phi_1+\phi_2) = -1.0337(10^{-3})\text{rads} \qquad \boxed{\gamma_A = -1034\mu\text{rads}}$$

2.31 A thin triangular plate ABC forms a right angle at point A. During deformation, point A moves horizontally by δ_A that is given in each problem. Determine the average shear strain at point A.

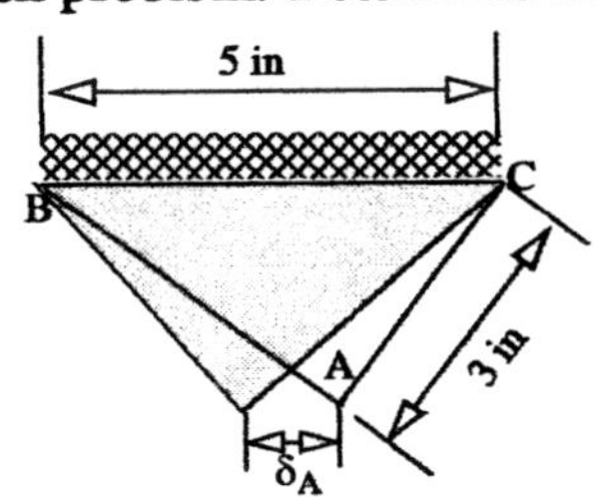

$\delta_A = 0.008$ in

Solution δ_A=0.008 γ_A= ?

The exaggerated deformed geometry can be drawn as shown below.

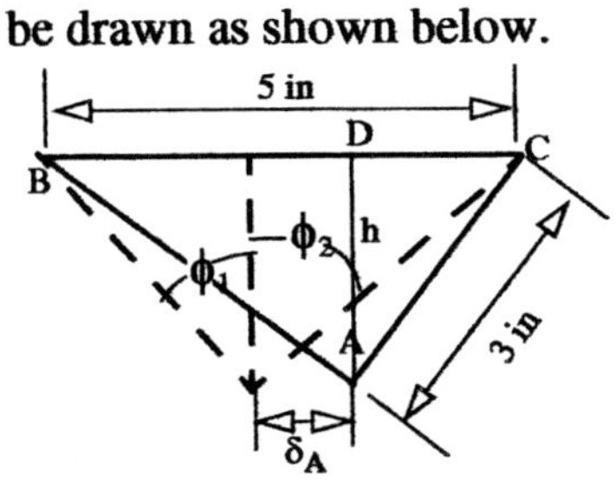

From Equations (1), (2) and (3) in Problem 2-28: BD = 3.2 h = 2.4 DC = 1.8

$$\tan\phi_1 = \frac{BD-\delta_A}{h} = \frac{3.2-0.008}{2.4} \quad \text{or} \quad \phi_1 = 0.92609\text{rads}$$

$$\tan\phi_2 = \frac{CD+\delta_A}{h} = \frac{1.8+0.008}{2.4} \quad \text{or} \quad \phi_2 = 0.64563\text{rads}$$

The decrease in the angle from the original right angle at A represents positive shear strain, i.e.,

$$\gamma_A = \frac{\pi}{2}-(\phi_1+\phi_2) = -0.9280(10^{-3})\text{rads} \qquad \boxed{\gamma_A = -928\mu\text{rad}}$$

2.32 A thin triangular plate ABC forms a right angle at point A. During deformation, point A moves horizontally by δ_A that is given in each problem. Determine the average shear strain at point A in each

problem

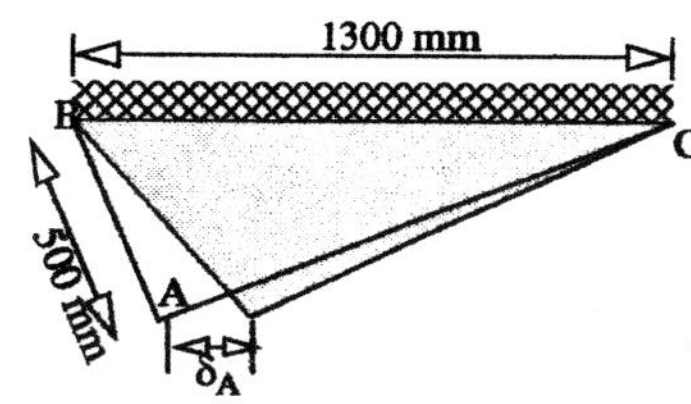

$\delta_A = 0.75$ mm

Solution $\delta_A=0.75$ $\gamma_A=?$

The exaggerated deformed geometry can be drawn as shown below.

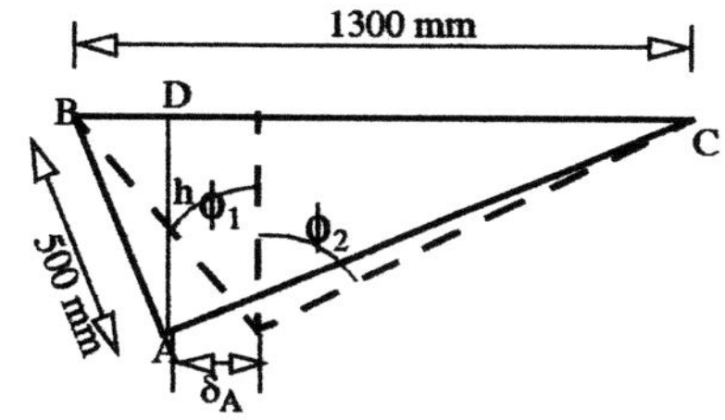

From Equations (1), (2) and (3) in problem 2-29: BD = 192.308 h = 461.538 DC = 1107.69

$$\tan\phi_1 = \frac{BD + \delta_A}{h} = \frac{192.308 + 0.75}{461.538} \quad \text{or} \quad \phi_1 = 0.39618\text{rads}$$

$$\tan\phi_2 = \frac{CD - \delta_A}{h} = \frac{1107.69 - 0.75}{461.538} \quad \text{or} \quad \phi_2 = 1.17576\text{rads}$$

The decrease in the angle from the original right angle at A represents positive shear strain, i.e.,

$$\gamma_A = \frac{\pi}{2} - (\phi_1 + \phi_2) = -1.1438(10^{-3})\text{rads} \quad \text{or} \quad \boxed{\gamma_A = -1144\mu\text{rad}}$$

2.33 The diagonals of two square form a right angle at point A. The two rectangles are pulled horizontally to a deformed shape shown by the colored lines. Determine the average shear strain at point A, if the displacements of points A and B are as given in each problem.

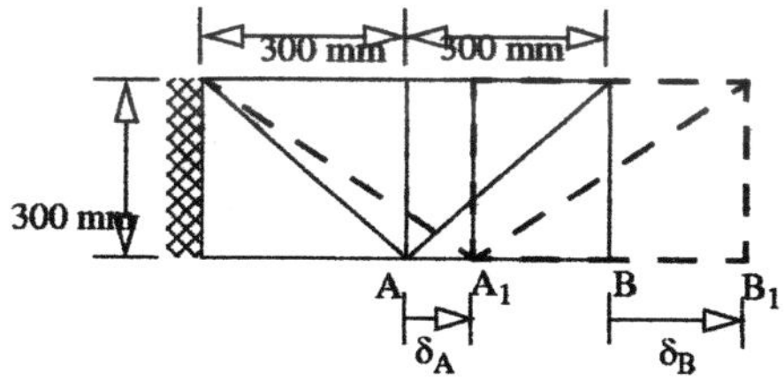

Solution $\delta_A=0.4$mm $\delta_B=0.8$mm $\gamma_A=?$

The exaggerated deformed geometry can be drawn as shown below.

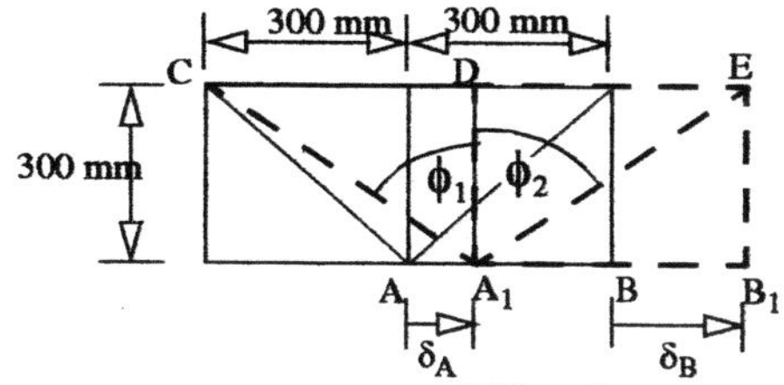

$$\tan\phi_1 = \frac{CD}{AD} = \frac{300 + \delta_A}{300} \qquad \textbf{1}$$

$$\tan\phi_2 = \frac{DE}{AD} = \frac{(CE - CD)}{AD} = \frac{(600 + \delta_B) - (300 + \delta_A)}{300} \quad \text{or}$$

$$\tan\phi_2 = \frac{300 + \delta_B - \delta_A}{300} \qquad \textbf{2}$$

Substituting $\delta_A = 0.4$ and $\delta_B = 0.8$ in Equations(1) and (2)

We obtain

$\tan\phi_1 = \frac{300.4}{300}$ or $\phi_1 = 0.78606\text{rads}$ and $\tan\phi_2 = \frac{300.4}{300}$ or $\phi_2 = 0.78606\text{rads}$

The decrease in the angle from the original right angle at A represents positive shear strain, i.e.,

$\gamma_A = \frac{\pi}{2} - (\phi_1 + \phi_2) = -1.332(10^{-3})\text{rads}$ $\boxed{\gamma_A = -1332\mu\text{rad}}$

2.34 The diagonals of two square form a right angle at point A. The two rectangles are pulled horizontally to a deformed shape shown by the colored lines. Determine the average shear strain at point A, if the displacements of points A and B are as given in each problem.

$\delta_A = 0.3$ mm and $\delta_B = 0.9$ mm.

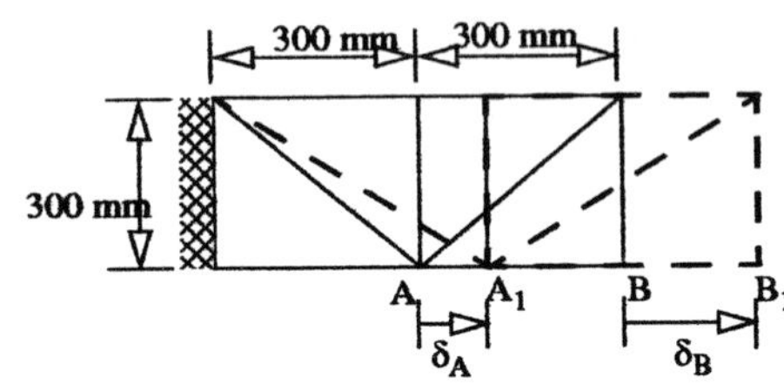

Solution δ_A=0.3 δ_B=0.9 γ_A= ?

Substituting $\delta_A = 0.3$ and $\delta_B = 0.8$ in Equations (1) and (2) in problem 2-33,

We obtain:

$\tan\phi_1 = \frac{300.3}{300}$ or $\phi_1 = 0.78590\text{rads}$ and $\tan\phi_2 = \frac{300.6}{300}$ or $\phi_2 = 0.78640\text{rads}$

The decrease in the angle from the original right angle at A represents positive shear strain, i.e.,

$\gamma_A = \frac{\pi}{2} - (\phi_1 + \phi_2) = -1.4987(10^{-3})\text{rads}$ $\boxed{\gamma_A = -1499\mu\text{rad}}$

2.35 The roller at P can slide only in the slot by the given amount. Determine the strains in bar AP by (a) finding the deformed length of AP without small-strain approximation, (b) using Equation 2.6, and (c) using Equation 2.7.

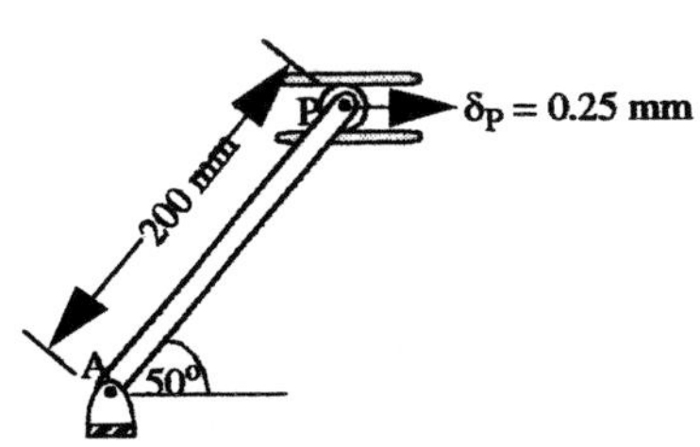

Solution $\delta_P = 0.25$ mm (a) ε_{AP}= ? using geometry

(b) ε_{AP}= ? using small strain scaler approach (c) ε_{AP}= ? using small strain vector approach.

We draw the exaggerated deformed shape as shown below.

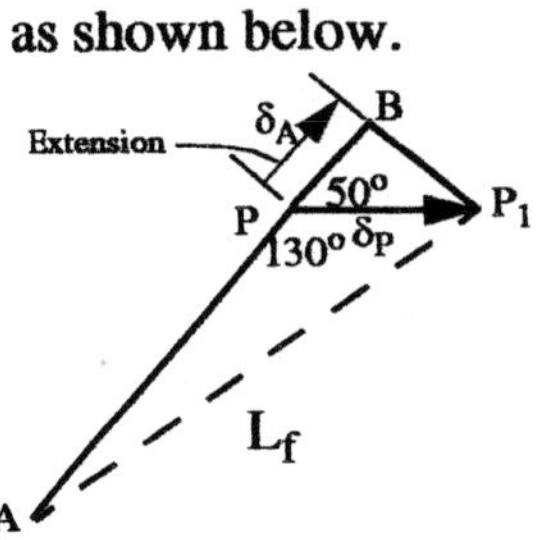

(a) From triangle APP_1, using cosine rule we obtain

$$L_f^2 = AP^2 + PP_1^2 - 2AP(PP_1)\cos 130 = 200^2 + 0.25^2 - 2(200)(0.25)\cos 130 = 200.16079$$

The average normal strain is $\varepsilon_{AP} = \frac{L_f - L_o}{L_o} = \frac{0.16079}{200} = 0.80394(10^{-3})$ or $\boxed{\varepsilon_{AP} = 803.9\mu\frac{mm}{mm}}$

(b) From triangle PBP_1

$$\delta_{AP} = PB = PP_1\cos 50 = 0.25\cos 50 = 0.1607$$

$$\varepsilon_{AP} = \frac{\delta_{AP}}{L_o} = \frac{0.16070}{200} = 0.80348(10^{-3}) \quad \text{or} \quad \boxed{\varepsilon_{AP} = 803.5\mu\frac{mm}{mm}}$$

(c) The deformation vector can be written as: $\bar{D} = 0.25\bar{i}$

The unit vector can be written as: $\bar{i}_{AP} = \cos 50\bar{i} + \sin 50\bar{j}$

The deformation of the bar AP can be calculated as: $\delta_{AP} = \bar{i}_{AP} \cdot \bar{D} = 0.25\cos 50 = 0.1607$

The average normal strain is $\varepsilon_{AP} = \frac{\delta_{AP}}{L_o} = \frac{0.16070}{200} = 0.80348(10^{-3})$ or $\boxed{\varepsilon_{AP} = 803.5\mu\frac{mm}{mm}}$

2.36 The roller at P can slide only in the slot by the given amount. Determine the strains in bar AP by (a) finding the deformed length of AP without small-strain approximation, (b) using Equation 2.6, and (c) using Equation 2.7.

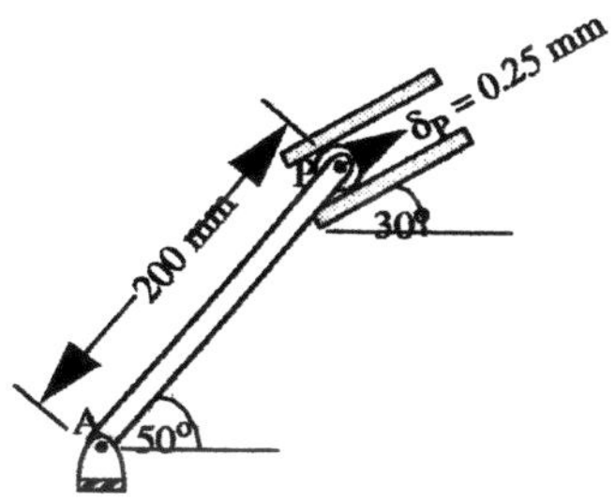

Solution $\delta_P = 0.25$ mm (a) $\varepsilon_{AP} = ?$ using geometry
(b) $\varepsilon_{AP} = ?$ using small strain scaler approach (c) $\varepsilon_{AP} = ?$ using small strain vector approach.

We draw the exaggerated deformed shape as shown below.

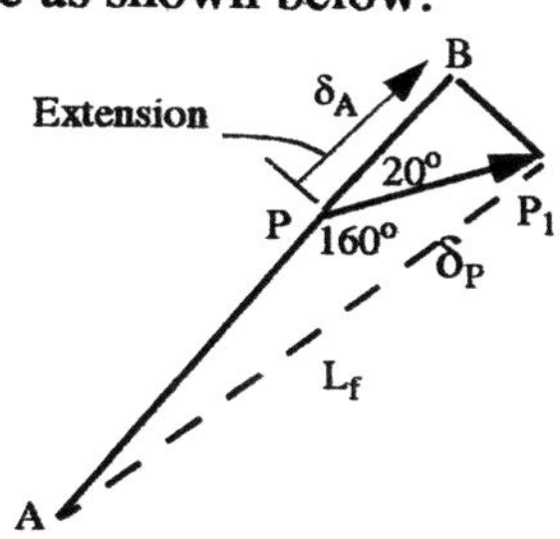

(a) From triangle APP_1, using cosine rule we obtain

$$L_f^2 = AP^2 + PP_1^2 - 2AP(PP_1)\cos 160 = 200^2 + 0.25^2 - 2(200)(0.25)\cos 160 = 200.23494$$

The average normal strain is: $\varepsilon_{AP} = \frac{L_f - L_o}{L_o} = \frac{0.23494}{200} = 1.1747(10^{-3})$ or $\boxed{\varepsilon_{AP} = 1174.7\mu\frac{mm}{mm}}$

(b) From triangle PBP_1: $\delta_{AP} = PB = PP_1\cos 20 = 0.25\cos 20 = 0.23492$

The average normal strain is: $\varepsilon_{AP} = \frac{\delta_{AP}}{L_o} = \frac{0.23492}{200} = 01.1746(10^{-3})$ or $\boxed{\varepsilon_{AP} = 1174.6\mu\frac{mm}{mm}}$

(c) The deformation vector can be written as: $\bar{D} = 0.25(\cos 30\bar{i} + \sin 30\bar{j})$

The unit vector can be written as $\bar{i}_{AP} = \cos 50\bar{i} + \sin 50\bar{j}$

$$\delta_{AP} = \bar{i}_{AP} \cdot \bar{D} = 0.25(\cos 30 \cos 50 + \sin 30 \sin 50) = 0.25\cos 20 = 0.23492$$

$$\varepsilon_{AP} = \frac{\delta_{AP}}{L_o} = \frac{0.23492}{200} = 01.1746(10^{-3})$$

$$\boxed{\varepsilon_{AP} = 1174.6\mu\frac{mm}{mm}}$$

2.37 A roller at P slides in a slot as shown in each problem. Determine the deformation in bar AP and bar BP using small strain approximation.

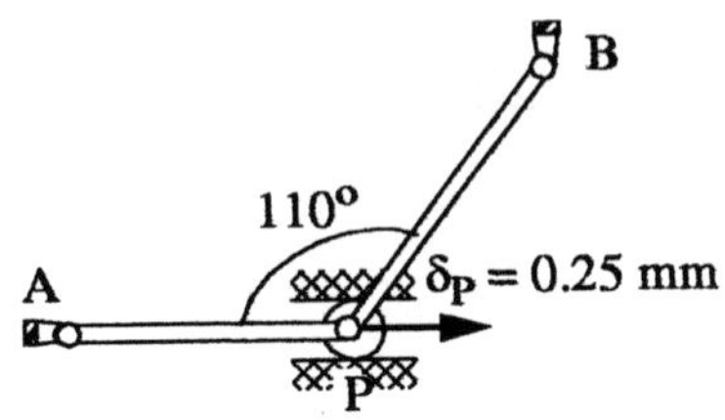

Solution $\delta_{AP} = ?$ $\delta_{BP} = ?$

We draw the exaggerated deformed shape as shown below.

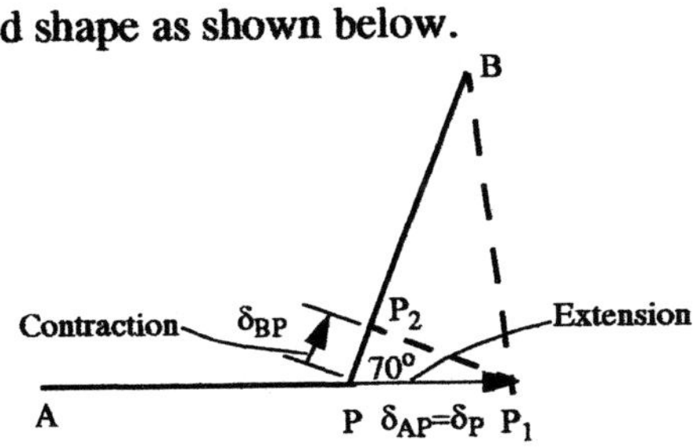

As per small strain approximation, we need component of PP_1 in the original direction of AP and BP, i.e., PP_1 represents the deformation of bar AP and PP_2 represents the deformation of bar BP and are calculated as shown below.

$\delta_{AP} = \delta_P = 0.25$ or $\boxed{\delta_{AP} = 0.25\text{mm extension}}$

$\delta_{BP} = \delta_P \cos 70 = 0.08551$ or $\boxed{\delta_{BP} = 0.0855\text{mm contraction}}$

2.38 A roller at P slides in a slot as shown in each problem. Determine the deformation in bar AP and bar BP using small strain approximation.

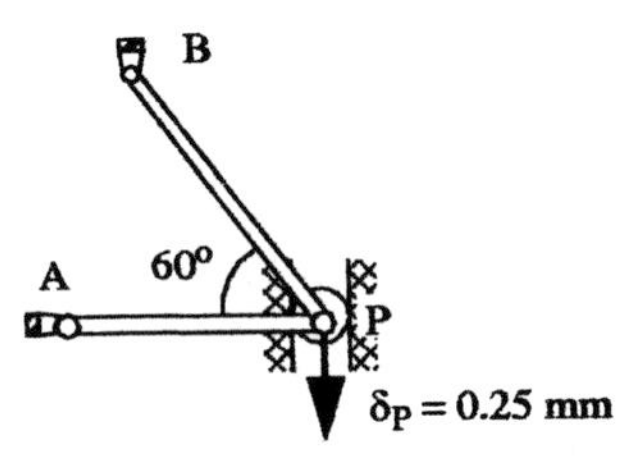

Solution $\delta_{AP} = ?$ $\delta_{BP} = ?$

We draw the exaggerated deformed shape as shown below.

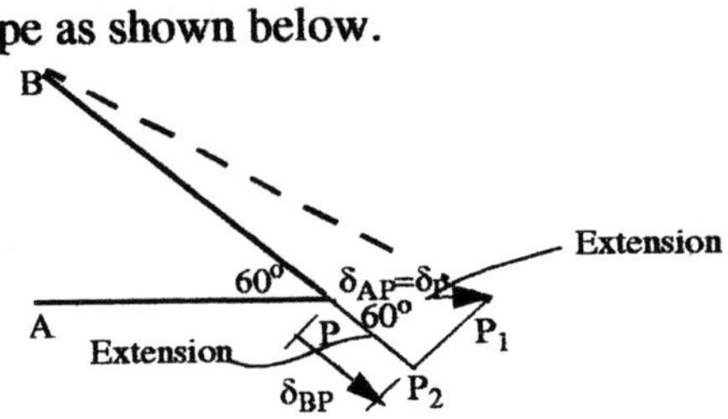

As per small strain approximation, we need component of PP_1 in the original direction of AP and BP, i.e., PP_1 represents the deformation of bar AP and PP_2 represents the deformation of bar BP and are calculated as shown below.

$\delta_{AP} = \delta_P = 0.25$ or $\boxed{\delta_{AP} = 0.25\text{mm extension}}$

$\delta_{BP} = \delta_P \cos 60 = 0.125$ or $\boxed{\delta_{BP} = 0.125\text{mm extension}}$

2.39 A roller at P slides in a slot as shown. Determine the deformation in bar AP and bar BP small strain approximation.

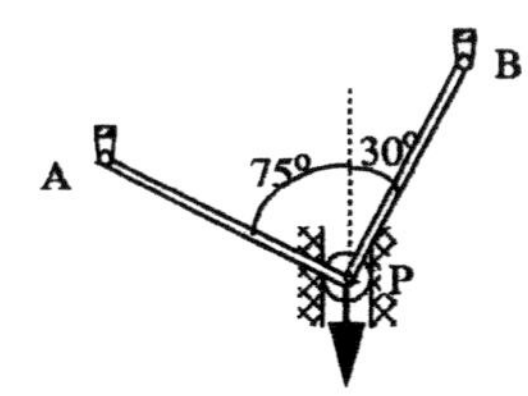

Fig. P2.39

Solution δ_{AP}= ? δ_{BP}= ?

We draw the exaggerated deformed shape as shown below.

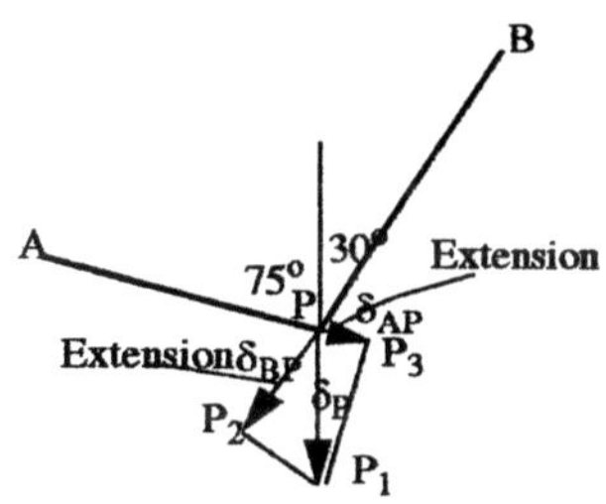

As per small strain approximation, we need component of PP_1 in the original direction of AP and BP, i.e., PP_1 represents the deformation of bar AP and PP_2 represents the deformation of bar BP and are calculated as shown below.

$\delta_{AP} = \delta_P \cos 75 = 0.06470$ or $\boxed{\delta_{AP} = 0.0647\text{mm extension}}$

$\delta_{BP} = \delta_P \cos 30 = 0.2165$ or $\boxed{\delta_{BP} = 0.2165\text{mm extension}}$

2.40 A roller at P slides in a slot as shown in each problem. Determine the deformation in bar AP and bar BP using small strain approximation.

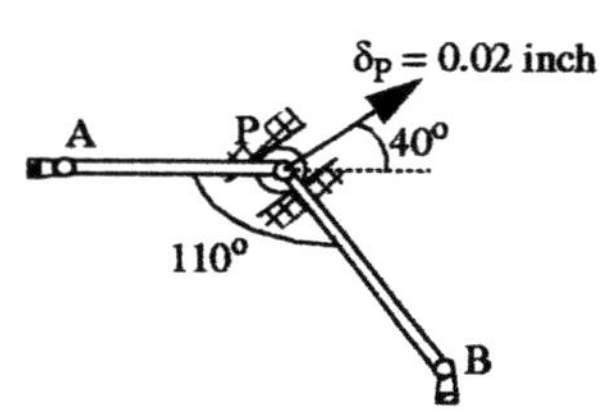

Fig. P2.40

Solution δ_{AP}= ? δ_{BP}= ?

We draw the exaggerated deformed shape as shown below.

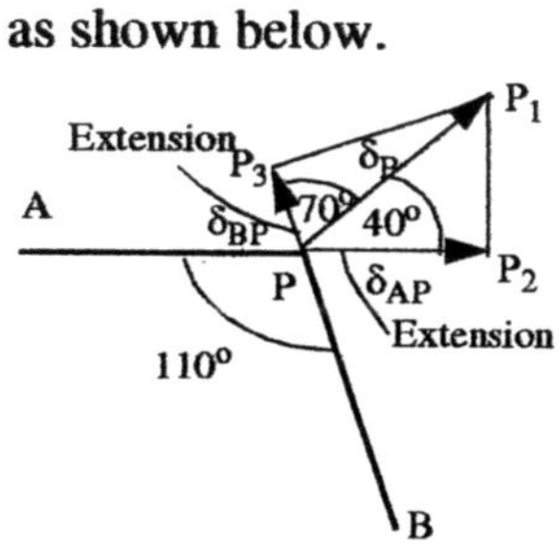

As per small strain approximation, we need component of PP_1 in the original direction of AP and BP, i.e., PP_2 represents the deformation of bar AP and PP_3 represents the deformation of bar BP and are calculated as shown below.

$\delta_{AP} = \delta_P \cos 40 = 0.01532$ or $\boxed{\delta_{AP} = 0.0153 \text{inch extension}}$

$\delta_{BP} = \delta_P \cos 70 = 0.00684$ or $\boxed{\delta_{BP} = 0.0068 \text{inch extension}}$

2.41 A roller at P slides in a slot as shown. Determine the deformation in bar AP and bar BP in each problem by using small strain approximation.

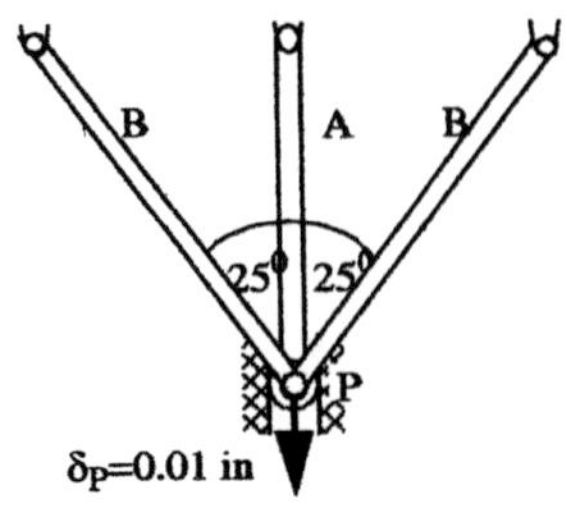

Fig. P2.41

Solution δ_{AP}= ? δ_{BP}= ?

We draw the exaggerated deformed shape as shown below.

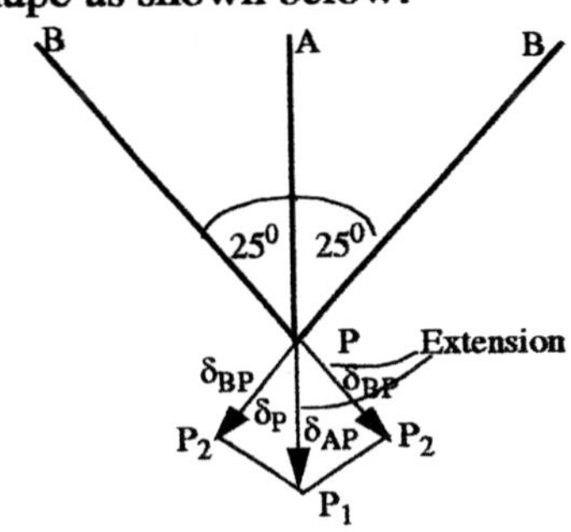

As per small strain approximation, we need component of PP_1 in the original direction of AP and BP, i.e., PP_1 represents the deformation of bar AP and PP_2 represents the deformation of bar BP and are calculated as shown below.

$\delta_{AP} = \delta_P = 0.01$ or $\boxed{\delta_{AP} = 0.01 \text{inch extension}}$

$\delta_{BP} = \delta_P \cos 25 = 0.00906$ or $\boxed{\delta_{BP} = 0.0091 \text{inch extension}}$

2.42 A roller at P slides in a slot as shown. Determine the deformation in bar AP and bar BP using small strain approximation.

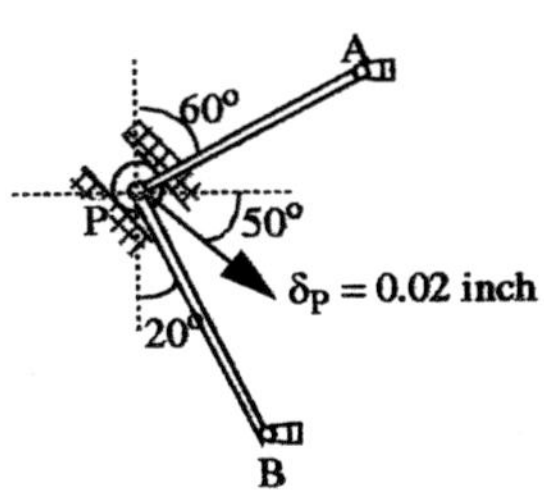

Fig. P2.42

Solution δ_{AP}= ? δ_{BP}= ?

We draw the exaggerated deformed shape as shown below.

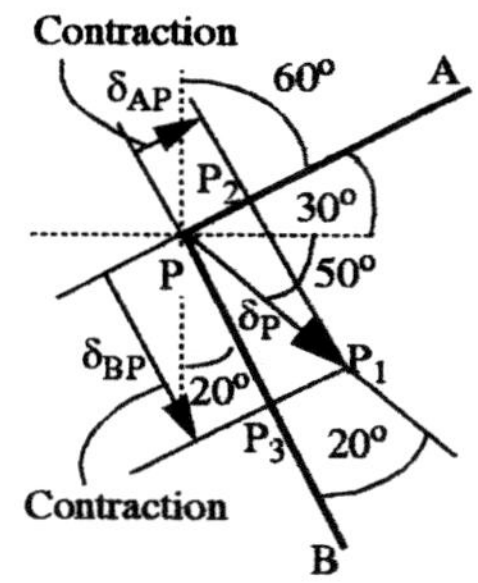

As per small strain approximation, we need component of PP_1 in the original direction of AP and BP, i.e., PP_2 represents the deformation of bar AP and PP_3 represents the deformation of bar BP and are calculated as shown below.

$\delta_{AP} = \delta_P \cos 80 = 0.00347$ or $\boxed{\delta_{AP} = 0.0035\text{inch contraction}}$

$\delta_{BP} = \delta_P \cos 20 = 0.01879$ or $\boxed{\delta_{BP} = 0.0188\text{inch contraction}}$

2.43 A gap of 0.004 of an inch exists between the rigid bar and bar A before the load P is applied. The rigid bar is hinged at point C. Due to force P, the strain in bar A was found to be - 500 μ. Determine the strain in bar B. The lengths of bars A and B are 30 and 50 inches respectively.

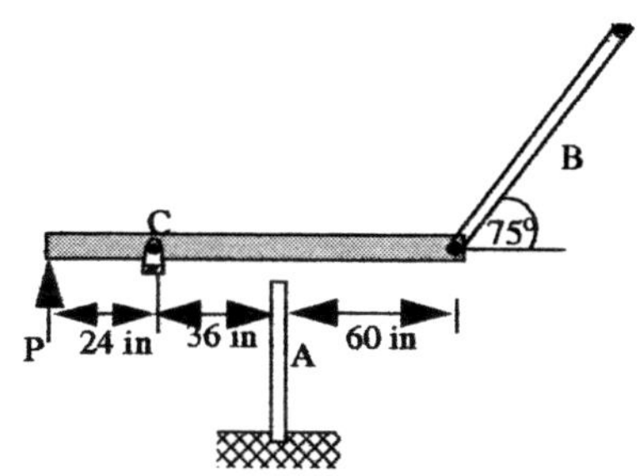

Fig. P2.43

Solution ε_A=-500m L_A=30in L_B=50in ε_B=?

The deformation of A is: $\delta_A = (500)(10^{-6})(30) = 0.015$ in contraction

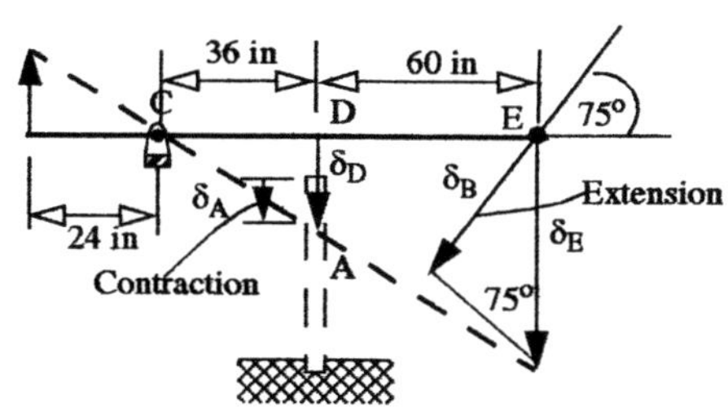

From geometry of the deformed shape we obtain: $\delta_D = \delta_A + \text{gap} = 0.015 + 0.004 = 0.019\text{in}$ and

$$\frac{\delta_D}{36} = \frac{\delta_E}{96} \qquad \therefore \delta_E = \frac{96}{36}(0.019) = 0.05067\text{in} \text{ and } \delta_B = \delta_E \cos 15 = 0.04894$$

The average normal strain in B is: $\varepsilon_B = \frac{\delta_B}{L_B} = \frac{0.04894}{50} = 0.97881(10^{-3})$ $\boxed{\varepsilon_B = 978.8\mu\frac{\text{in}}{\text{in}}}$

2.44 For the truss shown, the pin displacements in the x and y direction, given by u and v respectively were computed by FiniteElement Method and are as given. Determine the axial strain in members AB, BF,

FG, and GB.

$u_B = 12.6$ mm $\quad v_B = -24.48$ mm

$u_C = 21.0$ mm $\quad v_C = -69.97$ mm

$u_D = -16.8$ mm $\quad v_D = -119.65$ mm

$u_E = -12.6$ mm $\quad v_E = -69.97$ mm

$u_F = -8.4$ mm $\quad v_F = -28.68$ mm

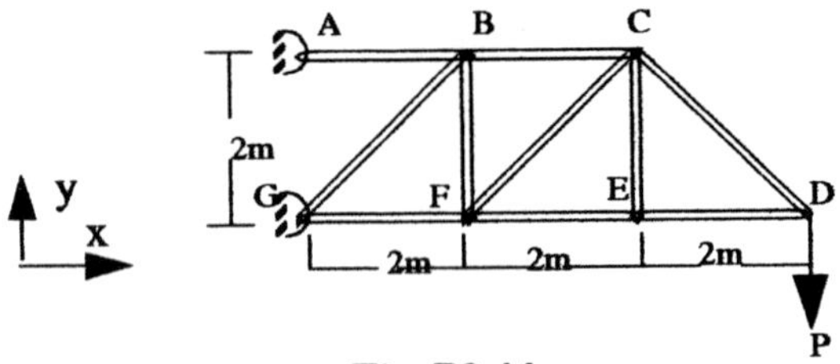

Fig. P2.44

Solution ε_{AB}=? $\quad\varepsilon_{BF}$=? $\quad\varepsilon_{FG}$=? $\quad\varepsilon_{GB}$=?

---

1. Deformation vector calculations

$$\bar{D}_{AB} = (u_B - u_A)\bar{i} + (v_B - v_A)\bar{j} = (12.6\bar{i} - 24.48\bar{j})\text{mm}$$

$$\bar{D}_{BF} = (u_F - u_B)\bar{i} + (v_F - v_B)\bar{j}$$

$$\bar{D}_{BF} = (-8.4 - 12.6)\bar{i} + (-28.68 + 24.48)\bar{j} = (-21\bar{i} - 4.2\bar{j})\text{mm}$$

$$\bar{D}_{FG} = (u_G - u_F)\bar{i} + (v_G - v_F)\bar{j} = (8.4\bar{i} + 28.68\bar{j})\text{mm}$$

$$\bar{D}_{GB} = (u_B - u_G)\bar{i} + (v_B - v_G)\bar{j} = (12.6\bar{i} - 24.48\bar{j})\text{mm}$$

2. Unit vector calculations

$$\bar{i}_{AB} = \bar{i}$$

$$\bar{i}_{BF} = -\bar{j}$$

$$\bar{i}_{FG} = -\bar{i}$$

$$\bar{i}_{GB} = \cos 45\bar{i} + \sin 45\bar{j} = 0.707\bar{i} + 0.707\bar{j}$$

3. Deformation calculation

$$\delta_{AB} = \bar{D}_{AB} \cdot \bar{i}_{AB} = 12.6\text{mm}$$

$$\delta_{BF} = \bar{D}_{BF} \cdot \bar{i}_{BF} = 4.2\text{mm}$$

$$\delta_{FG} = \bar{D}_{FG} \cdot \bar{i}_{FG} = -8.4\text{mm}$$

$$\delta_{GB} = (12.6)(0.707) - (24.48)(0.707) = -8.4004$$

4. Strain Calculation

$$\varepsilon_{AB} = \frac{\delta_{AB}}{L_{AB}} = \frac{12.6}{2(10^3)} = 6.3(10^{-3}) \qquad \boxed{\varepsilon_{AB} = 6300\mu\frac{\text{mm}}{\text{mm}}}$$

$$\varepsilon_{BF} = \frac{\delta_{BF}}{L_{BF}} = \frac{4.2}{2(10^3)} = 2.1(10^{-3}) \qquad \boxed{\varepsilon_{BF} = 2100\mu\frac{\text{mm}}{\text{mm}}}$$

$$\varepsilon_{FG} = \frac{\delta_{FG}}{L_{FG}} = \frac{-8.4}{2(10^3)} = -4.2(10^{-3}) \qquad \boxed{\varepsilon_{FG} = -4200\mu\frac{\text{mm}}{\text{mm}}}$$

$$\varepsilon_{GB} = \frac{\delta_{GB}}{L_{GB}} = \frac{-8.4004}{2\sqrt{2}(10^3)} = -2.9698(10^{-3}) \qquad \boxed{\varepsilon_{GB} = -2970\mu\frac{\text{mm}}{\text{mm}}}$$

2.45 For the truss shown, the pin displacements in the x and y direction given by u and v respectively were computed by FiniteElement Method and are as given. Determine the axial strain in members BC, CF, and FE.

$u_B = 12.6$ mm $v_B = -24.48$ mm
$u_C = 21.0$ mm $v_C = -69.97$ mm
$u_D = -16.8$ mm $v_D = -119.65$ mm
$u_E = -12.6$ mm $v_E = -69.97$ mm
$u_F = -8.4$ mm $v_F = -28.68$ mm

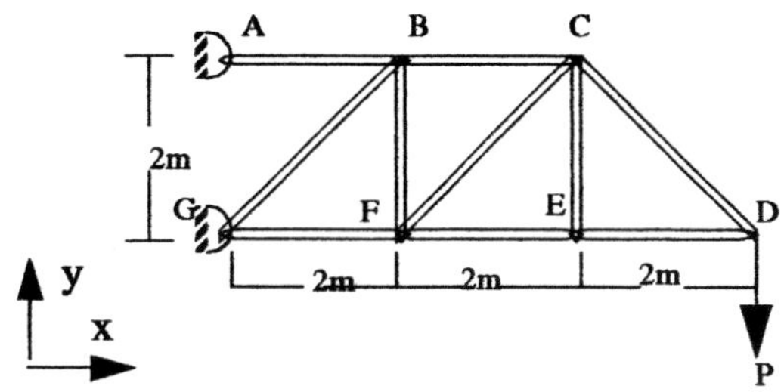

Solution $\varepsilon_{BC}=?$ $\varepsilon_{CF}=?$ $\varepsilon_{FE}=?$

---

1. Deformation vector calculation

$$\bar{D}_{BC} = (u_C - u_B)\bar{i} + (v_C - v_B)\bar{j} = (21 - 12.6)\bar{i} + (-69.97 + 24.48)\bar{j} = (8.4\bar{i} - 45.52\bar{j})\text{mm}$$

$$\bar{D}_{CF} = (u_F - u_C)\bar{i} + (v_F - v_C)\bar{j} = (-8.4 - 21.0)\bar{i} + (-28.68 + 69.97)\bar{j} = (29.4\bar{i} + 41.29\bar{j})\text{mm}$$

$$\bar{D}_{FE} = (u_E - u_F)\bar{i} + (v_E - v_F)\bar{j} = (-12.6 + 8.4)\bar{i} + (-69.97 + 28.68)\bar{j} = (-4.2\bar{i} - 41.29\bar{j})\text{mm}$$

2. Unit vector calculations

$$\bar{i}_{BC} = \bar{i}$$

$$\bar{i}_{CF} = -\cos 45\bar{i} - \sin 45\bar{j} = -0.707\bar{i} - 0.707\bar{j}$$

$$\bar{i}_{FE} = \bar{i}$$

3. Deformation calculation

$$\delta_{BC} = \bar{D}_{BC} \cdot \bar{i}_{BC} = 8.4\text{mm}$$

$$\delta_{CF} = \bar{D}_{CF} \cdot \bar{i}_{CF} = (29.4)(0.707) - (41.29)(0.707) = -8.408\text{mm}$$

$$\delta_{FE} = \bar{D}_{FE} \cdot \bar{i}_{FE} = -4.2\text{mm}$$

4. Strain Calculation

$$\varepsilon_{BC} = \frac{\delta_{BC}}{L_{BC}} = \frac{8.4}{2(10^3)} = 4.2(10^{-3}) \quad \text{or} \quad \boxed{\varepsilon_{BC} = 4200\mu\frac{\text{mm}}{\text{mm}}}$$

$$\varepsilon_{CF} = \frac{\delta_{CF}}{L_{CF}} = \frac{-8.408}{2\sqrt{2}(10^3)} = 2.973(10^{-3}) \quad \text{or} \quad \boxed{\varepsilon_{CF} = -2973\mu\frac{\text{mm}}{\text{mm}}}$$

$$\varepsilon_{FE} = \frac{\delta_{FE}}{L_{FE}} = \frac{-4.2}{2(10^3)} = -2.1(10^{-3}) \quad \text{or} \quad \boxed{\varepsilon_{FE} = -2100\mu\frac{\text{mm}}{\text{mm}}}$$

2.46 For the truss shown, the pin displacements in the x and y direction given by u and v respectively, were computed by FiniteElement Method and are as given. Determine the axial strain in members ED, DC, and CE.

$u_B = 12.6$ mm $v_B = -24.48$ mm
$u_C = 21.0$ mm $v_C = -69.97$ mm
$u_D = -16.8$ mm $v_D = -119.65$ mm
$u_E = -12.6$ mm $v_E = -69.97$ mm
$u_F = -8.4$ mm $v_F = -28.68$ mm

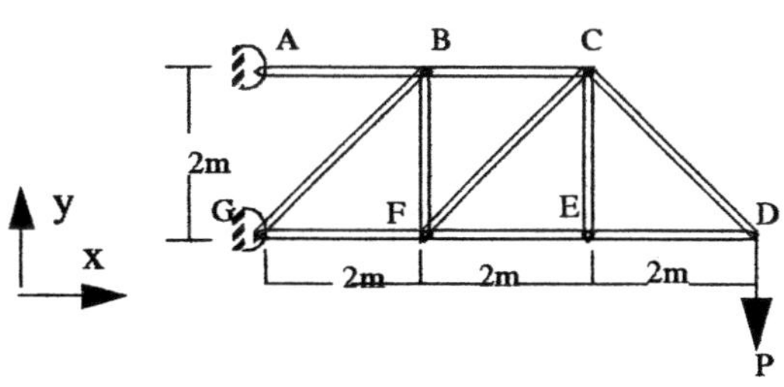

Solution ε_{ED}=? ε_{DC}=? ε_{CE}=?

1. Deformation vector calculation

$$\bar{D}_{ED} = (u_D - u_E)\bar{i} + (v_D - v_E)\bar{j} = (-16.8 + 12.6)\bar{i} + (-119.65 + 69.97)\bar{j} \text{ or } \bar{D}_{ED} = (-4.2\bar{i} - 49.98\bar{j})\text{mm}$$

$$\bar{D}_{DC} = (u_C - u_D)\bar{i} + (v_C - v_D)\bar{j} = (21.0 + 16.8)\bar{i} + (-69.97 + 119.65)\bar{j} \text{ or } \bar{D}_{DC} = (37.8\bar{i} + 49.68\bar{j})\text{mm}$$

$$\bar{D}_{CE} = (u_E - u_C)\bar{i} + (v_E - v_C)\bar{j} = (-12.6 - 21)\bar{i} + (-69.97 + 69.97)\bar{j} = -33.6\bar{i} \text{ mm}$$

2. Unit vector calculations

$$\bar{i}_{ED} = \bar{i}$$

$$\bar{i}_{DC} = -\cos 45\bar{i} + \sin 45\bar{j} = -0.707\bar{i} + 0.707\bar{j}$$

$$\bar{i}_{CE} = \bar{j}$$

3. Deformation calculation

$$\delta_{ED} = \bar{D}_{ED} \cdot \bar{i}_{ED} = -4.2\text{mm}$$

$$\delta_{DC} = \bar{D}_{DC} \cdot \bar{i}_{DC} = -(37.8)(0.707) + (49.68)(0.707) = 8.4004\text{mm}$$

$$\delta_{CE} = \bar{D}_{CE} \cdot \bar{i}_{CE} = 0$$

4. Strain Calculation

$$\varepsilon_{ED} = \frac{\delta_{ED}}{L_{ED}} = \frac{-4.2}{2(10^3)} = -2.1(10^{-3}) \quad \text{or} \quad \boxed{\varepsilon_{ED} = -2100\mu\frac{\text{mm}}{\text{mm}}}$$

$$\varepsilon_{DC} = \frac{\delta_{DC}}{L_{DC}} = \frac{8.4004}{2\sqrt{2}(10^3)} = 2.970(10^{-3}) \quad \text{or} \quad \boxed{\varepsilon_{DC} = 2970\mu\frac{\text{mm}}{\text{mm}}}$$

$$\boxed{\varepsilon_{CE} = 0}$$

2.47 For the truss shown, the pin displacements in the x and y directions given by u and v respectively, were computed by Finite Element Method and are as given. Determine the axial strains in members AB, BG, GA, and AH.

$u_B = 7.00$ mm $\quad v_B = 1.500$ mm

$u_C = 17.55$ mm $\quad v_C = 3.000$ mm

$u_D = 20.22$ mm $\quad v_D = -4.125$ mm

$u_E = 22.88$ mm $\quad v_E = -32.250$ mm

$u_F = 9.00$ mm $\quad v_F = -33.750$ mm

$u_G = 7.00$ mm $\quad v_G = -4.125$ mm

$u_H = 0 \quad v_H = 0$

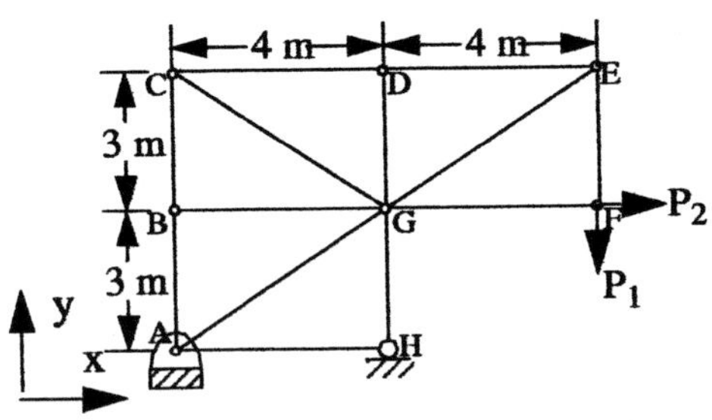

Fig. P2.47

Solution ε_{AB}=? ε_{BG}=? ε_{GA}=? ε_{AH}=?

1. Deformation vector calculations

$$\bar{D}_{AB} = (u_B - u_A)\bar{i} + (v_B - v_A)\bar{j} = (7\bar{i} + 1.5\bar{j})\text{mm}$$

$$\bar{D}_{BG} = (u_G - u_B)\bar{i} + (v_G - v_B)\bar{j}$$

$$\bar{D}_{BG} = (7 - 7)\bar{i} + (-4.125 - 1.5)\bar{j} = -5.625\bar{j}\text{mm}$$

$\bar{D}_{GA} = (u_A - u_G)\dot{i} + (v_A - v_G)\dot{j} = (-7\dot{i} + 4.125\dot{j})\text{mm}$

$\bar{D}_{AH} = (u_H - u_A)\dot{i} + (v_H - v_A)\dot{j} = 0$

2. Unit vector calculations

$\dot{i}_{AB} = \dot{j}$

$\dot{i}_{BG} = \dot{i}$

$$\dot{i}_{GA} = \frac{-4\dot{i} - 3\dot{j}}{\sqrt{3^2 + 4^2}} = -0.8\dot{i} - 0.6\dot{j}$$

$\dot{i}_{AH} = \dot{i}$

3. Deformation calculation

$\delta_{AB} = \bar{D}_{AB} \cdot \dot{i}_{AB} = 1.5\text{mm}$

$\delta_{BG} = \bar{D}_{BG} \cdot \dot{i}_{BG} = 0$

$\delta_{GA} = \bar{D}_{GA} \cdot \dot{i}_{GA} = (-7)(-0.8) + (4.125)(-0.6) = 3.125\text{mm}$

$\delta_{AH} = 0$

4. Strain Calculation

$$\varepsilon_{AB} = \frac{\delta_{AB}}{L_{AB}} = \frac{1.5}{3(10^3)} = 0.5(10^{-3}) \quad \text{or} \quad \boxed{\varepsilon_{AB} = 500\mu\frac{\text{mm}}{\text{mm}}}$$

$$\boxed{\varepsilon_{BG} = 0}$$

$$\varepsilon_{GA} = \frac{\delta_{GA}}{L_{GA}} = \frac{3.125}{5(10^3)} = 0.625(10^{-3}) \quad \text{or} \quad \boxed{\varepsilon_{GA} = 625\mu\frac{\text{mm}}{\text{mm}}}$$

$$\boxed{\varepsilon_{AH} = 0}$$

2.48 For the truss shown, the pin displacements in the x and y directions given by u and v respectively, were computed by Finite Element Method and are as given. Determine the axial strains in members BC, CG, GB, and CD.

$u_B = 7.00$ mm	$v_B = 1.500$ mm
$u_C = 17.55$ mm	$v_C = 3.000$ mm
$u_D = 20.22$ mm	$v_D = -4.125$ mm
$u_E = 22.88$ mm	$v_E = -32.250$ mm
$u_F = 9.00$ mm	$v_F = -33.750$ mm
$u_G = 7.00$ mm	$v_G = -4.125$ mm
$u_H = 0$	$v_H = 0$

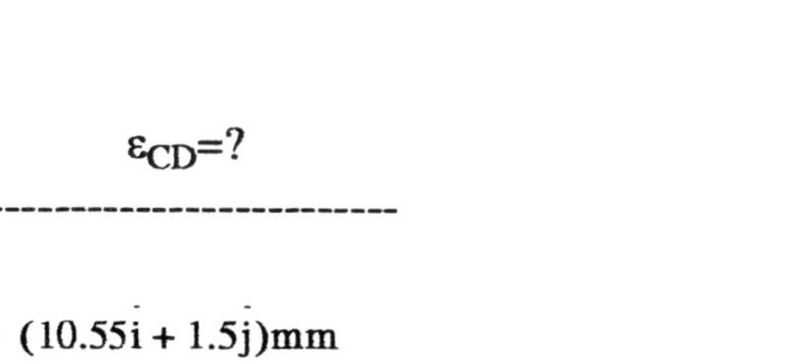

Solution ε_{BC}=? ε_{CG}=? ε_{GB}=? ε_{CD}=?

1. Deformation vector calculations

$\bar{D}_{BC} = (u_C - u_B)\dot{i} + (v_C - v_B)\dot{j} = (17.55 - 7)\dot{i} + (3 - 1.5)\dot{j} = (10.55\dot{i} + 1.5\dot{j})\text{mm}$

$\bar{D}_{CG} = (u_G - u_C)\dot{i} + (v_G - v_C)\dot{j} = (7 - 17.55)\dot{i} + (-4.125 - 3)\dot{j} = (-10.55\dot{i} - 7.125\dot{j})\text{mm}$

$\bar{D}_{GB} = (u_B - u_G)\dot{i} + (v_B - v_G)\dot{j} = (7-7)\dot{i} + (1.5 + 4.125)\dot{j} = (5.625\dot{j})mm$

$\bar{D}_{CD} = (u_D - u_C)\dot{i} + (v_D - v_C)\dot{j} = (20.22 - 17.55)\dot{i} + (-4.125 - 3)\dot{j} = (2.67\dot{i} - 7.125\dot{j})mm$

2. Unit vector calculations

$\dot{i}_{BC} = \dot{j}$

$\dot{i}_{CG} = \frac{4\dot{i} - 3\dot{j}}{5} = 0.8\dot{i} - 0.6\dot{j}$

$\dot{i}_{GB} = -\dot{i}$

$\dot{i}_{CD} = \dot{i}$

3. Deformation calculation

$\delta_{BC} = \bar{D}_{BC} \cdot \dot{i}_{BC} = 1.5mm$

$\delta_{CG} = \bar{D}_{CG} \cdot \dot{i}_{CG} = (0.8)(-10.55) + (-0.6)(-7.125) = -4.165mm$

$\delta_{GB} = \bar{D}_{GB} \cdot \dot{i}_{GB} = 0$

$\delta_{CD} = 2.67mm$

4. Strain Calculation

$\varepsilon_{BC} = \frac{\delta_{BC}}{L_{BC}} = \frac{1.5}{3(10^3)} = 0.5(10^{-3})$ or $\boxed{\varepsilon_{BC} = 500\mu\frac{mm}{mm}}$

$\varepsilon_{CG} = \frac{\delta_{CG}}{L_{CG}} = \frac{-4.165}{5(10^3)} = -0.833(10^{-3})$ or $\boxed{\varepsilon_{CG} = -833\mu\frac{mm}{mm}}$

$\varepsilon_{GB} = \frac{\delta_{GB}}{L_{GB}} = 0$ $\boxed{\varepsilon_{GB} = 0}$

$\varepsilon_{CD} = \frac{\delta_{CD}}{L_{CD}} = \frac{2.67}{4(10^3)} = 0.6675(10^{-3})$ or $\boxed{\varepsilon_{CD} = 667.5\mu\frac{mm}{mm}}$

2.49 For the truss shown, the pin displacements in the x and y directions given by u and v respectively, were computed by Finite Element Method and are as given. Determine the axial strains in members GF, FE, EG, and DE..

$u_B = 7.00$ mm	$v_B = 1.500$ mm
$u_C = 17.55$ mm	$v_C = 3.000$ mm
$u_D = 20.22$ mm	$v_D = -4.125$ mm
$u_E = 22.88$ mm	$v_E = -32.250$ mm
$u_F = 9.00$ mm	$v_F = -33.750$ mm
$u_G = 7.00$ mm	$v_G = -4.125$ mm
$u_H = 0$	$v_H = 0$

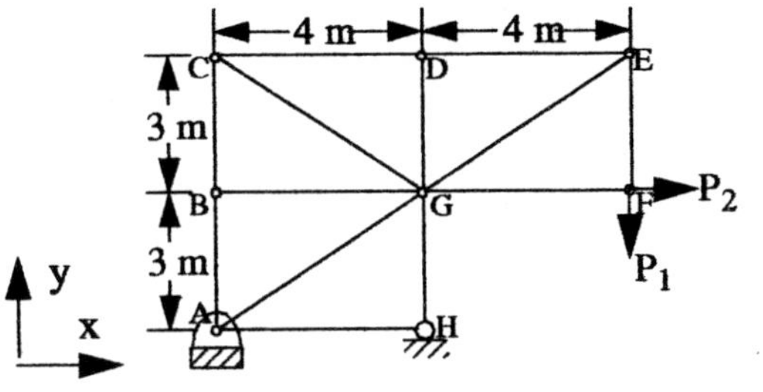

Solution ε_{GF}=? ε_{FE}=? ε_{EG}=? ε_{DE}=?

---

1. Deformation vector calculations

$\bar{D}_{GF} = (u_F - u_G)\dot{i} + (v_F - v_G)\dot{j} = (9-7)\dot{i} + (-33.75 + 4.125)\dot{j} = (2\dot{i} - 29.625\dot{j})mm$

$$\bar{D}_{FE} = (u_E - u_F)\hat{i} + (v_E - v_F)\hat{j} = (22.88 - 9)\hat{i} + (-32.25 + 33.75)\hat{j} = (13.88\hat{i} + 1.5\hat{j})\text{mm}$$

$$\bar{D}_{EG} = (u_G - u_E)\hat{i} + (v_G - v_E)\hat{j} = (7 - 22.88)\hat{i} + (-4.125 + 32.25)\hat{j} = (-15.88\hat{i} + 28.125\hat{j})\text{mm}$$

$$\bar{D}_{DE} = (u_E - u_D)\hat{i} + (v_E - v_D)\hat{j} = (22.88 - 20.22)\hat{i} + (-32.25 + 4.125)\hat{j} \text{ or } \bar{D}_{DE} = (2.66\hat{i} - 28.125\hat{j})\text{mm}$$

2. Unit vector calculations

$$\hat{i}_{GF} = \hat{i}$$

$$\hat{i}_{FE} = \hat{j}$$

$$\hat{i}_{EG} = \frac{-4\hat{i} - 3\hat{j}}{5} = -0.8\hat{i} - 0.6\hat{j}$$

$$\hat{i}_{DE} = \hat{i}$$

3. Deformation calculation

$$\delta_{GF} = \bar{D}_{GF} \cdot \hat{i}_{GF} = 2\text{mm}$$

$$\delta_{FE} = \bar{D}_{FE} \cdot \hat{i}_{FE} = 1.5\text{mm}$$

$$\delta_{EG} = \bar{D}_{EG} \cdot \hat{i}_{EG} = (-15.88)(-0.8) + (28.125)(-0.6) = -4.171\text{mm}$$

$$\delta_{DE} = 2.66\text{mm}$$

4. Strain Calculation

$$\varepsilon_{GF} = \frac{\delta_{GF}}{L_{GF}} = \frac{2}{4(10^3)} = 0.5(10^{-3}) \quad \text{or} \quad \boxed{\varepsilon_{GF} = 500\mu\frac{\text{mm}}{\text{mm}}}$$

$$\varepsilon_{FE} = \frac{\delta_{FE}}{L_{FE}} = \frac{1.5}{3(10^3)} = 0.5(10^{-3}) \quad \text{or} \quad \boxed{\varepsilon_{FE} = 500\mu\frac{\text{mm}}{\text{mm}}}$$

$$\varepsilon_{EG} = \frac{\delta_{EG}}{L_{EG}} = \frac{-4.171}{5(10^3)} = -0.8342(10^{-3}) \quad \text{or} \quad \boxed{\varepsilon_{EG} = -834\mu\frac{\text{mm}}{\text{mm}}}$$

$$\varepsilon_{DE} = \frac{\delta_{DE}}{L_{DE}} = \frac{2.66}{4(10^3)} = 0.665(10^{-3}) \quad \text{or} \quad \boxed{\varepsilon_{DE} = 665\mu\frac{\text{mm}}{\text{mm}}}$$

2.50 Three poles are pin connected to a ring at P and to the supports on the ground. The ring slides on a vertical rigid pole by 2 inches, as shown. The coordinates of the four points are as given. Determine the normal strain in each bar due to the movement of the ring.

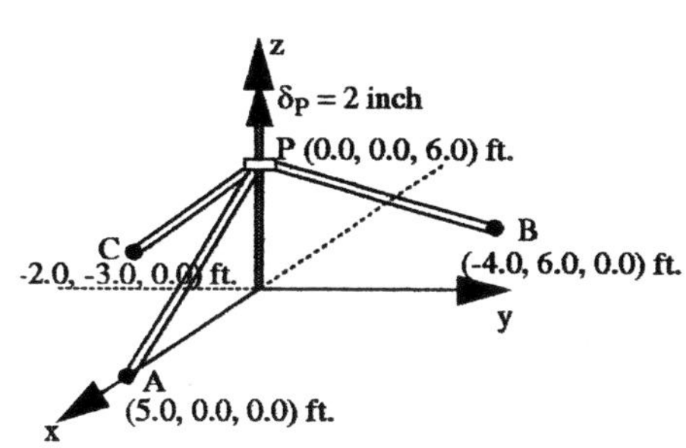

Fig. P2.50

1. Deformation vector calculations: Points A,B, and C are fixed. Point P moves vertically upwards. Thus, the deformation vectors for the three bars is given by:

$$\bar{D}_{AP} = 2\bar{k} \qquad \bar{D}_{BP} = 2\bar{k} \qquad \bar{D}_{CP} = 2\bar{k} \qquad (1)$$

2. Unit vector calcultions: The unit vectors in the directions of the rod can be found as shown below.

$$\hat{r}_{AP} = -5.0\hat{i} + 6\hat{k} \qquad |\hat{r}_{AP}| = \sqrt{5^2 + 6^2} = 7.81 \qquad (2)$$

$$\hat{i}_{AP} = \frac{\hat{r}_{AP}}{|\hat{r}_{AP}|} = -0.6402\hat{i} + 0.7682\hat{k}$$ 3

$$\hat{r}_{BP} = 4.0\hat{i} - 6\hat{j} + 6\hat{k} \qquad |\hat{r}_{BP}| = \sqrt{4^2 + 6^2 + 6^2} = 9.381$$ 4

$$\hat{i}_{BP} = \frac{\hat{r}_{BP}}{|\hat{r}_{BP}|} = 0.4264\hat{i} - 0.6396\hat{i} + 0.6396\hat{k}$$ 5

$$\hat{r}_{CP} = -2.0\hat{i} - 3\hat{j} + 6\hat{k} \qquad |\hat{r}_{BP}| = \sqrt{2^2 + 3^2 + 6^2} = 7.0\text{ft}$$ 6

$$\hat{i}_{CP} = \frac{\hat{r}_{CP}}{|\hat{r}_{CP}|} = 0.2857\hat{i} + 0.4286\hat{i} + 0.8571\hat{k}$$ 7

3. Deformation calculations

$$\delta_{AP} = \bar{D}_{AP} \bullet \hat{i}_{AP} = (2)(0.7682) = 1.5364 \text{ inch}$$ 8

$$\delta_{BP} = \bar{D}_{BP} \bullet \hat{i}_{BP} = (2)(0.6396) = 1.2792 \text{ inch}$$ 9

$$\delta_{CP} = \bar{D}_{CP} \bullet \hat{i}_{CP} = (2)(0.8571) = 1.7142 \text{ inch}$$ 10

4. Strain calcultions: The lengths of each bar are the magnitudes of the position vectors in Eqs. 2, 4, and 6. The normal strains can be found as shown below.

$$\varepsilon_{AP} = \frac{\delta_{AP}}{L_{AP}} = \frac{1.5364}{(12)(7.81)} = 16.393(10^{-3}) \qquad \boxed{\varepsilon_{AP} = 16393\mu \text{ in/in}}$$

$$\varepsilon_{BP} = \frac{\delta_{BP}}{L_{BP}} = \frac{1.2792}{(12)(9.381)} = 11.363(10^{-3}) \qquad \boxed{\varepsilon_{BP} = 11363\mu \text{ in/in}}$$

$$\varepsilon_{CP} = \frac{\delta_{CP}}{L_{CP}} = \frac{1.7142}{(12)(7)} = 20.407(10^{-3}) \qquad \boxed{\varepsilon_{CP} = 20407\mu \text{ in/in}}$$

Problem Set 2.2

2.51 A rectangle deforms into the colored shape shown in each problem. Determine the average values of strain components ε_{xx}, ε_{yy}, and γ_{xy} at point A.

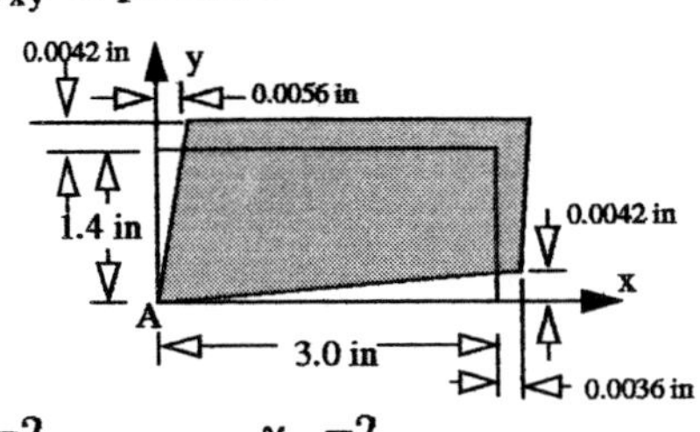

Solution ε_{xx}=? ε_{yy}=? γ_{xy}=?

The average strains can be found using Eqs.2.9a, 2.9b,and 2.9e as shown below.

$$\varepsilon_{xx} = \frac{\Delta u}{\Delta x} = \frac{0.0036}{3} = 0.0012 \qquad \boxed{\varepsilon_{xx} = 1200\mu\frac{\text{in}}{\text{in}}}$$

$$\varepsilon_{yy} = \frac{\Delta v}{\Delta y} = \frac{0.0042}{1.4} = 0.003 \qquad \boxed{\varepsilon_{yy} = 3000\mu\frac{\text{in}}{\text{in}}}$$

$$\gamma_{xy} = \frac{\Delta u}{\Delta y} + \frac{\Delta v}{\Delta x} = \frac{0.0056}{1.4} + \frac{0.0042}{3} = 0.0054 \qquad \boxed{\gamma_{xy} = 5400\mu\text{rad}}$$

2.52 A rectangle deforms into the colored shape shown in each problem. Determine the average values of strain components ε_{xx}, ε_{yy}, and γ_{xy} at point A.

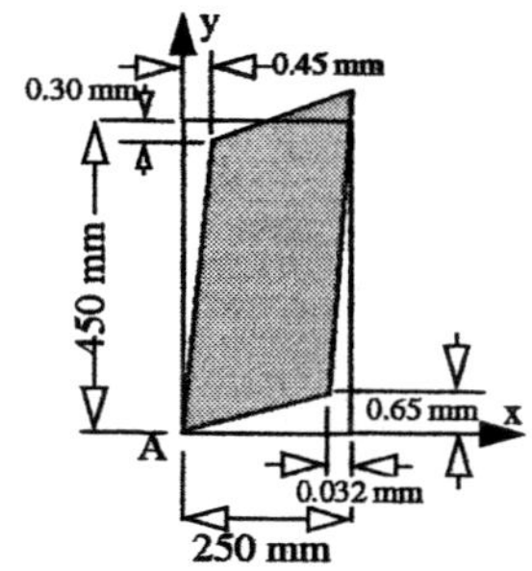

Solution ε_{xx}=? ε_{yy}=? γ_{xy}=?

The average strains can be found using Eqs.2.9a, 2.9b,and 2.9e as shown below.

$$\varepsilon_{xx} = \frac{\Delta u}{\Delta x} = \frac{-0.32}{250} = -0.128(10^{-3})$$

$$\boxed{\varepsilon_{xx} = -128\mu\frac{mm}{mm}}$$

$$\varepsilon_{yy} = \frac{\Delta v}{\Delta y} = \frac{-0.3}{450} = -0.6667(10^{-3})$$

$$\boxed{\varepsilon_{yy} = -666.7\mu\frac{mm}{mm}}$$

$$\gamma_{xy} = \frac{\Delta u}{\Delta y} + \frac{\Delta v}{\Delta x} = \frac{0.45}{450} + \frac{0.65}{250} = 3.6(10^{-3})$$

$$\boxed{\gamma_{xy} = 3600\mu rad}$$

2.53 A rectangle deforms into the colored shape shown in each problem. Determine the average values of strain components ε_{xx}, ε_{yy}, and γ_{xy} at point A.

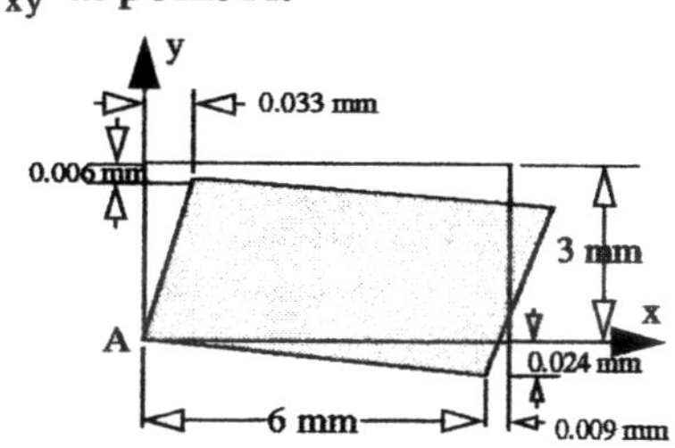

Solution ε_{xx}=? ε_{yy}=? γ_{xy}=?

The average strains can be found using Eqs.2.9a, 2.9b,and 2.9e as shown below.

$$\varepsilon_{xx} = \frac{\Delta u}{\Delta x} = \frac{-0.009}{6} = -1.5(10^{-3})$$

$$\boxed{\varepsilon_{xx} = -1500\mu\frac{mm}{mm}}$$

$$\varepsilon_{yy} = \frac{\Delta v}{\Delta y} = \frac{-0.006}{3} = -2.0(10^{-3})$$

$$\boxed{\varepsilon_{yy} = -2000\mu\frac{mm}{mm}}$$

$$\gamma_{xy} = \frac{\Delta u}{\Delta y} + \frac{\Delta v}{\Delta x} = \frac{0.033}{3} + \frac{(-0.024)}{6} = 7(10^{-3})$$

$$\boxed{\gamma_{xy} = 7000\mu rad}$$

2.54 Displacements u and v in the x and y directions respectively were measured by Moire Interferometry method at many points on a body. Displacements of four points on a body are as given. Determine the average values of strain components ε_{xx}, ε_{yy}, and γ_{xy} at point A shown.

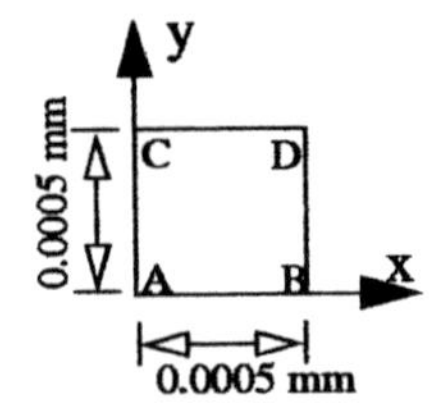

$u_A = 0$ $\quad v_A = 0$

$u_B = 0.625\mu mm$ $\quad v_B = -0.3125\mu mm$

$u_C = -0.500\mu mm$ $\quad v_C = -0.5625\mu mm$

$u_D = 0.250\mu mm$ $\quad v_D = -1.125\mu mm$

Solution $\quad \varepsilon_{xx}=?$ $\quad \varepsilon_{yy}=?$ $\quad \gamma_{xy}=?$

The average strains can be found using Eqs.2.9a, 2.9b,and 2.9e as shown below.

$$\varepsilon_{xx} = \frac{u_B - u_A}{x_B - x_A} = \frac{0.625(10^{-6})}{0.0005} = 1250(10^{-6}) \quad \text{or} \qquad \boxed{\varepsilon_{xx} = 1250\mu\frac{mm}{mm}}$$

$$\varepsilon_{yy} = \frac{v_C - v_A}{y_C - y_A} = \frac{-0.5625(10^{-6})}{0.0005} = -1125(10^{-6}) \quad \text{or} \qquad \boxed{\varepsilon_{yy} = -1125\mu\frac{mm}{mm}}$$

$$\gamma_{xy} = \frac{u_C - u_A}{y_C - y_A} + \frac{v_B - v_A}{x_B - x_A} = \frac{-0.5(10^{-6})}{0.0005} + \frac{(-0.3125)(10^{-6})}{0.0005} = -1625(10^{-6}) \text{ or}$$

$$\boxed{\gamma_{xy} = -1625\mu rad}$$

2.55 Displacements u and v in the x and y directions respectively were measured by Moire Interferometry method at many points on a body. Displacements of four points on a body are as given. Determine the average values of strain components ε_{xx}, ε_{yy}, and γ_{xy} at point A shown.

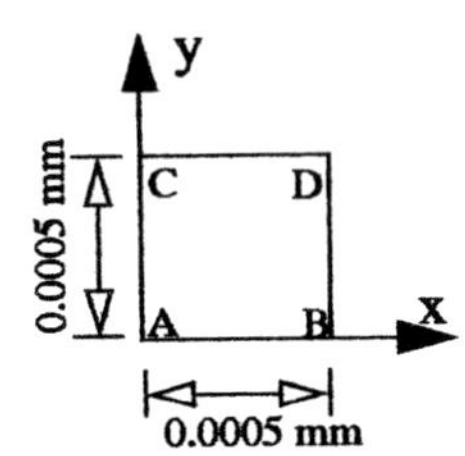

Fig. P2.55

$u_A = 0.625\mu mm$ $\quad v_A = -0.3125\mu mm$

$u_B = 1.500\mu mm$ $\quad v_B = -0.5000\mu mm$

$u_C = 0.250\mu mm$ $\quad v_C = -1.125\mu mm$

$u_D = 1.250\mu mm$ $\quad v_D = -1.5625\mu mm$

Solution $\quad \varepsilon_{xx}=?$ $\quad \varepsilon_{yy}=?$ $\quad \gamma_{xy}=?$

The average strains can be found using Eqs.2.9a, 2.9b,and 2.9e as shown below.

$$\varepsilon_{xx} = \frac{u_B - u_A}{x_B - x_A} = \frac{(1.5 - 0.625)(10^{-6})}{0.0005} = 1750(10^{-6}) \qquad \boxed{\varepsilon_{xx} = 1750\mu\frac{mm}{mm}}$$

$$\varepsilon_{yy} = \frac{v_C - v_A}{y_C - y_A} = \frac{(-1.125 + 0.3125)(10^{-6})}{0.0005} = -1625(10^{-6}) \qquad \boxed{\varepsilon_{yy} = -1625\mu\frac{mm}{mm}}$$

$$\gamma_{xy} = \frac{u_C - u_A}{y_C - y_A} + \frac{v_B - v_A}{x_B - x_A} = \frac{(0.25 - 0.625)(10^{-6})}{0.0005} + \frac{(-0.5 + 0.3125)(10^{-6})}{0.0005} \qquad \boxed{\gamma_{xy} = -1125\mu rad}$$

2.56 Displacements u and v in the x and y directions respectively were measured by Moire Interferometry method at many points on a body. Displacements of four points on a body are as given. Determine the average values of strain components ε_{xx}, ε_{yy}, and γ_{xy} at point A shown.

$u_A = -0.500\mu mm$ $v_A = -0.5625\mu mm$

$u_B = 0.250\mu mm$ $v_B = -1.125\mu mm$

$u_C = -1.250\mu mm$ $v_C = -1.250\mu mm$

$u_D = -0.375\mu mm$ $v_D = -2.0625\mu mm$

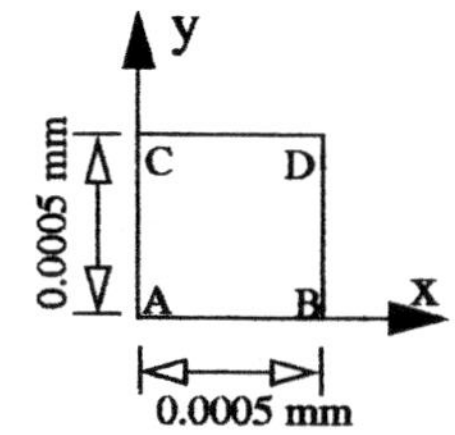

Fig. P2.56

Solution $\varepsilon_{xx}=?$ $\varepsilon_{yy}=?$ $\gamma_{xy}=?$

The average strains can be found using Eqs.2.9a, 2.9b,and 2.9e as shown below.

$$\varepsilon_{xx} = \frac{u_B - u_A}{x_B - x_A} = \frac{(0.25 + 0.5)(10^{-6})}{0.0005} = 1500(10^{-6})$$

$$\boxed{\varepsilon_{xx} = 1500\mu\frac{mm}{mm}}$$

$$\varepsilon_{yy} = \frac{v_C - v_A}{y_C - y_A} = \frac{(-1.25 + 0.5625)(10^{-6})}{0.0005} = -1375(10^{-6})$$

$$\boxed{\varepsilon_{yy} = -1375\mu\frac{mm}{mm}}$$

$$\gamma_{xy} = \frac{u_C - u_A}{y_C - y_A} + \frac{v_B - v_A}{x_B - x_A} = \frac{(-1.25 + 0.5)(10^{-6})}{0.0005} + \frac{(-1.125 + 0.5625)(10^{-6})}{0.0005} = -2625(10^{-6})$$

$$\boxed{\gamma_{xy} = -2625\mu rad}$$

2.57 Displacements u and v in the x and y directions respectively were measured by Moire Interferometry method at many points on a body. Displacements of four points on a body are as given. Determine the average values of strain components ε_{xx}, ε_{yy}, and γ_{xy} at point A shown.

$u_A = 0.250\mu mm$ $v_A = -1.125\mu mm$

$u_B = 1.250\mu mm$ $v_B = -1.5625\mu mm$

$u_C = -0.375\mu mm$ $v_C = -2.0625\mu mm$

$u_D = 0.750\mu mm$ $v_D = -2.7500\mu mm$

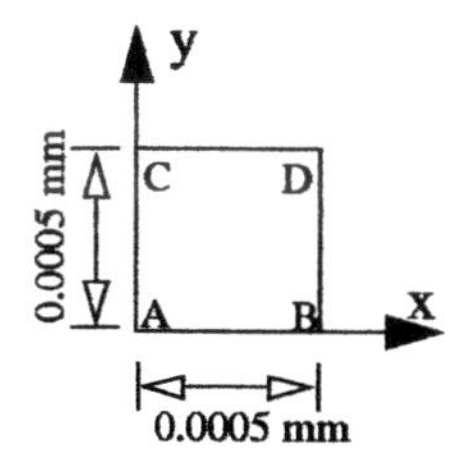

Fig. P2.57

Solution $\varepsilon_{xx}=?$ $\varepsilon_{yy}=?$ $\gamma_{xy}=?$

The average strains can be found using Eqs.2.9a, 2.9b,and 2.9e as shown below.

$$\varepsilon_{xx} = \frac{u_B - u_A}{x_B - x_A} = \frac{(1.25 - 0.25)(10^{-6})}{0.0005} = 2000(10^{-6})$$

$$\boxed{\varepsilon_{xx} = 2000\mu\frac{mm}{mm}}$$

$$\varepsilon_{yy} = \frac{v_C - v_A}{y_C - y_A} = \frac{(-2.0625 + 1.125)(10^{-6})}{0.0005} = -1875(10^{-6})$$

$$\boxed{\varepsilon_{yy} = -1875\mu\frac{mm}{mm}}$$

$$\gamma_{xy} = \frac{u_C - u_A}{y_C - y_A} + \frac{v_B - v_A}{x_B - x_A} = \frac{(-0.375 - 0.25)(10^{-6})}{0.0005} + \frac{(-1.5625 + 1.125)(10^{-6})}{0.0005} = -2125(10^{-6})$$

$$\boxed{\gamma_{xy} = -2125\mu rad}$$

2.58 In a tapered circular bar that is hanging vertically, the axial displacement due to its weight was found and is given below. Determine the axial strain ε_{xx} at x = 24 inches.

$$u(x) = \left[-19.44 + 1.44x - 0.01x^2 - \frac{933.12}{(72-x)}\right](10^{-3}) \text{ inches}$$

Solution $\varepsilon_{xx}(24)=?$

The strain at any point x can be found as shown below.

$$\varepsilon_{xx}(x) = \frac{du}{dx} = \left[1.44 - 0.02x - \frac{933.12}{(72-x)^2}\right](10^{-3})$$

$$\varepsilon_{xx}(24) = \frac{du}{dx} = \left[1.44 - 0.02(24) - \frac{933.12}{(72-24)^2}\right](10^{-3}) = 0.555(10^{-3}) \text{ or}$$

$$\boxed{\varepsilon_{xx}(24) = 555\mu\frac{in}{in}}$$

2.59 In a tapered rectangular bar that is hanging vertically, the axial displacement due to its weight was found and is given below. Determine the axial strain ε_{xx} at x = 150 mm

$$u(x) = [-50x + 20x^2 + 2.5\ln(1-0.8x)](10^{-6}) \text{ mm}$$

Solution $\varepsilon_{xx}(150)=?$

The strain at any point x can be found as shown below.

$$\varepsilon_{xx}(x) = \frac{du}{dx} = \left[-50 + 40x - \frac{2.5(0.8)}{(1-0.8x)}\right](10^{-6}) \quad \varepsilon_{xx}(150) = \frac{du}{dx} = \left[-50 + 40(150) - \frac{2}{(1-0.8(150))}\right](10^{-6})$$

$$\varepsilon_{xx}(150) = 5950(10^{-6}) \text{ or}$$

$$\boxed{\varepsilon_{xx}(150) = 5950\mu\frac{mm}{mm}}$$

2.60 The axial displacement in a quadratic one-dimensional finite element is as given below. Determine the strain at Node 2.

$$u(x) = \frac{u_1}{2a^2}(x-a)(x-2a) - \frac{u_2}{a^2}(x)(x-2a) + \frac{u_3}{2a^2}(x)(x-a)$$

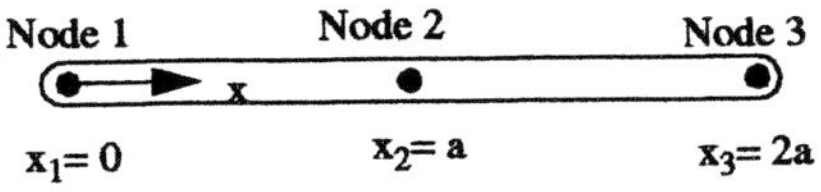

Solution $\varepsilon_{xx}(a)=?$

The strain at any point x can be found as shown below.

$$\varepsilon_{xx}(x) = \frac{du}{dx} = \frac{u_1}{2a^2}[(x-2a)+(x-a)] - \frac{u_2}{a^2}[(x-2a)+x] + \frac{u_3}{2a^2}[(x-a)+x]$$

$$\varepsilon_{xx}(x) = \frac{u_1}{2a^2}(2x-3a) - \frac{u_2}{a^2}(2x-2a) + \frac{u_3}{2a^2}(2x-a) = \frac{u_1}{2a^2}(-a) - \frac{u_2}{a^2}(0) + \frac{u_3}{2a^2}(a) \text{ or}$$

$$\boxed{\varepsilon_{xx}(a) = \frac{u_3 - u_1}{2a}}$$

2.61 Due to the applied load, the strain in the tapered bar due was found to be $\varepsilon_{xx} = \frac{0.2}{(40-x)^2}$. Determine the extension of the bar.

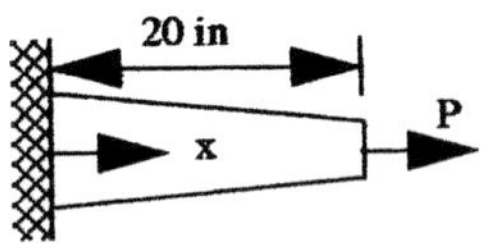

Solution u(20)=?

The strain at any point x can be written as shown below.

$$\varepsilon_{xx} = \frac{du}{dx} = \frac{0.2}{(40-x)^2}$$ 1

Method I : Integrating Eq. (1) we obtain: $u(x) = \frac{0.2}{(40-x)} + C_1$

Now $u(0) = 0$ therefore $C_1 = -\frac{0.2}{40} = -0.005$

$u(x) = \frac{0.2}{(40-x)} - 0.005$ or $u(20) = \frac{0.2}{(40-20)} - 0.005 = 0.005$ or $\boxed{u(20) = 0.005 in}$

Method II :Integrating Eq. (1) from x = 0 to x = 20 we obtain the following.

$$\int_0^{u(20)} du = \int_0^{20} \frac{0.2}{(40-x)^2}dx = \left.\frac{0.2}{(40-x)}\right|_0^{20} = \frac{0.2}{40-20} = 0.005 \text{ or } \boxed{u(20) = 0.005 in}$$

2.62 The axial strain in a bar of length L was found to be $\varepsilon_{xx} = \frac{KL}{(3L-2x)}$ $0 \le x \le L$ where K is a constant for a given material, loading, and cross-section dimension. Determine the total extension in terms of K and L.

Solution u(L)-u(0)=?

The strain at any point x can be written as shown below.

$$\varepsilon_{xx} = \frac{du}{dx} = \frac{KL}{(3L-2x)} \qquad 1$$

Integrating Eq. (1) from x = 0 to x = L, we obtain the following.

$$\int_{u(0)}^{u(L)} du = \int_0^L \frac{KL}{(3L-2x)}dx \text{ or } u(L)-u(0) = \left.\frac{KL}{(-2)}\ln(3L-2x)\right|_0^L = -\frac{KL}{2}[\ln(L)-\ln(3L)] = -\frac{KL}{2}(-\ln 3)$$

$$\boxed{u(L)-u(0) = 0.5493KL}$$

2.63 The axial strain in a bar due to its own weight of length L was found to be

$\varepsilon_{xx} = K\left[4L - 2x - \frac{8L^3}{(4L-2x)^2}\right]$ $0 \le x \le L$ where K is a constant for a given material and cross-section dimensions. Determine the total extension in terms of K and L.

Solution u(L)-u(0)=?

The strain at any point x can be written as shown below.

$$\varepsilon_{xx} = \frac{du}{dx} = K\left[4L - 2x - \frac{8L^3}{(4L-2x)^2}\right] \qquad 1$$

Integrating Eq. (1) from x = 0 to x = L, we obtain the following.

$$\int_{u(0)}^{u(L)} du = \int_0^L K\left[4L - 2x - \frac{8L^3}{(4L-2x)^2}\right]dx \text{ or}$$

$$u(L)-u(0) = K\left.\left[4Lx - \frac{2x^2}{2} - \frac{8L^3}{(4L-2x)}\right]\right|_0^L = K\left[4L^2 - L^2 - \frac{8L^3}{4L} + \frac{8L^3}{8L}\right] = 2KL^2 \qquad \boxed{u(L)-u(0) = 2KL^2}$$

2.64 A bar has a tapered and a uniform section securely fastened. Determine the total extension of the bar if the axial strain in each section is as given below.

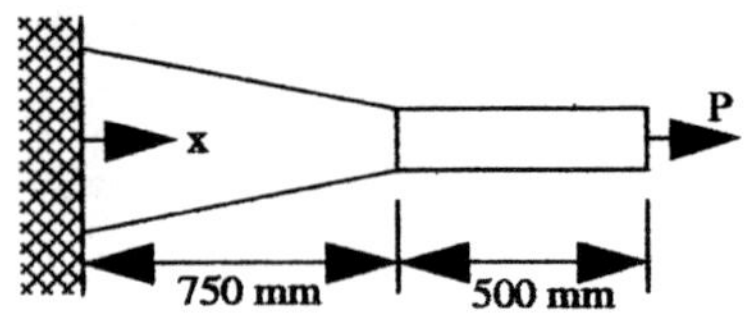

$$\varepsilon = \frac{1500(10^3)}{(1875-x)}\ \mu \qquad 0 \le x \le 750\ \text{mm}$$

$$\varepsilon = 1500\ \mu \qquad 750\ \text{mm} \le x \le 1250\ \text{mm}$$

Solution u(1250)=?

The strain at any point x can be written as shown below.

$$\varepsilon_{xx} = \frac{du}{dx} = \frac{1500(10^3)}{(1875-x)}\ \mu \qquad 0 \le x \le 750\ \text{mm} \qquad 1$$

$$\varepsilon_{xx} = \frac{du}{dx} = 1500\ \mu \qquad 750\ \text{mm} \le x \le 1250\ \text{mm} \qquad 2$$

Integrating Eq. (1), we obtain: $\int_{u(0)}^{u(750)} du = \int_0^{750} \frac{1500(10^3)}{(1875-x)} dx$ or

$$u(750) - u(0) = -1500(10^{-3})\ln(1875-x)\Big|_0^{750} = -1500(10^{-3})[\ln(1875-750) - \ln(1875)] = 766.24(10^{-3}) \qquad 3$$

$$\int_{u(750)}^{u(1250)} du = \int_{750}^{1250} (1500)(10^{-6})dx \quad \text{or} \quad u(1250) - u(750) = 1500(10^{-6})x\Big|_{750}^{1250} = 750(10^{-3}) \qquad 4$$

Adding Eq.(3) and Eq.(4) $u(1250) - u(0) = 1516.24(10^{-3})\text{m}$

Noting $u(0) = 0$ we obtain $\boxed{u(1250) = 1.516\text{mm}}$

2.65 N axial bars are securely fastened together. Determine the total extension of the composite bar if the strain in the ith section is as given.

$$\varepsilon_i = a_i \qquad x_{i-1} \le \quad \le x_i$$

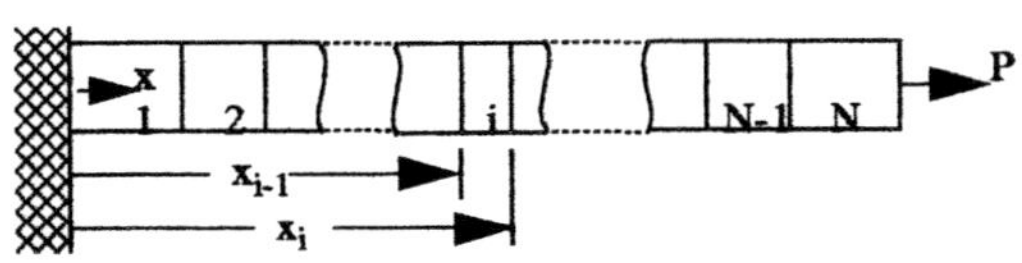

Solution u_N=?

The axial strain in the ith segment can be written as:

$$\varepsilon_i = \frac{du}{dx} = a_i \qquad x_{i-1} \le \quad \le x_i \qquad 1$$

Integraing Eq. (1) we obtain: $\int_{u_{i-1}}^{u_i} du = \int_{x_{i-1}}^{x_i} a_i dx$ or $u_i - u_{i-1} = a_i(x_i - x_{i-1}) \qquad 1 \le i \le N$

Adding the relative displacements over all segments, we obtain:

$$\sum_{i=1}^{N}(u_i - u_{i-1}) = \sum_{i=1}^{N} a_i(x_i - x_{i-1}) \text{ or } \sum_{i=1}^{N} u_i - \sum_{i=1}^{N} u_{i-1} = \sum_{i=1}^{N} a_i(x_i - x_{i-1}) \text{ or } u_N - u_0 = \sum_{i=1}^{N} a_i(x_i - x_{i-1})$$

Noting $u_0 = 0$ we obtain: $\boxed{u_N = \sum_{i=1}^{N} a_i(x_i - x_{i-1})}$

2.66 True strain (ε_T) is calculated from $d\varepsilon_T = (du)/(L_o + u)$, where u is the deformation at any given instance and L_o is the original undeformed length. Thus the increment in true strain is the ratio of change in length at any instance to the length at that given instance. If ε represents engineering strain show that at any instance the relationship between true strain and engineering strain is given by the following:

$$\varepsilon_T = \ln(1+\varepsilon) \qquad \textbf{2.12}$$

Solution u_N=?

Integrating $d\varepsilon_T = (du)/(L_o + u)$ we obtain: $\int_o^{\varepsilon_T} d\varepsilon_T = \int_0^{\mu} \frac{du}{(L_o+u)}$ or

$$\varepsilon_T = \ln(L_o+u)\Big|_0^u = \ln(L_o+u) - \ln(L_o) = \ln\left(\frac{L_o+u}{L_o}\right) = \ln\left(1+\frac{u}{L_o}\right) \qquad 1$$

By defination Engineering strain is given by: $\varepsilon = \frac{\text{Total Deformation}}{\text{Original Length}} = \frac{u}{L_o}$ Substituting this defination into Eq. (1), we obtain: $\boxed{\varepsilon_T = \ln(1+\varepsilon)}$

2.67 The displacements in a body are given by:

$u = [0.5(x^2 - y^2) + 0.5xy](10^{-3})\text{mm}$ $\qquad v = [0.25(x^2 - y^2) - xy]((10^{-3})\text{mm})$

Determine the strains ε_{xx}, ε_{yy}, and γ_{xy} at x = 5 mm and y = 7 mm.'

Solution $\{\varepsilon_{xx}=?$ $\qquad \varepsilon_{yy}=?$ $\qquad \gamma_{xy}=?\}$ at x =5mm and y=7 mm

---

The strains at a point can be found as shown below.

$$\varepsilon_{xx} = \frac{\partial u}{\partial x} = [0.5(2x) + 0.5y](10^{-3}) = (x + 0.5y)(10^{-3}) \qquad 1$$

$$\varepsilon_{yy} = \frac{\partial v}{\partial y} = [0.25(-2y) - x](10^{-3}) = (-0.5y - x)(10^{-3}) \qquad 2$$

$$\gamma_{xy} = \frac{\partial u}{\partial y} + \frac{\partial v}{\partial x} = [0.5(-2y) + 0.5x](10^{-3}) + [0.25(2x) - y](10^{-3}) = \gamma_{xy} = (x - 2y)(10^{-3}) \qquad 3$$

Substituting $x = 5$ and $y = 7$ in Eq.(1),(2) and (3). We obtain:

$\varepsilon_{xx} = (5 + 3.5)(10^{-3}) = 8.5(10^{-3})$ or $\boxed{\varepsilon_{xx} = 8500\mu\frac{\text{mm}}{\text{mm}}}$

$\varepsilon_{yy} = (-3.5 - 5)(10^{-3}) = -8.5(10^{-3})$ or $\boxed{\varepsilon_{yy} = -8500\mu\frac{\text{mm}}{\text{mm}}}$

$\gamma_{xy} = (5 - 14)(10^{-3}) = -9(10^{-3})$ or $\boxed{\gamma_{xy} = -9000\mu\text{rads}}$

2.68 A metal strip is to be pulled and bent to conform to a rigid surface such that the length of strip between OA fits the arc OB of the surface. The equation of the surface y=f(x) and the length OA are as given below. Determine the average normal strain in the metal strip.

$f(x) = 0.04x^{3/2}$ inches and length OA = 9 inches.

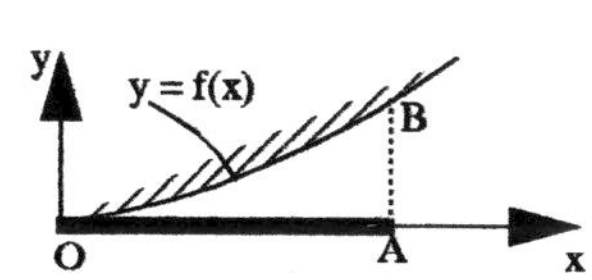

Solution $L_o = OA = 9$ $\qquad \varepsilon=?$

The first derivative of y can be written as: $\frac{dy}{dx} = (0.04)\left(\frac{3}{2}\right)x^{\frac{1}{2}} = 0.06x^{\frac{1}{2}}$

The length of the deformed rod can be written as

$$L_f = \int ds = \int_0^9 \sqrt{1+\left(\frac{dy}{dx}\right)^2}\,dx = \int_0^9 (\sqrt{1+0.0036x})dx = \frac{2}{3(0.0036)}(1+0.0036x)^{\frac{3}{2}}\Big|_0^9 \quad \text{or}$$

$$L_f = 185.19\left[(1+(0.0036)(9))^{\frac{3}{2}} - 1\right] = 9.07275$$

The average normal strain is: $\varepsilon = \frac{L_f - L_o}{L_f} = \frac{0.07275}{9} = 8.083(10^{-3})$

$$\boxed{\varepsilon = 8083\mu\frac{in}{in}}$$

2.69 A metal strip is to be pulled and bent to conform to a rigid surface such that the length of strip between OA fits the arc OB of the surface. The equation of the surface y=f(x) and the length OA are as given belo. Determine the average normal strain in the metal strip.

$f(x) = 625x^{3/2}$ μmm and length OA = 200 mm.

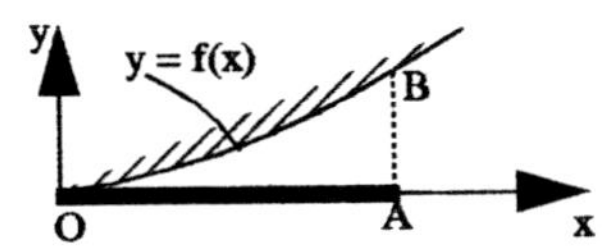

Solution $L_o = OA = 200mm$ $\varepsilon = ?$

The first derivative of y can be written as: $\frac{dy}{dx} = \left(\frac{3}{2}\right)(625)x^{\frac{1}{2}}(10^{-6}) = 0.9375(10^{-3})x^{\frac{1}{2}}$

The length of the deformed rod can be written as

$$L_f = \int ds = \int_0^{200} \sqrt{1+\left(\frac{dy}{dx}\right)^2}\,dx = \int_0^{200} \sqrt{1+(0.9375)^2(10^{-6})x}dx \quad \text{or}$$

$$L_f = \frac{2}{3(0.9375)^2(10^{-6})}(1+(0.9375)^2(10^{-6})x)^{\frac{3}{2}}\Big|_0^{200} \quad \text{or}$$

$$L_f = 711.11\left[(1+(0.9375)^2(10^{-6})200)^{\frac{3}{2}} - 1\right] = 200.00844 \text{ mm}$$

The average normal strain is: $\varepsilon = \frac{L_f - L_o}{L_f} = 42.19(10^{-6})$ or

$$\boxed{\varepsilon = 42.2\mu\frac{mm}{mm}}$$

2.70 A metal strip is to be pulled and bent to conform to a rigid surface such that the length of strip between OA fits the arc OB of the surface. The equation of the surface y=f(x) and the length OA are as given below. Determine the average normal strain in the metal strip.

$f(x) = (0.04x^{3/2} - 0.005x)$ inches and length OA = 9 inches.
Use numerical integration.

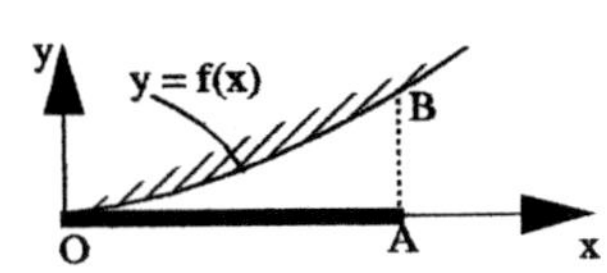

Solution $L_o = OA = 9$ $\varepsilon = ?$

The first derivative of y can be written as: $\frac{dy}{dx} = \frac{3}{2}(0.04)x^{\frac{1}{2}} - 0.005 = 0.06x^{\frac{1}{2}} - 0.005$

The length of the deformed rod can be written as:

$$L_f = \int ds = \int_0^9 \sqrt{1 + \left(\frac{dy}{dx}\right)^2}\, dx = \int_0^9 \sqrt{1 + \left(0.06x^{\frac{1}{2}} - 0.005\right)^2}\, dx$$

Let $f(x) = \sqrt{1 + \left(0.06x^{\frac{1}{2}} - 0.005\right)^2}$, thus $L_f = \int_0^9 f(x)dx$

Using numerical integration as described in Figure B2 with a Δx_i=0.25, we obtain the following the table below.

Numerical Integration Results from Spread Sheet.

xi	f(xi)	Integral value
0	1.0000125	0.250040619
0.25	1.000312451	0.500167191
0.5	1.000700123	0.750392467
0.75	1.001102085	1.000719147
1	1.001511358	1.251148721
1.25	1.001925237	1.501682168
1.5	1.002342333	1.752320187
1.75	1.002761823	2.003063311
2	1.00318317	2.253911957
2.25	1.003605998	2.504866462
2.5	1.004030038	2.755927102
2.75	1.004455082	3.007094109
3	1.004880973	3.258367678
3.25	1.005307582	3.509747977
3.5	1.005734807	3.761235148
3.75	1.006162564	4.012829316
4	1.006590781	4.264530589
4.25	1.007019398	4.516339059
4.5	1.007448365	4.768254809

xi	f(xi)	Integral value
4.75	1.007877637	5.020277911
5	1.008307175	5.272408426
5.25	1.008736947	5.524646409
5.5	1.009166921	5.776991909
5.75	1.009597073	6.029444965
6	1.010027379	6.282005615
6.25	1.010457817	6.534673888
6.5	1.010888369	6.787449811
6.75	1.011319017	7.040333407
7	1.011749746	7.293324693
7.25	1.012180542	7.546423685
7.5	1.012611393	7.799630394
7.75	1.013042285	8.052944831
8	1.013473208	8.306367001
8.25	1.013904153	8.559896909
8.5	1.01433511	8.813534556
8.75	1.01476607	9.067279943
9	1.015197025	

The deformed length from above table is: $L_f = 9.0672799$ in.

The average normal strain is: $\varepsilon = \frac{L_f - L_o}{L_f} = \frac{0.0672799}{9} = 7.4755(10^{-3})$ or $\boxed{\varepsilon = 7475.5\mu\frac{in}{in}}$

2.71 Measurements were made along the path of the stretch cord that is stretched over the canoe in problem 2.4 are shown in the table below. The y-coordinate was measured to the closest 1/32 inch. Between points A and B the cord path can be approximated by a straight line. Determine the average strain in the stretch cord if the original length of it is 40 inches. Use spread sheet and approximate each 2 inch x-interval by a straight line.

x_i	y_i
0	17
2	16 (30/32)
4	16 (29/32)
6	16 (19/32)
8	16 (3/32)

x_i	y_i
10	15 (16/32)
12	14 (24/32)
14	13 (28/32)
$x_B = 16$	$y_B = 12$
$x_A = 18$	$y_A = 0$

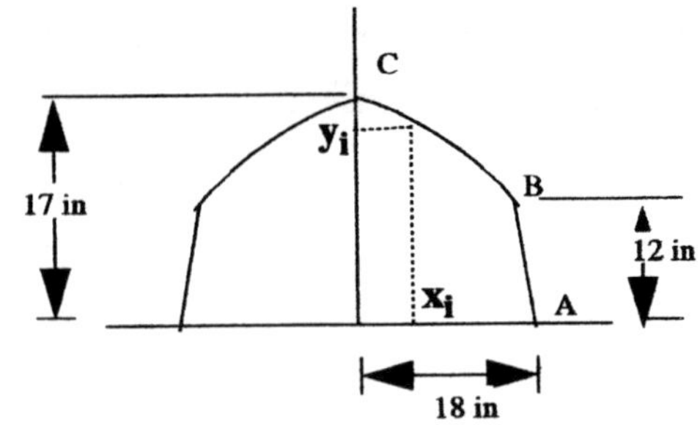

Solution $L_o = 40$ $\varepsilon = ?$

The deformed length of each 2 inch segment can be found using.

$$L_i = \sqrt{(x_i - x_{i-1})^2 + (y_i - y_{i-1})^2} \qquad 2 \le i \le 10$$

The final length of AC can be found as $AC = \sum_{i=2}^{10} L_i$. Using Spread sheet the following table can be constructed.

Results of Calculation of the total length using spread sheet.

x_i	y_i	L_i
0	17	
2	16.9375	2.000976324
4	16.90625	2.000244126
6	16.59375	2.024266843
8	16.09375	2.061552813
10	15.5	2.086273966
12	14.75	2.136000936
14	13.875	2.18303115
16	12	2.741464025
18	0	12.16552506
	Total Length	29.39933524

From the above table: $AC = 29.399335$

The average normal strain is: $\varepsilon = \dfrac{2(AC) - 40}{40} = 0.46997$ $\boxed{\varepsilon = 47\%}$

Problem Set 3.1

3.1—3.8 A tensile test specimen having a diameter of 10 mm and a gage length of 50 mm was tested to fracture. The stress-strain curve from the tension test is shown in Fig. P3.1. The lower plot is the expanded region OAB and associated with the strain values given in the lower scale. Solve the problems 3.1 through 3.8 using this graph

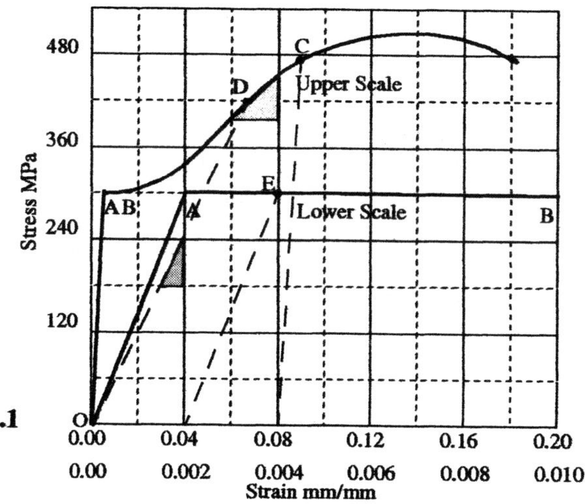

Fig. P3.1

3.1 Determine the axial force acting on the specimen when it is extended by (a) 0.2 mm (b) 4.0 mm.

Solution $d = 10$ mm; $L_o = 50$ mm; (a) P=? @ $\delta = 0.2$ mm (b) P=? @ $\delta = 4.0$ mm

---

The area of cross-section can be found as shown below.

$$A = \frac{\pi}{4}d^2 = 78.5398\,\text{mm}^2 \qquad (1)$$

(a) The strain in the specimen can be found as: $\varepsilon = \frac{\delta}{L_0} = \frac{0.2}{50} = 0.004$

From the lower plot in Fig. P3.8 $\sigma = 300$ MPa

$\sigma = \frac{P}{A}$ or $P = (300)(10^6)(78.5398)(10^{-6}) = 23561.9$ N or $\boxed{P = 23.56\text{kN}}$

(b) The strain in the specimen can be found as: $\varepsilon = \frac{\delta}{L_0} = \frac{4.0}{50} = 0.08$

From the upper plot in Fig. P3.8 $\sigma = 420 + 30 = 450$MPa

$\sigma = \frac{P}{A}$ or $P = (450)(10^6)(78.5398)(10^{-6}) = 35342.9$ N or $\boxed{P = 35.34\text{kN}}$

3.2 Determine the extension of the specimen when the axial force on the specimen is 33 kN.

Solution $d = 10$ mm $L_0 = 50$ $\delta = ?$ @$P = 33$ kN

The area of cross-section from Eq. 1 in problem 3.1 is $A = 78.5398$ mm^2.

The stress in the specimen can be found as: $\sigma = \frac{P}{A} = \frac{33(10^3)}{78.5398(10^{-6})} = 420.1(10^6)\frac{N}{m^2} = 420\text{MPa}$

From the upper plot in Fig. P3.8 $\varepsilon = (0.064)0.005 = 0.065$

$\varepsilon = \frac{\delta}{L_0}$ or $\delta = (0.065)(50) = 3.25$ or $\boxed{\delta = 3.25\text{mm}}$

3.3 Determine the (a) Ultimate Stress (b) Fracture Stress.

Solution σ_{ult}=? σ_{frac}=?

By inspection of Fig P3.8 σ_{ult}=480+30 $\boxed{\sigma_{ult} = 510MPa}$

σ_{frac}=480 $\boxed{\sigma_{frac} = 480MPa}$

3.4 Determine the (a) Modulus of Elasticity (b) proportional limit.

Solution E = ? δ_{prop}= ?

From the lower plot in Fig P3.8 $E = \frac{300(10^6)}{0.04} = 7.5(10^9)\frac{N}{m^2}$ or $\boxed{E = 7.5GPa}$

By inspection of the lower plot in Fig P3.8 $\boxed{\sigma_{prop} = 300MPa}$

3.5 Determine the total strain, the elastic strain, and the plastic strain when the axial force on the specimen is 33 kN.

Solution ε_{total} = ? ε_{elas}= ? ε_{plas}= ?@ P = 33kN

The area of cross-section from Eq. 1 in problem 3.1 is A = 78.5398 mm$^{2.}$ The normal stress in the specimen can be found as: $\sigma = \frac{P}{A} = \frac{33(10^6)}{78.5398(10^{-6})} = 420.2(10^6)\frac{N}{m^2} = 420.2MPa$

From upper plot in Fig. P3.8 ε_{total}= 0.06 + 0.005 = 0.065 $\boxed{\varepsilon_{total} = 0.065}$

From problem 3-4, we know E = 7.5 GPa, therefore, the elastic strain can be found as shown below.

$\varepsilon_{elas} = \frac{\sigma}{E} = \frac{420.2(10^6)}{7.5(10^9)} = 56.03(10^{-3})$ $\boxed{\varepsilon_{elas} = 0.056}$

The plastic strain can be found as shown below.

$\varepsilon_{plas} = \varepsilon_{total} - \varepsilon_{elas} = 65(10^{-3}) - 56.03(10^{-3}) = 8.97(10^{-3})$ $\boxed{\varepsilon_{plas} = 0.009}$

3.6 After the axial load was removed, the specimen was observed to have a length of 54 mm. What was the maximum axial load applied to the specimen.

Solution L_f = 54 mm P = ?

The plastic strain can be found as: $\varepsilon_{plas} = \frac{L_f - L_0}{L_0} = \frac{4}{50} = 0.08$

Starting from a strain value of 0.08, we draw a line parallel to OA on the upper plot, intersecting the stress-strain curve at point C. The stress at point C is approximated as $\sigma_c = 470MPa$. The area of cross-section from Eq. 1 in problem 3.1 is A = 78.5398 mm$^{2.}$ The applied force can be found as shown below

$P = \sigma_c A = (470)(10^{-6})(78.5398)(10^{-6}) = 36913.7N$ $\boxed{P = 36.9kN}$

3.7 Determine the offset yield stress at 0.2%.

Solution ε_{offset} = 0.2% σ_{yield}= ?

The offset strain is: $\varepsilon_{offset} = \frac{0.2}{100} = 0.002$ Starting from a strain value of 0.002, we draw a line parallel to OA on the lower plot in Fig. P3.1, intersecting the stress-strain curve at point E. The stress at point E is 300 MPA

$$\boxed{\sigma_{yield} = 300\text{MPa}}$$

3.8 At the stress level of 420 MPa, determine the (a) Tangent Modulus (b) Secant Modulus.

Solution $\sigma = 420$ MPa $E_s = ?$ $E_T = ?$

The stress value of 420 MPa corresponds to point D on the upper plot in Fig. P3.1. We draw a tangent line to the stress strain curve at point D and calculate the slope as shown below:

$$E_t = \frac{(450-400)(10^6)}{0.02} = 2.5(10^9)\frac{N}{m^2} \qquad \boxed{E_t = 2.5\text{GPa}}$$

To find the secant modulus, we draw a line from point D to point O in Fig. P3.1 and find the slope as shown below.

$$E_s = \frac{(420-0)(10^6)}{0.065-0} = 6.46(10^9)\frac{N}{m^2} \qquad \boxed{E_s = 6.5\text{GPa}}$$

3.9—3.16 A tensile test specimen having a diameter of 5/8 inch and a gage length of 2 inch was tested to fracture. The stress-strain curve from the tension test is shown in Fig. P3.9. The lower plot is the expanded region OAB and associated with the strain values given in the lower scale. Solve the problems 3.9 through 3.16 using this graph

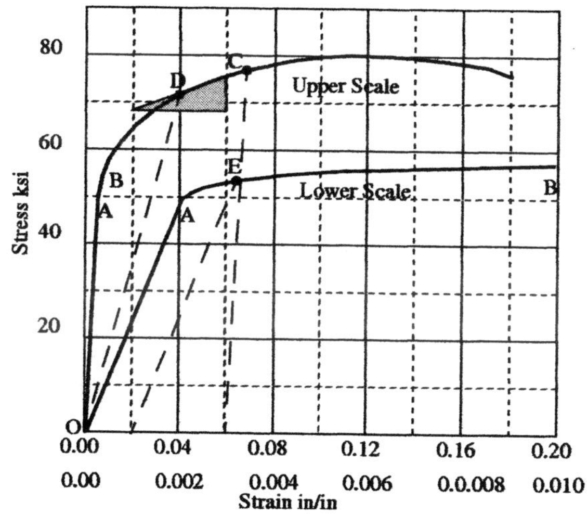

Fig. P3.9

3.9 Determine the axial force acting on the specimen when it is extended by (a) 0.006 inch (b) 0.120 inch.

Solution $d = \frac{5}{8}$inch $L_o = 2$ inch (a) P=? @ $\delta = 0.006$ inch (b) P =? @$\delta = 0.120$ inch

---

The area of cross-section can be found as shown below.

$$A = \frac{\pi}{4}d^2 = 0.3068\text{in}^2$$ 1

(a)The normal strain in specimen can be found as: $\varepsilon = \frac{\delta}{L_0} = \frac{0.006}{2} = 0.003$

From the lower plot in Fig. P3.9, the approximate value of stress is $\sigma = 53$ ksi.The applied load can be found as: $P = \sigma A = (53)(0.3068) = 16.26\text{kips}$ or $\boxed{P = 16.3\text{kips}}$

(b) The normal strain in specimen can be found as: $\varepsilon = \frac{\delta}{L_0} = \frac{0.120}{2} = 0.060$

From the upper plot in Fig. P3.9, the approximate value of stress is $\sigma = 75$ ksi. he applied load can be found as: $P = \sigma A = (75)(0.3068) = 23.01\text{kips}$ $\boxed{P = 23.0\text{kips}}$

3.10 Determine the extension of the specimen when the axial force on the specimen is (a) 10 kips (b) 20 kips.

Solution $d = \frac{5}{8}\text{inch}$ $L_0 = 2$ inch $\delta = ?$ @ P = 10 kips $\delta = ?$ @ P = 20 kips

---

The area of cross-section from Eq. 1 in problem 3.9 is: $A = 0.3068\ \text{in}^2$

(a)The normal stress in specimen can be found as: $\sigma = \frac{P}{A} = \frac{10}{0.3068} = 32.59\text{ksi}$

From the lower plot in Fig. P3.9, the approximate value of strain is $\varepsilon = 0.0013$. The deformation can be found as: $\delta = \varepsilon L_0 = (0.0013)(2) = 0.0026\text{in}$ or $\boxed{\delta = 0.0026\text{in}}$

(b) The normal stress in specimen can be found as: $\sigma = \frac{P}{A} = \frac{20}{0.3068} = 65.19\text{ksi}$

From the upper plot of Fig. P3.9, the approximate value of strain is $\varepsilon = 0.02$. The deformation can be found as: $\delta = \varepsilon L_0 = (0.02)(2) = 0.04\text{in}$ $\boxed{\delta = 0.04\text{in}}$

3.11 Determine the (a) Ultimate Stress (b) Fracture Stress.

Solution $\sigma_{ult} = ?$ $\sigma_{frac} = ?$

By inspection Fig. P3.9, we find: $\sigma_{ult} = 80$ ksi $\boxed{\sigma_{ult} = 80\text{ksi}}$

$\sigma_{frac} = 80 - 4 = 76$ ksi $\boxed{\sigma_{frac} = 76\text{ksi}}$

3.12 Determine the (a) Modulus of Elasticity (b) proportional limit.

Solution $E = ?$ $\delta_{prop} = ?$

From the lower plot in Fig. P3.9 $E = \frac{50 - 0}{0.002 - 0}$ $\boxed{E = 25,000\text{ksi}}$

By inspection of the lower plot in Fig. P3.9, we obtain $\boxed{\sigma_{prop} = 50\text{ksi}}$

3.13 Determine the total strain, the elastic strain, and the plastic strain when the axial force on the specimen is 20 kips.

Solution $\varepsilon_{total} = ?$ $\varepsilon_{elas} = ?$ $\varepsilon_{plas} = ?$ @ P = 20 kips

---

The area of cross-section from Eq. 1 in problem 3.9 is: $A = 0.3068\ \text{in}^2$. The normal stress in the specimen can be found as $\sigma = \frac{P}{A} = \frac{20}{0.3068} = 65.19\text{ksi}$

From upper plot in Fig. P3.9 we have: $\varepsilon_{total} = 0.02$ $\boxed{\varepsilon_{total} = 0.0200}$

From problem 3-12, we know E = 25,000 ksi, therefore the elastic strain can be found as shown below.

$\varepsilon_{elas} = \frac{\sigma}{E} = \frac{65.19}{25000} = 2.608(10^{-3})$ $\boxed{\varepsilon_{elas} = 0.0026}$

The plastic strain can be found as shown below.

$\varepsilon_{plas} = \varepsilon_{total} - \varepsilon_{elas} = 20(10^{-3}) - 2.608(10^{-3}) = 17.392(10^{-3})$ $\boxed{\varepsilon_{plas} = 0.0174}$

3.14 After the axial load was removed, the specimen was observed to have a length of 2.12 inch. What was the maximum axial load applied to the specimen.

Solution $L_f = 2.12$ inch $P = ?$

The plastic strain is: $\varepsilon_{plas} = \frac{L_f - L_0}{L_0} = \frac{0.12}{2} = 0.06$. Starting from a strain value of 0.06, we draw a line parallel to OA on the upper plot in Fig. P3.9, intersecting the stress-strain curve at point C. The stress at point C is approximated as $\sigma_c = 77\text{ksi}$. The area of cross-section from Eq. 1 in problem 3.9 is: A=0.3068 in^2. The applied load can be found as shown below.

$P = \sigma_c A = (77)(0.3068) = 23.62\text{kips}$ or $\boxed{P = 23.6\text{kips}}$

3.15 Determine the offset yield stress at (a) 0.1% (b) 0.2%.

Solution $\varepsilon_{offset} = 0.1\%$ $\sigma_{yield} = ?$

The offset strain is: $\varepsilon_{offset} = \frac{0.1}{100} = 0.001$. Starting from a strain value of 0.001, we draw a line parallel to OA on the lower plot in Fig. P3.9, intersecting the stress-strain curve at point E. The stress at point E is 54ksi. The yield stress is: $\boxed{\sigma_{yield} = 54\text{ksi}}$

3.16 At the stress level of 72 kip, determine the (a) Tangent Modulus (b) Secant Modulus.

Solution $\sigma = 72\text{ksi}$ $E_s = ?$ $E_T = ?$

The stress value of 72 ksi corresponds to point D. on the upper plot in Fig. P3.9, we draw a line to the stress strain curve at point D and calculate the slope as shown below:

$E_t = \frac{75-68}{0.06-0.02} = 325\text{ksi}$ or $\boxed{E_t = 325\text{ksi}}$

To find the secant modulus, we draw a line from point D to point O in Fig. P3.9, and find the slope as shown below:

$E_t = \frac{72-0}{0.04-0} = 1800\text{ksi}$ $\boxed{E_s = 1,800\text{ksi}}$

3.17 A 12 mm x 12 mm square metal alloy having a gage length of 50 mm was tested in tension. Draw the stress-strain curve and calculate the following quantities. (Use of Spread Sheet recommended).

(a) modulus of elasticity. (b) the proportional limit.
(c) the yield stress at 0.2 percent offset. (d) the tangent Modulus at stress level of 1400 MPa.
(e) the secant modulus at stress level of 1400 MPa.
(f) the plastic strain at stress level of 1400 MPa.

Load (KN)	Change in Length (mm)
0.00	0.00
17.32	0.02
60.62	0.07
112.58	0.13
147.22	0.17
161.18	0.53
168.27	1.10

Load (KN)	Change in Length (mm)
176.03	1.96
182.80	2.79
190.75	4.00
193.29	4.71
200.01	5.80
204.65	7.15
209.99	8.88

Load (KN)	Change in Length (mm)
212.06	9.99
212.17	11.01
208.64	11.63
204.99	12.03
199.34	12.31
192.15	12.47
185.46	12.63
break	

Solution $L_o = 50$ mm cross-section is 12 mm x 12 mm

The area of cross-section is $A = (12)(12) = 144\ mm^2 = 144(10^{-6})\ m^2$. Dividing the column of load by the area A and the column of change of length by the gage length Lo, we can obtain the columns of stress and strains on a spread sheet as shown in the table below. The stresses and strain can be plotted to obtain the stress-strain graph in figure (a). The data corresponding to first 5 points can be also plotted to get an enlarged view of region OAB as shown in figure (b).

Stress-Strain values in problem 3.17.

Strain	Stress (MPa)
0	0
0.0004	120.28
0.0014	420.7
0.0026	781.36
0.0034	1022.36
0.0106	1119.31
0.022	1168.54
0.0392	1222.43
0.0558	1269.44
0.08	1324.65

Strain	Stress (MPa)
0.0942	1342.29
0.116	1388.96
0.143	1421.18
0.1776	1458.26
0.1998	1472.64
0.2202	1473.40
0.2326	1448.89
0.2406	1423.54
0.2462	1384.31
0.2494	1334.38
0.2526	1287.92

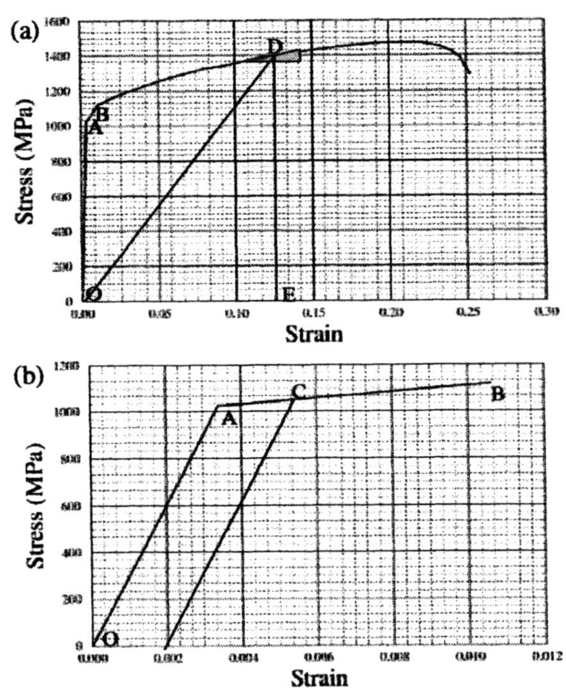

The modulus of elasticity is calculated as the average slope from the first three points as shown below:

$$E = \frac{1}{2}\left[\frac{120.28-0}{0.0004} + \frac{420.7-120.28}{0.0014-0.0004}\right] = 300560\text{MPa}$$ **E = 300Gpa**

From the figure (b) and the table: **$\sigma_{prop} = 1022\text{MPa}$**

The offset strain is: $\varepsilon_{offset} = \frac{0.2}{100} = 0.002$ On figure (b), starting at a strain value of 0.002, we draw a line parallel to OA to intersect the stress-strain curve at point C.The stress value at C is approximated as the yield stress **$\sigma_{yield} = 1060\text{MPa}$**

The stress value of 1400 MPa, corresponds to point D. in Figure (a). We draw tangent to the stress-strain curve at point D and calculate the slope as shown below:

$$E_t = \frac{1433 - 1367}{0.1467 - 0.1083} = 1719\text{MPa} \qquad \boxed{E_t = 1.72\text{GPa}}$$

To find the secant modulus we draw a line from point D to point O. and find the slope as

$$E_s = \frac{1400}{0.125} = 11200\text{MPa} \qquad \boxed{E_s = 11.2\text{GPa}}$$

The total strain at point D is $\varepsilon_{total} = 0.125$. The elastic strain at point D can be found as:

$$\varepsilon_{elas} = \frac{\sigma_D}{E} = \frac{1400(10^6)}{300(10^9)} = 4.667(10^{-3})$$ The plastic strain can be found as shown below.

$$\varepsilon_{plas} = \varepsilon_{total} - \varepsilon_{elas} = 0.125 - 0.004667 = 0.1203 \qquad \boxed{\varepsilon_{plas} = 0.1203}$$

3.18 A mild steel specimen of diameter 0.5 inches and a gage length of 2 inches was tested in tension. The results of the test are reported in the table below. Draw the stress-strain curve and calculate the following quantities. (Use of Spread Sheet recommended).

(a) Modulus of Elasticity. (b) the proportional limit.
(c) the yield stress at 0.05 percent offset. (d) the tangent Modulus at stress level of 50 ksi.
(e) the secant Modulus at stress level of 50 ksi.
(f) the plastic strain at stress level of 50 ksi.

Load (10^3)lbs	Change in Length (10^{-3})in
0.00	0.00
3.11	1.20
7.24	3.40
7.50	22.2
7.70	8.76
7.90	19.05
8.16	28.70
8.46	37.73
8.82	47.18
9.32	59.06
9.86	70.85
10.40	84.23
10.82	97.85
11.18	112.10
11.72	140.40
11.99	161.21
12.27	192.65
12.41	214.22
12.55	245.93
12.70	283.47
12.77	316.36
12.84	363.10
12.04	385.34
11.44	396.03
10.71	406.42
9.96	414.72
break	

Solution $L_o = 2$ inch $d = 0.5$ inch

The area of cross-section is $A = \pi d^2/4 = 0.1963\ \text{in}^2$. Dividing the column of load by the area A and the column of change of length by the gage length Lo, we can obtain the columns of stress and strains on a spread sheet as shown in the table below. The stress and strain can be plotted to obtain the stress-strain graph in figure (a). The data corresponding to first 5 points can be also plotted to get an enlarged view of region OAB as shown in figure (b).

The modulus of elasticity is calculated as the average slope from the first three points as shown below:

$$E = \frac{1}{2}\left[\frac{15.84 - 0}{0.00064} + \frac{36.87 - 15.84}{0.00148 - 0.00064}\right] = 24892.8\text{ksi} \qquad \boxed{E = 24893\text{ksi}}$$

From the figure (b) and the table: $\boxed{\sigma_{prop} = 36.9\text{ksi}}$

The offset strain is: $\varepsilon_{offset} = \frac{0.05}{100} = 0.0005$ on figure (b), starting at a strain value of 0.0005, we draw a line parallel to OA to intersect the stress-strain curve at point C. The stress value at C is approximated as

.. stress $\boxed{\sigma_{yield} = 37.5ksi}$

Stress-Strain values in problem 3.18

Strain	Stress (ksi)	Strain	Stress (ksi)
0.00000	0.00	0.05605	56.94
0.00064	15.84	0.07020	59.69
0.00148	36.87	0.08061	61.06
0.00153	38.25	0.09633	62.49
0.00438	39.22	0.10711	63.20
0.00953	40.23	0.12297	63.92
0.01435	41.56	0.14174	64.68
0.01887	43.09	0.15818	65.04
0.02359	44.92	0.18155	65.39
0.02953	47.47	0.19267	61.32
0.03543	50.22	0.19802	58.26
0.04212	52.97	0.20321	54.55
0.04893	55.11	0.20736	50.73

(a)

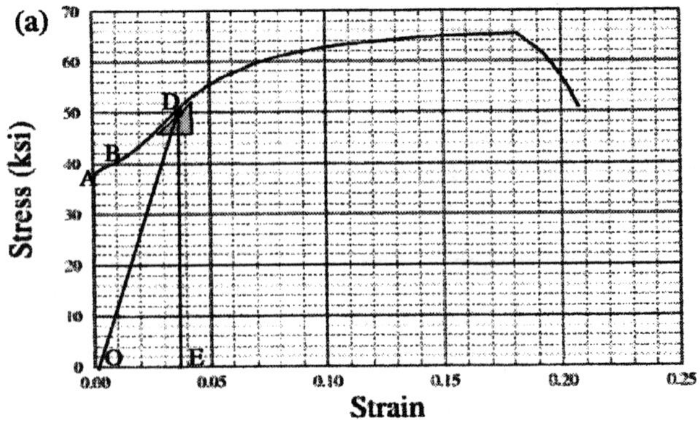

(b)

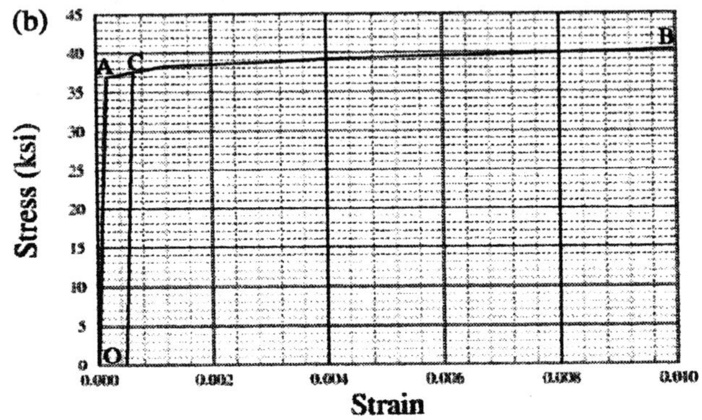

The stress value of 50ksi, corresponds to point D. in Figure (a). We draw tangent to the stress-strain curve at point D and calculate the slope as shown below:

$$E_t = \frac{52-46}{0.04-0.025} = 400ksi \qquad \boxed{E_t = 400ksi}$$

To find the secant modulus we draw a line from point D to point O. and find the slope as

$$E_s = \frac{50}{0.035} = 1428.57 \qquad \boxed{E_s = 1428.6ksi}$$

The total strain at point D is ε_{total}= 0.035. The elastic strain at point D is $\varepsilon_{elas} = \frac{\sigma_D}{E} = \frac{50}{24893} = 0.002$. The plastic strain can be found as shown below.

$$\varepsilon_{plas} = \varepsilon_{total} - \varepsilon_{elas} = 0.035 - 0.002 = 0.015 \qquad \boxed{\varepsilon_{plas} = 0.015}$$

3.19 A rectangular bar has an cross-sectional area of 2 in^2 and an undeformed length of five inches as shown in Fig. P3.19. When a load of P = 50,000 lbs is applied, the bar deforms to a position shown by the colored shape. Determine the Modulus of Elasticity and the Poisson's Ratio of the material.

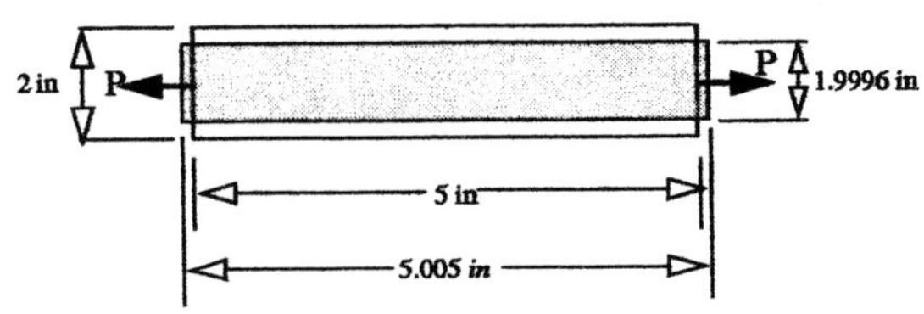

Fig. P3.19

Solution $A = 2\ in^2$ $L_0 = 5$ in $L_f = 5.005$ P = 50,000lbs E = ? ν = ?

The longitudinal (axial) strain is: $\varepsilon_{long} = \frac{L_f - L_0}{L_0} = \frac{0.005}{5} = 0.001$

The axial stress is: $\sigma = \frac{P}{A} = \frac{50000}{2} = 25,000\text{psi}$

The modulus of elasticity is: $E = \frac{\sigma}{\varepsilon_{long}} = \frac{25000}{0.001} = 25000(10^3)\text{psi}$ $\boxed{E = 25,000\text{ksi}}$

The transverse strain is: $\varepsilon_{trans} = -\frac{0.0004}{2} = -0.0002$

The Poisson's ratio is: $\nu = \frac{-\varepsilon_{trans}}{\varepsilon_{long}} = -\frac{(-0.0002)}{0.001} = 0.2$ $\boxed{\nu = 0.2}$

3.20 A force P= 20 kips is applied to the rigid plate that is attached to a square bar as shown in Fig. P3.20. If the plate moves an amount of 0.005 inch, determine the Modulus of Elasticity.

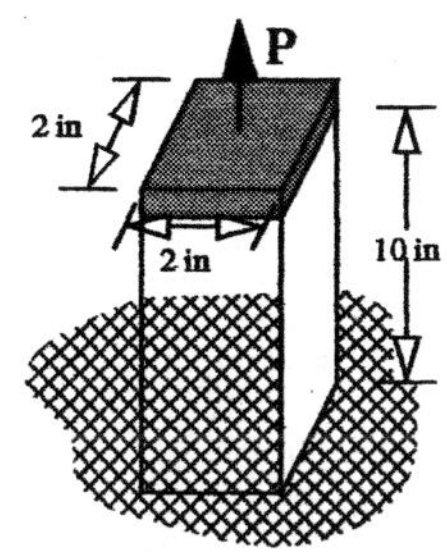

Fig. P3.20

Solution P = 20 kips δ=0.005 E = ?

The axial stress is: $\sigma = \frac{P}{A} = \frac{20}{4} = 5\text{ksi}$.

The axial strain is: $\varepsilon = \frac{0.005}{10} = 0.0005$

The modulus of elasticity is: $E = \frac{\sigma}{\varepsilon} = \frac{5}{0.0005} = 10,000\text{ksi}$ $\boxed{E = 10,000\text{ksi}}$

3.21 A force P= 20 kips is applied to the rigid plate that is attached to a square bar as shown in Fig. P3.21. If the plate moves an amount of 0.0125 inch, determine the Shear Modulus of Elasticity.

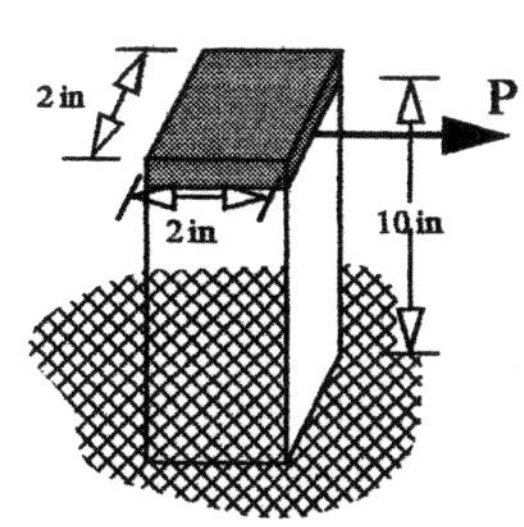

Fig. P3.21

Solution P = 20 ksi δ = 0.0125 A= 4 G = ?

The free body diagram of the rigid plate can be drawn as shown in figure a. By equilibrium of forces we obtain $V = P$. The shear stress on the rigid plate can be found as: $\tau = V/A = 20/4 = 5\text{ksi}$.
An exaggerated deformed geometry can be drawn as shown in figure b and the shear strain found as:
$\tan\gamma \approx \gamma = \delta/10 = 0.00125$

The shear modulus of elasticity can be found as: $G = \frac{\tau}{\gamma} = \frac{5}{0.00125} = 4000\text{ksi}$ $\boxed{G = 4000\text{ksi}}$

3.22 A circular bar of length 200 mm and diameter 20 mm is subjected to a tension test. Due to an axial force of 77 kN, the bar is seen to elongate by 4.5 mm and the diameter is seen to reduce by 0.162 mm. Determine the Modulus of Elasticity and the Shear Modulus of Elasticity.

Solution $L_0 = 200$ mm d = 20 mm P = 77 kN δ = 4-5mm Δd = -0.162 mm
E = ? G = ?

The area of cross-section is: $A = \frac{\pi}{4}d^2 = \frac{\pi}{4}(20^2) = 314.16\text{mm}^2 = 314.16(10^{-6})\text{m}^2$

The axial stress is: $\sigma = \frac{P}{A} = \frac{77(10^3)}{314.16(10^{-6})} = 245.1(10^6)\frac{N}{m^2}$

The longitudinal strain (axial) is: $\varepsilon_{long} = \frac{\delta}{L_0} = \frac{4.5}{200} = 0.0225$

The transverse strain is: $\varepsilon_{tran} = \frac{\Delta d}{d} = -\frac{0.162}{20} = -0.0081$

The modulus of elasticity is: $E = \frac{\sigma}{\varepsilon_{long}} = \frac{245.1(10^6)}{0.0225} = 10.893(10^9)$ $\boxed{E = 10.9\text{GPa}}$

The Poisson's ratio is: $\nu = \frac{-\varepsilon_{tran}}{\varepsilon_{long}} = -\frac{(-0.0081)}{(0.0225)} = 0.36$

The shear modulus of elasticity is: $G = \frac{E}{2(1+\nu)} = \frac{10.893}{2(1+0.36)} = 4.005\text{GPa}$ $\boxed{G = 4.0\text{GPa}}$

3.23 A circular bar of length 6 inch and diameter of 1 inch is made from a material with a Modulus of Elasticity of E=30,000 ksi and a Poisson's ratio of ν=1/3. Determine the change in length and diameter of the bar when a force of 20 kips is applied to the bar.

Solution $L_0 = 6$ d = 1 E= 30,000ksi ν=1/3 P = 20kips Δd = ? ΔL = ?

The area of cross-section is: $A = \frac{\pi}{4}d^2 = \frac{\pi}{4}(1^2) = 0.7854\text{in}^2$

The axial stress is: $\sigma = \frac{P}{A} = \frac{20}{0.7854} = 25.465\text{ksi}$

The longitudinal strain is: $\varepsilon_{long} = \frac{\sigma}{E} = \frac{25.465}{30,000} = 0.8488(10^{-3})$

The change of length is: $\Delta L = L\varepsilon_{long} = (6)(0.8488)(10^{-3}) = 5.093(10^{-3})$ $\boxed{\Delta L = 0.0051\text{in}}$

The transverse strain is: $\varepsilon_{tran} = -\nu\varepsilon_{long} = -[0.8488(10^{-3})]/3 = -0.2829(10^{-3})$

The change in diameter is $\Delta d = \varepsilon_{tran}d = (1)(-0.2829)(10^{-3})$ or $\boxed{\Delta d = -0.0002}$

3.24 A circular bar of length 15 inch and diameter of 0.75 inch is made from a material with a Modulus of Elasticity of E = 28,000 ksi and a Poisson's ratio of ν = 0.32. Due to a force the bar is seen to elongate by 0.04 inch. Determine the change in diameter and the applied force.

Solution $L_0 = 15$ $d = 0.75$ $E = 28{,}000\text{ksi}$ $\nu = 0.32$ $\delta = 0.04$ $\Delta d = ?$ $P = ?$

The longitudinal strain is: $\varepsilon_{long} = \frac{\delta}{L_0} = \frac{0.04}{15} = 2.667(10^{-3})$

The axial stress is: $\sigma = E\varepsilon_{long} = 28,000(2.667)(10^{-3}) = 74.667\text{ksi}$

The applied force is: $P = \sigma A = (74.667)\frac{\pi}{4}(0.75^2) = 32.987$ or $\boxed{P = 33\text{kips}}$

The transverse strain is: $\varepsilon_{tran} = -\nu\varepsilon_{long} = -(0.32)(2.667)(10^{-3}) = -0.8533(10^{-3})$

The change in diameter is: $\Delta d = (d)\varepsilon_{tran} = (0.75)(-0.8533)(10^{-3}) = -0.64(10^{-3})$ or

$\boxed{\Delta d = -0.00064\text{in}}$

3.25 A 1 inch x 1 inch square bar is 20 inches long and is made from a material that has a Poisson's ratio of 1/3. In a tension test, the bar is seen to elongate by 0.03 inch. Determine the percentage change in volume of the bar.

Solution $A = 1\text{ in}^2$ $L_0 = 20\text{in}$ $\nu = 1/3$ $\delta = 0.03$ % change in V= ?

The original volume is: $V_0 = AL_0 = 25\text{ in}^3$. Let Δa represent the change in width and thickness. The volume of the deformed bar is as given below.

$$V_f = \text{FinalVolume} = (1 + \Delta a)(1 + \Delta a)(L_0 + \delta) \quad \mathbf{1}$$

The longitudinal strain is: $\varepsilon_{long} = \frac{\delta}{L_0} = \frac{0.03}{20} = 0.0015$

The transverse strain is: $\varepsilon_{tran} = -\nu\varepsilon_{long} = -\frac{1}{3}(0.0015) = -0.0005$

We can find $\Delta a = (1)\varepsilon_{tran} = -0.0005$. Substituting δ and Δa into Eq. 1, we obtain:

$V_f = (1 - 0.0005)(1 - 0.0005)(20 + 0.3) = 20.00998$

The percentage change in volume is: $\frac{\Delta V}{V} \times 100 = \frac{V_f - V_0}{V_0} \times 100 = 0.04988$ or $\boxed{0.05\text{percent}}$

3.26 A circular bar of length 50 inch and diameter of 1 inch is made from a material with a Modulus of Elasticity of E = 28,000 ksi and a Poisson's ratio of 0.32. Determine the percentage change in volume of the bar when an axial force of 20 kips is applied to the bar.

Solution $d = 1\text{ in}$ $L_0 = 15\text{in}$ $\nu = 0.32$ $E = 28{,}000\text{ksi}$ % change in V @P = 20 kips

---

The axial stress is: $\sigma = \frac{P}{A} = \frac{20}{\pi(1^2)/4} = 25.465\text{ksi}$

The longitudinal strain is: $\varepsilon_{long} = \frac{\sigma}{E} = \frac{25.465}{28,000} = 0.9095(10^{-3})$

The deformation is: $\delta = L_0\varepsilon_{long} = (50)(0.9095)(10^{-3}) = 0.04547\text{in}$

The transverse strain is: $\varepsilon_{tran} = -\nu\varepsilon_{long} = -(0.32)(0.9095)(10^{-3}) = -0.2910(10^{-3})$

The change in diameter is: $\Delta d = (d)\varepsilon_{tran} = -(0.2910)(10^{-3})(1) = -0.2910(10^{-3})\text{in}$

The original volume is: $V_0 = AL_0 = (0.7854)(50) = 39.2699\text{in}^3$

The volume of the deformed bar is as shown below.

$$V_f = \frac{\pi}{4}(1-\Delta d)^2(L_0+\delta) = \frac{\pi}{4}(1-0.291(10^{-3}))^2(50+0.04547) = 39.2827\text{in}^3$$

The percentage change in volume is: $\frac{\Delta V}{V_0}\times 100 = \frac{V_f - V_0}{V_0}\times 100 = 0.0327$ **0.0327percent**

3.27 An aluminum rectangular bar has a cross-section of 25 mm x 50 mm and a length of 500 mm. The Modulus of Elasticity of E = 70 GPa and a Poisson's ratio of ν = 0.25. Determine the percentage change in the volume of the bar when an axial force of 300 kN is applied to the bar.

Solution $E = 70\text{GPa}$ $L_0 = 500\text{mm}$ $\nu = 0.25$ % change in V @ P = 300 kN

The axial stress is: $\sigma = \frac{P}{A} = \frac{300(10^3)}{(0.05)(0.025)} = 240(10^6)\frac{N}{m^2}$

The longitudinal strain is: $\varepsilon_{long} = \frac{\sigma}{E} = \frac{240(10^6)}{70(10^9)} = 3.4286(10^{-3})$

The transverse strain is: $\varepsilon_{tran} = -\nu\varepsilon_{long} = -(0.25)(3.4286)(10^{-3}) = -0.85714(10^{-3})$

The deformation of the bar is: $\delta = L_0\varepsilon_{long} = (500)(3.4286)(10^{-3}) = 1.7143\text{mm}$

The change in cross-sectional dimensions can be found as shown below.

$$\Delta a = a\varepsilon_{tran} = (25)(-0.85714)(10^{-3}) = -21.429(10^{-3})\text{mm}$$

$$\Delta b = b\varepsilon_{tran} = (50)(-0.85714)(10^{-3}) = -42.857(10^{-3})\text{mm}$$

The original volume is: $V_0 = AL_0 = (1250)(500) = 625(10^3)\text{mm}^3$

The volume of the deformed bar is: $V_f = (a+\Delta a)(b+\Delta b)(L_0+\delta) = 626.068(10^3)\text{mm}^3$

The percentage change in volume is: $\frac{\Delta V}{V_0}\times 100 = \frac{V_f - V_0}{V_0}\times 100 = 0.1709$ or **0.1709percent**

3.28 A circular bar of length L and diameter of d is made from a material with a Modulus of Elasticity of E and a Poisson's ratio of ν. Assuming small strain, show that the percentage change in the volume of the bar when an axial force P is applied is given by $400P(1-2\nu)/(E\pi d^2)$. Note the percentage change is zero when ν = 0.5?

Solution %change in V = f(L,d,E,ν,P)

The axial stress is: $\sigma = \frac{P}{A} = \frac{P}{\pi d^2/4} = \frac{4P}{\pi d^2}$

The longitudinal strain is: $\varepsilon_{long} = \frac{\sigma}{E} = \frac{4P}{E\pi d^2}$ and the transverse strain is: $\varepsilon_{tran} = -\nu\varepsilon_{long} = -\frac{4\nu P}{E\pi d^2}$

The change in length is: $\delta = L\varepsilon_{long} = \frac{4LP}{E\pi d^2}$ and the change in diameter is: $\Delta d = d\varepsilon_{tran} = -\frac{4\nu P}{E\pi d}$

The original volume is: $V_0 = \frac{\pi d^2}{4}L$. For small strain, the volume of the deformed bar can be found as shown below.

$$V_f = \frac{\pi}{4}(d+\Delta d)^2(L+\delta) = \frac{\pi}{4}d^2L\left(1+\frac{\Delta d}{d}\right)^2\left(1+\frac{\delta}{L}\right) \approx \frac{\pi}{4}d^2L\left(1+\frac{2\Delta d}{d}\right)\left(1+\frac{\delta}{L}\right) \approx \frac{\pi}{4}d^2L\left(1+\frac{2\Delta d}{d}+\frac{\delta}{L}\right)$$

The percentage change in volume is as shown below.

$$\left(\frac{\Delta V}{V}\right)(100) = \left(\frac{V_f - V_0}{V_0}\right)(100) = \left(2\frac{\Delta d}{d}+\frac{\delta}{L}\right)(100) = \left(-\frac{8\nu P}{E\pi d^2}+\frac{4P}{E\pi d^2}\right)(100) \text{ or } \boxed{\frac{400P}{E\pi d^2}(1-2\nu)}$$

3.29 A rectangular bar has a cross-sectional dimensions of a x b and a length L. The bar material has a Modulus of Elasticity of E and a Poisson's ratio of ν. Assuming small strain, show that the percentage change in the volume of the bar when an axial force P is applied is given by $100P(1-2\nu)/(Eab)$. Note the percentage change is zero when $\nu = 0.5$?

Solution %change in V = f(L,d,E,ν,P)

The axial stress is: $\sigma = \frac{P}{A} = \frac{P}{ab}$

The longitudinal strain is: $\varepsilon_{long} = \frac{\sigma}{E} = \frac{P}{Eab}$ o and the transverse strain is: $\varepsilon_{tran} = -\nu\varepsilon_{long} = -\frac{\nu P}{Eab}$

The change in length is: $\delta = L\varepsilon_{long} = \frac{\nu LP}{Eab}$. The change in cross-sectional dimensions can be found as shown below.

$$\Delta a = a\varepsilon_{tran} = -\frac{Pa\nu}{Eab} \qquad \Delta b = b\varepsilon_{tran} = -\frac{Pb\nu}{Eab}$$

The original volume is: $V_0 = abL$. The volume of deformed bar can be found as shown below.

$$V_f = (a+\Delta a)(b+\Delta b)(L+\delta) = abL\left(1+\frac{\Delta a}{a}\right)\left(1+\frac{\Delta b}{b}\right)\left(1+\frac{\delta}{L}\right) \approx (abL)\left(1+\frac{\Delta a}{a}+\frac{\Delta b}{b}\right)+\left(1+\frac{\delta}{L}\right)$$

Assuming small strain, the product terms of Δa, Δb, and δ can be neglected to obtain the following.

$V_f \approx (abL)\left(1+\frac{\Delta a}{a}+\frac{\Delta b}{b}+\frac{\delta}{L}\right)$. The percentage change in volume can be found as shown below.

$$\left(\frac{\Delta V}{V}\right)(100) = \left(\frac{V_f - V_0}{V_0}\right)(100) = \left(\frac{\Delta a}{a}+\frac{\Delta b}{b}+\frac{\delta}{L}\right)(100) = \left(-\frac{P\nu}{Eab}-\frac{P\nu}{Eab}+\frac{P}{Eab}\right)(100) \text{ or}$$

$$\boxed{(100)(1-2\nu)\frac{P}{Eab} \text{ percent}}$$

3.30 What is the strain energy in the bar of problem 3.19.

Solution P=50,000lbs A=2 in^2 L=5 in δ=0.005in U=?

The strain energy density for the bar can be written as:

$$U_0 = \frac{1}{2}\sigma\varepsilon = \frac{1}{2}\left(\frac{P}{A}\right)\frac{\delta}{L} = \frac{1}{2}\left(\frac{50,000}{2}\right)\left(\frac{0.005}{5}\right) = 12.5\text{in}-\text{lbs}/\text{in}^3$$

As the strain energy density is a constant, the total strain energy can be found as shown below:

$$U = \int U_0 dv = U_0 V = U_0 AL = (12.5)(2)(5) = 125 \text{ or } \boxed{U = 125\text{in}-\text{lbs}}$$

3.31 What is the strain energy in the bar of problem 3.20.

Solution P=20 kips A=4 in^2 L=10 in δ=0.005in U=?

The strain energy density for the bar can be written as:

$$U_0 = \frac{1}{2}\sigma\varepsilon = \frac{1}{2}\left(\frac{P}{A}\right)\frac{\delta}{L} = \frac{1}{2}\left(\frac{20(10^3)}{4}\right)\left(\frac{0.005}{10}\right) = 1.25\text{in}-\text{lbs/in}^3$$

As the strain energy density is a constant, the total strain energy can be found as shown below:

$$U = \int U_0 dV = U_0 V = U_0 AL = (1.25)(4)(10) = 50 \qquad \boxed{U = 50\text{in}-\text{lbs}}$$

3.32 What is the strain energy in the bar of problem 3.21.

Solution P=20 kips A=4 in^2 L=10 in δ=0.0125in U=?

The strain energy density for the bar can be written as:

$$U_0 = \frac{1}{2}\sigma\varepsilon = \frac{1}{2}\left(\frac{P}{A}\right)\frac{\delta}{L} = \frac{1}{2}\left(\frac{20(10^3)}{4}\right)\left(\frac{0.0125}{10}\right) = 3.125\text{in}-\text{lbs/in}^3$$

As the strain energy density is a constant, the total strain energy can be found as shown below:

$$U = \int U_0 dv = U_0 V = U_0 AL = (3.125)(4)(10) = 125 \qquad \boxed{U = 125\text{in}-\text{lbs}}$$

3.33 A circular bar of length L and diameter of d is made from a material with a Modulus of Elasticity of E and a Poisson's ratio of ν. In terms of the given variables, what is the linear strain energy in the bar of problem when axial load P is applied to the bar.

Solution U=f(L,d,E,ν,p)=?

The strain energy density for the bar can be written as:

$$U_0 = \frac{1}{2}\sigma\varepsilon = \frac{1}{2}\left(\frac{\sigma^2}{E}\right) = \frac{1}{2}\left(\frac{P^2}{EA^2}\right) = \frac{1}{2}\frac{p^2}{E\left(\frac{\pi d^2}{4}\right)^2} = \frac{8}{\pi^2 E}$$

As the strain energy density is a constant, the total strain energy can be found as shown below:

$$U = \int U_0 dv = U_0 V = U_0 AL = \left(\frac{8P^2}{\pi^2 Ed^4}\right)\left(\frac{\pi d^2}{4}\right)L \qquad \boxed{U = \frac{2P^2L}{\pi Ed^2}}$$

3.34 A rectangular bar has a cross-sectional dimensions of a x b and a length L. The bar material has a Modulus of Elasticity of E and a Poisson's ratio of ν. In terms of the given variables, what is the linear strain energy in the bar of problem when axial load P is applied to the bar.

Solution U=f(L,a,b,E,ν,p)=?

The strain energy density for the bar can be written as:

$$U_0 = \frac{1}{2}\sigma\varepsilon = \frac{1}{2}\left(\frac{\sigma^2}{E}\right) = \frac{1}{2}\left(\frac{P^2}{EA^2}\right) = \frac{1}{2}\frac{p^2}{Ea^2b^2}$$

As the strain energy density is a constant, the total strain energy can be found as shown below:

$$U = \int U_0 dv = U_0 V = U_0 AL = \left(\frac{P^2}{2(Ea^2b^2)}\right)(ab)(L) \qquad \boxed{U = \frac{2P^2L}{2Eab}}$$

3.35 For the material that has the stress-strain curve shown in Fig. P3.1, determine
(a) Modulus of Resilience, use proportional limit as an approximation for yield point.
(b) Strain energy density at a stress level of 420 MPa.
(c) Complimentary strain energy density at a stress level of 420 MPa.
(d) Modulus of Toughness.

Solution Modulus of Resilience=? U_o= ? and $\bar{U}_o$= ? @ σ = 420 MPa
Modulus of Toughness=?

The proportional limit corresponds to point A. From problem 3.4, the stress at problem limit is $\sigma_{prop} = 300$ MPa and the Modulus of Elasticity is 150 GPa. Thus, the strain at proportional limit (at A_1) is $\varepsilon_{prop} = 300(10^6)/[(150)(10^9)] = 0.002$. All other points are approximate points chosen to represent the stress strain curve by a series of straight line. The areas under the stress strain curve are calculated as shown on the right of the figure.

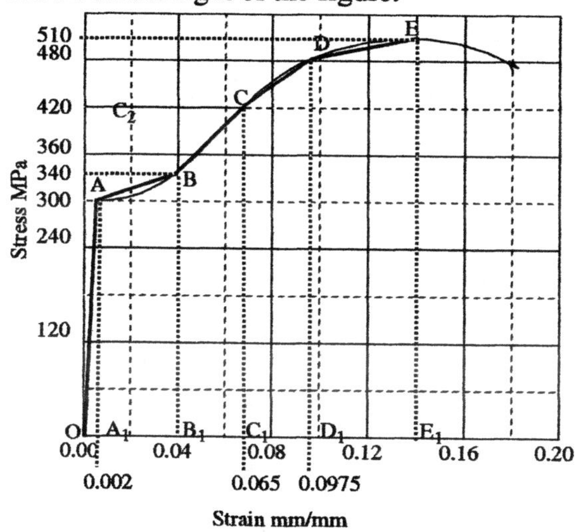

$$AOA_1 = \frac{(300)(10^6)(0.002)}{2} = 300(10^3)$$

$$AA_1BB_1 = \frac{(300+340)(10^6)(0.038)}{2} = 12160(10^3)$$

$$BB_1CC_1 = \frac{(340+420)(10^6)(0.025)}{2} = 9500(10^3)$$

$$CC_1DD_1 = \frac{(420+480)(10^6)(0.0325)}{2} = 14625(10^3)$$

$$DD_1EE_1 = \frac{(480+510)(10^6)(0.0425)}{2} = 21038(10^3)$$

The yield point corresponds to point A. The area of the triangle OAA_1 represents Modulus of Resilience.

$\boxed{\text{Modulus of Resilience is 300 kN-m/m}^3}$

Point C is at 420 MPa. The strain energy density at point C is the area AOA_1 plus areas AA_1BB_1 and BB_1CC_1. Thus the strain energy density at C is $U_O = (300 + 12160 + 9500)(10^3)\ \text{N} - \text{m/m}^3$ or

$$\boxed{U_O = 21,960\ \text{kN} - \text{m/m}^3}.$$

The complementary strain energy density at C can be found by subtracting U_O from the area of the rectangle OC_2CC_1. Thus, $\bar{U}o = (420)(10^6)(0.065) - 21,960\ (10^3) = 5340(10^3)\ \text{N} - \text{m/m}^3$ or

$$\boxed{\bar{U}o = 5,340\ \text{kN} - \text{m/m}^3}.$$

The ultimate stress corresponds to point E on the graph. The area underneath the curve can be calculated as the sum of the areas shown on the right of the graph. The total sum of the areas is: $57,623(10^3)$ N-m/m^3.

Thus, $\boxed{\text{Modulus of Toughness is 57,623 kN-m/m}^3}$

3.36 For the material that has the stress-strain curve shown in Fig. P3.9, determine

(a) Modulus of Resilience, use proportional limit as an approximation for yield point.
(b) Strain energy density at a stress level of 72 ksi.
(c) Complimentary strain energy density at a stress level of 72ksi.
(d) Modulus of Toughness.

Solution Modulus of Resilience=?U_o= ? $\bar{U}_o$= ? at 72 ksi
Modulus of Toughness=?

The proportional limit corresponds to point A. From problem 3.12, the stress at problem limit is $\sigma_{prop} = 50$ ksi and the Modulus of Elasticity is 25,000 ksi. Thus, the strain at proportional limit (at A_1) is $\varepsilon_{prop} = 50/25{,}000 = 0.002$. All other points are approximate points chosen to represent the stress strain curve by a series of straight line. The areas under the stress strain curve are calculated as shown on the right of the figure.

The area of the triangle OAA_1 represents Modulus of Resilience **Modulus of Resilience is 50 in-lbs/in³**

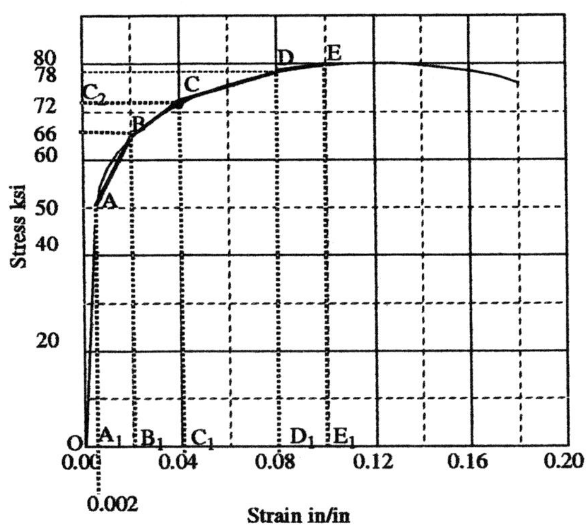

$$AOA_1 = \frac{(50)(10^3)(0.002)}{2} = 50$$

$$AA_1BB_1 = \frac{(50+66)(10^3)(0.018)}{2} = 1044$$

$$BB_1CC_1 = \frac{(66+72)(10^3)(0.02)}{2} = 1380$$

$$CC_1DD_1 = \frac{(72+78)(10^3)(0.04)}{2} = 3000$$

$$DD_1EE_1 = \frac{(78+80)(10^3)(0.02)}{2} = 1580$$

Point C is at 72 ksi. The strain energy density at point C is the area AOA_1 plus areas AA_1BB_1 and BB_1CC_1. Thus the strain energy density at C is $U_O = (50 + 1044 + 1380)\ \text{in-lbs/in}^3$ or

$$\boxed{U_O = 2474\ \text{in-lbs/in}^3}.$$

The complementary strain energy density at C can be found by subtracting U_O from the area of the rectangle OC_2CC_1. Thus, $\bar{U}_O = (72)(10^3)(0.04) - 2474 = 406\ \text{in-lbs/in}^3$ or $\boxed{\bar{U}_O = 406\ \text{in-lbs/in}^3}$.
The ultimate stress corresponds to point E on the graph. The area underneath the curve can be calculated as the sum of the areas shown on the right of the graph. The total sum of the areas is: 7054 in-lbs/in³. Thus,

Modulus of Toughness is 7054in-lbs/in³

3.37 For the metal alloy given in problem 3.17, determine

(a) Modulus of Resilience, use proportional limit as an approximation for yield point.
(b) Strain energy density at a stress level of 1400 MPa.
(c) Complimentary strain energy density at a stress level of 1400 MPa.
(d) Modulus of Toughness.

Solution Modulus of Resilience=? U_o= ? and $\bar{U}_o$= ? @σ = 1400 MPa

Modulus of Toughness=?

The proportional limit corresponds to point A in the figure below. From problem 3.17, the stress at problem limit is σ_{prop} = 1022 MPa and the Modulus of Elasticity is 300 GPa. Thus, the strain at proportional limit (at A_1) is $\varepsilon_{prop} = 1022(10^6)/[(300)(10^9)] = 0.0034$. All other points are approximate points chosen to represent the stress strain curve by a series of straight line. The areas under the stress strain curve are calculated as shown on the right of the figure.

The area of the triangle OAA_1 represents Modulus of Resilience. **Modulus of Resilience is 1734 kN-m/m³**
Point E is at 1400 MPa. The strain energy density at point E is the area AOA_1 through DD_1EE_1. Thus the strain energy density at E is $U_o = (1734 + 10775 + 43920 + 65850 + 34588)(10^3)\ \text{N}-\text{m}/\text{m}^3$ or

$$\boxed{U_o = 157\ \text{MN}-\text{m}/\text{m}^3}.$$

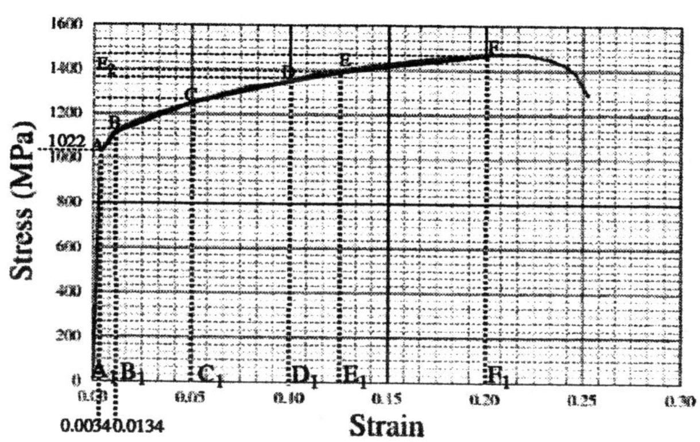

$$AOA_1 = \frac{(1022)(10^6)(0.0034)}{2} = 1734(10^3)$$

$$AA_1BB_1 = \frac{(1022 + 1133)(10^6)(0.01)}{2} = 10775(10^3)$$

$$BB_1CC_1 = \frac{(1133 + 1267)(10^6)(0.0366)}{2} = 43920(10^3)$$

$$CC_1DD_1 = \frac{(1267 + 1367)(10^6)(0.05)}{2} = 65850(10^3)$$

$$DD_1EE_1 = \frac{(1367 + 1400)(10^6)(0.025)}{2} = 34588(10^3)$$

$$EE_1FF_1 = \frac{(1400 + 1467)(10^6)(0.075)}{2} = 107512(10^3)$$

The complementary strain energy density at C can be found by subtracting U_o from the area of the rectangle OE_2EE_1. Thus, $\bar{U}o = (1400)(10^6)(0.125) - 157\ (10^6) = 18(10^6)\ \text{N}-\text{m}/\text{m}^3$ or

$$\boxed{\bar{U}o = 18\ \text{MN}-\text{m}/\text{m}^3}.$$

The ultimate stress corresponds to point E on the graph. The area underneath the curve can be calculated as the sum of the areas shown on the right of the graph. The total sum of the areas is: $264(10^6)$ N-m/m³.
Thus, **Modulus of Toughness is 264 MN-m/m³**

3.38 For the mild steel given in problem 3.18, determine

(a) Modulus of Resilience, use proportional limit as an approximation for yield point.
(b) Strain energy density at a stress level of 50 ksi.
(c) Complimentary strain energy density at a stress level of 50 ksi.
(d) Modulus of Toughness.

Solution Modulus of Resilience=? U_o= ? and $\bar{U}_o$= ? @ σ = 50 ksi
Modulus of Toughness=?

The proportional limit corresponds to point A in the figure below. From problem 3.18, the stress at problem limit is σ_{prop} = 36.9 ksi and the Modulus of Elasticity is 24,893 ksi. Thus, the strain at proportional limit (at A_1) is $\varepsilon_{prop} = 36.9/24{,}893 = 0.00148$. All other points are approximate points chosen to represent the stress strain curve by a series of straight line. The areas under the stress strain curve are calculated as shown on the right of the figure.

The area of the triangle OAA_1 represents Modulus of Resilience. **Modulus of Resilience is 27.3 in-lbs/in³**

Point B is at 50 ksi. The strain energy density at point B is the area AOA_1 plus area AA_1BB_1. Thus the strain energy density at B is $U_O = (27.31 + 1521)\ \text{in-lbs/in}^3$ or $\boxed{U_O = 1548.3\ \text{in-lbs/in}^3}$.

The complementary strain energy density at B can be found by subtracting U_O from the area of the rectangle OB_2BB_1. Thus, $\bar{U}_O = (50)(10^3)(0.035) - 1548.3 = 201.7\ \text{in-lbs/in}^3$ or

$$\boxed{\bar{U}_O = 201.7\ \text{in-lbs/in}^3}.$$

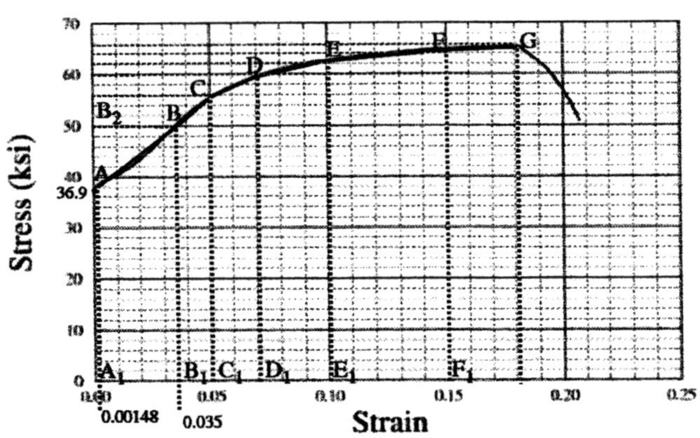

$$AOA_1 = \frac{(36.9)(10^3)(0.00148)}{2} = 27.31$$

$$AA_1BB_1 = \frac{(36.9 + 50)(10^3)(0.035)}{2} = 1521$$

$$BB_1CC_1 = \frac{(50 + 56)(10^3)(0.05)}{2} = 2650$$

$$CC_1DD_1 = \frac{(56 + 60)(10^3)(0.02)}{2} = 1160$$

$$DD_1EE_1 = \frac{(60 + 64)(10^3)(0.05)}{2} = 3100$$

$$DD_1EE_1 = \frac{(64 + 66)(10^3)(0.05)}{2} = 3250$$

The ultimate stress corresponds to point G on the graph. The area underneath the curve can be calculated as the sum of the areas shown on the right of the graph. The total sum of the areas is: 11708.3 in-lbs/in³. Thus,

Modulus of Toughness is 11,708 in-lbs/in³

3.39 The roller at P slides in the slot by an amount $\delta_P = 0.25$ mm due to the force F. Member AP has an area of cross-section of $A = 100\ \text{mm}^2$ and a Modulus of Elasticity $E = 200$ GPa. If the roller moves by the amount given, determine the force F.

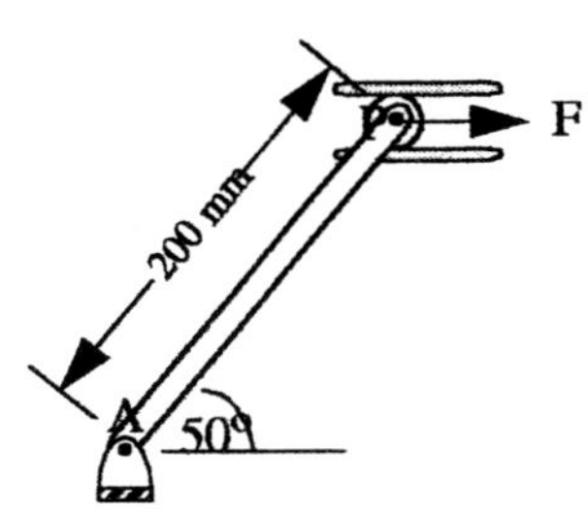

Fig. P3.39

Solution $\delta_P = 0.25$ mm $E = 200$ GPa $A = 100\ \text{mm}^2$ F=?

1.Strain Calculation: Assuming small strain, the average normal strain of bar AP is given by Eq.(2) in problem 2.35 $\varepsilon_{AP} = 803.5\mu\ \text{mm/mm}$

2.Stress Calculation: By Hooke's law we can find the normal stress as shown below.

$$\sigma_{AP} = E\varepsilon_{AP} = (200)(10^9)(803.5)(10^{-6}) = 160.7(10^6)\text{N/m}^2\ \text{(T)}$$

3.Internal Force Calculation: The internal normal force can be found as shown below.

$N_{AP} = A\sigma_{AP} = (100)(10^{-6})(160.7)(10^6) = 16.07(10^3)N = 16.07kN \;\; (T)$

4. External Force Calculation: We draw the free body diagram of the rigid bar as shown below.

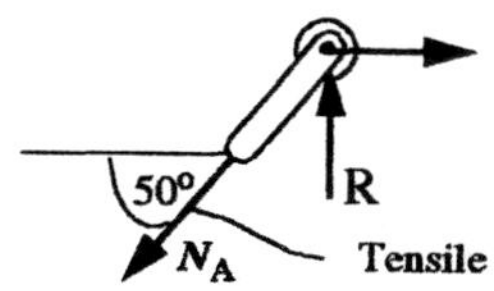

By force equilibrium in the x-direction, we obtain the following.

$F = N_{AP}\cos 50 = (16.07)\cos 50 = 10.33$ or $\boxed{F = 10.3kN}$

3.40 The roller at P slides in the slot by an amount $\delta_P = 0.25$ mm due to the force F. Member AP has an area of cross-section of A = 100 mm^2 and a Modulus of Elasticity E = 200 GPa. If the roller moves by the amount given, determine the force F.

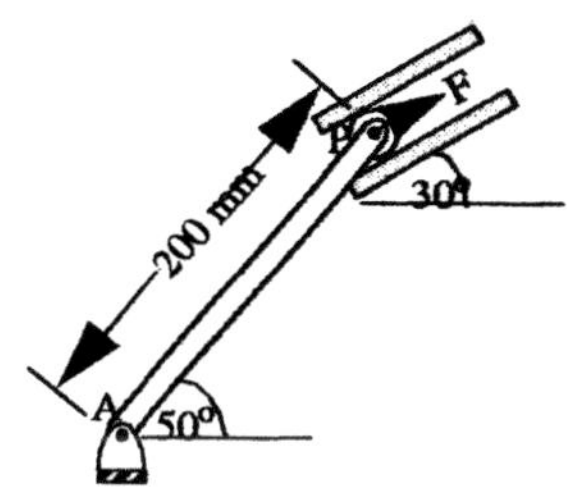

Fig. P3.40

Solution $\delta_P = 0.25$ mm E = 200 GPa A = 100 mm^2 F=?

1.Strain Calculation: Assuming small strain, the average normal strain of bar AP is given by E q.(2) in problem 2.36 as $\varepsilon_{AP} = 1174.6\mu \;\; mm/mm$

2.Stress Calculation: By Hooke's law we can find the normal stress as shown below.

$\sigma_{AP} = E\varepsilon_{AP} = (200)(10^9)(1174.6)(10^{-6}) = 234.9(10^6)N/m^2 \;\; (T)$

3.Internal Force Calculation: The internal normal force can be found as shown below.

$N_{AP} = A\sigma_{AP} = (100)(10^{-6})(234.9)(10^6) = 23.49(10^3)N = 23.49kN$

4. External Force Calculation: We draw the free body diagram of the rigid bar as shown below.

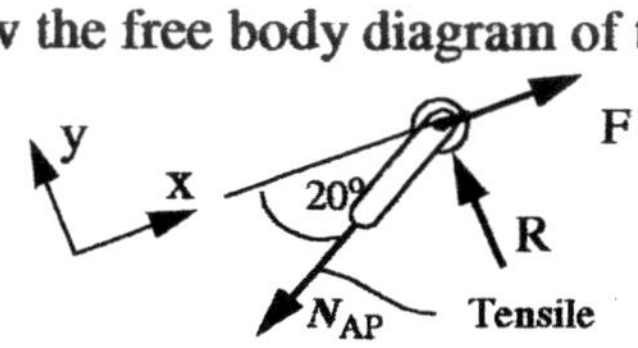

By force equilibrium in the x-direction in Fig.(a), we obtain the following.

$F = N_{AP}\cos 20 = (23.49)\cos 20 = 22.073$ $\boxed{F = 22.1kN}$

3.41 A roller slides in a slot by the amount $\delta_P = 0.25$ mm in the direction of the force F. Both bars have an area of cross-section of A = 100 mm^2 and a Modulus of Elasticity E = 200 GPa. Bar AP and BP have lengths of L_{AP}= 200 mm and L_{BP}= 250 mm respectively. Determine the applied force F.

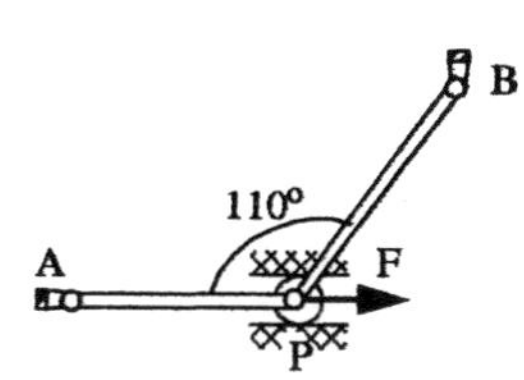

Fig. P3.41

Solution $\delta_P = 0.25$ mm E = 200 GPa A = 100 mm^2
L_{AP}= 200 mm L_{BP}= 250 mm F=?

1.Strain Calculation: Assuming small strain, the deformation of the bars can be found as in problem 2-37.

$\delta_{AP} = 0.25mm$ extension $\delta_{BP} = 0.0855mm$ contraction

$$\varepsilon_{AP} = \frac{\delta_{AP}}{L_{AP}} = \frac{0.25}{200} = 1.25(10^{-3})\text{extension and } \varepsilon_{BP} = \frac{\delta_{BP}}{L_{BP}} = \frac{0.0855}{250} = 0.342(10^{-3})\text{contraction}$$

2.Stress calculation:

$$\sigma_{AP} = E\varepsilon_{AP} = (200)(10^9)(1.25)(10^{-3}) = 250(10^6)\text{N/m}^2\text{(T)}$$

$$\sigma_{BP} = E\varepsilon_{BP} = (200)(10^9)(0.342)(10^{-3}) = 68.4(10^6)\text{N/m}^2\text{(C)}$$

3.Internal force Calculation:

$$N_{AP} = A\sigma_{AP} = (250)(10^6)(100)(10^{-6}) = 25(10^3)\text{N} = 25\text{kN(T)}$$

$$N_{BP} = A\sigma_{BP} = (68.4)(10^6)(100)(10^{-6}) = 6.84(10^3)\text{N} = 6.84\text{kN(C)}$$

4. External force Calculation: We draw the free body diagram of the rigid bar as shown below.

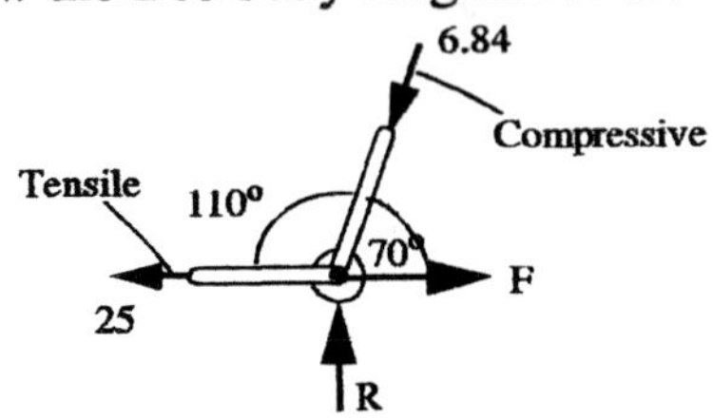

By force equilibrium in the x-direction we obtain the following.

$F - 25 - 6.84\cos 70 = 0$ or $\boxed{F = 27.34\text{kN}}$

3.42 A roller slides in a slot by the amount $\delta_P = 0.25$ mm in the direction of the force F. Both bars have an area of cross-section of $A = 100$ mm^2 and a Modulus of Elasticity $E = 200$ GPa. Bar AP and BP have lengths of L_{AP}= 200 mm and L_{BP}= 250 mm respectively. Determine the applied force F.

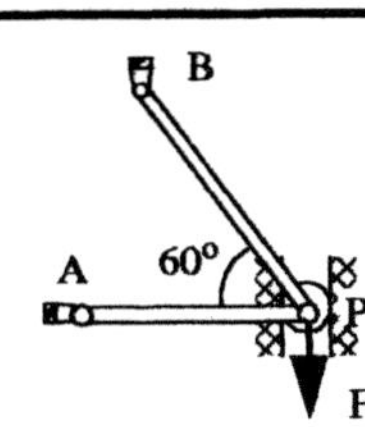

Fig. P3.42

Solution $\delta_P = 0.25$ mm $A = 100$ mm^2 L_{AP}= 200 mm L_{BP}= 250 mm F=?

1.Strain Calculation: Assuming small strain, the deformation of the bars can be found as in problem 2-38.

$\delta_{AP} = 0$ and $\delta_{BP} = 0.2165\text{mm extension}$

$$\varepsilon_{AP} = 0 \quad \text{and} \quad \varepsilon_{BP} = \frac{\delta_{BP}}{LBP} = \frac{0.2165}{250} = 0.866(10^{-3}) \text{ extension}$$

2. Stress calculation: By Hooke's law we can find the normal stress as shown below

$$\sigma_{AP} = 0 \text{ and } \sigma_{BP} = E\varepsilon_{BP} = (200)(10^9)(0.866)(10^{-3}) = 173.2(10^6)\text{N/m}^2\text{(T)}$$

3. Internal force Calculation:

$$N_{AP} = 0 \text{ and } N_{BP} = A\sigma_{BP} = (173.2)(10^6)(100)(10^{-6}) = 17.32(10^3)\text{N} = 17.32\text{kN(T)}$$

4. External Force Calculation We draw the free body diagram as shown below.

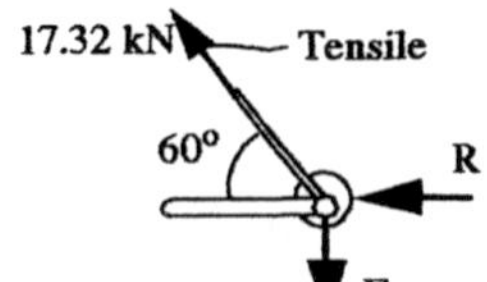

By force equilibrium in the y-direction, we obtain the $F = 17.32\sin 60 = 14.999\text{kN}$ $\boxed{F = 15\text{kN}}$

3.43 A roller slides in a slot by the amount $\delta_P = 0.25$ mm in the direction of the force F. Both bars have an area of cross-section of $A = 100\ \text{mm}^2$ and a Modulus of Elasticity E = 200 GPa. Bar AP and BP have lengths of L_{AP}= 200 mm and L_{BP}= 250 mm respectively. Determine the applied force F.

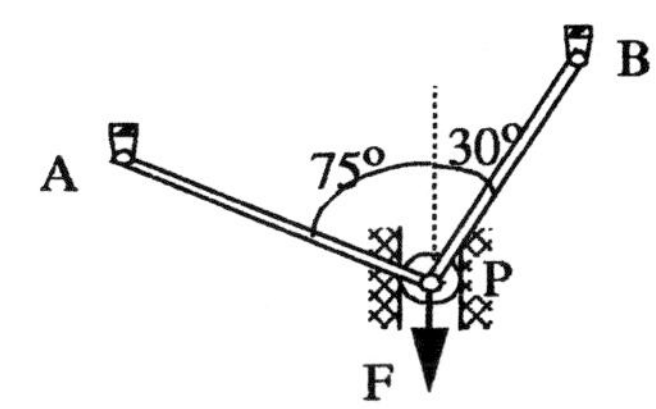

Fig. P3.43

Solution $\delta_P = 0.25$ mm E = 200 GPa $A = 100\ \text{mm}^2$
L_{AP}= 200 mm L_{BP}= 250 mm F=?

1.Strain Calculation: Assuming small strain, the deformation of the bars can be found as in problem 2-39.
$\delta_{AP} = 0.0647\text{mm}$ extension and $\delta_{BP} = 0.21651\text{mm}$ extension

$$\varepsilon_{AP} = \frac{\delta_{AP}}{LAP} = \frac{0.0647}{200} = 0.3235(10^{-3}) \text{ extension and } \varepsilon_{BP} = \frac{\delta_{BP}}{LBP} = \frac{0.21651}{250} = 0.866(10^{-3}) \text{ extension}$$

2.Stress calculation: By Hooke's law we can find the normal stress as shown below

$$\sigma_{AP} = E\varepsilon_{AP} = (200)(10^9)(0.3235)(10^{-3}) = 64.7(10^6)\text{N/m}^2(\text{T})$$

$$\sigma_{BP} = E\varepsilon_{BP} = (200)(10^9)(0.866)(10^{-3}) = 173.2(10^6)\text{N/m}^2(\text{T})$$

3.Internal force Calculation:

$$N_{AP} = A\sigma_{AP} = (64.7)(10^6)(100)(10^{-6}) = 6.47(10^3)\text{N} = 6.47\text{kN(T)}$$

$$N_{BP} = A\sigma_{BP} = (173.2)(10^6)(100)(10^{-6}) = 17.32(10^3)\text{N} = 17.32\text{kN(T)}$$

4. External Force Calculation: We draw the free body diagram of the roller as shown below.

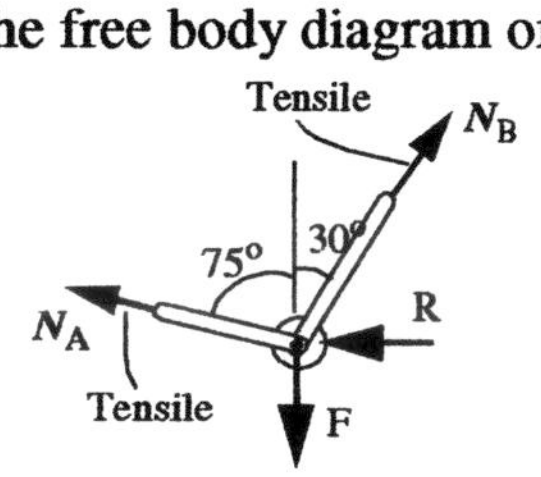

By force equilibrium in the y-direction, we obtain the following.

$F = 17.32\cos 30 + 6.47\cos 75 = 12.674$ or $\boxed{F = 16.7\text{kN}}$

3.44 A little boy is shooting paper darts at his friends using a rubber band that has an un-stretched length of 7 inches. The piece of rubber band between point A and B is pulled to form two sides AC and CB of a triangle as shown. Assume the same normal strain in AC and CB, and the rubber band around the thumb and forefinger is a total of one inch. The area of cross-section of the band is (1/128) in^2 and the rubber has Modulus of Elasticity E= 150 psi. Determine the approximate force F and the angle θ at which the paper dart leaves the boy's hand.

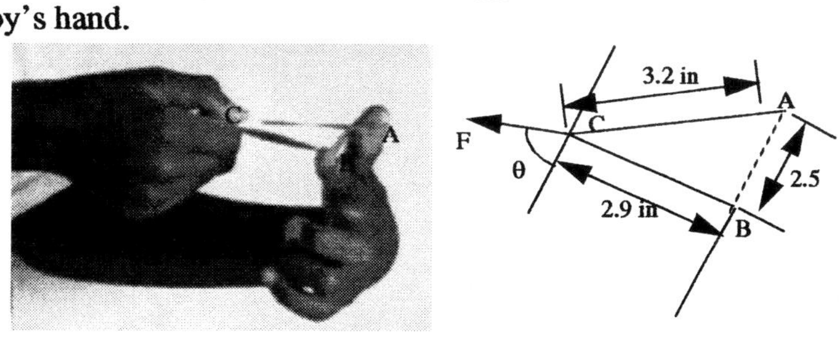

Fig. P3.44

Solution E =150 psi A = 1/128 in L_0= 7 in θ = ? F=?

1.Strain Calculation: The stretched length is: $L_f = 3.2 + 2.9 + 2.5 + 1 = 9.6\text{in}$

$$\varepsilon = \frac{L_f - L_0}{L_0} = \frac{9.6-7}{7} = 0.37143 \text{ extension}$$

2.Stress calculation: By Hooke's law: $\sigma = E\varepsilon = (150)(0.37143) = 55.71\text{psi(T)}$

3.Internal force Calculation: $N = A\sigma = (55.71)\frac{1}{128} = 0.43527\text{lbs(T)}$

4. External Force Calculation.The following free body diagram of the dart can be created

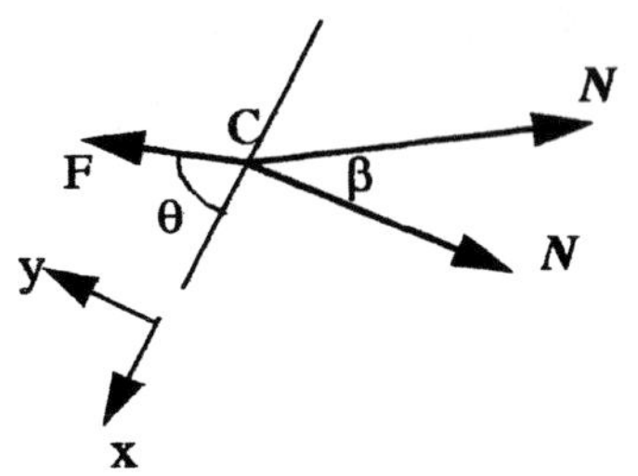

$$\cos\beta = \frac{3.2^2 + 2.9^2 - 2.5^2}{2(3.2)(2.9)} \quad \text{or} \quad \beta = 48.08^0$$

By equilibrium of forces in the x-direction $F\cos\theta - N\sin\beta = 0$ or

$$F\cos\theta = 0.43527(\sin 48.08) = 0.32387 \quad (1)$$

By equilibrium of forces in the y-direction. $F\sin\theta - N - N\cos\beta = 0$

$$F\sin\theta = 0.43527(1 + \cos 48.08) = 0.72608 \quad (2)$$

Dividing Eq(2) by Eq(1) we obtain $\tan\theta = 2.24187$ or $\boxed{\theta = 65.96^\circ}$

From Eq(1) $F = \dfrac{0.32387}{\cos 65.96} = 0.795\text{lb}$ $\boxed{F = 0.795\text{lb}}$

3.45 Three poles are pin connected to a ring at P and to the supports on the ground. The coordinates of the four points are as given. All poles have an area of cross-section of A=1 in^2. and a Modulus of Elasticity E= 10,000 ksi. If under the action of force F the ring at P moves vertically by an amount δ_P = 2 inch determine the force F.

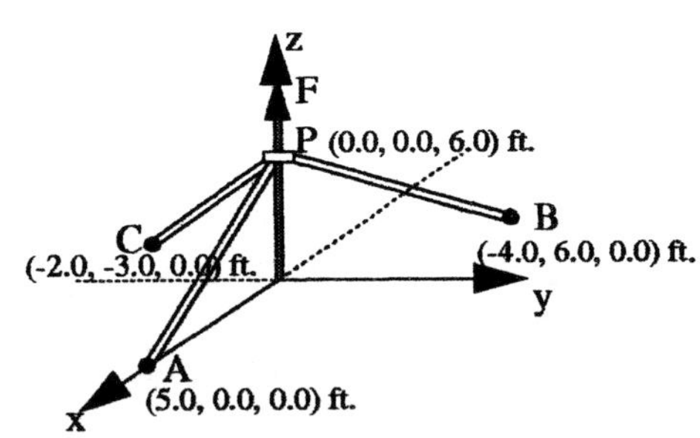

Fig. P3.45

Solution δ_P= 2 in E =10,000ksi A = 1in^2 F=?

1. Strain Calculation: From the solution of problem 2-50, we have:

$\varepsilon_{AP} = 16393\mu\text{in/in}$ $\varepsilon_{BP} = 11364\mu\text{in/in}$ $\varepsilon_{CP} = 20407\mu\text{in/in}$

2.Stress calculation:

$$\sigma_{AP} = E\varepsilon_{AP} = (10000)(16393)(10^{-6}) = 163.93\text{ksi(T)}$$

$$\sigma_{BP} = E\varepsilon_{BP} = (10000)(11364)(10^{-6}) = 113.64\text{ksi(T)}$$

$$\sigma_{CP} = E\varepsilon_{CP} = (10000)(20407)(10^{-6}) = 204.07\text{ksi(T)}$$

3.Internal force Calculation: $N_{AP} = A\sigma_{AP} = 163.93\text{kips(T)}$

$N_{BP} = A\sigma_{BP} = 113.64\text{kips(T)}$ and $N_{CP} = A\sigma_{CP} = 204.07\text{kips(T)}$

4. External Force Calculation. The free body diagram of the collar can be drawn as shown below.

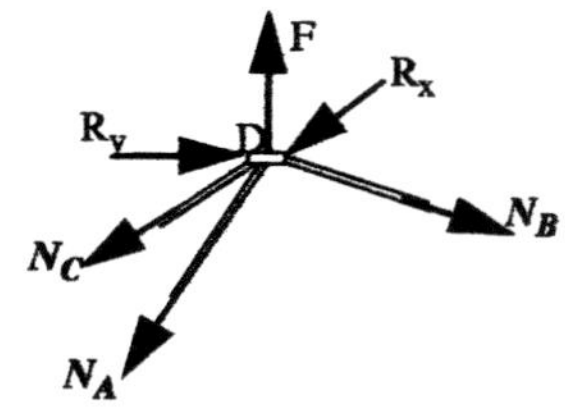

$$\overline{N_{AP}} + \overline{N_{BP}} + \overline{N_{CP}} + F\bar{k} + \bar{R} = 0 \qquad 1$$

Noting that the unit vectors in problem 2-50 were directed towards P and the internal forces are directed away from P, we can write

$$\overline{N_{AP}} = -N_{AP}\bar{i}_{AP} = (-163.93)(-0.6402\hat{i} + 0.7682\hat{k}) \qquad 2$$

$$\overline{N_{BP}} = -N_{BP}\bar{i}_{BP} = (-113.64)(-0.4264\hat{i} - 0.6396\hat{j} + 0.6396\hat{k}) \qquad 3$$

$$\overline{N_{CP}} = -N_{CP}\bar{i}_{CP} = (-204.07)(0.2857\hat{i} + 0.4286\hat{j} + 0.8571\hat{k}) \qquad 4$$

$$\bar{R} = R_x\hat{i} + R_y\hat{j} \qquad 5$$

Substituting Eq's (2),(3),(4) and (5) into Eq(1) and considering the coefficient of unit vector $\bar{k}$ we obtain:

$-163.93(0.7682) - 113.64(0.6396) - 204.07(0.8571) + F = 0$ or $\boxed{F = 373.52\text{kips}}$

3.46 A gap of 0.004 inch exists between the rigid bar and bar A before the force F is applied. The rigid bar is hinged at point C. Due to force F the strain in bar A was found to be - 500 μ in/in. The lengths of bar A and B are 30 and 50 inches respectively. Both bars have an area of cross-section A= 1 in^2 and Modulus of Elasticity E = 30,000 ksi. Determine the applied force F.

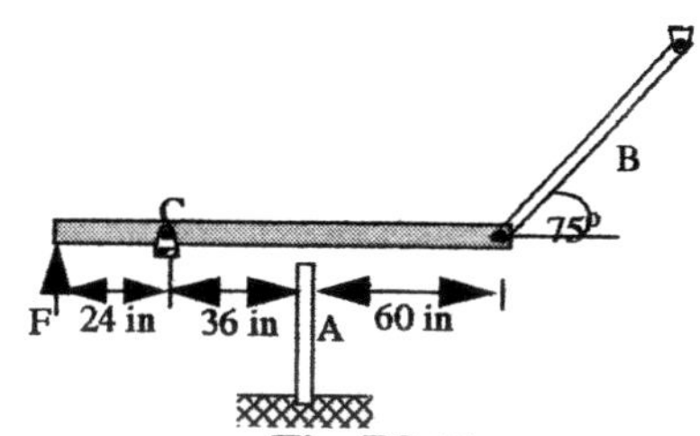

Fig. P3.46

Solution gap= 0.004 E_A=E_B =30,000ksi L_A= 30in

L_B= 50in ε_A=-500μ in/in A_A=A_B= 1in^2

1.Strain Calculation: From the solution of problem 2-43, we have: $\varepsilon_B = 978.8\mu$

2.Stress calculation: $\sigma_A = E\varepsilon_A = (30000)(-500)(10^{-6}) = -15\text{ksi}$ $\sigma_A = 15\text{ksi(C)}$

$\sigma_B = E\varepsilon_B = (30000)(978.8)(10^{-6}) = 29.364\text{ksi}$ $\sigma_B = 29.364\text{ksi(C)}$

3.Internal force Calculation: $N_A = A\sigma_A = 15\text{kips(C)}$ $N_B = A\sigma_B = 29.364\text{kips(T)}$

4. External Force Calculation: The free body diagram of the rigid bar can be drawn as shown below.

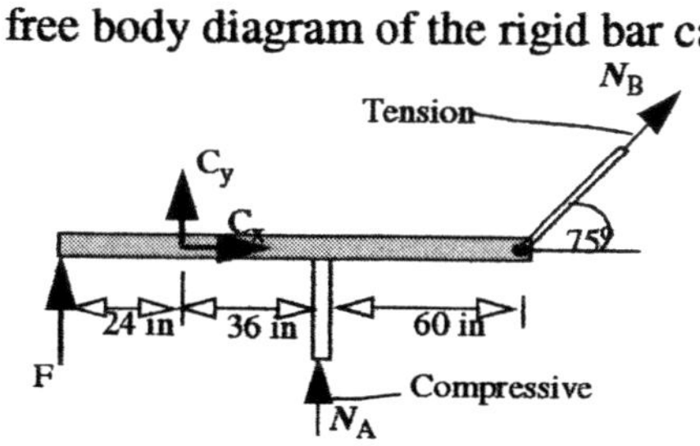

By moment equilibrium about point C, we obtain: $24F = (36)(15) + (29.364)(\sin 75)(96)$ or $\boxed{F = 136\text{kips}}$

3.47 The pins in the truss shown are displaced by u and v in the x and y direction respectively, as given. All rods in the truss have an area of cross-section A= 100 mm^2 and a Modulus of Elasticity E= 200 GPa. Determine the external forces P_1 and P_2.

$u_A = -4.6765$ mm $\quad v_A = 0$

$u_B = -3.3775$ mm $\quad v_B = -8.8793$ mm

$u_C = -2.0785$ mm $\quad v_C = -9.7657$ mm

$u_D = -1.0392$ mm $\quad v_D = -8.4118$ mm

$u_E = 0.0000$ mm $\quad v_E = 0.0000$ mm

$u_F = -3.2600$ mm $\quad v_F = -8.4118$ mm

$u_G = -2.5382$ mm $\quad v_G = -9.2461$ mm

$u_H = -1.5500$ mm $\quad v_H = -8.8793$ mm

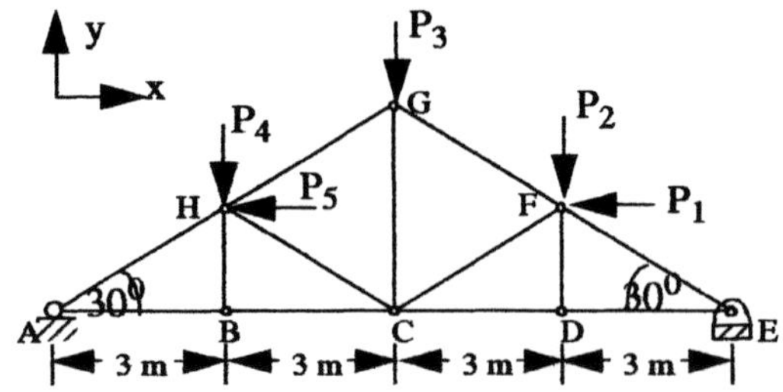

Fig. P3.47

Solution $\quad$ A=100 mm^2 $\quad$ E = 200GPa $\quad$ P_1 = ? P_2 = ?

---

1. Strain Calculation in all members at joint F.
Deformation vector calculation.

$$\bar{D}_{FC} = (u_C - u_F)\hat{i} + (v_C - v_F)\hat{j} = (-2.0785 + 3.26)\hat{i} + (-9.7657 + 8.4118)\hat{j} = (1.1815\hat{i} - 1.3539\hat{j})\text{ mm}$$

$$\bar{D}_{FG} = (u_G - u_F)\hat{i} + (v_G - v_F)\hat{j} = (-2.5382 + 3.26)\hat{i} + (-9.2461 + 8.4118)\hat{j} = (0.7218\hat{i} - 0.8343\hat{j})\text{mm}$$

$$\bar{D}_{FE} = (u_E - u_F)\hat{i} + (v_E - v_F)\hat{j} = (3.26\hat{i} - 8.4118\hat{j})\text{mm}$$

$$\bar{D}_{FD} = (u_D - u_F)\hat{i} + (v_D - v_F)\hat{j} = (-1.0392 + 3.26)\hat{i} + (-8.4118 + 8.4118)\hat{j} = (2.2208\hat{i})\text{mm}$$

Unit vector calculation.

$$\hat{i}_{FC} = -\cos 30\hat{i} - \sin 30\hat{j} = -0.866\hat{i} - 0.5\hat{j}$$

$$\hat{i}_{FG} = -\cos 30\hat{i} - \sin 30\hat{j} = -0.866\hat{i} - 0.5\hat{j}$$

$$\hat{i}_{FE} = -\cos 30\hat{i} - \sin 30\hat{j} = -0.866\hat{i} - 0.5\hat{j}$$

$$\hat{i}_{FD} = -\hat{j}$$

Deformation calculation.

$$\delta_{FC} = \hat{i}_{FC} \cdot \bar{D}_{FC} = (-0.866)(1.1815) + (-0.5)(-1.3539) = -0.34626\text{mm}$$

$$\delta_{FG} = \hat{i}_{FG} \cdot \bar{D}_{FG} = (-0.866)(0.7218) + (0.5)(-0.8343) = -1.04225\text{mm}$$

$$\delta_{FE} = \hat{i}_{FE} \cdot \bar{D}_{FE} = (-0.866)(3.26) + (-0.5)(8.4118) = -1.38274\text{mm}$$

$$\delta_{FD} = \hat{i}_{FD} \cdot \bar{D}_{FD} = 0$$

Strains.

$$L_{FC} = L_{FG} = L_{FE} = 31\cos 30 = 3.4641\text{m}$$

$$\varepsilon_{FC} = -\frac{0.34626}{3.4641(10^3)} = -0.1(10^{-3}) \qquad \varepsilon_{FG} = -\frac{1.04225}{3.4641(10^3)} = -0.3(10^{-3})$$

$$\varepsilon_{FE} = -\frac{1.38274}{3.4641(10^3)} = -0.4(10^{-3}) \qquad \varepsilon_{FD} = 0$$

2. Stress Calculation:

$\sigma_{FC} = E\varepsilon_{FC} = (200)(10^9)(-0.1)(10^{-3}) = -20(10^6) \text{or} \sigma_{FC} = 20(10^6)\frac{N}{m^2}(C)$

$\sigma_{FG} = E\varepsilon_{FG} = (200)(10^9)(-0.3)(10^{-3}) = -60(10^6) \text{or} \sigma_{FC} = 60(10^6)\frac{N}{m^2}(C)$

$\sigma_{FE} = E\varepsilon_{FE} = (200)(10^9)(-0.4)(10^{-3}) = -80(10^6) \text{or} \sigma_{FC} = 80(10^6)\frac{N}{m^2}(C)$

$\sigma_{FD} = 0$

3.Internal Force calculation.

$N_{FC} = \sigma_{FC}A = (20)(10^6)(100)(10^{-6}) = 2000N = 2kN(C)$

$N_{FG} = \sigma_{FG}A = (60)(10^6)(100)(10^{-6}) = 6000N = 6kN(C)$

$N_{FE} = \sigma_{FE}A = (80)(10^6)(100)(10^{-6}) = 8000N = 8kN(C)$

$N_{FD} = 0$

4.External Force Calculation. We draw the free body diagram of joint F as shown below.

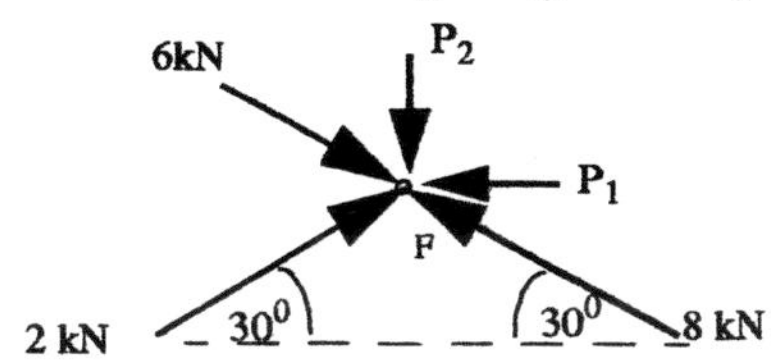

By force equilibrium in the x-direction, we obtain $-P_1 - 8\cos 30 + 2\cos 30 + 6\cos 30 = 0$ or $\boxed{P_1 = 0}$

By force equilibrium in the y-direction, we obtain $-P_2 + 2\sin 30 + 8\sin 30 - 6\sin 30 = 0$ or $\boxed{P_2 = 2kN}$

3.48 The pins in the truss shown are displaced by u and v in the x and y direction respectively, as given. All rods in the truss have an area of cross-section A= 100 mm² and a Modulus of Elasticity E= 200 GPa.Determine the external force P_3

$u_A = -4.6765$ mm $\quad v_A = 0$

$u_B = -3.3775$ mm $\quad v_B = -8.8793$ mm

$u_C = -2.0785$ mm $\quad v_C = -9.7657$ mm

$u_D = -1.0392$ mm $\quad v_D = -8.4118$ mm

$u_E = 0.0000$ mm $\quad v_E = 0.0000$ mm

$u_F = -3.2600$ mm $\quad v_F = -8.4118$ mm

$u_G = -2.5382$ mm $\quad v_G = -9.2461$ mm

$u_H = -1.5500$ mm $\quad v_H = -8.8793$ mm

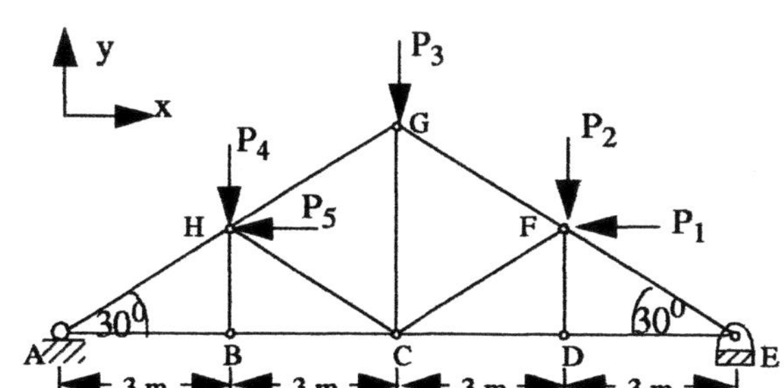

Fig. P3.48

Solution $\quad A=100\ mm^2 \quad E = 200GPa \quad P_3 = ?$

---

1. Strain Calculation in all members at joint G.
(i) Deformation vector calculation.

$\bar{D}_{GF} = (u_F - u_G)\hat{i} + (v_F - v_G)\hat{j} = (-3.26 + 2.5382)\hat{i} + (-8.4118 + 9.2461)\hat{j} = (-0.7218\hat{i} + 0.8343\hat{j})mm$

$\bar{D}_{GH} = (u_H - u_G)\hat{i} + (v_H - v_G)\hat{j} = (-1.55 + 2.5382)\hat{i} + (-8.8793 + 9.2461)\hat{j} = (0.98820\hat{i} + 0.3668\hat{j})mm$

$\bar{D}_{GC} = (u_C - u_G)\hat{i} + (v_C - v_G)\hat{j} = (-2.0785 + 2.5382)\hat{i} + (-9.7657 + 9.2461)\hat{j} = (0.4597\hat{i} - 0.5196\hat{j})mm$

(ii) Unit vector calculation.

$\hat{i}_{GF} = \cos 30\hat{i} - \sin 30\hat{j} = 0.866\hat{i} - 0.5\hat{j}$

$\hat{i}_{GH} = -\cos 30\hat{i} - \sin 30\hat{j} = -0.866\hat{i} - 0.5\hat{j}$

$\hat{i}_{GC} = -\hat{j}$

(iii) Deformation calculation.

$\delta_{GF} = \bar{i}_{GF} \cdot \bar{D}_{GF} = (0.866)(-0.7218) + (-0.5)(0.8343) = -1.0423\text{mm}$

$\delta_{GH} = \bar{i}_{GH} \cdot \bar{D}_{GH} = (-0.866)(0.9882) + (-0.5)(0.3668) = -1.03918\text{mm}$

$\delta_{GC} = \bar{i}_{GC} \cdot \bar{D}_{GC} = 0.5196\text{mm}$

(iv) Strains.

$L_{GF} = L_{GN} = L_{GC} = 3/(\cos 30) = 3.4641\text{m}$

$$\varepsilon_{GF} = -\frac{1.0423}{3.4641(10^3)} = -0.3(10^{-3})$$

$$\varepsilon_{GH} = -\frac{1.03918}{3.4641(10^3)} = -0.3(10^{-3})$$

$$\varepsilon_{GC} = -\frac{0.5196}{3.4641(10^3)} = 0.15(10^{-3})$$

2.Stress Calculation:

$$\sigma_{GF} = E\varepsilon_{GF} = (200)(10^9)(-0.3)(10^{-3}) = -60(10^6) \text{or} \sigma_{GF} = 60(10^6)\frac{N}{m^2}(C)$$

$$\sigma_{GH} = E\varepsilon_{GH} = (200)(10^9)(-0.3)(10^{-3}) = -60(10^6) \text{or} \sigma_{GH} = 60(10^6)\frac{N}{m^2}(C)$$

$$\sigma_{GC} = E\varepsilon_{GC} = (200)(10^9)(0.15)(10^{-3}) = -30(10^6) \text{or} \sigma_{GC} = 30(10^6)\frac{N}{m^2}(T)$$

3.Internal Force calculation.

$N_{GF} = \sigma_{GF}A = (60)(10^6)(100)(10^{-6}) = 6000\text{N} = 6\text{kN(C)}$

$N_{GH} = \sigma_{GH}A = (60)(10^6)(100)(10^{-6}) = 6000\text{N} = 6\text{kN(C)}$

$N_{GC} = \sigma_{GC}A = (30)(10^6)(100)(10^{-6}) = 3000\text{N} = 3\text{kN(T)}$

4. External Force Calculation. We draw the free body diagram of the joint G as shown below.

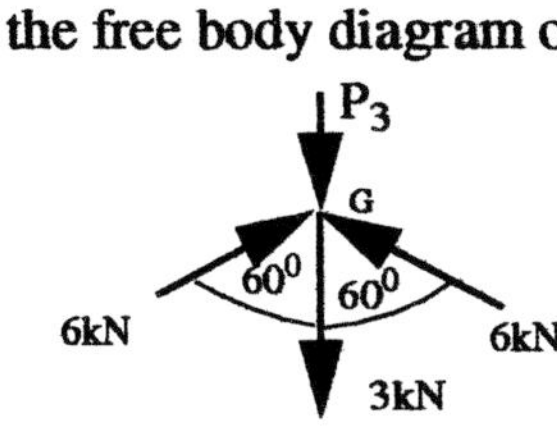

By equilibrium of forces in y-direction $-P_3 + 6\cos 60 + 6\cos 60 - 3 = 0$ or $\boxed{P_3 = 3\text{kN}}$

3.49 The pins in the truss shown are displaced by u and v in the x and y direction respectively, as given. All rods in the truss have an area of cross-section A= 100 mm² and a Modulus of Elasticity E=

200 GPa.Determine the external force P_4 and P_5.

$u_A = -4.6765$ mm $v_A = 0$
$u_B = -3.3775$ mm $v_B = -8.8793$ mm
$u_C = -2.0785$ mm $v_C = -9.7657$ mm
$u_D = -1.0392$ mm $v_D = -8.4118$ mm
$u_E = 0.0000$ mm $v_E = 0.0000$ mm
$u_F = -3.2600$ mm $v_F = -8.4118$ mm
$u_G = -2.5382$ mm $v_G = -9.2461$ mm
$u_H = -1.5500$ mm $v_H = -8.8793$ mm

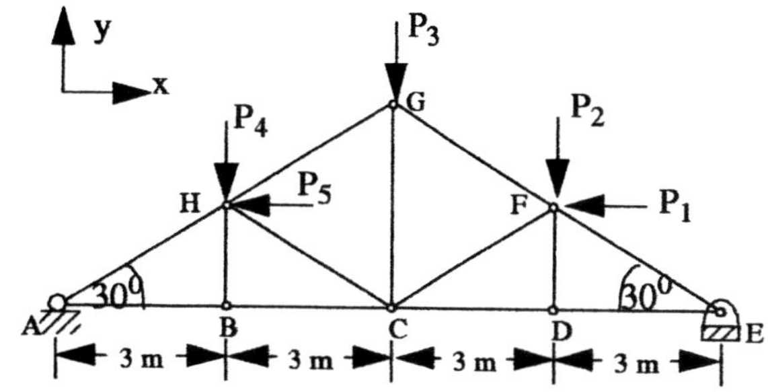

Fig. P3.49

Solution A=100 mm^2 E = 200GPa $P_4 = ?$ $P_5 = ?$

---

1. Strain Calculation in all members at joint H.
(i) Deformation vector calculation.

$$\bar{D}_{HA} = (u_A - u_H)\hat{i} + (v_A - v_H)\hat{j} = (-4.6765 + 1.55)\hat{i} + (8.8793)\hat{j} = (-3.1265\hat{i} + 8.8793\hat{j})$$

$$\bar{D}_{HG} = (u_G - u_H)\hat{i} + (v_G - v_H)\hat{j} = (-2.5382 + 1.55)\hat{i} + (-9.2461 + 8.8793)\hat{j} \text{ or}$$

$$\bar{D}_{HG} = (-0.9882\hat{i} - 0.3668\hat{j})$$

$$\bar{D}_{HC} = (u_C - u_H)\hat{i} + (v_C - v_H)\hat{j} = (-2.0785 + 1.55)\hat{i} + (-9.7657 + 8.8793)\hat{j} = (-0.5285\hat{i} - 0.8864\hat{j})$$

$$\bar{D}_{HB} = (u_B - u_H)\hat{i} + (v_B - v_H)\hat{j} = (-3.375 + 1.55)\hat{i} + (-8.8793 + 8.8793)\hat{j} = -1.825\hat{i}$$

(ii) Unit vector calculation.

$$\hat{i}_{HA} = -\cos 30\hat{i} - \sin 30\hat{j} = -0.866\hat{i} - 0.5\hat{j}$$

$$\hat{i}_{HG} = \cos 30\hat{i} + \sin 30\hat{j} = 0.866\hat{i} + 0.5\hat{j}$$

$$\hat{i}_{HC} = \cos 30\hat{i} - \sin 30\hat{j} = 0.866\hat{i} - 0.5\hat{j}$$

$$\hat{i}_{HB} = -\hat{j}$$

(iii) Deformation calculation.

$$\delta_{HA} = \bar{i}_{HA} \cdot \bar{D}_{HA} = (-0.866)(-3.1265) + (-0.5)(8.8793) = -1.73202\text{mm}$$

$$\delta_{HG} = \bar{i}_{HG} \cdot \bar{D}_{HG} = (0.866)(-0.9882) + (0.5)(-0.3668) = -1.03918\text{mm}$$

$$\delta_{HC} = \bar{i}_{HC} \cdot \bar{D}_{HC} = (0.866)(-0.5285) + (-0.5)(-0.8864) = -0.01449\text{mm}$$

$$\delta_{HB} = \bar{i}_{HB} \cdot \bar{D}_{HB} = 0$$

(iv) Strains.

$$L_{HA} = L_{HG} = L_{HC} = 3/\cos 30 = 3.4641\text{m}$$

$$\varepsilon_{HA} = \frac{\delta_{HA}}{L_{HA}} = -\frac{1.73202}{3.4641(10^3)} = -0.5(10^{-3}) \qquad \varepsilon_{HG} = \frac{\delta_{HG}}{L_{HG}} = -\frac{1.03918}{3.4641(10^3)} = -0.3(10^{-3})$$

$$\varepsilon_{HC} = \frac{\delta_{HC}}{L_{HC}} = -\frac{0.01449}{3.4641(10^3)} = -0.004(10^{-3}) \qquad \varepsilon_{HD} = 0$$

2.Stress Calculation:

$$\sigma_{HA} = E\varepsilon_{HA} = (200)(10^9)(-0.5)(10^{-3}) = -100(10^6) \text{ or} \qquad \sigma_{HA} = 100(10^6)\frac{N}{m^2}(C)$$

$\sigma_{HG} = E\varepsilon_{HG} = (200)(10^9)(-0.3)(10^{-3}) = -60(10^6)$ or $\qquad \sigma_{HG} = 60(10^6)\frac{N}{m^2}(C)$

$\sigma_{HC} = E\varepsilon_{HC} = (200)(10^9)(-0.004)(10^{-3}) = -0.8(10^6)$ or $\qquad \sigma_{FC} = 0.8(10^6)\frac{N}{m^2}(C)$

$\sigma_{HB} = 0$

3.Internal Force calculation.

$N_{HA} = \sigma_{HA}A = (100)(10^6)(100)(10^{-6}) = 10000N = 10kN(C)$

$N_{HG} = \sigma_{HG}A = (60)(10^6)(100)(10^{-6}) = 6000N = 6kN(C)$

$N_{HC} = \sigma_{HC}A = (0.8)(10^6)(100)(10^{-6}) = 80N = 0.08kN(C)$

$N_{HB} = 0$

4. External Force Calculation. We draw the free body diagram of the joint H as shown below.

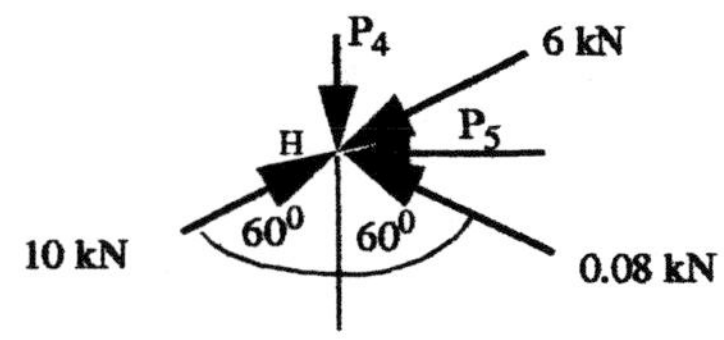

By equilibrium of forces in x-direction: $-P_5 - 6\sin 60 + 10\sin 60 - 0.8\sin 60 = 0$ or $\boxed{P_5 = 3.39kN}$

By equilibrium of forces in y-direction $-P_4 + 10\cos 60 - 6\cos 60 + 0.08\cos(60) = 0$ or $\boxed{P_4 = 2.04kN}$

3.50 A circular rod of diameter 15 mm is acted upon by a distributed force p(x) that has the units of kN / m as shown. The Modulus of Elasticity of the rod is 70 GPa. Determine the distributed force p(x) if the displacement u(x) in the x-direction is as given below. x is measured in meters. $u(x) = 30(x - x^2)(10^{-6})m$

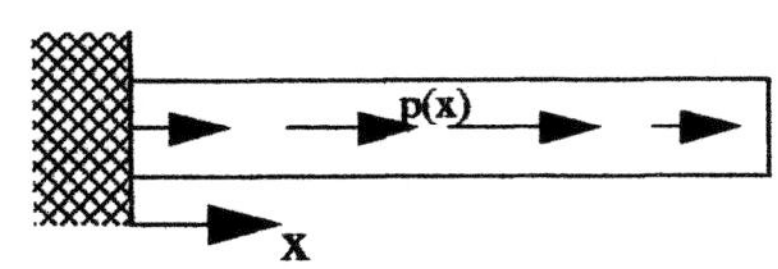

Fig. P3.50

Solution $\qquad d = 15$ mm $\qquad E = 70$ GPa $\qquad p(x) = ?$

1.Strain Calculation: $\varepsilon_{xx} = \frac{du}{dx} = 30(1 - 2x)(10^{-6})$

2.Stress Calculation: $\sigma_{xx} = E\varepsilon_{xx} = (70)(10^9)(30)(1 - 2x)(10^{-6}) = 2.1(1 - 2x)(10^6)\frac{N}{m^2}$

3.Internal Force Calculation:

$N = \sigma_{xx}A = 7.0(10^6)(1 - 2x)\left(\frac{\pi}{4}\right)(0.015)^2 = 371.1(1 - 2x)\text{Newtons}$

4. External Force Calculations: A free body diagram of a differential element can be drawn as shown below.

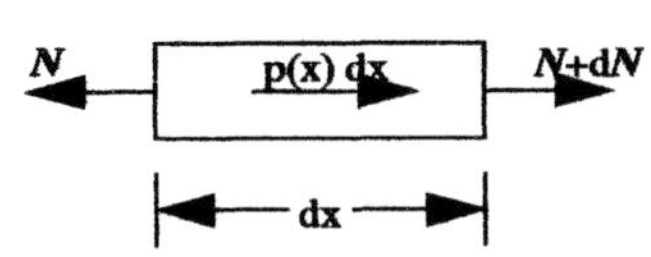

By balancing forces in the x -direction we obtain: $(N + dN) + p(x)dx - N = 0$ or

$p(x) = -\left(\frac{dN}{dx}\right) = -371.1(-2)$ or $\qquad \boxed{p(x) = 742.2\ N/m}$

3.51 A circular rod of diameter 15 mm is acted upon by a distributed force p(x) that has the units of kN / m as shown. The Modulus of Elasticity of the rod is 70 GPa. Determine the distributed force p(x) if the displacement u(x) in the x-direction is as given below. x is measured in meters. $u(x) = 50(x^2 - 2x^3)(10^{-6})m$

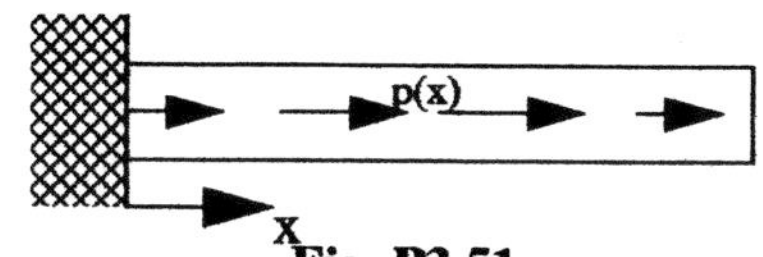

Fig. P3.51

Solution d =15mm E = 70G Pa p(x) = ?

1.Strain Calculation: $\varepsilon_{xx} = \frac{du}{dx} = 50(2x - 6x^2)(10^{-6})$

2.Stress Calculation: $\sigma_{xx} = E\varepsilon_{xx} = (70)(10^9)(50)(2x - 6x^2)(10^{-6}) = 7.0(10^6)(x - 3x^2)N/m^2$

3.Internal Force Calculation: $N = \sigma_{xx}A = 7.0(10^6)(x - 3x^2)\left(\frac{\pi}{4}\right)(0.015)^2 = 1237(x - 3x^2)\text{Newtons}$

4.External Force Calculation: A free body diagram of a differential element can be drawn as shown below

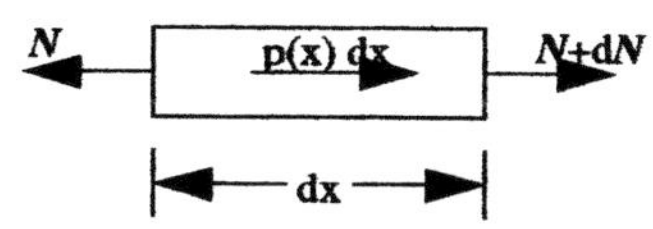

By balancing forces in the x -direction we obtain: $(N + dN) + p(x)dx - N = 0$ or

$$p(x) = -\left(\frac{dN}{dx}\right) = -1237(1 - 6x)\text{Newtons/m}$$

$$\boxed{p(x) = -1.237(1 - 6x)\frac{kN}{m}}$$

3.52 Due to the action of the forces, the displacement in the x-direction was found to be $u = \frac{(60x + 80xy - x^2y)}{180000}$ in. The Modulus of Elasticity of the beam is 30,000 ksi. Determine the statically equivalent internal normal force N and the internal bending moment M_z acting at point O at a section at x = 20 inch. The cross-section of the beam is shown on the right. Assume some unknown shear stress is present at the cross-section.

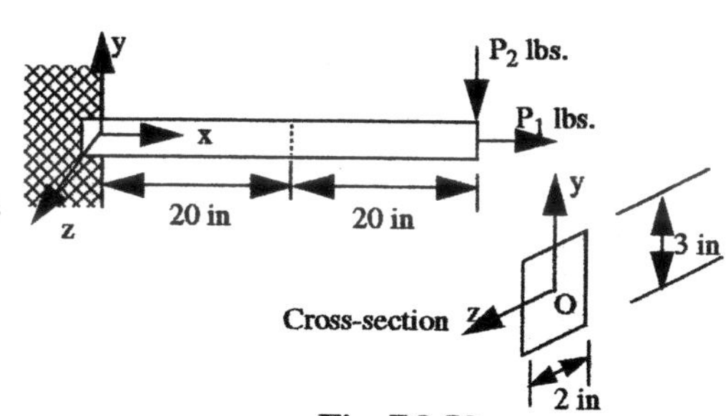

Fig. P3.52

Solution E=30000ksi N=? and M_z=? @ x=20

1.Strain Calculation: $\varepsilon_{xx} = \frac{du}{dx} = \frac{(60 + 80y - 2xy)}{180000}$

2.Stress Calculation: $\sigma_{xx} = E\varepsilon_{xx} = 30(10^3)\frac{(60 + 80y - 2xy)}{180000} = \frac{(30 + 40y - xy)}{3}\text{ksi}$

at x=20, $\sigma_{xx} = \frac{(30 + 40y - 20y)}{3} = \frac{30 + 20y}{3}\text{ksi}$

3. Internal forces and moments: The stress distribution can be replaced by an equivalent force at the centroid of the distribution and then replaced by an equivalent force and moment at O.

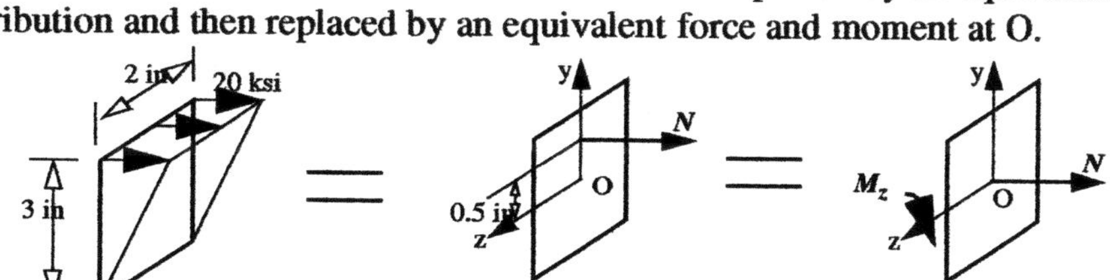

$N = \frac{1}{2}(20)(3)(2) = 60\text{kips}$ or $\boxed{N = 60\text{kips}}$

$M_z = N(0.5) = 30\text{in}-\text{kips}$ or $\boxed{M_z = 30\text{in}-\text{kips}}$

3.53 Using numerical integration and the data in Problem 3.17 determine the strain energy density when the specimen shows an average strain of 0.18 mm / mm.

Solution $\varepsilon = 0.18$ $U_o = ?$

The strain energy density is given by the equation $U_o = \int_0^{0.18} \sigma d\varepsilon$. Using the results of stresses and strains in problem 3.17 and the numerical integration described in Appendix B, we obtain the value of the integral on a spread sheet as shown below.
The stress value at a strain of 0.18 is obtained by linear interpolation of the known values of stresses and strain before and after the point.

Strain x_i	Stress (MPa) $f(x_i)$	U_o
0.0000	0.000	0.02
0.0004	120.280	0.29
0.0014	420.970	1.02
0.0026	781.810	1.74
0.0034	1022.360	9.45

Strain x_i	Stress (MPa) $f(x_i)$	U_o
0.0106	1119.310	22.49
0.0220	1168.540	43.05
0.0392	1222.430	63.73
0.0558	1269.440	95.12
0.0800	1324.650	114.06

Strain x_i	Stress (MPa) $f(x_i)$	U_o
0.0942	1342.290	143.83
0.1160	1388.960	181.76
0.1430	1421.180	231.58
0.1776	1458.260	235.08
0.1800	1459.815	← Interpolated Value.
0.1998	1472.640	

$\boxed{U_o = 235.1\text{MJ}/\text{m}^3}$

3.54 Using numerical integration determine the Modulus of Toughness for the material described in Problem 3.17.

Solution Modulus of Toughness = U_T=?

The Modulus of Toughness is the strain energy up to the ultimate stress.Using the results of stresses and strains in problem 3.17 and the numerical integration described in Appendix B, we obtain the value of the integral on a spread sheet as shown below. Note we neglect all values after the ultimate stress in the calculation.

Strain x_i	Stress (MPa) f_i	U_T
0.0000	0.00	0.02
0.0004	120.28	0.29
0.0014	420.97	1.02
0.0026	781.81	1.74
0.0034	1022.36	9.45
0.0106	1119.31	22.49

Strain x_i	Stress (MPa) f_i	U_T
0.0220	1168.54	43.05
0.0392	1222.43	63.73
0.0558	1269.44	95.12
0.0800	1324.65	114.06
0.0942	1342.29	143.83
0.1160	1388.96	181.76

Strain x_i	Stress (MPa) f_i	U_T
0.1430	1421.18	231.58
0.1776	1458.26	264.11
0.1998	1472.64	294.16
0.2202	1473.40	
0.2326	1448.89	

$\boxed{U_T = 294.2\text{MJ}/\text{m}^3}$

3.55 Assume that the stress-strain curve *after* yield stress in Problem 3.17 is described by the quadratic equation $\sigma = a + b\varepsilon + c\varepsilon^2$. (a) Determine the coefficients a, b, and c by Least Square Method. (b) Find the strain energy density when the specimen shows an average strain of 0.18 mm / mm. (c) Find the Tangent Modulus of Elasticity at a stress level of 1400 MPa.

Solution $a = ?$ $b = ?$ $c = ?$ $U_o = ?\ @\varepsilon = 0.18$ $E_T = ?\ @\sigma = 1400$ MPa

---

Using the results of stresses and strains in problem 3.17 after the proportional limit and the Least Square Method described in Appendix B, we obtain the value of the values of constants a, b, and c on a spread sheet as shown in the table below.

	Strain x_i	Stress (MPa) f_i	x_i^2	x_i^3	x_i^4	x_i*f_i	$x_i^2*f_i$
1	0.0106	1119.31	0.0001	0.0000	0.0000	11.86	0.13
2	0.0220	1168.54	0.0005	0.0000	0.0000	25.71	0.57
3	0.0392	1222.43	0.0015	0.0001	0.0000	47.92	1.88
4	0.0558	1269.44	0.0031	0.0002	0.0000	70.83	3.95
5	0.0800	1324.65	0.0064	0.0005	0.0000	105.97	8.48
6	0.0942	1342.29	0.0089	0.0008	0.0001	126.44	11.91
7	0.1160	1388.96	0.0135	0.0016	0.0002	161.12	18.69
8	0.1430	1421.18	0.0204	0.0029	0.0004	203.23	29.06
9	0.1776	1458.26	0.0315	0.0056	0.0010	258.99	46.00
10	0.1998	1472.64	0.0399	0.0080	0.0016	294.23	58.79
11	0.2202	1473.40	0.0485	0.0107	0.0024	324.44	71.44
12	0.2326	1448.89	0.0541	0.0126	0.0029	337.01	78.39
13	0.2406	1423.54	0.0579	0.0139	0.0034	342.50	82.41
14	0.2462	1384.31	0.0606	0.0149	0.0037	340.82	83.91
15	0.2494	1334.38	0.0622	0.0155	0.0039	332.79	83.00
16	0.2526	1287.92	0.0638	0.0161	0.0041	325.33	82.18
b_{ij} & r_i	2.3798	21540.14	0.4730	0.1034	0.0236	3309.21	660.77
C_{ij}	0.0005	0.0072	0.0224	0.1533	0.5288	1.90	
D	0.000770195						
a_i	1062.06	4493.16	-12993.12				

The values of the constants are: $a = 1062.1\text{MPa}$ $b = 4493.3\text{MPa}$ $c = -12993.1\text{MPa}$.

The strain energy density at $\varepsilon = 0.18$ can be written as the sum strain energy density up to and after the proportional limit. The strain energy density at the proportional limit is $U_{prop} = (\sigma_{prop})^2/2E$. We can also substitute the quadratic stress expression in the integral after the proportional limit as shown in Eq. 1.

$$U_o = \int_0^{0.18} \sigma d\varepsilon = \int_0^{\varepsilon_{prop}} \sigma d\varepsilon + \int_{\varepsilon_{prop}}^{0.18} \sigma d\varepsilon = \frac{\sigma_{prop}^2}{2E} + \int_{\varepsilon_{prop}}^{0.18} (a + b\varepsilon + c\varepsilon^2) d\varepsilon \quad (1)$$

From problem 3.17 $\sigma_{prop} = 1022$ MPa and E = 300 GPa. Substituting these value, we obtain $U_{prop} = 1.741\text{MJ}/\text{m}^3$ and $\varepsilon_{prop} = \sigma_{prop}/E = 0.0034$. Integrating and substituting the values of a, b, and c, we obtain:

$$U_o = 1.741 + \left(a\varepsilon + \frac{b\varepsilon^2}{2} + \frac{c\varepsilon^3}{3}\right)\Bigg|_{0.0034}^{0.18} = 236.8 \qquad \boxed{U_o = 236.8\text{MJ}/\text{m}^3}$$

The Tangent Modulus is the slope of the stress-strain curve in the non-linear region and can be written as $E_T = \frac{d\sigma}{d\varepsilon} = b + c\varepsilon$. At $\sigma = 1400$ MPa, the strain from the plot in problem 3.17 is $\varepsilon_{1400} = 0.125$. Substituting this value and the values of b and c, we obtain the Tangent Modulus as given below.

$E_T = 4493.3 + (-12993.1)(0.125) = 2869.2\text{MPa}$ $\boxed{E_T = 2.87\ \text{GPa}}$

3.56 Using numerical integration and the data in Problem 3.18, determine the strain energy density when the specimen shows an average strain of 0.15 in / in.

Solution $\varepsilon = 0.15$ $U_o = ?$

The strain energy density is given by the equation $U_o = \int_0^{0.15} \sigma d\varepsilon$. Using the results of stresses and strains in problem 3.18 and the numerical integration described in Appendix B, we obtain the value of the integral on a spread sheet as shown below.

The stress value at a strain of 0.15 is obtained by linear interpolation of the known values of stresses and strain before and after the point.

Strain x_i	Stress $f(x_i)$ (ksi)	U_o
0	0	0.005
0.00064	15.84	0.027
0.00148	36.87	0.029
0.00153	38.2	0.139
0.00438	39.22	0.344
0.00953	40.23	0.541
0.01435	41.56	0.732

Strain x_i	Stress $f(x_i)$ (ksi)	U_o
0.01887	43.09	0.940
0.02359	44.92	1.215
0.02953	47.47	1.503
0.03543	50.22	1.848
0.04212	52.97	2.216
0.04893	55.11	2.615
0.05605	56.94	3.440

Strain x_i	Stress $f(x_i)$ (ksi)	U_o
0.0702	59.69	4.068
0.08061	61.06	5.040
0.09633	62.49	5.717
0.10711	63.2	6.725
0.12297	63.92	7.932
0.14174	64.68	8.467
0.15	64.86 ← Interpolated Value.	
0.15818	65.04	

$\boxed{U_o = 8.5\ \text{in}-\text{kips/in}^3}$

3.57 Using numerical integration determine the Modulus of Toughness for the material described in Problem 3.18.

Solution Modulus of Toughness = U_T=?

The Modulus of Toughness is the strain energy up to the ultimate stress.Using the results of stresses and strains in problem 3.18 and the numerical integration described in Appendix B, we obtain the value of the integral on a spread sheet as shown below. Note we neglect all values after the ultimate stress in the calculation.

Strain x_i	Stress f_i (ksi)	U_T
0	0	0.005
0.00064	15.84	0.027
0.00148	36.87	0.029
0.00153	38.2	0.139
0.00438	39.22	0.344
0.00953	40.23	0.541
0.01435	41.56	0.732
0.01887	43.09	0.940

Strain x_i	Stress f_i (ksi)	U_T
0.02359	44.92	1.215
0.02953	47.47	1.503
0.03543	50.22	1.848
0.04212	52.97	2.216
0.04893	55.11	2.615
0.05605	56.94	3.440
0.0702	59.69	4.068
0.08061	61.06	5.040

Strain x_i	Stress f_i (ksi)	U_T
0.09633	62.49	5.717
0.10711	63.2	6.725
0.12297	63.92	7.932
0.14174	64.68	8.998
0.15818	65.04	10.522
0.18155	65.39	
0.19267	61.32	

$\boxed{U_T = 10.5\ \text{in}-\text{kips/in}^3}$

3.58 Assume that the stress-strain curve *after* yield stress in Problem 3.18 is described by the quadratic equation $\sigma = a + b\varepsilon + c\varepsilon^2$. (a) Determine the coefficients a, b, and c by Least Square Method. (b) Find the strain energy density when the specimen shows an average strain of 0.15 in / in. (c) Find the Tangent Modulus of Elasticity at a stress level of 50 ksi.

Solution $a = ?$ $b = ?$ $c = ?$ $U_o = ?$ @ $\varepsilon = 0.15$ $E_T = ?$ @ $\sigma = 50$ ksi

---

Using the results of stresses and strains in problem 3.18 after the proportional limit and the Least Square Method described in Appendix B, we obtain the value of the values of constants a, b, and c on a spread sheet as shown in the table below.

	Strain x_i	Stress (ksi) f_i	x_i^2	x_i^3	x_i^4	x_i*f_i	$x_i^2*f_i$
1	0.00438	39.22	0.000	0.000	0.000	0.172	0.001
2	0.00953	40.23	0.000	0.000	0.000	0.383	0.004
3	0.01435	41.56	0.000	0.000	0.000	0.596	0.009
4	0.01887	43.09	0.000	0.000	0.000	0.813	0.015
5	0.02359	44.92	0.001	0.000	0.000	1.060	0.025
6	0.02953	47.47	0.001	0.000	0.000	1.402	0.041
7	0.03543	50.22	0.001	0.000	0.000	1.779	0.063
8	0.04212	52.97	0.002	0.000	0.000	2.231	0.094
9	0.04893	55.11	0.002	0.000	0.000	2.697	0.132
10	0.05605	56.94	0.003	0.000	0.000	3.191	0.179
11	0.0702	59.69	0.005	0.000	0.000	4.190	0.294
12	0.08061	61.06	0.006	0.001	0.000	4.922	0.397
13	0.09633	62.49	0.009	0.001	0.000	6.020	0.580
14	0.10711	63.2	0.011	0.001	0.000	6.769	0.725
15	0.12297	63.92	0.015	0.002	0.000	7.860	0.967
16	0.14174	64.68	0.020	0.003	0.000	9.168	1.299
17	0.15818	65.04	0.025	0.004	0.001	10.288	1.627
18	0.18155	65.39	0.033	0.006	0.001	11.872	2.155
19	0.19267	61.32	0.037	0.007	0.001	11.815	2.276
20	0.19802	58.26	0.039	0.008	0.002	11.537	2.284
21	0.20321	54.55	0.041	0.008	0.002	11.085	2.253
22	0.20736	50.73	0.043	0.009	0.002	10.519	2.181
b_{ij} & r_i	2.04273	1202.06	0.29666296	0.050327768	0.009120184	120.3690455	17.60174158
C_{ij}	0.000	0.004	0.015	0.113	0.501	2.354	
D	0.001						
a_i	35.995	455.956	-1756.978				

The values of the constants are: $a = 36.0\text{ksi}$ $b = 456.0\text{ksi}$ $c = -1757.0\text{ksi}$.

The strain energy density at $\varepsilon = 0.15$ can be written as the sum strain energy density up to and after the proportional limit. The strain energy density at the proportional limit is $U_{prop} = (\sigma_{prop})^2/2E$. We can also substitute the quadratic stress expression in the integral after the proportional limit as shown in Eq. 1.

$$U_o = \int_0^{0.15} \sigma d\varepsilon = \int_0^{\varepsilon_{prop}} \sigma d\varepsilon + \int_{\varepsilon_{prop}}^{0.15} \sigma d\varepsilon = \frac{\sigma_{prop}^2}{2E} + \int_{\varepsilon_{prop}}^{0.15} (a + b\varepsilon + c\varepsilon^2) d\varepsilon \qquad 1$$

From problem 3.18, $\sigma_{prop} = 36.9$ ksi and $E = 24893$ ksi. Substituting these value, we obtain $U_{prop} = 0.0273\text{in-kips/in}^3$ and $\varepsilon_{prop} = \sigma_{prop}/E = 0.00148$. Integrating and substituting the values of a, b, and c, we obtain:

$$U_o = 0.0273 + \left(a\varepsilon + \frac{b\varepsilon^2}{2} + \frac{c\varepsilon^3}{3}\right)\Bigg|_{0.00148}^{0.15} = 8.526 \qquad \boxed{U_o = 8.53\ \text{in}-\text{kips/in}^3}$$

The Tangent Modulus is the slope of the stress-strain curve in the non-linear region and can be written as $E_T = \frac{d\sigma}{d\varepsilon} = b + c\varepsilon$. At $\sigma = 50$ ksi, the strain from the plot in problem 3.18 is $\varepsilon_{50} = 0.035$. Substituting this value and the values of b and c, we obtain the Tangent Modulus as given below.

$$E_T = 455.96 + (-1756.96)(0.035) = 394.46\text{ksi} \qquad \boxed{E_T = 394.5\text{ksi}}$$

3.59—3.61 Marks were made on the cord used for tying the canoe on top of the car in Example 3.5. These marks were made every two inches to produce a total of 20 segments. The stretch cord is symmetric with respect to the top of the canoe. The starting point of the first segment is on the carrier rail of the car and the end point of the tenth segment is on the top of the canoe. The measured length of each segment is as shown in the adjoining table. Using this information and the data given in each of the problems 3.59 through 3.61, determine

(a) the tension in the cord of each segment.

(b) the force exerted by the cord on the carrier of the car.

Segment Number	Deformed Length inches
1	3.4
2	3.4
3	3.4
4	3.4
5	3.4
6	3.4
7	3.1
8	2.7
9	2.3
10	2.2

3.59 Use Modulus of Elasticity of E = 510 psi and the diameter of the stretch cord as (1/2) inch.

Solution $L_o = 2$ in $\quad d_o = 1/2$ in $\quad E = 510$ psi $\quad T_i = ?$ $\quad P = ?$

The strain ε_i in the ith segment of column three in table below is given by: $\varepsilon_i = (\Delta L_i)/L_0$, where ΔL_i is the deformed length in column two and $L_0 = 2$ in.

Segment Number	Deformed Length. ΔL_i (inches)	Strain ε_i	Stress σ_i (psi)	Internal Tension T_i (lbs)
1	3.4	0.70	357	70.10
2	3.4	0.70	357	70.10
3	3.4	0.70	357	70.10
4	3.4	0.70	357	70.10
5	3.4	0.70	357	70.10
6	3.4	0.70	357	70.10
7	3.1	0.55	280.5	55.08
8	2.7	0.35	178.5	35.05
9	2.3	0.15	76.5	15.02
10	2.2	0.10	51	10.01

The stress σ_i in the ith segment in column four is given by $\sigma_i = E\varepsilon_i$. Multiplying the column of strain by E = 510 psi, we obtain the column of stress. The area of each segment A_i is $A_i = (\pi 0.5^2)/4 = 0.19635\ \text{in}^2$. The internal tension T_i in the ith segment is given by: $T_i = \sigma_i A_i$. Multiplying the column of stress by the area A_i we obtain the column of internal tension.

The force P exerted on the car carrier by the cord is given by the tension in segment 1. $\boxed{P = 70.1\ \text{lb}}$

3.60 Use the diameter of the stretch cord as (1/2) inch and the following equation for stress-strain curve:

$$\sigma = \begin{cases} (1020\varepsilon - 1020\varepsilon^2)\ \text{psi} & \varepsilon < 0.5 \\ 255\ \text{psi} & \varepsilon \geq 0.5 \end{cases}$$

Solution $L_o = 2$ in $d_o = 1/2$ in $T_i = ?$ $P = ?$

The strain calculation is as in problem 3.59 and shown in the table below.

The stress σ_i in the for segments 8,9, and 10 is found using the equation $\sigma_i = 1020\varepsilon_i - 1020\varepsilon_i^2$, as the strains in these segments is less than 0.5. For segments 1 through 7, the stress is 255 psi as the strain is greater than 0.5.Multiplying the column of strain by E = 510 psi, we obtain the column of stress.

The area of each segment A_i is $A_i = (\pi 0.5^2)/4 = 0.19635\ \text{in}^2$. The internal tension T_i in the ith segment is given by: $T_i = \sigma_i A_i$. Multiplying the column of stress by the area A_i we obtain the column of internal tension.

Segment Number	Deformed Length. ΔL_i (inches)	Strain ε_i	Stress σ_i (psi)	Internal Tension T_i (lbs)
1	3.4	0.70	255.00	50.07
2	3.4	0.70	255.00	50.07
3	3.4	0.70	255.00	50.07
4	3.4	0.70	255.00	50.07
5	3.4	0.70	255.00	50.07
6	3.4	0.70	255.00	50.07
7	3.1	0.55	255.00	50.07
8	2.7	0.35	232.05	45.56
9	2.3	0.15	130.05	25.54
10	2.2	0.10	91.80	18.02

The force P exerted on the car carrier by the cord is given by the tension in segment 1. $\boxed{P = 50.1\ \text{lbs}}$

3.61 Use Poisson's ratio of ν=1/2 and initial diameter of 1/2 inch and calculate the diameter in deformed position for each segment. Use stress-strain relationship given in problem 3.60.

Solution $L_o = 2$ in $d_o = 1/2$ in $\nu = 1/2$ $T_i = ?$ $P = ?$

The longitudinal strain ε_i and the corresponding stress si are calculated as in problem 3.60 and are shown in the table below.

The transverse strain is $-\nu\varepsilon_i$. The change in diameter is thus $-\nu\varepsilon_i d_o$. The deformed diameter d_i of the ith segment is $d_i = d_0 - \nu\varepsilon_i d_0$. Substituting the values of ν=1/2 and $d_o = 1/2$ in. the entries in column five can be found using $d_i = (1 - 0.5\varepsilon_i)/2$.

The area of each segment A_i is $A_i = \pi d_i^2/4$. The internal tension T_i in the ith segment is given by: $T_i = \sigma_i A_i$. Multiplying the column of stress by the area A_i we obtain the column of internal tension.

Segment Number	Deformed Length. ΔL_i (inches)	Strain ε_i	Stress σ_i (psi)	Deformed Diameter d_i (inches)	Internal Tension T_i (lbs)
1	3.4	0.70	255.00	0.33	21.15
2	3.4	0.70	255.00	0.33	21.15
3	3.4	0.70	255.00	0.33	21.15
4	3.4	0.70	255.00	0.33	21.15
5	3.4	0.70	255.00	0.33	21.15
6	3.4	0.70	255.00	0.33	21.15
7	3.1	0.55	255.00	0.36	26.32
8	2.7	0.35	232.05	0.41	31.01
9	2.3	0.15	130.05	0.46	21.85
10	2.2	0.10	91.80	0.48	16.27

The force P exerted on the car carrier by the cord is given by the tension in segment 1. $\boxed{P = 21.2\ \text{lbs}}$

Problem Set 3.2

3.62 Write the Generalized Hooke's Law for isotropic material in cylindrical coordinates (r, θ, z).

Solution

The Generalized Hooke's Law for isotropic material in cylindrical coordinates (r, θ, z) can be written as shown below.

$$\varepsilon_{rr} = [\sigma_{rr} - \nu(\sigma_{\theta\theta} + \sigma_{zz})]/E \qquad \gamma_{r\theta} = \tau_{r\theta}/G$$

$$\varepsilon_{\theta\theta} = [\sigma_{\theta\theta} - \nu(\sigma_{zz} + \sigma_{rr})]/E \qquad \gamma_{\theta z} = \tau_{\theta z}/G$$

$$\varepsilon_{zz} = [\sigma_{zz} - \nu(\sigma_{rr} + \sigma_{\theta\theta})]/E \qquad \gamma_{zr} = \tau_{zr}/G$$

3.63 Write the Generalized Hooke's Law for isotropic material in spherical coordinates (r, θ, ϕ).

Solution

The Generalized Hooke's Law for isotropic material in cylindrical coordinates (r, θ, ϕ) can be written as shown below.

$$\varepsilon_{rr} = [\sigma_{rr} - \nu(\sigma_{\theta\theta} + \sigma_{\phi\phi})]/E \qquad \gamma_{r\theta} = \tau_{r\theta}/G$$

$$\varepsilon_{\theta\theta} = [\sigma_{\theta\theta} - \nu(\sigma_{\phi\phi} + \sigma_{rr})]/E \qquad \gamma_{\theta\phi} = \tau_{\theta\phi}/G$$

$$\varepsilon_{\phi\phi} = [\sigma_{\phi\phi} - \nu(\sigma_{rr} + \sigma_{\theta\theta})]/E \qquad \gamma_{\phi r} = \tau_{\phi r}/G$$

3.64 The stresses and two material constants are given, calculate ε_{xx}, ε_{yy}, γ_{xy}, ε_{zz}, and σ_{zz} (a) assuming plane stress, and (b) assuming plane strain.

$\sigma_{xx} = 100\ \text{MPa(T)}$ $\quad \sigma_{yy} = 150\ \text{MPa(T)}$ $\quad \tau_{xy} = -125\ \text{MPa}$ $\quad E = 200\ \text{GPa}$ $\quad \nu = 0.32$

Solution $\quad \varepsilon_{xx}=?$ $\varepsilon_{yy}=?$ $\gamma_{xy}=?$ $\varepsilon_{zz}=?$ $\sigma_{zz}=?$ (a) plane stress (b) plane strain

---

$$G = \frac{E}{2(1+\nu)} = \frac{200}{2(1.32)} = 75.76\text{GPa}$$

(a) Assuming Plane Stress $\boxed{\sigma_{zz} = 0}$

$$\varepsilon_{xx} = \frac{\sigma_{xx} - \nu(\sigma_{yy} + \sigma_{zz})}{E} = \frac{[100 - 0.32(150)](10^6)}{200(10^9)} = \frac{52(10^6)}{200(10^9)} = 0.26(10^{-3}) \qquad \boxed{\varepsilon_{xx} = 260\mu}$$

$$\varepsilon_{yy} = \frac{\sigma_{yy} - \nu(\sigma_{xx})}{E} = \frac{[150 - 0.32(100)](10^6)}{200(10^9)} = \frac{118(10^6)}{200(10^9)} = 0.59(10^{-3}) \qquad \boxed{\varepsilon_{yy} = 590\mu}$$

$$\gamma_{xy} = \frac{\tau_{xy}}{G} = \frac{-125(10^6)}{75.76(10^9)} = -1.65(10^{-3}) \qquad \boxed{\gamma_{xy} = -1650\mu}$$

$$\varepsilon_{zz} = \frac{\sigma_{zz} - \nu(\sigma_{xx} + \sigma_{yy})}{E} = -\frac{0.32(250)(10^6)}{200(10^9)} = \frac{80(10^6)}{200(10^9)} = -0.4(10^{-3}) \qquad \boxed{\varepsilon_{zz} = -400\mu}$$

(b) Assuming Plane Strain $\boxed{\varepsilon_{zz} = 0}$

$$0 = \frac{\sigma_{zz} - \nu(\sigma_{xx} + \sigma_{yy})}{E}$$

$$\therefore \sigma_{zz} = \nu(\sigma_{xx} + \sigma_{yy}) = 0.32(250)(10^6) = 80(10^6) \qquad \boxed{\sigma_{zz} = 80\text{MPa(T)}}$$

$$\varepsilon_{xx} = \frac{\sigma_{xx} - \nu(\sigma_{yy} + \sigma_{zz})}{E} = \frac{[100 - 0.32(230)](10^6)}{200(10^9)} = 0.132(10^{-3}) \qquad \boxed{\varepsilon_{xx} = 132\mu}$$

$$\varepsilon_{yy} = \frac{\sigma_{yy} - \nu(\sigma_{xx} + \sigma_{zz})}{E} = \frac{[150 - 0.32(180)](10^6)}{200(10^9)} = 0.462(10^{-3}) \qquad \boxed{\varepsilon_{yy} = 462\mu}$$

Same as in part (a) $\boxed{\gamma_{xy} = -1650\mu}$

3.65 The stresses and two material constants are given, calculate ε_{xx}, ε_{yy}, γ_{xy}, ε_{zz}, and σ_{zz} (a) assuming plane stress, and (b) assuming plane strain.

$\sigma_{xx} = 225$ MPa(C) $\quad \sigma_{yy} = 125$ MPa(T) $\quad \tau_{xy} = 150$ MPa $\quad E = 70$ GPa $\quad G = 28$ GPa

Solution $\quad \varepsilon_{xx}=?\quad \varepsilon_{yy}=?\quad \gamma_{xy}=?\quad \varepsilon_{zz}=?\quad \sigma_{zz}=?\quad$ (a) plane stress $\quad$ (b) plane strain

---

$$G = \frac{E}{2(1+\nu)} \qquad \therefore 28(10^9) = \frac{70(10^9)}{2(1+\nu)} \qquad \nu = 0.25$$

(a) Assuming Plane Stress $\boxed{\sigma_{zz} = 0}$

$$\varepsilon_{xx} = \frac{\sigma_{xx} - \nu(\sigma_{yy} + \sigma_{zz})}{E} = \frac{[-225 - 0.25(125)](10^6)}{70(10^9)} = -3.661(10^{-3}) \qquad \boxed{\varepsilon_{xx} = -3661\mu}$$

$$\varepsilon_{yy} = \frac{\sigma_{yy} - \nu(\sigma_{xx})}{E} = \frac{[125 - 0.25(-225)](10^6)}{70(10^9)} = 2.589(10^{-3}) \qquad \boxed{\varepsilon_{yy} = 2589\mu}$$

$$\gamma_{xy} = \frac{\tau_{xy}}{G} = \frac{150(10^6)}{28(10^9)} = 5.357(10^{-3}) \qquad \boxed{\gamma_{xy} = 5357\mu\text{rad}}$$

$$\varepsilon_{zz} = \frac{\sigma_{zz} - \nu(\sigma_{xx} + \sigma_{yy})}{E} = -\frac{0.25(-225 + 125)(10^6)}{70(10^9)} = 0.357(10^{-3}) \qquad \boxed{\varepsilon_{zz} = 357\mu}$$

(b) Assuming Plane Strain $\boxed{\varepsilon_{zz} = 0}$

$$0 = \frac{\sigma_{zz} - \nu(\sigma_{xx} + \sigma_{yy})}{E}$$

$$\therefore \sigma_{zz} = \nu(\sigma_{xx} + \sigma_{yy}) = 0.25(-225 + 125)(10^6) = -25(10^6)$$ $\boxed{\sigma_{zz} = 25\text{MPa(C)}}$

$$\varepsilon_{xx} = \frac{\sigma_{xx} - \nu(\sigma_{yy} + \sigma_{zz})}{E} = \frac{[-225 - 0.25(125 - 25)](10^6)}{70(10^9)} = -3.571(10^{-3})$$ $\boxed{\varepsilon_{xx} = -3571\mu}$

$$\varepsilon_{yy} = \frac{\sigma_{yy} - \nu(\sigma_{xx} + \sigma_{zz})}{E} = \frac{[125 - 0.25(-225 - 25)](10^6)}{70(10^9)} = 2.679(10^{-3})$$ $\boxed{\varepsilon_{yy} = 2679\mu}$

Same as in part (a) $\boxed{\gamma_{xy} = 5357\mu\text{rad}}$

3.66 The stresses and two material constants are given, calculate ε_{xx}, ε_{yy}, γ_{xy}, ε_{zz}, and σ_{zz} (a) assuming plane stress, and (b) assuming plane strain.

$\sigma_{xx} = 22\ \text{ksi(C)}$ $\sigma_{yy} = 25\ \text{ksi(C)}$ $\tau_{xy} = -15\ \text{ksi}$ $E = 30,000\ \text{ksi}$ $\nu = 0.3$

Solution $\varepsilon_{xx}=?$ $\varepsilon_{yy}=?$ $\gamma_{xy}=?$ $\varepsilon_{zz}=?$ $\sigma_{zz}=?$ (a) plane stress (b) plane strain

---

$$G = \frac{E}{2(1+\nu)} = \frac{30(10^6)}{2(1+0.3)} = 11.54(10^6)\text{psi}$$

(a) Assuming Plane Stress $\boxed{\sigma_{zz} = 0}$

$$\varepsilon_{xx} = \frac{\sigma_{xx} - \nu(\sigma_{yy} + \sigma_{zz})}{E} = \frac{[-22 - 0.3(-25)](10^3)}{30(10^6)} = -0.483(10^{-3})$$ $\boxed{\varepsilon_{xx} = -483\mu}$

$$\varepsilon_{yy} = \frac{\sigma_{yy} - \nu(\sigma_{xx})}{E} = \frac{[-25 - 0.3(-22)](10^3)}{30(10^6)} = -0.613(10^{-3})$$ $\boxed{\varepsilon_{yy} = -613\mu}$

$$\gamma_{xy} = \frac{\tau_{xy}}{G} = \frac{-15(10^3)}{11.54(10^6)} = -1.3(10^{-3})$$ $\boxed{\gamma_{xy} = -1300\mu}$

$$\varepsilon_{zz} = \frac{\sigma_{zz} - \nu(\sigma_{xx} + \sigma_{yy})}{E} = -\frac{0.3(-22 - 25)(10^3)}{30(10^6)} = 0.47(10^{-3})$$ $\boxed{\varepsilon_{zz} = 470\mu}$

(b) Assuming Plane Strain $\boxed{\varepsilon_{zz} = 0}$

$$0 = \frac{\sigma_{zz} - \nu(\sigma_{xx} + \sigma_{yy})}{E}$$

$$\therefore \sigma_{zz} = \nu(\sigma_{xx} + \sigma_{yy}) = 0.3(-22 - 25)(10^3) = (-14.1)(10^3)$$ $\boxed{\sigma_{zz} = 14.1\text{ksi(C)}}$

$$\varepsilon_{xx} = \frac{\sigma_{xx} - \nu(\sigma_{yy} + \sigma_{zz})}{E} = \frac{[-22 - 0.3(-25 + 14.1)](10^3)}{30(10^6)} = -0.342(10^{-3})$$ $\boxed{\varepsilon_{xx} = -342\mu}$

$$\varepsilon_{yy} = \frac{\sigma_{yy} - \nu(\sigma_{xx} + \sigma_{zz})}{E} = \frac{[-25 - 0.3(-22 + 14.1)](10^3)}{30(10^6)} = -0.472(10^{-3})$$ $\boxed{\varepsilon_{yy} = -472\mu}$

Same as in part (a) $\boxed{\gamma_{xy} = -1300\mu}$

3.67 The stresses and two material constants are given, calculate ε_{xx}, ε_{yy}, γ_{xy}, ε_{zz}, and σ_{zz} (a) assuming plane stress, and (b) assuming plane strain.

$\sigma_{xx} = 15$ ksi(T) $\sigma_{yy} = 12$ ksi(C) $\tau_{xy} = -10$ ksi $E = 10,000$ ksi $G = 3900$ ksi

Solution $\varepsilon_{xx}=?$ $\varepsilon_{yy}=?$ $\gamma_{xy}=?$ $\varepsilon_{zz}=?$ $\sigma_{zz}=?$ (a) plane stress (b) plane strain

$$G = \frac{E}{2(1+\nu)} \quad \therefore 3900(10^3) = \frac{10000(10^3)}{2(1+\nu)} \quad \nu = 0.282$$

(a) Assuming Plane Stress $\boxed{\sigma_{zz} = 0}$

$$\varepsilon_{xx} = \frac{\sigma_{xx} - \nu(\sigma_{yy} + \sigma_{zz})}{E} = \frac{[15-0.282(-12)](10^3)}{10(10^6)} = 1.838(10^{-3}) \qquad \boxed{\varepsilon_{xx} = 1838\mu}$$

$$\varepsilon_{yy} = \frac{\sigma_{yy} - \nu(\sigma_{xx})}{E} = \frac{[-12-0.282(-15)](10^3)}{10(10^6)} = -1.623(10^{-3}) \qquad \boxed{\varepsilon_{yy} = -1623\mu}$$

$$\gamma_{xy} = \frac{\tau_{xy}}{G} = \frac{-10(10^3)}{3.9(10^6)} = -2.564(10^{-3}) \qquad \boxed{\gamma_{xy} = -2564\mu}$$

$$\varepsilon_{zz} = \frac{\sigma_{zz} - \nu(\sigma_{xx} + \sigma_{yy})}{E} = -\frac{0.282(-15-12)(10^3)}{30(10^6)} = -0.085(10^{-3}) \qquad \boxed{\varepsilon_{zz} = -85\mu}$$

(b) Assuming Plane Strain $\boxed{\varepsilon_{zz} = 0}$

$$0 = \frac{\sigma_{zz} - \nu(\sigma_{xx} + \sigma_{yy})}{E}$$

$$\therefore \sigma_{zz} = \nu(\sigma_{xx} + \sigma_{yy}) = 0.282(15-12)(10^3) = 846 \qquad \boxed{\sigma_{zz} = 846\text{psi(T)}}$$

$$\varepsilon_{xx} = \frac{\sigma_{xx} - \nu(\sigma_{yy} + \sigma_{zz})}{E} = \frac{[15-0.282(-12+0.846)](10^3)}{10(10^6)} = 1.815(10^{-3}) \qquad \boxed{\varepsilon_{xx} = 1815\mu}$$

$$\varepsilon_{yy} = \frac{\sigma_{yy} - \nu(\sigma_{xx} + \sigma_{zz})}{E} = \frac{[-12-0.282(15+0.846)](10^3)}{10(10^6)} = -1.647(10^{-3}) \qquad \boxed{\varepsilon_{yy} = -1647\mu}$$

Same as in part (a) $\boxed{\gamma_{xy} = -2564\mu}$

3.68 The stresses and two material constants are given, calculate ε_{xx}, ε_{yy}, γ_{xy}, ε_{zz}, and σ_{zz} (a) assuming plane stress, and (b) assuming plane strain.

$\sigma_{xx} = 300$ MPa(C) $\sigma_{yy} = 300$ MPa(T) $\tau_{xy} = 150$ MPa $G = 15$ GPa $\nu = 0.2$

Solution $\varepsilon_{xx}=?$ $\varepsilon_{yy}=?$ $\gamma_{xy}=?$ $\varepsilon_{zz}=?$ $\sigma_{zz}=?$ (a) plane stress (b) plane strain

$$G = \frac{E}{2(1+\nu)} \quad \therefore 15(10^9) = \frac{E}{2(1+0.2)} \quad E = 36(10^9) \quad \text{i.e., } E = 36\text{GPa}$$

(a) Assuming Plane Stress $\boxed{\sigma_{zz} = 0}$

$$\varepsilon_{xx} = \frac{\sigma_{xx} - \nu(\sigma_{yy} + \sigma_{zz})}{E} = \frac{[-300-0.2(300)](10^6)}{36(10^9)} = -10(10^{-3}) \qquad \boxed{\varepsilon_{xx} = -10000\mu}$$

$$\varepsilon_{yy} = \frac{\sigma_{yy} - \nu(\sigma_{xx} + \sigma_{zz})}{E} = \frac{[300-0.2(-300)](10^6)}{36(10^9)} = 10(10^{-3}) \qquad \boxed{\varepsilon_{yy} = 10000\mu}$$

$$\gamma_{xy} = \frac{\tau_{xy}}{G} = \frac{150(10^6)}{15(10^9)} = 10(10^{-3})$$ $\boxed{\gamma_{xy} = 10000\mu rad}$

$$\varepsilon_{zz} = \frac{\sigma_{zz} - \nu(\sigma_{xx} + \sigma_{yy})}{E} = -\frac{0.2(300-300)(10^6)}{36(10^9)} = 0$$ $\boxed{\varepsilon_{zz} = 0}$

(b) Assuming Plane Strain $\boxed{\varepsilon_{zz} = 0}$

$$0 = \frac{\sigma_{zz} - \nu(\sigma_{xx} + \sigma_{yy})}{E}$$

$$\therefore \sigma_{zz} = \nu(\sigma_{xx} + \sigma_{yy}) = 0.2(300-300)(10^6) = 0$$ $\boxed{\sigma_{zz} = 0}$

As σ_{zz} is zero, the results are the same as in plane stress.

3.69 The stresses and two material constants are given, calculate ε_{xx}, ε_{yy}, γ_{xy}, ε_{zz}, and σ_{zz} (a) assuming plane stress, and (b) assuming plane strain.

$\sigma_{xx} = 100$ psi(T) $\quad \sigma_{yy} = 125$ psi(T) $\quad \tau_{xy} = -50$ psi $\quad E = 2000$ psi $\quad G = 800$ psi

Solution $\varepsilon_{xx}=?$ $\varepsilon_{yy}=?$ $\gamma_{xy}=?$ $\varepsilon_{zz}=?$ $\sigma_{zz}=?$ (a) plane stress (b) plane strain

---

$$G = \frac{E}{2(1+\nu)} \quad \therefore 800 = \frac{2000}{2(1+\nu)} \qquad \nu = 0.25$$

(a) Assuming Plane Stress $\boxed{\sigma_{zz} = 0}$

$$\varepsilon_{xx} = \frac{\sigma_{xx} - \nu(\sigma_{yy} + \sigma_{zz})}{E} = \frac{100-0.25(125)}{2000} = 0.0344$$ $\boxed{\varepsilon_{xx} = 0.0344}$

$$\varepsilon_{yy} = \frac{\sigma_{yy} - \nu(\sigma_{xx} + \sigma_{zz})}{E} = \frac{125-0.25(100)}{2000} = 0.05$$ $\boxed{\varepsilon_{yy} = 0.05}$

$$\gamma_{xy} = \frac{\tau_{xy}}{G} = \frac{-50}{800} = -0.0625$$ $\boxed{\gamma_{xy} = -0.0625}$

$$\varepsilon_{zz} = \frac{\sigma_{zz} - \nu(\sigma_{xx} + \sigma_{yy})}{E} = -\frac{0.25(100+125)}{2000} = -0.0281$$ $\boxed{\varepsilon_{zz} = -0.0281}$

(b) Assuming Plane Strain $\boxed{\varepsilon_{zz} = 0}$

$$0 = \frac{\sigma_{zz} - \nu(\sigma_{xx} + \sigma_{yy})}{E}$$

$$\therefore \sigma_{zz} = \nu(\sigma_{xx} + \sigma_{yy}) = 0.25(100+125) = 56.25$$ $\boxed{\sigma_{zz} = 56.25 psi(T)}$

$$\varepsilon_{xx} = \frac{\sigma_{xx} - \nu(\sigma_{yy} + \sigma_{zz})}{E} = \frac{100-0.25(125+56.25)}{2000} = 0.0273$$ $\boxed{\varepsilon_{xx} = 0.0273}$

$$\varepsilon_{yy} = \frac{\sigma_{yy} - \nu(\sigma_{xx} + \sigma_{zz})}{E} = \frac{125-0.25(100+56.25)}{2000} = 0.0430$$ $\boxed{\varepsilon_{yy} = 0.043}$

Same as in part (a) $\boxed{\gamma_{xy} = -0.0625}$

3.70 The strains and two material constants are given. Calculate σ_{xx}, σ_{yy}, τ_{xy}, σ_{zz}, and ε_{zz} assuming the point is in plane stress.

$\varepsilon_{xx} = 500\ \mu \qquad \varepsilon_{yy} = 400\ \mu \qquad \gamma_{xy} = -300\ \mu \qquad E = 200\ \text{GPa} \qquad \nu = 0.32$

Solution $\sigma_{xx}=?\ \sigma_{yy}=?\ \tau_{xy}=?\ \sigma_{zz}=?\ \varepsilon_{zz}=?$ plane stress

---

$$G = \frac{E}{2(1+\nu)} \qquad G = \frac{15(10^9)}{2(1.32)} = 75.76(10^9)$$

Assuming Plane Stress $\boxed{\sigma_{zz} = 0}$

$$\varepsilon_{xx} = \frac{\sigma_{xx} - \nu(\sigma_{yy} + \sigma_{zz})}{E} \text{ or } \sigma_{xx} - \nu(\sigma_{yy} + \sigma_{zz}) = E\varepsilon_{xx} = 200(10^9)500(10^{-6}) = 100000(10^3)\frac{N}{m^2}$$

$$\therefore \sigma_{xx} - \nu(\sigma_{yy} + \sigma_{zz}) = 100\text{MPa} \qquad 1$$

Similarly

$$\varepsilon_{yy} = \frac{\sigma_{yy} - \nu(\sigma_{xx} + \sigma_{zz})}{E} \text{ or } \sigma_{yy} - \nu(\sigma_{xx} + \sigma_{zz}) = E\varepsilon_{yy} = 200(10^9)400(10^{-6}) = 80000(10^3)\frac{N}{m^2}$$

$$\sigma_{yy} - \nu(\sigma_{xx} + \sigma_{zz}) = 80\text{MPa} \qquad 2$$

Multiplying Eq.(2) by ν and adding to Eq.(1), we obtain: $\sigma_{xx}(1-\nu^2) = 100 + \nu 80$

$$\therefore \sigma_{xx} = \frac{100 + 0.32(80)}{1 - 0.32^2} = \frac{125.6}{0.8976} = 139.929 \qquad 3$$

$\boxed{\sigma_{xx} = 139.93\text{MPa(T)}}$

Substituting Eq.(3) in Eq.(2)

$\sigma_{yy} = 80 + \nu(139.93) = 124.78$ $\boxed{\sigma_{yy} = 124.78\text{MPa(T)}}$

$$\varepsilon_{zz} = \frac{\sigma_{zz} - \nu(\sigma_{xx} + \sigma_{yy})}{E} = -\frac{0.32}{200(10^9)}(139.9 + 124.8)(10^6) = -0.423(10^{-3})$$ $\boxed{\varepsilon_{zz} = -423\mu}$

$$\tau_{xy} = G\gamma_{xy} = 75.76(10^9)(-300)(10^{-6}) = 22727(10^3)$$ $\boxed{\tau_{xy} = -22.7\text{MPa}}$

3.71 The strains and two material constants are given. Calculate σ_{xx}, σ_{yy}, τ_{xy}, σ_{zz}, and ε_{zz} assuming the point is in plane stress.

$\varepsilon_{xx} = -3000\ \mu \qquad \varepsilon_{yy} = 1500\ \mu \qquad \gamma_{xy} = 2000\ \mu \qquad E = 70\ \text{GPa} \qquad G = 28\ \text{GPa}$

Solution $\sigma_{xx}=?\ \sigma_{yy}=?\ \tau_{xy}=?\ \sigma_{zz}=?\ \varepsilon_{zz}=?$ plane stress

---

$$G = \frac{E}{2(1+\nu)} \qquad 28(10^9) = \frac{70(10^9)}{2(1+\nu)} \qquad \nu = 0.25$$

Assuming Plane Stress $\boxed{\sigma_{zz} = 0}$

$$\varepsilon_{xx} = \frac{\sigma_{xx} - \nu(\sigma_{yy} + \sigma_{zz})}{E} \text{ or } \sigma_{xx} - \nu(\sigma_{yy} + \sigma_{zz}) = E\varepsilon_{xx} = 70(10^9)(-3000)(10^{-6}) = -210(10^6)\frac{N}{m^2} \text{ or}$$

$$\sigma_{xx} - \nu(\sigma_{yy} + \sigma_{zz}) = -210\text{MPa} \qquad 1$$

Similarly

$$\varepsilon_{yy} = \frac{\sigma_{yy} - \nu(\sigma_{xx} + \sigma_{zz})}{E} \text{ or } \sigma_{yy} - \nu(\sigma_{xx} + \sigma_{zz}) = E\varepsilon_{yy} = 70(10^9)1500(10^{-6}) = 105(10^6)\frac{N}{m^2} \text{ or}$$

$$\sigma_{yy} - \nu(\sigma_{xx} + \sigma_{zz}) = 105\text{MPa} \qquad 2$$

Multiplying Eq.(1) by ν and adding to Eq.(2) we obtain: $\sigma_{yy}(1-\nu^2) = 10.5+\nu(-21)$ or

$$\sigma_{yy} = \frac{105+0.25(-210)}{1-0.25^2} = \frac{52.5}{0.9375} = 56.0 \qquad 3$$

$\boxed{\sigma_{yy} = 56\text{MPa(T)}}$

Substituting Eq.(3) into Eq.(1)

$\sigma_{xx} = -210+\nu(56) = -196$ $\boxed{\sigma_{xx} = 196\text{MPa(C)}}$

$$\varepsilon_{zz} = \frac{\sigma_{zz}-\nu(\sigma_{xx}+\sigma_{yy})}{E} = -\frac{0.25}{70(10^9)}(-196+56)(10^6) = -0.5(10^{-3})$$ $\boxed{\varepsilon_{zz} = 500\mu}$

$\tau_{xy} = G\gamma_{xy} = 28(10^9)(2000)(10^{-6}) = 56(10^6)$ $\boxed{\tau_{xy} = 56\text{MPa}}$

3.72 The strains and two material constants are given. Calculate σ_{xx}, σ_{yy}, τ_{xy}, σ_{zz}, and ε_{zz} assuming the point is in plane stress.

$\varepsilon_{xx} = -800\ \mu$ $\varepsilon_{yy} = -1000\ \mu$ $\gamma_{xy} = -500\ \mu$ $E = 30,000\ \text{ksi}$ $\nu = 0.3$

Solution σ_{xx}=? σ_{yy}=? τ_{xy}=? σ_{zz}=? ε_{zz}=? plane stress

$$G = \frac{E}{2(1+\nu)} = \frac{30000}{2(1.3)} = 11538\text{ksi}$$

Assuming Plane Stress $\boxed{\sigma_{zz} = 0}$

$$\varepsilon_{xx} = \frac{\sigma_{xx}-\nu(\sigma_{yy}+\sigma_{zz})}{E} \text{ or } \sigma_{xx}-\nu(\sigma_{yy}+\sigma_{zz}) = E\varepsilon_{xx} = 30(10^3)(-800)(10^{-6}) = -24\text{ksi}$$

$$\therefore \sigma_{xx}-\nu(\sigma_{yy}+\sigma_{zz}) = -24\text{ksi} \qquad 1$$

Similarly

$$\varepsilon_{yy} = \frac{\sigma_{yy}-\nu(\sigma_{xx}+\sigma_{zz})}{E} \text{ or } \sigma_{yy}-\nu(\sigma_{xx}+\sigma_{zz}) = E\varepsilon_{yy} = 30(10^3)(-1000)(10^{-6}) = -30\text{ksi}$$

$$\sigma_{yy}-\nu(\sigma_{xx}+\sigma_{zz}) = -30\text{ksi} \qquad 2$$

Multiplying Eq.(1) by ν and adding to Eq.(2) we obtain: $\sigma_{yy}(1-\nu^2) = -24\nu-30$

$$\therefore \sigma_{yy} = \frac{-0.3(24)-30}{1-0.3^2} = -40.88\text{ksi} \qquad 3$$

$\boxed{\sigma_{yy} = 40.9\text{ksi(C)}}$

Substituting Eq.(3) in Eq.(1)

$\sigma_{xx} = -24+\nu(-40.88) = -36.26$ $\boxed{\sigma_{xx} = 36.26\text{ksi(C)}}$

$$\varepsilon_{zz} = \frac{\sigma_{zz}-\nu(\sigma_{xx}+\sigma_{yy})}{E} = -\frac{0.3}{30(10^3)}(-36.36-40.88) = 0.771(10^{-3})$$ $\boxed{\varepsilon_{zz} = 771\mu\frac{\text{in}}{\text{in}}}$

$\tau_{xy} = G\gamma_{xy} = 11538(-500)(10^{-6}) = -5.769$ $\boxed{\tau_{xy} = -5.77\text{ksi}}$

3.73 The strains and two material constants are given. Calculate σ_{xx}, σ_{yy}, τ_{xy}, σ_{zz}, and ε_{zz} assuming the point is in plane stress.

$\varepsilon_{xx} = 1500\ \mu$ $\varepsilon_{yy} = -1200\ \mu$ $\gamma_{xy} = -1000\ \mu$ $E = 10,000\ \text{ksi}$ $G = 3900\ \text{ksi}$

Solution σ_{xx}=? σ_{yy}=? τ_{xy}=? σ_{zz}=? ε_{zz}=? plane stress

$$G = \frac{E}{2(1+\nu)} \text{ or } 3900 = \frac{10000}{2(1+\nu)} \qquad \therefore \nu = 0.282$$

Assuming Plane Stress $\boxed{\sigma_{zz} = 0}$

$$\varepsilon_{xx} = \frac{\sigma_{xx} - \nu(\sigma_{yy} + \sigma_{zz})}{E} \text{ or } \sigma_{xx} - \nu(\sigma_{yy} + \sigma_{zz}) = E\varepsilon_{xx} = 10(10^3)(1500)(10^{-6}) = 15\text{ksi}$$

$$\therefore \sigma_{xx} - \nu(\sigma_{yy} + \sigma_{zz}) = 15\text{ksi} \qquad 1$$

Similarly

$$\varepsilon_{yy} = \frac{\sigma_{yy} - \nu(\sigma_{xx} + \sigma_{zz})}{E} \text{ or } \sigma_{yy} - \nu(\sigma_{xx} + \sigma_{zz}) = E\varepsilon_{yy} = 10(10^3)(-1200)(10^{-6}) = -12\text{ksi}$$

$$\sigma_{yy} - \nu(\sigma_{xx} + \sigma_{zz}) = -12\text{ksi} \qquad 2$$

Multiplying Eq.(1) by ν and adding to Eq.(2) we obtain: $\sigma_{yy}(1 - \nu^2) = 15\nu - 12$

$$\therefore \sigma_{yy} = \frac{0.282(15) - 12}{1 - (0.282^2)} = -8.44 \qquad 3$$

$\boxed{\sigma_{yy} = 8.44\text{ksi(C)}}$

Substituting Eq.(3) in Eq.(1)

$\sigma_{xx} = 15 + \nu(-8.44) = 15 + 0.282(-8.44) = 12.619$ $\boxed{\sigma_{xx} = 12.62\text{ksi(T)}}$

$$\varepsilon_{zz} = \frac{\sigma_{zz} - \nu(\sigma_{xx} + \sigma_{yy})}{E} = -\frac{0.282}{10(10^3)}(12.619-8.44) = -0.1178 \qquad \boxed{\varepsilon_{zz} = -118\mu\frac{\text{in}}{\text{in}}}$$

$\tau_{xy} = G\gamma_{xy} = 3900(-1000)(10^{-6}) = -3.9$ $\boxed{\tau_{xy} = -3.9\text{ksi}}$

3.74 The strains and two material constants are given. Calculate σ_{xx}, σ_{yy}, τ_{xy}, σ_{zz}, and ε_{zz} assuming the point is in plane stress.

$\varepsilon_{xx} = -2000\ \mu$ $\quad \varepsilon_{yy} = 2000\ \mu$ $\quad \gamma_{xy} = 1200\ \mu$ $\quad G = 15\ \text{GPa}$ $\quad \nu = 0.2$

Solution $\quad \sigma_{xx}$=? σ_{yy}=? τ_{xy}=? σ_{zz}=? ε_{zz}=? $\qquad$ plane stress

---

$$G = \frac{E}{2(1+\nu)} \text{ or } 15 = \frac{E}{2(1+0.2)} \text{ or } E = 36.0\text{GPa}$$

Assuming Plane Stress $\boxed{\sigma_{zz} = 0}$

$$\varepsilon_{xx} = \frac{\sigma_{xx} - \nu(\sigma_{yy} + \sigma_{zz})}{E} \text{ or } \sigma_{xx} - \nu(\sigma_{yy} + \sigma_{zz}) = E\varepsilon_{xx} = 36(10^9)(-2000)(10^{-6}) = -72\text{MPa or}$$

$$\sigma_{xx} - \nu(\sigma_{yy} + \sigma_{zz}) = -72\text{MPa} \qquad 1$$

Similarly

$$\varepsilon_{yy} = \frac{\sigma_{yy} - \nu(\sigma_{xx} + \sigma_{zz})}{E} \text{ or } \sigma_{yy} - \nu(\sigma_{xx} + \sigma_{zz}) = E\varepsilon_{yy} = 36(10^9)(2000)(10^{-6}) = 72\text{MPa or}$$

$$\sigma_{yy} - \nu(\sigma_{xx} + \sigma_{zz}) = 72\text{MPa} \qquad 2$$

Multiplying Eq.(1) by ν and adding to Eq.(2) $\sigma_{yy}(1 - \nu^2) = -72\nu - 72$ or

$$\sigma_{yy} = \frac{-0.2(72) + 72}{1 - 0.2^2} = 60.1 \text{ or } \boxed{\sigma_{yy} = 60\text{MPa(T)}} \qquad 3$$

Substituting Eq.(3) in Eq.(1)

$\sigma_{xx} = -72 + \nu(\sigma_{yy}) = -72 + 0.2(60.1) = -59.98$ or $\boxed{\sigma_{xx} = 60\text{MPa(C)}}$

$$\varepsilon_{zz} = \frac{\sigma_{zz} - \nu(\sigma_{xx} + \sigma_{yy})}{E} = -\frac{0.2}{36(10^9)}(-60 + 60) = 0 \qquad \boxed{\varepsilon_{zz} = 0}$$

$$\tau_{xy} = G\gamma_{xy} = 15(10^9)(1200)(10^{-6}) = 18(10^6) \qquad \boxed{\tau_{xy} = 18\text{MPa}}$$

3.75 The strains and two material constants are given. Calculate σ_{xx}, σ_{yy}, τ_{xy}, σ_{zz}, and ε_{zz} assuming the point is in plane stress.

$\varepsilon_{xx} = 50\ \mu$ $\quad \varepsilon_{yy} = 75\ \mu$ $\quad \gamma_{xy} = -25\ \mu$ $\quad E = 2000\ \text{psi}$ $\quad G = 800\ \text{psi}$

Solution σ_{xx}=? σ_{yy}=? τ_{xy}=? σ_{zz}=? ε_{zz}=? plane stress

---

$$G = \frac{E}{2(1+\nu)} \text{ or } 800 = \frac{2000}{2(1+\nu)} \qquad \therefore \nu = 0.25$$

Assuming Plane Stress $\boxed{\sigma_{zz} = 0}$

$$\varepsilon_{xx} = \frac{\sigma_{xx} - \nu(\sigma_{yy} + \sigma_{zz})}{E} \text{ or } \sigma_{xx} - \nu(\sigma_{yy} + \sigma_{zz}) = E\varepsilon_{xx} = 2000(50)(10^{-6}) = 0.1\text{psi}$$

$$\therefore \sigma_{xx} - \nu(\sigma_{yy} + \sigma_{zz}) = 0.1\text{psi} \qquad (1)$$

Similarly

$$\varepsilon_{yy} = \frac{\sigma_{yy} - \nu(\sigma_{xx} + \sigma_{zz})}{E} \text{ or } \sigma_{yy} - \nu(\sigma_{xx} + \sigma_{zz}) = E\varepsilon_{yy} = 2000(75)(10^{-6}) = 0.15\text{psi}$$

$$\sigma_{yy} - \nu(\sigma_{xx} + \sigma_{zz}) = 0.15\text{psi} \qquad (2)$$

Multiplying Eq.(1) by ν and adding to Eq.(2) we obtain: $\sigma_{yy}(1 - \nu^2) = 0.1\nu + 0.15$

$$\therefore \sigma_{yy} = \frac{0.25(0.1) + 0.15}{1 - 0.25^2} = 0.1867 \qquad (3)$$

$$\boxed{\sigma_{yy} = 0.187\text{psi(T)}}$$

Substituting Eq.(3) in Eq.(1)

$$\sigma_{xx} = 0.1 + \nu(0.1867) = 0.1 + 0.25(0.1867) = 0.1467 \qquad \boxed{\sigma_{xx} = 0.147\text{psi(T)}}$$

$$\varepsilon_{zz} = \frac{\sigma_{zz} - \nu(\sigma_{xx} + \sigma_{yy})}{E} = -\frac{0.25}{2000}(0.1467 + 0.1867) = -41.7(10^{-6}) \qquad \boxed{\varepsilon_{zz} = -41.7\mu\frac{\text{in}}{\text{in}}}$$

$$\tau_{xy} = G\gamma_{xy} = 800(-25)(10^{-6}) = -0.02 \qquad \boxed{\tau_{xy} = -0.02\text{psi}}$$

3.76 The cross-section of the wooden piece that is visible in the picture is 40 mm x 25 mm. The clamped length of the wooden piece in the vice is 125 mm. The Modulus of Elasticity of wood is E = 14 GPa and a Poisson's ratio of 0.3. The jaws of the vice exert a uniform pressure of 3.2 MPa on the wood. Determine the average change of length of the wood.

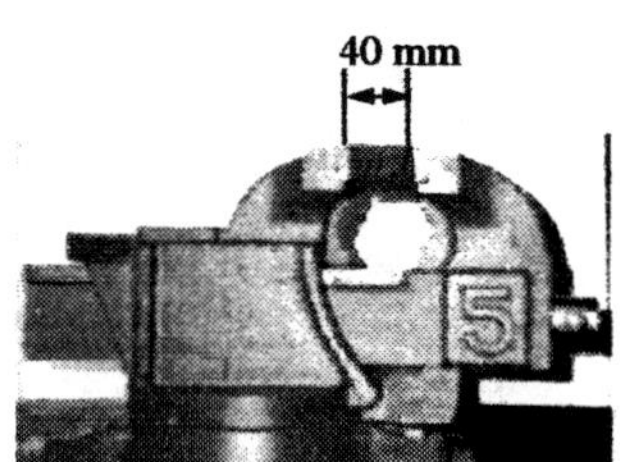

Fig. P3.76

Solution E=14GPa $\quad\nu$=0.3 $\quad\sigma_{xx}$=3.2MPa(C) $\quad\Delta L$=?

We assume $\sigma_{yy} = 0$ and $\sigma_{zz} = 0$

$$\varepsilon_{zz} = \frac{\sigma_{zz} - \nu(\sigma_{xx} + \sigma_{yy})}{E} = \frac{-\nu\sigma_{xx}}{E}$$

$$\varepsilon_{zz} = -\frac{(0.3)(-3.2)(10^6)}{14(10^9)} = 0.06857(10^{-3})$$

$$\Delta L = \varepsilon_{zz}L = 0.06857(10^{-3})(125) = 8.571(10^3)$$ $\boxed{\Delta L = 0.0086\text{mm}}$

3.77 A thin plate (E=30,000 ksi; ν = 0.25) under the action of uniform forces deforms to the shaded position. Assuming plane stress determine the average normal stresses in the x and y direction.

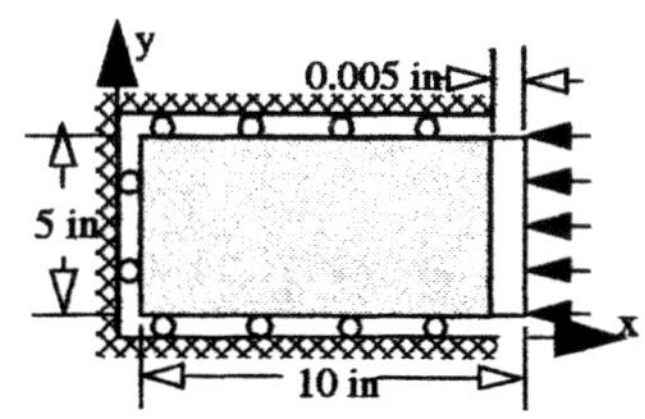

Fig. P3.77

Solution E=30000ksi ν=0.25 σ_{xx}=? σ_{yy}=?

Assuming plane stress $\sigma_{zz} = 0$

$$\varepsilon_{xx} = -\frac{0.005}{10} = -0.0005 \text{ or } \sigma_{xx} - \nu\sigma_{yy} = E\varepsilon_{xx} = 30000(-0.0005)$$

$$\sigma_{xx} - \nu\sigma_{yy} = E\varepsilon_{xx} = -15\text{ksi}$$ **1**

$$\varepsilon_{yy} = 0 \quad \text{or} \quad \sigma_{yy} - \nu\sigma_{xx} = 0$$ **2**

Substituting Eq.(2) into Eq.(1) we obtain: $\sigma_{xx}(1 - \nu^2) = -15\text{psi}$ or

$$\sigma_{xx} = \frac{-15}{1 - 0.25^2} = -16\text{ksi}$$ $\boxed{\sigma_{xx} = 16\text{ksi(C)}}$

From Eq.(2) we obtain: $\sigma_{yy} = 0.25(-16) = -4\text{ksi}$ $\boxed{\sigma_{yy} = 4\text{ksi(C)}}$

3.78 A 2in x 2 in square with a circle inscribed is stressed as shown Fig. P3.78. The plate material has a Modulus of Elasticity of E = 10,000 ksi and a Poisson's ratio ν = 0.25. Assuming plane stress, determine the major and minor axis of the ellipse formed due to deformation.

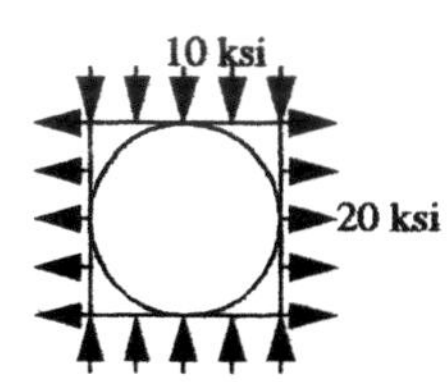

Fig. P3.78

Solution E=10000ksi ν=0.25 a = major axis=? b = minor axis=?

We are given: $\sigma_{xx} = 20\text{ksi}$ and $\sigma_{yy} = -10\text{ksi}$ and Plane stress $\therefore \sigma_{zz} = 0$

$$\varepsilon_{xx} = \frac{\sigma_{xx} - \nu(\sigma_{yy} + \sigma_{zz})}{E} = \frac{20 - 0.25(-10)}{10000} = 2.25(10^{-3})$$

$$\varepsilon_{yy} = \frac{\sigma_{yy} - \nu(\sigma_{xx} + \sigma_{zz})}{E} = \frac{-10 - 0.25(20)}{10000} = -1.5(10^{-3})$$

The circle extends in x direction and contracts in y direction

$\Delta a = \varepsilon_{xx}(2) = 4.5(10^{-3})$ $a = 2 + \Delta a = 2.0045\text{in}$ $\boxed{a = 2.0045\text{in}}$

$\Delta b = \varepsilon_{yy}(2) = -3(10^{-3})$ $b = 2 + \Delta b = 1.9970\text{in}$ $\boxed{b = 1.9770\text{in}}$

3.79 A 2in x 2 in square with a circle inscribed is stressed as shown Fig. P3.79. The plate material has a Modulus of Elasticity of E = 10,000 ksi and a Poisson's ratio $\nu = 0.25$. Assuming plane stress, determine the major and minor axis of the ellipse formed due to deformation.

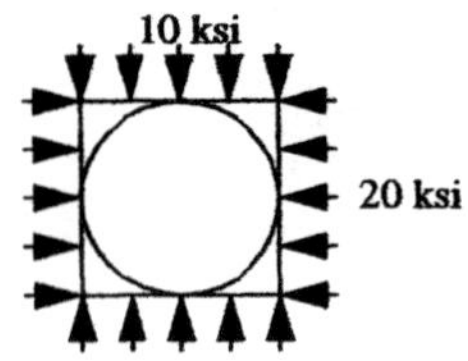

Fig. P3.79

Solution E=10000ksi ν=0.25 a = major axis =? b = minor axis =?

We are given: $\sigma_{xx} = -20\text{ksi}$ and $\sigma_{yy} = -10\text{ksi}$ and plane stress $\therefore \sigma_{zz} = 0$

$$\varepsilon_{xx} = \frac{\sigma_{xx} - \nu(\sigma_{yy} + \sigma_{zz})}{E} = \frac{-20 - 0.25(-10)}{10000} = -1.75(10^{-3})$$

$$\varepsilon_{yy} = \frac{\sigma_{yy} - \nu(\sigma_{xx} + \sigma_{zz})}{E} = \frac{-10 - 0.25(-20)}{10000} = -0.5(10^{-3})$$

The circle contracts more in x direction than in y direction

$\Delta a = \varepsilon_{xx}(2) = -3.5(10^{-3})$ $a = 2 + \Delta a = 1.9965\text{in}$ $\boxed{a = 1.9965\text{in}}$

$\Delta b = \varepsilon_{yy}(2) = -1(10^{-3})$ $b = 2 + \Delta b = 1.999\text{in}$ $\boxed{b = 1.999\text{in}}$

3.80 A 50 mm x 50 mm square with a circle inscribed is stressed as shown Fig. P3.80. The plate material has a Modulus of Elasticity of E = 70 GPa and a Poisson's ratio $\nu = 0.25$. Assuming plane stress, determine the major and minor axis of the ellipse formed due to deformation.

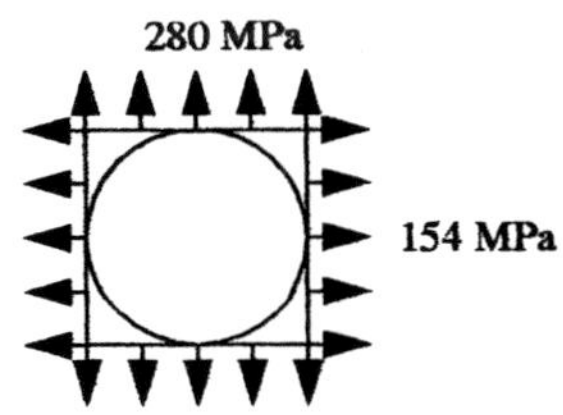

Fig. P3.80

Solution E=70GPa ν=0.25 a= major axis =? b = minor axis =?

We are given: $\sigma_{xx} = 154\text{MPa}$ and $\sigma_{yy} = 280\text{MPa}$ and plane stress $\therefore \sigma_{zz} = 0$

$$\varepsilon_{xx} = \frac{\sigma_{xx} - \nu(\sigma_{yy} + \sigma_{zz})}{E} = \frac{[154-0.25(280)](10^6)}{70(10^9)} = 1.2(10^{-3})$$

$$\varepsilon_{yy} = \frac{\sigma_{yy} - \nu(\sigma_{xx} + \sigma_{zz})}{E} = \frac{[280-0.25(154)](10^6)}{70(10^9)} = 3.45(10^{-3})$$

The circle elongates more in x direction than in y direction

$\Delta a = \varepsilon_{xx}(50) = 60(10^{-3})$ $a = 50 + \Delta a = 50.06\text{mm}$ $\boxed{a = 50.06\text{mm}}$

$\Delta b = \varepsilon_{yy}(50) = 172.5(10^{-3})$ $b = 50 + \Delta b = 50.1725\text{mm}$ $\boxed{b = 50.1725\text{mm}}$

3.81 Derive the following relation of normal stresses in terms of normal strain from the generalized Hooke's Law.

$$\sigma_{xx} = [(1-\nu)\varepsilon_{xx} + \nu\varepsilon_{yy} + \nu\varepsilon_{zz}]\frac{E}{(1-2\nu)(1+\nu)}$$

$$\sigma_{yy} = [(1-\nu)\varepsilon_{yy} + \nu\varepsilon_{zz} + \nu\varepsilon_{xx}]\frac{E}{(1-2\nu)(1+\nu)} \qquad \textbf{3.14}$$

$$\sigma_{zz} = [(1-\nu)\varepsilon_{zz} + \nu\varepsilon_{xx} + \nu\varepsilon_{yy}]\frac{E}{(1-2\nu)(1+\nu)}$$

Solution

Adding the three equations for normal strains in the generalized Hooke's Law.

$$\varepsilon_{xx} + \varepsilon_{yy} + \varepsilon_{zz} = \frac{1-2\nu}{E}(\sigma_{xx} + \sigma_{yy} + \sigma_{zz})$$

$$\therefore(\sigma_{xx} + \sigma_{yy} + \sigma_{zz}) = \left(\frac{E}{1-2\nu}\right)(\varepsilon_{xx} + \varepsilon_{yy} + \varepsilon_{zz}) \qquad 1$$

From generalized Hooke's Law.

$$\varepsilon_{xx} = \frac{\sigma_{xx}}{E} - \frac{\nu}{E}(\sigma_{yy} + \sigma_{zz}) = \sigma_{xx}\left(\frac{1+\nu}{E}\right) - \frac{\nu}{E}(\sigma_{xx} + \sigma_{yy} + \sigma_{zz}) \qquad 2$$

Substituting Eq. (1) in Eq. (2) we obtain

$$\varepsilon_{xx} = \sigma_{xx}\left(\frac{1+\nu}{E}\right) - \left(\frac{\nu}{E}\right)\left(\frac{E}{1-2\nu}\right)(\varepsilon_{xx} + \varepsilon_{yy} + \varepsilon_{zz})$$

$$\therefore\sigma_{xx}\left(\frac{1+\nu}{E}\right) = \varepsilon_{xx}\left(1 + \frac{\nu}{1-2\nu}\right) + \frac{\nu}{1-2\nu}\varepsilon_{yy} + \frac{\nu}{1-2\nu}\varepsilon_{zz}$$

$$\boxed{\therefore\sigma_{xx} = \frac{E}{(1+\nu)(1-2\nu)}[\varepsilon_{xx}(1-\nu) + \nu\varepsilon_{yy} + \nu\varepsilon_{zz}]} \qquad 3$$

From generalized Hooke's Law.

$$\varepsilon_{yy} = \frac{\sigma_{yy}}{E} - \frac{\nu}{E}(\sigma_{xx} + \sigma_{zz}) = \sigma_{yy}\left(\frac{1+\nu}{E}\right) - \frac{\nu}{E}(\sigma_{xx} + \sigma_{yy} + \sigma_{zz}) \qquad 4$$

Substituting Eq. (1) in Eq. (4) we obtain

$$\varepsilon_{yy} = \sigma_{yy}\left(\frac{1+\nu}{E}\right) - \left(\frac{\nu}{E}\right)\left(\frac{E}{1-2\nu}\right)(\varepsilon_{xx} + \varepsilon_{yy} + \varepsilon_{zz}) \text{ or } \sigma_{yy}\left(\frac{1+\nu}{E}\right) = \frac{\nu}{1-2\nu}\varepsilon_{xx} + \varepsilon_{yy}\left(1 + \frac{\nu}{1-2\nu}\right) + \frac{\nu}{1-2\nu}\varepsilon_{zz}$$

$$\boxed{\therefore\sigma_{yy} = \frac{E}{(1+\nu)(1-2\nu)}[\nu\varepsilon_{xx} + (1-\nu)\varepsilon_{yy} + \nu\varepsilon_{zz}]} \qquad 5$$

From generalized Hooke's Law.

$$\varepsilon_{zz} = \frac{\sigma_{zz}}{E} - \frac{\nu}{E}(\sigma_{xx} + \sigma_{yy}) = \sigma_{zz}\left(\frac{1+\nu}{E}\right) - \frac{\nu}{E}(\sigma_{xx} + \sigma_{yy} + \sigma_{zz}) \qquad 6$$

Substituting Eq. (1) in Eq. (6) we obtain

$$\varepsilon_{zz} = \sigma_{zz}\left(\frac{1+\nu}{E}\right) - \left(\frac{\nu}{E}\right)\left(\frac{E}{1-2\nu}\right)(\varepsilon_{xx} + \varepsilon_{yy} + \varepsilon_{zz})$$

$$\therefore\sigma_{zz}\left(\frac{1+\nu}{E}\right) = \frac{\nu}{1-2\nu}\varepsilon_{xx} + \frac{\nu}{1-2\nu}\varepsilon_{yy} + \left(1 + \frac{\nu}{1-2\nu}\right)\varepsilon_{zz}$$

$$\boxed{\therefore\sigma_{zz} = \frac{E}{(1+\nu)(1-2\nu)}[\nu\varepsilon_{xx} + \nu\varepsilon_{yy} + (1-\nu)\varepsilon_{zz}]} \qquad 7$$

Eqs. (3), (5), and (7) are the same as Eq. (3.14)

3.82 For a point in plane stress show

$$\sigma_{xx} = [\varepsilon_{xx} + \nu\varepsilon_{yy}]\frac{E}{(1-\nu^2)} \qquad \sigma_{yy} = [\varepsilon_{yy} + \nu\varepsilon_{xx}]\frac{E}{(1-\nu^2)} \qquad 3.15$$

Solution

Plane Stress $\therefore\sigma_{zz} = 0$

From generalized Hooke's Law.

$$\varepsilon_{xx} = \frac{\sigma_{xx}}{E} - \frac{\nu}{E}(\sigma_{yy} + \sigma_{zz}) \quad \text{or} \quad \sigma_{xx} - \nu\sigma_{yy} = E\varepsilon_{xx} \qquad 1$$

$$\varepsilon_{yy} = \frac{\sigma_{yy}}{E} - \frac{\nu}{E}(\sigma_{xx} + \sigma_{zz}) \quad \text{or} \quad \sigma_{yy} - \nu\sigma_{xx} = E\varepsilon_{yy} \qquad 2$$

Multiplying Eq. (2) by ν and adding to Eq. (1) we obtain: $\sigma_{xx}(1 - \nu^2) = E(\varepsilon_{xx} + \nu\varepsilon_{yy})$

$$\boxed{\therefore \sigma_{xx} = \frac{E}{(1-\nu^2)}(\varepsilon_{xx} + \nu\varepsilon_{yy})} \qquad 3$$

Multiplying Eq. (1) by ν and adding to Eq. (2) we obtain: $\sigma_{yy}(1 - \nu^2) = E(\varepsilon_{yy} + \nu\varepsilon_{xx})$

$$\boxed{\therefore \sigma_{yy} = \frac{E}{(1-\nu^2)}(\varepsilon_{yy} + \nu\varepsilon_{xx})} \qquad 4$$

Eq. (3) and (4) are the same as given in Eq. 3.15

3.83 For a point in plane stress show

$$\varepsilon_{zz} = -\left(\frac{\nu}{1-\nu}\right)(\varepsilon_{xx} + \varepsilon_{yy}) \qquad \textbf{3.16}$$

Solution

Plane Stress $\therefore \sigma_{zz} = 0$

Adding Eq. (1) and (2) in Problem 3.82 we obtain: $\sigma_{xx}(1-\nu) + \sigma_{yy}(1-\nu) = E(\varepsilon_{xx} + \varepsilon_{yy})$

$$\therefore \sigma_{xx} + \sigma_{yy} = \frac{E}{(1-\nu)}(\varepsilon_{xx} + \varepsilon_{yy}) \qquad 1$$

From generalized Hooke's Law

$$\varepsilon_{zz} = \frac{\sigma_{zz}}{E} - \frac{\nu}{E}(\sigma_{xx} + \sigma_{yy}) = 0 - \left(\frac{\nu}{E}\right)\left(\frac{E}{(1-\nu)}\right)(\varepsilon_{xx} + \varepsilon_{yy})$$

$$\boxed{\therefore \varepsilon_{zz} = -\left(\frac{\nu}{(1-\nu)}\right)(\varepsilon_{xx} + \varepsilon_{yy})} \qquad 2$$

Eq. (2) is the same as given in Eq. 3.16

3.84 Using Eqs. 3.15 and 3.16 solve for σ_{xx}, σ_{yy}, and ε_{zz} in Problem 3.70.

Solution σ_{xx}=? σ_{yy}=? and ε_{zz}=?

Substituting $\varepsilon_{xx} = 500(10^{-6})$, $\varepsilon_{yy} = 400(10^{-6})$, $E = 200(10^9)\frac{N}{m^2}$ and $\nu = 0.32$ into Eqs. 3.15 and 3.16 we obtain the following.

$$\sigma_{xx} = [500 + (0.32)(400)]\frac{(10^{-6})200(10^9)}{1 - 0.32^2} = 139.9(10^6)\frac{N}{m^2} \qquad \boxed{\sigma_{xx} = 139.9\text{MPa(T)}}$$

$$\sigma_{yy} = [400 + (0.32)(500)]\frac{(10^{-6})200(10^9)}{1 - 0.32^2} = 124.77(10^6)\frac{N}{m^2} \qquad \boxed{\sigma_{yy} = 124.8\text{MPa(T)}}$$

$$\varepsilon_{zz} = -\frac{0.25}{1 - 0.25}(500 + 400)(10^{-6}) = -423.5(10^{-6}) \qquad \boxed{\varepsilon_{zz} = -423.5\mu\frac{mm}{mm}}$$

3.85 Using Eqs. 3.15 and 3.16 solve for σ_{xx}, σ_{yy}, and ε_{zz} in Problem 3.71.

Solution σ_{xx}=? σ_{yy}=? and ε_{zz}=?

Substituting $\varepsilon_{xx} = -3000(10^{-6})$, $\varepsilon_{yy} = 1500(10^{-6})$, $E = 70(10^9)\frac{N}{m^2}$ and $\nu = 0.25$ into Eqs. 3.15 and 3.16 we obtain the following.

$$\sigma_{xx} = [-3000+(0.25)(1500)]\frac{(10^{-6})70(10^9)}{1-0.25^2} = -196(10^6)\frac{N}{m^2}$$ $\boxed{\sigma_{xx} = 196\text{MPa(C)}}$

$$\sigma_{yy} = [1500+(0.25)(-3000)]\frac{(10^{-6})70(10^9)}{1-0.25^2} = 56(10^6)\frac{N}{m^2}$$ $\boxed{\sigma_{yy} = 56\text{MPa(T)}}$

$$\varepsilon_{zz} = -\frac{0.25}{1-0.25}(-3000+1500)(10^{-6}) = 500(10^{-6})$$ $\boxed{\varepsilon_{zz} = 500\mu}$

3.86 Using Eqs. 3.15 and 3.16 solve for σ_{xx}, σ_{yy}, and ε_{zz} in Problem 3.72.

Solution σ_{xx}=? σ_{yy}=? and ε_{zz}=?

Substituting $\varepsilon_{xx} = -800(10^{-6})$, $\varepsilon_{yy} = -1000(10^{-6})$, $E = 30000\text{ksi}$ and $\nu = 0.3$ into Eqs. 3.15 and 3.16 we obtain the following.

$$\sigma_{xx} = [-800+(0.3)(-1000)]\frac{(10^{-6})30000}{1-0.3^2} = -36.26\text{ksi}$$ $\boxed{\sigma_{xx} = 36.3\text{ksi(C)}}$

$$\sigma_{yy} = [-1000+(0.3)(-800)]\frac{(10^{-6})30000}{1-0.3^2} = -40.88\text{ksi}$$ $\boxed{\sigma_{yy} = 40.9\text{ksi(C)}}$

$$\varepsilon_{zz} = -\frac{0.3}{1-0.3}(-800-1000)(10^{-6}) = 771.4(10^{-6})$$ $\boxed{\varepsilon_{zz} = 771.4\mu}$

3.87 Using Eqs. 3.15 and 3.16 solve for σ_{xx}, σ_{yy}, and ε_{zz} in Problem 3.73.

Solution σ_{xx}=? σ_{yy}=? and ε_{zz}=?

Substituting $\varepsilon_{xx} = 1500(10^{-6})$, $\varepsilon_{yy} = -1200(10^{-6})$, $E = 10000\text{ksi}$ and $\nu = 0.282$ into Eqs. 3.15 and 3.16 we obtain the following.

$$\sigma_{xx} = [1500+(0.282)(-1200)]\frac{(10^{-6})10000}{1-0.282^2} = 12.62\text{ksi}$$ $\boxed{\sigma_{xx} = 12.6\text{ksi(T)}}$

$$\sigma_{yy} = [-1200+(0.282)(1500)]\frac{(10^{-6})10000}{1-0.282^2} = -8.44\text{ksi}$$ $\boxed{\sigma_{yy} = 8.44\text{ksi(C)}}$

$$\varepsilon_{zz} = -\frac{0.282}{1-0.282}(1500+(-1200))(10^{-6}) = -117.8(10^{-6})$$ $\boxed{\varepsilon_{zz} = -118\mu}$

3.88 Using Eqs. 3.15 and 3.16 solve for σ_{xx}, σ_{yy}, and ε_{zz} in Problem 3.74.

Solution σ_{xx}=? σ_{yy}=? and ε_{zz}=?

Substituting $\varepsilon_{xx} = -2000(10^{-6})$, $\varepsilon_{yy} = 2000(10^{-6})$, $E = 36(10^9)\text{N/m}^2$ and $\nu = 0.2$ into Eqs. 3.15 and

3.16 we obtain the following.

$$\sigma_{xx} = [-2000 + (0.2)(2000)]\frac{(10^{-6})(36)(10^{9})}{1-0.2^2} = -60(10^6)\frac{N}{m^2}$$ $\boxed{\sigma_{xx} = 60\text{MPa(C)}}$

$$\sigma_{yy} = [2000 + (0.2)(-2000)]\frac{(10^{-6})(36)(10^{9})}{1-0.2^2} = 60(10^6)\frac{N}{m^2}$$ $\boxed{\sigma_{yy} = 60\text{MPa(T)}}$

$$\varepsilon_{zz} = -\frac{0.2}{1-0.2}(-2000+2000)(10^{-6}) = 0$$ $\boxed{\varepsilon_{zz} = 0}$

3.89 Using Eqs. 3.15 and 3.16 solve for σ_{xx}, σ_{yy}, and ε_{zz} in Problem 3.75.

Solution σ_{xx}=? σ_{yy}=? and ε_{zz}=?

Substituting $\varepsilon_{xx} = 50(10^{-6})$, $\varepsilon_{yy} = 75(10^{-6})$, $E = 2000\text{psi}$ and $\nu = 0.25$ into Eqs. 3.15 and 3.16 we obtain the following.

$$\sigma_{xx} = [50 + (0.25)(75)]\frac{(10^{-6})2000}{1-0.25^2} = 0.1467\text{psi}$$ $\boxed{\sigma_{xx} = 0.147\text{psi(T)}}$

$$\sigma_{yy} = [75 + (0.25)(50)]\frac{(10^{-6})2000}{1-0.25^2} = 0.187\text{psi}$$ $\boxed{\sigma_{yy} = 0.187\text{psi(T)}}$

$$\varepsilon_{zz} = -\frac{0.25}{1-0.25}(50+75)(10^{-6}) = -41.67(10^{-6})$$ $\boxed{\varepsilon_{zz} = -41.67\mu}$

3.90 For a point in plane strain show

$$\varepsilon_{xx} = [(1-\nu)\sigma_{xx} - \nu\sigma_{yy}]\frac{(1+\nu)}{E} \qquad \varepsilon_{yy} = [(1-\nu)\sigma_{yy} - \nu\sigma_{xx}]\frac{(1+\nu)}{E}$$ **3.17**

Solution

Plane strain $\therefore \varepsilon_{zz} = 0$

From Generalized Hooke's Law: $\varepsilon_{zz} = \frac{\sigma_{zz}}{E} - \frac{\nu}{E}(\sigma_{xx} + \sigma_{yy}) = 0$ or

$$\boxed{\therefore \sigma_{zz} = \nu(\sigma_{xx} + \sigma_{yy})}$$ **1**

From Generalized Hooke's Law

$$\varepsilon_{xx} = \frac{\sigma_{xx}}{E} - \frac{\nu}{E}(\sigma_{yy} + \sigma_{zz}) = \frac{\sigma_{xx}}{E} - \frac{\nu}{E}(\sigma_{yy} + \nu\sigma_{xx} + \nu\sigma_{yy}) = \frac{(1-\nu^2)}{E}\sigma_{xx} - \frac{\nu(1+\nu)}{E}\sigma_{yy}$$

$$\boxed{\therefore \varepsilon_{xx} = ((1-\nu)\sigma_{xx} - \nu\sigma_{yy})\frac{(1+\nu)}{E}}$$ **2**

From Generalized Hooke's Law

$$\varepsilon_{yy} = \frac{\sigma_{yy}}{E} - \frac{\nu}{E}(\sigma_{xx} + \sigma_{zz}) = \frac{\sigma_{yy}}{E} - \frac{\nu}{E}(\sigma_{xx} + \nu\sigma_{xx} + \nu\sigma_{yy}) = \frac{(1-\nu^2)}{E}\sigma_{yy} - \frac{\nu(1+\nu)}{E}\sigma_{xx}$$

$$\boxed{\therefore \varepsilon_{yy} = ((1-\nu)\sigma_{yy} - \nu\sigma_{xx})\frac{(1+\nu)}{E}}$$ **3**

Eqs. (2) and (3) are same as Eq. 3.17

3.91 For a point in plane strain show

$$\sigma_{xx} = [(1-\nu)\varepsilon_{xx} + \nu\varepsilon_{yy}]\frac{E}{(1-2\nu)(1+\nu)} \qquad \sigma_{yy} = [(1-\nu)\varepsilon_{yy} + \nu\varepsilon_{xx}]\frac{E}{(1-2\nu)(1+\nu)} \qquad \textbf{3.18}$$

Solution

Plane strain $\therefore \varepsilon_{zz} = 0$

Multiplying Eq. (1) in Problem 3.89 by $(1-\nu)$ and Eq. (2) by ν and adding

$$(1-\nu)\varepsilon_{xx} + \nu\varepsilon_{yy} = ((1-\nu)^2 - \nu^2)\sigma_{xx}\frac{(1+\nu)}{E} = (1 - 2\nu + \nu^2 - \nu^2)\sigma_{xx}\frac{(1+\nu)}{E}$$

$$\boxed{\therefore \sigma_{xx} = [(1-\nu)\varepsilon_{xx} + \nu\varepsilon_{yy}]\frac{E}{(1-2\nu)(1+\nu)}} \qquad 1$$

Multiplying Eq. (1) in Problem 3.89 by ν and Eq. (2) by $(1-\nu)$ and adding

$$\nu\varepsilon_{xx} + (1-\nu)\varepsilon_{yy} = ((1-\nu)^2 - \nu^2)\sigma_{yy}\frac{(1+\nu)}{E} = (1 - 2\nu + \nu^2 - \nu^2)\sigma_{yy}\frac{(1+\nu)}{E}$$

$$\boxed{\therefore \sigma_{yy} = [\nu\varepsilon_{xx} + (1-\nu)\varepsilon_{yy}]\frac{E}{(1-2\nu)(1+\nu)}} \qquad 2$$

Eq. (1) and (2) are the same as Eq. 3.18

3.92 A differential element subjected to only normal strains is shown in Fig. P3.92. The ratio of change in volume (ΔV) to the original volume (V) is called the volumetric strain (ε_V) or *dilation*. For small strain, prove

$$\varepsilon_V = \Delta V/V = \varepsilon_{xx} + \varepsilon_{yy} + \varepsilon_{zz} \qquad \textbf{3.19}$$

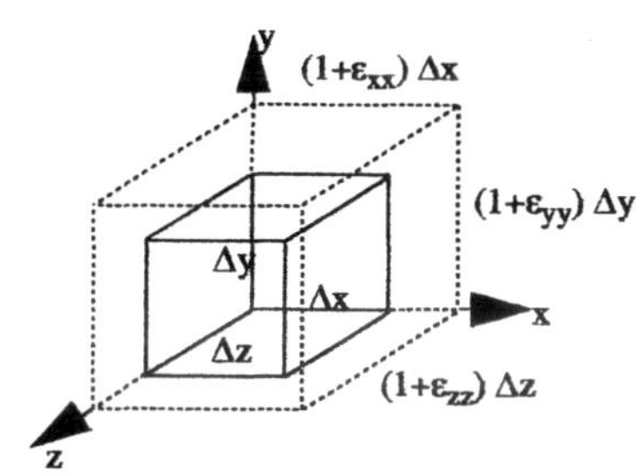

Fig. P3.92

Solution

The original volume is: $V = (\Delta x)(\Delta y)(\Delta z)$

The change of volume is: $\Delta V = (1+\varepsilon_{xx})\Delta x(1+\varepsilon_{yy})\Delta y(1+\varepsilon_{zz})\Delta(z) - (\Delta x\Delta y\Delta z)$

$$\therefore \Delta V = [1 + \varepsilon_{xx} + \varepsilon_{yy} + \varepsilon_{zz} + \varepsilon_{xx}\varepsilon_{yy} + \varepsilon_{yy}\varepsilon_{zz} + \varepsilon_{zz}\varepsilon_{xx} + \varepsilon_{xx}\varepsilon_{yy}\varepsilon_{zz} - 1](\Delta x\Delta y\Delta z)$$

For small strain, the quadratic and the cubic terms can be neglected i.e., $\Delta V = [\varepsilon_{xx} + \varepsilon_{yy} + \varepsilon_{zz}](\Delta x\Delta y\Delta z)$

$$\boxed{\varepsilon_v = \frac{\Delta V}{V} = \varepsilon_{xx} + \varepsilon_{yy} + \varepsilon_{zz}} \qquad 1$$

Eq. 1 is same as Eq. 3.19

3.93 Prove

$$p = -K\varepsilon_v \qquad p = -\left(\frac{\sigma_{xx} + \sigma_{yy} + \sigma_{zz}}{3}\right) \qquad K = \frac{E}{3(1-2\nu)} \qquad \textbf{3.20}$$

where, K is called the *bulk modulus* and p is called the hydrostatic pressure because at a point in fluid the normal stresses in all directions are equal to -p. Note that at ν = 1/2 there is no change in volume regardless of the value of stresses. Such materials are called *incompressible materials*.

Solution

Adding the three equations for normal strain in generalized Hooke's law we obtain

$$\varepsilon_{xx} + \varepsilon_{yy} + \varepsilon_{zz} = \frac{(1-2\nu)}{E}(\sigma_{xx} + \sigma_{yy} + \sigma_{zz})$$

Substituting $\sigma_{xx} + \sigma_{yy} + \sigma_{zz} = -3p$ and $\varepsilon_{xx} + \varepsilon_{yy} + \varepsilon_{zz} = \varepsilon_v$ we obtain: $\varepsilon_v = \frac{-3(1-2\nu)p}{E}$

$$\boxed{\therefore p = \frac{-E}{3(1-2\nu)}\varepsilon_v} \quad \mathbf{1}$$

Eq. (1) is the same as Eq. 3.20

3.94 The stresses at a point on free surface of an orthotropic material are as given. Also given are the material constants. Using Equation 3.21, solve for the strains ε_{xx}, ε_{yy}, and γ_{xy}.

$\sigma_{xx} = 4\text{ksi(T)}$ $\quad \sigma_{yy} = 10\text{ksi(C)}$ $\quad \tau_{xy} = 4\text{ksi}$

$E_x = 7500\text{ksi}$ $\quad E_y = 2500\text{ksi}$ $\quad \nu_{xy} = 0.3$ $\quad G_{xy} = 1250\text{ksi}$

Solution $\quad \varepsilon_{xx} = ?$ $\quad \varepsilon_{yy} = ?$ $\quad \gamma_{xy} = ?$

$$\frac{\nu_{xy}}{E_x} = \frac{0.3}{7500} = \frac{\nu_{yx}}{E_y}$$

Substituting the given value of stresses and material constants in Eq. 3.21, we obtain the following

$$\varepsilon_{xx} = \frac{4}{7500} - \frac{0.3}{7500}(-10) = 0.933(10^{-3}) \qquad \boxed{\varepsilon_{xx} = 933\mu}$$

$$\varepsilon_{yy} = \frac{(-10)}{2500} - \frac{0.3}{7500}(4) = -4.16(10^{-3}) \qquad \boxed{\varepsilon_{yy} = -4160\mu}$$

$$\gamma_{xy} = \frac{4}{1250} = 3.2(10^{-3}) \qquad \boxed{\gamma_{xy} = 3200\mu}$$

3.95 The stresses at a point on free surface of an orthotropic material are as given. Also given are the material constants. Using Equation 3.21, solve for the strains ε_{xx}, ε_{yy}, and γ_{xy}.

$\sigma_{xx} = 25\text{ksi(C)}$ $\quad \sigma_{yy} = 5\text{ksi(C)}$ $\quad \tau_{xy} = -8\text{ksi}$

$E_x = 25000\text{ksi}$ $\quad E_y = 2000\text{ksi}$ $\quad \nu_{xy} = 0.32$ $\quad G_{xy} = 1500\text{ksi}$

Solution $\quad \varepsilon_{xx} = ?$ $\quad \varepsilon_{yy} = ?$ $\quad \gamma_{xy} = ?$

$$\frac{\nu_{xy}}{E_x} = \frac{0.32}{25000} = \frac{\nu_{yx}}{E_y}$$

Substituting the given value of stresses and material constants in Eq. 3.21, we obtain the following

$$\varepsilon_{xx} = \frac{(-25)}{25000} - \frac{0.32}{25000}(-5) = -0.936(10^{-3}) \qquad \boxed{\varepsilon_{xx} = -936\mu}$$

$$\varepsilon_{yy} = \frac{(-5)}{2000} - \frac{0.3}{25000}(-25) = -2.18(10^{-3}) \qquad \boxed{\varepsilon_{yy} = -2180\mu}$$

$$\gamma_{xy} = \frac{-8}{1500} = -5.333(10^{-3}) \qquad \boxed{\gamma_{xy} = -5333\mu}$$

3.96 The stresses at a point on free surface of an orthotropic material are as given. Also given are the material constants. Using Equation 3.21, solve for the strains ε_{xx}, ε_{yy}, and γ_{xy}.

$\sigma_{xx} = 200$ MPa(C) $\quad \sigma_{yy} = 80$ MPa(C) $\quad \tau_{xy} = -54$ MPa

$E_x = 53$ GPa $\quad E_y = 18$ GPa $\quad \nu_{xy} = 0.25 \quad G_{xy} = 9$ GPa

Solution $\quad \varepsilon_{xx} = ? \quad \varepsilon_{yy} = ? \quad \gamma_{xy} = ?$

$$\frac{\nu_{xy}}{E_x} = \frac{0.25}{53(10^9)} = \frac{\nu_{yx}}{E_y}$$

Substituting the given value of stresses and material constants in Eq. 3.21, we obtain the following

$$\varepsilon_{xx} = \frac{(-200)(10^6)}{53(10^9)} - \frac{0.25}{53(10^9)}(-80)(10^6) = -3.396(10^{-3}) \qquad \boxed{\varepsilon_{xx} = -3396\mu}$$

$$\varepsilon_{yy} = \frac{(-80)(10^6)}{18(10^9)} - \frac{0.25}{53(10^9)}(-200)(10^6) = -3.501(10^{-3}) \qquad \boxed{\varepsilon_{yy} = -3501\mu}$$

$$\gamma_{xy} = \frac{-54(10^6)}{9(10^9)} = -6(10^{-3}) \qquad \boxed{\gamma_{xy} = -6000\mu}$$

3.97 The stresses at a point on free surface of an orthotropic material are as given. Also given are the material constants. Using Equation 3.21, solve for the strains ε_{xx}, ε_{yy}, and γ_{xy}.

$\sigma_{xx} = 300$ MPa(T) $\quad \sigma_{yy} = 50$ MPa(T) $\quad \tau_{xy} = 60$ MPa

$E_x = 180$ GPa $\quad E_y = 15$ GPa $\quad \nu_{xy} = 0.28 \quad G_{xy} = 11$ GPa

Solution $\quad \varepsilon_{xx} = ? \quad \varepsilon_{yy} = ? \quad \gamma_{xy} = ?$

$$\frac{\nu_{xy}}{E_x} = \frac{0.28}{180(10^9)} = \frac{\nu_{yx}}{E_y}$$

Substituting the given value of stresses and material constants in Eq. 3.21, we obtain the following

$$\varepsilon_{xx} = \frac{300(10^6)}{180(10^9)} - \frac{0.28}{180(10^9)}(50)(10^6) = 1.589(10^{-3}) \qquad \boxed{\varepsilon_{xx} = 1589\mu}$$

$$\varepsilon_{yy} = \frac{(50)(10^6)}{15(10^9)} - \frac{0.28}{180(10^9)}(300)(10^6) = 2.867(10^{-3}) \qquad \boxed{\varepsilon_{yy} = 2867\mu}$$

$$\gamma_{xy} = \frac{60(10^6)}{11(10^9)} = 5.454(10^{-3}) \qquad \boxed{\gamma_{xy} = 5454\mu}$$

3.98 The strains at a point on free surface of an orthotropic material is given in each problem. Also given are the material constants. Using Equation 3.21, solve for the stresses σ_{xx}, σ_{yy}, and τ_{xy}.

$\varepsilon_{xx} = -1000\mu \quad \varepsilon_{yy} = 500\mu \quad \gamma_{xy} = -250\mu$

$E_x = 7500$ksi $\quad E_y = 2500$ksi $\quad \nu_{xy} = 0.3 \quad G_{xy} = 1250$ksi

Solution $\quad \sigma_{xx} = ? \quad \sigma_{yy} = ? \quad \tau_{xy} = ?$

$$\frac{\nu_{xy}}{E_x} = \frac{\nu_{yx}}{E_y} = \frac{0.3}{7500} = 40(10^{-6})$$

$$\frac{1}{E_x} = \frac{1}{7500} = 133.33(10^{-6}) \text{ and } \frac{1}{E_y} = \frac{1}{2500} = 400(10^{-6})$$

Substituting the given strain values and the constants above in Eqs. 3.21 we obtain the following

$$\varepsilon_{xx} = 133.33(10^{-6})\sigma_{xx} - 40(10^{-6})\sigma_{yy} = -1000(10^{-6}) \quad \text{or} \quad 133.33\sigma_{xx} - 40\sigma_{yy} = -1000 \qquad 1$$

$$\varepsilon_{yy} = 400(10^{-6})\sigma_{yy} - 40(10^{-6})\sigma_{xx} = 500(10^{-6}) \quad \text{or} \quad 400\sigma_{yy} - 40\sigma_{xx} = 500 \qquad 2$$

Solving Eqs. (1) and (2), we obtain

$\sigma_{xx} = -7.34\text{ksi}$ $\boxed{\sigma_{xx} = 7.34\text{ksi(C)}}$

$\sigma_{yy} = 0.515\text{ksi}$ $\boxed{\sigma_{yy} = 0.515\text{ksi(T)}}$

$\tau_{xy} = (1250)(-250)(10^{-6}) = -0.3125\text{ksi}$ $\boxed{\tau_{xy} = -0.3125\text{ksi}}$

3.99 The strains at a point on free surface of an orthotropic material is given in each problem. Also given are the material constants. Using Equation 3.21, solve for the stresses σ_{xx}, σ_{yy}, and τ_{xy}.

$\varepsilon_{xx} = -750\mu$ $\varepsilon_{yy} = -250\mu$ $\gamma_{xy} = 400\mu$

$E_x = 25000\text{ksi}$ $E_y = 2000\text{ksi}$ $\nu_{xy} = 0.32$ $G_{xy} = 1500\text{ksi}$

Solution $\sigma_{xx} = ?$ $\sigma_{yy} = ?$ $\tau_{xy} = ?$

$$\frac{\nu_{xy}}{E_x} = \frac{\nu_{yx}}{E_y} = \frac{0.32}{25000} = 12.8(10^{-6})$$

$$\frac{1}{E_x} = \frac{1}{25000} = 40(10^{-6}) \text{ and } \frac{1}{E_y} = \frac{1}{2000} = 500(10^{-6})$$

Substituting the given strain values and the constants above in Eqs. 3.21 we obtain the following

$$\varepsilon_{xx} = 40(10^{-6})\sigma_{xx} - 12.8(10^{-6})\sigma_{yy} = -750(10^{-6}) \quad \text{or} \quad 40\sigma_{xx} - 12.8\sigma_{yy} = -750 \qquad 1$$

$$\varepsilon_{yy} = 500(10^{-6})\sigma_{yy} - 12.8(10^{-6})\sigma_{xx} = -250(10^{-6}) \quad \text{or} \quad 500\sigma_{yy} - 12.8\sigma_{xx} = -250 \qquad 2$$

Solving Eqs. (1) and (2), we obtain

$\sigma_{xx} = -19.07\text{ksi}$ $\boxed{\sigma_{xx} = 19.07\text{ksi(C)}}$

and $\sigma_{yy} = -0.988\text{ksi}$ $\boxed{\sigma_{yy} = 0.99\text{ksi(C)}}$

$\tau_{xy} = (1500)(400)(10^{-6}) = 0.6\text{ksi}$ $\boxed{\tau_{xy} = 0.6\text{ksi}}$

3.100 The strains at a point on free surface of an orthotropic material is given in each problem. Also given are the material constants. Using Equation 3.21, solve for the stresses σ_{xx}, σ_{yy}, and τ_{xy}.

$\varepsilon_{xx} = 2000\mu$ $\varepsilon_{yy} = -800\mu$ $\gamma_{xy} = 300\mu$

$E_x = 53\ \text{GPa}$ $E_y = 18\ \text{GPa}$ $\nu_{xy} = 0.25$ $G_{xy} = 9\ \text{GPa}$

Solution $\sigma_{xx} = ?$ $\sigma_{yy} = ?$ $\tau_{xy} = ?$

$$\frac{\nu_{xy}}{E_x} = \frac{\nu_{yx}}{E_y} = \frac{0.25}{53(10^9)} = 4.717(10^{-12})$$

$$\frac{1}{E_x} = \frac{1}{53(10^9)} = 18.868(10^{-12}) \text{ and } \frac{1}{E_y} = \frac{1}{18(10^9)} = 55.556(10^{-12})$$

Substituting the given strain values and the constants above in Eqs. 3.21 we obtain the following

$\varepsilon_{xx} = 18.868(10^{-12})\sigma_{xx} - 4.717(10^{-12})\sigma_{yy} = 2000(10^{-6})$ or $18.868\sigma_{xx} - 4.717\sigma_{yy} = 2000\text{MPa}$ **1**

$\varepsilon_{yy} = 55.556(10^{-12})\sigma_{yy} - 4.717(10^{-12})\sigma_{xx} = -800(10^{-6})$ or $55.556\sigma_{yy} - 4.717\sigma_{xx} = -800\text{MPa}$ **2**

Solving Eqs. (1) and (2), we obtain

$\sigma_{xx} = 104.62\text{MPa}$ $\boxed{\sigma_{xx} = 104.6\text{MPa(T)}}$

and $\sigma_{yy} = -5.52\text{MPa}$ $\boxed{\sigma_{yy} = 5.52\text{MPa(C)}}$

$\tau_{xy} = 9(10^9)(300)(10^{-6}) = 2700(10^3)\frac{N}{m^2}$ $\boxed{\tau_{xy} = 2.7\text{MPa}}$

3.101 The strains at a point on free surface of an orthotropic material is given in each problem. Also given are the material constants. Using Equation 3.21, solve for the stresses σ_{xx}, σ_{yy}, and τ_{xy}.

$\varepsilon_{xx} = 1500\mu$ $\varepsilon_{yy} = -750\mu$ $\gamma_{xy} = -450\mu$

$E_x = 180\ \text{GPa}$ $E_y = 15\ \text{GPa}$ $\nu_{xy} = 0.28$ $G_{xy} = 11\ \text{GPa}$

Solution $\sigma_{xx} = ?$ $\sigma_{yy} = ?$ $\tau_{xy} = ?$

$$\frac{\nu_{xy}}{E_x} = \frac{\nu_{yx}}{E_y} = \frac{0.28}{180(10^9)} = 1.555(10^{-12})$$

$$\frac{1}{E_x} = \frac{1}{180(10^9)} = 5.555(10^{-12}) \text{ and } \frac{1}{E_y} = \frac{1}{15(10^9)} = 66.667(10^{-12})$$

Substituting the given strain values and the constants above in Eqs. 3.21 we obtain the following

$\varepsilon_{xx} = 5.555(10^{-12})\sigma_{xx} - 1.555(10^{-12})\sigma_{yy} = 1500(10^{-6})$ or $5.555\sigma_{xx} - 1.555\sigma_{yy} = 1500\text{MPa}$ **1**

$\varepsilon_{yy} = 66.667(10^{-12})\sigma_{yy} - 1.555(10^{-12})\sigma_{xx} = -750(10^{-6})$ or $66.667\sigma_{yy} - 1.555\sigma_{xx} = -750\text{MPa}$ **2**

Solving Eqs. (1) and (2), we obtain

$\sigma_{xx} = 268.60\text{MPa}$ $\boxed{\sigma_{xx} = 268.6\text{MPa(T)}}$

and $\sigma_{yy} = -4.98\text{MPa}$ $\boxed{\sigma_{yy} = 4.98\text{MPa(C)}}$

$\tau_{xy} = 11(10^9)(-450)(10^{-6}) = -4950(10^3)\frac{N}{m^2}$ $\boxed{\tau_{xy} = -4.95\text{MPa}}$

3.102 Using Equation 3.21, show that on a free surface of an orthotropic material, the following relationships can be written.

$$\sigma_{xx} = \frac{E_x(\varepsilon_{xx} + \nu_{yx}\varepsilon_{yy})}{(1 - \nu_{yx}\nu_{xy})} \qquad \sigma_{yy} = \frac{E_y(\varepsilon_{yy} + \nu_{xy}\varepsilon_{xx})}{(1 - \nu_{yx}\nu_{xy})} \qquad \textbf{3.22}$$

Solution

Eq. 3.21 in matrix form can be written as:

$$\begin{bmatrix} \frac{1}{E_x} & -\frac{\nu_{yx}}{E_y} \\ -\frac{\nu_{xy}}{E_x} & \frac{1}{E_y} \end{bmatrix} \begin{bmatrix} \sigma_{xx} \\ \sigma_{yy} \end{bmatrix} = \begin{bmatrix} \varepsilon_{xx} \\ \varepsilon_{yy} \end{bmatrix}$$

The determinant (D) of the matrix on the left hand side yields: $D = \frac{1}{E_x E_y} - \frac{\nu_{xy}\nu_{yx}}{E_x E_y} = \frac{(1-\nu_{xy}\nu_{yx})}{E_x E_y}$

Using Kramer's rule, we can write

$$\sigma_{xx} = \frac{\begin{vmatrix} \varepsilon_{xx} & -\frac{\nu_{yx}}{E_y} \\ \varepsilon_{yy} & \frac{1}{E_y} \end{vmatrix}}{D} \quad \text{or} \quad \sigma_{xx} = \left(\frac{\varepsilon_{xx}}{E_y} + \frac{\nu_{yx}}{E_y}\varepsilon_{yy}\right)\frac{E_x E_y}{1-\nu_{xy}\nu_{yx}} \quad \text{or}$$

$$\boxed{\sigma_{xx} = \frac{E_x}{1-\nu_{xy}\nu_{yx}}(\varepsilon_{xx} + \nu_{yx}\varepsilon_{yy})}$$

Similarly

$$\sigma_{yy} = \frac{\begin{vmatrix} \frac{1}{E_x} & \varepsilon_{xx} \\ -\frac{\nu_{xy}}{E_x} & \varepsilon_{yy} \end{vmatrix}}{D} \quad \text{or} \quad \sigma_{yy} = \left(\frac{\varepsilon_{yy}}{E_x} + \frac{\nu_{xy}}{E_x}\varepsilon_{xx}\right)\frac{E_x E_y}{1-\nu_{xy}\nu_{yx}} \quad \text{or}$$

$$\boxed{\sigma_{yy} = \frac{E_y}{1-\nu_{xy}\nu_{yx}}(\varepsilon_{yy} + \nu_{xy}\varepsilon_{xx})}$$

Problem Set 3.3

3.103 A joint in a wooden structure is to be designed for a factor of safety of 3. If the average failure stress in shear on surface BCD is 1.5 ksi and if the average failure bearing stress on surface BEF is 6 ksi, determine the smallest dimensions h and d to the nearest 1/16th of an inch.

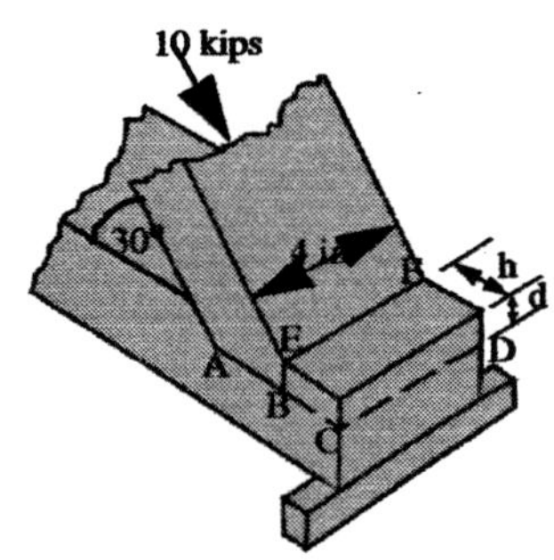

Fig. P3.103

Solution K=3 $(\tau_f)_{BCD}$=1.5ksi $(\sigma_f)_{BEF}$=6ksi h=? d=? to the nearest 1/16 inch

---

The following free body diagrams can be constructed.

(a)

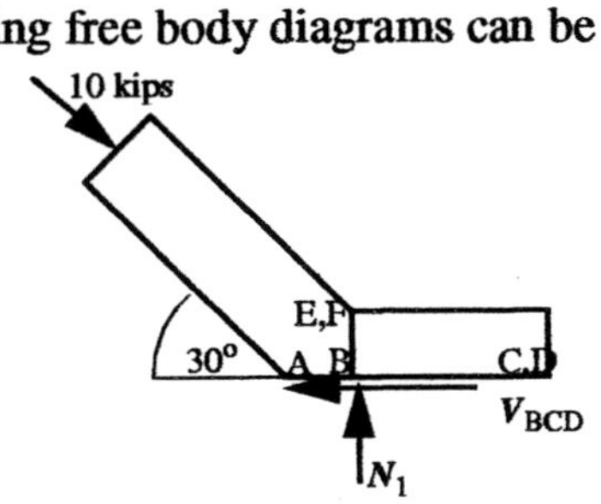

(b)

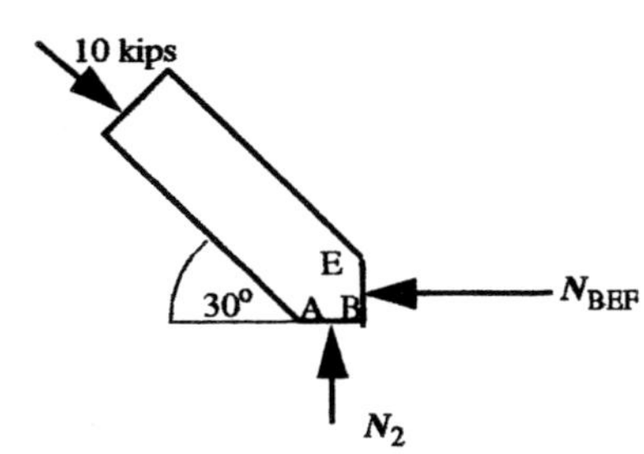

From Fig.(a), by force equilibrium in the x-direction we obtain:

$-V_{BCD} + 10\cos 30 = 0$ or $V_{BCD} = 10\cos 30 = 8.66\text{kips}$

The average shear stress is: $\tau_{BCD} = \frac{V_{BCD}}{A_{BCD}} = \frac{8.66}{4h}$

$$K = \frac{(\tau_f)_{BCD}}{\tau_{BCD}} \quad \text{or} \quad \frac{1.5}{(8.66/4h)} \quad \text{or} \quad h = 4.33$$

$$\boxed{h = 4\frac{3}{8}\text{in}}$$

From Fig.(b), by force equilibrium in the x-direction we obtain:

$10\cos 30 - N_{BEF} = 0$ or $N_{BEF} = 8.66\text{kips (C)}$

The average normal stress is: $\sigma_{BEF} = \dfrac{N_{BEF}}{A_{BEF}} = \dfrac{8.66}{4d}$ (C)

$$K = \frac{(\sigma_f)_{BEF}}{\sigma_{BEF}} \quad \text{or} \quad \frac{6}{(8.66/4d)} = 3 \quad \text{or} \quad d = 1.0825 \text{ or} \quad \boxed{d = 1\tfrac{1}{8}\text{in}}$$

3.104 A 300-lb light is hanging from a ceiling by a chain. The links of the chain are loops made from a thick wire. Determine the minimum diameter of the wire to the nearest 1/16th of an inch for a factor of safety of 4. The normal failure stress for the wire is 28 ksi.

Solution W=300lb K=4 σ_{fail}=28ksi d=? nearest 1/16 inch

The following free body diagram can be drawn by making an imaginary cut through the two wires in a link of the chain.

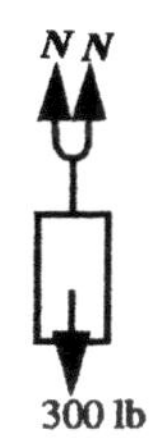

By force equilibrium $2N = 300\text{lb}$ or $N = 150\text{lbs}$

The average normal stress is: $\sigma = \dfrac{N}{A} = \dfrac{150}{(\pi d^2/4)} = \dfrac{600}{\pi d^2}$

$$K = \frac{\sigma_{fail}}{\sigma} = \frac{28(10^3)}{(600/(\pi d^2))} = 4 \text{ or } d^2 = \frac{2400}{28\pi(10^3)} = 0.02728 \quad d = 0.165 \text{ or} \quad \boxed{d = \tfrac{3}{16}\text{inch}}$$

3.105 Determine the maximum weight W that can be suspended using cables as shown for a factor of safety of 1.2. The cable's fracture stress is 200 MPa and the diameter is 10 mm.

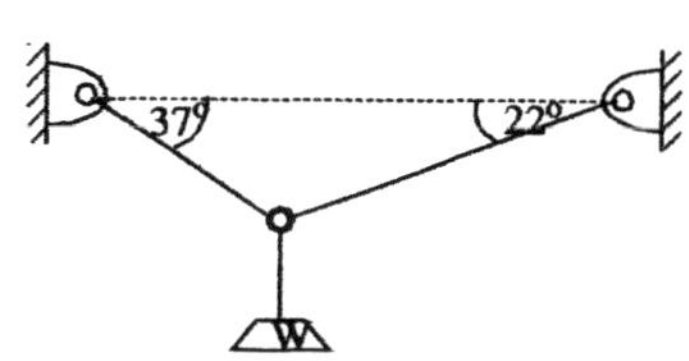

Fig. P3.105

Solution K=1.2 σ_{fail}=200MPa d=10mm W_{max}=?

By making imaginary cuts through the cables, the following free body diagram can be drawn.

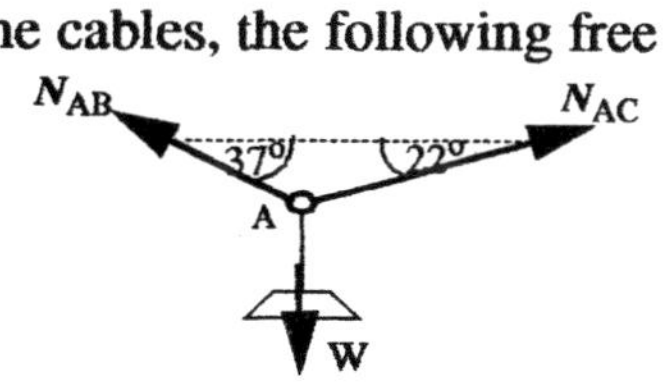

By force equilibrium in the x-direction: $-N_{AB}\cos 37 + N_{AC}\cos 22 = 0$ or

$$N_{AB} = 1.16096 N_{BC} \qquad (1)$$

By force equilibrium in the y-direction

$N_{AB}\sin 37 + N_{BC}\sin 22 - W = 0$ or $(1.16096\sin 37 + \sin 22)N_{BC} = W$ or

$$N_{BC} = 0.9317W \qquad (2)$$

From Eq. (1)

$$N_{AB} = 1.0817W \qquad (3)$$

The area of cross-section is: $A_{AB} = A_{BC} = \frac{\pi}{4}(10)^2 = 78.54\text{mm}^2 = 78.54(10^{-6})\text{m}^2$

The average normal stress is:

$$\sigma_{AB} = \frac{N_{AB}}{A_{AB}} = \frac{1.0817W}{78.54(10^{-6})} = 0.01377W(10^{-6}) \le \frac{\sigma_{fail}}{K} \text{ or } 0.01377W(10^6) \le \frac{200(10^6)}{1.2}$$

$$W \le 12.103(10^3)\text{N} \qquad (4)$$

$$\sigma_{BC} = \frac{N_{BC}}{A_{BC}} = \frac{0.9317W}{78.54(10^{-6})} = 0.01186W(10^6) \le \frac{\sigma_{fail}}{K} \text{ or } 0.01186W(10^6) \le \frac{200(10^6)}{1.2} \text{ or}$$

$$W \le 14.049(10^3)\text{N} \qquad (5)$$

The maximum weight must satisfy the inequalities in Eq(4) and (5) i.e., $\boxed{W_{max} = 12.1\text{kN}}$

3.106 An adhesively bonded joint in wood is fabricated as shown. For a factor of safety of 1.25, determine the minimum overlap length L and dimension h to the nearest 1/8th inch. The shear strength of adhesive is 400 psi and the wood strength is 6 ksi in tension.

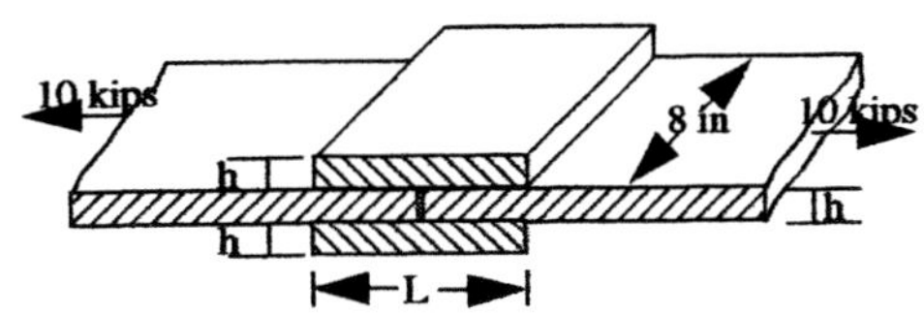

Fig. P3.106

Solution $K = 1.25$ $\tau_{fail} = 400\text{psi}$ $\sigma_{fail} = 6\text{ksi}$

$L_{min} = ?$ nearest 1/8 inch, $h_{min} = ?$ nearest 1/8 inch

---

The following free body diagrams can be drawn.

(a)

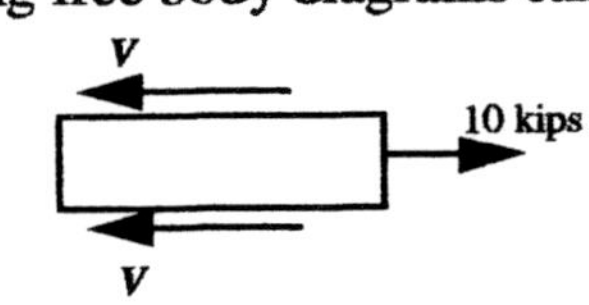

(b)

From Fig.(a): $2V = 10$ or $V = 5\text{kips}$ From Fig.(b) $N = 10\text{kips}$

The average normal stress is: $\sigma = \frac{N}{8h} = \frac{10}{8h} \le \frac{\sigma_{fail}}{K}$ or $\frac{10}{8h} \le \frac{6}{1.25}$ or $h \ge 0.2604$ or

$$\boxed{h_{min} = \frac{3}{8}\text{inch}}$$

The average shear stress is: $\tau = \frac{V}{8(L/2)} = \frac{5}{4L} \le \frac{\tau_{fail}}{K}$ or $\frac{5}{4L} \le \frac{0.4}{1.25}$ or $h \ge 3.906$

$$\boxed{L_{min} = 4\text{inches}}$$

3.107 A joint in a truss has the configuration shown. Determine the minimum diameter of the pin to the nearest millimeter for a factor of safety of 2.0. The pin's failure stress in shear is 300 MPa.

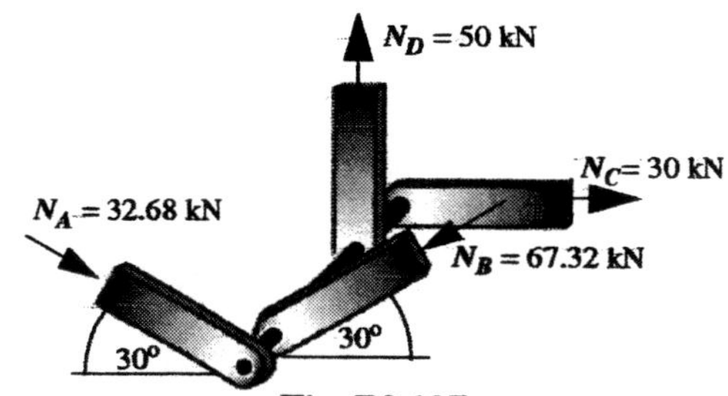

Fig. P3.107

Solution K=2 τ_{fail}=300MPa d_{min}=? nearest mm

By making imaginary cuts through the pin between the truss members, the following free body diagrams can be obtained.

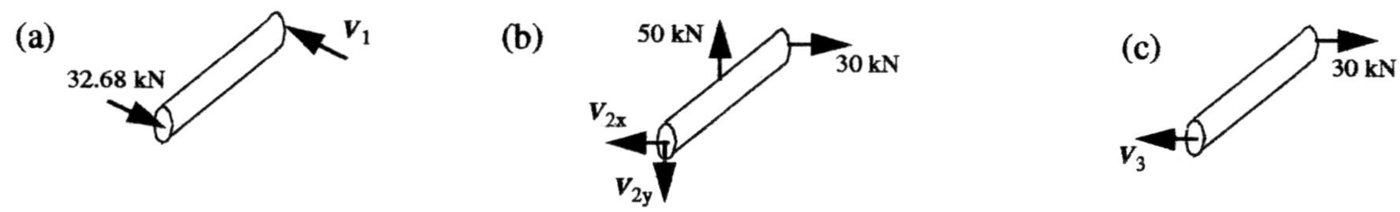

From Fig.(a) $V_1 = 32.68\text{kN}$

From Fig.(b) $V_{2x} = 30\text{kN}$ $V_{2y} = 50\text{kN}$ $V_2 = \sqrt{30^2 + 50^2} = 583\text{kN}$

From Fig.(c) $V_3 = 30\text{kN}$

The maximum shear force is V_2. Thus, the maximum shear stress is: $\tau_{max} = \frac{V_2}{A} = \frac{5831(10^3)}{(\pi d^2/4)} \le \frac{\tau_{fail}}{K}$ or

$\frac{(4)(58.31)(10^3)}{\pi d^2} \le \frac{300(10^6)}{2}$ or $d^2 \ge 0.4949(10^{-3})$ or $d \ge 0.02225(10^{-3})\text{m}$ or $\boxed{d_{min} = 23\text{mm}}$

3.108 The shear stress on the cross-section of the wire of a helical spring is given by $\tau = K(8PC/(\pi d^2))$ where, P is the force on the spring, d is the diameter of the wire from which the spring is constructed, C is called the spring index given by the ratio C = D / d, D is the diameter of coiled spring, and K is called the Wahl factor given by $K = \frac{4C-1}{4C-4} + \frac{0.615}{C}$. The spring is to be designed to resist a maximum force of 1200 N and must have a factor of safety of 1.1 in yield. The shear stress in yield is 350 MPa. Make a table for admissible values of D for values of d that range $8\text{mm} \le d \le 16\text{ mm}$ in steps of 2 mm.

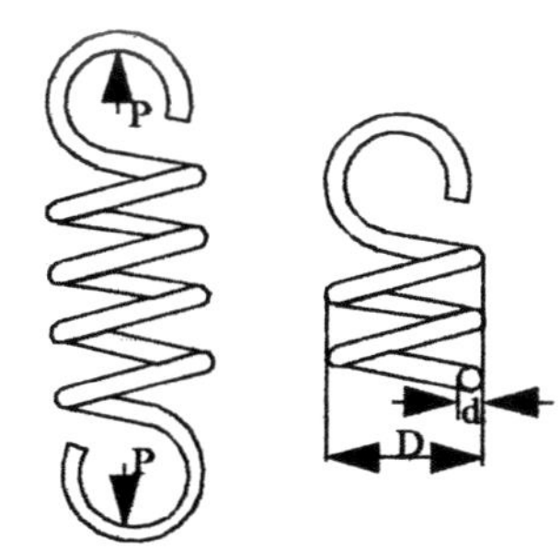

Fig. P3.108

Solution P=1200N K=1.1 τ_{yield}=350MPa D=? for ($8\text{mm} \le d \le 16\text{mm}$)

The allowable maximum stress is: $\tau_{max} = \frac{\tau_{yield}}{K} = \frac{350}{1.1} = 318.18\text{MPa}$

From the given expression of shear stress: $\tau = \frac{(KC)(8)(1200)}{\pi d^2} = 318.18(10^6)$

or $KC = 0.1041d^2(10^6)$, where d is in meters or $KC = 0.1041d^2$, where d is in millimeters

let $a = 0.1041d^2$

1

Substituting for K: $\left(\frac{4C-1}{4C-4}\right)C + 0.615 = a$ or $C^2 + (0.365 - a)C - (0.615 - a) = 0$ or

$$C = \frac{(a-0.365) + \sqrt{(a-0.365)^2 + 4(0.615-a)}}{2}$$

Let C_1 and C_2 be the two roots. Then $D_1 = C_1 d$ and $D_2 = C_2 d$. The values of D_1 & D_2 are tabulated below.

d (mm)	a	C_1	C_2	D_1 (mm)	D_2 (mm)
8	6.662	5.115	1.182	40.9	9.5
10	10.410	8.951	1.094	89.5	10.9
12	14.990	13.566	1.060	162.8	12.7
14	20.404	18.997	1.042	266.0	14.6
16	26.650	25.254	1.031	404.1	16.5

3.109 A steel bar is axially loaded as shown. Determine the factor of safety for the bar if yielding is to be avoided. The normal yield stress for steel is 30 ksi. Use stress concentration factor chart in Appendix C.

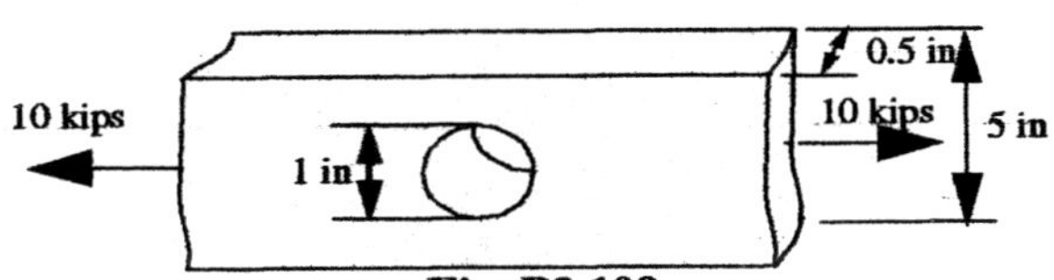

Fig. P3.109

Solution $\sigma_{yield} = 30 ksi$ $K = ?$

The nominal axial stress is: $\sigma_{nom} = \frac{10}{(5)(0.5)} = 4 ksi$

From Figure C.1, for the nominal stress and $\frac{d}{h} = \frac{1}{5} = 0.2$ the approximate value of stress concentration factor is $K_{gross} = \frac{(3+3.25)}{2} = 3.125$

$\sigma_{max} = K_{gross}\sigma_{nom} = (3.125)(4) = 12.5$

The factor of safety can be found as shown below

$K = \frac{\sigma_{yield}}{\sigma_{max}} = \frac{30}{12.5} = 2.4$ $\boxed{K = 2.4}$

3.110 The stress concentration factor for a stepped flat tension bar with shoulder fillets was determined as:

$$K_{conc} = 1.970 - 0.384\left(\frac{2r}{H}\right) - 1.018\left(\frac{2r}{H}\right)^2 + 0.430\left(\frac{2r}{H}\right)^3$$

The above equation is valid only if $\left(\frac{H}{d}\right) > \left(1 + 2\frac{r}{d}\right)$ and $\left(\frac{L}{H}\right) > \left(5.784 - 1.89\frac{r}{d}\right)$.

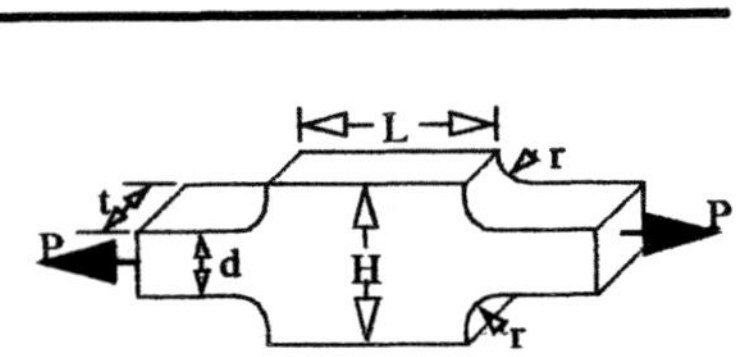

Fig. P3.110

The nominal stress is P/(dt). Make a chart for the stress concentration factor vs. (H/d) for the following values of (r/d): 0.2, 0.4, 0.6, 0.8, 1.0. Use of spread sheet is recommended.

Solution

The equation for K_{conc} can be re-written as:

$$K_{conc} = 1.970 - 0.384\left(\frac{2(r/d)}{(H/d)}\right) - 1.018\left(\frac{2(r/d)}{(H/d)}\right)^2 + 0.430\left(\frac{2(r/d)}{(H/d)}\right)^3 \quad 1$$

Using r/d as a parameter and Eq.1, the values of K_{conc} can be found for various values of H/d on a spread sheet. Those values for which the inequality $\left(\frac{H}{d}\right) > \left(1 + 2\frac{r}{d}\right)$ does not hold can be eliminated as shown in the table below.

H/d	r/d=0.2	r/d=0.4	r/d=0.6	r/d=0.8	r/d=1.0
1.4	1.787				
1.6	1.817				
1.8	1.839	1.636			
2.0	1.856	1.681			
2.2	1.869	1.716	1.527		
2.4	1.880	1.745	1.577		
2.6	1.888	1.768	1.618	1.448	
2.8	1.896	1.787	1.652	1.498	
3.0	1.902	1.803	1.681	1.541	1.389
3.2	1.907	1.817	1.706	1.577	1.437
3.4	1.911	1.829	1.727	1.609	1.479
3.6	1.915	1.839	1.745	1.636	1.516
3.8	1.919	1.848	1.761	1.660	1.549
4.0	1.922	1.856	1.775	1.681	1.577
4.2	1.925	1.863	1.787	1.700	1.603
4.4	1.927	1.869	1.798	1.716	1.626
4.6	1.929	1.875	1.808	1.731	1.646

H/d	r/d=0.2	r/d=0.4	r/d=0.6	r/d=0.8	r/d=1.0
4.8	1.931	1.880	1.817	1.745	1.664
5.0	1.933	1.884	1.825	1.757	1.681
5.2	1.935	1.888	1.832	1.768	1.696
5.4	1.936	1.892	1.839	1.778	1.710
5.6	1.938	1.896	1.845	1.787	1.723
5.8	1.939	1.899	1.851	1.796	1.734
6.0	1.940	1.902	1.856	1.803	1.745
6.2	1.941	1.904	1.861	1.810	1.755
6.4	1.942	1.907	1.865	1.817	1.764
6.6	1.943	1.909	1.869	1.823	1.772
6.8	1.944	1.911	1.873	1.829	1.780
7.0	1.945	1.913	1.876	1.834	1.787
7.2	1.946	1.915	1.880	1.839	1.794
7.4	1.946	1.917	1.883	1.844	1.800
7.6	1.947	1.919	1.886	1.848	1.806
7.8	1.948	1.920	1.888	1.852	1.812
8.0	1.948	1.922	1.891	1.856	1.817

The values in the table can be plotted as shown in the graph below.

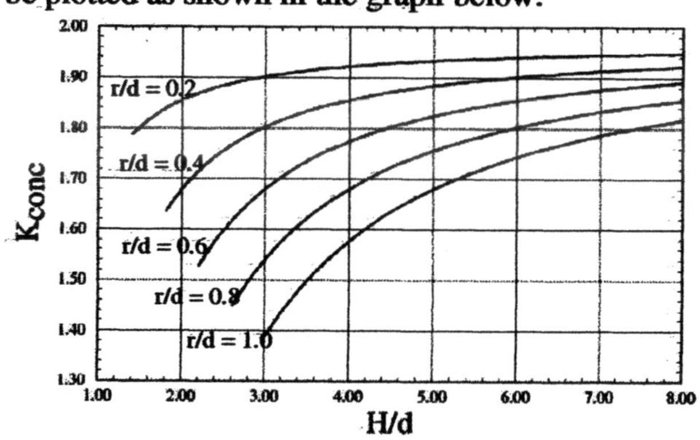

3.111 Determine the maximum normal stress in the stepped flat tension bar shown in Fig. P3.110 for the following data: P = 9 kips, H= 8 inches, d = 3 inches, t = 0.125 inches, and r = 0.625 inches.

Solution $\sigma_{max} = ?$

The nominal stress is: $\sigma_{nom} = \frac{P}{dt} = \frac{9}{(3)(0.125)} = 24\text{ksi}$ and $\frac{r}{H} = \frac{0.625}{8} = 0.07813$

Substituting (r/H) in the expression for K_{conc} given in problem 3.110, we obtain

$$K_{conc} = 1.97 - 0.384[(2)(0.07813)] - 1.018[(2)(0.07813)]^2 + 0.430[(2)(0.07813)]^3 = 1.8868$$

$\sigma_{max} = K_{conc}\sigma_{nom} = (1.8868)(24) = 45.283$ or $\boxed{\sigma_{max} = 45.3\text{ksi}}$

3.112 An aluminum stepped tension bar is to carry a load P = 56 kN. The normal yield stress of aluminum is 160 MPa. The bar has H = 300 mm, d = 100 mm, t = 10 mm in Fig. P3.110. For a factor of safety of 1.6, determine the minimum value r of the fillet radius if yielding is to be avoided.

Solution P=56kN σ_{yield}=160MPa H=300mm d=100mm
t=10mm K=1.6 r=?

The allowable maximum stress is: $\sigma_{max} = \frac{\sigma_{yield}}{12} = \frac{160}{1.6} = 100\text{MPa}$

The nominal stress is: $\sigma_{nom} = \frac{P}{dt} = \frac{56(10^3)}{(0.1)(0.01)} = 56(10^6)\frac{N}{m^2} = 56\text{MPa}$

The stress concentration factor is: $K_{cone} = \frac{\sigma_{max}}{\sigma_{nom}} = \frac{100}{56} = 1.785$ and $\frac{H}{d} = 3$

From the graphs in problem 3.110, for H/d=3 and K_{conc}=1.785, the approximate value of r/d=0.4

Thus $r = 0.4d = (0.4)(100) = 40\text{mm}$ or $\boxed{r = 40\text{mm}}$

3.113 The stress concentration factor for a flat tension bar with U-shaped notches was determined as:

$$K_{conc} = 3.857 - 5.066\left(\frac{4r}{H}\right) + 2.469\left(\frac{4r}{H}\right)^2 - 0.258\left(\frac{4r}{H}\right)^3$$

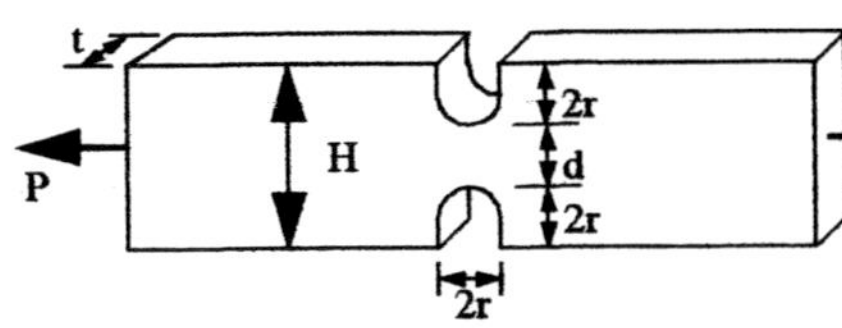

Fig. P3.113

The nominal stress is P/(Ht). Make a chart for the stress concentration factor vs. (r/d) for the following values of (H/d): 1.25, 1.50, 1.75, 2.0. Use of spread sheet is recommended.

Solution

The equation for K_{conc} can be re-written as:

$$K_{conc} = 3.857 - 5.066\left(\frac{4(r/d)}{(H/d)}\right) + 2.469\left(\frac{4(r/d)}{(H/d)}\right)^2 - 0.258\left(\frac{4(r/d)}{(H/d)}\right)^3 \quad 1$$

Using H/d as a parameter and Eq.1, the values of K_{conc} can be found for various values of r/d on a spread sheet as shown in the table below.

r/d	H/d=1.25	H/d=1.50	H/d=1.75	H/d=2.00
0.05	3.109	3.225	3.310	3.375
0.06	2.974	3.109	3.208	3.284
0.07	2.843	2.996	3.109	3.195
0.08	2.718	2.886	3.012	3.109
0.09	2.597	2.780	2.917	3.024
0.10	2.480	2.677	2.825	2.940
0.11	2.368	2.577	2.735	2.859
0.12	2.261	2.480	2.648	2.780
0.13	2.158	2.387	2.563	2.702
0.14	2.060	2.296	2.480	2.626
0.15	1.966	2.209	2.400	2.552
0.16	1.876	2.125	2.322	2.480
0.17	1.790	2.044	2.246	2.410

r/d	H/d=1.25	H/d=1.50	H/d=1.75	H/d=2.00
0.18	1.709	1.966	2.173	2.341
0.19	1.632	1.890	2.101	2.274
0.20	1.558	1.818	2.032	2.209
0.21	1.489	1.749	1.966	2.146
0.22	1.424	1.683	1.901	2.084
0.23	1.363	1.619	1.839	2.024
0.24	1.306	1.558	1.778	1.966
0.25	1.252	1.501	1.720	1.909
0.26	1.203	1.445	1.664	1.854
0.27	1.157	1.393	1.610	1.801
0.28	1.114	1.343	1.558	1.749
0.29	1.076	1.297	1.509	1.699
0.30	1.041	1.252	1.461	1.651

The values in the table can be plotted as shown in the graph below.

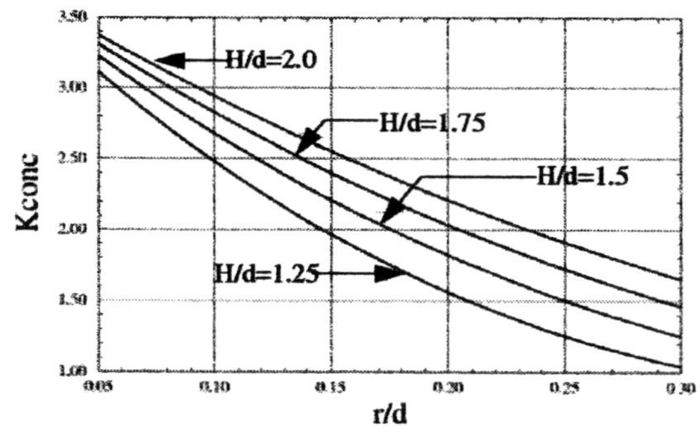

3.114 Determine the maximum normal stress in the flat tension bar shown in Fig. P3.113 for the following data: P =150 kN , H = 300 mm, r = 15 mm, t = 5 mm.

Solution σ_{max}=?

The nominal stress is: $\sigma_{nom} = \frac{P}{Ht} = \frac{150(10^3)}{(0.3)(0.005)} = 100(10^6)\frac{N}{m^2} = 100MPa$

$\frac{r}{H} = \frac{15}{300} = 0.05$ or $4\frac{r}{H} = 0.2$

Substituting(4r/H) in the expression for K_{conc} given in problem 3.113, we obtain

$K_{conc} = 3.857 - 5.066(0.2) + 2.469(0.2)^2 - 0.258(0.2)^3 = 2.9405$

$\sigma_{max} = K_{conc}\sigma_{nom} = (2.9405)(100) = 294.05$ or $\boxed{\sigma_{max} = 294MPa}$

3.115 A steel tension bar with U-shaped notches of the type shown in Fig. P3.113, is to carry a load P = 18 kips. The yield stress of steel is 30 ksi. The bar has H = 9 inches mm, t =0.25 inches. For a factor of safety of 1.4, determine the value of r if yielding is to be avoided.

Solution P=18kips σ_{yield}=30ksi H=9in t=0.25in d=6in, K=1.4 r=?

$\sigma_{max} = \frac{\sigma_{yield}}{K} = \frac{30}{1.4} = 21.43ksi$

$\sigma_{nom} = \frac{P}{Ht} = \frac{18}{(9)(0.25)} = 8ksi$

$K_{conc} = \frac{\sigma_{max}}{\sigma_{nom}} = \frac{21.43}{8} = 2.68$

$\frac{H}{d} = \frac{9}{6} = 1.5$

From the graph in problem 3.113, for H/d=1.5 and K_{conc}=2.68, the approximate value of r/d=0.1

$\therefore r = 0.1d = 0.6in$ or $\boxed{r = 0.6inch}$

3.116 An iron rim (α = 6.5 μ/°F) of diameter 35.98 inches is to be placed on a wooden cask of diameter 36 inches. Determine the minimum temperature increase needed to slip the rim onto the cask.

Solution α=6.5 μ/°F d_i=35.98 in d_f=36in ΔT=?

All stresses are zero. Thus, the total radial normal strain = thermal strain

$$\varepsilon = \frac{(d_f/2)-(d_i/2)}{(d_i/2)} = \alpha\Delta T \text{ or } \frac{36-35.98}{35.98} = (6.5)(10^{-6})\Delta T \text{ or } \Delta T = 85.518°F \text{ or } \boxed{\Delta T = 85.52°F}$$

3.117 The temperature in steel (E_s = 200 GPa, and α_s = 12.0 μ/°C) aluminum (E = 72 GPa, and α = 23.0 μ/°C is increased by 60°C. Determine the angle by which the pointer rotates from the vertical position.

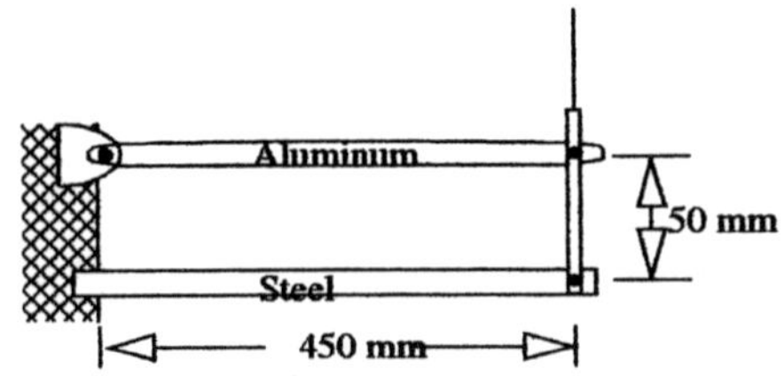

Fig. P3.117

Solution E_s=200GPa α_s=12.0 μ/°C E_{al}=72GPa α_{al}=23.0 μ/°C ΔT=60 θ=?

The thermal strain and the corresponding deformation in each rod can be found as shown below

$$\varepsilon_{al} = \alpha_{al}\Delta T = (23)(60)(10^{-6}) = 1.38(10^{-3}) \text{ and } \delta_{al} = \varepsilon_{al}L_{al} = (1.38)(10^{-3})(450) = 0.621\text{mm}$$

$$\varepsilon_s = \alpha_s\Delta T = (12)(60)(10^{-6}) = 0.72(10^{-3}) \text{ and } \delta_s = \varepsilon_s L_s = (0.72)(10^{-3})(450) = 0.324\text{mm}$$

The deformed geometry of the assembly can be drawn as shown below.

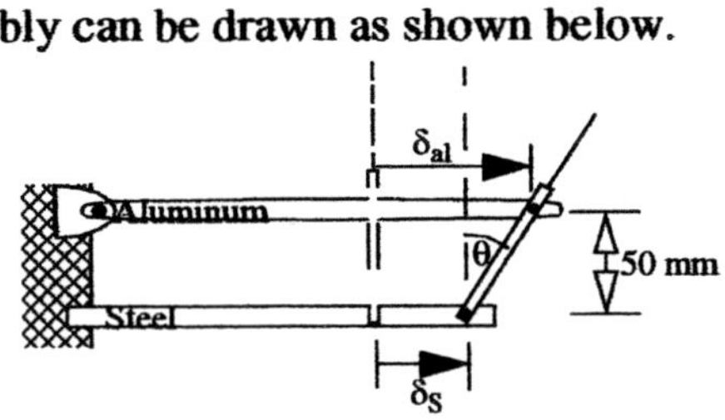

$$\therefore \tan\theta = \frac{\delta_{al}-\delta_s}{50} = \frac{0.621-0.324}{50} \text{ or } \boxed{\theta = 0.34°}$$

3.118 Solve Problem 3.64 if the temperature decrease is 25° C. Use α = 11.7 μ/°C

Solution ΔT=-25°C α=11.7 μ/°C
{ε_{xx}=?, ε_{yy}=?, γ_{xy}=?, ε_{zz}=?, σ_{zz}=?} for plane stress and plane strain

---

(a) Plane stress $\boxed{\sigma_{zz} = 0}$

Substituting $\sigma_{xx} = 100(10^6)\frac{N}{m^2}$, $\sigma_{yy} = 150(10^6)\frac{N}{m^2}$, $\sigma_{zz} = 0$, $E = 200(10^9)\frac{N}{m^2}$ and $\nu = 0.32$ into Eq's(3-27a), (3-27b), and(3-27c) we obtain the following:

$$\varepsilon_{xx} = \frac{[100-0.32(150)](10^6)}{200(10^9)} + (11.7)(-25)(10^{-6}) = -32.5(10^{-6}) \text{ or } \boxed{\varepsilon_{xx} = -32.5\mu}$$

$$\varepsilon_{yy} = \frac{[150-0.32(100)](10^6)}{200(10^9)} + 11.7(-25)(10^{-6}) = 297.5(10^{-6}) \text{ or } \boxed{\varepsilon_{yy} = 297.5\mu}$$

$$\varepsilon_{zz} = \frac{-0.32(100+150)(10^6)}{200(10^9)} + 11.7(-25)(10^{-6}) = -692.5(10^{-6}) \text{ or } \boxed{\varepsilon_{zz} = -692.5\mu}$$

Shear strain is same as in problem 3.63 $\boxed{\gamma_{xy} = -1650\mu}$

(b) Plane strain $\boxed{\varepsilon_{zz} = 0}$

From Eq(3-27c) $0 = \dfrac{\sigma_{zz} - 0.32(100 + 150)(10^6)}{200(10^9)} + 11.7(-25)(10^{-6})$ or $\sigma_{zz} = 138.5(10^6)\dfrac{N}{m^2}$ or

$\boxed{\sigma_{zz} = 138.5\text{MPa(T)}}$

From Eq's(3-27a) and(3-27b)

$\varepsilon_{xx} = \dfrac{[100 - 0.32(150 + 138.5)](10^6)}{200(10^9)} + (11.7)(-25)(10^{-6}) = -254.1(10^{-6})$ or $\boxed{\varepsilon_{xx} = -254.1\mu}$

$\varepsilon_{yy} = \dfrac{[150 - 0.32(100 + 138.5)](10^6)}{200(10^9)} + (11.5)(-25)(10^{-6}) = 75.9(10^{-6})$ or $\boxed{\varepsilon_{yy} = 75.9\mu}$

Shear strain value as in part(a) $\boxed{\gamma_{xy} = -1650\mu}$

3.119 Solve Problem 3.65 if the temperature increase is 50° C. Use $\alpha = 23.6\ \mu/^\circ C$

Solution $\Delta T = 50^\circ$ C. $\alpha = 23.6\mu$

$\{\varepsilon_{xx}=?, \varepsilon_{yy}=?, \gamma_{xy}=?, \varepsilon_{zz}=?, \sigma_{zz}=?\}$ for plane stress and plane strain

---

(a) Plane stress $\boxed{\sigma_{zz} = 0}$

Substituting $\sigma_{xx} = -225(10^6)\dfrac{N}{m^2}$, $\sigma_{yy} = 125(10^6)\dfrac{N}{m^2}$, $\sigma_{zz} = 0$, $E = 70(10^9)\dfrac{N}{m^2}$ and $\nu = 0.25$ into Eq's(3-27a), (3-27b), and(3-27c) we obtain the following:

$\varepsilon_{xx} = \dfrac{[-225 - 0.25(125)](10^6)}{70(10^9)} + (23.6)(50)(10^{-6}) = -2481(10^{-6})$ or $\boxed{\varepsilon_{xx} = -2481\mu}$

$\varepsilon_{yy} = \dfrac{[125 - 0.25(225)](10^6)}{70(10^9)} + 23.6(50)(10^{-6}) = 3769(10^{-6})$ or $\boxed{\varepsilon_{yy} = 3769\mu}$

$\varepsilon_{zz} = \dfrac{-0.25(-225 + 125)(10^6)}{70(10^9)} + 23.6(50)(10^{-6}) = 1537(10^{-6})$ or $\boxed{\varepsilon_{zz} = 1537\mu}$

Shear strain is same as in problem 3.64 $\boxed{\gamma_{xy} = 5357\mu}$

(b) Plane strain $\boxed{\varepsilon_{zz} = 0}$

From Eq(3-27c): $0 = \dfrac{\sigma_{zz} - 0.25(-225 + 125)(10^6)}{70(10^9)} + 23.6(-50)(10^{-6})$ or $\sigma_{zz} = -107.6(10^6)\dfrac{N}{m^2}$ or

$\boxed{\sigma_{zz} = 107.6\text{MPa(C)}}$

From Eq's(3-27a) and(3-27b)

$\varepsilon_{xx} = \dfrac{[-225 - 0.25(125 + 107.6)](10^6)}{70(10^9)} + (23.6)(50)(10^{-6}) = -2096.4(10^{-6})$ or $\boxed{\varepsilon_{xx} = -2096\mu}$

$\varepsilon_{yy} = \dfrac{[125 - 0.25(-225 - 107.6)](10^6)}{70(10^9)} + (23.6)(50)(10^{-6}) = 4153.6(10^{-6})$ or $\boxed{\varepsilon_{yy} = 4154\mu}$

Shear strain value as in part(a) $\boxed{\gamma_{xy} = 5357\mu}$

3.120 Solve problem 3.72 if the temperature increase is 40° F-. Use $\alpha = 6.5\ \mu/^\circ F$

Solution $\Delta T = 40^\circ F$ $\alpha = 6.5\mu/^\circ F$ $\sigma_{xx}=?, \sigma_{yy}=?, \tau_{xy}=?, \varepsilon_{zz}=?$ for plane stress

---

For Plane stress: $\sigma_{zz} = 0$. Eqs. (3-27a) and (3-27b) can be rewritten with $\sigma_{zz}=0$

$\sigma_{xx} - \nu\sigma_{yy} = E(\varepsilon_{xx} - \alpha\Delta T) = (30,000)(-800 - 6.5(40))(10^{-6})$ or

$$\sigma_{xx} - \nu\sigma_{yy} = -31.8\text{ksi} \qquad 1$$

$\sigma_{yy} - \nu\sigma_{xx} = E(\varepsilon_{yy} - \alpha\Delta T) = (30,000)(-1000 - 6.5(40))(10^{-6})$ or

$$\sigma_{yy} - \nu\sigma_{xx} = -37.8\text{ksi} \qquad 2$$

Solving Eq(1) and (2) with ν=0.3, we obtain: $\sigma_{xx} = -47.4\text{ksi}$ and $\sigma_{yy} = -52.02\text{ksi}$ or

$\boxed{\sigma_{xx} = 47.4\text{ksi(C)}}$ and $\boxed{\sigma_{yy} = 52.02\text{ksi(C)}}$

From Eq(3-27c): $\varepsilon_{zz} = \dfrac{0 - 0.3(-47.4 - 52.02)}{30,000} + (40)(6.5)(10^{-6}) = 1254.2$ or $\boxed{\varepsilon_{zz} = 1254\mu}$

Shear stress is same as in problem 3.72. $\boxed{\tau_{xy} = -5.77\text{ksi}}$

3.121 Solve problem 3.73 if the temperature decrease is 100° F. Use α = 12.8 μ/°F

Solution ΔT=-100 α=12.8μ/°F σ_{xx}=?, σ_{yy}=?, τ_{xy}=?, ε_{zz}=? for plane stress

---

For plane stress $\sigma_{zz} = 0$ Eqs. (3-27a) and (3-27b) can be rewritten with σ_{zz}=0

$\sigma_{xx} - \nu\sigma_{yy} = E(\varepsilon_{xx} - \alpha\Delta T) = (10,000)(1500 - 12.8(-100))(10^{-6})$ or

$$\sigma_{xx} - \nu\sigma_{yy} = 27.8\text{ksi} \qquad 1$$

$\sigma_{yy} - \nu\sigma_{xx} = E(\varepsilon_{yy} - \alpha\Delta T) = 10,000(-1200 - 12.8(-100))(10^{-6})$ or

$$\sigma_{yy} - \nu\sigma_{xx} = 0.8\text{ksi} \qquad 2$$

Solving Eq's(1) and (2) with ν=0.282 we obtain: $\boxed{\sigma_{xx} = 30.45\text{ks(T)}}$ and $\boxed{\sigma_{yy} = 9.39\text{ksi(T)}}$

From Eq(3-27c): $\varepsilon_{zz} = \dfrac{0 - 0.282(30.45 + 9.39)}{(10,000)} + 12.8(-100)(10^{-6}) = -2403(10^{-6})$ or $\boxed{\varepsilon_{zz} = -2403\mu}$

Shear stress is same as in problem 3.73 $\boxed{\tau_{xy} = -3.9\text{ksi}}$

3.122 Solve problem 3.74 if the temperature decrease is 75 ° C. Use α = 26.0 μ/°C

Solution ΔT=-75°F α=26 μ/°C σ_{xx}=? σ_{yy}=? τ_{xy}=? ε_{zz}=? for plane stress

---

For plane stress $\sigma_{zz} = 0$

Eq's(3-27a) and (3-27b) can be rewritten with σ_{zz}=0

$\sigma_{xx} - \nu\sigma_{yy} = E(\varepsilon_{xx} - \alpha\Delta T) = (36)(10^9)(-2000 - 26(-75))(10^{-6})$ or $\sigma_{xx} - \nu\sigma_{yy} = -1.8\text{MPa}$ 1

$\sigma_{yy} - \nu\sigma_{xx} = E(\varepsilon_{yy} - \alpha\Delta T) = 36(10^9)(-2000 - 26(-75))(10^{-6})$ or $\sigma_{yy} - \nu\sigma_{xx} = 142.2\text{MPa}$ 2

Solving Eqs. (1) and (2) with ν=0.2 we obtain: $\sigma_{xx} = 27.75\text{MPa}$ and $\sigma_{yy} = 147.75\text{MPa}$

$\boxed{\sigma_{xx} = 27.8\text{MPa(T)}}$ and $\boxed{\sigma_{yy} = 147.8\text{MPa(T)}}$

From Eq(3-27c) $\varepsilon_{zz} = \dfrac{0 - 0.2(27.75 + 147.75)(10^6)}{(36)(10^9)} + 26(-75)(10^{-6}) = -2925(10^{-6})$ or $\boxed{\varepsilon_{zz} = -2925\mu}$

Shear stress is same as in problem 3.74 $\boxed{\tau_{xy} = 18\text{MPa}}$

3.123 A plate (E=30,000ksi; $\nu = 0.25$; $\alpha = 6.5\times10^{-6}/^{o}F$) cannot expand in the y direction and can expand at most by 0.005 inch in the x direction as shown. Assuming plane stress determine the average normal stresses in the x and y direction due to a uniform temperature increase of 100°F.

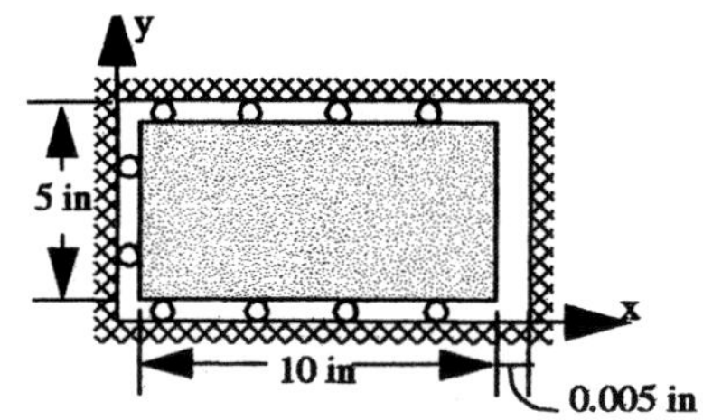

Fig. P3.123

Solution E=30,000ksi ν=0.25 $\alpha=6.5\ \mu/^{o}F$ $\varepsilon_{yy}=0$,

$\delta_x=0.005$ $\Delta T=100^{o}F$ $\sigma_{xx}=?$ $\sigma_{yy}=?$

The normal strain in the x-direction is: $\varepsilon_{xx} = \frac{\delta_x}{10} = \frac{0.005}{10} = 0.005 = 500(10^{-6})$

Assuming plane stress $\sigma_{zz} = 0$, Eq's(3-27a) and (3-27b) can be rewritten as

$$\sigma_{xx} - \nu\sigma_{yy} = E(\varepsilon_{xx} - \alpha\Delta T) = (30,000)[500 - (6.5)(100)](10^{-6}) \quad \text{or} \quad \sigma_{xx} - 0.25\sigma_{yy} = -4.5\text{ksi} \qquad \mathbf{1}$$

$$\sigma_{yy} - \nu\sigma_{xx} = E(\varepsilon_{yy} - \alpha\Delta T) = 30,000[0 - (6.5)(100)](10^{-6}) \quad \text{or} \quad \sigma_{yy} - \nu\sigma_{xx} = -19.5\text{ksi} \qquad \mathbf{2}$$

Solving Eqs. (1) and (2) we obtain: $\sigma_{xx} = -10\text{ksi}$ and $\sigma_{yy} = -22\text{ksi}$ or

$$\boxed{\sigma_{xx} = 10\text{ksi(C)}} \text{ and } \boxed{\sigma_{yy} = 22\text{ksi(C)}}$$

3.124 Derive the following relation of normal stresses in terms of normal strain from Equations 3.27a, 3.27b, and 3.27c.

$$\sigma_{xx} = [(1-\nu)\varepsilon_{xx} + \nu\varepsilon_{yy} + \nu\varepsilon_{zz}]\frac{E}{(1-2\nu)(1+\nu)} - \frac{E\alpha\Delta T}{(1-2\nu)}$$

$$\sigma_{yy} = [(1-\nu)\varepsilon_{yy} + \nu\varepsilon_{zz} + \nu\varepsilon_{xx}]\frac{E}{(1-2\nu)(1+\nu)} - \frac{E\alpha\Delta T}{(1-2\nu)} \qquad \mathbf{3.31}$$

$$\sigma_{zz} = [(1-\nu)\varepsilon_{zz} + \nu\varepsilon_{xx} + \nu\varepsilon_{yy}]\frac{E}{(1-2\nu)(1+\nu)} - \frac{E\alpha\Delta T}{(1-2\nu)}$$

Solution

Adding Equations (3-27a),(3-27b) and (3-27c) we obtain:

$$\varepsilon_{xx} + \varepsilon_{yy} + \varepsilon_{zz} = \frac{1-2\nu}{E}(\sigma_{xx} + \sigma_{yy} + \sigma_{zz}) + 3\alpha\Delta T$$

$$\therefore \sigma_{xx} + \sigma_{yy} + \sigma_{zz} = \frac{E}{1-2\nu}(\varepsilon_{xx} + \varepsilon_{yy} + \varepsilon_{zz}) - \frac{3(E\alpha\Delta T)}{1-2\nu} \qquad \mathbf{1}$$

Eq (3-27a) can be written as $\varepsilon_{xx} = (1+\nu)\frac{\sigma_{xx}}{E} - \frac{\nu}{E}(\sigma_{xx} + \sigma_{yy} + \sigma_{zz}) + \alpha\Delta T$. Substituting Eq(1) we obtain:

$$\varepsilon_{xx} = \frac{1+\nu}{E}\sigma_{xx} - \frac{\nu}{1-2\nu}(\varepsilon_{xx} + \varepsilon_{yy} + \varepsilon_{zz}) + \frac{3\nu}{1-2\nu}\alpha\Delta T + \alpha\Delta T \text{ or}$$

$$\frac{1+\nu}{E}\sigma_{xx} = \left[\varepsilon_{xx}\left(1 + \frac{\nu}{1-2\nu}\right) + \frac{\nu}{1-2\nu}\varepsilon_{yy} + \frac{\nu}{1-2\nu}\varepsilon_{zz}\right] - \alpha\Delta T\left(\frac{3\nu + 1 - 2\nu}{1-2\nu}\right) \text{ or}$$

$$\sigma_{xx}\frac{E}{(1+\nu)(1-2\nu)}[(1-\nu)\varepsilon_{xx} + \nu\varepsilon_{yy} + \nu\varepsilon_{zz}] - \frac{E\alpha\Delta T}{1-2\nu} \qquad \mathbf{2}$$

Eq(3-27b) can be written as: $\varepsilon_{yy} = \frac{1+\nu}{E}\sigma_{yy} - \frac{\nu}{E}(\sigma_{xx} + \sigma_{yy} + \sigma_{zz}) + \alpha\Delta T$. Substituting Eq.(1) and solving for σ_{yy} we obtain:

$$\sigma_{yy} = \frac{E}{(1+\nu)(1-2\nu)}[\nu\varepsilon_{xx} + (1-\nu)\varepsilon_{yy} + \nu\varepsilon_{zz}] - \frac{E\alpha\Delta T}{1-2\nu} \qquad 3$$

Eq(3-27c) can be written as $\varepsilon_{zz} = \frac{1+\nu}{E}\sigma_{zz} - \frac{\nu}{E}(\sigma_{xx} + \sigma_{yy} + \sigma_{zz}) + \alpha\Delta T$. Substituting Eq.(1) and solving for σ_{zz} we obtain

$$\sigma_{zz} = \frac{E}{(1+\nu)(1-2\nu)}[\nu\varepsilon_{xx} + \nu\varepsilon_{yy} + (1-\nu)\varepsilon_{zz}] - \frac{E\alpha\Delta T}{1-2\nu} \qquad 4$$

Eqs. (2),(3) and (4) are the same as Eq.(3-31)

3.125 For a point in plane stress show

$$\sigma_{xx} = [\varepsilon_{xx} + \nu\varepsilon_{yy}]\frac{E}{(1-\nu^2)} - \frac{E\alpha\Delta T}{(1-\nu)} \qquad \sigma_{yy} = [\varepsilon_{yy} + \nu\varepsilon_{xx}]\frac{E}{(1-\nu^2)} - \frac{E\alpha\Delta T}{(1-\nu)} \qquad 3.32$$

Solution

For plane stress(σ_{zz}=0) Eqs. (3-27a) and (3-27b) can be written as

$$\sigma_{xx} - \nu\sigma_{yy} = E(\varepsilon_{xx} - \alpha\Delta T) \qquad 1$$

$$\sigma_{yy} - \nu\sigma_{xx} = E(\varepsilon_{yy} - \alpha\Delta T) \qquad 2$$

Multiplying Eq.(2) by ν and adding it to Eq.(1) we obtain: $\sigma_{xx}(1-\nu^2) = E[\varepsilon_{xx} + \nu\varepsilon_{yy} - \alpha\Delta T(1+\nu)]$ or

$$\sigma_{xx} = \frac{E}{1-\nu^2}(\varepsilon_{xx} + \nu\varepsilon_{yy}) - \frac{E\alpha\Delta T}{1-\nu} \qquad 3$$

Multiplying Eq.(1) by ν and adding to Eq.(2) we obtain: $\sigma_{yy}(1-\nu^2) = E[\nu\varepsilon_{xx} + \varepsilon_{yy} - \alpha\Delta T(1+\nu)]$ or

$$\sigma_{yy} = \frac{E}{1-\nu^2}(\varepsilon_{yy} + \nu\varepsilon_{xx}) - \frac{E\alpha\Delta T}{1-\nu} \qquad 4$$

Eqs.(3) and (4) are same as Eq(3-32)

3.126 For a point in plane stress show

$$\varepsilon_{zz} = -\left(\frac{\nu}{1-\nu}\right)(\varepsilon_{xx} + \varepsilon_{yy}) + \frac{(1+\nu)}{(1-\nu)}\alpha\Delta T \qquad 3.33$$

Solution

Adding Eqs.(3) and(4) an problem 3.124, we obtain

$$\sigma_{xx} + \sigma_{yy} = \frac{E}{1-\nu^2}[(1+\nu)\varepsilon_{xx} + (1+\nu)\varepsilon_{yy}] - \frac{2(E\alpha\Delta T)}{1-\nu} \text{ or}$$

$$\sigma_{xx} + \sigma_{yy} = \frac{E}{1-\nu}(\varepsilon_{xx} + \varepsilon_{yy}) - \frac{2(E\alpha\Delta T)}{1-\nu} \qquad 5$$

Eq.(3-27c) for plane stress can be written as: $\varepsilon_{zz} = -\frac{\nu}{E}(\sigma_{xx} + \sigma_{yy}) + \alpha\Delta T$. Substituting Eq.(1)

$$\varepsilon_{zz} = -\frac{\nu}{1-\nu}(\varepsilon_{xx} + \varepsilon_{yy}) + \left(\frac{2\nu}{1-\nu} + 1\right)\alpha\Delta T \text{ or}$$

$$\varepsilon_{zz} = -\frac{\nu}{1-\nu}(\varepsilon_{xx} + \varepsilon_{yy}) + \frac{(1+\nu)}{1-\nu}\alpha\Delta T \qquad 6$$

Eq.(2) is the same as Eq.(3-33)

3.127 A machine component is made from a steel alloy that has an S-N curve shown in Figure 3.34. Estimate the service life of the component if the peak stress is reversed at the rate shown.

(a) 40 ksi at 200 cycles per minute.
(b) 36 ksi at 250 cycles per minute.
(c) 32 ksi at 300 cycles per minute.

Solution T=? for the three cases

(a) From figure 3.34 for steel at 40ksi, the number of cycles to failure is n=400,000

$T = \frac{400,000}{200} = 2,000$ minutes or $\boxed{T = 33.33\text{hours}}$

(b) At 36ksi, the number of cycles to failure is n=2(10^6)

$T = \frac{2(10^6)}{250} = 8000$ minutes $\boxed{T = 133.33\text{hours}}$

(c) The stress value of 32 ksi is below the endurance limit and the component won't fail. $\boxed{T = \infty}$

3.128 A machine component is made from a aluminum alloy that has an S-N curve shown in Figure 3.34. What should be the maximum permissible peak stress in MPa for the following situations:

(a) 17 hours of service at 100 cycles per minute.
(b) 40 hours of service at 50 cycles per minute.
(c) 80 hours of service at 20 cycles per minute.

Solution σ_{max}= ? for the three cases

(a) T=17 hours = 1020 minutes and n = number of cycles=(100)(1020)=102,000
From figure 3-34 for aluminum, the peak stress value at 100,000 is 182 MPa and the peak stress value at 200,000 is 168 MPa. By linear interpolation we obtain a peak stress= $182 - \frac{2000}{100,000}(14) = 181.7$ or

$\boxed{\sigma_{max} = 182\text{MPa}}$

(b) T=40 hours =2400 minutes
n=(2400)(50)=120,000

By linear interpolation, the peak stress value= $182 - \frac{20,000}{100,000}(14) = 177.2$ or $\boxed{\sigma_{max} = 177\text{MPa}}$

(c)T=80 hours =4800 min and n=(4800)(20)=96,000 cycles
The peak stress value between 96,000 cycles &100,000 cycles changes slightly and is approximated at 183MPa. $\boxed{\sigma_{max} = 183\text{MPa}}$

3.129 A uni-axial stress acts on an aluminum plate with a hole as shown. The aluminum has an S-N curve shown in Figure 3.34. Predict the number of cycles the plate could be used if d = 3.2 inches and the far field stress σ = 6 ksi.

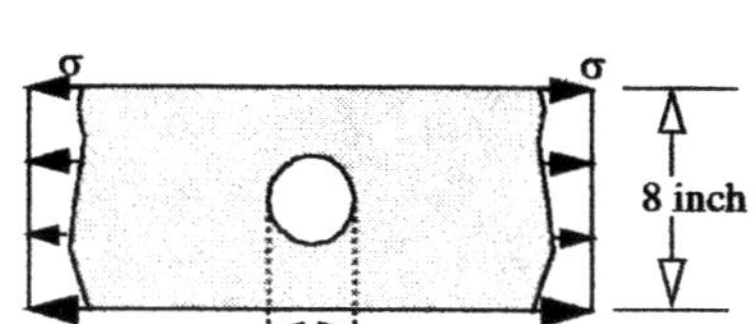

Fig. P3.129

Solution σ=6ksi d=3.2 H=8 n=?

From Figure C.1, for $\frac{d}{H} = \frac{3.2}{8} = 0.4$, the stress concentration factor K_{gross}=3.75

The maximum normal stress is: $\sigma_{max} = K_{gross}\sigma = (3.75)(6) = 22.5\text{ksi}$

From figure 3.34 for alumina, the number of cycles at 22.5ksi are estimated at 400,000

$\boxed{n = 400,000\text{cycles}}$

3.130 A uni-axial stress acts on an aluminum plate with a hole as shown. The aluminum has an S-N curve shown in Figure 3.34. Determine the maximum diameter of the hole to the nearest 1/8th of an inch, if the predicted service life of half-million cycle is desired for a uniform far field stress σ=6 ksi.

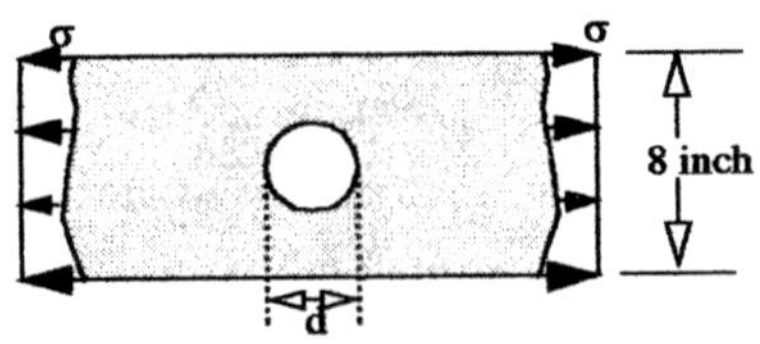

Fig. P3.130

Solution n=500,000 cycles σ= 6 ksi H=8 d_{max}= ? to the nearest 1/8 inch

---

From figure 3.34 for aluminum, the peak stress at 500,000 cycles is $\sigma_{max} = 22\text{ksi}$

The stress concentration factor is: $K_{gross} = \dfrac{\sigma_{max}}{\sigma} = \dfrac{22}{6} = 3.67$

From figure C.1, $\dfrac{d}{H} = 0.34$ for K_{gross}=3.5 and $\dfrac{d}{H} = 0.4$ for K_{gross}=3.75, by linear interpolation the value

of $\dfrac{d}{H} = 0.34 + \dfrac{0.4-0.34}{3.75-3.5}(3.67-3.5)$ or $\dfrac{d}{H} = 0.38$ or $d = (0.38)(8) = 3.04$ or $\boxed{d_{max} = 3.0\text{inches}}$

3.131 A uni-axial stress acts on an aluminum plate with a hole as shown. The aluminum has an S-N curve shown in Figure 3.34. Determine the maximum far field stress σ if the diameter of the hole is 2.4 inch and a predicted service life of three-quarters of a million is desired.

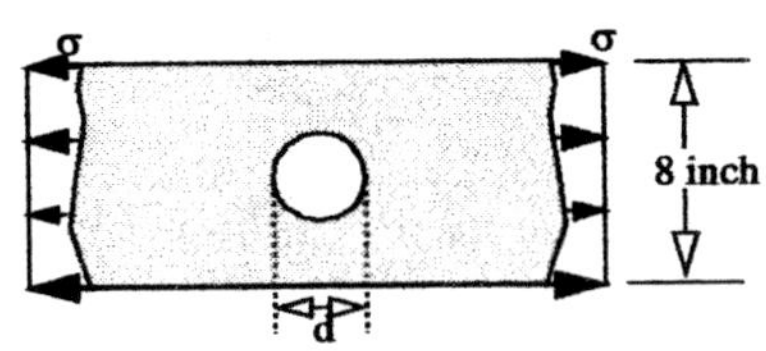

Fig. P3.131

Solution d=2.4 H=8 n=750,000 cycles σ = ?

From figure 3.34 for aluminum, the peak stress at 750,000 cycles is estimated as σ_{max}=21ksi

From Figure C.1, for $\dfrac{d}{H} = 0.3$ the stress concentration factor is estimated $K_{gross} = \left(\dfrac{3.25+3.5}{2}\right) = 3.375$

The nominal normal stress is: $\sigma = \dfrac{\sigma_{max}}{K_{gross}} = \dfrac{21}{3.375} = 6.222$ $\boxed{\sigma = 6.2\text{ksi}}$

3.132 Bronze has a yield stress of $\sigma_{yield} = 18$ ksi in tension, yield strain of $\varepsilon_{yield} = 0.0012$, ultimate stress of $\sigma_{ult} = 50$ ksi and the corresponding ultimate strain of $\varepsilon_{ult} = 0.50$. Determine the material constants and plot the resulting stress-strain curve for (a) elastic-perfectly plastic model (b) linear strain hardening model (c) and the non-linear power law model.

Solution $\sigma_{yield} = 18$ ksi $\varepsilon_{yield} = 0.0012$ $\sigma_{ult} = 50$ ksi $\varepsilon_{ult} = 0.50$

Constants in the three material models=?

We have coordinates of three points on the curve: P_o ($\sigma_0 = 0.00$, $\varepsilon_0 = 0.000$), P_1 ($\sigma_1 = 18.0$, $\varepsilon_1 = 0.0012$), and P_2 ($\sigma_2 = 50$, $\varepsilon_2 = 0.5$). Using this data we can find the various constants in the material models.

(a) The modulus of elasticity E is slope between points P_o and P_1 and can be found as shown below.

$$E_1 = \frac{\sigma_1 - \sigma_0}{\varepsilon_1 - \varepsilon_0} = \frac{18}{0.0012} = 15,000 \text{ ksi}$$

After yield stress, the stress is a constant. The stress strain behavior can be written as shown below.

$$\sigma = \begin{cases} 15,000\varepsilon \text{ ksi} & |\varepsilon| \le 0.0012 \\ 18 \text{ ksi} & 0.0012 \le |\varepsilon| \le 0.5 \end{cases} \quad (1)$$

(b) In the linear strain hardening model the slope of the straight line before yield stress is as calculated in part (a). After the yield stress, the slope of the line can be found from the coordinates of point P_1 and P_2.

$$E_2 = \frac{\sigma_2 - \sigma_1}{\varepsilon_2 - \varepsilon_1} = \frac{50 - 18}{0.5 - 0.0012} = 64.15 \text{ ksi}$$

The stress strain behavior can be written as shown below.

$$\sigma = \begin{cases} 15,000\varepsilon \text{ ksi} & |\varepsilon| \le 0.0012 \\ (18 + 64.15(\varepsilon - 0.0012)) \text{ ksi} & 0.0012 \le |\varepsilon| \le 0.5 \end{cases} \quad (2)$$

(c) The two constants E and n in $\sigma = E\varepsilon^n$ can be found by substituting the coordinates of the two point P_1 and P_2, to generate the following two equations:

$$18 = E(0.0012)^n \quad (3)$$

$$50 = E(0.5)^n \quad (4)$$

Dividing Eq. 4 by Eq. 3 and taking the logarithm of both sides, we can solve for 'n' as shown below:

$$\ln\left(\frac{0.5}{0.0012}\right)^n = \ln\left(\frac{50}{18}\right) \quad \text{or} \quad n \ln(416.67) = \ln(2.778) \quad \text{or} \quad n = 0.1694$$

Substituting the value of n in Eq. 4, we can obtain the value of E as $E = \frac{50}{(0.5)^{0.1694}} = 56.2 \text{ ksi}$

We can now write the stress-strain equations for the power law model as shown below:

$$\sigma = \begin{cases} 56.2\varepsilon^{0.1694} \text{ ksi} & \varepsilon \ge 0 \\ -56.2(-\varepsilon)^{0.1694} \text{ ksi} & \varepsilon < 0 \end{cases} \quad (5)$$

Stresses at different strains can be found plotted as shown below.

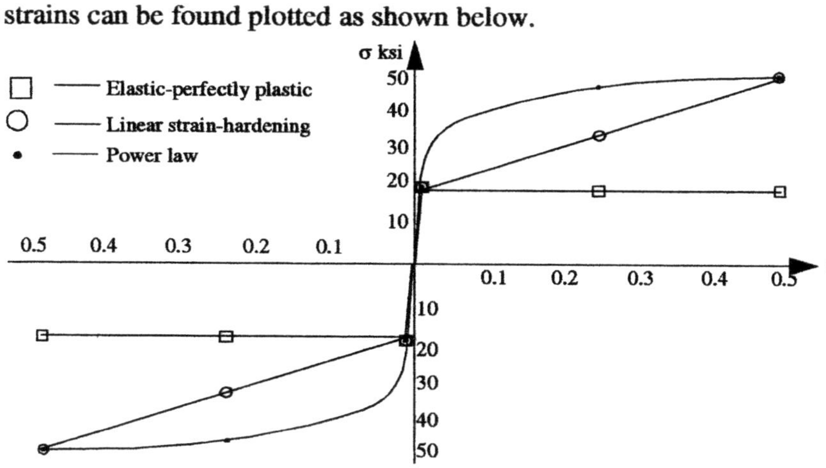

3.133 Cast iron has a yield stress of $\sigma_{yield} = 220$ MPa in tension, yield strain of $\varepsilon_{yield} = 0.00125$, ultimate stress of $\sigma_{ult} = 340$ MPa and the corresponding ultimate strain of $\varepsilon_{ult} = 0.20$. Determine the material constants and plot the resulting stress-strain curve for (a) elastic-perfectly plastic model (b) linear strain hardening model (c) and the non-linear power law model.

Solution $\sigma_{yield} = 220$ MPa $\varepsilon_{yield} = 0.00125$ $\sigma_{ult} = 340$ MPa $\varepsilon_{ult} = 0.20$.
Constants in the three material models=?

We have coordinates of three points on the curve: P_o ($\sigma_0 = 0.00$, $\varepsilon_0 = 0.000$), P_1 ($\sigma_1 = 220$, $\varepsilon_1 = 0.00125$), and P_2 ($\sigma_2 = 340$, $\varepsilon_2 = 0.2$). Using this data we can find the various constants in the material models.
(a) The modulus of elasticity E is slope between points P_o and P_1 and can be found as shown below.

$$E_1 = \frac{\sigma_1 - \sigma_0}{\varepsilon_1 - \varepsilon_0} = \frac{220}{0.00125} = 176,000 \text{ MPa}$$

After yield stress, the stress is a constant. The stress strain behavior can be written as shown below.

$$\sigma = \begin{cases} 176,000\varepsilon \text{ MPa} & |\varepsilon| \le 0.00125 \\ 220 \text{ MPa} & 0.00125 \le |\varepsilon| \le 0.2 \end{cases} \quad \mathbf{1}$$

(b) In the linear strain hardening model the slope of the straight line before yield stress is as calculated in part (a). After the yield stress, the slope of the line can be found from the coordinates of point P_1 and P_2

$$E_2 = \frac{\sigma_2 - \sigma_1}{\varepsilon_2 - \varepsilon_1} = \frac{340 - 220}{0.2 - 0.00125} = 603.8\text{MPa}$$ The stress strain behavior can be written as shown below.

$$\sigma = \begin{cases} 176,000\varepsilon \text{ MPa} & |\varepsilon| \le 0.00125 \\ 220 + 603.8(\varepsilon - 0.00125) \text{ MPa} & 0.00125 \le |\varepsilon| \le 0.2 \end{cases} \quad \mathbf{2}$$

(c) The two constants E and n in $\sigma = E\varepsilon^n$ can be found by substituting the coordinates of the two point P_1 and P_2, to generate the following two equations:

$$220 = E(0.00125)^n \quad \mathbf{3}$$
$$340 = E(0.2)^n \quad \mathbf{4}$$

Dividing Eq. 4 by 3 and taking the logarithm of both sides, we can solve for 'n' as shown below:

$$\ln\left(\frac{0.2}{0.00125}\right)^n = \ln\left(\frac{340}{220}\right) \quad \text{or} \quad n \ln(160) = \ln(1.545) \quad \text{or} \quad n = 0.0858$$

Substituting the value of n in Eq. 4, we can obtain the value of E as $E = \dfrac{340}{(0.2)^{0.0858}} = 390.3 \text{ MPa}$

We can now write the stress-strain equations for the power law model as shown below:

$$\sigma = \begin{cases} 390.3\varepsilon^{0.0858} \text{ MPa} & \varepsilon \ge 0 \\ -390.3(-\varepsilon)^{0.0858} \text{ MPa} & \varepsilon < 0 \end{cases} \quad \mathbf{5}$$

Stresses at different strains can be found plotted as shown below.

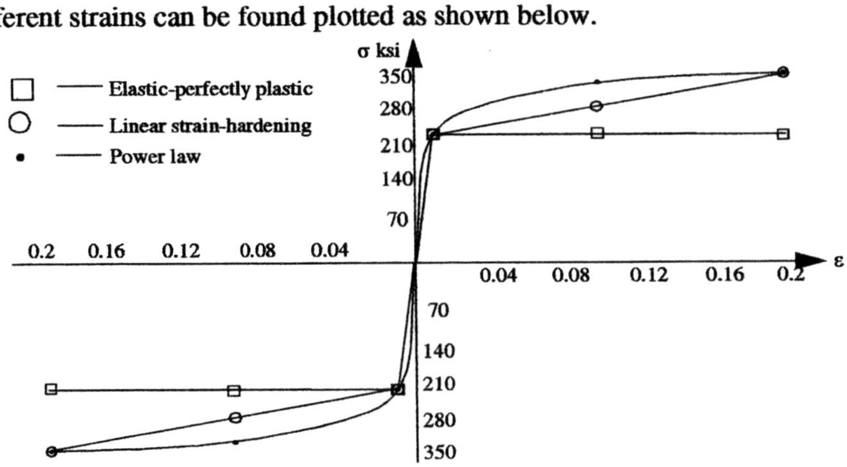

3.134 A solid circular shaft of diameter 3 inches has a shear strain at a section in polar coordinates was found to be $\gamma_{x\theta} = 2\rho(10^{-3})$ where ρ is the radial coordinate measured in inches. Write an expressions for $\tau_{x\theta}$ as a function of ρ and plot the shear strain $\gamma_{x\theta}$ and shear stress $\tau_{x\theta}$ distribution across the cross-section.The shaft is made from elastic-perfectly plastic material that has a yield stress of τ_{yield}= 18 ksi, and a Shear modulus G = 12,000 ksi.

Solution $\gamma_{x\theta}=2\rho(10^{-3})$ $\tau_{yield}=18\text{ksi}$ G=12,000 ksi $\tau_{x\theta}=f(\rho)$ = ? Elastic perfectly plastic

---

The strain at yield point is: $\gamma_{yield} = \frac{18}{12,000} = 0.0015 = 2\rho_{yield}(10^{-3})$ or $\rho_{yield} = 0.75\text{inch}$ where, ρ_{yield} is the radial coordinate of elastic-plastic beam For $0 < \rho < 0.75$, $\tau_{x\theta} = G\gamma_{x\theta} = (12000)(2\rho)(10^{-3}) = 24\rho$. The stress distribution can be written as:

$$\tau_{x\theta} = \begin{cases} 24\rho \text{ ksi} & 0 < \rho < 0.75\text{in} \\ 18 \text{ ksi} & 0.75\text{in} < \rho < 1.5\text{in} \end{cases}$$ 1

The plot of shear strain and shear stress are shown below.

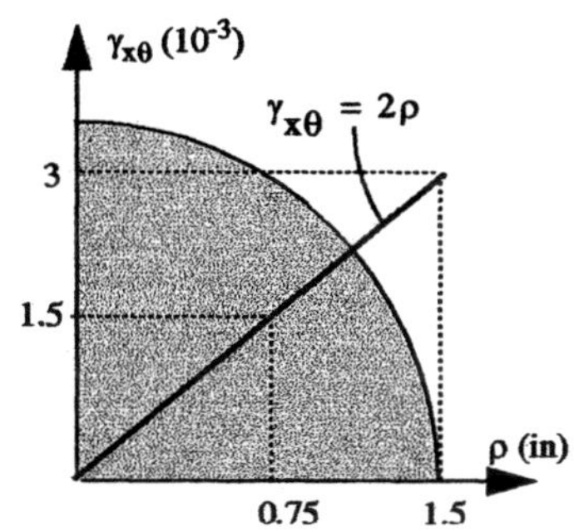

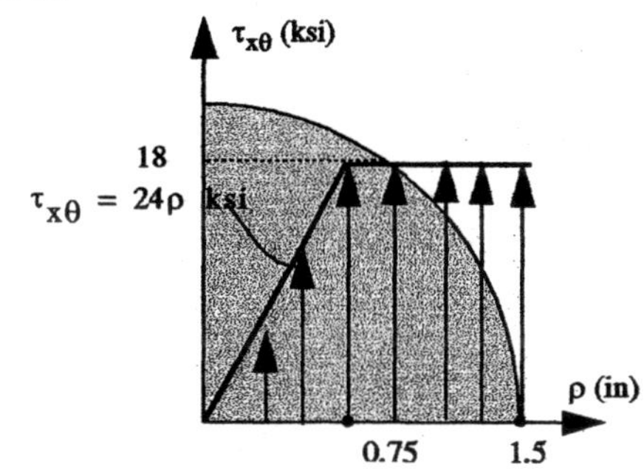

3.135 A solid circular shaft of diameter 3 inches has a shear strain at a section in polar coordinates was found to be $\gamma_{x\theta} = 0.002\rho$ where ρ is the radial coordinate measured in inches. Write an expressions for $\tau_{x\theta}$ as a function of ρ and plot the shear strain $\gamma_{x\theta}$ and shear stress $\tau_{x\theta}$ distribution across the cross-section.The shaft is made form a bi-linear material shown in Figure 3.37. The material has a yield stress of τ_{yield}= 18 ksi, Shear modulus G_1 = 12,000 ksi and G_2= 4,800 ksi.

Solution $\gamma_{x\theta}=2\rho(10^{-3})$ $\tau_{yield}=18\text{ksi}$ $G_1=12{,}000\text{ksi}$ $G_2=4{,}800\text{ksi}$ $\tau_{x\theta}=f(\rho)=?$

Before yield point we have: $\tau_1=G_1\gamma_{x\theta}=(12{,}000)(2\rho)(10^{-3})=24\rho$

The strain at yield point is: $\gamma_{yield} = \frac{\tau_{yield}}{G_1} = \frac{18}{12,000} = 0.0015$

After yield point we can write: $\tau_2 = \tau_{yield} + G_2(\gamma_{x\theta} - \gamma_{yield}) = 18 + 4800(2\rho(10^{-3}) - 0.0015) = 10.8 + 9.6\rho$
From problem 3.134 ρ_{yield}=0.75 inch. The stress distribution can be written as:

$$\tau_{x\theta} = \begin{cases} 24\rho \text{ ksi} & 0 < \rho < 0.75\text{in} \\ (10.8 + 9.6\rho)\text{ksi} & 0.75\text{in} < \rho < 1.5\text{in} \end{cases}$$ 1

The plot of shear strain and shear stress are given below.

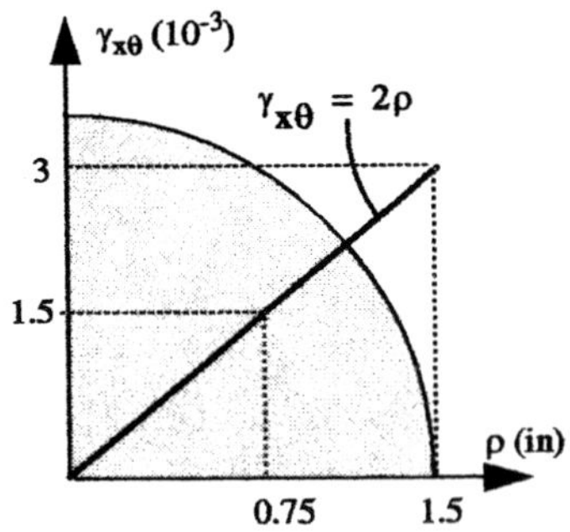

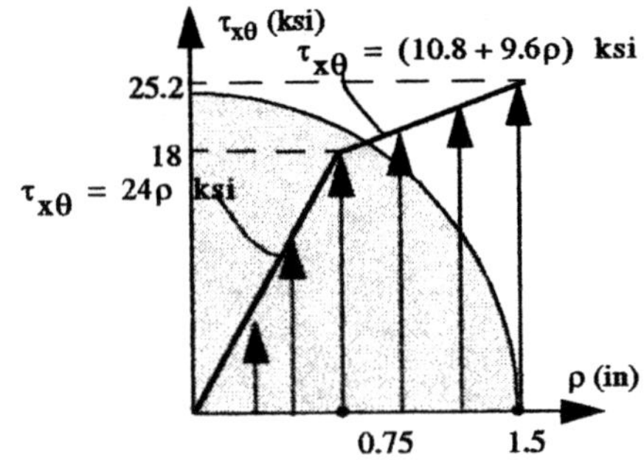

3.136 A solid circular shaft of diameter 3 inches has a shear strain at a section in polar coordinates was found to be $\gamma_{x\theta} = 2\rho(10^{-3})$ where ρ is the radial coordinate measured in inches. Write an expressions for $\tau_{x\theta}$ as a function of ρ and plot the shear strain $\gamma_{x\theta}$ and shear stress $\tau_{x\theta}$ distribution across the cross-section.The shaft material has a stress strain relationship given by $\tau = 243\gamma^{0.4}$ ksi

Solution $\gamma_{x\theta} = 2\rho(10^{-3})$ $\tau_{x\theta} = f(\rho) = ?$

The stress distribution is: $\tau_{x\theta} = 243\gamma_{x\theta} = 243(2\rho(10^{-3}))^{0.4}$ or $\boxed{\tau_{x\theta} = 20.23\rho^{0.4}\text{ksi} \qquad 0 \le \rho \le 1.5}$

The plots of shear strain and shear stress are given below.

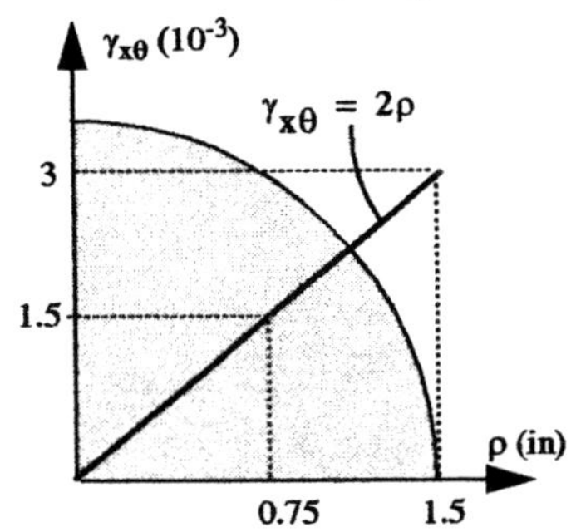

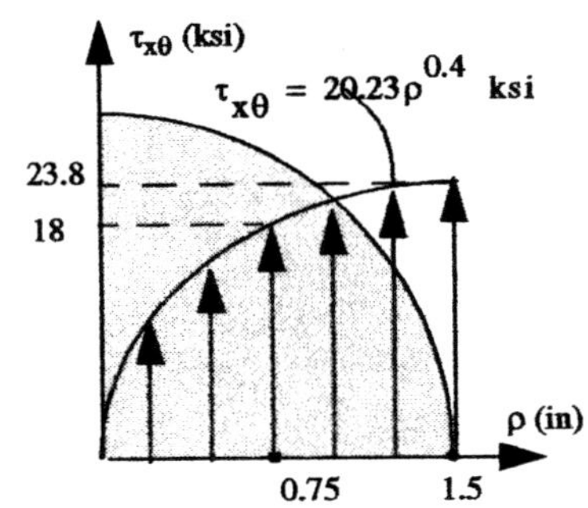

3.137 A solid circular shaft of diameter 3 inches has a shear strain at a section in polar coordinates was found to be $\gamma_{x\theta} = 0.2\rho$ where ρ is the radial coordinate measured in meters. Write an expressions for $\tau_{x\theta}$ as a function of ρ and plot the shear strain $\gamma_{x\theta}$ and shear stress $\tau_{x\theta}$ distribution across the cross-section.The shaft material has a stress strain relationship given by $\tau = (12,000\gamma - 120,000\gamma^2)$ ksi

Solution $\gamma_{x\theta} = 2\rho(10^{-3})$ $\tau_{x\theta} = f(\rho) = ?$

The shear stress distribution is:

$$\tau_{x\theta} = (12,000\gamma_{x\theta} - 120,000\gamma_{x\theta}^2) = (12,000)(2\rho)(10^{-3}) - 120,000[2\rho(10^{-3})]^2 \text{ or}$$

$$\boxed{\tau_{x\theta} = (24\rho - 0.48\rho^2)\text{ksi} \qquad 0 \le \rho \le 1.5}$$

The plots of shear strain and shear stress are given below.

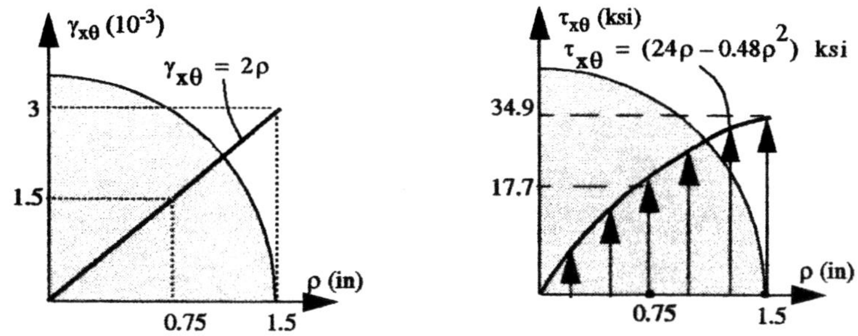

3.138 A hollow circular shaft has an inner diameter of 50 mm and an outside diameter of 100 mm. The shear strain at a section in polar coordinates was found to be $\gamma_{x\theta} = 0.2\rho$ where ρ is the radial coordinate measured in meters. Write an expressions for $\tau_{x\theta}$ as a function of ρ and plot the shear strain $\gamma_{x\theta}$ and shear stress $\tau_{x\theta}$ distribution across the cross-section. The shaft is made from elastic-perfectly plastic material that has a shear yield stress of $\tau_{yield} = 175$ MPa, and a Shear modulus G = 26 GPa.

Solution $\gamma_{x\theta}=0.2\rho$ ρ in meters $\tau_{yield}=175MP$ G=26GPa $d_i=50mm$ $d_0=150mm$

$\tau_{x\theta} = f(\rho) = ?$ Elastic-perfectly plastic

The strain at yield point is: $\gamma_{yield} = \dfrac{\tau_{yield}}{G} = \dfrac{175(10^6)}{26(10^9)} = 6.7308(10^{-3}) = 0.2\rho_{yield}$ or

$\rho_{yield} = 33.65(10^{-3})m$ where ρ_{yield} is the radial coordinate of elastic-plastic boundary

In the elastic region: $\tau_{x\theta} = G\gamma_{z\theta} = 26(10^9)(0.2\rho) = 5200\rho(10^6)N/m^2$ where ρ is in meters. The stress distribution can be written as:

$$\tau_{x\theta} = \begin{cases} 5200\rho MPa & 0.025m \le \rho \le 0.03365m \\ 175MPa & 0.03365m \le \rho \le 0.05m \end{cases}$$ 1

The plot of shear strain and shear stress are given below

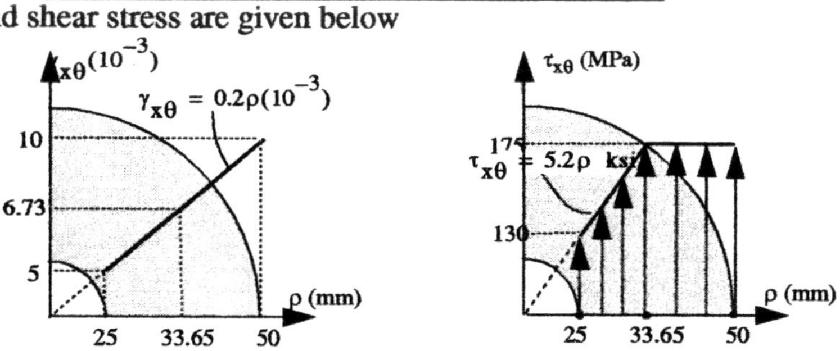

3.139 A hollow circular shaft has an inner diameter of 50 mm and an outside diameter of 100 mm. The shear strain at a section in polar coordinates was found to be $\gamma_{x\theta} = 0.2\rho$ where ρ is the radial coordinate measured in meters. Write an expressions for $\tau_{x\theta}$ as a function of ρ and plot the shear strain $\gamma_{x\theta}$ and shear stress $\tau_{x\theta}$ distribution across the cross-section. The shaft is made from a bi-linear material shown in Figure 3.37. The material has a shear yield stress of τ_{yield}= 175 MPa, Shear modulus G_1 = 26 GPa and G_2= 14 GPa.

Solution $\gamma_{yield}=0.2\rho$ where ρ is in meters $\tau_{yield}=175MPa$

$G_1=26GPa$ $G_2=14GPa$ $\tau_{x\theta} = f(\rho) = ?$

The stress before yield point is:

$$\tau_1 = G_1\gamma_{x\theta} = 26(10^9)(0.2\rho) = 5200\rho \text{MPa}$$ 1

From problem 3.138: $\gamma_{yield} = 6.7308(10^{-3})$ and $\rho_{yield} = 33.65(10^{-3})\text{m}$

The stress after yield point is: $\tau_2 = \tau_{yield} + G_2(\gamma_{x\theta} - \gamma_{yield}) = 175(10^6) + 14(10^9)(0.2\rho - 6.7308(10^{-3}))$

$$\tau_2 = 80.77(10^6) + 2800\rho(10^6)$$ 2

In Eqs.(1) &(2), ρ is in meters.The stress distribution can be written as:

$$\tau_{x\theta} = \begin{cases} 5200\rho \text{MPa} & 0.025\text{m} \le \rho \le 0.03365\text{m} \\ (80.77 + 2800\rho)\text{MPa} & 0.03365\text{m} \le \rho \le 0.05\text{m} \end{cases}$$ 1

The plot of shear strain and shear stress are given below.

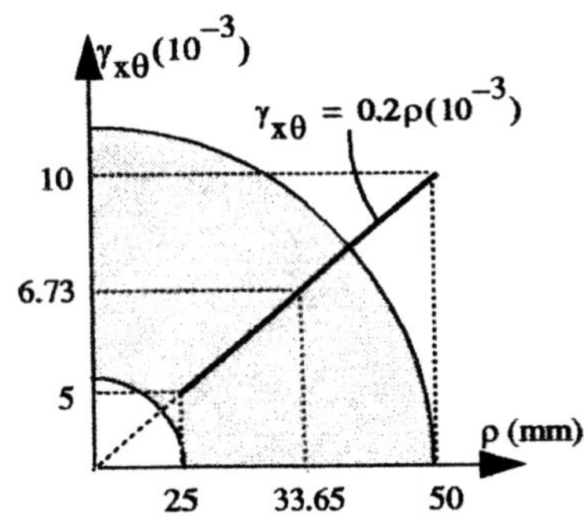

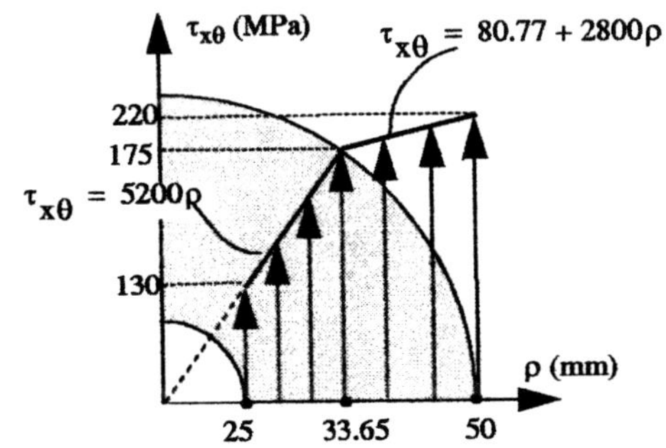

3.140 A hollow circular shaft has an inner diameter of 50 mm and an outside diameter of 100 mm. The shear strain at a section in polar coordinates was found to be $\gamma_{x\theta} = 0.2\rho$ where ρ is the radial coordinate measured in meters. Write an expressions for $\tau_{x\theta}$ as a function of ρ and plot the shear strain $\gamma_{x\theta}$ and shear stress $\tau_{x\theta}$ distribution across the cross-section. The shaft material has a stress strain relationship given by $\tau = 3435\gamma^{0.6}$ MPa

Solution $\gamma_{x\theta}=0.2\rho$ ρ in meters $\tau_{x\theta} = f(\rho) = ?$

The stress distribution is: $\tau_{x\theta} = 3435\gamma^{0.6} = 3435(0.2\rho)^{0.6} = 1307.8\rho^{0.6}$ or

$$\boxed{\tau_{x\theta} = 1307.8\rho^{0.6}\text{MPa} \qquad 0.025\text{m} \le \rho \le 0.05\text{m}}$$

The plot of shear strain and shear stress are given below.

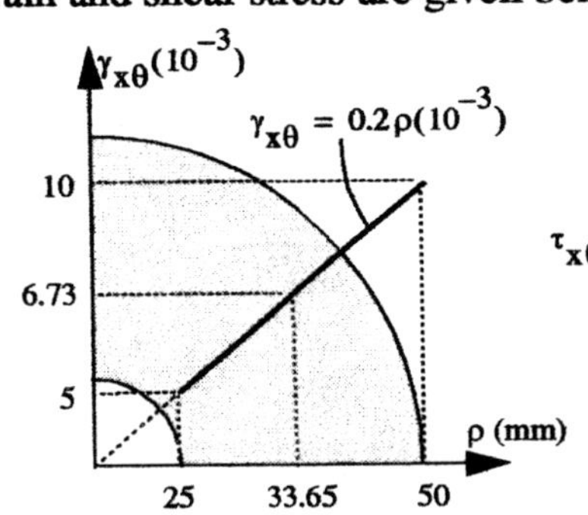

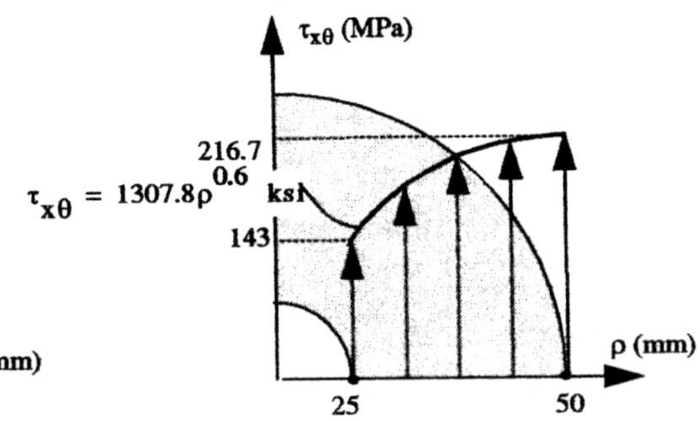

3.141 A hollow circular shaft has an inner diameter of 50 mm and an outside diameter of 100 mm. The shear strain at a section in polar coordinates was found to be $\gamma_{x\theta} = 0.2\rho$ where ρ is the radial coordinate measured in meters. Write an expressions for $\tau_{x\theta}$ as a function of ρ and plot the shear strain $\gamma_{x\theta}$ and shear stress $\tau_{x\theta}$ distribution across the cross-section. The shaft material has a stress strain relationship given by

$\tau = (26,000\gamma - 208,000\gamma^2)$ MPa

Solution $\gamma_{x\theta}=0.2\rho$ where ρ is in meters $\tau_{x\theta} = f(\rho) = ?$

The shear stress distribution is: $\tau_{x\theta} = (26,000\gamma - 208,000\gamma^2) = 26,000(0.2\rho) - 208,000(0.2\rho)^2$

$$\boxed{\tau_{x\theta} = (5,200\rho - 8,320\rho^2)\text{MPa} \qquad 0.025\text{m} \le \rho \le 0.05\text{m}}$$ 1

The plots of shear strain and stress are shown below.

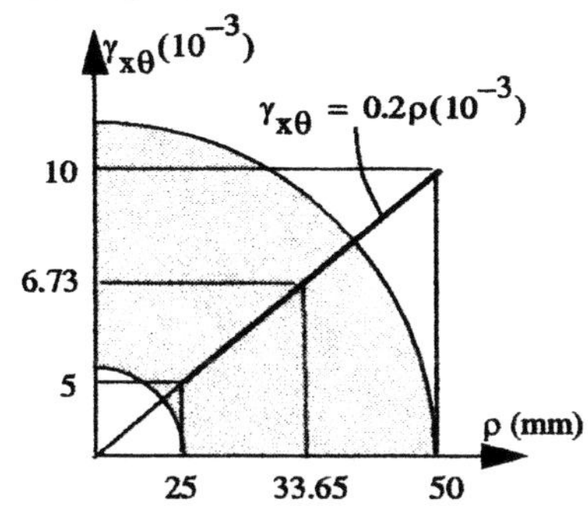

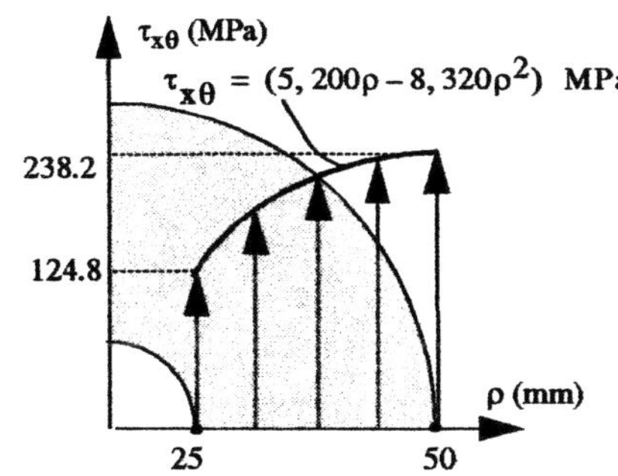

3.142 A rectangular beam has the dimensions as shown in Fig. P3.142. The normal strain due to bending about the z-axis was found to vary as $\varepsilon_{xx} = -0.01y$ with y measured in meters. For the given material model in the problem. Write an expressions for normal stress σ_{xx} as a function of y and plot the σ_{xx} distribution across the cross-section. The beam is made from elastic-perfectly plastic material that has a yield stress of σ_{yield}= 250 MPa, and a Modulus of Elasticity E = 200 GPa.

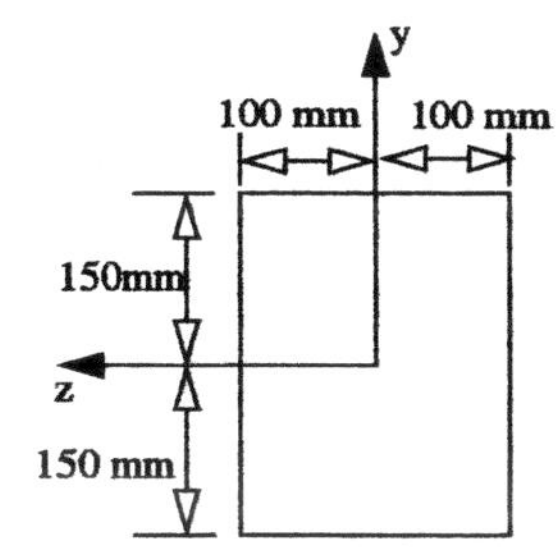

Fig. P3.142

Solution $\varepsilon_{xx}=-0.01y$ y in meters σ_{yield}=250MPa E=200GPa

σ_{xx}= f(y) Elastic-perfectly plastic

The strain at yield point is: $\varepsilon_{yield} = \dfrac{\sigma_{yield}}{E} = \dfrac{250(10^6)}{200(10^6)} = 1.25(10^{-3})$

For positive values of y we have contraction, and for negative values of y we have tension. With geometry being symmetric in y, the elastic-plastic boundary will be on both sides

$\varepsilon_{yield} = \pm 1.25(10^{-3}) = -0.01y_{yield}$ or

$$y_{yield} = \pm 0.125\text{m} = \pm 125\text{mm}$$ 1

In the linear region

$\sigma_{xx} = E\varepsilon_{xx} = (200)(10^9)(-0.01y) = -2000y$ y in meters.

$$\sigma_{xx} = \begin{cases} -250\text{MPa} & 0.125\text{m} < y < 0.150\text{m} \\ -2000y\text{MPa} & -0.125\text{m} < y < 0.125\text{m} \\ 250\text{MPa} & -0.150\text{m} < y < -0.125\text{m} \end{cases}$$ 2

The plots of normal strain and stress are shown below.

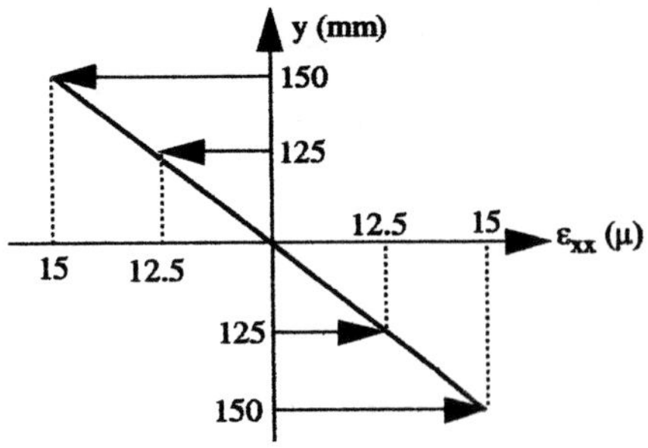

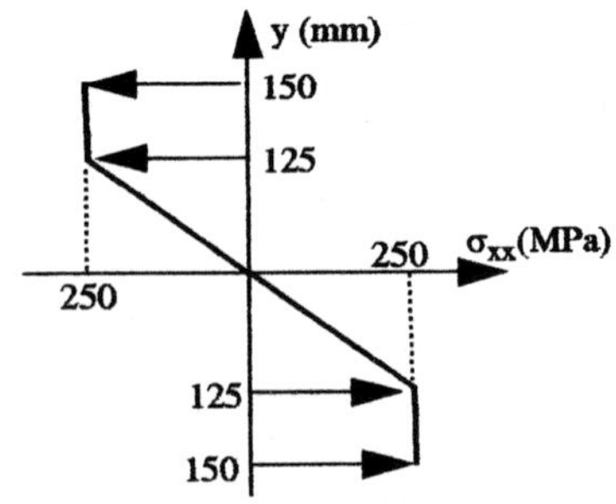

3.143 A rectangular beam has the dimensions as shown in Fig. P3.142. The normal strain due to bending about the z-axis was found to vary as $\varepsilon_{xx} = -0.01y$ with y measured in meters. For the given material model in the problem. Write an expressions for normal stress σ_{xx} as a function of y and plot the σ_{xx} distribution across the cross-section. The beam is made from a bi-linear material shown in Figure 3.37. The material has a yield stress of σ_{yield}= 250 MPa, Modulus of Elasticity E_1 = 200 GPa and E_2= 80 GPa.

Solution ε_{xx}=-0.01y where y is in meters σ_{yield}=250 MPa

E_1 =200 GPa E_2=80 GPa $\sigma_{xx} = f(y) = ?$

From problem 3-142: $\varepsilon_{yield} = 1.25(10^{-3})$ and $y_{yield} = \pm 125\text{mm}$.

In linear region: $\sigma_1 = E_1\varepsilon_1 = (200)(10^9)(-0.01y) = -2000y \text{ MPa}$.

For $y < -0.125\text{m}$ i.e., on the tensile side

$$\sigma_2 = \sigma_{yield} + E_2(\varepsilon_{xx} - \varepsilon_{yield}) = 250(10^6) + (80)(10^9)[-0.01y - 1.25(10^{-3})] = (150 - 800y)\text{MPa}$$

For $y > 0.125\text{m}$ i.e., on the compressive side

$$\sigma_2 = -\sigma_{yield} + E_2(\varepsilon_{xx} - (-\varepsilon_{yield})) = (-250)(10^6) + (80)(10^9)[-0.01y + 1.25(10^{-3})] = (-150 - 800y)\text{MPa}$$

The stress distribution can be written as:

$$\sigma_{xx} = \begin{cases} (-150 - 800y)\text{MPa} & 0.125\text{m} < y < 0.150\text{m} \\ -2000y\text{MPa} & -0.125\text{m} < y < 0.125\text{m} \\ (150 - 800y)\text{MPa} & -0.150\text{m} < y < -0.125\text{m} \end{cases}$$

1

The plots of normal strain and stress are shown below.

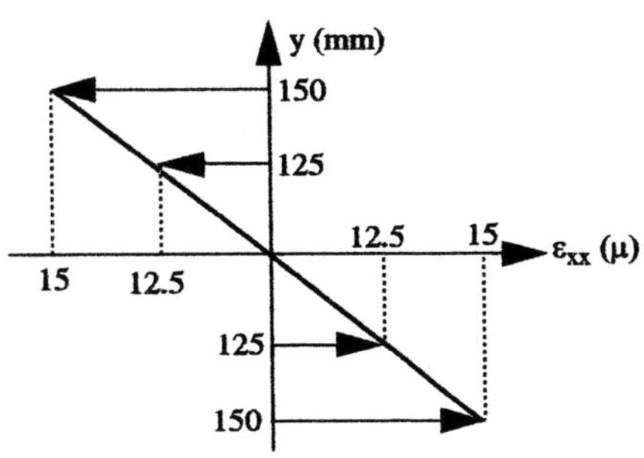

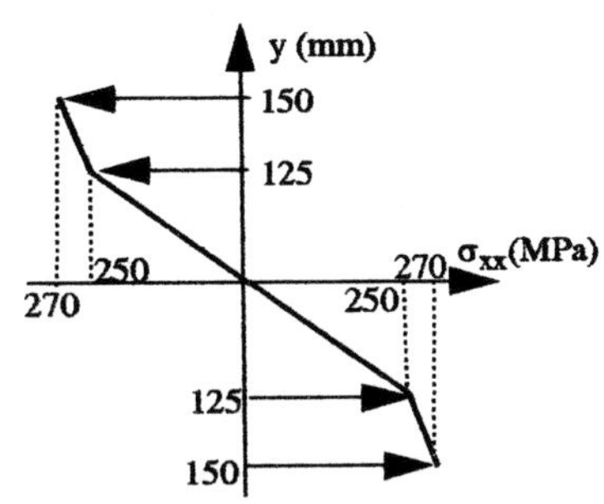

3.144 A rectangular beam has the dimensions as shown in Fig. P3.142. The normal strain due to bending about the z-axis was found to vary as $\varepsilon_{xx} = -0.01y$ with y measured in meters. For the given material model in the problem. Write an expressions for normal stress σ_{xx} as a function of y and plot the σ_{xx} distribution across the cross-section. The beam material has a stress strain relationship given by $\sigma = 952\varepsilon^{0.2}$ MPa

Solution ε_{xx}=-0.01y where y is in meters σ_{xx} = f(y) = ?

As the material behavior in tension and compression is the same, we have:

$$\sigma_{xx} = \begin{cases} 952\varepsilon^{0.2}\text{MPa} & \varepsilon > 0 \\ -952(-\varepsilon)^{0.2}\text{MPa} & \varepsilon < 0 \end{cases}$$

Noting that for positive y we have contraction and for negative y we have extension, the stress distribution can be written as shown below.

$$\sigma_{xx} = \begin{cases} 952(-0.01y)^{0.2}\text{MPa} & y < 0 \\ -952(0.01y)^{0.2}\text{MPa} & y > 0 \end{cases} \quad \text{or} \quad \boxed{\sigma_{xx} = \begin{cases} 379(-y)^{0.2}\text{MPa} & y < 0 \\ -379(y)^{0.2}\text{MPa} & y > 0 \end{cases}}$$

The plots of normal strain and stress are as shown below.

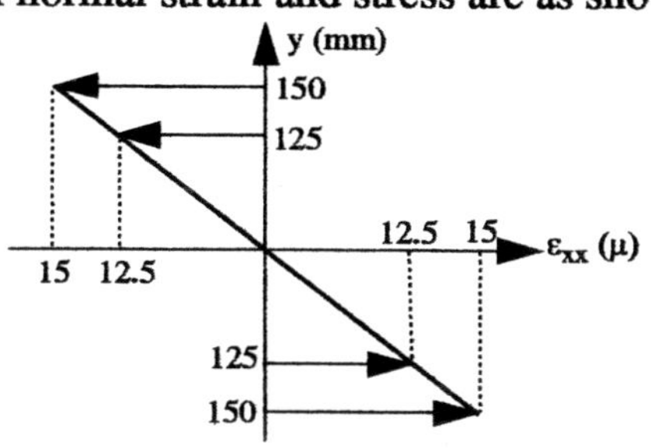

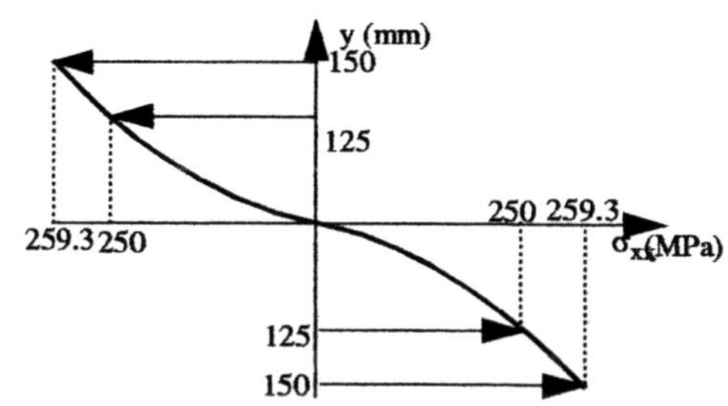

3.145 A rectangular beam has the dimensions as shown in Fig. P3.142. The normal strain due to bending about the z-axis was found to vary as $\varepsilon_{xx} = -0.01y$ with y measured in meters. For the given material model in the problem. Write an expressions for normal stress σ_{xx} as a function of y and plot the σ_{xx} distribution across the cross-section. The beam material has a stress strain relationship given by

$\sigma = (200\varepsilon - 2,000\varepsilon^2)$ MPa

Solution ε_{xx}=-0.01y where y is in meters σ_{xx} = f(y) = ?

As the material behavior in tension and compression is the same, we have:

$$\sigma_{xx} = \begin{cases} (200\varepsilon - 2,000\varepsilon^2)\text{MPa} & \varepsilon > 0 \\ (200\varepsilon + 2,000\varepsilon^2)\text{MPa} & \varepsilon < 0 \end{cases}$$

Noting that for positive y we have contraction and for negative y we have extension, the stress distribution can be written as shown below.

$$\sigma_{xx} = \begin{cases} [200(-0.01y) - 2,000(-0.01y)^2]\text{MPa} & y < 0 \\ [200(-0.01y) + 2,000(-0.01y)^2]\text{MPa} & y > 0 \end{cases} \quad \text{or} \quad \boxed{\sigma_{xx} = \begin{cases} (-2y - 0.2y^2)\text{MPa} & y < 0 \\ (-2y + 0.2y^2)\text{MPa} & y > 0 \end{cases}}$$

The plots of normal strain and stress are as shown below

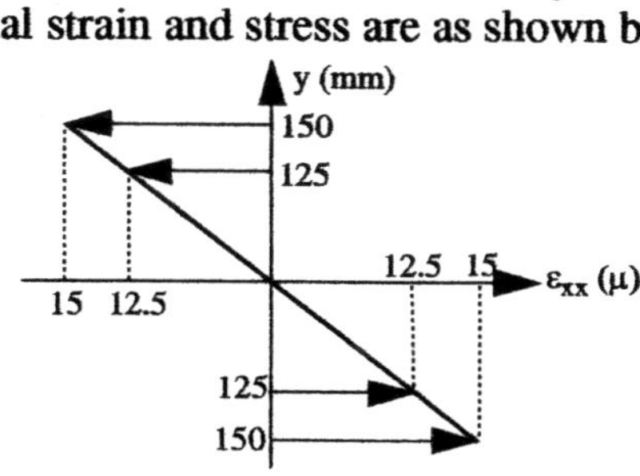

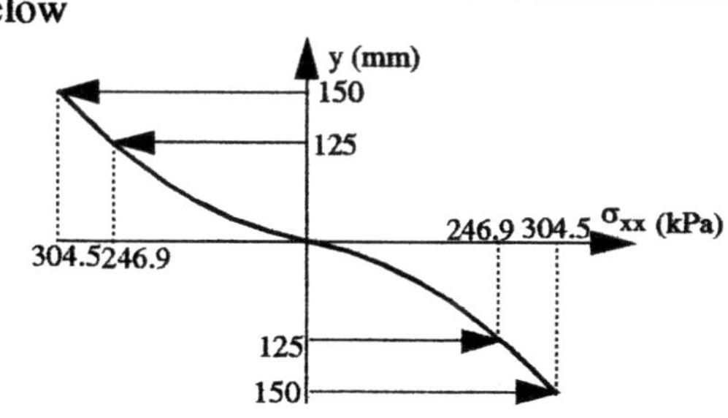

Problem Set 4.1

4.1 Aluminum (E = 30,000 ksi) bars are welded to rigid plates as shown. All bars have an area of cross-section of 0.5 in^2. Due to the applied forces the rigid plates at A, B, C, and D displaced in the x-direction without rotating by the amount given below. Determine the applied forces F_1, F_2, F_3, and F_4. (See Example 2.1)

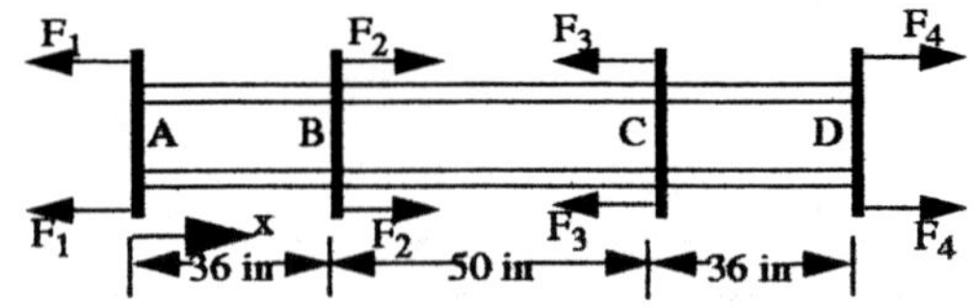

Fig. P4.1

$u_A = -0.0100$ in $\quad u_B = 0.0080$ in

$u_C = -0.0045$ in $\quad u_D = 0.0075$ in

Solution $\quad E = 30{,}000$ ksi $\quad A = 0.5\text{in}^2 \quad F_1 = ? \quad F_2 = ? \quad F_3 = ? \quad F_4 = ?$

---

1. Deformation calculations

$u_D - u_C = 0.0075 - (-0.0045) = 0.0120$ in

$u_C - u_B = -0.0045 - 0.0080 = -0.0125$ in

$u_B - u_A = 0.0080 - (-0.0100) = 0.0180$ in

2. Strain calculations

$$\varepsilon_{DC} = \frac{u_D - u_C}{x_D - x_C} = \frac{0.012}{36} = 0.3333(10^{-3})$$

$$\varepsilon_{CB} = \frac{u_C - u_B}{x_C - x_B} = \frac{-0.0125}{50} = -0.2500(10^{-3})$$

$$\varepsilon_{BA} = \frac{u_B - u_A}{x_B - x_A} = \frac{0.0180}{36} = 0.5000(10^{-3})$$

3. Stress calculations

$\sigma_{DC} = (30)(10^3)(0.3333(10^{-3})) = 10$ or $\sigma_{DC} = 10$ ksi(T)

$\sigma_{CB} = (30)(10^3)(-0.25)(10^{-3}) = -7.5$ or $\sigma_{CB} = 7.5$ ksi(C)

$\sigma_{BA} = (30)(10^3)(0.50)(10^{-3}) = 15$ or $\sigma_{BA} = 15$ ksi(T)

4. Internal force calculations

$N_{DC} = (10)(0.5) = 5$ kips(T)

$N_{CB} = (7.5)(0.5) = 3.75$ kips(C)

$N_{BA} = (15)(0.5) = 7.5$ kips(T)

5. External force calculations

By making imaginary cuts through segments AB, BC, and CD the following free body diagrams can be obtained.

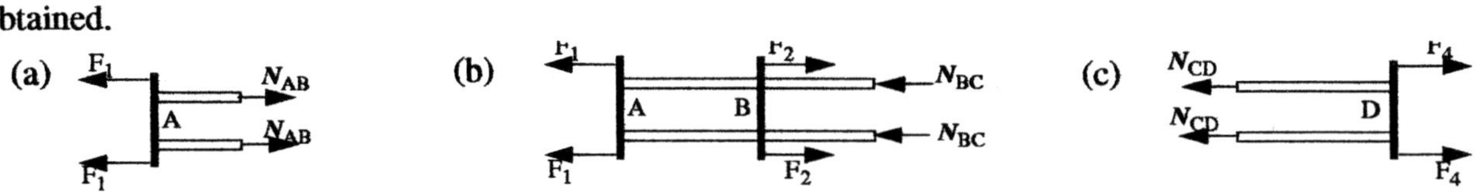

By equilibrium of forces in figure (a)

$2F_1 = 2N_{BA}$ $\quad \boxed{F_1 = 7.5 \text{ kips}}$

By equilibrium of forces in figure (b)

$-2F_1 + 2F_2 - 2N_{CB} = 0$ or $F_2 = F_1 + N_{CD} = 7.5 + 3.75$ $\quad \boxed{F_2 = 11.25 \text{ kips}}$

By equilibrium of forces in figure (c)

$2F_4 = 2N_{DC}$ $\boxed{F_4 = 5 \text{ kips}}$

By equilibrium of forces on the entire bar

$-2F_1 + 2F_2 - 2F_3 + 2F_4 = 0$ or $F_3 = F_2 + F_4 - F_1$ $\boxed{F_3 = 8.75 \text{ kips}}$

4.2 Brass bars between section A and B, aluminum bars between B and C, and steel bars between section C and D, are welded to rigid plates. The rigid plates displace in the x-direction without rotating by the following amounts: u_B= -1.8 mm, u_C = 0.7 mm, and u_D = 3.7 mm. Determine the external forces F_1, F_2 and F_3using the properties given in the adjoining Table.

	Brass	Aluminum	Steel
Modulus of Elasticity	70 GPa	100 GPa	200 GPa
Diameter	30 mm	25 mm	20 mm

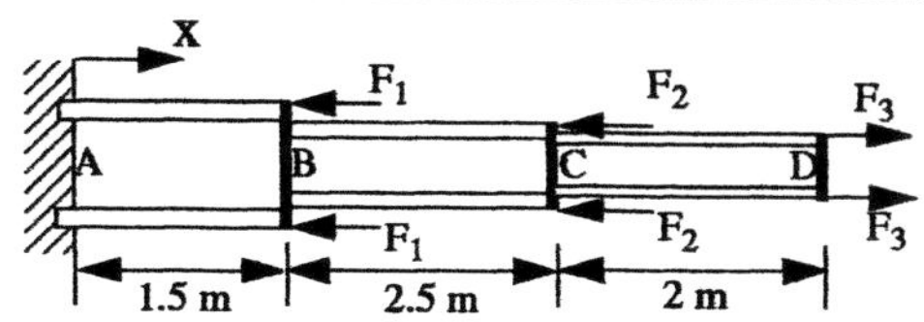

Fig. P4.2

Solution $F_1 = ?$ $F_2 = ?$ $F_3 = ?$

---

1. Deformation calculations

$u_D - u_C = 3.7 - 0.7 = 3.0 \text{ mm}$

$u_C - u_B = 0.7 + 1.8 = 2.5 \text{ mm}$

$u_B - u_A = -1.8 - 0 = -1.8 \text{ mm}$

2. Strain calculations

$$\varepsilon_{DC} = \frac{u_D - u_C}{x_D - x_C} = \frac{3}{2(10^3)} = 1.5(10^{-3})$$

$$\varepsilon_{CB} = \frac{u_C - u_B}{x_C - x_B} = \frac{2.5}{2.5(10^3)} = 1.0(10^{-3})$$

$$\varepsilon_{BD} = \frac{u_B - u_D}{x_B - x_D} = \frac{-1.8}{1.5(10^{-3})} = -1.2(10^{-3})$$

3. Stresses

$$\sigma_{DC} = E_{DC}\varepsilon_{DC} = (200)(10^9)(1.5)(10^{-3}) = 300(10^6) \text{ N/m}^2 \text{ (T)}$$

$$\sigma_{BC} = E_{BC}\varepsilon_{BC} = (100)(10^9)(1.0)(10^{-3}) = 100(10^6) \text{ N/m}^2 \text{ (T)}$$

$$\sigma_{AB} = E_{AB}\varepsilon_{AB} = (70)(10^9)(-1.2)(10^{-3}) = -84(10^6) \quad \text{or} \quad \sigma_{AB} = 84(10^6) \text{ (N/m}^2\text{) (C)}$$

4. Internal forces

$$N_{DC} = \sigma_{DC}A_{DC} = (300)(10^6)\left[\frac{\pi}{4}(0.03)^2\right] = 94.25(10^3) \text{ N} = 94.25 \text{ kN (T)}$$

$$N_{BC} = \sigma_{BC}A_{BC} = (100)(10^6)\left[\frac{\pi}{4}(0.025)^2\right] = 49.09(10^3) \text{ N} = 49.09 \text{ kN (T)}$$

$$N_{AB} = \sigma_{AB}A_{AB} = (84)(10^6)\left[\frac{\pi}{4}(0.020)^2\right] = 59.38(10^3) \text{ N} = 59.38 \text{ kN (C)}$$

5. External forces

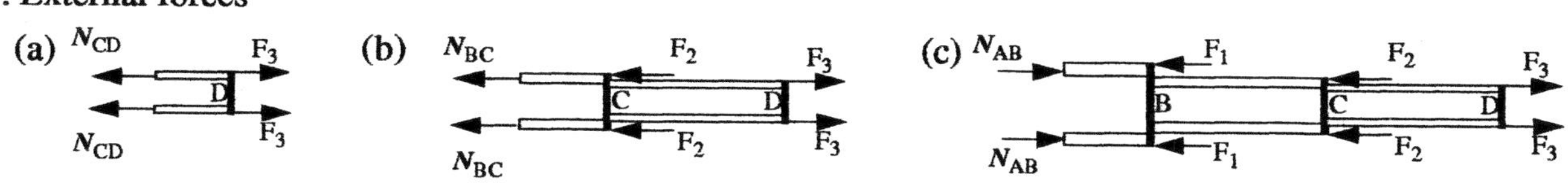

By equilibrium of forces in figure (a)

$2F_3 = 2N_{CD}$ $\boxed{F_3 = 94.3 \text{ kN}}$

By equilibrium of forces in figure (b)

$2F_3 - 2F_2 - 2N_{BC} = 0$ or $F_2 = F_3 - N_{BC} = 94.25 - 49.09 = 45.16$

$\boxed{F_2 = 45.2 \text{ kN}}$

By equilibrium of forces in figure (c)

$2F_3 - 2F_2 - 2F_1 + 2N_{AB} = 0$ or $F_1 = F_3 - F_2 + N_{AB} = 94.25 - 45.16 + 59.38 = 108.47$

$\boxed{F_1 = 108.5 \text{ kN}}$

4.3 Ends of four circular steel (E = 200 GPa) bars are welded to a rigid plate. The other ends of the bars are built into the wall. Due to the action of the external force F the rigid plate moves to the right by 0.1 mm without rotating. If the bars have a diameter of 10 mm, determine the applied force F.

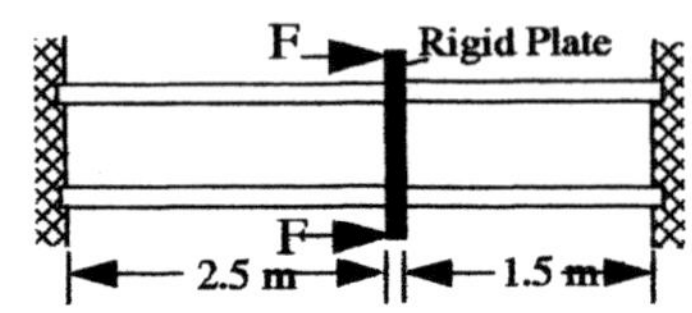

Fig. P4.3

Solution E = 200 GPa d = 10 mm δ = 0.1 mm to the right F = ?

1. Deformation calculation. Fig. (a) shows an exaggerated deformed geometry.

(a)

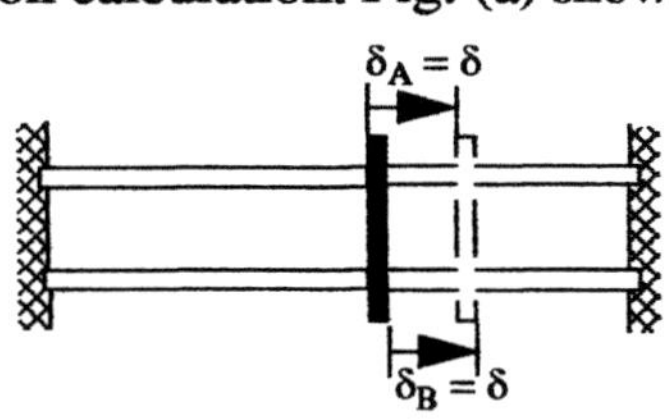

(b)

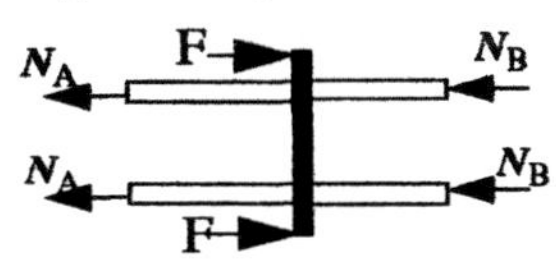

From the deformed geometry in figure (a) we obtain the following.

$\delta_A = \delta = 0.1$ mm extension

$\delta_B = \delta = 0.1$ mm contraction

2. Strain calculation

$$\varepsilon_A = \frac{\delta_A}{L_A} = \frac{0.1}{2.5(10^3)} = 40(10^{-6}) \text{ extension}$$

$$\varepsilon_B = \frac{\delta_B}{L_B} = \frac{0.1}{1.5(10^3)} = 66.67(10^{-6}) \text{ contraction}$$

3. Stress calculation

$$\sigma_A = E_A\varepsilon_A = (200)(10^9)(40)(10^{-6}) = 8.0(10^6) \text{ (N/m}^2\text{) (T)}$$

$$\sigma_B = E_B\varepsilon_B = (200)(10^9)(66.67)(10^{-6}) = 13.333(10^6) \text{ (N/m}^2\text{) (C)}$$

4. Internal force calculation

$$A_A = A_B = \frac{\pi}{4}(0.01)^2 = 78.54(10^{-6}) \text{ m}^2$$

$$N_A = \sigma_A A_A = (8.0)(10^6)(78.54)(10^{-6}) = 628.31 \text{ N(T)}$$

$$N_B = \sigma_B A_B = (13.333)(10^6)(78.54)(10^{-6}) = 1047.2 \text{ N(T)}$$

5. External force calculation. Fig. (b) shows the free body diagram of the rigid plate.

By force equilibrium $2F - 2N_A - 2N_B = 0$ or $F = N_A + N_B = 1675.5$ $\boxed{F = 1676 \text{ N}}$

4.4 Rigid plates are securely fastened to bars A and B as shown. A gap of 0.02 inch exists between the rigid plates before the forces are applied. After the application of the forces the normal strain in bar A was found to be 500 μ. The area of cross-section and Modulus of Elasticity for the each bar are as follows:

A_A= 1 in^2, E_A=10,000 ksi, A_B= 0.5 in^2, E_B=30,000 ksi.

Determine the applied forces F, assuming that the rigid plates do not rotate.

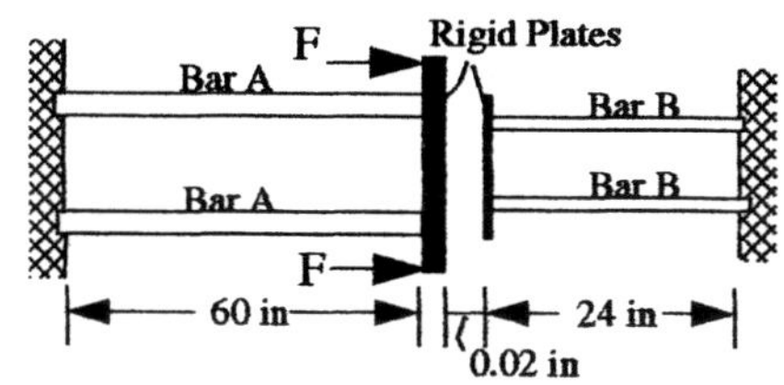

Fig. P4.4

Solution $A_A = 1\text{ in}^2$ $E_A = 10{,}000\text{ksi}$ $A_B = 0.5\text{in}^2$ $E_B = 30{,}000\text{ksi}$

$\varepsilon_A = 500\mu$ $F = ?$

---

1. Deformation calculation. Fig. (a) shows an exaggerated deformed geometry.

(a)

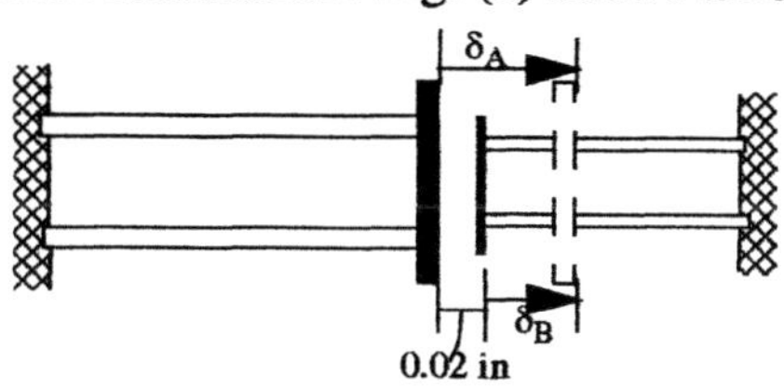

(b)

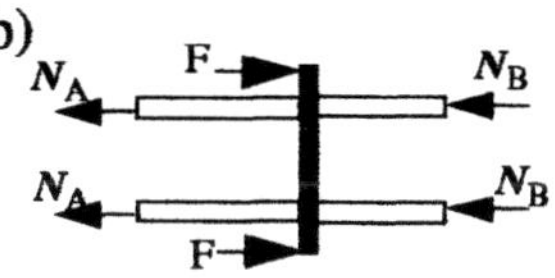

$\delta_A = \varepsilon_A L_A = (500)(10^{-6})(60) = 0.03 \text{ in}$

From deformed shape in Fig. (a)

$\delta_B = \delta_A - 0.02 = 0.01 \text{ in contraction}$

2. Strain calculation

$\varepsilon_A = 500(10^{-6}) \text{ extension}$

$$\varepsilon_B = \frac{\delta_B}{L_B} = \frac{0.01}{24} = 416.7(10^{-6}) \text{ contraction}$$

3. Stress calculations

$\sigma_A = E_A\varepsilon_A = 10(10^3)(500)(10^{-6}) = 5 \text{ ksi (T)}$

$\sigma_B = E_B\varepsilon_B = 30(10^3)(416.7)(10^{-6}) = 12.5 \text{ ksi (C)}$

4. Internal force calculations

$N_A = \sigma_A A_A = (5)(1) = 5.0 \text{ kips (T)}$

$N_B = \sigma_B A_B = (12.5)(0.5) = 6.25 \text{ kips (C)}$

5. External force calculation. Fig.(b) shows the free body diagram of the rigid plate. By force equilibrium

$2F - 2N_A - 2N_B = 0$ or $F = N_A + N_B = 5.0 + 6.25$ or $\boxed{F = 11.25 \text{ kips}}$

4.5 A crane is lifting a mass of 1000 kg. The mass of the iron ball at B is 25 kg. A single cable runs between A and B that has a diameter of 25 mm. Two cables are between the B and C, each cable has a diameter of 10 mm. Determine the axial stresses in the cables.

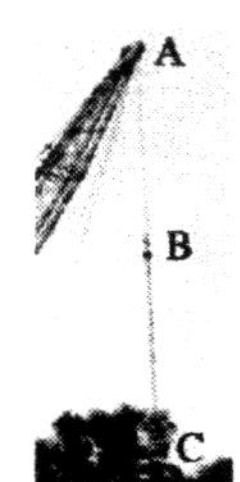

Fig. P4.5

Solution $m_1 = 1000kg$ $m_2 = 25kg$ $d_{AB} = 25mm$ $d_{BC} = 10mm$
$\sigma_{AB}=?$ $\sigma_{BC}=?$

By making imaginary cuts between A and B and between B and C, we obtain the following free body diagrams.

(a)

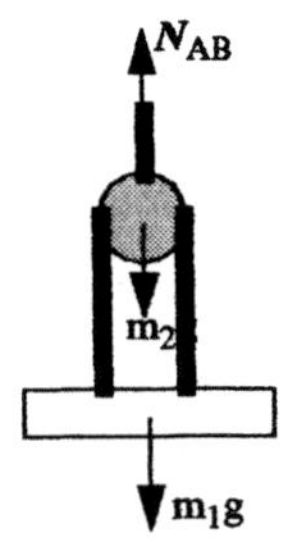

(b)

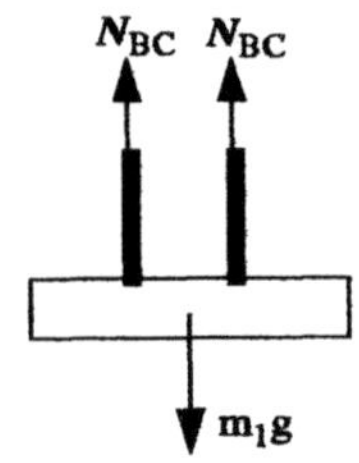

By force equilibrium in figure (a)

$$N_{AB} = m_1g + m_2g = (1025)(9.81) = 10.055(10^3)\ \text{N}$$

$$\sigma_{AB} = \frac{N_{AD}}{A_{AD}} = \frac{10.055(10^3)}{\pi(0.025)^2/4} = 20.483(10^6)\ \text{N/m}^2 \text{ or } \qquad \boxed{\sigma_{AB} = 20.5\ \text{MP}_a\ \text{(T)}}$$

By force equilibrium in figure (b)

$$2N_{BC} = m_1g \text{ or } N_{BC} = \frac{(1000)(9.81)}{2} = 4.905(10^3)\ \text{N}$$

$$\sigma_{BC} = \frac{N_{BC}}{A_{BC}} = \frac{4.905(10^3)}{\pi(0.01)^2/4} = 62045(10^6)\ (\text{N/m}^2) \text{ or } \qquad \boxed{\sigma_{BC} = 62.5\ \mu\text{P}_a\ \text{(T)}}$$

4.6 The counter-weight in a lift bridge has 12 cables on the left and 12 cables on the right. Each cable has an effective diameter of 0.75 inch, a length of 50 feet, Modulus of Elasticity of 30,000 ksi, and an ultimate strength of 60 ksi. (a) If the counter-weight is100 kips, determine the factor of safety for cable. (b) What is the extension of each cable when the bridge is being lifted?

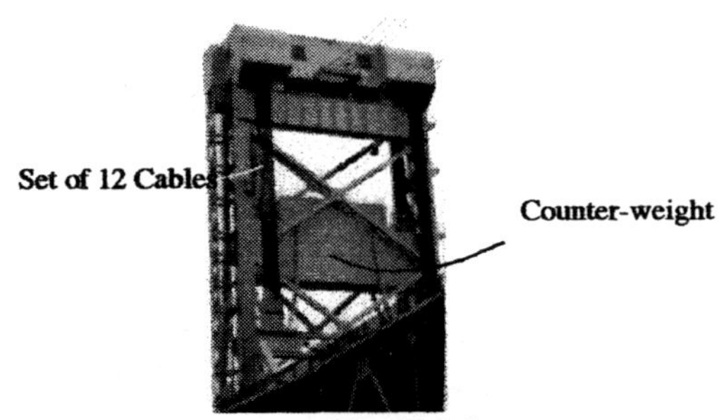

Fig. P4.6

Solution $d = 0.75in$ 2 sets of 12 cables $L = soft$ $E = 30,000ksi$
$\sigma_{ult} = 60\ \text{ksi}$ $W = 100\ ksi$ $k_{saftey} = ?$

The free body diagram of the counter-weight is shown below.

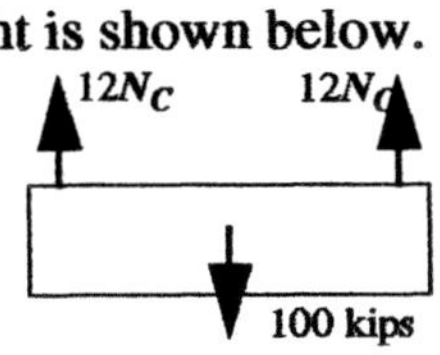

By force equilibrium $24N_C = 100$ or $\qquad N_C = 4.167\ \text{kips}$

The axial stress is: $\sigma_C = \frac{N_C}{A_C} = \frac{4.167}{\pi(0.75/2)^2} = 9.43\ \text{ksi}$

The factor of safety is: $k_{saftey} = \frac{\sigma_{ult}}{\sigma_c} = \frac{60}{9.43} = 6.361$ or $\boxed{k_{saftey} = 6.36}$

The axial strain is: $\varepsilon_C = \frac{\sigma_c}{E} = \frac{9.43}{30000} = 0.3143(10^{-3})$

The deformation of the cable is: $\delta_C = \varepsilon_C L = (0.3143)(10^{-3})(50)(12) = 0.1886$ in or $\boxed{\delta_C = 0.19 \text{ in}}$

4.7 Draw the axial force diagrams and then check your results by finding internal forces in each segment AB, BC, and CD by making imaginary cut and drawing free body diagrams.

Solution $N_{AB} = ?$ $N_{BC} = ?$ $N_{CD} = ?$

We draw a template as shown in figure (a) and write the template equation.

(a)

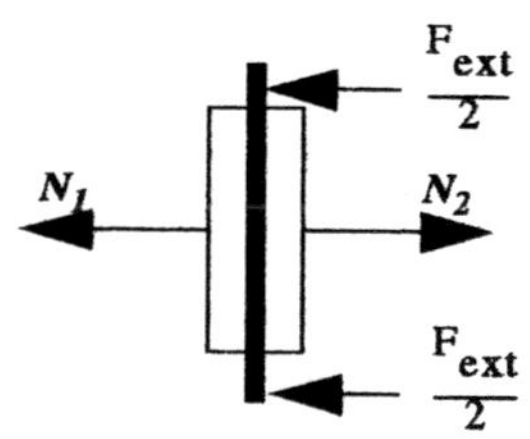

Template Equation $N_2 = N_1 + F_{ext}$

(b)

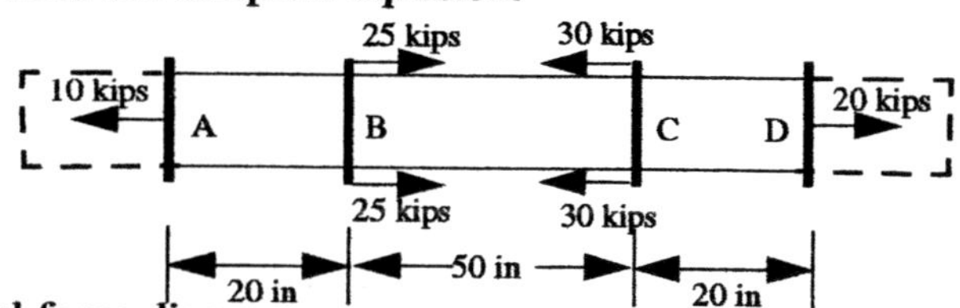

(c) Axial force diagram

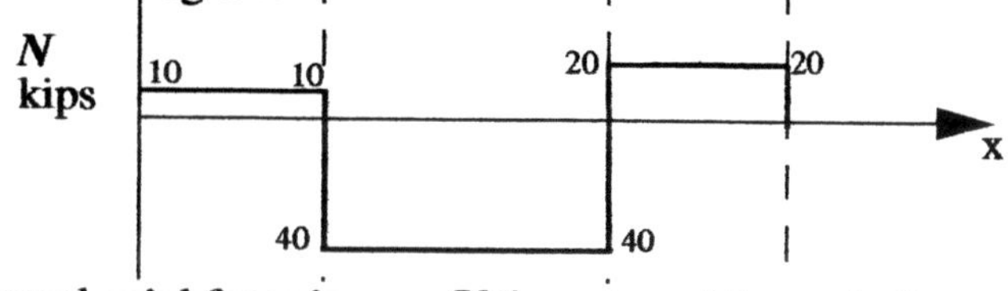

We start in the imaginary extension where the internal axial force is zero. If the external forces in figure (b) are in the direction shown on the template in figure (a), then we add the value as per the template equation, otherwise we subtract. In this manner we obtain the axial force diagram shown in figure (c).
By making imaginary cuts between A and B, between B and C, and between C and D, we obtain the free body diagrams shown in figures (d), (e), and (f), respectively.

(d)

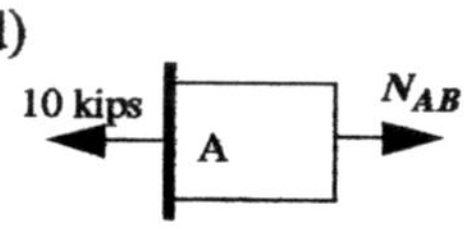

(e)

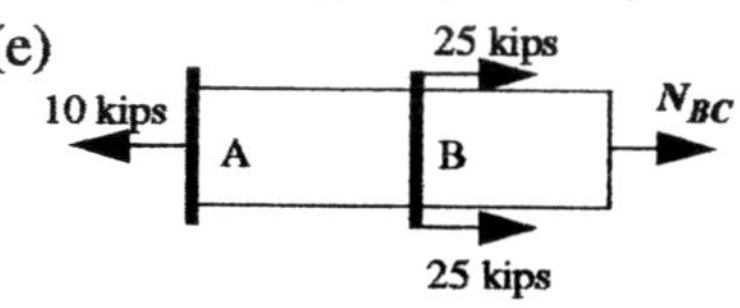

(f)

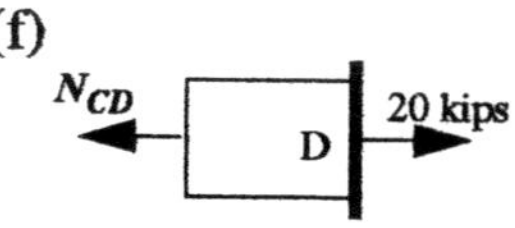

By equilibrium of forces in figure (d), we obtain: $\boxed{N_{AB} = 10 \text{ kips}}$ (1)

By equilibrium of forces in figure (d), we obtain: $\boxed{N_{BC} = -40 \text{ kips}}$ (2)

By equilibrium of forces in figure (d), we obtain: $\boxed{N_{CD} = 20 \text{ kips}}$ (3)

The value of internal forces in Eq's (1), (2) and (3) are the same as obtained from the axial force diagram in figure (c).

4.8 Draw the axial force diagrams and then check your results by finding internal forces in each segment AB, BC, and CD by making imaginary cut and drawing free body diagrams.

Solution $N_{AB} = ?$ $N_{BC} = ?$ $N_{CD} = ?$

By force equilibrium of the entire axial member, we obtain the reaction force at A as shown below.

$$R_A - 150 + 90 - 70 = 0 \qquad \text{or} \qquad R_A = 130 \text{ kN}$$

We draw a template as shown in figure (a) and write the template equation.

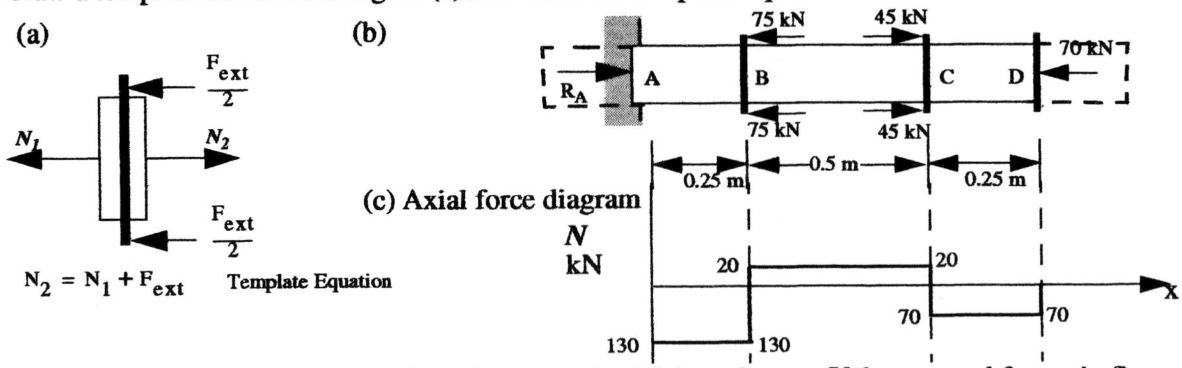

We start in the imaginary extension where the internal axial force is zero. If the external forces in figure (b) are in the direction shown on the template in figure (a), then we add the value as per the template equation, otherwise we subtract. In this manner we obtain the axial force diagram shown in figure (c).

By making imaginary cuts between A and B, between B and C, and between C and D, we obtain the free body diagrams shown in figures (d), (e), and (f), respectively.

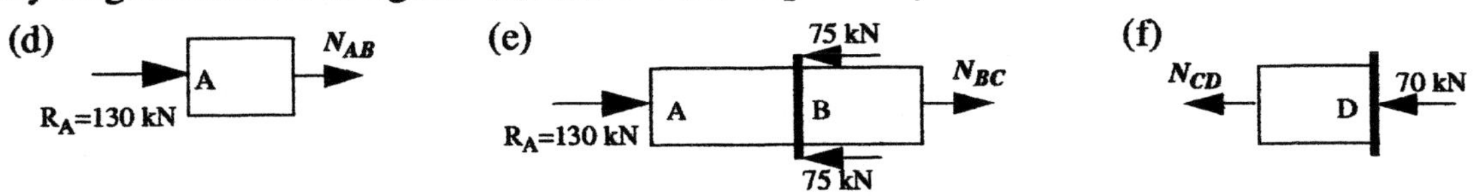

By equilibrium of forces in figure (d), we obtain: $\boxed{N_{AB} = -130 \text{ kN}}$ (1)

By equilibrium of forces in figure (d), we obtain: $\boxed{N_{BC} = 20 \text{ kN}}$ (2)

By equilibrium of forces in figure (d), we obtain: $\boxed{N_{CD} = -70 \text{ kN}}$ (3)

The value of internal forces in Eq's (1), (2) and (3) are the same as obtained from the axial force diagram in figure (c).

4.9 Draw the axial force diagrams and then check your results by finding internal forces in each segment AB, BC, and CD by making imaginary cut and drawing free body diagrams

Solution $N_{AB} = ?$ $N_{BC} = ?$ $N_{CD} = ?$

By force equilibrium of the entire axial member, we obtain the reaction force at D as shown below.

$$1.5 + 4 - 8 + R_D = 0 \qquad \text{or} \qquad R_D = 2.5 \text{ kips}$$

We draw a template as shown in figure (a) and write the template equation

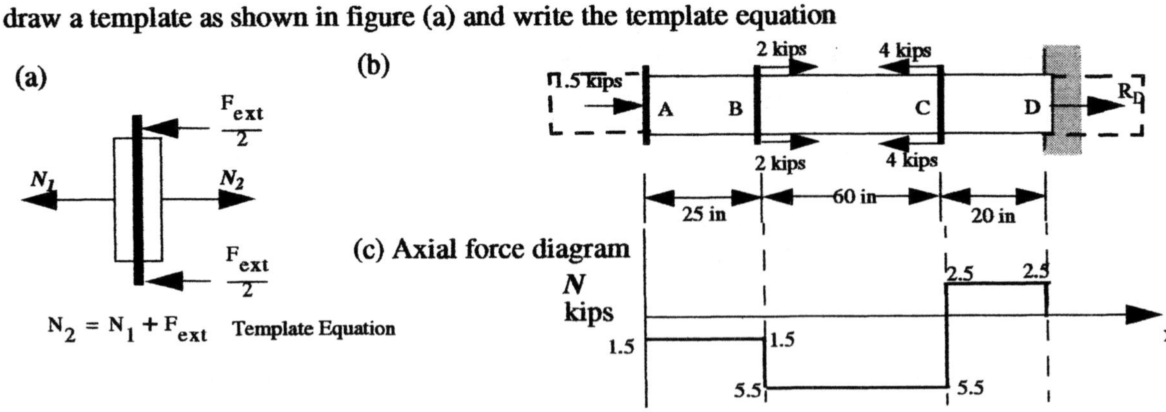

We start in the imaginary extension where the internal axial force is zero. If the external forces in figure (b) are in the direction shown on the template in figure (a), then we add the value as per the template equation,

otherwise we subtract. In this manner we obtain the axial force diagram shown in figure (c).
By making imaginary cuts between A and B, between B and C, and between C and D, we obtain the free body diagrams shown in figures (d), (e), and (f), respectively.

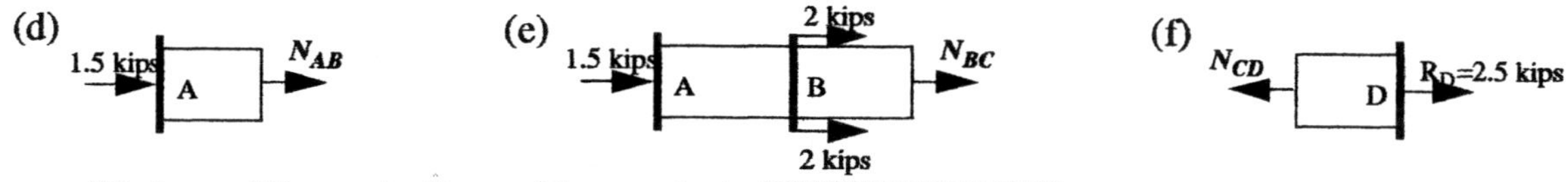

By equilibrium of forces in figure (d), we obtain: $\boxed{N_{AB} = -1.5 \text{ kips}}$ (1)

By equilibrium of forces in figure (d), we obtain: $\boxed{N_{BC} = -5.5 \text{ kips}}$ (2)

By equilibrium of forces in figure (d), we obtain: $\boxed{N_{CD} = 2.5 \text{ kips}}$ (3)

The value of internal forces in Eq's (1), (2) and (3) are the same as obtained from the axial force diagram in figure (c).

4.10 Draw the axial force diagrams and then check your results by finding internal forces in each segment AB, BC, and CD by making imaginary cut and drawing free body diagrams

Solution $N_{AB} = ?$ $N_{BC} = ?$ $N_{CD} = ?$

We draw a template as shown in figure (a) and write the template equation,

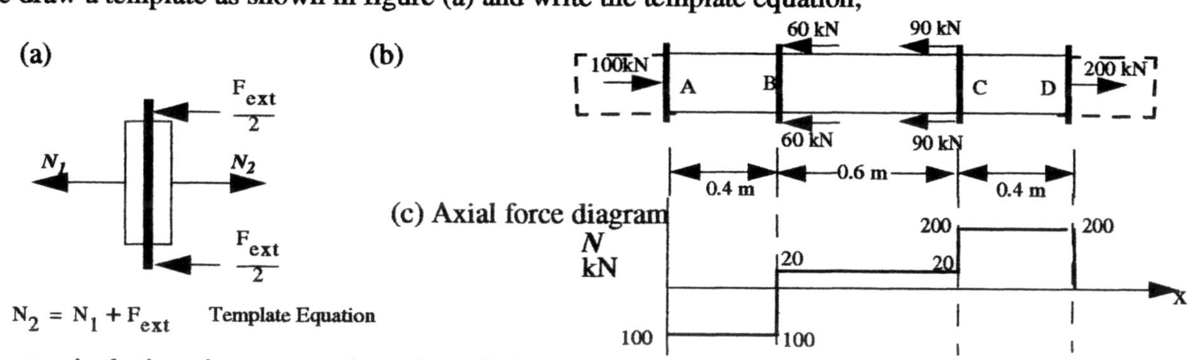

We start in the imaginary extension where the internal axial force is zero. If the external forces in figure (b) are in the direction shown on the template in figure (a), then we add the value as per the template equation, otherwise we subtract. In this manner we obtain the axial force diagram shown in figure (c).
By making imaginary cuts between A and B, between B and C, and between C and D, we obtain the free body diagrams shown in figures (d), (e), and (f), respectively.

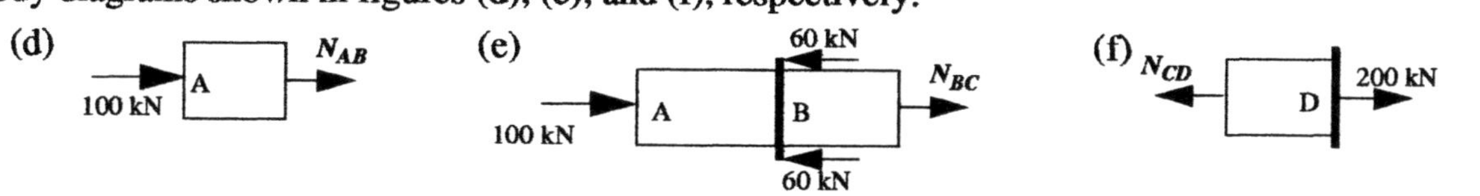

By equilibrium of forces in figure (d), we obtain: $\boxed{N_{AB} = -100 \text{ kN}}$ (1)

By equilibrium of forces in figure (d), we obtain: $\boxed{N_{BC} = 20 \text{ kN}}$ (2)

By equilibrium of forces in figure (d), we obtain: $\boxed{N_{CD} = 200 \text{ kN}}$ (3)

The value of internal forces in Eq's (1), (2) and (3) are the same as obtained from the axial force diagram in figure (c).

4.11 The axial rigidity of the bar in problem 4.7 is EA = 8,000 kips. Determine the movement of section at D with respect to section at A.

Solution EA = 8,000 kips $u_D - u_A = ?$

From problem 4.7 we have: $N_{AB} = 10 \text{ kips}$ $N_{BC} = -40 \text{ kips}$ $N_{CD} = 20 \text{ kips}$

$$u_B - u_A = \frac{N_{AB}(x_B - x_A)}{EA} = \frac{(10)(20)}{8000} = 0.025 \text{ in} \quad 1$$

$$u_C - u_B = \frac{N_{BC}(x_C - x_B)}{EA} = \frac{(-40)(50)}{8000} = -0.250 \text{ in} \quad 2$$

$$u_D - u_C = \frac{N_{CD}(x_D - x_C)}{EA} = \frac{(20)(20)}{8000} = 0.050 \text{ in} \quad 3$$

Adding Eq's (1), (2), and (3), we obtain:

$u_D - u_A = 0.025 - 0.25 + 0.05 = -0.175 \text{ in}$ or $\boxed{u_D - u_A = -0.175 \text{ in}}$

4.12 The axial rigidity of the bar in problem 4.8 is EA = 80,000 kN. Determine the movement of section at C with respect to section at A.

Solution EA = 80,000 kN $u_C = ?$

From problem 4.8 we have: $N_{AB} = -130 \text{ kN}$ $N_{BC} = 20 \text{ kN}$ $N_{CD} = -70 \text{ kN}$

$$u_B - u_A = \frac{N_{AB}(x_B - x_A)}{EA} = \frac{(-130)(0.25)}{80,000} = -0.4063(10^{-3}) \text{ m} \quad 1$$

$$u_C - u_B = \frac{N_{BC}(x_C - x_B)}{EA} = \frac{(20)(0.5)}{80,000} = 0.125(10^{-3}) \text{ m} \quad 2$$

Adding Eq's (1) and (2) we obtain: $u_C - u_A = (-0.4063 + 0.125)(10^{-3}) = -0.28125((10^{-3}) \text{ m})$

Point A is fixed to the wall, i.e., $u_A = 0$, thus we obtain: $\boxed{u_C = -0.281 \text{ mm}}$

4.13 The axial rigidity of the bar in problem 4.9 is EA = 2,000 kips. Determine the movement of section at B.

Solution EA = 2,000 kips $u_B = ?$

From problem 4.9 we have: $N_{AB} = -1.5 \text{ kips}$ $N_{BC} = -5.5 \text{ kips}$ $N_{CD} = 2.5 \text{ kips}$

$$u_C - u_B = \frac{N_{BC}(x_C - x_B)}{EA} = \frac{(-5.5)(60)}{2000} = -0.165 \text{ in} \quad 1$$

$$u_D - u_C = \frac{N_{CD}(x_D - x_C)}{EA} = \frac{(2.5)(20)}{2000} = 0.025 \text{ in} \quad 2$$

Adding Eq's (1) and (2) we obtain: $u_D - u_B = -0.165 + 0.025 = -0.140 \text{ in}$

Point D is fixed to the wall,i.e., $u_D = 0$, thus we obtain: $\boxed{u_B = 0.14 \text{ in}}$

4.14 The axial rigidity of the bar in problem 4.10 is EA = 50,000 kN. Determine the movement of section at D with respect to section at A.

Solution EA = 50,000 kN $u_D - u_A = ?$

From problem 4.10 we have: $N_{AB} = -100 \text{ kN}$ $N_{BC} = 20 \text{ kN}$ $N_{CD} = 200 \text{ kN}$

$$u_B - u_A = \frac{N_{AB}(x_B - x_A)}{EA} = \frac{(-100)(0.4)}{50,000} = -0.80(10^{-3}) \text{ m} \quad 1$$

$$u_C - u_B = \frac{N_{BC}(x_C - x_B)}{EA} = \frac{(20)(0.6)}{50,000} = 0.24(10^{-3}) \text{ m} \quad 2$$

$$u_D - u_C = \frac{N_{CD}(x_D - x_C)}{EA} = \frac{(200)(0.4)}{50,000} = 1.60(10^{-3})\ \text{m} \qquad 3$$

Adding Eq's (1),(2), and (3) we obtain:

$u_C - u_A = (-0.80 + 0.24 + 1.60)(10^{-3}) = 1.04((10^{-3})\ \text{m})$ or $\boxed{u_C = 1.04\ \text{mm}}$

4.15 Three segments of 4 in x 2 in rectangular wooden bars (E = 1600 ksi) are secured together with rigid plates and subjected to axial forces as shown. Determine: (a) the movement of rigid plate at D with respect to plate at A. (b) the maximum axial stress.

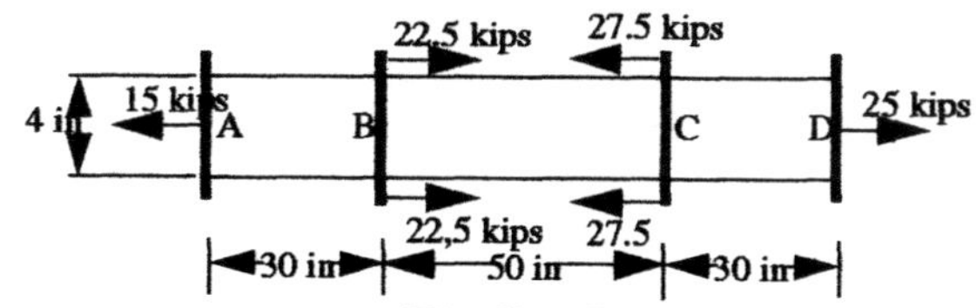

Fig. P4.15

Solution E = 1600 ksi $A = (4)(2) = 8\ \text{in}^2$ $u_D - u_A = ?$ $\sigma_{max} = ?$

We draw a template as shown in figure (a) and write the template equation. Using the template and template equation, we draw the axial force diagram as shown in figure (b)

(a)

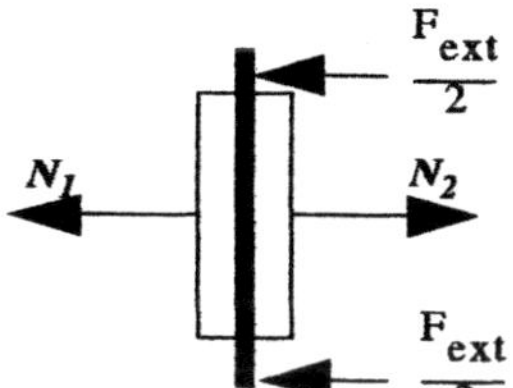

Template Equation $N_2 = N_1 + F_{ext}$

(b) Axial force diagram

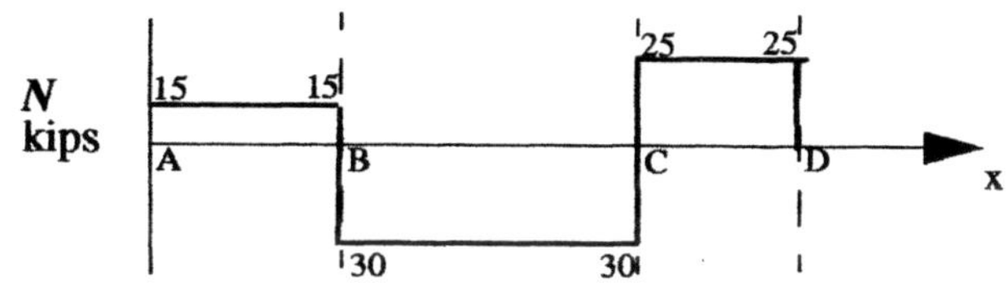

From figure (b) the internal axial forces are: $N_{AB} = 15\ \text{kips}$, $N_{BC} = -30\ \text{kips}$, and $N_{CD} = 25\ \text{kips}$.

$$u_B - u_A = \frac{N_{AB}(x_B - x_A)}{EA} = \frac{(15)(30)}{(1600)(8)} = 0.0352\ \text{in} \qquad 1$$

$$u_C - u_B = \frac{N_{BC}(x_C - x_B)}{EA} = \frac{(-30)(50)}{(1600)(8)} = -0.1172\ \text{in} \qquad 2$$

$$u_D - u_C = \frac{N_{CD}(x_D - x_C)}{EA} = \frac{(25)(30)}{(1600)(8)} = 0.0586\ \text{in} \qquad 3$$

Adding Eq's (1), (2), and (3), we obtain:

$u_D - u_A = 0.0352 - 0.1172 + 0.0586 = -0.0234\ \text{in}$ or $\boxed{u_D - u_A = -0.0234\ \text{in}}$

With all segments having the same area of cross-section, the maximum axial stress will be in the segment with the maximum internal axial force, i.e., segment BC. The axial stress in segment BC can be found as shown below.

$\sigma_{max} = \frac{N_{BC}}{A} = \frac{-30}{8} = -3.75\ \text{ksi}$ $\boxed{\sigma_{max} = 3.75\ \text{ksi (C)}}$

4.16 Two circular steel bars (E=30,000 ksi, ν = 0.3) of diameter 2 inch are securely connected to an aluminum bar (E=10,000 ksi, ν = 0.33) of diameter 1.5 inch. Determine (a) the displacement of section at C with respect to the wall. (b) the maximum change in the diameter.

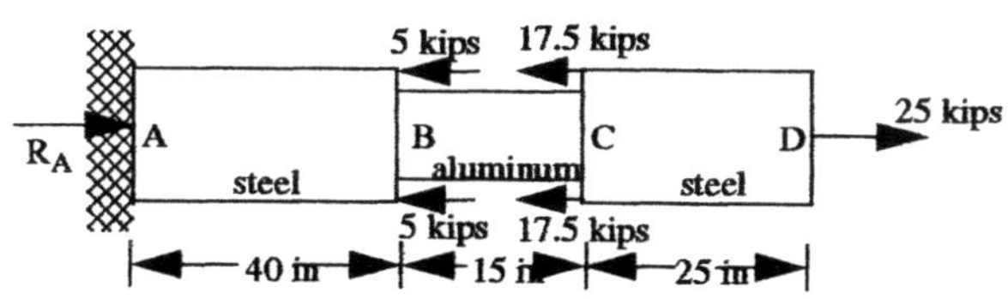

Fig. P4.16

Solution $E_{AB} = E_{CD} = 30,000$ ksi $E_{BC} = 10,000$ ksi $\nu_{AB} = \nu_{CD} = 0.3$ $\nu_{BC} = 0.33$

$d_{AB} = d_{CD} = 2.0$ in $d_{BC} = 1.5$ in $u_C - u_A = ?$ $\Delta d_{max} = ?$

By force equilibrium of the entire axial member, we obtain the reaction force at A as shown below.

$$R_A - 10 - 35 + 25 = 0 \qquad \text{or} \qquad R_A = 20 \text{ kips}$$

We draw a template as shown in figure (a) and write the template equation. Using the template and template equation, we draw the axial force diagram as shown in figure (b)

(a)

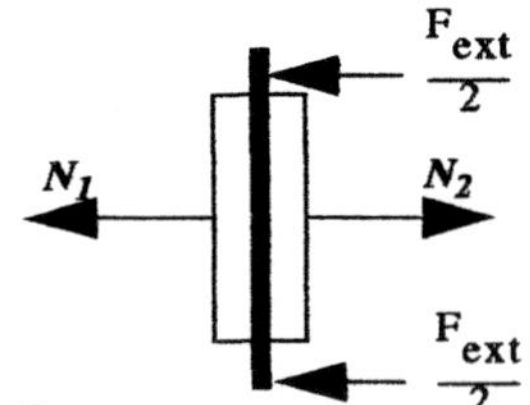

$N_2 = N_1 + F_{ext}$ Template Equation

(b) Axial force diagram

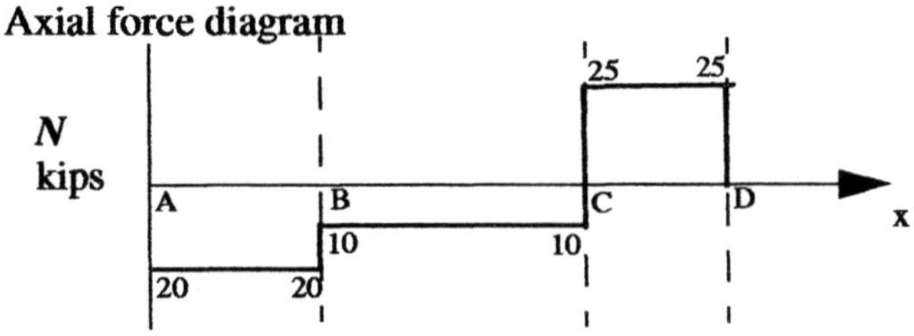

From figure (b) the internal axial forces are: $N_{AB} = -20 \text{ kips}$, $N_{BC} = -10 \text{ kips}$, and $N_{CD} = 25 \text{ kips}$.

The area of cross-sections are: $A_{AB} = A_{CD} = \pi(2^2)/4 = 3.142 \text{ in}^2$, and $A_{BC} = \pi(1.5^2)/4 = 1.767 \text{ in}^2$

$$u_B - u_A = \frac{N_{AB}(x_B - x_A)}{EA} = \frac{(-20)(40)}{(30,000)(3.142)} = -8.488(10^{-3}) \text{ in} \qquad (1)$$

$$u_C - u_B = \frac{N_{BC}(x_C - x_B)}{EA} = \frac{(-10)(15)}{(10,000)(1.767)} = -8.488(10^{-3}) \text{ in} \qquad (2)$$

Adding Eq's (1) and (2)

$u_C - u_A = (-8.488 - 8.488)(10^{-3}) = (-16.98)(10^{-3}) \text{ in}$ or $\boxed{u_C - u_A = -0.017 \text{ in}}$

The maximum change in diameter will occur in that segment in which the axial stress is maximum. As area of cross-sections for AB and CD are the same, we conclude that CD will have a larger stress as the axial force is larger. We calculate the axial stress in BC and CD as shown below:

$$\sigma_{BC} = \frac{N_{BC}}{A_{BC}} = \frac{(-10)}{1.767} = -5.659 \text{ ksi} \text{ and } \sigma_{CD} = \frac{N_{CD}}{A_{CD}} = \frac{(25)}{3.142} = 7.958 \text{ ksi}$$

The axial strain in BC and CD is:

$$\varepsilon_{BC} = \frac{\sigma_{BC}}{E_{BC}} = \frac{(-5.659)}{10,000} = (-0.5659)(10^{-3}) \text{ and } \varepsilon_{CD} = \frac{\sigma_{CD}}{E_{CD}} = \frac{(7.958)}{30,000} = 0.2653(10^{-3})$$

The transverse strain in BC and CD is:

$(\varepsilon_{BC})_T = -\nu_{BC}\varepsilon_{BC} = -(0.3)(-0.5659)(10^{-3}) = 0.1698(10^{-3})$ and

$(\varepsilon_{CD})_T = -\nu_{CD}\varepsilon_{CD} = -(0.33)(0.2653)(10^{-3}) = -0.0875(10^{-3})$.

The change in diameter in BC and CD is:

$\Delta d_{BC} = (1.5)(0.1698)(10^{-3}) = (0.0257)(10^{-3}) \text{ in}$ and $\Delta d_{CD} = (2)(-0.0875)(10^{-3}) = (-0.175)(10^{-3}) \text{ in}$

The maximum change in diameter occurs in BC. It's value is: or $\boxed{\Delta d_{max} = 0.00025 \text{ in}}$

4.17 Two cast iron pipes (E = 100 GPa) are adhesively bonded together as shown. The outer diameters of the two pipes are 50 mm and 70 mm and wall thickness of each pipe is 10 mm. Determine the displacement of end B with respect to end A.

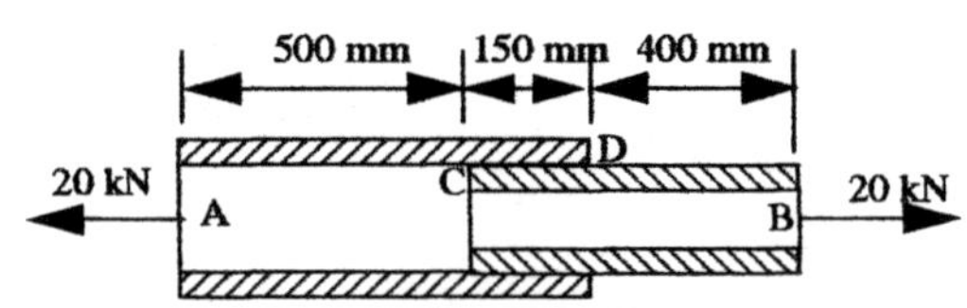

Fig. P4.17

Solution $E = 100GP_a$ $(d_{AC})_o = 70$ mm $(d_{AC})_i = 50$ mm $(d_{DB})_o = 50$ mm

$(d_{DB})_i = 30$ mm $(d_{CD})_o = 70$ mm $(d_{CD})_i = 30$ mm $u_B - u_A = ?$

The internal normal force in each segment is: $N_{AC} = N_{CD} = N_{DB} = 20\ \text{kN}$. The area of material cross-sections are as shown below.

$$A_{AC} = \frac{\pi[(0.07)^2 - (0.05)^2]}{4} = 1.885(10^{-3})\ \text{m}^2$$

$$A_{DB} = \frac{\pi[(0.05)^2 - (0.03)^2]}{4} = 1.257(10^{-3})\ \text{m}^2$$

$$A_{CD} = \frac{\pi[(0.07)^2 - (0.03)^2]}{4} = 3.142(10^{-3})\ \text{m}^2$$

The relative displacement of the ends of each segment can be written as shown below.

$$u_C - u_A = \frac{N_{AC}(x_C - x_A)}{E_{AC}A_{AC}} = \frac{(20)(10^3)(0.5)}{(100)(10^9)(1.885)(10^{-3})} = 0.05305(10^{-3})\ \text{m} \quad 1$$

$$u_D - u_C = \frac{N_{CD}(x_D - x_C)}{E_{CD}A_{CD}} = \frac{(20)(10^3)(0.15)}{(100)(10^9)(3.142)(10^{-3})} = 0.00955\ (10^{-3})\ \text{m} \quad 2$$

$$u_B - u_D = \frac{N_{BD}(x_B - x_D)}{E_{BD}A_{BD}} = \frac{(20)(10^3)(0.4)}{(100)(10^9)(1.257)(10^{-3})} = 0.06364\ (10^{-3})\ \text{m} \quad 3$$

Adding Eq's (1), (2), and (3)

$u_B - u_A = (0.05305 + 0.00955 + 0.06364)(10^{-3}) = 0.12624(10^{-3})$ m or $\boxed{u_B - u_A = 0.126\ \text{mm}}$

4.18 The tapered bar shown in Fig. P4.18 has a cross-sectional area that varies with x as $A = K(2\ L - 0.25\ x)^2$. Determine the elongation of the bar in terms of P, L, E and K.

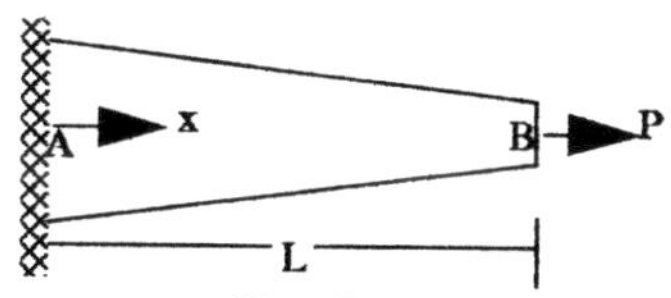

Fig. P4.18

Solution $u_B - u_A = ?$

From Eq. 4.8 we have $\frac{du}{dx} = \frac{N_{AB}}{E_{AB}A_{AB}} = \frac{P}{E[k(2L - 0.25x)^2]}$.

Integrating we obtain: $\int_0^L du = \frac{P}{Ek}\int_0^L \frac{dx}{(2L - 0.25x)^2}$

$u = \frac{P}{Ek}\left[\frac{1}{0.25(2L - 0.25x)}\bigg|_0^L\right]$ or $u = \frac{P}{Ek}\left[\frac{1}{0.4375L} - \frac{1}{0.5L}\right]$ or $\boxed{u = 0.2857\frac{P}{EkL}}$

4.19 The tapered bar shown in Fig. P4.18 has a cross-sectional area that varies with x as $A = K(4\ L - 3\ x)$. Determine the elongation of the bar in terms of P, L, E and K.

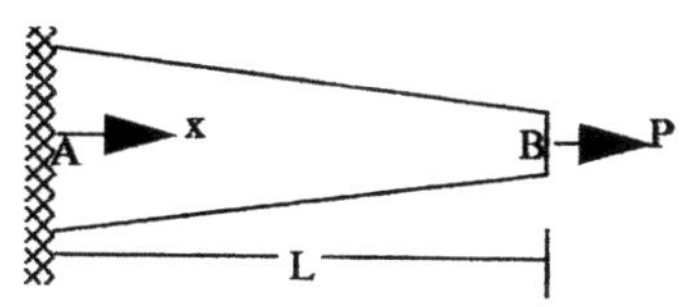

Solution $u_B - u_A = ?$

From Eq. 4.8 we have $\frac{du}{dx} = \frac{N_{AB}}{E_{AB}A_{AB}} = \frac{P}{EK(4L - 3x)}$. Integrating we obtain: $\int_0^L du = \int_0^L \frac{P}{EK}\cdot\frac{dx}{(4L - 3x)}$

or $u = \frac{P}{3EK}[\ln|4L - 3x|\,|_0^L]$ or $u = -\frac{P}{3EK}[\ln L - \ln 4L]$ or $u = -\frac{P}{3EK}\left[\ln\frac{1}{4}\right]$ or $\boxed{u = 0.4621\frac{P}{EK}}$

4.20 A tapered and un-tapered solid circular steel bars (E=30,000 ksi) are securely fastened to a solid circular aluminum bar (E=10,000 ksi). The un-tapered steel bar has a diameter of 2 inches. The aluminum bar has a diameter of 1.5 inches. The tapered bars diameter varies from 1.5 inches to 2 inches.(a) Determine the displacement of section at C with respect to the section at A (b) Determine the maximum axial stress in the bar.

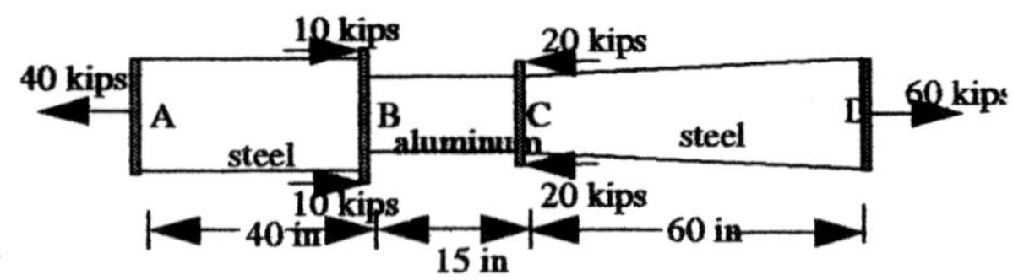

Fig. P4.20

Solution $E_{AB} = E_{CD} = 30,000$ ksi $E_{BC} = 10,000$ ksi

$d_{AB} = 2.0$ in $d_{BC} = 1.5$ in d_{CD} varies from 1.5 in to 2 in

$u_C - u_A = ?$ $\sigma_{max} = ?$

We draw a template as shown in figure (a) and write the template equation. Using the template and template equation, we draw the axial force diagram as shown in figure (b)

(a)

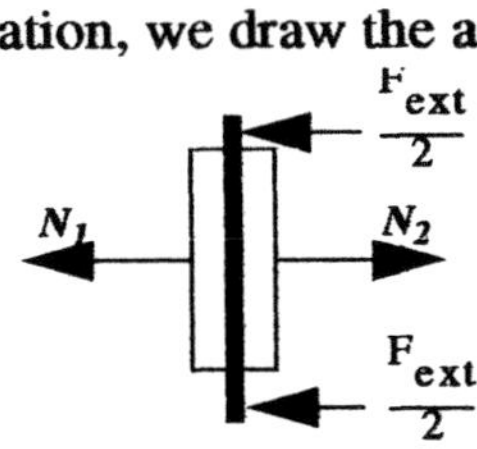

Template Equation

$N_2 = N_1 + F_{ext}$

(b) Axial force diagram

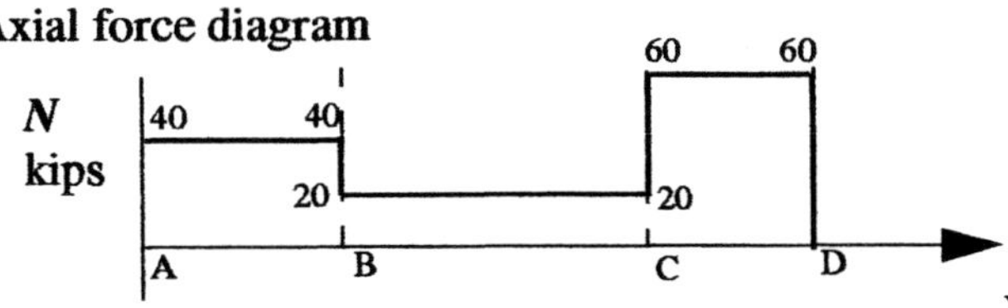

From figure (b) the internal axial forces are: $N_{AB} = 40$ kips , $N_{BC} = 20$ kips , and $N_{CD} = 60$ kips .

$$u_B - u_A = \frac{N_{AB}(x_B - x_A)}{E_{AB}A_{AB}} = \frac{(40)(40)}{(30,000)(\pi 2^2/4)} = 16.976(10^{-3}) \text{ in} \quad 1$$

$$u_C - u_B = \frac{N_{BC}(x_C - x_B)}{E_{BC}A_{BC}} = \frac{(20)(15)}{(10,000)(\pi 1.5^2/4)} = 16.976(10^{-3}) \text{ in} \quad 2$$

Adding Eq's (1), (2), and (3), we obtain:

$u_C - u_A = (16.976 + 16.976)(10^{-3}) = 33.952(10^{-3})$in or $\boxed{u_C - u_A = 0.034 \text{ in}}$

The axial stress in CD will be maximum where the area of cross-section is minimum. Thus, the maximum stress in CD will be just after C, where the diameter is 1.5 in.This stress will be greater than the stress in BC as the diameter there is also 1.5 but axial force is smaller.

$$\sigma_{max} = \frac{N_{CD}}{A} = \frac{60}{(\pi 1.5^2/4)} = 33.95$$ $\boxed{\sigma_{max} = 33.95 \text{ ksi (T)}}$

4.21 The column shown has a length L, Modulus of elasticity E, specific weight γ, and radius a. Determine the contraction of the column in terms of L, E, γ, and a.

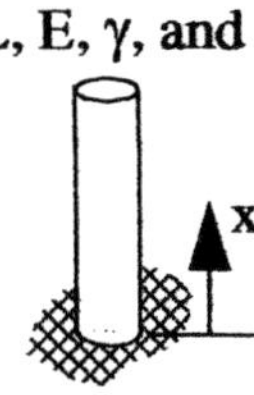

Solution $u_B - u_A = ?$

The distributed force is product of specific weight and the area of cross-section and it acts in the negative x-direction: $p = -\gamma A = -\gamma\pi a^2$

From Eq. 4.14 we have: $\frac{dN}{dx} = -p = \gamma\pi a^2$. Integrating we obtain: $N = \gamma\pi a^2 x + c_1$

at $x = L$ $\qquad N = 0$ We obtain: $c_1 = -\gamma\pi a^2 L$ and $N = \gamma\pi a^2 x - \gamma\pi a^2 L$

From Eq. 4.8 we have: $\frac{du}{dx} = \frac{N}{EA} = \frac{\gamma\pi a^2}{E\pi a^2}(x-L) = \frac{\gamma}{E}(x-L)$

Integrating we obtain: $\int_{u_A}^{u_B} du = \frac{\gamma}{E}\int_0^L (x-L)dx$ or $u_B = \frac{\gamma}{E}\left(\frac{x^2}{2} - Lx\right)\Bigg|_0^L$ $\qquad \boxed{u_B = \frac{-\gamma L^2}{2E}}$

4.22 The column shown has a length L, Modulus of elasticity E, specific weight γ, and length a as the side of an equilateral triangle. Determine the contraction of the column in terms of L, E, γ, and a.

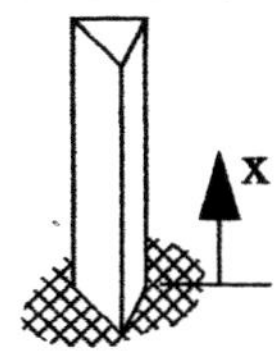

Solution $\qquad u_B - u_A = ?$

The area of cross-section is: $A = \frac{1}{2}ab = \frac{\sqrt{3}}{4}a^2$. The distributed force is product of specific weight and the area of cross-section and it acts in the negative x-direction: $p = -\gamma\left(\frac{\sqrt{3}}{4}a^2\right)$

From Eq. 4.14 we have: $\frac{dN}{dx} = -p = \frac{\sqrt{3}}{4}\gamma a^2$. Integrating we obtain: $N = \frac{\sqrt{3}}{4}\gamma a^2 x + c_1$

at $x = L$ $\qquad N = 0$ we obtain: $c_1 = -\frac{\sqrt{3}}{4}\gamma a^2 L$ and $N = \frac{\sqrt{3}}{4}\gamma a^2(x-L)$

From Eq. 4.8 we have: $\frac{du}{dx} = \left[\frac{N}{EA}\right] = \frac{\sqrt{3}}{4}\frac{\gamma\pi a^2}{E\frac{\sqrt{3}}{4}a^2}(x-L) = \frac{\gamma}{E}(x-L)$

Integrating we obtain: $\int_{u_A}^{u_B} du = \frac{\gamma}{E}\int_0^L (x-L)dx$ or $u_B = \frac{\gamma}{E}\left(\frac{x^2}{2} - Lx\right)\Bigg|_0^L$ $\qquad \boxed{u_B = \frac{-\gamma L^2}{2E}}$

4.23 The columns shown has a length L, Modulus of elasticity E, specific weight γ, and cross-sectional area A. Determine the contraction of the column in terms of L, E, γ, and A.

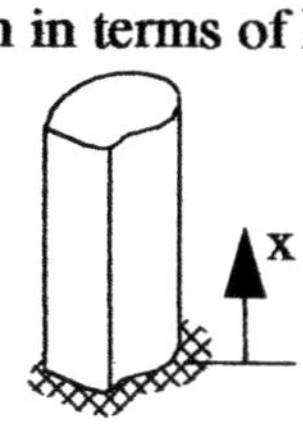

Solution $\qquad u_B - u_A = ?$

The distributed force is product of specific weight and the area of cross-section and it acts in the negative x-direction: $p = -\gamma A$

From Eq. 4.14 we have: $\frac{dN}{dx} = -p = \gamma A$. Integrating we obtain: $N = \gamma A x + c_1$

at $x = L$ $\qquad N = 0$ we obtain: $c_1 = -\gamma A L$ and $N = \gamma A(x-L)$

From Eq. 4.8 we have: $\frac{du}{dx} = \left[\frac{N}{EA}\right] = \gamma A(x-L)$

Integrating we obtain: $\int_{u_A}^{u_B} du = \gamma A\int_0^L (x-L)dx$ or $u_B - u_A = \frac{\gamma A}{EA}\left(\frac{x^2}{2} - Lx\right)\Big|_0^L = \frac{\gamma}{2E}\left(\frac{-L}{2}\right)$ or $\boxed{u_B = -\frac{\gamma L^2}{2E}}$

4.24 On the truncated cone of Example 4.6 a force of $P = (\gamma\pi r^2 L/5)$ is also applied as shown. Determine the total elongation of the cone due to its weight and the applied force. (Hint: use superposition)

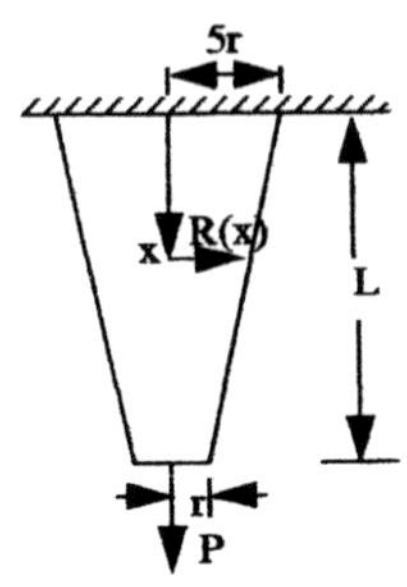

Fig. P4.24

Solution $\delta = ?$

Let the total elongation $\delta = \delta_W + \delta_P$, where δ_W is the elongation due to weight and δ_P is the elongation due to the force P. From example 4.6 we have:

$$\delta_W = \frac{7}{30}\frac{\gamma L^2}{E} \quad (1)$$

and $R(x) = \frac{r}{L}(5L - 4x)$ or $A(x) = \pi\frac{r^2}{L^2}(5L-4x)^2$ (2)

The free body diagram at any location x is:

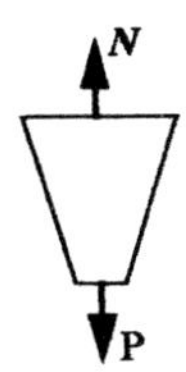

By equilibrium of forces we obtain: $N = P = \frac{\gamma\pi r^2 L}{5}$

From Eq. 4.14 we have: $\frac{du}{dx} = \frac{N}{EA} = \frac{\gamma\pi r^2 L}{5E\pi\frac{r^2}{L^2}(5L-4x)^2} = \frac{\gamma L^3}{5E(5L-4x)^2}$

Integrating from x = 0 to x = L we can obtain δ_P as shown below

$$\delta_P = \int_{u_0}^{u_L} du = \int_0^L \frac{\gamma L^3}{5E(5L-4x)^2}dx = \frac{\gamma L^3}{5E}\left[\frac{1}{-(-4)}\cdot\frac{1}{(5L-4x)}\right] \text{ or}$$

$$\delta_P = \frac{\gamma L^3}{20E}\left[\frac{1}{L} - \frac{1}{5L}\right] = \frac{\gamma L^2}{25E} \quad (3)$$

Adding equation (1) and (3), we obtain the total elongation as shown below

$$\delta = \frac{7}{30}\frac{\gamma L^2}{E} + \frac{\gamma L^2}{25E} = \frac{41}{150}\frac{\gamma L^2}{E} \text{ or } \quad \boxed{\delta = 0.273\frac{\gamma L^2}{E}}$$

4.25 A 20 feet tall, thin-hollow tapered tube of uniform wall thickness 1/8 inch is used for a light pole in a parking lot. The mean diameter at the bottom is 8 inches and at the top it is 2 inches. The weight of the lights on top of the pole is 80 lbs. The pole is made of aluminum alloy with a specific weight of 0.1 lb/in^3 and Modulus of Elasticity E = 11,000 ksi. Determine (a) the maximum axial stress. (b) the contraction of the pole.

(Hint: Approximate the area of cross-section as $A = \pi Dt$, where D is the mean diameter and t is the thickness of the thin-walled pole.)

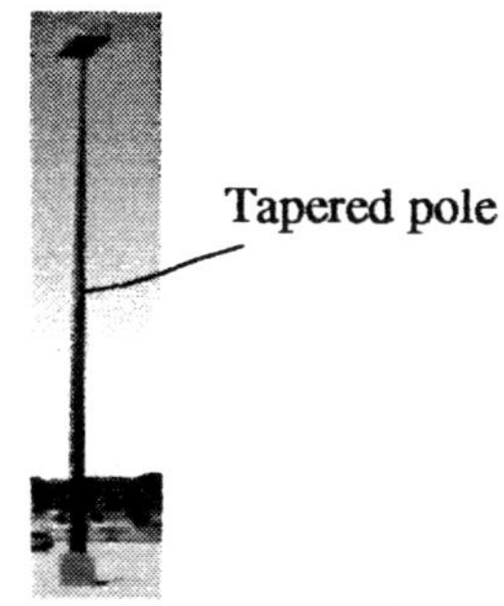

Fig. P4.25

Solution $L = 20ft$ $t = 118in$ $d_b = 8in$ $d_t = 2in$ $W_{Light} = 80lbs$

$\gamma = 0.1\ lb/in^3$ $E = 11,000ksi$ $G = 8000ksi$ $\sigma_{max} =$ $\delta = ?$

The geometry off the pole in two-dimension is as shown below.

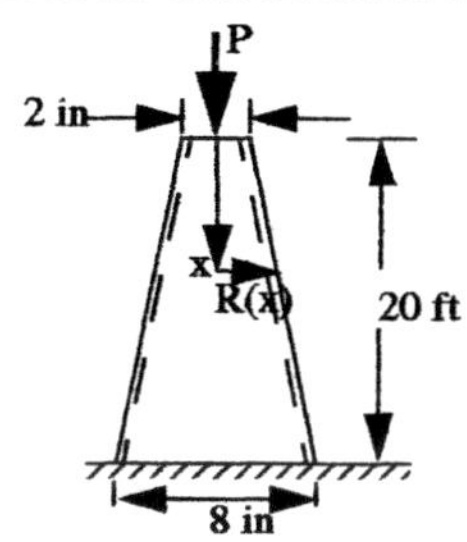

Let R(x) be the outer radius of the pole at any x.

$$R(x) = a + bx$$ 1

At x = 0 R(x) = 1 inch . Thus: a = 1

At x = 20 ft = 240 in R(x) = 4 inch, thus $4 = 1 + b(240)$ or $b = \frac{1}{80}$

$$R(x) = \frac{(80 + x)}{80}\ in$$ 2

For thin-walled tubes the area of cross-section can be approximated as $A = 2\pi Rt$ or

$$A(x) = (2\pi)\frac{(80 + x)}{80}\left(\frac{1}{8}\right) = 9.817(10^3)(80 + x)$$ 3

With gravity being the same direction as x

$$p(x) = \gamma A = 0.9817(10^{-3})(80 + x)$$ 4

From Eq. 4.14 we have the following.

$$\frac{dN}{dx} = -p = -0.9817(10^{-3})(80 + x)$$ 5

Integrating equation (5): $N = -0.4909(10^{-3})(80 + x)^2 + c_1$

At x = 0 N = -80lbs, thus $-80 = -0.4909(10^{-3})(80)^2 + c_1$ or $c_1 = -76.858$

$$N = -0.4909(10^{-3})(80 + x)^2 - 76.858$$ 6

The maximum axial force will be at the bottom of the pole i.e., x = 240 in

$$N_{max} = -0.4909(10^{-3})(320)^2 - 76.858 = -127.12lbs$$ 7

From Eq. (3) the area of cross-section at the bottom is $A_b = 9.817(10^3)(320) = 3.1416\ in^2$

The maximum axial force is: $\sigma_{max} = \frac{N_{max}}{A_b} = \frac{-127.12}{3.1416} = -54.78$ $\boxed{\sigma_{max} = 54.8\ psi(C)}$

From Eq. 4.8 we have: $\frac{du}{dx} = \frac{N}{EA} = \frac{-0.4909(10^{-3})(80+x)^2 - 76.858}{11(10^6)L9.817(10^{-3})(80+x)}$ or

$$\frac{du}{dx} = -0.4545(10^{-9})(80+x) - \frac{0.7117(10^{-3})}{80+x}$$

Integrating from x = 0 to x = 240

$$\int_{u_0}^{u_{240}} du = -0.4545(10^{-9})\int_0^{240}(80+x)dx - 0.7117(10^{-3})\int_0^{240}\frac{dx}{80+x} \quad \text{or}$$

$$u_{240} - u_0 = -0.4545(10^{-9})\frac{(80+x)^2}{2}\Bigg|_0^{240} - 0.7117(10^{-3})\ln(80+x)\Big|_0^{240} = -1.205(10^{-3})$$

Now $u_{240} = 0$ as the point is fixed to the ground and we obtain the contraction $\delta = u_0 = 1.205(10^{-3})$

$$\boxed{\delta = 0.0012 \text{ in}}$$

4.26 Determine the contraction of a column due to its own weight. The specific weight γ, the Modulus of Elasticity E, the length L, and the radius R are as given.

$\gamma = 0.28\ \text{lb/in}^3 \qquad E = 3,600\ \text{ksi}$

$L = 120\ \text{in} \qquad R = \sqrt{240 - x}$

where, R and x are in inches.

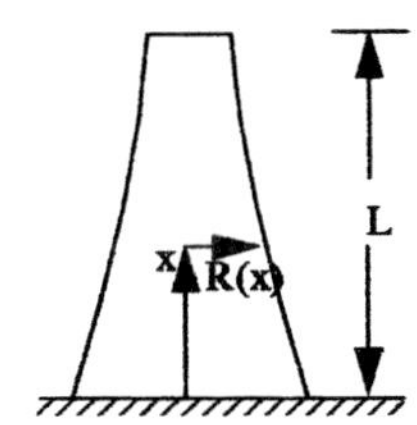

Fig. P4.26

Solution $\delta = ?$

Gravity is in the negative x-direction hence $p = -\gamma A$ or $p = -\gamma\pi R^2 = -0.28\pi(240 - x)$

From Eq. 4.14 we have the following.

$$\frac{dN}{dx} = -p = 0.28\pi(240 - x) \qquad 1$$

Integrating from x = L = 120, where N = 0 to any local x we obtain

$$\int_0^N dN = \int_{120}^x 0.2817(240-x)dx \text{ or } N = -\frac{(0.28\pi)}{2}(240-x)^2\Big|_{120}^x = -0.14\pi[(240-x)^2 - 120^2]$$

From Eq. 4.14 we have the following.

$$\frac{du}{dx} = \frac{N}{EA} = \frac{-0.14\pi[(240-x)^2 - 120^2]}{(3\cdot 6)(10^6)\pi(240-x)} \quad \text{or} \quad \frac{du}{dx} = -3.889(10^{-6})\left[(240-x) - \frac{120^2}{240-x}\right] \qquad 2$$

Integrating from x = 0 to x = 120

$$\int_{u_0}^{u_{120}} du = -3.889(10^{-6})\int_0^{120}\left[(240-x) - \frac{120^2}{(240-x)}\right]dx \quad \text{or}$$

$$u_{120} - u_0 = -3.889(10^{-6})\left[\frac{-(240-x)^2}{2} + 120^2\ln(240-x)\right]$$

Now point at x = 0 is fixed to the ground, i.e., $u_0 = 0$ and we obtain the following.

$$u_{120} = -3.889(10^{-6})\left[\frac{-120^2}{2} + \frac{240^2}{2} + 120^2\ln(240-x)\right] \quad \text{or} \quad u_{120} = -45.18(10^{-3})\ \text{inches}$$

The top point moves in negative x-direction, hence it is in contraction $\boxed{\delta = 0.045 \text{ in}}$

4.27 Determine the contraction of a column due to its own weight. The specific weight γ, the Modulus of Elasticity E, the length L, and the radius R are as given.

$$\gamma = 24\ \text{kN/m}^3 \qquad E = 25\ \text{GPa}$$
$$L = 10\ \text{m} \qquad R = 0.5e^{-0.07x}$$

where, R and x are in meters.

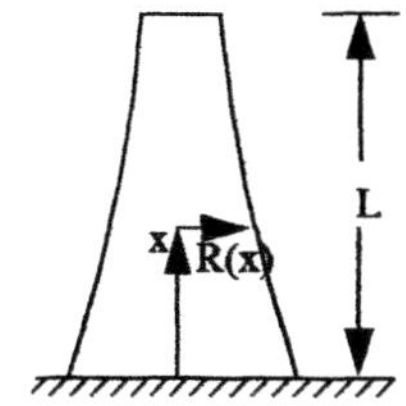

Fig. P4.27

Solution $\delta = ?$

Gravity is in the negative x-direction, hence $p = -\gamma A = -\gamma\pi R^2 = -24(10^3)\pi(0.5e^{-0.07x})^2$ or $p = -18.85(10^3)e^{0.14x}$. From Eq. 4.14 we have the following.

$$\frac{dN}{dx} = -p = 18.85(10^3)e^{-0.14x} \qquad 1$$

Integrating from x = L = 10 m, where $N = 0$ to any location x, we obtain

$$\int_0^N dN = \int_{10}^x (18.85\pi)(10^3)e^{-0.14x}dx \text{ or } N = -18.85\pi(10^3)\left(\frac{e^{-0.14x}}{0.14}\right)\Bigg|_{10}^{x} \text{ or}$$

$$N = -134.64(10^3)(e^{-0.14x} - e^{-1.4}) \qquad 2$$

From Eq. 4.14 we have the following:

$$\frac{du}{dx} = \frac{N}{EA} = \frac{-134.64(10^3)(e^{-0.14} - 0.2466)}{25(10^9)\pi[(0.5)^2e^{-0.14x}]} \text{ or } \frac{du}{dx} = -6.857(10^{-6})(1 - 0.2466e^{0.14x})$$

Integrating from x = 0 to x = 10 m

$$\int_{u_0}^{u_{10}} du = -6.857(10^{-6})\int_0^{10}(1 - 0.2466e^{0.14})dx \quad \text{or } u_{10} - u_0 = -6.857(10^{-6})\left(x - \frac{0.2466e^{0.14x}}{0.14}\right)\Bigg|_0^{10}$$

Point x = 0 is fixed to the ground i.e., $u_0 = 0$

$$u_{10} = -6.857(10^{-6})[10 - 1.761(e^{1.4} - 1)] = -31.67(10^{-6})$$

The top point moves in negative x-direction, hence it is contraction $\boxed{\delta = 0.317\ \text{mm}}$

4.28 The frictional force per unit length on a cast iron pipe being pulled from the ground varies as a quadratic function as shown. Determine the force F needed to pull the pipe out of ground and the elongation of the pipe before the pipe slips in terms of the modulus of elasticity E, area of cross-section A, length L and the maximum value of frictional force f_{max}.

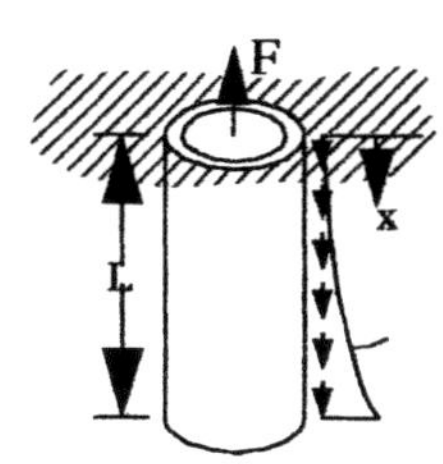

Fig. P4.28

Solution $F = ?$ $\delta = ?$

The distributed force is: $p(x) = f_{max}(x^2/L^2)$. From Eq. 4.14 we have the following.

$$\frac{dN}{dx} = -p(x) = -f_{max}(x^2/L^2) \qquad 1$$

Integrating from x = L, where N = 0, to any location x, we obtain $\int_0^N dN = -\int_L^x f_w - x\frac{x^2}{L^2}dx$ or

$$N = \left.\frac{-f_{max}x^3}{3L^2}\right|_L^x = \frac{f_{max}}{3L^2}(L^3 - x^3) \qquad 2$$

At $x = 0$ $\quad N = F$; Thus, $\boxed{F = \frac{f_{max}L}{3}}$

From Eq. 4.14 we have: $\frac{du}{dx} = \frac{N}{EA} = \frac{f_{max}}{3EAL^2}(L^3 - x^3)$ Integrating from $x = 0$ to $x = L$

$$\int_{u_0}^{u_L} du = \frac{f_{max}}{3EAL^2}\int_0^L (L^3 - x^3)dx \text{ or } u_L - u_0 = \left.\frac{f_{max}}{3EAL^2}\left(L^3x - \frac{x^4}{4}\right)\right|_0^L = \frac{f_{max}}{3EAL^2}\left(L^4 - \frac{L^4}{4}\right) \text{ or}$$

$$\boxed{u_L - u_0 = \frac{f_{max}L^2}{4EA}}$$

4.29 The spare wheel in an automobile is stored under the vehicle and raised and lowered by a cable. The wheel has a mass of 25 kg. The ultimate strength of the cable is 300 MPa and an effective Modulus of Elasticity of E = 180 GPa. The length of the hanging cable length is 36 cm. (a) For a factor of safety of four, determine to the nearest millimeter, the minimum diameter of the cable if failure due to rupture is to be avoided. (b) What is the maximum extension of the cable for the answer in part (a).

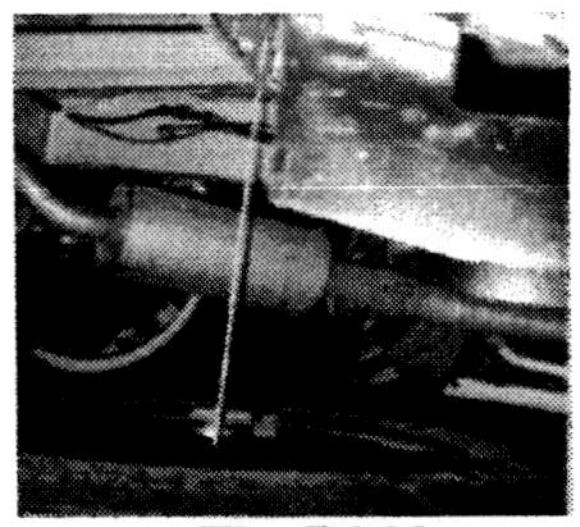

Fig. P4.29

Solution $\quad m = 25$ kg $\quad \sigma_{ult} = 300$ MPa $\quad E = 180$ GPa $\quad L_0 = 36$ cm

---

The free body diagram of the tire is as shown below.

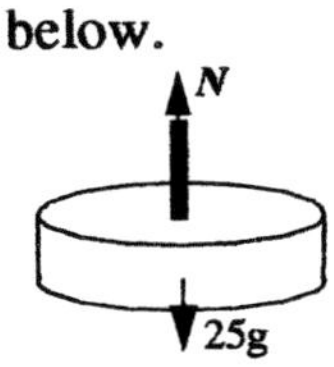

$$N = 25g = 245.25 \text{ Newtons} \qquad 1$$

The maximum allowable axial stress is: $\sigma_{max} = \frac{\sigma_{ult}}{k_{saftey}} = \frac{300}{4} = 75\text{MPa}$.

The axial stress can be written as: $\sigma = \frac{N}{A} = \frac{245.25}{(\pi d^2/4)} \le \sigma_{max}$ or $\frac{312.26}{d^2} \le 75(10^6)$ or $d^2 \ge 4.163(10^{-6})$ or

$$d \ge 2.04(10^{-3}) \text{ m} \qquad 2$$

To the nearest millimeter, that satisfies the inequality in Eq (2) is $\boxed{d_{min} = 3 \text{ mm}}$

$$\delta = \frac{NL}{EA} = \frac{(245.25)(0.36)}{(180)(\pi 0.003^2/4)} = 0.0694(10^{-3}) \text{ m} \qquad \boxed{\delta = 0.069 \text{ mm}}$$

4.30 A adhesively bonded joint in wood (E = 1800 ksi) is fabricated as shown. If the total elongation of the joint between A and D is to be limited to 0.05 inches, determine the maximum axial force F that can be applied.

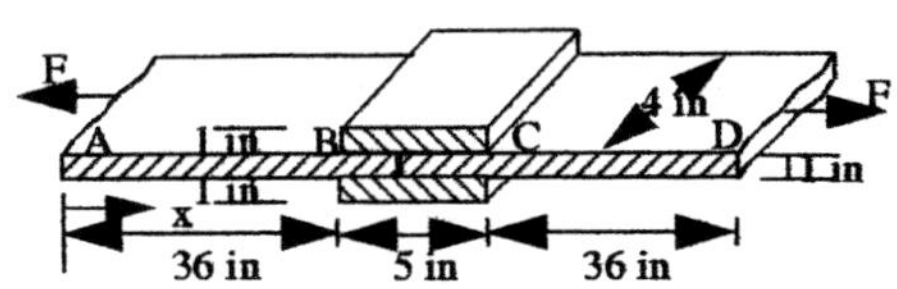

Fig. P4.30

Solution $E = 1800$ ksi $\delta \le 0.05$ $F_{max} = ?$

The internal axial force in all segments of the joint is $N = F$. The area of cross-sections of the segments are:

$A_{AB} = (4)(1) = 4\ in^2$ $A_{BC} = 3(4)(1) = 12\ in^2$ $A_{CD} = (4)(1) = 4\ in^2$

The relative displacement of the ends of each segment can be written as shown below.

$$u_B - u_A = \frac{N_{AB}(x_B - x_A)}{EA_{AB}} = \frac{F(36)}{(1800)(4)} = 5(10^{-3})F \quad (1)$$

$$u_C - u_B = \frac{N_{BC}(x_C - x_B)}{EA_{BC}} = \frac{F(5)}{(1800)(12)} = 0.2318(10^{-3})F \quad (2)$$

$$u_D - u_C = \frac{N_{CD}(x_D - x_C)}{EA_{CD}} = \frac{F(36)}{(1800)(4)} = 5(10^{-3})F \quad (3)$$

Adding equations (1), (2) and (3)

$u_D - u_A = F(5 + 0.2318 + 5)(10^{-3}) = 10.2318(10^{-3}) \le 0.05$ or $F \le 4.8869$ kips or $\boxed{F_{max} = 4886\ lb}$

4.31 A 5 feet long hollow rod is to transmit an axial force of 30 kips. The outer diameter of the rod must be 6 inches to fit existing attachments. The relative displacement of the two ends of the shaft is limited to 0.024 inches. The axial rod can be made of steel or aluminum. The Modulus of Elasticity E, the allowable axial stress σ_{allow}, and the specific weight γ, are given in Table 4.1. Determine the maximum inner diameter in increments of 1/8 inch of the lightest rod that can be used for transmitting the axial force and the corresponding weight.

Table 4.1.Material Properties in Problem 4.31

Material	E ksi	σ_{allow} ksi	γ lbs./in^3
Steel	30,000	24	0.285
Aluminum	10,000	14	0.100

Solution $L = 5$ ft $P = 30$ kips $d_o = 6$ in $\Delta u = 0.024$

$d_i = ?$ nearest 1/8th inch, $W = ?$ for lightest rod

We calculate the area of cross-section for steel to satisfy the stress and deformation requirements.

$$(\Delta u)_S = \frac{(30)(60)}{(30)(10^3)A_S} \le 0.024 \quad \text{or} \quad A_S \ge 2.25\ in^2 \quad (1)$$

$$\sigma_S = \frac{(30)}{A_S} \le 24 \quad \text{or} \quad A_S \ge 1.25\ in^2 \quad (2)$$

Using similar calculations for Aluminum we obtain the following limits on A_{Al}.

$$(\Delta u)_{Al} = \frac{(30)(60)}{10(10^3)A_{Al}} \le 0.024 \quad \text{or} \quad A_{Al} \ge 7.50\ in^2 \quad (3)$$

$$\sigma_{Al} = \frac{(30)}{A_{Al}} \le 14 \quad \text{or} \quad A_{Al} \ge 2.143\ in^2 \quad (4)$$

Thus if $A_S \ge 2.25\ in^2$ then it will meet both conditions in Eq's (1) and (2). Similarly if $A_{Al} \ge 7.50\ in^2$ then it will meet both conditions in Eq's (3) and (4). The internal diameters D_S and D_{Al} can be found as shown below:

$$A_S = \frac{\pi}{4}(6^2 - D_S^2) \geq 2.25 \qquad D_S \leq 5.756 \text{ in}$$

$$A_{Al} = \frac{\pi}{4}(6^2 - D_{Al}^2) \geq 7.5 \qquad D_{Al} \leq 5.143 \text{ in}$$ 5

Rounding downwards to the closest 1/8th inch, we obtain: $D_S = 5.75 \text{ in} \qquad D_{Al} = 5.125 \text{ in}$

We can find the weight of each material by taking the product of the specific weight and the volume of a hollow cylinder as shown below:

$$W_S = (0.285)\frac{\pi}{4}(6^2 - 5.75^2)(60) = 39.45 \text{ lbs} \qquad W_{Al} = (0.1)\frac{\pi}{4}(6^2 - 5.125^2)(60) = 45.87 \text{ lbs}$$

The rod should be of steel.

The inside diameter should be $\boxed{d_i = 5\frac{3}{4} \text{ inch}}$. The corresponding weight is $\boxed{W_S = 39.5 \text{ lbs}}$

4.32 A hitch for an automobile is to be designed for pulling a maximum load of 3,600 lbs. A solid-square-bar fits into a square-tube, and is held in place by a pin as shown. The allowable axial stress in the bar is 6 ksi, the allowable shear stress in the pin is 10 ksi, and the allowable axial stress in the steel tube is 12 ksi. To the nearest 1/16th of an inch, determine the minimum cross-sectional dimensions of the pin, the bar and the tube. Negect stress concentration.(Note: Pin is in double shear)

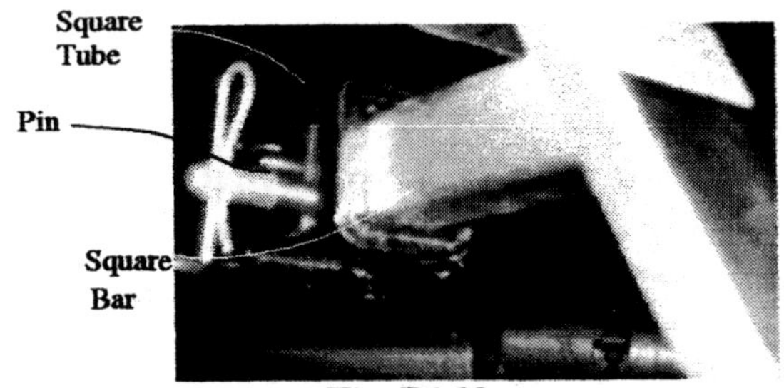

Fig. P4.32

Solution $P = 3{,}600\text{lbs} \qquad \sigma_b \leq 6 \text{ ksi} \qquad \tau_p = 10 \text{ ksi}$ double shear

$\sigma_s = 12 \text{ ksi} \qquad d_p = ? \qquad a_b = ? \qquad b_s = ?$ nearest 1116 inch

---

The two-dimensional picture of the hitch and the cross-section of the tube are shown below.

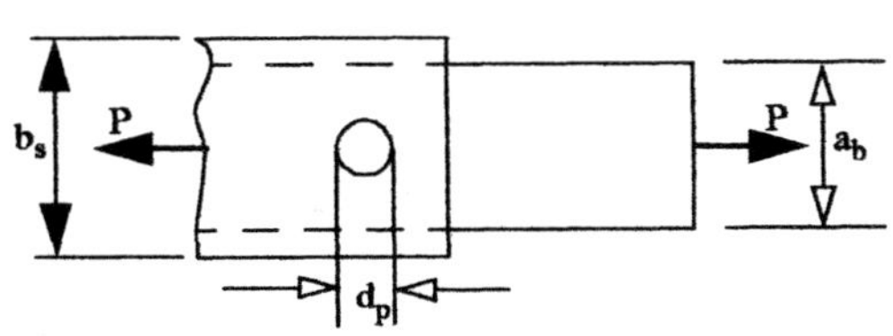

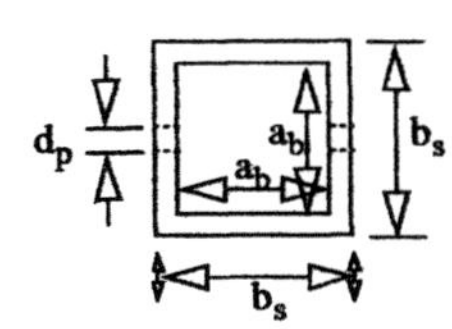

As pin is in double shear: $2V = P$ or $V = 180$

The shear stress is: $\tau_p \geq \frac{V}{A_p} = \frac{1800}{(\pi dp^2/4)} \leq 10,000 \quad d_p^2 \geq 0.2292 \quad d_p \geq 0.4787 \text{ inch} \qquad \boxed{d_p = 0.5 \text{ in}}$

The axial force in bar is $N = P = 3600$. The minimum area of cross-section in the bar is is at the pin hole.

The axial stress is: $\sigma_b = \frac{3600}{a_b(a_b - d_p)} \leq 6000$ or $a_b^2 - 0.5a_b \geq 0.6$ or $a_b \geq 1.063 \qquad \boxed{a_b = 1\frac{1}{8} \text{ in}}$

The axial force in the steel tube is $N_s = P = 3600$. The minimum area of cross-section in the tube is at the pin hole.

The axial stress is: $\sigma_s = \frac{3600}{[(b_s)(b_s - d_p) - (a_b)(a_b - d_p)]} \leq 12,000$ or $b_s^2 - 0.5b_s - (1.125)(1.125 - 0.5) \geq 0.3$ or

$b_s^2 - 0.5b_s - 1.003125 \geq 0 \qquad b_s \geq 1.282 \qquad \boxed{b_s = 1\frac{5}{16} \text{ in}}$

4.33 An axial rod has a constant axial rigidity EA and is acted upon by a distributed axial force p(x). *If* at section A the *internal axial force is zero* show that the relative displacement of section at B with respect to displacement of section at A is given by

$$u_B - u_A = \frac{1}{EA}\left[\int_{x_A}^{x_B}(x - x_B)p(x)dx\right] \qquad \textbf{4.15}$$

Solution $N(x_A) = 0$ $u_B - u_A = ?$

From Eq. 4.14 we have the following.

$$\frac{dN}{dx} = -p(x) \qquad 1$$

Integrating from $x = x_A$ to any location x, we obtain the following.

$$\int_{N_A=0}^{N} dN = -\int_{x_A}^{x} p(x)dx \quad \text{or} \quad N(x) = -\int_{x_A}^{x} p(x)dx \qquad 2$$

Integrating the equation $\frac{du}{dx} = \frac{N(x)}{EA}$ from x_A to x_B, we obtain the following.

$$\int_{u_A}^{u_B} du = \int_{x_A}^{x_A}\frac{N(x)}{EA}dx \quad \text{or} \quad u_B - u_A = \frac{1}{EA}\int_{x_A}^{x} N(x)dx \qquad 3$$

Integrating by parts, the above equation can be written as

$$u_B - u_A = \frac{1}{EA}\left[xN(x)\Big|_{x_A}^{x_B} - \int_{x_A}^{x} x\frac{dN}{dx}dx\right] \qquad 4$$

Imposing the limits and substituting Eq. (1), we obtain the following.

$$u_B - u_A = \frac{1}{EA}\left[x_B N(x_B) - N(x_A) - \int_{x_A}^{x} x(-p(x))dx\right] \qquad 5$$

Substituting Eq. (2) at $x = x_B$ and noting that $N(x_A) = 0$, we obtain

$$u_B - u_A = \frac{1}{EA}\left[x_B\left\{-\int_{x_A}^{x_B} p(x)dx\right\} - 0 + \int_{x_A}^{x_B} xp(x)dx\right] \qquad 6$$

Noting that x_B is a constant and can be taken inside the integral, we obtain:

$$u_B - u_A = \frac{1}{EA}\left[\int_{x_A}^{x_B}(x - x_B)p(x)dx\right] \qquad 7$$

Eq. (7) is same as Eq. 4.15.

4.34 The stress-strain relationship for a non-linear material is given by the power law $\sigma = E\varepsilon^n$. If all assumptions except Hooke's law are valid, show

$$\Delta u = \left(\frac{N}{EA}\right)^{\frac{1}{n}} L \qquad \textbf{4.16}$$

and the axial stress σ_{xx} is given by Equation 4.9, where N is the internal axial force, L is the length of the rod, A is the area of cross-section, and Δu is the relative displacement of the rod.

Solution $\sigma = E\varepsilon^n$ $\sigma_{xx} = ?$ $\Delta u = ?$

From kinematics we have $\varepsilon_{xx} = \frac{du}{dx}(x)$. Substituting this in the given stress-strain relationship, we obtain.

$$\sigma_{xx} = E\left[\frac{du}{dx}(x)\right]^n \qquad 1$$

The normal stress and internal normal force are related as

$$N = \int_A \sigma_{xx} dA \qquad 2$$

Substituting Eq. (1) into Eq. (2) and assuming homogenous material we obtain the following:

$$N = \int_A E\left[\frac{du}{dx}(x)\right]^n dA = E\left[\frac{du}{dx}(x)\right]^n \int_0 dA = EA\left[\frac{du}{dx}(x)\right]^n \text{ or}$$

$$\left[\frac{du}{dx}(x)\right]^n = \frac{N}{EA} \qquad 3$$

Substituting Eq. (3) into Eq. (1), we obtain: $\boxed{\sigma_{xx} = \frac{N}{A}}$

Eq. (2) can be re-written as:

$$\frac{du}{dx}(x) = \left[\frac{N}{EA}\right]^{\frac{1}{n}} \qquad 4$$

Assuming all quantities on the right in Eq (4) are constant over the length of the shaft, then $\frac{du}{dx} = \frac{\Delta u}{L}$.

Substituting this in Eq. (4) we obtain: $\boxed{\Delta u = \left[\frac{N}{EA}\right]^{\frac{1}{n}} L}$

4.35 Determine the elongation of a rotating bar in terms of the rotating speed ω, density ρ, length L, area of cross-section A, and Modulus of Elasticity E. (Hint: the body force per unit volume is $\rho\omega^2 x$).

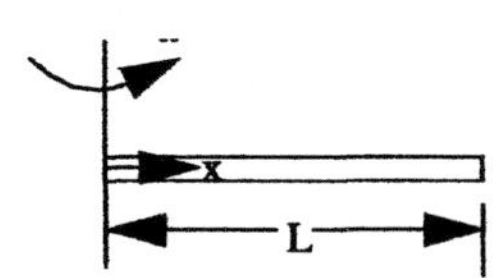

Fig. P4.35

Solution $\Delta u = f(\omega, \rho, L, A, E) = ?$

The free body diagram of a differential element is shown below.

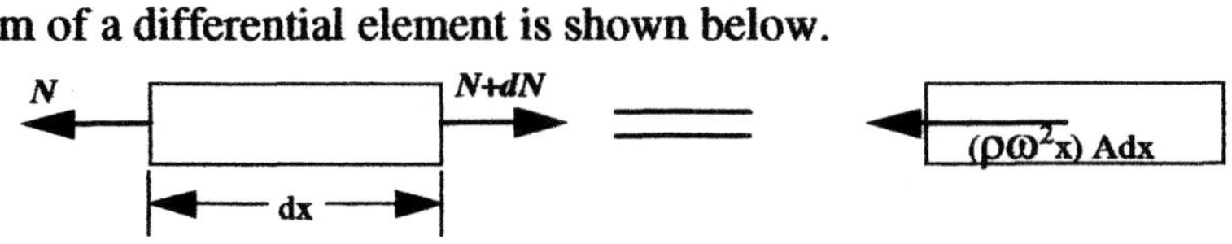

By equilibrium $N + dN - N = -\rho w^2 x A dx$ or

$$dN = \rho w^2 A x dx \qquad 1$$

At $x = L$ $N = 0$. Integrating Eq (1) from x = L to any location x we obtain

$$\int_0^N dN = \int_L^x -\rho w^2 A x dx \text{ or } N - 0 = \rho w^2 A \frac{x^2}{2}\bigg|_L^x \text{ or}$$

$$N = \frac{\rho w^2}{2}A(L^2 - x^2) \qquad 2$$

From Eq. 4.14 we have the following.

$$\frac{du}{dx} = \frac{N}{EA} = \frac{\rho w^2}{2E}(L^2 - x^2) \qquad 3$$

Integrating Eq (3) from x = 0 to x = L we obtain: $\int_{u_0}^{u_L} du = \frac{\rho w^2}{2E}\int_0^L (L^2 - x^2)dx$ or

$$\Delta u = u_L - u_0 = \frac{\rho w^2}{2E}\left(L^2 x - \frac{x^3}{3}\right)\Bigg|_0^L \quad \text{or} \quad \Delta u = \frac{\rho w^2}{2E}\left(L^3 - \frac{L^3}{3}\right) \qquad \boxed{\Delta u = \frac{\rho w^2 L^3}{3E}}$$

4.36 Consider the dynamic equilibrium of differential element shown in Fig. P4.36. Where N is the internal force, ρ is the density, A is the area of cross-section and $\frac{\partial^2 u}{\partial t^2}$ is acceleration.

Fig. P4.36

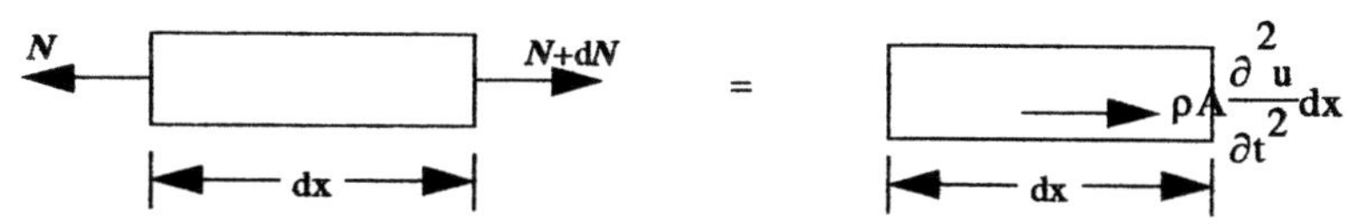

By substituting for N from Equation 4.8 into the dynamic equilibrium equation derive the following *wave equation:*

$$\frac{\partial^2 u}{\partial t^2} = c^2 \frac{\partial^2 u}{\partial x^2} \qquad c = \sqrt{\frac{E}{\rho}} \qquad 4.17$$

The material constant c is the velocity of propagation of sound in the material.

Solution

By equilibrium of forces in Fig. P4.36 we obtain: $N + dN - N = \rho A\frac{\partial^2 u}{\partial t^2}$ or

$$\frac{dN}{dx} = \rho A\frac{\partial^2 u}{\partial t^2} \qquad 1$$

Eq. 4.8 can be written as shown below.

$$N = EA\frac{\partial u}{\partial x} \qquad 2$$

Substituting Eq(2) into Eq(1), we obtain $\frac{\partial}{\partial x}\left(EA\frac{\partial u}{\partial x}\right) = \rho A\frac{\partial^2 u}{\partial t^2}$

Assuming EA is not a function of x we obtain the following. $EA\frac{\partial^2 u}{\partial x^2} = \rho A\frac{\partial^2 u}{\partial t^2}$ or $\frac{\partial^2 u}{\partial t^2} = c^2\frac{\partial^2 u}{\partial x^2}$

Above equation is same as Eq(4-17)

4.37 Show by substitution that the functions f(x - ct) and g(x + ct) satisfy the wave Equation 4.17.

Solution

Consider the solution: $u = f(x - ct)$. Substituting $y = x - ct$ we obtain: $\frac{\partial u}{\partial x} = \frac{\partial f}{\partial y}\frac{\partial y}{\partial x} = \frac{\partial f}{\partial y}$. We obtain the following:

$$\frac{\partial^2 u}{\partial x^2} = \frac{\partial}{\partial x}\left(\frac{\partial f}{\partial y}\right) = \frac{\partial}{\partial y}\left(\frac{\partial f}{\partial y}\right)\frac{\partial y}{\partial x} = \frac{\partial^2 f}{\partial y^2} \qquad 3$$

We can also write: $\frac{\partial u}{\partial t} = \frac{\partial f}{\partial y}\frac{\partial y}{\partial t} = -c\frac{\partial f}{\partial y}$. We obtain the following.

$$\frac{\partial^2 u}{\partial t^2} = \frac{\partial}{\partial t}\left(-c\frac{\partial f}{\partial y}\right) = \frac{\partial}{\partial y}\left(-c\frac{\partial f}{\partial y}\right)(-c) = c^2\frac{\partial^2 f}{\partial y^2} \qquad 4$$

Substituting Eq.(2) and (3) in Eq. 4-17, we observe that the equation is identically satisfied, hence f is a solution of Eq. 4-17.

Consider now $u = g(x+ct)$. Substituting $z = x+ct$ we obtain: $\frac{\partial^2 u}{\partial x^2} = \frac{\partial}{\partial x}\left(\frac{\partial g}{\partial z}\right) = \frac{\partial}{\partial z}\left(\frac{\partial g}{\partial z}\right)\frac{\partial z}{\partial x} = \frac{\partial^2 g}{\partial z^2}$ and

$\frac{\partial^2 u}{\partial t^2} = \frac{\partial}{\partial t}\left(c\frac{\partial g}{\partial z}\right) = \frac{\partial}{\partial z}\left(c\frac{\partial g}{\partial z}\right)(c) = c^2\frac{\partial^2 g}{\partial z^2}$ which on substitution satisfies Eq 4-17. Hence g(x + ct) is a solution of Eq 4-17.

4.38 The strain displacement relationship for large axial strain is given by

$$\varepsilon_{xx} = \frac{du}{dx} + \frac{1}{2}\left(\frac{du}{dx}\right)^2 \qquad \textbf{4.18}$$

where we recognize that as u is only a function of x and hence the strain from Equation 4.18 is uniform across the cross-section. For linear, elastic, homogenous material show:

$$\frac{du}{dx} = \sqrt{1 + \frac{2N}{EA}} - 1 \qquad \textbf{4.19}$$

and the axial stress σ_{xx} is given by Equation 4.1.

Solution

By Hooke's Law

$$\sigma_{xx} = E\varepsilon_{xx} = E\left[\frac{du}{dx} + \frac{1}{2}\left(\frac{du}{dx}\right)^2\right] \qquad 1$$

By static equivalency: $N = \int_A \sigma_{xx} dA = \int_A E\left[\frac{du}{dx} + \frac{1}{2}\left(\frac{du}{dx}\right)^2\right] dA$

Assuming material is homogenous across the cross-section i.e. E is a constant, and nothing that $\frac{du}{dx}$ is a function of x and does not change across the cross-section, we obtain. $N = E\left[\frac{du}{dx} + \frac{1}{2}\left(\frac{du}{dx}\right)^2\right]A$ or

$$\frac{du}{dx} + \frac{1}{2}\left(\frac{du}{dx}\right)^2 = \frac{N}{EA} \qquad 2$$

Substituting Eq(2) into Eq(1) we obtain $\sigma_{xx} = E\left(\frac{N}{EA}\right) = \frac{N}{A}$ as given by Eq. 4-9.

Eq. (3) can be re-written as $\left(\frac{du}{dx}\right)^2 + 2\left(\frac{du}{dx}\right) - 2\frac{N}{EA} = 0$.

Solving the quadratic, we obtain $\frac{du}{dx} = \frac{1}{2}\left(-2 \pm \sqrt{4 + 4\left(\frac{2N}{EA}\right)}\right)$ or $\frac{du}{dx} = \pm\sqrt{1 + 2\left(\frac{N}{EA}\right)} - 1$

The positive root is the admissible root, because with the negative root $\frac{du}{dx}$ is always negative and less that -1 irrespective of the value of N i.e. applied forces with a positive root we obtain Eq 4-19.

4.39 Table 4.2shows the measured radii of a solid tapered rod at several point along the axis. The rod is made of aluminum (E = 100 GPa) and has a length of 1.5 meters. Determine (a) elongation of the rod using numerical integration. (b) the maximum axial stress in the rod.

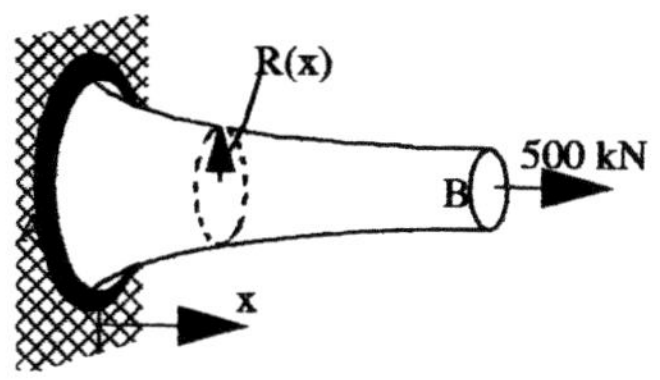

Fig. P4.39

Table 4.2.Data for Problem 4.39.

x (m)	R(x) (mm)	x (m)	R(x) (mm)
0.0	100.6	0.8	60.1
0.1	92.7	0.9	60.3
0.2	82.6	1.0	59.1
0.3	79.6	1.1	54.0
0.4	75.9	1.2	54.8
0.5	68.8	1.3	54.1
0.6	68.0	1.4	49.4
0.7	65.9	1.5	50.6

Solution E = 100 GPa $x_A = 0$ $x_B = 1.5$ m $\Delta u = u_B - u_A = ?$ $\sigma_{max} = ?$

The internal axial force is N = 500 kN. From Eq. 4.14 we have the following.

$$\frac{du}{dx} = \frac{N}{EA} = \frac{(500)(10^3)}{(100)(10^9)(\pi R^2(x))} = \frac{5(10^{-6})}{\pi R^2(x)} \qquad 1$$

In Eq. (1), R(x) is measured in meters. Eq. (1) can be rewritten as $\frac{du}{dx} = \frac{5}{\pi R^2(x)}$, where now R(x) is in millimeters. Integrating from $x_A = 0$ to $x_B = 1.5$, we obtain the following:

$$\int_{u_A}^{u_B} du = \int_{x_A=0}^{x_B=1.5} \left[\frac{5}{\pi R^2(x)}\right]dx \quad \text{or} \quad \Delta u = \int_0^{1.5} \left[\frac{5}{\pi R^2(x)}\right]dx \qquad 2$$

Representing $f(x) = 1.5915/[R^2(x)]$ and using numerical integration as described in Appendix B, we obtain the value of the integral on a spread sheet as shown in the table below.

From the results in the table we see the elongation of the bar is: $\boxed{\Delta u = 0.60 \text{ mm}}$

The maximum stress will occur at the cross-section where area is the smallest, i.e., just before point B and can be found as shown below.

$$\sigma_{max} = \frac{N}{A_B} = \frac{500(10^3)}{\pi(0.0506^2)} = 62.2(10^6)\ \text{N/m}^2 \text{ or} \qquad \boxed{\sigma_{max} = 62.2\ \text{MPa(T)}}$$

x (m)	R(x) (mm)	f(x)	Δu (10^{-3})(m)
0.00	100.60	0.1573	0.0171
0.10	92.70	0.1852	0.0380
0.20	82.60	0.2333	0.0623
0.30	79.60	0.2512	0.0886
0.40	75.90	0.2763	0.1193
0.50	68.80	0.3362	0.1533
0.60	68.00	0.3442	0.1888
0.70	65.90	0.3665	0.2292

x (m)	R(x) (mm)	f(x)	Δu (10^{-3})(m)
0.80	60.10	0.4406	0.2731
0.90	60.30	0.4377	0.3178
1.00	59.10	0.4557	0.3678
1.10	54.00	0.5458	0.4216
1.20	54.80	0.5300	0.4753
1.30	54.10	0.5438	0.5351
1.40	49.40	0.6522	0.5988
1.50	50.60	0.6216	

4.40 Let the radius of the tapered rod in problem 4.39 be represented by the equation $R(x) = a + bx$. Using the data in Table 4.2 determine the constant a and b by the least-square method and then find the elongation of the rod by analytical integration.

Solution $E = 100$ GPa $a = ?$ $b = ?$ $x_A = 0$ $x_B = 1.5$ m $\Delta u = u_B - u_A = ?$

We develop the equations for the least square method for a linear representation as described in Appendix B. This difference between the radius value and the value obtained by substituting $x = x_i$ in the equation $R(x) = a + bx$ is the error e_i that can be written as given below.

$$e_i = R_i - R(x_i) = R_i - (a + bx_i) \qquad 1$$

We define the error E as $E = \sum_{i=1}^{N} e_i^2$. The error E is minimized with respect to coefficients a, and b to generate a set of linear algebraic equations as shown below.

$$\frac{\partial E}{\partial a} = 0 \quad \text{or} \quad \sum 2e_i \frac{\partial e_i}{\partial a} = 0 \quad \text{or} \quad \sum 2[R_i - (a + bx_i)][-1] = 0$$

$$\frac{\partial E}{\partial b} = 0 \quad \text{or} \quad \sum 2e_i \frac{\partial e_i}{\partial b} = 0 \quad \text{or} \quad \sum 2[R_i - (a + bx_i)][-x_i] = 0$$

The above equations on the right can be rearranged and written in matrix form as shown below:

$$\begin{bmatrix} N & \sum x_i \\ \sum x_i & \sum x_i^2 \end{bmatrix} \begin{Bmatrix} a \\ b \end{Bmatrix} = \begin{Bmatrix} \sum R_i \\ \sum x_i R_i \end{Bmatrix} \quad \text{or} \quad \begin{bmatrix} b_{11} & b_{12} \\ b_{21} & b_{22} \end{bmatrix} \begin{Bmatrix} a \\ b \end{Bmatrix} = \begin{Bmatrix} r_1 \\ r_2 \end{Bmatrix} \qquad 2$$

The coefficients of the b-matrix and the r-vector can be determined by comparison to the matrix form of the equations on the left. The coefficients a and b can be determined by Cramer's rule. Let D represent the determinant of the b matrix. By Cramer's rule, we replace the first column in the matrix of b's by the right hand side and find the determinant of the so constructed matrix and divide by D. Thus, the coefficients a, and b can be written as shown below.

$$D = b_{11}b_{22} - b_{12}b_{21} \qquad 3$$

$$a = \left|\begin{bmatrix} r_1 & b_{12} \\ r_2 & b_{22} \end{bmatrix}\right| / D = \frac{r_1 b_{22} - r_2 b_{12}}{D} \qquad b = \left|\begin{bmatrix} b_{11} & r_1 \\ b_{21} & r_2 \end{bmatrix}\right| / D = \frac{r_2 b_{11} - r_1 b_{21}}{D} \qquad 4$$

The given data and Eq. (2) through (4) can be put in a spread sheet and the coefficients a and b can be found as shown in the table below as: $a = 90.226$ mm $b = -30.593$

	x_i (m)	R_i (mm)	x_i^2	$x_i{*}R_i$
1	0.00	100.60	0.0000	0.000
2	0.10	92.70	0.0100	9.270
3	0.20	82.60	0.0400	16.520
4	0.30	79.60	0.0900	23.880
5	0.40	75.90	0.1600	30.360
6	0.50	68.80	0.2500	34.400
7	0.60	68.00	0.3600	40.800
8	0.70	65.90	0.4900	46.130
9	0.80	60.10	0.6400	48.080

	x_i (m)	R_i (mm)	x_i^2	$x_i{*}R_i$
10	0.90	60.30	0.8100	54.270
11	1.00	59.10	1.0000	59.100
12	1.10	54.00	1.2100	59.400
13	1.20	54.80	1.4400	65.760
14	1.30	54.10	1.6900	70.330
15	1.40	49.40	1.9600	69.160
16	1.50	50.60	2.2500	75.900
b_{ij} & r_i	12.0000	1076.50	12.4000	703.360
D	54.4			
a and b	90.226	-30.593		

Thus R(x) = 90.226-30.593x, where x is in meters and R is in millimeters. Substituting this in Eq. (2) of problem 4.39 and integrating, we obtain the elongation as shown below.

$$\Delta u = \int_0^{1.5}\left[\frac{5}{\pi(90.226-30.593x)^2}\right]dx = \left(\frac{5}{\pi}\right)\left[\frac{1}{(30.593)(90.226-30.593x)}\right]_0^{1.5} = 0.5968(10^{-3})\ \text{m}$$

Thus the elongation of the bar is: $\boxed{\Delta u = 0.60\ \text{mm}}$

4.41 Table 4.4 shows the value of distributed axial force at several point along the axis of a hollow steel (E = 30,000 ksi) rod. The rod has a length of 36 inches, an outside diameter of 1 inch, and an inside diameter of 0.875 inch. Determine (a) the displacement of the end A using numerical integration. (b) the maximum axial stress in the rod.

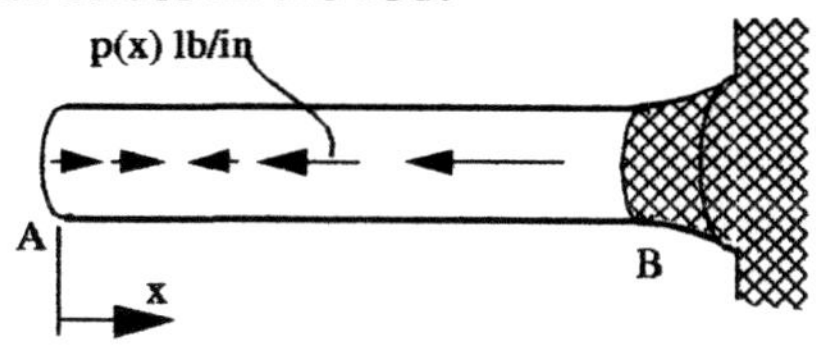

Fig. P4.41

Table 4.4.

x (inches)	p (x) lbs./in	x (inches)	p (x) lbs./in
0	260	21	-471
3	106	24	-598
6	32	27	-645
9	40	30	-880
12	-142	33	-1035
15	-243	36	-1108
18	-262		

Solution E = 30,000 ksi d_o= 1 inch d_i = 0.875 inch u_A = ? σ_{max}= ?

The equilibrium equation $\frac{dN}{dx} + p(x) = 0$ can be integrated from point A, where $N = 0$ to any location x, to obtain the internal axial force as a function of x as shown below.

$$N(x) = \int_{N_A=0}^{N(x)} dN = -\int_{x_A=0}^{x} p(x)dx \tag{1}$$

The area of cross-section is $A = \pi(d_o^2 - d_i^2)/4 = 0.1841\text{in}^2$.

$$\frac{du}{dx} = \frac{N(x)}{EA} = \frac{N(x)}{(30)(10^6)(0.1841)} = 181.08(10^{-9})N(x) \tag{2}$$

Integrating from point A i.e., x_A =0 to point B, i.e., x_B =36. We note that point B is fixed to the wall, hence u_B= 0. we obtain the following integral:

$$\int_{u_A}^{u_B=0} du = \int_{x_A=0}^{x_B=36} [181.08(10^{-9})N(x)]dx \quad \text{or} \quad u_A = -\int_0^{36}[181.08(10^{-9})N(x)]dx \tag{3}$$

The axial stress σ_{xx} can be found as shown below:

$$\sigma_{xx} = \frac{N}{A} = \frac{N}{0.1841} \tag{4}$$

Eq. (1) can be numerically integrated on a spread sheet to obtain $N(x_i)$, the value of internal force at any x_i. Then, Eq (3) can be numerically integrated to obtain the elongation. Eq. (4) can be used to find the axial stress at various x_i and the maximum value chosen by inspection. These calculations can be done on a spread sheet as shown in the table below.

The negative sign for the displacement u_A implies it moves in the negative x-direction, which means it displaces to the left. The answers are: $\boxed{u_A = 0.0176\text{inch to the left}}$ and $\boxed{\sigma_{max} = 73.7\text{ksi(T)}}$

x_i (inches)	$p(x_i)$ (lbs./in)	$N(x_i)$ (lbs.)	$[u_A-u(x_{i+1})](10^{-3})$ (inches)	$\sigma_{xx}(x_i)$ (ksi)
0	260	0.00	0.149	0.000
3	105.5	-548.25	0.503	-2.978
6	32.0	-754.50	0.942	-4.098
9	39.5	-861.75	1.368	-4.681
12	-142.0	-708.00	1.596	-3.846
15	-242.5	-131.25	1.462	-0.713
18	-262.0	625.50	0.824	3.398
21	-470.5	1724.25	-0.548	9.366
24	-598.0	3327.00	-2.862	18.072
27	-644.5	5190.75	-6.303	28.195
30	-880.0	7477.50	-11.145	40.617
33	-1034.5	10349.25	-17.640	56.215
36	-1108.0	13563.00		73.672

4.42 Let the distributed force p(x) in problem 4.41 be represented by the equation $p(x) = a + bx + cx^2$. Using the data in Table 4.4 determine the constant a, b, and c by the least-square method and then find the displacement of section at A by analytical integration.

Solution $a = ?$ $b = ?$ $c = ?$ $u_A = ?$

The equilibrium equation $\frac{dN}{dx} + p(x) = 0$ can be integrated from point A, where $N = 0$ to any location x, to obtain the internal axial force as a function of x as shown below.

$$N(x) = \int_{N_A=0}^{N(x)} dN = -\int_{x_A=0}^{x} p(x)dx = -\int_{x_A=0}^{x} [a+bx+cx^2]dx = -\left[ax + \frac{bx^2}{2} + \frac{cx^3}{3}\right] \qquad 1$$

The area of cross-section is $A = \pi(d_o^2 - d_i^2)/4 = 0.1841 in^2$.

$$\frac{du}{dx} = \frac{N(x)}{EA} = \frac{N(x)}{(30)(10^6)(0.1841)} = (-181.08)(10^{-9})\left[ax + \frac{bx^2}{2} + \frac{cx^3}{3}\right] \qquad 2$$

Integrating from point A i.e., $x_A = 0$ to point B, i.e., $x_B = 36$. We note that point B is fixed to the wall, hence $u_B = 0$. we obtain the following integral: $\int_{u_A}^{u_B=0} du = \int_{x_A=0}^{x_B=36} (-181.08)(10^{-9})\left[ax + \frac{bx^2}{2} + \frac{cx^3}{3}\right]dx$ or

$$u_A = (181.08)(10^{-9})\left(\frac{ax^2}{2} + \frac{bx^3}{6} + \frac{cx^4}{12}\right)\Bigg|_0^{36} = (181.08)(10^{-9})\left[\frac{a(36)^2}{2} + \frac{b(36)^3}{6} + \frac{c(36)^4}{12}\right] \qquad 3$$

Using the Least Square Method described in Appendix B, we obtain the value of the values of constants a, b, and c on a spread sheet as shown in the table below.

The values of the constants are: $a = 224.40$ $b = -23.60$ $c = -0.40$

Substituting the values of the above constants in Eq. (3), we obtain the displacement of point A

$$u_A = (181.08)(10^{-9})\left[\frac{(224)(36)^2}{2} + \frac{(-23.60)(36)^3}{6} + \frac{(-0.40)(36)^4}{12}\right] = -0.017 \text{ or } u_A = 0.017 \text{in to the left}$$

	x_i	$p(x_i)$	x_i^2	x_i^3	x_i^4	$x*p_i$	$x_i^2*p_i$
1	0.0	260.0	0.0	0.000E+00	0.000E+00	0.000E+00	0.000E+00
2	3.0	105.5	9.0	2.700E+01	8.100E+01	3.165E+02	9.495E+02
3	6.0	32.0	36.0	2.160E+02	1.296E+03	1.920E+02	1.152E+03
4	9.0	39.5	81.0	7.290E+02	6.561E+03	3.555E+02	3.200E+03
5	12.0	-142.0	144.0	1.728E+03	2.074E+04	-1.704E+03	-2.045E+04
6	15.0	-242.5	225.0	3.375E+03	5.063E+04	-3.638E+03	-5.456E+04
7	18.0	-262.0	324.0	5.832E+03	1.050E+05	-4.716E+03	-8.489E+04
8	21.0	-470.5	441.0	9.261E+03	1.945E+05	-9.881E+03	-2.075E+05
9	24.0	-598.0	576.0	1.382E+04	3.318E+05	-1.435E+04	-3.444E+05
10	27.0	-644.5	729.0	1.968E+04	5.314E+05	-1.740E+04	-4.698E+05
11	30.0	-880.0	900.0	2.700E+04	8.100E+05	-2.640E+04	-7.920E+05
12	33.0	-1034.5	1089.0	3.594E+04	1.186E+06	-3.414E+04	-1.127E+06
13	36.0	-1108.0	1296.0	4.666E+04	1.680E+06	-3.989E+04	-1.436E+06
b_{ij} & r_i	234.0	-4945.0	5850.0	1.643E+05	4.918E+06	-1.513E+05	-4.531E+06
C_{ij}	1.783E+09	1.897E+08	4.216E+06	2.971E+07	7.666E+05	2.129E+04	
D	3.453E+09						
a_i	224.40	-23.60	-0.40				

Problem Set 4.2

4.43 The strain at a cross-section of an axial rod is assumed to have a uniform value of $\varepsilon_{xx} = 200\ \mu$. (a) Plot the stress distribution across the laminated cross-section. (b) Determine the equivalent internal axial force N and its location from the bottom of the cross-section. Use $E_{aluminium} = 100$ GPa, $E_{wood} = 10$ GPa, and $E_{steel} = 200$ GPa.

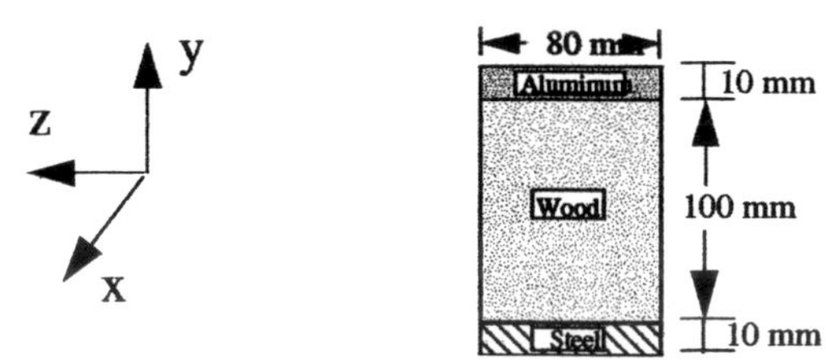

Fig. P4.43

Solution $\varepsilon_{xx} = 2000\mu$ $E_{al} = 100\text{GPa}$ $E_w = 10\text{GPa}$ $E_s = 200\text{GPa}$

$\sigma_{xx} = ?$ $N = ?$ $y_N = ?$

The normal stress in steel, wood and aluminum can be found using Hooke's Law as shown below.

$$(\sigma_{xx})_s = E_s\varepsilon_{xx} = (200)(10^9)(200)(10^{-6}) = 40(10^6)(\text{N/m}^2) = 40\ \text{MPa}$$

$$(\sigma_{xx})_w = E_s\varepsilon_{xx} = (10)(10^9)(200)(10^{-6}) = 2(10^6)(\text{N/m}^2) = 2\ \text{MPa}$$

$$(\sigma_{xx})_{al} = E_s\varepsilon_{xx} = (100)(10^9)(200)(10^{-6}) = 20(10^6)(\text{N/m}^2) = 20\ \text{MPa}$$

The stress distribution across the cross-section is as shown below.

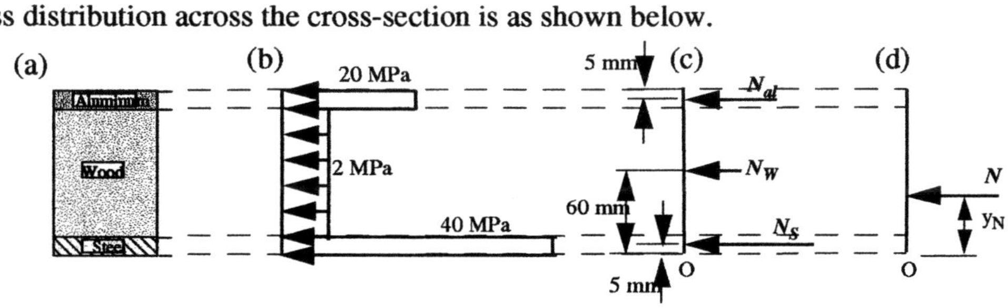

The stress distribution in figure (b) can be replaced by equivalent internal normal forces shown in figure (c) that act at the centroid of each distribution. The internal normal forces, can be found as shown below.

$N_s = (\sigma_{xx})_s A_s = 40(10^6)(0.08)(0.01) = 32(10^3)\text{Newtons} = 32\text{kN}$

$N_w = (\sigma_{xx})_w A_w = 2(10^6)(0.08)(0.1) = 16(10^3)\text{Newtons} = 16\text{kN}$

$N_{A1} = (\sigma_{xx})_{A1} A_{A1} = 20(10^6)(0.08)(0.01) = 16(10^3)\text{Newtons} = 16\text{kN}$

Equating the forces in figure (c) and (d) we obtain $N = N_s + N_w + N_{A1} = 64\text{kN}$

$\boxed{N = 64\text{kN(T)}}$

Equating moment about point 0 in figure (c) and (d), we obtain

$N(y_N) = N_s(0.005) + N_w(0.06) + (16)(0.115)$ or $(64)y_N = (32)(0.005) + (16)(0.006) + (16)(0.115)$

$y_N = \frac{2.96}{64} = 46.25(10^{-3})\ \text{m}$ $\boxed{y_N = 46.25\ \text{mm}}$

4.44 A reinforced concrete bar is constructed by embedding 2 in x 2 in square iron rods in concrete as shown. Assuming uniform strain of $\varepsilon_{xx} = -1500\ \mu$ in the cross-section (a) Plot the stress distribution across the cross-section (b) Determine the equivalent internal axial force N. Use $E_{iron} = 25{,}000$ ksi and $E_{concrete} = 3{,}000$ ksi.

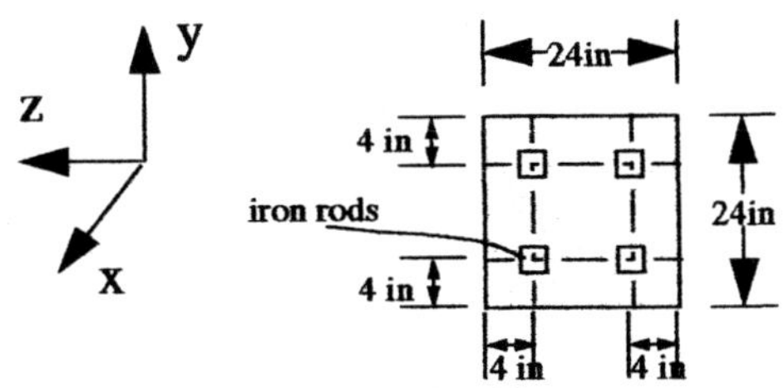

Fig. P4.44

Solution $\varepsilon_{xx} = -1500\mu$ $E_i = 25{,}000\text{ksi}$ $E_c = 3{,}000\text{ksi}$ $\sigma_{xx} = ?$ $N = ?$

The normal stress in iron and concrete can be found using Hooke's Law as shown below.

$(\sigma_{xx})_i = E_i\varepsilon_{xx} = 25(10^3)(-1500)(10^{-6}) = -37.5\ \text{ksi} = 37.5\ \text{ksi}$

$(\sigma_{xx})_c = E_c\varepsilon_{xx} = 3.0(10^3)(-1500)(10^{-6}) = -4.5\ \text{ksi} = 4.5\ \text{ksi}$

The normal stress distribution across the cross-section is:

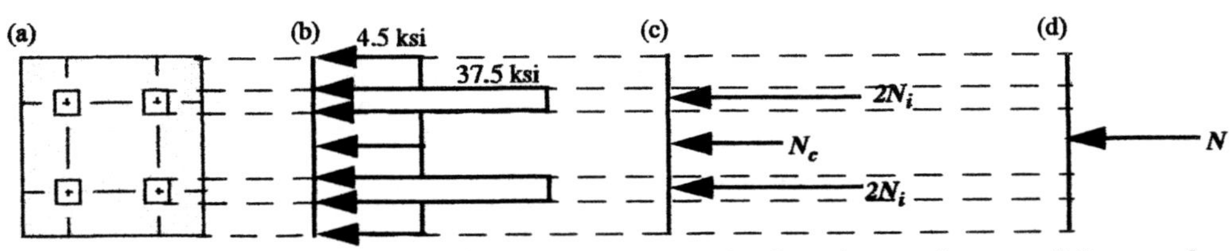

The stress distribution in figure (b) can be replaced by equivalent internal normal forces shown in figure (c) that act at the centroid of each distribution. The internal normal forces can be found as shown below.

$A_i = (2)(2) = 4\ \text{in}^2$ and $A_c = (24)(24) - 4(4) = 560\ \text{in}^2$

$N_i = (\sigma_{xx})_i A_i = (37.5)(4) = 150\ \text{kips (C)}$ and $N_c = (\sigma_{xx})_c A_c = (4.5)(560) = 2520\ \text{kips (C)}$

Equating forces in figure (c) and (d) we obtain

$N = 4N_i + N_c = 4(150) + 2520 = 3120\ \text{kips}$ $\boxed{N = 3120\ \text{kips}}$

4.45 A wooden rod ($E_W = 2000$ ksi) and steel strip ($E_s = 30{,}000$ ksi) are fastened securely to each other and to the rigid plates as shown. Determine (a) the location h of the line along which the external forces must act to produce no bending. (b) the maximum axial stress in steel and wood.

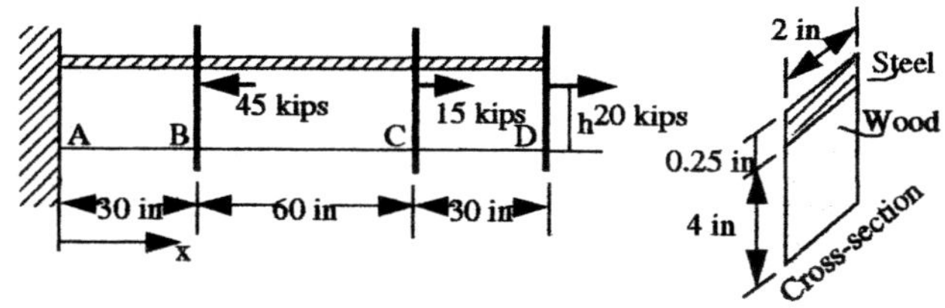

Fig. P4.45

Solution $E_W = 2000\text{ksi}$ $E_s = 30{,}000\text{ksi}$ $h = ?$ $(\sigma_{max})_s = ?$ $(\sigma_{max})_W = ?$

The geometry of the cross-section used for finding the origin is shown below.

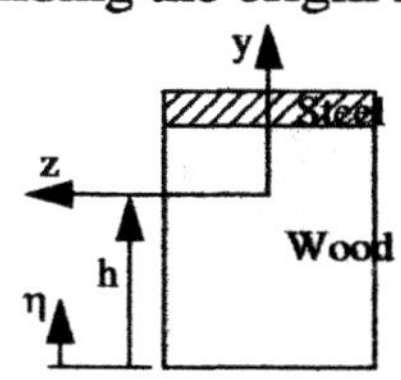

The location of the origin can be found using Eq. 4-23

$$h = \frac{\Sigma \eta_{ci} E_i A_i}{\Sigma E_i A_i} = \frac{\eta_{cw} E_w A_w + \eta_{cs} E_s A_s}{E_w A_w + E_s A_s} = \frac{(2)(2000)(4)(2) + (4.125)(30,000)(2)(0.25)}{(2000)(4)(2) + (30,000)(2)(0.25)}$$

$$\text{or } h = \frac{32,000 + 61.875}{10,000 + 15,000} = 3.028$$

$$\boxed{h = 3.3 \text{ inch}}$$

Making imaginary cuts in segments AB, BC and CD and taking right side of the cut we obtain the following free body diagrams.

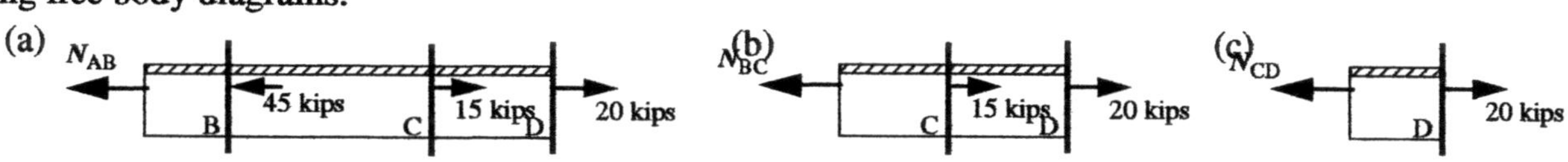

By force equilibrium we obtain the internal normal forces as shown below.

$$20 + 15 - 45 - N_{AD} = 0 \quad \text{or} \quad N_{AB} = -10 \text{ kips}$$

$$20 + 15 - N_{BC} = 0 \quad \text{or} \quad N_{BC} = 35 \text{ kips}$$

$$20 - N_{CD} = 0 \quad \text{or} \quad N_{CD} = 20\text{kips}$$

The axial rigidity of the cross-section is

$$\Sigma E_i A_i = E_w A_w + E_s A_s = (2000)(4)(2) + (30,000)(2)(0.25) = 31,000 \text{ kips}$$

The maximum stresses in each material will occur in the segment where internal normal force is maximum, i. e., in BC using Eq. 4-21, the maximum stress in each material can be found as shown below.

$$(\sigma_{max})_s = E_s \frac{N_{BC}}{\Sigma E_i A_i} = \frac{(30,000)(35)}{31,000} = 33.870 \text{ ksi}$$

$$\boxed{(\sigma_{max})_s = 33.9 \text{ ksi (T)}}$$

$$(\sigma_{max})_w = E_w \frac{N_{BC}}{\Sigma E_i A_i} = \frac{(2000)(35)}{31,000} = 2.258 \text{ ksi}$$

$$\boxed{(\sigma_{max})_w = 2.3 \text{ ksi (T)}}$$

4.46 A solid steel ($E_{steel} = 200 \text{ GPa}$) shaft circular bar of diameter 80 mm is 4 m long which extends through and is attached to a hollow brass ($E_{brass} = 100 \text{ GPa}$) bar of 120 mm outside diameter. Determine (a) the displacement of point C with respect to the wall. (b) the maximum axial stress in steel and brass.

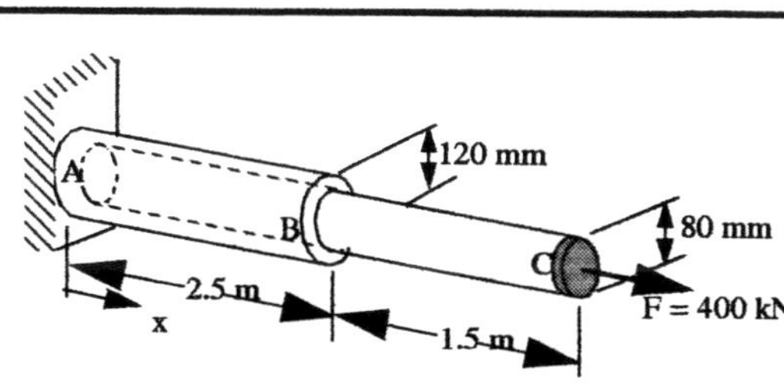

Fig. P4.46

Solution $E_s = 200GPa$ $d_s = 80$ mm $E_b = 100GPa$ $(d_b)_0$ 120 mm $(d_b)_i = 80$ mm
u_c-u_a = ? $(\sigma_{max})_s$ = ? $(\sigma_{max})_b$ = ?

The internal normal force of any cross-section of the bar is $N = 400KN$. The cross-sectioned areas of each material can be found as

$A_s = \frac{\pi}{4}(0.08)^2 = 5.026(10^{-3})\ m^2$ and $A_b = \frac{\pi}{4}[(0.12)^2 - (0.08)^2] = 6.283(10^{-3})\ m^2$

The axial rigidities of cross-sections AB and BC can be found as shown below

$(\Sigma E_i A_i)_{AB} = (200)(10^9)(5.026)(10^{-3}) + (100)(10^9)(6.283)(10^{-3}) = 1633.6(10^6)$ Newtons

$(EA)_{BC} = (200)(10^9)(5.026)(10^{-3}) = 1005.3(10^6)$ Newtons

The relative displacement of the end of each segment can be found as shown below

$$u_B - u_A = \frac{N(x_B - x_A)}{(\Sigma E_i A_i)_{AB}} = \frac{(400)(10^3)(2.5)}{1633.6(10^6)} = 0.612(10^{-3})\ m \quad 1$$

$$u_C - u_B = \frac{N(x_C - x_B)}{(EA)_{BC}} = \frac{(400)(10^3)(1.5)}{(1005.3)(10^6)} = 0.597(10^{-3})\ m \quad 2$$

Adding Eq(1) and (2) we obtain

$u_C - u_A = (0.612 + 0.597)(10^{-3}) = 1.209(10^{-3})\ m$ $\boxed{u_C - u_A = 1.209\ mm}$

In segment BC, the axial stress is

$$\sigma_{BC} = \frac{N}{A_{BC}} = \frac{400(10^3)}{5.026(10^{-3})} = 79.58(10^6)\ N/m^2 \quad 3$$

In segment AB, the axial stress in steel and bronze can be found as shown below

$$(\sigma_{AB})_s = \frac{NE_s}{(\Sigma E_i A_i)_{AB}} = \frac{400(10^3)(200)(10^9)}{1633.6(10^6)} = 48.97(10^6)\ N/m^2 \quad 4$$

$$(\sigma_{AB})_b = \frac{NE_b}{(\Sigma E_i A_i)_{AB}} = \frac{400(10^3)(100)(10^9)}{1633.6(10^6)} = 24.48(10^6)\ N/m^2 \quad 5$$

By comparing Eq's (3) and (4), we see that the maximum stress in steel is given by Eq(3) and maximum stress in bronze is given by Eq (5) $\boxed{(\sigma_{max})_s = 79.6\ MPa\ (T)}$ and $\boxed{(\sigma_{max})_b = 24.5MPa\ (T)}$

4.47 A concrete column is reinforced using nine iron circular bars of diameter 1 inch. The modulus of elasticity for concrete and iron are E_c= 4,500 ksi and E_i = 25,000 ksi. Determine (a) the maximum axial stress in concrete and iron. (b) the contraction of the column.

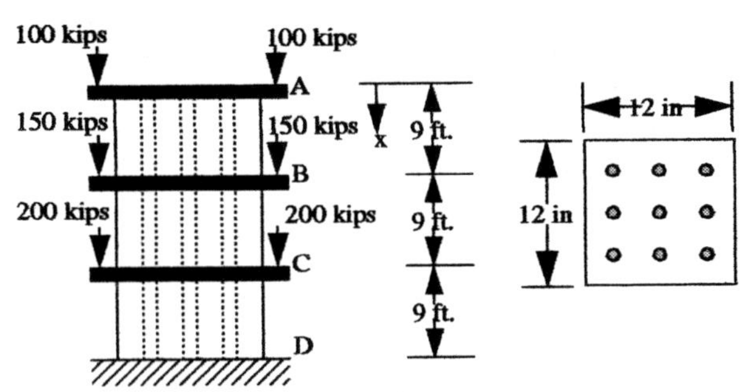

Fig. P4.47

Solution $n = 9$ $d = 1$ in $E_c = 4500$ ksi $E_i = 25{,}000$ ksi $(\sigma_i)_{max}$ = ? $(\sigma_c)_{max}$ = ?

The cross-sectional areas of each material can be found as:

$A_i = \frac{\pi}{4}(1)^2(9) = 7.069\ in^2$ and $A_c = 144 - 7.069 = 136.93\ in^2$.

The axial rigidity of the cross-section is: $\Sigma E_j A_j = 25,000(7.069) + (4500)(136.93) = 792.91(10^3)$ kips

Using the template shown in fig. (a) we draw the axial force diagram shown in fig. (b). The internal axial forces are:

$N_{AB} = -200$ kips $\quad N_{BC} = -500$ kips $\quad N_{CD} = -900$ kips

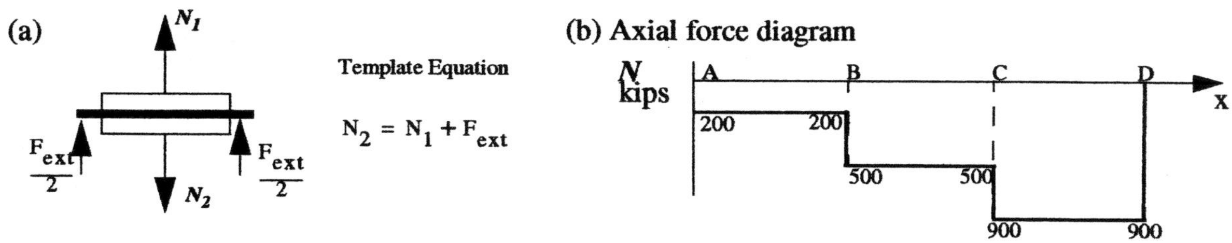

As the axial rigidity is the same for all cross-sections, the maximum axial stress will be in the segment is the maximum axial force, i.e., in segment CD.

$$(\sigma_i)_{max} = \frac{E_i N_{CD}}{\Sigma E_j A_j} = \frac{25(10^3)(-900)}{792.91(10^3)} = -28.38(10^3) \qquad \boxed{(\sigma_i)_{max} = 28.4 \text{ ksi (C)}}$$

$$(\sigma_c)_{max} = \frac{E_c N_{CD}}{\Sigma E_j A_j} = \frac{4.5(10^3)(-900)}{792.91(10^3)} = -5.108(10^3) \qquad \boxed{(\sigma_c)_{max} = 5.1 \text{ ksi (C)}}$$

The relative displacement of the ends of each segment can be written as:

$$u_B - u_A = \frac{N_{AB}(x_B - x_A)}{\Sigma E_j A_j} = \frac{(-200)(108)}{792.91} = -27.24(10^{-3}) \tag{1}$$

$$u_C - u_B = \frac{N_{BC}(x_C - x_B)}{\Sigma E_j A_j} = \frac{(-500)(108)}{792.91} = -68.1(10^{-3}) \tag{2}$$

$$u_D - u_C = \frac{N_{DC}(x_D - x_C)}{\Sigma E_j A_j} = \frac{(-900)(108)}{792.91} = -122.6(10^{-3}) \tag{3}$$

Adding Eq's (1), (2) and (3)

$u_D - u_C = (-27.24 - 68.1 - 122.6)(10^{-3}) = -217.94(10^{-3})$ in

Nothing that point D is fixed to the ground i.e. $u_D = 0$ we obtain the contraction of the column

$$\boxed{u_A = 0.218 \text{ inches}}$$

4.48 A cast iron pipe ($E_i = 100$ GPa) and a copper pipe ($E_c = 130$ GPa) are adhesively bonded together as shown. The outer diameters of the two pipes are 50 mm and 70 mm and wall thickness of each pipe is 10 mm. Determine (a) the maximum axial stress in the region BC. (b) the displacement of end D with respect to end A.

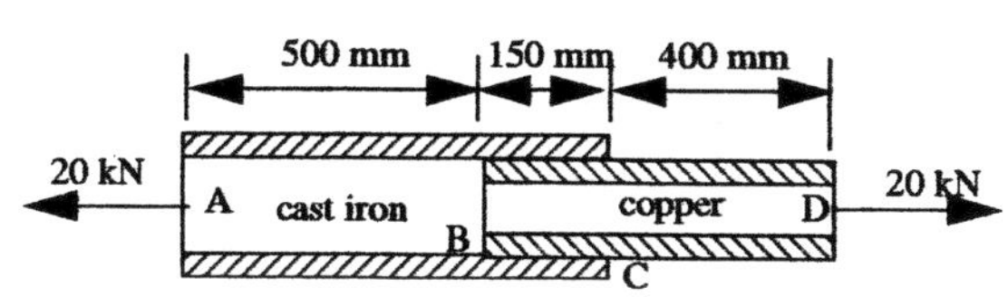

Fig. P4.48

Solution $\quad E_i = 100$ GPa $\quad E_c = 130$GPa $\quad (d_i)_o = 70$ mm $\quad (d_i)_i = 50$ mm

$(d_c)_o = 50$ mm $\quad (d_c)_i = 30$ mm $\quad (\sigma_{BC})_{max} = ?$ $\quad u_D - u_A = ?$

The internal forces are: $N_{AB} = N_{BC} = N_{CD} = 20$kN. The area of cross-section of each material is:

$A_i = \frac{\pi}{4}[(0.07) - (0.05)^2] = 1.885(10^{-3})\text{m}^2$ and $A_c = \frac{\pi}{4}[(0.05)^2 - (0.03)^2] = 1.257(10^{-3})\text{m}^2$

The axial rigidity can be found as shown below:

$E_i A_i = (100)(10^9)(1.885)(10^{-3}) = 188.5(10^6)$ N/m^2

$E_cA_c = (130)(10^9)(1.257)(10^{-3}) = 163.4(10^6)\ N/m^2$

$\Sigma E_jA_j = E_iA_i + E_cA_c = 351.9(10^6)\ N/m^2$

The maximum axial stress in BC will be in the copper because $E_c > E_i$ and N_{BC} and ΣE_jA_j are the constant across the cross-section.

$$(\sigma_{BC})_{max} = \frac{E_cN_{CD}}{\Sigma E_jA_j} = \frac{(130)(10^9)(20)(10^3)}{351.9(10^6)} = 7.39(10^6)\ N/m^2 \qquad \boxed{(\sigma_{BC})_{max} = 7.4MPa(T)}$$

The relative displacements of the ends of each segment can be written as shown below.

$$u_B - u_A = \frac{N_{AB}(x_B - x_A)}{E_iA_i} = \frac{(20)(10^3)(0.5)}{188.5(10^6)} = 53.05(10^{-6})m \qquad 1$$

$$u_C - u_B = \frac{N_{BC}(x_C - x_B)}{\Sigma E_jA_j} = \frac{(20)(10^3)(0.15)}{351.9(10^6)} = 8.53(10^{-6})m \qquad 2$$

$$u_D - u_C = \frac{N_{CD}(x_D - x_C)}{E_cA_c} = \frac{(20)(10^3)(0.4)}{163.4(10^6)} = 48.96(10^{-6})m \qquad 3$$

Adding Eq (1), (2), and (3) we obtain

$$u_D - u_A = (53.08 + 8.53 + 48.96)(10^{-6}) = 110.54(10^{-6})m \qquad \boxed{u_D - u_A = 0.11\ mm}$$

4.49 A cross-section of a bar is made from two materials. Assume parallel sections remain parallel i.e $\varepsilon_{xx} = \frac{du}{dx}(x)$. In terms of the variables P, E and h determine (a)the location (y_P) of force P on the cross-section so that there is only axial deformation and no bending. (b) the axial stress at point A.

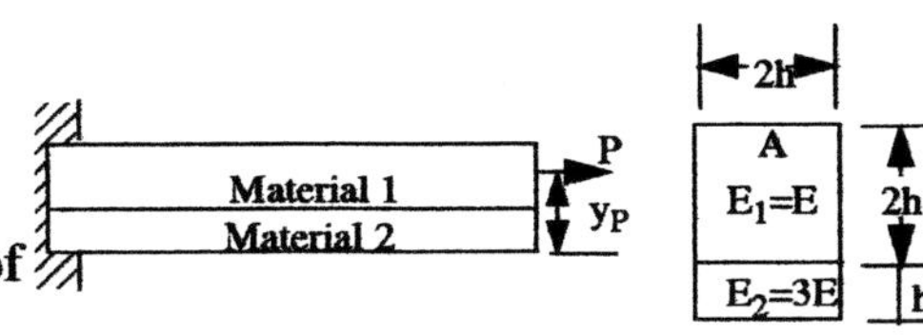

Fig. P4.49

Solution $y_p = ?$ $(\sigma_{xx})_A = ?$

The location of the origin can be found as shown below.

$$y_p = \frac{\Sigma E_iA_i\eta_{ci}}{\Sigma E_iA_i} = \frac{(3E)(2h^2)(h/2) + E(4h^2)(2h)}{(3E)(2h^2) + E(4h^2)} = \frac{11Eh^3}{10Eh^2} = 1.1h \qquad \boxed{y_p = 1.1h}$$

The internal axial force is: $N = P$. The axial rigidity is: $E_iA_i = 3E(2h^2) + E(4h^2) = 10Eh^2$.

The axial stress at A is: $$(\sigma_{xx})_A = \frac{E_AN}{\Sigma E_iA_i} = \frac{EP}{10Eh^2} = \frac{P}{10h^2} \qquad \boxed{(\sigma_{xx})_A = 0.1P/h^2}$$

4.50 Two 1 in thick wood (E_W = 1800 ksi) are joined together using two 1/8 in thick aluminum (E_{Al} = 10,000 ksi) sheets as shown. The allowable stress in wood is 1.5 ksi, the allowable stress in aluminum is 12 ksi and the total elongation of the joint length AD is to be limited to 0.05 inches. Determine the maximum axial force F that can be applied.

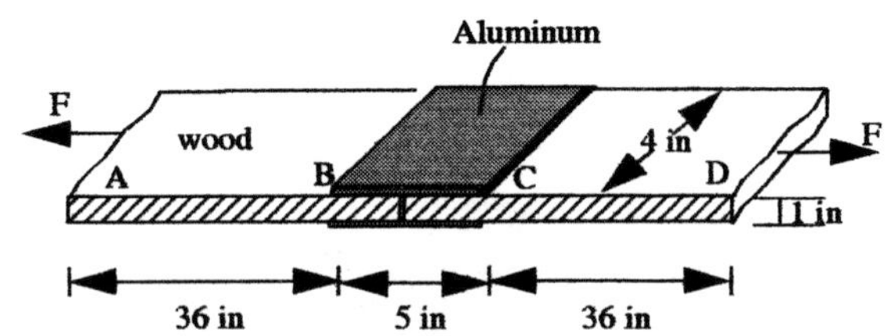

Fig. P4.50

Solution $E_W = 1800ksi$ $E_{AL} = 10{,}00ksi$ $t = 1/8$ inch

$(\sigma_W)_{max} \le 1.5\ ksi$ $(\sigma_{AL})_{max} \le 12\ ksi$ $u_D - u_A \le 0.05$ $F_{max} = ?$

The area of cross-section of each material are: $A_W = (4)(1) = 4\ in^2$ and $A_{AL} = (4)(1/8) = 0.5\ in^2$

The axial rigidities are: $E_W A_W = (1800)(4) = 7200\ \text{kips}$

$\Sigma E_i A_i = E_w A_w + 2E_{AL}A_{AL} = (7200) + 2(10,000)(0.5) = 17,200\text{kips}$

The internal axial force is: $N = F$. The relative displacements of the ends can be written as shown blow.

$$u_B - u_A = \frac{N(x_B - x_A)}{E_w A_w} = \frac{F(36)}{7200} = 5F(10^{-3}) \quad 1$$

$$u_C - u_B = \frac{N(x_C - x_B)}{E_i A_i} = \frac{F(5)}{17,200} = 0.2907F(10^{-3}) \quad 2$$

$$u_D - u_C = \frac{N(x_D - x_C)}{E_w A_w} = \frac{F(36)}{7,200} = 5F(10^{-3}) \quad 3$$

Adding Eq's (1), (2), and (3) $u_D - u_A = (5 + 0.2907 + 5)F(10^{-3}) = 10.2907F(10^{-3}) \le 0.05$ or

$$F \le 4.859\ \text{kips} \quad 4$$

The maximum stress in wood can be in either segment AB or in BC and can be found as shown below.

$$(\sigma_w)_{AB} = \frac{N}{A_w} = \frac{F}{4} \le 1.5 \quad \text{or} \quad F \le 6\ \text{kips} \quad 5$$

$$(\sigma_w)_{BC} = \frac{NE_w}{\Sigma E_i A_i} = \frac{F(1800)}{17,200} \le 1.5 \quad \text{or} \quad F \le 14.33\ \text{kips} \quad 6$$

The maximum axial stress in aluminum can be found as shown below

$$(\sigma_{AL})_{max} = \frac{NE_{AL}}{\Sigma E_i A_i} = \frac{F(10,000)}{17,200} \le 12 \quad \text{or} \quad F \le 20.6\ \text{kips} \quad 7$$

The maximum value of F that meets the inequalities in Eq's (4), (5), (6), and (7) is $\boxed{F_{max} = 4.8\ \text{kips}}$

4.51 The average stress for a composite Lamina shown in Fig. P4.51 can be obtained by dividing the internal axial force in Equation 4.20 by the total area of cross-section A. The tension test for a composite Lamina would yield the Hooke's law as $\sigma_{average} = E_{effective}\varepsilon$, where $E_{effective}$ is the effective Modulus of Elasticity for the lamina. Show that $E_{effective}$ is given by Equation 4.26.

$$E_{effective} = \frac{E_{fiber}V_{fiber} + E_{resin}V_{resin}}{V_{fiber} + V_{resin}} \quad \textbf{4.26}$$

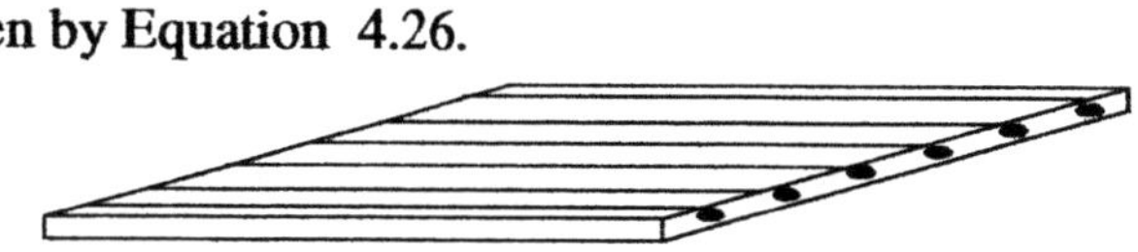

Fig. P4.51

where, E_{fiber} and E_{resin} are the Modulus of Elasticity of fiber and resin, respectively. V_{fiber} and V_{resin} is are the volume of fiber and volume of resin in the Lamina, respectively. Equation 4.26 is known as the *Rule of Mixtures.*

Solution

Noting that the axial strain is $\varepsilon = \frac{du}{dx}$, Eq 4-20 can be re-written as $N = [E_{fiber}A_{fiber} + E_{resin}\ A_{resin}]\varepsilon$

Dividing the above equation by the total area $A = A_{fiber} + A_{resin}$ we obtain the average stress as:

$\sigma_{org} = \frac{N}{A} = \frac{E_{fiber}A_{fiber} + E_{resin}\ A_{resin}}{A_{fiber} + A_{resin}}\varepsilon$. Multiplying the numerator and denominator by L, the length of lamina and noting that $V_{fiber} = LA_{fiber}$ and $V_{resin} = LA_{resin}$ we obtain

$$\sigma_{avg} = \frac{E_{fiber}V_{fiber} + E_{resin}\,V_{resin}}{V_{fiber} + V_{resin}}\varepsilon \qquad 1$$

Comparing Eq (1) with $\sigma_{avg} = E_{effective}\varepsilon$ we obtain the equation for $E_{effective}$ given in Eq 4.26.

4.52 An axial rod is made from two materials. The stress-strain relation in material one is given by: $\sigma = E_1\,\varepsilon^{0.5}$ and in material 2 the relation is $\sigma = E_2\,\varepsilon^{0.5}$. Let A_1 and A_2 be the area of cross-section of material 1 and material 2 respectively. Determine the elongation of the bar in terms of F, A_1, A_2, E_1, E_2 and L.

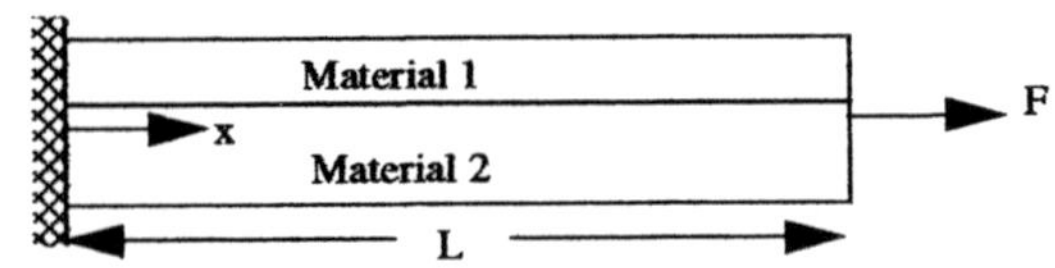

Fig. P4.52

Solution $\sigma_1 = E_1\varepsilon_1^{0.5}$ $\sigma_2 = E_2\varepsilon_2^{0.5}$ $\Delta u = f(F, A_1, A_2, E_1, E_2, \text{and } L)$

The strain at a cross section is $\varepsilon = \frac{du}{dx}(x)$ Thus, the stress can be written as

$$\sigma_{xx} = \begin{cases} E_1\left(\frac{du}{dx}\right)^{0.5} & \text{in material 1} \\ E_2\left(\frac{du}{dx}\right)^{0.5} & \text{in material 2} \end{cases} \qquad 1$$

The internal axial force is: $N = \int_A \sigma_{xx}dA = \int_{A_1} E_1\left(\frac{du}{dx}\right)^{0.5}dA + \int_{A_2} E_2\left(\frac{du}{dx}\right)^{0.5}dA$

Noting that $\frac{du}{dx}$ does not change across the cross-section and E_1 and E_2 are constants, we obtain

$N = \left(\frac{du}{dx}\right)^{0.5}(E_1A_1 + E_2A_2) = F$ or

$$\left(\frac{du}{dx}\right)^{0.5} = \frac{F}{E_1A_1 + E_2A_2} \qquad 2$$

All variables on the left are constants, hence, $\frac{du}{dx}$ is constant and can be written as $\frac{du}{dx} = \frac{\Delta u}{L}$, Eq(2) can be squared and rewritten as

$$\boxed{\Delta u = \left(\frac{F}{E_1A_1 + E_2A_2}\right)^2 L}$$

Problem Set 4.3

4.53 A rigid bar is hinged at C. The Modulus of Elasticity of the bar A is E= 30,000 ksi, the area of cross-section is A = 1.25 in^2, and the length is 24 inches. Determine the applied force F if point B moves upward by 0.002 inches.

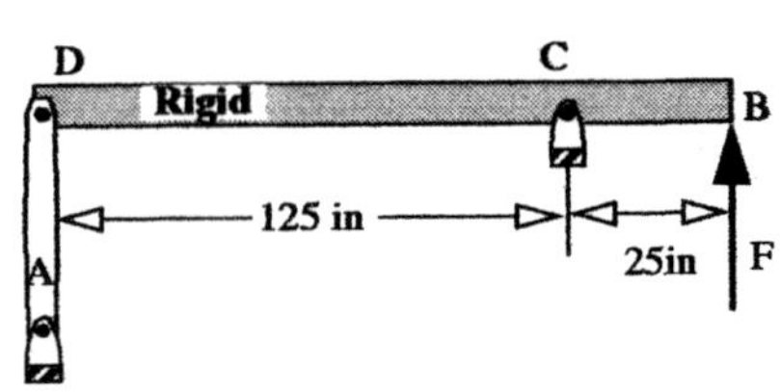

Solution E = 30,000 ksi A = 1.25 in^2 L = 24 inch $\delta_B = 0.002$ in

---

We draw the free body diagram of the rigid bar as shown in Fig. (a)

(a)

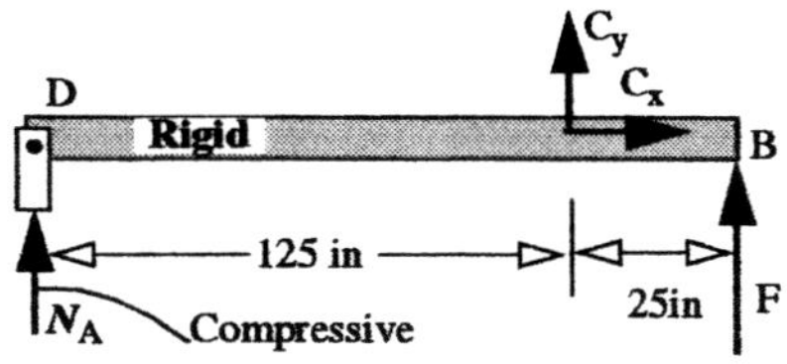

(b)

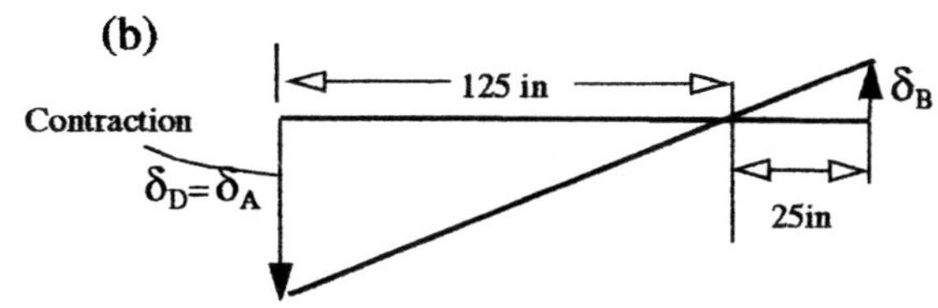

By taking moment equilibrium about point C in Fig.(a), we obtain the following.

$$125N_A = 25F \qquad \text{or} \qquad F = 5N_A$$ 1

We draw the exaggerated deformed shape as shown in Fig. (b) and write the deformation equations using similar triangles.

$$\frac{\delta_A}{125} = \frac{\delta_B}{25} \qquad \text{or} \qquad \delta_A = 5\delta_B = 5(0.002) = 0.01\text{in}$$ 2

From Eq 4.27, we have the following:

$$\delta_A = \frac{N_A L}{EA} = \frac{N_A(24)}{(30,000)(1.25)} = 0.01 \qquad \text{or} \qquad N_A = 15.625 \text{ kips}$$ 3

Substituting Eq. (3) into Eq. (1), we obtain the force F as shown below.

$F = 5(15.625) = 78.125\text{kips}$ or $\boxed{F = 78.1 \text{ kips}}$

4.54 A rigid bar is hinged at C. The Modulus of Elasticity of the bar A is E= 30,000 ksi, the area of cross-section is A = 1.25 in^2, and the length is 24 inches. Determine the applied force F if point B moves upward by 0.002 inches.

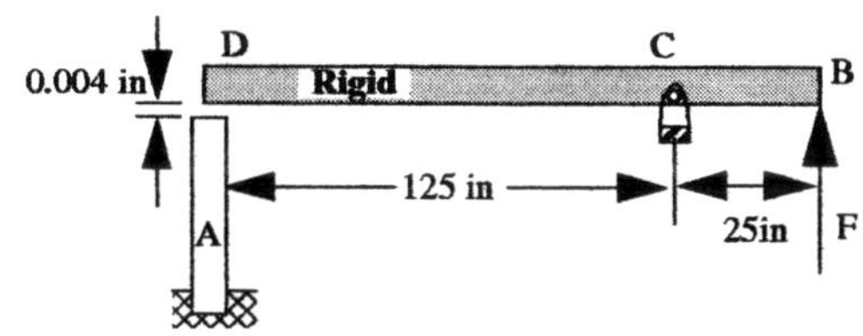

Solution E = 30,000 ksi A = 1.25 in^2 L = 24 inch $\delta_B = 0.002$ in

---

We draw the free body diagram of the rigid bar as shown in Fig. (a)

(a)

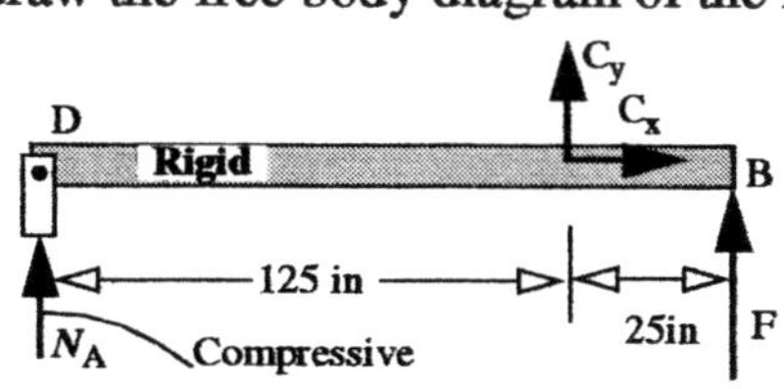

(b)

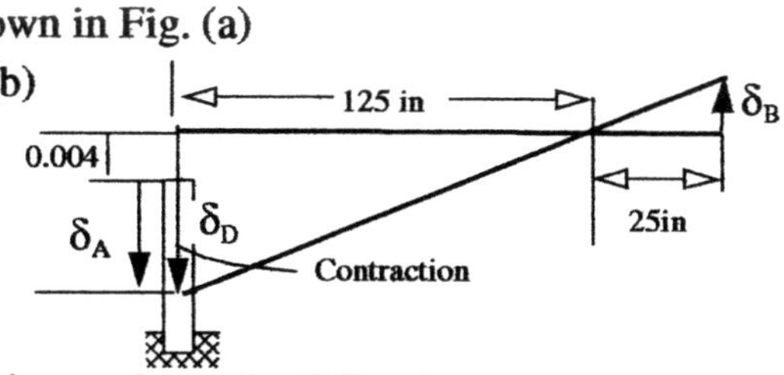

By taking moment equilibrium about point C in Fig.(a), we obtain the following.

$$125N_A = 25F \qquad \text{or} \qquad F = 5N_A$$ 1

We draw the exaggerated deformed shape as shown in Fig. (b) and write the deformation equations.

$$\delta_D = \delta_A + 0.04 \qquad \frac{\delta_D}{125} = \frac{\delta_B}{25} \qquad \delta_A = 5\delta_B - 0.004 = 5(0.002) - 0.004 = 0.006\text{in}$$ 2

From Eq 4.27, we have the following:

$$\delta_A = \frac{N_A L}{EA} = \frac{N_A(24)}{(30,000)(1.25)} = 0.006 \qquad \text{or} \qquad N_A = 9.375 \text{ kips}$$ 3

Substituting Eq. (3) into Eq. (1), we obtain the force F as shown below.

$F = 5(9.375) = 46.875\text{kips}$ or $\boxed{F = 46.9 \text{ kips}}$

4.55 A rigid bar is hinged at C. The Modulus of Elasticity of the bar A is E= 100 GPa, the area of cross-section is A = 15 mm^2, and the length is 1.2 m. Determine the applied force F if point B moves to the left by 0.75 mm.

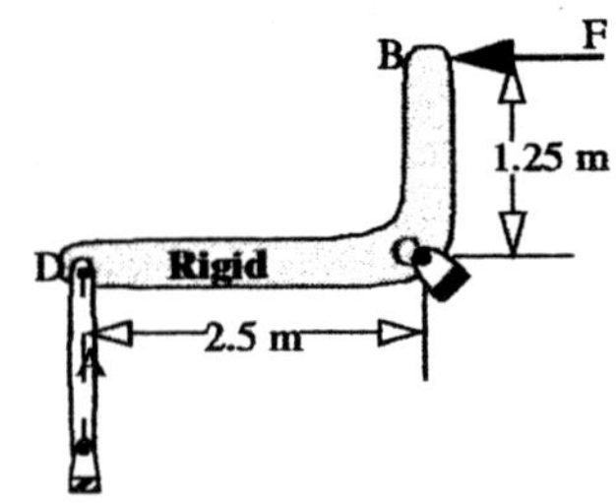

Solution E = 100 GPa A = 15 mm^2 L = 1.2 m δ_B = 0.75 mm

We draw the free body diagram of the rigid bar as shown in Fig. (a)

(a)

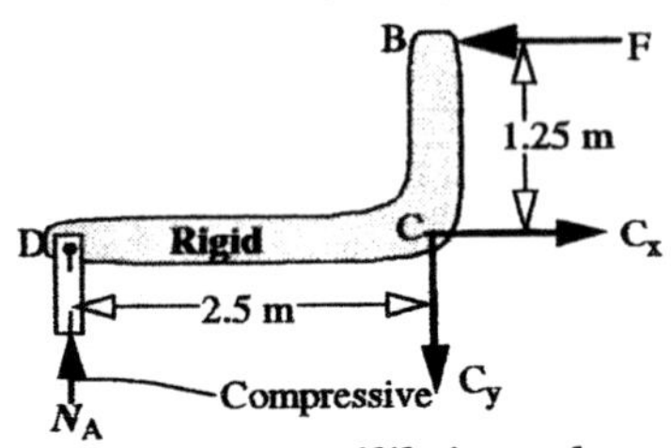

(b)

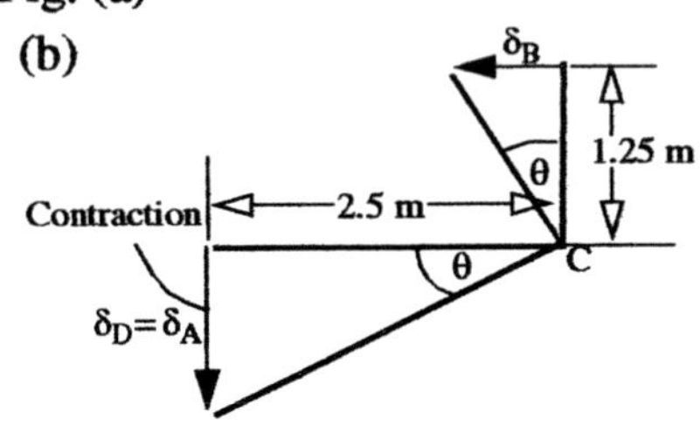

By taking moment equilibrium about point C in Fig.(a), we obtain the following.

$$2.5N_A = 1.25F \qquad \text{or} \qquad F = 2N_A \tag{1}$$

We draw the exaggerated deformed shape as shown in Fig. (b) and write the deformation equations.

$$\frac{\delta_A}{2.5} = \frac{\delta_B}{1.25} \qquad \text{or} \qquad \delta_A = 2\delta_B = 2(0.75) = 1.5\text{mm} = 1.5(10^{-3})\text{m} \tag{2}$$

From Eq 4.27, we have the following:

$$\delta_A = \frac{N_A L}{EA} = \frac{N_A(1.2)}{(100)(10^9)(15)(10^{-6})} = 1.5(10^{-3}) \qquad \text{or} \qquad N_A = 1.875(10^3)\ \text{Newtons} = 1.875\text{kN} \tag{3}$$

Substituting Eq. (3) into Eq. (1), we obtain the force F as shown below.

$F = 2(1.875) = 3.75\text{kN}$ or $\boxed{F = 3.75\ \text{kN}}$

4.56 A rigid bar is hinged at C. The Modulus of Elasticity of the bar A is E= 100 GPa, the area of cross-section is A = 15 mm^2, and the length is 1.2 m. Determine the applied force F if point B moves to the left by 0.75 mm.

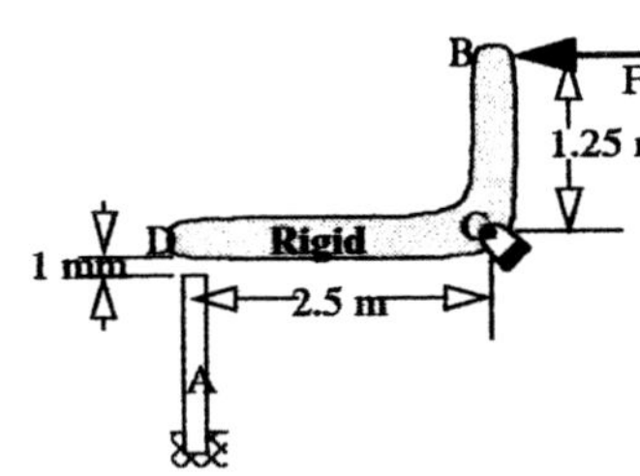

Solution E = 100 GPa A = 15 mm^2 L = 1.2 m δ_B = 0.75 mm

We draw the free body diagram of the rigid bar as shown in Fig. (a)

(a)

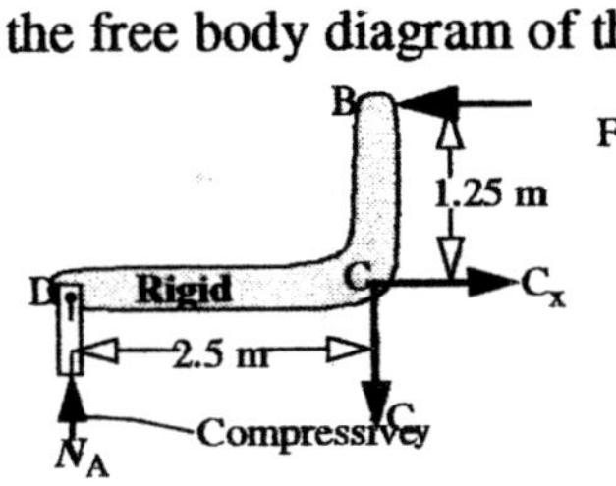

(b)

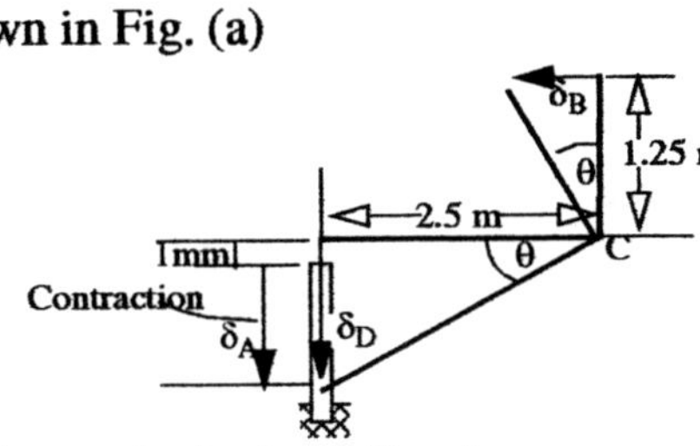

By taking moment equilibrium about point C in Fig.(a), we obtain the following.

$$2.5N_A = 1.25F \qquad \text{or} \qquad F = 2N_A \tag{1}$$

We draw the exaggerated deformed shape as shown in Fig. (b) and write the deformation equations.

$$\delta_D = \delta_A + 1 \qquad \frac{\delta_D}{2.5} = \frac{\delta_B}{1.25} \qquad \text{or} \qquad \delta_A = 2\delta_B - 1 = 2(0.75) - 1 = 0.5\text{mm} = 0.5(10^{-3})\text{m}$$ 2

From Eq 4.27, we have the following:

$$\delta_A = \frac{N_A L}{EA} = \frac{N_A(1.2)}{(100)(10^9)(15)(10^{-6})} = 0.5(10^{-3}) \qquad \text{or} \qquad N_A = 0.625(10^3)\ \text{Newtons} = 0.625\text{kN}$$ 3

Substituting Eq. (3) into Eq. (1), we obtain the force F as shown below.

$F = 2(0.625) = 1.25\text{kN}$ or $\boxed{F = 1.25\ \text{kN}}$

4.57 The roller at P slides in the slot due to the force F = 20 kN. Member AP has an area of cross-section of A = 100 mm^2 and a Modulus of Elasticity E = 200 GPa. Determine the displacement of the roller.

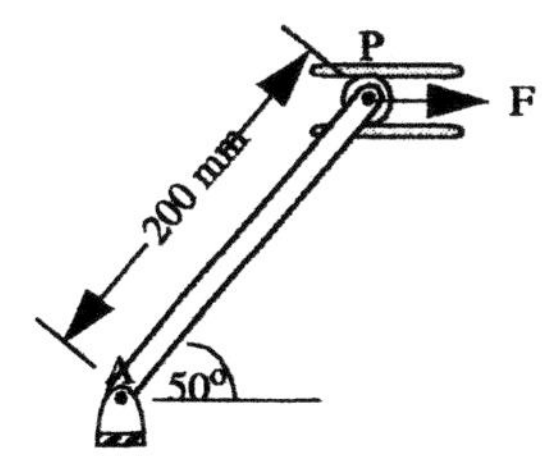

Solution E = 200 GPa A = 100 mm^2 F=20 kN $\delta_P = ?$

We draw the free body diagram of the rigid bar as shown in Fig. (a)

(a)

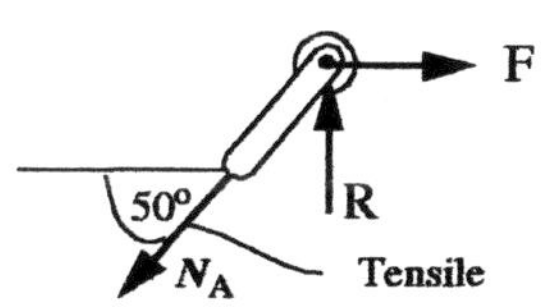

(b)

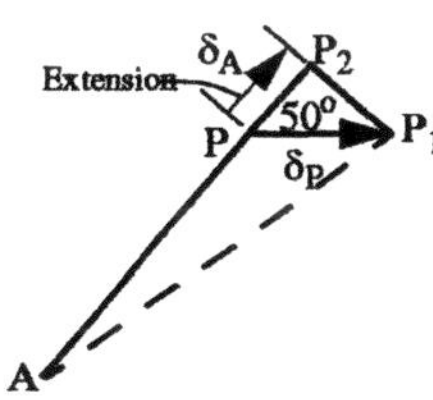

By force equilibrium in the x-direction in Fig.(a), we obtain the following.

$$N_A \cos 50 = F \qquad \text{or} \qquad N_A = 31.11\text{kN}$$ 1

We draw the exaggerated deformed shape as shown in Fig. (b). As per small strain approximation, we need component of PP_1 in the original direction of AP, i.e., PP_2 represents the deformation of bar AP and is calculated as shown below.

$$\delta_A = \delta_P \cos 50 \qquad \text{or} \qquad \delta_P = 1.556\delta_A$$ 2

From Eq 4.27, we have the following:

$$\delta_A = \frac{N_A L}{EA} = \frac{(31.11)(10^3)(0.2)}{(200)(10^9)(100)(10^{-6})} = 0.311(10^{-3})\text{m} = 0.311\text{mm}$$ 3

Substituting Eq. (3) into Eq. (2), we obtain the displacement of roller as shown below.

$\delta_P = 1.556\delta_A = (1.556)(0.311) = 0.484\text{mm}$ or $\boxed{\delta_P = 0.48\text{mm}}$

4.58 The roller at P slides in the slot due to the force F = 20 kN. Member AP has an area of cross-section of A = 100 mm^2 and a Modulus of Elasticity E = 200 GPa. Determine the displacement of the roller.

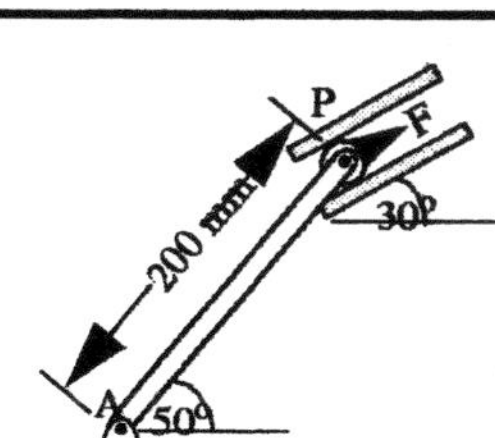

Solution E = 200 GPa A = 100 mm^2 F=20 kN $\delta_P = ?$

We draw the free body diagram of the rigid bar as shown in Fig. (a)

(a)

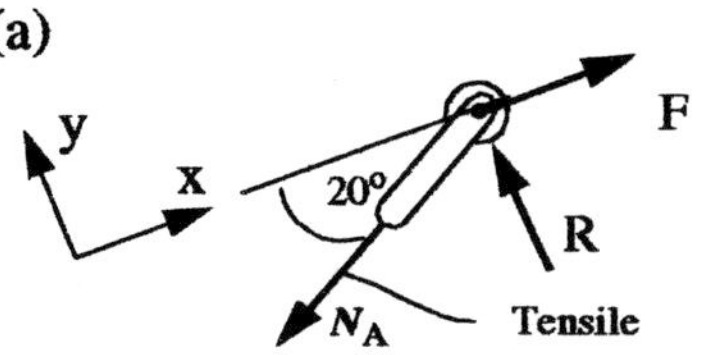

(b)

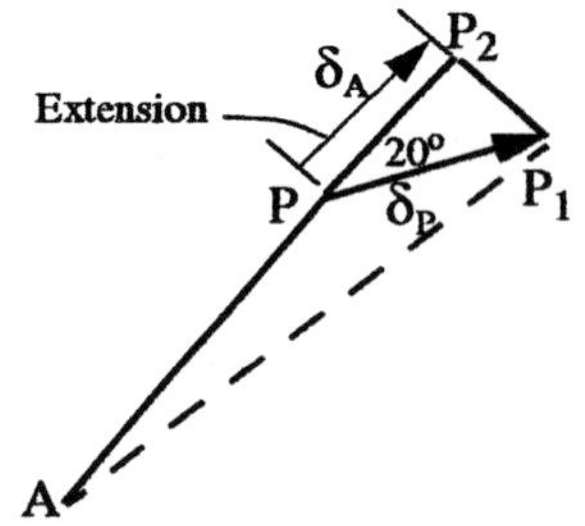

By force equilibrium in the x-direction in Fig.(a), we obtain the following.

$$N_A \cos 20 = F \qquad \text{or} \qquad N_A = 21.28\text{kN}$$ 1

We draw the exaggerated deformed shape as shown in Fig. (b). As per small strain approximation, we need component of PP_1 in the original direction of AP, i.e., PP_2 represents the deformation of bar AP and is calculated as shown below.

$$\delta_A = \delta_P \cos 20 \qquad \text{or} \qquad \delta_P = 1.064\delta_A$$ 2

From Eq 4.27, we have the following:

$$\delta_A = \frac{N_A L}{EA} = \frac{(21.28)(10^3)(0.2)}{(200)(10^9)(100)(10^{-6})} = 0.2128(10^{-3})\text{m} = 0.2128\text{mm}$$ 3

Substituting Eq. (3) into Eq. (2), we obtain the displacement of roller as shown below.

$\delta_P = 1.064\delta_A = (1.064)(0.2128) = 0.2264\text{mm}$ or $\boxed{\delta_P = 0.23\text{mm}}$

4.59 A rigid bar is hinged at C. The Modulus of Elasticity of the bar A is E= 30,000 ksi, the area of cross-section is A = 1.25 in^2 and the length is 24 inches. Determine the axial stress in bar A and the displacement of point B on the rigid bar.

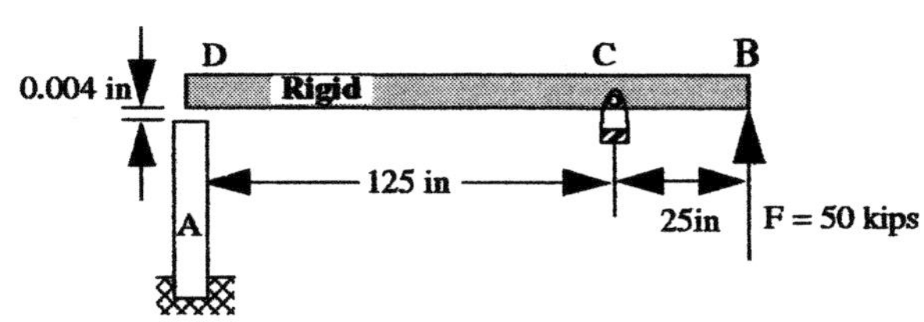

Solution E = 30,000 ksi A = 1.25 in^2
L = 24 inch $\sigma_A = ?$ $\delta_D = ?$

We draw the free body diagram of the rigid bar as shown in Fig. (a)

(a)

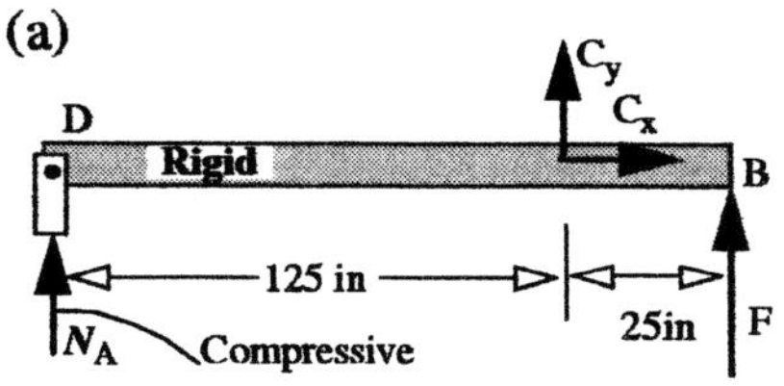

(b)

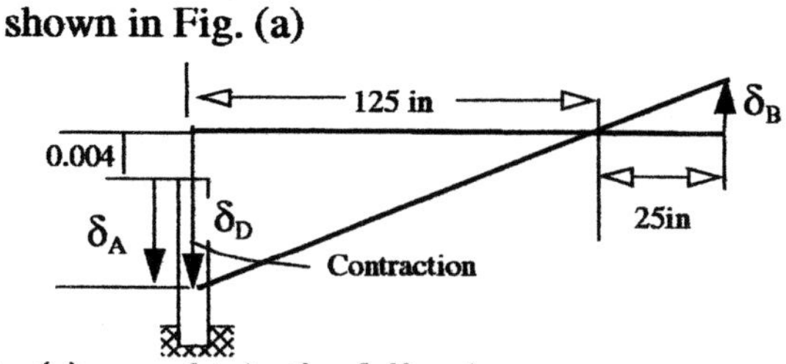

By taking moment equilibrium about point C in Fig.(a), we obtain the following.

$$125N_A = 25F \qquad \text{or} \qquad N_A = F/5 = 10\text{kips}$$ 1

The axial stress in member A is: $\sigma_A = \frac{N_A}{A} = \frac{10}{1.25} = 8.0 \text{ ksi}$ or $\boxed{\sigma_A = 8.0 \text{ ksi(C)}}$

We draw the exaggerated deformed shape as shown in Fig. (b) and write the deformation equations.

$$\delta_D = \delta_A + 0.004 \qquad \frac{\delta_D}{125} = \frac{\delta_B}{25} \qquad \text{or} \qquad \delta_B = \frac{(\delta_A + 0.004)}{5}$$ 2

From Eq 4.27, we have the following:

$$\delta_A = \frac{N_A L}{EA} = \frac{(10)(24)}{(30,000)(1.25)} = 0.0064 \text{ in} \qquad 3$$

Substituting Eq. (3) into Eq. (2), we obtain the displacement δ_B as shown below.

$$\delta_B = \frac{(0.0064 + 0.004)}{5} = 0.00208\text{in} \quad \text{or} \quad \boxed{\delta_B = 0.0021 \text{ in}}$$

4.60 A rigid bar is hinged at C. The Modulus of Elasticity of the bar A is E= 30,000 ksi, the area of cross-section is A = 1.25 in^2 and the length is 24 inches. Determine the axial stress in bar A and the displacement of point B on the rigid bar.

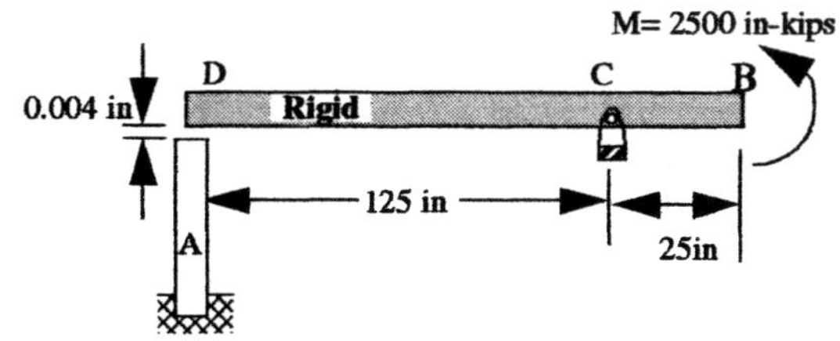

Solution E = 30,000 ksi A = 1.25 in^2 L = 24 inch σ_A = ? δ_D = ?

We draw the free body diagram of the rigid bar as shown in Fig. (a)

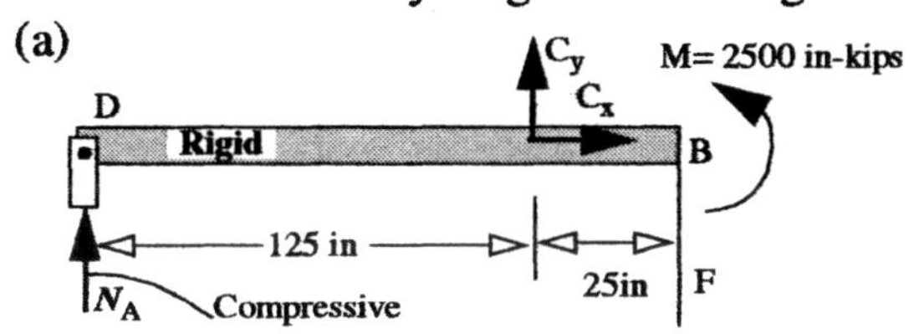

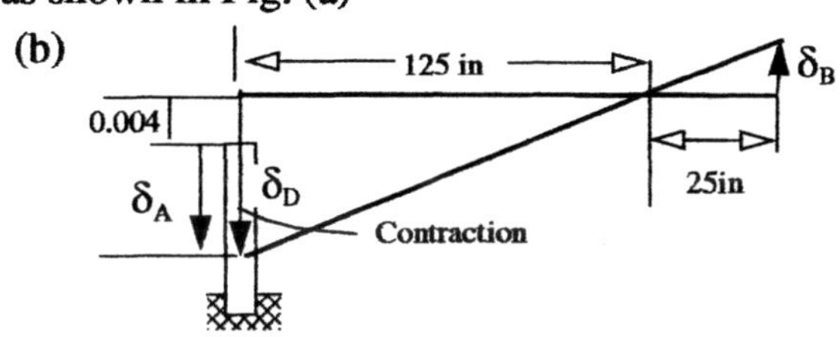

By taking moment equilibrium about point C in Fig.(a), we obtain the following.

$$125N_A = 2500 \quad \text{or} \quad N_A = 20\text{kips} \qquad 1$$

The axial stress in member A is: $\sigma_A = \frac{N_A}{A} = \frac{20}{1.25} = 16.0 \text{ ksi}$ or $\boxed{\sigma_A = 16.0 \text{ ksi(C)}}$

We draw the exaggerated deformed shape as shown in Fig. (b) and write the deformation equations.

$$\delta_D = \delta_A + 0.004 \qquad \frac{\delta_D}{125} = \frac{\delta_B}{25} \quad \text{or} \quad \delta_B = \frac{(\delta_A + 0.004)}{5} \qquad 2$$

From Eq 4.27, we have the following:

$$\delta_A = \frac{N_A L}{EA} = \frac{(20)(24)}{(30,000)(1.25)} = 0.0128 \text{ in} \qquad 3$$

Substituting Eq. (3) into Eq. (2), we obtain the displacement δ_B as shown below.

$$\delta_B = \frac{(0.0128 + 0.004)}{5} = 0.00336\text{in} \quad \text{or} \quad \boxed{\delta_B = 0.0034 \text{ in}}$$

4.61 A force F= 20 kN is applied to the roller that slides inside a slot. Both bars have an area of cross-section of A = 100 mm^2 and a Modulus of Elasticity E = 200 GPa. Bar AP and BP have lengths of L_{AP}= 200 mm and L_{BP}= 250 mm respectively. Determine the displacement of the roller and axial stress in bar A

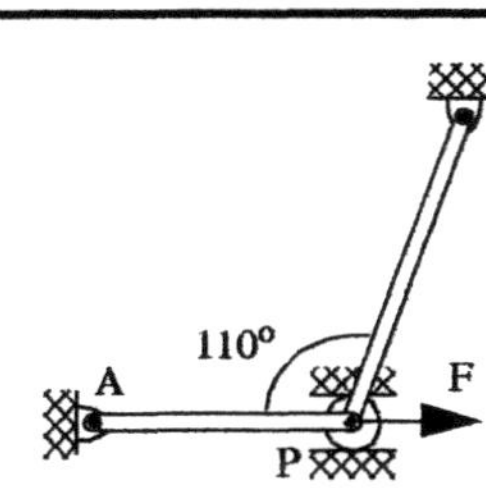

Solution F=20 kN E = 200 GPa A = 100 mm^2
L_{AP}= 200 mm L_{BP}= 250 mm δ_P = ? σ_A = ?

We draw the free body diagram of the rigid bar as shown in Fig. (a)

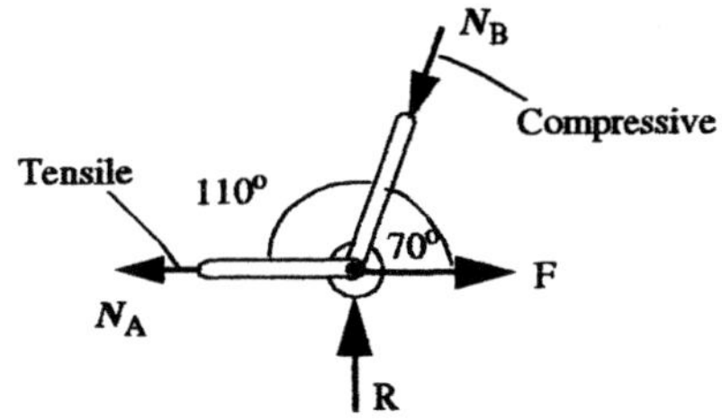

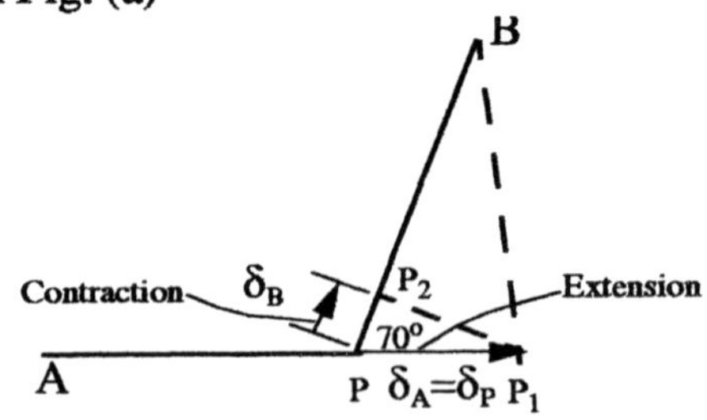

By force equilibrium in the x-direction in Fig.(a), we obtain the following.

$$N_A + N_B \cos 70 = F \qquad \text{or} \qquad N_A + 0.342 N_B = 20(10^3) \tag{1}$$

We draw the exaggerated deformed shape as shown in Fig. (b). As per small strain approximation, we need component of PP_1 in the original direction of AP and BP, i.e., PP_1 represents the deformation of bar AP and PP_2 represents the deformation of bar BP and are calculated as shown below.

$$\delta_A = \delta_P \tag{2}$$

$$\delta_B = \delta_P \cos 70 = 0.342\delta_P \tag{3}$$

From Eq 4.27, we have the following:

$$\delta_A = \frac{N_A L_A}{EA} = \frac{N_A(0.2)}{(200)(10^9)(100)(10^{-6})} = \delta_P \qquad \text{or} \qquad N_A = 100\delta_P(10^6) \tag{4}$$

$$\delta_B = \frac{N_B L_B}{EA} = \frac{N_B(0.25)}{(200)(10^9)(100)(10^{-6})} = 0.342\delta_P \qquad \text{or} \qquad N_B = 27.36\delta_P(10^6) \tag{5}$$

Substituting Eqs. (4) and (5) into Eq. (1), we obtain the displacement δ_P as shown below.

$$[(100\delta_P) + 0.342(27.36\delta_P)](10^6) = 20(10^3) \qquad \text{or} \qquad \delta_P = 0.1829(10^{-3})\text{m} \tag{6}$$

$$\boxed{\delta_P = 0.18 \text{ mm}}$$

Substituting Eq.(6) into Eq. (4) we obtain: $N_A = (100)(0.1829)(10^{-3})(10^6) = 18.29(10^3)$ Newtons

The axial stress in member A is: $\sigma_A = \dfrac{N_A}{A} = \dfrac{18.29(10^3)}{(100)(10^{-6})} = 182.9(10^6)$ or $\boxed{\sigma_A = 183\text{MPa(T)}}$

4.62 A force F= 20 kN is applied to the roller that slides inside a slot. Both bars have an area of cross-section of A = 100 mm^2 and a Modulus of Elasticity E = 200 GPa. Bar AP and BP have lengths of L_{AP}= 200 mm and L_{BP}= 250 mm respectively. Determine the displacement of the roller and axial stress in bar A.

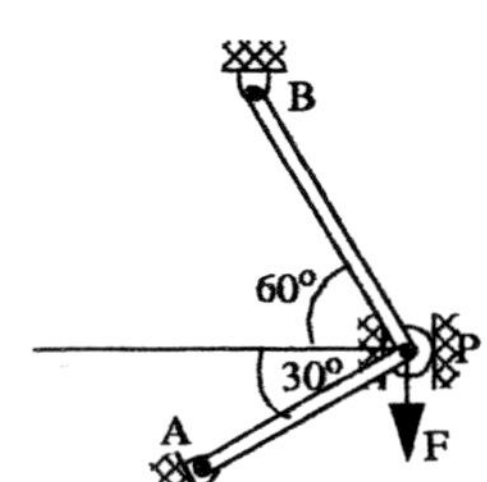

Solution F=20 kN E = 200 GPa A = 100 mm^2

L_{AP}= 200 mm L_{BP}= 250 mm δ_P = ? σ_A = ?

We draw the free body diagram of the rigid bar as shown in Fig. (a)
By force equilibrium in the y-direction in Fig.(a), we obtain the following.

$$N_A \sin 30 + N_B \sin 60 = F \qquad \text{or} \qquad 0.5N_A + 0.866N_B = 20(10^3) \tag{1}$$

(a)

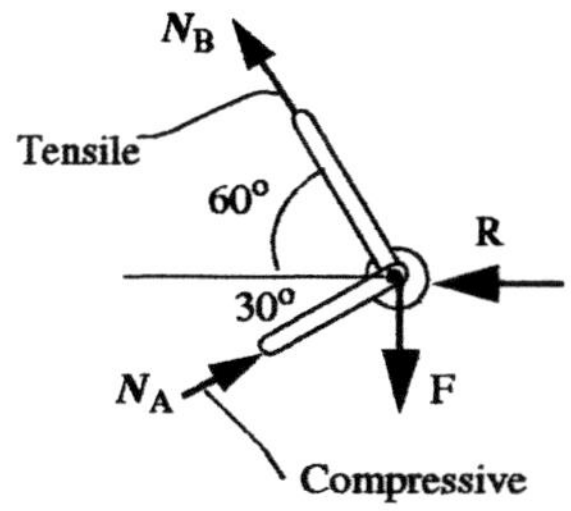

(b)

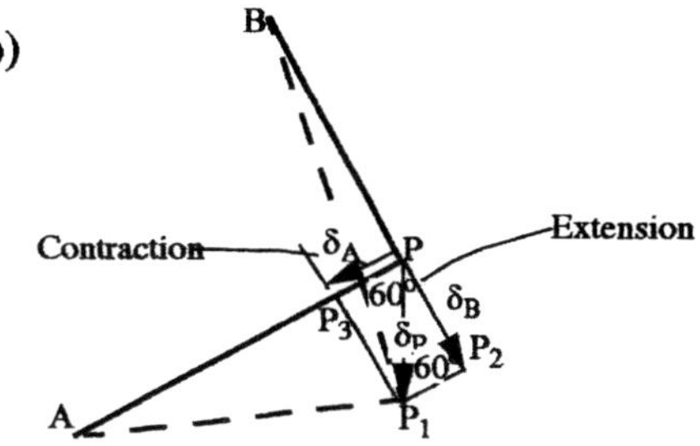

We draw the exaggerated deformed shape as shown in Fig. (b) As per small strain approximation, we need component of PP_1 in the original direction of AP and BP, i.e., PP_3 represents the deformation of bar AP and PP_2 represents the deformation of bar BP and are calculated as shown below.

$$\delta_A = \delta_P \cos 60 = 0.5\delta_P \quad 2$$

$$\delta_B = \delta_P \sin 60 = 0.866\delta_P \quad 3$$

From Eq 4.27, we have the following:

$$\delta_A = \frac{N_A L_A}{EA} = \frac{N_A(0.2)}{(200)(10^9)(100)(10^{-6})} = 0.5\delta_P \quad \text{or} \quad N_A = 50\delta_P(10^6) \quad 4$$

$$\delta_B = \frac{N_B L_B}{EA} = \frac{N_B(0.25)}{(200)(10^9)(100)(10^{-6})} = 0.866\delta_P \quad \text{or} \quad N_B = 69.28\delta_P(10^6) \quad 5$$

Substituting Eqs. (4) and (5) into Eq. (1), we obtain the displacement δ_P as shown below.

$$[0.5(50\delta_P) + 0.866(69.2\delta_P)](10^6) = 20(10^3) \quad \text{or} \quad \delta_P = 0.2353(10^{-3})\text{m} \quad 6$$

$$\boxed{\delta_P = 0.24 \text{ mm}}$$

Substituting Eq.(6) into Eq. (4) we obtain: $N_A = (50)(0.2353)(10^{-3})(10^6) = 11.76(10^3)$ Newtons

The axial stress in member A is: $\sigma_A = \dfrac{N_A}{A} = \dfrac{11.76(10^3)}{(100)(10^{-6})} = 117.6(10^6)$ or $\boxed{\sigma_A = 118\text{MPa(C)}}$

4.63 A force F= 20 kN is applied to the roller that slides inside a slot. Both bars have an area of cross-section of A = 100 mm^2 and a Modulus of Elasticity E = 200 GPa. Bar AP and BP have lengths of L_{AP}= 200 mm and L_{BP}= 250 mm respectively. Determine the displacement of the roller and axial stress in bar A

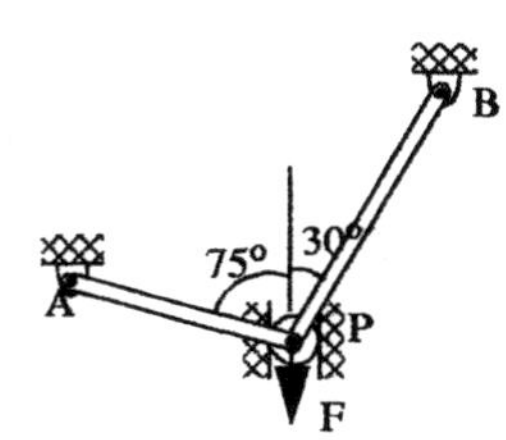

Solution F=20 kN E = 200 GPa A = 100 mm^2
L_{AP}= 200 mm L_{BP}= 250 mm δ_P = ? σ_A = ?

We draw the free body diagram of the rigid bar as shown in Fig. (a)

(a)

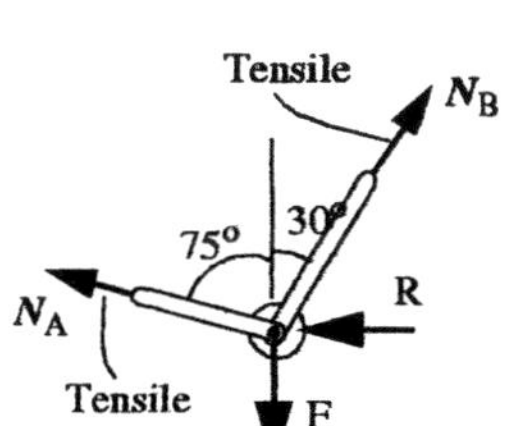

(b)

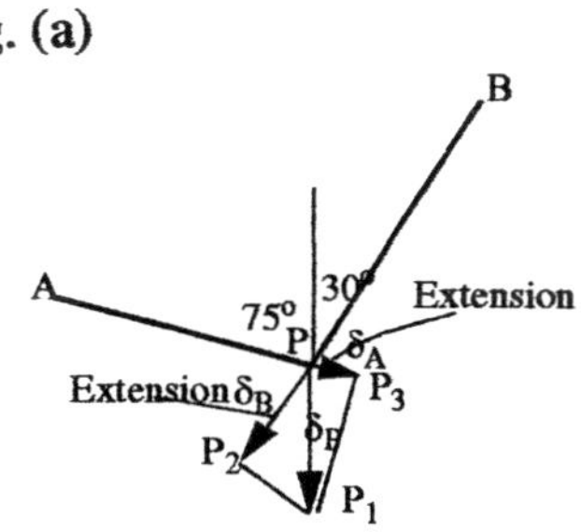

By force equilibrium in the y-direction in Fig.(a), we obtain the following.

$$N_A \cos 75 + N_B \cos 30 = \mathrm{F} \qquad \text{or} \qquad 0.2588 N_A + 0.8660 N_B = 20(10^3) \tag{1}$$

We draw the exaggerated deformed shape as shown in Fig. (b). As per small strain approximation, we need component of PP_1 in the original direction of AP and BP, i.e., PP_1 represents the deformation of bar AP and PP_2 represents the deformation of bar BP and are calculated as shown below.

$$\delta_A = \delta_P \cos 75 = 0.2588\delta_P \tag{2}$$

$$\delta_B = \delta_P \cos 30 = 0.8660\delta_P \tag{3}$$

From Eq 4.27, we have the following:

$$\delta_A = \frac{N_A L_A}{EA} = \frac{N_A(0.2)}{(200)(10^9)(100)(10^{-6})} = 0.2588\delta_P \qquad \text{or} \qquad N_A = 25.88\delta_P(10^6) \tag{4}$$

$$\delta_B = \frac{N_B L_B}{EA} = \frac{N_B(0.25)}{(200)(10^9)(100)(10^{-6})} = 0.866\delta_P \qquad \text{or} \qquad N_B = 69.28\delta_P(10^6) \tag{5}$$

Substituting Eqs. (4) and (5) into Eq. (1), we obtain the displacement δ_P as shown below.

$$[0.2588(25.88\delta_P) + 0.866(69.28\delta_P)](10^6) = 20(10^3) \quad \text{or} \quad \delta_P = 0.3002(10^{-3})\text{m} \tag{6}$$

$$\boxed{\delta_P = 0.30 \ \text{mm}}$$

Substituting Eq.(6) into Eq. (4) we obtain: $N_A = (25.88)(0.3002)(10^{-3})(10^6) = 7.769(10^3)$ Newtons

The axial stress in member A is: $\sigma_A = \dfrac{N_A}{A} = \dfrac{7.769(10^3)}{(100)(10^{-6})} = 77.69(10^6)$ or $\boxed{\sigma_A = 77.7\text{MPa(T)}}$

4.64 An aluminum (E_{al}=10,000 ksi and ν_{al}= 0.25) and steel (E_s=30,000 ksi and ν_s= 0.28) hollow cylinder are securely fastened to a rigid plate as shown in Fig. P4.65. Both cylinders are made from sheet metal thickness 1/8 inch. The outer diameters of aluminum and steel cylinders are 4 inch and 3 inch respectively. The load P = 20 kips in Fig. P4.65. Determine (a) the displacement of the rigid plate. (b) the change in diameters of each cylinder.

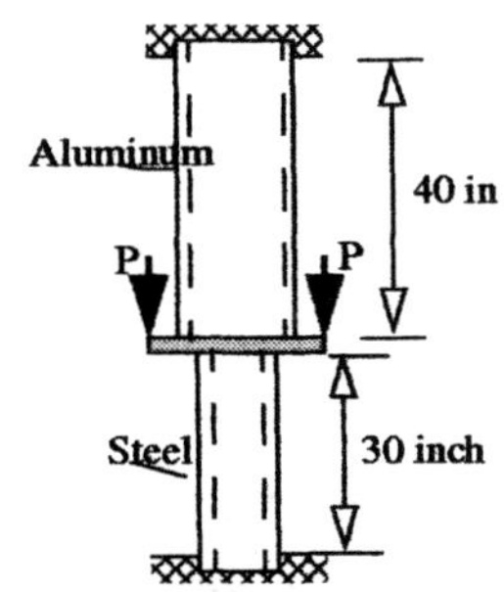

Fig. P4.64

Solution E_{al} = 10,000ksi ν_{al} = 0.25 E_s = 30,000ksi

ν_s = 0.28 t = 1/8 in $(d_{al})_0$ = 4in $(d_s)_0$ = 3in

P = 20 kips δp = ? Δd_s = ? Δd_{al} = ?

We draw the free body diagram of the rigid plate as shown in Fig.(a).

(a)

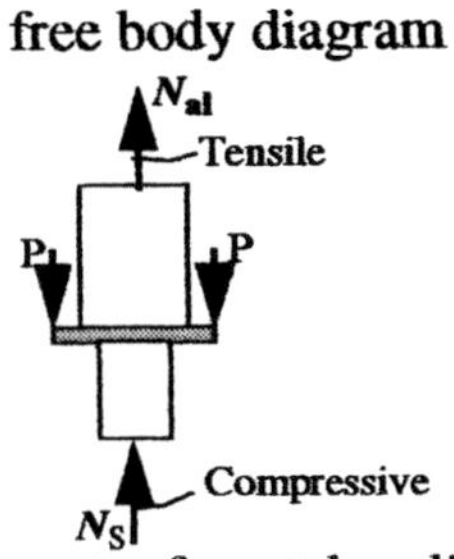

(b)

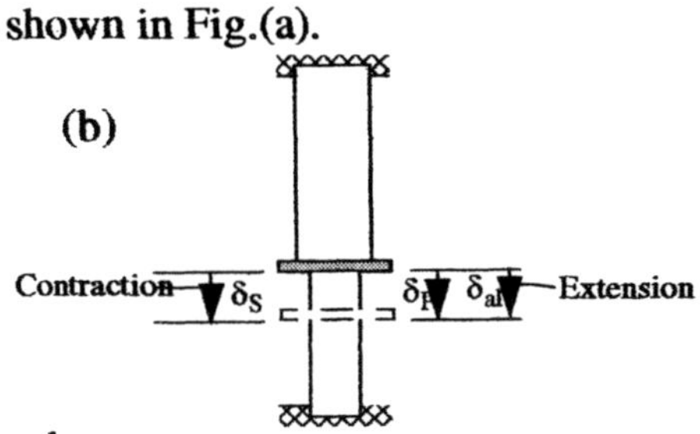

The inner diameter for each cylinder can be found as shown:

$(d_s)_i = (d_s)_0 - 2t = 2.75\text{in}$ and $(d_{al})_i = (d_{al})_0 - 2t = 3.75\text{in}$

The area of the cross section can be found as

$$A_s = \frac{\pi}{4}(3^2 - 2.75^2) = 1.129\text{in}^2 \qquad 1$$

$$A_{al} = \frac{\pi}{4}(4^2 - 3.75^2) = 1.522\text{in}^2 \qquad 2$$

We draw the free body diagram of the rigid plate as shown in Fig (a). By force equilibrium in Fig. (a) we obtain the following.

$$N_{al} + N_s = 2P = 40\text{kips} \qquad 3$$

We draw the exaggerated deformed shape as shown in Fig.(b) and write the deformation equations as shown below.

$$\delta_s = \delta_p \qquad 4$$

$$\delta_{al} = \delta_p \qquad 5$$

From Eq 4.27, we have the following

$$\delta_{al} = \frac{N_{al}L_{al}}{E_{al}A_{al}} = \frac{N_{al}(40)}{10(10^3)(1.522)} = 2.628(10^{-3})N_{al} = \delta_p \quad \text{or} \quad N_{al} = 0.3804\delta_p(10^3) \qquad 6$$

$$\delta_s = \frac{N_sL_s}{E_sA_s} = \frac{N_s(30)}{30(10^3)(1.129)} = 0.8857(10^{-3})N_s = \delta_p \quad \text{or} \quad N_s = 1.129\delta_p(10^3) \qquad 7$$

Substituting Eq.(6) and Eq. (7) into Eq. (3) we obtain the displacement of the plate δ_p as shown below

$(0.3804 + 1.129)\delta_p(10^3) = 40$ or $\delta_p = 26.5(10^{-3})\text{in}$ or $\boxed{\delta_p = 0.0265\text{in}}$

From Eq's(4) and (5) we obtain $\delta_{al} = 26.5(10^{-3})\text{in}$ and $\delta_s = 26.5(10^{-3})\text{in}$ .The axial strains are

$$\varepsilon_s = \frac{\delta_s}{L_s} = \frac{26.5(10^{-3})}{30} = 0.8833(10^{-3}) \text{ and } \varepsilon_{al} = \frac{\delta_{al}}{L_{al}} = \frac{26.5(10^{-3})}{40} = 0.6625(10^{-3})$$

Noting that steel cylinder contracts and Aluminum cylinder extends in the axial direction, we obtain the change in diameter as shown below.

$\Delta d_s = (d_s)_0[-\nu_s(-0.8833)(10^{-3})] = 0.7420(10^{-3})$ or $\boxed{\Delta d_s = 0.00074\text{in}}$

$\Delta d_{al} = (d_{al})_0[-\nu_{al}(0.6625)(10^{-3})] = -0.6625(10^{-3})$ or $\boxed{\Delta d_{al} = -0.00066\text{in}}$

4.65 An aluminum (E_{al}=10,000 ksi and ν_{al}= 0.25) and steel (E_s=30,000 ksi and ν_s= 0.28) hollow cylinder are securely fastened to a rigid plate as shown in Fig. P4.65. Both cylinders are made from sheet metal thickness 1/8 inch. The outer diameters of aluminum and steel cylinders are 4 inch and 3 inch respectively. The allowable stresses in aluminum and steel are 10 ksi and 25 ksi, determine the maximum force P that can be applied to the assembly shown in Fig. P4.65.

Solution $E_{al} = 10{,}000\text{ksi}$ $\quad \nu_{al} = 0.25$ $\quad E_s = 30{,}000\text{ksi}$

$\nu_s = 0.28$ $\quad t = 1/8\text{ in}$ $\quad (d_{al})_0 = 4\text{in}$ $\quad (d_s)_0 = 3\text{in}$

$\sigma_{al} \le 10\text{ksi}$ $\quad \sigma_s \le 25\text{ksi}$ $\quad P_{max} = ?$

From Eq. (3), (6) and (7) in problem 4-64, we have

$$N_{al} + N_s = 2P \qquad 8$$

$$N_{al} = 0.3804\delta_p(10^3) \qquad 9$$

$$N_s = 1.129\delta_p(10^3) \qquad 10$$

Substituting Eq.(2) and (3) into Eq. (1) we obtain: $[0.3804 + 1.129]\delta_p(10^3) = 2P$ or

$$\delta_p = 1.325P(10^{-3}) \qquad 11$$

We obtain the internal forces as $N_{al} = 0.504P$ and $N_s = 1.496P$

The axial stresses in each cylinder can be found using the areas in Eq.(1) and (2) in the problem 4-64 and maximum value of P calculated as shown below

$$\sigma_{al} = \frac{N_{al}}{A_{al}} = \frac{0.504P}{1.522} \le 10 \quad \text{or} \quad P \le 30.19\text{kips} \qquad 12$$

$$\sigma_s = \frac{N_s}{A_s} = \frac{1.496P}{1.129} \le 25 \quad \text{or} \quad P \le 18.87\text{kips} \qquad 13$$

The maximum value of P that satisfies Eq's (5) and (6) is $\boxed{P_{max} = 18.8\text{kips}}$

4.66 A gap of 0.004 inch exists between the rigid bar and bar A before the force P is applied. The rigid bar is hinged at point C. The lengths of bar A and B are 30 and 50 inches respectively. Both bars have an area of cross-section A= 1 in^2 and Modulus of Elasticity E = 30,000 ksi. If P = 100 kips in Fig. P4.66, determine the axial stresses in bars A and B.

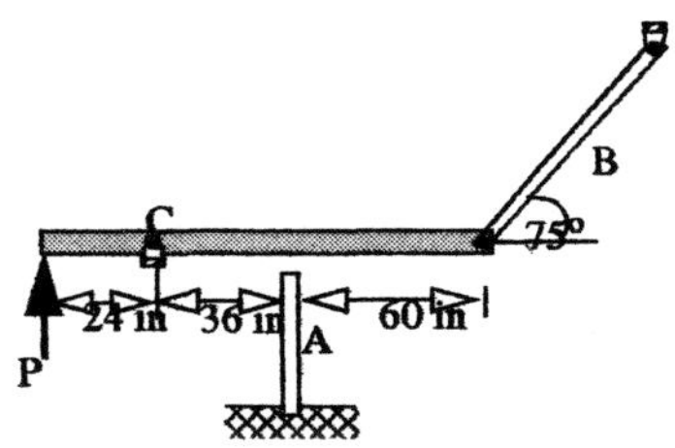

Fig. P4.66

Solution $E_A = E_B = 30{,}000\text{ksi}$ $L_A = 30$ in $L_B = 50$ in
$A_A = A_B = 1\text{in}^2$ $P = 100$ kips $\sigma_A = ?$ $\sigma_B = ?$

Assume gap closes at Equilibrium. The free body diagram of the rigid body as shown in Fig(a).

(a)

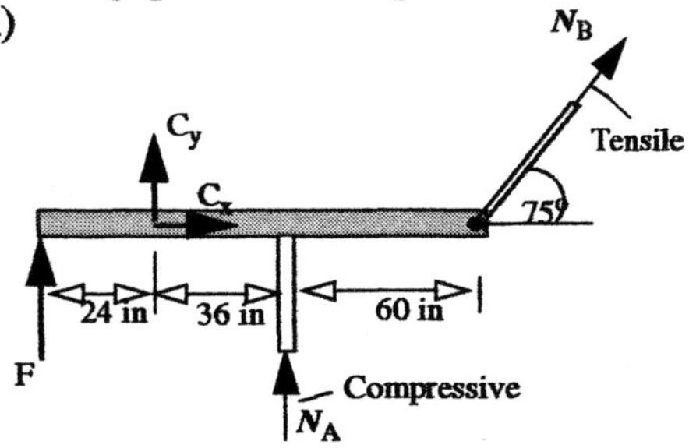

(b)

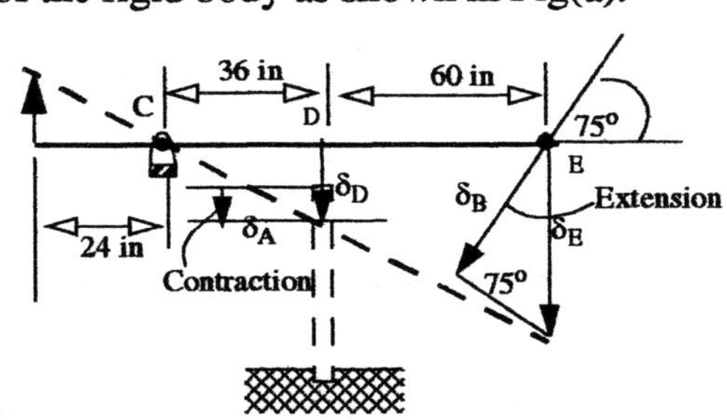

By moment equilibrium about point C in Fig.(a) we obtain: $F(24) - N_A(36) - N_B\sin 75(96) = 0$ or

$$36N_A + 92.73N_B = 24P = 2400 \qquad 1$$

We draw the exaggerated deformed shape as shown in Fig (b) and write deformation equations

$$\frac{\delta_D}{36} = \frac{\delta_E}{96} \qquad 2$$

$$\delta_A = \delta_D - 0.004 \qquad 3$$

$$\delta_B = \delta_E \sin 75 \qquad 4$$

From Eq.4.27

$$\delta_A = \frac{N_A L_A}{E_A A_A} = \frac{N_A(30)}{(30)(10^3)(1)} = N_A(10^{-3}) \qquad 5$$

$$\delta_B = \frac{N_B L_B}{E_B A_B} = \frac{N_B(50)}{(30)(10^3)(1)} = 1.667N_B(10^{-3}) \qquad 6$$

From Eq's (5) and (3) we obtain.

$$N_A(10^{-3}) = \delta_D - 0.004 \qquad \text{or} \qquad N_A = 10^3\delta_D - 4 \qquad 7$$

From Eq's (2),(3) and (6) we obtain.

$$\delta_B = 1.667N_B(10^{-3}) = \left(\frac{96}{36}\delta_D\right)\sin 75 \qquad \text{or} \qquad N_B = 0.647\delta_D(10^3) \qquad 8$$

Substituting Eq's (7) and (8) into Eq. (1), we obtain the following

$$36[10^3(\delta_D) - 4] + 92.73[0.647\delta_D(10^3)] = 2400 \qquad \text{or} \qquad 96\delta_D(10^3) = 2544 \text{ or}$$

$$\delta_D = 0.0265\text{in} \qquad 9$$

As δ_D is greater than the gap, the assumption of gap closing is correct. From Eq.(7) and (8) we obtain.

$N_A = (10^3)(0.0265) - 4 = 22.5\text{kips}$ and $N_B = (0.647)(0.0265)(10^3) = 17.15\text{kips}$

The axial stress can be found as

$$\sigma_A = \frac{N_A}{A_A} = \frac{22.5}{1} \text{ or} \qquad \boxed{\sigma_A = 22.5\text{ksi(C)}}$$

$$\sigma_B = \frac{N_B}{A_B} = \frac{17.15}{1} \text{ or} \qquad \boxed{\sigma_B = 17.2\text{ksi(T)}}$$

4.67 A gap of 0.004 inch exists between the rigid bar and bar A before the force P is applied in Fig. P4.67. The rigid bar is hinged at point C. The lengths of bars A and B are 30 and 50 inches respectively. Both bars have an area of cross-section A= 1 in^2 and Modulus of Elasticity E = 30,000 ksi. If the allowable normal stress in the bars is 20 ksi in tension or compression, determine the maximum force P that can be applied to the assembly shown in Fig. P4.67.

Solution $E_A = E_B = 30{,}000\text{ksi}$ $L_A = 30$ in $L_B = 50$ in

$A_A = A_B = 1\text{in}^2$ $\sigma_A \le 20\text{ksi}$ $\sigma_B \le 20\text{ksi}$ $P_{max} = ?$

From Eq's (1), (7) and (8) in problem 4-66, we have the following

$$36N_A + 92.73N_B = 24P \qquad 1$$

$$N_A = (10^3)\delta_D - 4 \qquad 2$$

$$N_B = 0.647\delta_D(10^3) \qquad 3$$

Substituting Eq.(2) and (3) into Eq.(1), we obtain the following

$$36[(10^3)\delta_D - 4] + 92.73[0.647\delta_D(10^3)] = 24P = 96(10^3)\delta_D = 24P + 144 \text{ or}$$

$$\delta_D = (0.25P + 1.5)(10^{-3}) \qquad 4$$

From Eq.(2) and (3) we obtain.

$$N_A = 0.25P - 2.5 \qquad 5$$

$$N_B = 0.1618P + 0.9705 \qquad 6$$

The axial stresses can be found and using the limiting values, the limit on P can be found as shown below

$$\sigma_A = \frac{N_A}{A_A} = (0.25P - 2.5) \le 20 \qquad \text{or} \qquad P \le 90\text{kips} \qquad 7$$

$$\sigma_B = \frac{N_B}{A_B} = (0.1618P + 0.9705) \le 20 \qquad \text{or} \qquad P \le 117.6\text{kips} \qquad 8$$

The maximum value of P that satisfies Eq's(7) and (8) is $\boxed{P_{max} = 90\text{kips}}$

4.68 In Fig. P4.68, a gap exists between the rigid bar and rod A before the force F is applied. The rigid bar is hinged at point C. The lengths of bar A and B are 1 m and 1.5 m respectively and the diameters are 50 mm and 30 mm respectively. The bars are made of steel with a Modulus of Elasticity E = 200 GPa and Poisson's ratio is 0.28. If F=75 kN in Fig. P4.68, determine (a)the deformation of the two bars. (b) the change in the diameters of the two bars.

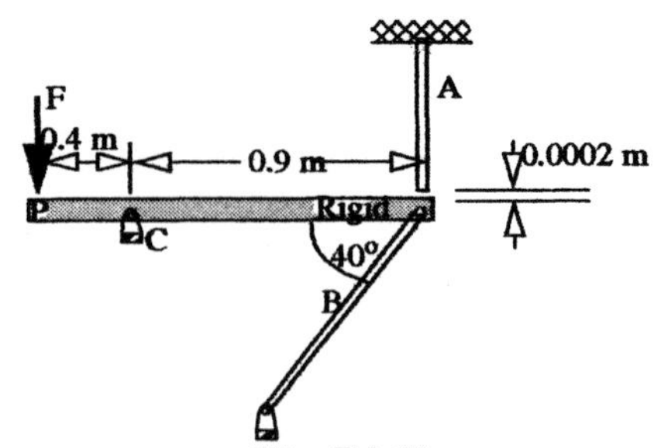

Fig. P4.68

Solution $E_s = 200\text{GPa}$ $\quad \nu = 0.28$ $\quad L_A = 1\text{ m}$ $\quad L_B = 1.5\text{m}$ $\quad F = 75\text{ kN}$

$d_A = 50\text{ mm}$ $\quad d_B = 30\text{mm}$ $\quad \delta_A = ?$ $\quad \delta_B = ?$ $\quad \Delta d_A = ?$ $\quad \Delta d_B = ?$

The area of cross-sections are.

$$A_A = \frac{\pi}{4}(0.05)^2 = 1.963(10^{-3})\text{m}^2 \tag{1}$$

$$A_B = \frac{\pi}{4}(0.03)^2 = 0.7068(10^{-3})\text{m}^2 \tag{2}$$

We assume the gap is closed at equilibrium. We can draw the free body diagram of bar as shown in Fig. (a)

(a)

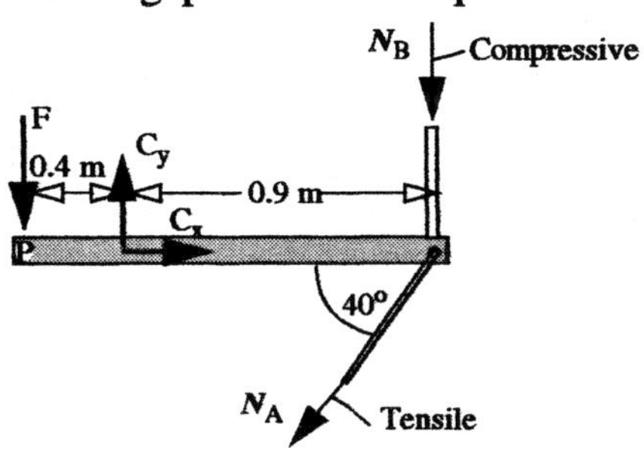

(b)

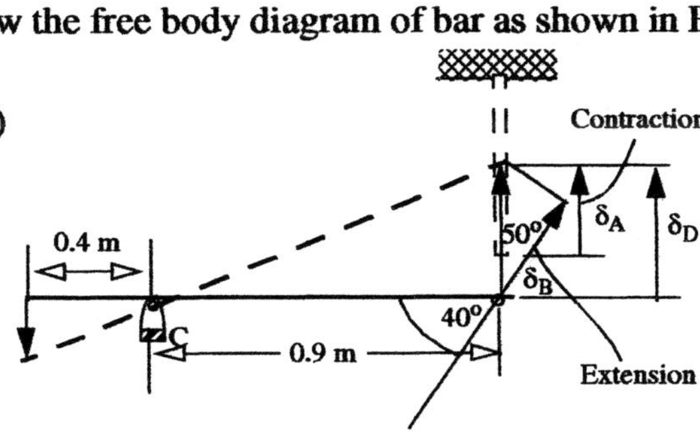

By moment equilibrium about point C, we obtain: $F(0.4) - N_A(0.9) - (N_B \sin 40)(0.9) = 0$ or

$$0.9N_A + 0.5785N_B = 0.4F = 30(10^3) \tag{3}$$

We draw the exaggerated deformed shape and write the deformation equations as shown below

$$\delta_A = \delta_D - 0.0002 \tag{4}$$

$$\delta_B = \delta_D \cos 50 = 0.6428\delta_D \tag{5}$$

From Eq. (4.27)

$$\delta_A = \frac{N_A L_A}{E_A A_A} = \frac{N_A(1)}{(200)(10^9)(1.963)(10^{-3})} = 2.547N_A(10^{-9}) \tag{6}$$

$$\delta_B = \frac{N_B L_B}{E_B A_B} = \frac{N_B(1.5)}{(200)(10^9)(0.7068)(10^{-3})} = 10.611N_B(10^{-9}) \tag{7}$$

From Eq's (6) and (4) we obtain.

$$N_A = (392.6\delta_D - 0.0785)(10^6) \tag{8}$$

From Eq's (5) and (7) we obtain.

$$N_B = 63.581\delta_D(10^6) \tag{9}$$

Substituting Eq's (8) and (9) into Eq (3), we obtain

$$0.9(392.6\delta_D - 0.0785)(10^6) + 0.5785(63.581)\delta_D(10^6) = 30(10^3) \text{ or}$$

$390\delta_D = 0.10065$ or, $\delta_D = 0.2581(10^{-3})$m

As δ_D is greater than the gap, the assumption of gap closing is correct, From Eq's (4) and (5) we obtain

$\delta_A = 0.2581(10^{-3}) - 0.2(10^{-3}) = 0.0581(10^{-3})$m $\boxed{\delta_A = 0.058\text{mm}}$

$\delta_B = 0.2581(0.6428)(10^{-3}) = 0.1659(10^{-3})$m $\boxed{\delta_B = 0.166\text{mm}}$

The axial strains in A and B can be found as

$$\varepsilon_A = \frac{\delta_A}{L_A} = 0.0581(10^{-3}) \text{ and } \varepsilon_B = \frac{\delta_B}{L_B} = \frac{0.166(10^{-3})}{1.5} = 0.1106(10^{-3})$$

Noting that A contracts and B extends in the axial direction, we obtain the change in the diameter as shown below

$\Delta d_A = d_A[-\nu(-0.0581)(10^{-3})] = 0.813(10^{-3})$mm $\boxed{\Delta d_A = 0.00081\text{mm}}$

$\Delta d_B = d_B[-\nu(0.1106)(10^{-3})] = -0.929(10^{-3})$mm $\boxed{\Delta d_B = -0.00093\text{mm}}$

4.69 In Fig. P4.68, a gap exists between the rigid bar and rod A before the force F is applied. The rigid bar is hinged at point C. The lengths of bar A and B are 1 m and 1.5 m respectively and the diameters are 50 mm and 30 mm respectively. The bars are made of steel with a Modulus of Elasticity E = 200 GPa and Poisson's ratio is 0.28. In Fig. P4.68, the allowable axial stresses in bars A and B are 110 MPa and 125 MPa, respectively. Determine the maximum force F that can be applied.

Solution $E_s = 200$ GPa $L_A = 1$ m $L_B = 1.5$ m $d_A = 50$ mm $d_B = 30$ mm

$\nu = 0.28$ $\sigma_A \le 110\text{MPa}$ $\sigma_B \le 125\text{MPa}$ $F_{max} = ?$

From Eq's (3), (8) and (9) in problem (4-68) we have the following.

$$0.9N_A + 0.5785N_B = 0.4F \quad (1)$$

$$N_A = (392.6\delta_D - 0.0785)(10^6) \quad (2)$$

$$N_B = 63.581\delta_D(10^6) \quad (3)$$

Substituting Eq's (2) and (3) into Eq. (1) we obtain.

$$[0.9(392.6\delta_D - 0.0785) + (0.5785)(63.581)\delta_D](10^6) = 0.4F + 390.1\delta_D = 0.4F(10^{-6}) + 0.07065 \text{ or}$$

$$\delta_D = 1.025F(10^{-9}) + 0.1811(10^{-3}) \quad (4)$$

From Eq. (2) and (3) we obtain.

$$N_A = 402.4F(10^{-3}) - 7.400(10^3) \quad (5)$$

$$N_B = 65.17F(10^{-3}) - 11.514(10^3) \quad (6)$$

Dividing the internal force in Eq's (5) and (6) by the areas given by Eq's (1) and (2) in the problem 4-68 we obtain the following.

$$\sigma_A = \frac{N_A}{A_A} = \frac{402.4F(10^{-3}) - 7.4(10^3)}{1.963(10^{-3})} \le 110(10^6) \quad \text{or} \quad F \le 555.0(10^3)\text{Newtons} \quad (7)$$

$$\sigma_B = \frac{N_B}{A_B} = \frac{65.17F(10^{-3}) + 11.514(10^3)}{0.7068(10^{-3})} \le 125(10^6) \quad \text{or} \quad F \le 1179(10^3)\text{Newtons} \quad (8)$$

The maximum value of F that satisfies Eq's (7) and (8) is $\boxed{F_{max} = 555\text{kN}}$

4.70 A rectangular aluminum bar (E = 10,000 ksi) a steel bar (E = 30,000 ksi), and a brass bar (E = 15,000 ksi) form a structure as shown in Fig. P4.70. All bars have the same thickness of 0.5 inch. A gap of 0.02 inch exists before the load of P = 15 kips is applied to the rigid plate. Assume that the rigid plate does not rotate. If P = 15 kips in Fig. P4.70, determine (a) the axial stress in steel. (b) the displacement of the rigid plate with respect to the right wall.

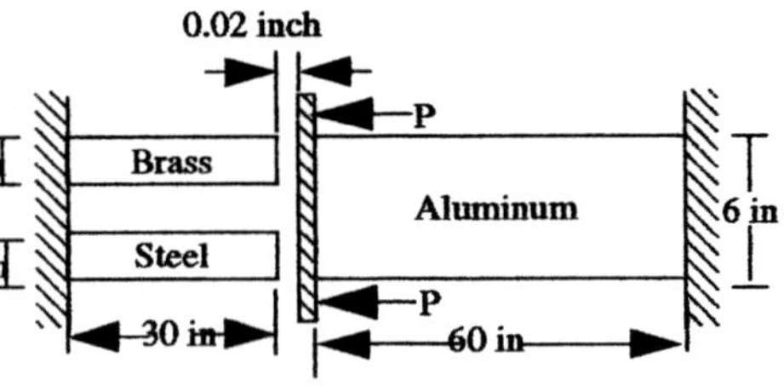

Fig. P4.70

Solution $E_{al} = 10{,}000\text{ksi}$ $E_s = 30{,}000\text{ksi}$ $E_{bl} = 15{,}000\text{ksi}$

$t = 0.5$ in $P = 15$ kips $\sigma_s =$ $\delta_p = ?$

The areas of cross sections of various material can be found as shown below.

$A_{br} = (2)(0.5) = 1\text{in}^2$ $A_s = (2)(0.5) = 1\text{in}^2$ $A_{al} = (6)(0.5) = 3\text{in}^2$

Assume that the gap closes at equilibrium. We can draw the free body diagram of the rigid plate as shown in Fig. (a)

(a)

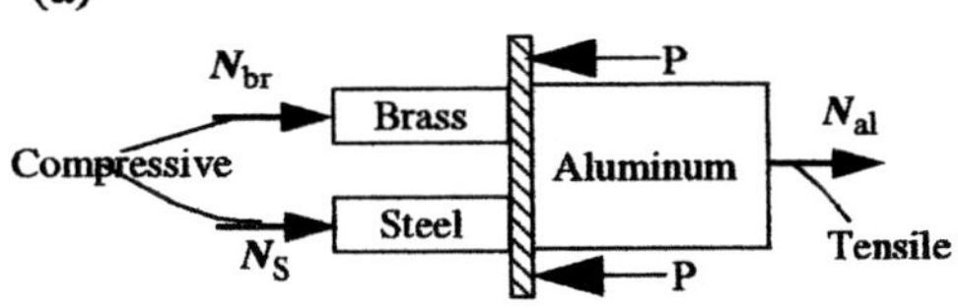

(b)

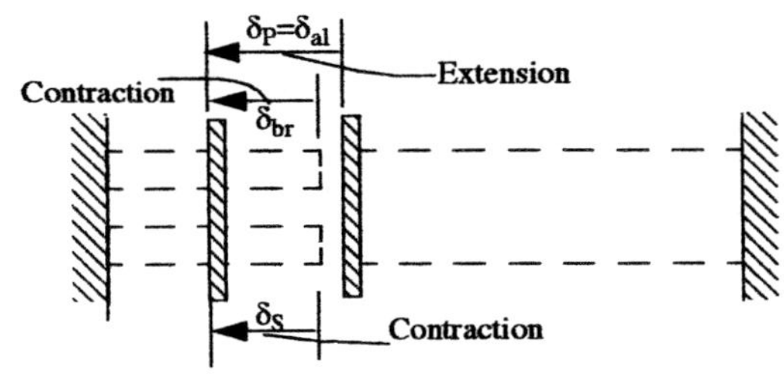

By force equilibrium in Fig (a)

$$N_{al} + N_{br} + N_s = 2P = 30\text{kips} \quad (1)$$

We draw the exaggerated deformed shape as shown in Fig. (b) and write the deformation equation as shown below.

$$\delta_{al} = \delta_p \quad (2)$$

$$\delta_s = \delta_p - 0.02 \quad (3)$$

$$\delta_{br} = \delta_p - 0.02 \quad (4)$$

From Eq. (4-27)

$$\delta_{al} = \frac{N_{al}L_{al}}{E_{al}A_{al}} = \frac{N_{al}(60)}{(10)(10^3)(3)} = 2N_{al}(10^{-3}) = \delta_p \quad \text{or} \quad N_{al} = 0.5\delta_p(10^3) \quad (5)$$

$$\delta_s = \frac{N_sL_s}{E_sA_s} = \frac{N_s(30)}{(30)(10^3)(1)} = N_s(10^{-3}) = \delta_p - 0.02 \quad \text{or} \quad N_s = (10^3)\delta_p - 20 \quad (6)$$

$$\delta_{br} = \frac{N_{br}L_{br}}{E_{br}A_{br}} = \frac{N_{br}(30)}{(15)(10^3)(1)} = 2N_{br}(10^{-3}) = \delta_p - 0.02 \quad \text{or} \quad N_{br} = 0.5(10^3)\delta_p - 10 \quad (7)$$

Substituting Eq. (5), (6) and (7) into Eq. (1)

$$0.5\delta_p(10^3) + (10^3)\delta_p - 20 + 0.5(10^3)\delta_p - 10 = 30 \text{ or } 2.0\delta_p(10^3) = 60 \quad \text{or} \quad \delta_p = 30(10^{-3})\text{inch}$$

As $\delta_p > 0.02$ the assumption gap closes is correct. $\boxed{\delta_p = 0.03\text{inch}}$

From Eq. (6) $N_s = (10^3)(30)(10^{-3}) - 20 = 10\text{kips}$

The axial stress in steel is $\sigma_s = \dfrac{N_s}{A_s} = \dfrac{10}{1}$ $\boxed{\sigma_s = 10\text{ksi(C)}}$

4.71 A rectangular aluminum bar (E = 10,000 ksi) a steel bar (E = 30,000 ksi), and a brass bar (E = 15,000 ksi) form a structure as shown in Fig. P4.70. All bars have the same thickness of 0.5 inch. A gap of 0.02 inch exists before the load of P = 15 kips is applied to the rigid plate. Assume that the rigid plate does not rotate. Determine the maximum load P in Fig. P4.70, if the allowable axial stresses in brass, steel, and aluminum are 8 ksi, 15 ksi, and 10 ksi, respectively.

Solution $E_{al} = 10{,}000\text{ksi}$ $E_s = 30{,}000\text{ksi}$ $E_{bl} = 15{,}000\text{ksi}$

$t = 0.5 \text{ in}$ $\sigma_{br} \le 8\text{ksi}$ $\sigma_s \le 15\text{ksi}$ $\sigma_{al} \le 10\text{ksi}$

From Eq's (1), (5), (6) and (7) in problem 4-70, we have

$$N_{al} + N_s + N_{br} = 2P \quad (1)$$

$$N_{al} = 0.5(10^3)\delta_p \quad (2)$$

$$N_s = (10^3)\delta_p - 20 \quad (3)$$

$$N_{br} = 0.5(10^3)\delta_p - 10 \quad (4)$$

Substituting Eq. (2), (3) and (4) into Eq(1) we obtain the following.

$0.5(10^3)\delta_p + (10^3)\delta_p - 20 + 0.5(10^3)\delta_p - 10 = 2P$ or

$$2.0\delta_p(10^3) = 2P + 30, \text{ or, } \delta_p = P(10^{-3}) + 0.015 \quad (5)$$

Substituting Eq. (5) into Eq(2), (3) and (4) we obtain

$N_{al} = 0.5P + 7.5$ $N_s = P - 5$ $N_{br} = 0.5P - 2.5$

The axial stress in each material can be found and the limits on P establish as shown below.

$$\sigma_{al} = \frac{N_{al}}{A_{al}} = \frac{0.5P + 7.5}{3} \le 10\text{ksi} \quad \text{or} \quad P \le 45\text{kips} \quad (6)$$

$$\sigma_s = \frac{N_s}{A_s} = \frac{P - 5}{1} \le 15\text{ksi} \quad \text{or} \quad P \le 20\text{kips} \quad (7)$$

$$\sigma_{br} = \frac{N_{br}}{A_{br}} = \frac{0.5P - 2.5}{1} \le 8\text{ksi} \quad \text{or} \quad P \le 21\text{kips} \quad (8)$$

The maximum value of P that satisfies the three inequalities is $\boxed{P_{max} = 20\text{kips}}$

4.72 Bar A and bar B have a cross-sectional area of 400 mm^2 and a Modulus of Elasticity E = 200 GPa. A gap exists between bar A and the rigid bar before the force F is applied as shown in Fig. P4.72. The applied force F= 10 kN in, determine (a) the axial stress in bar B and (b) the deformation of bar A.

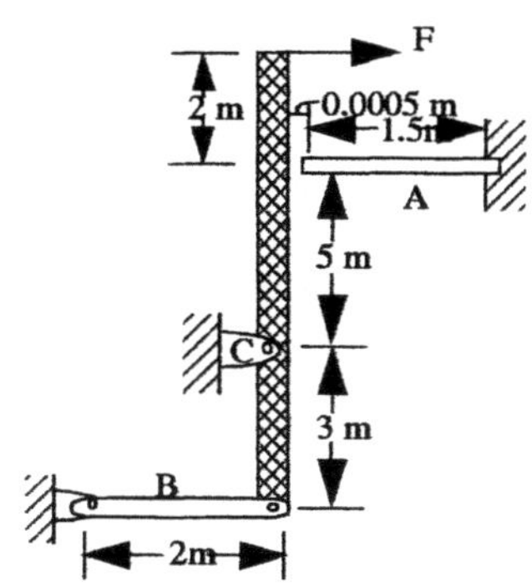

Fig. P4.72

Solution $A_A = A_B = 400 \text{ mm}^2$ $E_A = E_B = 200 \text{ GPa}$ $F = 10 \text{ kN}$ $\delta_A = ?$ $\sigma_B =$

Assume the gap closes at equilibriums, we draw the free body diagram of the rigid bar as shown in Fig(a)

(a)

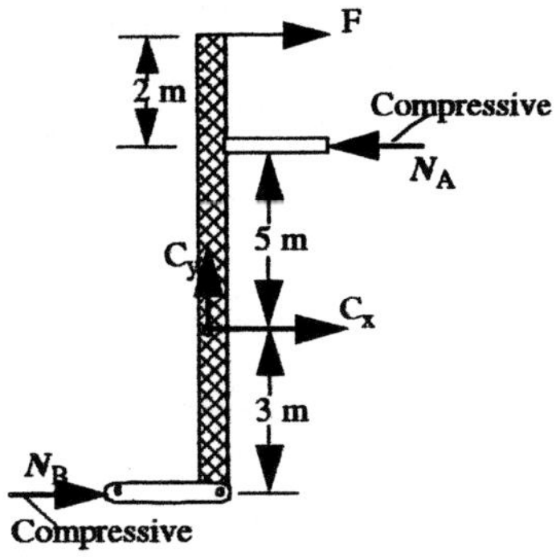

(b)

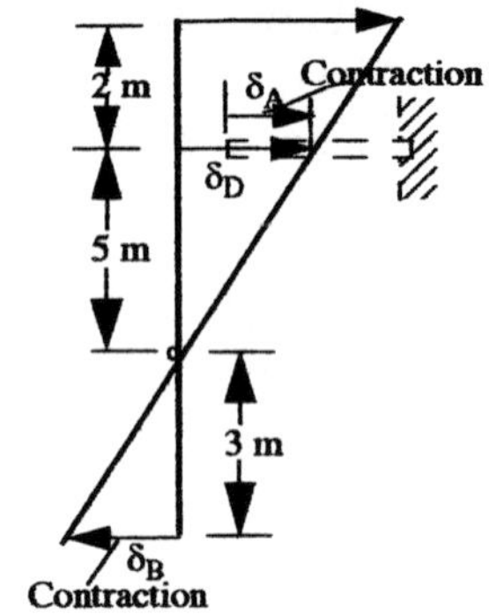

From moment equilibrium about point C, we obtain.

$$N_A(5) + N_B(3) - F(7) = 0 \quad \text{or} \quad 5N_A + 3N_B = 7F = 70(10^3) \qquad 1$$

We draw an exaggerated deformed shape as shown in Fig(b) and write the deformation equations

$$\frac{\delta_D}{5} = \frac{\delta_B}{3} \text{or, } \delta_B = \frac{3}{5}\delta_D \qquad 2$$

$$\delta_A = \delta_D - 0.0005 \qquad 3$$

From Eq. 4-27 we have $\delta_A = \frac{N_A L_A}{E_A A_A} = \frac{N_A(1.5)}{(200)(10^9)(400)(10^{-6})} = 18.75N_A(10^{-9}) = \delta_D - 0.0005$ or

$$N_A = (53.33\delta_D - 0.02667)(10^6) \qquad 4$$

$$\delta_B = \frac{N_B L_B}{E_B A_B} = \frac{N_B(2)}{(200)(10^9)(400)(10^{-6})} = 25N_B(10^{-9}) = \frac{3}{5}\delta_D \quad \text{or} \quad N_B = 24\delta_D(10^6) \qquad 5$$

Substituting Eq's (4) and (5) into Eq. (1) we obtain

$(5)(53.33\delta_D - 0.02667)(10^6) + (3)(24)(\delta_D)(10^6) = 70(10^3)$ or

$$338.65\delta_D(10^6) = 203.33 \quad \text{or} \quad \delta_D = 0.6004(10^{-3})\text{m} \qquad 6$$

As $\delta_D > 0.0005$ the assumption that gap closes is correct. By substituting Eq.(6) into Eq. (3) we obtain the deformation of A as $\delta_A = (0.6004 - 0.5)(10^{-3}) = 0.1004(10^{-3})\text{m}$ or $\boxed{\delta_A = 0.1\text{mm}}$

Substituting Eq. (6) into Eq. (5) we obtain $N_B = (24)(0.6004)(10^{-3})(10^6) = 14.41(10^3)\text{Newtons}$

The axial stress in B is $\sigma_B = \frac{N_B}{A_B} = \frac{14.41(10^3)}{400(10^{-6})} = 36.03(10^6)\text{N/m}^2$ or $\boxed{\sigma_B = 36\text{MPa(C)}}$

4.73 Bar A and bar B have a cross-sectional area of 400 mm^2 and a Modulus of Elasticity E = 200 GPa. A gap exists between bar A and the rigid bar before the force F is applied as shown in Fig. P4.72. Determine the maximum force F that can be applied if the allowable stress in member B is 120 MPa(C) and the allowable deformation of bar A is 0.25 mm.

Solution $A_A = A_B = 400\text{ mm}^2$ $E_A = E_B = 200\text{ GPa}$ $P_{max} = ?$ $\sigma_B \le 120\text{MPa(C)}$ $\delta_A \le 0.25\text{mm}$

From Eq's (1), (4) and (5) in problem 4-72, we have

$$5N_A + 3N_B = 7F \qquad 1$$

$$N_A = (53.33\delta_D - 0.02667)(10^6) \qquad 2$$

$$N_B = 24\delta_D(10^6)$$ **3**

Substituting Eq. (6) and (2) into Eq (1).

$$5(53.33\delta_D - 0.02667)(10^6) + (3)(24)\delta_D(10^6) = 7F \quad \text{or} \quad 338.65\delta_D = 7F(10^{-6}) + 0.1333 \text{ or}$$

$$\delta_D = 20.67F(10^{-9}) + 0.3936(10^{-3})$$ **4**

From Eq. (3) in problem 4-72

$$\delta_A = \delta_D - 0.5(10^{-3}) = 20.67F(10^{-9}) - 0.1064(10^{-3}) \le 0.25(10^{-3}) \quad \text{or} \quad F \le 17.24(10^3)\text{newtons}$$ **5**

Substituting Eq. (4) into Eq. (3): $N_B = 0.496F + 9.446(10^3)$

The axial stress in B is

$$\sigma_B = \frac{N_B}{A_B} = \frac{0.496F + 9.446(10^3)}{400(10^{-6})} \le 120(10^6) \quad \text{or} \quad F \le 77.73$$ **6**

The maximum value of F that satisfies Eq(5) and (6) is $\boxed{F_{max} = 17.2\text{kN}}$

4.74 A rectangular steel (E = 30,000 ksi, ν=0.25) bar of 0.5 inch thickness has a gap of 0.01 inch between the section at D and a rigid wall before the forces are applied. Assuming that the applied forces are sufficient to close the gap determine: (a) the movement of section at C with respect to the left wall.(b) the change in the depth d of segment CD.

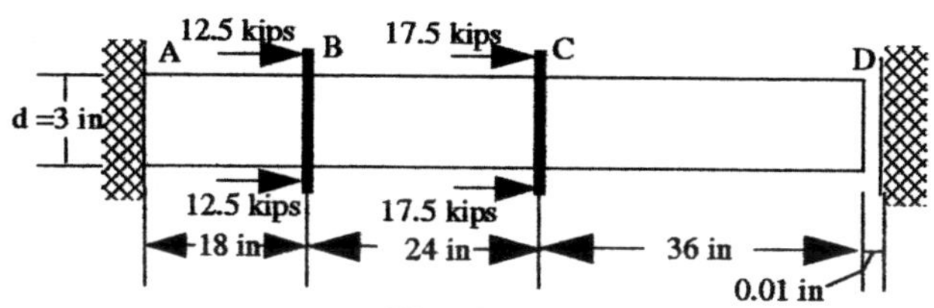

Fig. P4.74

Solution E = 30,000 ν = 0.25 t = 0.5 δc = ? Δd_{CD}

Assume gap is closed at equilibrium. We make imaginary cuts through segments AB, BC and CD and draw free body diagram as shown below. By force equilibrium in Figs. (a), (b) and (c) we obtain the following.

$$N_{AB} = R_A$$ **1**

$$N_{BC} = R_A - 25$$ **2**

$$N_{CD} = 60 - R_A$$ **3**

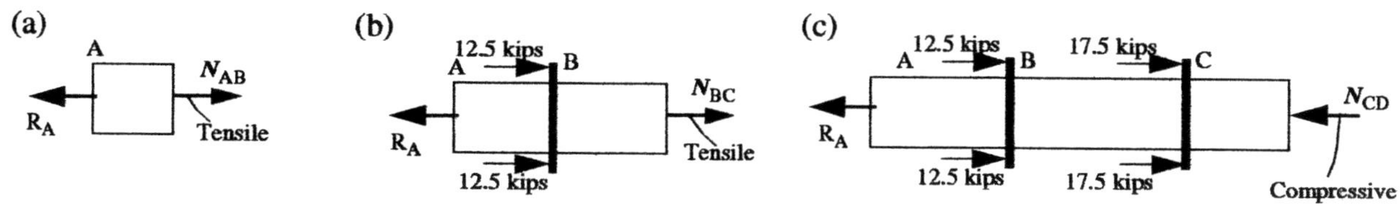

The total extension of the bar is equal to the gap. Noting that segment AB and BC will extend and segment CD will contract, we obtain the equation for total extension as

$$\delta_{AB} + \delta_{BC} - \delta_{CD} = 0.01$$ **4**

From Eq (4-27)

$$\delta_{AB} = \frac{N_{AB}L_{AB}}{EA} = \frac{(R_A)(18)}{(30)(10^3)(0.5)(3)} = 0.40R_A(10^{-3})$$ **5**

$$\delta_{BC} = \frac{N_{BC}L_{BC}}{EA} = \frac{(R_A - 25)(24)}{(30)(10^3)(0.5)(3)} = 0.533(R_A - 25)(10^{-3})$$ **6**

$$\delta_{CD} = \frac{N_{CD}L_{CD}}{EA} = \frac{(60 - R_A)(36)}{(30)(10^3)(0.5)(3)} = 0.80(60 - R_A)(10^{-3}) \quad 7$$

Substituting Eq. (5), (6) and (7) into Eq. (4): $[0.4R_A + 0.533(R_A - 25) - 0.8(60 - R_A)](10^{-3}) = 0.01$ or

$$1.733R_A = 71.325 \quad \text{or} \quad R_A = 41.157\text{kips} \quad 8$$

The movement of rigid plate at C is the extension of segment,

$\delta_C = \delta_{AB} + \delta_{BC} = [(0.4)(41.157 - 25)](10^{-3}) = 25.07(10^{-3})\text{inch}$ or $\boxed{\delta_C = 0.025\text{inch}}$

The axial strain in CD is $\varepsilon_{CD} = \frac{\delta_{CD}}{L_{CD}} = \frac{0.8(60 - 41.157)(10^{-3})}{36} = 0.4187(10^{-3})$

Noting that ε_{CD} is contraction, the change in the depth can be found as shown below.

$\Delta d_{CD} = -\nu(-0.4187)(10^{-3})(3) = 0.314(10^{-3})$ or $\boxed{\Delta d_{CD} = 0.00031\text{inch}}$

4.75 Three plastic members of equal cross-section are shown in Fig. P4.75. Member B is smaller than member A by 0.5 mm. A distributed force is applied to the rigid plate, which moves downwards without rotating. The modulus of elasticity for members A and B are 1.5 GPa and 2.0 GPa, respectively. The distributed force w = 20 MPa in Fig. P4.75. Determine the axial stress in each member.

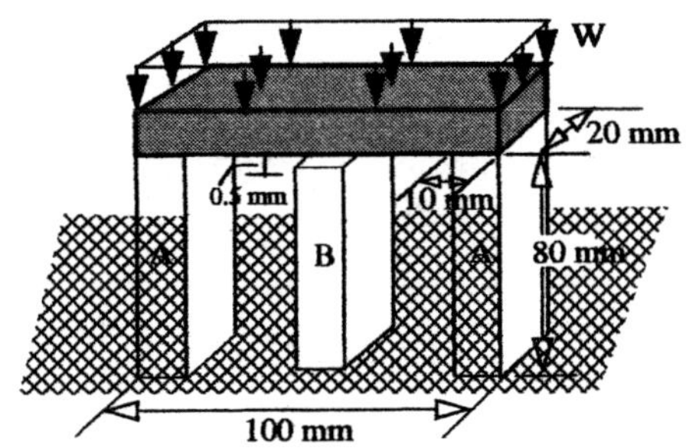

Fig. P4.75

Solution $E_A = 1.5\text{GPa}$ $E_B = 2.0\text{GPa}$ $w = 20\text{ MPa}$ $\sigma_A =$ $\sigma_B =$

Assume gap closes at equilibrium, we can draw the free body diagram of the rigid body after replacing the distributed force by an equivalent force as shown in Fig (a). By equilibrium of forces in Fig (a), we obtain.

$$2N_A + N_B = 0.002w = 40(10^3)\text{N/m}^2 \quad 1$$

(a)

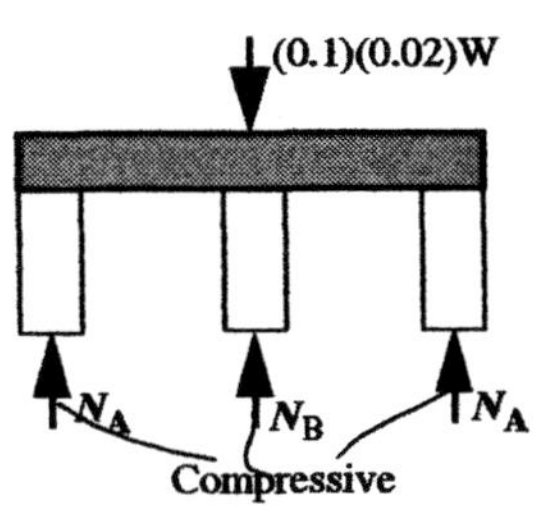

(b)

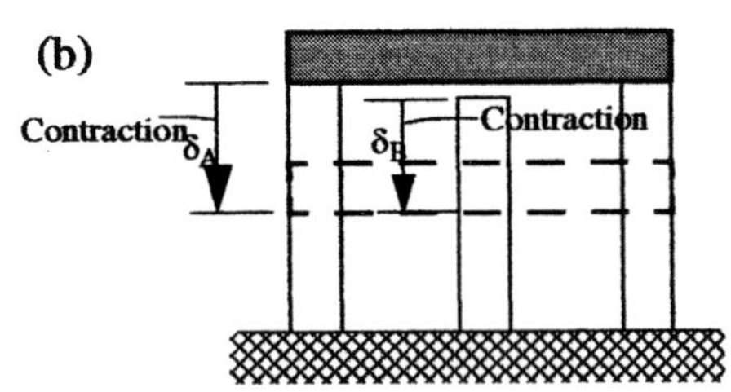

We can draw an exaggerated deformed geometry as shown in Fig (b) and write the deformation equations

$$\delta_A = \delta_B + 0.0005 \quad 2$$

From Eq. (4-27)

$$\delta_A = \frac{N_A L_A}{E_A A_A} = \frac{N_A(0.08)}{(1.5)(10^9)(0.01)(0.02)} = 0.2667(10^{-6})N_A \quad 3$$

$$\delta_B = \frac{N_B L_B}{E_B A_B} = \frac{N_B(0.08)}{(2.0)(10^9)(0.01)(0.02)} = 0.2000(10^{-6})N_B \quad 4$$

Equations (3) and (4) can be written in terms of δ_A as shown below

$$N_A = 3.75(10^6)\delta_A \quad 5$$

$$N_B = 5.00(10^6)(\delta_A - (0.0005)) \qquad 6$$

Substituting Eq. (5) and (6) into Eq (1), we obtain $[2(3.75\delta_A) + 5(\delta_A - 0.0005)](10^6) = 40(10^3)$ or

$$12.5\delta_A = 0.0425 \quad \text{or} \quad \delta_A = 0.0034 \qquad 7$$

The internal forces in A and B are

$N_A = (3.75)(0.0034)(10^6) = 12.75(10^3)\text{N/m}^2$ and $N_B = (5.0)(0.0034 - 0.0005)(10^6) = 14.5(10^3)\text{N/m}^2$

The axial stress in the bar A and B are

$$\sigma_A = \frac{N_A}{A_A} = \frac{(12.75)(10^3)}{(0.01)(0.02)} = 63.75(10^6)\text{N/m}^2 \qquad \boxed{\sigma_A = 63.75\text{MPa(C)}}$$

$$\sigma_B = \frac{N_B}{A_B} = \frac{(14.5)(10^3)}{(0.01)(0.02)} = 72.5(10^6)\text{N/m}^2 \qquad \boxed{\sigma_B = 72.5\text{MPa(C)}}$$

4.76 Three plastic members of equal cross-section are shown in Fig. P4.75. Member B is smaller than member A by 0.5 mm. A distributed force is applied to the rigid plate, which moves downwards without rotating. The modulus of elasticity for members A and B are 1.5 GPa and 2.0 GPa, respectively.The allowable stresses in members A and B in Fig. P4.75 are 50 MPa and 30 MPa. Determine the maximum intensity of the distributed force that can be applied to the rigid plate.

Solution $E_A = 1.5\text{GPa}$ $E_B = 2.0\text{GPa}$

$w_{max} = ?$ $\sigma_A \le 50\text{MPa}$ $\sigma_B = 30\text{MPa}$

From Eq's (1), (5) and (6) in problem 4-75, we have

$$2N_A + N_B = 0.002w \qquad 1$$

$$N_A = 3.75(10^6)\delta_A \qquad 2$$

$$N_B = 5.0(10^6)(\delta_A - 0.0005) \qquad 3$$

Substituting Eq. (2) and (3) into Eq(1) we obtain

$$[2(3.75)\delta_A + 5(\delta_A - 0.0005)](10^6) = 0.002w \quad \text{or} \quad 12.5\delta_A = 2w(10^{-9}) + 2.5(10^{-3})$$

$$\delta_A = 0.16w(10^{-9}) + 0.2(10^{-3}) \qquad 4$$

Substituting Eq. (4) into Eq's (2) and (3), we obtain

$$N_A = 0.6w(10^{-3}) + 0.75(10^3) \qquad 5$$

$$N_B = 0.8w(10^{-3}) - 1.5(10^3) \qquad 6$$

The axial stress can be found and limits on N obtained as shown below

$$\sigma_A = \frac{N_A}{A_A} = \frac{0.6w(10^{-3}) + 0.75(10^3)}{(0.01)(0.02)} \le 50(10^6)\text{N/m}^2 \quad \text{or} \quad w \le 17.92(10^6)\text{N/m}^2 \qquad 7$$

$$\sigma_B = \frac{N_B}{A_B} = \frac{0.8w(10^{-3}) - 1.5(10^3)}{(0.01)(0.02)} \le 30(10^6)\text{N/m}^2 \quad \text{or} \quad w \le 9.375(10^6)\text{N/m}^2 \qquad 8$$

The maximum value of w that satisfies Eq.(7) and (8) is $\boxed{w_{max} = 9.4\text{MPa}}$

4.77 An aluminum (E_{Al} = 70 GPa) circular bar of and a steel (E = 200 GPa) tapered circular bar are securely attached to a rigid plate on which axial forces are applied as shown in Fig. P4.77. Determine (a) the displacement of the rigid plate (b) the maximum axial stress in steel.

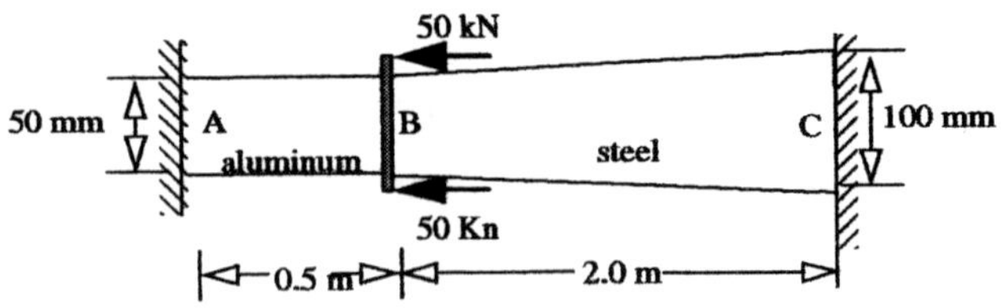

Fig. P4.77

Solution $E_{al} = 70GPa$ $E_s = 200GPa$ $\delta_B = ?$ $(\sigma_S)_{max} =$

The radius of the bar BC varies linearly and

$$R(x) = a + bx \quad 1$$

At x=0.5m; R=0.025m and x=2.5m; R=0.05. Using these values we obtain the following equations

$$a + b(0.5) = 0.025 \quad 2$$

$$a + b(2.5) = 0.050 \quad 3$$

Solving the above two equations, we obtain.

$a = 18.75(10^{-3})$and$b = 12.5(10^{-3})$

Thus, the area of cross-section in segment be varies as

$$A = \pi(18.75 + 12.5x)^2(10^{-6}) \quad 4$$

We make imaginary cuts in segment AB and BC to obtain free body diagrams shown below

(a) 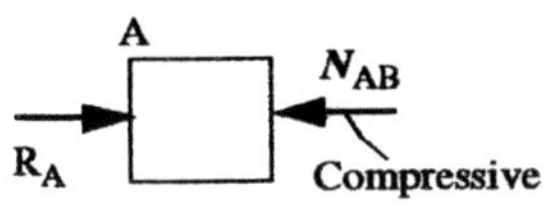

(b) 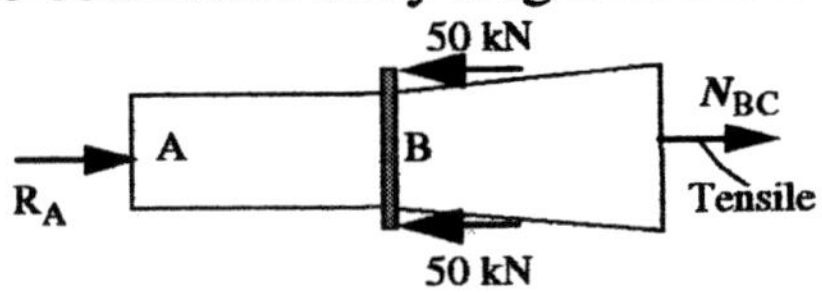

By force equilibrium we obtain.

$$N_{AB} = R_A \quad 5$$

$$N_{BC} = 100(10^3) - R_A \quad 6$$

The total extension of the bar is zero. Thus, the contraction of segment AB is equal to extension of segment BC

$$\delta_{AB} = \delta_{BC} \quad 7$$

From Eq. (4-27) we have

$$\delta_{AB} = \frac{N_{AB}L_{AB}}{E_{AB}A_{AB}} = \frac{R_A(0.5)}{(70)(10^9)\pi(0.025)^2} = 3.638R_A(10^{-9}) \quad 8$$

As the area of cross section varies, we use Eq. 4.8 to find the extension of bar BC as shown below.

$$\frac{du}{dx} = \frac{N_{BC}}{E_{BC}A_{BC}} = \frac{[100(10^3) - R_A]}{(200)(10^9)\pi(18.75 + 12.5x)^2(10^{-6})}$$

Integrating from B to C, we obtain the extension as shown below

$$\delta_{BC} = \int_{u_B}^{u_A} du = \frac{[100(10^3) - R_A]}{(200\pi)(10^3)}\int_{0.5}^{2.5}\frac{dx}{(18.75 + 12.5x)^2} = \frac{[100(10^3) - R_A]}{(200\pi)(10^3)}\frac{1}{(-12.5)}\frac{1}{(18.75 + 12.5x)}\Bigg|_{0.5}^{2.5} \text{ or}$$

$$\delta_{BC} = 2.546[(100)(10^3) - R_A](10^{-9}) \quad 9$$

Substituting Eq's (8) and (9) into Eq. (7), we obtain the following.

$$3.638R_A(10^{-9}) = 2.546[100(10^3) - R_A](10^{-9}) \quad \text{or} \quad R_A = 41.17(10^3)\text{Newtons} \quad 10$$

The movement of the rigid plate is the same as contraction of segment AB. From Eq. (8) we obtain.

$\delta_B = (3.638)(41.17)(10^3)(10^{-9}) = 0.1498(10^{-3})\text{m}$ $\boxed{\delta_B = 0.15\text{mm to the left}}$

From Eq. (6) and (10): $N_{BC} = 58.83(10^3)\text{Newtons}$. The maximum axial stress in the segment BC will be just after B, as the area of the cross-section is smallest there.

$$(\sigma_S)_{max} = \frac{58.83(10^3)}{\pi(0.025)^2} = 29.96(10^6)\text{N/m}^2 \qquad \boxed{(\sigma_S)_{max} = 30\text{MPa(T)}}$$

4.78 A 2 in x 1 in wooden bar is reinforced with 1/4 in steel bars to create laminated bars. Two pieces of these laminated bars are securely fastened to a rigid plate loaded as shown. The modulus of elasticity of steel and wood are 30,000 ksi and 8,000 ksi, respectively. Determine the movement of the rigid plate and the maximum stress in steel and wood

Fig. P4.78

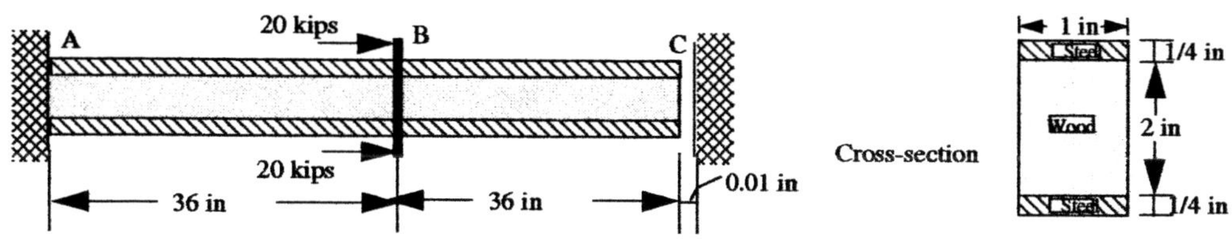

Solution $E_W = 800$ ksi $E_s = 30000$ $A_N = 2\text{ in}^2$ $A_S = 1/4\text{ in}^2$

$\delta_B = ?$ $(\sigma_S)_{max} =$ $(\sigma_w)_{max} =$

Assume gap closes at equilibrium. We make imaginary cuts section AB and BC and draw free body diagram as shown below.

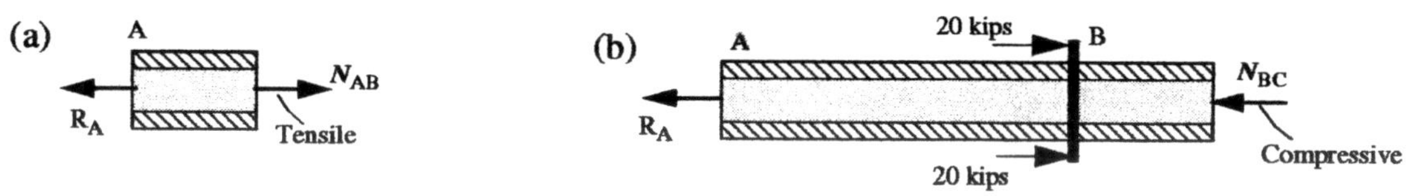

By force equilibrium we obtain.

$$N_{AB} = R_A \qquad 1$$

$$N_{BC} = 40 - R_A \qquad 2$$

The total extension of the bar is equal to the gap. Noting that AB extends and BC contracts, we obtain

$$\delta_{AB} - \delta_{BC} = 0.01 \qquad 3$$

The axial rigidity of the cross-section can be found as shown below

$$\sum E_j A_j = E_w A_w + E_s A_s = (8000)(2) + (30000)\left(\frac{1}{4}\right)(2) = 31000 \qquad 4$$

From Eq. 4-22

$$\delta_{AB} = u_B - u_A = \frac{N_{AB}(x_B - x_A)}{\sum E_j A_j} = \frac{R_A(36)}{31000} \qquad 5$$

$$\delta_{BC} = u_C - u_B = \frac{N_{BC}(x_C - x_B)}{\sum E_j A_j} = \frac{(40 - R_A)(36)}{31000} \qquad 6$$

Substituting Eq. (5) and (6) into Eq. (3),

$$\frac{R_A(36)}{31000} - \frac{(40 - R_A)(36)}{31000} = 0.01 \quad \text{or} \quad 72R_A = 310 + 1440 \quad \text{or} \quad R_A = 24.31\text{kips} \qquad 7$$

Noting that movment of the plate at B is equal to extension of AB, we obtain from Eq. (5)

$\delta_B = \delta_{AB} = 28.23(10^{-3})\text{inch}$ or $\boxed{\delta_B = 0.0282\text{inch}}$

As both segment has same axial rigidity, the maximum axial stress in each material will be in AB where internal force is greater. From Eq. (4-21)

$$(\sigma_S)_{max} = \frac{E_S N_{AB}}{\sum E_j A_j} = \frac{(30000)(24.31)}{31000} = 23.53 \text{ or } \boxed{(\sigma_S)_{max} = 23.5\text{ksi}}$$

$$(\sigma_W)_{max} = \frac{E_W N_{AB}}{\sum E_j A_j} = \frac{(8000)(24.31)}{31000} = 6.27 \text{ or } \boxed{(\sigma_W)_{max} = 6.27\text{ksi}}$$

4.79 A concrete column is reinforced using nine iron circular bars of diameter 1 inch. The modulus of elasticity for concrete and iron are E_c= 4,500 ksi and E_i = 25,000 ksi. Determine the maximum axial stress in concrete and iron.

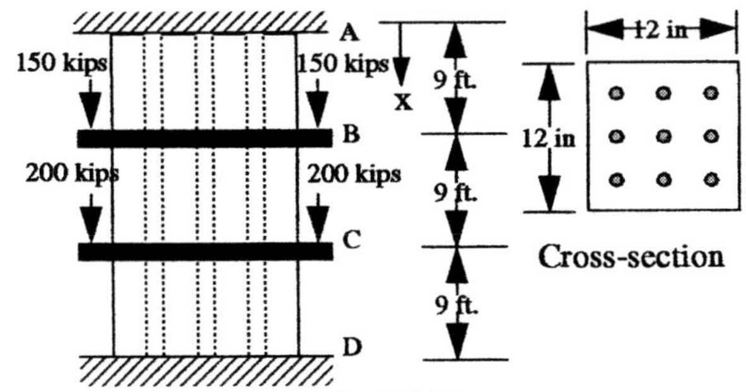

Fig. P4.79

Solution $E_C = 4500\text{ksi}$ $E_i = 25000$ $d = 1\text{in}$ $(\sigma_i)_{max} =$ $(\sigma_c)_{max} =$

From problem 4-47: $A_i = 7.069\text{in}^2$ and $A_C = 136.93\text{in}^2$ and $\sum E_j A_j = 792.91(10^3)\text{kips}$

We make imaginary cuts in segments AB, BC and CD and draw free body diagram as shown.

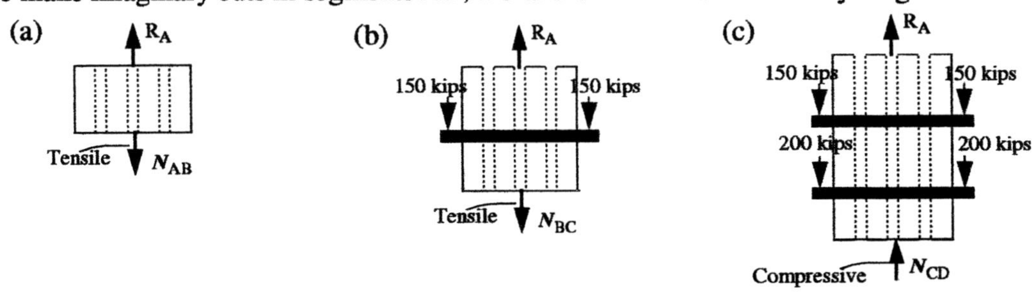

By force equilibrium, we obtain.

$$N_{AB} = R_A \quad 1$$

$$N_{BC} = R_A - 300 \quad 2$$

$$N_{CD} = 700 - R_A \quad 3$$

The total extension of the column is zero. Noting that AB and BC extends (we assumed) and CD contracts, we obtain.

$$\delta_{AB} + \delta_{BC} - \delta_{CD} = 0 \quad 4$$

From Eq. 4-21

$$\delta_{AB} = \frac{N_{AB}(x_B - x_A)}{\sum E_j A_j} = \frac{R_A(108)}{792.91(10^3)} \quad 5$$

$$\delta_{BC} = \frac{N_{BC}(x_C - x_B)}{\sum E_j A_j} = \frac{(R_A - 300)(108)}{792.91(10^3)} \quad 6$$

$$\delta_{CD} = \frac{N_{CD}(x_D - x_C)}{\sum E_j A_j} = \frac{(700 - R_A)(108)}{792.91(10^3)} \quad 7$$

Substituting Eq. (5), (6) and (7) into Eq. (4)

$$[R_A + (R_A - 300) - (700 - R_A)]\frac{(108)}{792.9(10^3)} = 0 \qquad \text{or} \qquad R_A = 333.3\text{kips}$$

From Eq. (1), (2) and (3) we obtain. $N_{AB} = 333.3\text{kips}$ $N_{BC} = 33.3\text{kips}$ and $N_{CD} = 366.67\text{kips}$

As axial rigidity is same throughout the column, the maximum axial stress will be in segment CD where internal axial force is maximum.

$$(\sigma_i)_{max} = \frac{E_i N_{CD}}{\sum E_j A_j} = \frac{(25000)(366.67)}{792.91(10^3)} = 11.56 \qquad \boxed{(\sigma_i)_{max} = 11.6\text{ksi(C)}}$$

$$(\sigma_C)_{max} = \frac{E_C N_{CD}}{\sum E_j A_j} = \frac{(4500)(366.67)}{792.91(10^3)} = 2.08 \qquad \boxed{(\sigma_C)_{max} = 2.1\text{ksi(C)}}$$

4.80 A rigid bar hinged at point O has a force P applied to it as shown. Bars A and B are made of steel (E = 30,000 ksi). The cross-sectional areas of the bars A and B are $A_A = 1$ in^2 and $A_B = 2$ in^2 respectively. If the allowable deflection at point C is 0.01 inch and the allowable stress in the bars is 25 ksi, determine the maximum force P that can be applied.

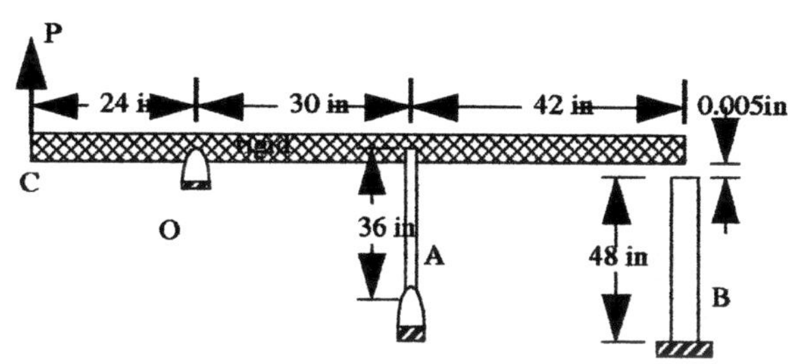

Fig. P4.80

Solution $E = 30000$ $A_A = 1\text{in}^2$ $A_B = 2\text{ in}^2$

$\delta_C \le 0.01\text{in}$ $\sigma_A \le 25\text{ksi}$ $\sigma_B \le 25\text{ksi}$ $P_{max} = ?$

Assume gap closes. We draw the free body diagram of the rigid bar as shown in Fig (a).

(a)

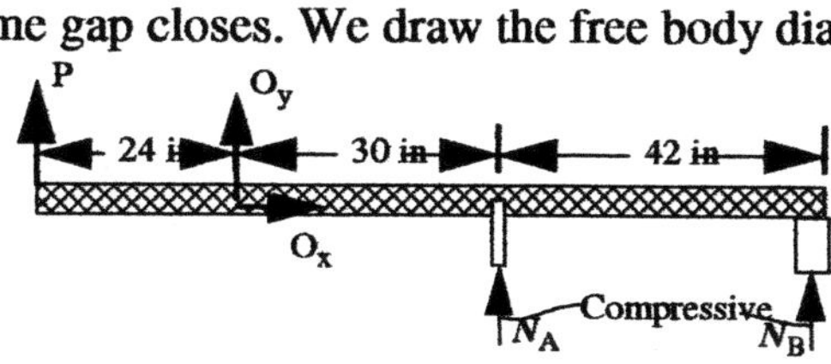

(b)

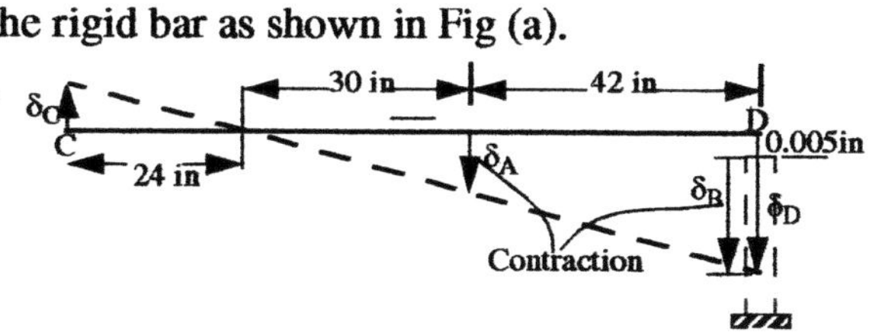

By moment equilibrium about point O, we obtain.

$$N_A(30) + N_B(72) - P(24) = 0 \qquad \text{or} \qquad 30N_A + 72N_B = 24P \quad 1$$

We draw an exaggerated deformed shape as shown in Fig. (b) and write the deformation Equations.

$$\frac{\delta_C}{24} = \frac{\delta_A}{30} = \frac{\delta_D}{72} \quad 2$$

$$\delta_D = \delta_B + 0.005 \quad 3$$

Eq. (2) and (3) can be rewritten as

$$\delta_A = \frac{30}{24}\delta_C = \frac{5}{4}\delta_C \quad 4$$

$$\delta_B = \frac{72}{24}\delta_C - 0.005 = 3\delta_C - 0.005 \qquad 5$$

From Eq. (4-27)

$$\delta_A = \frac{N_A L_A}{E_A A_A} = \frac{N_A(36)}{(30000)(1)} = \frac{5}{4}\delta_C \quad \text{or} \quad N_A = 1.0417\delta_C(10^3) \qquad 6$$

$$\delta_B = \frac{N_B L_B}{E_B A_B} = \frac{N_B(48)}{(30000)(2)} = 3\delta_C - 0.005 \quad \text{or} \quad N_B = 3.75\delta_C(10^3) - 6.25 \qquad 7$$

Substituting Eq. (6) and (7) into Eq (1). We obtain

$$30(1.0417\delta_C)(10^3) + 72(3.75\delta_C(10^3) - 6.25) = 24P \quad \text{or} \quad 301.25(10^3)\delta_C = 24P + 450$$

$$\delta_C = (0.0791P + 1.4938)(10^{-3}) \qquad 8$$

Noting the limit on δ_C, we obtain one limit on P as

$$(0.0797P + 1.4938)(10^{-3}) \le 0.01 \quad \text{or} \quad P \le 106.73\text{kips} \qquad 9$$

From Eq. (6) and (7). we obtain: $N_A = (0.083P + 1.556)\text{kips}$ and $N_B = (0.2988P - 0.6483)\text{kips}$

$$\sigma_A = \frac{N_A}{A_A} = \frac{0.083P + 1.556}{1} \le 25 \quad \text{or} \quad P \le 282.5 \qquad 10$$

$$\sigma_B = \frac{N_B}{A_B} = \frac{0.2988P - 0.6483}{2} \le 25 \quad \text{or} \quad P \le 169.5 \qquad 11$$

The maximum value of P that satisfies Eq. (9), (11) and (12) is $\boxed{P_{max} = 106.7\text{kips}}$

4.81 The structure at the base of a crane is modeled by the pin connected structure shown. The allowable axial stresses in members AC and BC are 15 ksi, and the Modulus of Elasticity is 30,000 ksi. To ensure adequate stiffness at the base, the displacement of pin C in the vertical direction to be limited to 0.1 inches. Determine the minimum area of cross-sections for members AC and BC.

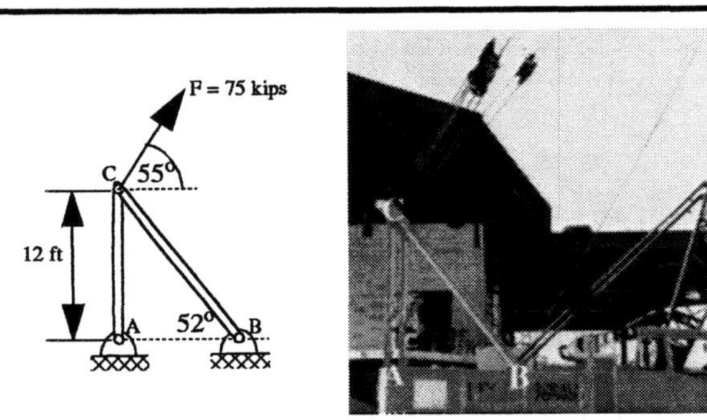

Fig. P4.81

Solution $\sigma_{AC} \le 15\text{ksi}$ $\sigma_{BC} = 15\text{ksi}$ $v_C \le 0.1\text{in}$ $A_{AC} = ?$ $A_{BC} = ?$

We can draw the free body diagram as shown in Fig.(a)

(a)

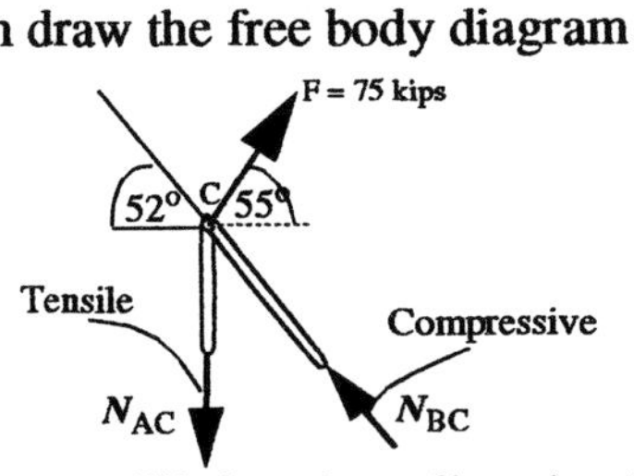

(b)

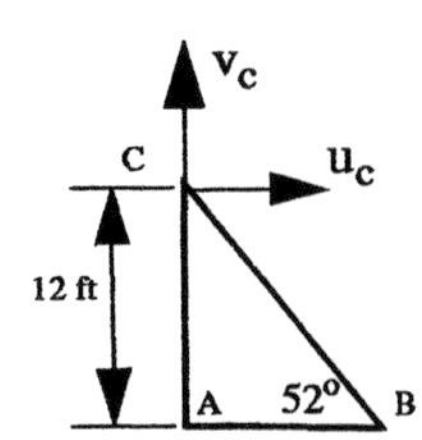

By force equilibrium in x-direction in Fig (a)

$$N_{BC}\cos 52 = 75\cos 55 \quad \text{or} \quad N_{BC} = 69.87\text{kips} \qquad 1$$

$$\sigma_{BC} = \frac{N_{BC}}{A_{BC}} = \frac{69.87}{A_{BC}} \le 15 \quad \text{or} \quad A_{BC} \ge 4.658\text{in}^2 \qquad 2$$

By force equilibrium in y-direction in Fig (b)

$$N_{AC} = N_{BC}\sin 52 + 75\sin 55 \qquad \text{or} \qquad N_{AC} = 116.49\text{kips} \qquad (3)$$

$$\sigma_{AC} = \frac{N_{AC}}{A_{AC}} = \frac{116.49}{A_{AC}} \le 15 \qquad \text{or} \qquad A_{AC} \ge 7.766\text{in}^2 \qquad (4)$$

The length L_{BC} = 12/sin52 = 15.23 ft. The deformation of each bar can be found as shown below

$$\delta_{AC} = \frac{N_{AC}L_{AC}}{EA_{AC}} = \frac{(69.87)(12)(12)}{(30000)A_{AC}} = \frac{0.3354}{A_{AC}}\text{in}^2 \qquad (5)$$

$$\delta_{BC} = \frac{N_{BC}L_{BC}}{EA_{BC}} = \frac{(116.49)(15.23)(12)}{(30000)A_{BC}} = \frac{0.7097}{A_{BC}}\text{in}^2 \qquad (6)$$

Let the pin move by u_C and v_C in the x and y-direction, respectively. Assuming small strain.

$$\delta_{AC} = v_C \qquad (7)$$

Substituting Eq. (5) into Eq. (7). we obtain

$$v_C = \frac{0.3354}{A_{AC}} \le 0.1 \qquad A_{AC} \ge 3.35\text{in}^2 \qquad (8)$$

The maximum area of cross section that satisfies Eq.(2), (4) and (8) are

$$\boxed{A_{AC} = 7.77 \text{ in}^2} \qquad \text{and} \qquad \boxed{A_{BC} = 4.66 \text{ in}^2}$$

4.82 The landing wheel of the plane is modeled as shown to the right. Pin at C is in double shear and has an allowable shear stress of 12 ksi The allowable axial stress for link BC is 30 ksi. Determine the diameter of pin C and the effective area of cross-section of link BC.

(Note the following: Attachments at A and B are approximated by pins to simplify analysis. There are two links represented by BC, one on either side of the hydraulic cylinder that we are modeling as a single link with an effective area cross-section that is to be determined so that the free body diagram is two-dimensional)

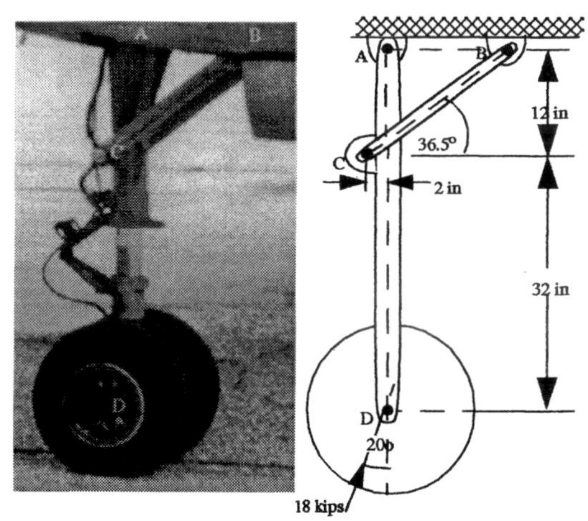

Fig. P4.82

Solution $\sigma_{BC} \le 30\text{ksi}$ $\qquad \tau_C \le 12\text{ksi}$ $\qquad d_c = ?$ $\qquad A_{BC} = ?$

We draw the free body diagram of bar ACD. as shown in Fig. (a)

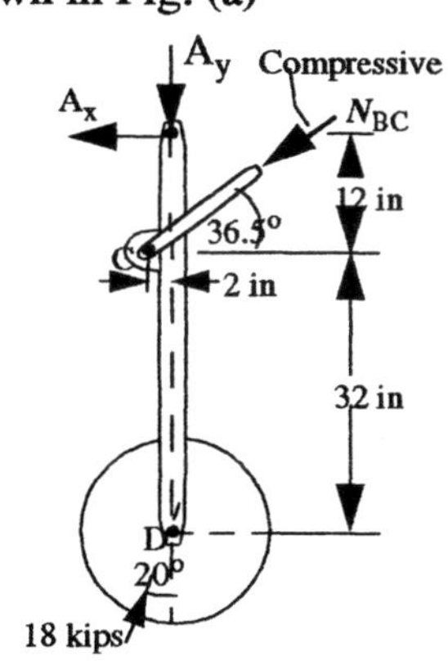

By moment equilibrium about point A.

$$18\sin 20(44) - (N_{BC}\cos 36.5)(12) + (N_{BC}\sin 36.5)(2) = 0 \quad \text{or} \quad N_{BC} = \frac{270.9}{8.457} = 32.03\text{kips}$$

The axial stress in BC is: $\sigma_{BC} = \frac{N_{BC}}{A_{BC}} = \frac{32.03}{A_{BC}} \leq 30$ or $A_{BC} \geq 1.068\text{in}^2$ or $\boxed{A_{BC} = 1.1\text{in}^2}$

The pin being in double sheer will have a shear or force $V_c = N_{BC}/2 = 16.015$

The shear stress in pin C is: $\tau_C = \frac{V_C}{A_C} = \frac{16.05}{(\pi d^2/4)} \leq 12$ or $d \geq 1.703\text{in}$ or $\boxed{d = 1.71\text{in}}$

4.83 The allowable normal stress in the stepped axial rod is 20 ksi. If F = 10kips, determine the smallest fillet radius that can be used at section B. Use stress concentration graphs given in Appendix C.

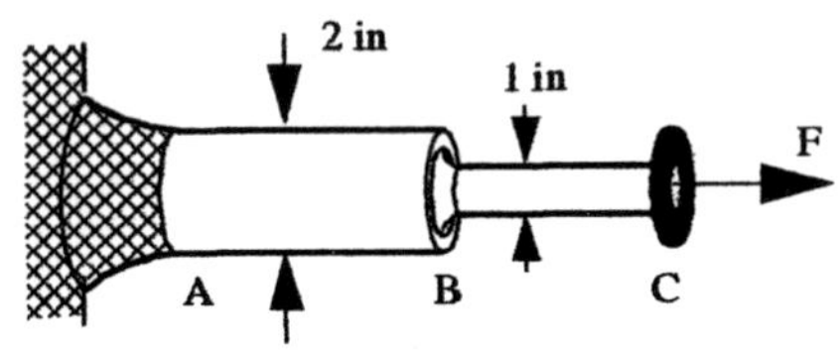

Fig. P4.83

Solution $\sigma_{max} \leq 20\text{ksi}$ $r = ?$

The axial stress in BC can be found as: $\sigma_{BC} = \frac{10}{\pi(1^2)/4} = 12.73\text{ksi}$

$$\sigma_{max} = k_{conc}\sigma_{BC} = 12.73k_{conc} \leq 20 \quad \text{or} \quad k_{conc} \leq 1.57$$ 1

From Fig. C.2, we obtain the approximate value of r/d corresponding to D/d = 2 and $k_{conc} = 1.5$

$\frac{r}{d} = 0.26$ or $r = (0.26)(1) = 0.26\text{inch}$ $\boxed{r = 0.26\text{inch}}$

4.84 The fillet radius in the stepped circular rod shown is 6 mm. Determine the maximum axial force F that can act on the rigid wheel if the allowable axial stress is 120 MPa, and Modulus of Elasticity is 70 GPa.Use stress concentration graphs given in Appendix C.

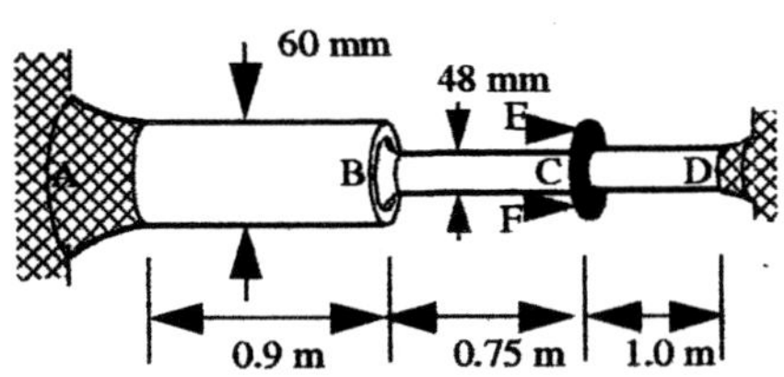

Fig. P4.84

Solution $\sigma_{max} \leq 120$ MPa $r = 6$ mm $E = 70$GPa $F_{max} = ?$

The area of cross-sections of various segments can be found as shown below.

$$A_{AB} = \frac{\pi}{4}(0.06)^2 = 2.83(10^{-3})\text{m}^2$$ 1

$$A_{BC} = A_{CD} = \frac{\pi}{4}(0.048)^2 = 1.81(10^{-3})\text{m}^2$$ 2

The ratio's $\frac{r}{d} = \frac{6}{48} = 0.125$ and $\frac{D}{d} = \frac{60}{48} = 1.25$

From Fig. C-2. we obtain the stress concentration factor as

$$K_{conc} = 1.7$$ 3

We made imaginary cuts in segments AB, BC and CD and draw free body diagram as shown below

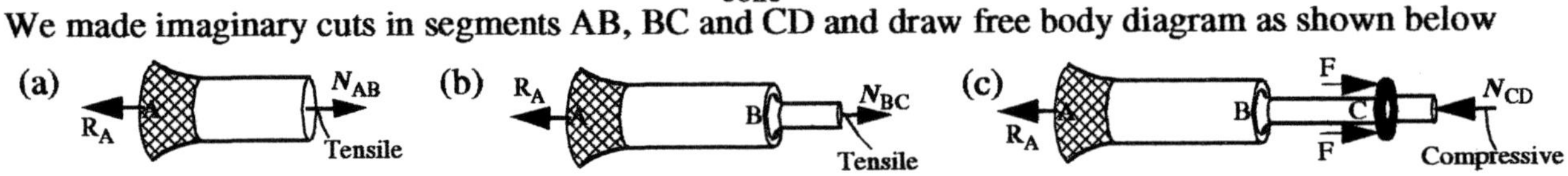

By force equilibrium we obtain

$$N_{AB} = R_A \quad (1)$$

$$N_{BC} = R_A \quad (2)$$

$$N_{CD} = 2F - R_A \quad (3)$$

The total extension of the bar is zero.

$$\delta_{AB} + \delta_{BC} - \delta_{CD} = 0 \quad (4)$$

From Eq. (4-27) we have

$$\delta_{AB} = \frac{N_{AB}L_{AB}}{E_{AB}A_{AB}} = \frac{R_A(0.9)}{70(10^9)(2.83)(10^{-3})} = 4.543R_A(10^{-6}) \quad (5)$$

$$\delta_{BC} = \frac{N_{BC}L_{BC}}{E_{BC}A_{BC}} = \frac{R_A(0.75)}{70(10^9)(1.81)(10^{-3})} = 5.919R_A(10^{-6}) \quad (6)$$

$$\delta_{CD} = \frac{N_{CD}L_{CD}}{E_{CD}A_{CD}} = \frac{(2F - R_A)(1)}{70(10^9)(1.81)(10^{-3})} = 7.893(2F - R_A)(10^{-6}) \quad (7)$$

Substituting Eq's (5), (6) and (7) into Eq. (4)

$$[4.543R_A + 5.919R_A - 7.893(2F - R_A)](10^{-6}) = 0 \qquad \text{or} \qquad R_A = \frac{15.786F}{18.355} = 0.86F$$

The nominal axial stress in each member can be found as

$$\sigma_{AB} = \frac{N_{AB}}{A_{AB}} = \frac{0.86F}{2.83(10^{-3})} = 303.9F \quad (8)$$

$$\sigma_{BC} = \frac{N_{BC}}{A_{BC}} = \frac{0.86F}{1.81(10^{-3})} = 475.1F \quad (9)$$

$$\sigma_{CD} = \frac{N_{CD}}{A_{CD}} = \frac{(2 - 0.86)F}{2.83(10^{-3})} = 629.8F \quad (10)$$

The maximum stress in BC is

$$(\sigma_{BC})_{max} = k_{conc}\sigma_{BC} = (1.7)(475.1F) = 807.7F \quad (11)$$

As the maximum stress in BC is the largest, we use to find maximum F as shown below.

$807.7F \le 120(10^6)$ or $F \le 148.6(10^3)$Newtons or $\boxed{F_{max} = 148.6kN}$

4.85 The fillet radius is 0.2 mm in the stepped steel circular rod shown in Fig. P4.83. What should be the peak value of the cyclic load F to ensure a service life of half-million cycles. Use the SN-curve shown in Figure 3.34.

Solution $r = 0.2$ $n = 0.5(10^6)$ steel $F_{max} = ?$

From Fig. 3.34 the peak stress for half a million cycles is $\sigma_{max} = 40ksi$

For r/d = 0.2 and d/D = 2, the stress concentration factor from Fig C-2 is $K_{conc} = 1.65$

The nominal stress in BC is as shown below.

$$\sigma_{BC} = \frac{\sigma_{max}}{K_{conc}} = \frac{40}{1.65} = \frac{F}{\pi(1)^2/4} \qquad \text{or} \qquad F = 19.04kips \text{ or } \boxed{F_{max} = 19kips}$$

4.86 The aluminum axial rods in Fig. P4.84 are subjected to a cyclic force F. Determine the peak value of the force F to ensure a service life of one-million cycle. Use the SN-curves shown in Figure 3.34 and material data shown in Table 4.6.

Solution $n = 10^6$ Aluminum $F_{max} = ?$

From Fig 3-34, the peak stress for a million cycles in Aluminum is approximately $\sigma_{max} = 147\text{MPa}$

From Eq. (11) in problem 4-84, we have

$$(\sigma_{BC})_{max} = 807.7F \le 147(10^6) \quad \text{or} \quad F \le 181.998(10^3)\text{Newtons} \qquad \boxed{F_{max} = 181.9\text{kN}}$$

Problem Set 4.4

4.87 During assembly of the structure, a misfit between bar A and the attachment of the rigid bar was found as shown. If bar A is pulled and attached, determine the initial stress introduced due to the misfit. The modulus of elasticity of the circular bars A and B is E= 10,000 ksi and the diameter is 1 inch.

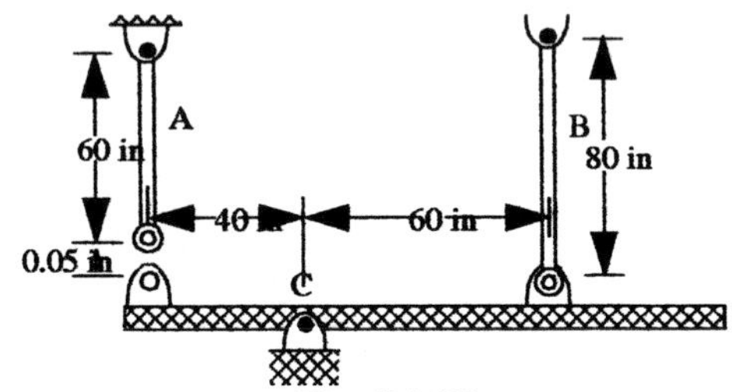

Fig. P4.87

Solution E = 10,000ksi d = 1inch $\sigma_A =$ $\sigma_B =$

We draw the free body diagram of rigid bar as shown in Fig. (a)

(a)

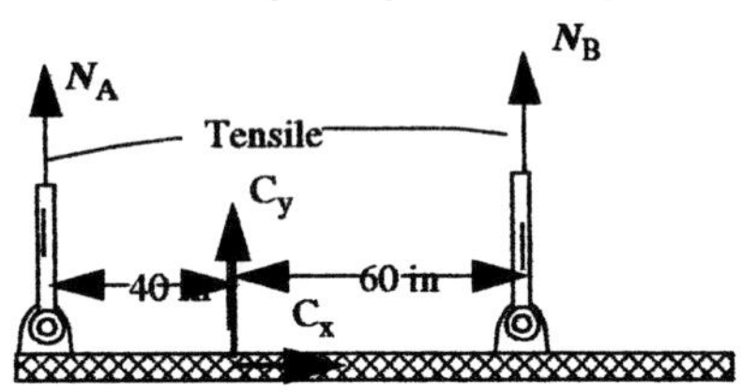

(b)

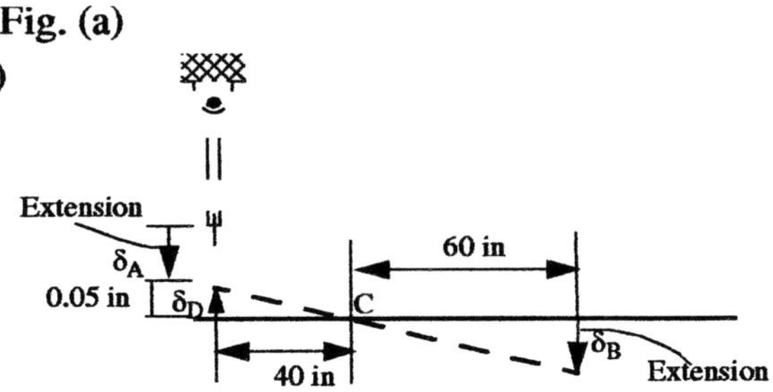

By moment equilibrium about point C, we obtain

$$0.4N_A = 0.6N_B \qquad 1$$

We can draw the exaggerated deformed geometry as shown in Fig.(b), and write the deformation equations

$$\frac{\delta_D}{0.4} = \frac{\delta_B}{0.4} \quad \text{or} \quad \delta_B = 1.5\delta_D \qquad 2$$

$$\delta_A = 0.05 - \delta_D \qquad 3$$

From Eq. 4-27

$$\delta_A = \frac{N_A L_A}{E_A A_A} = \frac{N_A(60)}{(10,000)\left(\frac{\pi}{4}\right)(1)^2} = 7.639N_A(10^{-3}) = 0.05 - \delta_D \quad \text{or} \quad N_A = 6.545 - 130.91\delta_D \qquad 4$$

$$\delta_B = \frac{N_B L_B}{E_B A_B} = \frac{N_B(80)}{(10,000)(\pi(1)^2/4)} = 10.186N_B(10^{-3}) = 1.5\delta_D \quad \text{or} \quad N_B = 147.26\delta_D \qquad 5$$

Substituting Eq. (4) and (5) into Eq. (1)

$$0.4(6.545 - 130.91\delta_D) = 0.6(147.26\delta_D) \quad \text{or} \quad \delta_D = \frac{2.618}{140.72} = 18.6(10^{-3}) \qquad 6$$

From Eq. (4) and (5) we obtain: $N_A = 4.11\text{kips}$ and $N_B = 2.74\text{kips}$

The initial axial stresses are

$$\sigma_A = \frac{N_A}{A_A} = \frac{4.11}{(\pi(1)^2/4)} = 5.233\text{ksi} \qquad \boxed{\sigma_A = 5.2\text{ksi(T)}}$$

$$\sigma_B = \frac{N_B}{A_B} = \frac{2.74}{(\pi(1)^2/4)} = 3.488\text{ksi} \qquad \boxed{\sigma_B = 3.5\text{ksi(T)}}$$

4.88 Bar A was manufactured 2 mm less than bar B due to an error. The attachment of these bars to the rigid bar would cause a misfit of 2 mm. Calculate the initial stress in each assembly. Which of the two assembly configuration you would recommend? Use modulus of elasticity of E = 70 GPa and diameter of the circular bars as 25 mm.

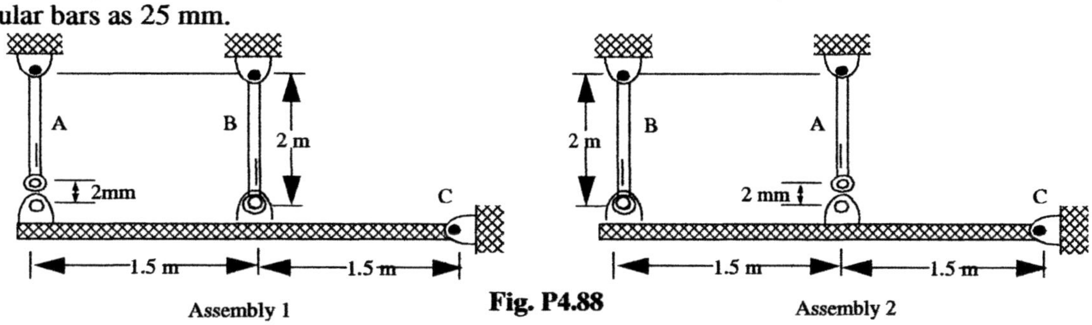

Fig. P4.88

Solution E = 70GPa d = 25mm assembly = ? σ_A = σ_B =

The area cross-section is $A = \frac{\pi}{4}(0.025)^2 = 0.4909(10^{-3})$

Assembly 1 Analysis

We can draw the free body diagram of rigid bar as shown in Fig.(a)

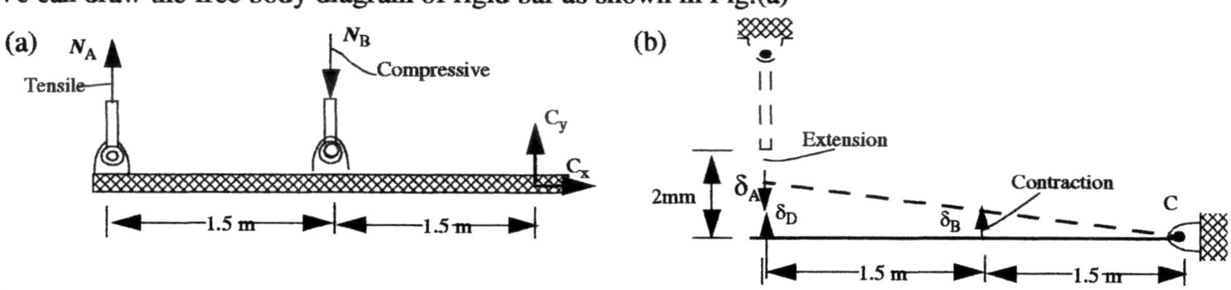

By moment equilibrium about point C

$$N_A(3) = N_B(1.5) \qquad \text{or} \qquad 2N_A = N_B(1) \tag{1}$$

We can draw an exaggerated deformed geometry and write deformation equations

$$\frac{\delta_D}{3} = \frac{\delta_B}{1.5} \qquad \text{or} \qquad \delta_B = \frac{\delta_D}{2} \tag{2}$$

$$\delta_A = 0.002 - \delta_D \tag{3}$$

From Eq. (4-27)

$$\delta_A = \frac{N_A L_A}{E_A A_A} = \frac{N_A(2)}{(70)(10^9)(0.4909)(10^{-3})} = 58.2N_A(10^{-9}) \tag{4}$$

Substituting eq. (3) into Eq. (4), we obtain

$$N_A = 34.36(10^3) - 17.18\delta_D(10^6) \tag{5}$$

$$\delta_B = \frac{N_B L_B}{E_B A_B} = 58.2N_B(10^{-9}) \tag{6}$$

Substituting eq. (2) into Eq. (6), we obtain

$$N_B = 8.591\delta_D(10^6) \qquad 7$$

Substituting Eq. (5) and (7)into Eq. (1), we obtain

$$2[34 - 36(10^3) - 17.18\delta_D(10^6)] = 8.591\delta_D(10^6) \quad \text{or} \quad \delta_D = \frac{68.72}{42.95}(10^{-3}) = 1.60(10^{-3}) \qquad 8$$

From Eq. (4) and (5): $N_A = 6.872(10^3)$ and $N_B = 13.75(10^3)$

The initial axial stresses are

$$\sigma_A = \frac{N_A}{A_A} = \frac{6.872(10^3)}{(0.4909)(10^{-3})} = 14.0(10^6) \qquad \boxed{\sigma_A = 14.0\text{MPa(T)}}$$

$$\sigma_B = \frac{N_B}{A_B} = \frac{13.75(10^3)}{(0.4909)(10^{-3})} = 28.0(10^6) \qquad \boxed{\sigma_B = 28.0\text{MPa(C)}}$$

Assembly 2 Analysis

We can draw the free body diagram of rigid bar as shown in Fig.(c)

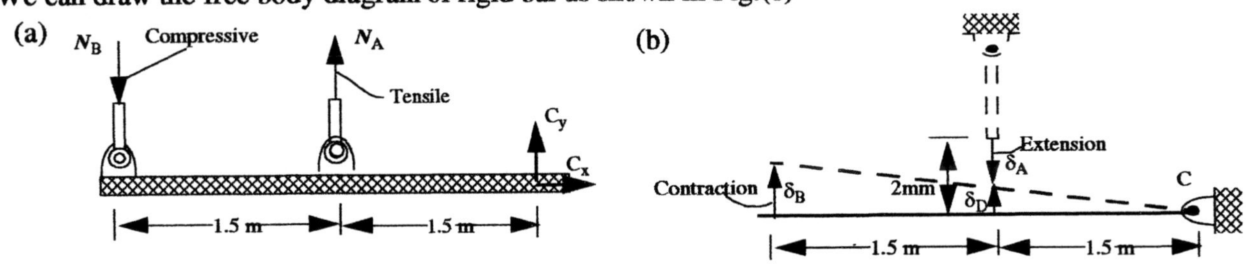

By moment equilibrium about point C

$$N_B(3) = 1.5N_A \quad \text{or} \quad N_A = 2N_B \qquad 9$$

We can draw an exaggerated deformed geometry and write deformation equations

$$\frac{\delta_B}{3} = \frac{\delta_D}{1.5} \quad \text{or} \quad \delta_B = 2\delta_D \qquad 10$$

$$\delta_A = 0.002 - \delta_D \qquad 11$$

Substituting Eq. (11) into Eq. (4) we obtain

$$N_A = 34.36(10^3) - 17.18\delta_D(10^6) \qquad 12$$

Substituting Eq. (8) into Eq. (6) we obtain

$$N_B = 34.36\delta_D(10^6) \qquad 13$$

Substituting Eq. (12) and (13) into Eq. (7) we obtain

$$[34.36(10^3) - 17.18\delta_D(10^6)] = 2(34.36)\delta_D(10^6) \quad \text{or} \quad \delta_D = \frac{34.36(10^3)}{85.908(10^6)} = 0.4(10^{-3})$$

From Eq. (12) and (13) we obtain: $N_A = 27.49(10^3)$ or $N_B = 13.74(10^3)$

The axial stresses are

$$\sigma_A = \frac{N_A}{A_A} = \frac{27.49(10^3)}{0.4909(10^{-3})} = 56(10^6) \qquad \boxed{\sigma_A = 56\text{MPa(T)}}$$

$$\sigma_B = \frac{N_B}{A_B} = \frac{13.74(10^3)}{0.4909(10^{-3})} = 27.989 \qquad \boxed{\sigma_B = 28\text{MPa(C)}}$$

Assembly 1 is preferred as the initial stress in member A is significantly smaller.

4.89 A steel bolt is passes through a aluminum sleeve. After assembling the unit by finger tightening the nut (no deformation) the nut is given 1/4 turn. If the pitch of the threads is 0.3 mm, determine the initial axial stress developed in the sleeve and the bolt. The modulus of elasticity for steel and copper are $E_{st} = 200$ GPa, $E_{al} = 70$ GPa, respectively the area of cross-sections are$A_{st} = 500\ mm^2$ $A_{al} = 1100\ mm^2$.

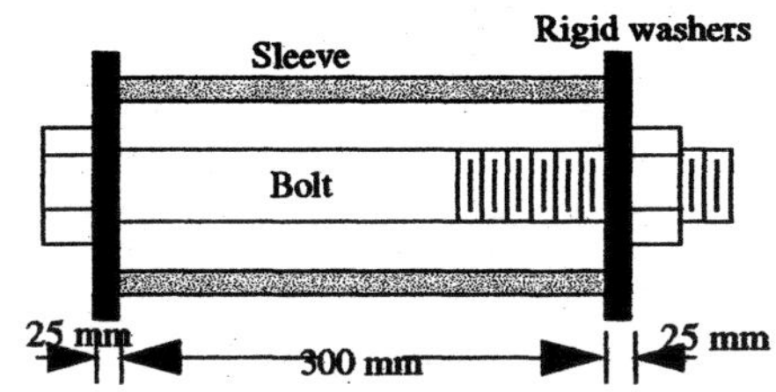

Fig. P4.89

Solution $p = 0.3$ mm $n = 1/4$ $E_{st} = 200$ GPa $E_{al} = 70$ GPa

$A_{st} = 500\ mm^2$ $A_{al} = 1100\ mm^2$

We can make an imaginary cut through the sleeve and the bolt and draw the free body diagram shown in Fig. (a).

(a)

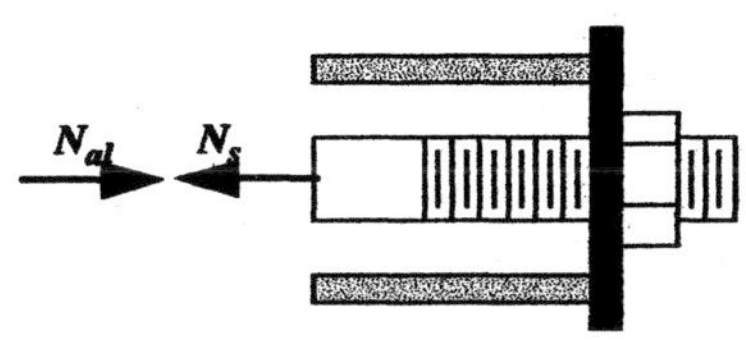

(b)

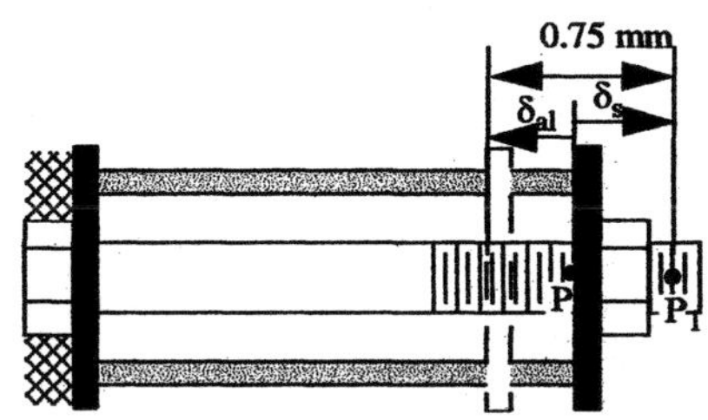

N_{al} represent the equivalent axial force that acts on the entire circulation cross-section of the sleeve

By force equilibrium in Fig(a)

$$N_{al} = N_s \qquad 1$$

Consider point P on the bolt. As the nut is tightened, point P moves to point P_1, and the distance PP_1, represents the extension of the bolt. Relative to point P_1(bolt) the nuts moves by an amount (n)(p) = 0.75mm. The deformation equation is

$$\delta_{al} + \delta_s = 0.75(10^{-3})\text{m} \qquad 2$$

From Eq. (4-27).

$$\delta_{al} = \frac{N_{al}L_{al}}{E_{al}A_{al}} = \frac{N_{al}(0.3)}{(70)(10^9)(1100)(10^{-6})} = 3.896N_{al}(10^{-9}) \qquad 3$$

$$\delta_s = \frac{N_sL_s}{E_sA_s} = \frac{N_s(0.3 + 0.025 - 0.025)}{(200)(10^9)(500)(10^{-6})} = 3.5N_s(10^{-9}) \qquad 4$$

Substituting Eq. (3),(4) and (1) into Eq. (2) we obtain

$$(3.896 + 3.5)N_{al}(10^{-9}) = 0.75(10^{-3}) \quad \text{or} \quad N_{al} = 101.4(10^3)\text{Newtons and } N_s = 101.4(10^3)\text{Newtons}$$

The axial stresses are

$$\sigma_{al} = \frac{N_{al}}{A_{al}} = \frac{101.4(10^3)}{1100(10^{-6})} = 92.2(10^6) \qquad \boxed{\sigma_{al} = 92.2\text{MPa(C)}}$$

$$\sigma_s = \frac{N_s}{A_s} = \frac{101.4(10^3)}{500(10^{-6})} = 202.8(10^6) \qquad \boxed{\sigma_s = 202.8\text{MPa(T)}}$$

4.90 The rigid bar is horizontal when the unit is put together by finger tightening the nut. The pitch of the threads is 0.125 inch. Develop a table in steps of quarter turn of the nut that can be used for prescribing

the pre-tension in bar B. Maximum number of quarter turns is limited by the yield stress.

	Bar A	Bar B
Modulus of Elasticity	10,000 ksi	30,000 ksi
Yield Stress	24 ksi	30 ksi
Area of cross-section	0.5 in^2	0.75 in^2

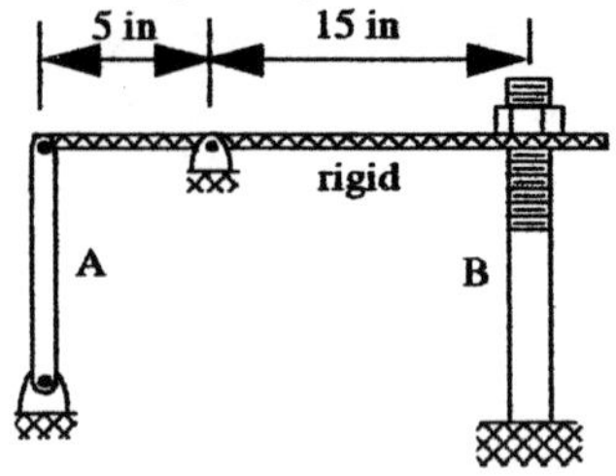

Fig. P4.90

Solution $\quad p = 0.125$ inch $\quad n$ = number of quarter turns = ? $\quad \sigma_A = ?$ $\quad \sigma_B = ?$

We can draw the free body diagram of the rigid plate as shown in Fig. (a).

(a)

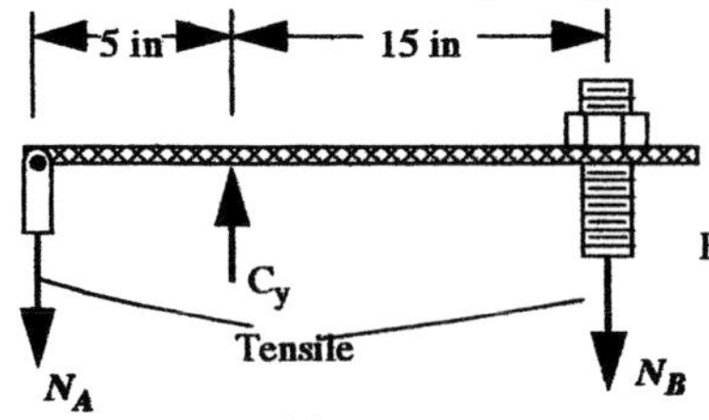

(b)

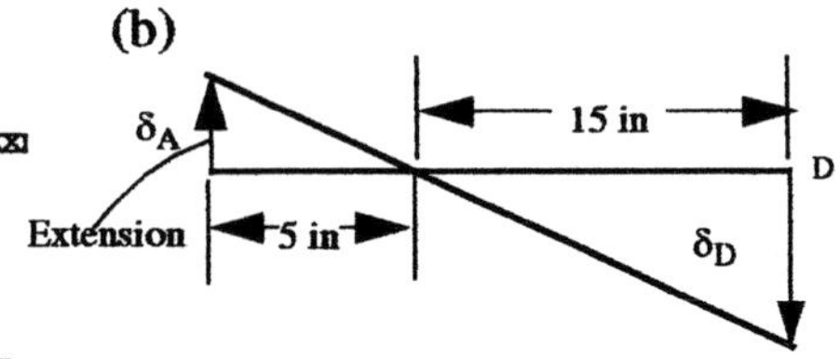

(c)

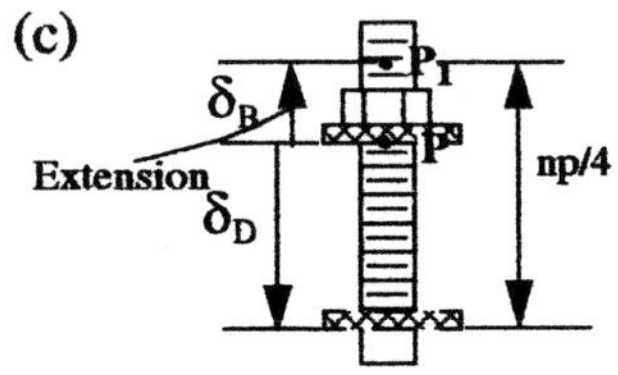

By moment equilibrium about point C

$$N_A(5) = 15N_B \quad \text{or} \quad N_A = 3N_B \tag{1}$$

Consider point D on the rigid plate. Using deformed geometry in Fig(b), we can write

$$\frac{\delta_A}{5} = \frac{\delta_D}{15} \quad \text{or} \quad \delta_A = \frac{\delta_D}{3} \tag{2}$$

Point P on the bolt moves to point P_1 as the bolt material is pulled through the nut. The distance PP_1 in the Fig (c) represents the extension of the bolt. The nut moves relative to the bolt. i.e. with respect to point P_1 by the amount nP/4. The deformed equation is thus

$$\delta_B + \delta_D = \frac{nP}{4} \quad \text{or} \quad \delta_B = 0.03125n - \delta_D \tag{3}$$

From Eq. (4-27)

$$\delta_A = \frac{N_A L_A}{E_A A_A} = \frac{N_A(50)}{(10)(10^3)(0.5)} = 0.01N_A = \frac{\delta_D}{3} \tag{4}$$

$$\text{or} \quad N_A = 33.33\delta_D \tag{5}$$

$$\delta_B = \frac{N_B L_B}{E_B A_B} = \frac{N_B(50)}{(30)(10^3)(0.75)} = 2.222(10^{-3})N_B = 0.03125n - \delta_D \tag{6}$$

$$\text{or} \quad N_B = 14.0625n - 450\delta_D \tag{7}$$

Substituting Eq. (5) and (7) into eq. (1) we obtain.

$$33.33\delta_D = 3(14.0625n - 450\delta_D) \quad \text{or} \quad \delta_D = \frac{42.1875n}{1383.33} = 0.0305n \tag{8}$$

From Eq. (5) and (7) we obtain: $N_A = 1.0166n\text{kips}$ and $N_B = 0.3388n$

The axial stresses are

$$\sigma_A = \frac{N_A}{A_A} = \frac{1.0166n}{0.5} = 2.033n \tag{9}$$

$$\sigma_B = \frac{N_B}{A_B} = \frac{0.3388n}{0.75} = 0.4518n \qquad \textbf{10}$$

The stress in A exceeds material yield stress when n = 12. The initial stress for 11 quarter turns are given in the table below.

n	σ_A	σ_B
1	2.03	0.45
2	4.07	0.90
3	6.10	1.36
4	8.13	1.81
5	10.17	2.26
6	12.20	2.71
7	14.23	3.16
8	16.26	3.61
9	18.30	4.07
10	20.33	4.52
11	22.36	4.97

4.91 The increase in temperature varies as $\Delta T = T_L(x^2/L^2)$. Determine the axial stress and movement of a point at x=L/2 in terms of length L, Modulus of elasticity E, area of cross-section A, coefficient of thermal expansion α and the increase of temperature at the end T_L

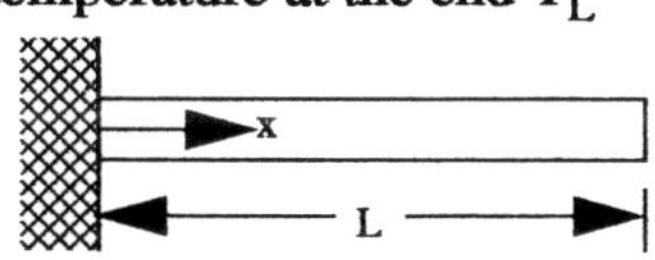

Solution $\quad \Delta T = T_L\left(\frac{x^2}{L^2}\right) \qquad \sigma_{xx} = \qquad u\left(\frac{L}{2}\right) =$

As the right side is free to expand and there are no external forces. $\boxed{\sigma_{xx} = 0}$

The axial strain is: $\varepsilon_{xx} = \frac{du}{dx} = \alpha\Delta T = \alpha T_L\frac{x^2}{L^2}$

Integrating from x=0, where displacement is zero to x=L/2 we obtain

$$\int_0^{u\left(\frac{L}{2}\right)} du = \int_0^{\frac{L}{2}} \alpha T_L\frac{x^2}{L^2}dx \quad \text{or} \quad u\left(\frac{L}{2}\right) = \frac{\alpha T_L}{3L^2}x^3\Big|_0^{\frac{L}{2}} = \frac{\alpha T_L L^3}{24L^2} \qquad \boxed{u\left(\frac{L}{2}\right) = \frac{\alpha T_L L}{24}}$$

4.92 The increase in temperature varies as $\Delta T = T_L(x^2/L^2)$. Determine the axial stress and movement of a point at x=L/2 in terms of length L, Modulus of elasticity E, area of cross-section A, coefficient of thermal expansion α and the increase of temperature at the end T_L

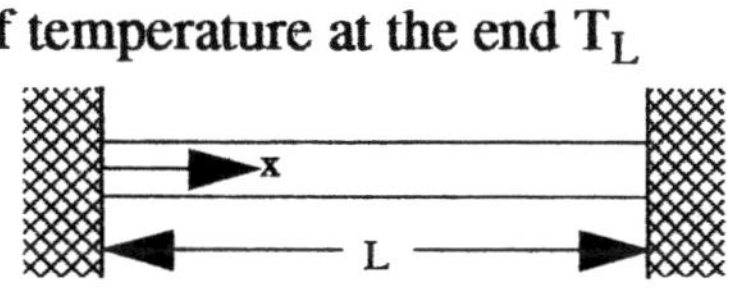

Solution $\quad \Delta T = T_L(x^2/L^2) \qquad \sigma_{xx} = \qquad u(L/2) =$

We make an imaginary cut and draw free body diagram as shown below.

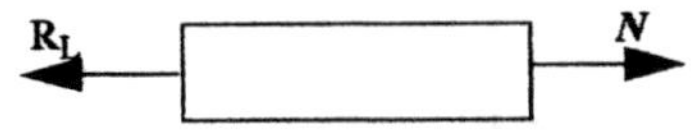

By force equilibrium we obtain

$$N = R_L \quad (1)$$

$$\sigma_{xx} = \frac{N}{A} = \frac{R_L}{A} \quad (2)$$

The axial strain is: $\varepsilon_{xx} = \frac{du}{dx} = \frac{\sigma_{xx}}{E} + \alpha\Delta T$ or $\frac{du}{dx} = \frac{R_L}{EA} + \alpha T_L(x^2/L^2)$ Integrating we obtain.

$$u(x) = \frac{R_L}{EA}x + \frac{\alpha T_L x^3}{3L^2} + C_1 \quad (3)$$

Noting $u(0) = 0$. $C_1 = 0$. Noting that $u(L) = 0$, we obtain the following.

$$u(L) = \frac{R_L L}{EA} + \frac{\alpha T_L L^3}{3L^2} = 0 \quad \text{or} \quad R_L = -\frac{\alpha T_L EA}{3} \quad (4)$$

From Eq. (2) we obtain.

$$\boxed{\sigma_{xx} = \frac{E\alpha T_L}{3}(C)}$$

Substituting Eq. (4) and $C_1 = 0$ in Eq (3): $u(x) = -\frac{\alpha T_L}{3}x + \frac{\alpha T_L x^3}{3L^2}$ at x=L/2, we obtain

$$u\left(\frac{L}{2}\right) = -\frac{\alpha T_L L}{6} + \frac{\alpha T_L L^3}{24L^2} = -\frac{\alpha T_L L}{8} \qquad \boxed{u\left(\frac{L}{2}\right) = -\frac{\alpha T_L L}{8}}$$

4.93 The tapered bar shown has a cross-sectional area that varies with x as $A = K(L-0.5\ x)^2$. If the increase in temperature of the bar varies as $\Delta T = T_L(x^2/L^2)$. Determine the axial stress at mid point in terms of the length L, Modulus of elasticity E, area of cross-section parameter K, coefficient of thermal expansion α, and the increase of temperature at the end T_L

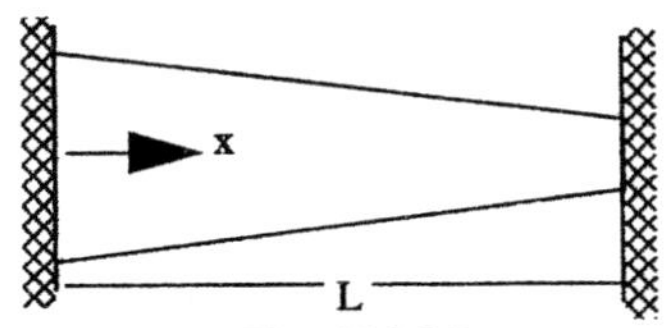

Fig. P4.93

Solution $A = K(L-0.5x)^2$ $\Delta T = T_L(x^2/L^2)$ $\sigma_{xx}(L/2) =$

We make an imaginary cut and draw a free body diagram as shown below.

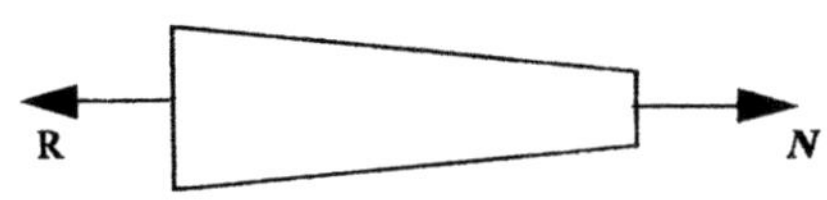

By force equilibrium

$$N = R \quad (1)$$

The axial stress and strain can be written as shown below.

$$\sigma_{xx} = \frac{N}{A} = \frac{R}{K(L-0.5x)^2} \quad (2)$$

$$\varepsilon_{xx} = \frac{du}{dx} = \frac{\sigma_{xx}}{E} + \alpha\Delta T = \frac{R}{EK(L-0.5x)^2} + \alpha T_L \frac{x^2}{L^2} \quad (3)$$

Integrating from x=0 to x=L and noting that displacement at both points are zero, we obtain the following

$$\int_{u_0=0}^{u_L=0} du = \int_0^L\left(\frac{R}{EK(L-0.5x)^2}+\frac{\alpha T_L x^2}{L^2}\right)dx = \left(\frac{R}{(0.5EK)(L-0.5x)}+\frac{\alpha T_L x^3}{3L^2}\right)\Bigg|_0^L = 0 \text{ or}$$

$$\frac{R}{(0.5EK)}\left[\frac{1}{0.5L}-\frac{1}{L}\right]+\frac{\alpha T_L L}{3} = 0$$

$$\frac{2R}{EKL} = \frac{\alpha T_L L}{3} \qquad \text{or} \qquad R = -\frac{EK\alpha T_L L^2}{6} \tag{4}$$

Substituting Eq. (4) in Eq. (2) and evaluating the stress at x=L/2

$$\sigma_{xx} = \frac{R}{K\left(L-0.5\frac{L}{2}\right)^2} = \frac{R}{K(0.75L)^2} = -\frac{E\alpha T_L}{6(0.75)^2} \qquad \boxed{\sigma_{xx} = \frac{8}{27}E\alpha T_L(C)}$$

4.94 Three metallic rods are attached to a rigid plate as shown. The temperature of the rods is lowered by 100 °F after the forces are applied. Assuming the rigid plate does not rotate, determine the movement of the rigid plate

	Area in²	E ksi	α 10⁻⁶/°F
Aluminum	4	10,000	12.5
Steel-1	4	30,000	6.6
Steel-2	12	30,000	6.6

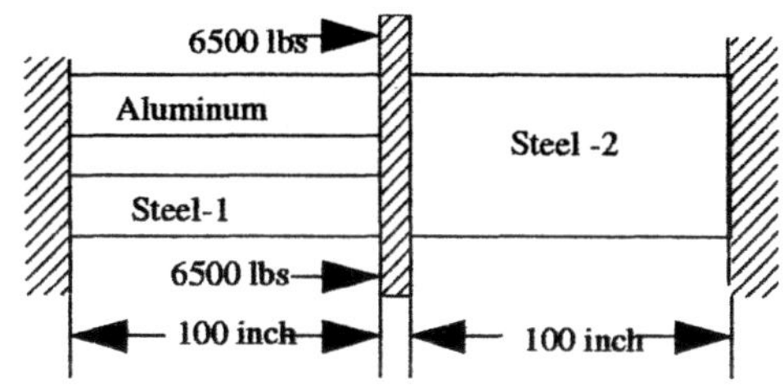

Fig. P4.94

Solution $\Delta T = -100°F$ $\delta_P =$

We can draw the free body diagram of the rigid plate as shown in Fig.(a).

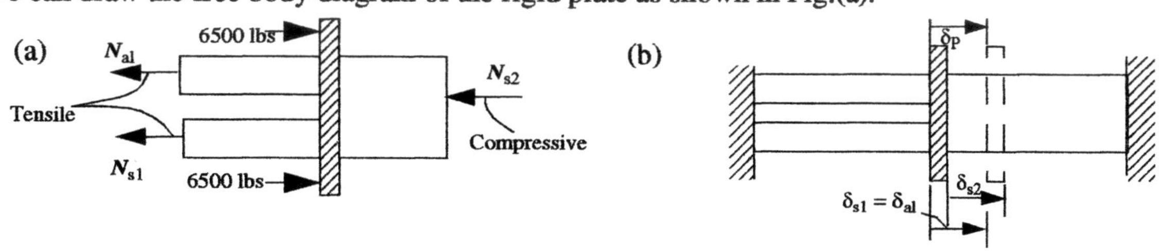

By force equilibrium in Fig (a)

$$N_{a1}+N_{s1}+N_{s2} = 13000\text{lbs} = 13\text{kips} \tag{1}$$

We can draw an exaggerated deformed geometry as shown in Fig.(b) and write the deformation equations

$$\delta_{a1} = \delta_P \qquad \delta_{s1} = \delta_P \qquad \delta_{s2} = \delta_P \tag{2}$$

Noting that the deformation shown in Fig (b) are considered positive we have the following

$$\delta_{a1} = \frac{N_{a1}(100)}{10(10^3)(4)}-(100)(12.5)(10^{-6})(100) = \delta_P \qquad \text{or} \qquad N_{a1} = 400\delta_P+50 \tag{3}$$

$$\delta_{s1} = \frac{N_{s1}(100)}{30(10^3)(4)}-(100)(6.6)(10^{-6})(100) = \delta_P \qquad \text{or} \qquad N_{s1} = 1200\delta_P+79.2 \tag{4}$$

$$\delta_{s2} = \frac{N_{s2}(100)}{30(10^3)(12)}+(100)(6.6)(10^{-6})(100) = \delta_P \qquad \text{or} \qquad N_{s2} = 3600\delta_P-237.6 \tag{5}$$

Substituting Eq's (3), (4) and (5) into Eq. (1) we obtain

$$(400 + 1200 + 3600)\delta_P + 50 + 79.2 - 237.6 = 13 \quad \text{or} \quad \delta_P = \frac{121.4}{5200} = 23.35(10^{-3})$$

$$\boxed{\delta_P = 0.0233 \text{ inch to the right}}$$

4.95 Solve problem 4.89 assuming that in addition to turning the nut the temperature of the assembled unit is raised by 40°C. The coefficients of thermal expansion for steel and aluminum are $\alpha_{st} = 12\ \mu/°C$, $\alpha_{al} = 22.5\ \mu/°C$.

Solution $\Delta T = 40°C$ $\quad\alpha_{st} = 12\mu/°C$ $\quad\alpha_{al} = 22.5\mu/°C$

From Eq. (1) and (2) in problem 4-89

$$N_{al} = N_s \qquad 1$$

$$\delta_{al} + \delta_s = 0.75(10^{-3})\text{m} \qquad 2$$

Noting that the deformation in Eq. (2) assumes for steel as positive in extension and for aluminum as positive in contraction, we can modify the deformation equations (3) and (4) of problem 4-89 as shown below.

$$\delta_{al} = 3.896N_{al}(10^{-9}) - (40)(22.5)(10^{-6})(0.3) \quad \text{or} \quad \delta_{al} = 3.896N_{al}(10^{-9}) - 270(10^{-6}) \qquad 3$$

$$\delta_s = 3.5N_s(10^{-9}) - (40)(12)(10^{-6})(0.3 + 0.025 + 0.025) \quad \text{or} \quad \delta_s = 3.5N_s(10^{-9}) - 168(10^{-6}) \qquad 4$$

Substituting Eq's (3), (4) and (1) into Eq. (2)

$$(3.896 + 3.5)(N_{al})(10^{-9}) + (-270 + 168)(10^{-6}) = 0.75(10^{-3}) \text{ or}$$

$$N_{al} = \frac{0.852(10^{-3})}{7.396(10^{-9})} = 115.2(10^3)\text{Newtons} \qquad 5$$

$$N_s = 115.2(10^3)\text{Newtons} \qquad 6$$

The axial stresses are

$$\sigma_{al} = \frac{N_{al}}{A_{al}} = \frac{115.2(10^3)}{1100(10^{-6})} = 104.7(10^6) \qquad \boxed{\sigma_{al} = 104.7\text{MPa(C)}}$$

$$\sigma_s = \frac{N_s}{A_s} = \frac{115.2(10^3)}{500(10^{-6})} = 230.4(10^6) \qquad \boxed{\sigma_s = 230.4\text{MPa(T)}}$$

4.96 Determine the axial stress in bar A of problem 4.90 assuming that the nut is turned 1 full turn and the temperature of bar A is decreased by 80°F. The coefficients of thermal expansion for bar A is $\alpha_{st} = 22.5\ \mu/°F$.

Solution $p = 0.125$ inch $\quad n = 4$ quarter turns $\quad \Delta T_A = -80°F \quad \alpha_{st} = 22.5\ \mu/°F$

$\sigma_A = ?$ $\quad \sigma_B = ?$

From Eqs. (1) (2) and (3) in problem 4.90 we have:

$$N_A(5) = 15N_B \quad \text{or} \quad N_A = 3N_B \qquad 1$$

$$\frac{\delta_A}{5} = \frac{\delta_D}{15} \quad \text{or} \quad \delta_A = \frac{\delta_D}{3} \qquad 2$$

$$\delta_B + \delta_D = \frac{nP}{4} \qquad \text{or} \qquad \delta_B = 0.03125n - \delta_D \qquad 3$$

Noting that the deformation in Eq. (4) of problem 4.90 assumes for bar A as positive in extension, we can modify the deformation in Eq. (4) of problem 4.90 as shown below.

$$\delta_A = \frac{N_A(50)}{(10)(10^3)(0.5)} - (80)(22.5)(10^{-6})(50) = 0.01N_A - 90(10^{-3}) = \frac{\delta_D}{3} \qquad 4$$

$$\text{or} \qquad N_A = 33.33\delta_D + 9 \qquad 5$$

There is no temperature change in bar B, hence from Eq. (7) of problem 4.90 we have the following with n = 4.

$$\text{or} \qquad N_B = 14.0625(4) - 450\delta_D = 56.25 - 450\delta_D \qquad 6$$

Substituting Eq. (5) and (6) into Eq. (1) we obtain.

$$33.33\delta_D + 9 = 3(56.25 - 450\delta_D) \qquad \text{or} \qquad \delta_D = \frac{159.75}{1383.33} = 0.1155 \qquad 7$$

From Eq. (5) we obtain: $N_A = 12.849\text{kips}$

The axial stress in A is: $\sigma_A = \frac{N_A}{A_A} = \frac{12.849}{0.5} = 25.70\text{ksi}$ $\boxed{\sigma_A = 25.70\text{ksi(T)}}$

Problem Set 4.5

4.97 A force F is applied to the roller that slides inside a slot. Both bars have an area of cross-section of A = 100 mm^2, Modulus of Elasticity E = 200 GPa, and a yield stress of 250 MPa. Bar AP and BP have lengths of L_{AP}= 200 mm and L_{BP}= 250 mm respectively. Draw the load deflection curve and determine the collapse load.

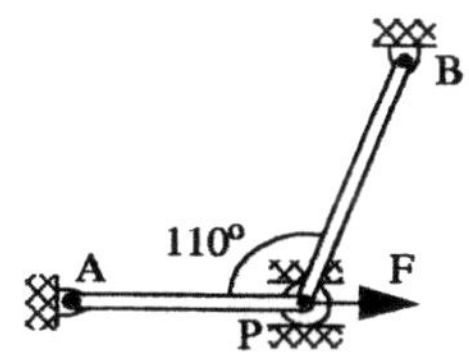

Fig. P4.97

Solution E = 200 GPa A = 100 mm^2 L_{AP}= 200 mm L_{BP}= 250 mm

σ_{yield} = 250 MPa F vs. δ_P = ? $F_{collapse}$=?

We draw the free body diagram of the rigid bar as shown in Fig. (a)

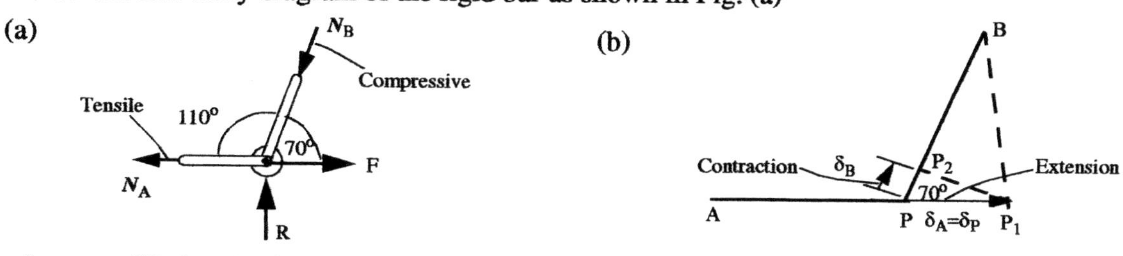

By force equilibrium in the x-direction in Fig.(a), we obtain the following.

$$N_A + N_B\cos 70 = F \qquad \text{or} \qquad N_A + 0.342N_B = F \qquad 1$$

We draw the exaggerated deformed shape as shown in Fig. (b). As per small strain approximation, we need component of PP_1 in the original direction of AP and BP, i.e., PP_1 represents the deformation of bar AP and PP_2 represents the deformation of bar BP and are calculated as shown below.

$$\delta_A = \delta_P \qquad 2$$

$$\delta_B = \delta_P\cos 70 = 0.342\delta_P \qquad 3$$

From Eq 4.27, we have the following:

$$\delta_A = \frac{N_A L_A}{EA} = \frac{N_A(0.2)}{(200)(10^9)(100)(10^{-6})} = \delta_P \qquad N_A = 100\delta_P(10^6) \qquad (4)$$

$$\delta_B = \frac{N_B L_B}{EA} = \frac{N_B(0.25)}{(200)(10^9)(100)(10^{-6})} = 0.342\delta_P \qquad N_B = 27.36\delta_P(10^6) \qquad (5)$$

First Member Yields: As area of cross-sections and yield stress of both bars are same, we conclude from Eqs (4) and (5) that bar A will reach yield stress first. When A is just at yield stress:

$$N_A = (25,000)(100)(10^{-6}) = 25(10^3) \text{ Newtons} \qquad (6)$$

Let F_1 and δ_1 be the values of the force and deformation when A reaches yield stress. Substituting Eq. (6) into Eq. (4) we obtain: $100\delta_1(10^6) = 25,000$ or

$$\delta_1 = 0.25(10^{-3}) \text{ m} \qquad (7)$$

Substituting Eq. (7) into Eq. (5), we obtain:

$$N_B = (27.36)(0.25)(10^{-3})(10^6) = 6.84(10^3) \text{ Newtons} \qquad (8)$$

Substituting Eq. (8) and (6) into Eq. (1) we obtain:

$$F_1 = (25)(10^3) + (0.342)(6.84)(10^3) = 27.34(10^3) \text{ Newtons} \qquad (9)$$

Second Member Yields: When B reaches yield stress, the structure collapses. When B is at yield stress, the internal axial force in member B is:

$$N_B = (25,000)(100)(10^{-6}) = 25(10^3) \text{ Newtons} \qquad (10)$$

Let $F_{collapse}$ and $\delta_{collapse}$ be the values of the force and deformation when B reaches yield stress. Substituting Eq. (10) into Eq. (5) we obtain: $27.36\delta_{collapse}(10^6) = 25,000$ or

$$\delta_{collapse} = 0.914(10^{-3}) \text{ m} \qquad (11)$$

Substituting Eq. (6) and (10) into Eq. (1) we obtain:

$F_{collapse} = (25)(10^3) + (0.342)(25)(10^3) = 33.55(10^3)$ Newtons or $\boxed{F_{collapse} = 33.6 \text{ kN}}$

The load deflection curve is given below:

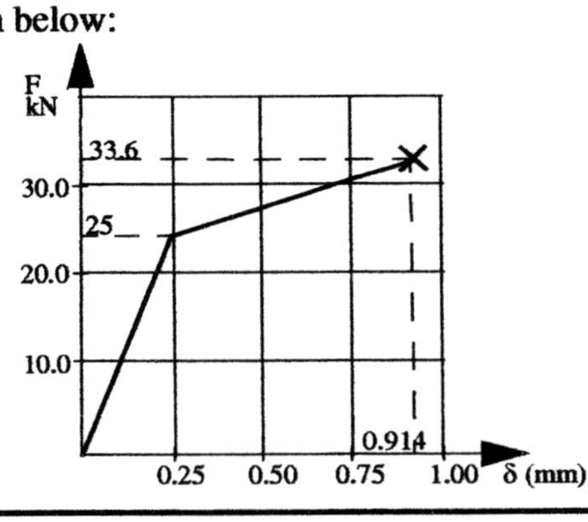

4.98 A force F is applied to the roller that slides inside a slot. Both bars have an area of cross-section of A = 100 mm², Modulus of Elasticity E = 200 GPa, and a yield stress of 250 MPa. Bar AP and BP have lengths of L_{AP}= 200 mm and L_{BP}= 250 mm respectively. Draw the load deflection curve and determine the collapse load.

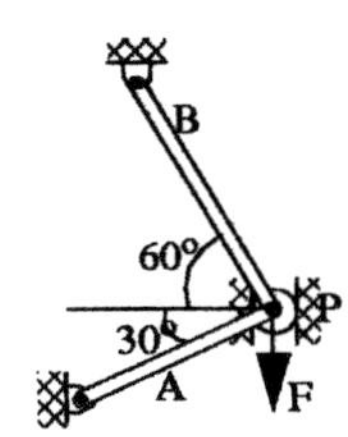

Fig. P4.98

Solution $E = 200$ GPa $A = 100\ mm^2$ $L_{AP} = 200$ mm $L_{BP} = 250$ mm
$\sigma_{yield} = 250$ MPa F vs. $\delta_P = ?$ $F_{collapse} = ?$

We draw the free body diagram of the rigid bar as shown in Fig. (a) below.

(a)

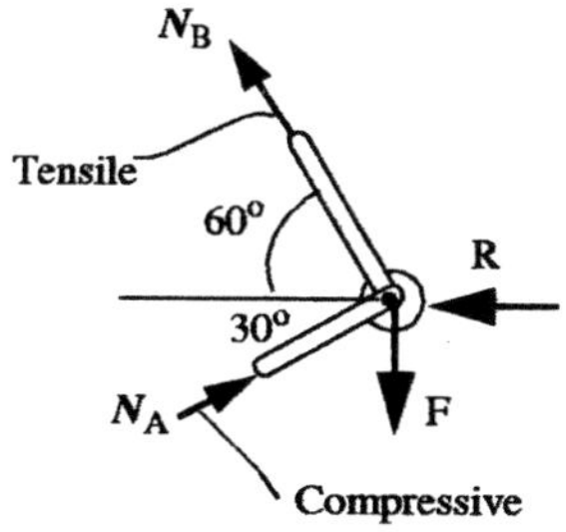

(b)

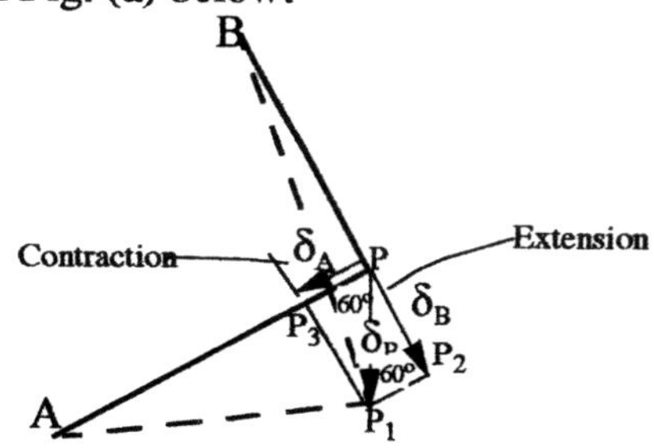

By force equilibrium in the y-direction in Fig.(a), we obtain the following.

$$N_A \sin 30 + N_B \sin 60 = F \qquad \text{or} \qquad 0.5N_A + 0.866N_B = F \tag{1}$$

We draw the exaggerated deformed shape as shown in Fig. (b). As per small strain approximation, we need component of PP_1 in the original direction of AP and BP, i.e., PP_3 represents the deformation of bar AP and PP_2 represents the deformation of bar BP and are calculated as shown below.

$$\delta_A = \delta_P \cos 60 = 0.5\delta_P \tag{2}$$

$$\delta_B = \delta_P \sin 60 = 0.866\delta_P \tag{3}$$

From Eq 4.27, we have the following:

$$\delta_A = \frac{N_A L_A}{EA} = \frac{N_A(0.2)}{(200)(10^9)(100)(10^{-6})} = 0.5\delta_P \qquad \text{or} \qquad N_A = 50\delta_P(10^6) \tag{4}$$

$$\delta_B = \frac{N_B L_B}{EA} = \frac{N_B(0.25)}{(200)(10^9)(100)(10^{-6})} = 0.866\delta_P \qquad \text{or} \qquad N_B = 69.28\delta_P(10^6) \tag{5}$$

First Member Yields: As area of cross-sections and yield stress of both bars are same, we conclude from Eqs (4) and (5) that bar B will reach yield stress first. When B is just at yield stress:

$$N_B = (25{,}000)(100)(10^{-6}) = 25(10^3)\ \text{Newtons} \tag{6}$$

Let F_1 and δ_1 be the values of the force and deformation when B reaches yield stress. Substituting Eq. (6) into Eq. (4) we obtain: $69.28\delta_1(10^6) = 25{,}000$ or

$$\delta_1 = 0.3609(10^{-3})\ \text{m} \tag{7}$$

Substituting (7) into Eq. (4), we obtain:

$$N_A = (50)(0.3609)(10^{-3})(10^6) = 18.04(10^3)\ \text{Newtons} \tag{8}$$

Substituting Eq. (8) and (6) into Eq. (1) we obtain:

$$F_1 = (0.5)(18.04)(10^3) + (0.866)(25)(10^3) = 30.67(10^3)\ \text{Newtons} \tag{9}$$

Second Member Yields: When member A reaches yield stress, the structure will collapses. When A is at yield stress, the internal axial force in member A is:

$$N_A = (25{,}000)(100)(10^{-6}) = 25(10^3)\ \text{Newtons} \tag{10}$$

Let $F_{collapse}$ and $\delta_{collapse}$ be the values of the force and deformation when B reaches yield stress. Substituting Eq. (10) into Eq. (4) we obtain: $50\delta_{collapse}(10^6) = 25{,}000$ or

$$\delta_{collapse} = 0.50(10^{-3})\ \text{m} \tag{11}$$

Substituting Eq. (8) and (10) into Eq. (1) we obtain:

$$F_{collapse} = (0.5)(25)(10^3) + (0.866)(25)(10^3) = 34.15(10^3) \text{ Newtons} \quad \text{or} \quad \boxed{F_{collapse} = 34.15 \text{ kN}}$$

The load deflection curve is given below:

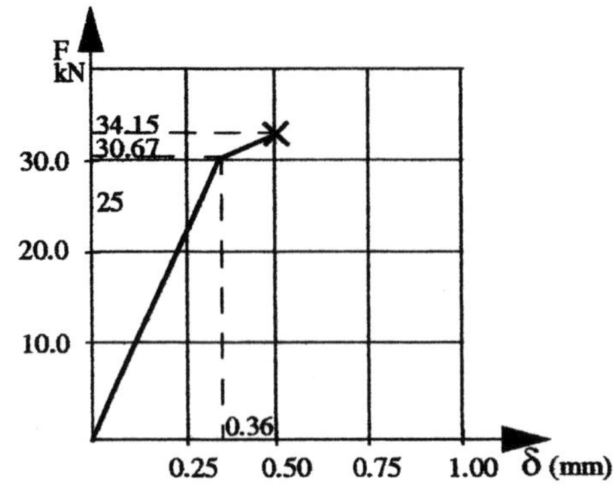

4.99 A force F is applied to the roller that slides inside a slot. Both bars have an area of cross-section of A = 100 mm^2, Modulus of Elasticity E = 200 GPa, and a yield stress of 250 MPa. Bar AP and BP have lengths of L_{AP}= 200 mm and L_{BP}= 250 mm respectively. Draw the load deflection curve and determine the collapse load.

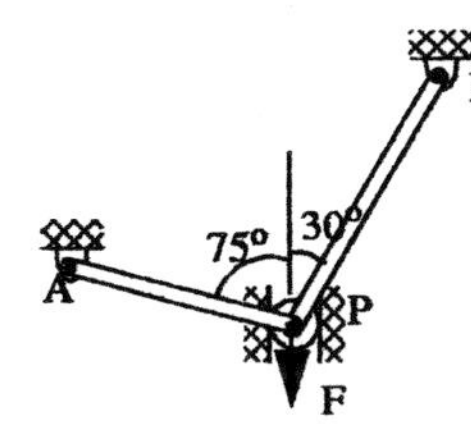

Fig. P4.99

Solution E = 200 GPa A = 100 mm^2 L_{AP}= 200 mm L_{BP}= 250 mm

σ_{yield} = 250 MPa F vs. δ_P = ? $F_{collapse}$=?

We draw the free body diagram of the rigid bar as shown in Fig. (a)

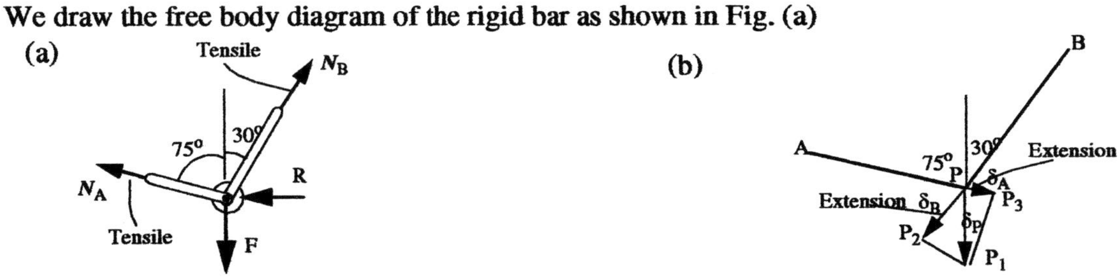

By force equilibrium in the y-direction in Fig.(a), we obtain the following.

$$N_A \cos 75 + N_B \cos 30 = F \quad \text{or} \quad 0.2588 N_A + 0.8660 N_B = F \tag{1}$$

We draw the exaggerated deformed shape as shown in Fig. (b). As per small strain approximation, we need component of PP_1 in the original direction of AP and BP, i.e., PP_1 represents the deformation of bar AP and PP_2 represents the deformation of bar BP and are calculated as shown below.

$$\delta_A = \delta_P \cos 75 = 0.2588 \delta_P \tag{2}$$

$$\delta_B = \delta_P \cos 30 = 0.8660 \delta_P \tag{3}$$

From Eq 4.27, we have the following:

$$\delta_A = \frac{N_A L_A}{EA} = \frac{N_A(0.2)}{(200)(10^9)(100)(10^{-6})} = 0.2588 \delta_P \quad \text{or} \quad N_A = 25.88 \delta_P (10^6) \tag{4}$$

$$\delta_B = \frac{N_B L_B}{EA} = \frac{N_B(0.25)}{(200)(10^9)(100)(10^{-6})} = 0.866 \delta_P \quad \text{or} \quad N_B = 69.28 \delta_P (10^6) \tag{5}$$

First Member Yields: As area of cross-sections and yield stress of both bars are same, we conclude from Eqs (4) and (5) that bar B will reach yield stress first. When B is just at yield stress:

$$N_B = (25,000)(100)(10^{-6}) = 25(10^3) \text{ Newtons} \qquad 6$$

Let F_1 and δ_1 be the values of the force and deformation when B reaches yield stress. Substituting Eq. (6) into Eq. (4) we obtain: $69.28\delta_1(10^6) = 25,000$ or

$$\delta_1 = 0.3609(10^{-3}) \text{ m} \qquad 7$$

Substituting (7) into Eq. (4), we obtain:

$$N_A = (25.88)(0.3609)(10^{-3})(10^6) = 9.34(10^3) \text{ Newtons} \qquad 8$$

Substituting Eq. (8) and (6) into Eq. (1) we obtain:

$$F_1 = (0.2588)(9.34)(10^3) + (0.866)(25)(10^3) = 24.07(10^3) \text{ Newtons} \qquad 9$$

Second Member Yields: When member A reaches yield stress, the structure will collapses. When A is at yield stress, the internal axial force in member A is:

$$N_A = (25,000)(100)(10^{-6}) = 25(10^3) \text{ Newtons} \qquad 10$$

Let $F_{collapse}$ and $\delta_{collapse}$ be the values of the force and deformation when B reaches yield stress. Substituting Eq. (10) into Eq. (4) we obtain: $(25.88)\delta_{collapse}(10^6) = 25,000$ or

$$\delta_{collapse} = (0.966)(10^{-3}) \text{ m} \qquad 11$$

Substituting Eq. (8) and (10) into Eq. (1) we obtain:

$$F_{collapse} = (0.2588)(25)(10^3) + (0.866)(25)(10^3) = 28.12(10^3) \text{ Newtons or } \boxed{F_{collapse} = 28.12 \text{ kN}}$$

The load deflection curve is given below:

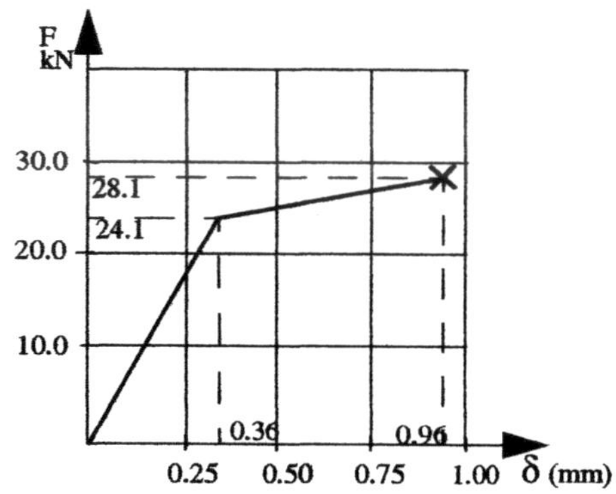

4.100 Three steel members shown have an area of cross-section of A = 100 mm^2, a modulus of elasticity E = 200 GPa and a yield stress 250 MPa. The three members are attached to a roller that can slide in the slot shown. Draw the load deflection curve and determine the collapse load.

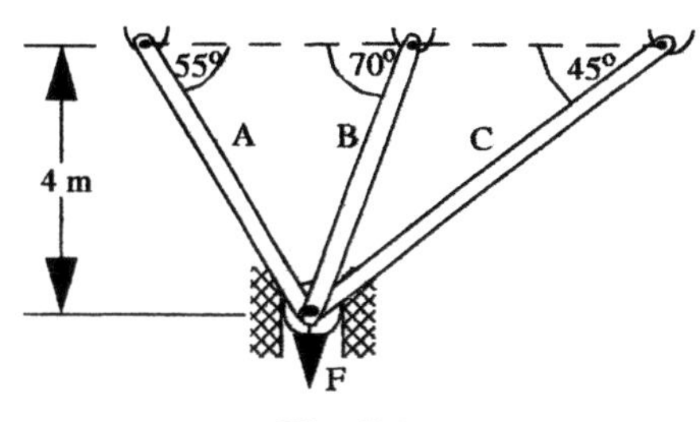

Fig. P4.100

Solution $E = 200$ GPa $\qquad A = 100 \text{ mm}^2$

$\sigma_{yield} = 250$ MPa $\qquad F$ vs. $\delta_F = ?$ $\qquad F_{collapse} = ?$

The length of each member can be found from geometry as shown below.

$$L_A = \frac{4}{\sin 55} = 4.8831\,\text{m} \qquad L_B = \frac{4}{\sin 70} = 4.257\,\text{m} \qquad L_C = \frac{4}{\sin 45} = 5.657\,\text{m}$$

We can draw the free body diagram of the roller as shown in Fig. (a).

(a)

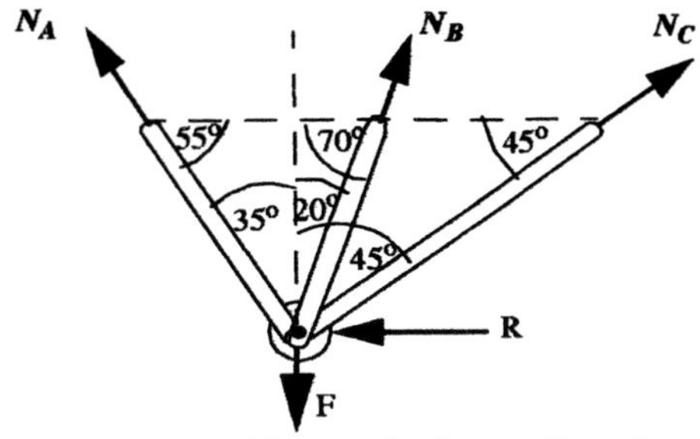

(b)

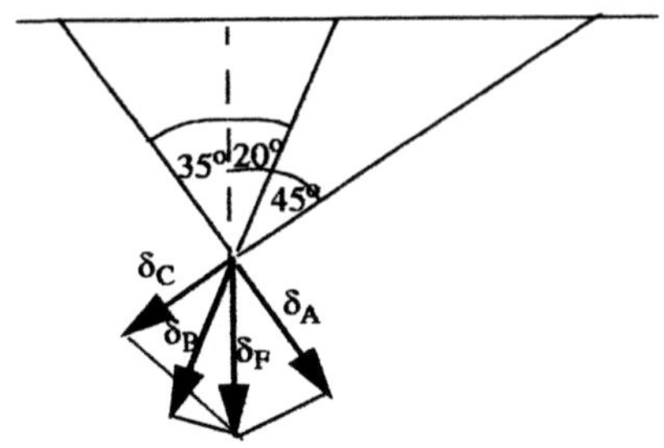

By equilibrium of forces in the y-direction in Fig (a) we obtain the following.

$$N_A\cos 35 + N_B\cos 20 + N_C\cos 45 - F = 0 \qquad \text{or} \qquad 0.8192N_A + 0.9397N_B + 0.7071N_C = F \tag{1}$$

As per small strain approximation, we need component of δ_F in the original directions of members A, B, and C as shown in Fig (b). The deformation equations can be written as.

$$\delta_A = \delta_F\cos 35 = 0.8192\delta_F \tag{2}$$

$$\delta_B = \delta_F\cos 20 = 0.9397\delta_F \tag{3}$$

$$\delta_C = \delta_F\cos 45 = 0.7071\delta_F \tag{4}$$

$$\delta_A = \frac{N_A L_A}{EA} = \frac{N_A(4.883)}{(200)(10^9)(100)(10^{-6})} = 0.8192\delta_F \qquad \text{or} \qquad N_A = 3.3551\delta_F(10^6) \tag{5}$$

$$\delta_B = \frac{N_B L_B}{EA} = \frac{N_B(4.257)}{(200)(10^9)(100)(10^{-6})} = 0.9397\delta_F \qquad \text{or} \qquad N_B = 4.4148\delta_F(10^6) \tag{6}$$

$$\delta_C = \frac{N_C L_C}{EA} = \frac{N_C(5.657)}{(200)(10^9)(100)(10^{-6})} = 0.7071\delta_F \qquad \text{or} \qquad N_C = 2.500\delta_F(10^6) \tag{7}$$

First Member Yields: As area of cross-sections and yield stress of all bars are same, we conclude from Eqs (4), (5), and (6) that bar B will reach yield stress first. When bar B is just at yield stress the internal axial force is:

$$N_B = (250)(10^6)(100)(10^{-6}) = 25000 \tag{8}$$

Let F_1 and δ_1 be the values of the force and deformation when bar B reaches yield stress. Substituting Eq. (8) into Eq. (6) we obtain: $4.4148\delta_F(10^6) = 25000$ or

$$\delta_1 = 5.663(10^{-3})\text{m} \tag{9}$$

Substituting (9) into Eq. (5) and (7) we obtain:

$$N_A = 3.3551(5.663)(10^{-3})(10^6) = 18999 \tag{10}$$

$$N_C = 2.500(5.663)(10^{-3})(10^6) = 14157 \tag{11}$$

Substituting Eqs. (8),(10) and (11) into Eq. (1) we obtain:

$$F_1 = (0.8192)(18999) + (0.9397)(25000) + (0.7071)(14157) = 49067\ \text{Newtons} \tag{12}$$

Second Member Yields: From Eqs. (5) and (7), we conclude that bar A will reach yield stress next. When bar A is at yield stress, the internal axial force is:

$$N_A = (250)(10^6)(100)(10^{-6}) = 25000 \tag{13}$$

Let F_2 and δ_2 be the values of the force and deformation when bar A reaches yield stress. Substituting Eq. (13) into Eq. (5) we obtain: $3.3551\delta_2(10^6) = 25000$ or

$$\delta_2 = 7.451(10^{-3})\text{m} \qquad \textbf{14}$$

Substituting (14) into Eq. (6) we obtain:

$$N_C = 2.500(7.451)(10^{-3})(10^6) = 18628 \qquad \textbf{15}$$

Substituting Eqs. (8),(13) and (15) into Eq. (7) we obtain:

$$F_2 = (0.8192)(25000) + (0.9397)(25000) + (0.7071)(18628) = 57144 \text{ Newtons} \qquad \textbf{16}$$

Third Member Yields: When bar C reaches yield the structure will collapse.When bar C is at yield stress, the internal axial force in bar C is:

$$N_B = (250)(10^6)(100)(10^{-6}) = 25000 \qquad \textbf{17}$$

Let $F_{collapse}$ and $\delta_{collapse}$ be the values of the force and deformation when bar C reaches yield stress. Substituting Eq. (17) into Eq. (7) we obtain: $2.500\delta_{collapse}(10^6) = 25000$ or

$$\delta_{collapse} = 10(10^{-3})\text{m} \qquad \textbf{18}$$

Substituting Eqs. (8),(13) and (18) into Eq. (7) we obtain:

$$F_{collapse} = (0.8192)(25000) + (0.9397)(25000) + (0.7071)(25000) = 61650 \text{ Newtons} \qquad \textbf{19}$$

or $\boxed{F_{collapse} = 61.65\text{kN}}$

The load deflection curve is given below.

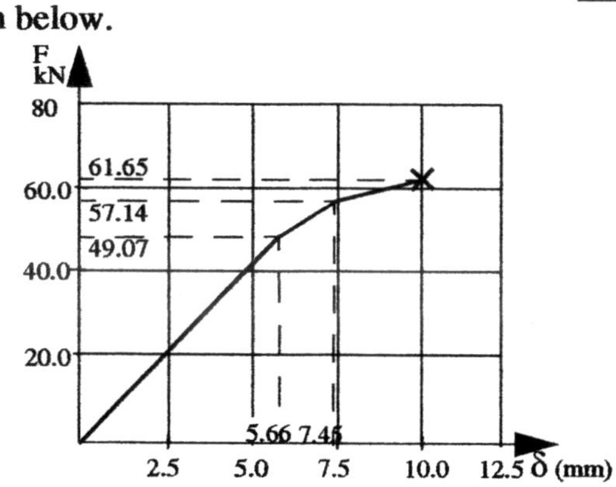

4.101 Three steel (E = 200 GPa, σ_{yield} = 200 MPa) bars A, B, and C have lengths of L_A=4 m, L_B=3 m and L_C= 2 m respectively. All bars have the same cross-sectional area of 500 mm^2. Draw the load deflection curve for the structure and determine the collapse load.

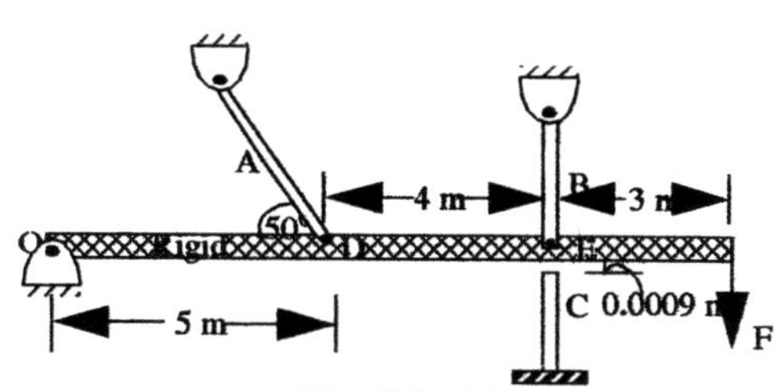

Fig. P4.101

Solution E = 200 GPa σ_{yield} = 200 MPa A = 500 mm^2 F vs. δ_P = ? L_A= 4 m $F_{collapse}$=? L_B= 3 m L_C= 2 m

---

We can draw the free body diagram of the rigid bar as shown in Fig. (a)
By moment equilibrium about point O we have the following.

$$N_A \sin 50(5) + N_B(9) + N_C(9) - F(12) = 0 \qquad 3.83N_A + 9N_B + 9N_C = 12F \qquad \textbf{1}$$

(a)

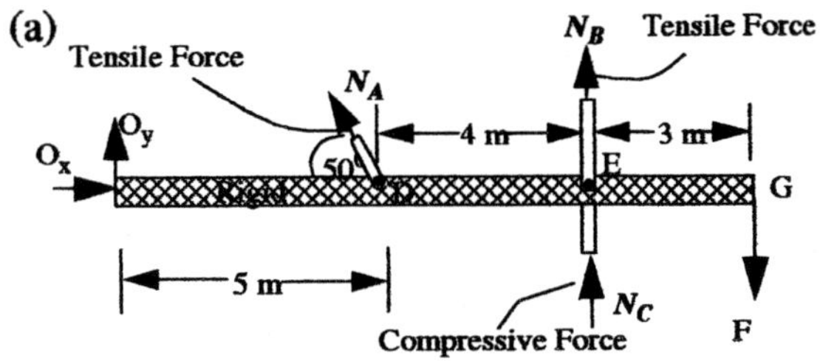

(b)

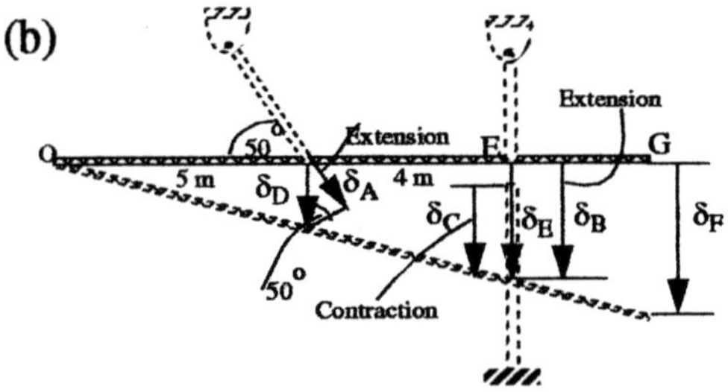

We consider points D, E, and G as points on the rigid bar, draw an exaggerated deformed geometry as shown in Fig. (b) and write the deformation equations as shown below.

$$\frac{\delta_D}{5} = \frac{\delta_E}{9} = \frac{\delta_F}{12} \text{ and } \delta_B = \delta_E \qquad \delta_C = \delta_E - 0.0009 \qquad \delta_A = \delta_D \sin 50$$

Writing the above deformation equations in term of δ_F, we obtain:

$$\delta_A = \left(\frac{5}{12}\right)\delta_F \sin 50 = 0.3192\delta_F \qquad \mathbf{2}$$

$$\delta_B = \left(\frac{9}{12}\right)\delta_F = 0.75\delta_F \qquad \mathbf{3}$$

$$\delta_C = \left(\frac{9}{12}\right)\delta_F - 0.0009 = 0.75\delta_F - 0.0009 \qquad \mathbf{4}$$

From Eq 4.27, we have the following:

$$\delta_A = \frac{N_A L_A}{E_A A_A} = \frac{N_A(4)}{(200)(10^9)(500)(10^{-6})} = 0.3192\delta_F \quad \text{or} \quad N_A = 7.980\delta_F(10^6) \qquad \mathbf{5}$$

$$\delta_B = \frac{N_B L_B}{E_B A_B} = \frac{N_B(3)}{(200)(10^9)(500)(10^{-6})} = 0.75\delta_F \quad \text{or} \quad N_B = 25\delta_F(10^6) \qquad \mathbf{6}$$

$$\delta_C = \frac{N_C L_C}{E_C A_C} = \frac{N_C(2)}{(200)(10^9)(500)(10^{-6})} = 0.75\delta_F - 0.0009 \quad \text{or} \quad N_C = (37.5\delta_F - 0.045)(10^6) \qquad \mathbf{7}$$

First Member Yields: As area of cross-sections and yield stress of all bars are same, we conclude from Eqs (5), (6), and (7) that bar C will reach yield stress first. When bar C is just at yield stress the internal axial force is:

$$N_C = (200)(10^6)(500)(10^{-6}) = 100,000 \qquad \mathbf{8}$$

Let F_1 and δ_1 be the values of the force and deformation when bar B reaches yield stress. Substituting Eq. (8) into Eq. (6) we obtain: $(37.5\delta_1 - 0.045)(10^6) = 100,000$ or

$$\delta_1 = 3.8667(10^{-3})\text{m} \qquad \mathbf{9}$$

Substituting (9) into Eq. (5) and (6) we obtain:

$$N_A = 7.980(3.8667)(10^{-3})(10^6) = 30856 \qquad \mathbf{10}$$

$$N_B = 25(3.8667)(10^{-3})(10^6) = 96667 \qquad \mathbf{11}$$

Substituting Eqs. (8),(10) and (11) into Eq. (1) we obtain:

$$F_1 = \frac{3.83(30856) + 9(96667) + 9(100,000)}{12} = 157.3(10^3) \text{ Newtons} \qquad \mathbf{12}$$

Second Member Yields: From Eqs. (5) and (7), we conclude that bar B will reach yield stress next. When bar B is at yield stress, the internal axial force is:

$$N_B = (200)(10^6)(500)(10^{-6}) = 100,000 \qquad \mathbf{13}$$

Let F_2 and δ_2 be the values of the force and deformation when bar A reaches yield stress. Substituting Eq. (13) into Eq. (5) we obtain: $25\delta_2(10^6) = 100,000$ or

$$\delta_2 = 4.0(10^{-3})\text{m} \qquad \textbf{14}$$

Substituting (14) into Eq. (5) we obtain:

$$N_A = 7.980(4)(10^{-3})(10^6) = 31920 \qquad \textbf{15}$$

Substituting Eqs. (8),(13) and (15) into Eq. (7) we obtain:

$$F_2 = \frac{3.83(31920) + 9(100,000) + 9(100,000)}{12} = 160.2(10^3) \text{ Newtons} \qquad \textbf{16}$$

Third Member Yields: When bar A reaches yield the structure will collapse.When bar A is at yield stress, the internal axial force in bar C is:

$$N_A = (200)(10^6)(500)(10^{-6}) = 100,000 \qquad \textbf{17}$$

Let $F_{collapse}$ and $\delta_{collapse}$ be the values of the force and deformation when bar C reaches yield stress. Substituting Eq. (17) into Eq. (7) we obtain: $7.980\delta_{collapse}(10^6) = 100,000$ or

$$\delta_{collapse} = 12.53(10^{-3})\text{m} \qquad \textbf{18}$$

Substituting Eqs. (8),(13) and (18) into Eq. (7) we obtain:

$$F_{collapse} = \frac{3.83(100,000) + 9(100,000) + 9(100,000)}{12} = 181.9(10^3) \text{ Newtons} \qquad \textbf{19}$$

or $\boxed{F_{collapse} = 181.9\text{kN}}$

The load deflection curve is given below.

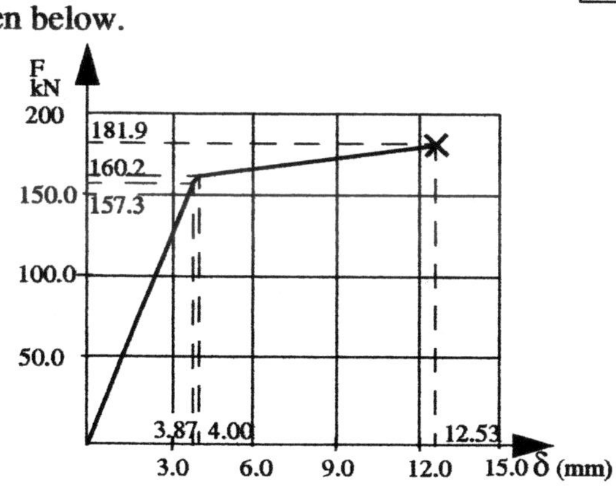

4.102 Draw the load deflection curve and determine the collapse load for the two bar linkage shown. Use the following data

	Bar A	Bar B
Modulus of Elasticity	200 GPa	170 GPa
Yield Stress	250 MPa	220 MPa
Area of cross-section	100 mm^2	75 mm^2
Length	1.5 m	1.5 m

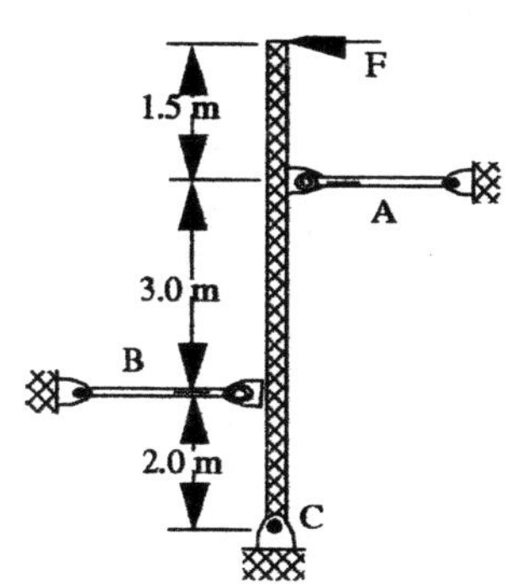

Fig. P4.102

Solution F vs. δ_F $F_{collapse}$ = ?

We can draw the free body diagram of the rigid bar as shown in Fig. (a).

(a)

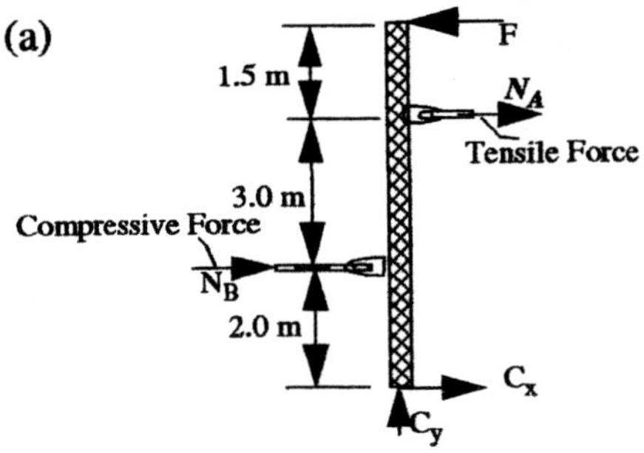

(b)

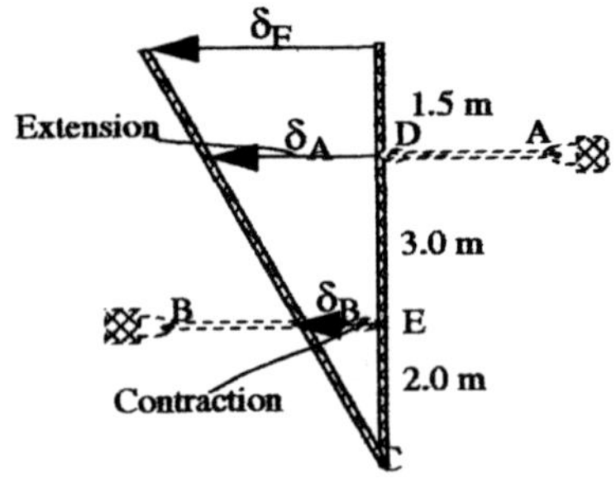

By moment equilibrium about point C we obtain the following.

$$5N_A + 2N_B = 6.5F \qquad (1)$$

We can draw the exaggerated deformed shape as shown in Fig. (b) and obtain the following deformation equation: $\frac{\delta_A}{5} = \frac{\delta_B}{2} = \frac{\delta_F}{6.5}$ or

$$\delta_A = \left(\frac{5}{6.5}\right)\delta_F \qquad (2)$$

$$\delta_B = \left(\frac{2}{6.5}\right)\delta_F \qquad (3)$$

From Eq 4.27, we have the following:

$$\delta_A = \frac{N_A L_A}{E_A A_A} = \frac{N_A(1.5)}{(200)(10^9)(100)(10^{-6})} = \left(\frac{5}{6.5}\right)\delta_F \quad \text{or} \quad N_A = 10.256\delta_F(10^6) \qquad (4)$$

$$\delta_B = \frac{N_B L_B}{E_B A_B} = \frac{N_B(1.5)}{(170)(10^9)(75)(10^{-6})} = \left(\frac{2}{6.5}\right)\delta_F \quad \text{or} \quad N_B = 2.615\delta_F(10^6) \qquad (5)$$

At yield stress, the value of internal forces in members A and B will be as shown below:

$$(N_A)_{yield} = (250)(10^6)(100)(10^{-6}) = 25,000\text{Newtons} \qquad (6)$$

$$(N_B)_{yield} = (220)(10^6)(75)(10^{-6}) = 16,500\text{Newtons} \qquad (7)$$

First Member Yields: Let F_1 and δ_1 be the values of the force and deformation when first member reaches yield stress. Substituting Eq. (6) into Eq. (4) we obtain: $10.256\delta_1(10^6) = 25,000$ or

$$\delta_1 = 2.438(10^{-3})\ \text{m} \qquad (8)$$

Substituting Eq. (8) into Eq. (5), we obtain:

$$N_B = 2.615(2.438)(10^{-3})(10^6) = 6374\ \text{Newtons} \qquad (9)$$

As the value of N_B in Eq. (9) is less that the value in Eq. (7), it implies that bar A yields first. Substituting Eq. (9) and (6) into Eq. (1) we obtain:

$$F_1 = \frac{5(25000) + 2(6374)}{6.5} = 21192\ \text{Newtons} \qquad (10)$$

Second Member Yields: When bar B reaches yield stress, the structure collapses. When bar B is at yield stress, the internal axial force in bar B is given by Eq. (7). Let $F_{collapse}$ and $\delta_{collapse}$ be the values of the force and deformation when bar B reaches yield stress. Substituting Eq. (7) into Eq. (5) we obtain:

$2.615\delta_{collapse}(10^6) = 16,500$ or

$$\delta_{collapse} = 6.310(10^{-3})\ \text{m} \qquad (11)$$

Substituting Eq. (6) and (7) into Eq. (1) we obtain:

$$F_{collapse} = \frac{5(25000)+2(16500)}{6.5} = 24308 \text{ Newtons or } \boxed{F_{collapse} = 24.3 \text{ kN}}$$

The load deflection curve is given below:

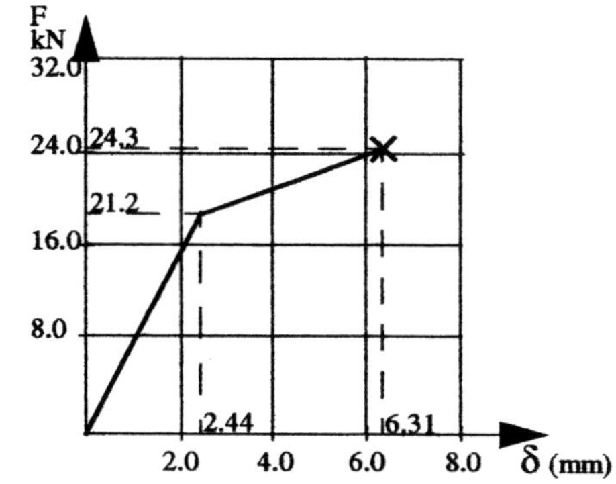

4.103 Three poles are pin connected to a ring at D and to the supports on the ground. The ring slides on a vertical rigid pole. The coordinates of the four points are as given. All poles have an area of cross-section of A =1 in^2. and a Modulus of Elasticity E = 10,000 ksi, and a yield stress of $\sigma_{yield} = 40$ ksi. Draw the load deflection curve and determine the collapse load.

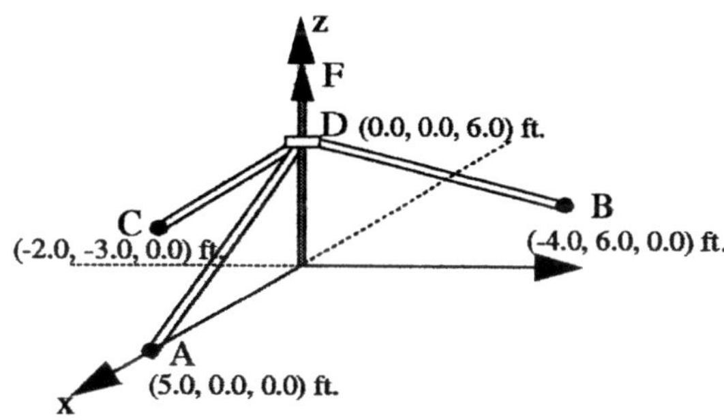

Fig. P4.103

Solution $A = 1 \text{ in}^2$ $E = 10,000$ ksi $\sigma_{yield} = 40$ ksi F vs. δ_F $F_{collapse} = ?$

The unit vectors in the directions of the rod can be found as shown below.

$$\bar{r}_{AD} = -5.0\bar{i} + 6\bar{k} \qquad |\bar{r}_{AD}| = \sqrt{5^2+6^2} = 7.81 \tag{1}$$

$$\bar{i}_{AD} = \frac{\bar{r}_{AD}}{|\bar{r}_{AD}|} = -0.6402\bar{i} + 0.7682\bar{k} \tag{2}$$

$$\bar{r}_{BD} = 4.0\bar{i} - 6\bar{j} + 6\bar{k} \qquad |\bar{r}_{BD}| = \sqrt{4^2+6^2+6^2} = 9.381 \tag{3}$$

$$\bar{i}_{BD} = \frac{\bar{r}_{BD}}{|\bar{r}_{BD}|} = 0.4264\bar{i} - 0.6396\bar{i} + 0.6396\bar{k} \tag{4}$$

$$\bar{r}_{CD} = 2.0\bar{i} + 3\bar{j} + 6\bar{k} \qquad |\bar{r}_{BD}| = \sqrt{2^2+3^2+6^2} = 7.0\text{ft} \tag{5}$$

$$\bar{i}_{CD} = \frac{\bar{r}_{CD}}{|\bar{r}_{CD}|} = 0.2857\bar{i} + 0.4286\bar{i} + 0.8571\bar{k} \tag{6}$$

We can draw the free body diagram the collar as shown below.

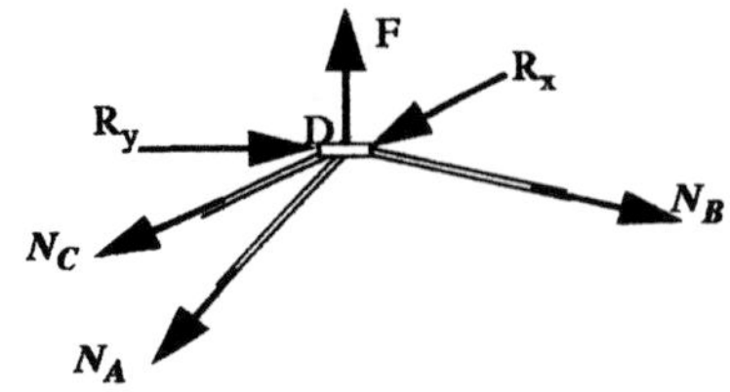

By equilibrium of forces in the z-direction (using the coefficient of the unit vectors in the k-direction)

$$0.7682N_A + 0.6396N_B + 0.8571N_C = F \qquad \mathbf{7}$$

Let δ_F be the movement of the collar. As points A, B, and C are fixed to the ground, the deformation vector for all three rods is the same and equal to $\vec{D} = \delta_F\hat{k}$. Using small strain approximation, the deformation of each rod can be found as shown below.

$$\delta_{AD} = \hat{i}_{AD} \bullet \vec{D} = 0.7682\delta_F \qquad \mathbf{8}$$

$$\delta_{BD} = \hat{i}_{BD} \bullet \vec{D} = 0.6396\delta_F \qquad \mathbf{9}$$

$$\delta_{CD} = \hat{i}_{CD} \bullet \vec{D} = 0.8571\delta_F \qquad \mathbf{10}$$

The lengths of each rod are the magnitude of the position vectors in Eqs. (2), (4), and (6). From Eq. (4-27)

$$\delta_{AD} = \frac{N_A L_{AD}}{EA} = \frac{N_A(7.81)}{(10,000)(1)} = 0.7682\delta_F \quad \text{or} \quad N_A = 0.9836\delta_F(10^3) \qquad \mathbf{11}$$

$$\delta_{BD} = \frac{N_B L_{BD}}{EA} = \frac{N_B(9.381)}{(10,000)(1)} = 0.6396\delta_F \quad \text{or} \quad N_B = 0.6818\delta_F(10^3) \qquad \mathbf{12}$$

$$\delta_{CD} = \frac{N_C L_{CD}}{EA} = \frac{N_C(7)}{(10,000)(1)} = 0.8571\delta_F \quad \text{or} \quad N_C = 1.2244\delta_F(10^3) \qquad \mathbf{13}$$

First Member Yields: As area of cross-sections and yield stress of all bars are same, we conclude from Eqs (11), (12), and (13) that bar CD will reach yield stress first. When CD is just at yield stress:

$$N_C = (40)(1) = 40 \text{ kips} \qquad \mathbf{14}$$

Let F_1 and δ_1 be the values of the force and deformation when CD reaches yield stress. Substituting Eq. (13) into Eq. (14) we obtain: $1.2244\delta_1(10^3) = 40$ or

$$\delta_1 = 0.0327 \text{ inch} \qquad \mathbf{15}$$

Substituting (15) into Eq. (11) and (12) we obtain:

$$N_A = (0.9836)(0.0327)(10^3) = 32.133\text{kips} \qquad \mathbf{16}$$

$$N_B = (0.6818)(0.0327)(10^3) = 21.92\text{kips} \qquad \mathbf{17}$$

Substituting Eqs. (14),(16) and (17) into Eq. (7) we obtain:

$$F_1 = (0.7682)(32.133) + (0.6396)(21.92) + (0.8571)(40) = 72.99\text{kips} \qquad \mathbf{18}$$

Second Member Yields: From Eqs. (11) and (12), we conclude that member AD will reach yield stress next. When A is at yield stress, the internal axial force in member AD is:

$$N_A = (40)(1) = 40 \text{ kips} \qquad \mathbf{19}$$

Let F_2 and δ_2 be the values of the force and deformation when AD reaches yield stress. Substituting Eq. (19) into Eq. (11) we obtain: $0.9836\delta_2(10^3) = 40$ or

$$\delta_2 = 0.0407 \text{ inch} \qquad \mathbf{20}$$

Substituting (20) into Eq. (12) we obtain:

$$N_B = (0.6818)(0.0407)(10^3) = 27.73\text{kips} \quad \textbf{21}$$

Substituting Eqs. (14),(19) and (22) into Eq. (7) we obtain:

$$F_2 = (0.7682)(40) + (0.6396)(27.73) + (0.8571)(40) = 82.75\text{kips} \quad \textbf{22}$$

Third Member Yields: When member BD reaches yield the structure will collapse.When BD is at yield stress, the internal axial force in member B is:

$$N_B = (40)(1) = 40 \text{ kips} \quad \textbf{23}$$

Let $F_{collapse}$ and $\delta_{collapse}$ be the values of the force and deformation when member BD reaches yield stress.

Substituting Eq. (12) into Eq. (24) we obtain: $0.6818\delta_{collapse}(10^3) = 40$ kips or

$$\delta_{collapse} = 0.0587\text{inch} \quad \textbf{24}$$

Substituting Eqs. (14),(19) and (24) into Eq. (7) we obtain:

$$F_{collapse} = (0.7682)(40) + (0.6396)(40) + (0.8571)(40) = 90.60\text{kips} \quad \textbf{25}$$

or $\boxed{F_{collapse} = 90.6\text{kips}}$

The load deflection curve is given below:

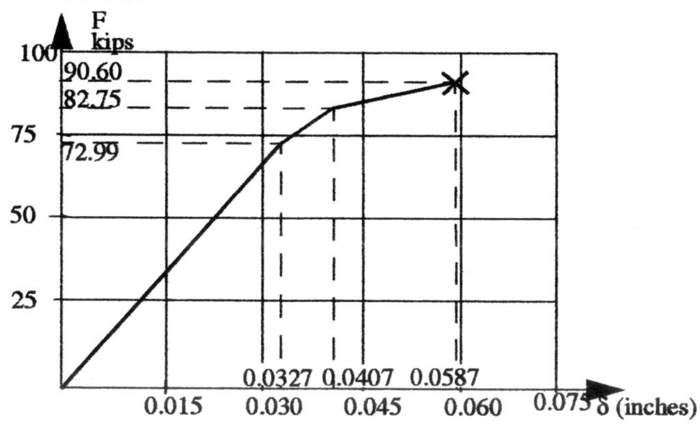

4.104 Fifty rivets of 10 mm diameter are used to attach caps at each end on a 1000 mm diameter cylinder. The wall of the cylinder is 10 mm thick and the gas pressure is 200 kPa. Determine the hoop stress and the axial stress in the cylinder and the shear stress in each rivet.

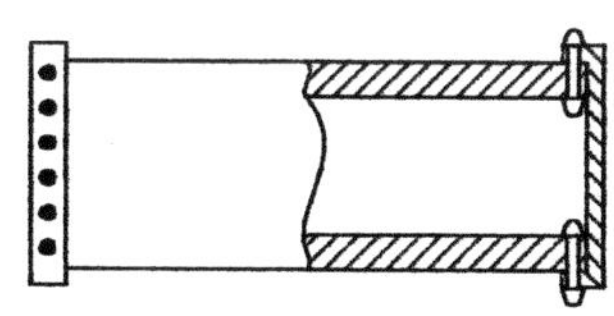

Fig. P4.104

Solution $n = 50$ $d_r = 10$ mm $d = 1000$ mm $t = 10$ mm $p = 200$kPa $\sigma_{qq} = ?$ $\tau_r = ?$

The hoop stress can be found as shown below:

$$\sigma_{\theta\theta} = \frac{pr}{t} = \frac{(200)(10^3)(0.500)}{(0.01)} = 10(10^6)\ \text{N/m}^2 \qquad \boxed{\sigma_{\theta\theta} = 10\text{MPa(T)}}$$

The shear force on each rivet is $V_r = \tau_r A_r$, where Ar is the area of cross-section of each rivet. The free body diagram of the cap can be drawn as shown below.

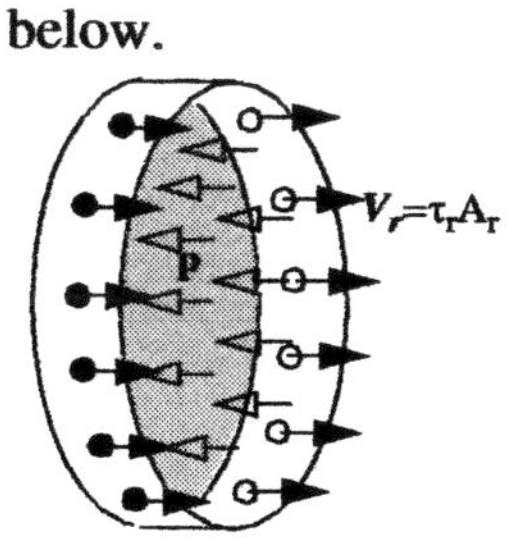

By force equilibrium we obtain:

$p(\pi r^2) = (50)(\tau_r A_r)$ or $\tau_r = \dfrac{(200)(10^3)\pi(0.5)^2}{(50)\pi(0.01/2)^2} = 40(10^6)\ \text{N/m}^2$ or $\boxed{\tau_r = 40\text{MPa}}$

4.105 A pressure tank 15 feet long and 40 inch diameter is to be fabricated from a 1/2 inch thick sheet. A 15 feet long, 8 inch wide and 1/2 inch thick plate is bonded onto the tank to seal the gap. What is the shear stress in the adhesive when the pressure in the tank is 500 psi? Assume uniform shear stress over the entire inner surface of the attaching plate.

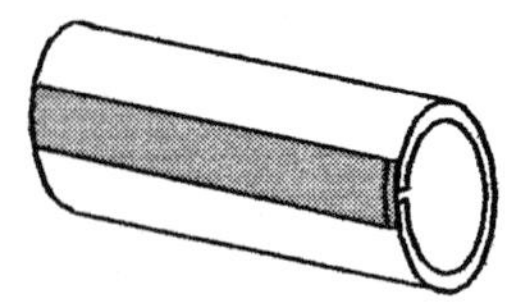

Fig. P4.105

Solution $L = 15$ ft. $d = 40$ inch $t = 0.5$ inch $W = 8$ inch $p = 500$ psi $\tau_a = ?$

The hoop stress can be found as $\sigma_{\theta\theta} = \dfrac{pr}{t} = \dfrac{(500)(20)}{(0.5)} = 20000\text{psi}$

We make two imaginary cuts, one through the adhesive and another one through the tank on the other side to produce the free body diagram shown below.

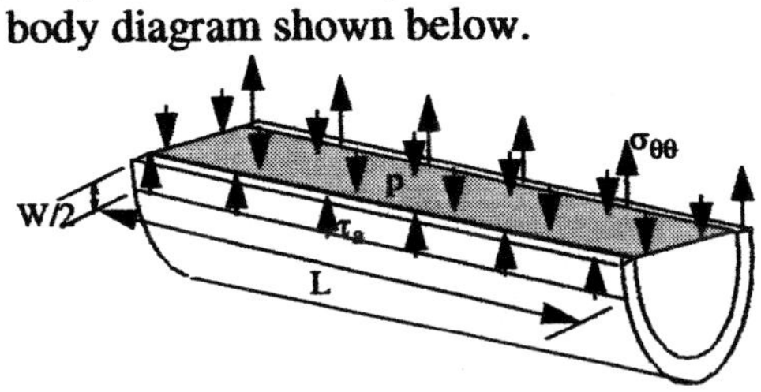

By force equilibrium we obtain: $\tau_a(W/2)(L) + \sigma_{\theta\theta}(L)(t) - (p)(L)(d) = 0$

$\tau_a = \dfrac{(p)(d) - \sigma_{\theta\theta}(t)}{(W/2)} = \dfrac{(500)(40) - (20000)(0.5)}{(8/2)} = 2500\text{psi}$ $\boxed{\tau_a = 2.5\text{ksi}}$

4.106 A 5 feet diameter spherical tank has a wall thickness of 3/4 inch. If the maximum normal stress is not to exceed 10 ksi, determine the maximum permissible pressure.

Solution $d = 5$ ft. $t = 0.75$ inch $\sigma_{max} \le 10$ ksi $p_{max} = ?$

The maximum pressure can be found as shown below.

$\sigma = \dfrac{pr}{t} = \dfrac{p(5)(12)}{0.75} \le 10$ or $p \le 0.125\text{ksi}$ or $\boxed{p_{max} = 125\text{psi}}$

4.107 In a spherical tank of radius 500 mm and a thickness of 40 mm, a hole of 50 mm diameter is drilled and then plugged using adhesive of shear strength 1.2 MPa to form a safety pressure release mechanism.Determine the maximum allowable pressure and the corresponding hoop stress in the tank material.

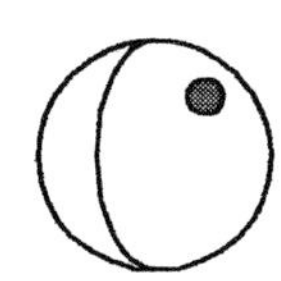

Fig. P4.107

Solution $r = 500$ mm $t = 40$ mm $d_{plug} = 50$ mm $\tau_a \le 1.2\text{MPa}$ $p_{max} = ?$ $\sigma = ?$

We can draw the free body diagram of the plug as shown below.

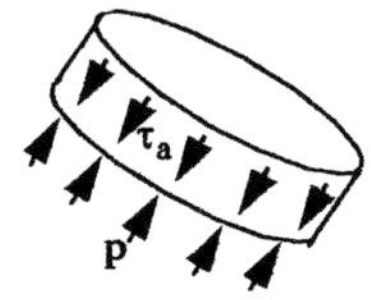

By force equilibrium we obtain $p(\pi d_{plug}^2/4) = (\tau_a)(\pi d_{plug})(t)$ or

$$(\tau_a) = \frac{pd_{plug}}{4t} = \frac{p(0.05)}{(4)(0.04)} \le 1.2(10^6) \quad \text{or} \quad p \le 3.84(10^6) \quad \text{or} \quad \boxed{p_{max} = 3.8\text{MPa}}$$

The hoop stress can be found as shown below.

$$\sigma = \frac{pr}{t} = \frac{3.84(10^6)(0.5)}{(0.04)} = 4.8(10^6)\ \text{N/m}^2 \qquad \boxed{\sigma = 4.8\text{MPa(T)}}$$

4.108 A 20 inch diameter pressure cooker is to be designed for a 15 psi pressure. The allowable normal stress in the cylindrical pressure cooker is to be limited to 3 ksi. Determine the minimum wall thickness of the pressure cooker. A 1/2 lbs. weight on top of the nozzle is used to control the pressure in the cooker. Determine the diameter 'd' of the nozzle.

Fig. P4.108

Solution $d = 20$ inch $p = 15$ psi $\sigma_{max} \le 3$ ksi $t_{min} = ?$ $W = 0.5$ lb $d_{noz} = ?$

The hoop stress in the cylinder is greater than the axial stress and is used to determine the minimum thickness as shown below.

$$\sigma_{\theta\theta} = \frac{pr}{t} = \frac{(15)(10)}{t} \le 3(10^3) \quad \text{or} \quad t \ge 0.05\ \text{in} \quad \text{or} \quad \boxed{t_{min} = 0.05\text{in}}$$

We can draw the free body diagram of the weight on the nozzle as shown below.

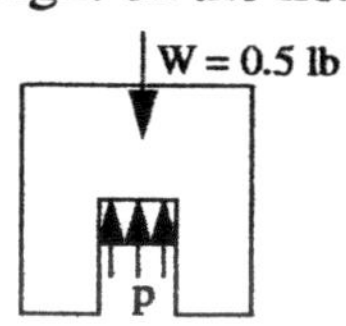

By force equilibrium we obtain: $p\pi(d_{noz}^2/4) = 0.5$ or

$$d_{noz} = \sqrt{\frac{2}{\pi p}} = \sqrt{\frac{2}{\pi(15)}} = 0.206 \text{ or} \qquad \boxed{d_{noz} = 0.206\ \text{in}}$$

4.109 The cylindrical gas tank shown is made from a sheet metal that is 8 mm thick and must be designed to sustain a maximum normal stress of 100 MPa. Develop a table of maximum permissible gas pressure and the corresponding diameter of the tank in steps of 100mm between diameter values of 400 mm to 900 mm.

Fig. P4.109

Solution $t = 8$ mm $\sigma_{max} \le 100\text{MPa}$ p vs. d

The hoop stress in the cylinder is greater than the axial stress and is used to determine the values of p and d. Assuming d is measured in millimeters and p in MPa, we can write the hoop stress expression as shown below:

$$\sigma_{\theta\theta} = \frac{pr}{t} = \frac{(p)(10^6)(d/2)(10^{-3})}{0.008} \le 100(10^6) \quad \text{or}$$

$$pd \le 1600$$

1

Eq. 1 can be used to determine the values of p and d as shown in the table below. All pressure values are rounded downwards,

d (mm	p (MPa)
400	4.00
500	3.20
600	2.66
700	2.28
800	2.00
900	1.77

4.110 A pressure tank 15 feet long and 40 inch diameter is to be fabricated from a 1/2 inch thick sheet. A 15 feet long, 8 inch wide and 1/2 inch thick plate is to be used for sealing the gap by using two rows of 90 rivets each. If the shear strength of the rivets is 36 ksi and the normal stress in the tank is to be limited to 20 ksi, determine the maximum pressure and the minimum diameter of the rivets that can be used.

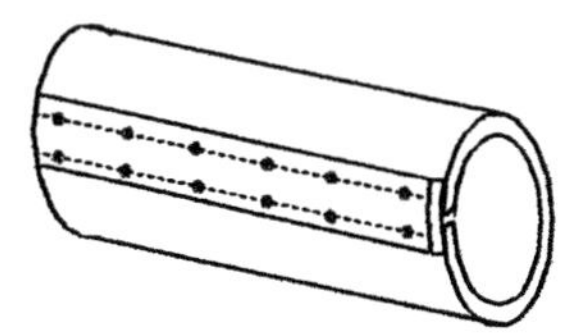

Fig. P4.110

Solution $L = 15$ ft. $d = 40$ inch $t = 0.5$ inch $W_{plate} = 8$ inch $t_{plate} = 0.5$ inch

$n_{rivets}=90$ $\sigma_{max} \le 20$ ksi $\tau_{riv} \le 36$ ksi $p_{max} = ?$ $(d_{riv})_{min} = ?$

The hoop stress in the cylinder is greater than the axial stress and is used to determine the maximum pressure as shown below.

$$\sigma_{\theta\theta} = \frac{pr}{t} = \frac{(p)(20)}{0.5} \le 20(10^3) \quad \text{or} \quad p \le 500\text{psi} \quad \text{or} \quad \boxed{p_{max} = 500\text{psi}}$$

We make imaginary cuts through the rivets and also cut through the tank on the other side to produce the free body diagram shown below.

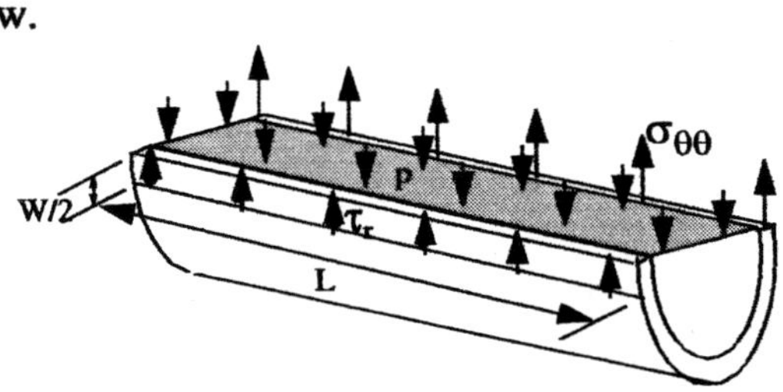

By force equilibrium we obtain: $n_{riv}\tau_r(\pi d_{riv}^2)/4 + \sigma_{\theta\theta}(L)(t) - (p)(L)(d) = 0$ or

$$\tau_{riv} = \frac{(p)(L)(d) - (\sigma_{\theta\theta})(L)(t)}{n_{riv}\pi(d_{riv}^2/4)} = \frac{(500)(15)(12)(40) - (20000)(15)(12)(0.5)}{(90)\pi(d_{riv}^2/4)} \le (36)(10^3) \text{ or}$$

$$d_{riv} \ge \sqrt{\frac{1800(10^3)}{810\pi(10^3)}} \text{ or } d_{riv} \ge 0.841\text{inch or} \quad \boxed{(d_{riv})_{min} = 0.85\text{in}}$$

Problem Set 5.1

5.1 A pair of 48 inch long bars and a pair of 60 inch long bars are symmetrically attached to a rigid disc at a radius of 2 inch at one end and built into the wall at the other end. The shear strain at point A due to a twist of the rigid disc was found to be 3000 μ rads. Determine the magnitude shear strain at point D.

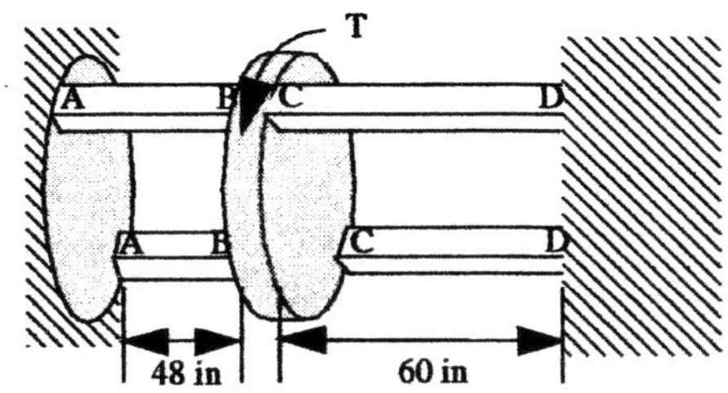

Fig. P5.1

Solution $\gamma_A = 3000\ \mu$ r= 2 inch. $\gamma_D = ?$

We can draw an exaggerated deformed geometry as shown below.

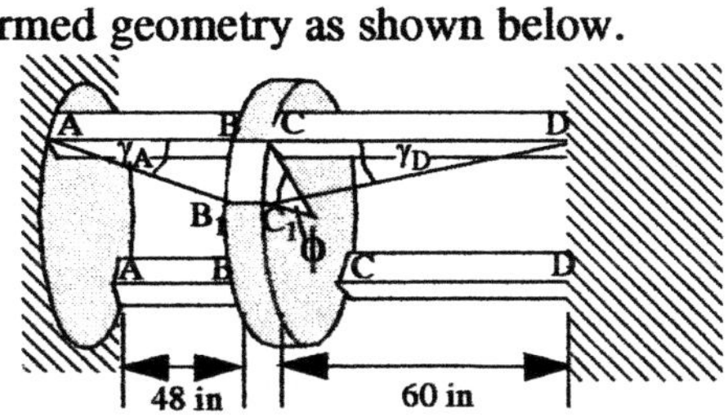

From triangle ABB_1 and small strain approximation, we have the following:

$$\tan\gamma_A = \frac{BB_1}{AB} \quad \text{or} \quad \gamma_A = \frac{BB_1}{48} = 3000(10^{-6}) \quad \text{or} \quad BB_1 = 0.144 \text{ in} \qquad 1$$

Note that $BB_1 = CC_1$. From triangle DCC_1 we have the following:

$$\tan\gamma_D = \frac{CC_1}{CD} = \frac{BB_1}{CD} \quad \text{or} \quad \gamma_D = \frac{0.144}{60} = 2.400(10^{-3}) \text{ or} \qquad \boxed{\gamma_D = 2400\ \mu\text{rad}}$$

5.2 If the four bars in problem 5.1 are made from material that has a Shear Modulus of 12,000 ksi., determine the applied torque T on the rigid disc.The area of cross-section of all bars is 0.25 in^2.

Solution G = 12,000 ksi A = 0.25in^2 T = ?

From problem 5.1 $\gamma_A = 3.000(10^{-3})$ and $\gamma_D = 2.400(10^{-3})$.

The shear stresses in bars AB and CD can be found from Hooke's Law as shown below.

$$\tau_{AB} = G\gamma_A = (12,000)(3.000)(10^{-3}) = 36 \text{ ksi} \qquad 1$$

$$\tau_{CD} = G\gamma_D = (12,000)(2.400)(10^{-3}) = 28.8 \text{ ksi} \qquad 2$$

The internal shear forces in the bars can be found as shown below:

$$V_{AB} = \tau_{AB}A = (36)(0.25) = 9.0 \text{ kips} \qquad 3$$

$$V_{CD} = \tau_{CD}A = (28.8)(0.25) = 7.2 \text{ kips} \qquad 4$$

We can make imaginary cuts through the bars and draw the free body diagram of the rigid disc as shown below.

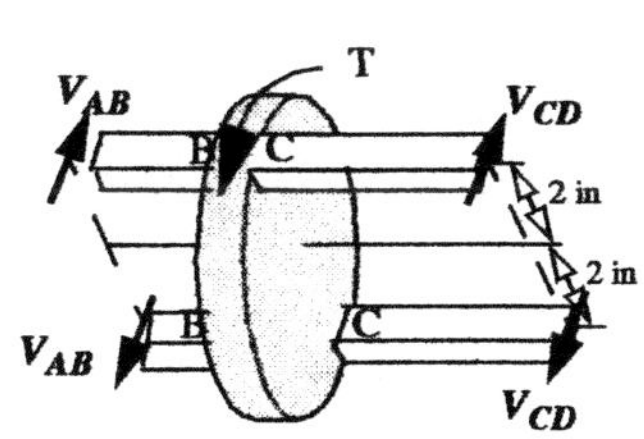

By equilibrium of moment about the axis of the rigid disc, we obtain:

$$T = (2)(V_{AB})(2) + (2)(V_{CD})(2) = (4)(V_{AB}) + (4)(V_{CD}) \quad \text{or} \qquad 5$$

$T = (4)(9) + (4)(7.2) = 64.8$ or $\boxed{T = 64.8 \text{ in} - \text{kips}}$

5.3 If the bars AB in problem 5.1 are made of aluminum with a Shear Modulus of $G_{al} = 4{,}000$ ksi and bars CD are made from bronze with a Shear Modulus of G_{br}=6,500 ksi, determine the applied torque T on the rigid disc.The area of cross-section of all bars is 0.25 in^2

Solution G_{AB}= 4,000 ksi G_{CD}= 6,500 ksi $A = 0.25\text{in}^2$ T = ?

From problem 5.1 $\gamma_A = 3.000(10^{-3})$ and $\gamma_D = 2.400(10^{-3})$.

The shear stresses in bars AB and CD can be found from Hooke's Law as shown below.

$$\tau_{AB} = G_{AB}\gamma_A = (4,000)(3.000)(10^{-3}) = 12.0 \text{ ksi} \qquad 1$$

$$\tau_{CD} = G_{CD}\gamma_D = (6,500)(2.400)(10^{-3}) = 15.6 \text{ ksi} \qquad 2$$

The internal shear forces in the bars can be found as shown below:

$$V_{AB} = \tau_{AB}A = (12.0)(0.25) = 3.0 \text{ kips} \qquad 3$$

$$V_{CD} = \tau_{CD}A = (15.6)(0.25) = 3.9 \text{ kips} \qquad 4$$

From Eq. (5) in problem 5.2, we have:

$T = (4)(V_{AB}) + (4)(V_{CD}) = (4)(3) + (4)(3.9) = 27.6$ or $\boxed{T = 27.6 \text{ in} - \text{kips}}$

5.4 Three pairs of bars are symmetrically attached to rigid discs at the radii shown. The discs were observed to rotate by angles $\phi_1 = 1.5°$, $\phi_2 = 3.0°$, and $\phi_3 = 2.5°$ in the direction of the applied torques T_1, T_2, and T_3 respectively. The shear modulus of the bars is 40 ksi and the area of cross-section is 0.04 in^2. Determine the shear strains in bars AB, CD, and EF.

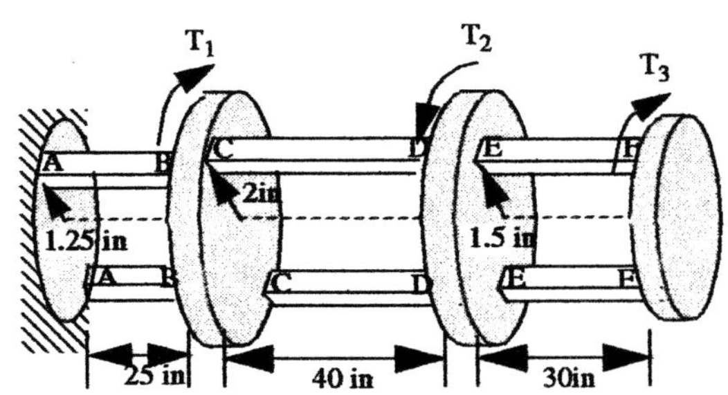

Fig. P5.4

Solution $\phi_1 = 1.5°$ $\phi_2 = 3.0°$ $\phi_3 = 2.5°$

G = 40 ksi $A = 0.04 \text{ in}^2$ $T_1 = ?$ $T_2 = ?$ $T_3 = ?$

Converting the given angles from degrees to radians we obtain:

$$\phi_1 = \left(\frac{1.5°}{180}\right)\pi = 0.0262\text{rads} \qquad \phi_2 = \left(\frac{3.0°}{180}\right)\pi = 0.0524\text{rads} \qquad \phi_2 = \left(\frac{2.5°}{180}\right)\pi = 0.0436\text{rads}$$

We can draw an exaggerated deformed geometry as shown below.

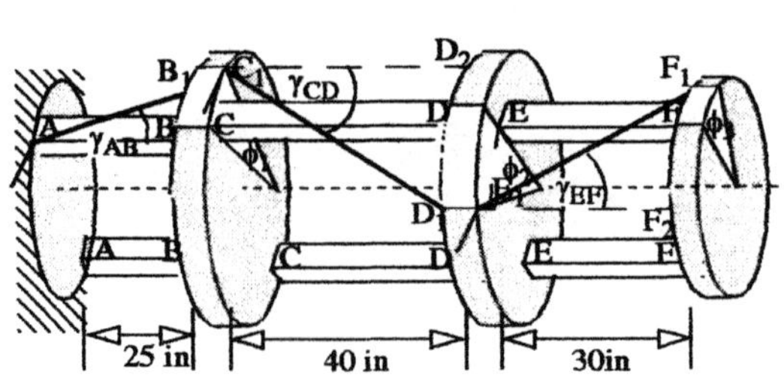

The magnitude of shear strains can be found using small strain approximation as shown below.

$$\tan|\gamma_{AB}| = \frac{BB_1}{AB} \quad \text{or} \quad |\gamma_{AB}| = \frac{1.25\phi_1}{25} = \frac{1.25(0.0262)}{25} = 1.31(10^{-3})$$

$$\tan|\gamma_{CD}| = \frac{D_2D_1}{C_1D} \quad \text{or} \quad |\gamma_{CD}| = \frac{D_2D + DD_1}{CD} = \frac{2\phi_1 + 2\phi_2}{CD} = \frac{2(0.0262 + 0.0524)}{40} = 3.93(10^{-3})$$

$$\tan|\gamma_{EF}| = \frac{F_2F_1}{E_1F_2} \quad \text{or} \quad |\gamma_{EF}| = \frac{F_2F + FF_1}{EF} = \frac{1.5\phi_2 + 1.5\phi_3}{EF} = \frac{2(0.0524 + 0.0436)}{30} = 4.8(10^{-3})$$

We note that the angle between the axial direction and the tangent direction, increases for AB and EF and decreases for CD. Using the sign convention that positive shear strain result in decrease in angle from right angle, we write the shear strains with the following signs.

$$\gamma_{AB} = -1.31(10^{-3}) \qquad \gamma_{CD} = 3.93(10^{-3}) \qquad \gamma_{EF} = -4.8(10^{-3}) \qquad (1)$$

The shear stress in each bar can be found using Hooke's law.

$$\tau_{AB} = G\gamma_{AB} = (40)(-1.31)(10^{-3}) = -52.4(10^{-3})\text{ksi} = -52.4 \text{ psi} \qquad (2)$$

$$\tau_{CD} = G\gamma_{CD} = (40)(3.93)(10^{-3}) = 157.2(10^{-3})\text{ksi} = 157.2 \text{ psi} \qquad (3)$$

$$\tau_{EF} = G\gamma_{EF} = (40)(-4.80)(10^{-3}) = -192.0(10^{-3})\text{ksi} = -192.0 \text{ psi} \qquad (4)$$

The internal shear forces in the bars can be found as shown below:

$$V_{AB} = \tau_{AB}A = (-52.4)(0.04) = -2.096 \text{ lbs} \qquad (5)$$

$$V_{CD} = \tau_{CD}A = (157.2)(0.04) = 6.288 \text{ lbs} \qquad (6)$$

$$V_{EF} = \tau_{EF}A = (-192.0)(0.04) = -7.680 \text{ lbs} \qquad (7)$$

We can make imaginary cuts through the bars and draw free body diagrams as shown. The direction of V_{EF} is determined by inspection to oppose the torque T_3. The direction of V_{CD} is opposite to V_{EF} and direction of V_{AB} is in the same direction as V_{EF} as per the signs obtained above.

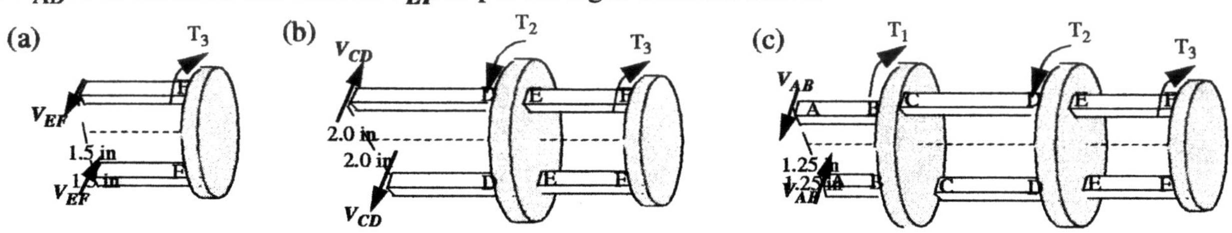

From moment equilibrium in Fig. (a), we obtain the following.

$$T_3 = (2)(V_{AB})(1.5) = (3)(7.628) = 23.04 \text{ or} \qquad \boxed{T_3 = 23.0 \text{ in} - \text{lbs}}$$

From moment equilibrium in Fig. (b), we obtain the following.

$$T_2 - T_3 - (2)(V_{CD})(2.0) = 0 \quad \text{or} \quad T_2 = 23.04 + (4)(6.288) = 48.192 \text{ or} \qquad \boxed{T_2 = 48.2 \text{ in} - \text{lbs}}$$

From moment equilibrium in Fig. (c), we obtain the following.

$$T_1 - T_2 + T_3 + (2)(V_{AB})(1.25) = 0 \text{ or } T_1 = 48.2 - 23.04 + (2.5)(2.096) = 30.4 \text{ or} \boxed{T_1 = 30.4 \text{ in} - \text{lbs}}$$

5.5 A circular shaft of radius r and length Δx has to rigid discs attached at each end. If the rigid discs are rotated as shown. Determine the shear strain γ at point A in terms of r, Δx, and $\Delta\phi$ assuming line AB remains straight, where. $\Delta\phi = \phi_2 - \phi_1$.

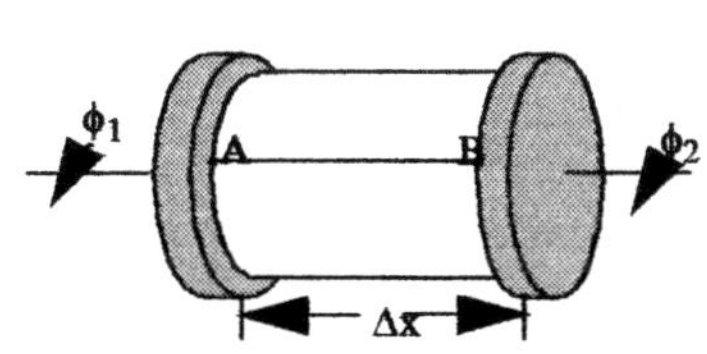

Fig. P5.5

Solution $\gamma = f(r, \Delta x, \Delta\phi) = ?$

We can draw an exaggerated deformed geometry as shown below.

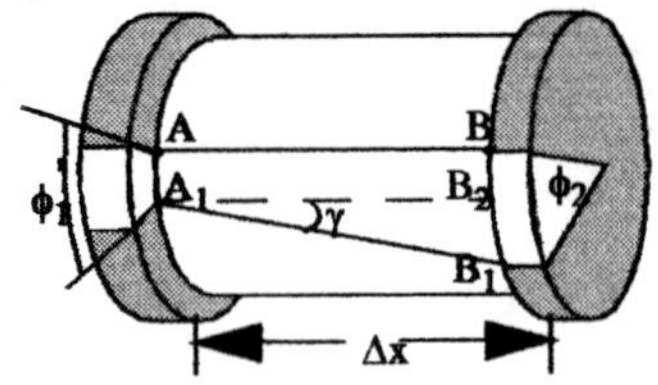

From triangle AB_2B_1 and small strain approximation, we have the following:

$$\tan\gamma = \frac{B_2B_1}{A_1B_2} \quad \text{or} \quad \gamma = \frac{BB_1 - BB_2}{AB} = \frac{BB_1 - AA_1}{AB} = \frac{r\phi_2 - r\phi_1}{\Delta x} \quad \text{or} \quad \boxed{\gamma = \frac{r\Delta\phi}{\Delta x}}$$

5.6 A hollow circular shaft made from hard rubber has an outer diameter of 4 inch and an inner diameter of 1.5 inch. The shaft is fixed to the wall on the left end and the rigid disc on the right hand is twisted as shown. The shear strain at point A which is on the outside surface was found to be 4000 μ rads. Determine the shear strain at point C which is on the inside surface and the angle of rotation. Assume lines AB and CD remain straight during deformation.

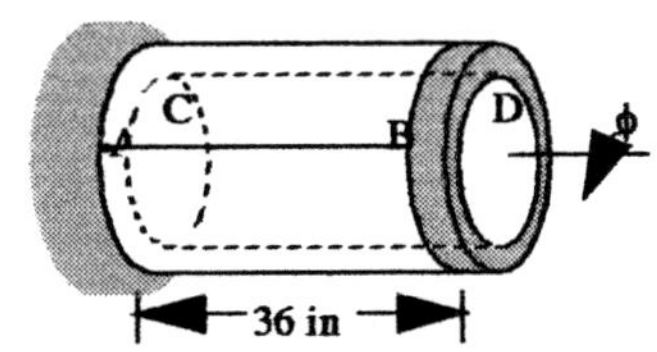

Fig. P5.6

Solution $d_o = 4$ inch $d_i = 1.5$ inch $\gamma_A = 4000\ \mu$ rads $\gamma_C = ?$

We can draw the exaggerated deformed geometry as shown below.

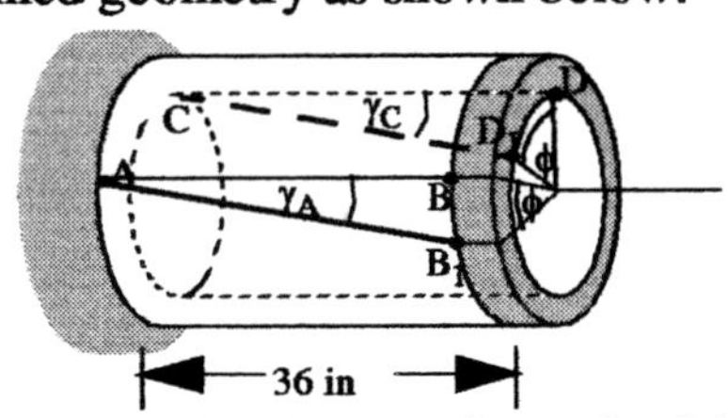

From triangle ABB_1 and small strain approximation, we have the following:

$$\tan\gamma_A \approx \gamma_A = \frac{BB_1}{AB} = \frac{(d_o/2)\phi}{AB} \quad \text{or} \quad 4000(10^{-6}) = \frac{(4/2)\phi}{36} \quad \text{or} \quad \phi = 0.072 \text{ rads}$$

From triangle CDD_1 and small strain approximation, we have the following:

$$\tan\gamma_C \approx \gamma_C = \frac{DD_1}{CD} = \frac{(d_i/2)\phi}{AB} = \frac{(1.5/2)(0.072)}{36} = 1.500(10^{-3}) \quad \text{or} \quad \boxed{\gamma_C = 1500\ \mu \text{ rad}}$$

5.7 The magnitude of shear strain in each of the segment of the stepped shaft shown were found to be $\gamma_{AB} = 3000\ \mu\text{rad}$, $\gamma_{CD} = 2500\ \mu\text{rad}$, and $\gamma_{EF} = 6000\ \mu\text{rad}$. The radius of section AB is 150 mm, of section CD is 70 mm, and section EF is 60 mm. Determine the angle by which each of the rigid discs was rotated.

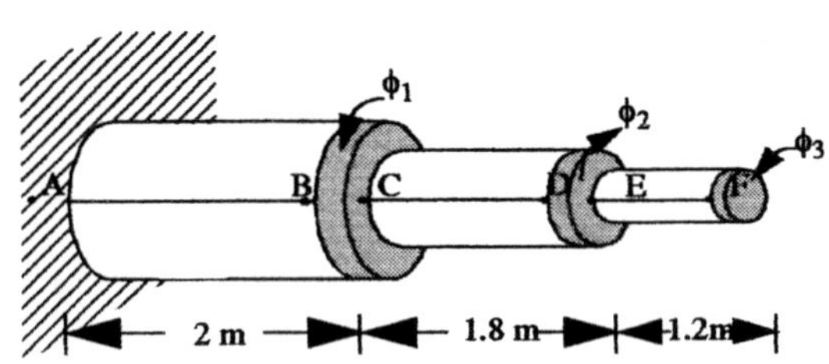

Fig. P5.7

Solution

$\gamma_{AB} = 3000\ \mu\text{rads}$ $\gamma_{CD} = 2500\ \mu\text{rads}$ $\gamma_{EF} = 6000\ \mu\text{rads}$

$r_{AB} = 150$ mm $r_{CD} = 70$ mm $r_{EF} = 60$ mm

$\phi_1 = ?$ $\phi_2 = ?$ $\phi_3 = ?$

We can draw the exaggerated deformed geometry as shown below.

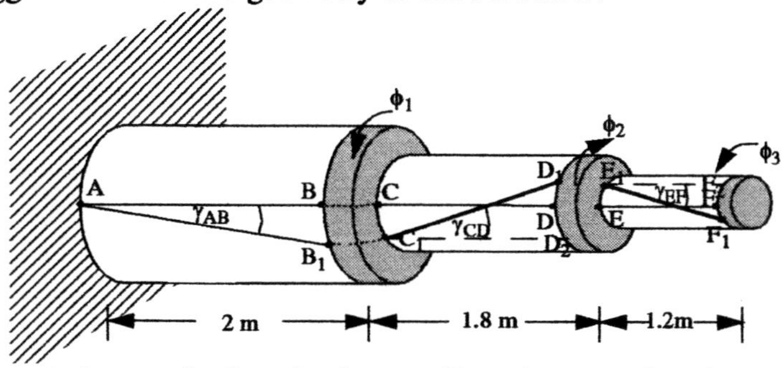

The magnitude of shear strains can be found using small strain approximation as shown below.

$$\tan\gamma_{AB} \approx \gamma_{AB} = \frac{BB_1}{AB} = \frac{r_{AB}\phi_1}{AB} = \frac{(0.150)\phi_1}{2} = 3000(10^{-6}) \text{ or } \quad \boxed{\phi_1 = 0.0400 \text{ rad}}$$

$$\tan\gamma_{CD} \approx \gamma_{CD} = \frac{D_2D_1}{C_1D_2} = \frac{D_2D + DD_1}{CD} = \frac{r_{CD}\phi_1 + r_{CD}\phi_2}{CD} = \frac{0.07(0.04 + \phi_2)}{1.8} = 2500(10^{-6})$$

or $\boxed{\phi_2 = 0.0243 \text{ rad}}$

$$\tan\gamma_{CD} \approx \gamma_{CD} = \frac{F_2F_1}{E_1F_2} = \frac{F_2F + FF_1}{EF} = \frac{r_{EF}\phi_2 + r_{EF}\phi_3}{EF} = \frac{0.06(0.0243 + \phi_3)}{1.2} = 6000(10^{-6})$$

or $\boxed{\phi_3 = 0.0957 \text{ rad}}$

5.8 A hollow brass (G_B = 6500 ksi) shaft and a solid steel (G_S = 13,000 ksi) shaft are securely fastened to form a composite shaft as shown. The shear strain $\gamma_{x\theta}$ in polar coordinates at the section and the dimensions of the cross-section are as given. Determine the equivalent internal torque acting at the cross-section.

$\gamma_{x\theta} = 0.001\rho$ where ρ is in inches. Use d_B= 4 in and d_S = 2 in.

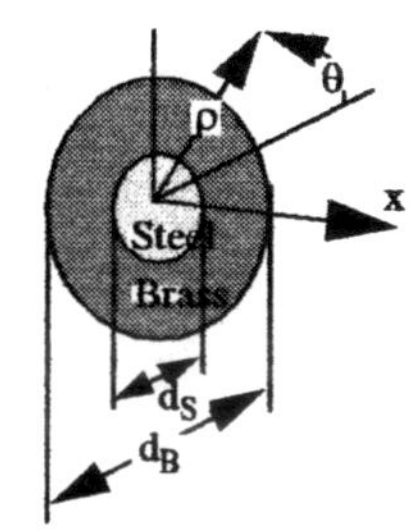

Fig. P5.8

Solution $\quad G_B = 6500$ ksi $\quad G_S = 13{,}000$ ksi)

$\gamma_{x\theta} = 0.001\rho \quad d_B$= 4 in $\quad d_S$ = 2 in. $\quad$ ***T*** = ?

From Hooke's law we can write:

$$(\tau_{x\theta})_{steel} = G_S\gamma_{x\theta} = (13,000)(0.001\rho) = 13\rho \text{ ksi} \qquad 1$$

$$(\tau_{x\theta})_{brass} = G_B\gamma_{x\theta} = (6,500)(0.001\rho) = 6.5\rho \text{ ksi} \qquad 2$$

The shear stress across the cross-section can be written as shown below.

$$\tau_{x\theta} = \begin{cases} 13\rho \text{ ksi} & 0.0 \le \rho < 1.0 \\ 6.5\rho \text{ ksi} & 1.0 < \rho \le 2.0 \end{cases} \qquad 3$$

The differential area dA is the area of a ring of radius ρ and thickness dρ i.e., $dA = (2\pi\rho)d\rho$. Equation 5.1 can be written as a sum of integrals over steel and brass and evaluated as shown.

$$T = \int_0^2 \rho\tau_{x\theta}(2\pi\rho)d\rho = \int_0^1 \rho(13\rho)(2\pi\rho)d\rho + \int_1^2 \rho(6.5\rho)(2\pi\rho)d\rho = (26\pi)\frac{\rho^4}{4}\bigg|_0^1 + (13\pi)\frac{\rho^4}{4}\bigg|_1^2$$

or $T = \left(\frac{26\pi}{4}\right) + \left(\frac{13\pi}{4}\right)(2^4 - 1^4) = 173.57 \text{ in-kips}$ or $\boxed{T = 173.6 \text{ in-kips}}$

5.9 A hollow brass ($G_B = 6500$ ksi) shaft and a solid steel ($G_S = 13{,}000$ ksi) shaft are securely fastened to form a composite shaft as shown. The shear strain $\gamma_{x\theta}$ in polar coordinates at the section and the dimensions of the cross-section are as given. Determine the equivalent internal torque acting at the cross-section.

$\gamma_{x\theta} = -0.0005\rho$ where ρ is in inches. Use $d_B = 6$ in and $d_S = 4$ in.

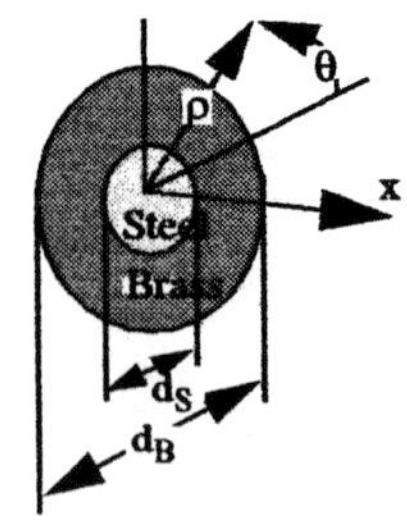

Fig. P5.9

Solution $G_B = 6500$ ksi $G_S = 13{,}000$ ksi) $\gamma_{x\theta} = -0.0005\rho$ $d_B = 6$ in $d_S = 4$ in.

From Hooke's law we can write:

$$(\tau_{x\theta})_{brass} = G_B\gamma_{x\theta} = (6,500)(-0.0005\rho) = -3.25\rho \text{ ksi} \qquad 1$$

$$(\tau_{x\theta})_{steel} = G_S\gamma_{x\theta} = (13,000)(-0.0005\rho) = -6.50\rho \text{ ksi} \qquad 2$$

The shear stress across the cross-section can be written as shown below.

$$\tau_{x\theta} = \begin{cases} -3.25\rho \text{ ksi} & 0.0 \le \rho < 2.0 \\ -6.50\rho \text{ ksi} & 2.0 < \rho \le 3.0 \end{cases} \qquad 3$$

The differential area dA is the area of a ring of radius ρ and thickness $d\rho$ i.e., $dA = (2\pi\rho)d\rho$. Equation 5.1 can be written as a sum of integrals over steel and brass and evaluated as shown.

$$T = \int_0^3 \rho\tau_{x\theta}(2\pi\rho)d\rho = \int_0^2 \rho(-6.50\rho)(2\pi\rho)d\rho + \int_2^3 \rho(-3.25\rho)(2\pi\rho)d\rho = (-13\pi)\frac{\rho^4}{4}\bigg|_0^2 + (-6.5\pi)\frac{\rho^4}{4}\bigg|_2^3 \text{ or}$$

$T = \left(\frac{-13\pi}{4}\right)(2^4 - 0) + \left(\frac{-6.5\pi}{4}\right)(3^4 - 2^4) = -495.19 \text{ in-kips}$ or $\boxed{T = -495.2 \text{ in-kips}}$

5.10 A hollow brass ($G_B = 6500$ ksi) shaft and a solid steel ($G_S = 13{,}000$ ksi) shaft are securely fastened to form a composite shaft as shown. The shear strain $\gamma_{x\theta}$ in polar coordinates at the section and the dimensions of the cross-section are as given. Determine the equivalent internal torque acting at the cross-section.

$\gamma_{x\theta} = 0.002\rho$ where ρ is in inches. Use $d_B = 3$ in and $d_S = 1$ in.

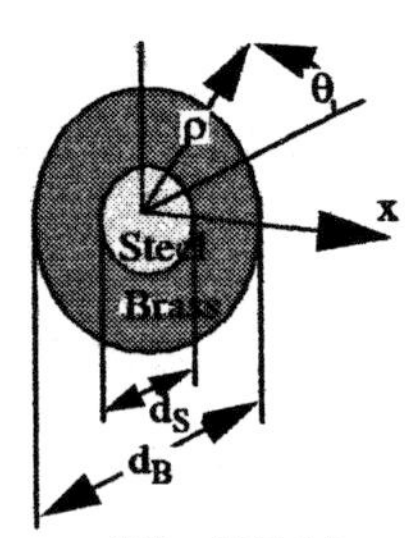

Fig. P5.10

Solution $G_B = 6500$ ksi $G_S = 13{,}000$ ksi) $\gamma_{x\theta} = 0.002\rho$ $d_B = 3$ in $d_S = 1$ in. $T = ?$

From Hooke's law we can write:

$$(\tau_{x\theta})_{steel} = G_S\gamma_{x\theta} = (13,000)(0.002\rho) = 26\rho \text{ ksi} \qquad 1$$

$$(\tau_{x\theta})_{brass} = G_B\gamma_{x\theta} = (6,500)(0.002\rho) = 13\rho \text{ ksi} \qquad 2$$

The shear stress across the cross-section can be written as shown below.

$$\tau_{x\theta} = \begin{cases} 26\rho \text{ ksi} & 0.0 \le \rho < 0.5 \\ 13\rho \text{ ksi} & 0.5 < \rho \le 1.5 \end{cases} \qquad 3$$

The differential area dA is the area of a ring of radius ρ and thickness dρ i.e., $dA = (2\pi\rho)d\rho$. Equation 5.1 can be written as a sum of integrals over steel and brass and evaluated as shown.

$$T = \int_0^{1.5} \rho\tau_{x\theta}(2\pi\rho)d\rho = \int_0^{0.5} \rho(26\rho)(2\pi\rho)d\rho + \int_{0.5}^{1.5} \rho(13\rho)(2\pi\rho)d\rho = (52\pi)\frac{\rho^4}{4}\bigg|_0^{0.5} + (26\pi)\frac{\rho^4}{4}\bigg|_{0.5}^{1.5} \text{ or}$$

$$T = \left(\frac{52\pi}{4}\right)(0.5^4 - 0) + \left(\frac{26\pi}{4}\right)(1.5^4 - 0.5^4) = 104.65 \text{ in-kips or}$$

$$\boxed{T = 104.65 \text{ in-kips}}$$

5.11 A hollow titanium (G_{Ti} = 36 GPa) shaft and a hollow Aluminum (G_{Al} = 26 GPa) shaft are securely fastened to form a composite shaft as shown. The shear strain $\gamma_{x\theta}$ in polar coordinates at the section is $\gamma_{x\theta} = 0.05\rho$ where ρ is in meters. Determine the equivalent internal torque acting at the cross-section. Use d_i= 40 mm, d_{Al}= 80 mm and d_{Ti} = 120 mm.

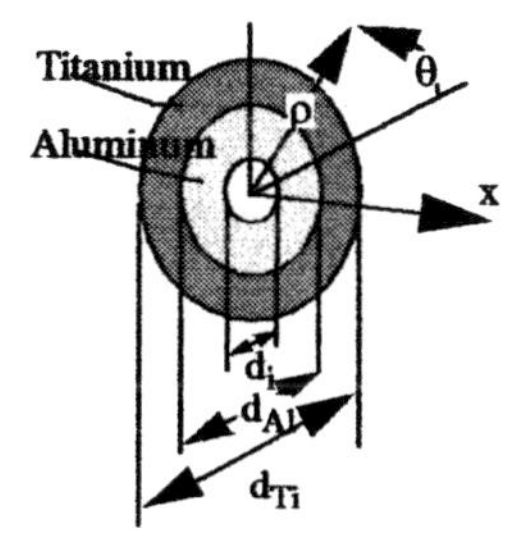

Solution G_{Ti} = 36 GPa, G_{Al} = 26 GPa, $\gamma_{x\theta} = 0.05\rho$, d_i= 40 mm, d_{Al}= 80 mm, d_{Ti} = 120 mm, T = ?

From Hooke's law we can write:

$$(\tau_{x\theta})_{Al} = G_{Al}\gamma_{x\theta} = (26)(10^9)(0.05\rho) = 1.3\rho(10^9) \text{ N/m}^2 \qquad 1$$

$$(\tau_{x\theta})_{Ti} = G_{Ti}\gamma_{x\theta} = (36)(10^9)(0.05\rho) = 1.8\rho(10^9) \text{ N/m}^2 \qquad 2$$

The shear stress across the cross-section can be written as shown below.

$$\tau_{x\theta} = \begin{cases} 1.3\rho(10^9)(\text{ N/m}^2) & 0.02 \le \rho < 0.04 \\ 1.8\rho(10^9) \text{ N/m}^2 & 0.04 < \rho \le 0.06 \end{cases} \qquad 3$$

The differential area dA is the area of a ring of radius ρ and thickness dρ i.e., $dA = (2\pi\rho)d\rho$. Equation 5.1 can be written as a sum of integrals over steel and brass and evaluated as shown.

$$T = \int_{0.02}^{0.06} \rho\tau_{x\theta}(2\pi\rho)d\rho = \int_{0.02}^{0.04} \rho(1.3\rho)(10^9)(2\pi\rho)d\rho + \int_{0.04}^{0.06} \rho(1.8\rho)(10^9)(2\pi\rho)d\rho \text{ or}$$

$$T = \left[(2.6\pi)\frac{\rho^4}{4}\bigg|_{0.02}^{0.04} + (3.6\pi)\frac{\rho^4}{4}\bigg|_{0.04}^{0.06}\right](10^9) = \left[\left(\frac{2.6\pi}{4}\right)(0.04^4 - 0.02^4) + \left(\frac{3.6\pi}{4}\right)(0.06^4 - 0.04^4)\right](10^9) \text{ or}$$

$$T = 34.306(10^3) \text{ N-m or}$$

$$\boxed{T = 34.3 \text{ kN-m}}$$

5.12 A hollow titanium (G_{Ti} = 36 GPa) shaft and a hollow Aluminum (G_{Al} = 26 GPa) shaft are securely fastened to form a composite shaft as shown. The shear strain $\gamma_{x\theta}$ in polar coordinates at the section is $\gamma_{x\theta} = 0.04\rho$ where ρ is in meters. Determine the equivalent internal torque acting at the cross-section Use d_i= 50 mm, d_{Al}= 90 mm and d_{Ti} = 100 mm.

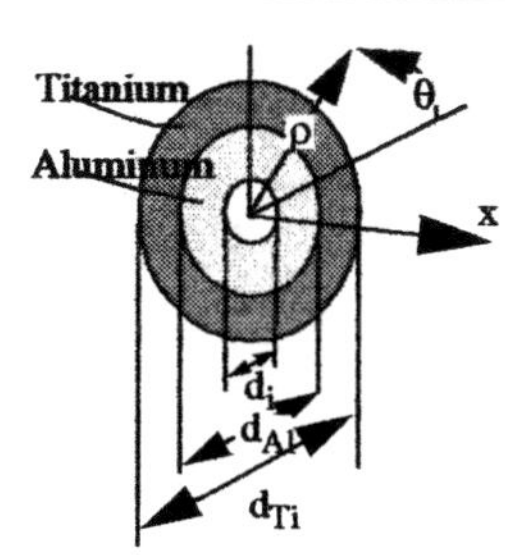

Solution $G_{Ti} = 36$ GPa $G_{Al} = 26$ GPa $\gamma_{x\theta} = 0.04\rho$
$d_i = 50$ mm $d_{Al} = 90$ mm $d_{Ti} = 100$ mm $T = ?$

From Hooke's law we can write:

$$(\tau_{x\theta})_{Al} = G_{Al}\gamma_{x\theta} = (26)(10^9)(0.04\rho) = 1.04\rho(10^9)\ \text{N/m}^2 \quad (1)$$

$$(\tau_{x\theta})_{Ti} = G_{Ti}\gamma_{x\theta} = (36)(10^9)(0.04\rho) = 1.44\rho(10^9)\ \text{N/m}^2 \quad (2)$$

The shear stress across the cross-section can be written as shown below.

$$\tau_{x\theta} = \begin{cases} 1.04\rho(10^9)(\text{ N/m}^2) & 0.025 \le \rho < 0.045 \\ 1.44\rho(10^9)\ \text{N/m}^2 & 0.045 < \rho \le 0.050 \end{cases} \quad (3)$$

The differential area dA is the area of a ring of radius ρ and thickness $d\rho$ i.e., $dA = (2\pi\rho)d\rho$. Equation 5.1 can be written as a sum of integrals over steel and brass and evaluated as shown.

$$T = \int_{0.025}^{0.050} \rho\tau_{x\theta}(2\pi\rho)d\rho = \int_{0.025}^{0.045} \rho(1.04\rho)(10^9)(2\pi\rho)d\rho + \int_{0.045}^{0.05} \rho(1.44\rho)(10^9)(2\pi\rho)d\rho \text{ or}$$

$$T = \left[(2.08\pi)\frac{\rho^4}{4}\Bigg|_{0.025}^{0.045} + (2.88\pi)\frac{\rho^4}{4}\Bigg|_{0.045}^{0.050}\right](10^9) \text{ or}$$

$$T = \left[\left(\frac{2.08\pi}{4}\right)(0.045^4 - 0.025^4) + \left(\frac{2.88\pi}{4}\right)(0.05^4 - 0.045^4)\right](10^9) = 10.92(10^3)\ \text{N-m}$$

or $\boxed{T = 10.9\ \text{kN-m}}$

5.13 A hollow titanium ($G_{Ti} = 36$ GPa) shaft and a hollow Aluminum ($G_{Al} = 26$ GPa) shaft are securely fastened to form a composite shaft as shown. The shear strain $\gamma_{x\theta}$ in polar coordinates at the section is $\gamma_{x\theta} = -0.06\rho$ where ρ is in meters. Determine the equivalent internal torque acting at the cross-section. Use $d_i = 30$ mm, $d_{Al} = 40$ mm and $d_{Ti} = 50$ mm.

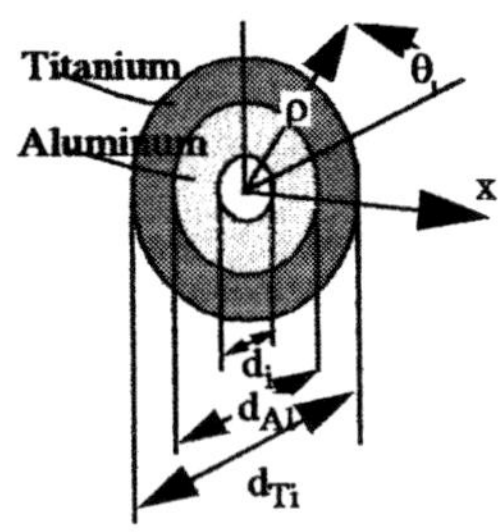

Solution $G_{Ti} = 36$ GPa $G_{Al} = 26$ GPa
$\gamma_{x\theta} = -0.06\rho$ $d_i = 30$ mm $d_{Al} = 40$ mm $d_{Ti} = 50$ mm $T = ?$

From Hooke's law we can write:

$$(\tau_{x\theta})_{Al} = G_{Al}\gamma_{x\theta} = (26)(10^9)(-0.06\rho) = -1.56\rho(10^9)\ \text{N/m}^2 \quad (1)$$

$$(\tau_{x\theta})_{Ti} = G_{Ti}\gamma_{x\theta} = (36)(10^9)(-0.06\rho) = -2.16\rho(10^9)\ \text{N/m}^2 \quad (2)$$

The shear stress across the cross-section can be written as shown below.

$$\tau_{x\theta} = \begin{cases} -1.56\rho(10^9)(\text{ N/m}^2) & 0.015 \le \rho < 0.020 \\ -2.16\rho(10^9)\ \text{N/m}^2 & 0.020 < \rho \le 0.025 \end{cases} \quad (3)$$

The differential area dA is the area of a ring of radius ρ and thickness $d\rho$ i.e., $dA = (2\pi\rho)d\rho$. Equation 5.1 can be written as a sum of integrals over steel and brass and evaluated as shown.

$$T = \int_{0.015}^{0.025} \rho\tau_{x\theta}(2\pi\rho)d\rho = \int_{0.015}^{0.020} \rho(-1.56\rho)(10^9)(2\pi\rho)d\rho + \int_{0.020}^{0.025} \rho(-2.16\rho)(10^9)(2\pi\rho)d\rho \text{ or}$$

$$T = \left[(-3.12\pi)\frac{\rho^4}{4}\Big|_{0.015}^{0.020} + (-4.32\pi)\frac{\rho^4}{4}\Big|_{0.020}^{0.025}\right](10^9) \text{ or}$$

$$T = \left[\left(\frac{-3.12\pi}{4}\right)(0.020^4 - 0.015^4) + \left(\frac{-4.32\pi}{4}\right)(0.025^4 - 0.020^4)\right](10^9) \text{ or} \qquad \boxed{T = -1050 \text{ N-m}}$$

Problem Set 5.2

5.14 Determine the direction of shear stress at points A and B (a) by inspection, and (b) by using the sign convention for internal torque and the subscripts. Report your answer as a positive or negative τ_{xy}.

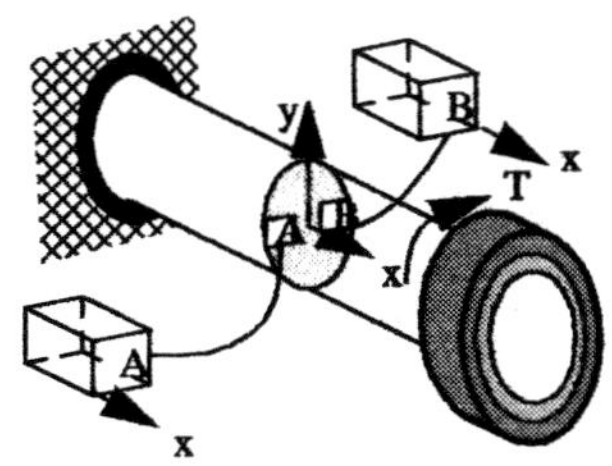

Solution

By Inspection: The reaction torque at the wall will be counter-clockwise with respect to the x-axis. Thus, the segment of the shaft near the wall would rotate counter-clockwise, that is point A would move downwards and point B upwards. To oppose this imaginary motion, the shear stress at point A will be upward and at point B it will be downwards as shown in Figs (a) and (b). The shear stress on the rest of the surfaces can be drawn using the observation that a symmetric pair of shear stress points towards the corner or away from the corner.

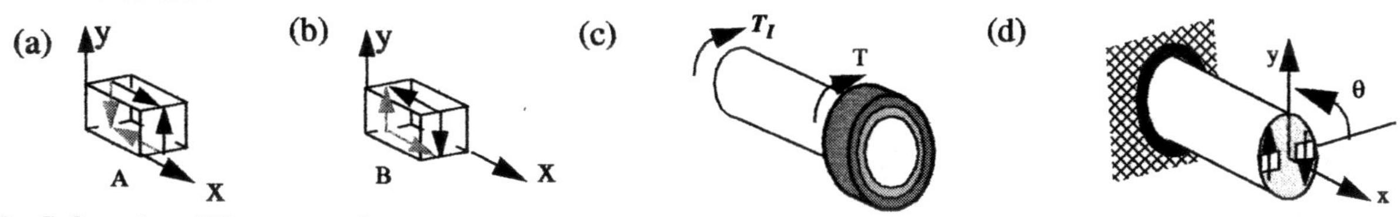

By Subscripts: We can make an imaginary cut through the cross-section containing points A and B and draw the free body diagram as shown in Fig. (c). the internal torque is drawn as per our sign convention. By moment equilibrium we obtain $T_I = -T$. Thus from the torsional stress formula we obtain $\tau_{x\theta} < 0$. Noting that the outward normal of the cross-section is in the positive x-direction, a negative $\tau_{x\theta}$ requires the stress to be in the negative θ-direction as shown in Fig. (d). In the x-y coordinate system we obtain:

$\boxed{(\tau_{xy})_A > 0}$ $\boxed{(\tau_{xy})_B < 0}$

5.15 Determine the direction of shear stress at points A and B (a) by inspection, and (b) by using the sign convention for internal torque and the subscripts. Report your answer as a positive or negative τ_{xy}

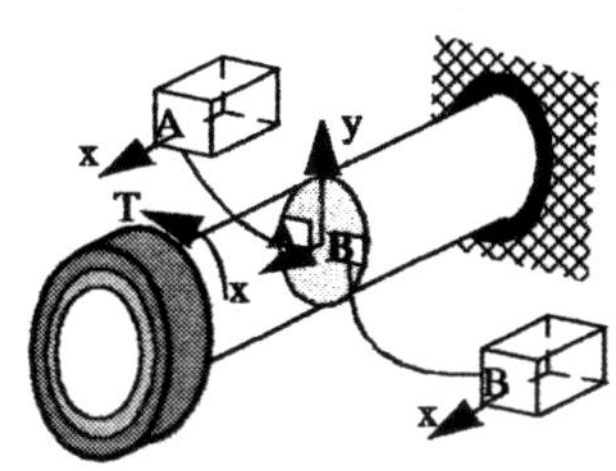

Solution

By Inspection: The reaction torque at the wall will be counter-clockwise with respect to the x-axis. Thus, the segment of the shaft near the wall would move counter-clockwise, that is point A would move upwards and point B would downwards. To oppose this imaginary motion, the shear stress at point A will be downwards and at point B it will be upwards as shown in Figs (a) and (b). The shear stress on the rest of the surfaces can be drawn using the observation that a symmetric pair of shear stress points towards the corner or away from the corner.

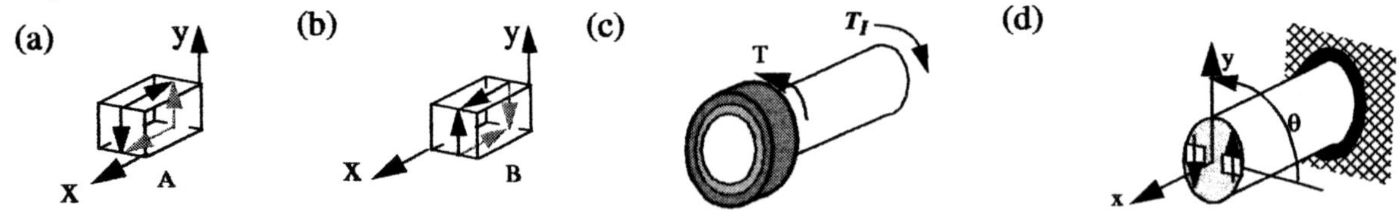

By Subscripts: We can make an imaginary cut through the cross-section containing points A and B and draw the free body diagram as shown in Fig. (c). the internal torque is drawn as per our sign convention. By moment equilibrium we obtain $T_I = T$. Thus from the torsional stress formula we obtain $\tau_{x\theta} > 0$. Noting that the outward normal of the cross-section is in the positive x-direction, a positive $\tau_{x\theta}$ requires the stress to be in the negative θ-direction as shown in Fig. (d). In the x-y coordinate system we obtain:

$(\tau_{xy})_A < 0$ $(\tau_{xy})_B > 0$

5.16 Determine the direction of shear stress at points A and B (a) by inspection, and (b) by using the sign convention for internal torque and the subscripts. Report your answer as a positive or negative τ_{xy}

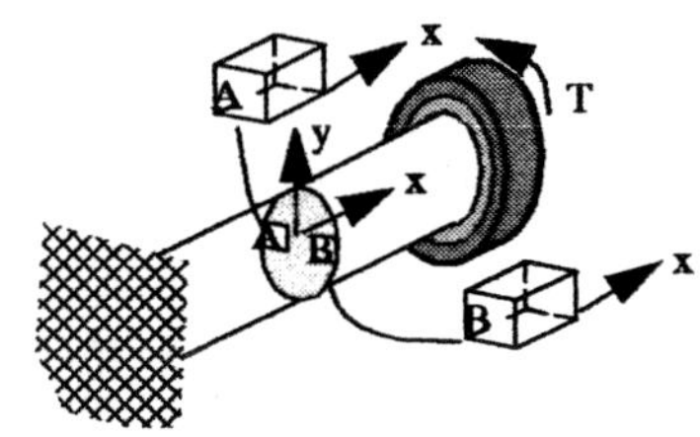

Solution

By Inspection: Due to the applied torque, the segment of the shaft containing the points A and B would rotate clockwise, that is point A would move downwards and point B upwards. To oppose this imaginary motion, the shear stress at point A will be upward and at point B it will be downwards as shown in Figs (a) and (b). The shear stress on the rest of the surfaces can be drawn using the observation that a symmetric pair of shear stress points towards the corner or away from the corner.

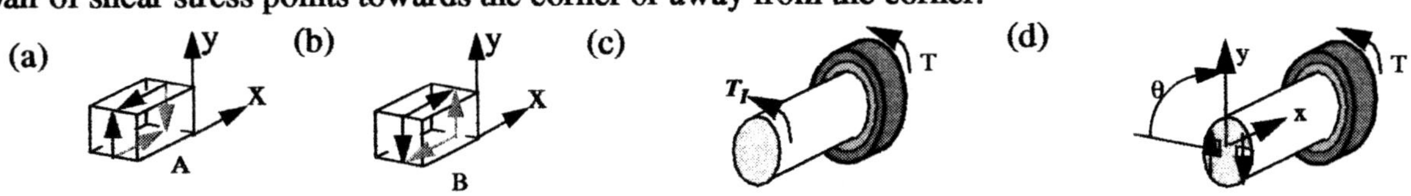

By Subscripts: We can make an imaginary cut through the cross-section containing points A and B and draw the free body diagram as shown in Fig. (c). the internal torque is drawn as per our sign convention. By moment equilibrium we obtain $T_I = -T$. Thus from the torsional stress formula we obtain $\tau_{x\theta} < 0$. Noting that the outward normal of the cross-section is in the negative x-direction, a negative $\tau_{x\theta}$ requires the stress to be in the positive θ-direction as shown in Fig. (d). In the x-y coordinate system we obtain:

$(\tau_{xy})_A < 0$ $(\tau_{xy})_B > 0$

5.17 Determine the direction of shear stress at points A and B (a) by inspection, and (b) by using the sign convention for internal torque and the subscripts. Report your answer as a positive or negative τ_{xy}

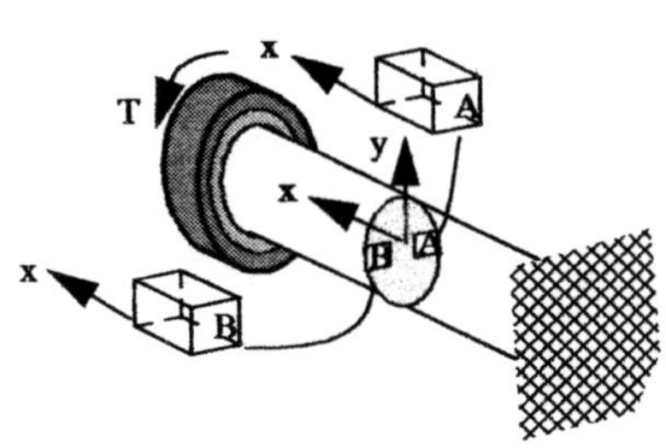

Solution

By Inspection: Due to the applied torque, the segment of the shaft containing the points A and B would rotate in the direction of the torque, that is point A would move upwards and point B would move downwards. To oppose this imaginary motion, the shear stress at point A will be downward and at point B it will be upwards as shown in Figs (a) and (b). The shear stress on the rest of the surfaces can be drawn using the observation that a symmetric pair of shear stress points towards the corner or away from the corner.

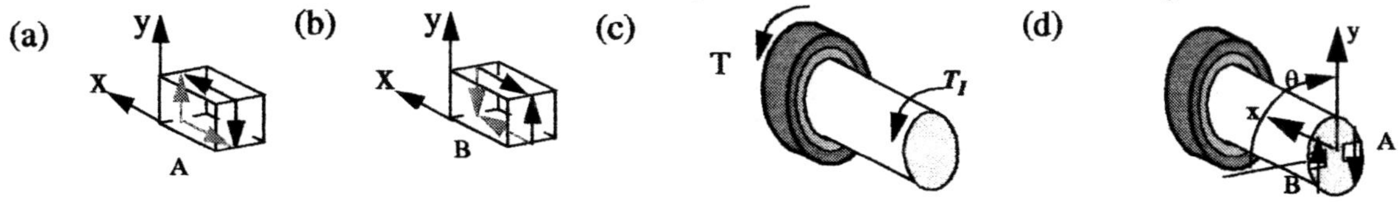

By Subscripts: We can make an imaginary cut through the cross-section containing points A and B and draw the free body diagram as shown in Fig. (c). The internal torque is drawn as per our sign convention. By moment equilibrium we obtain $T_I = -T$. Thus from the torsional stress formula we obtain $\tau_{x\theta} < 0$. Noting that the outward normal of the cross-section is in the negative x-direction, a negative $\tau_{x\theta}$ requires the stress to be in the positive θ-direction as shown in Fig. (d). In the x-y coordinate system we obtain:

$$\boxed{(\tau_{xy})_A > 0} \qquad \boxed{(\tau_{xy})_B < 0}$$

5.18 The two shaft shown have the same cross-sectional area A, show that the ratio of polar moment of inertia of the hollow shaft to the solid shaft is as given below:

$$\frac{J_{hollow}}{J_{Solid}} = \frac{\alpha^2 + 1}{\alpha^2 - 1}$$

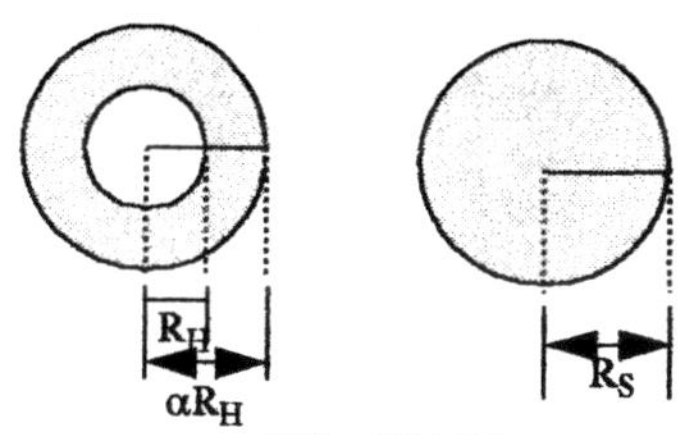

Fig. P5.18

Solution

We can calculate the radius R_H and R_S in terms of the cross-sectional area A as shown below

$$A_H = \pi[(\alpha R_H)^2 - R_H^2] = A \qquad \text{or} \qquad R_H^2 = \frac{A}{\pi(\alpha^2 - 1)} \qquad 1$$

$$A_S = \pi R_S^2 = A \qquad \text{or} \qquad R_S^2 = \frac{A}{\pi} \qquad 2$$

The polar area moment of inertia for a hollow shaft with inside radius R_i and outside radius R_o is $J = \frac{\pi}{2}(R_o^4 - R_i^4)$. For the hollow shaft $R_o = \alpha R_H$ and $R_i = R_H$, while for solid shaft $R_o = R_S$ and $R_i = 0$. Substituting these values and using Eqs (1) and (2), we obtain the two polar area moments shown below:

$$J_H = \frac{\pi}{2}[(\alpha R_H)^4 - (R_H)^4] = \frac{\pi}{2}(\alpha^4 - 1)(R_H)^4 = \frac{\pi}{2}(\alpha^4 - 1)\left(\frac{A}{\pi(\alpha^2 - 1)}\right)^2 = \left(\frac{\alpha^2 + 1}{\alpha^2 - 1}\right)\left(\frac{A^2}{2\pi}\right) \qquad 3$$

$$J_S = \frac{\pi}{2}R_S^4 = \frac{\pi}{2}\left(\frac{A}{\pi}\right)^2 = \frac{A^2}{2\pi} \qquad 4$$

Dividing Eq. (3) by Eq. (4) we obtain:

$$\frac{J_H}{J_S} = \frac{\left(\frac{\alpha^2 + 1}{\alpha^2 - 1}\right)\left(\frac{A^2}{2\pi}\right)}{A^2/(2\pi)} \quad \text{or}$$

$$\boxed{\frac{J_H}{J_S} = \left(\frac{\alpha^2 + 1}{\alpha^2 - 1}\right)}$$

5.19 Show that for a thin tube of thickness t and center-line radius R, the polar moment of inertia can be approximated by $J = 2\pi R^3 t$. By thin tube we imply $t < (R/10)$

Solution

The polar moment of inertia of a hollow shaft in terms of outer radius R_o and inner radius R_i can be written as shown below:

$$J = \frac{\pi}{2}(R_o^4 - R_i^4) = \frac{\pi}{2}(R_o^2 + R_i^2)(R_o^2 - R_i^2) = \frac{\pi}{2}(R_o^2 + R_i^2)(R_o + R_i)(R_o - R_i) \qquad 1$$

Substituting $R_o = R + t/2$ and $R_i = R - t/2$ in the equation above we obtain:

$$J = \frac{\pi}{2}\left(R^2 + Rt + \frac{t^2}{4} + R^2 - Rt + \frac{t^2}{4}\right)(R + t/2 + R - t/2)(R + t/2 - R + t/2) = \frac{\pi}{2}\left(2R^2 + \frac{t^2}{2}\right)(2R)(t) \qquad 2$$

Noting that the t^2 term is two orders of magnitude smaller than the R^2 term and hence can be neglected to obtain the following approximated expression:

$J \approx \frac{\pi}{2}(2R^2)(2R)(t)$ or $\boxed{J = 2\pi R^3 t}$

5.20 For the shaft shown draw the torque diagrams. Check the values of internal torque by making imaginary cuts and drawing free body diagrams.

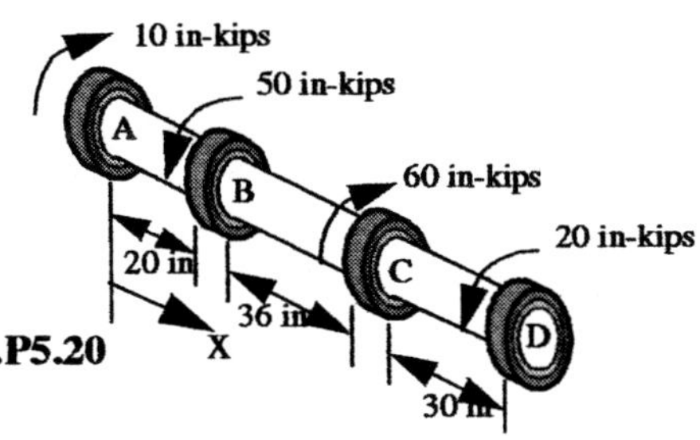

Fig.P5.20

Solution

The torque diagram can be drawn using the template and template equation shown.

(a) Template

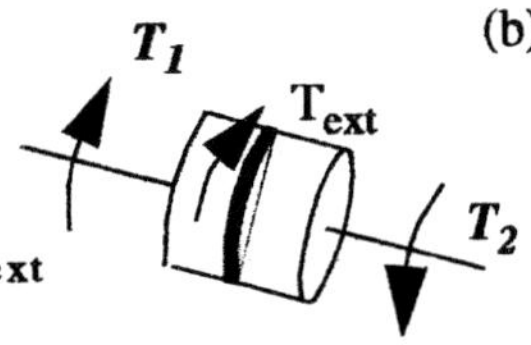

$T_2 = T_1 + T_{ext}$

(b) Torque Diagram

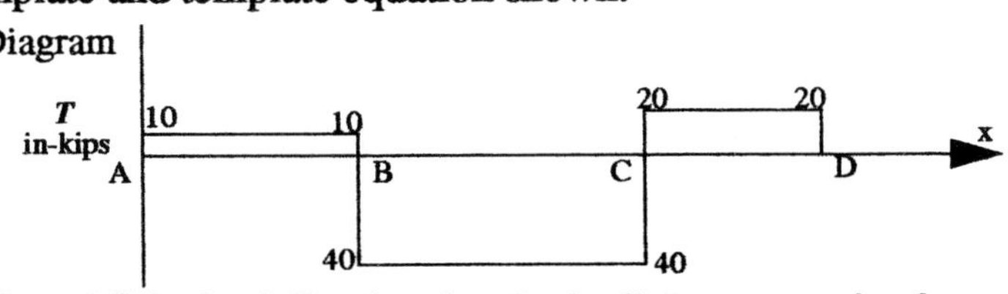

By making imaginary cuts in segments AB, BC, and CD, the following free body diagrams can be drawn. The internal torque is drawn as per our sign convention.

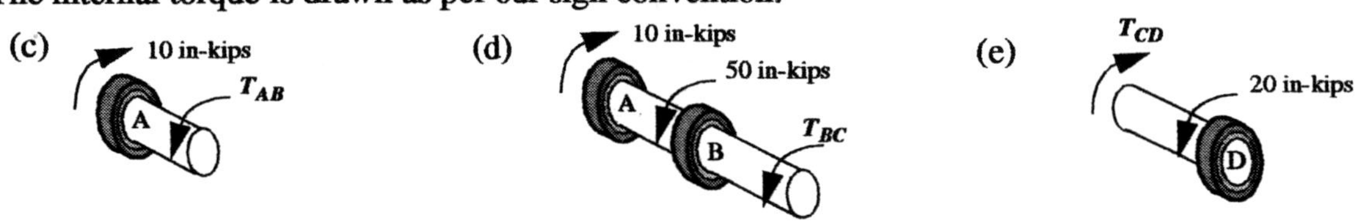

By equilibrium of moment about the shaft axis in Figs. (c), (d), and (e), we obtain:

$$T_{AB} = 10 \text{in} - \text{kips} \qquad 1$$

$$T_{BC} - 10 + 50 = 0 \quad \text{or} \quad T_{BC} = -40 \text{in} - \text{kips} \qquad 2$$

$$T_{CD} = 20 \text{in} - \text{kips} \qquad 3$$

The value of internal torques in Eqs. (1), (2), and (3) are the same as on the torque diagram in Fig. (b).

5.21 For the shaft shown draw the torque diagrams. Check the values of internal torque by making imaginary cuts and drawing free body diagrams.

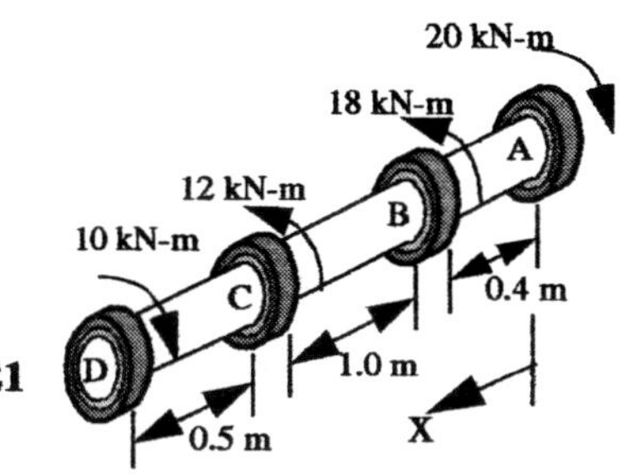

Fig.P5.21

Solution

The torque diagram can be drawn using the template and template equation shown.

(a) Template

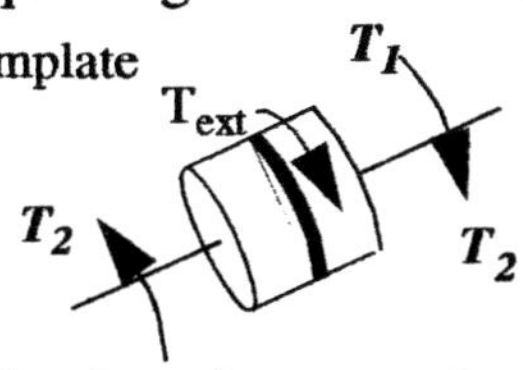

$$T_2 = T_1 + T_{ext}$$

(b) Torque Diagram

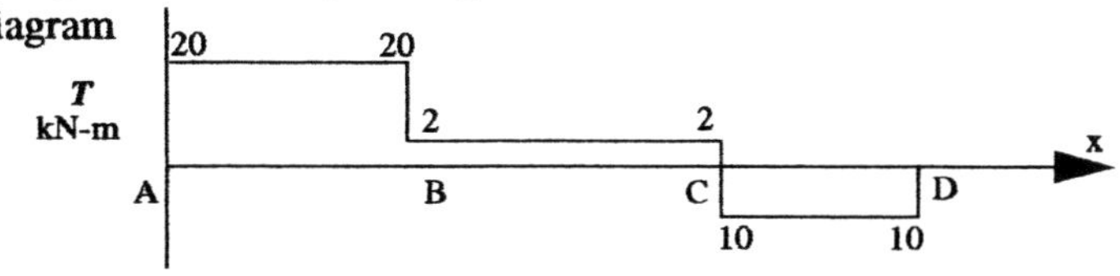

By making imaginary cuts in segments AB, BC, and CD, the following free body diagrams can be drawn. The internal torque is drawn as per our sign convention.

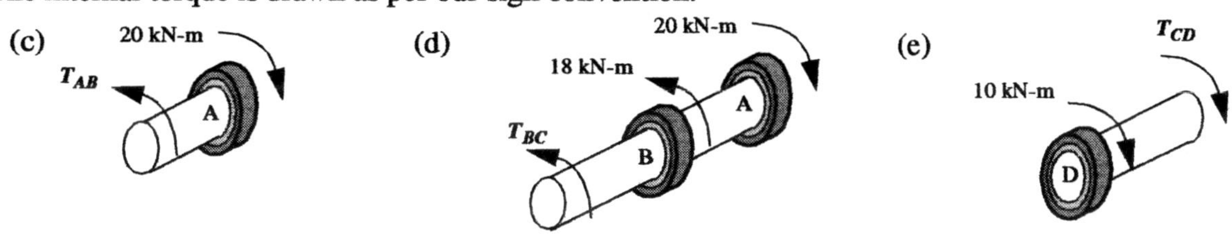

By equilibrium of moment about the shaft axis in Figs. (c), (d), and (e), we obtain:

$$T_{AB} = 20 \text{ kN}-\text{m} \tag{1}$$

$$T_{BC} - 20 + 18 = 0 \qquad \text{or} \qquad T_{BC} = 2 \text{ kN}-\text{m} \tag{2}$$

$$T_{CD} = -10 \text{ kN}-\text{m} \tag{3}$$

The value of internal torques in Eqs. (1), (2), and (3) are the same as on the torque diagram in Fig. (b).

5.22 For the shaft shown draw the torque diagrams. Check the values of internal torque by making imaginary cuts and drawing free body diagrams.

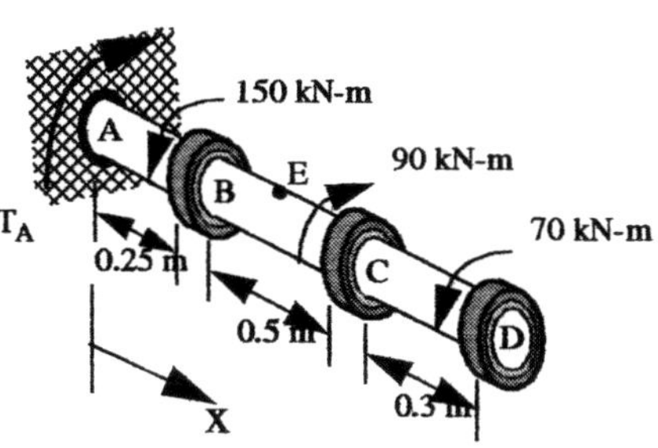

Fig.P5.22

Solution

The wall reaction torque T_A can be found from the free body diagram of the entire shaft in Fig. 5.22. By equilibrium of moments about the shaft axis, we obtain:

$$T_A - 150 + 90 - 70 = 0 \qquad \text{or} \qquad T_A = 130 \text{ kN}-\text{m} \tag{1}$$

The torque diagram can be drawn using the template and template equation shown.

(a) Template

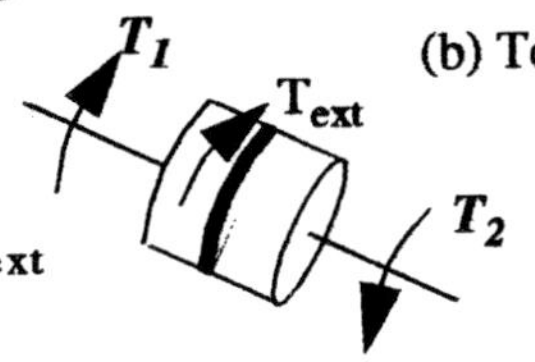

$T_2 = T_1 + T_{ext}$

(b) Torque Diagram

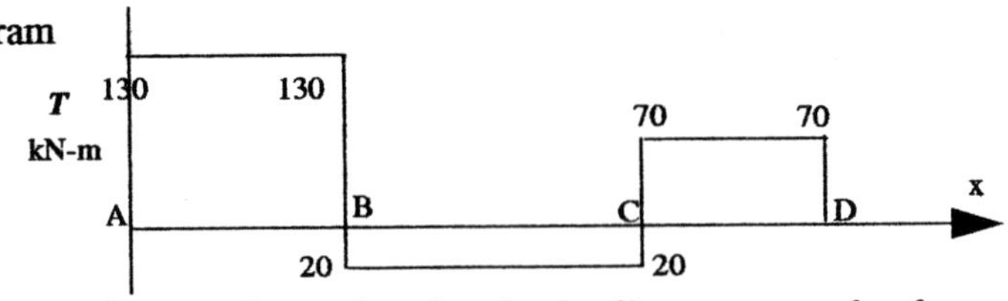

By making imaginary cuts in segments AB, BC, and CD, the following free body diagrams can be drawn. The internal torque is drawn as per our sign convention.

(c)

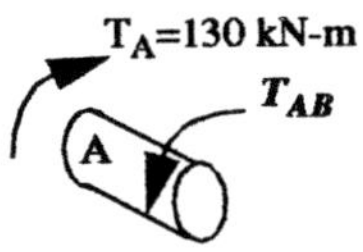

(d)

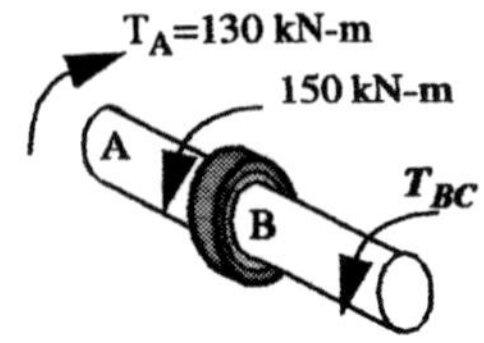

(e)

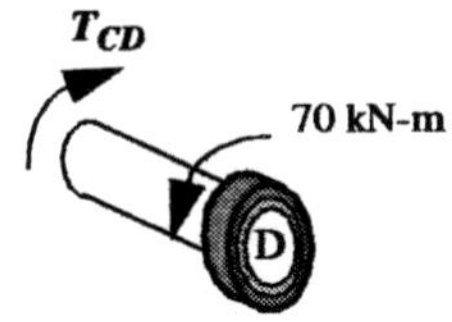

By equilibrium of moment about the shaft axis in Figs. (c), (d), and (e), we obtain:

$$T_{AB} = 130 \text{ kN-m}$$ 2

$$T_{BC} - 130 + 150 = 0 \qquad \text{or} \qquad T_{BC} = -20 \text{ kN-m}$$ 3

$$T_{CD} = 70 \text{ kN-m}$$ 4

The value of internal torques in Eqs. (2), (3), and (4) are the same as on the torque diagram in Fig. (b).

5.23 For the shaft shown draw the torque diagrams. Check the values of internal torque by making imaginary cuts and drawing free body diagrams.

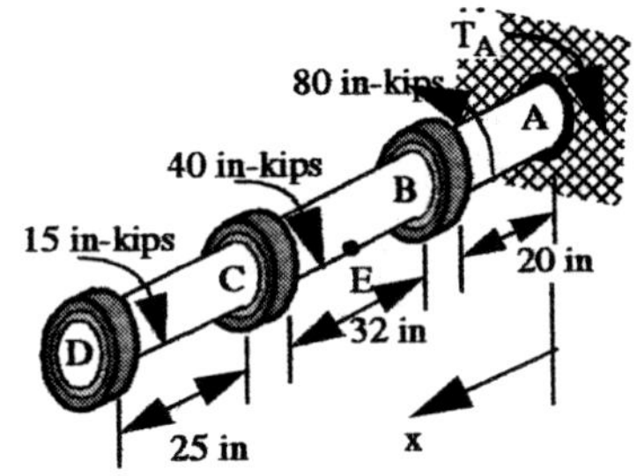

Fig.P5.23

Solution

The wall reaction torque T_A can be found from the free body diagram of the entire shaft in Fig. 5.23. By equilibrium of moments about the shaft axis, we obtain:

$$T_A - 80 + 40 + 15 = 0 \qquad \text{or} \qquad T_A = 25 \text{ in-kips}$$ 1

The torque diagram can be drawn using the template and template equation shown.

(a) Template

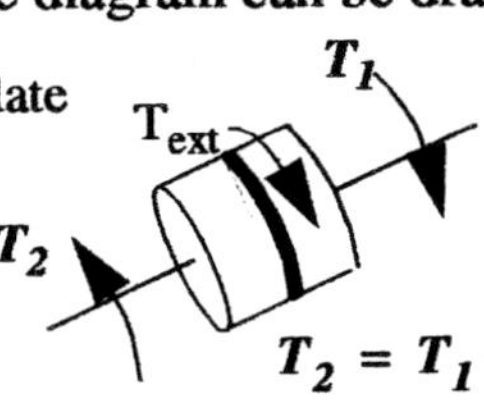

$T_2 = T_1 + T_{ext}$

(b) Torque Diagram

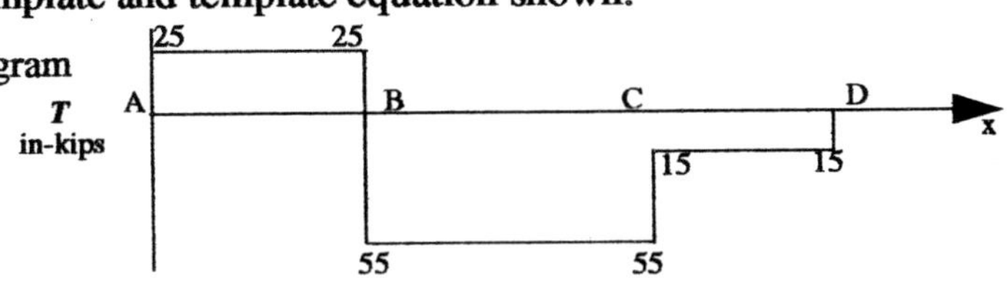

By making imaginary cuts in segments AB, BC, and CD, the following free body diagrams can be drawn. The internal torque is drawn as per our sign convention.

(c)

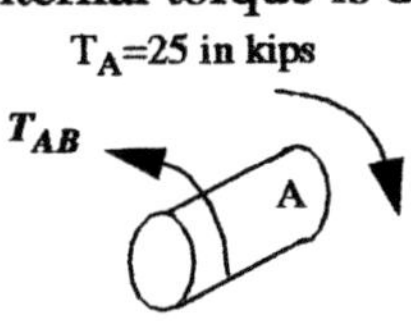

(d)

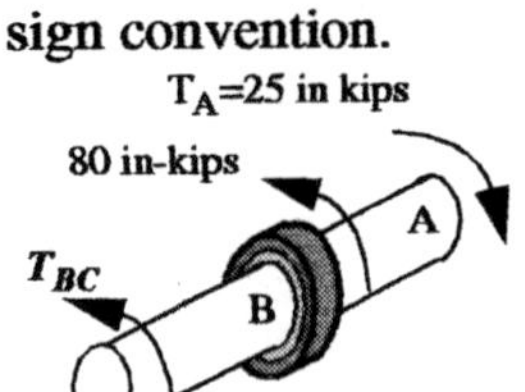

(e)

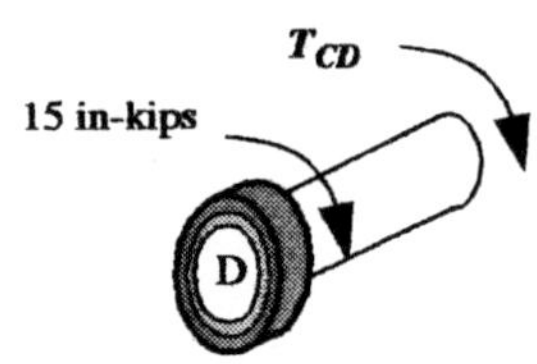

By equilibrium of moment about the shaft axis in Figs. (c), (d), and (e), we obtain:

$$T_{AB} = 25 \text{ in-kips} \quad (2)$$

$$T_{BC} - 25 + 80 = 0 \quad \text{or} \quad T_{BC} = -55 \text{ in-kips} \quad (3)$$

$$T_{CD} = -15 \text{ in-kips} \quad (4)$$

The value of internal torques in Eqs. (2), (3), and (4) are the same as on the torque diagram in Fig. (b).

5.24 The torsional rigidity of the shaft in Problem 5.20 is 90,000 kips-in^2. Determine the rotation of the rigid wheel D with respect to the rigid wheel A.

Solution GJ = 90,000 kips-in^2 $\phi_D - \phi_A = ?$

From Eq. (1), (2), and (3) in problem 5.20, we have:

$T_{AB} = 10\text{in-kips}$ $T_{BC} = -40\text{in-kips}$ $T_{CD} = 20\text{in-kips}$

The relative rotation of the ends of each segment can be found as shown below:

$$\phi_B - \phi_A = \frac{T_{AB}(x_B - x_A)}{GJ} = \frac{(10)(20)}{90,000} = 2.222(10^{-3}) \text{ rads} \quad (1)$$

$$\phi_C - \phi_B = \frac{T_{BC}(x_C - x_B)}{GJ} = \frac{(-40)(36)}{90,000} = -16.0(10^{-3}) \text{ rads} \quad (2)$$

$$\phi_D - \phi_C = \frac{T_{CD}(x_C - x_B)}{GJ} = \frac{(20)(30)}{90,000} = 6.667(10^{-3}) \text{ rads} \quad (3)$$

Adding Eqs. (1), (2), and (3) we obtain:

$$\phi_D - \phi_A = (2.222 - 16.0 + 6.667)(10^{-3}) = -7.111(10^{-3}) \text{ rads or } \boxed{\phi_D - \phi_A = 0.00711 \text{ rads CW}}$$

5.25 The torsional rigidity of the shaft in Problem 5.21 is 1,270 kN-m^2. Determine the rotation of the rigid wheel D with respect to the rigid wheel A.

Solution GJ = 1,270 kN-m^2 $\phi_D - \phi_A = ?$

From Eq. (1), (2), and (3) in problem 5.21, we have:

$T_{AB} = 20 \text{ kN-m}$ $T_{BC} = 2 \text{ kN-m}$ $T_{CD} = -10 \text{ kN-m}$

The relative rotation of the ends of each segment can be found as shown below:

$$\phi_B - \phi_A = \frac{T_{AB}(x_B - x_A)}{GJ} = \frac{(20)(0.4)}{1,270} = 6.299(10^{-3}) \text{ rads} \quad (1)$$

$$\phi_C - \phi_B = \frac{T_{BC}(x_C - x_B)}{GJ} = \frac{(2)(1)}{1,270} = 1.575(10^{-3}) \text{ rads} \quad (2)$$

$$\phi_D - \phi_C = \frac{T_{CD}(x_C - x_B)}{GJ} = \frac{(-10)(0.5)}{1,270} = -3.937(10^{-3}) \text{ rads} \quad (3)$$

Adding Eqs. (1), (2), and (3) we obtain:

$$\phi_D - \phi_A = (6.299 + 1.575 - 3.937)(10^{-3}) = 3.937(10^{-3}) \text{ rads or } \boxed{\phi_D - \phi_A = 0.00394 \text{ rads CCW}}$$

5.26 The shaft in Problem 5.22 is made of steel (G = 80 GPa) and has a diameter of 150 mm. Determine: (a) the rotation of the rigid wheel D, (b) the magnitude of torsional shear stress at point E and show it on a stress cube. Point E is on the top surface of the shaft, and (c) the magnitude of maximum torsional shear *strain* in the shaft.

Solution $G = 80$ GPa $d = 150$ mm $\phi_D = ?$ $(\tau_{x\theta})_E = ?$ $\gamma_{max} = ?$

The polar moment of the shaft cross-section can be found as:

$$J = \frac{\pi}{32}(0.15^4) = 49.701(10^{-6})\ \text{m}^4 \qquad 1$$

The torsional rigidity can be found as:

$$GJ = (80)(10^9)(49.701)(10^{-6}) = 3.976(10^6)\text{N}-\text{m}^2 = 3.976(10^3)\text{kN}-\text{m}^2 \qquad 2$$

From Eq. (1), (2), and (3) in problem 5.22, we have:

$T_{AB} = 130\ \text{kN}-\text{m}$ $T_{BC} = -20\ \text{kN}-\text{m}$ $T_{CD} = 70\ \text{kN}-\text{m}$

The relative rotation of the ends of each segment can be found as shown below:

$$\phi_B - \phi_A = \frac{T_{AB}(x_B - x_A)}{GJ} = \frac{(130)(0.25)}{3.976(10^3)} = 8.174(10^{-3})\ \text{rads} \qquad 3$$

$$\phi_C - \phi_B = \frac{T_{BC}(x_C - x_B)}{GJ} = \frac{(-20)(0.5)}{3.976(10^3)} = -2.515(10^{-3})\ \text{rads} \qquad 4$$

$$\phi_D - \phi_C = \frac{T_{CD}(x_C - x_B)}{GJ} = \frac{(70)(0.3)}{3.976(10^3)} = 5.282(10^{-3})\ \text{rads} \qquad 5$$

Adding Eqs. (1), (2), and (3) we obtain:

$$\phi_D - \phi_A = (8.174 - 2.515 + 5.282)(10^{-3}) = 10.94(10^{-3})\ \text{rads}$$

Noting that point A is fixed to the wall, hence $\phi_A = 0$, we obtain:

$$\boxed{\phi_D = 0.0109\ \text{rads CCW}}$$

The shear stress at point E can be found as shown below.

$$(\tau_{x\theta})_E = \frac{T_{BC}\rho_E}{J} = \frac{(-20)(10^3)(0.075)}{49.701(10^{-6})} = -30.18(10^6)\ \text{N/m}^2 \text{ or}$$

$$\boxed{(\tau_{x\theta})_E = -30.2\ \text{MPa}}$$

The outward normal is in the positive x-direction, thus the shear stress must be in the negative θ-direction at point E as shown below.

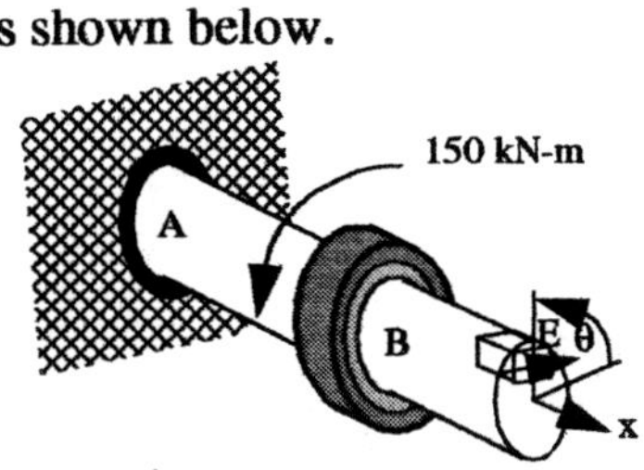

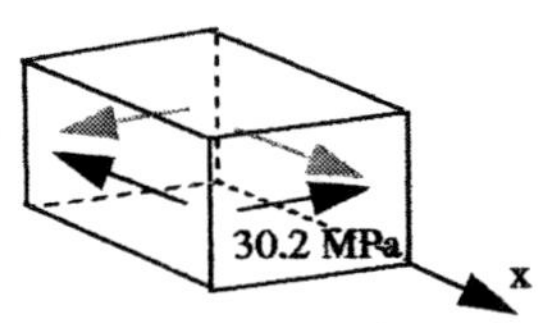

The maximum shear stress will be in segment AB as the internal torque is maximum and the diameter is the same for the entire shaft. The maximum shear stress can be found as shown below:

$$\tau_{max} = \frac{T_{AB}\rho_{max}}{J} = \frac{130(10^3)(0.075)}{49.701(10^{-6})} = 196.2(10^6)\ \text{N/m}^2.$$ The maximum shear strain can be found by using Hooke's law as shown below:

$$\gamma_{max} = \frac{\tau_{max}}{G} = \frac{196.2(10^6)}{80(10^9)} = 2.452(10^{-3}) \text{ or}$$

$$\boxed{\gamma_{max} = 2452\mu}$$

5.27 The shaft in Problem 5.23 is made of aluminum ($G = 4{,}000$ ksi) and has a diameter of 4 inches. Determine: (a) the rotation of the rigid wheel D, and (b) the magnitude of torsional shear stress at point E and show it on a stress cube. Point E is on the bottom surface of the shaft, and (c) the magnitude of maximum torsional shear *strain* in the shaft.

Solution $G = 4000$ ksi $d = 4$ in $\phi_D = ?$ $(\tau_{x\theta})_E = ?$ $\gamma_{max} = ?$

The polar moment of the shaft cross-section can be found as:

$$J = \frac{\pi}{32}(4^4) = 25.133 \text{ in}^4 \tag{1}$$

The torsional rigidity can be found as:

$$GJ = (4000)(25.133) = 100.53(10^3)\text{kips}-\text{in}^2 \tag{2}$$

From Eq. (1), (2), and (3) in problem 5.23, we have:

$T_{AB} = 25$ in – kips $T_{BC} = -55$ in – kips $T_{CD} = -15$ in – kips

The relative rotation of the ends of each segment can be found as shown below:

$$\phi_B - \phi_A = \frac{T_{AB}(x_B - x_A)}{GJ} = \frac{(25)(20)}{(100.53)(10^3)} = 4.974(10^{-3}) \text{ rads} \tag{3}$$

$$\phi_C - \phi_B = \frac{T_{BC}(x_C - x_B)}{GJ} = \frac{(-55)(32)}{(100.53)(10^3)} = -17.507(10^{-3}) \text{ rads} \tag{4}$$

$$\phi_D - \phi_C = \frac{T_{CD}(x_C - x_B)}{GJ} = \frac{(-15)(25)}{(100.53)(10^3)} = -3.730(10^{-3}) \text{ rads} \tag{5}$$

Adding Eqs. (1), (2), and (3) we obtain:

$\phi_D - \phi_A = (4.974 - 17.507 - 3.730)(10^{-3}) = -16.263(10^{-3})$ rads

Noting that point A is fixed to the wall, hence $\phi_A = 0$, we obtain: $\boxed{\phi_D = 0.0163 \text{ rads CW}}$

The shear stress at point E can be found as shown below.

$(\tau_{x\theta})_E = \frac{T_{BC}\rho_E}{J} = \frac{(-55)(2)}{25.133} = -4.377$ ksi or $\boxed{(\tau_{x\theta})_E = -4.4 \text{ ksi}}$

The outward normal is in the positive x-direction, thus the shear stress must be in the negative θ-direction at point E as shown below.

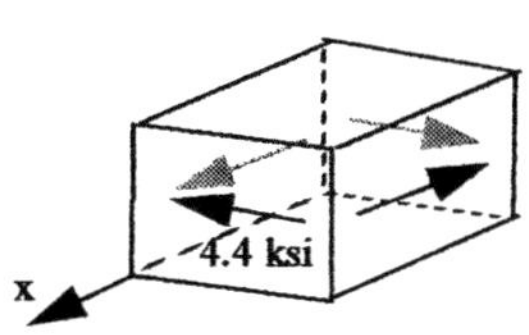

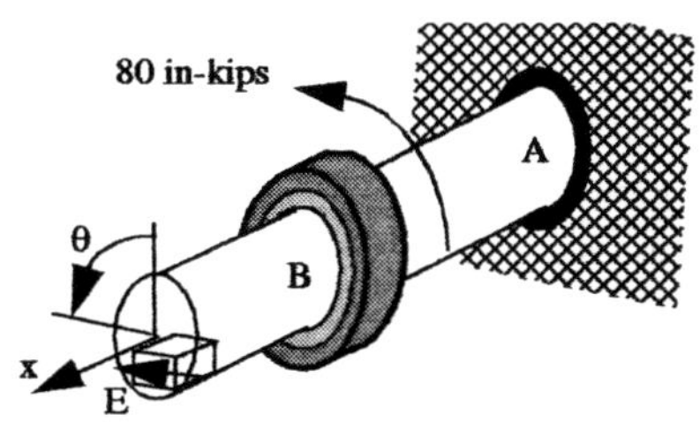

The maximum shear stress will be in segment BC as the internal torque is maximum and the diameter is the same for the entire shaft. The maximum shear stress value is the same as the shear stress at point E. The maximum shear strain can be found by using Hooke's law as shown below:

$\gamma_{max} = \frac{\tau_{max}}{G} = \frac{-4.377}{4,000} = -1.094(10^{-3})$ or $\boxed{\gamma_{max} = -1094\mu}$

5.28 A solid circular steel (G_s = 12,000 ksi) shaft BC is securely attached to two hollow steel shafts AB and CD as shown. Determine: (a) the angle of rotation of section at D with respect to section at A. (b) the magnitude of maximum torsional shear stress in the shaft (c) the torsional shear stress at point E and show it on a stress cube. Point E is on the inside bottom surface of CD.

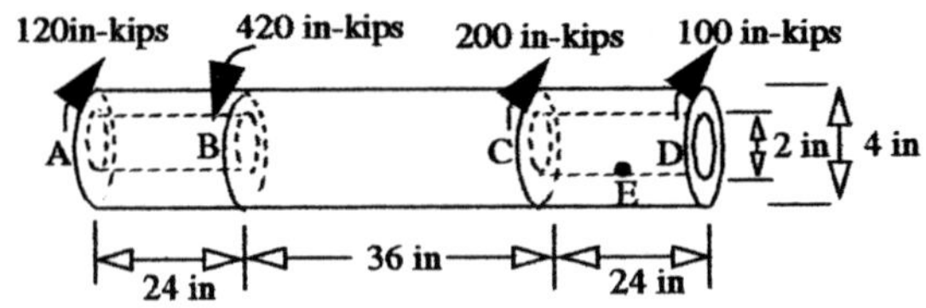

Fig. P5.28

Solution G = 12,000 ksi $\phi_D - \phi_A = ?$ $\tau_{max} = ?$ $(\tau_{x\theta})_E = ?$

The polar moment of the cross-sections can be found as shown below.

$$J_{AB} = J_{AB} = \frac{\pi}{32}(4^4 - 2^4) = 23.562\ \text{in}^4 \qquad J_{BC} = \frac{\pi}{32}(4^4) = 25.133\ \text{in}^4$$ 1

The torque diagram can be drawn using the template and template equation shown.

(a) Template

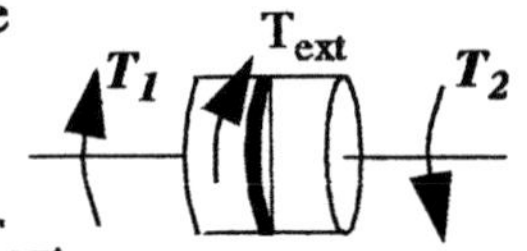

$T_2 = T_1 + T_{ext}$

(b) Torque Diagram

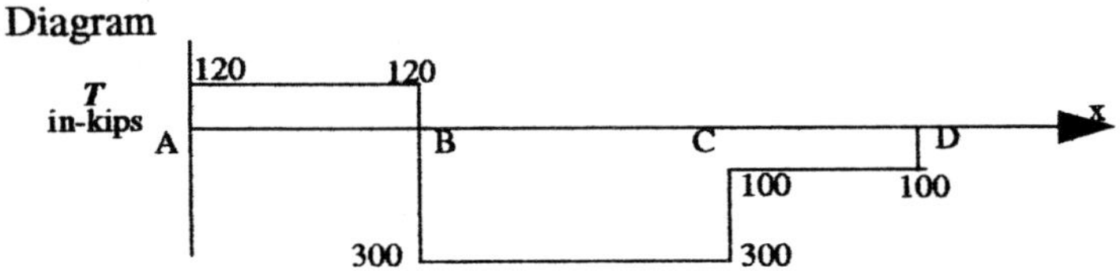

From Fig. (b) we have: $T_{AB} = 120\ \text{in-kips}\ \ T_{BC} = -300\ \text{in-kips}\ \ T_{CD} = -100\ \text{in-kips}$

The relative rotation of the ends of each segment can be found as shown below:

$$\phi_B - \phi_A = \frac{T_{AB}(x_B - x_A)}{G_{AB}J_{AB}} = \frac{(120)(24)}{(12,000)(23.562)} = 10.186(10^{-3})\ \text{rads}$$ 2

$$\phi_C - \phi_B = \frac{T_{BC}(x_C - x_B)}{G_{BC}J_{BC}} = \frac{(-300)(36)}{(12,000)(25.133)} = -35.810(10^{-3})\ \text{rads}$$ 3

$$\phi_D - \phi_C = \frac{T_{CD}(x_D - x_C)}{G_{CD}J_{CD}} = \frac{(-100)(24)}{(12,000)(23.562)} = -8.488(10^{-3})\ \text{rads}$$ 4

Adding Eqs. (2), (3), and (4) we obtain:

$$\phi_D - \phi_A = (10.186 - 35.810 - 8.488)(10^{-3}) = -34.11(10^{-3})\ \text{rads}$$

$$\boxed{\phi_D - \phi_A = 0.0341\ \text{rads CW}}$$

The maximum torsional shear stress will be in BC and can be found as shown below.

$$\tau_{max} = \left|\frac{T_{BC}\rho_{max}}{J_{BC}}\right| = \frac{(300)(2)}{25.133} = 23.87\ \text{ksi}$$

$$\boxed{\tau_{max} = 23.87\ \text{ksi}}$$

The torsional shear stress at point E can be found as shown below.

$$(\tau_{x\theta})_E = \frac{T_{CD}\rho_E}{J_{CD}} = \frac{(-100)(1)}{23.562} = -4.244\ \text{ksi or}$$

$$\boxed{(\tau_{x\theta})_E = -4.24\ \text{ksi}}$$

Figure below shows the shear stress on a stress cube.

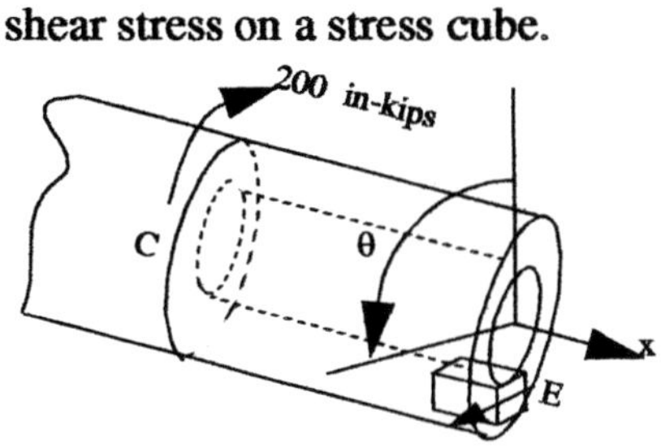

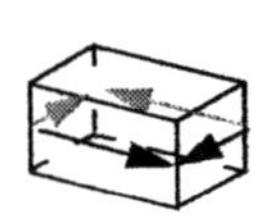

5.29 A steel shaft (G=80GPa) is subjected to the torques shown. Determine (a) the rotation of section A with respect to no load position. (b) the torsional shear stress at point E and show it on a stress cube. Point E is on the surface of the shaft.

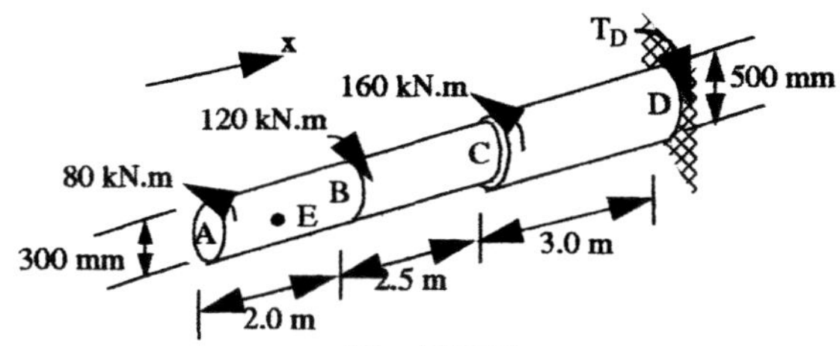

Fig. P5.29

Solution G=80GPa $\phi_A = ?$ $(\tau_{x\theta})_E = ?$

The polar moment of the cross-sections can be found as shown below.

$$J_{AB} = J_{BC} = \frac{\pi}{32}(0.3^4) = 0.7952(10^{-3})\ \text{mm}^4 \qquad J_{CD} = \frac{\pi}{32}(0.5^4) = 6.1359(10^{-3})\ \text{mm}^4$$ 1

The wall reaction torque T_D can be found from the free body diagram of the entire shaft in Fig. 5.23. By equilibrium of moments about the shaft axis, we obtain:

$$T_D - 80 + 120 - 160 = 0 \qquad \text{or} \qquad T_D = 120\ \text{kN} - \text{m}$$ 2

The torque diagram can be drawn using the template and template equation shown.

(a) Template

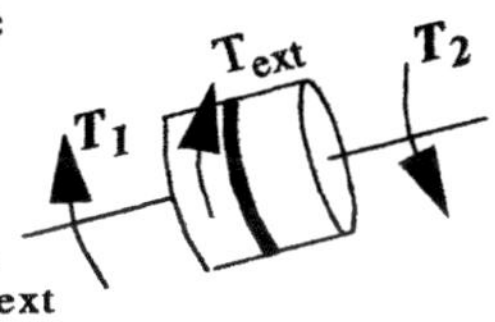

$T_2 = T_1 + T_{ext}$

(b) Torque Diagram

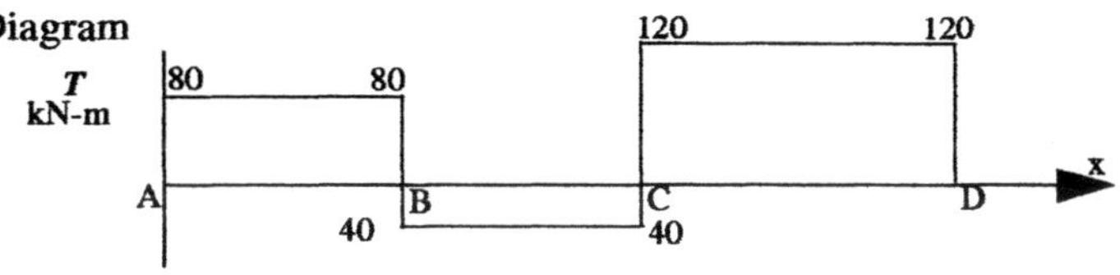

From Fig. (b) we have: $T_{AB} = 80\ \text{kN} - \text{m}$ $T_{BC} = -40\ \text{kN} - \text{m}$ $T_{CD} = 120\ \text{kN} - \text{m}$

The relative rotation of the ends of each segment can be found as shown below:

$$\phi_B - \phi_A = \frac{T_{AB}(x_B - x_A)}{G_{AB}J_{AB}} = \frac{(80)(10^3)(2)}{(80)(10^9)(0.7952)(10^{-3})} = 2.515(10^{-3})\ \text{rads}$$ 3

$$\phi_C - \phi_B = \frac{T_{BC}(x_C - x_B)}{G_{BC}J_{BC}} = \frac{(-40)(10^3)(2.5)}{(80)(10^9)(0.7952)(10^{-3})} = -1.572(10^{-3})\ \text{rads}$$ 4

$$\phi_D - \phi_C = \frac{T_{CD}(x_D - x_C)}{G_{CD}J_{CD}} = \frac{(120)(10^3)(3.0)}{(80)(10^9)(6.1359)(10^{-3})} = 0.7334(10^{-3})\ \text{rads}$$ 5

Adding Eqs. (3), (4), and (5) we obtain: $\phi_D - \phi_A = (2.515 - 1.572 + 0.7334)(10^{-3}) = 1.676(10^{-3})\ \text{rads}$

Note that point D is fixed to the wall, hence ϕ_D is zero. Thus, $\phi_A = -1.676(10^{-3})\ \text{rads}$

or $\boxed{\phi_A = 1676\ \mu\text{rads CW}}$

The torsional shear stress at point E can be found as shown below.

$$(\tau_{x\theta})_E = \frac{T_{AB}\rho_E}{J_{AB}} = \frac{(80)(10^3)(0.150)}{(0.7952)(10^{-3})} = 15.09(10^6)\ \text{N/m}^2 \text{ or}$$ $\boxed{(\tau_{x\theta})_E = 15.1\ \text{MPa}}$

Figure below shows the shear stress at point E on a stress cube.

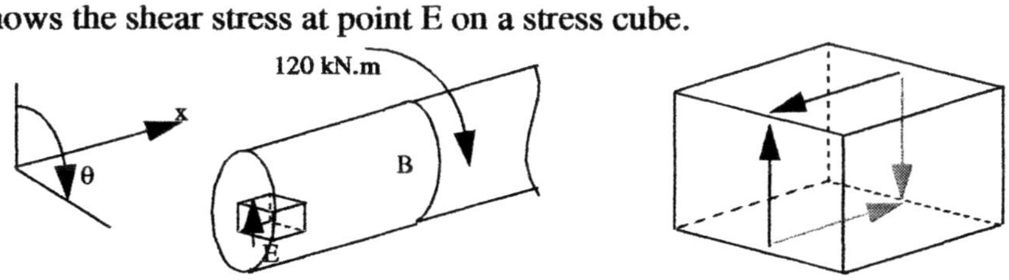

5.30 The radius of the tapered circular shaft varies from 200 mm at A to 50 mm at B. The shear modulus of the material is G = 40 GPa. Determine: (a) the angle of rotation of wheel C with respect to the fixed end. (b) the magnitude of maximum torsional shear strain in the shaft.

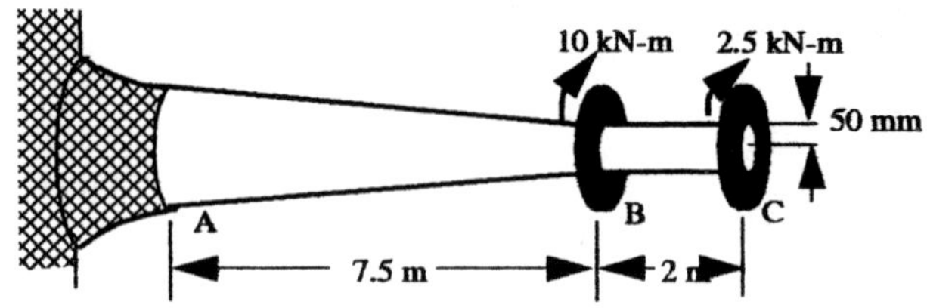

Fig. P5.30

Solution $R_A = 200$ mm $R_B = 50$ mm $G = 40$ GPa $\phi_C = ?$ $\gamma_{max} = ?$

We note that R in segment AB is a linear function of x i.e., $R(x) = a + bx$. The constants can be found as shown below.

$$R(x = 0) = R_A = a = 0.2 \quad \text{and} \quad R(x = 7.5) = R_B = 0.2 + b(7.5) = 0.05 \quad \text{or} \quad b = -0.02$$

$$R(x) = 0.2 - 0.02x \qquad 1$$

The polar moment of the cross-sections can be calculated as shown below.

$$J_{AB} = \frac{\pi}{2}(0.2 - 0.02x)^4 \text{ m}^4 \qquad J_{BC} = \frac{\pi}{2}(0.05)^4 = 9.817(10^{-6}) \text{ m}^4 \qquad 2$$

We can make imaginary cuts in segment AB and BC and obtain the free body diagrams shown in Fig.(a) and Fig. (b). The internal torques are drawn as per our sign convention.
From moment equilibrium in Figs. (a) and (b) we obtain:

$$T_{AB} = -12.5\text{kN-m} \qquad T_{BC} = -2.5\text{kN-m} \qquad 3$$

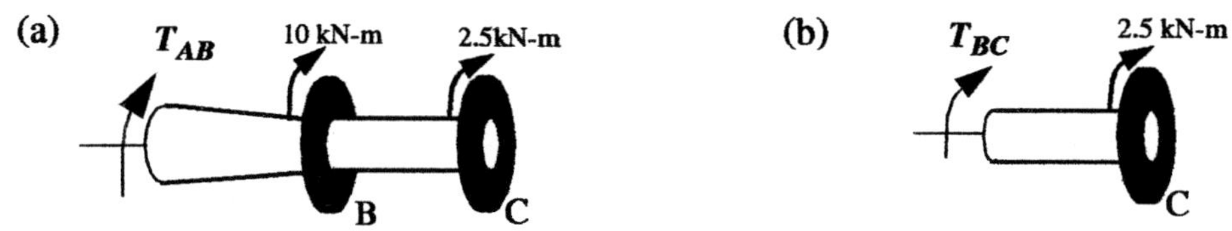

The relative rotation of the ends of segment AB can be found as shown below

$$\left(\frac{d\phi}{dx}\right)_{AB} = \frac{T_{AB}}{G_{AB}J_{AB}} = \frac{-(12.5)(10^3)}{(40)(10^9)\left[\frac{\pi}{2}(0.2 - 0.02x)^4\right]} \qquad 4$$

Integrating from point A to point B, we obtain:

$$\int_{\phi_A}^{\phi_B} d\phi = -\frac{25(10^{-6})}{40\pi}\int_0^{7.5}\frac{dx}{(0.2 - 0.02x)^4} = -\frac{0.1989(10^{-6})}{(-3)(-0.02)}\left[\frac{1}{(0.2 - 0.02x)^3}\right]\Bigg|_0^{7.5} \quad \text{or}$$

$$\phi_B - \phi_A = -(3.315)(10^{-6})\left[\frac{1}{(0.05)^3} - \frac{1}{(0.20)^3}\right] = -26.111(10^{-3}) \text{ rads} \qquad 5$$

The relative rotation

$$\phi_C - \phi_B = \frac{T_{BC}(x_C - x_B)}{G_{BC}J_{BC}} = \frac{(-2.5)(10^3)(2)}{(40)(10^9)(9.817)(10^{-6})} = -12.732(10^{-3})\text{rads} \qquad 6$$

Adding Eqs (5) and (6), we obtain the rotation of section at C with respect to section at A. Noting that the section at A is built into the wall we obtain the rotation of section at C.

$$\phi_C - \phi_A = (-26.111 - 12.732)(10^{-3}) = -38.843(10^{-3}) \text{ rads or} \qquad \boxed{\phi_C = 0.0388 \text{ rads CW}}$$

The maximum shear stress will be on a cross-section just left of B. Just left of section at B we have: $J_{AB} = 9.817(10^{-6}) \text{ m}^4$ and $\rho_{max} = 0.05\text{m}$. The magnitude of maximum torsional shear stress in the shaft can be found as shown below.

$$\tau_{max} = \left|\frac{T_{AB}\rho_{max}}{J_{AB}}\right| = \left|\frac{(12.5)(10^3)(0.05)}{9.817(10^{-6})}\right| = 63.66(10^6)\ \text{N/m}^2$$ 7

From Hooke's law the maximum shear strain can be found, as shown below.

$$\gamma_{max} = \frac{\tau_{max}}{G} = \frac{63.66(10^6)}{(40)(10^9)} = 1.591(10^{-3})$$

$\boxed{\gamma_{max} = 1591\ \mu}$

5.31 The radius of a tapered shaft varies as $R = \left(\frac{r}{L}\right)(2L - 0.25x)$. In terms of T_{ext}, L, G and r, determine: (a) the rotation of wheel B. (b) the magnitude maximum torsional shear stress in the shaft.

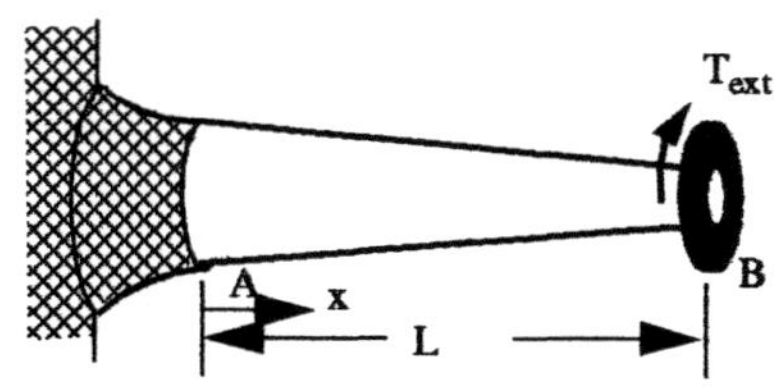

Fig. P5.31

Solution $R = \left(\frac{r}{L}\right)(2L - 0.25x)$ $R_B = 50$ mm $G = 40$ GPa

$\phi_B = f(T_{ext}, L, G, r) = ?$ $\tau_{max} = g(T_{ext}, L, G, r) = ?$

The polar moment of the cross-section can be calculated as shown below.

$$J_{AB} = \frac{\pi r^4}{2L^4}(2L - 0.25x)^4$$ 1

We can make imaginary cut in segment AB and obtain the following free body diagram. The internal torque is drawn as per our sign convention.

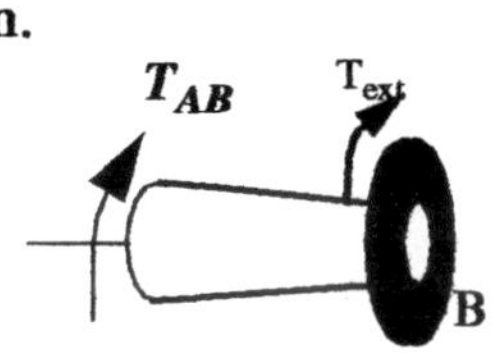

From moment equilibrium we obtain: $T_{AB} = -T_{ext}$

The relative rotation of the ends of segment AB can be found as shown below

$$\left(\frac{d\phi}{dx}\right)_{AB} = \frac{T_{AB}}{G_{AB}J_{AB}} = \frac{-T_{ext}}{G\left[\frac{\pi r^4}{2L^4}(2L - 0.25x)^4\right]} = \frac{-2T_{ext}L^4}{\pi G r^4(2L - 0.25x)^4}$$ 2

Integrating from point A to point B, we obtain:

$$\int_{\phi_A}^{\phi_B} d\phi = \left(\frac{-2T_{ext}L^4}{\pi G r^4}\right)\int_0^L \frac{dx}{(2L - 0.25x)^4} = -\left(\frac{2T_{ext}L^4}{\pi G r^4}\right)\frac{1}{(-3)(-0.02)}\left[\frac{1}{(2L - 0.25x)^3}\right]\Bigg|_0^L \quad \text{or}$$

$$\phi_B - \phi_A = -\left(\frac{2T_{ext}L^4}{\pi G r^4}\right)\left[\frac{1}{(1.75L)^3} - \frac{1}{(2L)^3}\right] = -0.0523\left(\frac{T_{ext}L}{Gr^4}\right) \quad \text{or}$$

$$\boxed{\phi_B = 0.0523\left(\frac{T_{ext}L}{Gr^4}\right)\ \text{CW}}$$

The maximum shear stress will be on a cross-section just left of B. Just left of section at B we have: $R = \left(\frac{r}{L}\right)(2L - 0.25L) = 1.75r$. The magnitude of maximum torsional shear stress in the shaft can be found as shown below.

$$\tau_{max} = \left|\frac{T_{AB}\rho_{max}}{J_{AB}}\right| = \frac{T_{ext}(1.75r)}{\pi(1.75r)^4/2} \quad \text{or}$$

$$\boxed{\tau_{max} = \frac{0.1188T_{ext}}{r^3}}$$

5.32 The external torque on a drill bit varies as a quadratic function to a maximum intensity of q in-lb/in as shown. If the drill bit diameter is d, its length L, and modulus of rigidity G, determine (a) the maximum shear stress on the drill bit. (b) the relative rotation of the end of the drill bit with respect to the chuck.

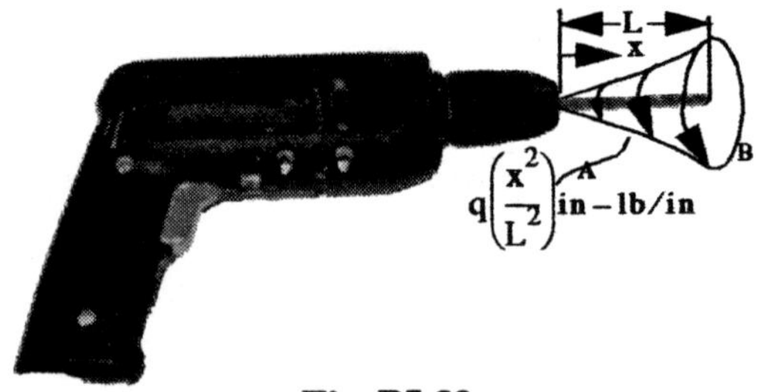

Fig. P5.32

Solution $\phi_B - \phi_A = f(q, L, G, r) = ?$ $\tau_{max} = g(q, L, G, r) = ?$

The distributed torque on the drill bit is counter-clockwise with respect to the x-axis, thus we can substitute

$t(x) = q\left(\frac{x^2}{L^2}\right)$ in the equilibrium equation and integrate to obtain the following:

$$\frac{dT}{dx} + q\left(\frac{x^2}{L^2}\right) = 0 \qquad \text{or} \qquad T = -q\left(\frac{x^3}{3L^2}\right) + c$$ 1

Where, c is an integration constant. At point B i.e., at x = L the internal torque should be zero as there is no concentrated applied torque at B. Substituting this in Eq. (1) we can find the integration constant as shown below.

$$T(x = L) = -q\left(\frac{L^3}{2L^2}\right) + c = 0 \qquad \text{or} \qquad c = \frac{qL}{3}$$ 2

Substituting Eq (2) into Eq (1) and simplifying, we obtain:

$$T = \frac{q}{3L^2}(L^3 - x^3)$$ 3

The maximum internal torque will exist at x = 0. From Eq. (3) we obtain: $T_{max} = \frac{qL}{3}$. The maximum shear stress can be found as shown below.

$$\tau_{max} = \left|\frac{T_{AB}\rho_{max}}{J}\right| = \frac{(qL/3)(d/2)}{[\pi d^4/32]} = \frac{16qL}{3\pi d^3} \quad \text{or}$$

$$\boxed{\tau_{max} = \frac{16qL}{3\pi d^3}}$$

The relative rotation can be found as shown below.

$$\frac{d\phi}{dx} = \frac{\frac{q}{3L^2}(L^3 - x^3)}{G(\pi d^4/32)}$$ 4

Integrating from A to B we obtain:

$$\int_{\phi_A}^{\phi_B} d\phi = \left(\frac{32q}{3\pi GL^2 d^4}\right)\int_{x_A = 0}^{x_B = L}(L^3 - x^3)dx \qquad \text{or} \qquad \phi_B - \phi_A = \left(\frac{32q}{3\pi GL^2 d^4}\right)\left(L^3 x - \frac{x^4}{4}\right)\Bigg|_0^L$$

or

$$\boxed{\phi_B - \phi_A = \left(\frac{8qL^2}{\pi G d^4}\right) \text{ CCW}}$$

5.33 A circular solid shaft is acted upon by torques as shown. Determine the rotation of the rigid wheel A with respect to the fixed end C in terms of q, L, G and J.

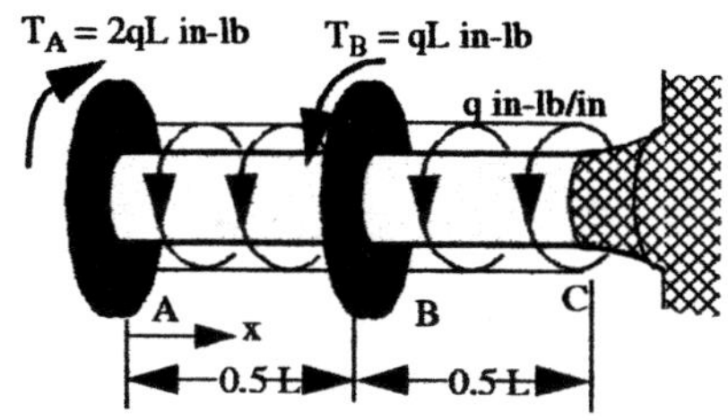

Fig. P5.33

Solution $\phi_A = f(q, L, G, J) = ?$

We can make imaginary cuts in segments AB and BC to obtain the following free body diagrams.

(a)

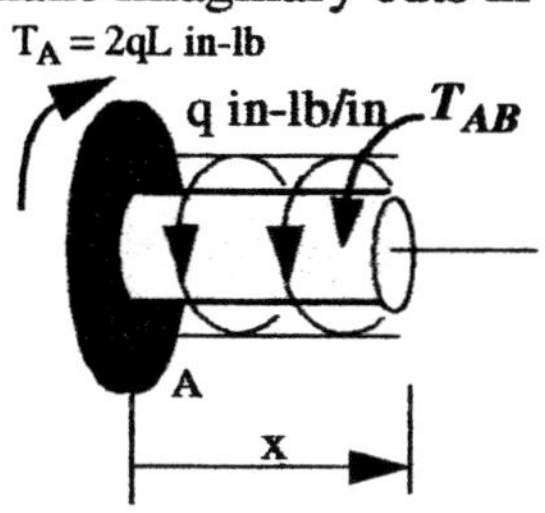

(b)

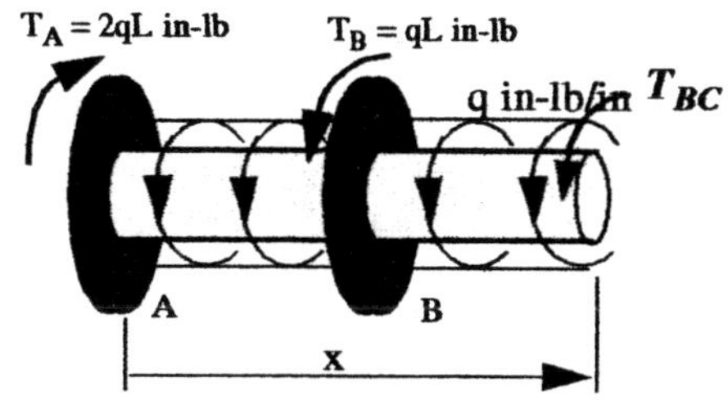

From moment equilibrium in Fig (a) we obtain:

$$T_{AB} - 2qL + qx = 0 \quad \text{or} \quad T_{AB} = 2qL - qx \qquad 1$$

From moment equilibrium in Fig (b) we obtain:

$$T_{BC} - 2qL + qx + qL = 0 \quad \text{or} \quad T_{BC} = qL - qx \qquad 2$$

The relative rotation for the ends of segments AB can be found by integrating from point A to point B as shown below

$$\left(\frac{d\phi}{dx}\right)_{AB} = \frac{T_{AB}}{G_{AB}J_{AB}} = \frac{2qL - qx}{GJ} \quad \text{or} \quad \int_{\phi_A}^{\phi_B} d\phi = \int_0^{0.5L} \left(\frac{2qL - qx}{GJ}\right) dx \quad \text{or}$$

$$\phi_B - \phi_A = \frac{1}{GJ}\left(2qLx - \frac{qx^2}{2}\right)\Bigg|_0^{0.5L} = \frac{7}{8}\left(\frac{qL^2}{GJ}\right) \qquad 3$$

The relative rotation for the ends of segments BC can be found by integrating from point A to point B as shown below.

$$\left(\frac{d\phi}{dx}\right)_{BC} = \frac{T_{BC}}{G_{BC}J_{BC}} = \frac{qL - qx}{GJ} \quad \text{or} \quad \int_{\phi_B}^{\phi_C} d\phi = \int_{0.5L}^{L} \left(\frac{qL - qx}{GJ}\right) dx \quad \text{or}$$

$$\phi_C - \phi_B = \frac{1}{GJ}\left(qLx - \frac{qx^2}{2}\right)\Bigg|_{0.5L}^{L} = \frac{1}{8}\left(\frac{qL^2}{GJ}\right) \qquad 4$$

Adding Eqs. (3) and (4), we obtain the rotation of section at C with respect to section at A. Noting that section at C is built into the wall we obtain the rotation of section at A as shown below.

$$\phi_C - \phi_A = \left(\frac{qL^2}{GJ}\right) \quad \text{or} \qquad \boxed{\phi_A = \left(\frac{qL^2}{GJ}\right) \text{ CW}}$$

5.34 A thin steel (G = 12,000 ksi) tube of 1/8 inch thickness has a mean diameter of 6 inches and a length of 36 inches. What is the maximum torque the tube can transmit if the allowable shear stress in is 10 ksi and the allowable relative rotation of the two ends is 0.015 rads?

Solution: $G = 12{,}000$ ksi $t = 1/8$ inch $d = 6$ inch $L = 36$ inch
$\tau_{max} \le 10\text{ksi}$ $\Delta\phi \le 0.015\text{rads}$ $T_{max} = ?$

The outer and inner diameters of the tube are: $d_o = 6 + (1/8) = 6.125$ in. and $d_i = 6 - (1/8) = 5.875$ in.

The polar moment of inertia is: $J = \frac{\pi}{32}[(6.125)^4 - (5.875)^4] = 21.215 \text{ in}^4$

The maximum stress is

$$\tau_{max} = \frac{T(d_o)/2}{J} = \frac{T(6.125/2)}{21.215} \le 10 \quad \text{or} \quad T \le 69.27\text{in} - \text{kips} \qquad 1$$

The relative rotation of the two ends is

$$\Delta\phi = \frac{TL}{GJ} = \frac{T(36)}{(12000)(21.211)} \le 0.015 \quad \text{or} \quad T \le 106.1\text{in} - \text{kips} \qquad 2$$

The maximum torque that satisfies Eq. (1) and (2) is $\boxed{T = 69.2\text{in} - \text{kips}}$

5.35 Determine the maximum torque that can be applied on a 2 inch diameter solid aluminum ($G = 4{,}000$ ksi) shaft if the allowable shear stress is 18 ksi and the relative rotation over 4 feet of the shaft is to be limited to 0.2 rads.

Solution: $d = 2$ inch. $G = 4{,}000$ ksi $L = 4$ feet
$\tau_{max} \le 18\text{ksi}$ $\Delta\phi \le 0.2\text{rads}$ $T_{max} = ?$

The polar moment of inertia is: $J = \frac{\pi}{32}(2)^4 = 1.5708 \text{ in}^4$

The maximum shear stress is

$$\tau_{max} = \frac{T(d/2)}{J} = \frac{T(1)}{1.5708} \le 18 \quad \text{or} \quad T \le 28.27\text{in} - \text{kips} \qquad 1$$

The relative rotation of the two ends is

$$\Delta\phi = \frac{TL}{GJ} = \frac{T(4)(42)}{(4000)(1.5708)} \le 0.2 \quad \text{or} \quad T \le 26.18\text{in} - \text{kips} \qquad 2$$

The maximum torque that satisfies Eq(1) and (2) is $\boxed{T = 28.1\text{in} - \text{kips}}$

5.36 A hollow steel ($G = 80$ GPa) shaft with an outside radius of 30 mm is to transmit a torque of 2700 N-m. The allowable shear stress is 120 MPa and the allowable relative rotation over 1 m is 0.1 rad. Determine the maximum permissible inner radius to the nearest millimeter.

Solution: $G = 80$ GPa $r_o = 3.0$ mm $T = 2700$ m
$L = 1$ m. $\tau_{max} \le 120\text{MPa}$ $\Delta\phi \le 0.1\text{rads}$ $(r_i)_{max} = ?$nearest mm.

The maximum shear stress is:

$$\tau_{max} = \frac{Tr_o}{J} = \frac{(2700)(30)(10^{-3})}{J} \le 120(10^6) \quad \text{or} \quad J \ge 675(10^{-9})\text{m}^4 \qquad 1$$

The relative rotation of the two ends is

$$\Delta\phi = \frac{TL}{GJ} = \frac{2700(1)}{(80)(10^9)(J)} \le 0.1 \quad \text{or} \quad J \ge 337.5(10^{-9})\text{m}^4 \qquad 2$$

If J is given by Eq(1) then it satisfies Eq(2).

Thus: $\frac{\pi}{2}(r_o^4 - r_i^4) \ge 675(10^{-9})$ or $r_i^4 \le (0.03)^4 - \frac{1350}{\pi}(10^{-9})$ or $r_i \le 0.0248\text{m}$

The maximum inner radius to the nearest millimeter is: $\boxed{(r_i)_{max} = 24\text{mm}}$

5.37 A 5 feet long hollow shaft is to transmit a torque of 200 in-kips. The outer diameter of the shaft must be 6 inches to fit existing attachments. The relative rotation of the two ends of the shaft is limited to 0.05 rads. The shaft can be made of steel or aluminum. The shear modulus of elasticity G, the allowable shear stress τ_{allow}, and the specific weight γ, are given in Table 5.1. Determine the maximum inner diameter to the nearest 1/8 inch of the lightest shaft that can be used for transmitting the torque and the corresponding weight.

Table 5.1.Material Properties in Problem 5.37.

Material	G ksi	τ_{allow} ksi	γ lbs./in^3
Steel	12,000	18	0.285
Aluminum	4,000	10	0.100

Solution L = 5 ft. T= 200 in-kips $d_o = 6$ inch $\Delta\phi \le 0.05$ rads

$(d_i)_{max}$ = ? nearest 1/8 inch W = ?

The relative rotation of the ends and maximum shear stress in steel can be found and the following limits on J_S, the polar moment of steel can be obtained.

$$(\Delta\phi)_S = \frac{(200)(5)(12)}{12(10^3)J_S} \le 0.05 \quad \text{or} \quad J_S \ge 20 \text{ in}^4 \tag{1}$$

$$(\tau_{max})_S = \frac{(200)(3)}{J_S} \le 18 \quad \text{or} \quad J_S \ge 33.33 \text{ in}^4 \tag{2}$$

Using similar calculations for Aluminum shaft we obtain the following limits on J_{Al}.

$$(\Delta\phi)_{Al} = \frac{(200)(5)(12)}{(4000)J_{Al}} \le 0.05 \quad \text{or} \quad J_{Al} \ge 60 \text{ in}^4 \tag{3}$$

$$(\tau_{max})_{Al} = \frac{(200)(3)}{J_{Al}} \le 10 \quad \text{or} \quad J_{Al} \ge 60 \text{ in}^4 \tag{4}$$

Thus if $J_S \ge 33.33$ in^4 it will meet both conditions in Eqs (1) and (2). Similarly if $J_{Al} \ge 60$ in^4 it will meet both conditions in Eqs. (3) and (4). The internal diameters d_S and d_{Al} can be found as shown below:

$$J_S = \frac{\pi}{32}(6^4 - d_S^4) \ge 33.33 \text{ in}^4 \quad \text{or} \quad d_S \le 5.561 \text{ in} \tag{5}$$

$$J_{Al} = \frac{\pi}{32}(6^4 - d_{Al}^4) \ge 60 \text{ in}^4 \quad \text{or} \quad d_{Al} \le 5.116 \text{ in} \tag{6}$$

Rounding downwards to the closest 1/8 inch we obtain:

$$d_S = 5.5 \text{ in} \quad \text{or} \quad d_{Al} = 5.0 \text{ in} \tag{7}$$

We can find the weight of each material by taking the product of the material density and the volume of a hollow cylinder as shown below:

$$W_S = (0.285)\left(\frac{\pi}{4}\right)(6^2 - 5.5^2)(60) = 77.224 \text{ lbs} \tag{8}$$

$$W_{Al} = (0.1)\left(\frac{\pi}{4}\right)(6^2 - 5.0^2)(60) = 51.8 \text{ lbs} \tag{9}$$

From Eqs 8 and 9 we see that the aluminum shaft is lighter. Thus, the answer is:

$$\boxed{(d_i)_{max} = 5.0 \text{ inch} \qquad W = 51.8 \text{ lbs}}$$

5.38 A 100-hp motor is driving a pulley and belt system as shown in Figure 5.38. If the system is to operate at 3,600 rpm, determine the minimum diameter of the solid shaft AB to the nearest 1/8 inch if the allowable torsional stress in the shaft is 10 ksi.

Fig. P5.38

Solution: P = 100 hpw = 3600 rpm$\tau_{max} \le 10\text{ksi}$ d_{min} = ? nearest 1/8 inch.

The rotational speed in rads/sec can be found as: $\omega = (3600)(2\pi)/60 = 376.99(\text{rad})/(\text{sec})$

The torque on the shaft can be found as shown below

$$T = \frac{P}{\omega} = \frac{(100)(6600)}{(376.99)} = 1751\text{in}-\text{lbs} \qquad 1$$

The maximum torsion and shear stress in terms of T can be found and from it the maximum torque can be obtained as shown below.

$$\tau_{max} = \frac{T(d/2)}{\frac{\pi}{32}d^4} = \frac{16}{\pi}\frac{1751}{d^3} \le 10(10^3) \qquad \text{or} \qquad d \ge 0.9625$$

The minimum diameter to the nearest inch 1/8 inch $\boxed{d_{min} = 1.0 \text{ in}}$

5.39 The bolts used in the coupling for transferring power in problem 5.38 have an allowable strength of 12 ksi. Determine the minimum number of 1/4 inch diameter bolts that must be placed at a radius of 5/8 inch.

Fig. P5.39

Solution: d_{bolt} = 1/4 inch $\tau_{bolt} \le 12\text{ksi}$ R = 5/8 inch n_{min} = ?

From Eq(1) in problem 5-38, we have T = 1751 in - kips

The area of cross section of a bolt is: $A = \frac{\pi}{4}(d_{bolt}^2) = 0.4909\text{in}^4$

The following force body diagram can be drawn

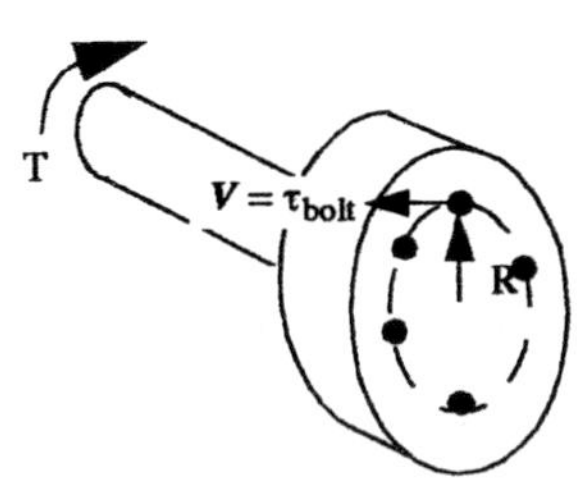

Each bolt resists a shear force of V. By moment equilibrium about the shaft axis we obtain

$$T = nVR = (nR)(\tau_{bolt})(A) \qquad \text{or} \qquad \tau_{bolt} = \frac{T}{nRA} = \frac{1751}{n(5/8)(0.04909)} \le 12(10^3) \qquad \text{or} \qquad n \ge 4.76$$

The minimum number of bolts needed are $\boxed{n_{min} = 5}$

5.40 A 20 kW motor drives 3 gears that are rotating at a frequency of 20 Hz. Gear A next to the motor transfers 8 kW of power. Gear B which is in the middle transfers 7 kW of power. Gear c which is at the far end from the motor transfers the remaining 5 kW of power. A single steel solid shaft connecting the motors to all three gears is to be used. The steel used has a yield strength in shear of 145 MPa. Assuming a factor of safety of 1.5, what is the minimum diameter of the shaft to the nearest millimeter that can be used if failure due to yielding is to be avoided? What is the magnitude of maximum torsional shear stress in the shaft segment between gears A and B.

Solution: $P = 20\text{ kW}$ $f = 20\text{ Hz.}$ $P_A = 8\text{kW}$ $P_B = 7\text{KN}$ $P_C = 5\text{ kW}$

$\tau_{yield} = 145\text{ MPa}$ $k = 1.5$ $d_{min} = ?$ to the nearest millimeter $\tau_{AB} = ?$

The torque delivered by the motor and the torques transferred by the gears can be found as shown below.

$$T = \frac{P}{2\pi f} = \frac{(20)(10^3)}{(2\pi)(20)} = 159.2\text{N}-\text{m} \qquad T_a = \frac{P_a}{2\pi f} = \frac{(8)(10^3)}{(2\pi)(20)} = 63.7\text{N}-\text{m}$$

$$T_b = \frac{P_b}{2\pi f} = \frac{(7)(10^3)}{(2\pi)(20)} = 55.7\text{N}-\text{m} \qquad T_c = \frac{P_c}{2\pi f} = \frac{(5)(10^3)}{(2\pi)(20)} = 39.8\text{N}-\text{m}$$

The shaft with the torques are shown in Fig (a) and the associated torque diagram shown in Fig(b). From the torque diagram the internal torques are:

$$T_{MA} = 159.2\text{N}-\text{m} \qquad T_{MB} = 95.5\text{N}-\text{m} \qquad T_{MB} = 39.8\text{N}-\text{m}$$

(a)

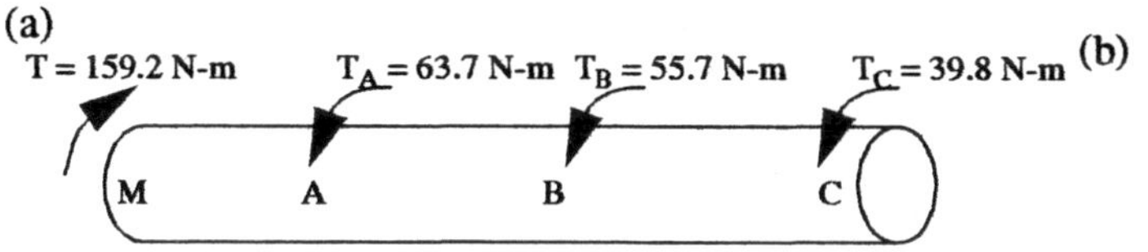

(b)

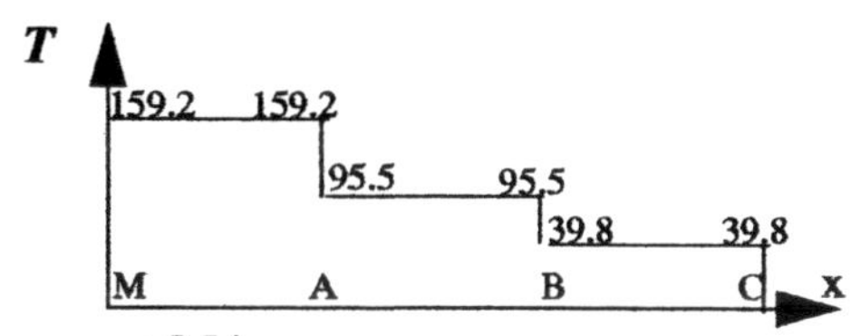

The maximum torsional shear stress in the shaft will be in segment MA

$$\tau_{max} = \frac{T_{MA}(d/2)}{(\pi d^4)/32} = \frac{16(159.2)}{\pi \quad d^3} \le \frac{145(10^6)}{1.5} \qquad \text{or} \qquad d \ge 0.0203\text{m}$$

The minimum diameter is $\boxed{d_{min} = 21\text{mm}}$

The maximum torsional shear stress in segment AB is

$$\tau_{AB} = \frac{T_{AB}(d/2)}{(\pi d^4)/32} = \frac{16 T_{AB}}{\pi \quad d^3} = \frac{16(95.5)}{\pi(0.021)^3} = 52.5(10^6) \text{ or} \qquad \boxed{\tau_{AB} = 52.5\text{MPa}}$$

5.41 A circular shaft has a constant torsional rigidity GJ and is acted upon by a distributed torque t(x). *If* at section A the *internal torque is zero* show that the relative rotation of section at B with respect to rotation of section at A is given by

$$\phi_B - \phi_A = \frac{1}{GJ}\left[\int_{x_A}^{x_B}(x - x_B)t(x)dx\right] \qquad \textbf{5.14}$$

Solution $T(x_A) = 0$ $\phi_B - \phi_A = ?$

From equilibrium equation, we have

$$\frac{dT}{dx} = -t(x) \qquad \textbf{1}$$

Integrating from $x = x_A$ to any location x, we obtain the following.

$$\int_{T_A=0}^{T} dT = -\int_{x_A}^{x} t(x)dx \qquad \text{or} \qquad T(x) = -\int_{x_A}^{x} t(x)dx \qquad \textbf{2}$$

Integrating the equation $\frac{d\phi}{dx} = \frac{T(x)}{GJ}$ from x_A to x_B, we obtain the following.

$$\int_{\phi_A}^{\phi_B} d\phi = \int_{x_A}^{x_A} \frac{T(x)}{GJ} dx \qquad \text{or} \qquad \phi_B - \phi_A = \frac{1}{GJ}\int_{x_A}^{x} T(x)dx \tag{3}$$

Integrating by parts, the above equation can be written as

$$\phi_B - \phi_A = \frac{1}{GJ}\left[xT(x)\Big|_{x_A}^{x_B} - \int_{x_A}^{x} x\frac{dT}{dx}dx \right] \tag{4}$$

Imposing the limits and substituting Eq. (1), we obtain the following.

$$\phi_B - \phi_A = \frac{1}{GJ}\left[x_B T(x_B) - T(x_A) - \int_{x_A}^{x} x(-t(x))dx \right] \tag{5}$$

Substituting Eq. (2) at $x = x_B$ and noting that $T(x_A) = 0$, we obtain

$$\phi_B - \phi_A = \frac{1}{GJ}\left[x_B\left\{ -\int_{x_A}^{x_B} t(x)dx \right\} - 0 + \int_{x_A}^{x_B} xt(x)dx \right] \tag{6}$$

Noting that x_B is a constant and can be taken inside the integral, we obtain:

$$\phi_B - \phi_A = \frac{1}{GJ}\left[\int_{x_A}^{x_B} (x - x_B)t(x)dx \right] \tag{7}$$

Eq. (7) is same as Eq. 5.14.

5.42 A solid circular shaft of radius R, and length L, is twisted by an applied torque T. The stress-strain relationship for a non-linear material is given by the power law $\tau = G\gamma^n$. If all assumptions except Hooke's law are applicable, show that the maximum shear stress in the shaft and the relative rotation of the two ends are as given below.

$$\tau_{max} = \frac{T(n+3)}{2\pi R^3} \qquad\qquad \Delta\phi = \left[\frac{(n+3)T}{2\pi G R^{(3+n)}}\right]^{\frac{1}{n}} L$$

Substitute n = 1 in the above formulas and show that we obtain the same results as from Equations 5.8 and 5.10.

Solution $\quad \tau = G\gamma^n \qquad \tau_{max} = ? \qquad \Delta\phi = ?$

From kinematics we have $\gamma_{x\theta} = \rho\frac{d\phi}{dx}(x)$. Substituting this in the given stress-strain relationship, we obtain.

$$\tau_{x\theta} = G\rho^n\left[\frac{d\phi}{dx}(x)\right]^n \tag{1}$$

The shear stress and internal torque are related as

$$T = \int_A \rho\tau_{x\theta} dA \tag{2}$$

Substituting Eq. (1) into Eq. (2), assuming homogenous material, and noting that $dA = 2\pi\rho\, d\rho$, we obtain the following:

$$T = \int_0^R \rho\left\{G\rho^n\left[\frac{d\phi}{dx}(x)\right]^n\right\}(2\pi\rho)d\rho = 2\pi G\left[\frac{d\phi}{dx}(x)\right]^n \int_0^R \rho^{n+2}d\rho = 2\pi G\left[\frac{d\phi}{dx}(x)\right]^n\left(\frac{R^{n+3}}{n+3}\right) \text{ or}$$

$$\left[\frac{d\phi}{dx}(x)\right]^n = \frac{T(n+3)}{2\pi G R^{n+3}} \qquad 3$$

Substituting Eq. (3) into Eq. (1), we obtain:

$$\tau_{x\theta} = G\rho^n\frac{T(n+3)}{2\pi G R^{n+3}} \qquad 4$$

The maximum torsional shear stress is at $\rho = R$: $\tau_{max} = GR^n\frac{T(n+3)}{2\pi G R^{n+3}}$ or $\boxed{\tau_{max} = \frac{T(n+3)}{2\pi R^3}}$

Eq. (2) can be re-written as:

$$\frac{d\phi}{dx} = \left[\frac{T(n+3)}{2\pi G R^{n+3}}\right]^{\frac{1}{n}} \qquad 5$$

Assuming all quantities on the right in Eq (5) are constant over the length of the shaft, then $\frac{d\phi}{dx} = \frac{\Delta\phi}{L}$.

Substituting this in Eq. (5) we obtain: $\boxed{\Delta\phi = \left[\frac{(n+3)T}{2\pi G R^{(3+n)}}\right]^{\frac{1}{n}} L}$

5.43 The internal torque T and the displacements of a point on a cross-section of a non-circular shaft are as given below.

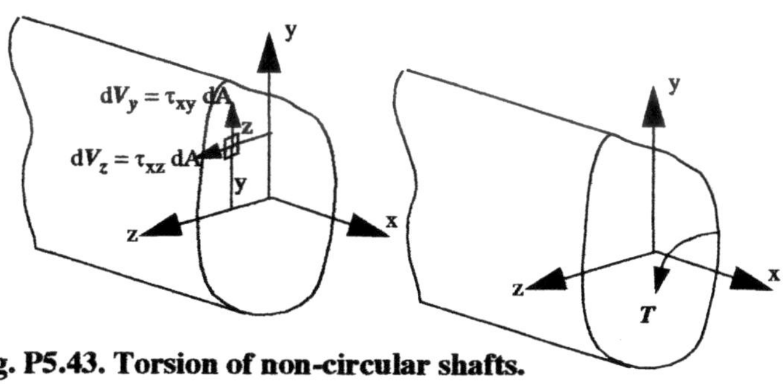

Fig. P5.43. Torsion of non-circular shafts.

$$u = \psi(y,z)\frac{d\phi}{dx} \qquad \textbf{5.15a}$$

$$v = -xz\frac{d\phi}{dx} \qquad \textbf{5.15b}$$

$$w = xy\frac{d\phi}{dx} \qquad \textbf{5.15c}$$

$$T = \int_A (y\tau_{xz} - z\tau_{xy})dA \qquad \textbf{5.16}$$

where u, v, and w are the displacements in the x, y, and z, respectively,

$\frac{d\phi}{dx}$ is the rate of twist and is considered *constant*,

$\psi(x, y)$ is called the warping function and describes the movement of points out of the plane of cross-section.

Using Equations 2.10d and 2.10f, and the Hooke's law show that the shear stresses for non-circular are given by:

$$\tau_{xy} = G\left(\frac{\partial\psi}{\partial y} - z\right)\frac{d\phi}{dx} \qquad \textbf{5.17a} \qquad\qquad \tau_{xz} = G\left(\frac{\partial\psi}{\partial z} + y\right)\frac{d\phi}{dx} \qquad \textbf{5.17b}$$

Solution

From Eq. 2.10d, we have $\gamma_{xy} = \frac{\partial u}{\partial y} + \frac{\partial v}{\partial x}$. Substituting Eqs. (5.15a) and (5.15b), we obtain the following.

$$\gamma_{xy} = \frac{\partial}{\partial y}\left[\psi(y,z)\frac{d\phi}{dx}\right] + \frac{\partial}{\partial x}\left[-xz\frac{d\phi}{dx}\right] = \frac{d\phi}{dx}\left[\frac{\partial\psi}{\partial y} - z\right] \qquad 1$$

From Hooke's law, we have $\tau_{xy} = G\gamma_{xy}$. Substituting Eq. (1), we obtain Eq. (5.17a).

From Eq. 2.10f, we have $\gamma_{xz} = \frac{\partial w}{\partial x} + \frac{\partial u}{\partial z}$. Substituting Eqs. (5.15a) and (5.15c), we obtain the following.

$$\gamma_{xz} = \frac{\partial}{\partial x}\left[xy\frac{d\phi}{dx}\right] + \frac{\partial}{\partial z}\left[\psi(y,z)\frac{d\phi}{dx}\right] = \frac{d\phi}{dx}\left[y + \frac{\partial\psi}{\partial z}\right] \qquad 2$$

From Hooke's law, we have $\tau_{xz} = G\gamma_{xz}$. Substituting Eq. (2), we obtain Eq. (5.17b).

5.44 Show that for circular shafts ($\psi(x, y) = 0$) the Equations in problem 5.43 reduce to Equation 5.7.

Solution

Substituting $\psi(x, y) = 0$ into Eqs. (5.17a) and (5.17b), we obtain:

$$\tau_{xy} = -Gz\frac{d\phi}{dx} \qquad \tau_{xz} = Gy\frac{d\phi}{dx} \qquad 1$$

Substituting Eq. (1) into Eq. 5.16, we obtain the following.

$$T = \iint_A \left[y\left(Gy\frac{d\phi}{dx}\right) - z\left(-Gz\frac{d\phi}{dx}\right)\right]dA = G\frac{d\phi}{dx}\int_A [y^2 + z^2]dA = G\frac{d\phi}{dx}\int_A [\rho^2]dA = GJ\frac{d\phi}{dx} \qquad 2$$

Eq. (2) is the same as Equation 5.7.

5.45 Consider the dynamic equilibrium of the differential element shown in Fig. P5.45, where T is the internal torque, ρ is the material density, J is the polar area moment of inertia, and $\frac{\partial^2\phi}{\partial t^2}$ is the angular acceleration. Show

$$\frac{\partial^2\phi}{\partial t^2} = c^2\frac{\partial^2\phi}{\partial x^2} \qquad \text{where } c^2 = \sqrt{\frac{G}{\rho}} \qquad \mathbf{5.18}$$

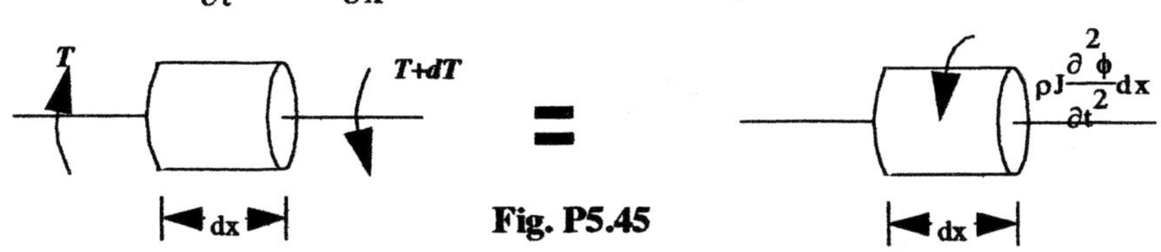

Fig. P5.45

Solution

By moment equilibrium in Fig. P5.45, we obtain the following.

$$T + dT - T = \rho J\frac{\partial^2\phi}{\partial t^2}dx \qquad \text{or} \qquad \frac{\partial T}{\partial x} = \rho J\frac{\partial^2\phi}{\partial t^2} \qquad 1$$

From Equation 5.7 we have $T = GJ\frac{\partial\phi}{\partial x}$. Substituting in Eq. (3) and assuming GJ as constant, we obtain

$$\frac{\partial}{\partial x}\left[GJ\frac{\partial\phi}{\partial x}\right] = \rho J\frac{\partial^2\phi}{\partial t^2} \qquad \text{or} \qquad GJ\left[\frac{\partial^2\phi}{\partial x^2}\right] = \rho J\frac{\partial^2\phi}{\partial t^2} \qquad \text{or} \qquad \frac{\partial^2\phi}{\partial t^2} = \left(\frac{G}{\rho}\right)\frac{\partial^2\phi}{\partial x^2} \qquad 2$$

Eq. (4) is same as Eq. 5.18.

5.46 Show by substitution that the solution below satisfies Equation 5.18.

$$\phi = \left(A\cos\frac{\omega x}{c} + B\sin\frac{\omega x}{c}\right)(C\cos\omega t + D\sin\omega t) \qquad \mathbf{5.19}$$

where A, B, C, and D are constants that are determined from the boundary conditions and the initial conditions, ω is the frequency of vibration.

Solution

The partial derivative of ϕ with respect to x can be found as shown below.

$$\frac{\partial\phi}{\partial x} = [C\cos\omega t + D\sin\omega t]\frac{\partial}{\partial x}\left[A\cos\frac{\omega x}{c} + B\sin\frac{\omega x}{c}\right] \text{ or}$$

$$\frac{\partial\phi}{\partial x} = [C\cos\omega t + D\sin\omega t]\left[-\left(\frac{\omega}{c}\right)A\sin\frac{\omega x}{c} + \left(\frac{\omega}{c}\right)B\cos\frac{\omega x}{c}\right] \qquad 1$$

The second partial derivative of ϕ with respect to x can be found as shown below

$$\frac{\partial^2\phi}{\partial x^2} = [C\cos\omega t + D\sin\omega t]\frac{\partial}{\partial x}\left[-\left(\frac{\omega}{c}\right)A\sin\frac{\omega x}{c} + \left(\frac{\omega}{c}\right)B\cos\frac{\omega x}{c}\right] \text{ or}$$

$$\frac{\partial^2\phi}{\partial x^2} = [C\cos\omega t + D\sin\omega t]\left[-\left(\frac{\omega}{c}\right)^2 A\cos\frac{\omega x}{c} - \left(\frac{\omega}{c}\right)^2 B\sin\frac{\omega x}{c}\right] \text{ or}$$

$$\frac{\partial^2\phi}{\partial x^2} = -\left(\frac{\omega}{c}\right)^2[C\cos\omega t + D\sin\omega t]\left(A\cos\frac{\omega x}{c} + B\sin\frac{\omega x}{c}\right) = -\left(\frac{\omega}{c}\right)^2\phi \qquad 2$$

The partial derivative of ϕ with respect to can be found as shown below.

$$\frac{\partial\phi}{\partial t} = \left[A\cos\frac{\omega x}{c} + B\sin\frac{\omega x}{c}\right]\frac{\partial}{\partial t}[C\cos\omega t + D\sin\omega t] \text{ or}$$

$$\frac{\partial\phi}{\partial t} = \left[A\cos\frac{\omega x}{c} + B\sin\frac{\omega x}{c}\right][-\omega C\sin\omega t + \omega D\cos\omega t] \qquad 3$$

The second partial derivative of ϕ with respect to t can be found as shown below

$$\frac{\partial^2\phi}{\partial t^2} = \left[A\cos\frac{\omega x}{c} + B\sin\frac{\omega x}{c}\right]\frac{\partial}{\partial t}[-\omega C\sin\omega t + \omega D\cos\omega t] \text{ or}$$

$$\frac{\partial^2\phi}{\partial t^2} = \left[A\cos\frac{\omega x}{c} + B\sin\frac{\omega x}{c}\right][-\omega^2 C\cos\omega t - \omega^2 B\sin\omega t] \text{ or}$$

$$\frac{\partial^2\phi}{\partial t^2} = -\omega^2\left[A\cos\frac{\omega x}{c} + B\sin\frac{\omega x}{c}\right][C\cos\omega t + D\sin\omega t] = -\omega^2\phi \qquad 4$$

Substituting Eqs. (2) and (4) into Eq. 5.18, we obtain an identity, proving Eq. 5.19 is the solution to Eq. 5.18.

5.47 A hollow aluminum shaft of 5 feet length is to carry a torque of 200 in-kips. The inner radius of the shaft is 1 inch. If the maximum torsional shear stress in the shaft is to be limited to 10 ksi, determine the minimum outer radius of the shaft to the nearest 1/8 inch.

Solution $L = 5$ ft $\quad T = 200$ in-kips $\quad R_i = 1$ inch

$\tau_{max} \le 10$ ksi $\quad R_o = ?$ nearest 1/8 inch

The maximum torsional shear stress will exist on the outer radius of the shaft and can be expressed in terms of R_o as shown below.

$$\tau_{max} = \frac{T\rho}{J} = \frac{(200)(R_o)}{\frac{\pi}{2}[R_o^4 - 1^4]} \le 10 \qquad \text{or} \qquad (R_o^4 - 12.732R_o - 1) \ge 0 \qquad 1$$

The minimum value of R_o corresponds to the root of the left hand side of Eq. 1, i.e.,

$f(R_o) = R_o^4 - 12.732R_o - 1 = 0$. We find the root on the spread sheet as described in Appendix B. The calculations are shown in the table below.

R_o	$f(R_o)$	R_o	$f(R_o)$
1.1	-13.542	2.3	-2.300
1.2	-14.205	2.31	-1.938
1.3	-14.696	2.32	-1.569
1.4	-14.984	2.33	-1.194
1.5	-15.036	2.34	-0.812
1.6	-14.818	2.35	-0.423
1.7	-14.293	2.36	-0.028
1.8	-13.421	2.37	0.374
1.9	-12.159	2.38	0.782
2	-10.465	2.39	1.198
2.1	-8.290	2.4	1.620
2.2	-5.586		
2.3	-2.300		
2.4	1.620		
2.5	6.232		

From the above table, we see that the root of $f(R_o)$ is between 2.36 and 2.37. The value of R_o to the nearest 1/8 that will satisfy Eq. (1) is 2.375. Thus, the minimum value of R_o is: $\boxed{R_o = 2\frac{3}{8} \text{ in}}$

5.48 A 4 feet long hollow shaft is to transmit a torque of 100 in-kips. The relative rotation of the two ends of the shaft is limited to 0.06 rads. The shaft can be made of steel or aluminum. The Shear modulus of rigidity G, the allowable shear stress τ_{allow}, and the specific weight γ, are given in Table 5.1. The inner radius of the shaft is to be 1 inch. Determine outer radius to the nearest 1/8 inch of the lightest shaft that can be used for transmitting the torque and the corresponding weight.

Solution $L = 4$ ft. $T = 100$ in-kips $\Delta\phi \le 0.06$ rads
$R_i = 1$ inch $R_o = ?$ to the nearest 1/8 inch $W = ?$

The relative rotation of the ends and maximum shear stress in steel can be found and the following limits on d_o can be obtained.

$$(\Delta\phi)_S = \frac{(100)(4)(12)}{12(10^3)\left(\frac{\pi}{2}\right)[R_o^4 - 1^4]} \le 0.06 \qquad \text{or} \qquad R_o \ge 1.513 \text{ in} \qquad 1$$

$$(\tau_{max})_S = \frac{(100)(R_o)}{\left(\frac{\pi}{2}\right)[R_o^4 - 1^4]} \le 18 \qquad \text{or} \qquad R_o^4 - 3.5368R_o - 1 \ge 0 \qquad 2$$

Substituting the value of R_o in Eq. 1 into the left hand side we obtain a negative value, which implies R_o must be greater than 1.513 inch to satisfy Eq. 2. The value of R_o corresponds to the root of the left hand side of Eq. 2, i.e., $f(R_o) = R_o^4 - 3.536R_o - 1 = 0$. We find the root on the spread sheet as described in Appendix B. The calculations are shown in the table below.

R_o	$f(R_o)$
1.5	-1.243
1.6	-0.105
1.7	1.340
1.8	3.131
1.9	5.312
2	7.926

R_o	$f(R_o)$
1.6	-0.105
1.61	0.025
1.62	0.158
1.63	0.294
1.64	0.434
1.65	0.576

From the above table, we see that the root of $f(R_o)$ is between 1.6 and 1.61. Rounding to the nearest 1/ 8 inch, we obtain for steel.

$$R_O = 1.625\text{inch} \qquad 3$$

Using similar calculations for Aluminum shaft we obtain the following.

$$(\Delta\phi)_{Al} = \frac{(100)(4)(12)}{(4000)\left(\frac{\pi}{2}\right)[r_o^4 - 1^4]} \le 0.06 \qquad \text{or} \qquad r_o \ge 1.925 \text{ in} \qquad 4$$

$$(\tau_{max})_{Al} = \frac{(100)(r_o)}{\left(\frac{\pi}{2}\right)[r_o^4 - 1^4]} \le 10 \qquad \text{or} \qquad r_o^4 - 6.366r_o - 1 \ge 0 \qquad 5$$

Substituting the value of r_o in Eq. (4) into the left hand side we obtain a positive value, which implies that $r_o = 1.925\text{inch}$ satisfies both Eqs. (4) and (5). Rounding to the nearest 1/8 inch, we obtain for aluminum.

$$r_o = 2.0\text{inch} \qquad 6$$

We can find the weight of each material by taking the product of the material density and the volume of a hollow cylinder as shown below:

$$W_S = (0.285)(\pi)(1.625^2 - 1^2)(48) = 70.51 \text{ lbs} \qquad 7$$

$$W_{Al} = (0.1)(\pi)(2^2 - 1^2)(48) = 45.24 \text{ lbs} \qquad 8$$

From Eqs 7 and 8 we see that the aluminum shaft is lighter. Thus, the answer is:

$$\boxed{r_o = 2.0\text{inch} \qquad W = 45.24 \text{ lbs}}$$

5.49 Table 5.2 shows the measured radii of a solid tapered shaft at several point along the axis of the shaft. The shaft is made of aluminum (G = 28 GPa) and has a length of 1.5 meters. Determine (a) the rotation of the free end with respect to the wall using numerical integration. (b) the maximum shear stress in the shaft.

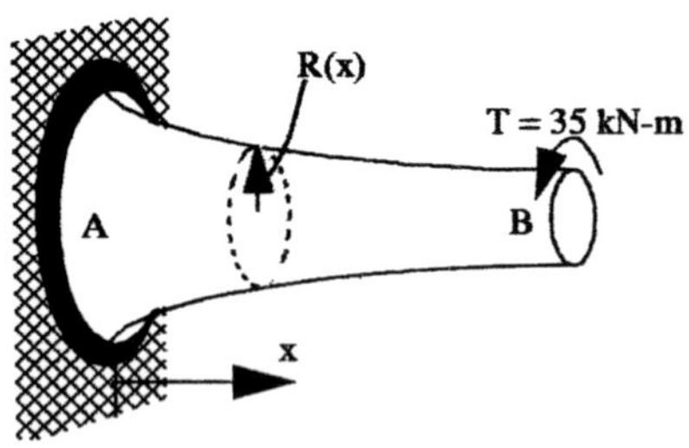

Fig. P5.49

Table 5.2.. Data for Problem 5.49.

x (m)	R(x) (mm)	x (m)	R(x) (mm)
0.0	100.6	0.8	60.1
0.1	92.7	0.9	60.3
0.2	82.6	1.0	59.1
0.3	79.6	1.1	54.0
0.4	75.9	1.2	54.8
0.5	68.8	1.3	54.1
0.6	68.0	1.4	49.4
0.7	65.9	1.5	50.6

Solution G = 28 GPa $x_A = 0$ $x_B = 1.5$ m $\Delta\phi = \phi_B - \phi_A = ?$ $\tau_{max} = ?$

The internal torque is T = 35 kN-m.

$$\frac{d\phi}{dx} = \frac{T}{GJ} = \frac{(35)(10^3)}{(28)(10^9)(\pi/2)R^4(x)} = \frac{2.5(10^{-6})}{\pi R^4(x)} \qquad 1$$

In Eq. (1), R(x) is measured in meters. Eq. (1) can be rewritten as $\frac{d\phi}{dx} = \frac{2.5(10^6)}{\pi R^4(x)}$, where now R(x) is in millimeters. Integrating from $x_A = 0$ to $x_B = 1.5$, we obtain the following:

$$\int_{\phi_A}^{\phi_B} d\phi = \int_{x_A=0}^{x_B=1.5} \left[\frac{2.5(10^6)}{\pi R^4(x)}\right] dx \quad \text{or} \quad \Delta\phi = \int_0^{1.5} \left[\frac{2.5(10^6)}{\pi R^4(x)}\right] dx \qquad 2$$

Representing $f(x) = \frac{2.5(10^6)}{\pi R^4(x)}$ and using numerical integration as described in Appendix B, we obtain the value of the integral on a spread sheet as shown below.

x	R(x)	f(x)	Integral
0.00	100.60	0.0078	0.0009
0.10	92.70	0.0108	0.0023
0.20	82.60	0.0171	0.0042
0.30	79.60	0.0198	0.0064
0.40	75.90	0.0240	0.0093
0.50	68.80	0.0355	0.0130
0.60	68.00	0.0372	0.0169
0.70	65.90	0.0422	0.0221

x	R(x)	f(x)	Integral
0.80	60.10	0.0610	0.0282
0.90	60.30	0.0602	0.0344
1.00	59.10	0.0652	0.0424
1.10	54.00	0.0936	0.0515
1.20	54.80	0.0882	0.0605
1.30	54.10	0.0929	0.0718
1.40	49.40	0.1336	0.0846
1.50	50.60	0.1214	

The rotation of the free end is: $\boxed{\Delta\phi = 0.085 \text{ rad}}$

The maximum shear stress will occur at the cross-section where radius is the smallest, i.e., just before point B and can be found as shown below.

$$\tau_{max} = \frac{T\rho}{J_B} = \frac{35(10^3)(0.0506)}{(\pi/2)(0.0506^4)} = 171.99(10^6)\ \text{N/m}^2 \text{ or} \qquad \boxed{\tau_{max} = 172 \text{ MPa}}$$

5.50 Let the radius of the tapered shaft in problem 5.49 be represented by the equation $R(x) = a + bx$. Using the data in Table 5.2 determine the constant a and b by the least-square method and then find the rotation of section at B by analytical integration.

Solution G = 28 GPa a = ? b = ? $x_A = 0$ $x_B = 1.5$ m $\Delta\phi = \phi_B - \phi_A = ?$

From results in problem 4-40, we obtain $\boxed{a = 90.226 \text{ mm} \quad b = -30.593}$

Thus R(x) = 90.226-30.593x, where x is in meters and R is in millimeters. Substituting this in Eq. (2) of problem 5.49 and integrating, we obtain the rotation as shown below.

$$\Delta\phi = \int_0^{1.5}\left[\frac{2.5(10^6)}{\pi(90.226-30.593x)^4}\right]dx = \left(\frac{2.5(10^6)}{\pi}\right)\left[\frac{1}{(3)(30.593)(90.226-30.593x)^3}\right]\Bigg|_0^{1.5} = 0.0877$$

Thus the rotation of the free end of the shaft is: $\boxed{\Delta\phi = 0.088 \text{ rad}}$

5.51 Table 5.3 shows the value of distributed torque at several point along the axis of a solid steel (G = 12,000 ksi) shaft. The shaft has a length of 36 inches and a diameter of 1 inch. Determine (a) the rotation of the end A with respect to the wall using numerical integration. (b) the maximum shear stress in the shaft.(Hint: Use Equation 5.14 for part (a))

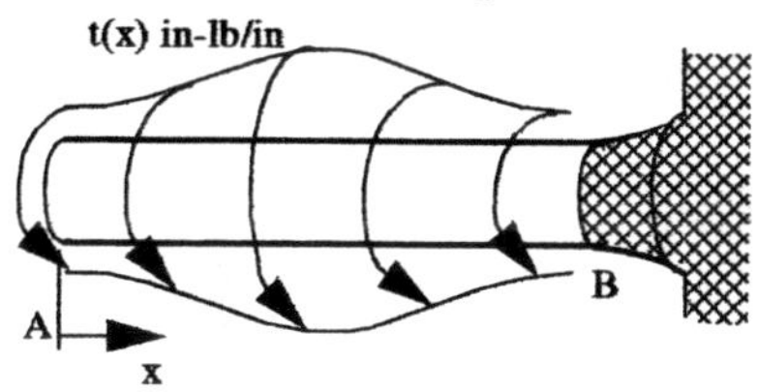

Fig. P5.51

Table 5.3.. Data for Problem 5.51.

x (inches)	t (x) in-lbs./in	x (inches)	t (x) in-lbs./in
0	93.0	21	588.8
3	146.0	24	700.1
6	214.1	27	789.6
9	260.4	30	907.4
12	335.0	33	1040.3
15	424.7	36	1151.4
18	492.6		

Solution G = 12,000 ksi d= 1 inch $u_A = ?$ $\sigma_{max} = ?$

The equilibrium equation $\frac{dT}{dx} + t(x) = 0$ can be integrated from point A, where $T = 0$ to any location x, to obtain the internal torque as a function of x as shown below.

$$T(x) = \int_{T_A=0}^{T(x)} dT = -\int_{x_A=0}^{x} t(x)dx \qquad 1$$

The polar moment of inertia of the cross-section is $J = \pi d^4/32 = 0.0981 \text{in}^4$.

$$\frac{d\phi}{dx} = \frac{T(x)}{GJ} = \frac{T(x)}{(12)(10^6)(0.0981)} = 848.8(10^{-9})T(x) \qquad 2$$

Integrating from point A i.e., $x_A = 0$ to point B, i.e., $x_B = 36$. We note that point B is fixed to the wall, hence $\phi_B = 0$. we obtain the following integral:

$$\int_{\phi_A}^{\phi_B=0} d\phi = \int_{x_A=0}^{x_B=36} [848.8(10^{-9})T(x)]dx \quad \text{or} \quad \phi_A = -\int_0^{36}[848.8(10^{-9})T(x)]dx \qquad 3$$

The maximum torsional shear stress $\tau_{x\theta}$ at a section can be found as shown below:

$$\tau_{x\theta} = \frac{T\rho}{J} = \frac{T(0.5)}{0.0981} = 5.0929T \qquad 4$$

Eq. (1) can be numerically integrated on a spread sheet to obtain $T(x_i)$, the value of internal torque at any x_i. Then, Eq (3) can be numerically integrated to obtain the rotation. Eq. (4) can be used to find the torsional shear stress at various x_i and the maximum value chosen by inspection. These calculations can be done on a spread sheet as shown below.

x_i (inches)	$t(x_i)$ (in-lbs./in)	$T(x_i)$ (in-lbs.)	$[\phi_A-\phi(x_{i+1})](10^{-3})$ (rads)	$\tau_{x\theta}(x_i)$ (ksi)
0	93	0.00	0.456	0.000
3	146.0	-358.50	2.057	-1.826
6	214.1	-898.65	5.251	-4.577
9	260.0	-1609.80	10.487	-8.199
12	335.0	-2502.30	18.310	-12.744
15	424.7	-3641.85	29.334	-18.548
18	492.0	-5016.90	44.174	-25.551
21	588.8	-6638.10	63.539	-33.808
24	700.1	-8571.45	88.211	-43.654
27	789.6	-10806.00	118.970	-55.034
30	907.4	-13351.50	156.689	-67.999
33	1040.3	-16273.05	202.314	-82.878
36	1151.4	-19560.60		-99.621

The rotation of section at A is counter-clockwise with respect to x-axis. $\boxed{\phi_A = 0.202 \text{ rads}}$

The magnitude of the maximum torsional shear stress is $\boxed{\tau_{max} = 99.6 \text{ ksi}}$

5.52 Let the distributed torque t(x) in problem 5.51 be represented by the equation $t(x) = a + bx + cx^2$. Using the data in Table 5.3 determine the constant a, b, and c by the least-square method and then find the rotation of section at B by analytical integration.

Solution $a = ?$ $b = ?$ $c =$ $\phi_A = ?$

The equilibrium equation $\frac{dT}{dx} + t(x) = 0$ can be integrated from point A, where $T = 0$ to any location x, to obtain the internal axial force as a function of x as shown below.

$$T(x) = \int_{T_A=0}^{T(x)} dT = -\int_{x_A=0}^{x} t(x)dx = -\int_{x_A=0}^{x} [a + bx + cx^2]dx = -\left[ax + \frac{bx^2}{2} + \frac{cx^3}{3}\right] \quad 1$$

The polar moment of inertia of the cross-section is $J = \pi d^4/32 = 0.0981 \text{in}^4$.

$$\frac{d\phi}{dx} = \frac{T(x)}{GJ} = \frac{T(x)}{(12)(10^6)(0.0981)} = -848.82(10^{-9})\left[ax + \frac{bx^2}{2} + \frac{cx^3}{3}\right] \quad 2$$

Integrating from point A i.e., $x_A = 0$ to point B, i.e., $x_B = 36$. We note that point B is fixed to the wall, hence $\phi_B = 0$. we obtain the following integral: $\int_{\phi_A}^{\phi_B = 0} d\phi = \int_{x_A=0}^{x_B=36} (-848.82)(10^{-9})\left[ax + \frac{bx^2}{2} + \frac{cx^3}{3}\right]dx$ or

$$\phi_A = (848.82)(10^{-9})\left(\frac{ax^2}{2} + \frac{bx^3}{6} + \frac{cx^4}{12}\right)\Bigg|_0^{36} = (848.82)(10^{-9})\left[\frac{a(36)^2}{2} + \frac{b(36)^3}{6} + \frac{c(36)^4}{12}\right] \quad 3$$

Using the Least Square Method described in Appendix B, we obtain the value of the values of constants a, b, and c on a spread sheet as shown in the table below.

	x_i	$t(x_i)$	x_i^2	x_i^3	x_i^4	$x*t_i$	$x_i^2*t_i$
1	0.0	93	0.0	0.000E+00	0.000E+00	0.000E+00	0.000E+00
2	3.0	146.0	9.0	2.700E+01	8.100E+01	4.380E+02	1.314E+03
3	6.0	214.1	36.0	2.160E+02	1.296E+03	1.285E+03	7.708E+03
4	9.0	260.0	81.0	7.290E+02	6.561E+03	2.340E+03	2.106E+04
5	12.0	335.0	144.0	1.728E+03	2.074E+04	4.020E+03	4.824E+04
6	15.0	424.7	225.0	3.375E+03	5.063E+04	6.371E+03	9.556E+04
7	18.0	492.0	324.0	5.832E+03	1.050E+05	8.856E+03	1.594E+05
8	21.0	588.8	441.0	9.261E+03	1.945E+05	1.236E+04	2.597E+05
9	24.0	700.1	576.0	1.382E+04	3.318E+05	1.680E+04	4.033E+05
10	27.0	789.6	729.0	1.968E+04	5.314E+05	2.132E+04	5.756E+05
11	30.0	907.4	900.0	2.700E+04	8.100E+05	2.722E+04	8.167E+05
12	33.0	1040.3	1089.0	3.594E+04	1.186E+06	3.433E+04	1.133E+06
13	36.0	1151.4	1296.0	4.666E+04	1.680E+06	4.145E+04	1.492E+06
b_{ij} &r_i	234.0	7142.4	5850.0	1.643E+05	4.918E+06	1.768E+05	5.014E+06
C_{ij}	1.783E+09	1.897E+08	4.216E+06	2.971E+07	7.666E+05	2.129E+04	
D	3.453E+09						
a_i	96.36	15.45	0.39				

The values of the constants are: $a = 96.36 \quad b = 15.45 \quad c = 0.39$

Substituting the values of the above constants in Eq. (3) we obtain:

$$\phi_A = (848.82)(10^{-9})\left[\frac{(96.36)(36)^2}{2} + \frac{(15.45)(36)^3}{6} + \frac{(0.39)(36)^4}{12}\right] = 0.2013 \text{ or } \quad \boxed{\phi_A = 0.2013 \text{ rads}}$$

Problem Set 5.3

5.53 A steel (G_s =12,000 ksi) and a bronze (G_{Br} = 5,600 ksi) shaft are securely connected at B as shown. Determine the maximum torsional shear stress in the shaft and rotation of section at B if the applied torque T = 50 in-kips.

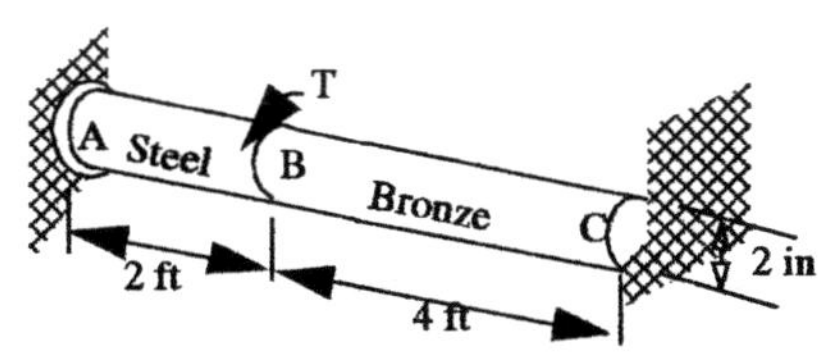

Fig. P5.53

Solution $G_{AB} = 12000$ ksi $\quad G_{BC} = 5600$ksi $\quad T = 50$in–kips τ_{max}=? ϕ_B=?

The polar moment of inertia for the shaft cross-section can be found as: $J = (\pi 2^4)/32 = 1.5708 \text{in}^4$

The following free body diagrams can be drawn with internal torques drawn as per our sign convention.

(a) 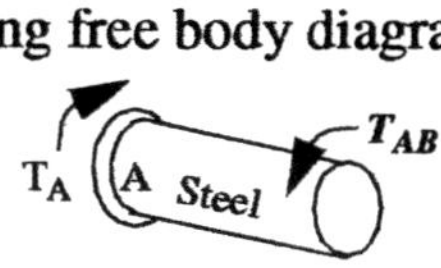

(b) 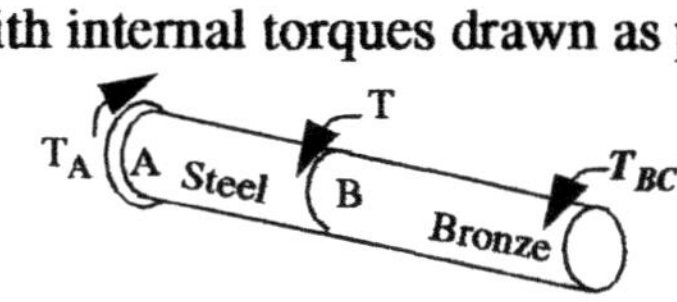

By moment equilibrium about the shaft axis we have the following equations.

$$T_{AB} = T_A \qquad T_{BC} = T_A - T \qquad (1)$$

The relative rotations of the ends of the each segment can be written as

$$\phi_B - \phi_A = \frac{T_{AB}(x_B - x_A)}{G_{AB}J} = \frac{T_A(2)(12)}{(12000)(1.5708)} = 1.273(10^{-3})T_A \qquad (2)$$

$$\phi_C - \phi_B = \frac{T_{BC}(x_C - x_B)}{G_{BC}J} = \frac{(T_A - 50)(4)(12)}{(5600)(1.5708)} = 5.456(10^{-3})(T_A - T) \qquad 3$$

Adding Eq's(2) and (3) and noting that points A and C are built into the wall, hence $\phi_C = 0$ and $\phi_A = 0$

$\phi_C - \phi_A = [1.273T_A + 5.456(T_A - T)](10^{-3}) = 0$ or

$$T_A = \frac{5.456}{(1.273 + 5.456)}T = (0.8108)(50) = 40.541\text{in} - \text{kips}$$

From Eq.(2) we obtain $\phi_B = 1.273(10^{-3})(40.541) = 51.608(10^{-3})\text{rads}$ or $\boxed{\phi_B = 0.0516 \text{ rads ccw}}$

From Eq.(1) , we obtain $T_{AB} = 40.541 \qquad T_{BC} = -9.459\text{in} - \text{kips}$

The magnitude of the maximum torsional shear stress will be in segment AB and can be found as

$$\tau_{max} = \left|\frac{T_{AB}\rho_{max}}{J}\right| = \frac{(40.541)(1)}{1.5708} \quad \text{or} \qquad \boxed{\tau_{max} = 25.8\text{ksi}}$$

5.54 A steel (G_s =12,000 ksi) and a bronze (G_{Br} = 5,600 ksi) shaft are securely connected at B as shown. Due to the torque T, the section at B rotates by an amount of 0.02 rads. Determine the maximum torsional shear strain and the applied torque T.

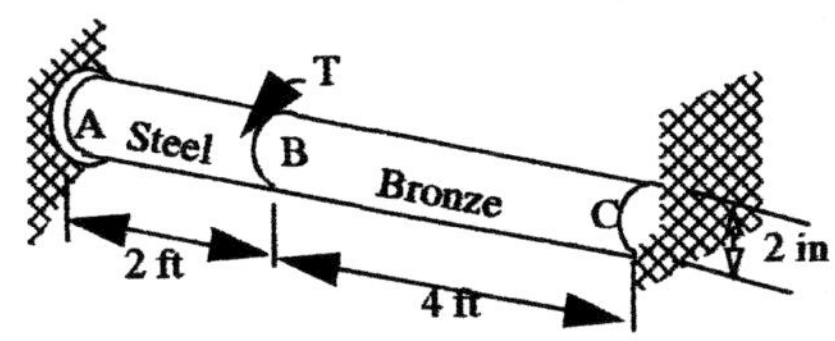

Fig. P5.54

Solution $\quad \phi_B = 0.02 \text{ rads} \qquad \gamma_{max} = ? \qquad T = ?$

The rate of twist in each segment can be found as shown below:

$$\left(\frac{d\phi}{dx}\right)_{AB} = \frac{\phi_B - \phi_A}{x_B - x_A} = \frac{0.02 - 0}{24} = 0.833(10^{-3})\ \frac{\text{rads}}{\text{in}} \qquad 1$$

$$\left(\frac{d\phi}{dx}\right)_{BC} = \frac{\phi_C - \phi_B}{x_C - x_B} = \frac{0 - 0.02}{48} = -0.4167(10^{-3})\ \frac{\text{rads}}{\text{in}} \qquad 2$$

The maximum torsional shear strain will be in AB and can be found as shown

$$\gamma_{max} = \rho_{max}\left(\frac{d\phi}{dx}\right)_{AB} = (1)(0.833)(10^{-3}) \quad \text{or} \qquad \boxed{\gamma_{max} = 833\ \mu\text{rads}}$$

Substituting $\phi_B = 0.02$ and $\phi_A = 0$ in Eq.(2) in problem 5.53 we obtain

$$T_A = \frac{0.02}{1.273(10^{-3})} = 15.711 \text{ in} - \text{kips}$$

Substituting $\phi_B = 0.02$ and $\phi_C = 0$ in Eq.(3) in problem 5.53 we obtain

$$(15.711 - T) = \frac{-0.02}{5.456(10^{-3})} \quad \text{or} \quad T = 19.377 \text{ in} - \text{kips} \qquad \boxed{T = 19.4 \text{ in} - \text{kips}}$$

5.55 Two hollow aluminum (G = 10,000 ksi) shafts are securely fastened to a solid aluminum shaft and loaded as shown. Point E is on the inner surface of the shaft. If T= 300 in-kips, Determine (a) the rotation of section at C with respect to rotation the wall at A. (b) the shear strain at point E.

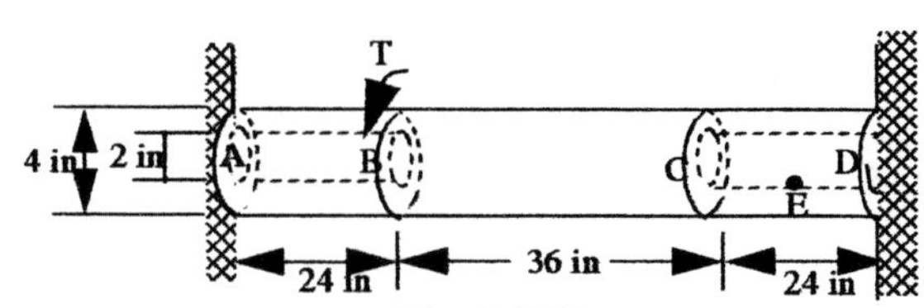

Fig. P5.55

Solution $\quad G = 10000 \text{ ksi} \qquad T = 300 \text{ in-kips} \qquad \phi_C - \phi_A = ? \qquad \gamma_E = ?$

The polar moment of cross - sections in various segments can be found as shown below

$$J_{AB} = J_{CD} = \frac{\pi}{32}(4^4 - 2^4) = 23.562 \text{ in}^4 \qquad 1$$

$$J_{BC} = \frac{\pi}{32}4^4 = 25.133 \text{ in}^4 \qquad 2$$

The following free body diagrams can be drawn with internal torques drawn as per our sign convention.

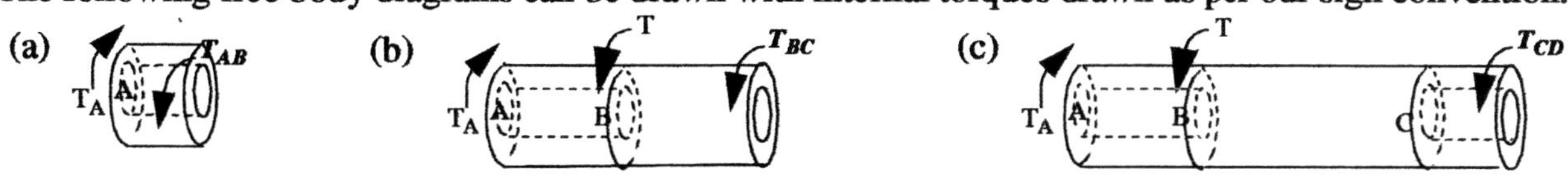

By moment equilibrium about the shaft axis we have the following equations.

$$T_{AB} = T_A \qquad T_{BC} = T_A - T \qquad T_{CD} = T_A - T \qquad 3$$

The relative rotation of the ends of each segment can be written as shown below

$$\phi_B - \phi_A = \frac{T_{AB}(x_B - x_A)}{GJ_{AB}} = \frac{(T_A)(24)}{(10000)(23.562)} = 0.1019(10^{-3})T_A \qquad 4$$

$$\phi_C - \phi_B = \frac{T_{BC}(x_C - x_B)}{GJ_{BC}} = \frac{(T_A - T)(36)}{(10000)(23.562)} = 0.1527(10^{-3})(T_A - T) \qquad 5$$

$$\phi_D - \phi_C = \frac{T_{CD}(x_D - x_C)}{GJ_{CD}} = \frac{(T_A - T)(24)}{(10000)(23.562)} = 0.1019(10^{-3})(T_A - T) \qquad 6$$

Adding Eqs. (4),(5) and(6) and noting that points A and C are built into wall, hence $\phi_D = 0$ and $\phi_A = 0$

$$\phi_D - \phi_A = [0.1019T_A + 0.1527(T_A - T) + 0.1019(T_A - T)](10^{-3}) = 0 \qquad \text{or}$$

$$T_A = \frac{(0.1527 + 0.1019)T}{0.1019 + 0.1527 + 0.1019} = 0.7141T = 214.27 \qquad 7$$

Substituting Eq.(7) into Eqs.(4) and (5) and adding, we obtain,

$\phi_C - \phi_A = [0.1019(214.27) + 0.1527(211.93 - 300)](10^{-3}) = 8.738(10^{-3})$ $\boxed{\phi_C - \phi_A = 0.00874 \text{ rads CCW}}$

From Eq.(3) and (7), we obtain: $T_{CD} = -85.75 \text{ in-kips}$

The shear stress at point E is: $\tau_E = \frac{T_{CD}\rho_E}{J_{CD}} = \frac{(-85.75)(1)}{23.562} = -3.639 \text{ ksi}$

The shear strain at point E is $\gamma_E = \frac{\tau_E}{G} = \frac{-3.639}{10000} = -0.3639(10^{-3})$ $\boxed{\gamma_E = -364 \ \mu\text{rads}}$

5.56 Two hollow aluminum (G = 10,000 ksi) shafts are securely fastened to a solid aluminum shaft and loaded as shown. Point E is on the inner surface of the shaft. If the torsional shear strain at point E is - 250 μ, determine the rotation of section C and the applied torque T that produced this shear strain.

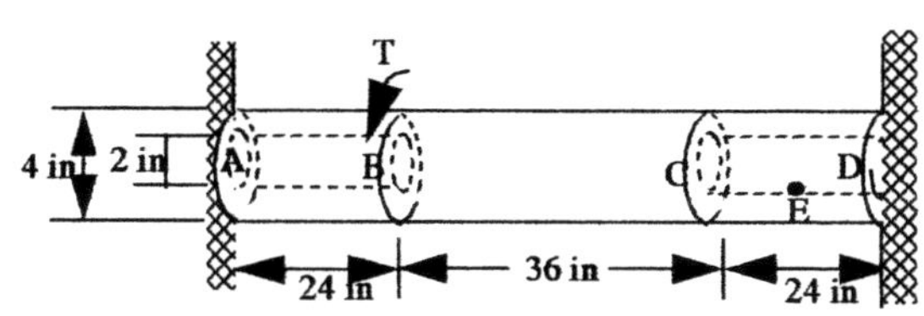

Fig. P5.56

Solution $G = 10000 \text{ ksi}$ $\gamma_E = -250 \ \mu\text{rads}$ $\phi_C = ?$ $T = ?$

The rotation of the section at C can be found as show below

$$\gamma_E = \rho_E\left(\frac{d\phi}{dx}\right)_{CD} \text{ or } \gamma_E = (-250)(10^{-6}) = (1)\frac{(\phi_D - \phi_C)}{x_D - x_C} \text{ or } (-250)(10^{-6}) = (1)\frac{(0 - \phi_C)}{24} \text{ or}$$

$\phi_C = 6(10^{-3})\text{rads or}$ $\boxed{\phi_C = 0.006 \text{ rads CCW}}$

Substituting $\phi_D = 0$ and $\phi_C = 0.006$ in Eq.(6) of problem 5.55, we obtain

$$T_A - T = -\frac{0.006}{0.1019(10^{-3})} = -58.88 \qquad 8$$

From Eq. (7) in problem 5-55 we have

$$T_A = 0.7141\ T \qquad 9$$

Substituting Eq.(3) into Eq.(2) we obtain: $(0.7141 - 1)T = -58.88$ or $\boxed{T = 205.9\ \text{in-kips}}$

5.57 Two solid circular steel (G_s= 80 GPa) shafts and a solid circular bronze (G_{br} = 40 GPa) are securely connected by a coupling at C. A torque of T =100 kN-m is applied to the rigid wheel B as shown. If the coupling plates cannot rotate relative to each other, determine the angle of rotation of the wheel B due to the applied torque.

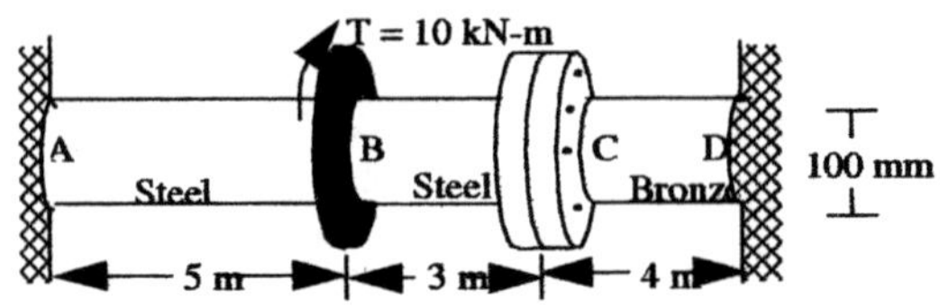

Fig. P5.57

Solution $G_{AB} = G_{BC} = 80\ \text{GPa}$ $\quad G_{CD} = 40\ \text{GPa}$ $\quad T = 10\ \text{kN-m}$ $\quad \phi_B = ?$

The polar moment of the cross-section can be found as shown below

$$J = \frac{\pi}{32}(0.1)^4 = 9.817(10^{-6})\ \text{m}^4 \qquad 1$$

The following free body diagrams can be drawn with internal torques drawn as per our sign convention.

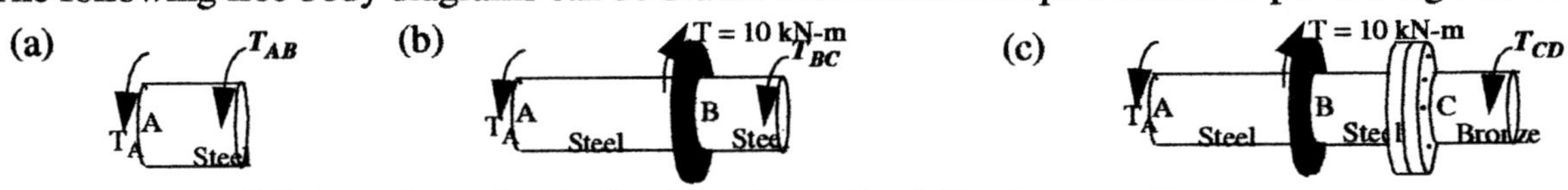

By moment equilibrium about the shaft axis we have the following equations.

$$T_{AB} = -T_A \qquad T_{BC} = T - T_A \qquad T_{CD} = T - T_A \qquad 2$$

The relative rotation of the ends of the each segment can be written as shown below.

$$\phi_B - \phi_A = \frac{T_{AB}(x_B - x_A)}{G_{AB}J} = \frac{T_A(5)}{(80)(10^9)(9.817)(10^{-6})} = 6.366(10^{-6})T_A \qquad 3$$

$$\phi_C - \phi_B = \frac{T_{BC}(x_C - x_B)}{G_{BC}J} = \frac{(T-T_A)(3)}{(80)(10^9)(9.817)(10^{-6})} = 3.8199(10^{-6})(T - T_A) \qquad 4$$

$$\phi_D - \phi_C = \frac{T_{CD}(x_D - x_C)}{G_{CD}J} = \frac{(T-T_A)(4)}{(40)(10^9)(9.817)(10^{-6})} = 10.1864(10^{-6})(T - T_A) \qquad 5$$

Adding Eqs. (3),(4) and (5) and noting that $\phi_D = 0$ and $\phi_A = 0$ we obtain the following.

$$\phi_D - \phi_A = [-6.3661T_A + 3.8199(T - T_A) + 10.1864(T - T_A)](10^{-6}) = 0 \qquad \text{or}$$

$$T_A = \frac{(3.8199 + 10.1864)T}{(6.3661 + 3.8199 + 10.1864)} = 0.6875T = 6.875\ \text{kN-m} \qquad 6$$

From Eq.(3) we obtain: $\phi_B = (-6.366)(10^{-6})(6.875)(10^3) = -43.77(10^{-3})$ $\boxed{\phi_B = 0.0438\ \text{rads CW}}$

5.58 Two solid circular steel (G_s= 80 GPa) shafts and a solid circular bronze (G_{br} = 40 GPa) are securely connected by a coupling at C. A torque of T =10 kN-m is applied to the rigid wheel B as shown.If the coupling plates can rotate relative to each other by 0.5° before engaging then what will be the angle of rotation of the wheel B.

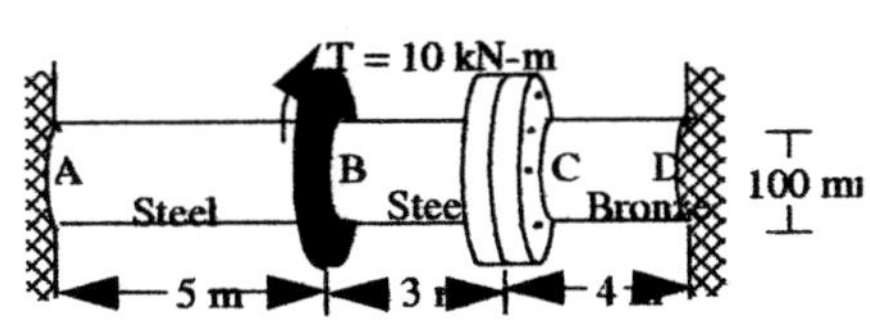

Fig. P5.58

Solution $G_{AB} = G_{BC} = 80\ \text{GPa}\ \ G_{CD} = 40\ \text{GPa}\ \ T = 10\ \text{kN-m}\ \ |\phi_{C_2} - \phi_{C_1}| = 0.5^\circ [\qquad \phi_B = ?$

The coupling plate C_1 will rotate clockwise relative to the coupling plate C_2, hence the relative rotation of plate C_2 with respect to plate C_1 will be positive. We can write the following

$$\phi_{C_2} - \phi_{C_1} = \frac{0.5\pi}{180} = 8.7267(10^{-3}) \qquad 1$$

Eqs (3) and(4) and (5) of problem 5-57 can be rewritten as

$$\phi_B - \phi_A = 6.366(10^{-6})T_A \qquad 2$$

$$\phi_{C1} - \phi_B = 3.8199(10^{-6})(T - T_A) \qquad 3$$

$$\phi_D - \phi_{C2} = 10.1864(10^{-6})(T - T_A) \qquad 4$$

Adding the above four equations and noting that $\phi_D = 0$ and $\phi_A = 0$, we obtain

$$\phi_D - \phi_A = [-6.3661 T_A + 3.8199(T - T_A) + 10.1864(T - T_A)](10^{-6}) + 8.7267(10^3) = 0 \text{ or}$$

$$T_A = \frac{(3.8199 + 10.1864)T + 8.7267(10^3)}{(6.3661 + 3.8199 + 10.1864)} \qquad 5$$

Substituting $T = 10(10^3)\ \text{N-m}$ in Eq.(5) $T_A = (0.6875)(10)(10^3) + (0.4283)(10^3) = 7.303(10^3)\ \text{N-m}$

From Eq. (2) we obtain,

$$\phi_B = (-6.366)(10^{-6})(7.303)(10^3) = -46.497(10^{-3})\ \text{rads}$$

$$\boxed{\phi_B = -0.0465\ \text{rads CW}}$$

5.59 A solid steel (G = 80 GPa) is securely fastened to a solid bronze (G = 40 GPa) shaft that is 2 m long as shown. If T = 10 kN-m in Fig. P5.59, determine (a) the magnitude of maximum torsional shear stress in the shaft. (b) the rotation of section at 1 m from the left wall.

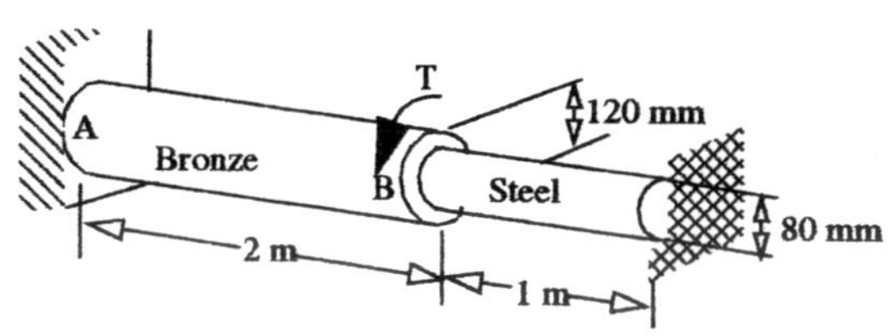

Fig. P5.59

Solution $G_{AB} = 40\text{GPa}\ \ G_{BC} = 80\ \text{GPa}\ \ T = 10\ \text{kN-m} \qquad \tau_{max} = ? \qquad \phi_1 = ?$

The polar moment of cross - sections in various segments can be found as shown below

$$J_{AB} = \frac{\pi}{32}(0.12^4) = 20.358(10^{-6})\ \text{m}^4 \qquad 1$$

$$J_{BC} = \frac{\pi}{32}(0.08)^4 = 4.021(10^{-6})\ \text{m}^4 \qquad 2$$

The following free body diagrams can be drawn. The internal torques are drawn as per our sign convention.

(a)

(b)

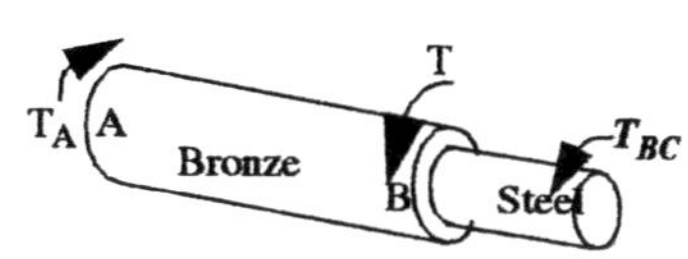

By moment equilibrium about the shaft axis we have the following equations.

$$T_{AB} = T_A \qquad T_{BC} = T_A - T \qquad 3$$

The relative rotation of the ends of the each segment can be written as shown below,

$$\phi_B - \phi_A = \frac{T_{AB}(x_B - x_A)}{G_{AB}J_{AB}} = \frac{T_A(2)}{(40)(10^9)(20.358)(10^{-6})} = 2.456(10^{-6})T_A \qquad 4$$

$$\phi_C - \phi_B = \frac{T_{BC}(x_C - x_B)}{G_{BC}J_{BC}} = \frac{(T - T_A)(1)}{(80)(10^9)(4.021)(10^{-6})} = 3.109(10^{-6})(T_A - T) \qquad 5$$

Adding Equations(4) and (5) and noting that $\phi_C = 0$ and $\phi_A = 0$ we obtain the following.

$$\phi_C - \phi_A = [2.456T_A + 3.109(T_A - T)](10^{-6}) = 0 \quad \text{or}$$

$$T_A = \frac{3.109T}{2.456 + 3.109} = 0.5586T = 5.586(10^3)\text{N} - \text{m} \qquad 6$$

From Eq.(3) $T_{AB} = 5.586(10^3)\text{N} - \text{m}$ $\quad T_{BC} = (-4.413(10^3))\text{N} - \text{m}$

The maximum torsional shear stress in each segment can be found as shown below

$$(\tau_{AB})_{max} = \frac{T_{AB}(\rho_{AB})_{max}}{J_{AB}} = \frac{(5.586)(10^3)(0.06)}{20.358(10^{-6})} = 16.46(10^6)\frac{\text{N}}{\text{m}^2}$$

$$(\tau_{BC})_{max} = \frac{T_{BC}(\rho_{BC})_{max}}{J_{BC}} = \frac{(-4.413)(10^3)(0.04)}{4.021(10^{-6})} = -43.9(10^6)\frac{\text{N}}{\text{m}^2}$$

The maximum torsional shear stress is $\boxed{\tau_{max} = 43.9 \text{ MPa}}$

The rotation at 1m from the wall of A can be found as shown below,

$$\phi_1 - \phi_A = \frac{T_{AB}(x_1 - x_A)}{G_{AB}J_{AB}} = \frac{(5.586)(10^3)(1)}{(40)(10^9)(20.358)(10^{-6})} = 6.86(10^{-3}) \text{ rads}$$

Noting $\phi_A = 0$, we obtain $\boxed{\phi_1 = 0.0069 \text{ rads CCW}}$

5.60 A solid steel (G = 80 GPa) is securely fastened to a solid bronze (G = 40 GPa) shaft that is 2 m long as shown. If the section at B rotates by 0.05 rads, determine (a) the maximum torsional shear strain in the shaft. (b) the applied torque T.

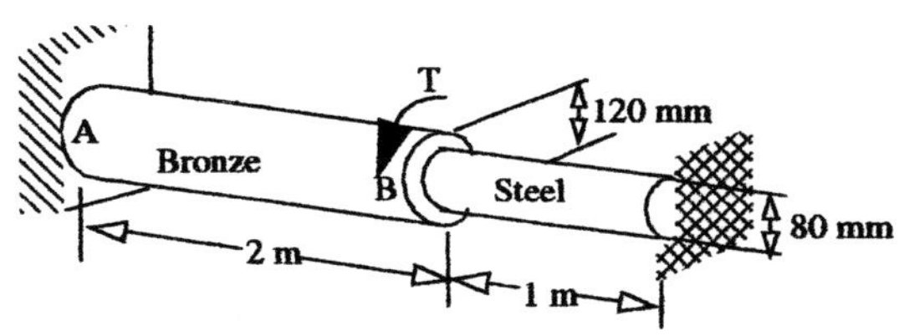

Fig. P5.60

Solution $\quad G_{AB} = 80\text{GPa} \quad G_{BC} = 40 \text{ GPa} \quad \phi_B = 0.05 \text{ rads} \qquad \tau_{max} = ? \qquad T = ?$

The rate of twist in the segment AB and BC can be found as

$$\left(\frac{d\phi}{dx}\right)_{AB} = \frac{\phi_B - \phi_A}{x_B - x_A} = \frac{0.05}{2} = 0.025 \frac{\text{rads}}{\text{m}} \quad \text{and} \quad \left(\frac{d\phi}{dx}\right)_{BC} = \frac{\phi_C - \phi_B}{x_C - x_B} = \frac{0 - 0.05}{1} = -0.05 \frac{\text{rads}}{\text{m}}$$

The maximum torsional shear strain in each segment can be found as:

$$(\gamma_{AB})_{max} = (\rho_{AB})_{max}\left(\frac{d\phi}{dx}\right)_{AB} = (0.06)(0.025) = 1.5(10^{-3})$$

$$(\gamma_{BC})_{max} = (\rho_{BC})_{max}\left(\frac{d\phi}{dx}\right)_{BC} = (0.04)(-0.05) = -2.0(10^{-3})$$

The maximum torsional shear strain is, $\boxed{\gamma_{max} = -2000 \ \mu\text{rads}}$

Substituting $\phi_B = 0.05$ and $\phi_A = 0$ in Eqs. (4) in problem (5.59) we obtain

$$T_A = \frac{0.05}{2.456(10^{-6})} = 20.36(10^3)\text{N} - \text{m}$$

From Eq.(6) of problem (5-59), we obtain,

$$T = \frac{T_A}{0.5586} = \frac{20.36(10^3)}{0.5586} = 36.44(10^3)\text{N} - \text{m} \qquad \boxed{T = 36.44 \text{ kN} - \text{m}}$$

5.61 An aluminum tube and a copper tube are securely fastened to two rigid bars as shown. The bars force the tubes to rotate by equal angles. The tubes are 5 mm thick and 1.5 m long. The mean diameters of the aluminum tube and copper tube are 125 mm and 50 mm, respectively. The shear modulus for aluminum and copper are $G_{al} = 28$ GPa and $G_{cu} = 40$ GPa, respectively.Under the action of the applied couple the section B of the two tubes rotate by an angle of 0.03 rads. Determine (a) the magnitude maximum torsional shear stress in aluminum and copper. (b) the magnitude of the couple that produced the given rotation.

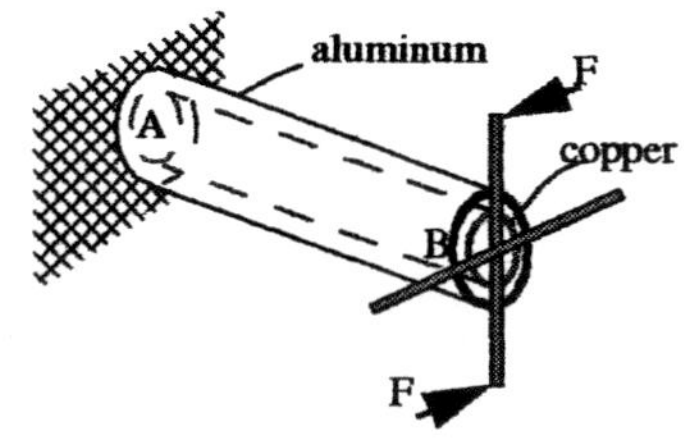

Fig. P5.61

Solution $t = 5\text{ mm}$ $d_{al} = 125\text{ mm}$ $d_{cu} = 50\text{ mm}$ $L = 1.5\text{ m}$ $G_{al} = 28\text{ GPa}$

$G_{cu} = 40\text{ GPa}$ $\phi_1 = 0.03\text{ rads}$ $(\tau_{al})_{max} = ?$ $(\tau_{cu})_{max} = ?$ $T = Fd = ?$

The polar moment of cross - section of each material can be found as shown below

$$J_{al} = \frac{\pi}{32}[(0.130)^4 - (0.120^4)] = 7.6820(10^{-6})\text{ m}^4 \quad (1)$$

$$J_{cu} = \frac{\pi}{32}[(0.055)^4 - (0.045^4)] = 0.4958(10^{-6})\text{ m}^4 \quad (2)$$

T_{al} and T_{cu} represents the internal torque on each material cross section. By equilibrium we have

$$T_{al} + T_{cu} = T \quad (3)$$

The relative rotation of the ends of the each segment can be written as shown below,

$$(\phi_B)_{al} - (\phi_A)_{al} = \frac{T_{al}(x_B - x_A)}{G_{al}J_{al}} = \frac{T_{al}(1.5)}{(28)(10^9)(7.682)(10^{-6})} = 6.973(10^{-6})T_{al} \quad (4)$$

$$(\phi_B)_{cu} - (\phi_A)_{cu} = \frac{T_{cu}(x_C - x_B)}{G_{cu}J_{cu}} = \frac{T_{cu}(1.5)}{(40)(10^9)(0.4958)(10^{-6})} = 75.63(10^{-6})T_{cu} \quad (5)$$

For both materials, ϕ_A is zero and $\phi_B = 0.03$ .From Eqs. (4) and (5) we obtain the following,

$$T_{al} = \frac{0.03}{6.973(10^{-6})} = 4.302(10^3)\text{ N-m} \quad (6)$$

$$T_{cu} = \frac{0.03}{75.63(10^{-6})} = 0.397(10^3)\text{ N-m} \quad (7)$$

Substituting Eq (6) and(7) into Eq.(3) we obtain

$$T = (4.302 + 0.397)(10^3) = 4.699(10^3)\text{ N-m}$$

$$\boxed{T = 4.699\text{ kN-m}}$$

The maximum torsional shear stress in each material can be found as shown below

$$(\tau_{al})_{max} = \frac{T_{al}(\rho_{al})_{max}}{J_{al}} = \frac{(4.302)(10^3)(0.065)}{7.682(10^{-6})} = 36.4(10^6)\text{ N/m}^2$$

$$\boxed{(\tau_{al})_{max} = 36.4\text{ MPa}}$$

$$(\tau_{cu})_{max} = \frac{T_{cu}(\rho_{cu})_{max}}{J_{cu}} = \frac{(0.397)(10^3)(0.0275)}{0.4958(10^{-6})} = 22.0(10^6)\text{ N/m}^2$$

$$\boxed{(\tau_{cu})_{max} = 22.0\text{MPa}}$$

5.62 An aluminum tube and a copper tube are securely fastened to two rigid bars as shown. The bars force the tubes to rotate by equal angles. The tubes are 5 mm thick and 1.5 m long. The mean diameters of the aluminum tube and copper tube are 125 mm and 50 mm, respectively. The shear modulus for aluminum and copper are $G_{al} = 28$ GPa and $G_{cu} = 40$ GPa, respectively. The applied couple on the tubes shown is 10 kN-m. Determine (a) the magnitude maximum torsional shear stress in aluminum and copper. (b) the rotation of section at B.

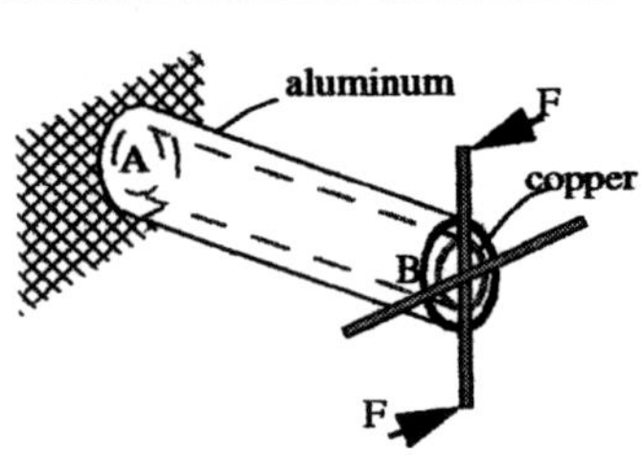

Fig. P5.62

Solution $t = 5 \text{ mm}$ $\quad d_{al} = 125 \text{ mm}$ $\quad d_{cu} = 50 \text{ mm}$ $\quad L = 1.5 \text{ m}$ $\quad G_{al} = 28 \text{ GPa}$

$G_{cu} = 40 \text{ GPa}$ $\quad T = Fd = 10\text{kN}-\text{m}$ $\quad (\tau_{al})_{max} = ?$ $\quad (\tau_{cu})_{max} = ?$ $\quad \phi_B = ?$

As ϕ_A is zero for both material and ϕ_B is same for both material, we can equate Eq.(4) and Eq.(5) of problem 5-61 to obtain $6.973(10^{-6})T_{al} = 75.63(10^{-6})T_{cu}$ or

$$T_{al} = 10.846T_{cu} \qquad 1$$

From Eq.(3) of the problem 5-61, we have

$$T_{al} + T_{cu} = 10(10^3) \text{ N}-\text{m} \qquad 2$$

Substituting Eq.(1) into Eq.(2) we obtain $(10.846 + 1)T_{cu} = 10(10^3)$ or

$$T_{cu} = 0.8442(10^3) \text{ N}-\text{m} \qquad 3$$

$$T_{al} = 9.1558(10^3) \text{ N}-\text{m} \qquad 4$$

From Eq.(4) we obtain $\phi_B = (6.973)(10^{-6})(9.1558)(10^{-6}) = 63.85(10^{-3})$ $\quad \boxed{\phi_B = 0.064 \text{ rads CCW}}$

The maximum torsional shear stress in each material can be found as shown below

$$(\tau_{al})_{max} = \frac{(9.1558)(10^3)(0.065)}{7.682(10^{-6})} = 77.47(10^6) \text{ N/m}^2 \qquad \boxed{(\tau_{al})_{max} = 77.5\text{MPa}}$$

$$(\tau_{cu})_{max} = \frac{(0.8442)(10^3)(0.065)}{0.4958(10^{-6})} = 46.8(10^6) \text{ N/m}^2 \qquad \boxed{(\tau_{cu})_{max} = 46.8\text{MPa}}$$

5.63 Two shafts with shear modulus $G_1 = G$ and $G_2 = 2G$ are securely fastened at section B. In terms of T, L, G, and d, determine the magnitude maximum torsional shear stress in the shaft and the rotation of section at B.

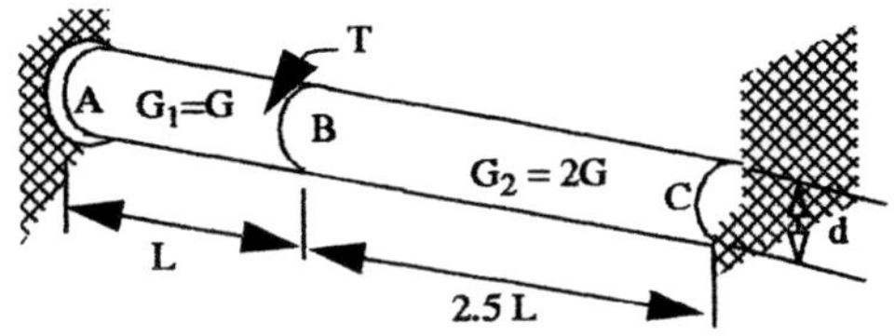

Fig. P5.63

Solution $\tau_{max} = f(T, L, G, d) = ?$ $\quad \phi_B = g(T, L, G, d) = ?$

The following free body diagrams can be drawn with internal torques drawn as per our sign convention.

(a)

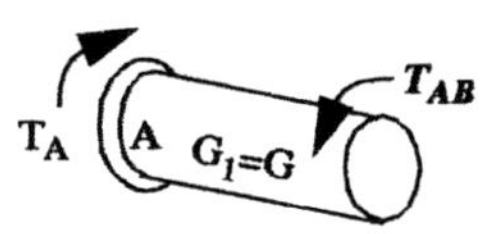

(b)

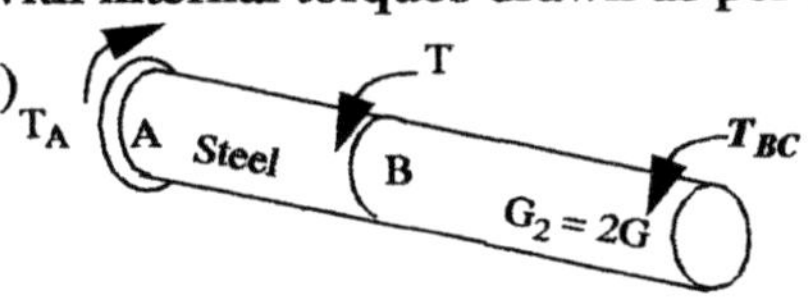

By moment equilibrium about the shaft axis we have the following equations

$$T_{AB} = T_A \qquad T_{BC} = T_A - T \qquad 1$$

The relative rotations of each segment can be written as shown below.

$$\phi_B - \phi_A = \frac{T_{AB}(x_B - x_A)}{G_{AB}J} = \frac{T_A(L)}{GJ} \quad (2)$$

$$\phi_C - \phi_B = \frac{T_{BC}(x_C - x_B)}{G_{BC}J} = \frac{(T_A - T)(2.5L)}{2GJ} = \frac{1.25}{GJ}(T_A - T)L \quad (3)$$

Adding Eq's(2) and (3) and noting that $\phi_C=0$ and $\phi_A=0$, we obtain the following

$$\phi_C - \phi_A = \frac{L}{GJ}[T_A + 1.25(T_A - T)] = 0 \quad \text{or} \quad T_A = \frac{1.25}{1.25+1}T = \frac{5}{9}T \quad (4)$$

From Eq.(2) we obtain

$$\boxed{\phi_B = \frac{5TL}{9GJ} \text{ CCW}}$$

Substituting Eq.(3) into Eq.(1), we obtain

$$T_{AB} = \frac{5}{9}T \qquad T_{BC} = -\frac{4}{9}T \quad (5)$$

The maximum torsional shear stress will be in segment AB and can be found as shown below

$$\tau_{max} = \frac{(5T/9)(d/2)}{(\pi d^3/32)} \qquad \boxed{\tau_{max} = \frac{80}{9\pi}\frac{T}{d^3}}$$

5.64 A uniform distributed torque of q in-lb/in is applied to the entire shaft as shown. In addition to the distributed torque a concentrated torque of T=3 q L in-lb is applied at section B. Let the shear modulus be G and radius of the shaft be r. Determine in terms of q, L, G, and r: (a) the rotation of section at B. (b) the magnitude maximum torsional shear stress in the shaft.

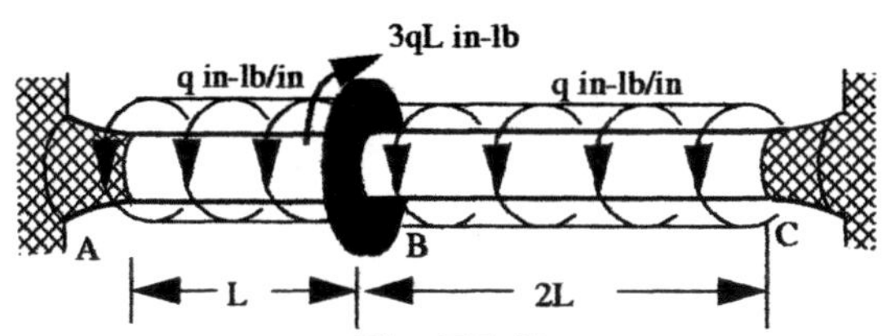

Fig. P5.64

Solution $\tau_{max} = f(q, L, G, r) = ?$ $\phi_B = g(q, L, G, r) = ?$

The following free body diagrams can be drawn with internal torques drawn as per our sign convention.

(a)

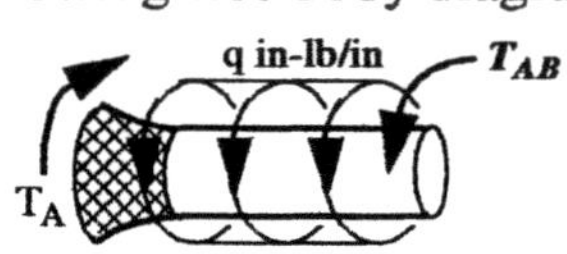

(b)

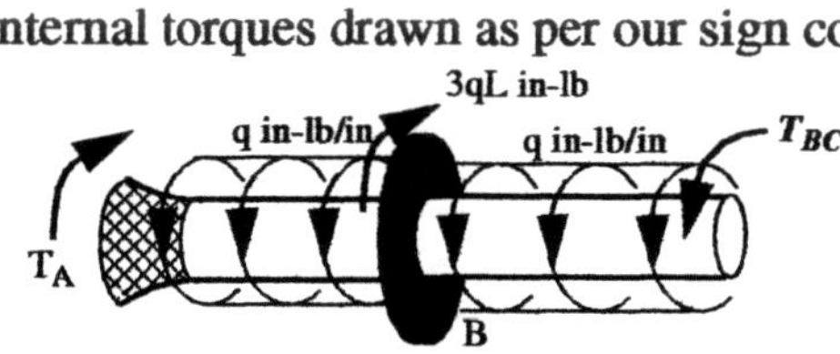

By moment equilibrium about the shaft axis we have the following equations.

$$T_{AB} + qx - T_A = 0 \quad \text{or} \quad T_{AB} = T_A - qx \quad (1)$$

$$T_{BC} + qx - T_A - 3qL = 0 \quad \text{or} \quad T_{BC} = T_A - qx + 3qL \quad (2)$$

The rate of twist of segment AB can be written as: $\left(\frac{d\phi}{dx}\right)_{AB} = \frac{T_{AB}}{GJ} = \frac{(T_A - qx)}{GJ}dx$

Integrating from x_A to x_B we obtain $\int_{\phi_A}^{\phi_B} d\phi = \int_{x_B=0}^{x_B=L} \frac{(T_A - qx)}{GJ}dx$ or

$$\phi_B - \phi_A = \frac{1}{GJ}\left(T_A x - \frac{q}{2}x^2\right)\Big|_0^L = \frac{1}{GJ}\left(T_A L - \frac{q}{2}L^2\right) \quad (3)$$

Similarly, the rate of twist of segment BC can be written as $\left(\frac{d\phi}{dx}\right)_{BC} = \frac{T_{BC}}{GJ} = \frac{(T_A - qx + 3qL)}{GJ}dx$

Integrating from x_B to x_C we obtain $\int_{\phi_B}^{\phi_C} d\phi = \int_{x_B = L}^{x_C = 3L} \frac{(T_A - qx + 3qL)}{GJ} dx$ or

$$\phi_c - \phi_B = \frac{1}{GJ}\left(T_A x - \frac{q}{2}x^2 + 3qLx\right)\Big|_L^{3L} = \frac{1}{GJ}(2T_A L + 2qL^2) \qquad 4$$

Adding Eqs. (3) and (4) and noting that ϕ_C=0 and ϕ_A=0, we obtain

$$\phi_C - \phi_A = \frac{L}{GJ}\left(T_A L - \frac{q}{2}L^2 + 2T_A L + 2qL^2\right) = 0 \quad \text{or} \quad T_A = -q\frac{L}{2} \qquad 5$$

From Eq. (3) we obtain $\phi_B = \frac{1}{GJ}\left(-\frac{q}{2}L^2 - \frac{q}{2}L^2\right) = \frac{-qL^2}{G(\pi r^4/2)}$ or $\boxed{\phi_B = \frac{2qL^4}{\pi G r^4} \text{ CW}}$

From Eq.(1)and (2), we obtain

$$T_{AB} = -\frac{q}{2}\left(x + \frac{L}{2}\right) \qquad T_{BC} = q\left(\frac{5}{2}L - x\right) \qquad 6$$

The maximum value of T_{AB} and T_{BC} is at x = L

$$(T_{AB})_{max} = -\frac{3}{2}qL \qquad (T_{BC})_{max} = \frac{3}{2}qL \qquad 7$$

The maximum torsional shear stress will be just after B and can be found as shown below

$$\tau_{max} = \frac{(T_{BC})_{max}(\rho_{BC})_{max}}{J_{BC}} = \frac{\left(\frac{3}{2}qL\right)r}{(\pi r^4/2)} \qquad \boxed{\tau_{max} = \frac{3qL}{\pi r^3}}$$

5.65 A steel ($G_s = 80$ GPa) and a bronze ($G_{Br} = 40$ GPa) shaft are securely connected at B as shown. The maximum torsional shear stresses in steel and bronze are to be limited to 160 MPa and 60 MPa, respectively and the rotation of section B is limited to 0.05 rads. (a) Determine the maximum allowable torque T to the nearest kN-m that can act on shaft in Fig. P5.65 if the diameter of the shaft is d = 100 mm. (b) What is the maximum torsional shear stress and maximum rotation in the shaft corresponding to the answer in part (a).

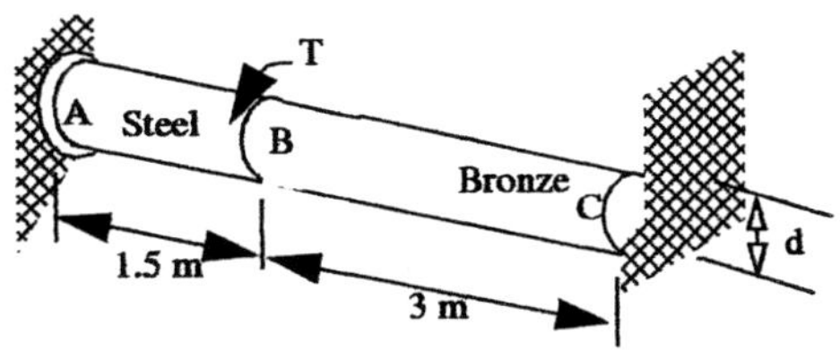

Fig. P5.65

Solution $G_{AB} = 80$ GPa $G_{BC} = 40$ GPa $(\tau_{AB})_{max} \le 160$ MPa $(\tau_{BC})_{max} \le 60$ MPa

$\phi_B \le 0.05$ rads d = 100 mm T_{max} = ? nearest kN τ_{max} = ? ϕ_B = ?

The following free body diagrams can be drawn with internal torques drawn as per our sign convention.

(a)

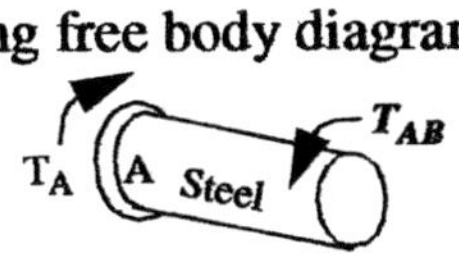

(b)

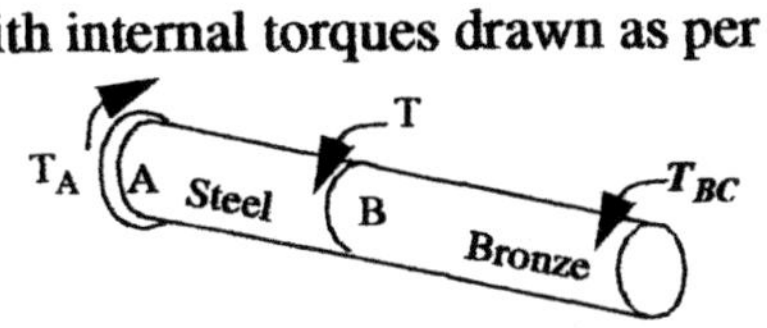

By moment equilibrium about the shaft axis we have the following equations.

$$T_{AB} = T_A \qquad T_{BC} = T_A - T \qquad 1$$

The relative rotation of the ends of each segment can be written as shown below.

$$\phi_B - \phi_A = \frac{T_{AB}(x_B - x_A)}{G_{AB}J} = \frac{1.5T_A}{80(10^9)J} \qquad 2$$

$$\phi_C - \phi_B = \frac{T_{BC}(x_C - x_B)}{G_{BC}J} = \frac{3(T_A - T)}{80(10^9)J} \quad 3$$

Adding Eq's. (2) and (3) and noting that ϕ_C=0 and ϕ_A=0, we obtain the following

$$\phi_C - \phi_A = \frac{1}{40(10^9)J}\left[\frac{1.5}{2}T_A + 3(T_A - T)\right] = 0 \text{ or}$$

$$T_A = \frac{3}{0.75+3}(9.1558)(10^{-6})T = 0.8T \quad 4$$

The polar moment of cross - section can be found as

$$J = \frac{\pi}{32}(0.1)^4 = 9.817(10^{-6}) \text{ m}^4 \quad 5$$

Substituting Eq.(4) and Eq.(5) into Eq.(2) we obtain

$$\phi_B = \frac{(1.5)(0.8T)}{(80)(10^9)(9.817)(10^{-6})} = 1.528T(10^{-6}) \le 0.05 \quad \text{or} \quad 6$$

$$T \le 32.72(10^3) \text{ N-m} \quad 7$$

Substituting Eq.(4) into Eq.(1) we obtain

$$T_{AB} = 0.8T \qquad T_{BC} = -0.2T \quad 8$$

The magnitude of the maximum torsional shear stress in each segment can be written as

$$(\tau_{AB})_{max} = \frac{(0.8T)(0.05)}{9.817(10^{-6})} = 4.074(10^3)T \le 160(10^6) \quad \text{or} \quad 9$$

$$T \le 39.27(10^3) \text{ N-m} \quad 10$$

$$(\tau_{BC})_{max} = \frac{(0.2T)(0.05)}{9.817(10^{-6})} = 1.018(10^3)T \le 60(10^6) \quad \text{or} \quad 11$$

$$T \le 58.90(10^3) \text{ N-m} \quad 12$$

The maximum value of T that meets all the requirements of Eqs.(7),(10) and (12) is T = 32.72 (10^3) N-m.To the nearest kN the maximum value is $\boxed{T_{max} = 32 \text{ kN-m}}$

Substituting T = 32 kN-m into Eq. (6) we obtain: $\phi_B = 48.896(10^{-3})$ rads $\boxed{\phi_B = 0.048 \text{ rads CCW}}$

Substituting T = 32 kN-m into Eqs. (9) and (11) we obtain the following:

$(\tau_{AB})_{max} = 130.37(10^6) \text{ N/m}^2$ and $(\tau_{BC})_{max} = 32.6(10^6) \text{ N/m}^2$

The maximum torsional shear stress is in steel and has a value: $\boxed{\tau_{max} = 130.4 \text{ MPa}}$

5.66 A steel (G_s = 80 GPa) and a bronze (G_{Br} = 40 GPa) shaft are securely connected at B as shown. The maximum torsional shear stresses in steel and bronze are to be limited to 160 MPa and 60 MPa, respectively and the rotation of section B is limited to 0.05 rads. Determine the minimum diameter 'd' of the shaft to the nearest millimeter, if the applied torque T = 20 kN-m. What is the maximum torsional shear stress and maximum rotation in the shaft corresponding to the answer in part (a).

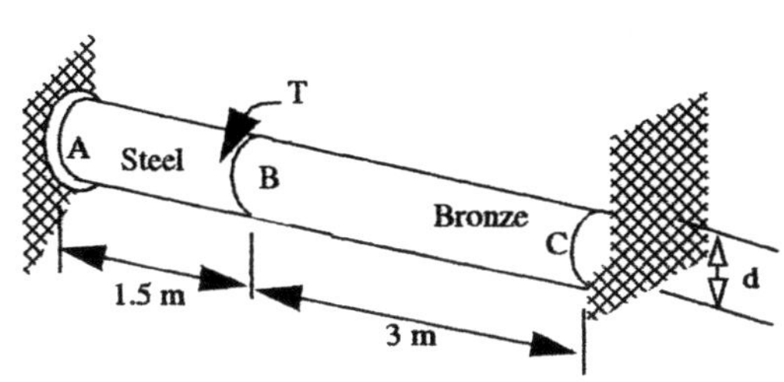

Fig. P5.66

Solution G_{AB} = 80 GPa G_{BC} = 40 GPa $(\tau_{AB})_{max} \le 160$ MPa

$(\tau_{BC})_{max} \leq 60$ MPa $\quad \phi_B \leq 0.05$ rads $\quad T = 20$ kN-m

$d_{min} = ?$ nearest mm $\quad T = ?$ nearest kN $\quad \tau_{max} = ?$ $\quad \phi_B = ?$

From Eq.(4) in problem 5-65

$$T_A = 0.8T = 16 \text{ kN}-\text{m}$$ 1

From Eq.(2) in problem 5-65

$$\phi_B = \frac{(1.5)(16)(10^3)}{(80)(10^9)\frac{\pi}{32}d^4} = \frac{3.056}{d^4}(10^{-6}) \leq 0.05 \qquad \text{or}$$ 2

$$d \geq 88.417(10^{-3}) \text{ m}$$ 3

Substituting T = 16 k N-m in Eq.(1) of problem 5-65

$$T_{AB} = 16 \text{ kN}-\text{m} \qquad T_{BC} = 4\text{kN}-\text{m}$$ 4

The magnitude of the maximum torsional shear stress in each segment can be written as

$$(\tau_{AB})_{max} = \frac{(16)(10^3)\left(\frac{d}{2}\right)}{(\pi d^4/32)} = \frac{81.487(10^3)}{d^3} \leq 160(10^6) \qquad \text{or}$$ 5

$$d \geq 79.86(10^{-3}) \text{ m}$$ 6

$$(\tau_{BC})_{max} = \frac{(4)(10^3)\left(\frac{d}{2}\right)}{(\pi d^4/32)} = \frac{20.37(10^3)}{d^3} \leq 60(10^6) \qquad \text{or}$$ 7

$$d \geq 69.76(10^{-3}) \text{ m}$$ 8

The minimum value of d that meets all the requirements of Eqs (3),(6) and (8) is d = 88.417 mm.To the nearest mm the minimum value is $\boxed{d_{min} = 89 \text{ mm}}$

Substituting d = 0.08 m into Eq.(2) we obtain: $\phi_B = \dfrac{3.056(10^{-6})}{(0.089)^4} = 48.71(10^{-3})$ or

$\boxed{\phi_B = 0.0487 \text{ rads CCW}}$

Substituting d = 0.089 m in Eqs. (5) and (8), we obtain the following.

$(\tau_{AB})_{max} = 115.6(10^6)\ \text{N/m}^2$ and $(\tau_{BC})_{max} = 39.8(10^6)\ \text{N/m}^2$

The maximum torsional shear stress is in steel and has a value: $\boxed{\tau_{max} = 116 \text{ MPa}}$

5.67 The solid steel shaft shown has a shear modulus of elasticity of G = 80 GPa and a allowable torsional shear stress of 60 MPa. The allowable rotation of any section is 0.03 rads. The applied torques on shaft are $T_1 = 10$ kN-m and $T_2 = 25$ kN-m. Determine (a) the minimum diameter 'd' of the shaft to the nearest millimeter. (b) the maximum torsional shear stress in the shaft and the maximum rotation of any section.

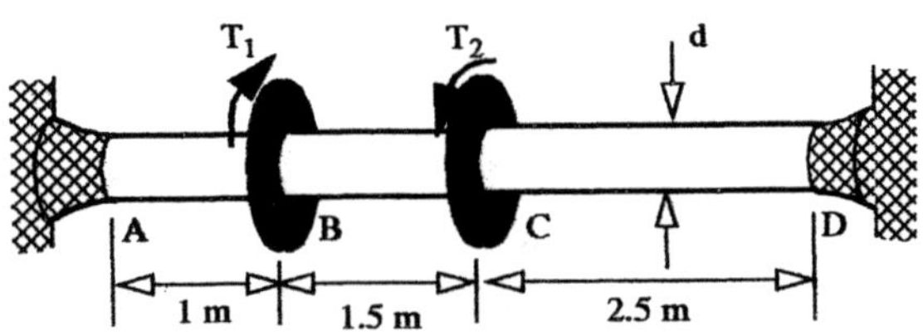

Fig. P5.67

Solution G = 80 GPa $\quad |\tau_{max}| \leq 60$ MPa $\quad |\phi_{max}| \leq 0.03$ rads $\quad T_1 = 10$ kN-m

$T_2 = 25$ kN-m $\quad d_{min} = ?$ nearest mm $\quad \tau_{max} = ?$ $\quad \phi_{max} = ?$

The following free body diagrams can be drawn. The internal torques are drawn as per our sign convention.

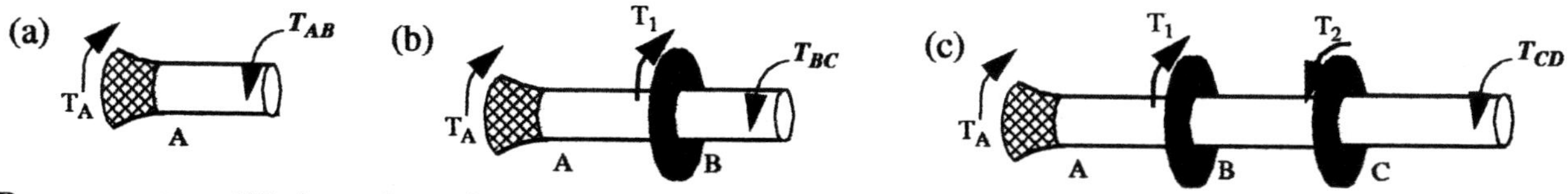

By moment equilibrium about the shaft axis we have the following equations.

$$T_{AB} = T_A \qquad T_{BC} = T_A + T_1 \qquad T_{CD} = T_A + T_1 - T_2 \tag{1}$$

The relative rotation of the ends of each segment can be written as shown below.

$$\phi_B - \phi_A = \frac{T_{AB}(x_B - x_A)}{GJ} = \frac{(1)T_A}{GJ} \tag{2}$$

$$\phi_C - \phi_B = \frac{T_{BC}(x_C - x_B)}{GJ} = \frac{(T_A + T_1)(1.5)}{GJ} \tag{3}$$

$$\phi_D - \phi_C = \frac{T_{CD}(x_D - x_C)}{GJ} = \frac{(T_A + T_1 - T_2)(2.5)}{GJ} \tag{4}$$

Adding Eq's(2), (3)and (4) and noting that ϕ_A=0 and ϕ_D=0, we obtain

$$\phi_D - \phi_A = \frac{1}{GJ}[T_A + 1.5(T_A + T_1) + 2.5(T_A + T_1 - T_2)] = 0 \quad \text{or} \quad T_A = \frac{2.5T_2 - 4T_1}{5} = 0.5T_2 - 0.8T_1 \tag{5}$$

From Eq. (1) we obtain

$$T_{AB} = 0.5T_2 - 0.8T_1 \tag{6}$$

$$T_{BC} = 0.5T_2 + 0.2T_1 \tag{7}$$

$$T_{CD} = -0.5T_2 + 0.2T_1 \tag{8}$$

Substituting $T_1 = 10$ kN-m and $T_2 = 25$kN-m we obtain

$$T_{AB} = 4.5 \text{ kN-m} \qquad T_{BC} = 14.5 \text{ kN-m} \qquad T_{CD} = -10.5\text{kN-m}$$

From the values of the internal torques, we conclude that maximum torsional shear stress will be in the signet BC and the maximum rotation will be of the cross-section at C

Substituting $\phi_D = 0$ and $\phi_C = \phi_{max}$ in Eq.(4) we obtain

$$-\phi_C = \frac{(-10.5)(10^3)(2.5)}{(80)(10^9)\frac{\pi}{32}d^4} = \frac{-3.342}{d^4}(10^{-6}) \quad \text{or} \quad \phi_{max} = \frac{3.342}{d^4}(10^{-6}) \le 0.03 \quad \text{or} \tag{9}$$

$$d \ge 0.1027 \text{ m} \tag{10}$$

The magnitude of the maximum torsional shear stress in segment BC can be written as

$$\tau_{max} = \frac{(14.5)(10^3)\left(\frac{d}{2}\right)}{\frac{\pi}{32}d^4} = \frac{73.847(10^3)}{d^3} \le 60(10^6) \quad \text{or} \tag{11}$$

$$d \ge 0.1072 \text{ m} \tag{12}$$

The minimum value of d that meets all the requirements of Eqs (10) and (12) is d = 107.2mm.The nearest mm that meets the requirement is

$$\boxed{d_{min} = 108 \text{ mm}}$$

Substituting d = 0.108 m in Eqs (9) and (10), we obtain the following.

$\phi_{max} = 0.0246$ rads

$$\boxed{\phi_B = 0.025 \text{ rads CCW}}$$

$\tau_{max} = 58.62(10^6)\ N/m^2$

$$\boxed{\tau_{max} = 58.62 \text{ MPa}}$$

5.68 The solid steel shaft shown has a shear modulus of elasticity of G = 80 GPa and a allowable torsional shear stress of 60 MPa. The allowable rotation of any section is 0.03 rads.The diameter of the shaft is d = 80 mm. Determine the maximum values of the torques T_1 and T_2 that can be applied to the shaft.

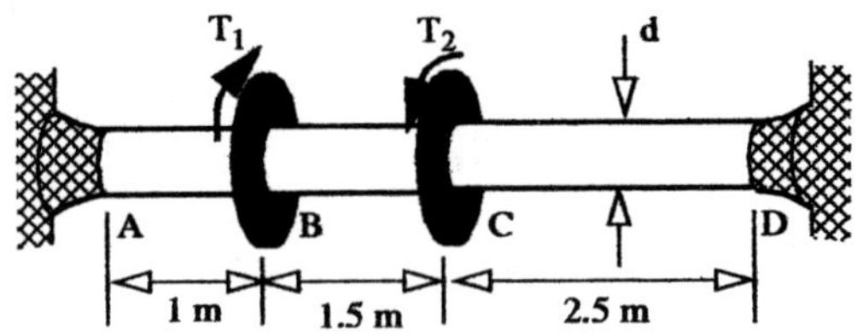

Fig. P5.68

Solution $G = 80$ GPa $\quad |\tau_{max}| \le 60$ MPa $\quad |\phi_{max}| \le 0.03$ rads $\quad d = 80$ mm

$T_1 = ?$ $\quad T_2 = ?$ nearest kN-m

The polar moment of cross - section can be found as

$$J = \frac{\pi}{32}(0.08)^4 = 4.021(10^{-6})\ \text{m}^4 \qquad 1$$

From Eqs. (6),(7) and (8) in problem 5-67, it is clear that for positive values of T_1 and T_2 the maximum internal torque will be in the segment BC. Hence the maximum torsional shear stress will be in segment BC and can be written as shown below.

$$\tau_{max} = \frac{(0.5T_2 + 0.2T_1)(0.04)}{4.021(10^{-6})} = 60(10^6) \quad \text{or} \quad 0.5T_2 + 0.2T_1 = 6.032(10^3)\ \text{N-m} \qquad 2$$

The rotation of her section at B and C can be written as shown below using Eqs (6) and (8) and noting that ϕ_A=0 and ϕ_D=0, we obtain

$$\phi_B - \phi_A = \frac{T_{AB}(1)}{(80)(10^9)(4.021)(10^{-6})} \quad \text{or} \quad \phi_B = (0.5T_2 - 0.8T_1)(3.109)(10^{-6}) \qquad 3$$

$$\phi_D - \phi_C = \frac{T_{CD}(2.5)}{(80)(10^9)(4.021)(10^{-6})} \quad \text{or} \quad \phi_C = -(0.2T_1 - 0.5T_2)(7.772)(10^{-6}) \qquad 4$$

Assuming ϕ_B is greater in magnitude than ϕ_C, and ϕ_B is clockwise, we obtain

$$(0.5T_2 - 0.8T_1)(3.109)(10^{-6}) = -0.03 \quad \text{or} \quad 0.5T_2 - 0.8T_1 = -9.649(10^3) \qquad 5$$

Solving Eqs. (2) and (5), we obtain

$$T_1 = 15.68(10^3)\ \text{N-m} \quad \text{and} \quad T_2 = 5.792(10^3)\ \text{N-m} \qquad 6$$

Substituting Eq.(6) into Eq. (4), we obtain $\phi_C = 0.0019$, which is less than 0.03, Hence our assumption that ϕ_B is greater in magnitude is correct, rounding downwards to the nearest kN, we obtain the maximum values of the torques $\boxed{T_1 = 15\ \text{k N-m}}$ and $\boxed{T_2 = 5.0\ \text{k N-m}}$

5.69 A composite shaft is constructed from aluminum (G_{al}= 4,000 ksi), bronze (G_{br} = 6,000 ksi) and steel (G_s = 12,000 ksi). (a) Determine the rotation of the free end with respect to the wall. (b) Plot the torsional shear strain and shear stress across the cross-section.

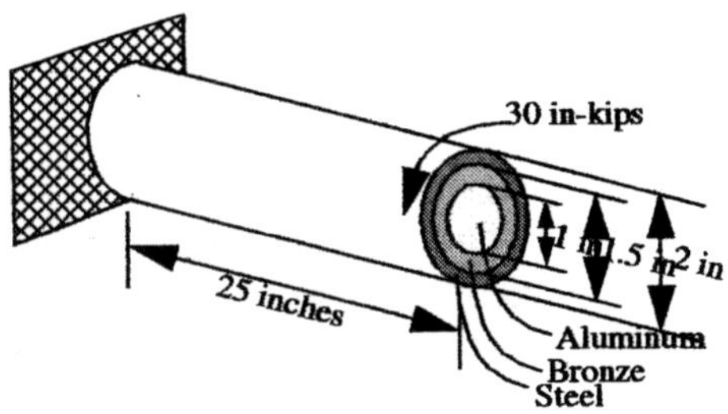

Fig. P5.69

Solution $G_{al} = 4000$ ksi $\quad G_{br} = 6000$ ksi $\quad G_s = 12000$ ksi

$\phi_B - \phi_A = ?$ $\quad \gamma_{x\theta}$ vs. $\rho = ?$ $\quad \tau_{x\theta}$ vs. $\rho = ?$

The polar moment of each material cross - section can be found as shown below.

$$J_s = \frac{\pi}{32}(2^4 - 1.5^4) = 1.0738 \text{ in}^4 \qquad J_{br} = \frac{\pi}{32}(1.5^4 - 1^4) = 0.3988 \text{ in}^4 \qquad J_{al} = \frac{\pi}{32}(1^4) = 0.0982 \text{ in}^4$$

The torsional rigidity of the cross-section is

$$\Sigma G_j J_j = (4000)(0.0982) + (6000)(0.3988) + (12000)(1.0738) = 15.671(10^3) \text{ kips} - \text{in}^2$$

The relative rotation of the ends of the shaft can be written as

$$\phi_B - \phi_A = \frac{T_{AB}(x_B - x_A)}{\Sigma G_j J_j} = \frac{(30)(25)}{15.671(10^3)} = 47.861(10^{-3})$$

Noting that ϕ_A=0, we obtain $\boxed{\phi_B = 0.0479 \text{ rads CCW}}$

The torsional shear stress and shear strainat any point on the cross section can be found as

$$\tau_{x\theta} = G_i \frac{T_{AB}\rho}{\Sigma G_j J_j} = 1.914(10^{-3}) G_i \rho \qquad \gamma_{x\theta} = \frac{\tau_{x\theta}}{G_i} = 1914(10^{-6})\rho$$ 1

The plots of torsional shear strain and stress are given below.

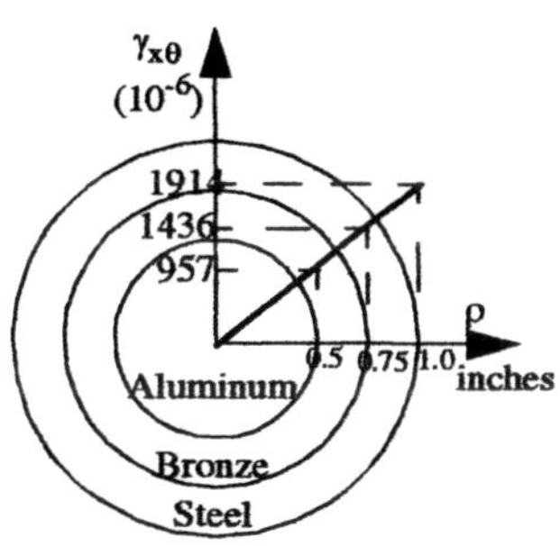

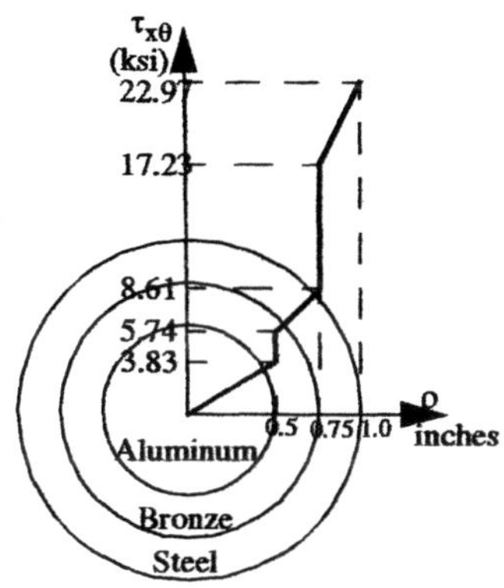

5.70 Solve problem 5.61 using Equations 5.21 and 5.22.

Solution $t = 5$ mm $\quad d_{al} = 125$ mm $\quad d_{cu} = 50$ mm $\quad L = 1.5$ m $\quad \phi_1 = 0.03$ rads

$G_{al} = 28$ GPa $\quad G_{cu} = 40$ GPa $\quad (\tau_{al})_{max} = ? \quad (\tau_{cu})_{max} = ? \quad T = Fd = ?$

From Eqs (1) and (2) in problem 5-61, we have

$$J_{al} = 7.6820(10^{-6}) \text{ m}^4 \qquad J_{cu} = 0.4958(10^{-6}) \text{ m}^4$$ 1

The torsional rigidity of the cross section is

$$\Sigma G_j J_j = (28)(10^9)7.6820(10^{-6}) + (40)(10^9)(0.4958)(10^{-6}) = 234.93(10^3) \text{ N} - \text{m}^2$$ 2

The relative rotation of the ends of the shaft can be written as

$$\phi_B - \phi_A = \frac{T_{AB}(x_B - x_A)}{\Sigma G_j J_j} = \frac{T(1.5)}{234.93(10^3)} = 0.03 \text{ or} T = 4.669(10^3) \text{ N} - \text{m}$$ $\boxed{T = 4.7 \text{ kN} - \text{m}}$

The torsional shear stress at any point on the cross section can be found as shown below

$$\tau_{x\theta} = G_i \frac{T_{AB}\rho}{\Sigma G_j J_j} = \frac{4.669(10^3)}{234.93(10^3)} G_i \rho = 0.02 G_i \rho$$ 3

The maximum torsional shear stress in copper will be at $\rho = (d_{cu}/2 + t/2) = 0.0275$ m. From Eq.(3) we obtain:

$$(\tau_{cu})_{max} = (0.02)(40)(10^9)(0.0275) = 22(10^6) \text{ N/m}^2$$ $\boxed{(\tau_{cu})_{max} = 22 \text{ MPa}}$

The maximum torsional shear stress in aluminum will be at $\rho = (d_{al}/2 + t/2) = 0.065 \text{ m}$. From Eq.(3) we obtain:

$(\tau_{al})_{max} = (0.02)(28)(10^9)(0.065) = 36.4(10^6) \text{ N/m}^2$ $\boxed{(\tau_{al})_{max} = 36.4 \text{ MPa}}$

5.71 Solve problem 5.62 using Equations 5.21 and 5.22.

Solution $t = 5 \text{ mm}$ $d_{al} = 125 \text{ mm}$ $d_{cu} = 50 \text{ mm}$ $L = 1.5 \text{ m}$

$G_{al} = 28 \text{ GPa}$ $G_{cu} = 40 \text{ GPa}$ $T = Fd = 10\text{kN} - \text{m}$

$(\tau_{al})_{max} = ?$ $(\tau_{cu})_{max} = ?$ $\phi_B = ?$

The torsional rigidity of the cross section is

$$\Sigma G_j J_j = 234.93(10^3) \text{ N} - \text{m}^2 \quad 1$$

The relative rotation of the ends of the shaft can be written as

$$\phi_B - \phi_A = \frac{T_{AB}(x_B - x_A)}{\Sigma G_j J_j} = \frac{(10)(10^3)(1.5)}{234.93(10^3)} = 63.85(10^{-3})$$

Noting that ϕ_A=0, we obtain $\boxed{\phi_B = 0.0639 \text{ rads CCW}}$

The torsional shear stress at any point on the cross section can be found as shown below

$$\tau_{x\theta} = G_i \frac{T_{AB}\rho}{\Sigma G_j J_j} = \frac{(10)(10^3)}{234.93(10^3)} G_i \rho = 0.0426 G_i \rho \quad 2$$

The maximum torsional shear stress in copper will be at $\rho = (d_{cu}/2 + t) = 0.03 \text{ m}$.

From Eq.(3) we obtain: $(\tau_{cu})_{max} = (0.0426)(40)(10^9)(0.0275) = 46.86(10^6) \frac{\text{N}}{\text{m}^2}$ or $\boxed{(\tau_{cu})_{max} = 46.9 \text{ MPa}}$

The maximum torsional shear stress in aluminum will be at $\rho = (d_{al}/2 + t) = 0.0675 \text{ m}$.

From Eq.(3) we obtain: $(\tau_{al})_{max} = (0.0426)(28)(10^9)(0.065) = 77.53(10^6) \frac{\text{N}}{\text{m}^2}$ or $\boxed{(\tau_{al})_{max} = 77.5 \text{ MPa}}$

5.72 A cast iron pipe (G_{ir} = 70 GPa) and a copper pipe (G_{cu} = 40 GPa) are securely bonded together as shown. The outer diameters of the two pipes are 50 mm and 70 mm and wall thickness of each pipe is 10 mm. If T = 1500 N-m, determine (a) the maximum torsional shear stress in cast iron and copper. (b) the rotation of section at D with respect to the section at A.

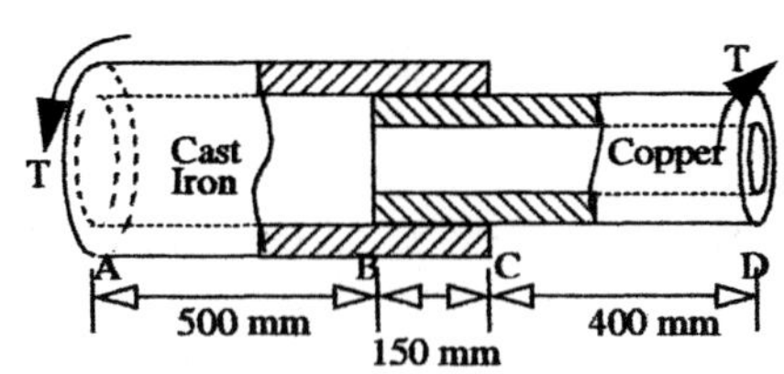

Fig. P5.72

Solution $G_{ir} = 70 \text{ GPa}$ $G_{cu} = 40 \text{ GPa}$ $(d_{ir})_o = 70 \text{ mm}$ $(d_{ir})_i = 50 \text{ mm}$

$(d_{cu})_o = 50 \text{ mm}$ $(d_{cu})_i = 30 \text{ mm}$ $T = 1500 \text{ N} - \text{m}$

$(\tau_{ir})_{max} = ?$ $(\tau_{cu})_{max} = ?$ $\phi_D - \phi_A = ?$

The polar moment of each material cross - section can be found as shown below

$$J_{ir} = \frac{\pi}{32}[(0.07)^4 - (0.05)^4] = 1.7436(10^{-6})\text{m}^4 \quad 1$$

$$J_{cu} = \frac{\pi}{32}[(0.05)^4 - (0.03)^4] = 0.5341(10^{-6})\text{m}^4 \quad 2$$

The torsional rigidity of the cross section in segment BC is

$$\Sigma G_j J_j = (70)(10^9)(1.7436)(10^{-6}) + (40)(10^9)(0.5341)(10^{-6}) = 143.41(10^3)\ \text{N}-\text{m}^2$$

The internal torque in all the segments is equal to -T.The relative rotation of the ends of each segment can be written as

$$\phi_B - \phi_A = \frac{T_{AB}(x_B - x_A)}{G_{ir}J_{ir}} = \frac{(-1500)(0.5)}{(70)(10^9)(1.7436)(10^{-6})} = -6.145(10^{-3}) \qquad 3$$

$$\phi_C - \phi_B = \frac{T_{BC}(x_C - x_B)}{\Sigma G_j J_j} = \frac{(-1500)(0.15)}{(143.4)(10^{-3})} = -1.569(10^{-3}) \qquad 4$$

$$\phi_D - \phi_C = \frac{T_{CD}(x_D - x_C)}{G_{cu}J_{cu}} = \frac{(-1500)(0.4)}{(40)(10^9)(0.5341)(10^{-6})} = -28.08(10^{-3}) \qquad 5$$

Adding Eqs. (3),(4) and (5), we obtain

$$\phi_D - \phi_A = -(0.145 + 1.569 + 28.08)(10^{-3}) = 35.798(10^{-3})$$

$$\boxed{\phi_D - \phi_A = 0.0358\ \text{rads CW}}$$

The magnitude of maximum torsional shear stress in segment AB and CD can be found as shown below

$$(\tau_{AB})_{max} = \left|\frac{T_{AB}(\rho_{AB})_{max}}{J_{ir}}\right| = \left|\frac{(-1500)(0.035)}{1.7436(10^{-6})}\right| = 30.1(10^6)\ \frac{\text{N}}{\text{m}^2} \qquad 6$$

$$(\tau_{CD})_{max} = \left|\frac{T_{CD}(\rho_{CD})_{max}}{J_{cu}}\right| = \left|\frac{(-1500)(0.025)}{0.5341(10^{-6})}\right| = 70.2(10^6)\ \frac{\text{N}}{\text{m}^2} \qquad 7$$

The magnitude of torsional shear stress at any point in segment BC is

$$\tau_{x\theta} = \left|G_i\frac{T_{BC}\rho}{\Sigma G_j J_j}\right| = \left|\frac{(-1500)}{143.41(10^3)}G_i\rho\right| = 10.46(10^{-3})G_i\rho \qquad 8$$

At ρ=0.035, G_i=G_{ir}, we obtain $\tau_{x\theta} = 25.62(10^{-6})$ which is less than in Eqs. (6). Thus, the maximum torsional shear stress in iron is

$$\boxed{(\tau_{ir})_{max} = 30.1\ \text{MPa}}$$

At ρ=0.025, G_i=G_{cu}, we obtain $\tau_{x\theta} = 10.46(10^{-6})$ which is less than in Eq.(7). Thus, the maximum torsional shear stress in copper is

$$\boxed{(\tau_{cu})_{max} = 70.2\ \text{MPa}}$$

5.73 A cast iron pipe (G_{ir} = 70 GPa) and a copper pipe (G_{cu} = 40 GPa) are securely bonded together as shown. The outer diameters of the two pipes are 50 mm and 70 mm and wall thickness of each pipe is 10 mm, The allowable rotation of cross-section at D with respect to that at A is 0.02 rads. Determine the maximum torque T to the nearest kN-m that can be applied to the shaft.

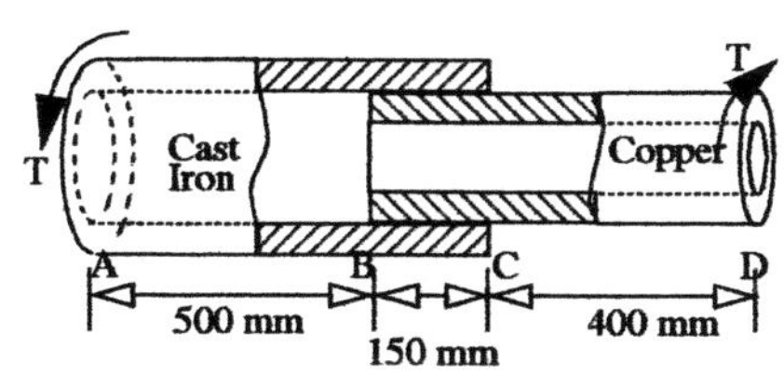

Fig. P5.73

Solution $G_{ir} = 70\ \text{GPa}$ $\quad G_{cu} = 40\ \text{GPa}$ $\quad (d_{ir})_o = 70\ \text{mm}$ $\quad (d_{ir})_i = 50\ \text{mm}$

$(d_{ir})_o = 50\ \text{mm}$ $\quad (d_{ir})_i = 30\ \text{mm}$ $\quad |\phi_D - \phi_A| \leq 0.2\ \text{rads}$ $\quad T_{max} = ?$

From problem 5-72, we have $J_{ir} = 1.7436(10^{-6})\ \text{m}^4$, $J_{cu} = 0.5341(10^{-6})\ \text{m}^4$ and $\Sigma G_j J_j = 143.41(10^3)\text{N}-\text{m}^2$

The internal torque in all the segments is equal to -T.The relative rotation of the ends of each segment can be found as shown below.

$$\phi_B - \phi_A = \frac{T_{AB}(x_B - x_A)}{G_{ir}J_{ir}} = \frac{(-T)(0.5)}{(70)(10^9)(1.7436)(10^{-6})} = -4.097(10^{-6})T \qquad 1$$

$$\phi_C - \phi_B = \frac{T_{BC}(x_C - x_B)}{\Sigma G_j J_j} = \frac{(-T)(0.15)}{(143.4)(10^{-3})} = -1.046(10^{-6})T \qquad 2$$

$$\phi_D - \phi_C = \frac{T_{CD}(x_D - x_C)}{G_{cu} J_{cu}} = \frac{(-T)(0.4)}{(40)(10^9)(0.5341)(10^{-6})} = -18.72(10^{-6})T \qquad 3$$

Adding Eqs.(1),(2) and (3), we obtain

$$\phi_D - \phi_A = -(4.097 + 1.046 + 18.72)(10^{-6})T = 23.867(10^{-6})T \qquad 4$$

$|\phi_D - \phi_A| = 23.867(10^{-6})T \le 0.02$ or $T \le 837.98$ N – m

The maximum torque to the nearest N-m is $\boxed{T_{max} = 837 \text{ N-m}}$

5.74 A solid steel (G_s = 80 GPa) shaft 3 m long is securely fastened to a hollow bronze (G_{br} = 40 GPa) shaft that is 2 m long as shown. The applied torque in is T = 10 kN-m. Determine (a) the magnitude of maximum torsional shear stress in the shaft. (b) the rotation of section at 1 m from the left wall.

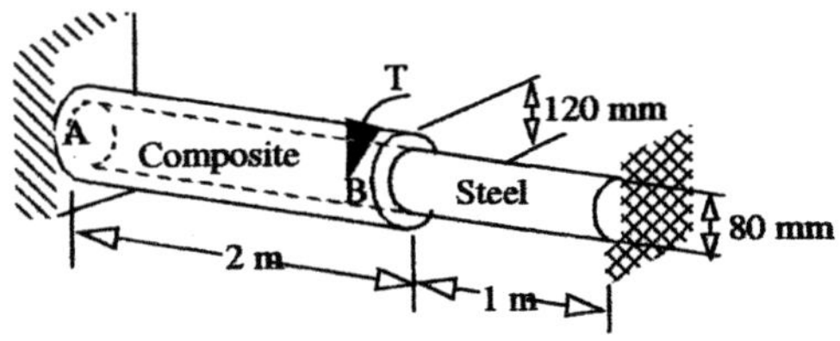

Fig. P5.74

Solution G_s= 80 GPa G_{br}= 40 GPa T = 10 kN-m τ_{max} = ? ϕ_1 = ?

The polar moment of cross-section of each material can be found as

$$J_{br} = \frac{\pi}{32}[(0.12)^4 - (0.08)^4] = 16.336(10^{-6}) \text{ m}^4 \qquad 1$$

$$J_s = \frac{\pi}{32}(0.08)^4 = 4.021(10^{-6}) \text{ m}^4 \qquad 2$$

The torsional rigidity of the cross section in segment AB is

$$\Sigma G_j J_j = (40)(10^9)(16.336)(10^{-6}) + (80)(10^9)(4.021)(10^{-6}) = 975.15(10^3) \text{ N-m}^2 \qquad 3$$

The following free body diagrams can be drawn with internal torques drawn as per our sign convention.

(a)

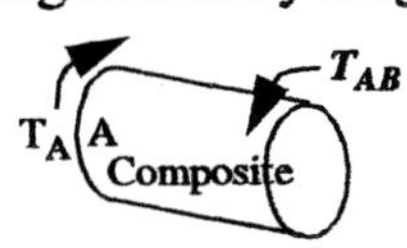

(b)

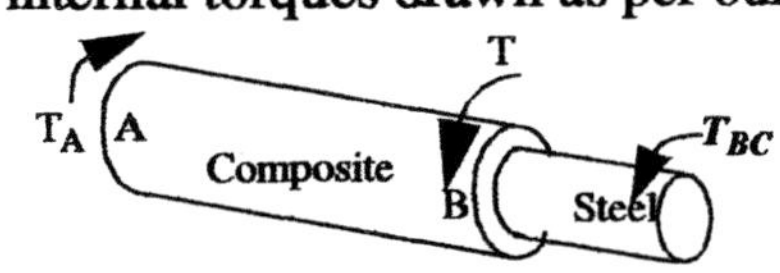

By moment equilibrium about the shaft axis we have the following equations.

$$T_{AB} = T_A \qquad T_{BC} = T_A - T \qquad 4$$

The relative rotation of the ends of each segment can be written as shown below

$$\phi_B - \phi_A = \frac{T_{AB}(x_B - x_A)}{\Sigma G_j J_j} = \frac{T_A(2)}{(975.15)(10^3)} = -2.051(10^{-6})T_A \qquad 5$$

$$\phi_C - \phi_B = \frac{T_{BC}(x_C - x_B)}{G_s J_s} = \frac{(T_A - T)(1)}{(80)(10^9)(4.021)(10^{-6})} = 3.109(10^{-6})(T_A - T) \qquad 6$$

Adding Eq's(5) and (6) and noting that ϕ_A=0 and ϕ_C=0, we obtain

$\phi_C - \phi_A = [2.051T_A + 3.109(T_A - T)](10^{-6}) = 0$ or

$$T_A = \frac{3.109T}{2.051 + 3.109} = 0.6025T = 6.025(10^3) \text{ N-m} \qquad 7$$

From Eq.(4) $T_{AB} = 6.025(10^3)$ N – m $\qquad T_{BC} = -3.9748(10^3)$ N – m

The maximum torsional shear stress in segment BC is

$$(\tau_{BC})_{max} = \frac{T_{BC}(\rho_{BC})_{max}}{J_s} = \frac{(-3.9748)(10^3)(0.04)}{4.021(10^{-6})} = -39.54(10^6)\ \frac{N}{m^2} \qquad 8$$

The torsional shear stress at any point of the cross -section in segment AB is

$$\tau_{x\theta} = G_i\frac{T_{AB}\rho}{\Sigma G_jJ_j} = \frac{6.025(10^3)}{975.15(10^3)}G_i\rho = 6.1787(10^{-3})G_i\rho \qquad 9$$

At the furthest point in the steel, in segment AB, $\rho = 0.04$ m $G_i = G_s$, we obtain $\tau_{x\theta} = 19.77(10^6)\ \frac{N}{m^2}$

which is smaller in magnitude than the shear stress value in Eq.(8).

At the furthest point in the bronze, in segment AB, $\rho = 0.06$m $G_i = G_{br}$, we obtain $\tau_{x\theta} = 14.83(10^6)\ \frac{N}{m^2}$

which is smaller in magnitude than the shear stress value in Eq.(8).The magnitude of the maximum torsional shear stress is $\boxed{\tau_{max} = 39.54\ \text{MPa}}$

The rotation at 1m from the wall of A can be found as shown below

$$\phi_1 - \phi_A = \frac{T_{AB}(x_1 - x_A)}{\Sigma G_jJ_j} = \frac{(6.025)(10^3)(1)}{(975.15)(10^3)} = 6.179(10^{-3})\ \text{rads}$$

Noting that ϕ_A=0, we obtain $\boxed{\phi_1 = 0.0062\ \text{rads CCW}}$

5.75 A solid steel (G_s = 80 GPa) shaft 3 m long is securely fastened to a hollow bronze (G_{br} = 40 GPa) shaft that is 2 m long as shown. If the section at B rotates by 0.05 rads, (a) determine the magnitude of maximum torsional shear *strain* in the shaft. (b) the applied torque T.

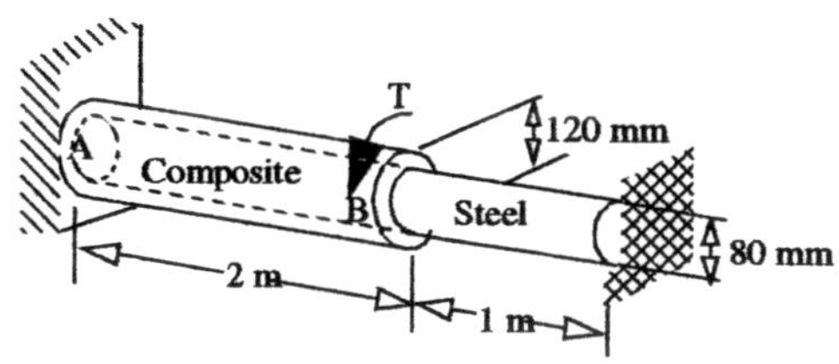

Fig. P5.75

Solution G_s = 80 GPa G_{br} = 40 GPa ϕ_B = 0.05 rads γ_{max} = ? T = ?

The calculation for maximum torsional shear strain is same as in problem 5-60 and is independent of the material values.Thus,from 5-60, we obtain $\boxed{\gamma_{max} = -2000\ \mu\text{rads}}$

Substituting ϕ_A=0 and ϕ_B=0.05 in Eq.(5) of problem 5-74, we obtain

$$T_A = \frac{0.5}{2.05(10^{-6})} = 24.39(10^3)\ \text{N}-\text{m}$$

From Eq.(7) of problem 5-74, we obtain $T = \frac{T_A}{0.6025} = \frac{24.39(10^3)}{0.6025} = 40.46(10^3)\text{N}-\text{m}$ or $\boxed{T = 40.5\text{kN}-\text{m}}$

5.76 A composite shaft of composite shaft is constructed from aluminum (G_{al}= 4,000 ksi) and steel (G_s = 12,000 ksi) as shown. The allowable shear stress is 18 ksi and 24 ksi in aluminum and steel, respectively. (a) Determine the maximum torque T that can be applied. (b) Determine the rotation of the free end with respect to the wall when the maximum torque is applied.

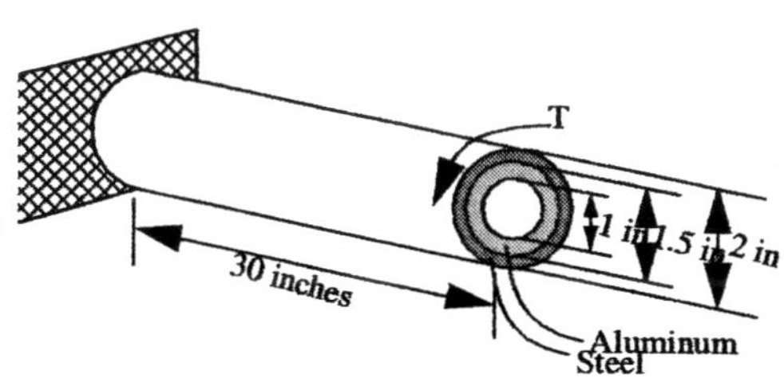

Fig. P5.76

Solution $G_s = 4{,}000$ ksi $G_{al} = 12{,}000$ ksi $(\tau_s)_{max} \le 24$ ksi $(\tau_{al})_{max} \le 18$ ksi

$T_{max} = ?$ $\phi_B = ?$

The polar moment of each material cross-section of each material can be found as

$$J_s = \frac{\pi}{32}[2^4 - 1.5^4] = 1.0738 \text{ in}^4 \text{ and } J_{al} = \frac{\pi}{32}[1.5^4 - 1^4] = 0.3988 \text{ in}^4$$

The torsional rigidity of the cross section is

$$\Sigma G_j J_j = (4000)(0.3988) + (12000)(1.0738) = 14.48(10^3) \text{ kips} - \text{in}^2$$

The magnitude of the maximum torsional shear stress in each material can be found as

$$(\tau_{al})_{max} = G_{al}\frac{T(\rho_{al})_{max}}{\Sigma G_j J_j} = \frac{(4000)(0.75)T}{14.48(10^3)} \le 18 \quad \text{or} \quad T \le 86.88 \text{ kips} - \text{in} \qquad 1$$

$$(\tau_s)_{max} = G_s\frac{T(\rho_s)_{max}}{\Sigma G_j J_j} = \frac{(12000)(1)T}{14.48(10^3)} \le 24 \quad \text{or} \quad T \le 28.96 \text{ kips} - \text{in} \qquad 2$$

The maximum value of T that satisfies Eqs(1) and (2) is $\boxed{T_{max} = 28.9 \text{ kips} - \text{in}}$

The relative rotation of the ends of shaft can be as shown below

$$\phi_B - \phi_A = \frac{T_{AB}(x_B - x_A)}{\Sigma G_j J_j} = \frac{(28.9)(30)}{(14.48)(10^3)} = 60.03(10^{-3})$$

Noting that $\phi_A = 0$, we obtain $\boxed{\phi_B = 0.06 \text{ rads CCW}}$

5.77 A 1-m long shaft is to be designed to transmit a torque of 3300 N-m. The outside diameter of the shaft must be 40 mm to fit existing attachments. The shaft can be all aluminum, all titanium, or a composite of the two material. The Shear Modulus of rigidity G, the allowable shear stress τ_{allow}, and the density γ of titanium and aluminum are given in Table 5.6. Determine the diameters to the nearest millimeter of the *lightest* shaft and the corresponding mass.

Table 5.6..Material Properties

Material	G GPa	τ_{allow} MPa	γ Mg./m^3
Titanium Alloy	36	300	4.4
Aluminum	28	180	2.8

Solution $L = m$ $T = 3300$ N-m $d_o = 40$ mm diameters = ? m = ?

Figure below shows the three possible material selection with the unknown variable diameters.

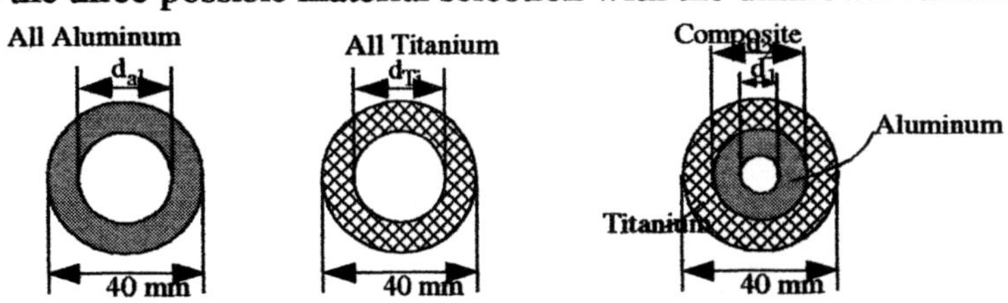

For homogenous cross -section, the maximum polar moment corresponds to solid shaft and its value is $J_{max} = 0.2513(10^{-6})$ m^4.

All Aluminum shaft: The maximum torsional shear stress can be written and compared to the limiting value as shown below

$$(\tau_{al})_{max} = \frac{T\rho_{max}}{J_{al}} = \frac{(3300)(0.02)}{J_{al}} \le 180(10^6) \text{ or } J_{al} \ge 0.3667(10^{-6}) \text{ m}^4$$ which is greater than J_{max}, hence the shaft cannot be all aluminum.

All Titanium shaft: The maximum torsional shear stress can be written and compared to the limiting value as shown below

$$(\tau_{ti})_{max} = \frac{T\rho_{max}}{J_{ti}} = \frac{(3300)(0.02)}{J_{ti}} \le 300(10^6) \text{ or } J_{al} \ge 0.22(10^{-6}) \text{ or } \frac{\pi}{32}[(0.04)^4 - d_{ti}^{\,4}] \ge 0.22(10^{-6}) \text{ or}$$

$$d_{ti} \le 23.77(10^{-3}) \text{ m} \qquad 1$$

The diameter to the nearest millimeter that satisfies Eq.(1) is $d_{ti} = 23.77$ mm .The corresponding mass is

$$m_{ti} = \gamma_{ti}AL = (4.4)(10^6)\frac{\pi}{4}[(0.04)^2 - (0.023)^2](1) = 3701 \text{ gms} \qquad 2$$

Composite Shaft: The maximum torsional shear stress in each material can be written and compared to the limiting value as shown below

$$(\tau_{ti})_{max} = \frac{TG_{ti}(\rho_{ti})_{max}}{\Sigma G_jJ_j} = \frac{(3300)(36)(10^9)(0.02)}{\Sigma G_jJ_j} \le 300(10^6) \quad \text{or} \quad \Sigma G_jJ_j \ge 7.92(10^3)\text{N} - \text{m}^2 \qquad 3$$

$$(\tau_{al})_{max} = \frac{TG_{al}(\rho_{al})_{max}}{\Sigma G_jJ_j} = \frac{(3300)(28)(10^9)(d_2/2)}{\Sigma G_jJ_j} \le 180(10^6) \qquad 4$$

Substituting Eq.(3) into (4), we obtain

$$d_2 \le 30.77(10^{-3}) \qquad 5$$

The diameter to the nearest millimeter that satisfies Eq.(5) is

$$d_2 = 30.77 \text{ mm} \qquad 6$$

The torsional rigidity of the cross section can be written and substituted in Eq.(3) to obtain the following

$$\Sigma G_jJ_j = (36)(10^9)\left(\frac{\pi}{32}[(0.04)^4 - (0.03)^4]\right) + (28)(10^9)\left(\frac{\pi}{32}[(0.03)^4 - d_1^{\,4}]\right) \ge 7.92(10^3) \text{ or}$$

$$(0.03)^4 - d_1^{\,4} \ge \frac{32}{\pi(28)(10^9)}[7.92 - 6.185](10^3) \quad \text{or} \quad d_1 \le 20.56(10^3)\text{m} \qquad 7$$

The diameter to the nearest millimeter that satisfies Eq.(7) is

$$d_1 = 20 \text{ mm} \qquad 8$$

The mass of the composite shaft can be found as shown below

$$m = \gamma_{ti}A_{ti}L + \gamma_{al}A_{al}L = (4.4)(10^6)\frac{\pi}{4}[(0.04)^2 - (0.03)^2](1) + (2.8)(10^6)\frac{\pi}{4}[(0.03)^2 - (0.02)^2](1) \text{ or}$$

$$m = 2419 + 1100 = 3519 \text{ gms} \qquad 9$$

From Eq.(9) and (2), we see that the mass of the composite shaft is less.Thus, the answer

$$\boxed{\text{CompositeShaft}; \quad d_1 = 20 \text{ mm} \quad d_2 = 30 \text{ mm} \quad W = 3519 \text{ gms}}$$

5.78 The allowable shear stress in the stepped shaft shown is 17 ksi. Determine the smallest fillet radius that can be used at section B. Use stress concentration graphs given in Appendix C.

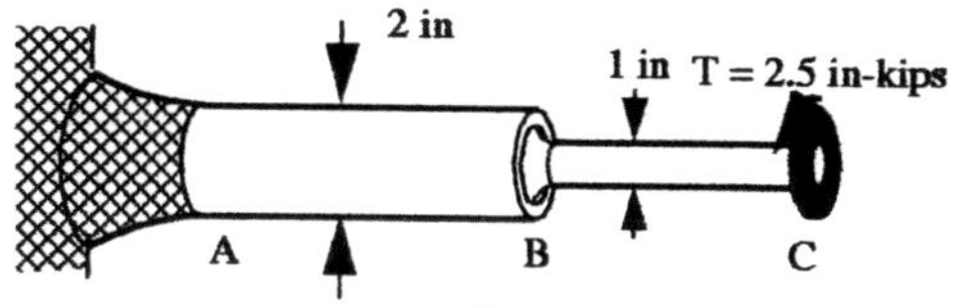

Fig. P5.78

Solution $\tau_{max} \le 17$ ksi $\qquad r = ?$

The maximum shear stress in section BC can be found as,

$$(\tau_{BC})_{max} = \frac{T\rho_{max}}{J_{BC}} = \frac{(2.5)(0.5)}{\pi(1)^4/32} = 12.73 \text{ ksi}$$

$\tau_{max} = K_{conc}(\tau_{BC})_{max} = 12.73K_{conc} \le 17 \therefore K_{conc} \le 1.335$

From Fig. C.3, we obtain the approximate value of r/d corresponding to D/d = 2 and $K_{conc} = 1.335$,

$\frac{r}{d} = 0.135$ or $r = (0.135)(1)$ **r = 0.135 inch**

5.79 The fillet radius in the stepped shaft shown is 6 mm. Determine the maximum torque that can act on the rigid wheel if the allowable shear stress is 80 MPa and a Modulus of Rigidity is 28 GPa. Use stress concentration graphs given in Appendix C.

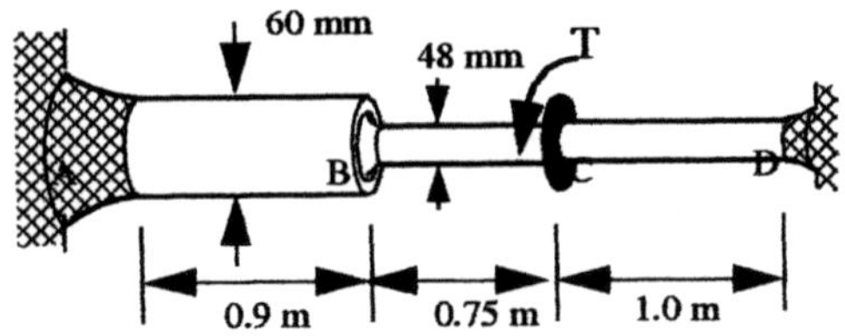

Fig. P5.79

Solution $r = 6$ mm $\tau_{max} \le 80$ MPa $G = 28$ MPa T_{max} ?

The polar moment of cross-sections in various segments can be found as

$$J_{AB} = \frac{\pi}{32}(0.06)^4 = 1.272(10^{-6})\ m^4 \qquad 1$$

$$J_{BC} = J_{CD} = \frac{\pi}{32}(0.048)^4 = 0.521(10^{-6})\ m^4 \qquad 2$$

The following free body diagrams can be drawn. The internal torques are drawn as per our sign convention.

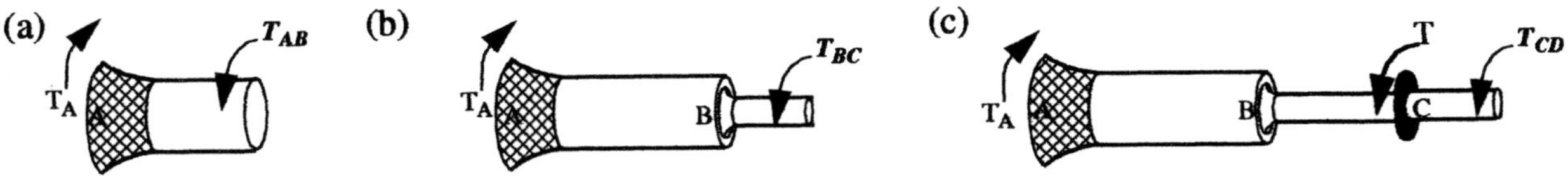

By moment equilibrium about the shaft axis we have the following equations.

$$T_{AB} = T_A \qquad T_{BC} = T_A \qquad T_{CD} = T_A - T \qquad 3$$

The relative rotation of the ends of each segment can be written as shown below

$$\phi_B - \phi_A = \frac{T_{AB}(x_B - x_A)}{G_{AB}J_{AB}} = \frac{T_A(0.9)}{(28)(10^9)(1.272(10^{-6}))} = 25.26(10^{-6})T_A \qquad 4$$

$$\phi_C - \phi_B = \frac{T_{BC}(x_C - x_B)}{G_{BC}J_{BC}} = \frac{(T_A)(0.75)}{(28)(10^9)(0.521)(10^{-6})} = 51.40(10^{-6})T_A \qquad 5$$

$$\phi_D - \phi_C = \frac{T_{CD}(x_D - x_C)}{G_{CD}J_{CD}} = \frac{(T_A - T)(1)}{(28)(10^9)(0.521)(10^{-6})} = 68.53(10^{-6})(T_A - T) \qquad 6$$

Adding Eq's(5) and (6) and noting that ϕ_A=0 and ϕ_D=0, we obtain

$$\phi_D - \phi_A = [25.26T_A + 51.40T_A + 68.53(T_A - T)] = 0 \quad \text{or} \quad T_A = \frac{68.53T}{25.26 + 51.40 + 68.53} = 0.472T$$

From Eq.(3), $T_{AB} = 0.472T \qquad T_{BC} = 0.472T \qquad T_{CD} = -0.528T$

The nominal torsional shear stresses in segment BC can be found as shown below

$$(\tau_{BC})_{max} = \frac{T_{BC}(\rho_{BC})_{max}}{J_{BC}} = \frac{(0.472T)(0.04)}{0.521(10^{-6})} = 21.74T(10^3)$$

Now $\frac{r}{d} = \frac{6}{48} = 0.125$ and $\frac{D}{d} = \frac{60}{48} = 1.25$ .From Figure C.3, we obtain the stress concentration factor as $K_{conc} = 1.29$.Thus, the maximum shear stress in segment BC is

$$(\tau_{max})_{BC} = 1.29(\tau_{BC})_{max} = (1.29)(21.74)T(10^3) \le 80(10^6) \quad \text{or} \quad T \le 2.853(10^3)\ \text{N-m} \qquad 7$$

The magnitude of maximum shear stress in CD is

$$(\tau_{CD})_{max} = \left|\frac{T_{CD}(\rho_{CD})_{max}}{J_{CD}}\right| = \frac{(0.528T)(0.024)}{0.521(10^{-6})} = 24.32T(10^3) \le 80(10^6) \text{ or } T \le 3.29(10^3)\ \text{N-m} \qquad 8$$

The maximum value of T that meets the requirements in Eq.(7) and (8) is $\boxed{T_{max} = 2.853\ \text{kN-m}}$

5.80 A composite circular shaft is made from two non-linear materials. The stress-strain relationship for material one is given by $\tau_1 = G\gamma^{0.5}$ and for material two it is given by $\tau_2 = 2G\gamma^{0.5}$. Determine the rotation of section at B in terms of T_{ext}, L, G, and R.

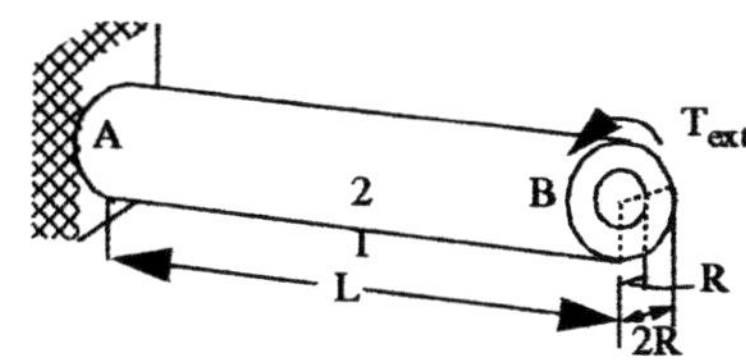

Fig. P5.80

Solution $\phi_B = f(T_{ext}, L, G, R) = ?$

Assuming all kinematic assumptions are valid, the shear strain is given by:

$$\gamma_{x\theta} = \rho\frac{d\phi}{dx} \qquad 1$$

By substituting Eq. (1) into the given stress-strain equations, we can obtain the following stress distribution across the cross-section.

$$\tau_{x\theta} = \begin{cases} G\rho^{0.5}\left(\frac{d\phi}{dx}\right)^{0.5} & 0 \le \rho < R \\ 2G\rho^{0.5}\left(\frac{d\phi}{dx}\right)^{0.5} & R < \rho < 2R \end{cases} \qquad 2$$

The equivalent internal torque is given by the equation: $T = \int_A \rho\tau_{x\theta}dA$

Writing the integral as the sum of integral over each material, noting that $\left(\frac{d\phi}{dx}\right)$ is only a function of x, and $dA = 2\pi\rho d\rho$, we obtain the following.

$$T = \int_0^R \rho\left[G\rho^{0.5}\left(\frac{d\phi}{dx}\right)^{0.5}\right](2\pi\rho)d\rho + \int_R^{2R} \rho\left[2G\rho^{0.5}\left(\frac{d\phi}{dx}\right)^{0.5}\right](2\pi\rho)d\rho = (2\pi G)\left(\frac{d\phi}{dx}\right)^{0.5}\left[\int_0^R \rho^{2.5}d\rho + 2\int_R^{2R}\rho^{2.5}d\rho\right] \quad \text{or}$$

$$T = (2\pi G)\left(\frac{d\phi}{dx}\right)^{0.5}\left[\frac{\rho^{3.5}}{3.5}\bigg|_0^R + 2\frac{\rho^{3.5}}{3.5}\bigg|_R^{2R}\right] = \frac{(2\pi G)}{3.5}\left(\frac{d\phi}{dx}\right)^{0.5}[R^{3.5} + 2(2R)^{3.5} - 2R^{3.5}] \quad \text{or}$$

$$\left(\frac{d\phi}{dx}\right)^{0.5} = \frac{T}{38.825GR^{3.5}} \quad \text{or} \quad \left(\frac{d\phi}{dx}\right) = \frac{T^2}{1507.4G^2R^7} \qquad 3$$

The internal torque is equal to external torque, and G and R are constant, thus $\frac{d\phi}{dx} = \frac{\phi_B - \phi_A}{x_B - x_A} = \frac{\phi_B}{L}$. Substituting in the above equation, we obtain.

$$\boxed{\phi_B = \frac{T_{ext}^2 L}{1507.4G^2R^7}}$$

Problem Set 5.4

5.81 A solid circular shaft of diameter 3 inches has a shear strain at a section in polar coordinates was found to be $\gamma_{x\theta} = 0.002\rho$ where ρ is the radial coordinate measured in inches. For the given material model, (a) Write expressions for $\tau_{x\theta}$ as a function of ρ and plot the shear strain $\gamma_{x\theta}$ and shear stress $\tau_{x\theta}$ distribution across the cross-section. (b) Determine the equivalent internal torque acting at the cross-section.The shaft is made from elastic-perfectly plastic material that has a yield stress of τ_{yield}= 18 ksi, and a Shear modulus G = 12,000 ksi.

Solution $\gamma_{x\theta}=0.002\rho$ $\tau_{yield}=18\text{ksi}$ G=12,000 ksi
$\tau_{x\theta}=f(\rho) = ?$ $T = ?$ Elastic perfectly plastic

---

(a) The strain at yield point is: $\gamma_{yield} = \frac{18}{12,000} = 0.0015 = 0.002\rho_{yield}$ or $\rho_{yield} = 0.75\text{inch}$ where, ρ_{yield} is the radial coordinate of elastic-plastic beam. For $0 < \rho < 0.75$, $\tau_{x\theta} = G\gamma_{x\theta} = (12000)(0.002\rho) = 24\rho$. The stress distribution can be written as:

$$\tau_{x\theta} = \begin{cases} 24\rho \text{ ksi} & 0 < \rho < 0.75\text{in} \\ 18 \text{ ksi} & 0.75\text{in} < \rho < 1.5\text{in} \end{cases} \quad (1)$$

The plot of shear strain and shear stress are shown below.

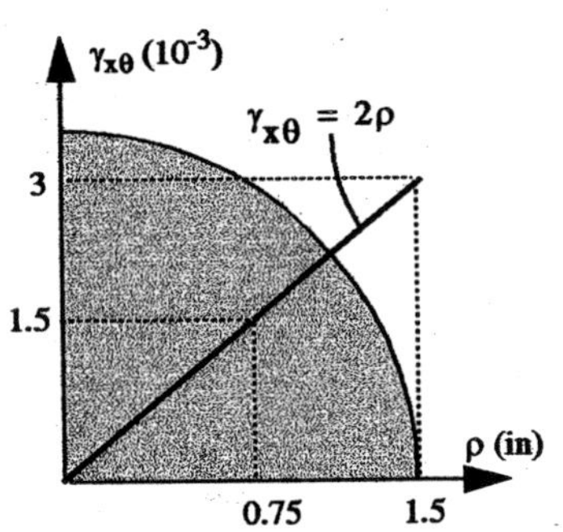

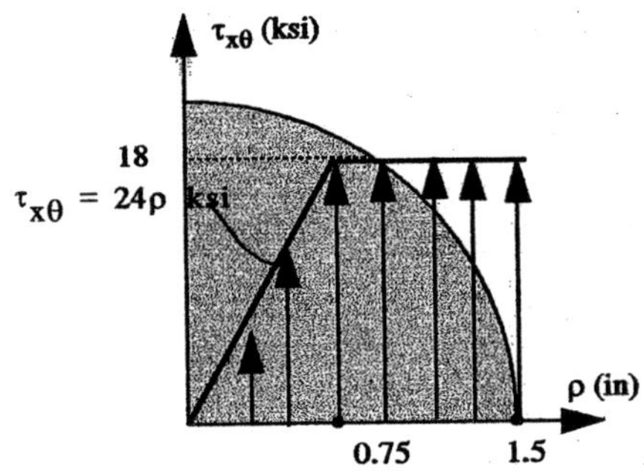

The internal torque can be found from Eq. 5.1 as shown below.

$$T = \int_A \rho\tau_{x\theta}dA = \underbrace{\int_{A_E} \rho\tau_{x\theta}dA}_{T_E} + \underbrace{\int_{A_P} \rho\tau_{x\theta}dA}_{T_P} \quad (2)$$

The differential area dA is the area of a ring of radius ρ and thickness dρ i.e. $dA = (2\pi\rho)d\rho$. Substituting the stress distribution from Eq. 1 into Eq. 2 and performing integration we obtain the internal torque

$$T_E = \int_0^{0.75} (\rho)(24\rho)(2\pi\rho)d\rho = (48\pi)\left.\frac{\rho^4}{4}\right|_0^{0.75} = 11.93 \text{ in} - \text{kips}$$

$$T_P = \int_{0.75}^{1.5} (\rho)(18)(2\pi\rho)d\rho = (36\pi)\left.\frac{\rho^3}{3}\right|_{0.75}^{1.5} = 111.33 \text{ in} - \text{kips}$$

$T = T_E + T_P = 123.26 \text{ in} - \text{kips}$ or $\boxed{T = 123.3 \text{ in} - \text{kips}}$

5.82 A solid circular shaft of diameter 3 inches has a shear strain at a section in polar coordinates was found to be $\gamma_{x\theta} = 0.002\rho$ where ρ is the radial coordinate measured in inches. For the given material model, (a) Write expressions for $\tau_{x\theta}$ as a function of ρ and plot the shear strain $\gamma_{x\theta}$ and shear stress $\tau_{x\theta}$ distribution across the cross-section. (b) Determine the equivalent internal torque acting at the cross-section. The shaft is made from a bi-linear material that has a yield stress of τ_{yield}= 18 ksi, Shear modulus G_1 = 12,000 ksi and G_2= 4,800 ksi.

Solution $\gamma_{x\theta}=0.002\rho$ $\tau_{yield}=18ksi$ $G_1=12{,}000ksi$ $G_2=4{,}800ksi$ $\tau_{x\theta}= f(\rho) = ?$ $T = ?$

Before yield point we have: $\tau_1=G_1\gamma_{x\theta}=(12{,}000)(0.002\rho)=24\rho$

The strain at yield point is: $\gamma_{yield} = \dfrac{\tau_{yield}}{G_1} = \dfrac{18}{12{,}000} = 0.0015$

After yield point we can write: $\tau_2 = \tau_{yield} + G_2(\gamma_{x\theta} - \gamma_{yield}) = 18 + 4800(2\rho(10^{-3}) - 0.0015) = 10.8 + 9.6\rho$

From problem 3.134 ρ_{yield}=0.75 inch. The stress distribution can be written as:

$$\tau_{x\theta} = \begin{cases} 24\rho \text{ ksi} & 0 < \rho < 0.75\text{in} \\ (10.8 + 9.6\rho)\text{ksi} & 0.75\text{in} < \rho < 1.5\text{in} \end{cases} \qquad (1)$$

The plot of shear strain and shear stress are given below.

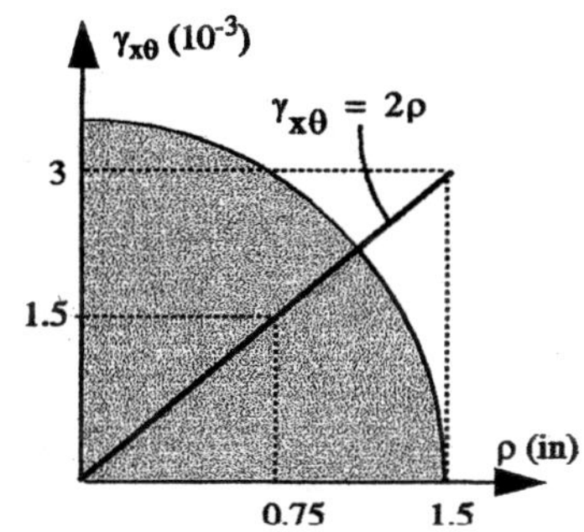

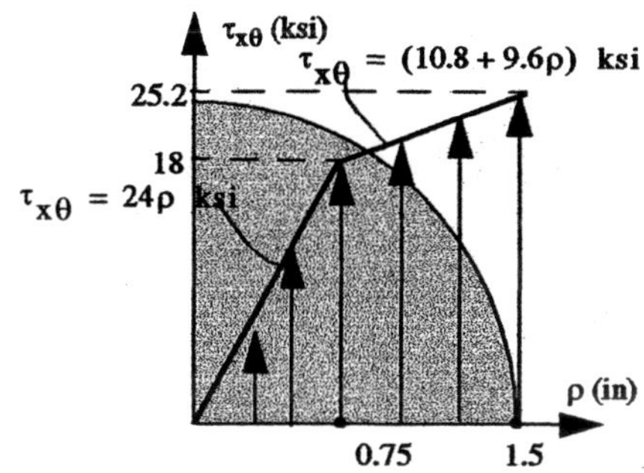

The internal torque can be found from Eq. 5.1 as shown below.

$$T = \int_A \rho\tau_{x\theta}dA = \underbrace{\int_{A_E} \rho\tau_{x\theta}dA}_{T_E} + \underbrace{\int_{A_P} \rho\tau_{x\theta}dA}_{T_P} \qquad (2)$$

The differential area dA is the area of a ring of radius ρ and thickness dρ i.e. $dA = (2\pi\rho)d\rho$. Substituting the stress distribution from Eq. 1 into Eq. 2 and performing integration we obtain the internal torque

$$T_E = \int_0^{0.75} (\rho)(24\rho)(2\pi\rho)d\rho = (48\pi)\frac{\rho^4}{4}\bigg|_0^{0.75} = 11.93 \text{ in-kips}$$

$$T_P = \int_{0.75}^{1.5} (\rho)(10.8 + 9.6\rho)(2\pi\rho)d\rho = (2\pi)\left(10.8\frac{\rho^3}{3} + 9.6\frac{\rho^4}{4}\right)\bigg|_{0.75}^{1.5} = 138.37 \text{ in-kips}$$

$T = T_E + T_P = 150.29$ in-kips or $\boxed{T = 150.3 \text{ in-kips}}$

5.83 A solid circular shaft of diameter 3 inches has a shear strain at a section in polar coordinates was found to be $\gamma_{x\theta} = 0.002\rho$ where ρ is the radial coordinate measured in inches. For the given material model, (a) Write expressions for $\tau_{x\theta}$ as a function of ρ and plot the shear strain $\gamma_{x\theta}$ and shear stress $\tau_{x\theta}$ distribution across the cross-section. (b) Determine the equivalent internal torque acting at the cross-section.The shaft material has a stress strain relationship given by $\tau = 243\gamma^{0.4}$ ksi.

Solution $\gamma_{x\theta}= 0.002\rho$ $\tau_{x\theta}= f(\rho) = ?$ $T = ?$

The stress distribution is: $\tau_{x\theta} = 243\gamma_{x\theta} = 243(0.002\rho)^{0.4}$ or $\boxed{\tau_{x\theta} = 20.23\rho^{0.4}\text{ksi} \qquad 0 \le \rho \le 1.5}$

The plots of shear strain and shear stress are given below.

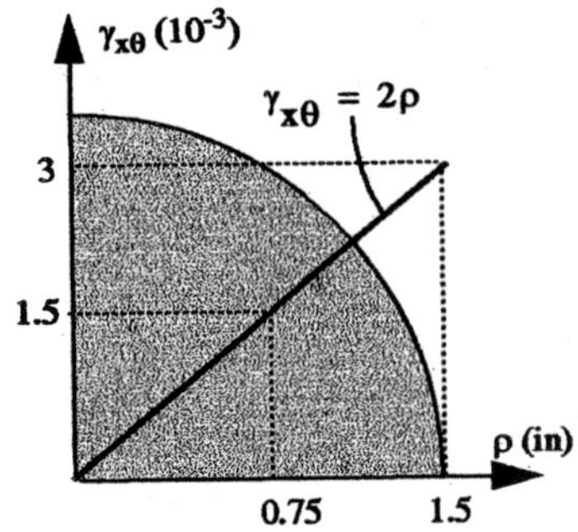

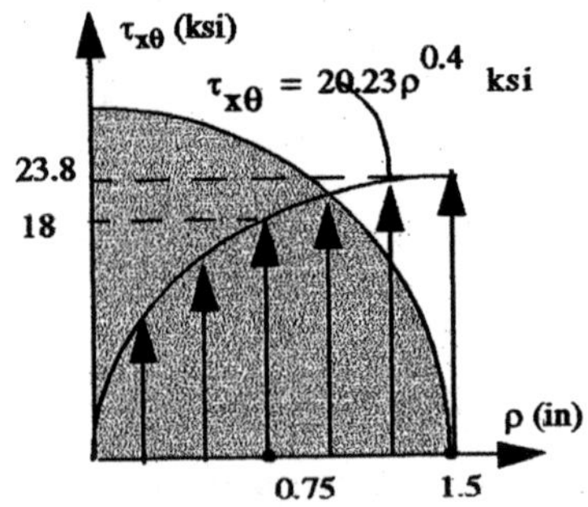

The differential area dA is the area of a ring of radius ρ and thickness dρ i.e. $dA = (2\pi\rho)d\rho$. The internal torque can be found from Equation 5.1 as shown below.

$$T = \int_A \rho\tau_{x\theta} dA = \int_0^{1.5} (\rho)(20.23\rho^{0.4})(2\pi\rho)d\rho = (40.46\pi)\frac{\rho^{3.4}}{3.4}\Bigg|_0^{1.5} = 148.39 \text{ in-kips or } \boxed{T = 148.4 \text{ in-kips}}$$

5.84 a solid circular shaft of diameter 3 inches has a shear strain at a section in polar coordinates was found to be $\gamma_{x\theta} = 0.002\rho$ where ρ is the radial coordinate measured in meters. For the given material model, (a) Write expressions for $\tau_{x\theta}$ as a function of ρ and plot the shear strain $\gamma_{x\theta}$ and shear stress $\tau_{x\theta}$ distribution across the cross-section. (b) Determine the equivalent internal torque acting at the cross-section.The shaft material has a stress strain relationship given by $\tau = (12,000\gamma - 120,000\gamma^2)$ ksi.

Solution $\gamma_{x\theta}=0.002\rho$ $\quad \tau = (12,000\gamma - 120,000\gamma^2)$ ksi $\quad \tau_{x\theta} = f(\rho) = ?$ $\quad T = ?$

The shear stress distribution is: $\tau_{x\theta} = (12,000\gamma_{x\theta} - 120,000\gamma_{x\theta}^2) = (12,000)(0.002\rho) - 120,000(0.002\rho)^2$ or

$$\boxed{\tau_{x\theta} = (24\rho - 0.48\rho^2)\text{ksi} \qquad 0 \le \rho \le 1.5}$$ 1

The plots of shear strain and shear stress are given below.

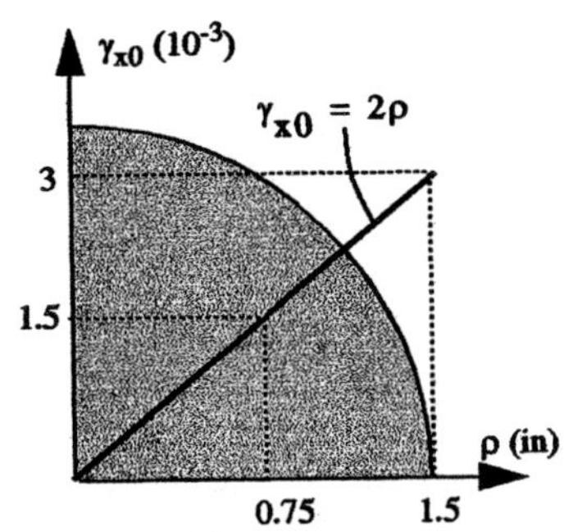

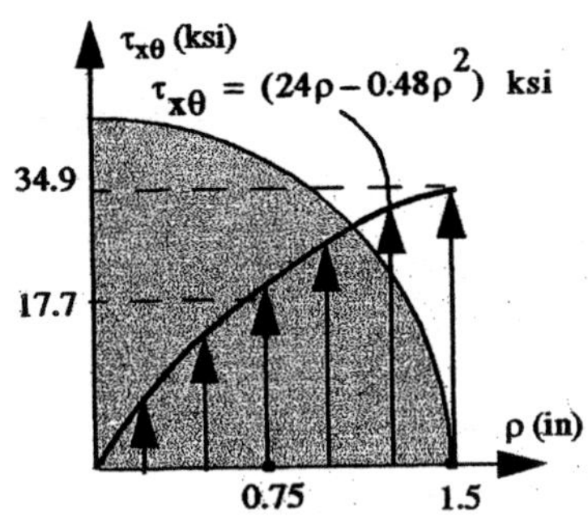

The differential area dA is the area of a ring of radius ρ and thickness dρ i.e. $dA = (2\pi\rho)d\rho$. The internal torque can be found from Eq. 5.1 as shown below.

$$T = \int_A \rho\tau_{x\theta} dA = \int_0^{1.5} (\rho)(24\rho - 0.48\rho^2)(2\pi\rho)d\rho = (2\pi)\left(24\frac{\rho^4}{4} - 0.48\frac{\rho^5}{5}\right)\Bigg|_0^{1.5} = 186.27 \text{ in-kips}$$

or $\boxed{T = 186.3 \text{ in-kips}}$

5.85 A hollow circular shaft has an inner diameter of 50 mm and an outside diameter of 100 mm. The shear strain at a section in polar coordinates was found to be $\gamma_{x\theta} = 0.2\rho$, where ρ is the radial coordinate measured in meters. For the given material model: (a) Write expressions for $\tau_{x\theta}$ as a function of ρ and plot the shear strain $\gamma_{x\theta}$ and shear stress $\tau_{x\theta}$ distribution across the cross-section. (b) Determine the equivalent internal torque acting at the cross-section. The shaft is made from elastic-perfectly plastic material that has a shear yield stress of $\tau_{yield} = 175$ MPa, and a Shear modulus G = 26 GPa.

Solution $\gamma_{x\theta}=0.2\rho$ ρ in meters $\tau_{yield}=175MP$ G=26GPa d_i=50mm d_0=100 mm
$\tau_{x\theta} = f(\rho) = ?$ $T = ?$ Elastic-perfectly plastic

---

The strain at yield point is: $\gamma_{yield} = \frac{\tau_{yield}}{G} = \frac{175(10^6)}{26(10^9)} = 6.7308(10^{-3}) = 0.2\rho_{yield}$ or

$\rho_{yield} = 33.65(10^{-3})$m where ρ_{yield} is the radial coordinate of elastic-plastic boundary

In the elastic region: $\tau_{x\theta} = G\gamma_{z\theta} = 26(10^9)(0.2\rho) = 5200\rho(10^6)\text{N/m}^2$ where ρ is in meters. The stress distribution can be written as:

$$\tau_{x\theta} = \begin{cases} 5200\rho\text{MPa} & 0.025\text{m} \le \rho \le 0.03365\text{m} \\ 175\text{MPa} & 0.03365\text{m} \le \rho \le 0.05\text{m} \end{cases}$$ 1

The plot of shear strain and shear stress are given below

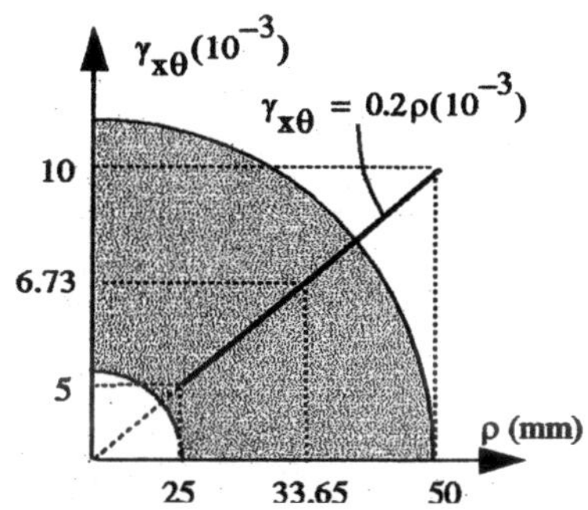

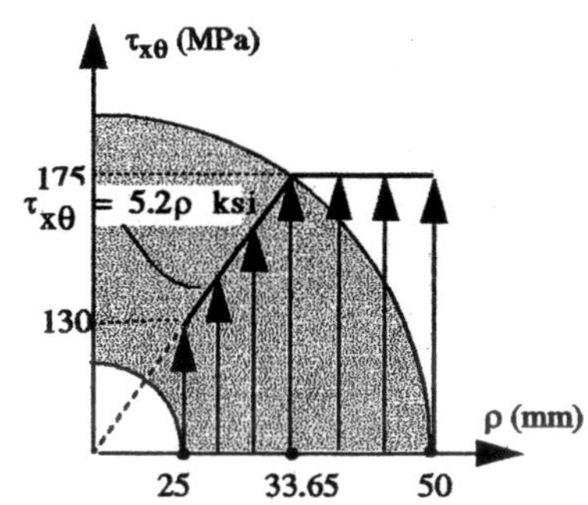

The internal torque can be found from Eq. 5.1 as shown below.

$$T = \int_A \rho\tau_{x\theta}dA = \underbrace{\int_{A_E} \rho\tau_{x\theta}dA}_{T_E} + \underbrace{\int_{A_P} \rho\tau_{x\theta}dA}_{T_P}$$ 2

The differential area dA is the area of a ring of radius ρ and thickness $d\rho$ i.e. $dA = (2\pi\rho)d\rho$. Substituting the stress distribution from Eq. 1 into Eq. 2 and performing integration we obtain the internal torque.

$$T_E = \int_{0.025}^{0.03365} (\rho)[5200\rho(10^6)](2\pi\rho)d\rho = (10400\pi)(10^6)\frac{\rho^4}{4}\Bigg|_{0.025}^{0.03365} = 7282.1 \text{ N-m}$$

$$T_P = \int_{0.03365}^{0.05} (\rho)[175(10^6)](2\pi\rho)d\rho = (350\pi)(10^6)\frac{\rho^3}{3}\Bigg|_{0.03365}^{0.05} = 31849.5 \text{ N-m}$$

$T = T_E + T_P = 39131.6$ N-m or $\boxed{T = 39.1 \text{ kN-m}}$

5.86 A hollow circular shaft has an inner diameter of 50 mm and an outside diameter of 100 mm. The shear strain at a section in polar coordinates was found to be $\gamma_{x\theta} = 0.2\rho$, where ρ is the radial coordinate measured in meters. For the given material model: (a) Write expressions for $\tau_{x\theta}$ as a function of ρ and plot the shear strain $\gamma_{x\theta}$ and shear stress $\tau_{x\theta}$ distribution across the cross-section. (b) Determine the equivalent internal torque acting at the cross-section. The shaft is made from a bi-linear material that has a shear yield stress of τ_{yield}= 175 MPa, Shear modulus G_1 = 26 GPa and G_2= 14 GPa.

Solution $\gamma_{yield}=0.2\rho$ where ρ is in meters $\tau_{yield}=175\text{MPa}$
$G_1=26\text{GPa}$ $G_2=14\text{GPa}$ $\tau_{x\theta} = f(\rho) = ?$ $T = ?$

The stress before yield point is:

$$\tau_1 = G_1\gamma_{x\theta} = 26(10^9)(0.2\rho) = 5200\rho\text{MPa} \tag{1}$$

From problem 3.138: $\gamma_{yield} = 6.7308(10^{-3})$ and $\rho_{yield} = 33.65(10^{-3})\text{m}$

The stress after yield point is: $\tau_2 = \tau_{yield} + G_2(\gamma_{x\theta} - \gamma_{yield}) = 175(10^6) + 14(10^9)(0.2\rho - 6.7308(10^{-3}))$

$$\tau_2 = 80.77(10^6) + 2800\rho(10^6) \tag{2}$$

In Eq(1) &(2) ρ is in meters. Eq's (1) and (2). The stress distribution can be written as:

$$\tau_{x\theta} = \begin{cases} 5200\rho\text{MPa} & 0.025\text{m} \le \rho \le 0.03365\text{m} \\ (80.77 + 2800\rho)\text{MPa} & 0.03365\text{m} \le \rho \le 0.05\text{m} \end{cases} \tag{3}$$

The plot of shear strain and shear stress are given below.

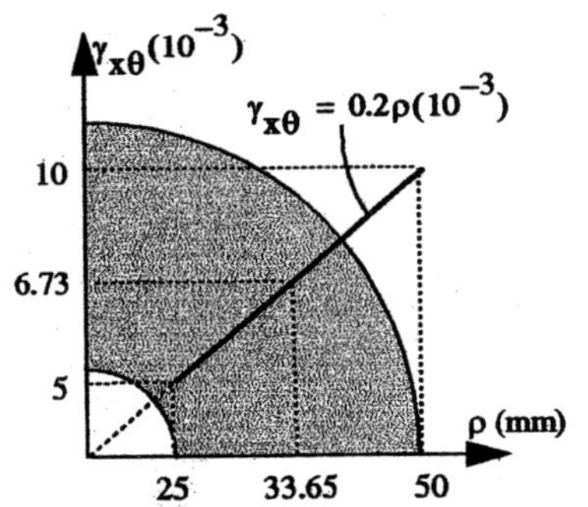

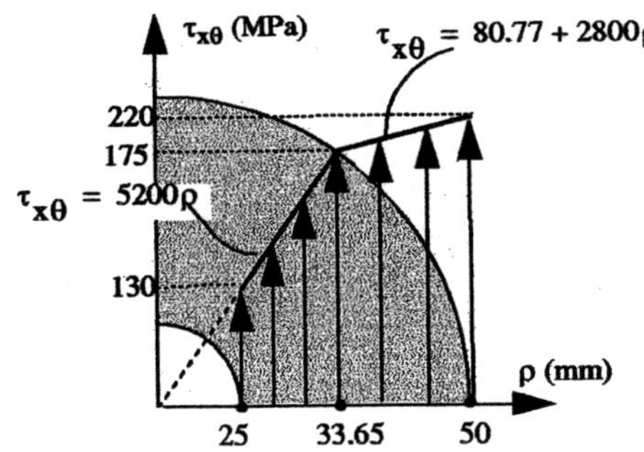

The internal torque can be found from Eq. 5.1 as shown below.

$$T = \int_A \rho\tau_{x\theta}dA = \underbrace{\int_{A_E} \rho\tau_{x\theta}dA}_{T_E} + \underbrace{\int_{A_P} \rho\tau_{x\theta}dA}_{T_P} \tag{4}$$

The differential area dA is the area of a ring of radius ρ and thickness dρ i.e. $dA = (2\pi\rho)d\rho$. Substituting the stress distribution from Eq. 1 into Eq. 2 and performing integration we obtain the internal torque.

$$T_E = \int_{0.025}^{0.03365} (\rho)[5200\rho(10^6)](2\pi\rho)d\rho = (10400\pi)(10^6)\frac{\rho^4}{4}\Bigg|_{0.025}^{0.03365} = 7282.1 \text{ N-m}$$

$$T_P = \int_{0.03365}^{0.05} (\rho)(80.77 + 2800\rho)(10^6)(2\pi\rho)d\rho = (2\pi)(10^6)\left(80.77\frac{\rho^3}{3} + 2800\frac{\rho^4}{4}\right)\Bigg|_{0.03365}^{0.05} = 36549.7 \text{ N-m}$$

$T = T_E + T_P = 43831.7 \text{ N-m}$ or $\boxed{T = 43.8 \text{ kN-m}}$

5.87 A hollow circular shaft has an inner diameter of 50 mm and an outside diameter of 100 mm. The shear strain at a section in polar coordinates was found to be $\gamma_{x\theta} = 0.2\rho$, where ρ is the radial coordinate measured in meters. For the given material model: (a) Write expressions for $\tau_{x\theta}$ as a function of ρ and plot the shear strain $\gamma_{x\theta}$ and shear stress $\tau_{x\theta}$ distribution across the cross-section. (b) Determine the equivalent internal torque acting at the cross-section. The shaft material has a stress strain relationship given by $\tau = 3435\gamma^{0.6}$ MPa.

Solution $\gamma_{x\theta}=0.2\rho$ where ρ is in meters $\tau_{x\theta} = f(\rho) = ?$ $T = ?$

The stress distribution is: $\tau_{x\theta} = 3435\gamma^{0.6} = 3435(0.2\rho)^{0.6} = 1307.8\rho^{0.6}$ or

$$\boxed{\tau_{x\theta} = 1307.8\rho^{0.6}\text{MPa} \qquad 0.025\text{m} \le \rho \le 0.05\text{m}}$$ 1

The plot of shear strain and shear stress are given below.

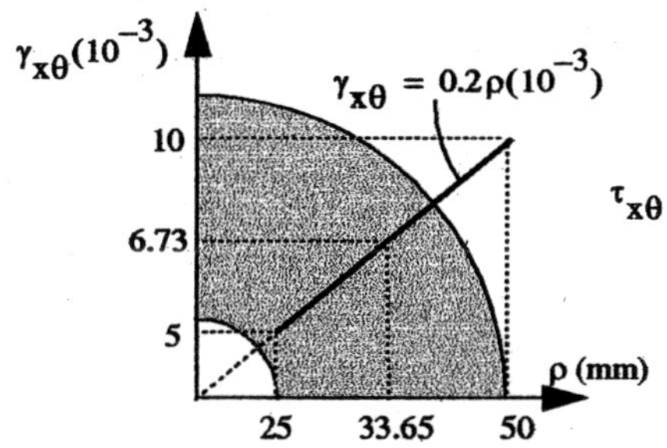

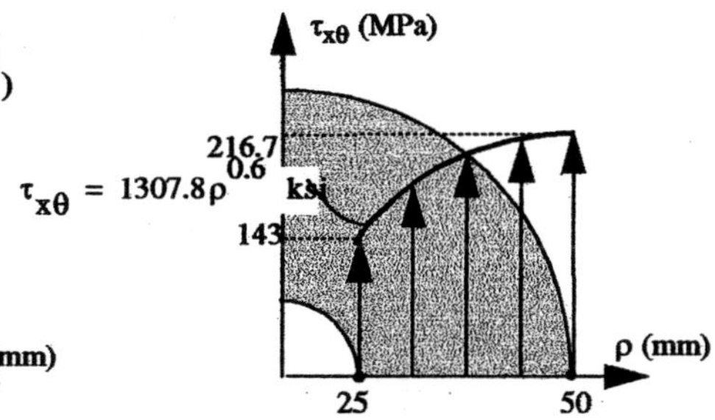

The differential area dA is the area of a ring of radius ρ and thickness dρ i.e. $dA = (2\pi\rho)d\rho$. The internal torque can be found from Eq. 5.1 as shown below.

$$T = \int_A \rho\tau_{x\theta}dA = \int_{0.025}^{0.05} (\rho)(1307.8\rho^{0.6})(10^6)(2\pi\rho)d\rho = (2615.6\pi)(10^6)\frac{\rho^{3.6}}{3.6}\bigg|_{0.025}^{0.05} = 43384.2 \text{ N}-\text{m}$$

or $$\boxed{T = 43.4 \text{ kN}-\text{m}}$$

5.88 A hollow circular shaft has an inner diameter of 50 mm and an outside diameter of 100 mm. The shear strain at a section in polar coordinates was found to be $\gamma_{x\theta} = 0.2\rho$ where ρ is the radial coordinate measured in meters. Write an expressions for $\tau_{x\theta}$ as a function of ρ and plot the shear strain $\gamma_{x\theta}$ and shear stress $\tau_{x\theta}$ distribution across the cross-section. The shaft material has a stress strain relationship given by $\tau = (26,000\gamma - 208,000\gamma^2)$ MPa

Solution $\gamma_{x\theta}=0.2\rho$ where ρ is in meters $\tau_{x\theta} = f(\rho) = ?$ $T = ?$

The shear stress distribution is: $\tau_{x\theta} = (26,000\gamma - 208,000\gamma^2) = 26,000(0.2\rho) - 208,000(0.2\rho)^2$

$$\boxed{\tau_{x\theta} = (5,200\rho - 8,320\rho^2)\text{MPa} \qquad 0.025\text{m} \le \rho \le 0.05\text{m}}$$ 1

The plots of shear strain and stress are shown below.

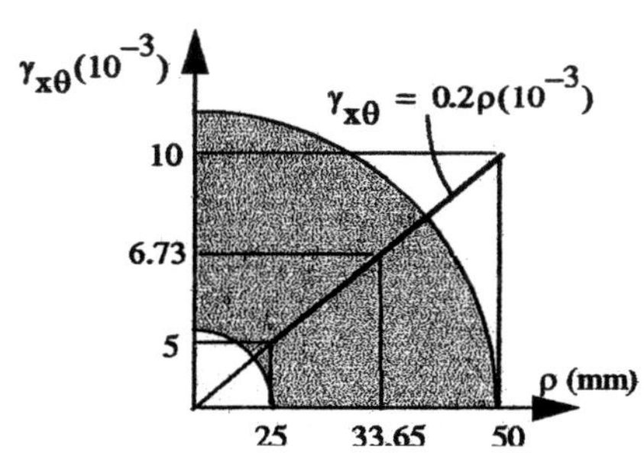

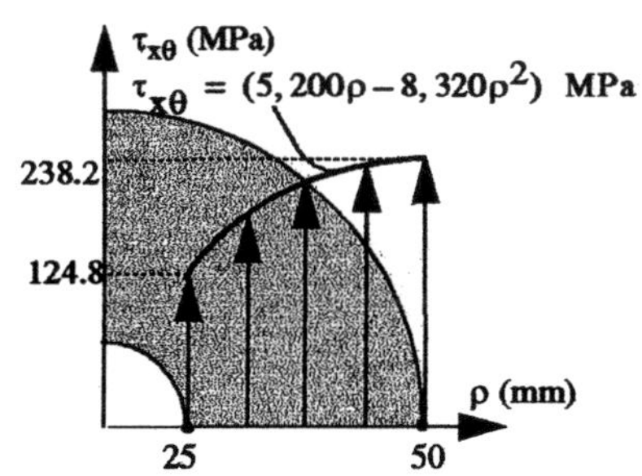

The differential area dA is the area of a ring of radius ρ and thickness $d\rho$ i.e. $dA = (2\pi\rho)d\rho$. The internal torque can be found from Eq. 5.1 as shown below.

$$T = \int_A \rho\tau_{x\theta}dA = \int_{0.025}^{0.05} (\rho)(5200\rho - 8320\rho^2)(10^6)(2\pi\rho)d\rho = (2\pi)(10^6)\left(5200\frac{\rho^4}{4} - 8320\frac{\rho^5}{5}\right)\Bigg|_{0.025}^{0.05} \text{ or}$$

$T = 44695.5 \text{ N-m}$ or $\boxed{T = 44.7 \text{ kN-m}}$

5.89 A 5 feet long hollow shaft with an outside diameter of 4 inches and an inside diameter of 2 inches is twisted through an angle of 12°. The shaft material has a yield stress of 24 ksi and a modulus of rigidity of 4000 ksi, Assuming elastic-perfectly plastic material behavior, determine (a) the magnitude of the applied torque. (b) the residual shear stress when the applied torque is removed. (c) Show that the residual stresses are self equilibrating, i.e., produce no internal torque.

Solution $L = 5$ ft. $d_o = 4$ in $d_i = 2$ in $\Delta\phi = 12°$
$\tau_{yield} = 24$ ksi $G = 4000$ ksi $T = ?$ $\tau_{residual} = ?$

The relative rotation is: $\Delta\phi = \frac{12}{180}\pi = 0.2094 \text{ rad}$

The torsional shear strain is: $\gamma = \rho\frac{\Delta\phi}{\Delta x} = \rho\frac{\Delta\phi}{L} = \rho\frac{0.2094}{60} = 3.491(10^{-3})\rho$.

At yield point: $\gamma_{yield} = \frac{\tau_{yield}}{G} = \frac{24}{4000} = 6(10^{-3}) = 3.491(10^{-3})\rho_{yield}$ or $\rho_{yield} = 1.718\text{in}$ where, ρ_{yield} is the radial coordinate of elastic-plastic beam.

For $1 < \rho < 1.718\text{in}$, $\tau_{x\theta} = G\gamma_{x\theta} = (4000)(3.491)(10^{-3})\rho = 13.964\rho \text{ ksi}$. The stress distribution can be written as:

$$\boxed{\tau_{x\theta} = \begin{cases} 13.964\rho \text{ ksi} & 1\text{in} < \rho < 1.718\text{in} \\ 24 \text{ ksi} & 1.718\text{in} < \rho < 2.0\text{in} \end{cases}} \quad (1)$$

The internal torque can be found from Eq. 5.1 as shown below.

$$T = \int_A \rho\tau_{x\theta}dA = \underbrace{\int_{A_E} \rho\tau_{x\theta}dA}_{T_E} + \underbrace{\int_{A_P} \rho\tau_{x\theta}dA}_{T_P} \quad (2)$$

The differential area dA is the area of a ring of radius ρ and thickness $d\rho$ i.e. $dA = (2\pi\rho)d\rho$. Substituting the stress distribution from Eq. 1 into Eq. 2 and performing integration we obtain the internal torque.

$$T_E = \int_1^{1.718} (\rho)(13.964\rho)(2\pi\rho)d\rho = (87.74\pi)\frac{\rho^4}{4}\Bigg|_1^{1.718} = 169.15 \text{ in-kips}$$

$$T_P = \int_{1.718}^{2.0} (\rho)(24)(2\pi\rho)d\rho = (48\pi)\frac{\rho^3}{3}\Bigg|_{1.718}^{2.0} = 147.24 \text{ in-kips}$$

$T = T_E + T_P = 316.39 \text{ in-kips}$ or $\boxed{T = 316.4 \text{ in-kips}}$

During unloading we can use Eq. 5.8 and obtain the elastic rebound shear stress. The polar moment for the shaft is $J = (\pi/2)(2^4 - 1^4) = 23.56 \text{ in}^4$. The elastic rebound shear stress is

$$\tau_{elastic} = \frac{T\rho}{J} = \frac{316.39}{23.56}\rho \qquad \tau_{elastic} = 13.43\rho \text{ ksi} \quad (3)$$

We subtract Eq.(3) from stresses in Eq.(1) to obtain the residual shear stresses i.e.,: $\tau_{residual} = \tau - \tau_{elastic}$

or

$$\tau_{residual} = \begin{cases} 0.536\rho \ \text{ksi} & 1\text{in} < \rho < 1.718\text{in} \\ (24 - 13.43\rho) \ \text{ksi} & 1.718\text{in} < \rho < 2.0\text{in} \end{cases}$$

5.90 The elastic-perfectly plastic shaft shown has a plastic zone 30 mm deep in section BC. The yield stress of shaft is 200 MPa and a shear modulus of G = 80 GPa. (a) Determine the magnitude of the applied torque T_{ext}. (b) The rotation of section A with respect to the wall.

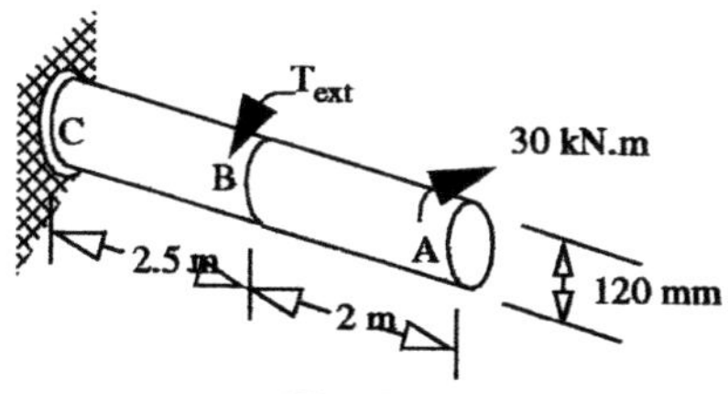

Fig. P5.90

Solution depth of plastic zone = 30 mm τ_{yield}=200 MPa G=80 GPa

$T_{ext} = ?$ $\phi_A - \phi_C = ?$

The elastic-plastic boundary is located at ρ_{yield} = 60 - 30 mm. The shear stress varies linearly from zero at the center to yield stress at the elastic plastic boundary. Thus for $0 \le \rho \le 0.02$ i

$$\tau_{x\theta} = \frac{\tau_{yield}\rho}{\rho_y} = \left[\frac{200(10^6)}{30(10^{-3})}\right]\rho = 6666.7(10^6)\rho$$

Beyond the elastic plastic boundary the shear stress is equal to yield stress. We can write the equation for shear stress as

$$\tau_{x\theta} = \begin{cases} 6666.7\rho \ \text{MPa} & 0 \le \rho \le 0.03 \\ 200 \ \text{MPa} & 0.03 \le \rho \le 0.06 \end{cases} \qquad 1$$

Substituting Eq.1 into Eq. 5.1, we can obtain the equivalent internal torque as shown below:

$$T_{BC} = \int_A \rho\tau_{x\theta}dA = \int_0^{0.03} \rho[6666.7(10^6)\rho](2\pi\rho)d\rho + \int_{0.03}^{0.06} \rho[200(10^6)](2\pi\rho)d\rho \ \text{ or}$$

$$T_{BC} = [13333.3\pi](10^6)\frac{\rho^4}{4}\Bigg|_0^{0.03} + [400\pi](10^6)\frac{\rho^3}{3}\Bigg|_{0.03}^{0.06} = 81868.1 \ \text{N-m} = 81.87\text{kN-m}$$

We draw the free body diagram after making an imaginary cut in AB and taking the right part to avoid calculating the wall reaction as shown in Fig. (a).

By equilibrium of torque in Fig. (a) we obtain: $T_{ext} = 81.87 + 30$ or $\boxed{T_{ext} = 111.9 \ \text{kN-m}}$

The strain at yield is $\gamma_{yield} = \frac{\tau_{yield}}{G} = \frac{200(10^6)}{80(10^9)} = 2.5(10^{-3})$ Using Eq. 5.24, we can find the relative rotation of section at B with respect to section at A as shown below.

$$\gamma_{yield} = \rho_{yield}\frac{d\phi}{dx} = 30(10^{-3})\frac{d\phi}{dx} = 2.5(10^{-3}) \quad \text{or} \quad \frac{d\phi}{dx} = 0.0833 = \frac{\phi_B - \phi_C}{(x_B - x_C)} \ \text{or}$$

$$\phi_B - \phi_C = 0.0833(x_B - x_C) = 0.0833(2.5) = 0.2083 \ \text{rad} \qquad 2$$

(a)

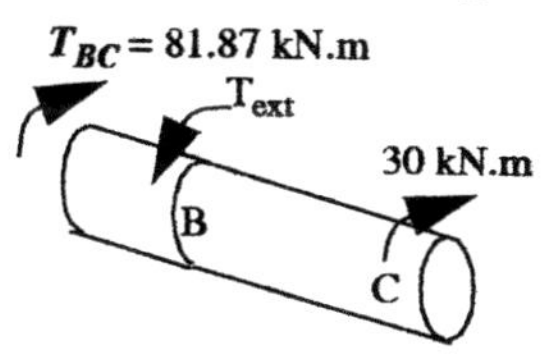

(b)

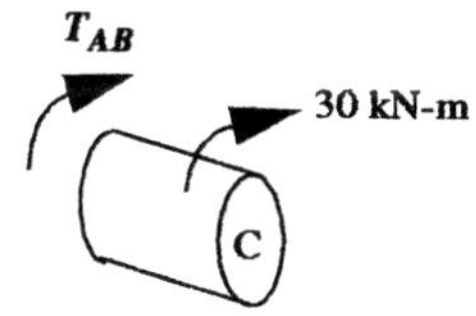

We draw the free body diagram after making an imaginary cut in AB as shown in Fig. (b). By equilibrium of torque we obtain: $T_{AB} = -30 \text{ kN}-\text{m}$. The polar moment in AB is:

$J_{AB} = \pi(0.12)^4/32 = 20.357(10^{-6})$. We find the maximum shear stress in section AB from Eq. 5.8 as:

$\tau_{max} = \dfrac{T_{BC}\rho_{max}}{J_{AB}} = \dfrac{-30(10^3)(60)(10^{-3})}{20.357(10^{-6})} = -88.42(10^6)\text{N/m}^2$. The magnitude of the maximum torsional shear stress in AB of 88.4 MPa does not exceed the yield stress of 200 MPa, hence we can use Eq. 5.10 to find the rotation of section at A with respect to B as shown below.

$$\phi_A - \phi_B = \frac{-30(10^3)(2)}{80(10^9)(20.357)(10^{-6})} = -0.0368 \text{ rad} \qquad 3$$

Adding Eq. 2 and 3 we obtain the rotation of section at C with respect to section at A as: $\phi_A - \phi_C = 0.2083 - 0.0368 = 0.1715 \text{ rad}$. Noting that the rotation at C is zero, we obtain the rotation at section at A as: $\boxed{\phi_A = 0.1715 \text{ rad ccw}}$

5.91 A 4 inch solid elastic-perfectly plastic shaft has a yield stress of 30 ksi and a shear modulus of G = 12,000 ksi. The plastic zone in AB is 0.5 inch deep. (a) Determine the magnitude of the applied torque T_{ext}.(b) The rotation of section C with respect to the wall

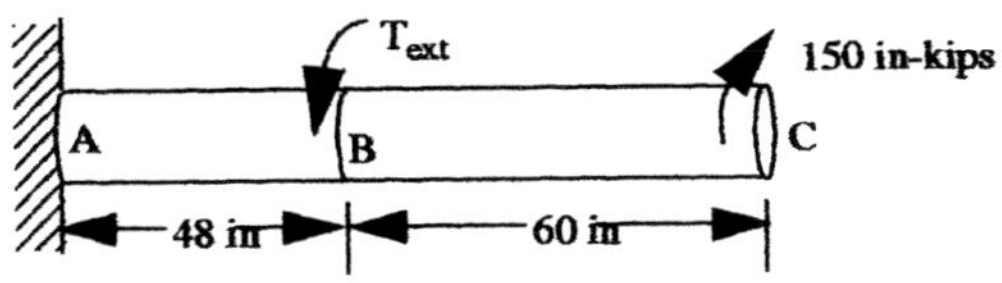

Fig. P5.91

Solution depth of plastic zone = 0.5in τ_{yield}=30 ksi G=12,000 ksi

T_{ext} = ? $\phi_A - \phi_C$ = ?

The elastic-plastic boundary is located at ρ_{yield} = 2 - 0.5= 1.5 in. The shear stress varies linearly from zero at the center to yield stress at the elastic plastic boundary. Thus for $0 \le \rho \le 1.5\text{in}$ the shear stress is:

$$\tau = \frac{\tau_{yield}\rho}{\rho_y} = \left[\frac{30}{1.5}\right]\rho = 20\rho \text{ ksi}$$

Beyond the elastic plastic boundary the shear stress is equal to yield stress. We can write the equation for shear stress as

$$\tau = \begin{cases} 20\rho \text{ ksi} & 0 \le \rho \le 1.5 \\ 30\text{ksi} & 1.5 \le \rho \le 2.0 \end{cases} \qquad 1$$

Substituting Eq.1 into Eq. 5.1, we can obtain the equivalent internal torque as shown below:

$$T_{AB} = \int_A \rho\tau_{x\theta}dA = \int_0^{1.5} \rho(20\rho)(2\pi\rho)d\rho + \int_{1.5}^{2.0} \rho(30)(2\pi\rho)d\rho = (40\pi)\frac{\rho^4}{4}\bigg|_0^{1.5} + (60\pi)\frac{\rho^3}{3}\bigg|_{1.5}^{2} = 449.6 \text{ in}-\text{kips}$$

We draw the free body diagram after making an imaginary cut in AB and taking the right part to avoid calculating the wall reaction as shown in Fig. (a).

(a)

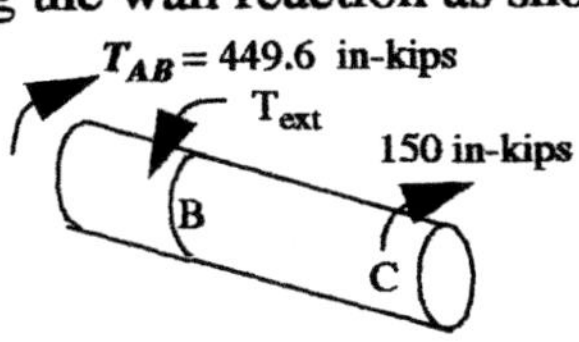

(b)

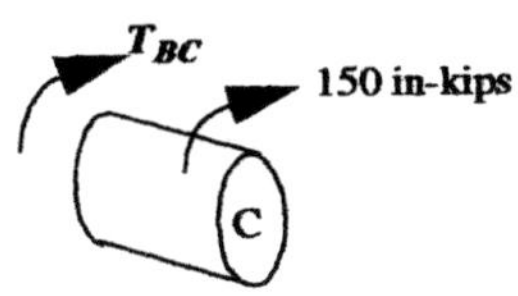

By equilibrium of torque in Fig. (a) we obtain: $T_{ext} = 449.6 + 150$ or $\boxed{T_{ext} = 599.6 \text{ in}-\text{kips}}$

The strain at yield is $\gamma_{yield} = \dfrac{\tau_{yield}}{G} = \dfrac{30}{12000} = 2.5(10^{-3})$ Using Eq. 5.24, we can find the relative rotation of section at B with respect to section at A as shown below.

$$\gamma_{yield} = \rho_{yield}\frac{d\phi}{dx} = 1.5\frac{d\phi}{dx} = 2.5(10^{-3}) \quad \text{or} \quad \frac{d\phi}{dx} = 1.667(10^{-3}) = \frac{\phi_B - \phi_A}{(x_B - x_A)} \quad \text{or}$$

$$\phi_B - \phi_A = 1.667(10^{-3})(x_B - x_A) = 1.667(10^{-3})(48) = 0.080 \text{ rad} \qquad 2$$

We draw the free body diagram after making an imaginary cut in BC as shown in Fig. (b). By equilibrium of torque we obtain: $T_{BC} = -150 \text{ in-kips}$. The polar moment in BC is: $J_{BC} = \pi(4)^4/32 = 25.13\text{in}^4$. We find the maximum shear stress in section BC from Eq. 5.8 as: $\tau_{max} = \frac{T_{BC}\rho_{max}}{J_{BC}} = \frac{(-150)(2)}{25.13} = -11.94\text{ksi}$.

The magnitude of the maximum torsional shear stress in BC of 11.94 ksi does not exceed the yield stress of 30 ksi hence we can use Eq. 5.10 to find the rotation of section at C with respect to B as shown below.

$$\phi_C - \phi_B = \frac{-(150)(60)}{(12000)(25.13)} = -0.0298 \text{ rad} \qquad 3$$

Adding Eq. 2 and 3 we obtain the rotation of section at C with respect to section at A as: $\phi_C - \phi_A = 0.080 - 0.0298 = 0.0502 \text{ rad}$. Noting that the rotation at A is zero, we obtain the rotation at section at C as:

$$\boxed{\phi_C = 0.05 \text{ rad ccw}}$$

5.92 A shaft made from elastic - perfectly plastic material has a yield stress of 200 MPa and a shear modulus of G = 80 GPa. The plastic zone in section AB is 25 mm deep (a) Determine the torque T_{ext} (b) Determine the rotation of section at B.

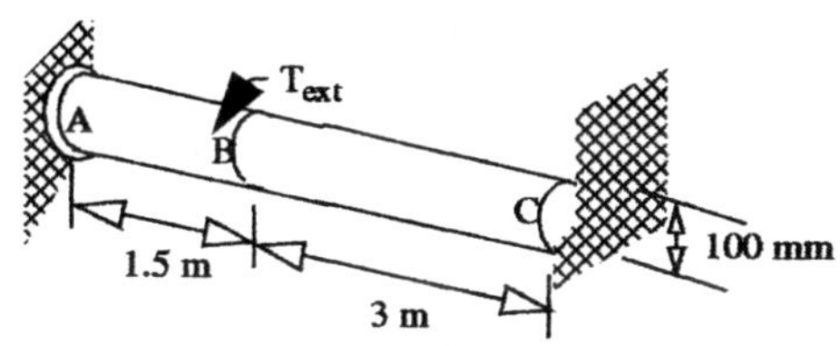

Fig. P5.92

Solution depth of plastic zone = 25 mm τ_{yield}=200 MPa G=80 GPa

$T_{ext} = ?$ $\phi_B = ?$

At yield point: $\gamma_{yield} = \frac{\tau_{yield}}{G} = \frac{200(10^6)}{80(10^9)} = 2.5(10^{-3})$

The elastic-plastic boundary is located at ρ_{yield} = 50 - 25= 25 mm = 0.025 m.

Using Eq. 5.24, we can find the relative rotation of section at B with respect to section at A as shown below.

$$\gamma_{yield} = \rho_{yield}\frac{d\phi}{dx} = 0.025\frac{\phi_B - \phi_A}{(x_B - x_A)} = 0.025\frac{\phi_B - \phi_A}{1.5} = 2.5(10^{-3}) \quad \text{or} \quad \phi_B - \phi_A = 0.15\text{rad}$$

Noting that rotation at A is zero, we obtain:

$$\boxed{\phi_B = 0.15\text{rad}}$$

We can find the maximum shear strain in BC as:

$$(\gamma_{BC})_{max} = \rho_{max}\left(\frac{d\phi}{dx}\right)_{BC} = 0.05\frac{\phi_C - \phi_B}{(x_C - x_B)} = 0.05\frac{0 - 0.15}{3.0} = (-2.5)(10^{-3})$$

The magnitude of the maximum shear strain is equal to the yield strain, i.e., the outer surface of the shaft is at yield while all points in the interior are less than yield, hence the cross-section in BC is fully elastic. The polar moment of inertia in BC is: $J_{BC} = \pi(0.1)^4/32 = 9.817(10^{-6})\text{m}^4$. We can use Eq. 5.10 to find the internal torque in BC as shown below.

$$\phi_C - \phi_B = \frac{T_{BC}(x_C - x_B)}{GJ_{BC}} = \frac{T_{BC}(3)}{80(10^9)(9.817)(10^{-6})} = -0.15 \quad \text{or}$$

$$T_{BC} = -39.27(10^3)\ \text{N-m} = -39.27\text{kN-m} \quad 4$$

In AB, the elastic-plastic boundary is located at $\rho_{yield} = 0.025$. The shear stress varies linearly from zero at the center to yield stress at the elastic plastic boundary. Thus for $0 \le \rho \le 0.02$ we have

$$\tau_{x\theta} = \frac{\tau_{yield}\rho}{\rho_y} = \left[\frac{200(10^6)}{0.025}\right]\rho = 800(10^6)\rho$$

Beyond the elastic plastic boundary the shear stress is equal to yield stress. We can write the equation for shear stress as

$$\tau_{x\theta} = \begin{cases} 8000\rho\ \text{MPa} & 0 \le \rho \le 0.025 \\ 200\ \text{MPa} & 0.025 \le \rho \le 0.05 \end{cases} \quad 5$$

Substituting Eq.5 into Eq. 5.1, we can obtain the equivalent internal torque as shown below:

$$T_{AB} = \int_A \rho\tau_{x\theta}dA = \int_0^{0.025} \rho[800(10^6)\rho](2\pi\rho)d\rho + \int_{0.025}^{0.05} \rho[200(10^6)](2\pi\rho)d\rho \quad \text{or}$$

$$T_{AB} = [16000\pi](10^6)\frac{\rho^4}{4}\bigg|_0^{0.025} + [400\pi](10^6)\frac{\rho^3}{3}\bigg|_{0.025}^{0.05} = 50723.6\ \text{N-m} = 50.72\text{kN-m}$$

We draw the free body diagram after making an imaginary cut in AB and BC as shown below.

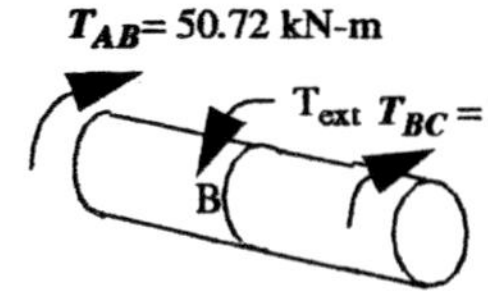

By equilibrium of torque in the above free body diagram we obtain: $T_{ext} = 50.72 + 39.27 = 89.99$ or

$$\boxed{T_{ext} = 90.0\ \text{kN-m}}$$

5.93 A circular solid shaft of radius R is made from a non-linear material that has a shear stress-shear strain relationship given by $\tau = K\gamma^{0.4}$. Show $\tau_{max} = 0.5411\frac{T}{R^3}$ $\quad \phi_2 - \phi_1 = 0.2154(x_2 - x_1)\left(\frac{T}{KR^{3.4}}\right)^{2.5}$, where τ_{max} is the maximum torsional shear stress at a section, T is the internal torque at the section, ϕ_2, and ϕ_1 are the rotation of section at x_1 and x_2.

Solution $\quad \tau = K\gamma^{0.4} \quad \tau_{max} = f(T, R) = ? \quad \phi_2 - \phi_1 = g(T, R, K) = ?$

Kinematics are unchanged, hence: $\gamma_{x\theta} = \rho\frac{d\phi}{dx}$. The torsional shear stress is as below.

$$\tau_{x\theta} = K(\gamma_{x\theta})^{0.4} = K\rho^{0.4}\left(\frac{d\phi}{dx}\right)^{0.4} \quad 1$$

Substituting Eq.1 into Eq. 5.1, and noting that $\frac{d\phi}{dx}$ does not change across the cross-section, we can obtain the equivalent internal torque as shown below:

$$T = \int_A \rho\tau_{x\theta}dA = \int_0^R \rho\left[K\rho^{0.4}\left(\frac{d\phi}{dx}\right)^{0.4}\right](2\pi\rho)d\rho = (2\pi K)\left(\frac{d\phi}{dx}\right)^{0.4}\int_0^R \rho(\rho^{0.4})(\rho)d\rho = (2\pi K)\left(\frac{d\phi}{dx}\right)^{0.4}\frac{\rho^{3.4}}{3.4}\bigg|_0^R \quad \text{or}$$

$$T = (2\pi K)\left(\frac{d\phi}{dx}\right)^{0.4}\frac{R^{3.4}}{3.4} \quad \text{or} \quad \left(\frac{d\phi}{dx}\right)^{0.4} = \frac{3.4T}{2\pi KR^{3.4}} = 0.5411\left(\frac{T}{KR^{3.4}}\right) \qquad 2$$

Substituting $\frac{d\phi}{dx}$ from Eq. (4) into Eq. 1 we obtain:

$$\tau_{x\theta} = K\rho^{0.4}\left[0.5411\left(\frac{T}{KR^{3.4}}\right)\right] = 0.5411\frac{T\rho^{0.4}}{R^{3.4}} \qquad 3$$

The maximum torsional shear stress will be at $\rho = R$. From Eq. 3 we obtain:

$$\tau_{max} = 0.5411\frac{TR^{0.4}}{R^{3.4}} \quad \text{or} \quad \boxed{\tau_{max} = 0.5411\frac{T}{R^3}}$$

Eq.2 can be re-written as: $\frac{d\phi}{dx} = \left[0.5411\left(\frac{T}{KR^{3.4}}\right)\right]^{\frac{1}{0.4}} = 0.2154\left(\frac{T}{KR^{3.4}}\right)^{2.5}$. The quantities on the right hand side are all constants and the derivative which is the slope can be written as: $\frac{d\phi}{dx} = \frac{\phi_2 - \phi_1}{x_2 - x_1}$ and we obtain the following:

$$\boxed{\phi_2 - \phi_1 = 0.2154\left(\frac{T}{KR^{3.4}}\right)^{2.5}(x_2 - x_1)}$$

5.94 A circular solid shaft of radius R is made from a non-linear material that has a shear stress-shear strain relationship given by $\tau = K\gamma^{0.6}$ Assuming small deformation determine(a) the rotation of section B with respect to the left wall.(b) the maximum torsional shear stress in the shaft.

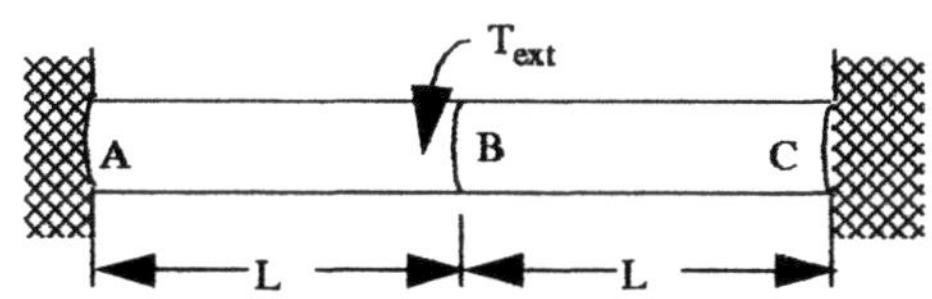

Fig. P5.94

Solution $\tau = K\gamma^{0.6}$ $\quad \tau_{max} = f(T_{ext}, R) = ?$ $\quad \phi_B = g(T_{ext}, R, K, L) = ?$

We first derive formulas for maximum stress and relative rotation in terms of the internal torque as in problem 5.93 and then solve the problem shown in Fig. P5.94.

Kinematics are unchanged, hence: $\gamma_{x\theta} = \rho\frac{d\phi}{dx}$. The torsional shear stress is as below.

$$\tau_{x\theta} = K(\gamma_{x\theta})^{0.6} = K\rho^{0.6}\left(\frac{d\phi}{dx}\right)^{0.6} \qquad 1$$

Substituting Eq.1 into Eq. 5.1, and noting that $\frac{d\phi}{dx}$ does not change across the cross-section, we can obtain the equivalent internal torque as shown below:

$$T = \int_A \rho\tau_{x\theta}dA = \int_0^R \rho\left[K\rho^{0.6}\left(\frac{d\phi}{dx}\right)^{0.6}\right](2\pi\rho)d\rho = (2\pi K)\left(\frac{d\phi}{dx}\right)^{0.6}\int_0^R \rho(\rho^{0.6})(\rho)d\rho = (2\pi K)\left(\frac{d\phi}{dx}\right)^{0.6}\frac{\rho^{3.6}}{3.6}\Bigg|_0^R \quad \text{or}$$

$$T = (2\pi K)\left(\frac{d\phi}{dx}\right)^{0.6}\frac{R^{3.6}}{3.6} \quad \text{or} \quad \left(\frac{d\phi}{dx}\right)^{0.6} = \frac{3.6T}{2\pi KR^{3.6}} = 0.5730\left(\frac{T}{KR^{3.6}}\right) \qquad 2$$

Substituting $\frac{d\phi}{dx}$ from Eq. (4) into Eq. 1 we obtain:

$$\tau_{x\theta} = K\rho^{0.6}\left[0.5730\left(\frac{T}{KR^{3.6}}\right)\right] = 0.5730\frac{T\rho^{0.6}}{R^{3.6}} \qquad 3$$

The maximum torsional shear stress will be at $\rho = R$. From Eq. 3 we obtain:

$$\tau_{max} = 0.5730\frac{TR^{0.6}}{R^{3.6}} \quad \text{or} \quad \tau_{max} = 0.5730\frac{T}{R^3} \qquad 4$$

Eq.2 can be re-written as: $\frac{d\phi}{dx} = \left[0.5730\left(\frac{T}{KR^{3.6}}\right)\right]^{\frac{1}{0.6}} = 0.3952\left(\frac{T}{KR^{3.6}}\right)^{\frac{5}{3}}$. The quantities on the right hand side are all constants and the derivative which is the slope can be written as: $\frac{d\phi}{dx} = \frac{\phi_2 - \phi_1}{x_2 - x_1}$ and we obtain the following:

$$\phi_2 - \phi_1 = 0.3952\left(\frac{T}{KR^{3.6}}\right)^{\frac{5}{3}}(x_2 - x_1) \qquad 5$$

From the symmetry in the problem, we conclude that the internal torques in segment AB and BC is the same. Hence the internal torque in each segment is $T = T_{ext}/2$. Substituting this in Eqs. 4 and 5 and noting that rotation at A is zero we obtain the following results.

$$\tau_{max} = 0.5730\frac{(T_{ext}/2)}{R^3} \quad \text{or} \quad \boxed{\tau_{max} = (0.2865)\frac{T_{ext}}{R^3}}$$

$$\text{and } \phi_B - \phi_A = 0.3952\left(\frac{T_{ext}/2}{KR^{3.6}}\right)^{\frac{5}{3}}(x_B - x_A) = 0.1245\left(\frac{T_{ext}}{KR^{3.6}}\right)^{\frac{5}{3}}L \qquad \boxed{\phi_B = 0.1245\left(\frac{T_{ext}}{KR^{3.6}}\right)^{\frac{5}{3}}L}$$

5.95 Calculate the magnitude of the maximum torsional shear stress in the cross-sections shown that is subjected to a torque of T=100 in-kips

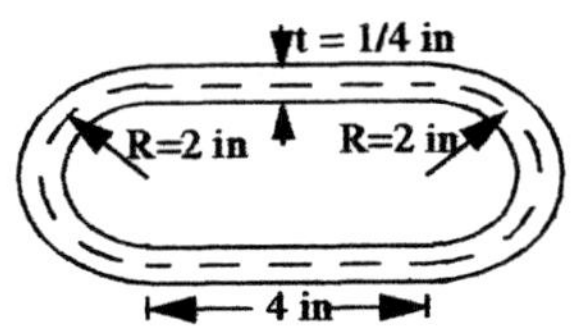

Fig. P5.95

Solution T=100 in-kips $\tau_{max} = ?$

The enclosed area is: $A_E = 2\left(\frac{1}{2}\pi 2^2\right) + (4)(4) = 28.566 \text{ in}^2$

As thickness is uniform the torsional shear stress is uniform across the cross-section and maximum value of it can be found using Equation 5.26 as shown below.

$$\tau_{max} = \frac{T}{2tA_E} = \frac{100}{2(1/4)(28.566)} = 7.001 \text{ ksi or} \qquad \boxed{\tau_{max} = 7.0 \text{ ksi}}$$

5.96 Calculate the maximum torsional shear stress in the cross-sections shown that is subjected to a torque of T = 900 N-m.

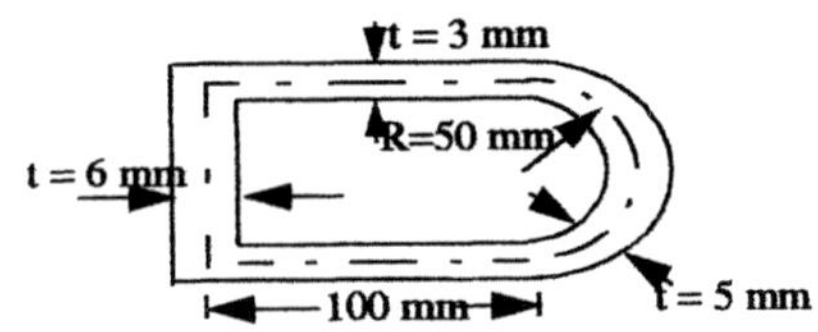

Fig. P5.96

Solution T = 900 N-m $\tau_{max} = ?$

The enclosed area is: $A_E = \frac{1}{2}\pi(0.05)^2 + (0.1)(0.1) = 0.0139\ m^2$

The maximum torsional shear stress will be where the thickness is minimum, that is t = 3 mm = 0.003 m. Its value can be found using Equation 5.26 as shown below.

$$\tau_{max} = \frac{T}{2tA_E} = \frac{900}{2(0.003)(0.0139)} = 10.77(10^6)\text{N/m}^2 \text{ or } \boxed{|\tau_{max}| = 10.8\ \text{MPa}}$$

5.97 Calculate the magnitude of the maximum torsional shear stress in the cross-sections shown that is subjected to a torque of T=15 kN-m

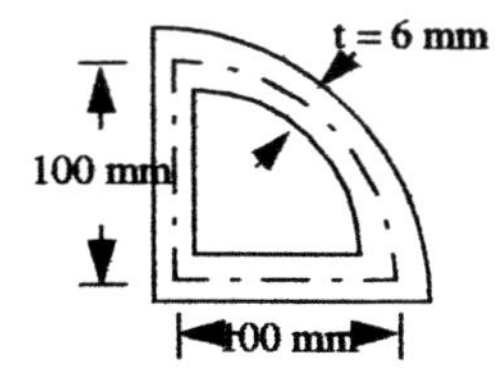

Fig. P5.97

Solution T = 15 kN-m $\tau_{max} = ?$

The enclosed area is: $A_E = \frac{1}{4}\pi(0.1)^2 = 7.854(10^{-3})\ m^2$

As thickness is uniform, the torsional shear stress is uniform across the cross-section and maximum value of it can be found using Equation 5.26 as shown below.

$$\tau_{max} = \frac{T}{2tA_E} = \frac{15(10^3)}{2(0.006)(7.854)(10^{-3})} = 159.1(10^6)\text{N/m}^2 \text{ or } \boxed{|\tau_{max}| = 159\ \text{MPa}}$$

5.98 A tube of uniform thickness t has a torque T applied to it. The cross-section of the tube is as shown. Determine the maximum torsional shear stress in terms of t, a, and T.

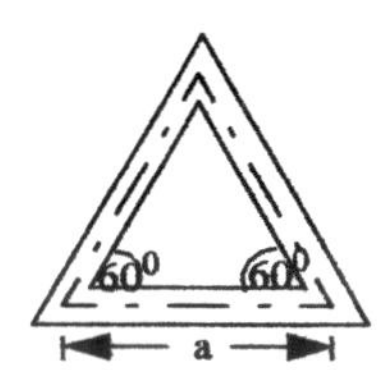

Fig. P5.98

Solution $\tau_{max} = f(t, a, T) = ?$

The enclosed area is: $A_E = \frac{1}{2}(a)\left(\frac{a}{2}\tan 60\right) = \frac{\sqrt{3}}{4}a^2$

As thickness is uniform the torsional shear stress is uniform across the cross-section and maximum value of it can be found using Equation 5.26 as shown below

$$\tau_{max} = \frac{T}{2tA_E} = \frac{T}{2(t)\left(\frac{\sqrt{3}}{4}a^2\right)} \text{ or } \boxed{\tau_{max} = \frac{2}{\sqrt{3}}\frac{T}{a^2 t}}$$

5.99 A tube of uniform thickness t has a torque T applied to it. The cross-section of the tube is as shown. Determine the maximum torsional shear stress in terms of t, a, and T.

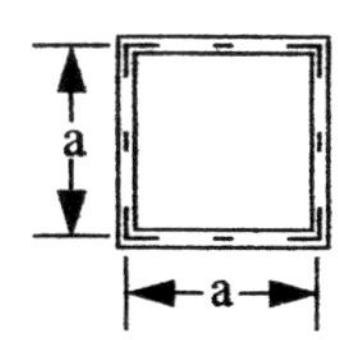

Fig. P5.99

Solution $\tau_{max} = f(t, a, T) = ?$

The enclosed area is: $A_E = (a)(a) = a^2$

As thickness is uniform the torsional shear stress is uniform across the cross-section and maximum value of it can be found using Equation 5.26 as shown below

$$\tau_{max} = \frac{T}{2tA_E} = \frac{T}{2(t)(a^2)} \text{ or } \boxed{\tau_{max} = \frac{T}{2a^2t}}$$

5.100 A tube of uniform thickness t has a torque T applied to it. The cross-section of the tube is as shown. Determine the maximum torsional shear stress in terms of t, a, and T.

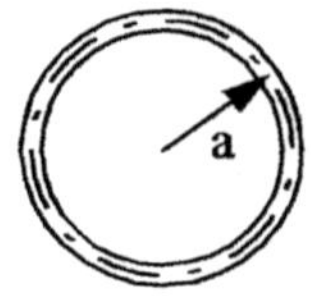

Fig. P5.100

Solution $\tau_{max} = f(t, a, T) = ?$

The enclosed area is: $A_E = \pi a^2$

As thickness is uniform the torsional shear stress is uniform across the cross-section and maximum value of it can be found using Equation 5.26 as shown below

$$\tau_{max} = \frac{T}{2tA_E} = \frac{T}{2(t)(\pi a^2)} \text{ or } \boxed{\tau_{max} = \frac{T}{2\pi a^2 t}}$$

5.101 A tube of uniform thickness t has a torque T applied to it. Determine the maximum torsional shear stress in terms of t, a, b, and T

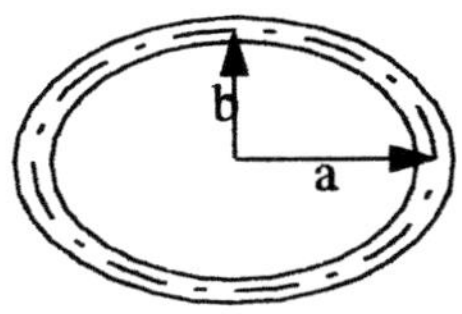

Fig. P5.101

Solution $\tau_{max} = f(t, a, b, T) = ?$

The enclosed area is: $A_E = \pi ab$

As thickness is uniform the torsional shear stress is uniform across the cross-section and maximum value of it can be found using Equation 5.26 as shown below

$$\tau_{max} = \frac{T}{2tA_E} = \frac{T}{2(t)(\pi ab)} \text{ or } \boxed{\tau_{max} = \frac{T}{2\pi abt}}$$

5.102 A hexagonal tube of uniform thickness is loaded as shown. Determine the magnitude of the maximum torsional shear stress in the tube.

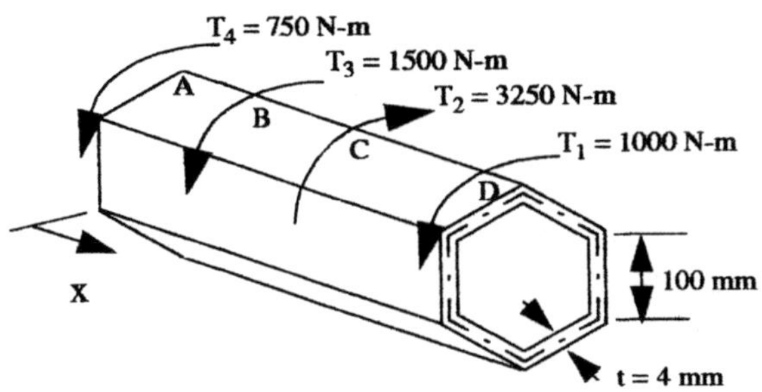

Fig. P5.102

Solution $\tau_{max} = ?$

The torque diagram can be drawn using the template and template equation shown.

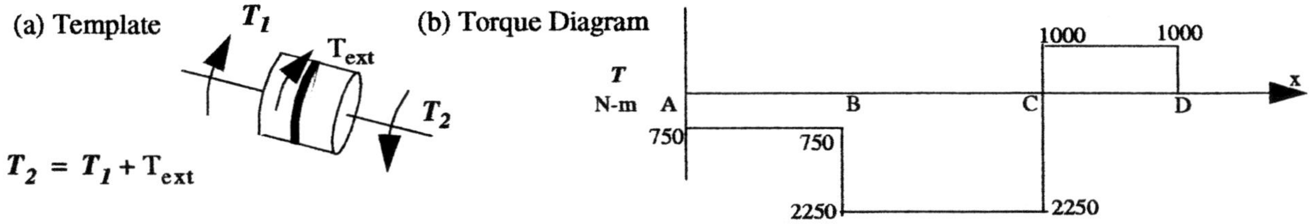

The maximum internal torque is in segment AB and its value is T_{max} = -2250 N-m.

The enclosed area is equal to the area of 6 equilateral triangles with sides of 0.1 m. Thus the total enclosed area is: $A_E = (6)\left[\frac{1}{2}(0.1)\left(\frac{0.1}{2}\tan 60\right)\right] = 0.0260\ \text{m}^2$

As thickness is uniform the torsional shear stress is uniform across the cross-section and the maximum value of it can be found using Equation 5.26 as shown below.

$$\tau_{max} = \frac{T_{max}}{2tA_E} = \frac{(-2250)}{2(0.004)(0.0260)} = -10.82(10^6)\text{N/m}^2 \text{ or} \qquad \boxed{|\tau_{max}| = 10.8\ \text{MPa}}$$

5.103 A rectangular tube is loaded as shown. Determine the maximum shear stress

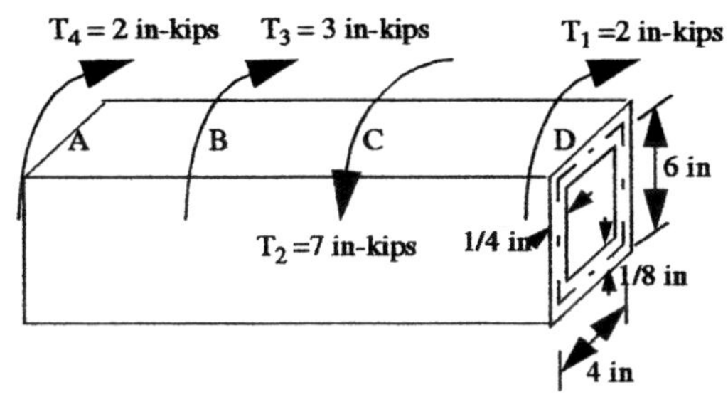

Fig. P5.103

Solution $\tau_{max} = ?$

The torque diagram can be drawn using the template and template equation shown.

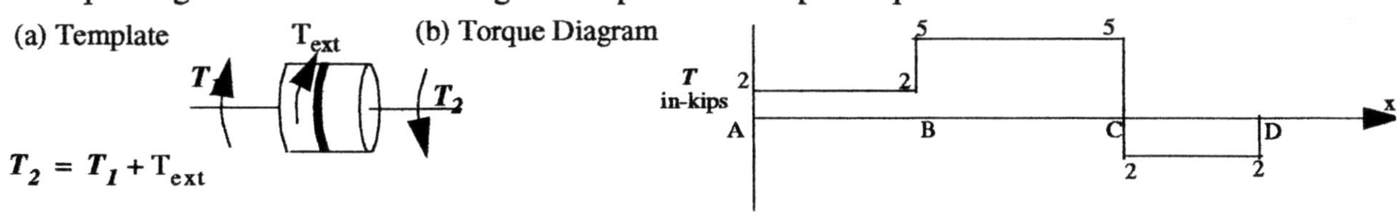

The maximum internal torque is in segment AB and its value is T_{max} = 5 in-kips.

The enclosed area is: $A_E = (4)(6) = 24\ \text{in}^2$

The maximum torsional shear stress will be where the thickness is minimum, that is t = 1/8 in. Its value can be found using Equation 5.26 as shown below.

$$\tau_{max} = \frac{T_{max}}{2tA_E} = \frac{(5)}{2(1/8)(24)} = 0.833\ \text{ksi or} \qquad \boxed{|\tau_{max}| = 833\ \text{psi}}$$

5.104 The three tubes shown in Problems 5.98 through 5.100 are to be compared for the maximum torque carrying capability, assuming that all tubes have the same thickness t, that the maximum shear stress in each tube can be τ and the amount of material used in the cross-section of each tube is A. Which shape would you use? What is the percentage torque carried by the remaining two shapes in terms of the most efficient structural shape?

Solution Shape= Triangle or Square or Circle = ? and percentage torque = ?

Let a_T, a_S and a_C represent the dimension of the triangle, square and circle shown in Figures P5.98, P5.99, P5.100, respectively. The material area in each shape is the product of the perimeter and the thickness. We can find the dimension of each shape in terms of the material area can be found and used in the results of Problems 5.98 through 5.100 as shown below.

For triangle: $A = (3a_T)t$ or $a_T = \frac{A}{3t}$. Substituting this into the result of problem 5.98 and representing the torque on triangle as T_T, we obtain the following.

$$\tau_{max} = \frac{2}{\sqrt{3}}\frac{T_T}{a_T^2 t} = \frac{2}{\sqrt{3}}\frac{T_T}{\left(\frac{A}{3t}\right)^2 t} = 10.392\frac{T_T}{A^2 t} \quad \text{or} \quad T_T = \frac{1}{10.392}A^2 t\tau_{max} \qquad 1$$

For square: $A = (4a_S)t$ or $a_S = \frac{A}{4t}$. Substituting this into the result of problem 5.99 and representing the torque on square as T_S, we obtain the following.

$$\tau_{max} = \frac{T_S}{2a_S^2 t} = \frac{T_S}{2\left(\frac{A}{4t}\right)^2 t} = 8.0\frac{T_S}{A^2 t} \quad \text{or} \quad T_S = \frac{1}{8.0}A^2 t\tau_{max} \qquad 2$$

For Circle: $A = (2\pi a_C)t$ or $a_C = \frac{A}{2\pi t}$. Substituting this into the result of problem 5.100 and representing the torque on circle as T_C, we obtain the following

$$\tau_{max} = \frac{T_C}{2\pi a_C^2 t} = \frac{T_C}{2\pi\left(\frac{A}{2\pi t}\right)^2 t} = 6.283\frac{T_C}{A^2 t} \quad \text{or} \quad T_C = \frac{1}{6.283}A^2 t\tau_{max} \qquad 3$$

By inspection of Equations (1), (2), and (3) we conclude that the circular shape will support the maximum internal torque. Thus our answer is: **Circular shape should be used.**

% torque in triangle relative to circle is: $= \left(\frac{T_T}{T_C}\right)(100) = \left(\frac{\frac{1}{8.0}A^2 t\tau_{max}}{\frac{1}{6.283}A^2 t\tau_{max}}\right)(100) = 60.46$ or

% torque in triangle relative to circle is 60.46%

% torque in square relative to circle is: $= \left(\frac{T_S}{T_C}\right)(100) = \left(\frac{\frac{1}{10.392}A^2 t\tau_{max}}{\frac{1}{6.283}A^2 t\tau_{max}}\right)(100) = 78.54$ or

% torque in square relative to circle is 78.54%

Problem Set 6.1

6.1 The rigid plate that is welded to the two bars is rotated about the z-axis causing the two bars to bend. The normal strains in bar 1 and 2 were found to be $\varepsilon_1 = 2000\ \mu\ \text{in/in}$ and $\varepsilon_2 = -1500\ \mu\ \text{in/in}$ respectively. Determine the angle of rotation ψ.

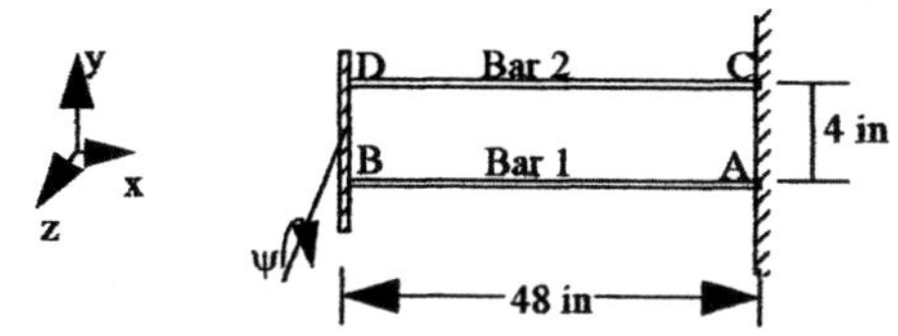

Solution $\varepsilon_1 = 2000\ \mu\ \text{in/in}$ $\varepsilon_2 = -1500\ \mu\ \text{in/in}$ $\psi = ?$

We can draw an exaggerated approximate deformed geometry as shown below.

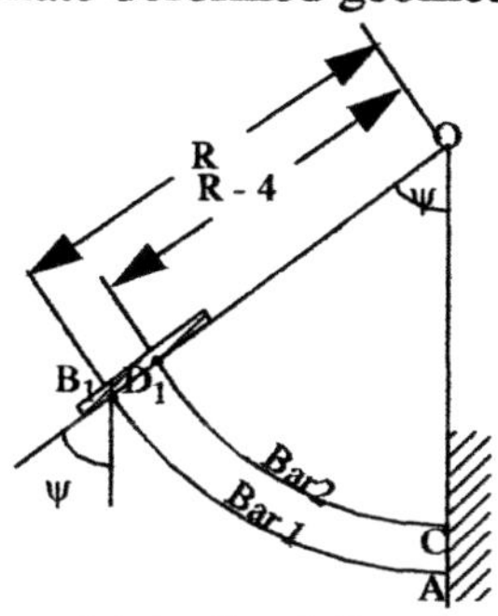

From the above deformed geometry we obtain the following.

$$\varepsilon_1 = \frac{AB_1 - AB}{AB} = \frac{R\psi - 48}{48} = 2000(10^{-6}) \qquad 1$$

$$\varepsilon_2 = \frac{CD_1 - CD}{CD} = \frac{(R-4)\psi - 48}{48} = -1500(10^{-6}) \qquad 2$$

Subtracting Eq. (2) from Eq. (1), we obtain: $\frac{4\psi}{48} = 3500(10^{-6})$ or $\psi = 0.042\ \text{rads}$ or $\boxed{\psi = 2.41^\circ}$

6.2 Determine 'h', the location at which a third bar in problem 6.1 must be placed so that there in no normal strain in the third bar.

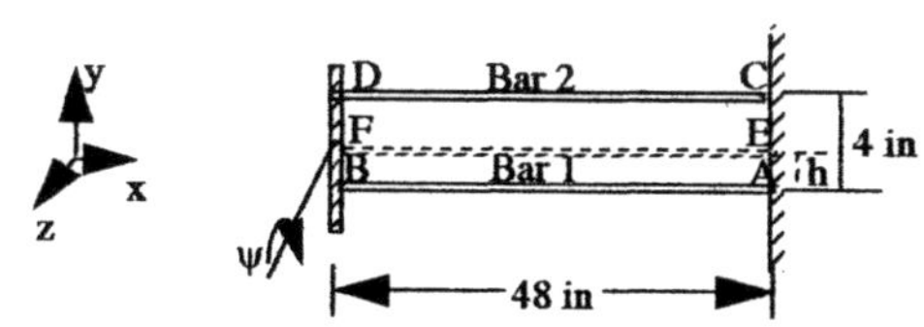

Solution $\varepsilon_1 = 2000\ \mu\ \text{in/in}$ $\varepsilon_2 = -1500\ \mu\ \text{in/in}$ $h = ?$ where $\varepsilon_3 = 0$

We draw an exaggerated approximate deformed geometry as shown below.

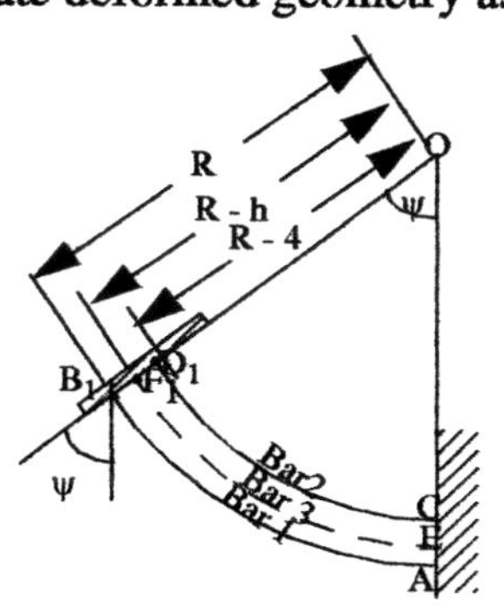

From the deformed geometry we obtain the following.

$$\varepsilon_3 = \frac{EF_1 - EF}{EF} = \frac{(R-h)\psi - 48}{48} = 0$$ 1

Subtracting the above equation from Eq. (1) in problem 6.1, we obtain: $\frac{h\psi}{48} = 2000(10^{-6})$ or $h = \frac{0.096}{\psi}$.

From the solution of problem 6.1 we have $\psi = 0.042$ rads, which on substitution yields

$h = \frac{0.096}{0.042} = 2.2857$ in or $\boxed{h = 2.29 \text{ in}}$

6.3 The two rigid plates that are welded to six bars are rotated about the z-axis causing the six bars to bend. The normal strains in bar 2 and 5 were found to be zero. What are the strains in the remaining bars?

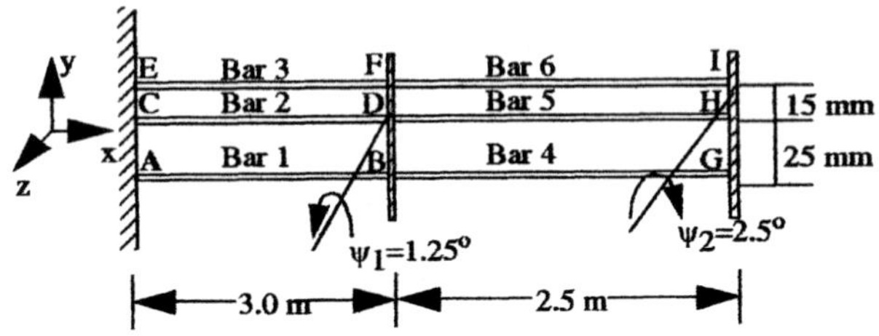

Solution $\varepsilon_2 = 0$ $\varepsilon_5 = 0$ $\varepsilon_1 = ?$ $\varepsilon_3 = ?$ $\varepsilon_4 = ?$ $\varepsilon_6 = ?$

The rotation of the rigid plates can be written in radians as shown below.

$$\psi_1 = \left(\frac{1.25}{180}\right)\pi = 0.02182 \text{ rads} \qquad \psi_2 = \left(\frac{2.5}{180}\right)\pi = 0.04363 \text{ rads}$$ 1

We can draw an exaggerated approximate deformed geometry as shown below

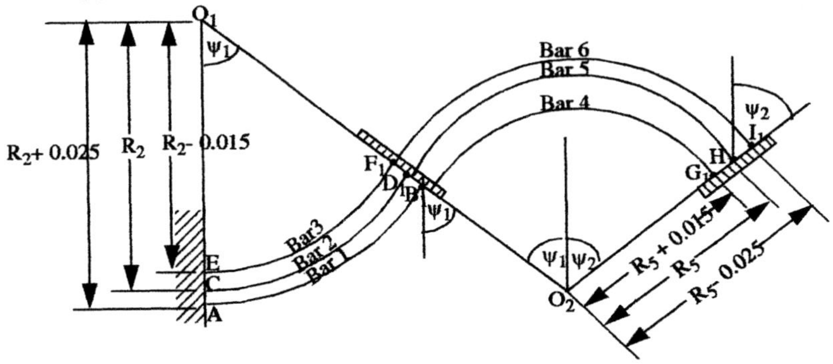

From the deformed geometry, we obtain the following.

$$\varepsilon_2 = \frac{CD_1 - CD}{CD} = \frac{R_2\psi_1 - 3}{3} = 0 \quad \text{or} \quad R_2\psi_1 = 3$$ 2

$$\varepsilon_1 = \frac{AB_1 - AB}{AB} = \frac{(R_2 + 0.025)\psi_1 - 3}{3} = \frac{R_2\psi_1 + (0.025)\psi_1 - 3}{3}$$ 3

Substituting Eq. (2) into Eq. (3) we obtain: $\varepsilon_1 = \frac{(0.025)(0.02182)}{3} = 181.8(10^{-6})$ or $\boxed{\varepsilon_1 = 182\mu \text{ m/m}}$

$$\varepsilon_3 = \frac{EF_1 - EF}{EF} = \frac{(R_2 - 0.015)\psi_1 - 3}{3} = \frac{R_2\psi_1 - (0.015)\psi_1 - 3}{3}$$ 4

Substituting Eq. (2) into Eq. (4) we obtain: $\varepsilon_3 = -\frac{(0.015)(0.02182)}{3} = -109.1(10^{-6})$ or

$\boxed{\varepsilon_3 = -109.1\mu \text{ m/m}}$

$$\varepsilon_5 = \frac{D_1H_1 - DH}{DH} = \frac{R_5(\psi_1 + \psi_2) - 2.5}{2.5} = 0 \quad \text{or} \quad R_5(\psi_1 + \psi_2) = 2.5$$ 5

$$\varepsilon_4 = \frac{B_1G_1 - BG}{BG} = \frac{(R_5 - 0.025)(\psi_1 + \psi_2) - 2.5}{2.5} = \frac{R_5(\psi_1 + \psi_2) - (0.025)(\psi_1 + \psi_2) - 2.5}{2.5} \quad 6$$

Substituting Eq. (5) into Eq. (6) we obtain: $\varepsilon_4 = -\frac{(0.025)(0.02182 + 0.04363)}{2.5} = -654.49(10^{-6})$ or

$\boxed{\varepsilon_4 = -654\ \mu\ m/m}$

$$\varepsilon_6 = \frac{F_1I_1 - FI}{FI} = \frac{(R_5 + 0.015)(\psi_1 + \psi_2) - 2.5}{2.5} = \frac{R_5(\psi_1 + \psi_2) + (0.015)(\psi_1 + \psi_2) - 2.5}{2.5} \quad 7$$

Substituting Eq. (5) into Eq. (7) we obtain: $\varepsilon_6 = \frac{(0.015)(0.02182 + 0.04363)}{2.5} = 392.7(10^{-6})$ or

$\boxed{\varepsilon_6 = 393\ \mu\ m/m}$

6.4 The strain in bar 1 and 3 were found to be $\varepsilon_1 = 800\ \mu$ and $\varepsilon_3 = 500\ \mu$. Determine the strains in the remaining bars.

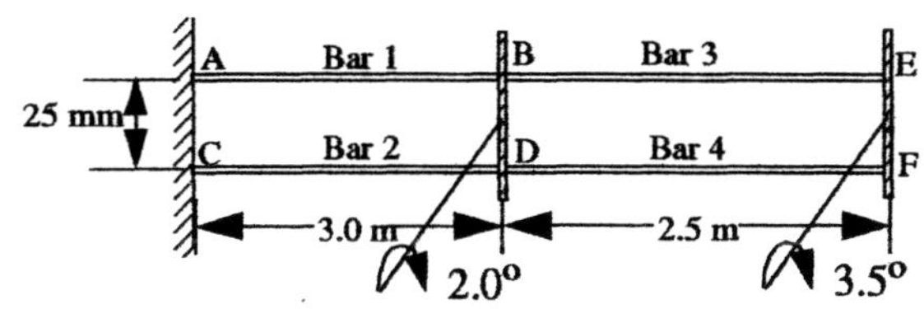

Fig. P6.4

Solution $\varepsilon_1 = 800\ \mu$ $\varepsilon_3 = 500\ \mu$ $\varepsilon_2 = ?$ $\varepsilon_4 = ?$

The rotation of the rigid plates can be written in radians as shown below.

$$\psi_1 = \left(\frac{2.0}{180}\right)\pi = 0.03491\ \text{rads} \qquad \psi_2 = \left(\frac{3.5}{180}\right)\pi = 0.06109\ \text{rads} \quad 1$$

We can draw an exaggerated approximate deformed geometry as shown below

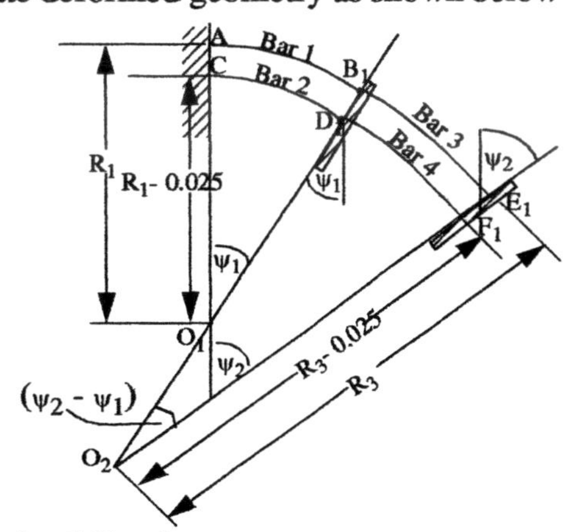

From the deformed geometry we obtain the following.

$$\varepsilon_1 = \frac{AB_1 - AB}{AB} = \frac{R_1\psi_1 - 3}{3} = 800(10^{-6}) \quad 2$$

$$\varepsilon_2 = \frac{CD_1 - CD}{CD} = \frac{(R_1 - 0.025)\psi_1 - 3}{3} = \frac{R_1\psi_1 - 3}{3} - \frac{0.025\psi_1}{3} \quad 3$$

Substituting Eqs. (1), (2) and ψ_1 into Eq. (3), we obtain the following.

$\varepsilon_2 = 800(10^{-6}) - \frac{0.025(0.03491)}{3} = 509.11(10^{-6})$ or $\boxed{\varepsilon_2 = 509\ \mu\ m/m}$

$$\varepsilon_3 = \frac{B_1E_1 - BE}{BE} = \frac{R_3(\psi_2 - \psi_1) - 2.5}{2.5} = 500(10^{-6}) \quad 4$$

$$\varepsilon_4 = \frac{D_1F_1 - DF}{DF} = \frac{(R_3 - 0.025)(\psi_2 - \psi_1) - 2.5}{2.5} = \frac{R_3(\psi_2 - \psi_1) - 2.5}{2.5} - \frac{0.025(\psi_2 - \psi_1)}{2.5} \quad 5$$

Substituting Eqs. (1),(4), ψ_1 and ψ_2 into Eq. (5), we obtain the following.

$$\varepsilon_4 = 500(10^{-6}) - \frac{0.025(0.06109 - 0.03491)}{2.5} = 238.2(10^{-6}) \text{ or } \quad \boxed{\varepsilon_4 = 238\ \mu\ m/m}$$

6.5 Due to the action of the external moment M_z and force P, the rigid plate was observed to rotate by 2° and the normal strain in bar 1 was found $\varepsilon_1 = 2000\ \mu\ in/in$. Both bars have an area of cross-section of $A = 1/2\ in^2$ and a Modulus of Elasticity of $E = 30{,}000$ ksi. Determine the applied moment M_z and force P.

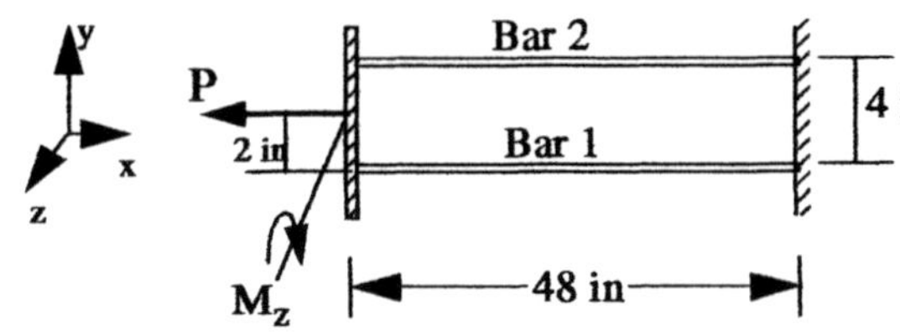

Fig. P6.5

Solution $\varepsilon_1 = 2000\ \mu\ in/in$ $\psi = 2°$ $E = 30{,}000$ ksi $A = 1/2\ in^2$ $M_z = ?$ $P = ?$

---

1. Strain calculations: The rotation of the rigid plates can be written in radians as shown below.

$$\psi = \left(\frac{2.0}{180}\right)\pi = 0.03491 \text{ rads} \quad 1$$

We can draw an exaggerated approximate deformed geometry as shown below.

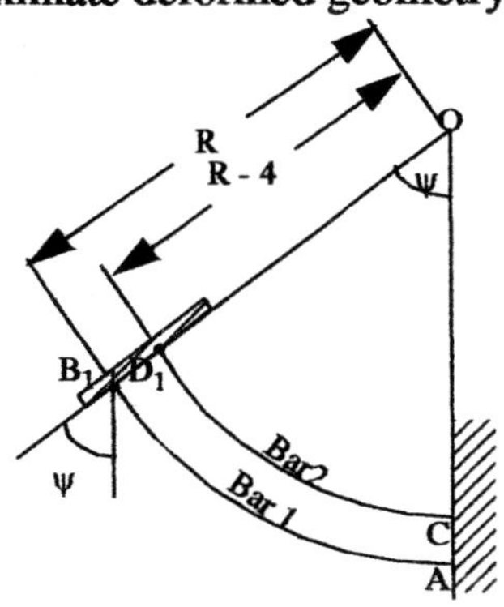

From the deformed geometry we obtain the following.

$$\varepsilon_1 = \frac{AB_1 - AB}{AB} = \frac{R\psi - 48}{48} = 2000(10^{-6}) \quad 2$$

$$\varepsilon_2 = \frac{CD_1 - CD}{CD} = \frac{(R-4)\psi - 48}{48} = \frac{R\psi - 48}{48} - \frac{4\psi}{48} \quad 3$$

Substituting Eqs. (1) and (2) into Eq. (3), we obtain:

$$\varepsilon_2 = 2000(10^{-6}) - \frac{4(0.03491)}{48} = -908.88(10^{-6}) \quad 4$$

2. Stress calculations: From Hooke's Law we can find the normal stresses in the bar as shown below.

$$\sigma_1 = E\varepsilon_1 = (30,000)(2000)(10^{-6}) = 60 \text{ ksi or } \sigma_1 = 60 \text{ ksi (T)}$$

$$\sigma_2 = E\varepsilon_2 = (30,000)(-908.88)(10^{-6}) = -27.27 \text{ ksi or } \sigma_2 = 27.27 \text{ ksi (C)}$$

3. Internal force calculations: The internal normal force in each bar can be found as shown below.

$N_1 = \sigma_1 A = (60)(1/2) = 30$ kips(T) $\qquad N_2 = \sigma_2 A = (27.27)(1/2) = 13.63$ kips(C)

4. External forces and moments calculations: The free body diagram of the rigid plate can be drawn by making imaginary cuts through the rods as shown below.

By equilibrium of forces we obtain: $P + 13.63 - 30 = 0$ or $\boxed{P = 16.37 \text{ kips}}$

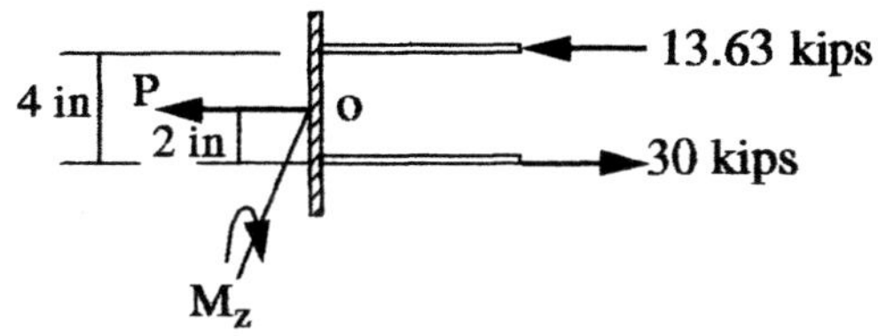

By equilibrium of moment about point O we obtain: $M_z - (13.63)(2) - (30)(2) = 0$ or

$$\boxed{M_z = 87.26 \text{ in} - \text{kips}}$$

6.6 The rigid plate was observed to rotate 1.25° due to the action of the external moment M_z and Force P. All three bars have an area of cross-section of A= 100 mm² and a Modulus of Elasticity E=200 GPa. If the strain in bar 2 was measured as zero determine the external moment M_z and Force P.

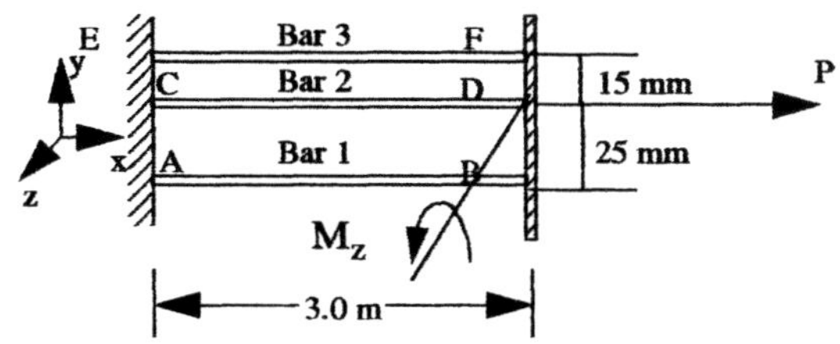

Solution $\varepsilon_2 = 0$ $\psi = 1.25^\circ$

E = 200 GPa A = 100 mm² M_z = ? P= ?

---

1. Strains calculations: The rotation of the rigid plates can be written in radians as shown below.

$$\psi = \left(\frac{1.25}{180}\right)\pi = 0.02182 \text{ rads} \quad 1$$

We can draw an exaggerated approximate deformed geometry as shown below.

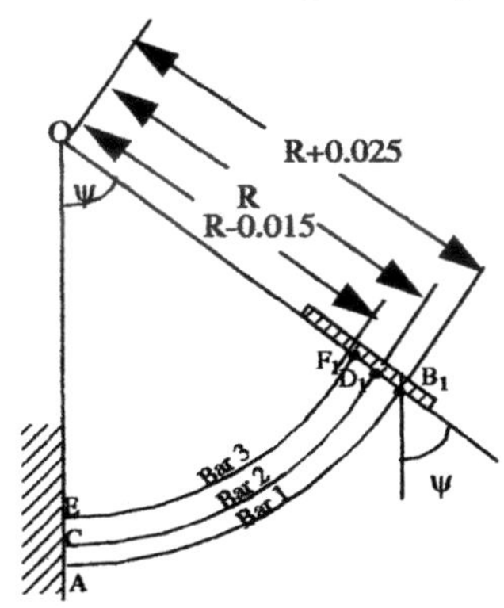

From the deformed geometry, we obtain the following.

$$\varepsilon_2 = \frac{CD_1 - CD}{CD} = \frac{R\psi - 3}{3} = 0 \quad \text{or} \quad R\psi = 3 \quad 2$$

$$\varepsilon_1 = \frac{AB_1 - AB}{AB} = \frac{(R + 0.025)\psi - 3}{3} = \frac{R\psi + (0.025)\psi - 3}{3} \quad 3$$

$$\varepsilon_3 = \frac{EF_1 - EF}{EF} = \frac{(R - 0.015)\psi - 3}{3} = \frac{R\psi - (0.015)\psi - 3}{3} \quad 4$$

Substituting Eqs. (1) and (2) into Eq. (3) we obtain: $\varepsilon_1 = \frac{(0.025)(0.02182)}{3} = 181.8(10^{-6})$

Substituting Eqs.(1) and (2) into Eq. (4) we obtain: $\varepsilon_3 = -\frac{(0.015)(0.02182)}{3} = -109.1(10^{-6})$

2. Stress calculations: From Hooke's Law we can find the normal stresses in the bar as shown below.

$\sigma_1 = E\varepsilon_1 = (200)(10^9)(181.8)(10^{-6}) = 36.36(10^6)\ N/m^2$ or $\sigma_1 = 36.36(10^6)\ N/m^2 (T)$

$\sigma_2 = E\varepsilon_2 = (200)(10^9)(-109.1)(10^{-6}) = -21.82(10^6) N/m^2$ or $\sigma_2 = 21.82(10^6)\ N/m^2 (C)$

3. Internal force calculations: The internal normal force in each bar can be found as shown below.

$N_1 = \sigma_1 A = (36.36)(10^6)(100)(10^{-6}) = 3636N\ (T)$

$N_2 = \sigma_2 A = (21.82)(10^6)(100)(10^{-6}) = 2182N\ (C)$

4. External forces and moments calculations: The free body diagram of the rigid plate can be drawn by making imaginary cuts through the rods as shown below.

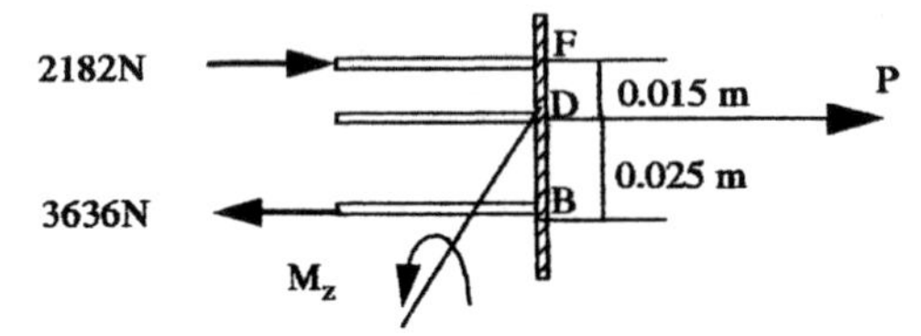

By equilibrium of forces we obtain: $P + 2182 - 3636 = 0$ or $\boxed{P = 1454\ N}$

By equilibrium of moment about point D we obtain: $M_z - (2182)(0.015) - (3636)(0.025) = 0$ or

$\boxed{M_z = 123.6\ N-m}$

6.7 The rigid plates BD and EF were observed to rotate by 2° and 3.5° in the direction of applied moments. All bars have an area of cross-section of A= 125 mm². Bars 1 and 3 are made of steel E_S= 200 GPa. and bars 2 and 4 are made of aluminum E_{al} = 70 GPa. If the strains in bar 1 and 3 were found to be $\varepsilon_1 = 800\ \mu$ and $\varepsilon_3 = 500\ \mu$ determine the applied moment M_1 and M_2 and the forces P_1 and P_2 that act at the center of the rigid plates.

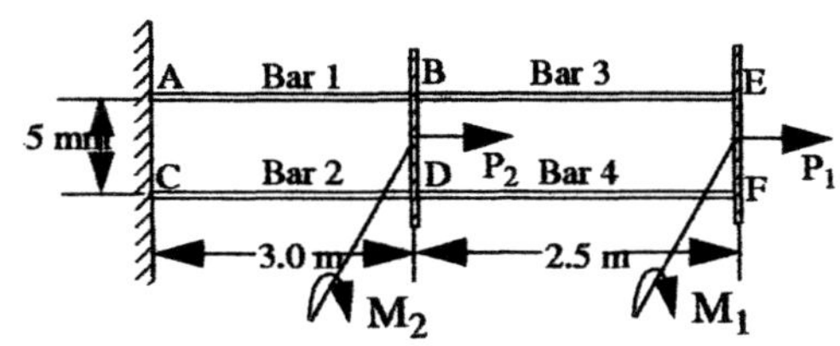

Fig. P6.7

Solution $\varepsilon_1 = 800\ \mu$ $\varepsilon_3 = 500\ \mu$ $\psi_1 = 2^o$ $\psi_1 = 3.5^o$ $E_1 = E_3 = 200$ GPa

$E_2 = E_4 = 70$ GPa $A = 125\ mm^2$ $M_1 = ?$ $P_1 = ?$ $M_2 = ?$ $P_2 = ?$

---

1. Strain calculations: From problem 6.4 we have: $\varepsilon_4 = 238.2(10^{-6})$ and $\varepsilon_2 = 509.11(10^{-6})$

2. Stress calculations: From Hooke's Law we can find the normal stresses in the bar as shown below.

$\sigma_1 = E_1\varepsilon_1 = (200)(10^9)(800)(10^{-6}) = 160(10^6)\ N/m^2\ (T)$

$\sigma_2 = E_2\varepsilon_2 = (70)(10^9)(509.11)(10^{-6}) = 35.64(10^6)\ N/m^2\ (T)$

$\sigma_3 = E_3\varepsilon_3 = (200)(10^9)(500)(10^{-6}) = 100(10^6)\ N/m^2\ (T)$

$\sigma_4 = E_4\varepsilon_4 = (70)(10^9)(238.2)(10^{-6}) = 16.67(10^6)\ N/m^2\ (T)$

3. Internal force calculations: The internal normal force in each bar can be found as shown below.

$N_1 = \sigma_1 A = (160)(10^6)(125)(10^{-6}) = 20(10^3)N = 20kN\ (T)$

$N_2 = \sigma_2 A = (35.64)(10^6)(125)(10^{-6}) = 4.46(10^3)N = 4.46kN\ (T)$

$N_3 = \sigma_3 A = (100)(10^6)(125)(10^{-6}) = 12.5(10^3)N = 12.5kN\ (T)$

$N_4 = \sigma_4 A = (16.67)(10^6)(125)(10^{-6}) = 2.08(10^3)N = 2.08kN \ (T)$

4. External forces and moments calculations: The free body diagram of the rigid plates can be drawn by making imaginary cuts through the rods as shown below.

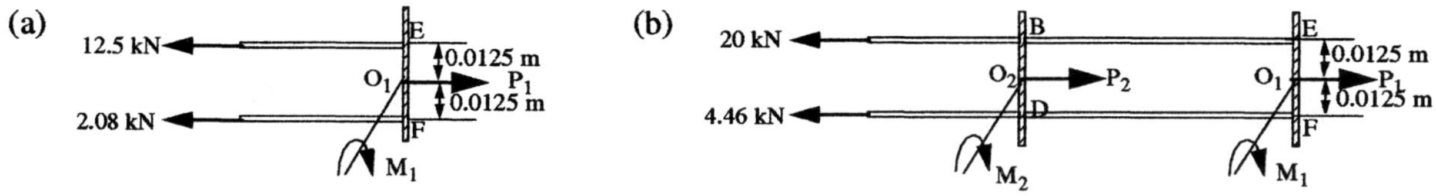

By equilibrium of forces in Fig. (a) we obtain: $P_1 - 12.5 - 2.08 = 0$ or $\boxed{P_1 = 14.58kN}$

By equilibrium of moment about point O_1 we obtain: $M_1 - (12.5)(0.0125) + (2.08)(0.0125) = 0$ or

$M_1 = 0.1303 \ \ kN-m$ $\boxed{M_1 = 130.3N-m}$

By equilibrium of forces in Fig. (b) we obtain: $P_2 + P_1 - 20 - 4.46 = 0$ or $\boxed{P_2 = 9.88kN}$

By equilibrium of moment about point O_1 we obtain: $M_2 + M_1 - (20)(0.0125) + (4.46)(0.0125) = 0$ or

$M_2 = 0.06395 \ \ kN-m$ $\boxed{M_2 = 64.0N-m}$

6.8 Steel ($E_{steel} = 30{,}000$ ksi) strips are securely attached to a wooden ($E_{wood} = 2{,}000$ ksi) beam as shown. The normal strain ε_{xx} at the cross-section due to bending about the z-axis, and the dimensions of the cross-section are as given below. Determine the equivalent internal moment M_z.

$\varepsilon_{xx} = -100y \ \mu$. where y is measured in inches. Use d =2 in, h_W =4 in and h_S= (1/8) in.

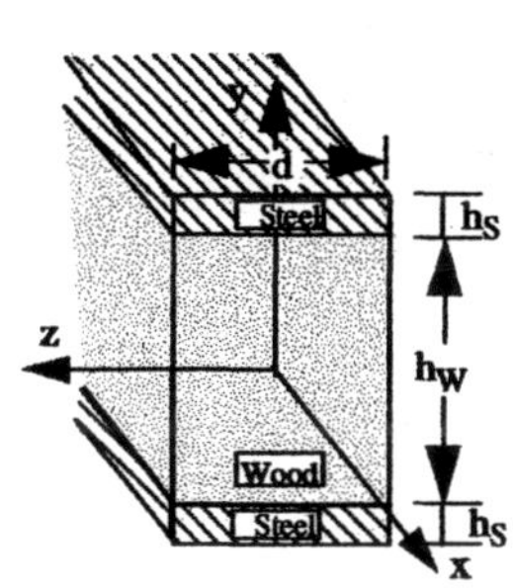

Solution $E_{steel} = 30{,}000$ ksi $E_{wood} = 2{,}000$ ksi $\varepsilon_{xx} = -100y \ \mu$.
d =2 in h_W =4 in h_S= (1/8) in. $M_z = ?$

From Hooke's law we can write:

$$(\sigma_{xx})_{steel} = E_{steel}\varepsilon_{xx} = (30000)(-100y)(10^{-6}) = -3.0y \ \ ksi \qquad 1$$

$$(\sigma_{xx})_{wood} = E_{wood}\varepsilon_{xx} = (2000)(-100y)(10^{-6}) = -0.2y \ \ ksi \qquad 2$$

The normal stress across the cross-section can be written as shown below.

$$\sigma_{xx} = \begin{cases} -3.0y \ \ ksi & 2.0 < y \le 2.125 \\ -0.2y \ \ ksi & -2.0 < y < 2.0 \\ -3.0y \ \ ksi & -2.125 \le y < -2.0 \end{cases} \qquad 3$$

The differential area dA is the area of a strip whose width is d = 2 inch and thickness dy i.e., $dA = 2dy$. Eq. 6.1 can be written as a sum of integrals over steel and wood and evaluated as shown.

$$M_z = -\int_A y\sigma_{xx}dA = -\left[\int_{-2.125}^{-2.0} y(-3.0y)(2)dy + \int_{-2.0}^{2.0} y(-0.2y)(2)dy + \int_{2.0}^{2.125} y(-3.0y)(2)dy\right] \text{ or}$$

$$M_z = -\left[-6\frac{y^3}{3}\Big|_{-2.125}^{-2.0} - 0.4\frac{y^3}{3}\Big|_{-2.0}^{2.0} - 6\frac{y^3}{3}\Big|_{2.0}^{2.125}\right] = -[-3.191 - 2.133 - 3.191] \text{ or } \quad \boxed{M_z = 8.52in-kips}$$

6.9 Steel (E_{steel} = 30,000 ksi) strips are securely attached to a wooden (E_{wood} = 2,000 ksi) beam as shown. The normal strain ε_{xx} at the cross-section due to bending about the z-axis, and the dimensions of the cross-section are as given below. Determine the equivalent internal moment M_z
ε_{xx} = - 50 y μ. where y is measured in inches. Use d =1 in, h_W =6 in and h_S= (1/4) in.

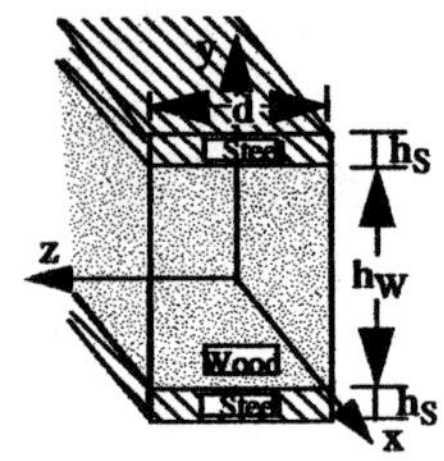

Solution E_{steel} = 30,000 ksi, d =1 in; E_{wood} = 2,000 ksi, h_W =6 in; ε_{xx} = - 50 y μ., h_S= (1/4) in. M_z = ?

From Hooke's law we can write:

$$(\sigma_{xx})_{steel} = E_{steel}\varepsilon_{xx} = (30000)(-50y)(10^{-6}) = -1.5y \text{ ksi} \quad 1$$

$$(\sigma_{xx})_{wood} = E_{wood}\varepsilon_{xx} = (2000)(-50y)(10^{-6}) = -0.1y \text{ ksi} \quad 2$$

The normal stress across the cross-section can be written as shown below.

$$\sigma_{xx} = \begin{cases} -1.5y \text{ ksi} & 3.0 < y \le 3.25 \\ -0.1y \text{ ksi} & -3.0 < y < 3.0 \\ -1.5y \text{ ksi} & -3.25 \le y < -3.0 \end{cases} \quad 3$$

The differential area dA is the area of a strip whose width is d = 1 inch and thickness dy i.e., $dA = (1)dy$. Eq. 6.1 can be written as a sum of integrals over steel and wood and evaluated as shown.

$$M_z = -\int_A y\sigma_{xx}dA = -\left[\int_{-3.25}^{-3.0} y(-1.5y)(1)dy + \int_{-3.0}^{3.0} y(-0.1y)(1)dy + \int_{3.0}^{3.25} y(-1.5y)(1)dy\right] \text{ or}$$

$$M_z = -\left[-1.5\frac{y^3}{3}\bigg|_{-3.25}^{-3.0} - 0.1\frac{y^3}{3}\bigg|_{-3.0}^{3.0} - 1.5\frac{y^3}{3}\bigg|_{3.0}^{3.25}\right] = -[-3.664 - 1.800 - 3.664] \text{ or } \boxed{M_z = 9.13\text{in}-\text{kips}}$$

6.10 Steel (E_{steel} = 30,000 ksi) strips are securely attached to a wooden (E_{wood} = 2,000 ksi) beam as shown. The normal strain ε_{xx} at the cross-section due to bending about the z-axis, and the dimensions of the cross-section are as given below.Determine the equivalent internal moment M_z
ε_{xx}= 200 y μ. where y is measured in inches. Use d =1 in, h_W =2 in and h_S= (1/16) in

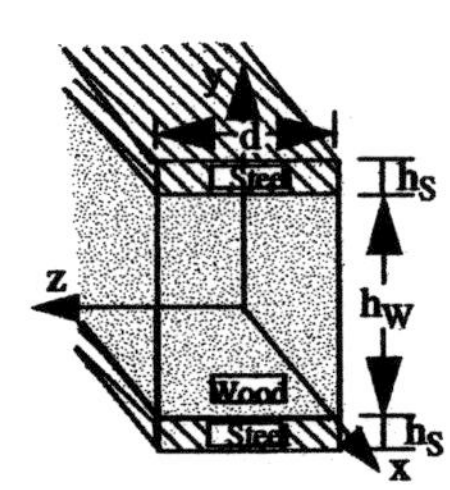

Solution E_{steel} = 30,000 ksi, d =1 in; E_{wood} = 2,000 ksi, h_W =2 in; ε_{xx} = 200 y μ., h_S= (1/16) in. M_z = ?

From Hooke's law we can write:

$$(\sigma_{xx})_{steel} = E_{steel}\varepsilon_{xx} = (30000)(200y)(10^{-6}) = 6y \text{ ksi} \quad 1$$

$$(\sigma_{xx})_{wood} = E_{wood}\varepsilon_{xx} = (2000)(200y)(10^{-6}) = 0.4y \text{ ksi} \quad 2$$

The normal stress across the cross-section can be written as shown below.

$$\sigma_{xx} = \begin{cases} 6y \text{ ksi} & 1.0 < y \le 1.0625 \\ 0.4y \text{ ksi} & -1.0 < y < 1.0 \\ 6y \text{ ksi} & -1.0625 \le y < -1.0 \end{cases} \quad 3$$

The differential area dA is the area of a strip whose width is d = 1 inch and thickness dy i.e., $dA = (1)dy$. Eq. 6.1 can be written as a sum of integrals over steel and wood and evaluated as shown.

$$M_z = -\int_A y\sigma_{xx}dA = -\left[\int_{-1.0625}^{-1.0} y(6y)(1)dy + \int_{-1.0}^{1.0} y(0.4y)(1)dy + \int_{1.0}^{1.0625} y(6y)(1)dy\right] \text{ or}$$

$$M_z = -\left[6\frac{y^3}{3}\Bigg|_{-1.0625}^{-1.0} + 0.4\frac{y^3}{3}\Bigg|_{-1.0}^{1.0} + 6\frac{y^3}{3}\Bigg|_{1.0}^{1.0625}\right] = -[0.3989 + 0.2667 + 0.3989] \text{ or} \quad \boxed{M_z = -1.06\text{in} - \text{kips}}$$

6.11 The flanges made from steel ($E_{steel} = 200$ GPa) are securely attached to the wooden ($E_{wood} = 10$ GPa) beam as shown. The normal strain ε_{xx} at the cross-section due to bending about the z-axis and the dimensions of the cross-section are as given below. Determine the equivalent internal moment M_z
$\varepsilon_{xx} = -0.012y$, where y is measured in meters. Use $t_W = 20$ mm in, $h_W = 250$ mm, $t_f = 20$ mm, and $d_f = 125$ mm.

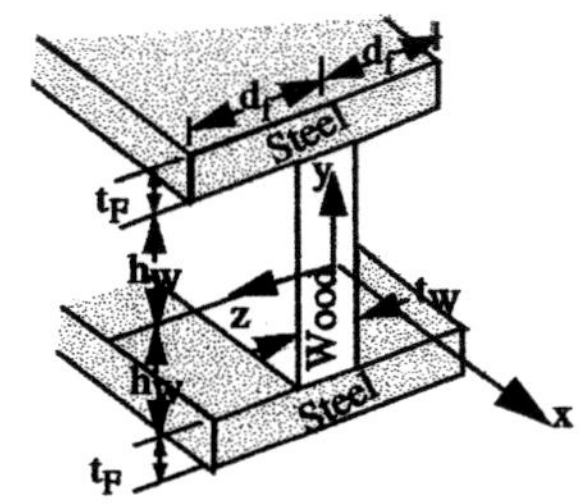

Solution $E_{steel} = 200$ GPa $\quad E_{wood} = 10$ GPa $\quad \varepsilon_{xx} = -0.012y$
$t_W = 20$ mm $\quad h_W = 250$ mm $\quad t_f = 20$ mm $\quad d_f = 125$ mm. $\quad M_z = ?$

From Hooke's law we can write:

$$(\sigma_{xx})_{steel} = E_{steel}\varepsilon_{xx} = (200)(10^9)(-0.012y) = -2400y(10^6) \text{ N/m}^2 \qquad 1$$

$$(\sigma_{xx})_{wood} = E_{wood}\varepsilon_{xx} = (10)(10^9)(-0.012y) = -120y(10^6) \text{ N/m}^2 \qquad 2$$

The differential area dA in the steel is the area of a strip whose width is $2d_f = 0.25$ m and thickness dy i.e., in steel $dA = 0.25dy$. The differential area dA in the wood is the area of a strip whose width is $t_W = 0.02$ m and thickness dy i.e., in wood $dA = 0.02dy$.
The normal stress and dA across the cross-section can be written as shown below.

$$\sigma_{xx} = \begin{cases} -2400y(10^6) \text{ N/m}^2 & dA = 0.25dy & 0.250 < y \le 0.270 \\ -120y(10^6) \text{ N/m}^2 & dA = 0.02dy & -0.250 < y < 0.250 \\ -2400y(10^6) \text{ N/m}^2 & dA = 0.25dy & -0.270 \le y < -0.250 \end{cases} \qquad 3$$

Noting the symmetry with respect to y, Eq. 6.1 can be written twice the sum of integrals over steel and wood in the upper half (y>0) of the cross-section.

$$M_z = -\int_A y\sigma_{xx}dA = -2\left[\int_0^{0.250} y(-120y)(0.02)dy + \int_{0.250}^{0.270} y(-2400y)(0.25)dy\right](10^6) \text{ or}$$

$$M_z = -2\left[-2.4\frac{y^3}{3}\Bigg|_0^{0.250} - 600\frac{y^3}{3}\Bigg|_{0.250}^{0.270}\right](10^6) = -2[-12.5 - 811.6](10^3) \text{ or} \quad \boxed{M_z = 1648\text{kN} - \text{m}}$$

6.12 The flanges made from steel ($E_{steel} = 200$ GPa) are securely attached to the wooden ($E_{wood} = 10$ GPa) beam as shown. The normal strain ε_{xx} at the cross-section due to bending about the z-axis and the dimensions of the cross-section are as given below. Determine the equivalent internal moment M_z
$\varepsilon_{xx} = -0.15y$, where y is measured in meters. Use $t_W = 10$ mm in, $h_W = 50$ mm, $t_f = 10$ mm, and $d_f = 25$ mm.

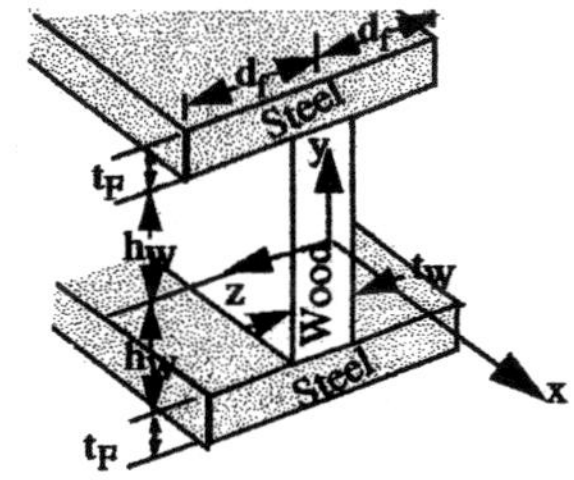

Solution $E_{steel} = 200$ GPa $E_{wood} = 10$ GPa $\varepsilon_{xx} = -0.15y$
$t_W = 10$ mm $h_W = 50$ mm $t_f = 10$ mm $d_f = 25$ mm. $M_z = ?$

From Hooke's law we can write:

$$(\sigma_{xx})_{steel} = E_{steel}\varepsilon_{xx} = (200)(10^9)(-0.15y) = -30y(10^9)\ \text{N/m}^2 \qquad 1$$

$$(\sigma_{xx})_{wood} = E_{wood}\varepsilon_{xx} = (10)(10^9)(-0.15y) = -1.5y(10^9)\ \text{N/m}^2 \qquad 2$$

The differential area dA in the steel is the area of a strip whose width is $2d_f = 0.05$ m and thickness dy i.e., in steel $dA = 0.05dy$. The differential area dA in the wood is the area of a strip whose width is $t_W = 0.01$ m and thickness dy i.e., in wood $dA = 0.01dy$. The normal stress and dA across the cross-section can be written as shown below.

$$\sigma_{xx} = \begin{cases} -30y(10^9)\ \text{N/m}^2 & dA = 0.05dy & 0.05 < y \le 0.06 \\ -1.5y(10^9)\ \text{N/m}^2 & dA = 0.01dy & -0.05 < y < 0.05 \\ -30y(10^9)\ \text{N/m}^2 & dA = 0.05dy & -0.06 \le y < -0.05 \end{cases} \qquad 3$$

Noting the symmetry with respect to y, Eq. 6.1 can be written twice the sum of integrals over steel and wood in the upper half (y>0) of the cross-section.

$$M_z = -\int_A y\sigma_{xx}dA = -2\left[\int_0^{0.05} y(-1.5y)(0.01)dy + \int_{0.05}^{0.06} y(-30y)(0.05)dy\right](10^9) \text{ or}$$

$$M_z = -2\left[-0.015\frac{y^3}{3}\bigg|_0^{0.05} - 1.5\frac{y^3}{3}\bigg|_{0.05}^{0.06}\right](10^9) = -2[-625 - 45500]\ \text{N-m} \text{ or } \boxed{M_z = 4214\text{kN-m}}$$

6.13 The flanges made from steel ($E_{steel} = 200$ GPa) are securely attached to the wooden ($E_{wood} = 10$ GPa) beam as shown. The normal strain ε_{xx} at the cross-section due to bending about the z-axis and the dimensions of the cross-section are as given below. Determine the equivalent internal moment M_z $\varepsilon_{xx} = 0.02y$, where y is measured in meters. Use $t_W = 15$ mm in, $h_W = 200$ mm, $t_f = 20$ mm, and $d_f = 150$ mm.

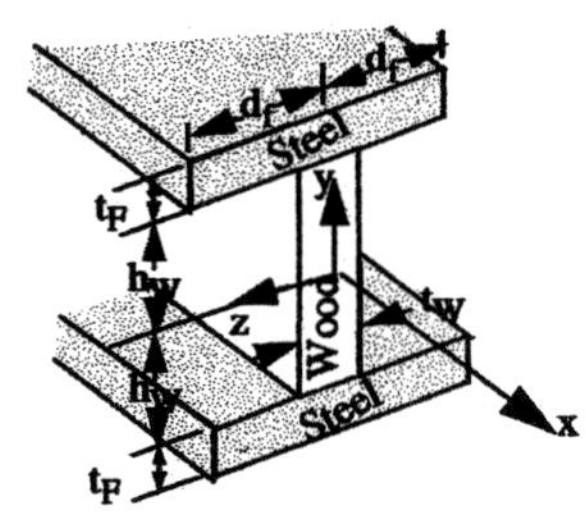

Solution $E_{steel} = 200$ GPa $E_{wood} = 10$ GPa $\varepsilon_{xx} = 0.02y$
$t_W = 15$ mm $h_W = 200$ mm $t_f = 20$ mm $d_f = 150$ mm. $M_z = ?$

From Hooke's law we can write:

$$(\sigma_{xx})_{steel} = E_{steel}\varepsilon_{xx} = (200)(10^9)(0.02y) = 4y(10^9)\ \text{N/m}^2 \qquad 1$$

$$(\sigma_{xx})_{wood} = E_{wood}\varepsilon_{xx} = (10)(10^9)(0.02y) = 0.2y(10^9)\ \text{N/m}^2 \qquad 2$$

The differential area dA in the steel is the area of a strip whose width is $2d_f = 0.3$ m and thickness dy i.e., in steel $dA = 0.3dy$. The differential area dA in the wood is the area of a strip whose width is $t_W = 0.015$ m and thickness dy i.e., in wood $dA = 0.015dy$. The normal stress and dA across the cross-section can be written as shown below.

$$\sigma_{xx} = \begin{cases} 4y(10^9)\ \text{N/m}^2 & dA = 0.3dy & 0.20 < y \le 0.22 \\ 0.2y(10^9)\ \text{N/m}^2 & dA = 0.015dy & -0.20 < y < 0.20 \\ 4y(10^9)\ \text{N/m}^2 & dA = 0.3dy & -0.22 \le y < -0.20 \end{cases} \qquad 3$$

Noting the symmetry with respect to y, Eq. 6.1 can be written twice the sum of integrals over steel and wood in the upper half (y>0) of the cross-section.

$$M_z = -\int_A y\sigma_{xx}dA = -2\left[\int_0^{0.20} y(0.2y)(0.015)dy + \int_{0.20}^{0.22} y(4y)(0.3)dy\right](10^9) \quad \text{or}$$

$$M_z = -2\left[0.003\frac{y^3}{3}\bigg|_0^{0.20} + 1.2\frac{y^3}{3}\bigg|_{0.20}^{0.22}\right](10^9) = -2[8 + 1059.2](10^3) \quad \text{or} \quad \boxed{M_z = -2134\text{kN}-\text{m}}$$

Problem Set 6.2

6.14 A solid and a hollow square beam have the same cross-sectional area A. Show that the ratio of second area moment of inertia for the hollow beam (I_H) to that of the solid beam (I_S) is as given below:

$$\frac{I_H}{I_S} = \frac{\alpha^2 + 1}{\alpha^2 - 1}$$

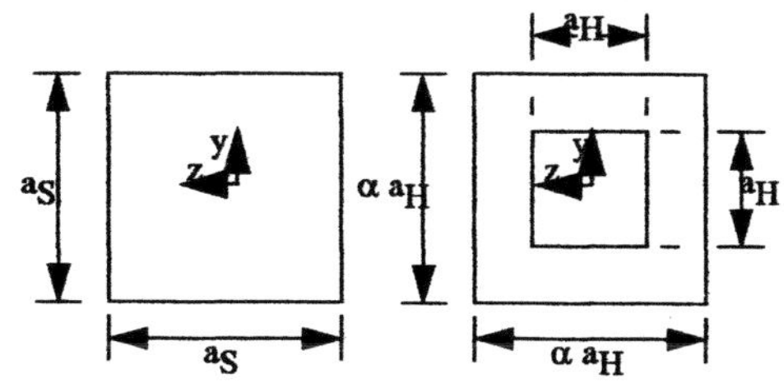

Solution

The lengths a_S and a_H in terms of area can be found as shown below:

$$A_S = a_S^2 = A \quad \text{or} \quad a_S = \sqrt{A}$$

$$A_H = (\alpha a_H)^2 - a_H^2 = (\alpha^2 - 1)a_H^2 = A \quad \text{or} \quad a_H = \sqrt{\frac{A}{(\alpha^2 - 1)}}$$

Let I_S and I_H represent the area moment of inertias about the z-axis for the solid cross-section and the hollow cross-section respectively. We can find I_S and I_H in terms of the area A as shown below using the formula for rectangles in Table A2.

$$I_S = \frac{1}{12}a_S a_S^3 = \frac{1}{12}A^2$$

$$I_H = \frac{1}{12}(\alpha a_H)(\alpha a_H)^3 - \frac{1}{12}a_H a_H^3 = \frac{(\alpha^4 - 1)}{12}a_H^4 = \frac{(\alpha^4 - 1)}{12}\left[\frac{A}{(\alpha^2 - 1)}\right]^2 = \frac{1}{12}\left(\frac{\alpha^2 + 1}{\alpha^2 - 1}\right)A^2$$

The ratio of the two area moment of inertias is: $\boxed{I_H/I_S = (\alpha^2 + 1)/(\alpha^2 - 1)}$

6.15 Figure 6.15(a) shows four separate wooden strips that bend independently about the neutral axis passing through the centroid of each strip. Figure 6.15(b) shows the four strips are glued together and bend as a unit about the centroid of the glued cross-section. (a) Show that $I_G = 16I_S$, where I_G is the area moment of inertias for the glued cross-section and I_S is the total area moment of inertia of the four separate beams. (b) Also show $\sigma_G = \sigma_S/4$, where σ_G and σ_S are the maximum bending normal stress at any cross-section for the glued and separate beams, respectively.

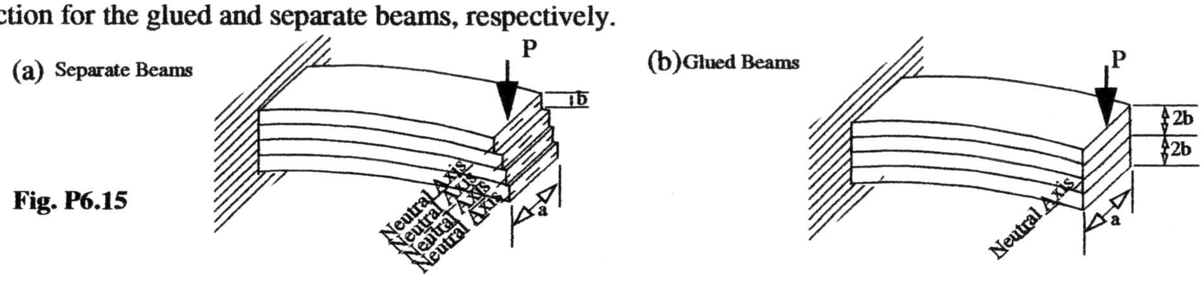

Fig. P6.15

Solution

The total area moment of inertia for separate beams is four times the area moment of inertia for the individual beam. Using the formula for rectangles in Table A2, I_s can be written as: $I_S = 4\left[\frac{1}{12}ab^3\right] = \frac{ab^3}{3}$

The glued beams act as a unit and I_G can be written as: $I_G = \frac{1}{12}a(4b)^3 = \frac{16ab^3}{3}$ or $\boxed{I_G = 16I_S}$

The maximum bending normal stress σ_S for separate beams will be at the top and bottom of each beam, i.e. $y_{max} = b/2$. From flexure stress formula, the magnitude of the maximum bending normal stress σ_S for separate beams can be written as:

$$\sigma_S = \frac{M_Z(b/2)}{I_S} = \frac{M_Z(b/2)}{(ab^3/3)} = \frac{3}{2}\left(\frac{M_Z}{ab^2}\right)$$

The maximum bending normal stress σ_G for glued beams will be at the top and bottom of the glued unit i.e., $y_{max} = 2b$. From flexure stress formula, the magnitude of the maximum bending normal stress σ_G for separate beams can be written as:

$$\sigma_G = \frac{M_Z(2b)}{I_G} = \frac{M_Z(2b)}{(16ab^3/3)} = \frac{3}{8}\left(\frac{M_Z}{ab^2}\right) \text{ or } \qquad \boxed{\sigma_G = \sigma_S/4}$$

6.16 The cross-section of the beam shown are constructed from thin sheet metal of thickness 't'.Assume that the thickness t<<a and obtain the second area moment of inertia about an axis passing through the centroid in terms of 'a' and 't'.

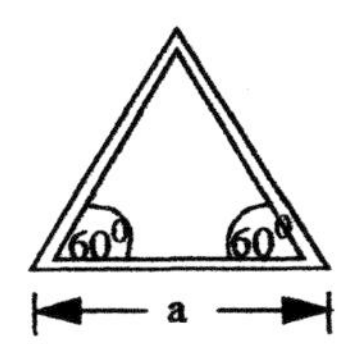

Fig. P6.16

Solution $I_T = ?$

The height of the triangle in terms of 'a' is $h = (a)\sin 60 = \left(\frac{\sqrt{3}}{2}\right)a$. The second area moment of inertia about an axis passing through the centroid of a triangle can be found using the formula in Table A2 as shown below.

$$I_T = \frac{1}{36}a\left[\left(\frac{\sqrt{3}}{2}\right)a\right]^3 - \frac{1}{36}(a-t)\left[\left(\frac{\sqrt{3}}{2}\right)(a-t)\right]^3 = \left(\frac{\sqrt{3}}{24}\right)[a^4-(a-t)^4] = \left(\frac{\sqrt{3}}{24}\right)[a^4-(a^2-2at+t^2)^2]$$

As t<<a, the t^2 term can be neglected and the remainder expanded as shown below.

$$I_T = \left(\frac{\sqrt{3}}{24}\right)[a^4-(a^2-2at)^2] \approx \left(\frac{\sqrt{3}}{24}\right)[a^4-(a^4-4a^3t+t^2)]$$

Once more neglecting the t^2 term we obtain: $I_T \approx \left(\frac{\sqrt{3}}{24}\right)[a^4 - a^4 + 4a^3t] \approx \frac{\sqrt{3}}{6}a^3t$ or $\boxed{I_T \approx \frac{a^3t}{2\sqrt{3}}}$

6.17 The cross-section of the beam shown are constructed from thin sheet metal of thickness 't'.Assume that the thickness t<<a and obtain the second area moment of inertia about an axis passing through the centroid in terms of 'a' and 't'.

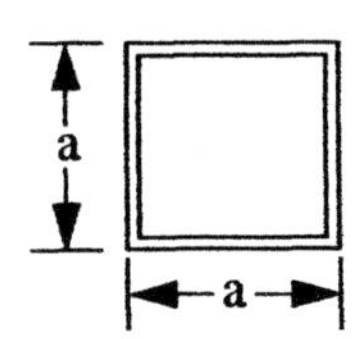

Fig. P6.17

Solution $I_S = ?$

The second area moment of inertia about an axis passing through the centroid of a rectangle can be found using the formula in Table A2 as shown below.

$$I_S = \frac{1}{12}aa^3 - \frac{1}{12}(a-t)(a-t)^3 = \frac{1}{12}[a^4 - (a-t)^4] = \frac{1}{12}[a^4 - (a^2 - 2at + t^2)^2]$$

As t<<a, the t^2 term can be neglected and the remainder expanded as shown below.

$$I_S = \frac{1}{12}[a^4 - (a^2 - 2at)^2] \approx \frac{1}{12}[a^4 - (a^4 - 4a^3t + 4a^2t^2)]$$

Once more neglecting the t^2 term we obtain: $I_S \approx \frac{1}{12}[a^4 - a^4 + 4a^3t] \approx \frac{1}{3}a^3t$ or $\boxed{I_S \approx \frac{a^3t}{3}}$

6.18 The cross-section of the beam shown are constructed from thin sheet metal of thickness 't'. Assume that the thickness t<<a and obtain the second area moment of inertia about an axis passing through the centroid in terms of 'a' and 't'.

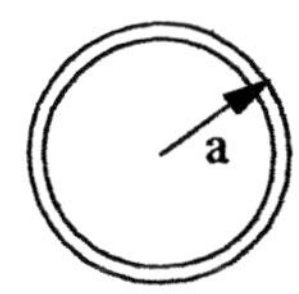

Fig. P6.18

Solution $I_C = ?$

The second area moment of inertia about an axis passing through the centroid of a circle can be found using the formula in Table A2 as shown below.

$$I_C = \frac{\pi}{4}(a)^4 - \frac{\pi}{4}(a-t)^4 = \frac{\pi}{4}[a^4 - (a^2 - 2at + t^2)^2]$$

As t<<a, the t^2 term can be neglected and the remainder expanded as shown below.

$$I_C = \frac{\pi}{4}[a^4 - (a^2 - 2at)^2] \approx \frac{\pi}{4}[a^4 - (a^4 - 4a^3t + t^2)]$$

Once more neglecting the t^2 term we obtain: $I_C \approx \frac{\pi}{4}[a^4 - a^4 + 4a^3t] \approx a^3t$ or $\boxed{I_C \approx \frac{\pi a^3 t}{3}}$

6.19 The same amount of material is used for constructing each cross-section shown in Fig. P6.16, Fig. P6.17, and Fig. P6.18. Let the maximum bending normal stress be σ_T, σ_S, and σ_C for the triangular, square, and circular cross-section. For the same moment carrying capability determine the proportional ratio of maximum bending normal stress, i.e., σ_T: σ_S:σ_C.

Solution

The material area for a thin-walled cross-section can be approximated as the product of the perimeter and the thickness. Let a_T, a_S, and a_C represent the dimension shown in Fig. P6.16, Fig. P6.17, and Fig. P6.18, respectively. a_T, a_S, and a_C can be found in terms of the material area as shown below.

$$(3a_T)t = A \quad \text{or} \quad a_T = \frac{A}{3t} \qquad (1)$$

$$(4a_S)t = A \quad \text{or} \quad a_S = \frac{A}{4t} \qquad (2)$$

$$(2\pi a_C)t = A \quad \text{or} \quad a_C = \frac{A}{2\pi t} \qquad (3)$$

The maximum bending normal stress exists at point furthest from the centroid. The maximum values of y for the three cases are: $y_T = 2a_T/3$, $y_S = a_S/2$, and $y_C = a_C$. The magnitude maximum bending normal stress for the three cases can be found using the flexure stress formula, the results of problems 6.16, 6.17, and 6.18, and the above three equations as shown below.

$$\sigma_T = \frac{M_z y_T}{I_T} = \frac{M_z(2a_T/3)}{(a_T^3 t/2\sqrt{3})} = \frac{4\sqrt{3}}{3}\frac{M_z}{(a_T^2 t)} = \frac{4\sqrt{3}}{3}\frac{M_z}{[(A/3t)^2 t]} = 12\sqrt{3}\frac{M_z t}{A^2} = 20.78\left(\frac{M_z t}{A^2}\right) \quad 4$$

$$\sigma_S = \frac{M_z y_S}{I_S} = \frac{M_z(a_S/2)}{(a_S^3 t/3)} = \frac{3}{2}\frac{M_z}{(a_S^2 t)} = \frac{3}{2}\frac{M_z}{[(A/4t)^2 t]} = 24\frac{M_z t}{A^2} \quad 5$$

$$\sigma_C = \frac{M_z y_C}{I_C} = \frac{M_z(a_C)}{(\pi a_C^2 t/3)} = \frac{3}{\pi}\frac{M_z}{(a_C^2 t)} = \frac{3}{\pi}\frac{M_z}{[(A/2\pi t)^2 t]} = 12\pi\frac{M_z t}{A^2} = 37.7\left(\frac{M_z t}{A^2}\right) \quad 6$$

The ratio of the maximum bending normal stresses is: $\sigma_T : \sigma_S : \sigma_C = 20.78 : 24 : 37.7$

6.20 The cross-section of a beam with a coordinate system that has an origin at the centroid C of the cross-section is shown. The normal strain at point A due to bending about the z-axis, and the Modulus of Elasticity are as given.

(a) Plot the stress distribution across the cross-section.
(b) Determine the maximum bending normal stress in the cross-section.
(c) Determine the equivalent internal bending moment M_z by using Equation 6.1.
(d) Determine the equivalent internal bending moment M_z by using Equation 6.12.

$\varepsilon_{xx} = 200\ \mu$

$E = 8000$ ksi

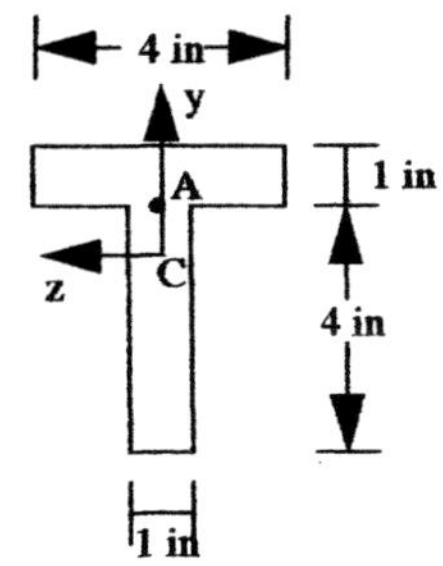

Fig. P6.20

Solution σ_{xx} vs. y $\sigma_{max} = ?$ $M_z = ?$ using Eq. 6.1 and 6.12

(a,b) The location of the centroid from the bottom of the cross-section shown in figure (a) can be found as shown below.

$$\eta_c = \frac{\sum_i \eta_i A_i}{\sum_i A_i} = \frac{(4)(1)(2) + (4)(1)(4.5)}{(4)(1) + (4)(1)} = 3.25 \text{ in} \quad 1$$

The bending normal stress at point A is: $\sigma_A = E\varepsilon_A = (8000)(200)(10^{-6}) = 1.6$ ksi(T). The bending normal stress is zero at the centroid C and is a linear function of y. Figure (b) below shows the stress distribution across the cross-section as a function of y.

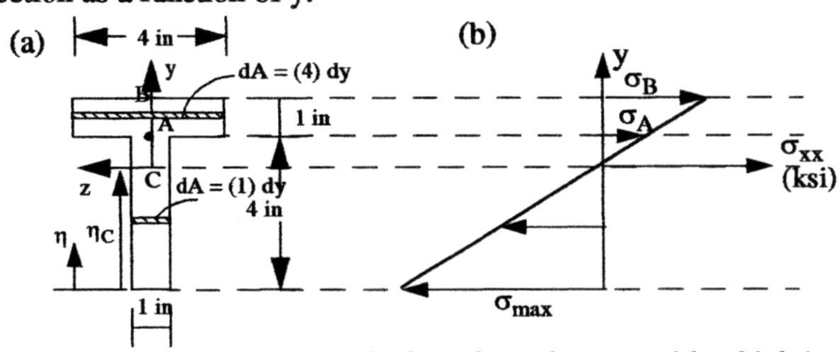

The maximum bending normal stress is at a point furthest from the centroid, which in this case is at the bottom of the beam as shown in figure (b). Using similar triangle, we obtain

$\frac{\sigma_B}{(5-3.25)} = \frac{\sigma_{max}}{3.25} = \frac{\sigma_A}{(4-3.25)}$ or $\sigma_B = 3.733$ ksi and $\sigma_{max} = 6.933$ ksi. Noting that point A is in tension, thus point B will also be in tension and the bottom of the beam will be in compression.

$$\boxed{\sigma_{max} = 6.93\ \text{ksi(C)}}$$

(c) The bending normal stress as a function of y can be written as: $\sigma_{xx} = \frac{\sigma_A}{(4-3.25)}y = 2.133y\ \text{ksi}$. The differential area of cross-section changes dA = (1) dy in the web to dA = (4) dy in the flange. Eq. 6.1 can be written as a sum of integrals over web and flange and evaluated as shown below.

$$M_z = -\int_A y\sigma_{xx}dA = -\left[\int_{-3.25}^{0.75} y(2.133y)(1)dy + \int_{0.75}^{1.75} y(2.133y)(4)dy\right] \text{ or}$$

$$M_z = -\left[2.133\frac{y^3}{3}\Bigg|_{-3.25}^{0.75} + 8.533\frac{y^3}{3}\Bigg|_{0.75}^{1.75}\right] = -[24.707 + 14.044] \text{ or} \qquad \boxed{M_z = -38.76\text{in}-\text{kips}}$$

(d) The second area moment of inertia about the z-axis can be found as shown below.

$$I_{zz} = \frac{1}{12}(1)(4)^3 + (1)(4)(3.25-2)^2 + \frac{1}{12}(4)(1)^3 + (4)(1)(4.5-3.25)^2 = 18.167\text{in}^4$$ 2

Substituting y_A= 0.75inch, σ_A= 1.6 ksi and I_{zz} = 18.167 in^4 in Eq. 6.12, we can find the internal bending moment as shown below.

$$\sigma_A = -\frac{M_z y_A}{I_{zz}} = -\frac{M_z(0.75)}{18.167} = 1.6 \text{ or} \qquad \boxed{M_z = -38.76\ \text{in}-\text{kips}}$$

6.21 The cross-section of a beam with a coordinate system that has an origin at the centroid C of the cross-section is shown. The normal strain at point A due to bending about the z-axis, and the Modulus of Elasticity are as given.

(a) Plot the stress distribution across the cross-section.
(b) Determine the maximum bending normal stress in the cross-section.
(c) Determine the equivalent internal bending moment M_z by using Equation 6.1.
(d) Determine the equivalent internal bending moment M_z by using Equation 6.12.

$\varepsilon_{xx} = -200\ \mu$

$E = 200\ \text{GPa}$

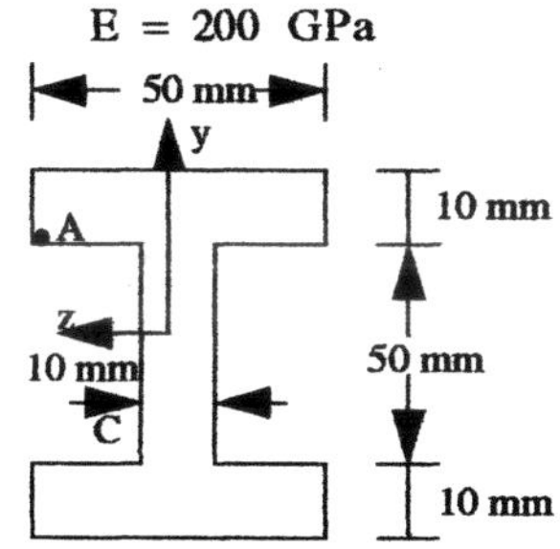

Fig. P6.21

Solution σ_{xx} vs. y σ_{max} = ? M_z = ? using Eq. 6.1 and 6.12

(a,b) From symmetry, the centroid is at 35 mm from the bottom.
The bending normal stress at point A is:

$\sigma_A = E\varepsilon_A = (200)(10^9)(-200)(10^{-6}) = -40(10^6)\ \text{N/m}^2 = 40\text{MPa(C)}$. The bending normal stress is zero at the centroid C and is a linear function of y. Figure (b) below shows the stress distribution across the cross-section as a function of y.

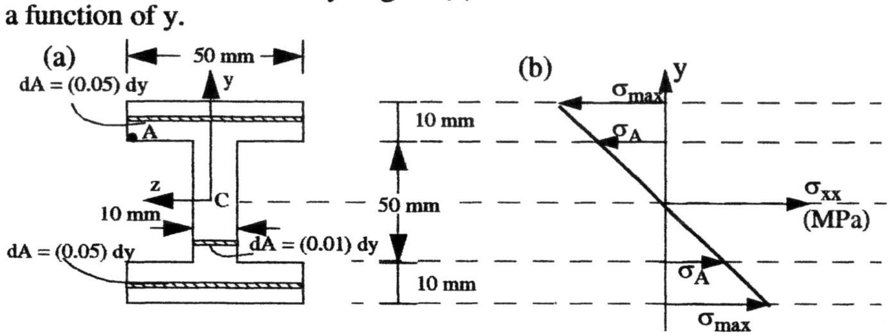

The maximum bending normal stress is at a point furthest from the centroid, which in this case is at the bottom of the beam as shown in figure (b). Using similar triangle, we obtain $\frac{\sigma_{max}}{0.035} = \frac{\sigma_A}{0.025}$ or and $\sigma_{max} = 56\ \text{MPa}$. The maximum bending normal stress will be compressive on top and tensile at the bottom.

$$\boxed{\sigma_{max} = 56\text{MPa (C)or(T)}}$$

(c) The bending normal stress as a function of y can be written as: $\sigma_{xx} = \frac{\sigma_A}{(0.025)}y = -1600y\ \text{MPa}$. The differential area of cross-section changes dA = (0.01) dy in the web to dA = (0.05) dy in the flange. Noting the symmetry with respect to y, Eq. 6.1 can be written as twice the sum of integrals over web and flange in the upper half (y>0) of the cross-section and evaluated as shown below.

$$M_z = -\int_A y\sigma_{xx}dA = -2(10^6)\left[\int_0^{0.025} y(-1600y)(0.01)dy + \int_{0.025}^{0.035} y(-1600y)(0.05)dy\right] \text{ or}$$

$$M_z = 2(10^6)\left[16\frac{y^3}{3}\bigg|_0^{0.025} + 80\frac{y^3}{3}\bigg|_{0.025}^{0.035}\right] = 2[83.333 + 726.667]\ \text{N-m or} \qquad \boxed{M_z = 1620\text{N-m}}$$

(d) The second area moment of inertia about the z-axis can be found as shown below.

$$I_{zz} = \frac{1}{12}(0.01)(0.05)^3 + 2\left[\frac{1}{12}(0.05)(0.01)^3 + (0.05)(0.01)(0.005 - 0.035)^2\right] = 1.0125(10^{-6})\text{m}^4$$ 1

Substituting y_A= 0.025 m, σ_A= -40.0(10^6) N/m^2 and I_{zz} = 1.0125 m^4 in Eq. 6.12, we can find the internal bending moment as shown below.

$$\sigma_A = -\frac{M_z y_A}{I_{zz}} = -\frac{M_z(0.025)}{1.0125(10^{-6})} = -40.0(10^6) \text{ or} \qquad \boxed{M_z = 1620\ \text{N-m}}$$

6.22 The cross-section of a beam with a coordinate system that has an origin at the centroid C of the cross-section is shown. The normal strain at point A due to bending about the z-axis, and the Modulus of Elasticity are as given.

(a) Plot the stress distribution across the cross-section.
(b) Determine the maximum bending normal stress in the cross-section.
(c) Determine the equivalent internal bending moment M_z by using Equation 6.1.
(d) Determine the equivalent internal bending moment M_z by using Equation 6.12.

$\varepsilon_{xx} = 300\ \mu$

$E = 30,000\ \text{ksi}$

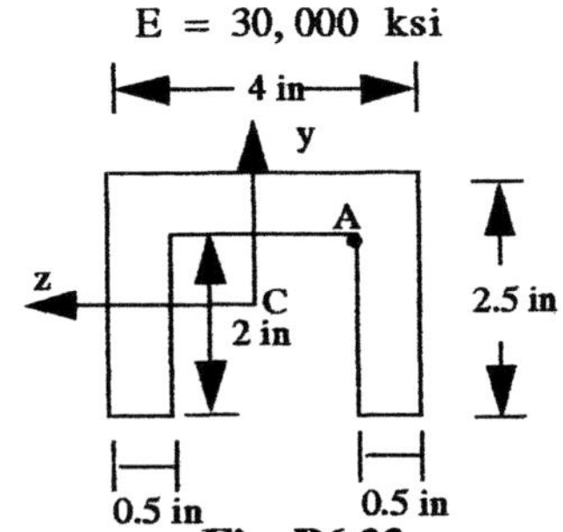

Fig. P6.22

Solution σ_{xx} vs. y $\quad \sigma_{max}$ = ? $\quad M_z$ = ? using Eq. 6.1 and 6.12

(a,b) The location of the centroid from the bottom of the cross-section shown in figure (a) can be found as shown below.

$$\eta_c = \frac{\sum_i \eta_i A_i}{\sum_i A_i} = \frac{2[(2.0)(0.5)(1.0)] + (4)(0.5)(2.25)}{2[(2.0)(0.5)] + (4)(0.5)} = 1.625\ \text{in}$$ 1

The bending normal stress at point A is: $\sigma_A = E\varepsilon_A = (30000)(300)(10^{-6}) = 9.0\ \text{ksi(T)}$. The bending normal stress is zero at the centroid C and is a linear function of y. Figure (b) below shows the stress distribution across the cross-section as a function of y.

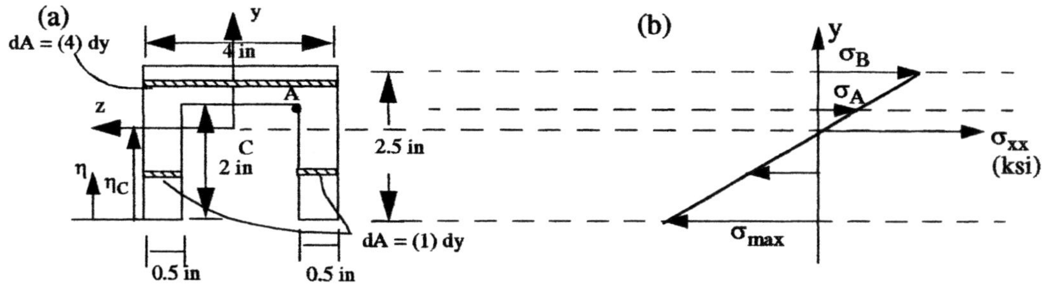

The maximum bending normal stress is at a point furthest from the centroid, which in this case is at the bottom of the beam as shown in figure (b). Using similar triangle, we obtain

$\frac{\sigma_B}{(2.25-1.625)} = \frac{\sigma_{max}}{1.625} = \frac{\sigma_A}{(2-1.625)}$ or $\sigma_B = 5.4$ ksi and $\sigma_{max} = 39.0$ ksi. Noting that point A is in tension, thus point B will also be in tension and the bottom of the beam will be in compression.

$\boxed{\sigma_{max} = 39.0 \text{ ksi(C)}}$

(c) The bending normal stress as a function of y can be written as: $\sigma_{xx} = \frac{\sigma_A}{(2-1.625)}y = 24y$ ksi. The differential area of cross-section changes dA = (1) dy in the web to dA = (4) dy in the flange. Eq. 6.1 can be written as a sum of integrals over web and flange and evaluated as shown below.

$$M_z = -\int_A y\sigma_{xx}dA = -\left[\int_{-1.625}^{0.375} y(24y)(1)dy + \int_{0.375}^{0.875} y(24y)(4)dy\right] \text{ or}$$

$$M_z = -\left[24\frac{y^3}{3}\Bigg|_{-1.625}^{0.375} + 96\frac{y^3}{3}\Bigg|_{0.375}^{0.875}\right] = -[34.75 + 19.75] \text{ or} \qquad \boxed{M_z = -54.5 \text{ in-kips}}$$

(d) The second area moment of inertia about the z-axis can be found as shown below.

$$I_{zz} = 2\left[\frac{1}{12}(0.5)(2.0)^3 + (0.5)(2)(0.625)^2\right] + \frac{1}{12}(4)(0.5)^3 + (4)(0.5)(0.625)^2 = 2.2708\text{in}^4$$ 2

Substituting y_A= 0.375inch, σ_A= 9.0 ksi and I_{zz} = 2.2708 in^4 in Eq. 6.12, we can find the internal bending moment as shown below.

$$\sigma_A = -\frac{M_z y_A}{I_{zz}} = -\frac{M_z(0.375)}{2.2708} = 9.0 \text{ or} \qquad \boxed{M_z = -54.5 \text{ in-kips}}$$

6.23 The cross-section of a beam with a coordinate system that has an origin at the centroid C of the cross-section is shown. The internal moment at a beam cross-section is as given. Determine the bending normal stresses at points A, B, and D

$M_z = 20$ in-kips

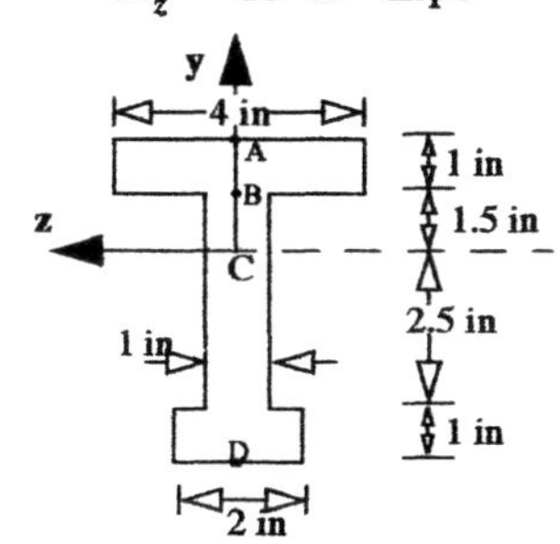

Fig. P6.23

Solution σ_A= ? σ_B= ? σ_D= ?

The second area moment of inertia about the z-axis can be found as shown below.

$$I_{zz} = \frac{1}{12}(2)(1)^3 + (2)(1)(3)^2 + \frac{1}{12}(1)(4)^3 + (1)(4)(0.5)^2 + \frac{1}{12}(4)(1)^3 + (4)(1)(2)^2 = 40.833\text{in}^4 \qquad 3$$

Substituting M_z = 20 in-kips, and I_{zz} = 40.833 in^4 in Eq. 6.12, we find bending normal stress as

$$\sigma_{xx} = -\frac{M_z y}{I_{zz}} = -\frac{(20)y}{40.833} = -0.4898y \qquad 1$$

Substituting y_A=2.5inch into Eq. 1, we obtain: $\sigma_A = -0.4898(2.5) = -1.224\text{ksi}$ or $\boxed{\sigma_A = 1224\text{psi(C)}}$

Substituting y_B=1.5inch into Eq. 1, we obtain: $\sigma_B = -0.4898(1.5) = -0.735\text{ksi}$ or $\boxed{\sigma_B = 735\text{psi(C)}}$

Substituting y_D=-3.5inch into Eq. 1, we obtain: $\sigma_A = -0.4898(-3.5) = 1.714\text{ksi}$ or $\boxed{\sigma_D = 1714\text{ksi(T)}}$

6.24 The cross-section of a beam with a coordinate system that has an origin at the centroid C of the cross-section is shown. The internal moment at a beam cross-section is given in each problem. Determine the bending normal stresses at points A, B, and D

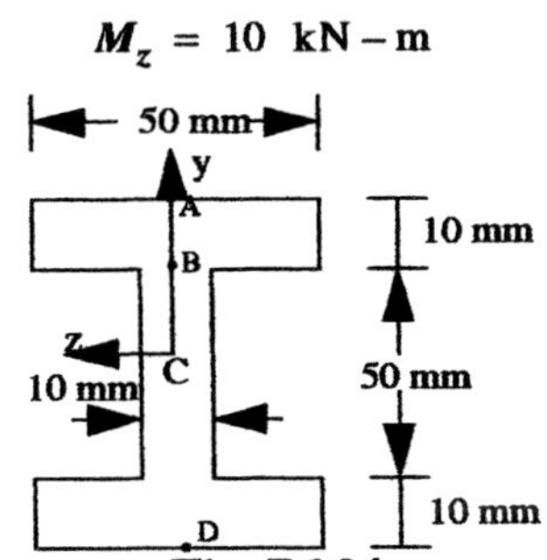

Fig. P6.24

Solution σ_A= ? σ_B= ? σ_D= ?

From symmetry, the centroid is at 35 mm from the bottom. From Eq.(1) in problem 6.21, the second area moment of inertia about the z-axis is: $I_{zz} = 1.0125(10^{-6})\text{m}^4$.

Substituting M_z = 10 (10^3) N-m, and I_{zz} = 1.0125(10^{-6}) m^4 in Eq. 6.12, we find bending normal stress as

$$\sigma_{xx} = -\frac{M_z y}{I_{zz}} = -\frac{(10)(10^3)y}{1.0125(10^{-6})} = -9.8765(10^9)y \qquad 1$$

Substituting y_A= 0.035 m into Eq. 1, we obtain: $\sigma_A = -9.8765(10^9)(0.035) = -345.7(10^6)\ \text{N/m}^2$ or

$\boxed{\sigma_A = 346\ \text{MPa(C)}}$

Substituting y_B= 0.025 m into Eq. 1, we obtain: $\sigma_B = -9.8765(10^9)(0.025) = -246.9(10^6)\ \text{N/m}^2$ or

$\boxed{\sigma_B = 247\ \text{MPa(C)}}$

Substituting y_D=-0.035 m into Eq. 1, we obtain $\sigma_D = -9.8765(10^9)(-0.035) = 345.7(10^6)\ \text{N/m}^2$ or

$\boxed{\sigma_D = 346\ \text{MPa(T)}}$

6.25 The cross-section of a beam with a coordinate system that has an origin at the centroid C of the cross-section is shown. The internal moment at a beam cross-section is given in each problem. Determine the bending normal stresses at points A, B, and D

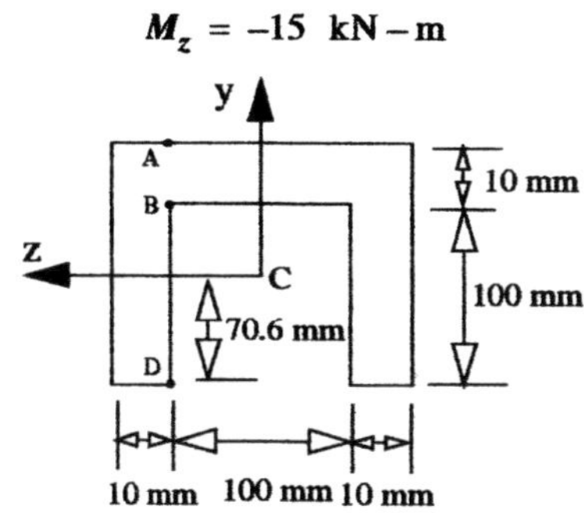

Fig. P6.25

Solution $\sigma_A = ?$ $\sigma_B = ?$ $\sigma_D = ?$

The second area moment of inertia about the z-axis can be found as shown below.

$$I_{zz} = 2\left[\frac{1}{12}(0.01)(0.10)^3 + (0.01)(0.1)(0.0206)^2\right] + \frac{1}{12}(0.12)(0.01)^3 + (0.12)(0.01)(0.0344)^2$$

or $I_{zz} = 3.9463(10^{-6})\text{m}^4$.

Substituting $M_z = -15(10^3)$ N-m, and $I_{zz} = 3.9463(10^{-6})$ m^4 in Eq. 6.12, we find bending normal stress as

$$\sigma_{xx} = -\frac{M_z y}{I_{zz}} = -\frac{(-15)(10^3)y}{3.9463(10^{-6})} = 3.801(10^9)y \quad (1)$$

Substituting y_A= (0.11-0.0706)= 0.0394 m into Eq. 1, we obtain:

$\sigma_A = 3.801(10^9)(0.0394) = 149.8(10^6)$ N/m^2 or $\boxed{\sigma_A = 150 \text{ MPa(T)}}$

Substituting y_B= (0.100-0.0706)= 0.0294 m into Eq. 1, we obtain:

$\sigma_B = 3.801(10^9)(0.0294) = 111.8(10^6)$ N/m^2 or $\boxed{\sigma_B = 112 \text{ MPa(T)}}$

Substituting y_D=-0.0706 m into Eq. 1, we obtain:

$\sigma_D = 3.801(10^9)(-0.0706) = 268.3(10^6)$ N/m^2 or $\boxed{\sigma_D = 268 \text{ MPa(C)}}$

6.26 A beam and loading in three different coordinate system is shown. Determine the internal shear force and bending moment at the section containing point A for the three cases shown using the sign convention described in Section 6.2.5.

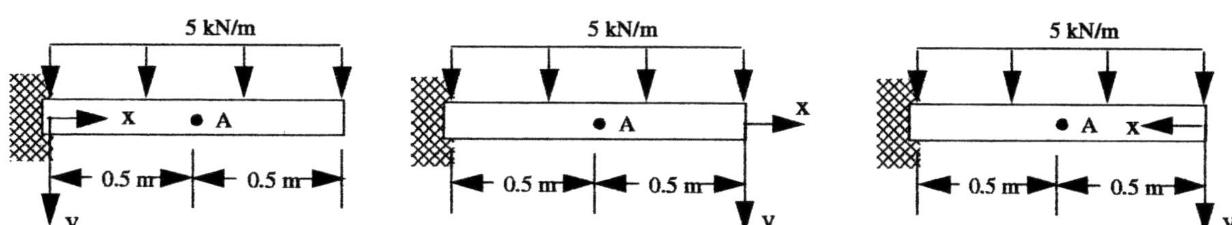

Fig. P6.26

Solution $V_y = ?$ $M_z = ?$ for the three cases

By making an imaginary cut through he section containing point A and taking the right part we obtain the free body diagrams shown on the top for each case. The internal shear force and internal bending moment are drawn as per the sign convention. The distributed loads can be replace by an equivalent load at the centroid of the distribution as shown on the bottom of each case below.

Case I: By equilibrium of forces in the y-direction we obtain: $\boxed{V_y = 0.25 \text{ kN}}$

By equilibrium of moment about point A we obtain: $\boxed{M_Z = 0.0625 \text{ kN}-\text{m}}$

Case 1 Case 2 Case 3

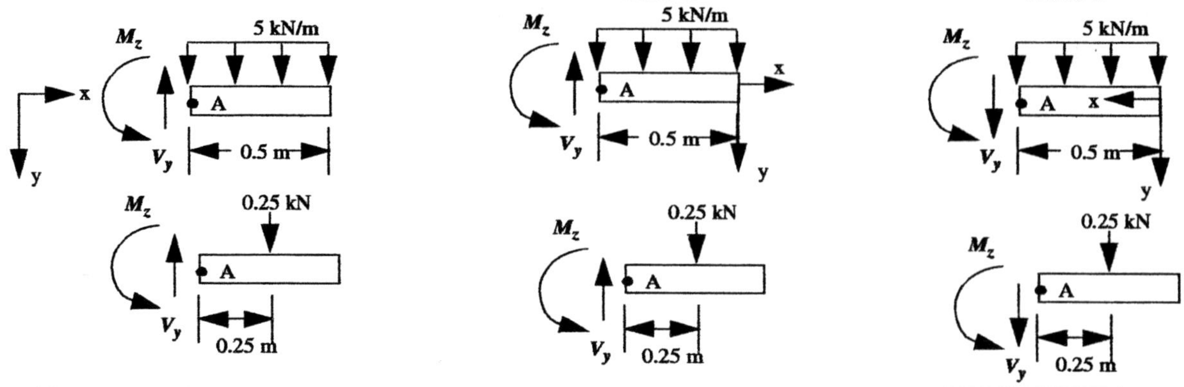

Case II: By equilibrium of forces in the y-direction we obtain: $\boxed{V_y = 0.25 \text{ kN}}$

By equilibrium of moment about point A we obtain: $\boxed{M_Z = 0.0625 \text{ kN}-\text{m}}$

Case III: By equilibrium of forces in the y-direction we obtain: $\boxed{V_y = -0.25 \text{ kN}}$

By equilibrium of moment about point A we obtain: $\boxed{M_Z = 0.0625 \text{ kN}-\text{m}}$

6.27 A beam and loading in three different coordinate system is shown. Determine the internal shear force and bending moment at the section containing point A for the three cases shown using the sign convention described in Section 6.2.5.

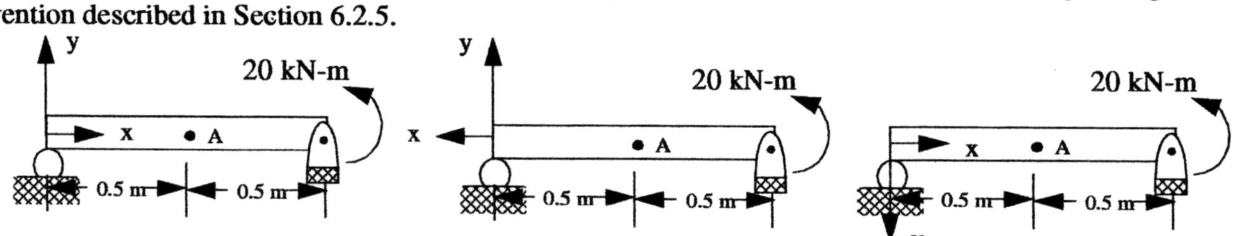

Fig. P6.27 Case 1 Case 2 Case 3

Solution $V_y = ?$ $M_z = ?$ for the three cases

The free body diagram of the entire beam is shown below.

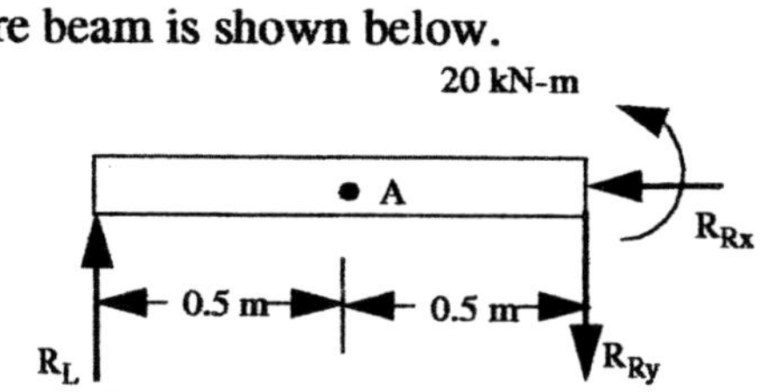

By moment equilibrium at the right end we obtain: $R_L(1) - 20 = 0$ or $R_L = 20 \text{ kN}$

By making an imaginary cut through the section containing point A and taking the left part we obtain the free body diagrams shown below. The internal shear force and internal bending moment are drawn as per the sign convention.

Case I: By equilibrium of forces in the y-direction we obtain: $\boxed{V_y = -20 \text{ kN}}$

By equilibrium of moment about point A we obtain: $\boxed{M_Z = 10 \text{ kN}-\text{m}}$

Case II: By equilibrium of forces in the y-direction we obtain: $\boxed{V_y = 20 \text{ kN}}$

By equilibrium of moment about point A we obtain: $\boxed{M_Z = 10 \text{ kN}-\text{m}}$

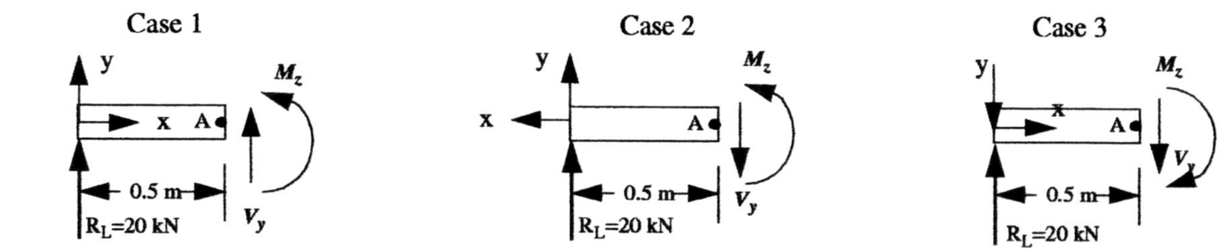

Case III: By equilibrium of forces in the y-direction we obtain: $\boxed{V_y = 20 \text{ kN}}$

By equilibrium of moment about point A we obtain: $\boxed{M_Z = -10 \text{ kN}-\text{m}}$

6.28 A beam and loading in three different coordinate system is shown. Determine the internal shear force and bending moment at the section containing point A for the three cases shown using the sign convention described in Section 6.2.5.

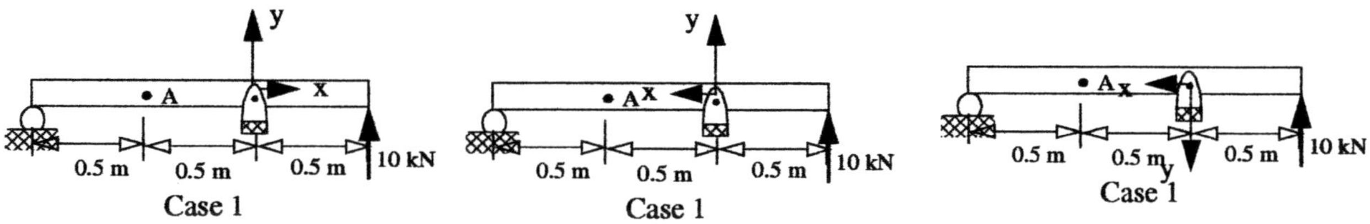

Fig. P6.28

Solution $V_y = ?$ $M_z = ?$ for the three cases

The free body diagram of the entire beam is shown below.

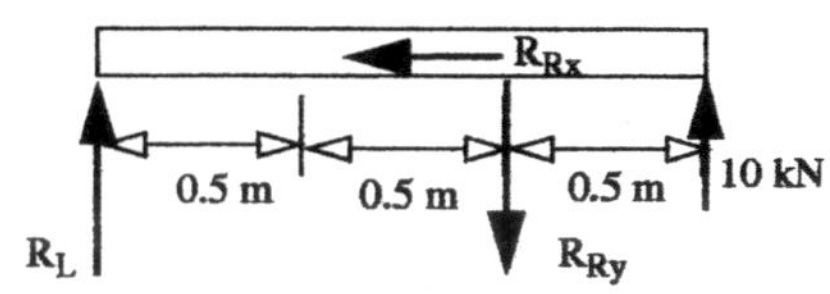

By moment equilibrium at the right support we obtain: $R_L(1)-(10)(0.5) = 0$ or $R_L = 5$ kN

By making an imaginary cut through the section containing point A and taking the left part we obtain the free body diagrams shown below. The internal shear force and internal bending moment are drawn as per the sign convention.

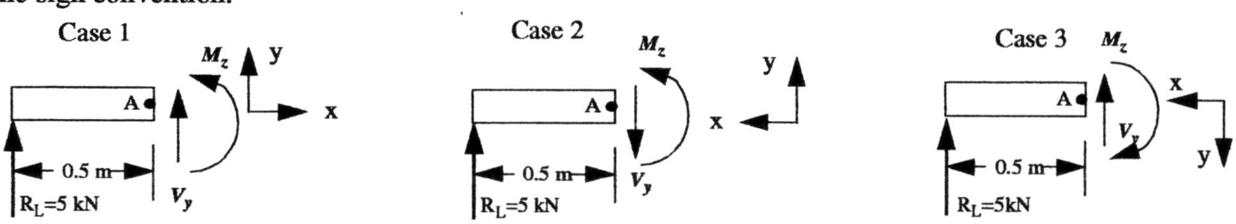

Case I: By equilibrium of forces in the y-direction we obtain: $\boxed{V_y = -5 \text{ kN}}$

By equilibrium of moment about point A we obtain: $\boxed{M_Z = 2.5 \text{ kN}-\text{m}}$

Case II: By equilibrium of forces in the y-direction we obtain: $\boxed{V_y = 5 \text{ kN}}$

By equilibrium of moment about point A we obtain: $\boxed{M_Z = 2.5 \text{ kN}-\text{m}}$

Case III: By equilibrium of forces in the y-direction we obtain: $\boxed{V_y = -5 \text{ kN}}$

By equilibrium of moment about point A we obtain: $\boxed{M_Z = -2.5 \text{ kN}-\text{m}}$

6.29 For the beam and loading shown, draw an approximate deformed shape of the beam. By inspection determine whether the bending normal stress is tensile or compressive at points A and B.

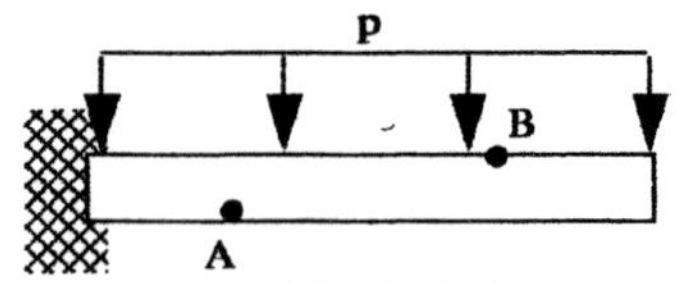

Fig. P6.29

Solution σ_A= (T) or (C) ? σ_B= (T) or (C) ?

An approximate deformed shape of the beam is shown below

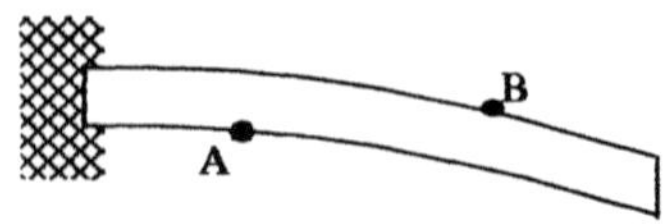

By inspection, the bending normal stresses are: σ_A is (C) and σ_B is (T)

6.30 For the beam and loading shown, draw an approximate deformed shape of the beam. By inspection determine whether the bending normal stress is tensile or compressive at points A and B.

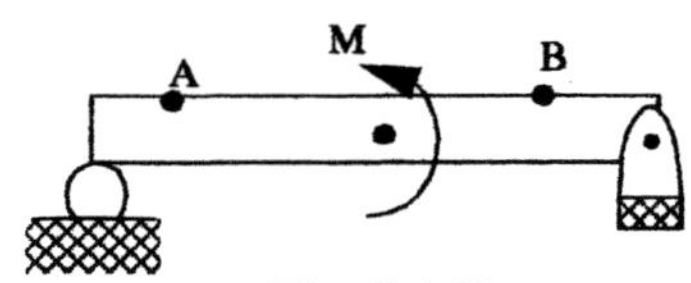

Fig. P6.30

Solution σ_A= (T) or (C) ? σ_B= (T) or (C) ?

An approximate deformed shape of the beam is shown below

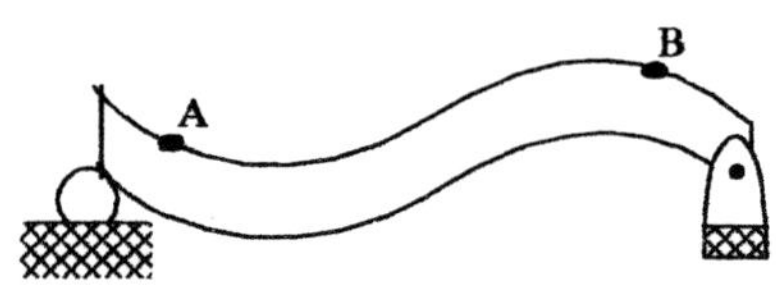

By inspection, the bending normal stresses are: σ_A is (C) and σ_B is (T)

6.31 For the beam and loading shown, draw an approximate deformed shape of the beam. By inspection determine whether the bending normal stress is tensile or compressive at points A and B.

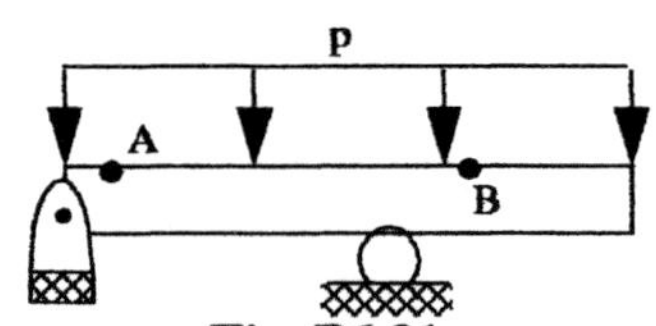

Fig. P6.31

Solution σ_A= (T) or (C) ? σ_B= (T) or (C) ?

There are two possibilities, depending upon the length of the overhang. Approximate deformed shape of the beam for both possibilities are shown below.

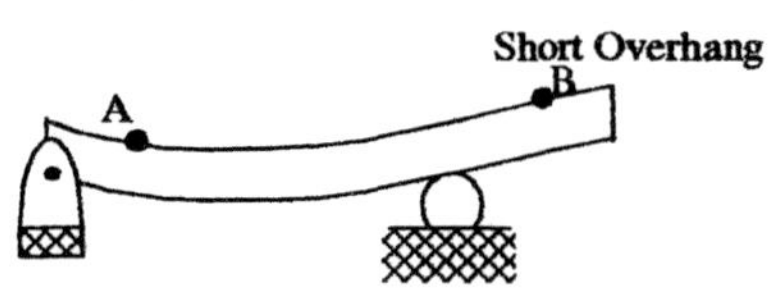

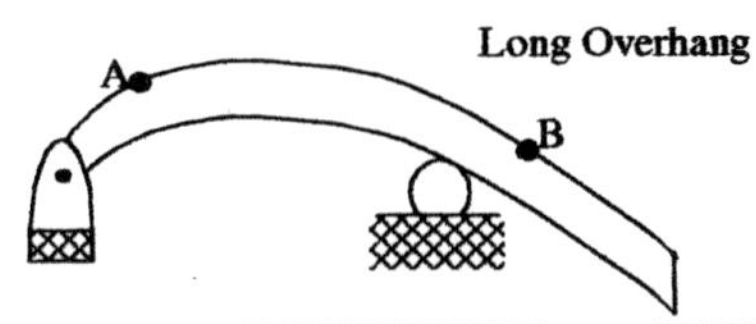

By inspection, the bending normal stresses for *short overhang* are: σ_A is (C) and σ_B is (T)

By inspection, the bending normal stresses for *long overhang* are: σ_A is (T) and σ_B is (T)

6.32 For the beam and loading shown, draw an approximate deformed shape of the beam. By inspection determine whether the bending normal stress is tensile or compressive at points A and B.

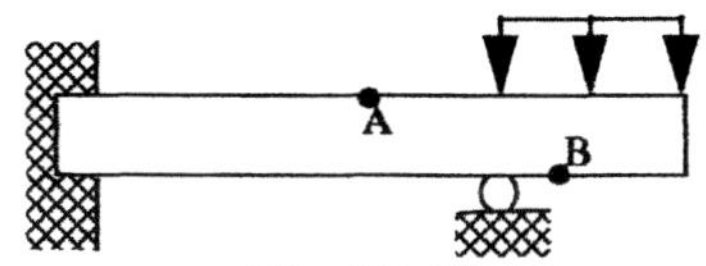

Fig. P6.32

Solution σ_A= (T) or (C) ? σ_B= (T) or (C) ?

An approximate deformed shape of the beam is shown below

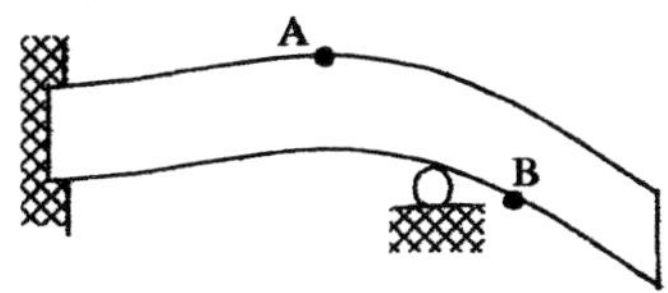

By inspection, the bending normal stresses are: σ_A is (T) and σ_B is (C)

6.33 For the beam and loading shown, draw an approximate deformed shape of the beam. By inspection determine whether the bending normal stress is tensile or compressive at points A and B.

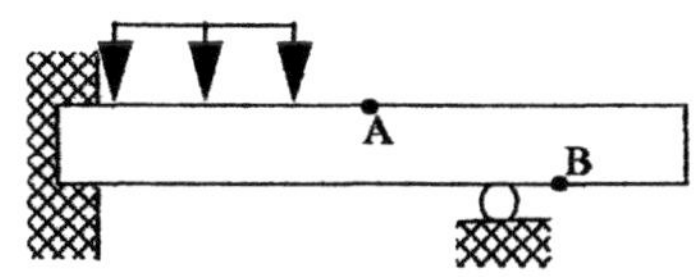

Fig. P6.33

Solution σ_A= (T) or (C) ? σ_B= (T) or (C) ?

An approximate deformed shape of the beam is shown below

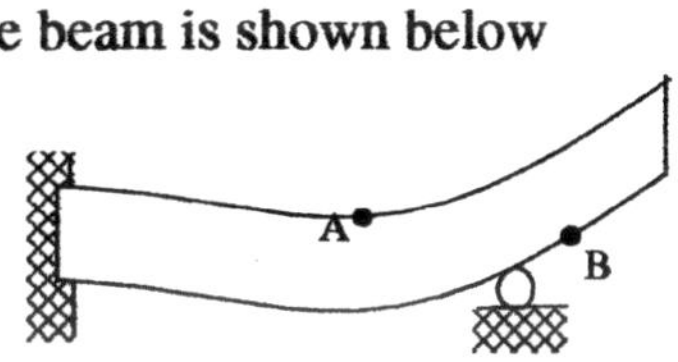

By inspection, the bending normal stresses are: σ_A is (C) and $\sigma_B = 0$

6.34 For the beam and loading shown, draw an approximate deformed shape of the beam. By inspection determine whether the bending normal stress is tensile or compressive at points A and B.

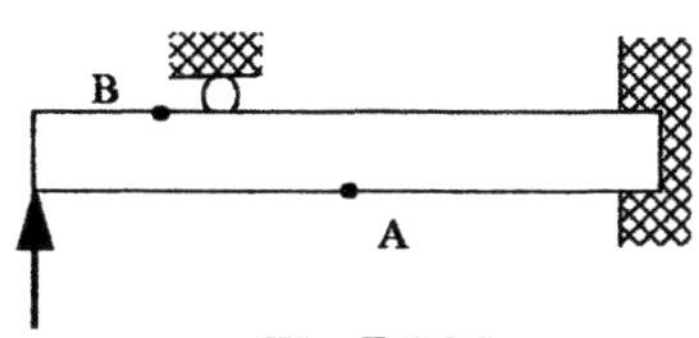

Fig. P6.34

Solution σ_A= (T) or (C) ? σ_B= (T) or (C) ?

An approximate deformed shape of the beam is shown below

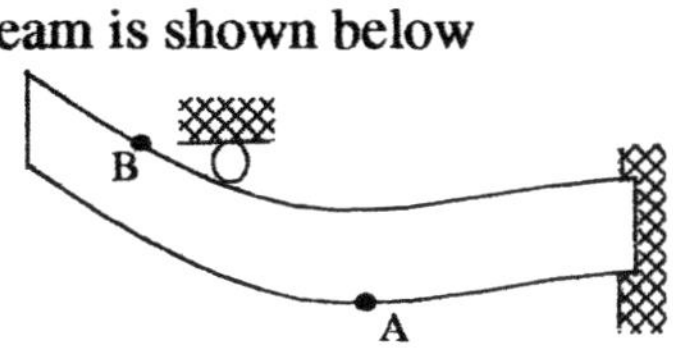

By inspection, the bending normal stresses are: σ_A is (T) and σ_B is (C)

6.35 A W 150 x 24 steel beam is simply supported over a length of 4 m and supports a distributed load of 2 kN/m. At the mid-section of the beam, determine (a) the bending normal stress at a point 40 mm above the bottom surface. (b) the maximum bending normal stress.

Solution W 150 x 24 L = 4 m p = 2 kN/m σ_{40} = ? and σ_{max} = ? @ x=L/2

A simply supported beam with uniform distributed load will have a reaction force at each end whose value is half of the total force acting on the beam, i.e., R = (pL)/2 = 4kN
Fig. (a) shows the free body diagram after making an imaginary cut at the mid-section. Internal shear force and bending moment are drawn as per our sign convention. Fig. (b) is the statically equivalent free body diagram.

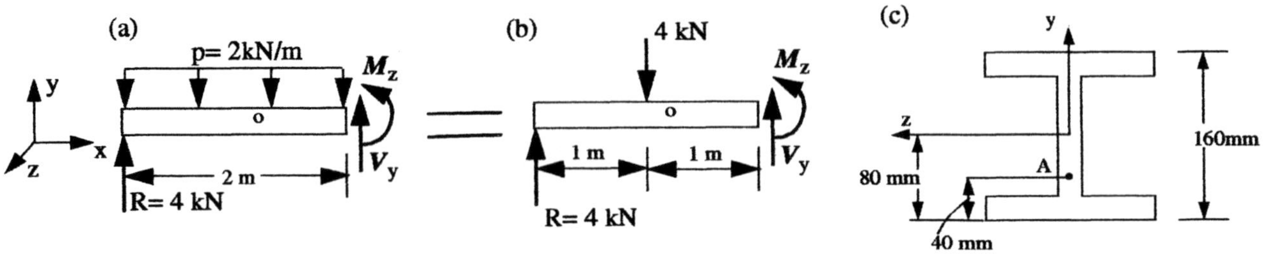

By equilibrium of moment about point O in Fig. (b) we obtain:

$$M_z - (4)(2) + (4)(1) = 0 \quad \text{or} \quad M_z = 4 \text{ kN-m} \qquad 1$$

From Appendix Table E.2, we obtain the geometry and dimension of W150 x 24 beam cross-section shown in Fig. (c) and $I_{zz} = 13.36(10^6)\ \text{mm}^4$. Substituting y= -(0.08 - 0.04)= - 0.04 m, Eq.(1) and $I_{zz} = 13.36(10^{-6})\ \text{m}^4$, into Eq. 6.12, we obtain the following.

$$\sigma_{40} = -\frac{M_z y_A}{I_{zz}} = -\frac{(4)(10^3)(-0.04)}{13.36(10^{-6})} = 11.98(10^6)\ \text{N/m}^2 \quad \text{or} \quad \boxed{\sigma_{40} = 12.0\ \text{MPa(T)}}$$

The maximum bending normal stress will be on the top and bottom surface i.e., $y_{max} = \pm 0.08$ m. From Eq. 6.12, we obtain:

$$\sigma_{max} = -\frac{M_z y_{max}}{I_{zz}} = -\frac{(4)(10^3)(\pm 0.08)}{13.36(10^{-6})} = \mp 23.95(10^6)\ \text{N/m}^2 \text{ or } \boxed{\sigma_{max} = 24.0\ \text{MPa (C) or (T)}}$$

Intuitive check: For a simply supported beam under a uniform distributed load, the visualization of the deformed shape shows that all points below the neutral axis will be in tension and all points above the neutral axis will be in compression, which is consistent with our answers.

6.36 A W10 x 30 steel beam is simply supported over a length of 10 ft and supports a distributed load of 1.5 kips /ft. At the mid-section of the beam, determine (a) the bending normal stress at a point 3 inches below the top surface. (b) the maximum bending normal stress.

Solution W10 x 30 L = 10 ft p = 1.5 kips /ft $\sigma_{3.0}$ = ? and σ_{max} = ? @ x=L/2

A simply supported beam with uniform distributed load will have a reaction force at each end whose value is half of the total force acting on the beam, i.e., R = (pL)/2 = 7.5 kips
Fig. (a) shows the free body diagram after making an imaginary cut at the mid-section. Internal shear force and bending moment are drawn as per our sign convention. Fig. (b) is the statically equivalent free body diagram.

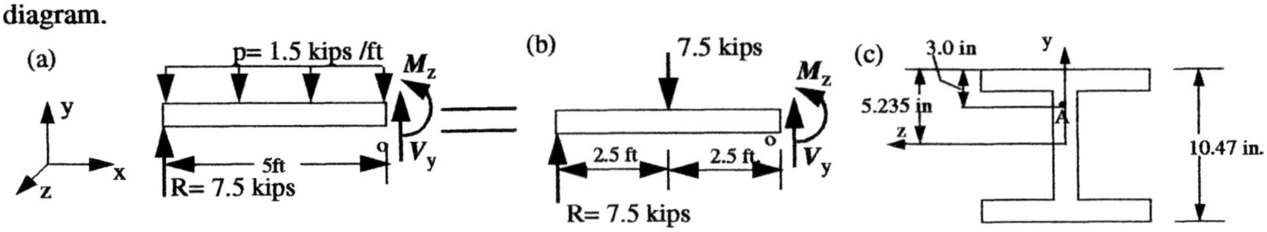

By equilibrium of moment about point O in Fig. (b) we obtain:

$$M_z - (7.5)(5) + (7.5)(2.5) = 0 \qquad \text{or} \qquad M_z = 18.75 \ \text{ft-kips} \qquad \mathbf{1}$$

From Appendix Table E.2, we obtain the geometry and dimension of W10 x 30 beam cross-section shown in Fig. (c) and $I_{zz} = 170 \ \text{in}^4$. Substituting y_A= (5.235 - 3)= 2.235 inch, I_{zz}, and $M_z = (18.75)(12) = 225 \text{in-lbs}$, into Eq. 6.12, we obtain the following.

$$\sigma_{3.0} = -\frac{M_z y_A}{I_{zz}} = -\frac{(225)(2.235)}{170} = -2.958 \ \text{ksi or} \qquad \boxed{\sigma_{3.0} = 2.96 \ \text{ksi(C)}}$$

The maximum bending normal stress will be on the top and bottom surface i.e., $y_{max} = \pm 5.235$ inch. From Eq. 6.12, we obtain:

$$\sigma_{max} = -\frac{M_z y_{max}}{I_{zz}} = -\frac{(225)(\pm 5.235)}{170} = \mp 6.928 \ \text{ksi or} \qquad \boxed{\sigma_{max} = 6.93 \ \text{ksi (C) or (T)}}$$

Intuitive check: For a simply supported beam under a uniform distributed load, the visualization of the deformed shape shows that all points below the neutral axis will be in tension and all points above the neutral axis will be in compression, which is consistent with our answers.

6.37 An S12 x 35 steel cantilever beam has a length of 20 ft. At the free end a force of 3 kips acts downwards. At the section near the built-in end, determine (a) the bending normal stress at a point 2 inches above the bottom surface. (b) the maximum bending normal stress.

Solution S12 x 35 L = 20 ft F = 3 kips $\sigma_{2.0}$ = ? and σ_{max} = ? @ x=0

Fig. (a) shows the free body diagram after making an imaginary cut just right of the built in end. Internal shear force and bending moment are drawn as per our sign convention.

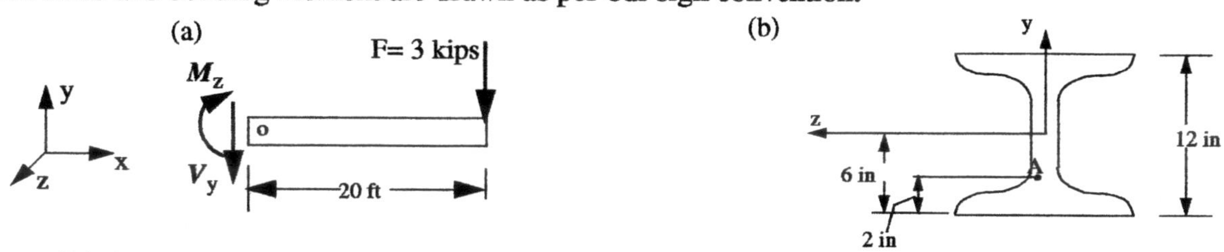

By equilibrium of moment about point O in Fig. (b) we obtain:

$$M_z + (3)(20) = 0 \qquad \text{or} \qquad M_z = -60 \ \text{ft-kips} \qquad \mathbf{1}$$

From Appendix Table E.2, we obtain the geometry and dimension of S12 x 35 beam cross-section shown in Fig. (b) and` $I_{zz} = 229 \ \text{in}^4$. Substituting y_A= - (6 - 2)= -4 inch, I_{zz}, and $M_z = (-60)(12) = -720 \text{in-lbs}$, into Eq. 6.12, we obtain the following.

$$\sigma_{2.0} = -\frac{M_z y_A}{I_{zz}} = -\frac{(-720)(-4)}{229} = -12.58 \ \text{ksi or} \qquad \boxed{\sigma_{2.0} = 12.58 \ \text{ksi(C)}}$$

The maximum bending normal stress will be on the top and bottom surface i.e., $y_{max} = \pm 6$ inch. From Eq. 6.12, we obtain:

$$\sigma_{max} = -\frac{M_z y_{max}}{I_{zz}} = -\frac{(-720)(\pm 6)}{229} = \pm 18.86 \ \text{ksi or} \qquad \boxed{\sigma_{max} = 18.86 \ \text{ksi (T) or (C)}}$$

Intuitive check: For a cantilever beam with a downward load at the free end, the visualization of the deformed shape shows that all points below the neutral axis will be in compression and all points above the neutral axis will be in tension, which is consistent with our answers.

6.38 An S250 x 52 steel cantilever beam has a length of 5 m. At the free end a force of 15 kN acts downwards. At the section near the built-in end, determine (a) the bending normal stress at a point 30 mm below the top surface. (b) the maximum bending normal stress.

Solution S250 x 52 L = 5 m F = 15 kN

σ_{30}= ? and σ_{max} = ? @ x=0

Fig. (a) shows the free body diagram after making an imaginary cut just right of the built in end. Internal shear force and bending moment are drawn as per our sign convention.

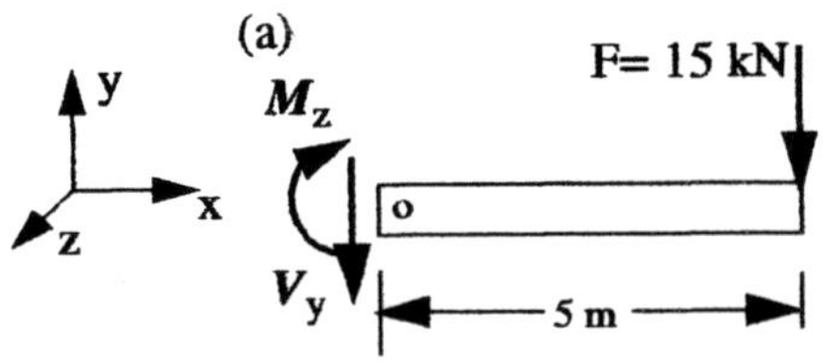

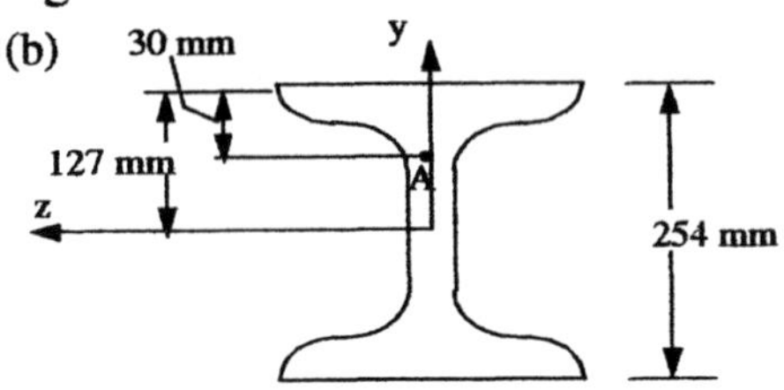

By equilibrium of moment about point O in Fig. (b) we obtain:

$$M_z + (5)(15) = 0 \quad \text{or} \quad M_z = -75 \text{ kN}-\text{m} \qquad 1$$

From Appendix Table E.2, we obtain the geometry and dimension of S250 x 52 beam cross-section shown in Fig. (b) and $I_{zz} = 61.20(10^6)\ \text{mm}^4$. Substituting y_A= (0.127 - 0.030)= 0.097 m,

$I_{zz} = 61.20(10^{-6})\ \text{m}^4$, and Eq.(1), into Eq. 6.12, we obtain the following.

$$\sigma_{30} = -\frac{M_z y_A}{I_{zz}} = -\frac{(-75)(10^3)(0.097)}{61.20(10^{-6})} = 118.87\ (10^6)\ \text{N/m}^2 \text{ or } \boxed{\sigma_{30} = 118.9 \text{ MPa (T)}}$$

The maximum bending normal stress will be on the top and bottom surface i.e., $y_{max} = \pm 0.127$ m. From Eq. 6.12, we obtain:

$$\sigma_{max} = -\frac{M_z y_{max}}{I_{zz}} = -\frac{(-75)(10^3)(\pm 0.127)}{61.20(10^{-6})} = \pm 155.63\ (10^6)\ \text{N/m}^2 \text{ or}$$

$$\boxed{\sigma_{max} = 155.6 \text{ MPa (T) or (C)}}$$

Intuitive check: For a cantilever beam with a downward load at the free end, the visualization of the deformed shape shows that all points below the neutral axis will be in compression and all points above the neutral axis will be in tension, which is consistent with our answers.

6.39 The beam, loading and the cross-section of the beam are as shown. Determine the bending normal stress at point A and the maximum bending normal stress in the section containing point A.

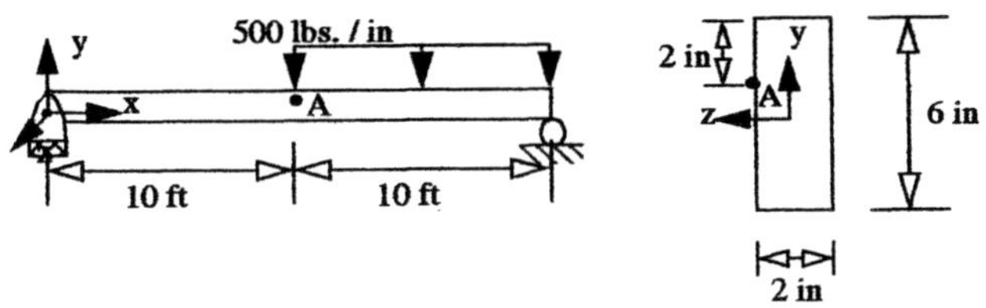

Fig. P6.39

Solution: σ_A = ? σ_{max} = ?

The second area moment of Inertia can be found out as

$$I_{zz} = \frac{1}{12}(2)(6^3) = 36 \text{ in}^4 \qquad 1$$

Fig.(a) shows the free body diagram of the entire beam with distributed load replaced by an equivalent load By equilibrium of moment about point C in Fig.(a) we obtain:

$R_B(20) - 5000(5) = 0$ or $R_B = 1250$ lbs

By making an imaginary cut through section containing point A, we obtain the free body diagram shown in Fig.(b). By equilibrium of moment about point A we obtain.

$$M_z = 10R_B = 12500\text{ft}-\text{lbs} = 150 \text{ in}-\text{kips} \quad (2)$$

Substituting Eqs. (1) and (2) and $y_A = 1$ inch into Eq. 6.12, we obtain.

$$\sigma_A = \frac{-M_z y_A}{I_{zz}} = -\frac{(150)(1)}{36} = -4.17 \text{ ksi(C)} \qquad \boxed{\sigma_A = 4.17 \text{ ksi(C)}}$$

Top and bottom surfaces are equally distant from the centroid. Substituting Eqs. (1), (2) and $y_{max} = \pm 3$ in Eq. 6-12, we obtain

$$\sigma_{max} = -\frac{(150)(\pm 3)}{36} = \mp 12.5 \text{ ksi(C)} \qquad \boxed{\sigma_{max} = 12.5 \text{ ksi (C) or (T)}}$$

Intuitive check: By visualizing the deformed shape of the beam, we can say that points below the neutral axis (bottom surface) will be in tension and the points above the neutral axis (top surface) will be in compression

---

6.40 The beam, loading and the cross-section of the beam are as shown. Determine the bending normal stress at point A and the maximum bending normal stress in the section containing point A.

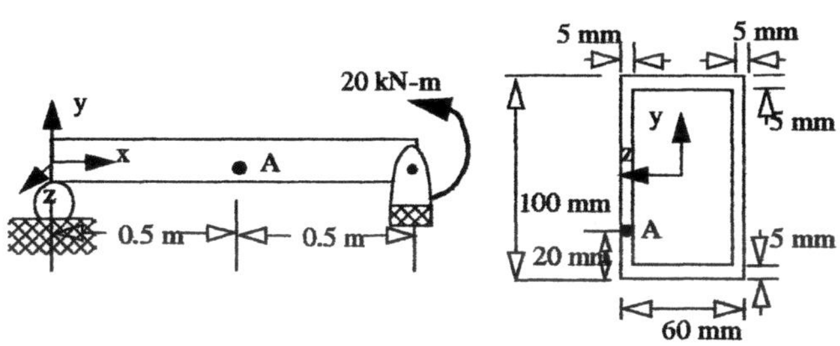

Fig. P6.40

Solution: $\sigma_A = ?$, $\sigma_{max} = ?$

The second area moment of Inertia can be found out as

$$I_{zz} = \frac{1}{12}(0.06)(0.1^3) - \frac{1}{12}(0.05)(0.09^3) = 1.9625 \text{ in}^4 \quad (1)$$

Fig.(a) shows the free body diagram of the entire beam.

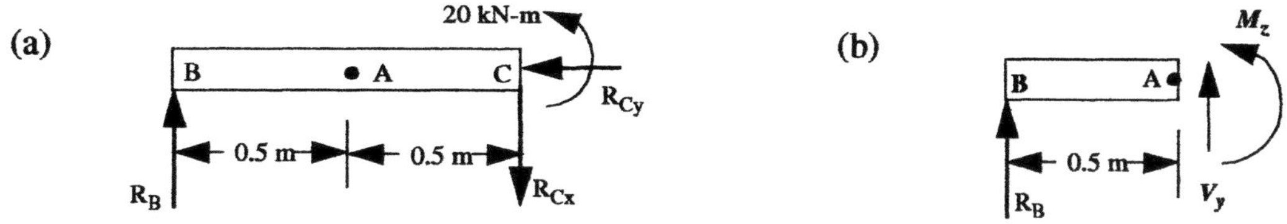

By equilibrium of moment about point C in Fig. (a) we obtain:

$R_B(1) - 20 = 0$ or $R_B = 20$ kN

By making an imaginary cut through section containing point A, we obtain the free body diagram shown in Fig.(b). By equilibrium of moment about point A we obtain.

$$M_z = (0.5)R_B = 10 \text{ kN}-\text{m} = 10{,}000 \text{ N}-\text{m} \quad (2)$$

Substituting Eqs. (1), (2) and $y_A = 0.03$ m, we obtain.

$$\sigma_A = \frac{-M_z y_A}{I_{zz}} = -\frac{(10000)(-0.03)}{1.9625(10^{-6})} = -152.87(10^6)\ (N/m^2) \text{ or } \qquad \boxed{\sigma_A = -152.9 \text{ MPa(T)}}$$

Top and bottom surfaces are equally distant from the centroid. Substituting Eqs. (1), (2) and $y_{max} = \pm 0.05$ in Eq. 6-12, we obtain

$$\sigma_{max} = -\frac{(10000)(\pm 0.05)}{1.9625(10^{-6})} = \mp 254.77(10^6)\ (N/m^2) \text{ or } \qquad \boxed{\sigma_{max} = -254.8 \text{ MPa(C) or (T)}}$$

Intuitive check: By visualizing the deformed shape of the beam, we can say that points below the neutral axis (bottom surface) will be in tension and the points above the neutral axis (top surface) will be in compression.

6.41 The beam, loading and the cross-section of the beam are as shown. Determine the bending normal stress at point A and the maximum bending normal stress in the section containing point A.

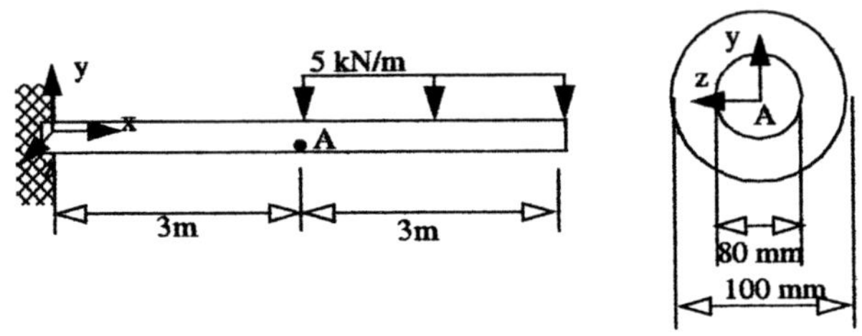

Fig. P6.41

Solution: $\sigma_A = ?$ $\qquad \sigma_{max} = ?$

The second area moment of Inertia can be found out as

$$I_{zz} = \frac{\pi}{64}[(0.1)^4 - (0.08^4)] = 2.8916(10^{-6})\ m^4 \qquad 1$$

By making an imaginary cut through section containing point A and replacing the distributed load by an equivalent force we obtain the free body diagram shown below.

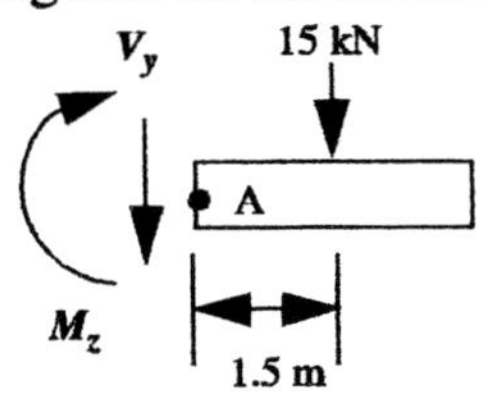

By equilibrium of moment about point A, we obtain.

$$M_z = -(15)(1.5) = -22.5KN - m = -22500\ N - m \qquad 2$$

Substituting Eqs. (1) and (2) and $y_A = -0.04$ m, we obtain

$$\sigma_A = \frac{-M_z y_A}{I_{zz}} = -\frac{(-22500)(-0.04)}{2.8916(10^{-6})} = -310.5(10^6)\ (N/m^2) \text{ or } \qquad \boxed{\sigma_A = 310.5 \text{ MPa(C)}}$$

Top and bottom surfaces are equally distant from the centroid. Substituting Eqs. (1), (2) and $y_{max} = \pm 0.05$ in Eq. 6-12, we obtain.

$$\sigma_{max} = -\frac{(-22500)(\pm 0.05)}{2.8916(10^{-6})} = \pm 388.18(10^6)\ (N/m^2) \text{ or } \qquad \boxed{\sigma_{max} = -388.2 \text{ MPa(C)or(T)}}$$

Intuitive check: By visualizing the deformed shape of the beam, we can say that points above the neutral axis (top surface) will be in tension and the points below the neutral axis (bottom surface) will be in compression.

6.42 The beam, loading and the cross-section of the beam are as shown. Determine the bending normal stress at point A and the maximum bending normal stress in the section containing point A.

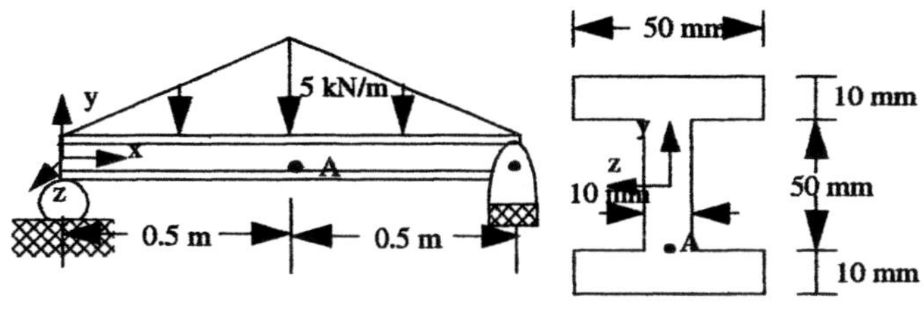

Fig. P6.42

Solution: $I_{zz} = 1.01(10^6)mm^4$ $\sigma_A = ?$ $\sigma_{max} = ?$

By symmetry the reaction force of each support is half the total force.

$$R_A = R_C = \frac{1}{2}\left[\frac{1}{2}(5)(1.0)\right] = 1.25kN$$ 1

The free body diagram after the imaginary cut through the section at A in Fig. (a). The distributed load is replaced by an equivalent load as shown in Fig. (b).

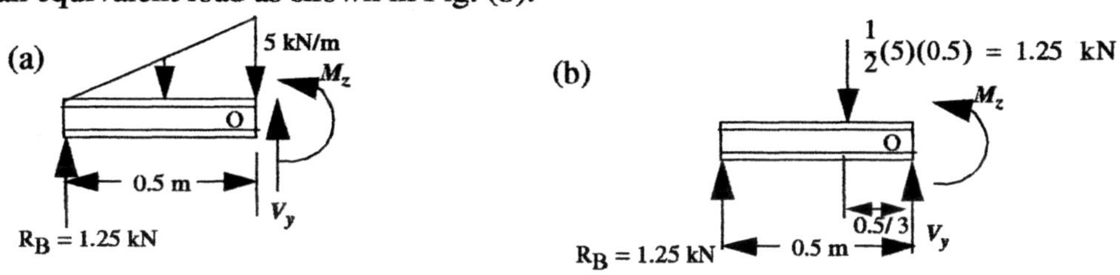

By equilibrium of moment about point O. $M_z - (1.25)(0.5) + (1.25)\left(\frac{0.5}{3}\right) = 0$ or $M_z = 0.4167kN-m$

The bending normal stress at A is

$$\sigma_A = -\frac{M_z y_A}{I_{zz}} = -\frac{(0.4167)(10^3)(-0.025)}{1.01(10^{-6})} = 10.31(10^6)\ \frac{N}{m^2} \text{ or}$$ $\boxed{\sigma_A = 10.3\ \text{MPa(T)}}$

The bending normal stress in the section is

$$\sigma_A = -\frac{M_z y_{max}}{I_{ZZ}} = -\frac{(0.4167)(10^3)(\pm 0.035)}{1.01(10^{-6})} = \mp 14.44(10^6)\ \frac{N}{m^2} \text{ or}$$ $\boxed{\sigma_{max} = 14.4\ \text{MPa(C) or (T)}}$

6.43 The beam, loading and the cross-section of the beam are as shown. Determine the bending normal stress at point A and the maximum bending normal stress in the section containing point A.

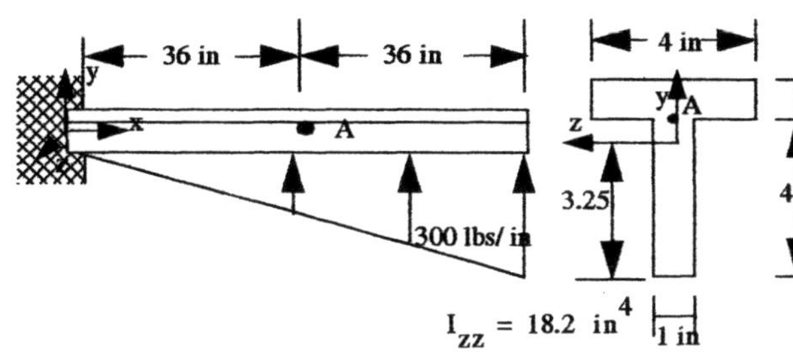

Fig. P6.43

Solution: $I_{zz} = 18.2(10^6)\ in^4$, $\sigma_a = ?$, $\sigma_{max} = ?$

Fig. (a) shows the free body diagram of the entire beam. By force and moment equilibrium in Fig. (a) we obtain: $R_B = 10,800$ lbs and $M_B = (48)(10800) = 518.4(10^3)$ in – lbs

Fig. (b) is the free body diagram after the imaginary cut through the section at A. The distributed load is replaced by equivalent load in Fig.(c). By equilibrium of moment about point O.

$$M_z - M_B + R_B(36) - (2700)(12) = 0 \quad \text{or} \quad M_z = 162000 \ \text{in} - \text{lbs}$$

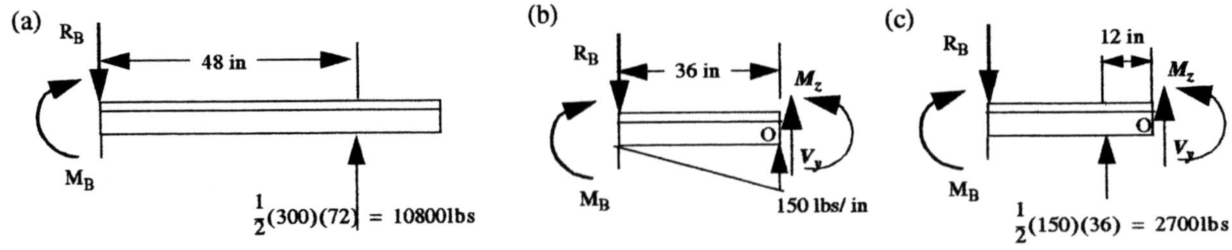

The bending normal stress at A is

$$\sigma_A = -\frac{M_t y_A}{I_{zz}} = -\frac{(162000)(1.75)}{18.2} = -15577 \ \text{psi or} \qquad \boxed{\sigma_A = 15.6 \ \text{ksi(C)}}$$

The maximum bending normal stress in the section is

$$\sigma_{max} = -\frac{M_{max} y_{max}}{I_{zz}} = -\frac{(162000)(-3.25)}{18.2} = 28929 \ \text{psi or} \qquad \boxed{\sigma_{max} = 28.9 \ \text{ksi(T)}}$$

6.44 The beam, loading and the cross-section of the beam are as shown. Determine the bending normal stress at point A and the maximum bending normal stress in the section containing point A.

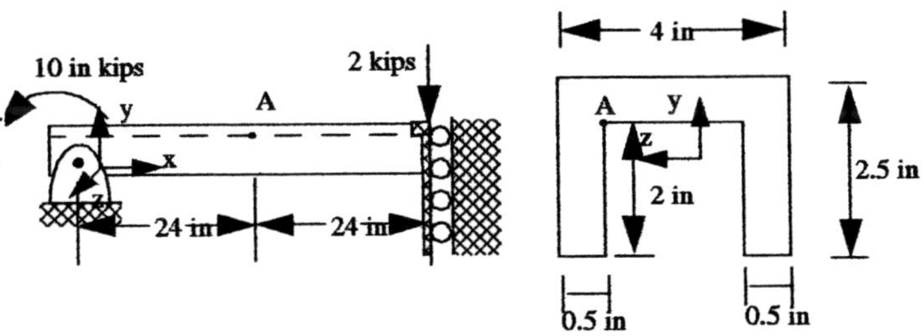

Fig. P6.44

Solution: $I_{zz} = 2.27\text{in}^4$ $\sigma_a = ?,$ $\sigma_{max} = ?$

Fig.(a) shows the free body diagram of the entire beam.

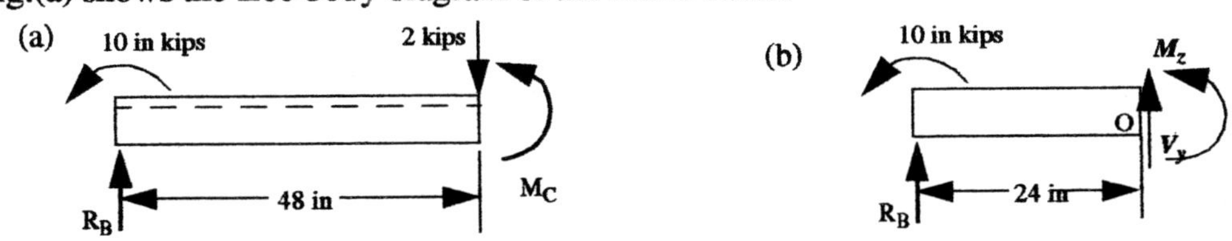

By equilibrium of force we obtain we obtain $R_B = 2$ kips

Fig. (b) shows the free body diagram after the imaginary cut through section containing point (a). By equilibrium of moment about point O we obtain: $M_z + 10 - (24)(R_B) = 0$ or $M_z = 38$ in – lbs

The bending normal stress at A is

$$\sigma_A = -\frac{M_z y_A}{I_{zz}} = -\frac{(38)(0.375)}{2.27} = -6.28 \ \text{ksi or} \qquad \boxed{\sigma_A = 6.3 \ \text{ksi(C)}}$$

The maximum bending normal stress in the section is

$$\sigma_{max} = -\frac{M_z y_{max}}{I_{zz}} = -\frac{(38)(-1.625)}{2.27} = 27.2 \ \text{ksi or} \qquad \boxed{\sigma_{max} = 27.2 \ \text{ksi(T)}}$$

6.45 A wooden (E = 10 GPa) rectangular beam, loading and cross-section are as shown. If the distributed force w = 5 kN/m , determine the normal strain ε_{xx} at point A.

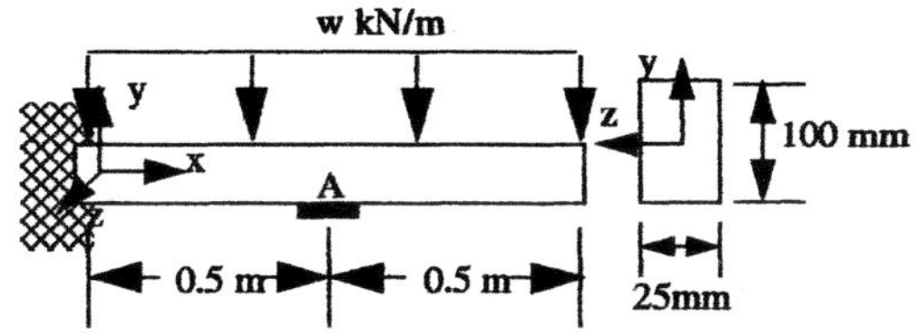

Fig. P6.45

Solution: E = 10 GPa w = 5 kN/m $\varepsilon_a = ?$

The area moment of Inertia is

$$I_{zz} = \frac{1}{12}(0.025)(0.1^3) = 2.083(10^{-6})\ m^4 \qquad 1$$

Fig. (a) shows the free body diagram of the beam after the imaginary cut. The distributed load is replaced by equivalent load in Fig. (b)

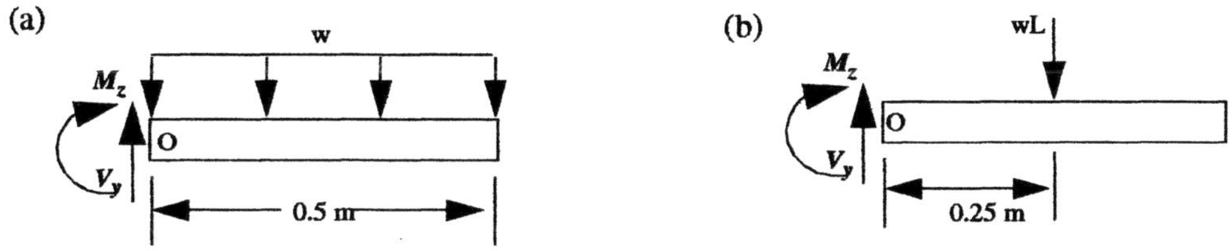

By equilibrium of moment about point O.

$$M_z + (0.25)(0.5w) = 0 \quad \text{or} \quad M_z = (-0.125)w \qquad 2$$

Substituting w = 5 kN/m we obtain $M_z = (-0.625)$ kN/m

The bending normal stress at A is

$$\sigma_A = -\frac{M_z y_A}{I_{zz}} = -\frac{(-625)(-0.05)}{2.083(10^{-6})} = -15(10^6)\ N/m^2$$

The bending normal strain at A is $\varepsilon_A = \frac{\sigma_A}{E} = -\frac{(15)(10^6)}{(10)(10^9)} = -1.5(10^{-3})$ or $\boxed{\varepsilon_A = -1500\ \mu}$

6.46 A wooden (E = 10 GPa) rectangular beam, loading and cross-section are as shown. The normal strain at point A was measured as $\varepsilon_{xx} = -500\ \mu$. Determine the distributed force w that is acting on the beam.

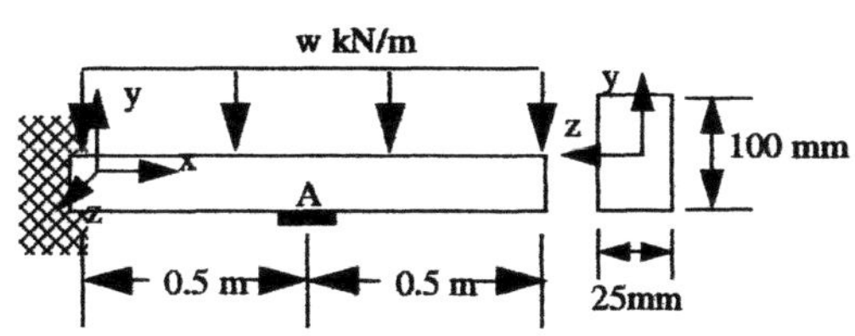

Fig. P6.46

Solution: E = 10 GPa $\varepsilon_A = -600\ \mu$ w = ?

From Eqs. (1) and (2) in problem 6.45 we have $I_{zz} = 2.083(10^{-6})\ m^4$ and $M_z = (-0.125w)$ kN – m

The bending normal stress of A is $\sigma_A = -\frac{M_z y_A}{I_{zz}} = -\frac{(-125w)(-0.05)}{2.083(10^{-6})} = -3.0w(10^6)\ N/m^2$

The bending normal strain at A is $\varepsilon_A = \frac{\sigma_A}{E} = -\frac{-3.0w(10^6)}{(10)(10^9)} = -600(10^{-6})$ or $\boxed{w = 2kN/m}$

6.47 A wooden beam (E = 8,000 ksi), loading, and the cross-section of the beam are as shown. If the applied load P = 6 kips in, determine the normal strain ε_{xx} at point A.

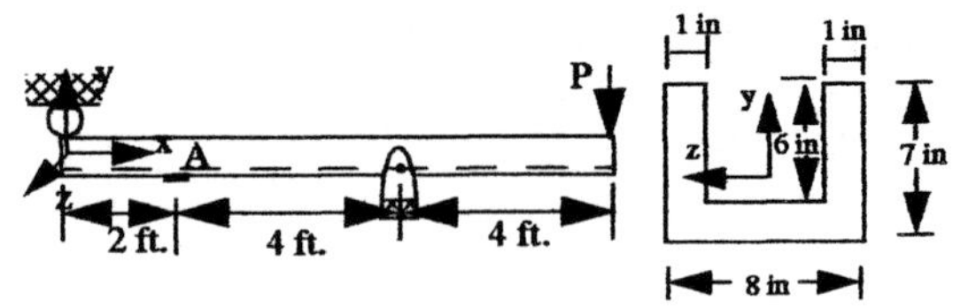

Fig. P6.47

Solution: $E = 8000\ \text{ksi}$ $\quad I_{zz} = 95.47\ \text{in}^4$ $\quad P = 6\ \text{kips}$ $\quad \varepsilon_a = ?$

The free body diagram of the entire beam is shown in Fig. (a).

(a)

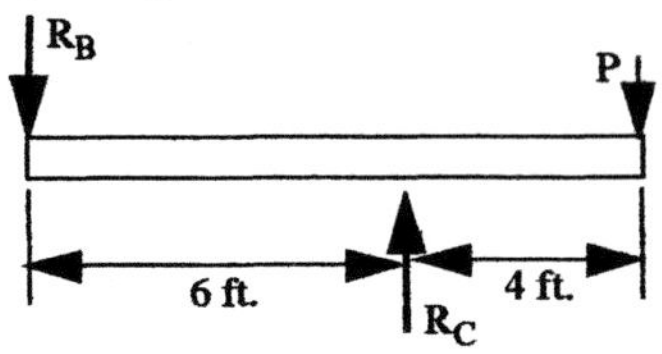

(b)

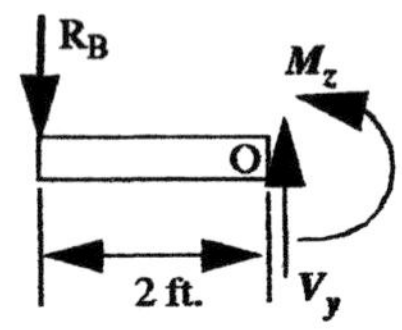

By equilibrium about point B, we obtain $R_B(6) - P(4) = 0$ or $R_B = 2P/3$

Fig. (b) shows the free body diagram of the beam after the imaginary cut through A. By equilibrium of moment about point O, we obtain,

$$M_z + (2)R_B = 0 \qquad \text{or} \qquad M_z = \frac{-4P}{3}\ \text{ft-kips} \tag{1}$$

Substituting P = 6 kips, we obtain $M_z = (-8)\ \text{ft-kips}$ or $M_z = -96\ \text{in-kips}$

The bending normal stress at A is $\sigma_A = -\left[\frac{M_z y_A}{I_{zz}}\right] = -\left[\frac{(-96)(-2.6)}{95.47}\right] = (-2.614)\ \text{ksi}$

The bending normal strain at A is $\varepsilon_A = \frac{\sigma_A}{E} = \frac{-2.614}{(8000)} = -0.3268(10^{-3})$ or $\boxed{\varepsilon_A = -327\ \mu}$

6.48 A wooden beam (E = 8,000 ksi), loading, and the cross-section of the beam are as shown. The normal strain at point A was measured as $\varepsilon_{xx} = -250\ \mu$. Determine the load P.

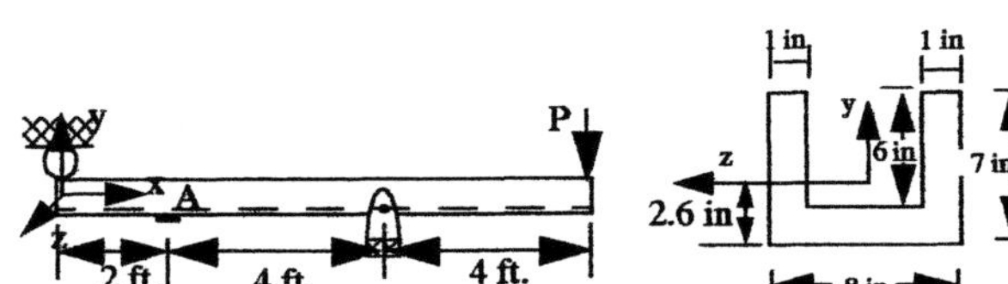

$I_{zz} = 95.47\ \text{in}^4$

Fig. P6.48

Solution: $E = 8000\ \text{ksi}$ $\quad I_{zz} = 95.47\ \text{in}^4$ $\quad \varepsilon_A = 250$ $\quad P = ?$

From Eq. (1) of problem 6.47 we have $M_z = \frac{-4P}{3}\ \text{ft-kips} = -16P\ \text{in-kips}$

The bending normal stress of A is $\sigma_A = -\frac{M_z y_A}{I_{zz}} = -\frac{(-16P)(-2.6)}{95.47} = -0.4357P$

The bending normal strain at A is $\varepsilon_A = \frac{\sigma_A}{E} = \frac{-0.4357P}{8000} = -250(10^{-6})$ or $\boxed{P = 4.59\ \text{kips}}$

Problem Set 6.3

6.49 (a) Write the equations for shear force and bending moments as a function of x for the entire beam. (b) Show your results satisfy Equations 6.13 and 6.14.

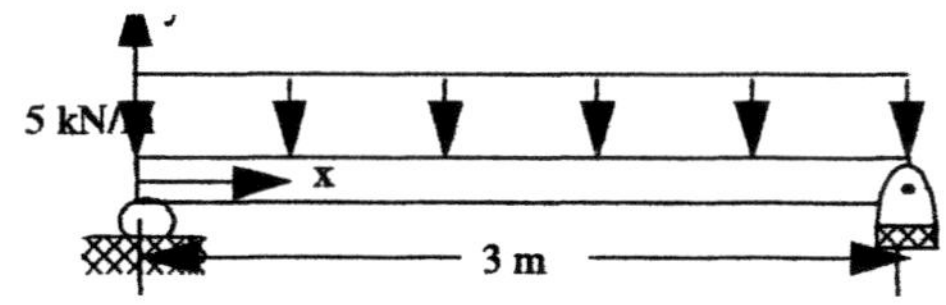

Fig. P6.49

Solution $V_y(x) = ?$ $M_z(x) = ?$

Fig.(a) shows the free body diagram of the entire beam with the distributed load replaced by an equivalent load.

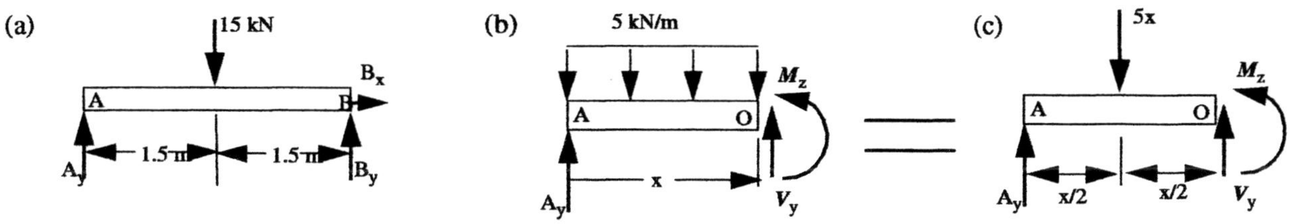

By equilibrium of moment about point B in Fig.(a) we obtain.

$$A_y(3) - 15(1.5) = 0 \quad \text{or} \quad A_y = 7.5 \text{ kN} \tag{1}$$

We can make an imaginary cut at a distance x and take the left part as a free body diagram. The internal shear force V_y and internal bending moment M_z are drawn as per our sign convention and shown in Fig. (b). The distributed load is replaced by equivalent load as shown in Fig. (c).

By equilibrium of forces in the y-direction in Fig. (c) we obtain: $V_y + A_y - 5x = 0$ or

$$\boxed{V_y = (5x - 7.5) \text{ kN}} \tag{2}$$

By equilibrium of moment about point O in Fig. (c) we obtain: $M_z - A_y(x) + (5x)\left(\frac{x}{2}\right) = 0$ or

$$\boxed{M_z = (7.5x - 2.5x^2) \text{ kN} - \text{m}} \tag{3}$$

Taking the first derivative of Eq. (2) we obtain:

$$\frac{dV_y}{dx} = 5 = -p_y \tag{4}$$

Taking the first derivative of Eq. (3) we obtain:

$$\frac{dM_z}{dx} = 7.5 - (2)(2.5)x = (7.5 - 5x) = -V_y \tag{5}$$

Eqs. (4) and (5) confirm that shear force and bending moment expressions of Eqs. (2) and (3) satisfy Eqs. 6.13 and 6.14, respectively.

6.50 (a) Write the equations for shear force and bending moments as a function of x for the entire beam. (b) Show your results satisfy Equations 6.13 and 6.14.

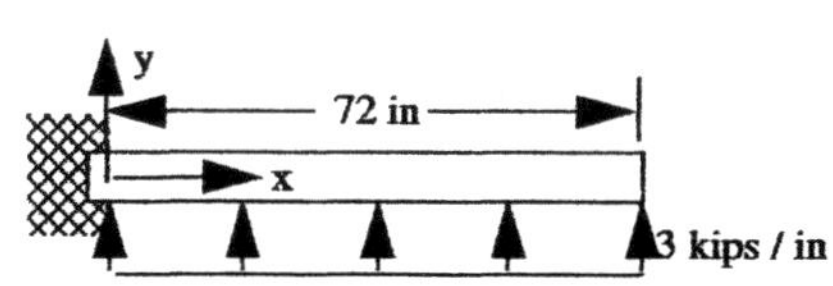

Fig. P6.50

Solution $V_y(x) = ?$ $M_z(x) = ?$

We can make an imaginary cut at a distance x and take the right part to avoid calculating the wall reactions.The internal shear force V_y and internal bending moment M_z are drawn as per our sign convention and shown in Fig. (a). The distributed load is replaced by equivalent load as shown in Fig. (b).

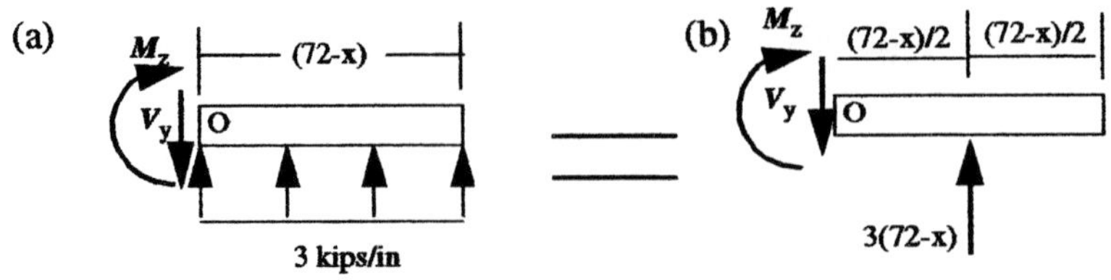

By equilibrium of forces in the y-direction in Fig. (a) we obtain: $-V_y + 3(72 - x) = 0$ or

$$\boxed{V_y = 3(72 - x) \text{ kips}} \qquad 1$$

By equilibrium of moment about point O in Fig. (b) we obtain: $M_z + 3(72 - x)\left(\frac{72 - x}{2}\right) = 0$ or

$$\boxed{M_z = -1.5(72 - x)^2 \text{ in} - \text{kips}} \qquad 2$$

Taking the first derivative of Eq. (1) we obtain:

$$\frac{dV_y}{dx} = 3(-1) = -3 = -p_y \qquad 3$$

Taking the first derivative of Eq. (2) we obtain:

$$\frac{dM_z}{dx} = (1.5)(2)(-1)(72 - x) = -3(72 - x) = -V_y \qquad 4$$

Eqs. (3) and (45) confirm that shear force and bending moment expressions of Eqs. (1) and (2) satisfy Eqs. 6.13 and 6.14, respectively.

6.51 (a) Write the equations for shear force and bending moments as a function of x for the entire beam. (b) Show your results satisfy Equations 6.13 and 6.14.

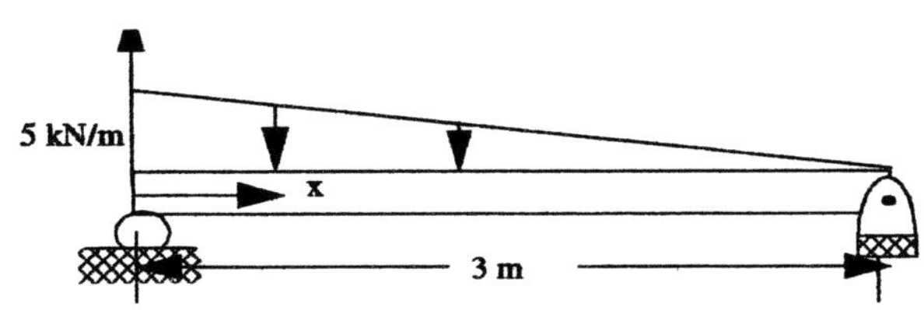

Fig. P6.51

Solution $V_y(x) = ?$ $M_z(x) = ?$

Fig.(a) shows the free body diagram of the entire beam with the distributed load replaced by an equivalent load.

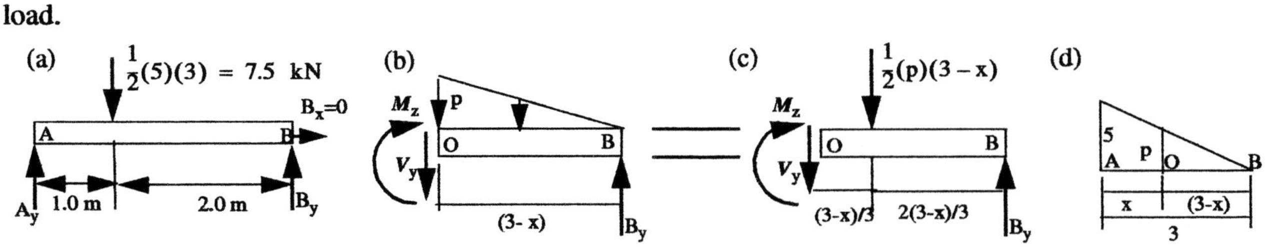

By equilibrium of moment about point A in Fig.(a) we obtain.

$$B_y(3) - (7.5)(1) = 0 \quad \text{or} \quad B_y = 2.5 \text{ kN} \qquad 1$$

We can make an imaginary cut at a distance x and take the right part as a free body diagram. The internal shear force V_y and internal bending moment M_z are drawn as per our sign convention and shown in Fig. (b). The distributed load is replaced by equivalent load as shown in Fig. (c). The intensity of distributed load at the imaginary cut (point O) can be found from similar triangle shown in Fig. (d) as:

$$\frac{p}{(3-x)} = \frac{5}{3} \quad \text{or} \quad p = \frac{5}{3}(3-x)$$ 2

By equilibrium of forces in the y-direction in Fig. (c) we obtain: $-\boldsymbol{V}_y + B_y - \frac{1}{2}(p)(3-x) = 0$ or

$$\boxed{\boldsymbol{V}_y = \left[2.5 - \frac{5}{6}(3-x)^2\right] \text{ kN}}$$ 3

By equilibrium of moment about point O in Fig. (c) we obtain:

$$\boldsymbol{M}_z - B_y(3-x) + \left[\frac{1}{2}(p)(3-x)\right]\left(\frac{3-x}{3}\right) = 0 \quad \text{or} \quad \boldsymbol{M}_z = B_y(3-x) - \frac{1}{6}p(3-x)^2 \text{ or}$$

$$\boxed{\boldsymbol{M}_z = \left[2.5(3-x) - \frac{5}{18}(3-x)^3\right] \text{ kN}-\text{m}}$$ 4

Taking the first derivative of Eq. (3) we obtain:

$$\frac{d\boldsymbol{V}_y}{dx} = -\left(\frac{5}{6}\right)(2)(-1)(3-x) = \frac{5}{3}(3-x) = -p_y$$ 5

Taking the first derivative of Eq. (4) we obtain:

$$\frac{d\boldsymbol{M}_z}{dx} = 2.5(-1) - \frac{5}{18}(3)(-1)(3-x)^2 = -\left[2.5 - \frac{5}{6}(3-x)^2\right] = -\boldsymbol{V}_y$$ 6

Eqs. (5) and (6) confirm that shear force and bending moment expressions of Eqs. (3) and (4) satisfy Eqs. 6.13 and 6.14, respectively.

6.52 (a) Write the equations for shear force and bending moments as a function of x for the entire beam. (b) Show your results satisfy Equations 6.13 and 6.14.

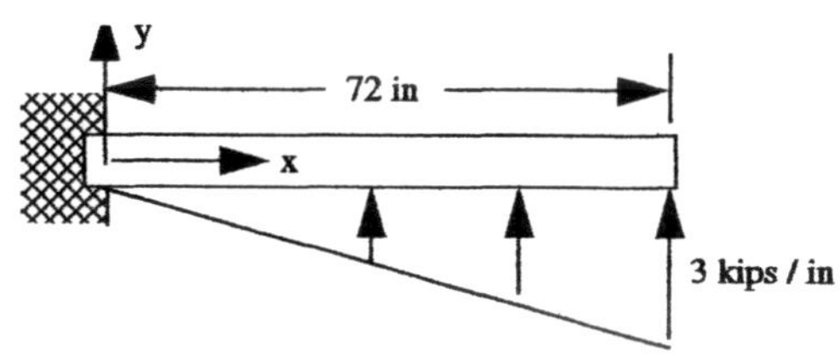

Fig. P6.52

Solution $\quad \boldsymbol{V}_y(x) = ? \qquad \boldsymbol{M}_z(x) = ?$

Fig.(a) shows the free body diagram of the entire beam with the distributed load replaced by an equivalent load.

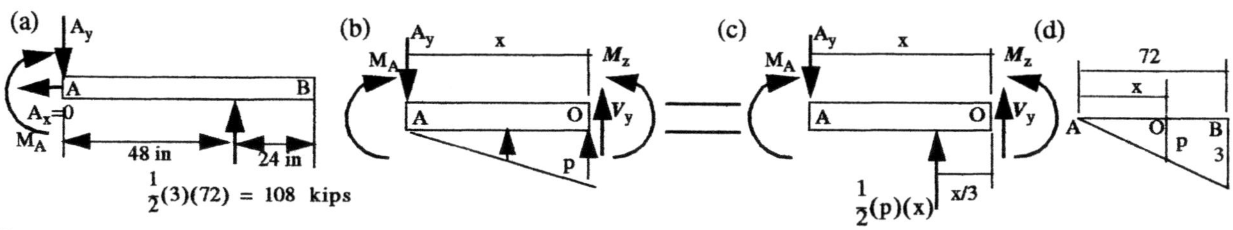

By equilibrium of forces in the y-direction in Fig. (a) we obtain

$$A_y = 108 \text{ kips}$$ 1

By equilibrium of moment about point A in Fig.(a) we obtain.

$$M_A - (108)(48) = 0 \quad \text{or} \quad M_A = 5184 \text{ in}-\text{kips}$$ 2

We can make an imaginary cut at a distance x and take the left part as a free body diagram. The internal shear force $\boldsymbol{V}_y$ and internal bending moment $\boldsymbol{M}_z$ are drawn as per our sign convention and shown in Fig. (b). The distributed load is replaced by equivalent load as shown in Fig. (c). The intensity of distributed load at the imaginary cut (point O) can be found from similar triangle shown in Fig. (d) as:

$$\frac{p}{x} = \frac{3}{72} \qquad \text{or} \qquad p = \frac{x}{24}$$ 3

By equilibrium of forces in the y-direction in Fig. (c) we obtain: $V_y - A_y - \frac{1}{2}(p)(x) = 0$ or

$$\boxed{V_y = \left[108 - \frac{1}{48}x^2\right] \text{ kips}}$$ 4

By equilibrium of moment about point O in Fig. (c) we obtain: $M_z + A_y(x) - \frac{1}{2}(p)(x)\left(\frac{x}{3}\right) - M_A = 0$ or

$$\boxed{M_z = \left[5184 - 108x + \frac{1}{144}x^3\right] \text{ in} - \text{kips}}$$ 5

Taking the first derivative of Eq. (3) we obtain:

$$\frac{dV_y}{dx} = -\left(\frac{1}{48}\right)(2)(x) = -\frac{1}{24}(x) = -p_y$$ 6

Taking the first derivative of Eq. (4) we obtain:

$$\frac{dM_z}{dx} = -108 + \frac{1}{144}(3)x^2 = -\left[108 - \frac{1}{48}x^2\right] = -V_y$$ 7

Eqs. (6) and (7) confirm that shear force and bending moment expressions of Eqs. (4) and (5) satisfy Eqs. 6.13 and 6.14, respectively.

6.53 (a) Write the equations for shear force and bending moments as a function of x for the entire beam. (b) Show your results satisfy Equations 6.13 and 6.14.

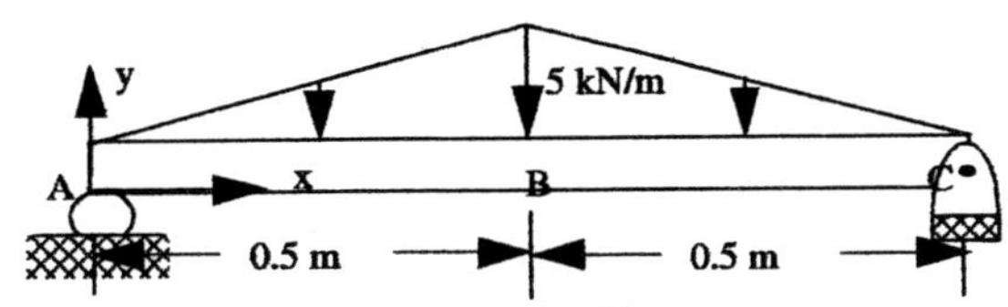

Fig. P6.53

Solution $V_y(x) = ?$ $M_z(x) = ?$

From the symmetry in the problem we conclude that the reaction forces at A and C in the y-direction are equal.

$$A_y = C_y = 1.25 \text{ kN}$$ 1

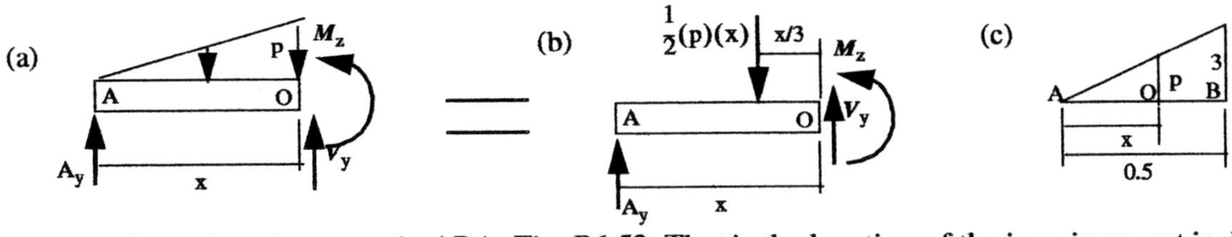

We can make an imaginary cut in AB in Fig. P6.53. That is the location of the imaginary cut is defined by: $0 \le x < 0.5$. We take the left part of the cut and draw the free body diagram. The internal shear force V_y and internal bending moment M_z are drawn as per our sign convention and shown in Fig. (a). The distributed load is replaced by equivalent load as shown in Fig. (b). The intensity of distributed load at the imaginary cut (point O) can be found from similar triangle shown in Fig. (c) as:

$$\frac{p}{x} = \frac{5}{0.5} \qquad \text{or} \qquad p = 10x$$ 2

By equilibrium of forces in the y-direction in Fig. (b) we obtain: $V_y + A_y - \frac{1}{2}(p)(x) = 0$ or

$$\boxed{V_y = [5x^2 - 1.25] \text{ kN} \qquad 0 \le x < 0.5}$$ 3

By equilibrium of moment about point O in Fig. (b) we obtain: $M_z + \frac{1}{2}(p)(x)\left(\frac{x}{3}\right) - A_y(x) = 0$ or

$$\boxed{M_z = \left[1.25x - \frac{5}{3}x^3\right] \text{ kN-m} \qquad 0 \le x < 0.5} \tag{4}$$

Taking the first derivative of Eq. (2) we obtain:

$$\frac{dV_y}{dx} = 10x = -p_y \tag{5}$$

Taking the first derivative of Eq. (3) we obtain:

$$\frac{dM_z}{dx} = 1.25 - \left(\frac{5}{3}\right)(3)x^2 = -V_y \tag{6}$$

Eqs. (5) and (6) confirm that shear force and bending moment expressions of Eqs. (3) and (4) satisfy Eqs. 6.13 and 6.14, respectively.To obtain the shear force and bending moment expressions in segment BC we can repeat the above procedure by making an imaginary cut and taking the right part for free body diagram. A simpler process is to recognize the symmetry of the problem and replace x by (1-x) in Eqs. (3) and (4) to obtain the following.

$$\boxed{V_y = [5(1-x)^2 - 1.25] \text{ kN} \qquad 0.5 < x \le 1} \tag{7}$$

$$\boxed{M_z = \left[1.25(1-x) - \frac{5}{3}(1-x)^3\right] \text{ kN-m} \qquad 0.5 < x \le 1} \tag{8}$$

6.54 (a) Write the equations for shear force and bending moments as a function of x for the entire beam. (b) Show your results satisfy Equations 6.13 and 6.14.

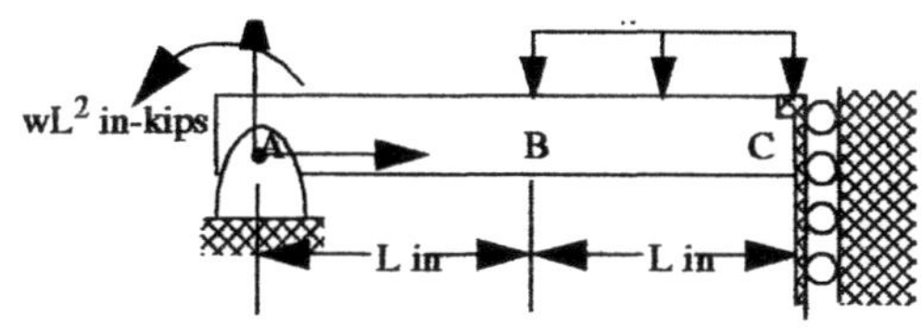

Fig. P6.54

Solution $V_y(x) = ?$ $M_z(x) = ?$

Fig.(a) shows the free body diagram of the beam with the distributed load replaced by an equivalent load.

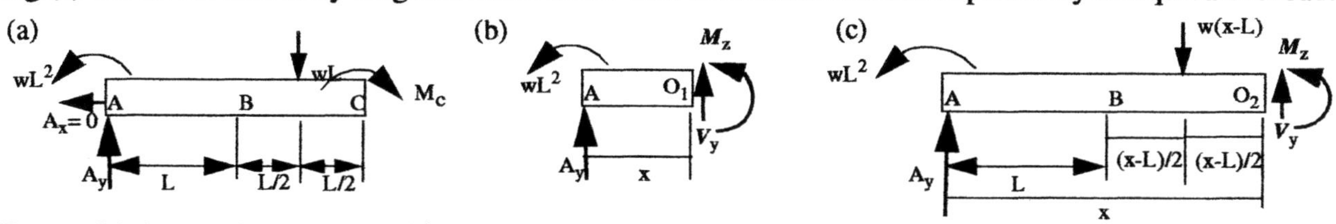

By equilibrium of forces in Fig.(a) we obtain.

$$A_y = wL \text{ kips} \tag{1}$$

We can make an imaginary cut in AB in Fig. P6.54. That is the location of the imaginary cut is defined by: $0 \le x < L$. We take the left part of the cut and draw the free body diagram. The internal shear force V_y and internal bending moment M_z are drawn as per our sign convention and shown in Fig. (b).

By equilibrium of forces in the y-direction in Fig. (b) we obtain: $V_y = -A_y$ or

$$\boxed{V_y = -wL \text{ kips} \qquad 0 \le x < L} \tag{2}$$

By equilibrium of moment about point O_1 in Fig. (b) we obtain: $M_z - A_y(x) = 0$ or

$$\boxed{M_z = wLx \text{ in-kips} \qquad 0 \le x < L} \tag{3}$$

Taking the first derivative of Eq. (2) we obtain.

$$\frac{dV_y}{dx} = 0 \qquad 4$$

Taking the first derivative of Eq. (3) we obtain:

$$\frac{dM_z}{dx} = wL = -V_y \qquad 5$$

Eqs. (4) and (5) confirm that shear force and bending moment expressions of Eqs. (2) and (3) satisfy Eqs. 6.13 and 6.14, respectively.

We can make an imaginary cut in BC in Fig. P6.54. That is the location of the imaginary cut is defined by: $L \le x < 2L$. We can take the left part of the cut and after replacing the distributed force, draw the free body diagram. The internal shear force V_y and internal bending moment M_z are drawn as per our sign convention and shown in Fig. (c).

By equilibrium of forces in the y-direction in Fig. (c) we obtain: $V_y + A_y - w(x-L) = 0$ or

$$\boxed{V_y = [w(x-L) - wL] \text{ kips} \qquad L < x \le 2L} \qquad 6$$

By equilibrium of moment about point O_2 in Fig. (c) we obtain: $M_z + (w)(x-L)\left(\frac{x-L}{2}\right) - A_y(x) = 0$ or

$$\boxed{M_z = \left[wLx - \frac{w}{2}(x-L)^2\right] \text{ in-kips} \qquad L < x \le 2L} \qquad 7$$

Taking the first derivative of Eq. (6) we obtain:

$$\frac{dV_y}{dx} = w = -p_y \qquad 8$$

Taking the first derivative of Eq. (7) we obtain:

$$\frac{dM_z}{dx} = wL - \frac{w}{2}(2)(x-L) = -V_y \qquad 9$$

Eqs. (8) and (9) confirm that shear force and bending moment expressions of Eqs. (6) and (7) satisfy Eqs. 6.13 and 6.14, respectively.

6.55 (a) Write the equations for shear force and bending moments as a function of x for the entire beam. (b) Show your results satisfy Equations 6.13 and 6.14.

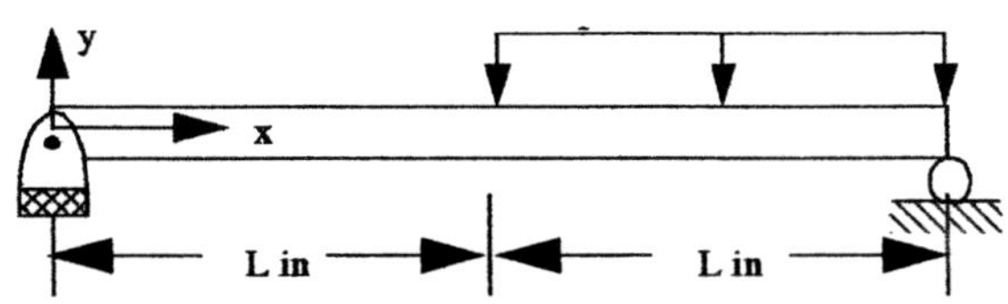

Fig. P6.55

Solution $V_y(x) = ?$ $M_z(x) = ?$

Fig.(a) shows the free body diagram of the beam with the distributed load replaced by an equivalent load. By equilibrium of moment about point C in Fig.(a) we obtain.

$$A_y(2L) - wL\left(\frac{L}{2}\right) = 0 \quad \text{or} \quad A_y = \frac{wL}{4} \text{ kips} \qquad 1$$

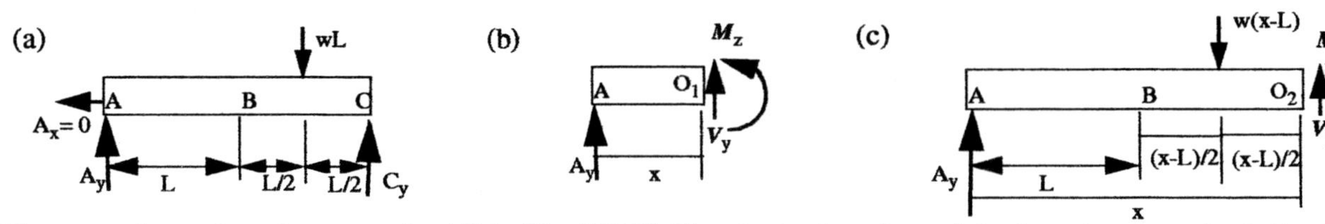

We can make an imaginary cut in AB in Fig. P6.55. That is the location of the imaginary cut is defined by:

$0 \le x < L$. We take the left part of the cut and draw the free body diagram. The internal shear force V_y and internal bending moment M_z are drawn as per our sign convention and shown in Fig. (b).
By equilibrium of forces in the y-direction in Fig. (b) we obtain: $V_y = -A_y$ or

$$\boxed{V_y = -\left(\frac{wL}{4}\right) \text{ kips} \qquad 0 \le x < L} \qquad (2)$$

By equilibrium of moment about point O_1 in Fig. (b) we obtain: $M_z - A_y(x) = 0$ or

$$\boxed{M_z = \frac{wL}{4}x \text{ in-kips} \qquad 0 \le x < L} \qquad (3)$$

Taking the first derivative of Eq. (2) we obtain.

$$\frac{dV_y}{dx} = 0 \qquad (4)$$

Taking the first derivative of Eq. (3) we obtain:

$$\frac{dM_z}{dx} = \frac{wL}{4} = -V_y \qquad (5)$$

Eqs. (4) and (5) confirm that shear force and bending moment expressions of Eqs. (2) and (3) satisfy Eqs. 6.13 and 6.14, respectively.
We can make an imaginary cut in BC in Fig. P6.55. That is the location of the imaginary cut is defined by: $L \le x < 2L$. We can take the left part of the cut and after replacing the distributed force, draw the free body diagram. The internal shear force V_y and internal bending moment M_z are drawn as per our sign convention and shown in Fig. (c).
By equilibrium of forces in the y-direction in Fig. (c) we obtain: $V_y + A_y - w(x-L) = 0$ or

$$\boxed{V_y = \left[w(x-L) - \frac{wL}{4}\right] \text{ kips} \qquad L < x \le 2L} \qquad (6)$$

By equilibrium of moment about point O_2 in Fig. (c) we obtain: $M_z + (w)(x-L)\left(\frac{x-L}{2}\right) - A_y(x) = 0$ or

$$\boxed{M_z = \left[\frac{wL}{4}x - \frac{w}{2}(x-L)^2\right] \text{ in-kips} \qquad L < x \le 2L} \qquad (7)$$

Taking the first derivative of Eq. (6) we obtain:

$$\frac{dV_y}{dx} = w = -p_y \qquad (8)$$

Taking the first derivative of Eq. (7) we obtain:

$$\frac{dM_z}{dx} = \frac{wL}{4} - \frac{w}{2}(2)(x-L) = -V_y \qquad (9)$$

Eqs. (8) and (9) confirm that shear force and bending moment expressions of Eqs. (6) and (7) satisfy Eqs. 6.13 and 6.14, respectively.

6.56 (a) Write the equations for shear force and bending moments as a function of x for the entire beam. (b) Show your results satisfy Equations 6.13 and 6.14.

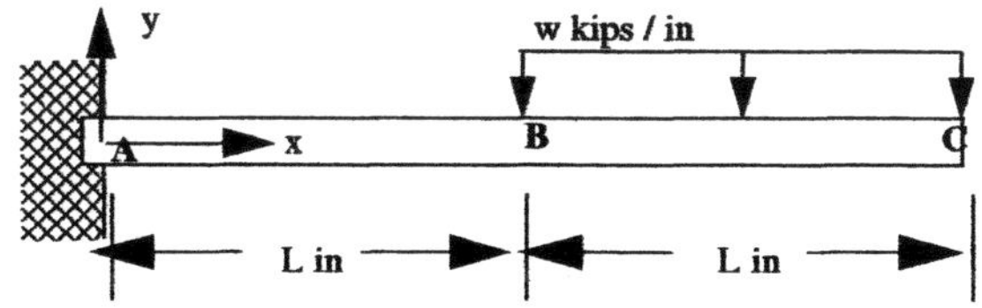

Fig. P6.56

Solution $\qquad V_y(x) = ? \qquad M_z(x) = ?$

Fig.(a) shows the free body diagram of the beam with the distributed load replaced by an equivalent load.

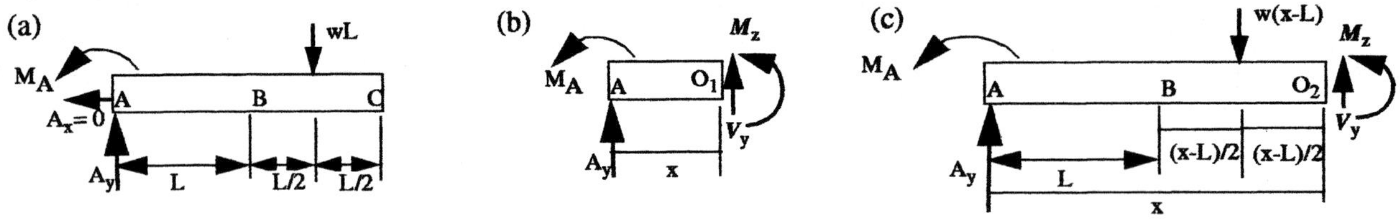

By equilibrium of forces in Fig.(a) we obtain.

$$A_y = wL \text{ kips} \qquad 1$$

By equilibrium of moment about point A in Fig. (a) we obtain: $M_A - wL\left(\frac{3L}{2}\right) = 0$ or

$$M_A = \frac{3}{2}wL^2 \qquad 2$$

We can make an imaginary cut in AB in Fig. P6.56. That is the location of the imaginary cut is defined by: $0 \le x < L$. We take the left part of the cut and draw the free body diagram. The internal shear force V_y and internal bending moment M_z are drawn as per our sign convention and shown in Fig. (b).

By equilibrium of forces in the y-direction in Fig. (b) we obtain: $V_y = -A_y$ or

$$\boxed{V_y = -wL \text{ kips} \qquad 0 \le x < L} \qquad 3$$

By equilibrium of moment about point O_1 in Fig. (b) we obtain: $M_z - A_y(x) + M_A = 0$ or

$$\boxed{M_z = \left(wLx - \frac{3}{2}wL^2\right) \text{ in-kips} \qquad 0 \le x < L} \qquad 4$$

Taking the first derivative of Eq. (3) we obtain.

$$\frac{dV_y}{dx} = 0 \qquad 5$$

Taking the first derivative of Eq. (4) we obtain:

$$\frac{dM_z}{dx} = wL = -V_y \qquad 6$$

Eqs. (5) and (6) confirm that shear force and bending moment expressions of Eqs. (3) and (4) satisfy Eqs. 6.13 and 6.14, respectively.

We can make an imaginary cut in BC in Fig. P6.56. That is the location of the imaginary cut is defined by: $L \le x < 2L$. We can take the left part of the cut and after replacing the distributed force, draw the free body diagram. The internal shear force V_y and internal bending moment M_z are drawn as per our sign convention and shown in Fig. (c).

By equilibrium of forces in the y-direction in Fig. (c) we obtain: $V_y + A_y - w(x-L) = 0$ or

$$\boxed{V_y = [w(x-L) - wL] \text{ kips} \qquad L < x \le 2L} \qquad 7$$

By equilibrium of moment about point O_2 in Fig. (c) we obtain:

$$M_z + (w)(x-L)\left(\frac{x-L}{2}\right) - A_y(x) + M_A = 0 \text{ or}$$

$$\boxed{M_z = \left[wLx - \frac{w}{2}(x-L)^2 - \frac{3}{2}wL^2\right] \text{ in-kips} \qquad L < x \le 2L} \qquad 8$$

Taking the first derivative of Eq. (7) we obtain:

$$\frac{dV_y}{dx} = w = -p_y \qquad 9$$

Taking the first derivative of Eq. (8) we obtain:

$$\frac{dM_z}{dx} = wL - \frac{w}{2}(2)(x - L) = -V_y \qquad \textbf{10}$$

Eqs. (9) and (10) confirm that shear force and bending moment expressions of Eqs. (7) and (8) satisfy Eqs. 6.13 and 6.14, respectively.

6.57 For the beam shown (a) Write the shear force and moment equation as a function of x in segment AB and segment BC. (b) Show that your results satisfy Equations 6.13 and 6.14. (c) What are the shear force and bending moment value just before and just after point B.

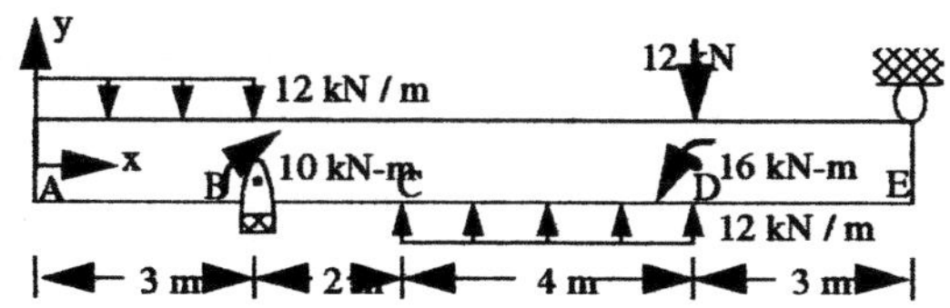

Fig. P6.57

Solution $V_y(x) = ?$ $M_z(x) = ?$

Fig.(a) shows the free body diagram of the beam with the distributed loads replaced by an equivalent loads.

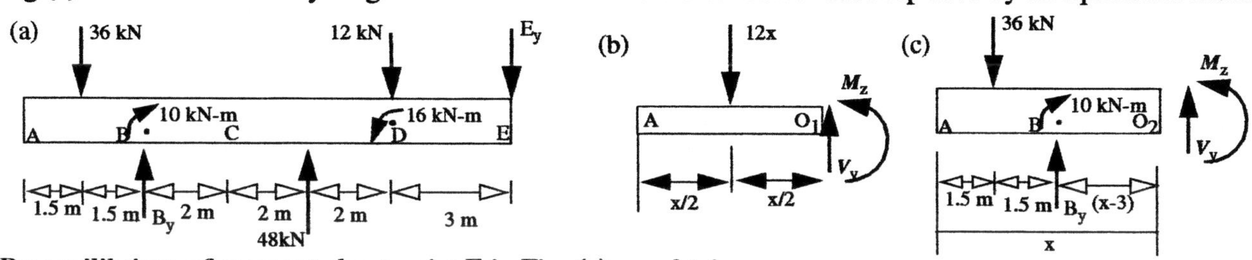

By equilibrium of moment about point E in Fig. (a) we obtain:

$B_y(9) - (36)(10.5) + 10 + (48)(5) - (12)(3) - 16 = 0$ or

$$B_y = 20 \text{ kN} \qquad \textbf{1}$$

By equilibrium of forces in Fig.(a) we obtain. $B_y - E_y - 36 + 48 - 12 = 0$ or

$$E_y = 20 \text{ kN} \qquad \textbf{2}$$

We can make an imaginary cut in AB in Fig. P6.57. That is the location of the imaginary cut is defined by: $0 \le x < 3\text{m}$. We take the left part of the cut and draw the free body diagram. The internal shear force V_y and internal bending moment M_z are drawn as per our sign convention and shown in Fig. (b).

By equilibrium of forces in the y-direction in Fig. (b) we obtain:

$$\boxed{V_y = 12x \text{ kN} \qquad 0 \le x < 3\text{m}} \qquad \textbf{3}$$

By equilibrium of moment about point O_1 in Fig. (b) we obtain: $M_z + (12x)\left(\frac{x}{2}\right) = 0$ or

$$\boxed{M_z = -6x^2 \text{ kN} - \text{m} \qquad 0 \le x < 3\text{m}} \qquad \textbf{4}$$

Taking the first derivative of Eq. (3) we obtain.

$$\frac{dV_y}{dx} = 12 = -p_y \qquad \textbf{5}$$

Taking the first derivative of Eq. (4) we obtain:

$$\frac{dM_z}{dx} = -12x = -V_y \qquad \textbf{6}$$

Eqs. (5) and (6) confirm that shear force and bending moment expressions of Eqs. (3) and (4) satisfy Eqs. 6.13 and 6.14, respectively.

We can make an imaginary cut in BC in Fig. P6.57. That is the location of the imaginary cut is defined by: $3\text{m} < x < 5\text{m}$. We can take the left part of the cut and after replacing the distributed force, draw the free body diagram. The internal shear force V_y and internal bending moment M_z are drawn as per our sign convention and shown in Fig. (c).

By equilibrium of forces in the y-direction in Fig. (c) we obtain: $V_y + B_y - 36 = 0$ or

$$\boxed{V_y = 16\ \text{kN} \qquad 3\text{m} < x < 5\text{m}} \tag{7}$$

By equilibrium of moment about point O_2 in Fig. (c) we obtain: $M_z + (36)(x - 1.5) - B_y(x - 3) - 10 = 0$ or

$$\boxed{M_z = (4 - 16x)\ \text{kN} - \text{m} \qquad 3\text{m} < x < 5\text{m}} \tag{8}$$

Taking the first derivative of Eq. (7) we obtain:

$$\frac{dV_y}{dx} = 0 \tag{9}$$

Taking the first derivative of Eq. (8) we obtain:

$$\frac{dM_z}{dx} = -16 = -V_y \tag{10}$$

Eqs. (9) and (10) confirm that shear force and bending moment expressions of Eqs. (7) and (8) satisfy Eqs. 6.13 and 6.14, respectively.

Point B is at x = 3m. Substituting x = 3 in Eqs. (3) and (4) we obtain the value of V_y and M_z just before point B as:

$$\boxed{V_y(3^-) = 36\ \text{kN}} \text{ and } \boxed{M_z(3^-) = -54\ \text{kN-m}}$$

where the superscript - refers to just before x=3. Substituting x=3 in Eqs, (7) and (8) we obtain the value of V_y and M_z just after point B as:

$$\boxed{V_y(3^+) = 16\ \text{kN}} \text{ and } \boxed{M_z(3^+) = -44\ \text{kN-m}}$$

where the superscript + refers to just after x=3. The differences in values for shear force and bending moment just before and after B are equal to the value of concentrated force By and the applied moment of 10 kN-m at B.

6.58 For the beam shown (a)write the shear force and moment equation as a function of x in segment CD and segment DE. (b) Show that your results satisfy Equations 6.13 and 6.14. (c) What are the shear force and bending moment value just before and just after point D.

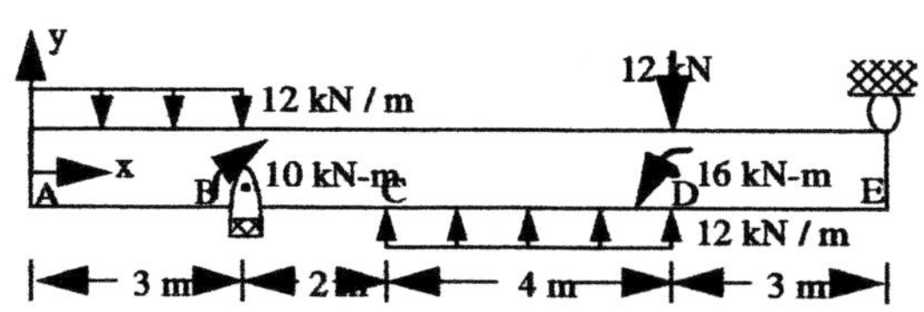

Fig. P6.58

Solution $V_y(x) = ?$ $M_z(x) = ?$

From Eqs. (1) and (2) in 6.57 we have the support reactions as:

$$B_y = 20\ \text{kN} \tag{1}$$

$$E_y = 20\ \text{kN} \tag{2}$$

We can make an imaginary cut in CD in Fig. P6.58. That is the location of the imaginary cut is defined by: $5\text{m} < x < 9\text{m}$. We take the right part of the cut and after replacing the distributed force, draw the free body diagram. The internal shear force V_y and internal bending moment M_z are drawn as per our sign convention and shown in Fig. (a).

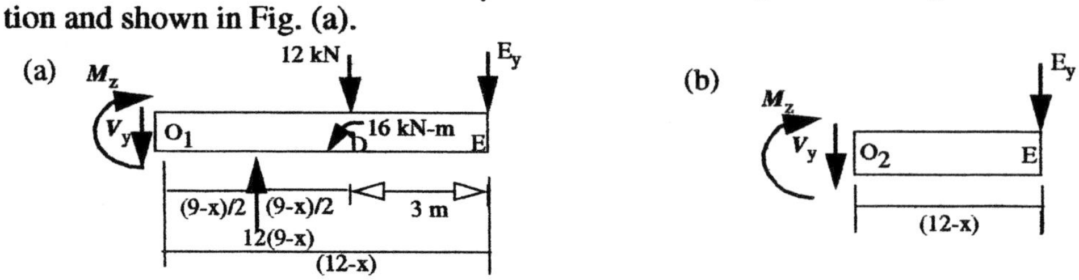

By equilibrium of forces in the y-direction in Fig. (a) we obtain: $-V_y - E_y - 12 + 12(9 - x) = 0$ or

$$\boxed{V_y = (76 - 12x)\ \text{kN} \qquad 5m < x < 9m} \tag{3}$$

By equilibrium of moment about point O_1 in Fig. (b) we obtain:

$$M_z - 12(9-x)\frac{(9-x)}{2} + 12(9-x) + E_y(12-x) - 16 = 0 \text{ or}$$

$$\boxed{M_z = (6x^2 - 76x + 154)\ \text{kN} - \text{m} \qquad 5m < x < 9m} \tag{4}$$

Taking the first derivative of Eq. (3) we obtain.

$$\frac{dV_y}{dx} = -12 = -p_y \tag{5}$$

Taking the first derivative of Eq. (4) we obtain:

$$\frac{dM_z}{dx} = (12x - 76) = -V_y \tag{6}$$

Eqs. (5) and (6) confirm that shear force and bending moment expressions of Eqs. (3) and (4) satisfy Eqs. 6.13 and 6.14, respectively.

We can make an imaginary cut in DE in Fig. P6.58. That is the location of the imaginary cut is defined by: $9m < x < 12m$. We can take the right part of the cut and draw the free body diagram. The internal shear force V_y and internal bending moment M_z are drawn as per our sign convention and shown in Fig. (b).

By equilibrium of forces in the y-direction in Fig. (b) we obtain: $V_y = -E_y$ or

$$\boxed{V_y = -20\ \text{kN} \qquad 9m < x < 12m} \tag{7}$$

By equilibrium of moment about point O_2 in Fig. (b) we obtain: $M_z + E_y(12 - x) = 0$ or

$$\boxed{M_z = (20x - 240)\ \text{kN} - \text{m} \qquad 9m < x < 12m} \tag{8}$$

Taking the first derivative of Eq. (7) we obtain:

$$\frac{dV_y}{dx} = 0 \tag{9}$$

Taking the first derivative of Eq. (8) we obtain:

$$\frac{dM_z}{dx} = 20 = -V_y \tag{10}$$

Eqs. (9) and (10) confirm that shear force and bending moment expressions of Eqs. (7) and (8) satisfy Eqs. 6.13 and 6.14, respectively.

Point D is at x = 9m. Substituting x = 9 in Eqs. (3) and (4) we obtain the value of V_y and M_z just before point D as:

$$\boxed{V_y(9^-) = -32\ \text{kN and } M_z(9^-) = -44\ \text{kN-m}}$$

where the superscript - refers to just before x=9. Substituting x=9 in Eqs, (7) and (8) we obtain the value of V_y and M_z just after point D as:

$$\boxed{V_y(9^+) = -20\ \text{kN and } M_z(9^+) = -60\ \text{kN-m}}$$

where the superscript + refers to just after x=9. The differences in values for shear force and bending moment just before and after D are equal to the value of concentrated force of 12 kN and the applied moment of 16 kN-m at D.

6.59 For the beam shown (a) write the shear force and moment equation as a function of x in segment AB and segment BC. (b) Show that your results satisfy Equations 6.13 and 6.14. (c) What are the shear force and bending moment value just before and just after point B.

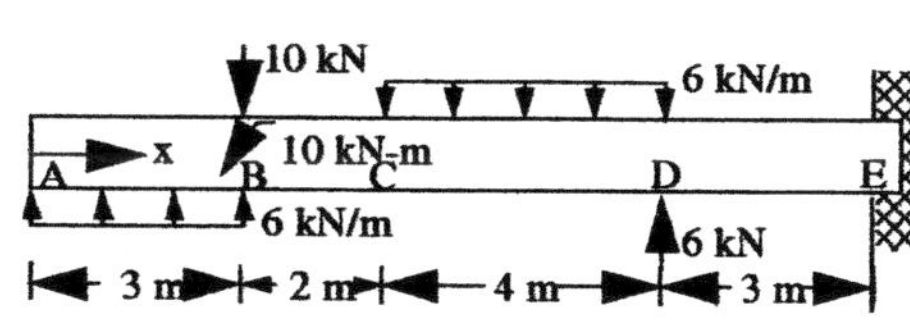

Fig. P6.59

Solution $V_y(x) = ?$ $M_z(x) = ?$

We can make an imaginary cut in AB in Fig. P6.59. That is the location of the imaginary cut is defined by: $0 \le x < 3\,m$. We take the left part of the cut and draw the free body diagram. The internal shear force V_y and internal bending moment M_z are drawn as per our sign convention and shown in Fig. (a).

(a)

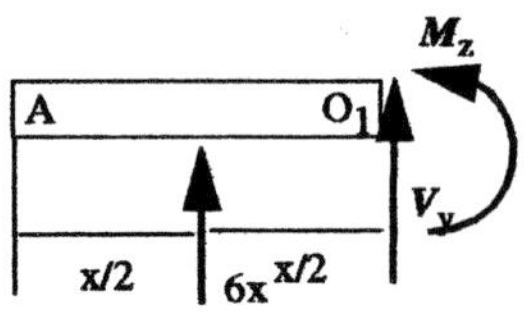

(b)

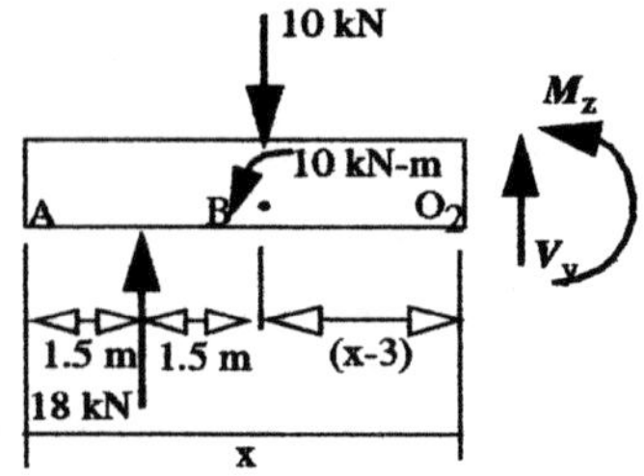

By equilibrium of forces in the y-direction in Fig. (a) we obtain:

$$\boxed{V_y = -6x \text{ kN} \qquad 0 \le x < 3\text{m}} \qquad 1$$

By equilibrium of moment about point O_1 in Fig. (a) we obtain: $M_z - (6x)\frac{x}{2} = 0$ or

$$\boxed{M_z = 3x^2 \text{ kN} - \text{m} \qquad 0 \le x < 3\text{m}} \qquad 2$$

Taking the first derivative of Eq. (1) we obtain.

$$\frac{dV_y}{dx} = -6 = -p_y \qquad 3$$

Taking the first derivative of Eq. (2) we obtain:

$$\frac{dM_z}{dx} = 6x = -V_y \qquad 4$$

Eqs. (3) and (4) confirm that shear force and bending moment expressions of Eqs. (1) and (2) satisfy Eqs. 6.13 and 6.14, respectively.

We can make an imaginary cut in BC in Fig. P6.59. That is the location of the imaginary cut is defined by: $3\,m < x < 5\,m$. We can take the left part of the cut and after replacing the distributed force, draw the free body diagram. The internal shear force V_y and internal bending moment M_z are drawn as per our sign convention and shown in Fig. (b).

By equilibrium of forces in the y-direction in Fig. (b) we obtain: $V_y + 18 - 10 = 0$ or

$$\boxed{V_y = -8 \text{ kN} \qquad 3\text{m} < x < 5\text{m}} \qquad 5$$

By equilibrium of moment about point O_2 in Fig. (c) we obtain: $M_z - (18)(x - 1.5) + 10(x - 3) + 10 = 0$ or

$$\boxed{M_z = (8x - 7) \text{ kN} - \text{m} \qquad 3\text{m} < x < 5\text{m}} \qquad 6$$

Taking the first derivative of Eq. (7) we obtain:

$$\frac{dV_y}{dx} = 0 \qquad 7$$

Taking the first derivative of Eq. (8) we obtain:

$$\frac{dM_z}{dx} = 8 = -V_y \qquad 8$$

Eqs. (7) and (8) confirm that shear force and bending moment expressions of Eqs. (5) and (6) satisfy Eqs. 6.13 and 6.14, respectively.

Point B is at x = 3m. Substituting x = 3 in Eqs. (1) and (2) we obtain the value of V_y and M_z just before point B as:

$\boxed{V_y(3^-) = -18 \text{ kN}}$ and $\boxed{M_z(3^-) = 27 \text{ kN-m}}$

where the superscript - refers to just before x=3. Substituting x=3 in Eqs, (5) and (6) we obtain the value of V_y and M_z just after point B as:

$\boxed{V_y(3^+) = -8 \text{ kN}}$ and $\boxed{M_z(3^+) = 17 \text{ kN-m}}$

where the superscript + refers to just after x=3. The differences in values for shear force and bending moment just before and after B are equal to the value of concentrated force of 10 kN and the applied moment of 10 kN-m at B.

6.60 For the beam shown (a) write the shear force and moment equation as a function of x in segment CD and segment DE. (b) Show that your results satisfy Equations 6.13 and 6.14. (c) What are the shear force and bending moment value just before and just after point D.

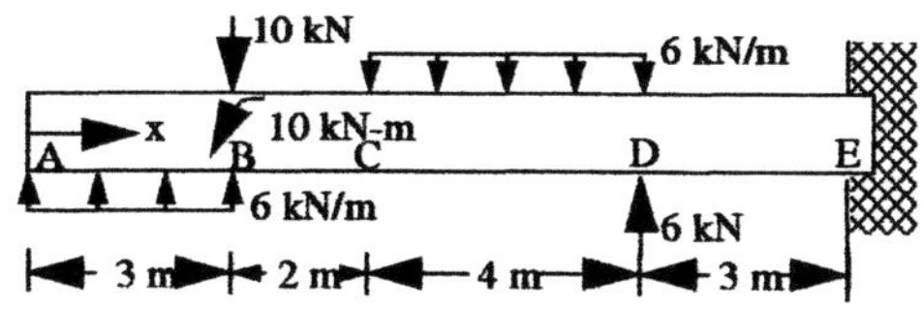

Fig. P6.60

Solution $V_y(x) = ?$ $M_z(x) = ?$

Fig.(a) shows the free body diagram of the entire beam with the distributed loads replaced by an equivalent loads.

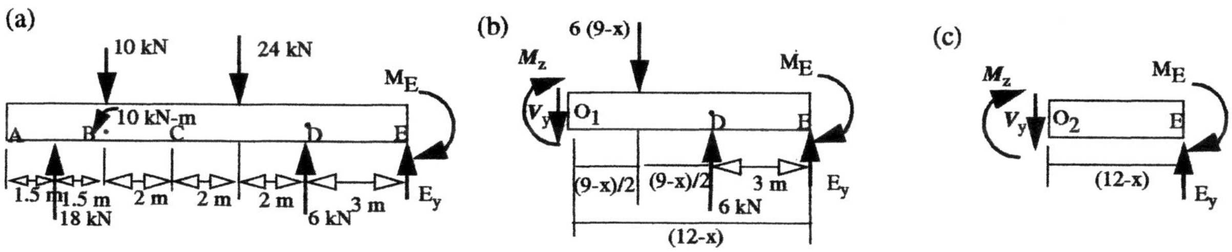

By equilibrium of forces in Fig.(a) we obtain. $E_y + 18 - 10 - 24 + 6 = 0$ or

$$E_y = 10 \text{ kN} \tag{1}$$

By equilibrium of moment about point E in Fig. (a) we obtain:

$M_E + (18)(10.5) - (10)(9) - 10 - (24)(5) + (6)(3) = 0$ or

$$M_E = 13 \text{ kN} - \text{m} \tag{2}$$

We can make an imaginary cut in CD in Fig. P6.60. That is the location of the imaginary cut is defined by: $5\text{m} < x < 9\text{m}$. We take the right part of the cut and after replacing the distributed force, draw the free body diagram. The internal shear force V_y and internal bending moment M_z are drawn as per our sign convention and shown in Fig. (b).

By equilibrium of forces in the y-direction in Fig. (a) we obtain: $-V_y + E_y - 6(9-x) + 6 = 0$ or

$$\boxed{V_y = (6x - 38)\text{kN} \qquad 5\text{m} < x < 9\text{m}} \tag{3}$$

By equilibrium of moment about point O_1 in Fig. (b) we obtain:

$M_z + 6(9-x)\dfrac{(9-x)}{2} - 6(9-x) - E_y(12-x) + M_E = 0$ or

$$\boxed{M_z = (-3x^2 + 38x - 82) \text{ kN} - \text{m} \qquad 5\text{m} < x < 9\text{m}} \tag{4}$$

Taking the first derivative of Eq. (3) we obtain.

$$\frac{dV_y}{dx} = 6 = -p_y \tag{5}$$

Taking the first derivative of Eq. (4) we obtain:

$$\frac{dM_z}{dx} = (-6x + 38) = -V_y \qquad 6$$

Eqs. (5) and (6) confirm that shear force and bending moment expressions of Eqs. (3) and (4) satisfy Eqs. 6.13 and 6.14, respectively.
We can make an imaginary cut in DE in Fig. P6.60. That is the location of the imaginary cut is defined by: $9m < x < 12m$. We can take the right part of the cut and draw the free body diagram. The internal shear force V_y and internal bending moment M_z are drawn as per our sign convention and shown in Fig. (b).
By equilibrium of forces in the y-direction in Fig. (b) we obtain: $V_y = E_y$ or

$$\boxed{V_y = 10 \text{ kN} \qquad 9m < x < 12m} \qquad 7$$

By equilibrium of moment about point O_2 in Fig. (b) we obtain: $M_z + M_E - E_y(12 - x) = 0$ or

$$\boxed{M_z = (107 - 10x) \text{ kN}-\text{m} \qquad 9m < x < 12m} \qquad 8$$

Taking the first derivative of Eq. (7) we obtain:

$$\frac{dV_y}{dx} = 0 \qquad 9$$

Taking the first derivative of Eq. (8) we obtain:

$$\frac{dM_z}{dx} = 10 = -V_y \qquad 10$$

Eqs. (9) and (10) confirm that shear force and bending moment expressions of Eqs. (7) and (8) satisfy Eqs. 6.13 and 6.14, respectively.
Point D is at x = 9m. Substituting x = 9 in Eqs. (3) and (4) we obtain the value of V_y and M_z just before point D as:

$$\boxed{V_y(9^-) = 16 \text{ kN}} \text{ and } \boxed{M_z(9^-) = 17 \text{ kN-m}}$$

where the superscript - refers to just before x=9. Substituting x=9 in Eqs, (7) and (8) we obtain the value of V_y and M_z just after point D as:

$$\boxed{V_y(9^+) = 10 \text{ kN}} \text{ and } \boxed{M_z(9^+) = 17 \text{ kN-m}}$$

where the superscript + refers to just after x=9. The difference in values for shear force just before and after D are equal to the value of concentrated force of 6 kN. As no moment is applied at D, the internal moment are continuous.

6.61 During skiing, the weight of a person is often all on one ski. The ground reaction p(x), when the weight is on one ski, is modeled as shown. (a) Find shear force and bending moment as a function of x across the ski. (b) The ski is 50 mm wide and the thickness of the ski varies as shown. Determine the maximum bending normal stress in the ski. Use of spread sheet recommended.

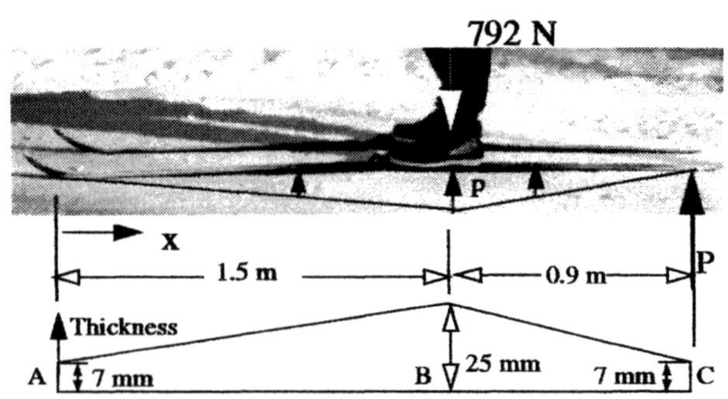

Fig. P6.61

Solution $V_y(x) = ?$ $M_z(x) = ?$ $\sigma_{max} = ?$

A schematic of the beam representing the ski and loading are shown in Fig. (a). The taper in the ski is not shown as it does not affect the bending moment distribution. Fig. (b) shows the free body diagram with distributed force replaced by an equivalent force. By moment equilibrium about point C in Fig. (b) we obtain:

$$\left[\frac{1}{2}(p)(1.5)\right](1.4) + \left[\frac{1}{2}(p)(0.9)\right]0.6 - (792)(0.9) = 0 \qquad \text{or} \qquad p = 540 \ \text{N/m} \qquad \mathbf{1}$$

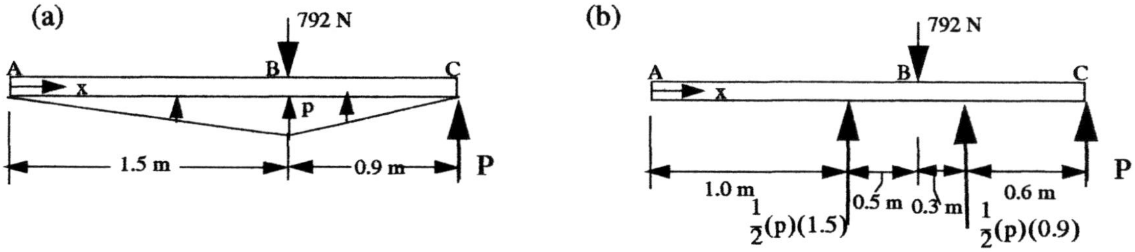

By force equilibrium in Fig. (b) we obtain:

$$P - 792 + \frac{1}{2}(p)(1.5) + \frac{1}{2}(p)(0.9) = 0 \qquad \text{or} \qquad P = 144 \ \text{N} \qquad \mathbf{2}$$

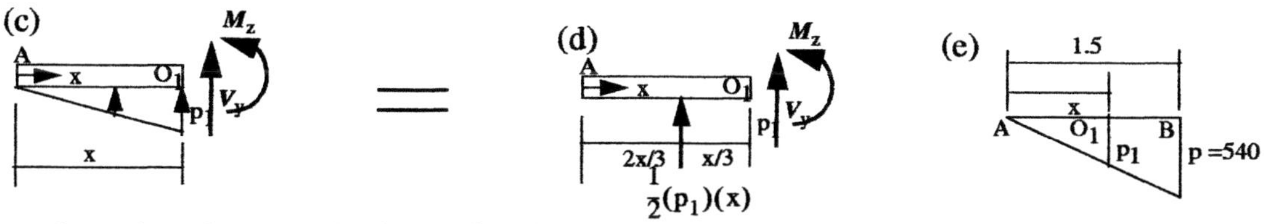

We can make an imaginary cut in AB in Fig. (a). That is the location of the imaginary cut is defined by: $0 \le x < 1.5$. We take the left part of the cut and draw the free body diagram.The internal shear force V_y and internal bending moment M_z are drawn as per our sign convention and shown in Fig. (c). The distributed load is replaced by equivalent load as shown in Fig. (d). The intensity of distributed load at the imaginary cut (point O_1) can be found from similar triangle shown in Fig. (e) as:

$$\frac{p_1}{x} = \frac{540}{1.5} \qquad \text{or} \qquad p_1 = 360x \ \text{N/m} \qquad \mathbf{3}$$

By equilibrium of forces in the y-direction in Fig. (d) we obtain: $V_y + \frac{1}{2}(p_1)(x) = 0$ or

$$\boxed{V_y = -180x^2 \ \text{N} \qquad 0 \le x < 1.5} \qquad \mathbf{4}$$

By equilibrium of moment about point O_1 in Fig. (d) we obtain: $M_z - \left[\frac{1}{2}(p_1)(x)\right]\left(\frac{x}{3}\right) = 0$ or

$$\boxed{M_z = 60x^3 \ \text{N-m} \qquad 0 \le x < 1.5} \qquad \mathbf{5}$$

We can make an imaginary cut in BC in Fig. (a). That is the location of the imaginary cut is defined by: $1.5 < x < 2.4$. We take the left part of the cut and draw the free body diagram. The internal shear force V_y and internal bending moment M_z are drawn as per our sign convention and shown in Fig. (f). The distributed load is replaced by equivalent load as shown in Fig. (g). The intensity of distributed load at the imaginary cut (point O_2) can be found from similar triangle shown in Fig. (h) as:

$$\frac{p_2}{(2.4 - x)} = \frac{540}{0.9} \qquad \text{or} \qquad p_2 = 600(2.4 - x) \qquad \mathbf{6}$$

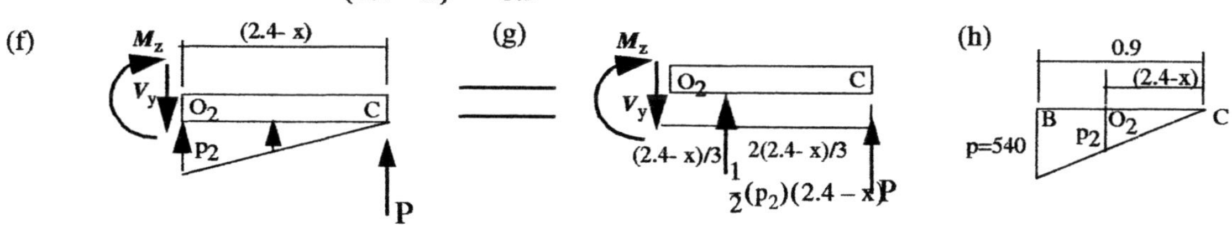

By equilibrium of forces in the y-direction in Fig. (f) we obtain: $-V_y + \frac{1}{2}(p_2)(2.4-x) + P = 0$ or

$$\boxed{V_y = [300(2.4-x)^2 - 144]\ \text{N} \qquad 1.5 < x < 2.4} \qquad 7$$

By equilibrium of moment about point O_2 in Fig. (g) we obtain:

$$M_z - \left[\frac{1}{2}(p_2)(2.4-x)\right]\left(\frac{2.4-x}{3}\right) - 144(2.4-x) = 0 \text{ or}$$

$$\boxed{M_z = [100(2.4-x)^3 + 144(2.4-x)]\ \text{N-m} \qquad 1.5 < x < 2.4} \qquad 8$$

The thickness t(x) of the ski is a two part linear function. Equation of straight line for each part can be found and written as shown below.

$$t(x) = \begin{cases} 0.007 + 0.012x & 0 \le x < 1.5 \\ 0.055 - 0.020x & 1.5 < x < 2.4 \end{cases} \qquad 9$$

The area moment of inertia for rectangular cross-section of width b = 0.05 m and thickness t(x) $I_{zz} = \frac{1}{12}b[t(x)]^3$. The maximum bending normal stress will exist at top and bottom surface, i.e., $y_{max} = t(x)/2$. Thus, the section modulus is

$$S(x) = \frac{I_{zz}}{y_{max}} = \frac{1}{6}b[t(x)]^2 = \frac{0.05}{6}[t(x)]^2 \qquad 10$$

The maximum bending normal stress at any section is: $\sigma = M_z/S(x)$. Substituting Eqs. 5, 8,9, and 10 we obtain the following expression for the maximum bending normal stress at any x as:

$$\sigma = \frac{60x^3}{\frac{0.05}{6}[0.007 + 0.012x]^2} = 7200\frac{x^3}{[0.007 + 0.012x]^2}\ \text{N/m}^2 \qquad 0 \le x < 1.5 \qquad 11$$

$$\sigma = \frac{100(2.4-x)^3 + 144(2.4-x)}{\frac{0.05}{6}[0.055 - 0.020x]^2} = \frac{12000(2.4-x)^3 + 17280(2.4-x)}{[0.055 - 0.020x]^2}\ \text{N/m}^2 \qquad 1.5 < x < 2.4 \qquad 12$$

Eqs. 11 and 12 are evaluated for various values of x on a spread sheet and the results shown in the table below.

x (m)	σ (N/m²)
0	0.000E+00
0.1	1.071E+05
0.2	6.519E+05
0.3	1.730E+06
0.4	3.309E+06
0.5	5.325E+06

x (m)	σ (N/m²)
0.6	7.713E+06
0.7	1.041E+07
0.8	1.338E+07
0.9	1.657E+07
1	1.994E+07
1.1	2.349E+07

x (m)	σ (N/m²)
1.2	2.717E+07
1.3	3.097E+07
1.4	3.488E+07
1.5	3.888E+07
1.6	1.452E+07
1.7	1.167E+07

x (m)	σ (N/m²)
1.8	8.975E+06
1.9	6.488E+06
2	4.267E+06
2.1	2.396E+06
2.2	9.917E+05
2.3	1.852E+05
2.4	0.000

The maximum bending normal stress is: $\boxed{\sigma_{max} = 38.9\ \text{MPa(T) or (C)}}$

6.62 Draw the shear and moment diagram for the beam and loading shown. Determine the maximum values of shear force V_y and bending moment M_z.

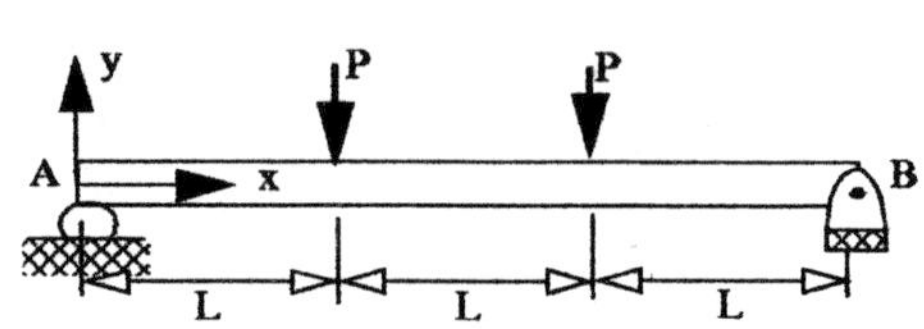

Fig. P6.62

Solution Plot of V_y and M_z $(V_y)_{max}$ = ? $(M_z)_{max}$ = ?

By symmetry the reactions at the support are equal. The beam with the reactions and imaginary extensions LA and BR are drawn in Fig. (a), the shear force diagram is drawn in Fig. (b) and bending moment diagram is drawn in Fig. (c).

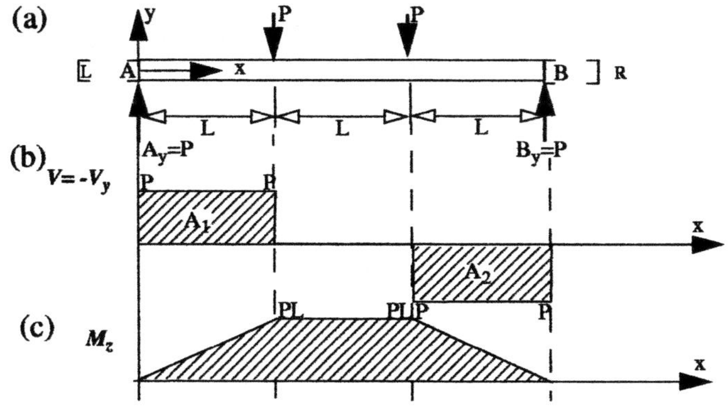

$$A_1 = A_2 = PL$$

The maximum value of shear force is: $(V_y)_{max} = \pm P$

The maximum value of bending moment is: $(M_z)_{max} = PL$

6.63 Draw the shear and moment diagram for the beam and loading shown. Determine the maximum values of shear force V_y and bending moment M_z.

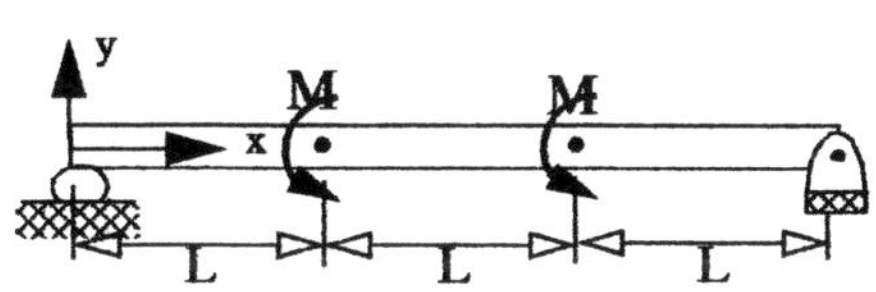

Fig. P6.63

Solution Plot of V_y and M_z $(V_y)_{max}$ = ? $(M_z)_{max}$ = ?

The free body diagram of the beam is shown in Fig. (a). By equilibrium of moment about point B we obtain:

$$3A_yL = 2M \quad \text{or} \quad A_y = 2M/3L \qquad 1$$

By force equilibrium in the y direction we obtain:

$$B_y = A_y = 2M/3L \qquad 2$$

The beam with the reactions and imaginary extensions LA and BR are drawn in Fig. (a), the shear force diagram is drawn in Fig. (b) and bending moment diagram is drawn in Fig. (c).

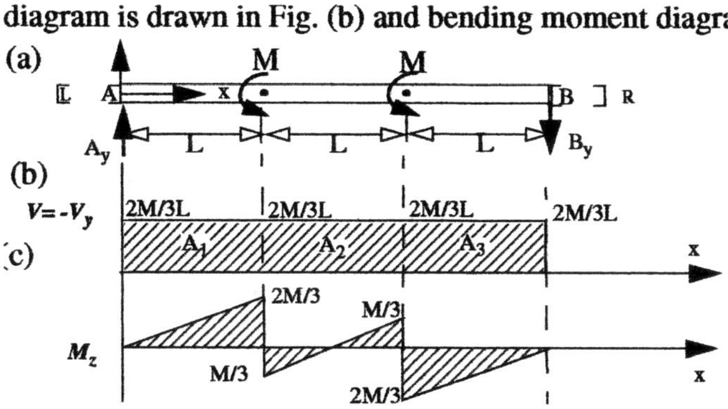

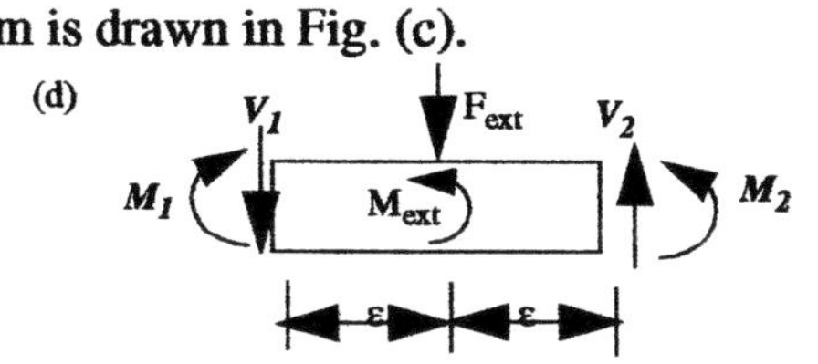

$$V_2 = V_1 + F_{ext}$$

Template Equations

$$M_2 = M_1 - M_{ext}$$

$$A_1 = A_2 = A_3 = 2M/3$$

The maximum value of shear force is: $(V_y)_{max} = 2M/3L$

The maximum value of bending moment is: $(M_z)_{max} = 2M/3$

6.64 For the beam shown in Fig. P6.49, draw the shear force bending moment diagram. Determine the maximum values of shear force V_y and bending moment M_z.

Solution Plot of V_y and M_z $(V_y)_{max} = ?$ $(M_z)_{max} = ?$

The reactions at the support were calculated in problem Fig. P6.49. The beam with the reactions and imaginary extensions LA and BR are drawn in Fig. (a), the shear force diagram is drawn in Fig. (b) and bending moment diagram is drawn in Fig. (c).

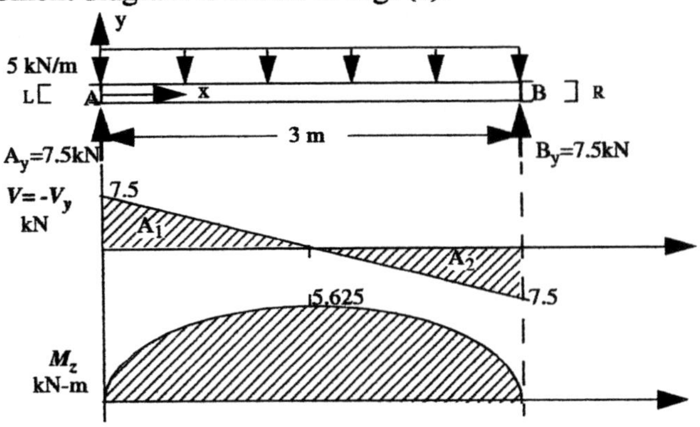

$$A_1 = A_2 = \frac{1}{2}(7.5)(1.5) = 5.625$$

The maximum value of shear force is: $(V_y)_{max} = \pm 7.5 \text{ kN}$

The maximum value of bending moment is: $(M_z)_{max} = 5.625 \text{ kN} - \text{m}$

6.65 For the beam shown in Fig. P6.50, draw the shear force bending moment diagram. Determine the maximum values of shear force V_y and bending moment M_z.

Solution Plot of V_y and M_z $(V_y)_{max} = ?$ $(M_z)_{max} = ?$

Fig.(a) shows free body diagram of the beam with distributed force replaced by an equivalent force. By equilibrium of forces in the y-direction we obtain: $A_y = 216 \text{ kips}$. By equilibrium of moment about point A, we obtain: $M_A = (216)(36) = 7776 \text{ in-kips}$

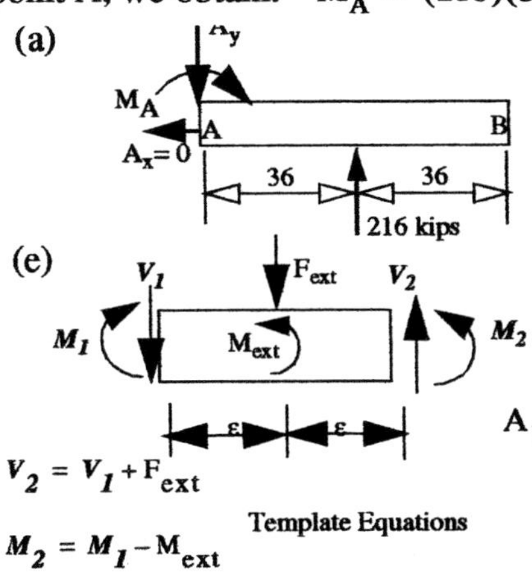

$$A_1 = \frac{1}{2}(216)(772) = 776$$

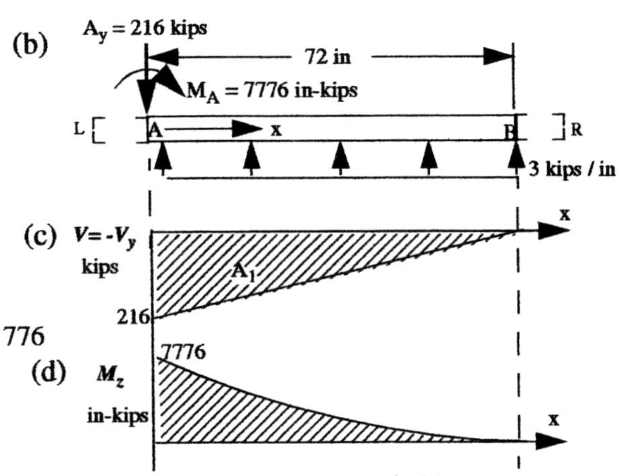

The beam with the reactions and imaginary extensions LA and BR are drawn in Fig. (a). Using the template shown in Fig. (e), the shear force diagram is drawn in Fig. (c) and bending moment diagram is drawn in Fig. (d).

The maximum value of shear force is: $(V_y)_{max} = 216 \text{ kips}$

The maximum value of bending moment is: $(M_z)_{max} = 7776 \text{ in} - \text{kips}$

6.66 Draw the shear and moment diagram for the beam and loading shown. Determine the maximum values of shear force V_y and bending moment M_z.

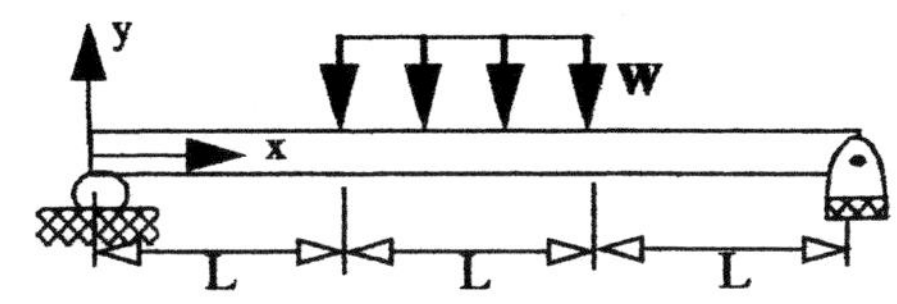

Fig. P6.66

Solution Plot of V_y and M_z $(V_y)_{max} = ?$ $(M_z)_{max} = ?$

Fig.(a) shows free body diagram of the beam with distributed force replaced by an equivalent force. By equilibrium of moment about point B, we obtain: $A_y(3L) - (wL)3(L/2) = 0$ or $A_y = wL/2$. By equilibrium of forces in the y-direction we obtain: $A_y + B_y - wL = 0$ or $B_y = wL/2$.

(a)

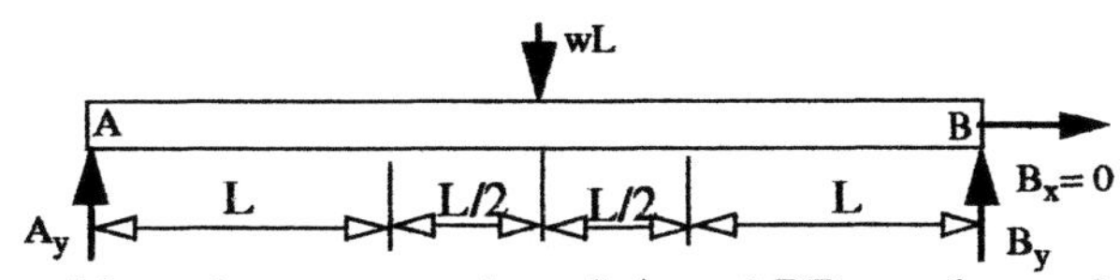

The beam with the reactions and imaginary extensions LA and BR are drawn in Fig. (b). The shear force diagram is drawn in Fig. (c) and bending moment diagram is drawn in Fig. (c).

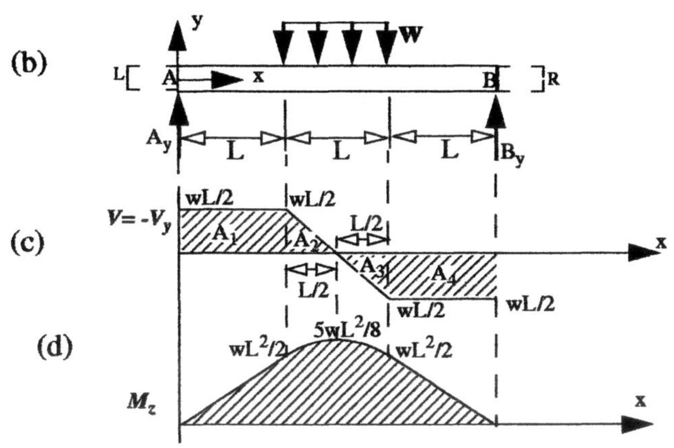

$$A_1 = A_4 = \frac{wL^2}{2}$$

$$A_2 = A_3 = \frac{1}{2}\left(\frac{wL}{2}\right)\left(\frac{L}{2}\right) = \frac{wL^2}{8}$$

The maximum value of shear force is: $(V_y)_{max} = \pm\frac{wL}{2}$

The maximum value of bending moment is: $(M_z)_{max} = \frac{5}{8}wL^2$

6.67 Draw the shear and moment diagram for the beam and loading shown. Determine the maximum values of shear force V_y and bending moment M_z.

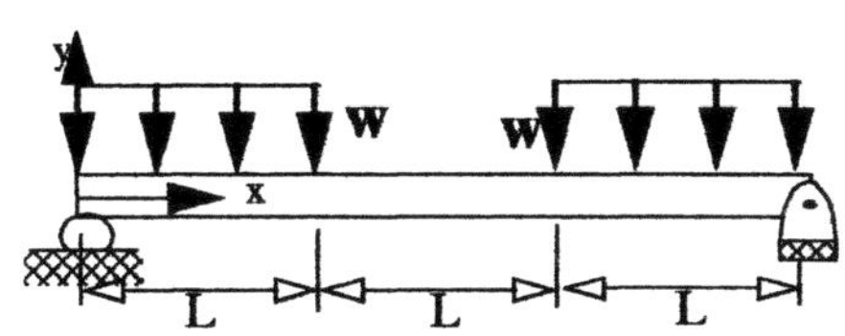

Fig. P6.67

Solution Plot of V_y and M_z $(V_y)_{max} = ?$ $(M_z)_{max} = ?$

Fig.(a) shows free body diagram of the beam with distributed force replaced by an equivalent force. By equilibrium of moment about point B, we obtain: $A_y(3L) - (wL)(5L/2) - (wL)(L/2) = 0$ or $A_y = wL$. By equilibrium of forces in the y-direction we obtain: $A_y + B_y - 2wL = 0$ or $B_y = wL$.
The beam with the reactions and imaginary extensions LA and BR are drawn in Fig. (a). The shear force diagram is drawn in Fig. (b) and bending moment diagram is drawn in Fig. (c).

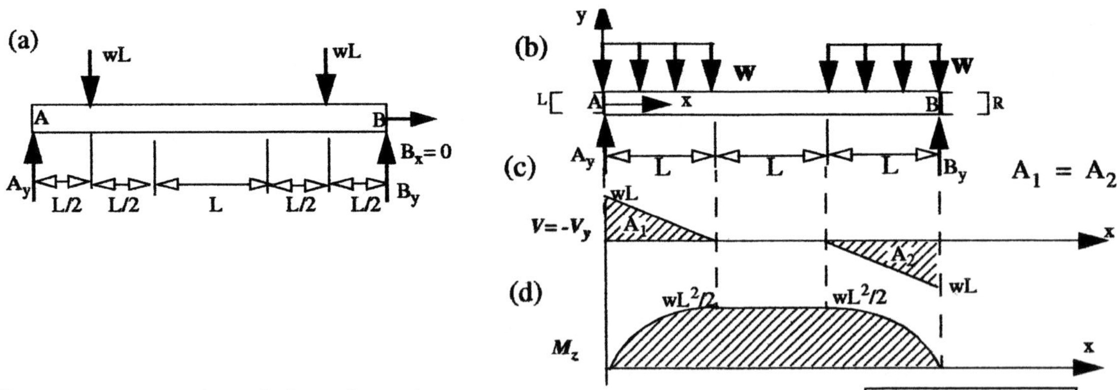

$$A_1 = A_2 = \frac{1}{2}wL^2$$

The maximum value of shear force is: $(V_y)_{max} = \pm wL$

The maximum value of bending moment is: $(M_z)_{max} = \frac{1}{2}wL^2$

6.68 Draw the shear and moment diagram for the beam shown in Fig. P6.57. Determine the maximum values of shear force V_y and bending moment M_z.

Solution Plot of V_y and M_z $(V_y)_{max} = ?$ $(M_z)_{max} = ?$

The reactions at the support were calculated in problem Fig. P6.57. The beam with the reactions and imaginary extensions LA and ER are drawn in Fig. (a), the shear force diagram is drawn in Fig. (b) and bending moment diagram is drawn in Fig. (c). The template and equations are shown in Fig. (d).

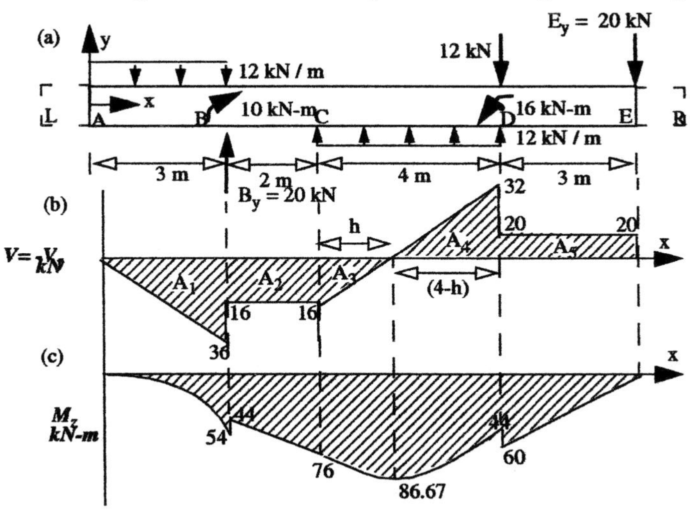

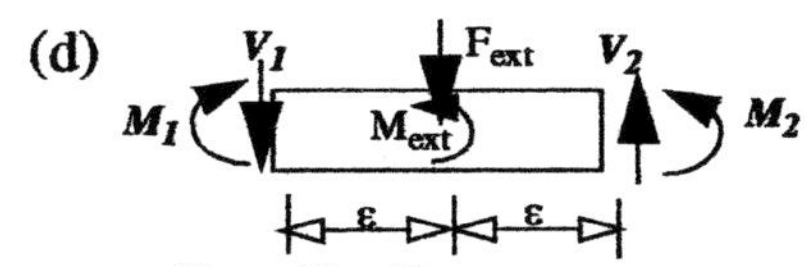

$$V_2 = V_1 + F_{ext}$$
$$M_2 = M_1 - M_{ext}$$

Template Equations

$$\frac{h}{16} = \frac{4-h}{32} \quad \text{or} \quad h = 1.333\,\text{m}$$

$$A_1 = \frac{1}{2}(36)(3) = 54$$

$$A_2 = (16)(2) = 32$$

$$A_3 = \frac{1}{2}(16)(h) = 10.67$$

$$A_4 = \frac{1}{2}(32)(4-h) = 42.67$$

$$A_5 = (20)(3) = 60$$

The maximum value of shear force is: $(V_y)_{max} = 36\ \text{kN}$

The maximum value of bending moment is: $(M_z)_{max} = -86.67\ \text{kN}-\text{m}$

6.69 Draw the shear and moment diagram for the beam shown in Fig. P6.60 determine the values of maximum shear force and bending moment.

Solution Plot of V_y and M_z $(V_y)_{max} = ?$ $(M_z)_{max} = ?$

The reactions at the support were calculated in problem Fig. P6.60. The beam with the reactions and imaginary extensions LA and ER are drawn in Fig. (a), the shear force diagram is drawn in Fig. (b) and bending moment diagram is drawn in Fig. (c). The template and equations are shown in Fig. (d).

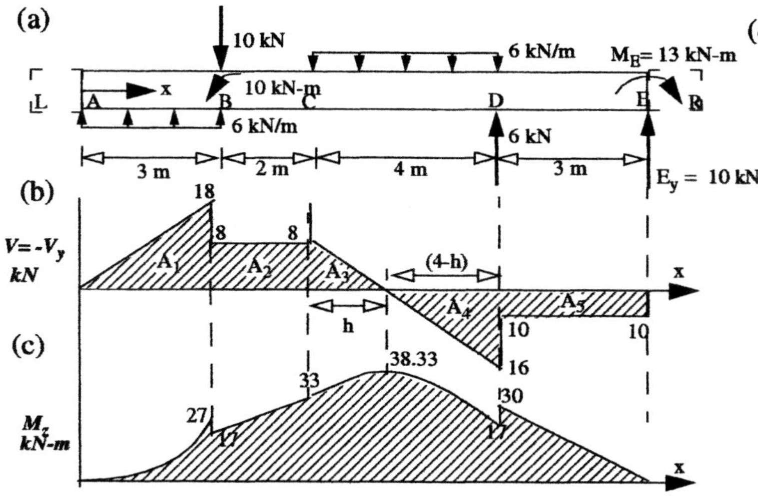

(d)

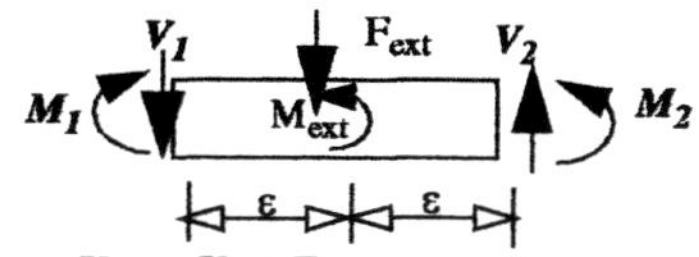

$V_2 = V_1 + F_{ext}$ Template Equations

$M_2 = M_1 - M_{ext}$

$\frac{h}{8} = \frac{4-h}{16}$ or $h = 1.333\text{m}$

$A_1 = \frac{1}{2}(18)(3) = 27$

$A_2 = (8)(2) = 16$

$A_3 = \frac{1}{2}(8)(h) = 5.333$

$A_4 = \frac{1}{2}(16)(4-h) = 21.333$

$A_5 = (10)(3) = 30$

The maximum value of shear force is: $(V_y)_{max} = -18 \text{ kN}$

The maximum value of bending moment is: $(M_z)_{max} = 38.33 \text{ kN}-\text{m}$

6.70 Draw the shear and moment diagram and determine the values of maximum shear force and bending moment.

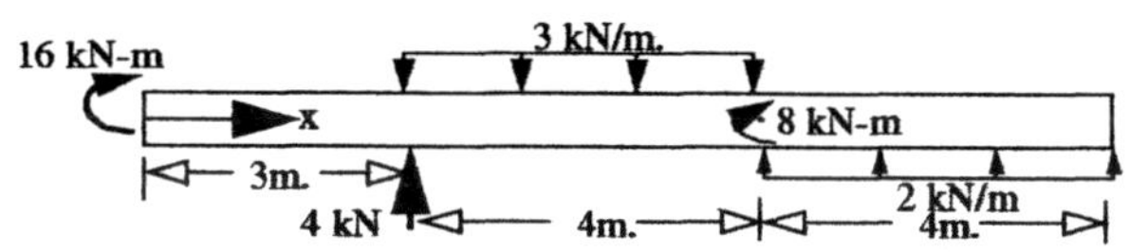

Fig. P6.70

Solution Plot of V_y and M_z $(V_y)_{max} = ?$ $(M_z)_{max} = ?$

The beam with imaginary extensions LA and ER are drawn in Fig. (a), the shear force diagram is drawn in Fig. (b) and bending moment diagram is drawn in Fig.(c). The template and equations are shown in Fig.(d).

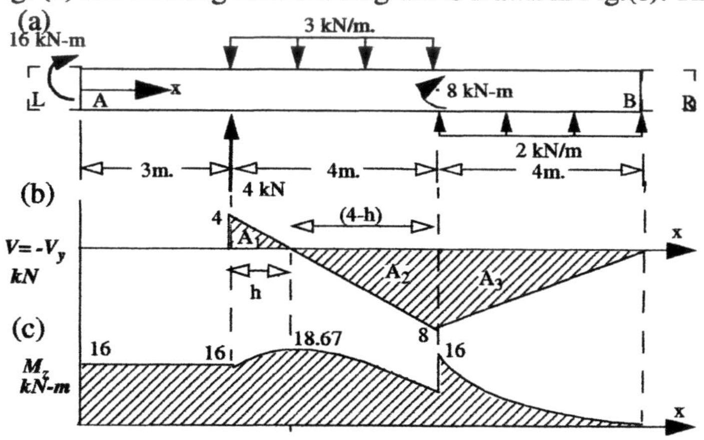

(d)

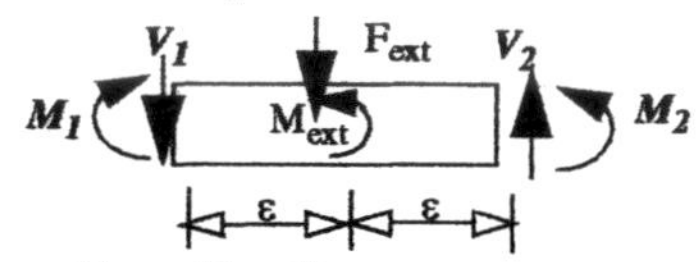

$V_2 = V_1 + F_{ext}$ Template Equations

$M_2 = M_1 - M_{ext}$

$\frac{h}{4} = \frac{4-h}{8}$ or $h = 1.333\text{m}$

$A_1 = \frac{1}{2}(4)(h) = 2.667$

$A_2 = \frac{1}{2}(8)(4-h) = 10.667$

$A_3 = \frac{1}{2}(8)(4) = 16$

The maximum value of shear force is: $(V_y)_{max} = 8 \text{ kN}$

The maximum value of bending moment is: $(M_z)_{max} = 18.67 \text{ kN}-\text{m}$

6.71 Draw the shear and moment diagram and determine the values of maximum shear force and bending moment.

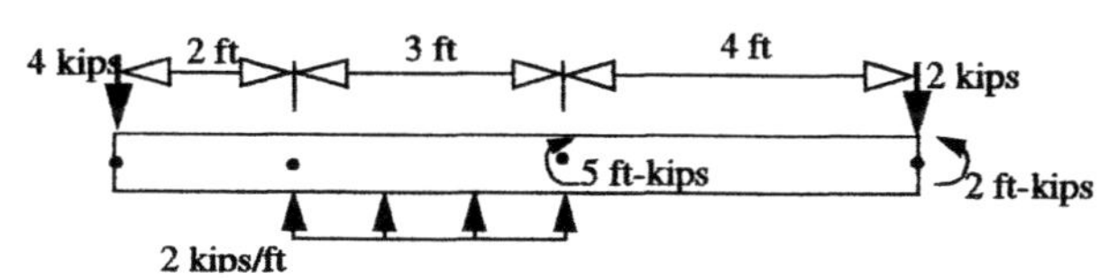

Fig. P6.71

Solution Plot of V_y and M_z $(V_y)_{max} = ?$ $(M_z)_{max} = ?$

The beam with the reactions and imaginary extensions LA and ER are drawn in Fig. (a), the shear force diagram is drawn in Fig. (b) and bending moment diagram is drawn in Fig. (c). The template and equations are shown in Fig. (d).

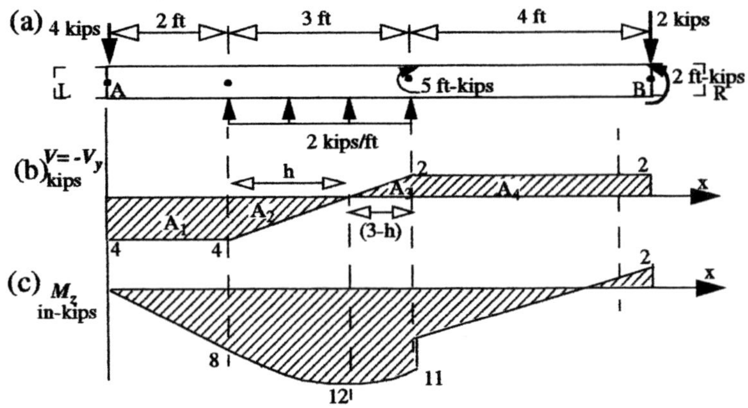

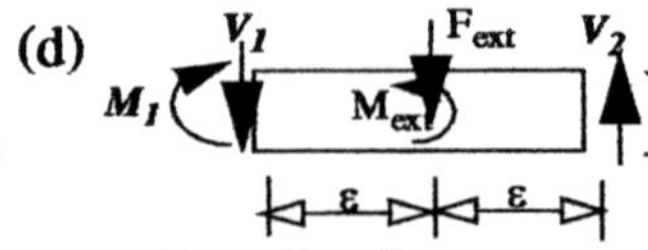

$V_2 = V_1 + F_{ext}$ Template Equations

$M_2 = M_1 - M_{ext}$

$\frac{h}{4} = \frac{3-h}{2}$ or $h = 2\text{ft}$

$A_1 = (4)(2) = 8$

$A_2 = \frac{1}{2}(4)(h) = 4$

$A_3 = \frac{1}{2}(2)(3-h) = 1$

$A_4 = (2)(4) = 8$

The maximum value of shear force is: $(V_y)_{max} = 4 \text{ kips}$

The maximum value of bending moment is: $(M_z)_{max} = -12 \text{ in-kips}$

6.72 Draw the shear and moment diagram and determine the values of maximum shear force and bending moment.

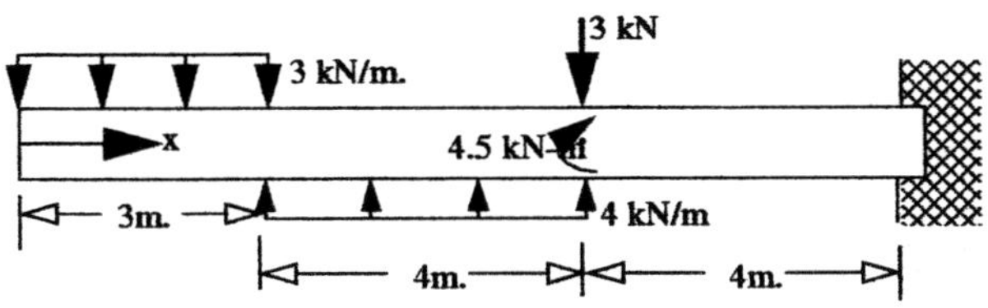

Fig. P6.72

Solution Plot of V_y and M_z $(V_y)_{max} = ?$ $(M_z)_{max} = ?$

The beam with the reactions and imaginary extensions LA and ER are drawn in Fig. (a), the shear force diagram is drawn in Fig. (b) and bending moment diagram is drawn in Fig. (c). The template and equations are shown in Fig. (d).

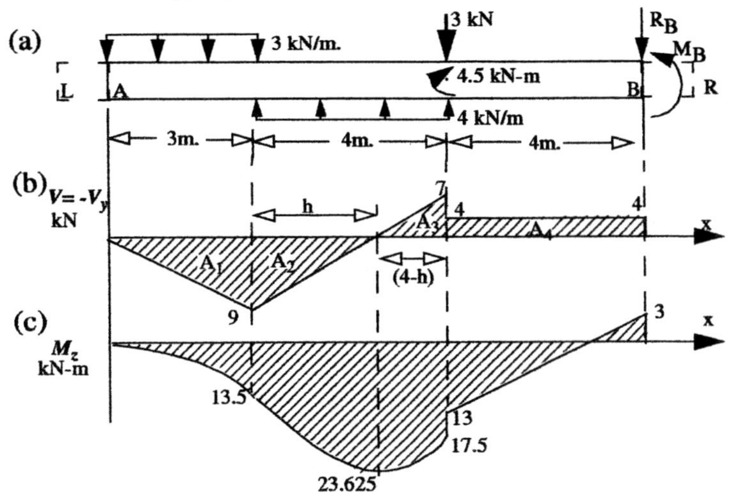

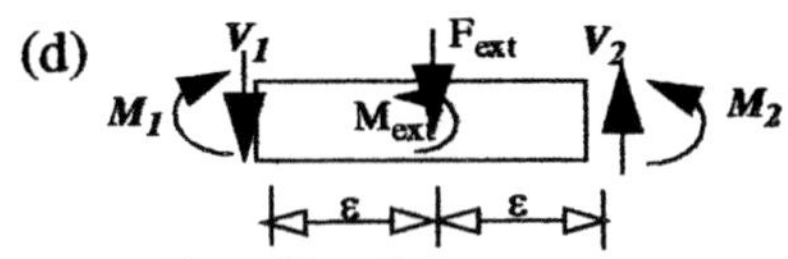

$V_2 = V_1 + F_{ext}$ Template Equations

$M_2 = M_1 - M_{ext}$

$\frac{h}{9} = \frac{4-h}{7}$ or $h = 2.25\text{m}$

$A_1 = \frac{1}{2}(9)(3) = 13.5$

$A_2 = \frac{1}{2}(9)(h) = 10.125$

$A_3 = \frac{1}{2}(7)(4-h) = 6.125$

$A_4 = (4)(4) = 16$

To check the results we draw the free body diagram of the entire beam with distributed forces replaced by equivalent forces as shown in Fig. (e).

By equilibrium of forces in the y direction we obtain: $-R_B - 9 + 16 - 3 = 0$ or $R_B = 4kN$.

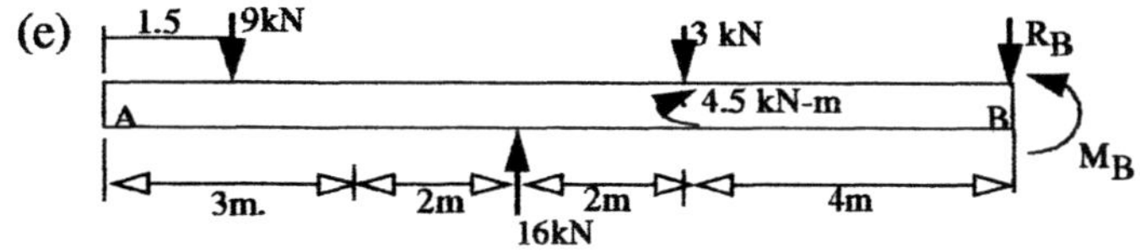

By moment equilibrium about point B we obtain: $M_B + (9)(9.5) - (16)(6) + (3)(4) - 4.5 = 0$ or $M_B = 3kN - m$. These values of R_B and M_B ensure that shear force and bending moment in the imaginary extension BR are zero.

The maximum value of shear force is: $(V_y)_{max} = 9\ kN$

The maximum value of bending moment is: $(M_z)_{max} = -23.625\ kN - m$

6.73 Draw the shear and moment diagram and determine the values of maximum shear force and bending moment.

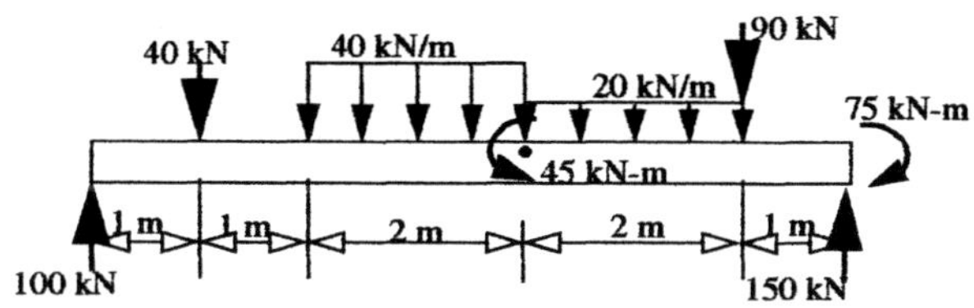

Fig. P6.73

Solution Plot of V_y and M_z $(V_y)_{max} = ?$ $(M_z)_{max} = ?$

The beam with the reactions and imaginary extensions LA and ER are drawn in Fig. (a), the shear force diagram is drawn in Fig. (b) and bending moment diagram is drawn in Fig. (c). The template and equations are shown in Fig. (d).

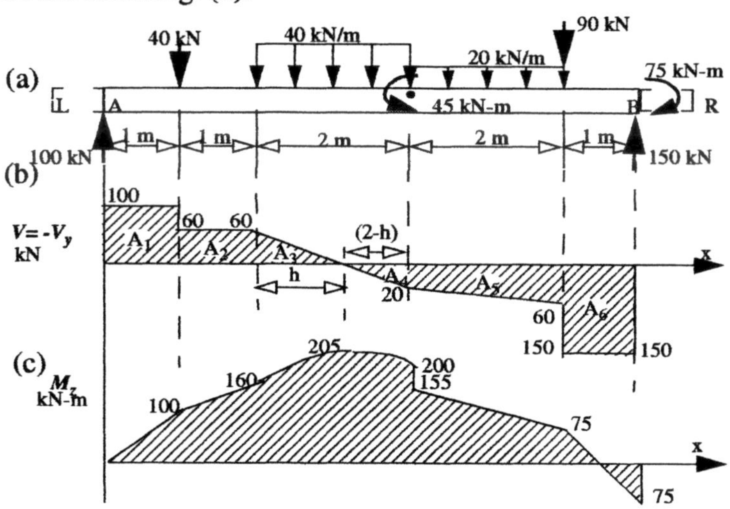

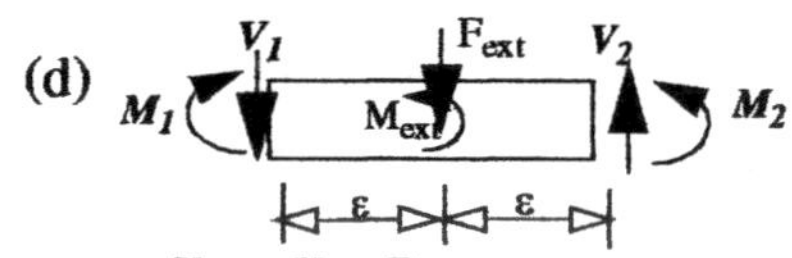

$V_2 = V_1 + F_{ext}$ Template Equations

$M_2 = M_1 - M_{ext}$

$$\frac{h}{60} = \frac{2-h}{20} \quad \text{or} \quad h = 1.5m$$

$A_1 = (100)(1) = 100$

$A_2 = (60)(1) = 60$

$A_3 = \frac{1}{2}(60)(h) = 45$

$A_4 = \frac{1}{2}(20)(2-h) = 5$

$A_5 = \frac{1}{2}(60+20)(2) = 80$

$A_6 = (150)(1) = 150$

The maximum value of shear force is: $(V_y)_{max} = 150\ kN$

The maximum value of bending moment is: $(M_z)_{max} = 205\ kN - m$

6.74 Draw the shear and moment diagram and determine the values of maximum shear force and bending moment.

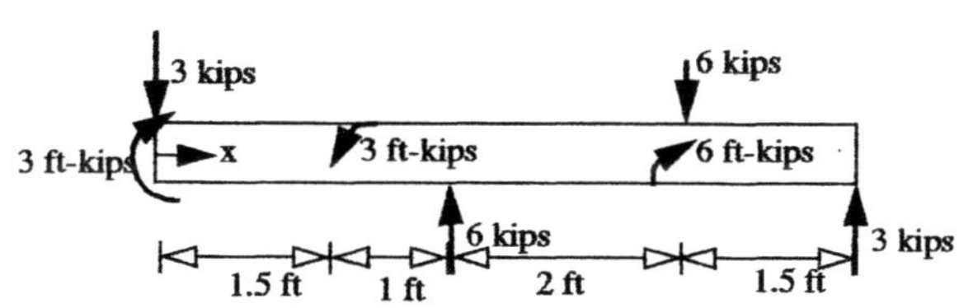

Fig. P6.74

Solution Plot of V_y and M_z $(V_y)_{max} = ?$ $(M_z)_{max} = ?$

The beam with the reactions and imaginary extensions LA and ER are drawn in Fig. (a), the shear force diagram is drawn in Fig. (b) and bending moment diagram is drawn in Fig. (c). The template and equations are shown in Fig. (d).

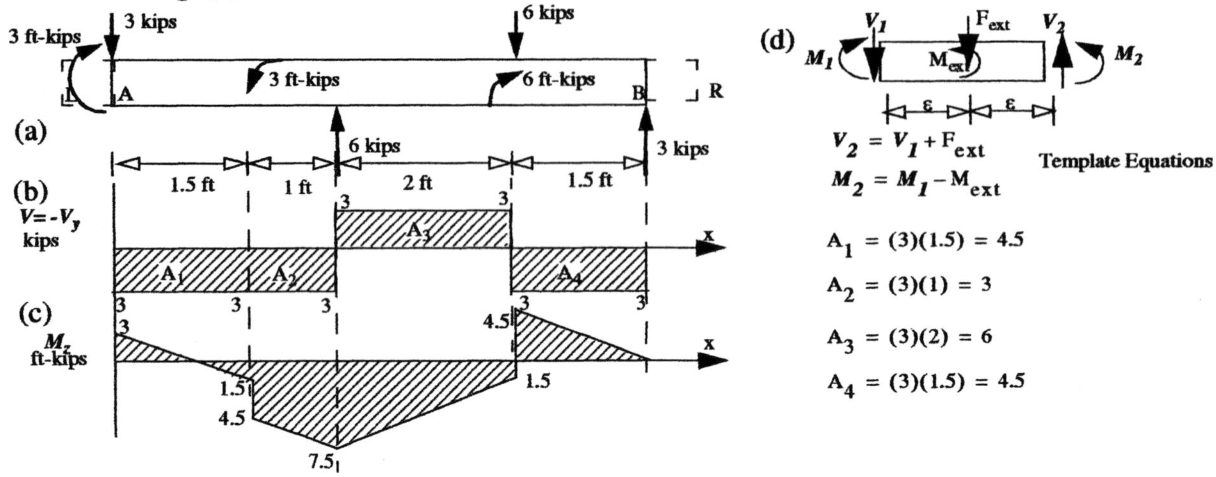

The maximum value of shear force is: $(V_y)_{max} = \pm 3 \text{ kips}$

The maximum value of bending moment is: $(M_z)_{max} = 7.5 \text{ in-kips}$

6.75 Draw the shear and moment diagram and determine the values of maximum shear force and bending moment.

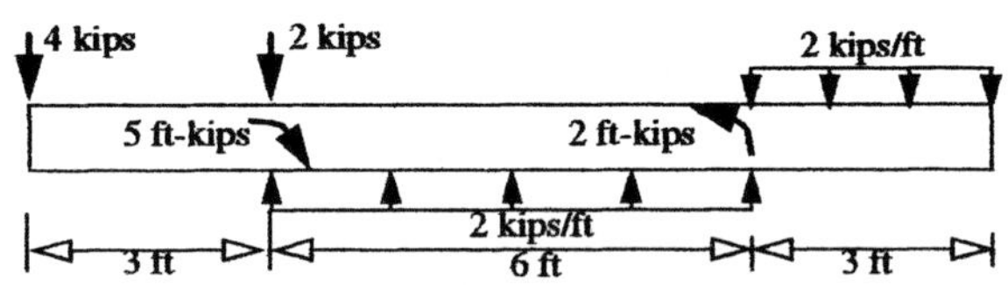

Fig. P6.75

Solution Plot of V_y and M_z $(V_y)_{max} = ?$ $(M_z)_{max} = ?$

The beam with the reactions and imaginary extensions LA and ER are drawn in Fig. (a), the shear force diagram is drawn in Fig. (b) and bending moment diagram is drawn in Fig. (c). The template and equations are shown in Fig. (d).

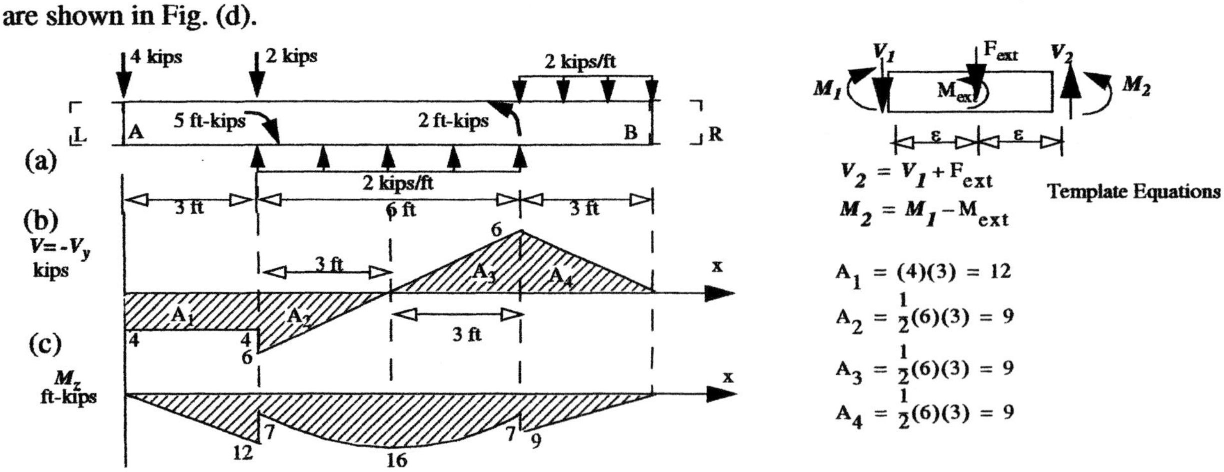

The maximum value of shear force is: $(V_y)_{max} = \pm 6 \text{ kips}$

The maximum value of bending moment is: $(M_z)_{max} = -16 \text{ in-kips}$

6.76 The beam, loading and the cross-section are as shown Fig. P6.76. Determine the intensity w of the distributed load, if the maximum bending normal stress is limited to 10 ksi (C) and 6 ksi (T). The second area moment of inertia is $I_{zz} = 47.73 in^4$.

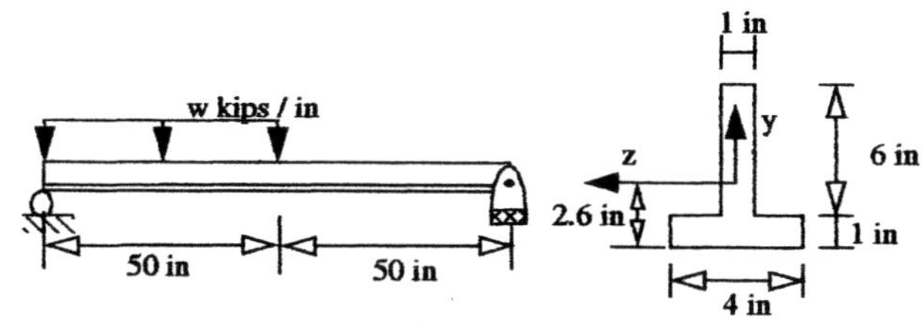

Fig. P6.76

Solution $\sigma_C \le 10 \text{ ksi(C)}$ $\sigma_T \le 6 \text{ ksi(T)}$ $w_{max} = ?$

Fig. (a) shows the free body diagram of the beam. By equilibrium of moment about point B in Fig.(a) we obtain.

$$A_y(100) - (50w)(75) = 0 \quad \text{or} \quad A_y = 37.5w \text{ kips} \qquad 1$$

By equilibrium of forces in the y direction we obtain:

$$A_y + B_y - (50w) = 0 \quad \text{or} \quad B_y = 12.5w \text{ kips} \qquad 2$$

The shear force and bending moment diagrams are shown in Fig. (b)

(a)

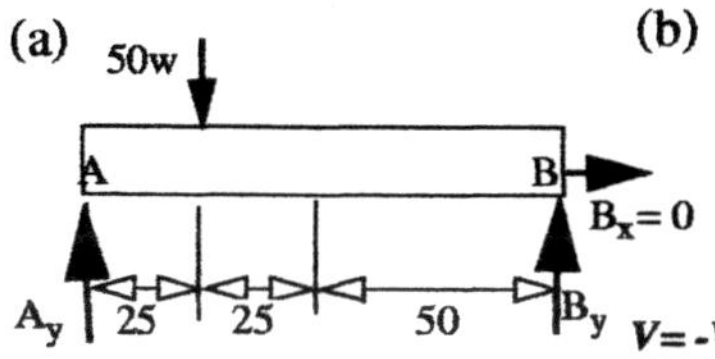

(b)

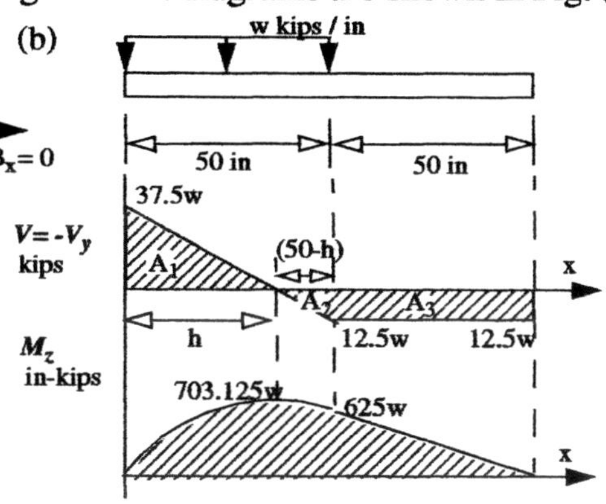

$$\frac{h}{37.5w} = \frac{50-h}{12.5w} \quad \text{or} \quad h = 37.5\text{in}$$

$$A_1 = \frac{1}{2}(37.5w)(h) = 703.125w$$

$$A_2 = \frac{1}{2}(12.5w)(50-h) = 78.125w$$

$$A_3 = (12.5w)(50) = 625$$

The maximum bending moment is $M_{max} = 703.125w$

The maximum bending normal stress at point D is $\sigma_D = \frac{-M_{max}y_D}{I_{zz}} = -\frac{(703.125w)(4.4)}{47.73} = -64.82w$ or

$$\sigma_D = 64.82w \text{ ksi(C)} \le 10 \text{ ksi(C)} \text{ or } w \le 0.1543 \text{ kips/(in)} \qquad 3$$

The maximum bending normal stress at point E is $\sigma_E = \frac{-M_{max}y_E}{I_{zz}} = -\frac{(703.125w)(-2.6)}{47.73} = -38.3w$ or

$$\sigma_E = 38.3w \text{ ksi(T)} \le 6\text{ksi(T)} \text{ or } w \le 0.1566\text{kips/in} \qquad 4$$

The maximum value of w that satisfies Eqs. (3) and (4) is $\boxed{w_{max} = 154.3\text{lb/in}}$

6.77 Two pieces of lumber are glued together to form the beam shown Fig. P6.76. Determine the intensity w of the distributed load, if the maximum tensile bending normal stress in the glue is limited to 800 psi (T) and maximum bending normal stress in wood is limited to 1200 psi.

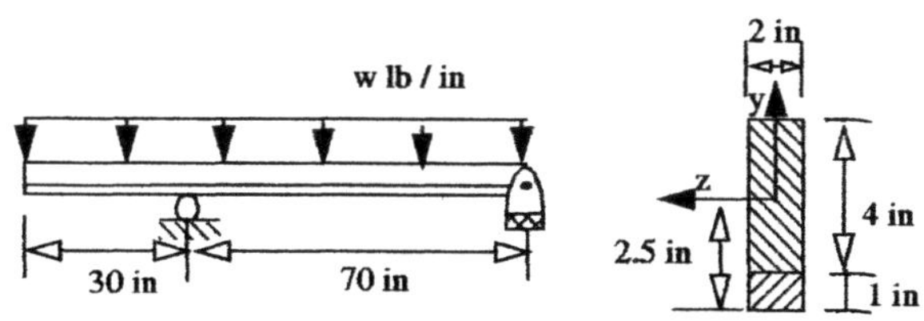

Fig. P6.77

Solution: $\sigma_G \le 800 \text{ psi(T)}$ $\sigma_w \le 1200 \text{ psi}$ $w_{max} = ?$

Fig. (a) shows the free body diagram of the beam. By equilibrium of moment about point B in Fig.(a) we obtain.

$$A_y(70) - (100w)(50) = 0 \quad \text{or} \quad A_y = 71.43w \text{ kips} \qquad 1$$

By equilibrium of forces in the y direction we obtain:

$$A_y + B_y - (100w) = 0 \quad \text{or} \quad B_y = 28.57w \text{ kips} \qquad 2$$

The shear force and bending moment diagrams are shown in Fig. (b)

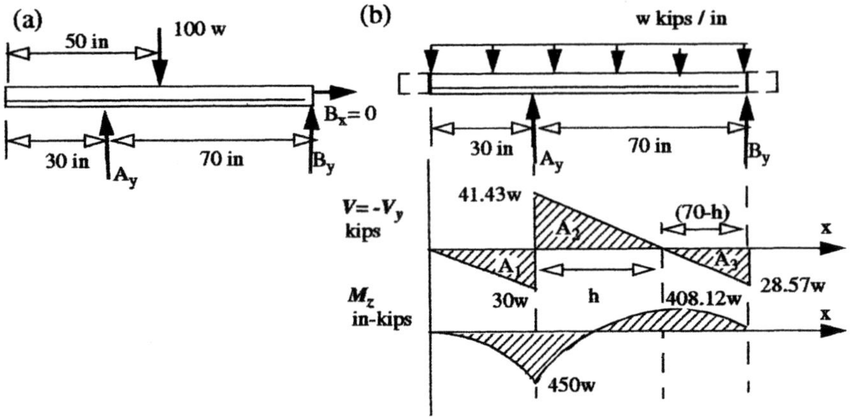

$$\frac{h}{41.43w} = \frac{70-h}{28.57w} \quad \text{or} \quad h = 41.43\text{in}$$

$$A_1 = \frac{1}{2}(30w)(30) = 450w$$

$$A_2 = \frac{1}{2}(41.43w)(h) = 858.22w$$

$$A_3 = \frac{1}{2}(28.57w)(70-h) = 408.12w$$

The maximum bending moment of $M_{max} = -450w$

The area moment of Inertia is $I_{zz} = \frac{1}{12}(2)(5)^3 = 20.83 \text{ in}^4$

The maximum bending normal stress in wood is $\sigma_W = \dfrac{-M_{max}y_{max}}{I_{zz}} = -\dfrac{(-450w)(\pm 2.5)}{20.83} = \pm 54w \text{ psi}$

The magnitude of the bending normal stress in wood should be less than 1200 psi i.e.,

$$|\sigma_w| = 54w \le 1200 \text{ or } w \le 22.2 \text{ lbs/in} \qquad 3$$

The maximum bending normal stress in glue is

$$\sigma_G = \frac{-M_{max}y_G}{I_{zz}} = -\frac{(-450w)(-2)}{20.83} = -43.2w \qquad 4$$

The normal stress in Eq.(4) is compressive and our limit is only on tensile normal stress. We consider the maximum positive moment i.e. M = 408.12 w. The maximum tensile stress in glue is

$$\sigma_G = \frac{-My_G}{I_{zz}} = \frac{(-408.12w)(-2)}{20.83} = 39.18w \text{ or } \sigma_G = 39.18w \text{ psi(T)} \le 800 \text{ psi(T) or } w \le 20.4 \text{ lbs/in} \qquad 5$$

The maximum value of w that satisfies Eqs. (4) and (5) is $\boxed{w_{max} = 20.4 \text{ lbs/in}}$

6.78 **The beam shown in Fig. P6.54 has a load of w = 25 lb/in and L = 72 inches. Select the lightest W or S shape beams from the ones given in Appendix E if the allowable bending normal stress is 21 ksi in tension and compression.**

Solution: $w = 25 \text{ lbs/in}$ $\quad L = 72$ $\quad |\sigma_{max}| \le 12 \text{ ksi}$ **Lightest W or S shape = ?**

The reaction forces were found out in problem 6.54. The shear and moment diagrams can be drawn as shown in Fig (a).

The maximum moment is $M_{max} = wL^2 = -(25)(72^2) = 129.6(10^3) \text{ in} - \text{lbs}$

The magnitude of maximum bending normal stress is:

$$|\sigma_{max}| = \left|\frac{M_{max}}{S}\right| = \frac{129.6(10^3)}{S} \le 12(10^3) \quad \text{or} \quad S \ge 10.8 \text{ in}^3 \qquad 1$$

(a)

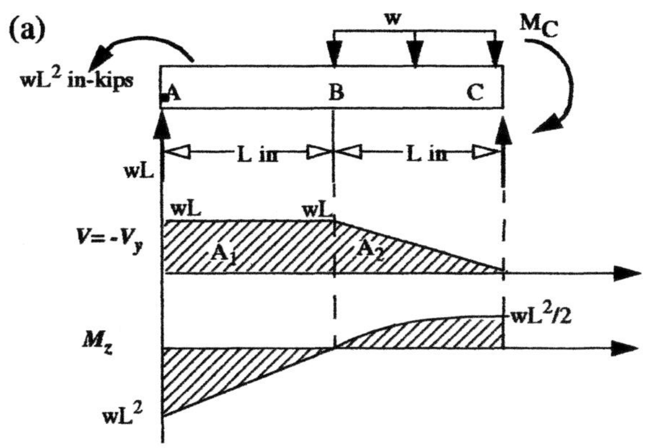

$$A_1 = wL(L) = wL^2$$

$$A_2 = \frac{1}{2}(wL)(L) = \frac{1}{2}(wL^2)$$

From appendix E, shapes with a section modulus just greater than 10.8 in^3 are W 8 x 15 and S 7 x20. Thus the lightest beam is **W8x15**

6.79 The beam shown in Fig. P6.55 has a load of 0.4 kips/inch and L = 48 inches. Select the lightest W or S shape beams from the ones given in Appendix E if the allowable bending normal stress is 16 ksi in tension and compression.

Solution: $w = 0.4$ (kips)/in , $L = 48$, $|\sigma_{max}| \le 16$ ksi Lightest W or S shape = ?

The reaction forces were found out in problem 6.54. The shear and moment diagrams can be drawn as shown in Fig (a).

The maximum bending moment is $M_{max} = \frac{9wL^2}{32} = \frac{9(0.4)(48^2)}{32} = 259.2$ in – kips

The magnitude of maximum bending normal stress is $|\sigma_{max}| = \left|\frac{M_{max}}{S}\right| = \frac{259.2}{S} \le 16$ or $S \ge 16.2$ in^3

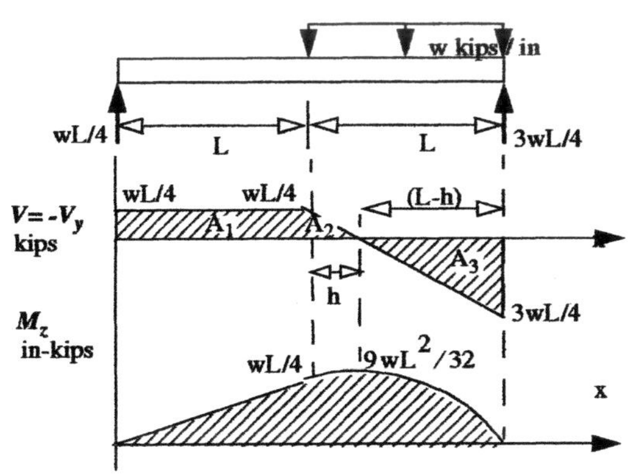

$$\frac{h}{(wL/4)} = \frac{L-h}{3wL/4} \quad \text{or} \quad h = L/4$$

$$A_1 = \left(\frac{wL}{4}\right)(L) = \frac{wL^2}{4}$$

$$A_2 = \frac{1}{2}\left(\frac{wL}{4}\right)\left(\frac{L}{4}\right) = \frac{wL^2}{32}$$

$$A_3 = \frac{1}{2}\left(\frac{3wL}{4}\right)\left(3\frac{L}{4}\right) = \frac{9wL^2}{32}$$

From appendix E, shapes with a section modulus just greater than 16.2 in^3 are W 10 x 22and S 8x23. Thus the lightest beam is **W10x22**

6.80 The beam shown in Fig. P6.56 has a load of w = 0.15 kips/in and L = 48 in. Select the lightest W or S shape beams from the ones given in Appendix E if the allowable bending normal stress is 21 ksi in tension and compression.

Solution: $w = 0.15$ (kips)/in , $L = 48$, $|\sigma_{max}| \le 21$ ksi Lightest W or S shape = ?

The maximum moment will be at the built in end and its magnitude equal to reaction moment. Calculate in problem 6.58 i.e. $M_{max} = \frac{3wL^2}{2} = \frac{3(0.15)(48^2)}{2} = 518.4 \text{ in-kips}$

The magnitude of maximum bending normal stress is

$$|\sigma_{max}| = \left|\frac{M_{max}}{S}\right| = \frac{518.4}{S} \le 21 \text{ or } S \ge 24.68 \text{ in}^3$$

From appendix E, shapes with a section modulus just greater than 24.68 in^3 are W 10 x 30 and S 10 x25.4. Thus the lightest beam is **S10x25.4**

6.81 Consider the beam shown in Fig. P6.58. Select the lightest W or S shape beams from the ones given in Appendix E if the allowable bending normal stress is 180 MPa in tension and compression.

Solution: $|\sigma_{max}| \le 180 \text{ MPa}$ Lightest W or S shape = ?

From the solution in problem 6.68 we have $M_{max} = -86.68 \text{ kN-m}$

The magnitude of maximum bending normal stress is

$$|\sigma_{max}| = \left|\frac{M_{max}}{S}\right| = \frac{86.67(10^3)}{S} \le 180(10^6) \text{ or } S \ge 0.4815(10^{-3}) \text{ m}^3 \text{ or } S \ge 481.5(10^3) \text{ mm}^3$$

From appendix E, shapes with a section modulus just greater than 481.5(10^3) mm^3 are W 250 x 44.8 and S 250 x 52. Thus the lightest beam is **W250x44.8**

6.82 Consider the beam shown in Fig. P6.60. Select the lightest W or S shape beams from the ones given in Appendix E if the allowable bending normal stress is 225 MPa in tension and compression.

Solution: $|\sigma_{max}| \le 225 \text{ MPa}$ Lightest W or S shape = ?

From the solution in problem 6.69 we have $M_{max} = 38.33 \text{ kN-m}$

The magnitude of maximum bending normal stress is

$$|\sigma_{max}| = \left|\frac{M_{max}}{S}\right| = \frac{38.33(10^3)}{S} \le 225(10^6) \text{ or } S \ge 0.1704(10^{-3}) \text{ m}^3 \text{ or } S \ge 170.4(10^3) \text{ mm}^3$$

From appendix E, shapes with a section modulus just greater than 170.4(10^3) mm^3 are W 150 x 29.8 and S 180 x 22.8. Thus the lightest beam is **W180x22.8**

6.83 The wind pressure on a sign post is approximated as a uniform pressure as shown Fig. P6.83. A similar sign post is to be designed using a hollow-square steel beam for the post. The outer dimension of the square is to be 12 inches. If the allowable bending normal stress is 24 ksi, and the pressure p = 33 lbs/ft.^{2}, determine the inner dimension of the lightest hollow beam to the nearest 1/8 in.

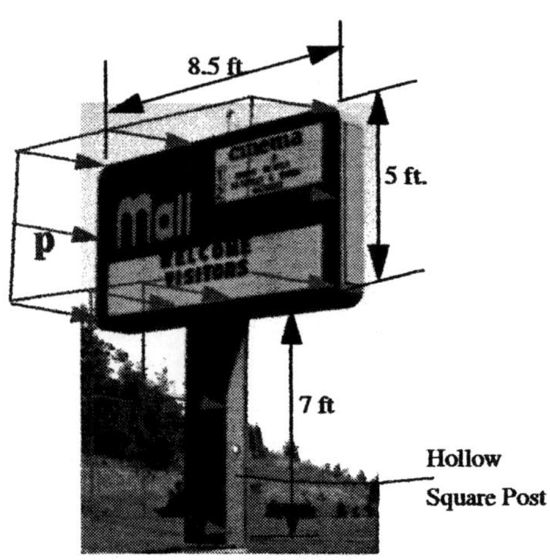

Fig. P6.83

Solution: $p = 33 \text{ lbs/(ft}^2)$, $|\sigma_{max}| \leq 24$ ksi , $a_0 = 12$ in , a_i = ?nearest 118 inch

---

The maximum bending moment will be at the base of the sign post. Fig (a) shows the free body diagram after an imaginary cut is made at the base. Internal shear force and moment are drawn as per our sign convention and the distributed forces are replaced by equivalent forces.

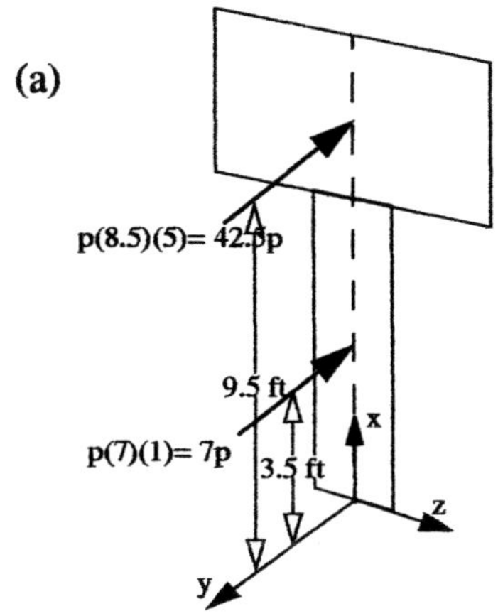

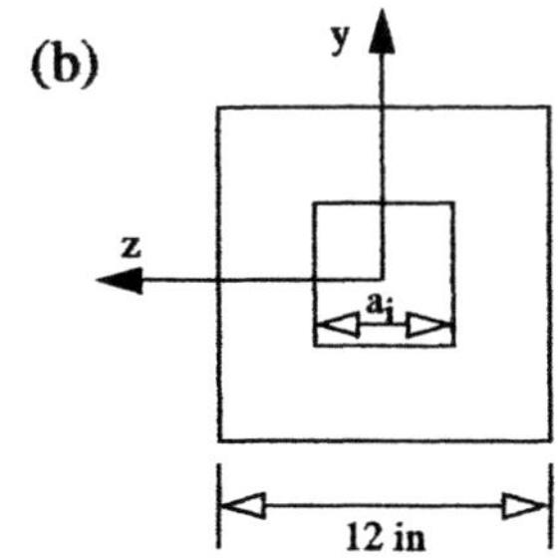

By equilibrium of moment about point O, we have $M_z + (42.5p)(9.5) + 7p(3.5) = 0$

or $M_z = (428.25)p = 14132.25 \text{ ft-lbs} = 169.6 \text{ in-kips}$

The magnitude of maximum bending normal stress is $|\sigma_{max}| = \left|\frac{M_{max}y_{max}}{I_{zz}}\right| = \frac{169.6(6)}{I_{zz}} \leq 24$ or

$$I_{zz} \geq 42.397 \qquad 1$$

The cross section of the beam is shown in Fig (b). The area moment of inertia is

$$I_{zz} = \frac{1}{12}(12)(12)^3 - \frac{1}{12}(a_i)(a_i)^3 \geq 42.397 \quad \text{or} \quad (a_i)^4 \leq (12)^4 - (12)(42.397) \quad \text{or} \quad (a_i) \leq 11.925 \qquad 2$$

The lightest hollow beam to the nearest 1/8th is $\boxed{a_i = 11\frac{7}{8} \text{ inch}}$

6.84 The allowable bending normal stress in the stepped circular beam shown is 200 MPa. If P = 200 N, determine the smallest fillet radius that can be used at section B. Use stress concentration graphs given in Appendix C.

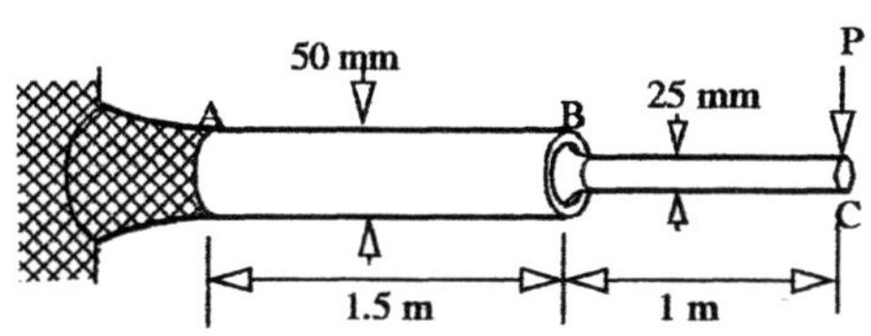

Fig. P6.84

Solution: $\sigma_{max} = 200$ MPa , $P = 200$ N , r = ?

An imaginary cut just right of B is made and the free body diagram drawn as shown in Fig (a). By moment equilibrium about point O, we obtain $M_z = -P(1) = -200$ N – m

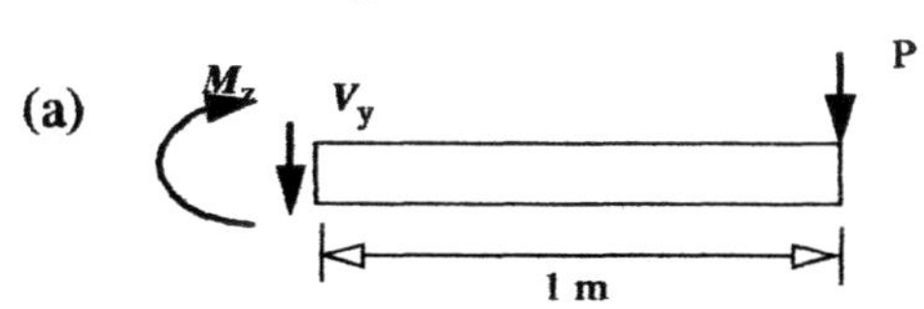

The area moment of inertia is $I_{zz} = \frac{\pi}{64}(0.025)^4 = 19.17(10^{-9}) \text{ m}^4$

The maximum bending normal stress is $|(\sigma_{BC})_{max}| = \left|\frac{-(-200)(12.5)(10^{-3})}{19.17(10^{-9})}\right| = 130.4(10^6)\ N/m^2$ or

$|(\sigma_{BC})_{max}| = 130.4\ MPa$. We have $|\sigma_{max}| = K_{conc}|(\sigma_{BC})_{max}| = 130.4K_{conc} \le 200$ or $K_{conc} \le 1.53$

From Fig C.4, we obtain the approximate value of r/d corresponding to D/d = 2 and $K_{cone} = 1.53$

$\frac{r}{d} = 0.15$ or $r = (0.15)(25)$ or $\boxed{r = 3.75\ mm}$

6.85 The allowable bending normal stress in the stepped circular beam shown is 48 ksi. Determine the maximum intensity of the distributed load w assuming the fillet radius is: (a) 0.3 inches (b) 0.5 inches. Use stress concentration graphs given in Appendix C.

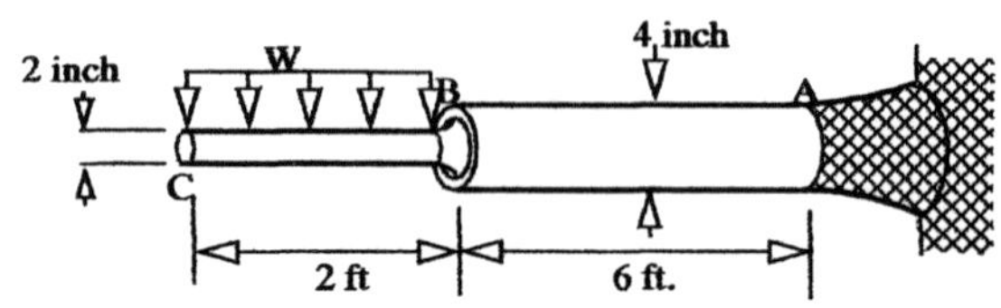

Fig. P6.85

Solution: $\sigma_{max} = 48\ ksi$ (a) r = 0.3 $w_{max} = ?$ (b) r = 0.5 $w_{max} = ?$

Fig (a) shows the free body diagram after an imaginary cut is made just left of B and the distributed force is replaced by an equivalent force.

By equilibrium of moment about point O we obtain

$$M_z + (24w)(12) = 0 \quad \text{or} \quad M_z = -288w\ \text{in-lb} \qquad 1$$

The area moment of Inertia is $I_{zz} = \frac{\pi}{64}(2)^4 = 0.7854\ in^4$

The magnitude of maximum bending normal stress is

$$|(\sigma_{BC})_{max}| = \left|\frac{-(-288w)(1)}{0.7854}\right| = 366.7w \qquad 2$$

(a) From Fig. C4, for $r/d = 0.3/2 = 0.15$ and $D/d = 2$ we obtain an approximate value of $K_{conc} = 1.55$.

Thus we have $|\sigma_{max}| = K_{conc}|(\sigma_{BC})_{max}| = (1.55)(366.7)w$ or $\sigma_{max} = 568.37w \le 48(10^3)$ psi

or $w \le 84.45$ lbs/in . The maximum value is $\boxed{w \le 84.45\ lbs/in}$

(b) From Fig. C4, for $r/d = 0.25$ and $D/d = 2$ we obtain an approximate value of $K_{conc} = 1.35$. Thus we have

$|\sigma_{max}| = K_{conc}|(\sigma_{BC})_{max}| = (1.35)(366.7)w$ or $\sigma_{max} = 495w \le 48(10^3)$ psi or $w \le 96.96$ lbs/in

The maximum value is $\boxed{w \le 96.96\ lbs/in}$

6.86 The fillet radius is 5 mm in the stepped aluminum (circular beam shown in Fig. P6.84. What should be the peak value of the cyclic load P to ensure a service life of half-million cycles. Use the SN-curve shown in Figure 3.34.

Solution: $r = 5\ mm$, $n = (0.5)(10)^6$, Aluminium $\sigma_{max} = 200\ MPa$

From Fig. 3.34 the peak stress for half a million cycles in aluminum is approximately $\sigma_{max} = 154\ MPa$.

For $r/d = 5/25 = 0.2$ and $D/d = 2$, the stress concentration factor is approximately $K_{conc} = (1.4 + 1.45)/2 = 1.425$

The nominal stress in BC can be written as $|(\sigma_{BC})_{max}| = \frac{\sigma_{max}}{K_{cone}} = \frac{154}{1.425} = 108.07\ MPa$

From problem 6.84 we have $M_z = -P$ and $I_{zz} = 19.17(10^{-9})\ m^4$

The nominal stress in BC can be written as

$$|(\sigma_{BC})_{max}| = \left|\frac{-(-P)(12.5)(10^{-3})}{19.17(10^{-9})}\right| = 0.652P(10^6) \le 108.07(10^6) \text{ or } P \le 165.74$$

The maximum value of P is $\boxed{P = 165.7\ N}$

6.87 The beam in Fig. P6.85 is made from a steel alloy that has the SN-curve shown in Figure 3.34. The peak intensity of the cyclic distributed load is w = 80lbs./in. and the fillet radius is 0.36. What is the predicted service life of the beam.

Solution: $r = 0.36in$, $w = 80\ lbs/in$, Steel , $n = ?$

From problem 6.85 we have $I_{zz} = 0.7854in^4$ and $M_z = -288w = -(288)(80) = -23040\ in-lb$

The nominal stress in BC is: $|(\sigma_{BC})_{max}| = \left|\frac{-(-23040)(1)}{0.7854}\right| = 29.34(10^3)psi = 29.34\ ksi$

From Fig. C.4 for $r/d = 0.36/2 = 0.18$ and $D/d = 2$, the stress concentration factor is approximately $K_{conc} = (1.4 + 1.5)/2 = 1.5$.

The maximum stress is $|\sigma_{max}| = K_{conc}|(\sigma_{BC})_{max}| = (1.475)(29.34) = 43.3\ ksi$

From Fig. 3.34, for a peak stress of 43.3 ksi in steel the approximate number of cycles is $n \approx 175,000$

$\boxed{n = 175,000 \text{ cycles}}$

Problem Set 6.4

6.88 Assuming a positive shear force V_y, (a) sketch the direction of the shear flow along the center-line on the thin cross-sections shown.(b) At points A, B, C, and D, determine if the stress component is τ_{xy} or τ_{xz} and if it is positive or negative.

Solution Shear flow sketch=? τ_{xy} or τ_{xz} at points A, B, C, and D =?

The shear flow is as shown.

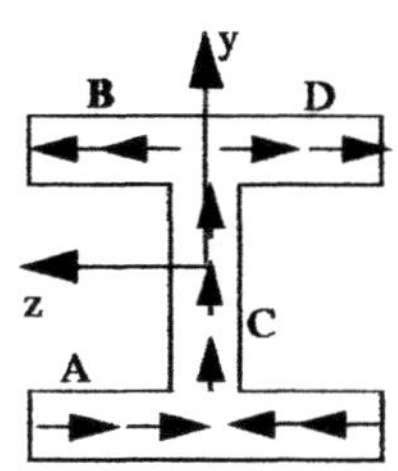

Point A: negative τ_{xz} | Point B: positive τ_{xz} | Point C: positive τ_{xy} | Point D: negative τ_{xz}

6.89 Assuming a positive shear force V_y, (a) sketch the direction of the shear flow along the center-line on the thin cross-sections shown.(b) At points A, B, C, and D, determine if the stress component is τ_{xy} or τ_{xz} and if it is positive or negative.

Solution Shear flow sketch=? τ_{xy} or τ_{xz} at points A, B, C, and D =?

The shear flow is as shown.

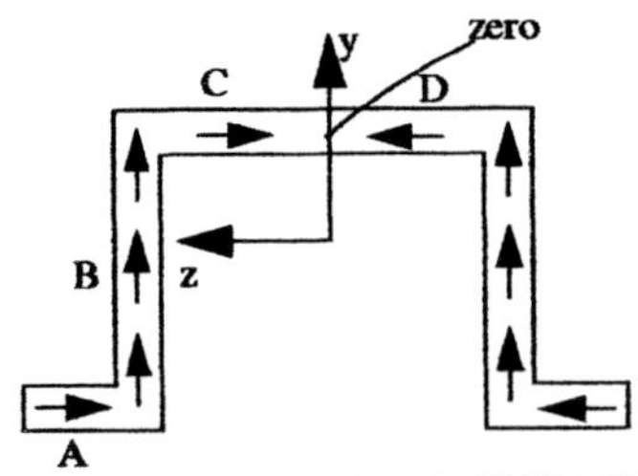

Point A: negative τ_{xz}	Point B: positive τ_{xy}	Point C: negative τ_{xz}	Point D: positive τ_{xz}

6.90 Assuming a positive shear force V_y, (a) sketch the direction of the shear flow along the center-line on the thin cross-sections shown.(b) At points A, B, C, and D, determine if the stress component is τ_{xy} or τ_{xz} and if it is positive or negative.

Solution Shear flow sketch=? τ_{xy} or τ_{xz} at points A, B, C, and D =?

The shear flow is as shown.

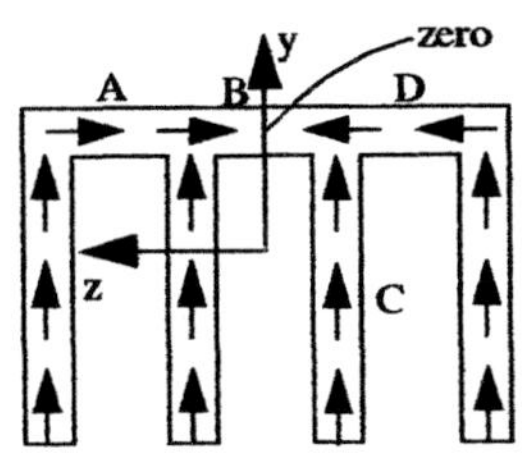

Point A: negative τ_{xz}	Point B: negative τ_{xz}	Point C: positive τ_{xy}	Point D: positive τ_{xz}

6.91 Assuming a positive shear force V_y, (a) sketch the direction of the shear flow along the center-line on the thin cross-sections shown.(b) At points A, B, C, and D, determine if the stress component is τ_{xy} or τ_{xz} and if it is positive or negative.

Solution Shear flow sketch=? τ_{xy} or τ_{xz} at points A, B, C, and D =?

The shear flow is as shown.

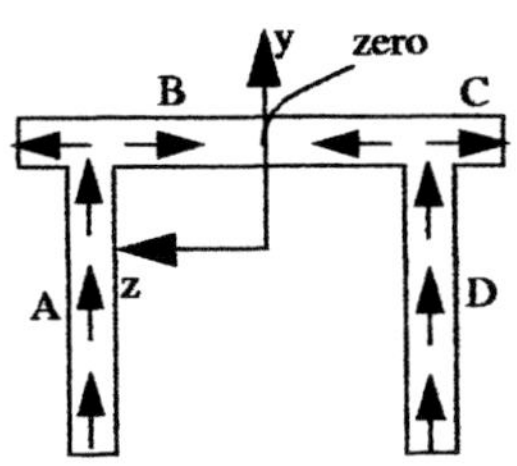

Point A: positive τ_{xy}	Point B: negative τ_{xz}	Point C: negative τ_{xz}	Point D: positive τ_{xy}

6.92 Assuming a positive shear force V_y, (a) sketch the direction of the shear flow along the center-line on the thin cross-sections shown.(b) At points A, B, C, and D, determine if the stress component is τ_{xy} or τ_{xz} and if it is positive or negative.

Solution Shear flow sketch=? τ_{xy} or τ_{xz} at points A, B, C, and D =?

The shear flow is as shown.

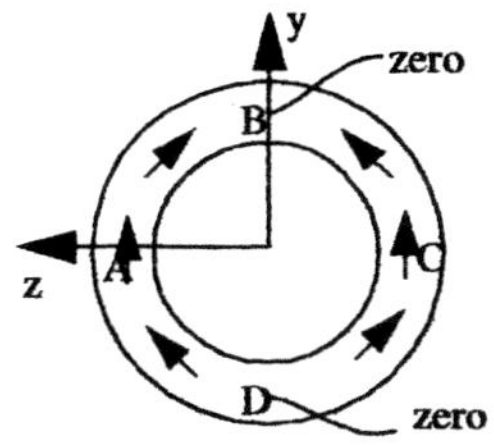

Point A: positive τ_{xy} | Point B: zero shear stress | Point C: positive τ_{xy} | Point D: zero shear stress

---

6.93 Assuming a positive shear force V_y, (a) sketch the direction of the shear flow along the center-line on the thin cross-sections shown.(b) At points A, B, C, and D, determine if the stress component is τ_{xy} or τ_{xz} and if it is positive or negative.

Solution Shear flow sketch=? τ_{xy} or τ_{xz} at points A, B, C, and D =?

The shear flow is as shown

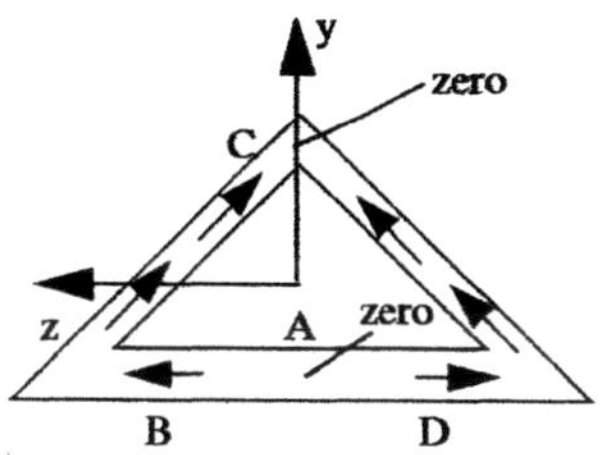

Point A: zero shear stress | Point B: positive τ_{xz} | Point C: zero shear stress | Point D: negative τ_{xz}

6.94 Assuming a positive shear force V_y, (a) sketch the direction of the shear flow along the center-line on the thin cross-sections shown.(b) At points A, B, C, and D, determine if the stress component is τ_{xy} or τ_{xz} and if it is positive or negative.

Solution Shear flow sketch=? τ_{xy} or τ_{xz} at points A, B, C, and D =?

The shear flow is as shown.

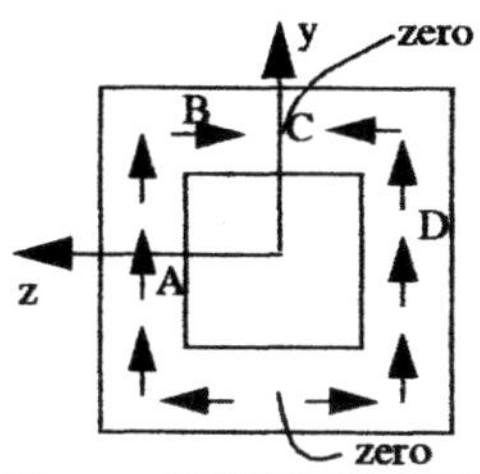

Point A: positive τ_{xy} | Point B: negative τ_{xz} | Point C: zero shear stress | Point D: positive τ_{xy}

6.95 Determine the magnitude of the maximum bending normal stress and shear stress in the beam.

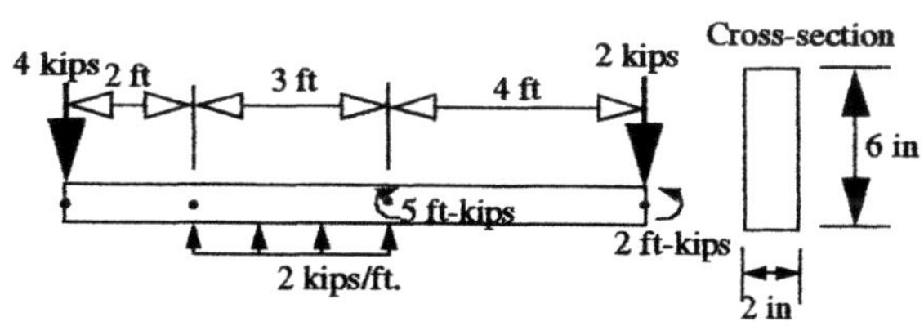

Fig. P6.95

Solution $|\sigma_{max}| = ?$ $|\tau_{max}| = ?$

The shear and moment diagrams are drawn as shown below

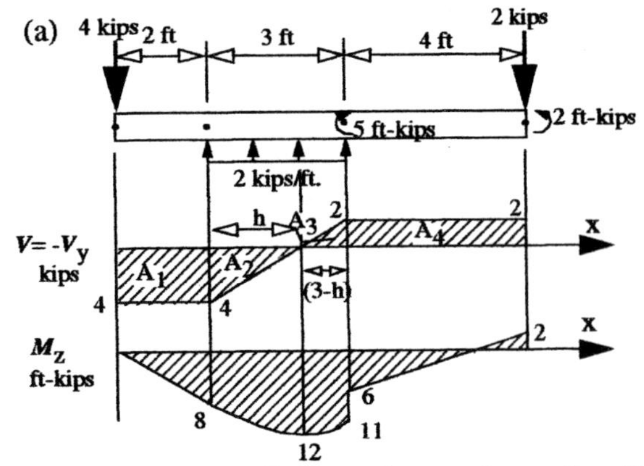

$$\frac{h}{4} = \frac{(3-h)}{2} \quad \text{or} \quad h = 2$$

$$A_1 = (4)(2) = 8$$

$$A_2 = \frac{1}{2}(4)(h) = 8$$

$$A_3 = \frac{1}{2}(4)(3-h) = 8$$

$$A_4 = (4)(2) = 8$$

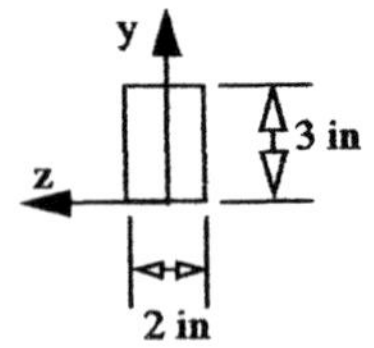

From the moment diagram the maximum bending moment is M_{max} = -12 ft-kips = -144 in-kips and maximum shear force is $(V_y)_{max} = 4$ kips. The area moment of inertia is: $I_{zz} = \frac{1}{12}(2)(6^3) = 36 \text{ in}^4$ The maximum bending normal stress in the section is: $\sigma_{max} = -\frac{M_{max}y_{max}}{I_{zz}} = -\frac{(-144)(\pm 3)}{36} = \pm 12\text{ksi}$ or

$$\boxed{|\sigma_{max}| = 12 \text{ ksi}}$$

Maximum Q_z is at the neutral axis. Fig (b) shows the area A_s that can be used for finding maximum Q_z at the neutral axis. The maximum Q_z is $Q_{max} = (3)(2)(1.5) = 9 \text{ in}^3$. The

The maximum shear stress is: $|\tau_{max}| = \left|\frac{(V_y)_{max}Q_{max}}{I_{zz}(t)}\right| = \frac{(4)(9)}{(36)(2)} = 0.5$ ksi or $\boxed{|\tau_{max}| = 500 \text{ psi}}$

6.96 For the beam, loading and cross-section shown, determine: (a) the maximum tensile and compressive bending normal stress. (b) the maximum shear bending stress. (c) the bending normal stress and the bending shear stress *at point A*. Point A is just below the flange on the cross-section in the middle of the beam. Show your result on a stress cube. The area moment of inertia for the beam was calculated to be $I_{zz} = 3.6\,(10^6)$ mm^4

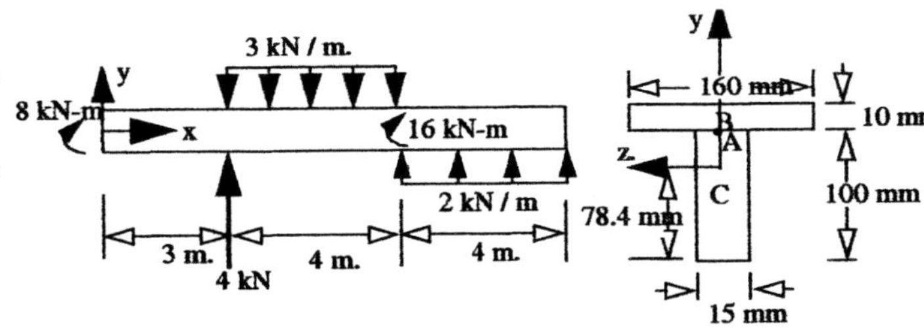

Fig. P6.96

Solution: $I_{zz} = 3.6(10^6)\text{mm}^4$, $|\tau_{max}| = ?$ $|\sigma_{max}| = ?$ $(\sigma_{xx})_A = ?$ $(\tau_{xy})_A = ?$

The shear and moment diagrams are drawn as shown in Fig. (a).

(a) The maximum moment is M_{max} = 16 kN-m. The maximum bending normal stress in the section is

$$\sigma_{max} = -\frac{M_{max}y_{max}}{I_{zz}} = -\frac{(16)(10^3)(-0.0784)}{3.6(10^{-6})} = 348.4(10^6)\frac{N}{m^2} \text{ or } \boxed{|\sigma_{max}| = 348.4\text{MPa}}$$

Fig (b) shows the area A_s that can be used for finding the maximum Q_z at the neutral axis.The maximum Q_z is $Q_{max} = (0.0784)(0.015)\left(-\frac{0.0784}{2}\right) = -46.10(10^{-6})\text{m}^3$

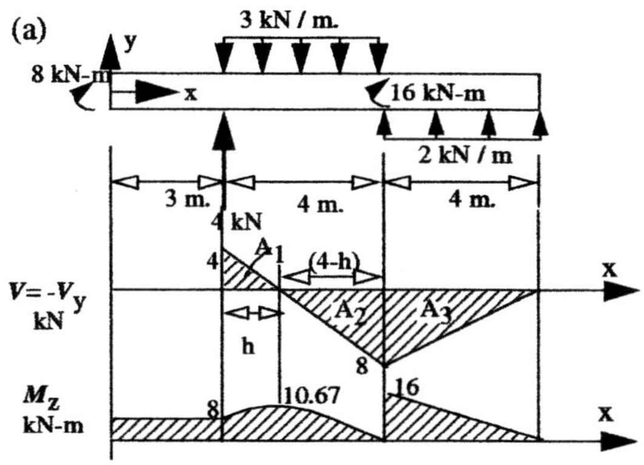

$$\frac{h}{4} = \frac{(4-h)}{8}$$

or $h = 1.333$

$$A_1 = \frac{1}{2}(4)(h) = 2.67$$

$$A_2 = \frac{1}{2}(8)(4-h) = 10.67$$

$$A_3 = \frac{1}{2}(8)(4) = 16$$

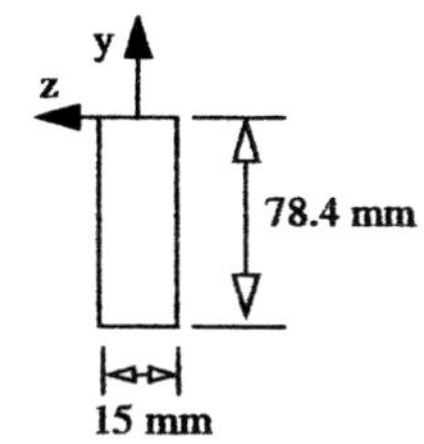

The maximum shear force is $(V_y)_{max} = 8\text{kN}$. The magnitude of maximum shear stress is

$$|\tau_{max}| = \left|-\frac{Q_{max}V_{ymax}}{I_{zz}\ t}\right| = \frac{8(10^3)(46.10)(10^{-6})}{3.6(10^{-6})(0.015)} = 6.84(10^6)\frac{N}{m^2} \text{ or } \boxed{|\tau_{max}| = 6.84\text{MPa}}$$

(b) The moment and shear force at the section containing point A are $M_A = 8\text{kN}-\text{m}$ and $(V_y)_A = -4\text{kN}$. The bending normal stress of point A is:

$$(\sigma_{xx})_A = -\frac{M_A y_A}{I_{zz}} = -\frac{(-8)(10^3)(0.0216)}{3.6(10^{-6})} = -48(10^6)\frac{N}{m^2} \text{ or } \boxed{(\sigma_{xx})_A = 48\text{MPa(C)}}$$

Fig (c) shows the area A_s that can be used in finding Q_z at point A. The value of Q_z at point A is $Q_A = (0.160)(0.010)(0.0266) = 42.56(10^{-6})m^3$. The bending shear stress at point A is

$$(\tau_{xs})_A = -\frac{Q_A(V_y)_A}{I_{zz}\ t} = -\frac{(-4)(10^3)42.56(10^{-6})}{3.6(10^{-6})(0.015)} = 3.15(10^6)\frac{N}{m^2}.$$ From Fig. (c) we note that s and y are opposite. Thus $(\tau_{xy})_A = -(\tau_{xs})_A = -3.15\text{MPa}$ $\boxed{(\tau_{xy})_A = -3.2\text{MPa}}$

The stress cube is shown in Fig. (d).

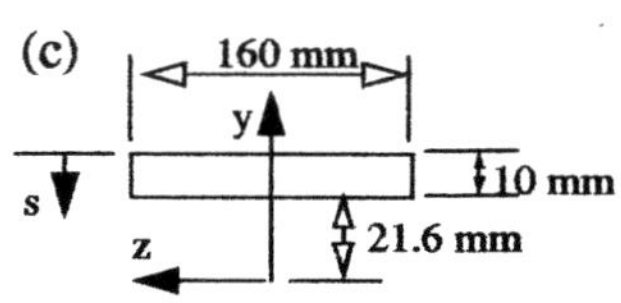

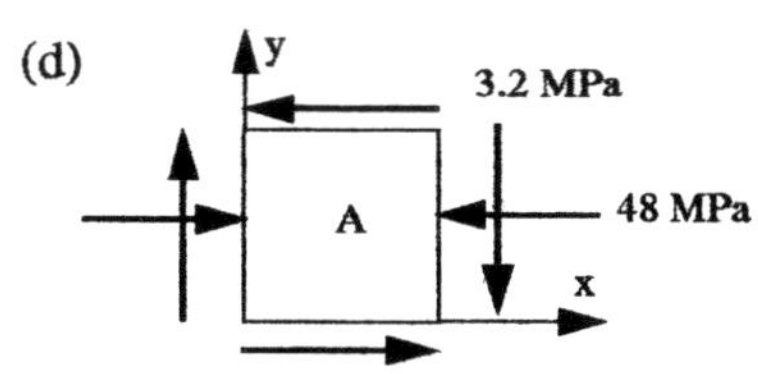

6.97 For the beam, loading and cross-section shown, determine: (a) the magnitude of the maximum bending normal stress and shear stress. (b) the bending normal stress and the bending shear stress *at point A*. Point A is on the cross-section 2 m from the right end. Show your result on a stress cube. The area moment of inertia for the beam was calculated to be $I_{zz} = 453\ (10^6)\ mm^4$.

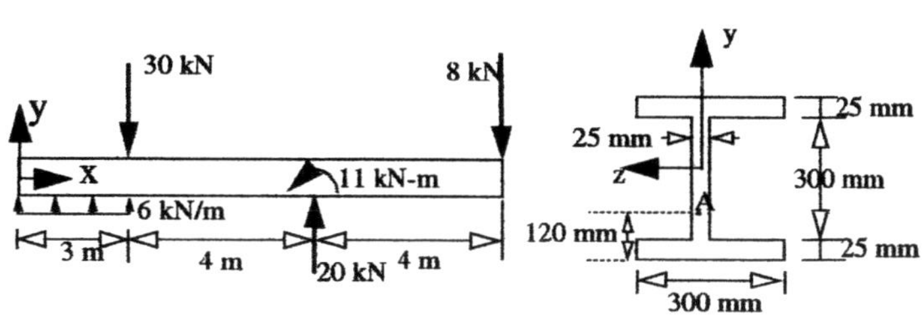

Fig. P6.97

Solution: $I_{zz} = 453(10^6)mm^4$, $|\tau_{max}| = ?$ $|\sigma_{max}| = ?$ $(\sigma_{xx})_A = ?$ $(\tau_{xy})_A = ?$

The shear and moment diagrams are drawn as shown below.

(a) The maximum moment is $M_{max} = -32$kN-m. The maximum bending normal stress in the section is

$$\sigma_{max} = -\frac{M_{max} y_{max}}{I_{zz}} = -\frac{(-32)(10^3)(\pm 0.175)}{453(10^{-6})} = 12.36(10^6)\frac{N}{m^2} \text{ or } \boxed{|\sigma_{max}| = 12.4\text{MPa}}$$

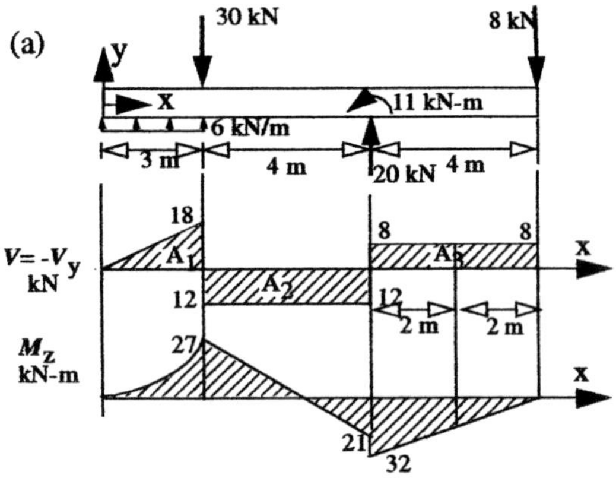

$A_1 = \frac{1}{2}(18)(3) = 27$

$A_2 = (12)(4) = 48$

$A_3 = (8)(4) = 32$

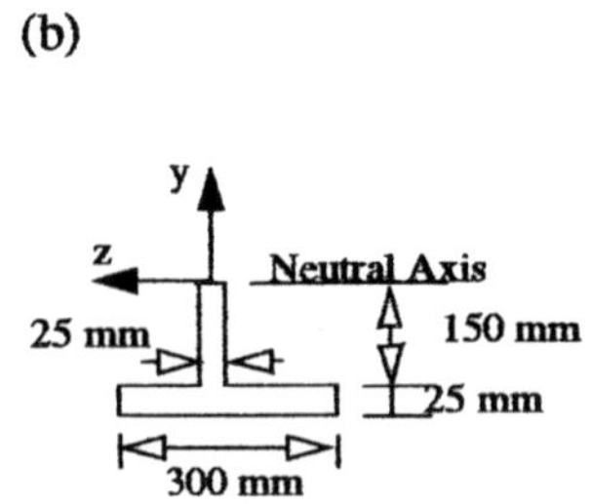

Fig. (b) shows the area A_s that can be used for finding the maximum Q_z at the neutral axis.

$$Q_{max} = (0.3)(0.025)(-0.1625) + (0.15)(0.025)(-0.075) = -1.5(10^{-3})m^3$$

The maximum shear force is $V_y = -18$kN . The magnitude of maximum shear stress is

$$|\tau_{max}| = \left|-\frac{Q_{max} V_{ymax}}{I_{zz}\, t}\right| = \frac{18(10^3)(1.5)(10^{-3})}{(453)(10^{-6})(0.025)} = 2.38(10^6)\frac{N}{m^2} \quad \boxed{|\tau_{max}| = 2.38\text{MPa}}$$

(b) The moment and shear force at the section containing point A are $M_A = -16\text{kN} - \text{m}$ and $(V_y)_A = -8$kN. The bending normal stress at point A is

$$(\sigma_{xx})_A = -\frac{M_A y_A}{I_{zz}} = -\frac{(-16)(10^3)(-0.055)}{(453)(10^{-6})} = -1.943(10^6)\frac{N}{m^2} \text{ or } \boxed{(\sigma_{xx})_A = 1.94\text{MPa(C)}}$$

Fig (c) shows the area A_s that can be used in finding Q_z at point A. The value of Q_z at point A is

$$Q_A = (0.3)(0.025)(-0.1625) + (0.095)(0.025)(-0.1025) = -1.46(10^{-3})m^3$$

The bending shear stress at point A is: $(\tau_{xs})_A = -\frac{Q_A (V_y)_A}{I_{zz}\, t} = -\frac{(-8)(10^3)(-1.46)(10^{-3})}{(453)(10^{-6})(0.025)} = -1.03(10^6)\frac{N}{m^2}$

From Fig. (c) we see that s is in the same direction as y. Thus $(\tau_{xy})_A = (\tau_{xs})_A$ or $\boxed{(\tau_{xy})_A = -1.03\text{MPa}}$

The stress cube is shown in Fig. (d) .

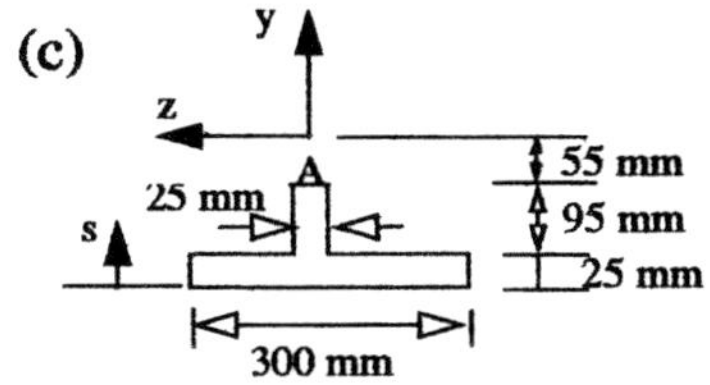

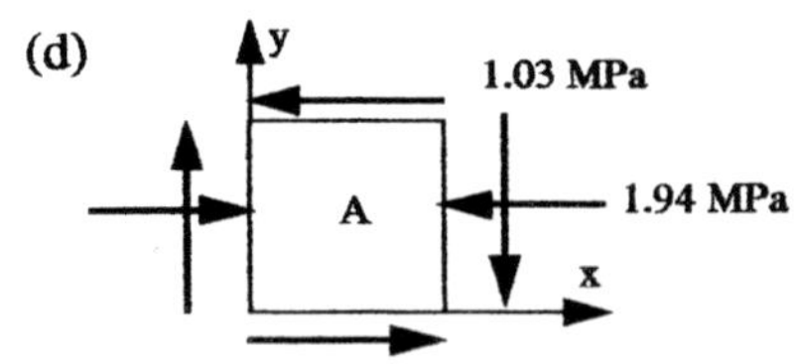

6.98 Two pieces of lumber are nailed together as shown. The nails are uniformly spaced 10 inches apart along the length. Determine the average shear force in each nail in segment AB and segment BC.

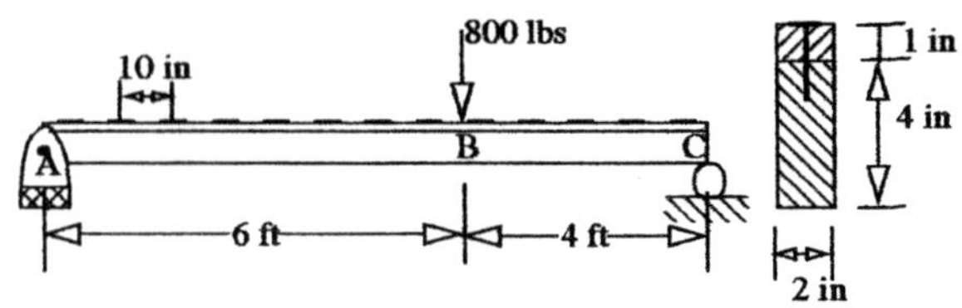

Fig. P6.98

Solution: $\Delta s = 10\text{in}$, $V_{AB} = ?$ $V_{BC} = ?$

Fig(a) shows the free body diagram of the entire beam. By equilibrium of moment about point C we obtain. $R_A(10) - 800(4) = 0$ or $R_A = 320\text{lb}$

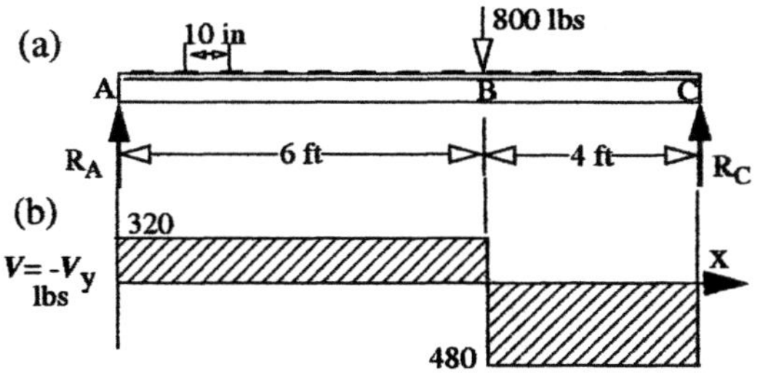

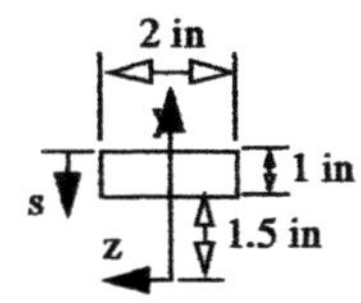

Fig (b) shows the shear force diagram. The shear forces in segments AB and BC are $(V_y)_{AB} = -320\text{lb}$, $(V_y)_{BC} = -480\text{lb}$. Fig. (c) shows the area A_s that can be used in finding Q_z at the line junction of the wood pieces. Its value is $Q_z = (2)(1)(2) = 4\text{in}^3$. The area moment of inertia is $I_{zz} = \frac{1}{12}(2)(5^3) = 20.83\text{in}^4$

The shear flow in each segment is

$$q_{AB} = \left|\frac{(V_y)_{AB}Q_z}{I_{zz}}\right| = \frac{(320)(4)}{20.83} = 61.44\frac{\text{lbs}}{\text{in}} \text{ and } q_{BC} = \left|\frac{(V_y)_{BC}Q_z}{I_{zz}}\right| = \frac{(480)(4)}{20.83} = 92.16\frac{\text{lbs}}{\text{in}}$$

The average shear forces in nails in each segment are

$V_{AB} = q_{AB}\Delta s = (61.44)(10)$ or $\boxed{V_{AB} = 614.4\text{lbs}}$

$V_{BC} = q_{BC}\Delta s = (92.16)(10)$ or $\boxed{V_{BC} = 921.6\text{lbs}}$

6.99 A cantilever beam is constructed by nailing three piece of lumber. The nails are uniformly spaced at an interval of 75 mm. Determine the average shear force in each nail.

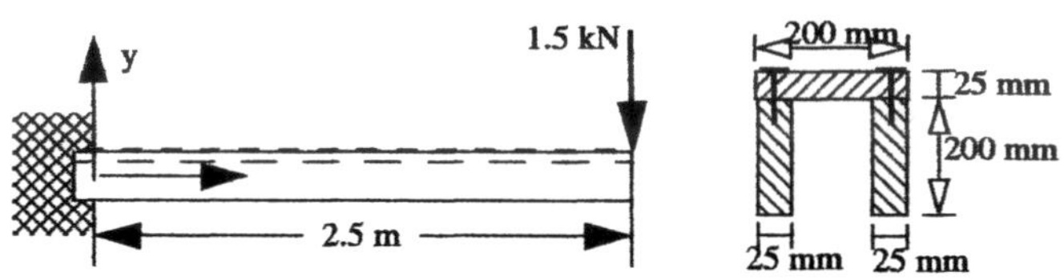

Fig. P6.99

Solution: $\Delta s = 75\text{mm}$, $V_{nail} = ?$

Fig (a) shows the free body diagram after an imaginary cut is made in the beam at any location. By equilibrium of forces $V_y = -1.5\text{kN}$

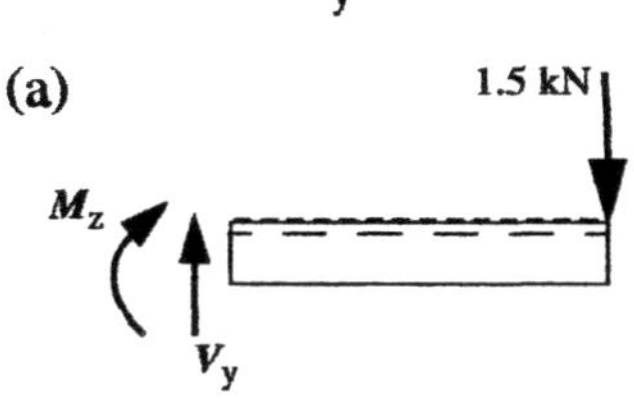

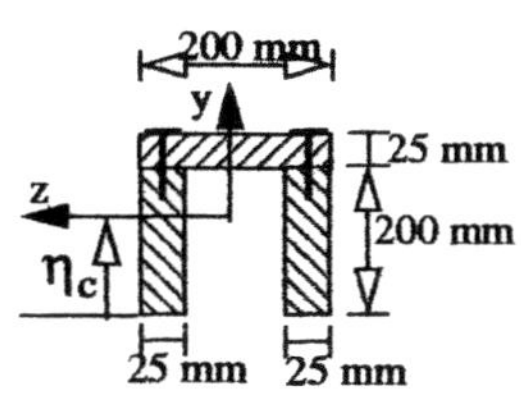

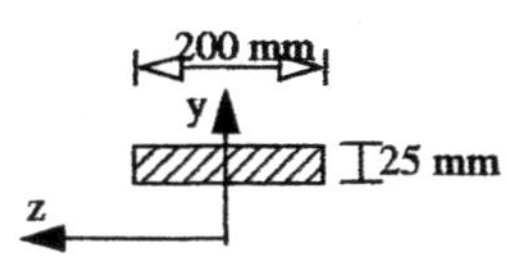

The location of the centroid is shown in Fig. (b) can be found out as shown below.

$$\eta_c = \frac{(2)(200)(25)(100) + (200)(25)(212.5)}{(2)(200)(25) + (200)(25)} = 137.5\text{mm}$$

The area moment of inertia is

$$I_{zz} = 2\left[\frac{1}{12}(25)(200^3) + (25)(200)(137.5 - 100)^2\right] + \left[\frac{1}{12}(200)(25^3) + (200)(25)(212.5 - 137.5)^2\right] \text{ or}$$

$I_{zz} = 75.78(10^6)\text{mm}^4$

Fig (c) shows the area A_s that can be used for finding Q_z at the junction of the wood pieces

$Q_t = (200)(25)(212.5 - 137.5) = 375(10)^3\text{mm}^3$

The magnitude of shear flow is $q = \left|\frac{V_y Q_t}{I_{zz}}\right| = \frac{1.5(10^3)(375)(10^{-6})}{(75.78)(10^{-6})} = 7.42(10^3)\text{N/m}$

The shear force is carried by two rows of nails. The average shear force in each nail.

$V_{nail} = \left(\frac{q}{2}\right)\Delta s = \frac{(7.42)}{2}(10^3)(0.075) = 278.25\text{N}$ or $\boxed{V_{nail} = 278\text{N}}$

6.100 A cantilever beam is constructed by nailing three piece of lumber. The nails are uniformly spaced at an interval of 75 mm. (a) Determine the average shear force in each nail. (b) Which is the better nailing method, the one shown in problem 6.99 or the one in this problem.

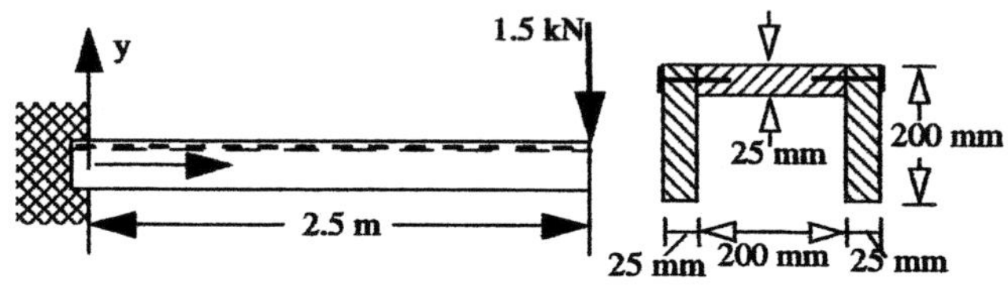

Fig. P6.100

Solution: $\Delta s = 75\text{mm}$, $V_{nail} = ?$

The location of the centroid is shown in Fig.(b) can be found out as shown below.

$$\eta_c = \frac{(2)(200)(75)(100) + (200)(25)(187.5)}{(2)(200)(25) + (200)(25)} = 129.17\text{mm}$$

The area moment of inertia is

$$I_{zz} = 2\left[\frac{1}{12}(25)(200^3) + (25)(200)(129.17 - 100)^2\right] + \left[\frac{1}{12}(200)(25^3) + (200)(25)(187.5 - 129.17)^2\right] \text{ or}$$

$I_{zz} = 45.66(10^6)\text{mm}^4$

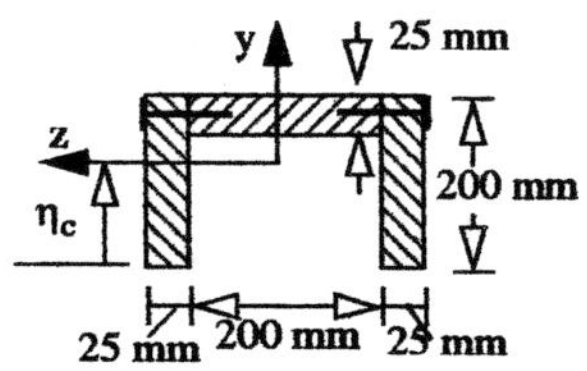

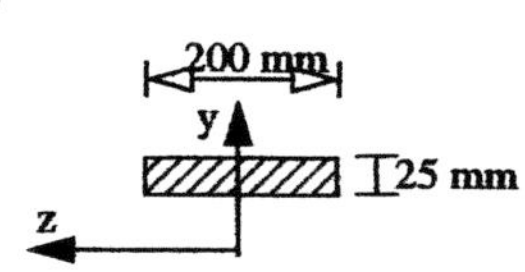

Fig (b) shows the area A_s that can be used for finding Q_t at the junction of the wood pieces

$Q_z = (200)(25)(187.5 - 129.17) = 291.65(10)^3\text{mm}^3$

The shear force value found in problem is $V_y = -1.5\text{kN}$

The magnitude of shear flow is $q = \left|\frac{V_y Q_t}{I_{zz}}\right| = \frac{(1.5)(10^3)(291.65)(10^{-6})}{115.66(10^{-6})} = 9.58(10^3)\text{N/m}$

The shear flow is carried by two rows of nails. The average shear force in each nail is

$V_{nail} = \left(\frac{q}{2}\right)\Delta s = \frac{(9.58)}{2}(10^3)(0.075) = 359.2\text{N}$ $\boxed{V_{nail} = 359\text{N}}$

As the force on the nail in problem 6.99 is smaller, the joining method of problem 6.99 is better.

6.101 The planks in a park bench are made from re-cycled plastic and are bolted to concrete supports as shown. For purpose of design the front plank is modeled as simply supported beam that caries all the weight of two individuals. Assume that each persons has a mass 100 kg. and the weight acts at one third the length of the plank as shown. The allowable bending normal stress for the re-cycled plastic is 10 MPa and allowable bending shear stress is 2 MPa. The width d of the planks that can be manufactured are in increments of 2 cm. from 12 cm to 20 cm. To design the lightest bench determine the corresponding values of the thickness t to the closest cm for the various values of d.

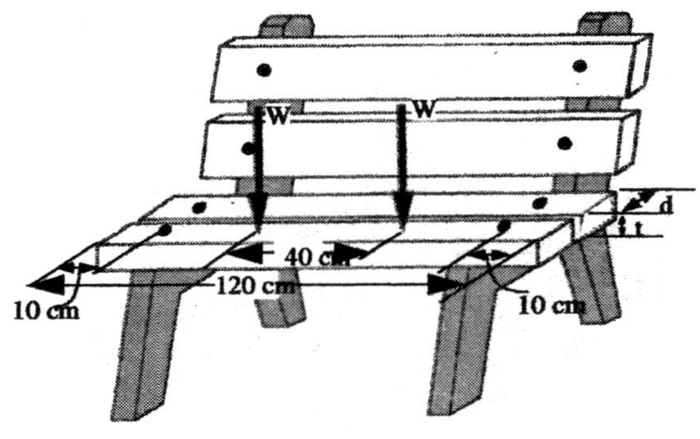

Fig. P6.101

Solution $m = 100$ kg $|\sigma_{max}| \le 10$MPa $|\tau_{max}| \le 2$MPa

$12\text{cm} \le d \le 20\text{cm}$ in steps of 2 cm. Lightest beam t =? nearest cm

---

The weight of each person is $W = (100)(9.81) = 981\ \text{N}$. The free body diagram of the beam is shown in Fig(a). By equilibrium of moment about point B, we obtain:

$R_A(0.1) - W(0.07) - W(0.03) = 0$ or $R_A = W$

Fig. (b) shows the shear and moment diagrams. The maximum value of shear force and bending moment is $(V_y)_{max} = \pm W = \pm 981\text{N}$ and $M_{max} = 0.03W = 29.43\text{N}-\text{m}$

The area moment of inertia is $I_{zz} = \frac{1}{12}dt^3$

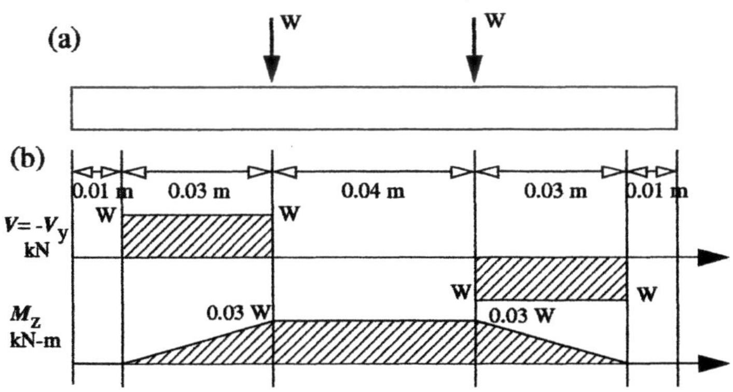

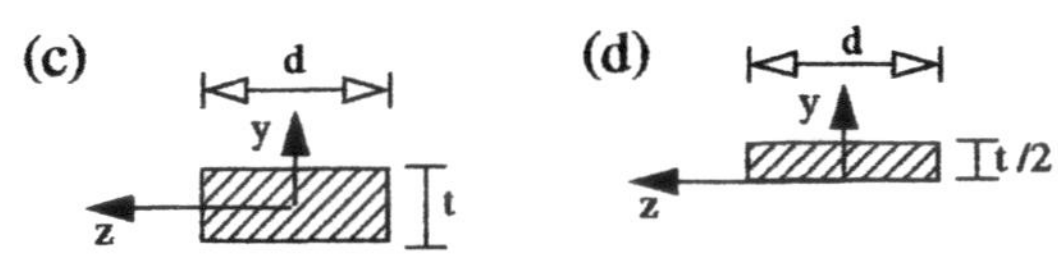

The magnitude of maximum bending normal stress is

$$|\sigma_{max}| = \left|\frac{M_{max}y_{max}}{I_{zz}}\right| = \frac{(-29.43)(t/2)}{\frac{1}{12}dt^3} = \frac{176.58}{dt^2} \le 10(10^6) \qquad \text{or} \qquad t \ge \sqrt{\frac{17.658(10^{-6})}{d}} \qquad \mathbf{1}$$

Fig. (d) shows the area A_s that can be used for finding maximum Q_z at the neutral axis. The value of maximum Q_z is $Q_{max} = (d)(t/2)(t/4) = dt^2/8$

The magnitude of maximum shear stress is:

$$|\tau_{max}| = \left|\frac{Q_{max}(V_y)_{max}}{I_{zz}\ t_{ma}}\right| = \frac{(981)(dt^2/8)}{(dt^3/12)(d)} = \frac{1471.5}{dt} \le 2000 \qquad \text{or} \qquad t \ge \left[\frac{735.75(10^{-6})}{d}\right] \qquad \mathbf{2}$$

In Eqs. (1) and (2), d and t are in meters. Table below shows the calculation for the value of t for each corresponding value of d. The minimum of t must satisfy Eqs. (1) and (2) and is rounded upward to the nearest cm.

d (m)	t from Eq. 1 (m)	t from Eq.2 m	t_{min} cm
0.12	0.0121	0.0061	2
0.14	0.0112	0.0053	2
0.16	0.0105	0.0046	2
0.18	0.0099	0.0041	1
0.20	0.0094	0.0037	1

6.102 Two piece of wood are glued together to form a beam. The allowable bending normal and shear stress in wood are 3 ksi and 1 ksi, respectively. The allowable bending normal and shear stress in the glue are 600 psi (T) and 250 psi, respectively. Determine the maximum moment M_{ext} that can be applied to the beam.

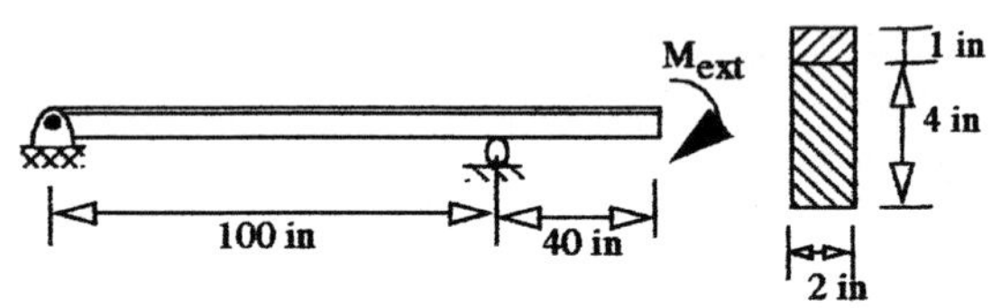

Fig. P6.102

Solution $|\sigma_w| \le 3\text{ksi}$ $|\tau_w| \le 1\text{ksi}$ $|\sigma_G| \le 600\text{psi(T)}$ $|\tau_G| \le 250\text{psi}$ $M_{ext} = ?$

Fig (a) shows the free body diagram of the beam. By equilibrium of moment about point B we obtain

$R_A(100) - M_{ext} = 0$ or $R_A = M_{ext}/100$

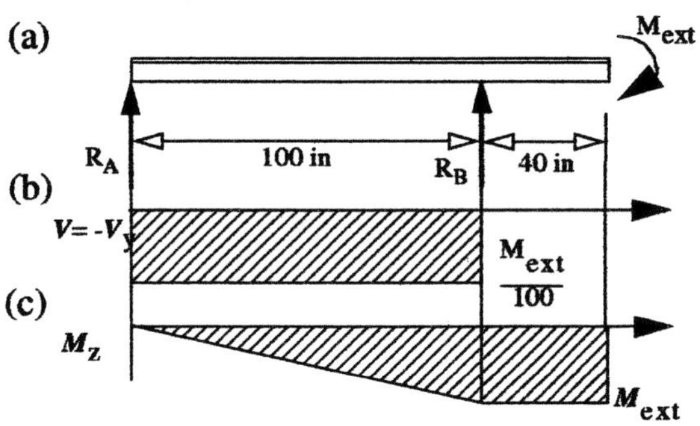

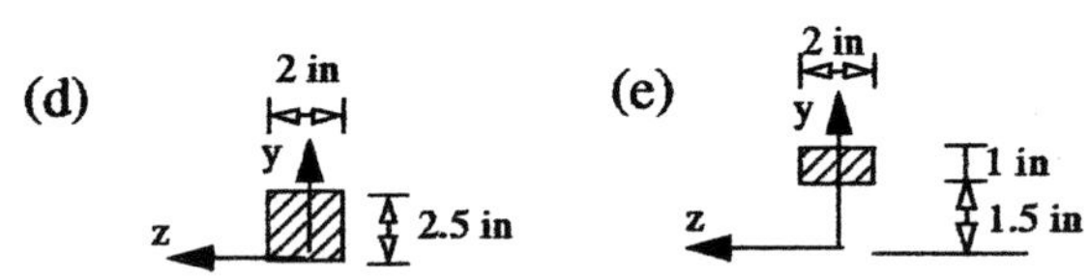

Figs. (b) and (c) show the shear and moment diagrams. The maximum values of shear force and bending moment are $(V_y)_{max} = M_{ext}/100$ and $M_{max} = -M_{ext}$

The area moment of inertia is $I_{zz} = \frac{1}{12}(2)(5^3) = 20.83\text{in}^4$

The magnitude of maximum bending stress in wood is

$$\sigma_w = \left|\frac{M_{max}y_{max}}{I_{zz}}\right| = \frac{(M_{ext})(2.5)}{20.83} \le 3000\text{psi} \quad \text{or} \quad M_{ext} \le 25000\text{in-lbs} \qquad 1$$

The maximum normal stress in glue is:

$$\sigma_G = \frac{-M_{max}y_G}{I_{zz}} = \frac{-(-M_{ext})(1.5)}{20.83} = \frac{M_{ext}(1.5)}{20.83} \le 600 \quad \text{or} \quad M_{ext} \le 8333.33\text{in-lbs} \qquad 2$$

Figs (d) and (e) show the area A_s that can be used for finding Q_z at the neutral axis and the glue as

$Q_{max} = (2.5)(2)(1.25) = 6.25\text{in}^3$ and $Q_G = (2)(1)(2) = 4\text{in}^3$

The magnitude of maximum shear stress in wood is

$$|\tau_w| = \left|\frac{Q_{max}(V_y)_{max}}{I_{zz}\,t}\right| = \frac{(M_{ext}/100)(6.25)}{(20.83)(2)} \le 1000 \quad \text{or} \quad M_{ext} \le 666.6(10^3)\text{in-lbs} \qquad 3$$

The magnitude of maximum shear stress in glue is

$$|\tau_G| = \left|\frac{Q_G(V_y)_{max}}{I_{zz}\ t}\right| = \frac{(M_{ext}/100)(4)}{(20.83)(2)} \leq 250 \qquad \text{or} \qquad M_{ext} \leq 260.3(10^3)\text{in} - \text{lbs} \qquad \textbf{4}$$

The maximum value of M_{ext} that satisfies Eqs.(1) through (4) is $\boxed{M_{ext} = 8333.33\text{in} - \text{lbs}}$

6.103 A wooden cantilever beam is to be constructed by nailing two pieces of lumber. The allowable bending normal and shear stress in the wood are 7 MPa and 1.5 MPa, respectively. The maximum force that the nail can support is 300 N. Determine the maximum value of load P to the nearest Newton and the spacing of the nails to the nearest centimeter.

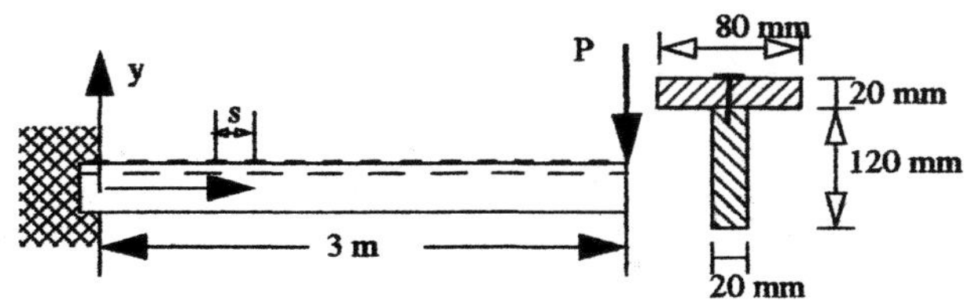

Fig. P6.103

Solution $|\sigma_w| \leq 7\text{MPa}$ $|\tau_w| \leq 1.5\text{MPa}$ $|V_{nail}| \leq 300$ $P_{max} = ?$ $\Delta s = ?$

The location of the centroid is shown in Fig. (b) can be found out as shown below.

$$\eta_c = \frac{(120)(20)(60) + (80)(20)(130)}{(120)(20) + (80)(20)} = 88\text{mm}$$

The area moment of inertia is

$$I_{zz} = \left[\frac{1}{12}(20)(120^3) + (20)(120)(88-60)^2\right] + \left[\frac{1}{12}(80)(20^3) + (80)(20)(130-88)^2\right] = 7.63(10^6)\text{mm}^4$$

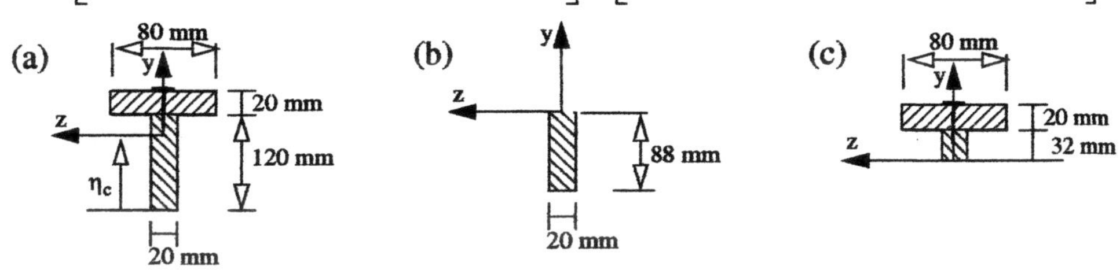

The maximum bending will be at the build in wall. The magnitude of this moment is $|M_{max}| = 3P$. The maximum bending normal stress in wood is

$$|\sigma_{max}| = \left|\frac{-M_{max}y_{max}}{I_{zz}}\right| = \frac{(3P)(-0.088)}{7.63(10^{-6})} = 34.6P(10^3) \leq 7(10^6) \qquad \text{or} \qquad P \leq 202.3\text{N} \qquad \textbf{1}$$

The shear force is uniform across the beam. Its magnitude is $|V_y| = P$. Figs. (b) and (c) show the area A_s that can be used for calculating Q_z at the neutral axis and the junction of the two wood pieces.

$$Q_{max} = (0.088)(0.02)\left(-\frac{0.088}{2}\right) = -77.44(10^{-6})\text{m}^3 \quad \text{and} \quad Q_{nail} = (0.08)(0.02)(-0.042) = 67.2(10^{-6})\text{m}^3$$

The magnitude of maximum shear stress in wood is

$$\tau_w = \left|\frac{-V_y Q_{max}}{I_{zz}t}\right| = \frac{(P)(77.44)(10^{-6})}{(7.63)(10^{-6})(0.02)} = 507.5P \leq 1.5(10^6) \qquad \text{or} \qquad P \leq 2956\text{N} \qquad \textbf{2}$$

The magnitude of the shear flow at the wood junction is

$$q = \left|\frac{Q_{nail}V_y}{I_{zz}}\right| = \frac{(P)(67.2)(10^{-6})}{(7.63)(10^{-6})} = 8.807P \leq 1.5(10^6)$$

The average shear force on each nail is $V_{nail} = q\Delta s = 8.807P\Delta s \leq 300$ or

$$P\Delta s \leq 34.06 \qquad \textbf{3}$$

The maximum value of P that satisfies Eqs. (1) and (2) is $\boxed{P_{max} = 202N}$

Substituting P = 202 N in Eq. (3), we obtain $\Delta s \le 0.168m$ or $\Delta s \le 16.8cm$

Rounding downwards we obtain $\boxed{\Delta s = 16cm}$

6.104 A wooden cantilever box beam is to be constructed by nailing four 1 inch x 6 inch pieces of lumber in one of the two ways shown. The allowable bending normal and shear stress in the wood are 750 psi and 150 psi, respectively. The maximum force that the nail can support is 100 lbs. Determine the maximum value of load P to the nearest pound, the spacing of the nails to the nearest half inch, and the preferred nailing method.

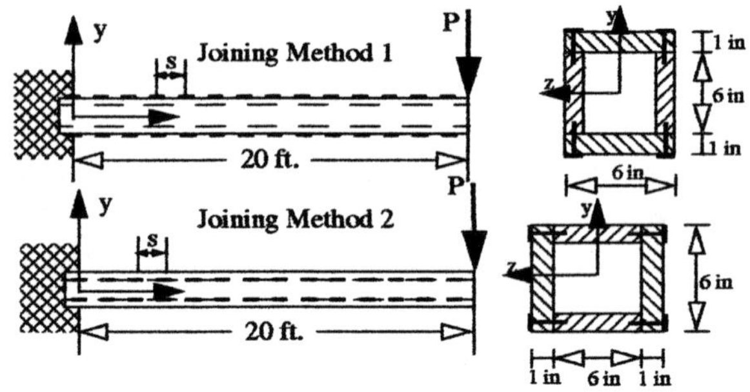

Fig. P6.104

Solution $|\sigma_w| \le 730psi$ $|\tau_w| \le 150$ psi $|V_{nail}| \le 100lbs$ P_{max} =? configuration =?

The configuration that gives the largest values of P_{max} and Δs is the better joining method. The area moment of inertias for the joining methods are

$$(I_{zz})_1 = \frac{1}{12}(6)(8^3) - \frac{1}{12}(4)(6^3) = 184in^4 \text{ and } (I_{zz})_2 = \frac{1}{12}(8)(6^3) - \frac{1}{12}(6)(4^3) = 112in^4$$

The section modulus for the two joining methods are

$$S_1 = \frac{(I_{zz})_1}{(y_{max})_1} = \frac{184}{4} = 46in^3 \text{ and } S_2 = \frac{(I_{zz})_2}{(y_{max})_2} = \frac{112}{4} = 37.33in^3$$

For the same bending moment and shear force, the stresses in configuration will be smaller because the section modulus and area moment of inertia are larger. $\boxed{\text{Use Joining method 1}}$

The maximum bending moment will be at the wall, its magnitude is M_{max} = 240P in-lbs. The magnitude of maximum bending normal stress in wood is

$$|\sigma_w| = \left|\frac{M_{max}}{S_1}\right| = \frac{240P}{46} = 5.218P \le 750 \quad \text{or} \quad P \le 143.7lbs \qquad 1$$

Figs (a) and (b) show the area A_s that can be used for finding the value of Q_z at the neutral axis and at the junction of the wood pieces. The values of these Q_z's are

$$Q_{max} = (2)(3)(1)(1.5) + (6)(1)(3.5) = 30in^3 \text{ and } Q_{Nail} = (6)(1)(3.5) = 21in^3$$

(a)

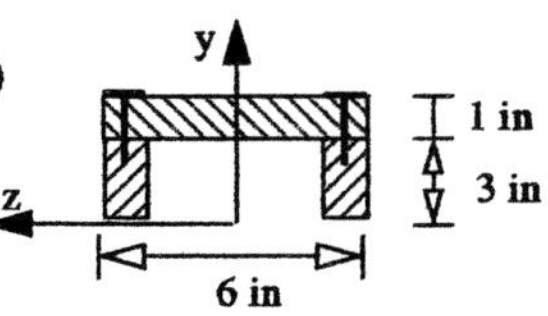

(b)

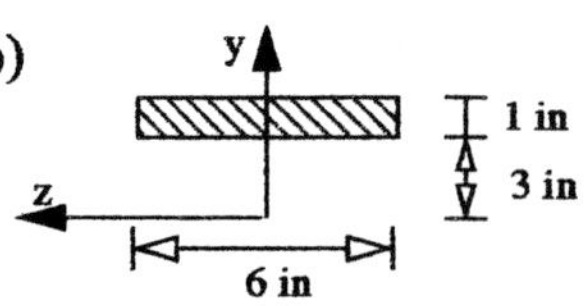

The shear force is uniform across the beam. Its magnitude is $|V_y| = P$

The magnitude of maximum shear stress in wood is

$$|\tau_W| = \left|\frac{Q_{max}V_y}{(I_{zz})_1\ t}\right| = \frac{P(30)}{(184)(2)} = 81.52P(10^{-3}) \le 150 \quad \text{or} \quad P \le 1840lbs \qquad 2$$

The magnitude of shear flow at the wood junction is $q = \left|\frac{Q_{Nail}V_y}{(I_{zz})_1}\right| = \frac{P(21)}{(184)} = 0.114P$

There are two rows of nails, each row supports half the shear flow. The average shear flow in each nail is

$$|V_{Nail}| = \left(\frac{q}{2}\right)\Delta S = 0.0570P\Delta S \le 100\text{lbs} \quad \text{or} \quad P\Delta S \le 1752 \qquad 3$$

The maximum value of P that satisfies Eqs. (1) and (2) is $\boxed{P_{max} = 143\text{lbs}}$

Substituting P = 143 in Eq.(3) we obtain $\Delta S \le \frac{1752}{143}$ or $\Delta S \le 12.25\text{inch}$

Rounding downwards we obtain $\boxed{\Delta S = 12\text{inch}}$

6.105 Leonardo da Vinci conducted experiments on simply supported beams and drew the following conclusion:

> "If a beam 2 barrccia long (L) supports 100 libbre (W), a beam 1 barrccia long (L/2) will support 200 libbre (2W). As many times as the shorter length is contained in the longer (L/α), so many times more weight (αW) will it support than the longer one."

Prove the above statement to be true by considering the two simply supported beams below and showing that W_2=αW for the same allowable bending normal stress.

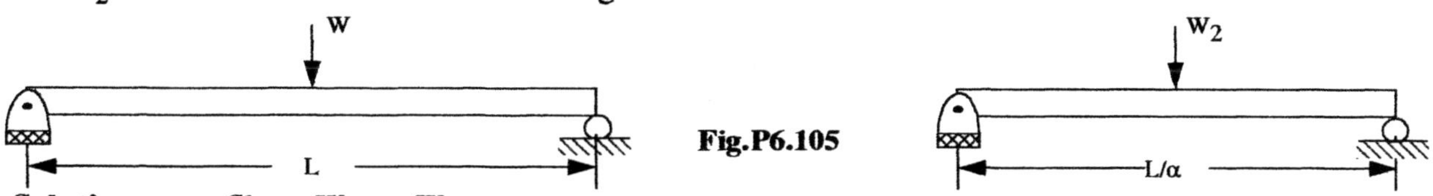

Fig.P6.105

Solution: Show $W_2 = \alpha W$

By symmetry each support reaction is half the load on the beam. The shear and moment diagrams can be drawn as shown below.

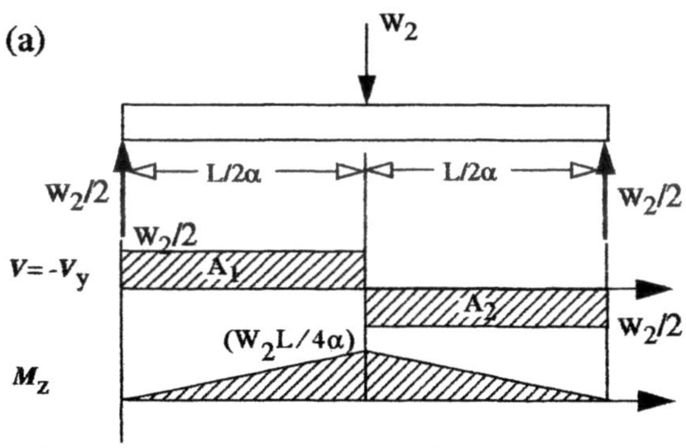

$$A_1 = A_2 = \left(\frac{W_2}{2}\right)\left(\frac{L}{2\alpha}\right) = \frac{W_2L}{4\alpha}$$

Let S be the section modulus of the beam. The maximum bending moment is $M_{max} = W_2L/4\alpha$. The maximum bending normal stress is

$$\sigma_{max} = \frac{M_{max}}{S} = \frac{W_2L}{4\alpha S} \qquad 1$$

When α = 1, W_2 = W and we obtain

$$\sigma_{max} = \frac{WL}{4S} \qquad 2$$

The maximum stress for the two beams is same. Equating Eqs.(1) and (2) we obtain

$\frac{W_2L}{4\alpha S} = \frac{WL}{4S}$ or $\boxed{W_2 = \alpha W}$

6.106 Galileo believed that a cantilever beam shown in Fig. P6.106 would break at point B which he considered to be a fulcrum point of a lever with AB and BC as the two arms. He believed that the material resistance (stress) was uniform across the cross-section. Show the stress value σ that Galileo obtained from

Fig. P6.106 is three times larger than the bending normal stress predicted by Equation 6.12.

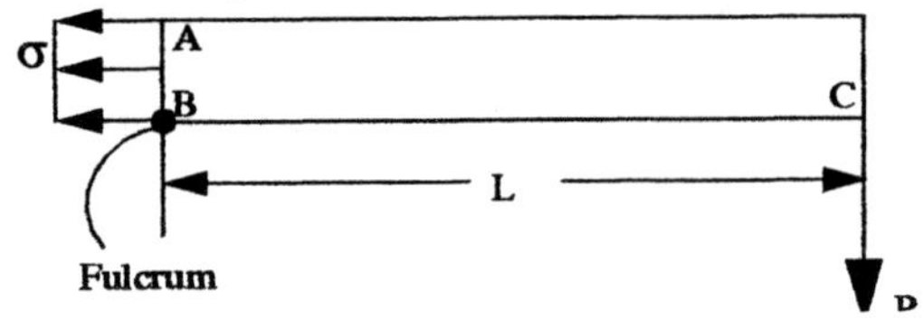

Fig. P6.106 Galileo's beam experiment.

Solution: Show $\sigma_{correct} = 3\sigma_{Galileo}$

Fig (a) shows the assumed rectangular cross section. Fig (b) shows the distributed forces in Galileo's beam experiment replaced by an equivalent load.

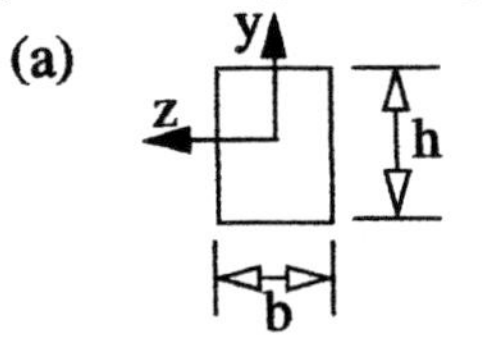

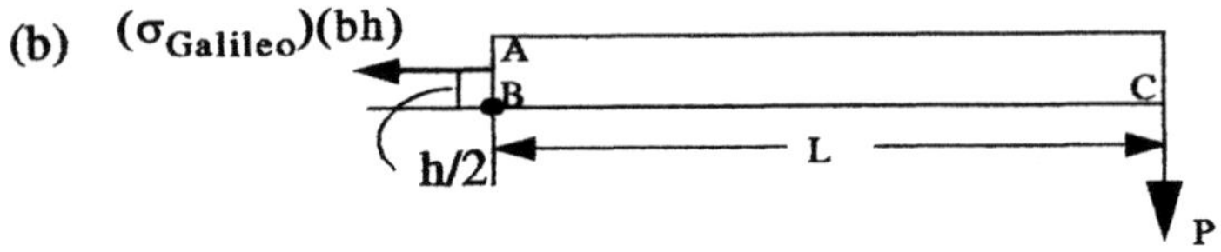

By moment equilibrium about point B, we obtain $(\sigma_{Galileo})(bh)\left(\frac{h}{2}\right) = (P)(L)$ or

$$\sigma_{Galileo} = \frac{2PL}{bh^2} \qquad 1$$

The area moment of inertia is $I_{zz} = \frac{1}{12}(b)(h^3)$. The moment of section AB is $|M_z| = PL$

The correct stress at B is

$$\sigma_{correct} = \left|\frac{y_B M_z}{I_{zz}}\right| = \frac{(PL)(h/2)}{\frac{1}{12}(b)(h^3)} = \frac{6PL}{bh^2} \qquad 2$$

Dividing Eqs. (1) by Eqs. (2) we obtain $\sigma_{Galileo}/\sigma_{correct} = 1/3$ or $\boxed{\sigma_{correct} = 3\sigma_{Galileo}}$

6.107 Galileo concluded that the bending moment due to its weight increases as the square of the length at the built in end of a cantilever beam. Show Galileo's statement is correct by deriving the bending moment at the built in end in a cantilever beam in terms of the specific weight g, area of cross-section A, and the length of the beam L.

Solution: $M_z = f(\gamma, A, L) = ?$

The distributed force due to gravity acts in the negative y- direction and its magnitude is γA force / length as shown in Fig (a).

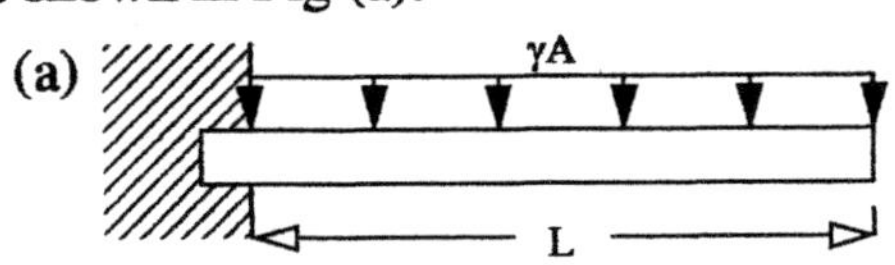

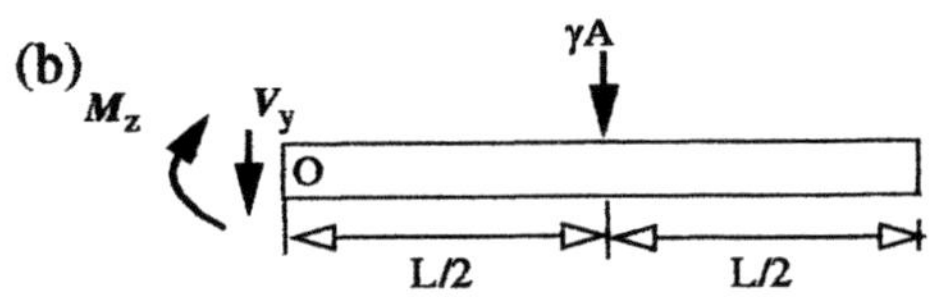

Fig (b) shows the free body diagram of the beam with distributed load replaced by an equivalent force and internal shear and bending moment drawn as per our sign convention. By equilibrium of moment about point O we obtain

$$M_z + (\gamma AL)\left(\frac{L}{2}\right) = 0 \text{ or } \qquad \boxed{M_z = -\left(\frac{\gamma AL^2}{2}\right)}$$

The moment varies with the square of the length as concluded by Galileo before Newton's law were formulated.

6.108 In the simply supported beam, Galileo determined that the bending moment is maximum at the applied load and its value is proportional to the product ab. He then concludes that to break the beam with the smallest load P, the load should be placed in the middle.
Prove Galileo's conclusions by drawing shear force bending moment diagrams and finding the value of maximum bending moment in terms of P, a, and b and then show that this value is largest when a = b.

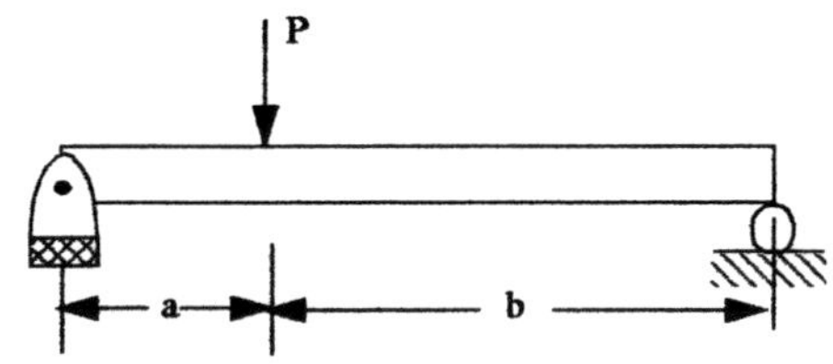

Fig. P6.108

Solution: Show $\boldsymbol{M}$ is maximum when a = b

The free body diagram of the beam is shown in Fig. (a). By equilibrium of moment about point C we obtain.

$$\boldsymbol{R}_A(a+b) - \boldsymbol{P}(b) = 0 \quad \text{or} \quad \boldsymbol{R}_A = \frac{b}{(a+b)}\boldsymbol{P}$$

The shear force and bending moment diagrams are drawn as shown below.

(a)

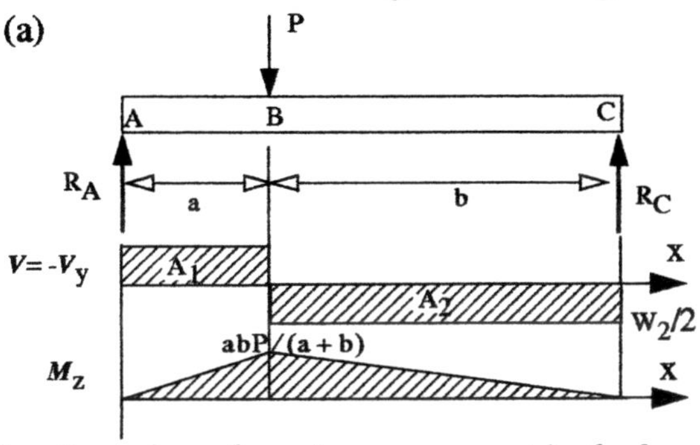

$$A_1 = \left(\frac{bP}{a+b}\right)(a) = \frac{abP}{a+b}$$

$$A_2 = \left(\frac{bP}{a+b}\right)(b) = \frac{abP}{a+b}$$

Thus the value of maximum moment in the beam is

$$\boldsymbol{M} = \frac{Pab}{(a+b)} \qquad 1$$

Let the length of the beam be L = a + b. Eqs. (1) can be written as

$$\boldsymbol{M} = \frac{\boldsymbol{P}a(L-a)}{L} = \frac{\boldsymbol{P}(aL-a^2)}{L} \qquad 2$$

Taking the first derivative of Eq. (2) with respect to a and setting it to zero we obtain

$$\frac{d\boldsymbol{M}}{da} = \frac{P(L-2a)}{L} = 0 \text{ or } a = (L/2)$$

Thus the bending moment is maximum when a = b = L/2 as concluded by Galileo before calculus.

6.109 Mariotte in attempt to correct Galileo's strength prediction, hypothesized that the stress varied in proportion to the distance from the fulcrum point B in Fig. P6.106, i.e., it varied linearly from point B. Show that the maximum bending stress value obtained by Mariotte is twice that predicted by Equation 6.12.

Solution: Show $\sigma_{correct} = 2\sigma_{Mariotte}$

In problem 6-106 the correct value of stress at the bottom of the beam was found as

$$\sigma_{correct} = \frac{6\boldsymbol{P}L}{bh^2} \qquad 1$$

Fig (a) shows Mariotte's assumed stress distribution. Fig. (b) shows the distributed forces replaced by an equivalent force.

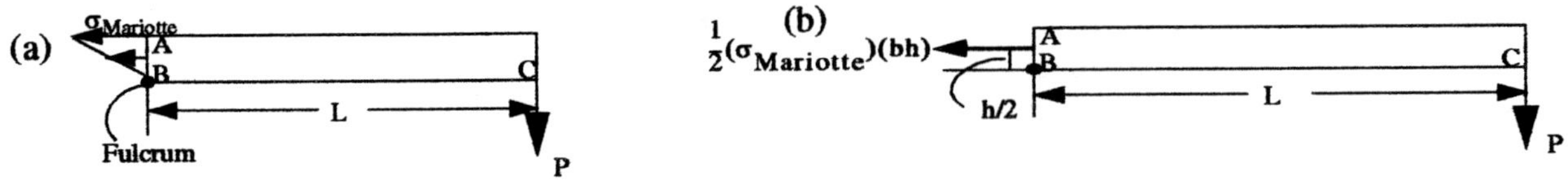

By equilibrium of moment about point B we obtain

$$\frac{1}{2}(\sigma_{Mariotte})(bh)\left(\frac{2h}{3}\right) = PL \quad \text{or} \quad \sigma_{Mariotte} = \frac{3PL}{bh^2} \qquad 2$$

Dividing Eq.(1) by Eq.(2) we obtain $\sigma_{correct}/\sigma_{Mariotte} = 2$ or $\boxed{\sigma_{correct} = 2\sigma_{Mariotte}}$

6.110 A beam is acted upon by a distributed load p(x). Let V_A and M_A represent the internal bending moment and shear force at A. Show that the internal moment at B is given by equation below.

$$M_B = M_A - V_A(x_B - x_A) + \int_{x_A}^{x_B}(x_B - x)p(x)dx \qquad 6.26$$

Solution

The equilibrium Eq. 6.13 is

$$\frac{dV_y}{dx} = -p(x) \qquad 1$$

Integrating Eq. 1 from x_A to x we obtain:

$$\int_{V_A}^{V_y} dV_y = -\int_{x_A}^{x} p(x)dx \quad \text{or} \quad V_y = V_A - \int_{x_A}^{x} p(x)dx$$

The equilibrium Eq. 6.14 is

$$\frac{dM_z}{dx} = -V_y \qquad 2$$

Integrating Eq. 2 from x_A to x_B we obtain:

$$\int_{M_A}^{M_B} dM_z = -\int_{x_A}^{x_B} V_y dx \quad \text{or} \quad M_B = M_A - \int_{x_A}^{x_B} V_y dx \qquad 3$$

Integrating Eq. 3 by parts we obtain:

$$M_B = M_A - (x - x_B)V_y\Big|_{x_A}^{x_B} + \int_{x_A}^{x_B}(x - x_B)\frac{dV_y}{dx}dx \qquad 4$$

Imposing the limits and substituting Eq. 1, we obtain:

$$M_B = M_A - [(x_B - x_B)V_B - (x_A - x_B)V_A] + \int_{x_A}^{x_B}(x - x_B)[-p(x)]dx \text{ or}$$

$$M_B = M_A + (x_A - x_B)V_A + \int_{x_A}^{x_B}(x_B - x)[p(x)]dx \qquad 5$$

Eq. 5 is same as Eq. 6.26.

6.111 The displacement in the x direction in a beam cross-section is given by $u = u_o(x) - y\frac{dv}{dx}(x)$. Assuming small strains and linear, elastic, isotropic, homogenous material with no inelastic strains, show:

$$N = EA\frac{du_o}{dx} - EAy_c\frac{d^2v}{dx^2} \qquad M_z = -EAy_c\frac{du_o}{dx} + EI_{zz}\frac{d^2v}{dx^2}$$

where y_c is the y coordinate of the centroid of the cross-section measured form some arbitrary origin, A is the area of cross-section, I_{zz} is the area moment of inertia about the z axis, N and M_z are the internal axial force and internal bending moment. Note that if y is measured from the centroid of the cross-section i.e., y_c=0, then the axial and bending problem de-couple. In such a case show $\sigma_{xx} = \frac{N}{A} - \frac{M_z y}{I_{zz}}$.

Solution

The normal strain in the x-direction is:

$$\varepsilon_{xx} = \frac{\partial u}{\partial x} = \frac{du_o}{dx} - y\frac{d^2v}{dx^2} \qquad 1$$

The normal stress in the x-direction is:

$$\sigma_{xx} = E\varepsilon_{xx} = E\left(\frac{du_o}{dx} - y\frac{d^2v}{dx^2}\right) \qquad 2$$

The internal normal force can be written as:

$$N = \int_A \sigma_{xx} dA = \int_A E\left(\frac{du_o}{dx} - y\frac{d^2v}{dx^2}\right) dA \qquad 3$$

Noting that material is homogenous across the cross-section and u_o and v are only functions of x and do not change across the cross-section we obtain:

$$N = E\left(\frac{du_o}{dx}\right)\int_A dA - E\left(\frac{d^2v}{dx^2}\right)\int_A y dA \qquad 4$$

In the above equation the first integral is the area of cross-section A and the second integral is the first moment of the area about the origin which by definition of centroid is $y_C A$. We thus obtain:

$$N = EA\left(\frac{du_o}{dx}\right) - Ey_c A\left(\frac{d^2v}{dx^2}\right) \qquad 5$$

The internal bending moment can be written as:

$$M_z = -\int_A y\sigma_{xx} dA = -\int_A Ey\left(\frac{du_o}{dx} - y\frac{d^2v}{dx^2}\right) dA = -\left[E\left(\frac{du_o}{dx}\right)\int_A y dA - E\left(\frac{d^2v}{dx^2}\right)\int_A y^2 dA\right] \qquad 6$$

Once more we note that the first integral in the last term of the equation above is $y_C A$ and the second integral in the last term is I_{zz}. We thus obtain:

$$M_z = -Ey_c A\left(\frac{du_o}{dx}\right) + EI_{zz}\left(\frac{d^2v}{dx^2}\right) \qquad 7$$

Eqs. 6 and 7 are the expressions of N and M_z that we were required to obtain.

6.112 Show that the bending normal stresses in a homogenous, linearly-elastic, isotropic symmetric-beam subject to a temperature change $\Delta T(x,y)$ is given by the equation below:

$$\sigma_{xx} = -\left(\frac{M_z y}{I_{zz}}\right) + \frac{M_T y}{I_{zz}} - E\alpha\Delta T(x, y) \qquad \textbf{6.27}$$

where, $M_T = E\alpha \int_A y\Delta T(x, y)dA$, α is coefficient of thermal expansion, and E is the Modulus of Elasticity.

Solution

By Eq. 6.5b the strain in bending is:

$$\varepsilon_{xx} = -y\frac{d^2v}{dx^2} \qquad 1$$

where, v is only a function of x. Including temperature effects the stress strain relationship is:

$$\varepsilon_{xx} = \frac{\sigma_{xx}}{E} + \alpha\Delta T(x, y) = -y\frac{d^2v}{dx^2} \quad \text{or} \quad \sigma_{xx} = -\left[Ey\frac{d^2v}{dx^2} + E\alpha\Delta T(x, y)\right] \qquad 2$$

The bending moment can be written as:

$$M_z = -\int_A y\sigma_{xx}dA = \int_A y\left[Ey\frac{d^2v}{dx^2} + E\alpha\Delta T(x, y)\right]dA \qquad 3$$

Noting that material is homogenous and v is only a function of x we obtain:

$$M_z = E\frac{d^2v}{dx^2}\int_A y^2 dA + E\alpha\int_A y\Delta T(x, y)dA = EI_{zz}\frac{d^2v}{dx^2} + M_T \quad \text{or} \qquad 4$$

$$\frac{d^2v}{dx^2} = \frac{(M_z - M_T)}{EI_{zz}} \qquad 5$$

Substituting Eq. 5 into Eq. 2, we obtain:

$$\sigma_{xx} = -\left[Ey\frac{(M_z - M_T)}{EI_{zz}} + E\alpha\Delta T(x, y)\right] = -\left[\frac{M_z y}{I_{zz}} - \frac{M_T y}{I_{zz}} + E\alpha\Delta T(x, y)\right] \qquad 6$$

Eq. 6 is same as Eq. 6.27.

6.113 In unsymmetrical bending of beams, under the assumption of plane sections remaining plane and perpendicular to the beam axis, the displacement u in the x-direction can be shown to be given by $u = -y\frac{dv}{dx} - z\frac{dw}{dx}$, where y and z are measured from the centroid of the cross-section, v and w are the deflection of the beam in the y and z direction respectively. Assume: small strain, a linear-elastic-isotropic-homogenous material, and no inelastic strain. Using Equations 1.5b and 1.5c, show

$$M_z = EI_{zz}\frac{d^2v}{dx^2} + EI_{yz}\frac{d^2w}{dx^2} \qquad M_y = EI_{yz}\frac{d^2v}{dx^2} + EI_{yy}\frac{d^2w}{dx^2} \quad \text{and} \qquad \textbf{6.28}$$

$$\sigma_{xx} = -\left(\frac{I_{yy}M_z - I_{yz}M_y}{I_{yy}I_{zz} - I_{yz}^2}\right)y - \left(\frac{I_{zz}M_y - I_{yz}M_z}{I_{yy}I_{zz} - I_{yz}^2}\right)z \qquad \textbf{6.29}$$

Note that if either y or z is a plane of symmetry, then $I_{yz} = 0$. From Equation 6.28, this implies that moment about z-axis causes deformation in the y-direction only and moment about the y-axis causes deformation in the z-direction only. In other words the bending problems about the y and the z-axis are de-coupled.

Solution

The normal strain in the x-direction is:

$$\varepsilon_{xx} = \frac{\partial u}{\partial x} = -y\frac{d^2v}{dx^2} - z\frac{d^2w}{dx^2} \qquad 1$$

The normal stress in the x-direction is:

$$\sigma_{xx} = E\varepsilon_{xx} = E\left(-y\frac{d^2v}{dx^2} - z\frac{d^2w}{dx^2}\right) \qquad 2$$

From Eq. 1.5b and 1.5c we have the following equivalent internal moments.

$$M_z = -\int_A y\sigma_{xx}dA = \int_A y\left[E\left(y\frac{d^2v}{dx^2} + z\frac{d^2w}{dx^2}\right)\right]dA = E\frac{d^2v}{dx^2}\int_A y^2dA + E\frac{d^2w}{dx^2}\int_A yzdA \text{ and} \qquad 3$$

$$M_y = -\int_A z\sigma_{xx}dA = \int_A z\left[E\left(y\frac{d^2v}{dx^2} + z\frac{d^2w}{dx^2}\right)\right]dA = E\frac{d^2v}{dx^2}\int_A yzdA + E\frac{d^2w}{dx^2}\int_A z^2dA \qquad 4$$

Substituting $I_{zz} = \int_A y^2dA$, $I_{yy} = \int_A z^2dA$, and $I_{yz} = \int_A yzdA$ in Eqs. (3) and (4) we obtain:

$$M_z = EI_{zz}\frac{d^2v}{dx^2} + EI_{yz}\frac{d^2w}{dx^2} \quad \text{and} \quad M_y = EI_{yz}\frac{d^2v}{dx^2} + EI_{yy}\frac{d^2w}{dx^2} \qquad 5$$

Eq. (5) is the same as Eq. 6.28.

Solving Eqs. (5) and (6) for $\frac{d^2v}{dx^2}$ and $\frac{d^2w}{dx^2}$ using Kramers rule we obtain the following.

$$\frac{d^2v}{dx^2} = \frac{1}{E}\left(\frac{I_{yy}M_z - I_{yz}M_y}{I_{yy}I_{zz} - I_{yz}^2}\right) \quad \text{and} \quad \frac{d^2w}{dx^2} = \frac{1}{E}\left(\frac{I_{zz}M_y - I_{yz}M_z}{I_{yy}I_{zz} - I_{yz}^2}\right) \qquad 6$$

Substituting Eq. (6) into Eq. (2) we obtain:

$$\sigma_{xx} = E\left[-\frac{y}{E}\left(\frac{I_{yy}M_z - I_{yz}M_y}{I_{yy}I_{zz} - I_{yz}^2}\right) - \frac{z}{E}\left(\frac{I_{zz}M_y - I_{yz}M_z}{I_{yy}I_{zz} - I_{yz}^2}\right)\right] \qquad 7$$

On cancelling E in the numerator and denominator of Eq. 7 we obtain Eq. 6.29.

6.114 The equation $\frac{\partial\sigma_{xx}}{\partial x} + \frac{\partial\tau_{yx}}{\partial y} = 0$ was derived in problem 1.98. Into this equation, substitute Equations 6.12 and 6.14 and integrate with y to show that:

$$\tau_{yx} = \frac{6V_y(b^2/4 - y^2)}{b^3t}$$

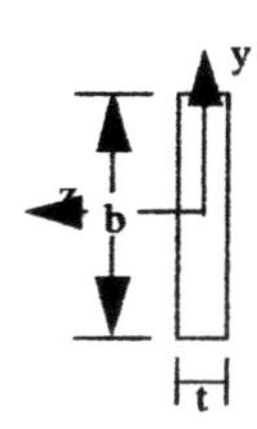

Fig. P6.114

Solution

Substituting $\sigma_{xx} = -\left(\frac{M_z y}{I_{zz}}\right)$ into $\frac{\partial\sigma_{xx}}{\partial x} + \frac{\partial\tau_{yx}}{\partial y} = 0$ we obtain the following:

$$-\frac{\partial}{\partial x}\left(\frac{M_z y}{I_{zz}}\right)+\frac{\partial \tau_{yx}}{\partial y}=0 \quad \text{or} \quad -\left(\frac{y}{I_{zz}}\right)\left[\frac{\partial M_z}{\partial x}\right]+\frac{\partial \tau_{yx}}{\partial y}=0 \qquad 1$$

Substituting $\frac{dM_z}{dx}=-V_y$ in the above equation we obtain:

$$\frac{\partial \tau_{yx}}{\partial y}=\left(\frac{y}{I_{zz}}\right)\left[\frac{\partial M_z}{\partial x}\right]=-\left(\frac{y}{I_{zz}}\right)V_y \qquad 2$$

We note that V_y and I_{zz} are not functions of y and the shear stress at the bottom is zero. Integrating from the bottom i.e., y = -b/2 to any point y we obtain:

$$\int_{\tau_{yx}=0}^{\tau_{yx}} d\tau_{yx}=-\left(\frac{V_y}{I_{zz}}\right)\int_{-\frac{b}{2}}^{y} y\,dy \quad \text{or} \quad \tau_{yx}=-\left(\frac{V_y}{I_{zz}}\right)\frac{y^2}{2}\Bigg|_{-\frac{b}{2}}^{y}=-\left(\frac{V_y}{2I_{zz}}\right)\left(y^2-\frac{b^2}{4}\right) \qquad 3$$

Substituting $I_{zz}=(tb^3)/12$ into Eq. (3) we obtain:

$$\tau_{yx}=\left(\frac{V_y}{2(tb^3)/12}\right)\left(-y^2+\frac{b^2}{4}\right) \quad \text{or} \quad \boxed{\tau_{yx}=\frac{6V_y(b^2/4-y^2)}{b^3t}}$$

6.115 A cantilever, hollow-circular aluminum beam of 5 feet length is to support a load of 1200-lbs. The inner radius of the beam is 1 inch. If the maximum bending normal stress is to be limited to 10 ksi, determine the minimum outer radius of the beam to the nearest 1/16th of an inch.

Solution $R_i = 1$ inch L = 5ft. P = 1200 lbs $\sigma_{max} \le 10$ksi R_o = ? nearest 1/16th in.

Fig. (a) shows the cantilever beam. The maximum moment will be at the built in wall. Its magnitude is:

$$M_{max} = (1200)(5) = 6000 \text{ ft-lbs} = 72,000 \text{ in-lbs} \qquad 1$$

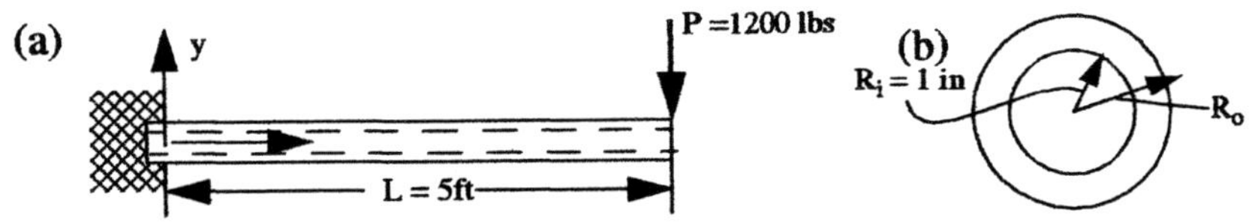

R_o	$f(R_o)$	R_o	$f(R_o)$
1.100	-9.620	2.100	-0.803
1.200	-9.927	2.105	-0.663
1.300	-10.061	2.110	-0.522
1.400	-9.993	2.115	-0.379
1.500	-9.688	2.120	-0.235
1.600	-9.114	2.125	-0.090
1.700	-8.232	2.130	0.057
1.800	-7.004	2.135	0.205
1.900	-5.386	2.140	0.355
2.000	-3.335	2.145	0.506
2.100	-0.803	2.150	0.658
2.200	2.257		
2.300	5.899		

Fig. (b) shows the cross-section of the beam. The area moment of inertia is $I_{zz} = \frac{\pi}{4}(R_o^4 - 1)$. The magnitude of maximum bending normal stress is:

$$\sigma_{max} = \left|\frac{M_{max} y_{max}}{I_{zz}}\right| = \frac{(72000)(R_o)}{\frac{\pi}{4}(R_o^4 - 1)} \le 10,000\text{psi} \quad \text{or} \quad R_o^4 - 1 - 9.167R_o \ge 0 \qquad 2$$

The value of R_o corresponds to the root of the left hand side of Eq. 2, i.e., $f(R_o) = R_o^4 - 9.167R_o - 1 = 0$. We find the root on the spreadsheet as described in Appendix B. The calculations are shown in the table above.

The value of R_o is between 2.125 and 2.130. Rounding upwards to the nearest 1/16 inch, we obtain the minimum value of R_o as

$$\boxed{R_O = 2\frac{3}{16}\text{inch}}$$

6.116 Table 6.1 shows the value of distributed load at several point along the axis of a rectangular beam. Determine the maximum bending normal and shear stress in the beam.

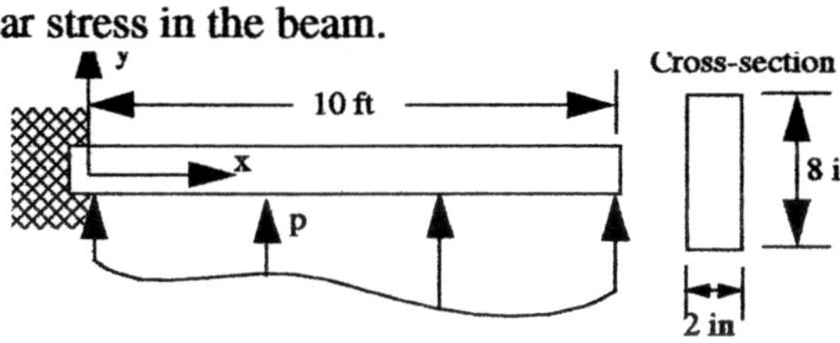

Fig. P6.116

Table 6.1. Data for Problem 6.116.

x ft.	p (x) lbs/ft.	x ft.	p (x) lbs/ft.
0	275	6	377
1	348	7	316
2	398	8	233
3	426	9	128
4	432	10	0
5	416		

Solution $|\sigma_{max}| = ?$ $|\tau_{max}| = ?$

Integrating the equilibrium equation $\frac{dV_y}{dx} = -p(x)$ from x = L to any point x and noting that the shear force at x = L is zero we obtain the following:

$$\int_0^{V_y} dV_y = -\int_L^x p(x)dx \quad \text{or} \quad V_y(x) = -\int_L^x p(x)dx \text{ lbs} \qquad 1$$

Integrating the equilibrium equation $\frac{dM_z}{dx} = -V_y$ from x = L to any point x and noting that the bending moment at x = L is zero we obtain the following:

$$\int_0^{M_z} dM_z = -\int_L^x V_y dx \quad \text{or} \quad M_z(x) = -\int_L^x V_y(x)dx \text{ ft-lbs} \qquad 2$$

The area moment of inertia is: $I_{zz} = \frac{(2)(8)^3}{12} = 85.33\text{in}^4$. The maximum Q_z is at the neutral axis. I_{zz} value is: $Q_{max} = (4)(2)(2) = 16\text{in}^3$. The magnitude of maximum bending normal stress in a section at any x_i can be written as:

$$\sigma_{xx}(x_i) = \left|\frac{M_z(x_i)y_{max}}{I_{zz}}\right| = \left|\frac{M_z(x_i)(12)(4)}{85.33}\right| = 0.562522|M_z(x_i)| \qquad 3$$

The magnitude of maximum bending shear stress in a section at any x_i can be written as:

$$\tau(x_i) = \left|\frac{V_y(x_i)Q_{max}}{I_{zz}t}\right| = \left|\frac{V_y(x_i)(16)}{(85.33)(2)}\right| = 0.093754|V_y(x_i)| \qquad 4$$

Eq. (1) can be numerically integrated on a spread sheet to obtain $V_y(x_i)$, the value of internal shear force at any x_i. Eq. (2) can be numerically integrated on a spread sheet to obtain $M_z(x_i)$, the value of internal bending moment at any x_i. Eq. (3) and Eq.(4) can be used to find the normal and shear stress at various x_i and the maximum values chosen by inspection. These calculations can be done on a spread sheet as shown below.The maximum bending stresses are at the built in end as expected.

x_i	$p(x_i)$	$-V_y(x_i)$ (lbs)	$M_z(x_i)$ ft-lbs	$\sigma_{xx}(x_i)$ psi	$\tau(x_i)$ psi
10	0	0.00	0.00	0.00	0.00
9	128	-180.50	90.25	50.77	16.92
8	233	-455.00	408.00	229.51	42.66
7	316	-801.50	1036.25	582.91	75.14
6	377	-1198.00	2036.00	1145.29	112.32
5	416	-1622.00	3446.00	1938.45	152.07
4	432	-2051.00	5282.50	2971.52	192.29
3	426	-2463.00	7539.50	4241.13	230.92
2	398	-2836.00	10189.00	5731.54	265.89
1	348	-3147.50	13180.75	7414.46	295.09
0	275	-3147.50	16328.25	9185.00	295.09

The magnitude of maximum bending normal stress is: $|\sigma_{max}| = 9185 \text{ psi}$

The magnitude of maximum bending shear stress is: $|\tau_{max}| = 295 \text{ psi}$

6.117 Let the distributed load p(x) in problem 6.116 be represented by the equation $p(x) = a + bx + cx^2$. Using the data in Table 6.1 determine the constant a, b, and c by the least-square method. Then find the maximum bending moment and maximum shear force by analytical integration and determine the maximum bending normal and shear stress.

Solution $a = ?$ $b = ?$ $c =$ $|\sigma_{max}| = ?$ $|\tau_{max}| = ?$

Integrating the equilibrium equation $\frac{dV_y}{dx} = -p(x)$ from x = L to any point x and noting that the shear force at x = L is zero we obtain:

$$\int_0^{V_y} dV_y = -\int_L^x p(x)dx \quad \text{or} \quad V_y(x) = -\int_L^x [a + bx + cx^2]dx = -\left[ax + \frac{bx^2}{2} + \frac{cx^3}{3}\right]_L^x \quad \text{or}$$

$$V_y(x) = -\left[a(x-L) + \frac{b}{2}(x^2 - L^2) + \frac{c}{3}(x^3 - L^3)\right] \qquad 1$$

Integrating the equilibrium equation $\frac{dM_z}{dx} = -V_y$ from x = L to any point x and noting that the bending moment at x = L is zero we obtain the following:

$$\int_0^{M_z} dM_z = -\int_L^x V_y dx \quad \text{or} \quad M_z(x) = \int_L^x \left[a(x-L) + \frac{b}{2}(x^2 - L^2) + \frac{c}{3}(x^3 - L^3)\right]dx \text{ ft-lbs or}$$

$$M_z(x) = a\left(\frac{x^2}{2} - Lx\right) + \frac{b}{2}\left(\frac{x^3}{3} - L^2x\right) + \frac{c}{3}\left(\frac{x^4}{4} - L^3x\right)\Bigg|_L^x \quad \text{or}$$

$$M_z(x) = a\left(\frac{x^2}{2} - Lx + \frac{L^2}{2}\right) + \frac{b}{2}\left(\frac{x^3}{3} - L^2x + \frac{2L^3}{3}\right) + \frac{c}{3}\left(\frac{x^4}{4} - L^3x + \frac{3L^4}{4}\right) \qquad 2$$

The maximum value of shear force and bending moment is at the wall i.e., at x = 0. From Eqs. (1) and (2) we obtain:

$$V_y(x=0) = \left(aL + \frac{bL^2}{2} + \frac{cL^3}{3}\right) \text{ lbs} \qquad 3$$

$$M_z(x=0) = \left(\frac{aL^2}{2} + \frac{bL^3}{3} + \frac{cL^4}{4}\right) \text{ ft} - \text{lbs} \qquad 4$$

Using the Least Square Method described in Appendix B, we obtain the value of the values of constants a, b, and c on a spread sheet as shown in the table below.

	x_i	$p(x_i)$	x_i^2	x_i^3	x_i^4	x^*p_i	$x_i^{2*}p_i$
1	10	0	100.0	1.000E+03	1.000E+04	0.000E+00	0.000E+00
2	9	128	81.0	7.290E+02	6.561E+03	1.152E+03	1.037E+04
3	8	233	64.0	5.120E+02	4.096E+03	1.864E+03	1.491E+04
4	7	316	49.0	3.430E+02	2.401E+03	2.212E+03	1.548E+04
5	6	377	36.0	2.160E+02	1.296E+03	2.262E+03	1.357E+04
6	5	416	25.0	1.250E+02	6.250E+02	2.080E+03	1.040E+04
7	4	432	16.0	6.400E+01	2.560E+02	1.728E+03	6.912E+03
8	3	426	9.0	2.700E+01	8.100E+01	1.278E+03	3.834E+03
9	2	398	4.0	8.000E+00	1.600E+01	7.960E+02	1.592E+03
10	1	348	1.0	1.000E+00	1.000E+00	3.480E+02	3.480E+02
11	0	275	0.0	0.000E+00	0.000E+00	0.000E+00	0.000E+00
Sums (b_{ij} &r_j)	55.0	3349.0	385.0	3025.0	25333.0	13720.0	77422.0
C_{ij}	6.026E+05	2.287E+05	1.815E+04	1.304E+05	1.210E+04	1.210E+03	
D	1.038E+06						
a_i	275.119	83.724	-11.122				

The values of the constants are: $a = 275.119 \quad b = 83.724 \quad c = -11.122$

Substituting the above values of a, b, and c and L = 10 in Eqs. (3) and (4) we obtain:

$$V_y(x=0) = 3230 \text{ lbs} \qquad 5$$

$$M_z(x=0) = 13858.95 \text{ ft} - \text{lbs} \qquad 6$$

Substituting Eqs. (5) and (6) into Eqs. (3) and (4) of problem 6.116 we obtain:

$$|\sigma_{max}| = \sigma_{xx}(x=0) = 0.562522|M_z(x=0)| = (0.562522)(13858.95) = 7795.96 \text{ psi}$$

$$|\tau_{max}| = \tau(x=0) = 0.093754|V_y(x=0)| = (0.093754)(3230) = 302.83 \text{ psi}$$

The magnitude of maximum bending normal stress is: $|\sigma_{max}| = 7796 \text{ psi}$

The magnitude of maximum bending shear stress is: $|\tau_{max}| = 303 \text{ psi}$

Problem Set 6.5

6.118 The cross-section of a composite beam with a coordinate system that has an origin at C is shown. The normal strain at point A due to bending about the z-axis, and the Modulus of Elasticity of each material are as given. In solving the problems use the fact that ε_{xx} varies linearly with y.

$\varepsilon_{xx} = -200\ \mu$
$E_1 = 8000$ ksi
$E_2 = 2000$ ksi

(a) Plot the stress distribution across the cross-section.
(b) Determine the maximum bending normal stress in each material.
(c) Determine the equivalent internal bending moment M_z by using Equation 6.1.
(d) Determine the equivalent internal bending moment M_z by using Equation 6.31.

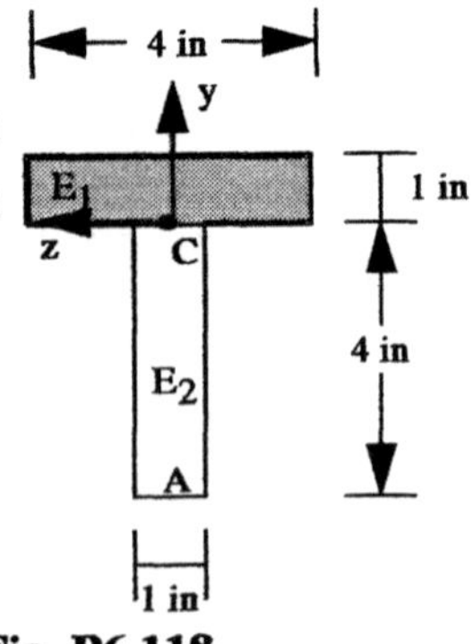

Fig. P6.118

Solution $(\varepsilon_{xx})_A = -200\mu$ $E_1 = 8000$ ksi $E_2 = 2000$ ksi σ_{xx} vs. y
$(\sigma_1)_{max} = ?$ $(\sigma_2)_{max} = ?$ $M_z = ?$ using Eq. 6.1 and Eq.6.31

At $y_A = -4$ in. the strain is -200μ and is zero at C. Noting the bending normal strain varies linearly with y, we obtain:

$$\varepsilon_{xx} = \frac{(-200)y}{(-4)} = 50y\ \mu \qquad (1)$$

The stress in each material is:

$$(\sigma_{xx})_1 = E_1\varepsilon_{xx} = (8000)(10^3)(50y)(10^{-6}) = 400y \text{ psi} \qquad (2)$$

$$(\sigma_{xx})_2 = E_2\varepsilon_{xx} = (2000)(10^3)(50y)(10^{-6}) = 100y \text{ psi} \qquad (3)$$

The bending normal stress can be written as:

$$\sigma_{xx} = \begin{cases} 100y \text{ psi} & -4 \le y < 0 \\ 400y \text{ psi} & 0 < y \le 1 \end{cases} \qquad (4)$$

(a) Fig, (a) shows the plot of bending normal stress

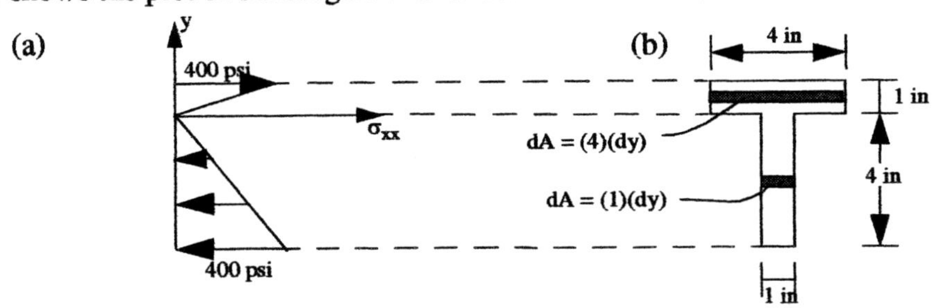

(b) The maximum bending normal stress in material 1 is at y = 1. Its value is: $\boxed{(\sigma_1)_{max} = 400 \text{ psi(T)}}$

The maximum bending normal stress in material 2 is at y = -4. Its value is: $\boxed{(\sigma_2)_{max} = 400 \text{ psi(C)}}$

(c) Fig. (b) shows the variation of differential area dA. Substituting Eq.(4) and the variation of dA in Eq. 6.1, we obtain:

$$M_z = -\int_A y\sigma_{xx}dA = -\left[\int_{-4}^{0} y(100y)dy + \int_{0}^{1} y(400y)(4)dy\right] = -\left[100\frac{y^3}{3}\bigg|_{-4}^{0} + 1600\frac{y^3}{3}\bigg|_{0}^{1}\right] = -2666.67 \text{ or}$$

$$\boxed{M_z = -2667 \text{ in} - \text{lb}}$$

(d) The area moment of inertias of each material cross-section can be found as:

$$(I_{zz})_1 = \frac{1}{12}(4)(1^3) + (4)(1)(0.5)^2 = 1.333 \text{ in}^4 \text{ and } (I_{zz})_2 = \frac{1}{12}(1)(4^3) + (1)(4)(2)^2 = 21.333 \text{ in}^4$$

The bending rigidity of the cross-section is:

$$\sum E_j(I_{zz})_j = (8000)(10^3)(1.333) + (2000)(10^3)(21.333) = 53.333(10^6) \text{ lb-in}^2$$

At y = -4in the stress σ_{xx} = -400 psi. From Eq. 6.31

$$\sigma_{xx} = -\frac{E y M_z}{\sum E_j(I_{zz})_j} = -\frac{(2000)(10^3)M_z(-4)}{53.333(10^6)} = (-400) \quad \text{or } M_z = -2666.66 \text{ or} \quad \boxed{M_z = -2667 \text{ in-lb}}$$

6.119 The cross-section of a composite beam with a coordinate system that has an origin at C is shown. The normal strain at point A due to bending about the z-axis, and the Modulus of Elasticity of each material are as given. In solving the problems use the fact that ε_{xx} varies linearly with y.

(a) Plot the stress distribution across the cross-section.
(b) Determine the maximum bending normal stress in each material.
(c) Determine the equivalent internal bending moment M_z by using Equation 6.1.
(d) Determine the equivalent internal bending moment M_z by using Equation 6.31.

$\varepsilon_{xx} = -200\ \mu$
$E_1 = 200$ GPa
$E_2 = 70$ GPa

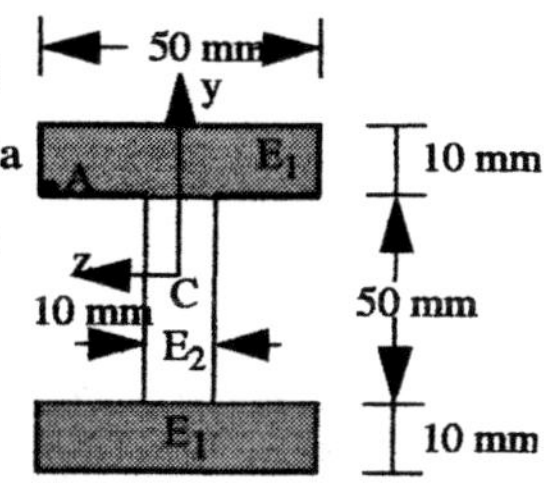

Fig. P6.119

Solution $(\varepsilon_{xx})_A = -200\mu$ $E_1 = 200$ GPa $E_2 = 70$ GPa σ_{xx} vs. y
$(\sigma_1)_{max} = ?$ $(\sigma_2)_{max} = ?$ $M_z = ?$ using Eq. 6.1 and Eq.6.31

At $y_A = 0.025$ m. the strain is -200μ and is zero at C. Noting the bending normal strain varies linearly with y, we obtain:

$$\varepsilon_{xx} = \frac{(-200)y}{(0.025)} = -8000y\ \mu \qquad 1$$

The stress in each material is:

$$(\sigma_{xx})_1 = E_1\varepsilon_{xx} = (200)(10^9)(-8000y)(10^{-6}) = -1600y(10^6)\text{N/m}^2 \qquad 2$$

$$(\sigma_{xx})_2 = E_2\varepsilon_{xx} = (70)(10^9)(-8000y)(10^{-6}) = -560y(10^6)\text{N/m}^2 \qquad 3$$

The bending normal stress can be written as:

$$\sigma_{xx} = \begin{cases} -1600y(10^6)\text{N/m}^2 & -0.035 \le y < -0.025 \\ -560y(10^6)\text{N/m}^2 & -0.025 < y < 0.025 \\ -1600y(10^6)\text{N/m}^2 & 0.025 < y < 0.035 \end{cases} \qquad 4$$

(a) Fig, (a) shows the plot of bending normal stress

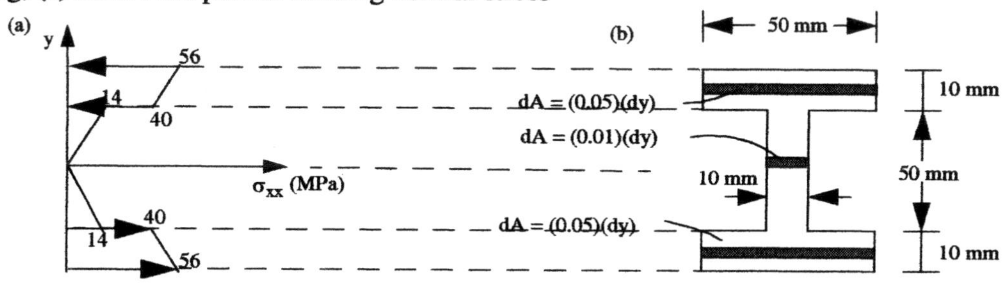

(b) The maximum bending normal stress in material 1 is at y = ± 0.035.

Its value is: $\boxed{(\sigma_1)_{max} = 56 \text{ MPa (T) (C)}}$

The maximum bending normal stress in material 2 is at y = ± 0.025.

Its value is: $\boxed{(\sigma_2)_{max} = 14 \text{ MPa (T) (C)}}$

(c) Fig. (b) shows the variation of differential area dA. Substituting Eq.(4) and the variation of dA in Eq. 6.1, we obtain:

$$M_z = -\int_A y\sigma_{xx} dA = (10^6)\left[\int_{-0.035}^{-0.025} y(1600y)(0.05dy) + \int_{-0.025}^{0.025} y(560y)(0.01)dy + \int_{0.025}^{0.035} y(1600y)(0.05dy)\right] \text{ or}$$

$$M_z = \left[80\frac{y^3}{3}\Bigg|_{-0.035}^{-0.025} + 5.6\frac{y^3}{3}\Bigg|_{-0.025}^{0.025} + 80\frac{y^3}{3}\Bigg|_{0.025}^{0.035}\right](10^6) = 1511.6 \text{ N-m or} \qquad \boxed{M_z = 1512 \text{ N-m}}$$

(d) The area moment of inertias of each material cross-section can be found as:

$$(I_{zz})_1 = 2\left[\frac{1}{12}(0.05, p)(0.01^3) + (0.05)(0.01)(0.03)^2\right] = 0.9083(10^{-6}) \text{ m}^4$$

$$(I_{zz})_2 = \frac{1}{12}(0.01)(0.05^3) = 0.1042(10^{-6}) \text{ m}^4$$

The bending rigidity of the cross-section is:

$$\sum E_j (I_{zz})_j = (200)(10^9)(0.9083)(10^{-6}) + (70)(10^9)(0.1042)(10^{-6}) = 188.95(10^3) \text{ N-m}^2$$

At y = 0.035 the modulus of elasticity is $E = 200(10^9)$ and the stress is stress $\sigma_{xx} = -56(10^6)$ N/m². From Eq. 6.31 we have

$$\sigma_{xx} = -\frac{EyM_z}{\sum E_j(I_{zz})_j} = -\frac{(200)(10^9)M_z(0.035)}{188.95(10^3)} = -56(10^6) \quad \text{or } M_z = 1511.6 \text{ or} \qquad \boxed{M_z = 1512 \text{ N-m}}$$

6.120 The cross-section of a composite beam with a coordinate system that has an origin at C is shown. The normal strain at point A due to bending about the z-axis, and the Modulus of Elasticity of each material are as given. In solving the problems use the fact that ε_{xx} varies linearly with y.

$\varepsilon_{xx} = 300\ \mu$

$E_1 = 30,000$ ksi

$E_2 = 20,000$ksi

(a) Plot the stress distribution across the cross-section.
(b) Determine the maximum bending normal stress in each material.
(c) Determine the equivalent internal bending moment M_z by using Equation 6.1.
(d) Determine the equivalent internal bending moment M_z by using Equation 6.31.

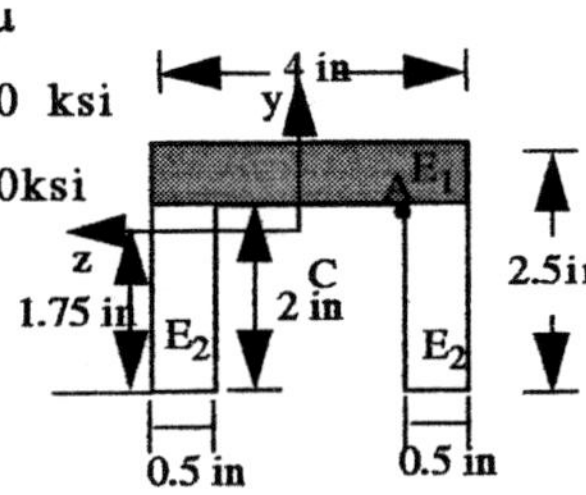

Fig. P6.120

Solution $(\varepsilon_{xx})_A = 300\mu$ $E_1 = 30{,}000$ ksi $E_2 = 20{,}000$ ksi σ_{xx} vs. y

$(\sigma_1)_{max} = ?$ $(\sigma_2)_{max} = ?$ $M_z = ?$ using Eq. 6.1 and Eq.6.31

At $y_A = 0.25$ in. the strain is 30μ and is zero at C. Noting the bending normal strain varies linearly with y, we obtain:

$$\varepsilon_{xx} = \frac{(300)y}{(0.25)} = 1200y\ \mu \qquad \textbf{1}$$

The stress in each material is:

$$(\sigma_{xx})_1 = E_1\varepsilon_{xx} = (30000)(10^3)(1200y)(10^{-6}) = 36000y \text{ psi} = 36y \text{ ksi} \qquad \textbf{2}$$

$$(\sigma_{xx})_2 = E_2\varepsilon_{xx} = (20000)(10^3)(1200y)(10^{-6}) = 24000y \text{ psi} = 24y \text{ ksi} \qquad 3$$

The bending normal stress can be written as:

$$\sigma_{xx} = \begin{cases} 24y \text{ ksi} & -1.75 \le y < 0.25 \\ 36y \text{ ksi} & 0.25 < y \le 0.75 \end{cases} \qquad 4$$

(a) Fig, (a) shows the plot of bending normal stress

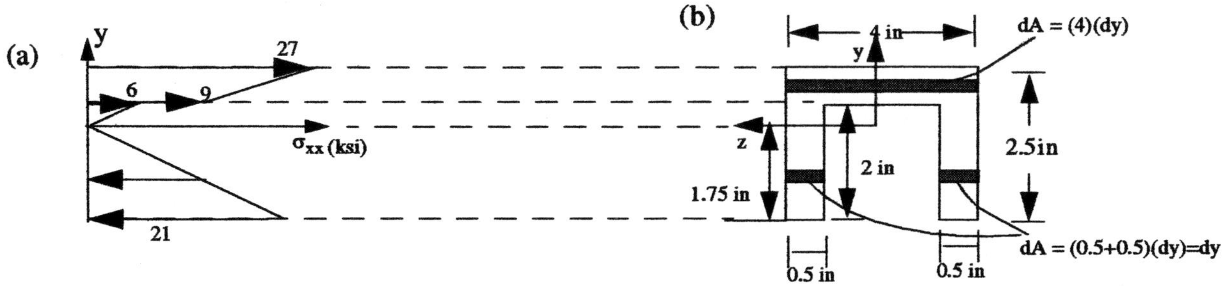

(b) The maximum bending normal stress in material 1 is at y = 0.75in. Its value is: $\boxed{(\sigma_1)_{max} = 27 \text{ ksi(T)}}$

The maximum bending normal stress in material 2 is at y = -1.75in. Its value is: $\boxed{(\sigma_2)_{max} = 42 \text{ ksi(C)}}$

(c) Fig. (b) shows the variation of differential area dA. Substituting Eq.(4) and the variation of dA in Eq. 6.1, we obtain:

$$M_z = -\int_A y\sigma_{xx}dA = -\left[\int_{-1.75}^{0.25} y(24y)dy + \int_{0.25}^{0.75} y(36y)(4)dy\right] = -\left[24\frac{y^3}{3}\Bigg|_{-1.75}^{0.25} + 144\frac{y^3}{3}\Bigg|_{0.25}^{0.75}\right] \text{ or}$$

$$\boxed{M_z = -62.5 \text{ in-kip}}$$

(d) The area moment of inertias of each material cross-section can be found as:

$$(I_{zz})_1 = \frac{1}{12}(4)(0.5^3) + (4)(0.5)(0.5)^2 = 0.5417 \text{ in}^4 \text{ and}$$

$$(I_{zz})_2 = 2\left[\frac{1}{12}(0.5)(2^3) + (0.5)(2)(0.75)^2\right] = 1.792 \text{ in}^4$$

The bending rigidity of the cross-section is:

$$\sum E_j(I_{zz})_j = (30000)(1.0104) + (20000)(1.792) = 52.09(10^3) \text{ kip-in}^2$$

At y = 0.75in the modulus of elasticity is E = 30,000 ksi and the stress is σ_{xx} = 27 ksi. From Eq. 6.31

$$\sigma_{xx} = -\frac{EyM_z}{\sum E_j(I_{zz})_j} = -\frac{(30000)M_z(0.75)}{52.09(10^3)} = 27 \text{ or} \qquad \boxed{M_z = -62.5 \text{ in-kip}}$$

6.121 A simply supported, 3 m long beam has a uniform distributed load of 10 kN/m over the entire length of the beam. If the beam has the composite cross-section shown, determine the maximum bending normal stress in each of the three materials. The modulus of elasticity for aluminum, wood, and steel is E_{al} = 70 GPa, E_w = 10 GPa, and E_s = 200 GPa, respectively.

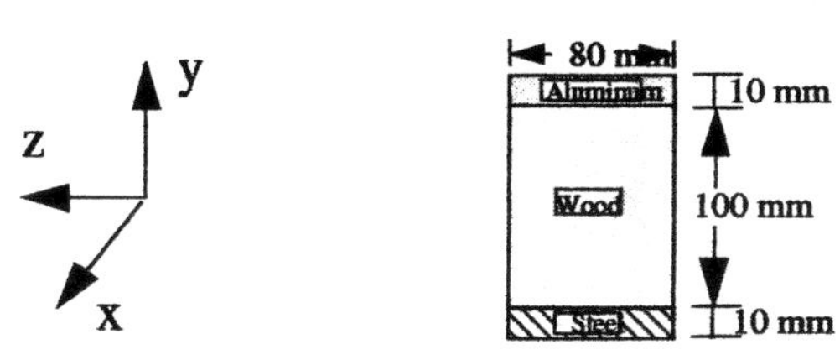

Fig. P6.121

Solution: L = 3m p = 10 kN/m E_{al} = 70 GPa E_s = 200 GPa

$(\sigma_{al})_{max}$ = ? $(\sigma_w)_{max}$ = ? $(\sigma_s)_{max}$ = ?

Fig (a) shows the cross section of the beam. The location of the neutral axis can be found as shown below.

$$\eta_c = \frac{\sum E_i A_i \eta_{ci}}{\sum E_i A_i} = \frac{(200)(10)(80)(5) + (10)(100)(80)(60) + (70)(10)(80)(115)}{(200)(10)(80) + (10)(100)(80) + (70)(10)(80)} = 40.7\text{mm}$$ 1

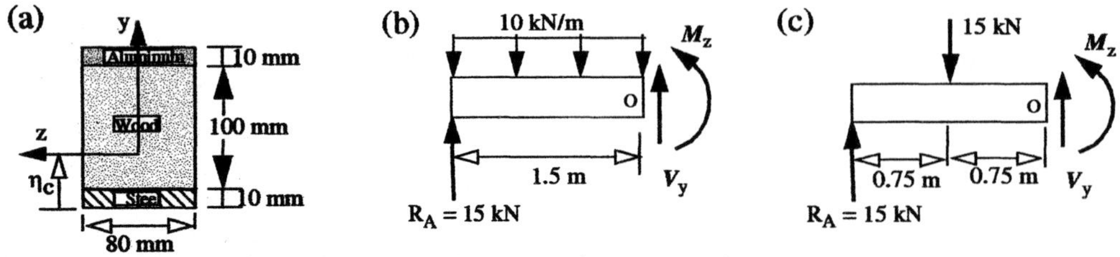

The area moment of inertias for each material cross section is:

$$(I_{zz})_{al} = \frac{1}{12}(0.08)(0.01)^3 + (0.08)(0.01)(0.115 - 0.0407)^2 = 4.426(10^{-6})\text{m}^4$$

$$(I_{zz})_w = \frac{1}{12}(0.08)(0.1)^3 + (0.08)(0.1)(0.06 - 0.0407)^2 = 9.654(10^{-6})\text{m}^4$$

$$(I_{zz})_s = \frac{1}{12}(0.08)(0.01)^3 + (0.08)(0.01)(0.0407 - 0.005)^2 = 1.025(10^{-6})\text{m}^4$$

The bending rigidity of the cross-section is

$$\sum E_j (I_{zz})_j = [(70)(4.426) + (10)(9.654) + (200)(1.025))](10^3) = 611.3(10^3)\text{N} - \text{m}^2$$ 2

By symmetry the reaction on each support of the simply supported beam is half the total load on the beam, i.e. $R_A = (3)(10)/(2) = 15\text{kN}$. The maximum moment will be at the center of the beam, Fig(b) shows the free body diagram of the beam after the imaginary cut has been made through the center of the beam. Fig(c) shows the free body diagram after the distributed force is replaced by an equivalent force. By equilibrium of moment about point O in Fig (c) we obtain,

$$M_{max} + 15(0.75) - 15(1.5) = 0 \quad \text{or} \quad M_{max} = 11.25\text{kN} - \text{m}$$ 3

The maximum bending normal stress in aluminium will be at $y = (0.12 - 0.0407) = 0.0793\text{m}$. Its value is

$$(\sigma_{al})_{max} = \frac{-E_{al}(M_{max})y}{\sum E_j (I_{zz})_j} = -\frac{(70)(10^9)(11.25)(10^3)(0.0793)}{611.3(10^3)} = -102.1(10^6)\frac{\text{N}}{\text{m}^2} \text{ or } \boxed{(\sigma_{al})_{max} = 102.1\text{MPa(C)}}$$

The maximum bending normal stress in wood will be at $y = (0.100 - 0.0407) = 0.0593\text{m}$. Its value is

$$(\sigma_w)_{max} = \frac{-E_w(M_{max})y}{\sum E_j (I_{zz})_j} = -\frac{(10)(10^9)(11.25)(10^3)(0.0593)}{611.3(10^3)} = -10.92(10^6)\frac{\text{N}}{\text{m}^2} \text{ or } \boxed{(\sigma_w)_{max} = 10.9\text{MPa(C)}}$$

The maximum bending normal stress in steel will be at y = -0.0407m. Its value is

$$(\sigma_s)_{max} = \frac{-E_s(M_{max})y}{\sum E_j (I_{zz})_j} = -\frac{(200)(10^9)(11.25)(10^3)(-0.0407)}{611.3(10^3)} = 149.7(10^6)\frac{\text{N}}{\text{m}^2} \text{ or } \boxed{(\sigma_s)_{max} = 149.7\text{MPa(T)}}$$

6.122 A wooden rod (E_W = 2000 ksi) and steel strip (E_s = 30,000 ksi) are fastened securely to rigid plates as shown. Determine (a) the maximum intensity of the load w, if the allowable bending normal stresses in steel and wood are 20 ksi, and 4 ksi.respectively. (b) the magnitude of the maximum shear stress in the beam corresponding to the load in part (a).

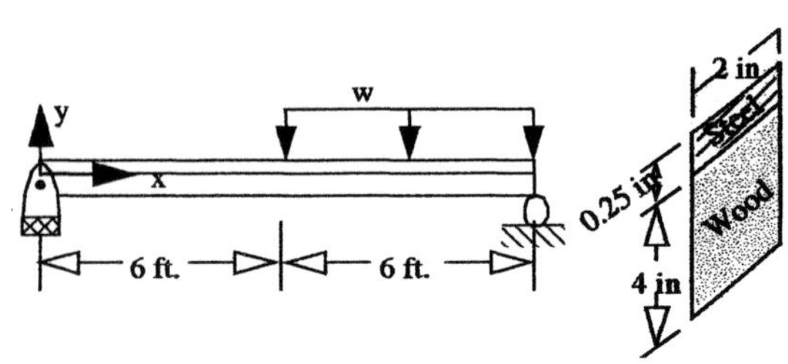

Fig. P6.122

Solution: $E_w = 2000$ ksi$E_s = 3000$ ksi $\quad |\sigma_s| \le 20\text{ksi} \quad |\sigma_w| \le 4\text{ksi} \quad |W_{max}| = ? \quad |\tau_{max}| = ?$

Fig (a) shows the cross section of the beam. The location of neutral axis can be found as shown below

$$\eta_c = \frac{\sum E_i A_i \eta_{ci}}{\sum E_i A_i} = \frac{(2000)(4)(2)(2) + (30000)(0.25)(2)(4.125)}{(2000)(4)(2) + (30000)(0.25)(2)} = 3.028\text{in} \qquad \mathbf{1}$$

The area moment of inertias for each material cross-section is:

$$(I_{zz})_s = \frac{1}{12}(2)(0.25)^3 + (2)(0.25)(4.125 - 3.028)^2 = 0.6043\text{in}^4$$

$$(I_{zz})_w = \frac{1}{12}(2)(4)^3 + (2)(4)(3.028 - 2)^2 = 19.12\text{in}^4$$

The bending rigidity of the cross-section is

$$\sum E_j (I_{zz})_j = [(30000)(0.6043) + (2000)(19.12)] = 56369\text{kip} - \text{in}^4 \qquad \mathbf{2}$$

Fig (b) shows the free body diagram of the entire beam with distributed load replaced by an equivalent force. By equilibrium of moment about point B we obtain $R_A(144) - (72w)(36) = 0$ or $R_A = 18w$

Fig (c) shows the shear force bending moment diagram. The maximum bending moment and shear force are

$$\boldsymbol{M}_{max} = 1458w \text{ and } (\boldsymbol{V}_y)_{max} = 54w \qquad \mathbf{3}$$

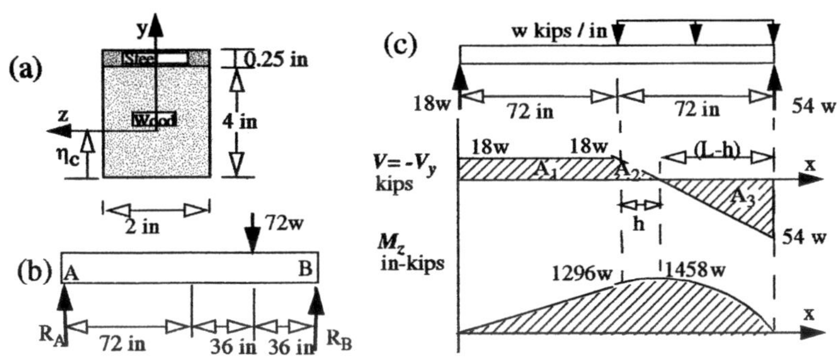

$$\frac{h}{(18w)} = \frac{72 - h}{54w} \quad \text{or} \quad h = 18$$

$$A_1 = (18w)(72) = 1296w$$

$$A_2 = \frac{1}{2}(18w)(18) = 162w$$

$$A_3 = \frac{1}{2}(54w)(54) = 1458w$$

(a) The maximum bending normal stress in steel is at y = 4.25 - 3.028 = 1.222 m. Its magnitude

$$|\sigma_s| = \left|\frac{-E_s(\boldsymbol{M}_{max})y}{\sum E_j (I_{zz})_j}\right| = \frac{(30000)(1458w)(1.222)}{56369} = 948.2w \le 200 \quad \text{or} \quad w \le 0.0211\text{kip/in} \qquad \mathbf{4}$$

The maximum bending normal stress in steel is at y = 3.028 in . Its magnitude is

$$|\sigma_W| = \left|\frac{-E_w(\boldsymbol{M}_{max})y}{\sum E_j (I_{zz})_j}\right| = \frac{(2000)(1458w)(3.028)}{56369} = 156.6w \le 4 \quad \text{or} \quad w \le 0.0255\text{kip/in} \qquad \mathbf{5}$$

The maximum value of w that satisfies Eq. 4 and 5 is w = 0.0211 kip/in or $\boxed{w = 21.1\text{lb/in}}$

(b) The maximum bending shear stress is at the neutral axis. Fig (d) shows the area A_s used in computing Q_z. As area A_s is that of only wood we obtain.

$$(Q_z)_w = (3.067)(2)\left(\frac{-3.067}{2}\right) = -9.406\text{in}^3 \text{ and } Q_{comp} = E_w(Q_z)_w = (2000)(-9.406) = (-18813)\text{kip} - \text{in}$$

Substituting w = 0.0211 into Eq. (5) we obtain $(\boldsymbol{V}_y)_{max} = (54)(0.0211) = 1.139\text{kip}$

The magnitude of maximum shear stress is

$$|\tau_{max}| = \left|\frac{-Q_{comp}(\boldsymbol{V}_y)_{max}}{\sum E_j (I_{zz})_j t}\right| = \frac{(18813)(1.139)}{(56369)(2)} = 0.1901\text{ksi or} \quad \boxed{|\tau_{max}| = 190.1\text{psi}}$$

6.123 A steel $E_{steel} = 200\ \text{GPa}$ strip is attached to a aluminum $E_{Al} = 70\ \text{GPa}$ beam to form a composite cantilever beam as shown. Determine (a) the maximum bending normal stress in steel and aluminum. (b) the magnitude of the maximum shear stress in the beam.

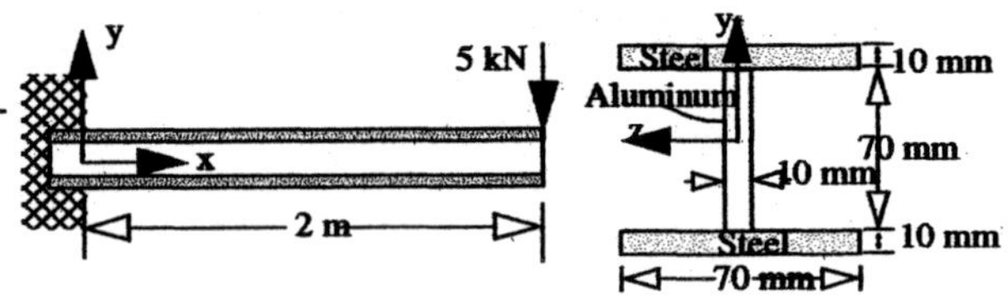

Fig. P6.123

Solution: $E_s = 200$ GPa $\quad E_{al} = 70$ GPa $\quad (\sigma_s)_{max} = ?$ $\quad (\sigma_{al})_{max} = ?$ $\quad |\tau_{max}| = ?$

The area moment of inertias of material cross section are:

$$(I_{zz})_s = 2\left[\frac{1}{12}(0.07)(0.01)^3 + (0.07)(0.01)(0.04)^2\right] = 2.252(10^{-6})\text{m}^4$$

$$(I_{zz})_{al} = \frac{1}{12}(0.01)(0.07)^3 = 0.2858(10^{-6})\text{m}^4$$

The bending rigidity of the cross-section is

$$\sum E_j(I_{zz})_j = [(200)(2.252) + (70)(0.2858)](10^3) = 470.5\text{N}-\text{m} \qquad 1$$

The bending moment is maximum at the wall. Its value is

$$M_{max} = (5)(2) = 10\text{kNm} \qquad 2$$

The maximum bending normal stress in steel is at the top and bottom i.e. $y = \pm 0.045$ m, its magnitude is.

$$|(\sigma_s)_{max}| = \left|\frac{-E_s(M_{max})y}{\sum E_j(I_{zz})_j}\right| = \frac{(200)(10^9)(100)(10^3)(0.045)}{470.5(10^3)} = 191.3(10^6)(\text{N/m}^2)$$

$$\boxed{|(\sigma_s)_{max}| = 191.3\text{MPa(T)or(C)}}$$

The maximum bending normal stress in aluminium is at the top and bottom i.e. $y = \pm 0.035$ m, its magnitude is.

$$|(\sigma_{al})_{max}| = \left|\frac{-E_{al}(M_{max})y}{\sum E_j(I_{zz})_j}\right| = \frac{(70)(10^9)(10)(10^3)(0.035)}{470.5(10^3)} = 52.08(10^6)(\text{N/m}^2)$$

$$\boxed{|(\sigma_{al})_{max}| = 52.1\text{MPa(T)or(C)}}$$

The shear force is uniform across the beam. The magnitude of the shear force is $|V_y| = 5\text{kN}$

Fig. (a) shows the cross section of the area. As found in calculation Q_z

(a)

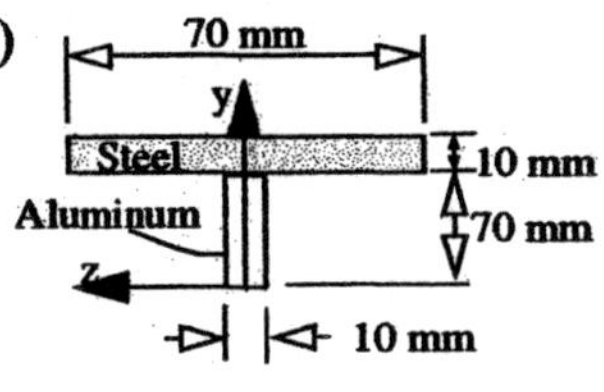

$$(Q_z)_s = (0.07)(0.01)(0.04) = 28(10^{-6})\text{m}^3 \text{ and } (Q_z)_{al} = (0.035)(0.01)\left(\frac{0.035}{2}\right) = 6.125(10^{-6})\text{m}^3$$

$$Q_{comp} = \sum E_j(Q_z)_j = [(200)(28) + (70)(6.125)](10^3) = 6028.75(10^3)\text{N}-\text{m}$$

The magnitude of maximum shear stress is

$$|\tau_{max}| = \left|\frac{-Q_{comp}(V_y)_{max}}{\sum E_j(I_{zz})_j t}\right| = \frac{6028.75(10^3)(5)(10^3)}{(470.5)(10^3)(0.01)} = 6.406(10^6)(\text{N/m}^2) \text{ or } \qquad \boxed{|\tau_{max}| = 6.41\text{MPa}}$$

6.124 A steel ($E_{steel} = 200$ GPa) tube of outside diameter of 240 mm is attached to a brass ($E_{brass} = 100$ GPa) tube to form the cross-section shown in Fig. P6.124. Determine (a) the maximum bending normal stress in steel and brass. (b) the magnitude of the maximum shear stress in the beam.

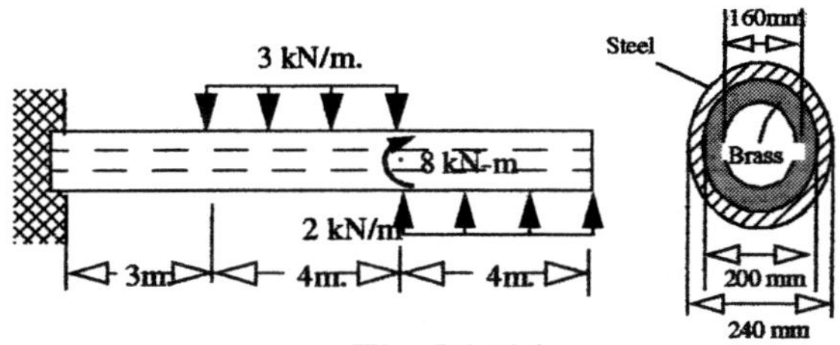

Fig. P6.124

Solution: $E_s = 200$ GPa $E_{br} = 100$ GPa $(\sigma_s)_{max} = ?$ $(\sigma_{br})_{max} = ?$ $|\tau|_{max} = ?$

The area moment of inertia of material cross section are:

$$(I_{zz})_s = \frac{\pi}{64}(0.24^2 - 0.20^2) = 84.32(10^{-6})m^4 \text{ and } (I_{zz})_{br} = \frac{\pi}{64}(0.20^2 - 0.16^2) = 46.37(10^{-6})m^4$$

The bending rigidity of the cross-section is

$$\sum E_j(I_{zz})_j = [(200)(84.32) + (100)(46.37)](10^3) = 21.5(10^6)N - m$$ 1

Fig (a) shows the free body diagram of the entire beam with distributed loads replaced by an equivalent loads. By equilibrium of forces in the y- direction we obtain $R_A = 12 - 8 = 4kN$. By equilibrium of moment about point A we obtain $M_A + (12)(5) + (8) - (8)(9) = 0$ or $M_A = 4kN - m$

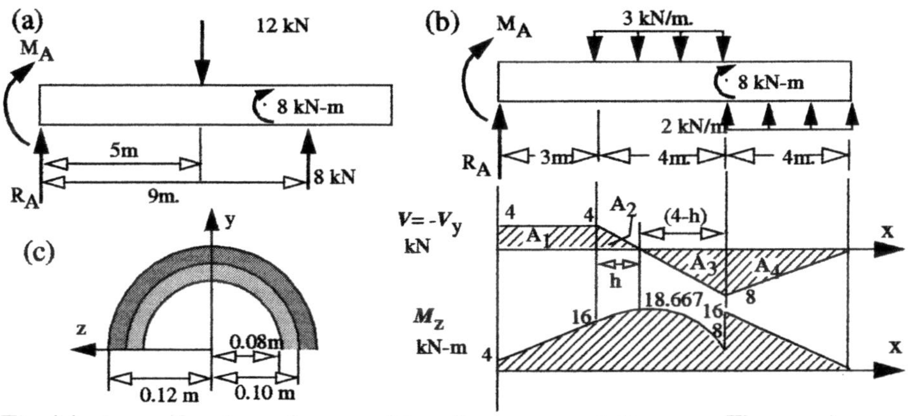

$$\frac{h}{4} = \frac{(4-h)}{8}$$

or $h = 1.333$

$$A_1 = (4)(3) = 12$$

$$A_2 = \frac{1}{2}(4)(h) = 2.667$$

$$A_3 = \frac{1}{2}(4)(4-h) = 10.667$$

$$A_4 = \frac{1}{2}(8)(4) = 16$$

Fig (b) shows the shear force and bending moment diagram. The maximum shear force and bending moment are $(V_y)_{max} = 8$ kN and $M_{max} = 18.667$ kN – m

The maximum bending normal stress in steel will be at y = ± 0.12 m, its magnitude is.

$$|(\sigma_s)_{max}| = \left|\frac{-E_s(M_{max})y}{\sum E_j(I_{zz})_j}\right| = \frac{(200)(10^9)(18.666)(10^3)(0.12)}{21.5(10^6)} = 21.5(10^6)(N/m^2)$$

$$\boxed{|(\sigma_s)_{max}| = 21.5\text{MPa (T) or (C)}}$$

The maximum bending normal stress in aluminium will be at y = ± 0.10 m, its magnitude is.

$$|(\sigma_{al})_{max}| = \left|\frac{-E_{al}(M_{max})y}{\sum E_j(I_{zz})_j}\right| = \frac{(100)(10^9)(18.666)(10^3)(0.10)}{21.5(10^6)} = 8.681(10^6)(N/m^2)$$

$$\boxed{|(\sigma_{al})_{max}| = 8.7\text{MPa (T) or (C)}}$$

Fig. (c) shows the cross section of the area A_s used in the calculation of Q_z.

$$(Q_z)_s = \left[\frac{\pi(0.12)^2}{2}\right]\left[\frac{(4)(0.12)}{3\pi}\right] - \left[\frac{\pi(0.1)^2}{2}\right]\left[\frac{(4)(0.1)}{3\pi}\right] = 0.4853(10^{-3})m^3$$

$(Q_z)_{br} = 0.3253(10^{-3})(m^3)$

$Q_{comp} = \sum E_j(Q_z)_j = [(200)(0.4853) + (100)(0.3253)](10^6) = 129.59(10^6)N - m$

The magnitude of maximum shear stress is

$$|\tau_{max}| = \left|\frac{-Q_{comp}(V_y)_{max}}{\sum E_j(I_{zz})_j t}\right| = \frac{(129.59)(10^6)(8)(10^3)}{(21.5)(10^6)(0.08)} = 602.7(10^3)N/m^2 \text{ or } \quad \boxed{|\tau_{max}| = 0.60MPa}$$

6.125 A reinforced concrete beam is constructed by embedding circular steel rods of 20 mm diameter. The maximum resisting (internal) moment in the beam was determined as 30 kN-m. Determine the maximum bending normal stress in the concrete and steel. Use modulus of elasticity for steel and concrete as $E_{st} = 200$ GPa and $E_c = 28$ GPa, respectively. Neglect the concrete capacity to support stresses in tension.

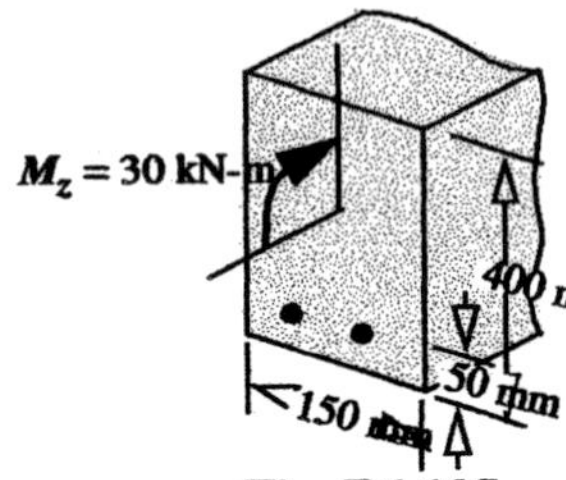

Fig. P6.125

Solution $d_s = 20$ mm $E_{st} = 200$ GPa $E_c = 28$ GPa $M_{max} = 30$ kN-m
$(\sigma_s)_{max} = ?$ $(\sigma_c)_{max} = ?$

We assume that the neutral axis is η_c from the top as shown in Fig. (a). The area of concrete in compression is $A_1 = 0.15\eta_c$. The total area of steel is $A_3 = 2\pi(0.01)^2 = 0.628(10^{-3})\ m^2$. Substituting these values and $E_1 = E_c = 28$ GPa, E_2=0, and $E_3 = E_s = 200$GPa in Eq. 6.33 we obtain a quadratic equation in η_c as shown below.

$$\eta_c = \frac{E_1A_1\eta_1 + E_2A_2\eta_3 + E_3A_3\eta_3}{E_1A_1 + E_2A_2 + E_3A_3} = \frac{(28)(0.15\eta_c)(\eta_c/2) + 0 + (200)(0.628)(10^{-3})(0.35)}{(28)(0.15\eta_c) + 0 + (200)(0.628)(10^{-3})} \text{ or}$$

$\eta_c^2 + 0.05981\eta_c - 0.02094 = 0$. Solving for the positive root in the quadratic equation we obtain:

$$\eta_c = 0.1179\ \ m \qquad 1$$

(a)

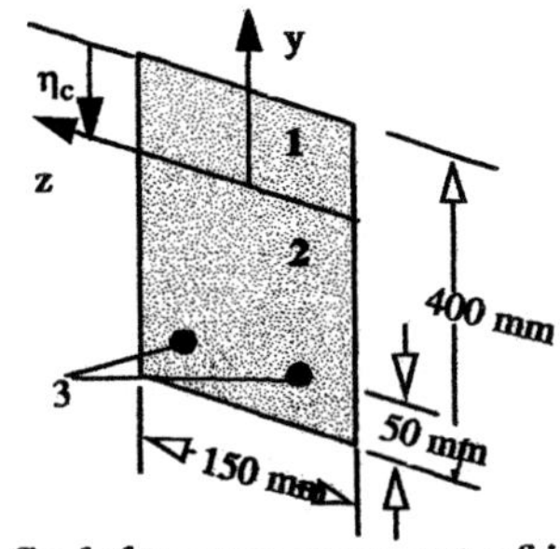

Using parallel axis theorem we can find the area moment of inertia about the z-axis as shown below:

$$(I_{zz})_1 = \frac{1}{12}(0.15)(0.1179)^3 + (0.15)(0.1179)(0.1179/2)^2 = 81.88(10^{-6})\ m^4$$

$$(I_{zz})_3 = (2)\left[\frac{\pi}{4}(0.01)^4 + \pi(0.01)^2(0.35 - 0.1179)^2\right] = 33.87(10^{-6})\ m^4$$

We can find the total bending rigidity as shown below:

$$\sum E_j(I_{zz})_j = E_1(I_{zz})_1 + E_2(I_{zz})_2 + E_3(I_{zz})_3 \text{ or}$$

$$\sum E_j(I_{zz})_j = (28)(10^9)(81.88)(10^{-6}) + 0 + (200)(10^9)(33.87)(10^{-6}) = 9.07(10^6)N - m^2 \qquad 2$$

The maximum compressive bending stress in concrete will occur at the top surface. Substituting $y = \eta_c = 0.1179\ m$, $M_z = 30$ kN-m, $E_1 = 28$GPa and Eq. (2) into Eq. 6.32 to obtain

$$(\sigma_c)_{max} = -E_1 y \frac{M_z}{\sum E_j (I_{zz})_j} = -(28)(10^9)(0.1179)\frac{(30)(10^3)}{9.07(10^6)} = -10.92(10^6)\frac{N}{m^2} \text{ or } \boxed{(\sigma_c)_{max} = 10.92\text{MPa(C)}}$$

The maximum tensile bending stress in steel will occur at the lowest point on the circular rod. Substituting $y = -(0.35 + 0.01 - \eta_c) = -0.2421\ m$, $M_z = 30$ kN-m, $E_3 = 200$GPa and Eq. (2) into Eq. 6.32 to obtain

$$(\sigma_s)_{max} = -E_3 y \frac{M_z}{\sum E_j (I_{zz})_j} = -(200)(10^9)(-0.2421)\frac{(30)(10^3)}{9.07(10^6)} = 160.2(10^6)\frac{N}{m^2} \text{ or } \boxed{(\sigma_s)_{max} = 160.2\text{MPa(T)}}$$

6.126 A reinforced concrete beam is constructed by embedding 2 in x 2 in square steel rods.The reinforced concrete beam is used as a 10 ft. cantilever beam, with a force of 36 kips applied at the free end in the negative y-direction. Determine the maximum compressive bending normal stress in the concrete. Use modulus of elasticity for steel and concrete as $E_{st} = 25{,}000$ ksi and $E_c = 4500$ ksi, respectively. Neglect the concrete capacity to support stresses in tension.

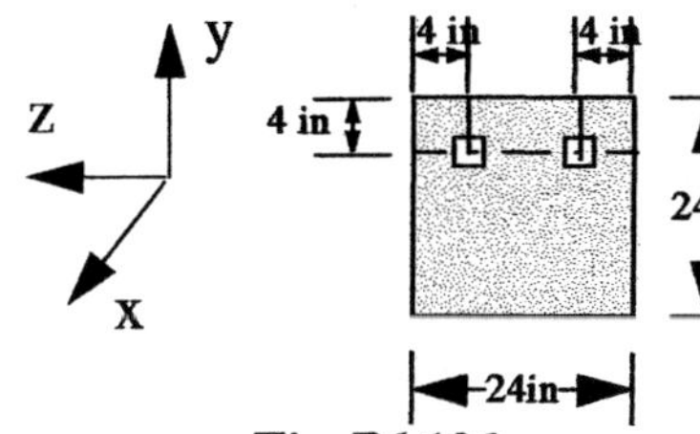

Fig. P6.126

Solution 2 in x 2 in $E_{st} = 25000$ ksi $E_c = 4500$ ksi $P = 36$ kips $L = 10$ ft

$(\sigma_s)_{max} = ?$ $(\sigma_c)_{max} = ?$

We assume that the neutral axis is η_c from the bottom as shown in Fig. (a). The area of concrete in compression is $A_1 = 24\eta_c$. The total area of steel is $A_3 = (2)(2)(2) = 8\ in^2$. Substituting these values and $E_1 = E_c = 4500$ ksi, $E_2 = 0$, and $E_3 = E_s = 25000$ ksi in Eq. 6.33 we obtain a quadratic equation in η_c as shown below.

$$\eta_c = \frac{E_1 A_1 \eta_1 + E_2 A_2 \eta_3 + E_3 A_3 \eta_3}{E_1 A_1 + E_2 A_2 + E_3 A_3} = \frac{(4500)(24\eta_c)(\eta_c/2) + 0 + (25000)(8)(20)}{(4500)(24\eta_c) + 0 + (25000)(8)} \text{ or}$$

$\eta_c^2 + 3.704\eta_c - 148.15 = 0$. Solving for the positive root in the quadratic equation we obtain:

$$\eta_c = 10.46\ in$$ **1**

(a)

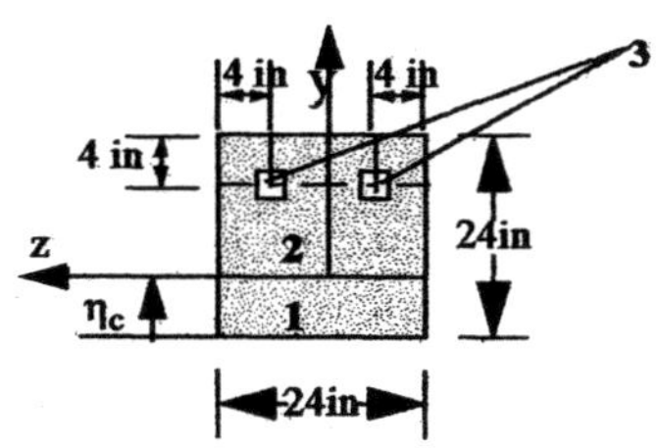

Using parallel axis theorem we can find the area moment of inertia about the z-axis as shown below:

$$(I_{zz})_1 = \frac{1}{12}(24)(10.46)^3 + (24)(10.46)(10.46/2)^2 = 9155.1\ in^4$$

$$(I_{zz})_3 = (2)\left[\frac{1}{12}(2)(2)^3 + (4)(20 - 10.46)^2\right] = 730.8\ in^4$$

We can find the total bending rigidity as shown below:

$$\sum E_j (I_{zz})_j = E_1 (I_{zz})_1 + E_2 (I_{zz})_2 + E_3 (I_{zz})_3 \text{ or}$$

$$\sum E_j (I_{zz})_j = (4500)(9155.1) + 0 + (25000)(730.8) = 59.47(10^6)\text{kip} - \text{in}^2$$ **2**

The maximum bending moment will occur at the wall in a cantilever beam. The magnitude of the maximum bending moment is $M_{max} = (36)(10) = 360$ ft-kips = 4320 in-kips.The maximum compressive bending stress in concrete will occur at the bottom surface. Substituting $y = -\eta_c = -10.46$ in , $M_z = 4320$ in-kips, $E_1 = 4.5$ ksi and Eq. (2) into Eq. 6.32 to obtain the magnitude of the bending stress in concrete as

$$|(\sigma_c)_{max}| = \left|-E_1 y \frac{M_z}{\sum E_j (I_{zz})_j}\right| = (4500)(10.46)\frac{4320}{59.47(10^6)} = 3.419 \text{ ksi or } \boxed{(\sigma_c)_{max} = 3.42 \text{ ksi(C)}}$$

The maximum tensile bending stress in steel will occur at the highest point on the square rod. Substituting $y = (20 + 1 - \eta_c) = 10.54$ in, $M_z = 4320$ in-kips, $E_3 = 25000$ ksi and Eq. (2) into Eq. 6.32 to obtain the magnitude of the bending stress in steel as

$$|(\sigma_s)_{max}| = \left|-E_3 y \frac{M_z}{\sum E_j (I_{zz})_j}\right| = (25000)(10.54)\frac{4320}{59.47(10^6)} = 19.14 \text{ ksi or } \boxed{(\sigma_s)_{max} = 19.14 \text{ ksi(T)}}$$

6.127 A reinforced concrete beam is constructed by embedding 20 mm x 20 mm square steel rods and loaded and supported as shown in Fig. P6.127. The allowable tensile bending normal stress is 160 MPa and the allowable compressive bending normal stress for concrete is 20 MPa. Determine the maximum intensity of the distributed load w. Use modulus of elasticity for steel and concrete as $E_{st} = 200$ GPa and $E_c = 28$ GPa, respectively. Neglect the concrete capacity to support stresses in tension.

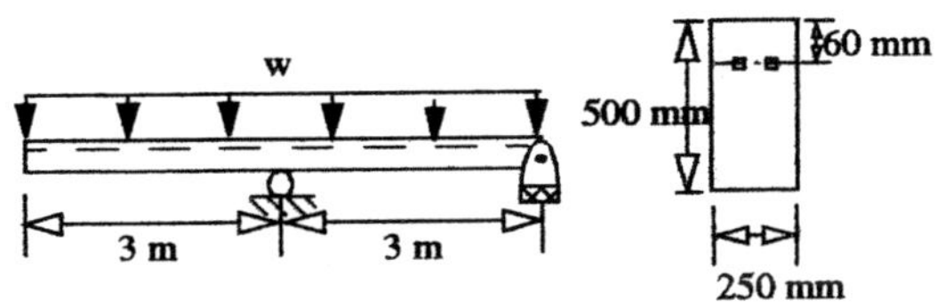

Fig. P6.127

Solution 20 mm x 20 mm $E_{st} = 200$ GPa $E_c = 28$ GPa $|\sigma_s| \le 160$ MPa $\sigma_C \le 20$ MPa(C)

We assume that the neutral axis is η_c from the bottom as shown in Fig. (a). The area of concrete in compression is $A_1 = 0.25\eta_c$. The total area of steel is $A_3 = (2)(0.02)(0.02) = 0.80(10^{-3})$ m^2. Substituting these values and $E_1 = E_c = 28$ GPa, $E_2 = 0$, and $E_3 = E_s = 200$GPa in Eq. 6.33 we obtain a quadratic equation in η_c as shown below.

$$\eta_c = \frac{E_1A_1\eta_1 + E_2A_2\eta_3 + E_3A_3\eta_3}{E_1A_1 + E_2A_2 + E_3A_3} = \frac{(28)(0.25\eta_c)(\eta_c/2) + 0 + (200)(0.80)(10^{-3})(0.44)}{(28)(0.25\eta_c) + 0 + (200)(0.80)(10^{-3})} \text{ or}$$

$\eta_c^2 + 0.0457\eta_c - 0.0201 = 0$. Solving for the positive root in the quadratic equation we obtain:

$$\eta_c = 0.1208 \text{ m}$$ 1

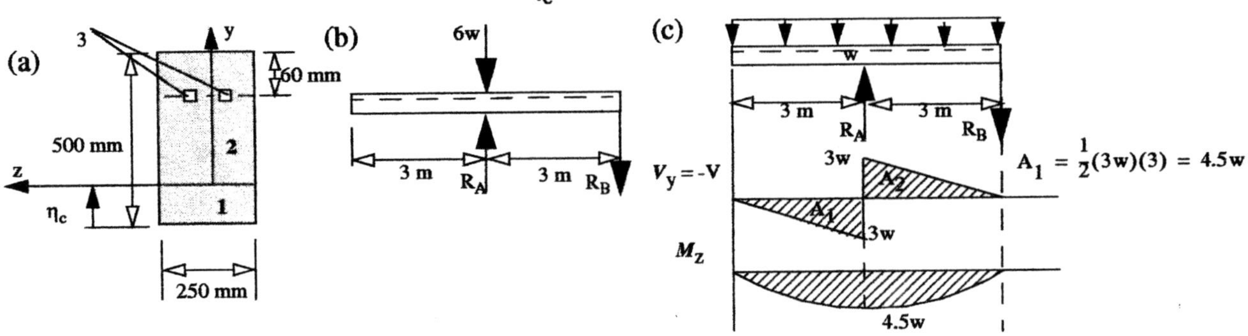

Using parallel axis theorem we can find the area moment of inertia about the z-axis as shown below:

$$(I_{zz})_1 = \frac{1}{12}(0.25)(0.1208)^3 + (0.25)(0.1208)(0.1208/2)^2 = 146.89(10^{-6})\ m^4$$

$$(I_{zz})_3 = (2)\left[\frac{1}{12}(0.02)(0.02)^3 + (0.02)(0.02)(0.44 - 0.1208)^2\right] = 81.54(10^{-6})\ m^4$$

We can find the total bending rigidity as shown below:

$$\sum E_j(I_{zz})_j = E_1(I_{zz})_1 + E_2(I_{zz})_2 + E_3(I_{zz})_3 \text{ or}$$

$$\sum E_j(I_{zz})_j = (28)(10^9)(146.89)(10^{-6}) + 0 + (200)(10^9)(81.54)(10^{-6}) = 20.42(10^6)N - m^2 \qquad 2$$

Fig. (b) shows the free body diagram of the entire beam with distributed force replaced by an equivalent force. By equilibrium of moment about point B we obtain: $R_A(3) = (3)(6w)$ or $R_A = 6w$. By equilibrium of forces in the y-direction $R_B = 0$. Fig. (c) shows the shear-moment diagram. The maximum moment is: $M_z = -4.5w$.

The maximum compressive bending stress in concrete will occur at the bottom surface. Substituting $y = -\eta_c = -0.1208$ m, $M_z = -4.5w$, $E_1 = 28GPa$ and Eq. (2) into Eq. 6.32 to obtain

$$(\sigma_c)_{max} = -E_1 y\frac{M_z}{\sum E_j(I_{zz})_j} = -(28)(10^9)(-0.1208)\frac{(-4.5w)}{20.42(10^6)} = (-0.7454w)(10^3)N/m^2 \text{ or}$$

$$(\sigma_c)_{max} = (0.7454w)(10^3)N/m^2(C) \le 20\ (10^6)(C) \quad \text{or} \quad w \le 26.83(10^3)\ N/m \qquad 3$$

The maximum tensile bending stress in steel will occur at the highest point on the square rod. Substituting $y = (0.44 + 0.01 - \eta_c) = 0.3292$ m, $M_z = -4.5w$, $E_3 = 200GPa$ and Eq. (2) into Eq. 6.32 to obtain

$$(\sigma_s)_{max} = -E_3 y\frac{M_z}{\sum E_j(I_{zz})_j} = -(200)(10^9)(0.3292w)\frac{(-4.5w)}{20.42(10^6)} = (14.509w)(10^3) \le (160)(10^6) \text{ or}$$

$$w \le 11.03(10^3)\ N/m \qquad 4$$

The maximum value of w that satisfies Eq. (3) and (4) is 11.03 (10^3) N/m $\boxed{w_{max} = 11.0\ kN/m}$

6.128 The bending normal stress on the cross-section shown is given by

$$\sigma_{xx} = \begin{cases} 24\ \text{ksi} & -5 \le y < -4 \\ -6y\ \text{ksi} & -4 < y < 4 \\ -24\ \text{ksi} & 4 < \ \le 5 \end{cases}$$

Determine the equivalent internal bending moment M_z.

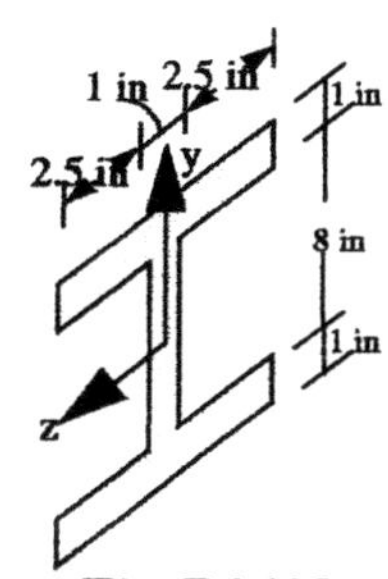

Fig. P6.128

Solution $M_z = ?$

Fig. (a) shows the variation of the differential area across the cross-section.

(a)

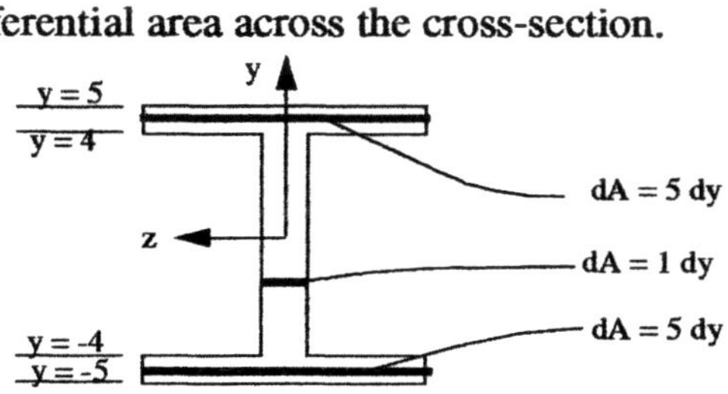

Substituting the given stress distribution and the variation of dA in Eq. 6.1, we obtain:

$$M_z = -\int_A y\sigma_{xx}dA = -\left[\int_{-5.0}^{-4.0} y(24)(5dy) + \int_{-4.0}^{4.0} y(-6y)(1)dy + \int_{4.0}^{5.0} y(-24)(5dy)\right] \text{ or}$$

$$M_z = -\left[144\frac{y^2}{2}\Bigg|_{-5.0}^{-4.0} - 6\frac{y^3}{3}\Bigg|_{-4.0}^{4.0} - 144\frac{y^2}{2}\Bigg|_{4.0}^{5.0}\right] = 1552 \text{ in-kip or} \qquad \boxed{M_z = 1552 \text{ in-kip}}$$

6.129 A rectangular beam has the dimensions as shown in Fig. P6.129. The normal strain due to bending about the z-axis was found to vary as $\varepsilon_{xx} = -0.01y$ with y measured in meters. Determine the equivalent internal moment that produced the given state of strain. The beam is made from elastic-perfectly plastic material that has a yield stress of σ_{yield}= 250 MPa, and a Modulus of Elasticity E = 200 GPa.

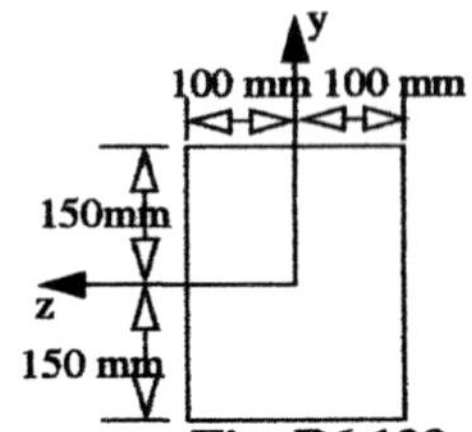

Fig. P6.129

Solution ε_{xx}=-0.01y y in meters σ_{yield}=250MPa E=200GPa M_z = ?

From problem 3.142, we have the following stress-distribution for the given state of strain and material.

$$\sigma_{xx} = \begin{cases} -250 \text{ MPa} & 0.125\text{m} < y < 0.150\text{m} \\ -2000y \text{ MPa} & -0.125\text{m} < y < 0.125\text{m} \\ 250 \text{ MPa} & -0.150\text{m} < y < -0.125\text{m} \end{cases}$$ 1

Substituting the stress distribution and dA = 0.2 dy in Eq. 6.1, we obtain:

$$M_z = -\int_A y\sigma_{xx}dA = -\left[\int_{-0.15}^{-0.125} y(250)(0.2dy) + \int_{-0.125}^{0.125} y(-2000y)(0.2dy) + \int_{0.125}^{0.15} y(-250)(0.2dy)\right](10^6) \text{ or}$$

$$M_z = -\left[50\frac{y^2}{2}\Bigg|_{-0.15}^{-0.125} - 400\frac{y^3}{3}\Bigg|_{-0.125}^{0.125} - 50\frac{y^2}{2}\Bigg|_{0.125}^{0.15}\right](10^6) = 864.5(10^3) \text{ N-m or} \qquad \boxed{M_z = 864.5 \text{ kN-m}}$$

6.130 A rectangular beam has the dimensions as shown in Fig. P6.129. The normal strain due to bending about the z-axis was found to vary as $\varepsilon_{xx} = -0.01y$ with y measured in meters. Determine the equivalent internal moment that would produce the given strain. The beam is made from a bi-linear material that has a yield stress of σ_{yield}= 200 MPa, Modulus of Elasticity E_1 = 250 GPa and E_2= 80 GPa.

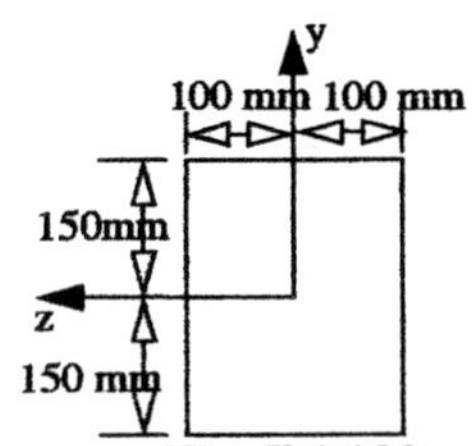

Fig. P6.130

Solution ε_{xx}=-0.01y y in meters σ_{yield}=250MPa E_1 = 200 GPa E_2= 80 GPa M_z = ?

From problem 3.143, we have the following stress-distribution for the given state of strain and material.

$$\sigma_{xx} = \begin{cases} (-150 - 800y)\text{MPa} & 0.125\text{m} < y < 0.150\text{m} \\ -2000y\text{MPa} & -0.125\text{m} < y < 0.125\text{m} \\ (150 - 800y)\text{MPa} & -0.150\text{m} < y < -0.125\text{m} \end{cases}$$ 1

Substituting the stress distribution and dA = 0.2 dy in Eq. 6.1, we obtain:

$$M_z = -\left[\int_{-0.15}^{-0.125} y(150-800y)(0.2dy) + \int_{-0.125}^{0.125} y(-2000y)(0.2dy) + \int_{0.125}^{0.15} y(-150-800y)(0.2dy)\right](10^6) \text{ or}$$

$$M_z = -\left[\left(30\frac{y^2}{2} - 40\frac{y^3}{3}\right)\Bigg|_{-0.15}^{-0.125} - 400\frac{y^3}{3}\Bigg|_{-0.125}^{0.125} + \left(-30\frac{y^2}{2} - 40\frac{y^3}{3}\right)\Bigg|_{0.125}^{0.15}\right](10^6) = 765.0(10^3) \text{ N-m or}$$

$$\boxed{M_z = 765 \text{ kN-m}}$$

6.131 A rectangular beam has the dimensions as shown in Fig. P6.129. The normal strain due to bending about the z-axis was found to vary as $\varepsilon_{xx} = -0.01y$ with y measured in meters. Determine the equivalent internal moment that would produce the given strain. The beam material has a stress strain relationship given by $\sigma = 952\varepsilon^{0.2}$ MPa.

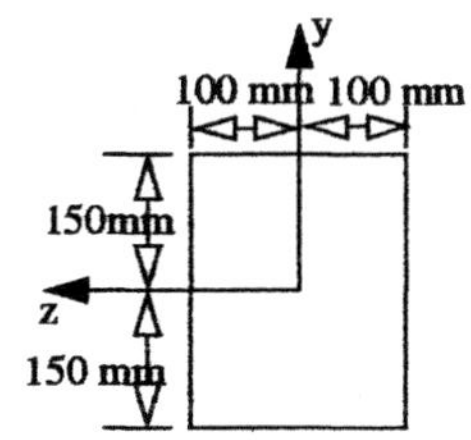

Fig. P6.131

Solution ε_{xx}=-0.01y y in meters $\sigma = 952\varepsilon^{0.2}$ MPa $M_z = ?$

From problem 3.144, we have the following stress-distribution for the given state of strain and material.

$$\sigma_{xx} = \begin{cases} 379(-y)^{0.2}\text{MPa} & y < 0 \\ -379(y)^{0.2}\text{MPa} & y > 0 \end{cases} \qquad 1$$

From the symmetry we conclude that the equivalent internal moment in the lower half ($y < 0$) is same as that from the upper half ($y > 0$). The integral in Eq. 6.1 can be written as twice the integral in the positive y half as shown below.

$$M_z = -\int_A y\sigma_{xx} dA = -\left[\int_{-0.15}^{0} y[\sigma_{xx}](0.2dy) + \int_0^{0.15} y[\sigma_{xx}](0.2dy)\right] = -2\int_0^{0.15} y[\sigma_{xx}](0.2dy)$$

Substituting the stress distribution we obtain:

$$M_z = -2\left[\int_0^{0.15} y[-379(y)^{0.2}](0.2dy)\right](10^6) = 151.6\left[\int_0^{0.15} y^{1.2}dy\right](10^6) = 151.6\left(\frac{y^{2.2}}{2.2}\right)\Bigg|_0^{0.15}(10^6) = 1061(10^3)$$

or $$\boxed{M_z = 1061 \text{ kN-m}}$$

6.132 A rectangular beam has the dimensions as shown in Fig. P6.129. The normal strain due to bending about the z-axis was found to vary as $\varepsilon_{xx} = -0.01y$ with y measured in meters. Determine the equivalent internal moment that would produce the given strain. The beam material has a stress strain relationship given by $\sigma = (200\varepsilon - 2,000\varepsilon^2)$ MPa

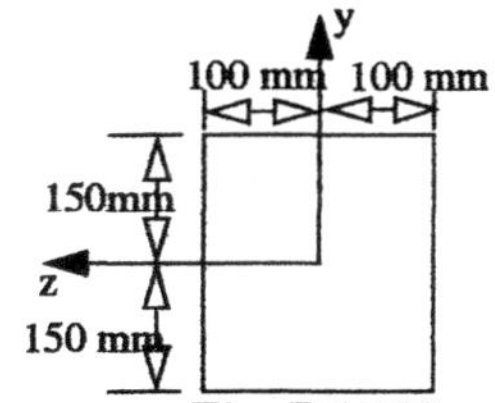

Fig. P6.132

Solution ε_{xx}=-0.01y y in meters $\sigma = (200\varepsilon - 2,000\varepsilon^2)$ MPa $M_z = ?$

From problem 3.145, we have the following stress-distribution for the given state of strain and material.

$$\sigma_{xx} = \begin{cases} (-2y - 0.2y^2)\text{MPa} & y < 0 \\ (-2y + 0.2y^2)\text{MPa} & y > 0 \end{cases}$$ 1

From the symmetry we conclude that the equivalent internal moment in the lower half ($y < 0$)is same as that from the upper half ($y > 0$). The integral in Eq. 6.1 can be written as twice the integral in the positive y half as shown below.

$$M_z = -\int_A y\sigma_{xx}dA = -\left[\int_{-0.15}^{0} y[\sigma_{xx}](0.2dy) + \int_{0}^{0.15} y[\sigma_{xx}](0.2dy)\right] = -2\int_{0}^{0.15} y[\sigma_{xx}](0.2dy)$$

Substituting the stress distribution we obtain:

$$M_z = -2\left[\int_{0}^{0.15} y[-2y + 0.2y^2](0.2dy)\right](10^6) = (-0.4)\left(-2\frac{y^3}{3} + 0.2\frac{y^4}{4}\right)\Bigg|_0^{0.15}(10^6) = 889.9 \text{ N-m}$$

or $\boxed{M_z = 890 \text{ N-m}}$

6.133 A wide flange beam has the dimensions shown in Fig. P6.133. The normal strain due to bending about the z-axis was found to vary as $\varepsilon_{xx} = -0.4y(10^{-3})$ with y measured in inches. Determine the equivalent internal moment that would produce the given strain. The beam is made from elastic-perfectly plastic material that has a yield stress of σ_{yield}= 40 ksi, and a Modulus of Elasticity E = 10,000 ksi.

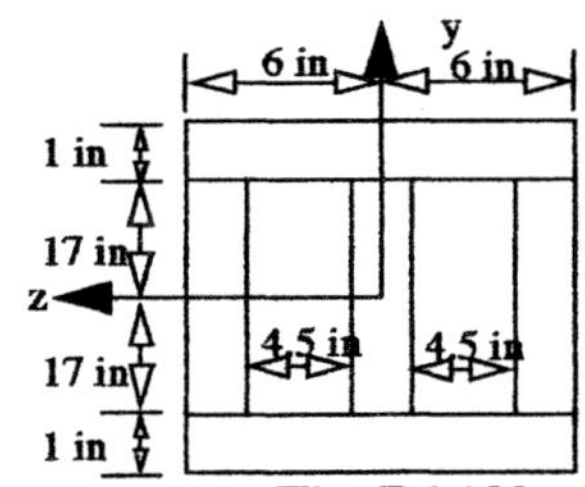

Fig. P6.133

Solution ε_{xx}=-0.0004y y in inches σ_{yield}=σ_{yield}= 40 ksi E = 10,000 ksi M_z = ?

The strain at yield point is: $\varepsilon_{yield} = \frac{\sigma_{yield}}{E} = \frac{40}{10000} = 0.004$

For positive values of y we have contraction, and for negative values of y we have tension. With geometry being symmetric in y, the elastic-plastic boundary will be on both sides

$$\varepsilon_{yield} = \pm 0.004 = -0.0004y_{yield} \quad \text{or} \quad y_{yield} = \pm 10 \text{ in}$$ 1

In the linear region the stress variation is given by: $\sigma_{xx} = E\varepsilon_{xx} = (10000)(-0.0004y) = -4y$. The stress distribution across the cross-section can be written as:

$$\sigma_{xx} = \begin{cases} 40 \text{ ksi} & -18 \text{ in} < y < -10 \text{ in} \\ -4y \text{ ksi} & -10 \text{ in} < y < 10 \text{ in} \\ -40 \text{ ksi} & 10 \text{ in} < y < 18 \text{ in} \end{cases}$$ 2

The variation of the differential area dA is shown in Fig. (a).

(a)

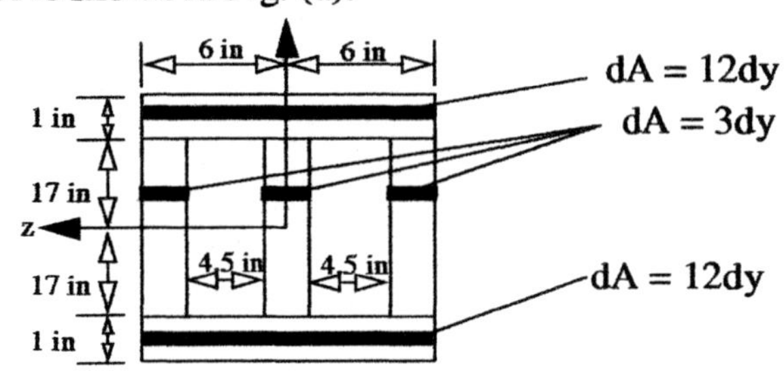

From the symmetry we conclude that the equivalent internal moment in the lower half ($y < 0$)is same as that from the upper half ($y > 0$). The integral in Eq. 6.1 can be written as twice the integral in the positive y half as shown below.

$$M_z = -\int_A y\sigma_{xx} dA = -2\left[\int_0^{10} y[\sigma_{xx}](3dy) + \int_{10}^{17} y[\sigma_{xx}](3dy) + \int_{17}^{18} y[\sigma_{xx}](12dy)\right]$$

Substituting the stress distribution we obtain:

$$M_z = -2\left[\int_0^{10} y[-4y](3dy) + \int_{10}^{17} y[-40](3dy) + \int_{17}^{18} y[-40](12dy)\right] \text{ or}$$

$$M_z = -2\left[-12\left(\frac{y^3}{3}\right)\Bigg|_0^{10} - 120\left(\frac{y^2}{2}\right)\Bigg|_{10}^{17} - 480\left(\frac{y^2}{2}\right)\Bigg|_{17}^{18}\right] = 2[4000 + 11340 + 8400] \qquad \boxed{M_z = 47480 \text{ in} - \text{kip}}$$

6.134 A beam has the dimensions shown in Fig. P6.134. The normal strain due to bending about the z-axis was found to vary as $\varepsilon_{xx} = -0.4y(10^{-3})$ with y measured in inches. Determine the equivalent internal moment that would produce the given strain. The beam is made from a bi-linear material that has a yield stress of σ_{yield} = 40 ksi, Modulus of Elasticity E_1 = 10,000 ksi and E_2 = 2,000 ksi.

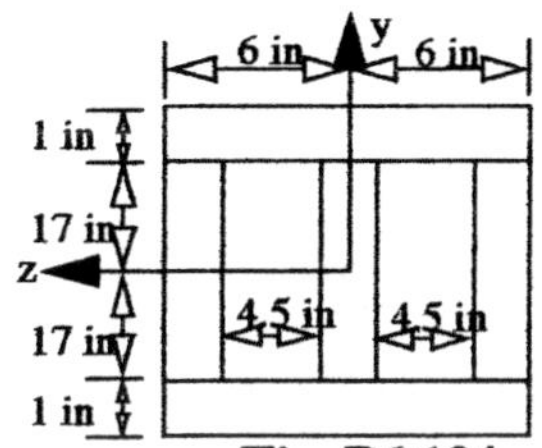

Fig. P6.134

Solution ε_{xx}=-0.0004y y in inches σ_{yield}= 40 ksi E_1 = 10,000 ksi E_2= 2,000 ksi. M_z = ?

From problem 6.133: $\varepsilon_{yield} = 0.004$ and $y_{yield} = \pm 10$ in .

In the linear region the stress variation is given by: $\sigma_{xx} = E\varepsilon_{xx} = (10000)(-0.0004y) = -4y$.

For $y < -10$ in i.e., on the tensile side

$$\sigma_2 = \sigma_{yield} + E_2(\varepsilon_{xx} - \varepsilon_{yield}) = 40 + (2000)[-0.0004y - 0.004] = (32 - 0.8y)\text{ksi}$$

For $y > 10$ in i.e., on the compressive side

$$\sigma_2 = -\sigma_{yield} + E_2(\varepsilon_{xx} - (-\varepsilon_{yield})) = (-40) + (2000)[-0.0004y + 0.004] = (-32 - 0.8y)\text{ksi}$$

The stress distribution can be written as:

$$\sigma_{xx} = \begin{cases} (32 - 0.8y) \text{ ksi} & -18 \text{ in} < y < -10 \text{ in} \\ -4y \text{ ksi} & -10 \text{ in} < y < 10 \text{ in} \\ (-32 - 0.8y) \text{ ksi} & 10 \text{ in} < y < 18 \text{ in} \end{cases} \qquad 1$$

The variation of the differential area dA is shown in Fig. (a) in problem 6.133. From the symmetry we conclude that the equivalent internal moment in the lower half ($y < 0$)is same as that in the upper half ($y > 0$). The integral in Eq. 6.1 can be written as twice the integral in the positive y half as shown below.

$$M_z = -\int_A y\sigma_{xx} dA = -2\left[\int_0^{10} y[\sigma_{xx}](3dy) + \int_{10}^{17} y[\sigma_{xx}](3dy) + \int_{17}^{18} y[\sigma_{xx}](12dy)\right]$$

Substituting the stress distribution we obtain:

$$M_z = -2\left[\int_0^{10} y[-4y](3dy) + \int_{10}^{17} y[-32 - 0.8y](3dy) + \int_{17}^{18} y[-32 - 0.8y](12dy)\right] \text{ or}$$

$$M_z = -2\left[-12\frac{y^3}{3}\Big|_0^{10} - 3\left(32\frac{y^2}{2} + 0.8\frac{y^3}{3}\right)\Big|_{10}^{17} - 12\left(32\frac{y^2}{2} + 0.8\frac{y^3}{3}\right)\Big|_{17}^{18}\right] = 2[4000 + 12202 + 9661]$$

$$\boxed{M_z = 51726 \text{ in} - \text{kip}}$$

6.135 A wide flange beam has the dimensions shown in Fig. P6.135. The normal strain due to bending about the z-axis was found to vary as $\varepsilon_{xx} = -0.4y(10^{-3})$ with y measured in inches. Determine the equivalent internal moment that would produce the given strain. The beam material has a stress strain relationship given by $\sigma = 70\varepsilon^{0.1}$ ksi

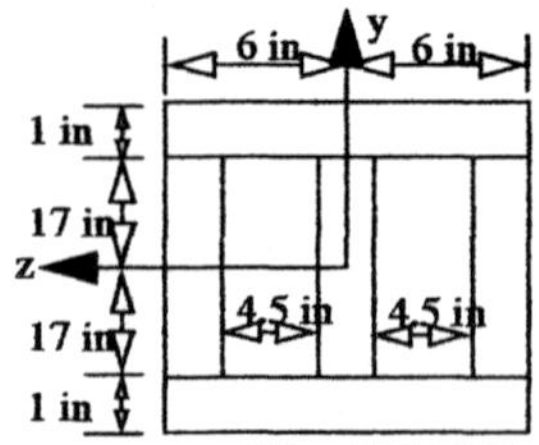

Fig. P6.135

Solution ε_{xx}=-0.0004y y in inches $\varepsilon_{xx} = -0.4y(10^{-3})$ $M_z = ?$

As the material behavior in tension and compression is the same, we have: $\sigma_{xx} = \begin{cases} 70\varepsilon^{0.1}\text{ksi} & \varepsilon > 0 \\ -70(-\varepsilon)^{0.1}\text{ksi} & \varepsilon < 0 \end{cases}$

Noting that for positive y we have contraction and for negative y we have extension, the stress distribution can be written as shown below.

$$\sigma_{xx} = \begin{cases} 70(-0.0004y)^{0.1}\text{ksi} & y < 0 \\ -70(0.0004y)^{0.1}\text{ksi} & y > 0 \end{cases} \quad \text{or} \quad \sigma_{xx} = \begin{cases} 32(-y)^{0.1}\text{ksi} & y < 0 \\ -32(y)^{0.1}\text{ksi} & y > 0 \end{cases} \qquad 1$$

The variation of the differential area dA is shown in Fig. (a) in problem 6.133. From the symmetry we conclude that the equivalent internal moment in the lower half ($y < 0$) is same as that in the upper half ($y > 0$). The integral in Eq. 6.1 can be written as twice the integral in the positive y half as shown below.

$$M_z = -\int_A y\sigma_{xx}dA = -2\left[\int_0^{17} y[\sigma_{xx}](3dy) + \int_{17}^{18} y[\sigma_{xx}](12dy)\right]$$

Substituting the stress distribution we obtain:

$$M_z = -2\left[\int_0^{17} y[-32(y)^{0.1}](3dy) + \int_{17}^{18} y[-32(y)^{0.1}](12dy)\right] \text{ or}$$

$$M_z = -2\left[-96\frac{y^{2.1}}{2.1}\Big|_0^{17} - 384\frac{y^{2.1}}{2.1}\Big|_{17}^{18}\right] = 2[17538.6 + 8947.0]$$

$$\boxed{M_z = 52971 \text{ in} - \text{kip}}$$

6.136 A wide flange beam has the dimensions shown in Fig. P6.136. The normal strain due to bending about the z-axis was found to vary as $\varepsilon_{xx} = -0.4y(10^{-3})$ with y measured in inches. Determine the equivalent internal moment that would produce the given strain. The beam material has a stress strain relationship given by $\sigma = (10,000\varepsilon - 90,000\varepsilon^2)$ ksi.

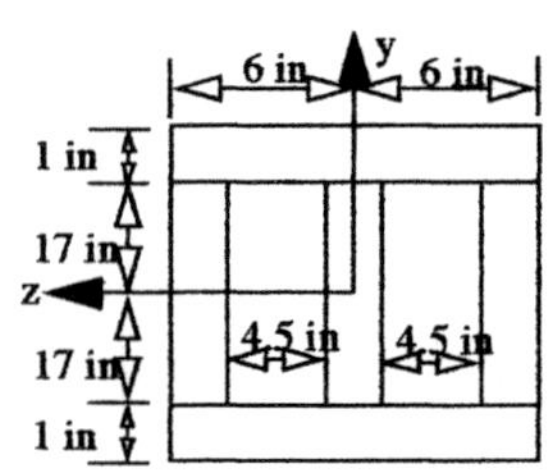

Fig. P6.136

Solution ε_{xx}=-0.0004y y in inches $\varepsilon_{xx} = -0.4y(10^{-3})$ $M_z = ?$

As the material behavior in tension and compression is the same, we have:

$$\sigma_{xx} = \begin{cases}(10,000\varepsilon - 90,000\varepsilon^2)\text{ksi} & \varepsilon > 0 \\ (10,000\varepsilon + 90,000\varepsilon^2)\text{ksi} & \varepsilon < 0\end{cases}$$

Noting that for positive y we have contraction and for negative y we have extension, the stress distribution can be written as shown below.

$$\sigma_{xx} = \begin{cases}[10,000(-0.0004y) - 90,000(-0.0004y)^2]\text{ksi} & y < 0 \\ [10,000(-0.0004y) + 90,000(-0.0004y)^2]\text{ksi} & y > 0\end{cases} \quad \text{or}$$

$$\sigma_{xx} = \begin{cases}[-4y - 0.0144y^2]\text{ksi} & y < 0 \\ [-4y + 0.0144y^2]\text{ksi} & y > 0\end{cases}$$ 2

The variation of the differential area dA is shown in Fig. (a) in problem 6.133. From the symmetry we conclude that the equivalent internal moment in the lower half $(y < 0)$is same as that in the upper half $(y > 0)$. The integral in Eq. 6.1 can be written as twice the integral in the positive y half as shown below.

$$M_z = -\int_A y\sigma_{xx} dA = -2\left[\int_0^{17} y[\sigma_{xx}](3dy) + \int_{17}^{18} y[\sigma_{xx}](12dy)\right]$$

Substituting the stress distribution we obtain:

$$M_z = -2\left[\int_0^{17} y[-4y + 0.0144y^2](3dy) + \int_{17}^{18} y[-4y + 0.0144y^2](12dy)\right] \quad \text{or}$$

$$M_z = -2\left[\left[-12\frac{y^3}{3} + 0.0432\frac{y^4}{4}\right]\Bigg|_0^{17} + \left[-48\frac{y^3}{3} + 0.1728\frac{y^4}{4}\right]\Bigg|_{17}^{18}\right] = 2[-18750 - 13777] \quad \boxed{M_z = 65054 \text{ in} - \text{kip}}$$

6.137 A beam made from elastic- perfectly plastic material has a yield stress of 30 ksi. Determine the internal bending moment if the elastic plastic boundary is half- inch from the top and bottom

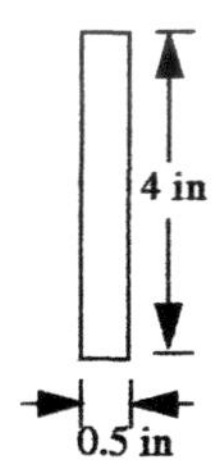

Fig. P6.137

Solution $\sigma_{yield} = 30$ ksi Elastic-plastic boundary 0.5 in from top and bottom. $M_z = ?$

Fig(a) shows the stress distribution across the cross-section with elastic-plastic boundary 1/2 in from the top and bottom.

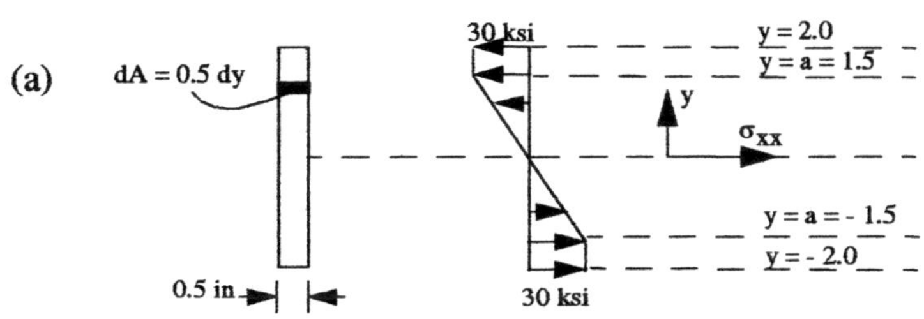

The stress distribution in the linear region is: $\sigma = -\frac{30y}{1.5} = -20y$ ksi. We can write the stress expression and value of dA for each interval as:

$$\sigma_{xx} = 30 \text{ ksi} \qquad dA = 0.5dy \qquad -2 \le y < -1.5$$

$$\sigma_{xx} = -20y \text{ ksi} \qquad dA = 0.5dy \qquad -1.5 < y < 1.5$$ 1

$$\sigma_{xx} = -30 \text{ ksi} \qquad dA = 0.5dy \qquad 1.5 < y < 2$$

From the symmetry we conclude that the equivalent internal moment in the lower half ($y < 0$)is same as that in the upper half ($y > 0$). The integral in Eq. 6.1 can be written as twice the integral in the positive y half as shown below.

$$M_z = -\int_A y\sigma_{xx}dA = -2\left[\int_0^{1.5} y[\sigma_{xx}](0.5dy) + \int_{1.5}^{2} y[\sigma_{xx}](0.5dy)\right]$$

Substituting the stress distribution we obtain:

$$M_z = -2\left[\int_0^{1.5} y[-20y](0.5dy) + \int_{1.5}^{2} y[-30](0.5dy)\right] = -2\left[-10\frac{y^3}{3}\bigg|_0^{1.5} - 15\frac{y^2}{2}\bigg|_{1.5}^{2}\right] = 2[11.25 + 13.125] \text{ or}$$

$$\boxed{M_z = 48.75 \text{ in} - \text{kip}}$$

6.138 Determine the shape factor for the rectangular cross-section shown in Fig. P6.137.

Solution $\sigma_{yield} = 30$ ksi Shape Factor = ?

Fig. (a) shows the stress distribution across the cross-section for a fully plastic moment.

(a)

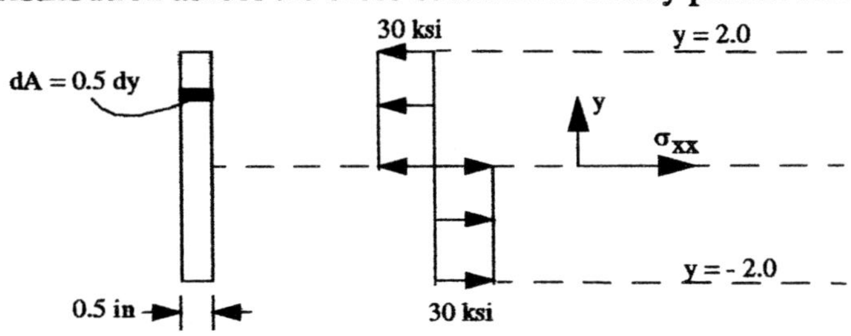

The stress distribution in the linear region is: $\sigma = -\frac{30y}{1.5} = -20y$ ksi. We can write the stress expression and value of dA for each interval as:

$$\sigma_{xx} = 30 \text{ ksi} \qquad dA = 0.5dy \qquad -2 \le y < 0$$ 1

$$\sigma_{xx} = -30 \text{ ksi} \qquad dA = 0.5dy \qquad 0 < y < 2$$

From the symmetry we conclude that the equivalent internal moment in the lower half ($y < 0$)is same as that in the upper half ($y > 0$). The integral in Eq. 6.1 can be written as twice the integral in the positive y half. Substituting the stress distribution we obtain the plastic moment.

$$M_P = -\int_A y\sigma_{xx}dA = -2\left[\int_0^{2} y[\sigma_{xx}](0.5dy)\right] = -2\left[\int_0^{2} y[-30](0.5dy)\right] = -2\left[-15\frac{y^2}{2}\bigg|_0^{2}\right] = 60 \text{ in} - \text{kips}$$ 2

The area moment of inertia of the cross-section is $I_{zz} = \frac{1}{12}(0.5)(4)^3 = 26.667 \text{ in}^4$.

We equate the magnitude of the maximum bending normal stress to yield stress and determine the elastic moment as shown below.

$$|\sigma_{xx}| = \left|\frac{M_e y_{max}}{I_{zz}}\right| = \frac{M_e(2)}{(26.667)} \qquad \text{or} \qquad M_e = 40 \text{ in} - \text{kips}$$ 3

The shape factor is $f = \dfrac{M_P}{M_e} = \dfrac{60}{40}$ or $\boxed{f = 1.5}$

6.139 An elastic - perfectly plastic material has a yield stress of $\sigma_{yield} = 40$ ksi. Point A in Fig. P6.139 is at yield stress due to bending of the beam. (a) Determine the location of the neutral axis assuming it is in the web.(b) The applied moment that produced the state of stress.

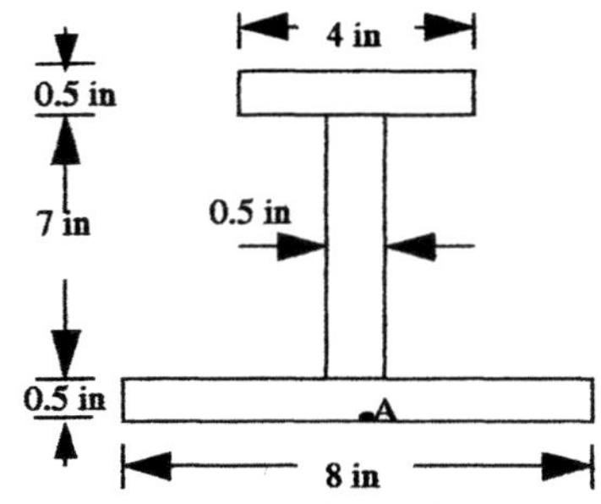

Fig. P6.139

Solution $\sigma_A = \sigma_{yield} = 40$ ksi $\quad a = ? \quad M_z = ?$

Fig(a) shows the stress distribution across the cross-section. We assume that the elastic-plastic boundary is in the web.

(a)

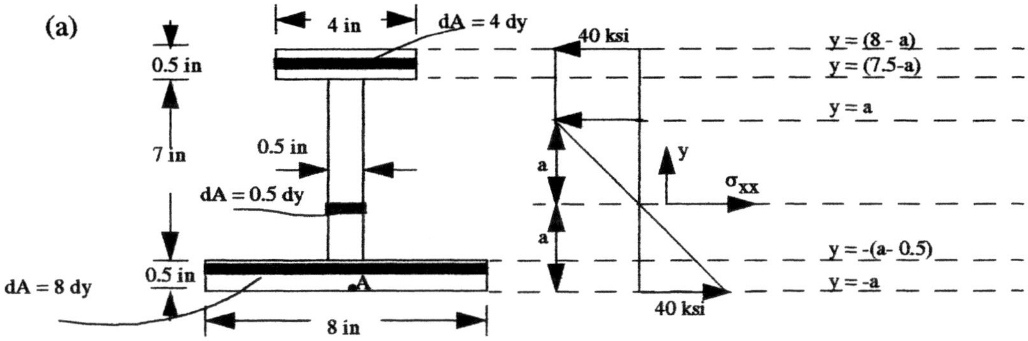

The stress distribution in the linear region is: $\sigma = -\dfrac{40y}{a}$. We can write the stress expression and value of dA for each interval as:

$$\sigma_{xx} = -\left(\frac{40y}{a}\right) \text{ ksi} \qquad dA = 8dy \qquad -a \le y < -(a-0.5)$$

$$\sigma_{xx} = -\left(\frac{40y}{a}\right) \text{ ksi} \qquad dA = 0.5dy \qquad -(a-0.5) < y < a \qquad (1)$$

$$\sigma_{xx} = -40 \text{ ksi} \qquad dA = 0.5dy \qquad a < y < (7.5-a)$$

$$\sigma_{xx} = -40 \text{ ksi} \qquad dA = 4dy \qquad (7.5-a) < y < (8-a)$$

Noting that the stress distribution must result in a zero axial force we obtain:

$$\int_A \sigma_{xx} dA = \int_{-a}^{-(a-0.5)} \left(-\frac{40y}{a}\right)(8dy) + \int_{-(a-0.5)}^{0} \left(-\frac{40y}{a}\right)(0.5dy) + \int_{0}^{(7.5-a)} (-40)(0.5dy) + \int_{(7.5-a)}^{(8-a)} (-40)(4dy) = 0 \text{ or}$$

$$(-40)\left[(8)\left(\frac{y^2}{2a}\right)\bigg|_{-a}^{-(a-0.5)} + (0.5)\left(\frac{y^2}{2a}\right)\bigg|_{-(a-0.5)}^{a} + (0.5)(y)\big|_a^{(7.5-a)} + (4)(y)\big|_{(7.5-a)}^{(8-a)}\right] = 0 \quad \text{or}$$

$$\left(\frac{4}{a}\right)[(a-0.5)^2 - a^2] + \left(\frac{0.25}{a}\right)[a^2 - (a-0.5)^2] + (0.5)[7.5 - a - a] + (4)[8 - a - 7.5 + a] = 0 \text{ or}$$

$a^2 - 2a - 0.9375 = 0$. The two roots of the quadratic equation are: a=2.3919 and a=- 0.3919. Only the positive root is admissible. As a = 2.3919 places the neutral axis in the web our starting assumption is correct. Thus the location of neutral axis is: $\boxed{a = 2.392 \text{ in}}$

Substituting the value of a = 2.392 in Eq. 1 we obtain the stress expression and value of dA for each interval as shown below.

$$\begin{array}{lll} \sigma_{xx} = -16.72y \text{ ksi} & dA = 8dy & -2.392 \le y < -1.892 \\ \sigma_{xx} = -16.72y \text{ ksi} & dA = 0.5dy & -1.892 < y < 2.392 \\ \sigma_{xx} = -40 \text{ ksi} & dA = 0.5dy & 2.392 < y < 5.108 \\ \sigma_{xx} = -40 \text{ ksi} & dA = 4dy & 5.108 < y \le 5.608 \end{array} \quad 2$$

The integral $M_z = -\int_A y\sigma_{xx} dA$ can be written as.

$$M_z = -\left[\int_{-2.392}^{-1.892} y(-16.72y)(8dy) + \int_{-1.892}^{2.392} y(-16.72y)(0.5dy) + \int_{2.392}^{5.108} y(-40)(0.5dy) + \int_{5.108}^{5.608} y(-40)(4dy)\right] \text{ or}$$

$$M_z = \left[133.76\frac{y^3}{3}\bigg|_{-2.392}^{-1.892} + 8.36\frac{y^3}{3}\bigg|_{-1.892}^{2.392} + 20\frac{y^2}{2}\bigg|_{2.392}^{5.108} + 160\frac{y^2}{2}\bigg|_{5.108}^{5.608}\right] = 308.25 + 57.012 + 203.7 + 428.64 \text{ or}$$

$$\boxed{M_z = 997.6 \text{ in} - \text{kip}}$$

6.140 For the cross-section shown in Fig. P6.139 determine the shape factor

Solution shape factor f = ? $\sigma_{yield} = 40$ ksi

Fig(a) shows the stress distribution across the cross-section for a fully plastic moment. We assume that the elastic-plastic boundary is in the web.

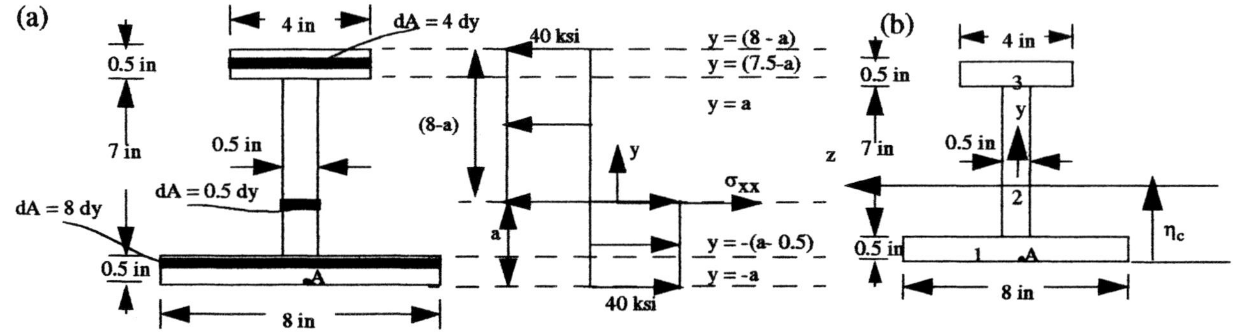

We can write the stress expression and value of dA for each interval.

$$\begin{array}{lll} \sigma_{xx} = 40 \text{ ksi} & dA = 8dy & -a \le y < -(a-0.5) \\ \sigma_{xx} = 40 \text{ ksi} & dA = 0.5dy & -(a-0.5) < y < 0 \\ \sigma_{xx} = -40 \text{ ksi} & dA = 0.5dy & 0 < y < (7.5-a) \\ \sigma_{xx} = -40 \text{ ksi} & dA = 4dy & (7.5-a) < y \le (8-a) \end{array} \quad 1$$

Noting that the stress distribution must result in a zero axial force we obtain

$$\int_A \sigma_{xx} dA = \int_{-a}^{-(a-0.5)} (40)(8dy) + \int_{-(a-0.5)}^{0} (40)(0.5dy) + \int_{0}^{(7.5-a)} (-40)(0.5dy) + \int_{(7.5-a)}^{(8-a)} (-40)(4dy) = 0 \text{ or}$$

$$(40)[(8)y|_{-a}^{-(a-0.5)} + (0.5)y|_{-(a-0.5)}^{a} - (0.5)(y)|_{a}^{(7.5-a)} - (4)(y)|_{(7.5-a)}^{(8-a)}] = 0 \text{ or}$$

$$(8)(0.5)+(0.5)(a-0.5)-(0.5)[7.5-a]+(4)(0.5) = 0 \qquad \text{or} \qquad a = 2 \text{ in}$$ 2

As a = 2.0 inch places the neutral axis in the web, our starting assumption is correct. The fully plastic moment can be written as:

$$M_P = -\int_A y\sigma_{xx} dA = -\left[\int_{-2}^{-1.5} y(40)(8dy) + \int_{-1.5}^{0} y(40)(0.5dy) + \int_{0}^{5.5} y(-40)(0.5dy) + \int_{5.5}^{6} y(-40)(4dy)\right] \text{ or}$$

$$M_P = -\left[320\frac{y^2}{2}\Big|_{-2}^{-1.5} + 20\frac{y^2}{2}\Big|_{-1.5}^{0} - 20\frac{y^2}{2}\Big|_{0}^{5.5} + 160\frac{y^2}{2}\Big|_{5.5}^{6}\right] = -[-280-22.5-302.5-460] = 1065 \text{ in-kip}$$ 3

To determine the elastic moment, we first determine the centroid of the cross-section shown in Fig (b). Location of the centroid from the bottom is:

$$\eta_c = \frac{(8)(0.5)(0.25)+(7)(0.5)(4)+(4)(0.5)(7.75)}{(8)(0.5)+(7)(0.5)+(4)(0.5)} = 3.158 \text{ in}$$

The area moment of inertia about the z-axis passing through the centroid for each section shown in Fig. (b) can be found as shown below.

$$(I_{zz})_1 = \left[\frac{1}{12}(8)(0.5)^3 + (8)(0.5)(3.158-0.25)^2\right] = 33.909 \text{ in}^4$$

$$(I_{zz})_2 = \left[\frac{1}{12}(0.5)(7)^3 + (0.5)(7)(3.158-4)^2\right] = 16.773 \text{ in}^4$$

$$(I_{zz})_3 = \left[\frac{1}{12}(4)(0.5)^3 + (4)(0.5)(7.75-3.158)^2\right] = 42.215 \text{ in}^4$$

The total area moment of inertia is the sum of the above three moment of inertias, i.e., $I_{zz} = 92.897 \text{ in}^4$. The maximum bending normal stress will be at the top surface where y = 8-3.158=4.842 in. The magnitude of the maximum bending stress can be equated to yield stress to obtain the elastic moment as shown below.

$$\sigma_{xx} = \left|\frac{M_e y_{max}}{I_{zz}}\right| = \frac{M_e(4.842)}{92.897} = 40 \qquad \text{or} \qquad M_e = 767.43 \text{ in-kips}$$ 4

The shape factor is $f = M_P/M_e = 1065/767.43$ or $\boxed{f = 1.388}$

6.141 A uniformly loaded simply supported beam is made of elastic-perfectly plastic material that has a yield stress of 30 ksi. The beam has the hollow square cross section shown. If point A is at yield stress, determine the intensity w of the uniform load

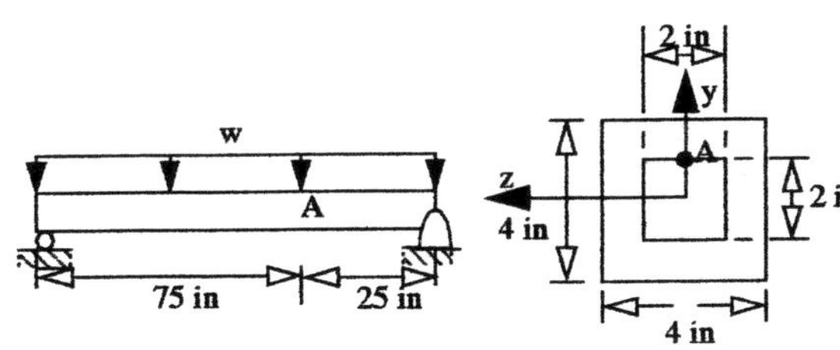

Fig. P6.141

Solution $\sigma_A = \sigma_{yield} = 30$ ksi $\qquad w = ?$

Fig(a) shows the stress distribution across the cross-section. We assume that the elastic-plastic boundary is in the web.The stress distribution in the linear region is: $\sigma = -(30y)/1 = -30y$. We can write the stress expression and value of dA for each interval as:

$$\sigma_{xx} = 30 \text{ ksi} \qquad dA = 4dy \qquad -2 \le y < -1$$

$$\sigma_{xx} = -30y \text{ ksi} \qquad dA = 2dy \qquad -1 < y < 1$$ 1

$$\sigma_{xx} = -30 \text{ ksi} \qquad dA = 4dy \qquad 1 < y < 2$$

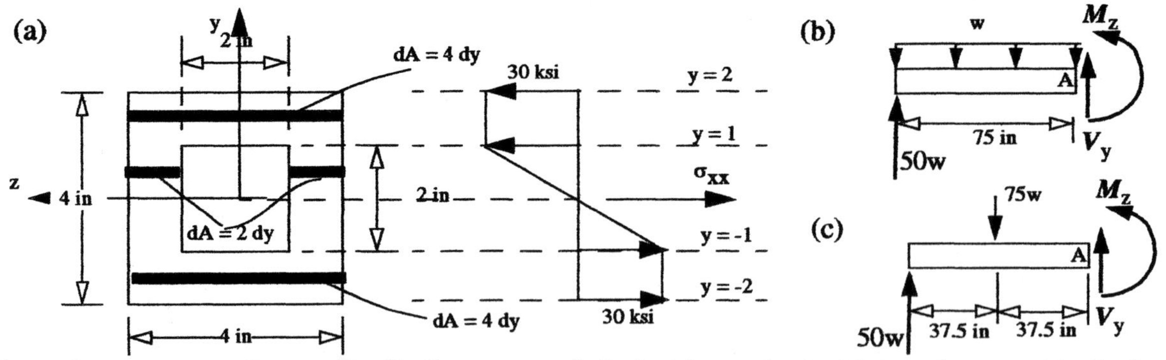

From the symmetry of stress distribution we conclude that the equivalent internal moment in the lower half ($y < 0$) is same as that in the upper half ($y > 0$). The integral in Eq. 6.1 can be written as twice the integral in the positive y half as shown below.

$$M_z = -\int_A y\sigma_{xx}dA = -2\left[\int_0^1 y[\sigma_{xx}](2dy) + \int_1^2 y[\sigma_{xx}](4dy)\right]$$

Substituting the stress distribution we obtain:

$$M_z = -2\left[\int_0^1 y[-30y](2dy) + \int_1^2 y[-30](4dy)\right] = -2\left[-60\frac{y^3}{3}\bigg|_0^1 - 120\frac{y^2}{2}\bigg|_1^2\right] = 400\text{in}-\text{kips} \qquad \mathbf{2}$$

From symmetry of loading we conclude that reaction at each support in Fig. P6.141 is half the load i.e., 50w. Fig. (b) shows the free body diagram after making an imaginary cut through point A. In Fig. (c) the distributed force is replaced by an equivalent force. By moment equilibrium about point A we obtain:

$$M_z - (50w)(75) + (75w)(37.5) = 0 \qquad \text{or} \qquad M_z = 937.5w \qquad \mathbf{3}$$

Equating Eqs. (2) and (3) we obtain: $937.5w = 400\text{in}-\text{kips}$ or $w = 0.4266\ \text{kips/in}$ or

$$\boxed{w = 426.7\ \text{lb/in}}$$

6.142 A beam of elastic-perfectly plastic material has a yield stress of 50 ksi and a cross-section shown. If point A just reaches yield stress, determine (a) the location of the neutral axis (b) the applied moment that produced the state of stress.

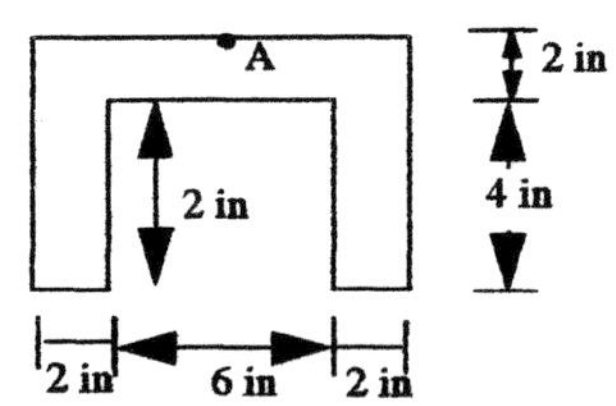

Fig. P6.142

Solution $\sigma_A = \sigma_{yield} = 500$ ksi $\quad a = ? \quad M_z = ?$

Fig(a) shows the stress distribution across the cross-section. We assume that the elastic-plastic boundary is in the web.

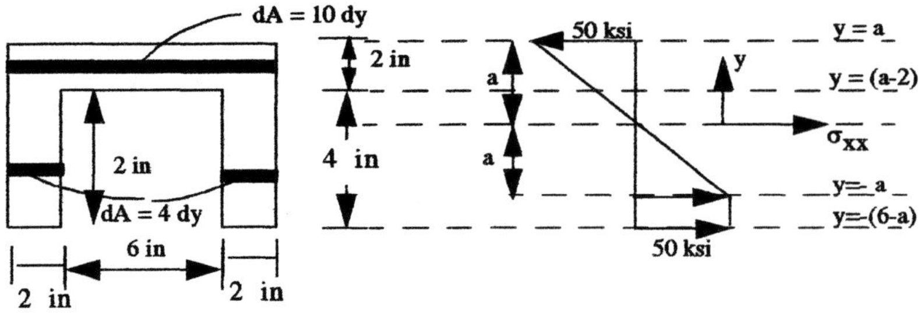

The stress distribution in the linear region is: $\sigma = -50(y/a)$. We can write the stress expression and value of dA for each interval as:

$$\sigma_{xx} = 50 \text{ ksi} \qquad dA = 4dy$$
$$\sigma_{xx} = -50(y/a) \text{ ksi}$$
$$\sigma_{xx} = -50(y/a) \text{ ksi} \qquad 1$$

Noting that the stress distribution must result in a zero axial force we obtain:

$$\int_A \sigma_{xx} dA = \int_{-(6-a)}^{-a} (50)(4dy) + \int_{-a}^{(a-2)} \left(-\frac{50y}{a}\right)(4dy) + \int_{(a-2)}^{a} \left(-\frac{50y}{a}\right)(10dy) = 0 \text{ or}$$

$$(50)\left[4y\Big|_{-(6-a)}^{-a} + (-4)\left(\frac{y^2}{2a}\right)\Big|_{-a}^{(a-2)} + (-10)\left(\frac{y^2}{2a}\right)\Big|_{(a-2)}^{a}\right] = 0 \text{ or}$$

$$4[-a+6-a] - \left(\frac{2}{a}\right)[(a-2)^2 - a^2] - \left(\frac{5}{a}\right)[a^2 - (a-2)^2] = 0 \text{ or } 2a^2 - 3a - 3 = 0$$

The two roots of the quadratic equation are: a=2.186 and a=- 0.686. Only the positive root is admissible. As a = 2.186 places the neutral axis in the web our starting assumption is correct. Thus the location of neutral axis is: $\boxed{a = 2.186 \text{ in}}$

Substituting the value of a = 2.186 in Eq. 1 we obtain the stress expression and value of dA for each interval as shown below.

$$\sigma_{xx} = 50 \text{ ksi} \qquad dA = 4dy \qquad -3.814 \le y < -2.186$$
$$\sigma_{xx} = -22.873y \text{ ksi} \qquad dA = 4dy \qquad -2.186 < y < 0.186 \qquad 2$$
$$\sigma_{xx} = -22.873y \text{ ksi} \qquad dA = 10dy \qquad 0.186 < y < 2.186$$

The integral $M_z = -\int_A y\sigma_{xx} dA$ can be written as.

$$M_z = -\left[\int_{-3.814}^{-2.186} y(50)(4dy) + \int_{-2.186}^{0.186} y(-22.873y)(4dy) + \int_{0.186}^{2.186} y(-22.873y)(10dy)\right] \text{ or}$$

$$M_z = \left[200\frac{y^2}{2}\Big|_{-3.814}^{-2.186} + (-91.491)\frac{y^3}{3}\Big|_{-2.186}^{0.186} + (-228.73)\frac{y^3}{3}\Big|_{0.186}^{2.186}\right] = -976.80 - 318.77 - 795.95 \text{ or}$$

$$\boxed{M_z = 2091 \text{ in-kip}}$$

6.143 A hollow-square beam is made from a material that has a stress -strain relation given by $\sigma = K\varepsilon^{0.4}$. Assume same behavior in tension and compression. In terms of K, L, a, M_{ext} determine the bending normal stress at point A.

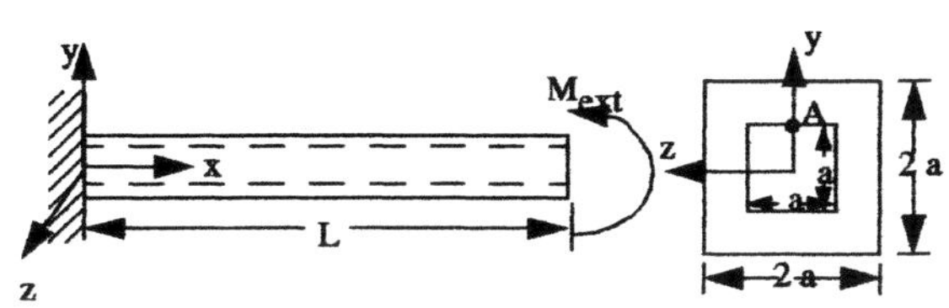

Fig. P6.143

Solution $\sigma = K\varepsilon^{0.4}$ $\qquad \sigma_A = f(K, L, a, M_{ext}) = ?$

The normal strain in bending is:

$$\varepsilon_{xx} = -y\frac{d^2v}{dx^2}(x) \qquad 1$$

As the material behavior in tension and compression is the same, we have:

$$\sigma_{xx} = \begin{cases} K\varepsilon_{xx}^{0.4} & \varepsilon_{xx} \ge 0 \\ -K(-\varepsilon_{xx})^{0.4} & \varepsilon_{xx} \le 0 \end{cases} \qquad 2$$

Noting that for positive y we have contraction and for negative y we have extension, the stress distribution can be written as shown below.

$$\sigma_{xx} = \begin{cases} K\left(-y\dfrac{d^2v}{dx^2}\right)^{0.4} & y < 0 \\ -K\left(y\dfrac{d^2v}{dx^2}\right)^{0.4} & y > 0 \end{cases} \qquad 3$$

From the symmetry we conclude that the equivalent internal moment in the lower half ($y < 0$)is same as that in the upper half ($y > 0$). The integral in Eq. 6.1 can be written as twice the integral in the positive y half as shown below.

$$M_z = -\int_A y\sigma_{xx}dA = -2\left[\int_0^{a/2} y[\sigma_{xx}](ady) + \int_{a/2}^{a} y[\sigma_{xx}](2ady)\right]$$

Substituting the stress distribution and noting that $\dfrac{d^2v}{dx^2}$ is only a function of x we obtain:

$$M_z = -\int_A y\sigma_{xx}dA = -2\left[\int_0^{a/2} y\left[-K\left(y\frac{d^2v}{dx^2}\right)^{0.4}\right](ady) + \int_{a/2}^{a} y\left[-K\left(y\frac{d^2v}{dx^2}\right)^{0.4}\right](2ady)\right] \text{ or}$$

$$M_z = 2aK\left(\frac{d^2v}{dx^2}\right)^{0.4}\left[\int_0^{a/2} y^{1.4}dy + 2\int_{a/2}^{a} y^{1.4}dy\right] = 2aK\left(\frac{d^2v}{dx^2}\right)^{0.4}\left[\frac{y^{2.4}}{2.4}\bigg|_0^{a/2} + 2\frac{y^{2.4}}{2.4}\bigg|_{a/2}^{a}\right] \text{ or}$$

$$= 2aK\left(\frac{d^2v}{dx^2}\right)^{0.4}\left[\frac{(a/2)^{2.4}}{2.4} + 2\left(\frac{(a)^{2.4}}{2.4} - \frac{(a/2)^{2.4}}{2.4}\right)\right] = 1.509a^{3.4}K\left(\frac{d^2v}{dx^2}\right) \text{ or}$$

$$\left(\frac{d^2v}{dx^2}\right)^{0.4} = \frac{M_z}{1.509a^{3.4}K} \qquad 4$$

Substituting Eq. (4) into Eq. (2) we obtain:

$$\sigma_{xx} = \begin{cases} \left(\dfrac{M_z}{1.509a^{3.4}}\right)(-y)^{0.4} & y < 0 \\ -\left(\dfrac{M_z}{1.509a^{3.4}}\right)(y)^{0.4} & y > 0 \end{cases} \qquad 5$$

The internal moment $M_z = M_{ext}$ across the beam and $y_A = a$. Substituting these values into Eq. 5 we obtain the stress at point A as:

$$\sigma_A = -\left(\frac{M_{ext}}{1.509a^{3.4}}(a)^{0.4}\right) = -\left(\frac{M_{ext}}{1.509a^3}\right) \text{ or}$$

$$\boxed{\sigma_A = \left(\frac{M_{ext}}{1.509a^3}\right) \text{ compression}}$$

6.144 The stress-strain curve in tension, for a material is given by $\sigma = K\,\varepsilon^{0.5}$. For a rectangular cross-section show that the bending normal stress is given by:

$$\sigma_{xx} = \begin{cases} \dfrac{-5\sqrt{2}}{bh^2}\left(\dfrac{y}{h}\right)^{0.5} M_z & y > 0 \\ \dfrac{5\sqrt{2}}{bh^2}\left(\dfrac{-y}{h}\right)^{0.5} M_z & y < 0 \end{cases}$$

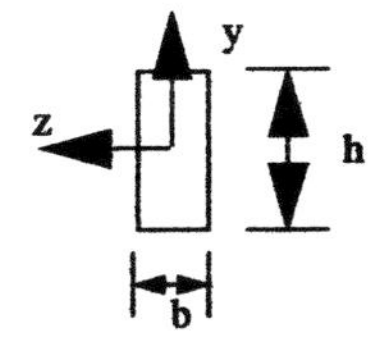

Fig. P6.144

Solution $\sigma = K\,\varepsilon^{0.5}$ $\qquad \sigma_{xx} = f(y, h, b, M_z) = ?$

The normal strain in bending is:

$$\varepsilon_{xx} = -y\frac{d^2v}{dx^2}(x) \qquad (1)$$

As the material behavior in tension and compression is the same, we have:

$$\sigma_{xx} = \begin{cases} K\varepsilon_{xx}^{0.5} & \varepsilon_{xx} \ge 0 \\ -K(-\varepsilon_{xx})^{0.5} & \varepsilon_{xx} \le 0 \end{cases} \qquad (2)$$

Noting that for positive y we have contraction and for negative y we have extension, the stress distribution can be written as shown below.

$$\sigma_{xx} = \begin{cases} K\left(-y\dfrac{d^2v}{dx^2}\right)^{0.5} & y < 0 \\ -K\left(y\dfrac{d^2v}{dx^2}\right)^{0.5} & y > 0 \end{cases} \qquad (3)$$

From the symmetry we conclude that the equivalent internal moment in the lower half ($y < 0$) is same as that in the upper half ($y > 0$). The integral in Eq. 6.1 can be written as twice the integral in the positive y half as shown below. Noting that $\frac{d^2v}{dx^2}$ is only a function of x we obtain:

$$M_z = -\int_A y\sigma_{xx}dA = -2\left\{\int_0^{h/2} y\left[-K\left(y\frac{d^2v}{dx^2}\right)^{0.5}\right](bdy)\right\} = 2bK\left(\frac{d^2v}{dx^2}\right)^{0.5}\int_0^{h/2} y^{1.5}dy = 2bK\left(\frac{d^2v}{dx^2}\right)^{0.5}\left.\frac{y^{2.5}}{2.5}\right|_0^{h/2} \quad \text{or}$$

$$M_z = 2bK\left(\frac{d^2v}{dx^2}\right)^{0.5}\frac{(h/2)^{2.5}}{2.5} = \frac{2}{2.5(2^{2.5})}bK\left(\frac{d^2v}{dx^2}\right)^{0.5}h^{2.5} = \frac{1}{5\sqrt{2}}bK\left(\frac{d^2v}{dx^2}\right)^{0.5}h^{2.5} \quad \text{or} \quad \left(\frac{d^2v}{dx^2}\right)^{0.5} = \frac{5\sqrt{2}M_z}{bh^{2.5}K} \qquad (4)$$

Substituting Eq. (4) into Eq. (2) we obtain:

$$\sigma_{xx} = \begin{cases} \left(\dfrac{5\sqrt{2}M_z}{bh^{2.5}K}\right)(-y)^{0.5} & y < 0 \\ -\left(\dfrac{5\sqrt{2}M_z}{bh^{2.5}K}\right)(y)^{0.5} & y > 0 \end{cases} \quad \text{or} \quad \boxed{\sigma_{xx} = \begin{cases} \left(\dfrac{5\sqrt{2}}{bh^2}\right)\left(\dfrac{-y}{h}\right)^{0.5} M_z & y < 0 \\ -\left(\dfrac{5\sqrt{2}}{bh^2}\right)\left(\dfrac{y}{h}\right)^{0.5} M_z & y > 0 \end{cases}}$$

6.145 A cantilever beam is made from a material that has a stress -strain relation given by $\sigma = K\varepsilon^{0.7}$ in tension In terms of K, w, a, and L determine the fiber stress at point A. Point A is just above the bottom flange.

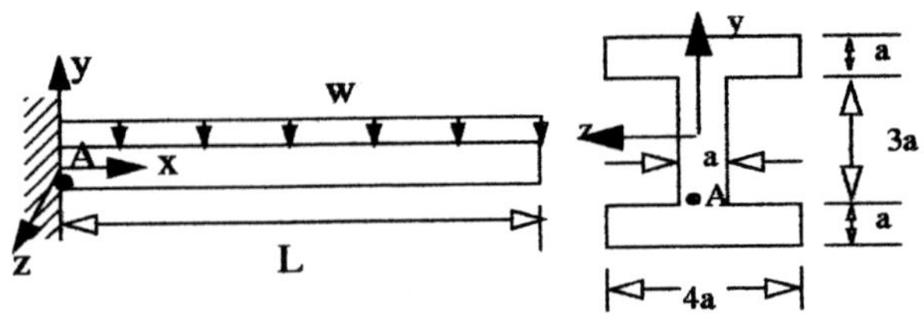

Fig. P6.145

Solution $\sigma = K\varepsilon^{0.4}$ $\sigma_A = f(K, L, a, w) = ?$

The normal strain in bending is:

$$\varepsilon_{xx} = -y\frac{d^2v}{dx^2}(x) \qquad (1)$$

As the material behavior in tension and compression is the same, we have:

$$\sigma_{xx} = \begin{cases} K\varepsilon_{xx}^{0.7} & \varepsilon_{xx} \ge 0 \\ -K(-\varepsilon_{xx})^{0.7} & \varepsilon_{xx} \le 0 \end{cases} \qquad (2)$$

Noting that for positive y we have contraction and for negative y we have extension, the stress distribution can be written as shown below.

$$\sigma_{xx} = \begin{cases} K\left(-y\frac{d^2v}{dx^2}\right)^{0.7} & y < 0 \\ -K\left(y\frac{d^2v}{dx^2}\right)^{0.7} & y > 0 \end{cases} \qquad (3)$$

From the symmetry we conclude that the equivalent internal moment in the lower half $(y < 0)$is same as that in the upper half $(y > 0)$. The integral in Eq. 6.1 can be written as twice the integral in the positive y half as shown below. Noting that $\frac{d^2v}{dx^2}$ is only a function of x we obtain:

$$M_z = -\int_A y\sigma_{xx}dA = -2\left\{\int_0^{1.5a} y\left[-K\left(y\frac{d^2v}{dx^2}\right)^{0.7}\right](ady) + \int_{1.5a}^{2.5a} y\left[-K\left(y\frac{d^2v}{dx^2}\right)^{0.7}\right](4ady)\right\} \text{ or}$$

$$M_z = 2aK\left(\frac{d^2v}{dx^2}\right)^{0.7}\left[\int_0^{1.5a} y^{1.7}dy + 4\int_{1.5a}^{2.5a} y^{1.7}dy\right] = 2aK\left(\frac{d^2v}{dx^2}\right)^{0.5}\left[\frac{y^{2.7}}{2.7}\bigg|_0^{1.5a} + 4\frac{y^{2.7}}{2.7}\bigg|_{1.5a}^{2.5a}\right] \text{ or}$$

$$M_z = aK\left(\frac{d^2v}{dx^2}\right)^{0.7}\left[\frac{(1.5a)^{2.7}}{2.7} + 4\frac{(2.5a)^{2.7} - (1.5a)^{2.7}}{2.7}\right] = a^{3.7}K\left(\frac{d^2v}{dx^2}\right)^{0.5}\left[\frac{1.5^{2.7}}{2.7} + 4\frac{2.5^{2.7} - 1.5^{2.7}}{2.7}\right] \text{ or}$$

$$M_z = 28.528a^{3.7}K\left(\frac{d^2v}{dx^2}\right)^{0.7} \quad \text{or} \quad \left(\frac{d^2v}{dx^2}\right)^{0.7} = \frac{M_z}{28.528a^{3.7}K} \qquad (4)$$

Substituting Eq. (4) into Eq. (2) we obtain:

$$\sigma_{xx} = \begin{cases} \left(\dfrac{M_z}{28.528a^{3.7}K}\right)(-y)^{0.7} & y < 0 \\ -\left(\dfrac{M_z}{28.528a^{3.7}K}\right)(y)^{0.7} & y > 0 \end{cases} \qquad 5$$

Fig. (a) shows the free body diagram of the beam after an imaginary cut has been made through section containing point A. Fig. (b) shows the equivalent free body diagram with distributed force replaced by an equivalent force.

(a)

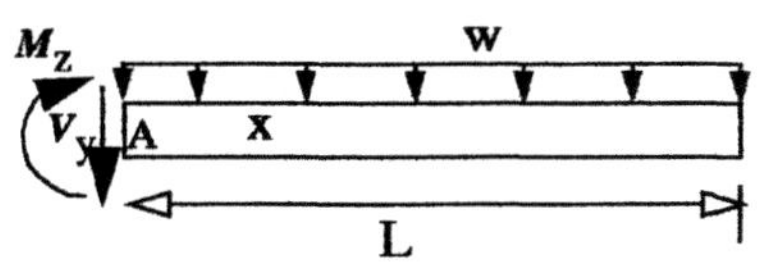

(b)

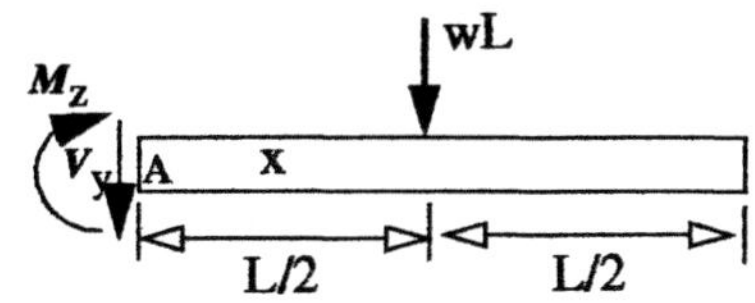

By equilibrium of moment about point A, we obtain;

$$M_z = -\left(\frac{wL^2}{2}\right) \qquad 6$$

Substituting y = -1.5a and Eq. (6) into Eq. (5) we obtain:

$$\sigma_A = \frac{-\left(\frac{wL^2}{2}\right)}{28.528a^{3.7}K}(1.5a)^{0.7} = -\frac{(1.5)^{0.7}}{(2)(28.528)}\left(\frac{wL^2}{Ka^3}\right) = -0.0233\left(\frac{wL^2}{Ka^3}\right) \text{ or } \boxed{\sigma_A = 0.0233\left(\frac{wL^2}{Ka^3}\right) \text{ compression}}$$

Problem Set 7.1

7.1 In terms of w, P, L, E, and I determine (a) equation of the elastic curve. (b) the deflection of the beam at point A

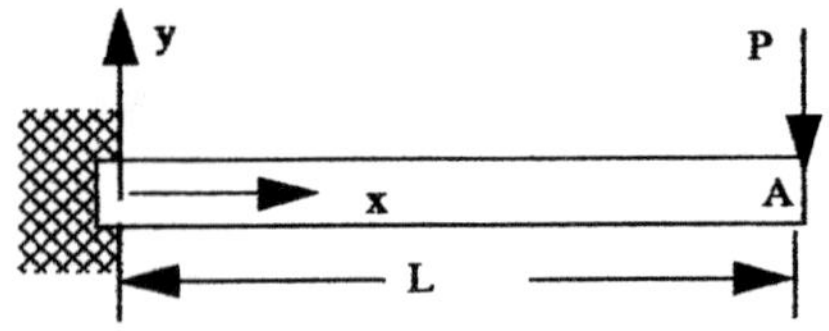

Fig. P7.1

Solution $v(x) = ?$ $v(L) = ?$

Fig.(a) shows the free body diagram after on imaginary cut is made at the same location x.

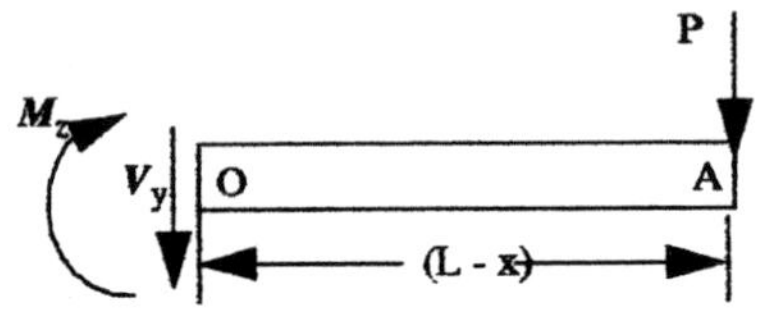

By equilibrium of moment about point O, we obtain: $M_z + P(L-x) = 0$ or $M_z = -P(L-x)$

The boundary value problem statement is

$$EI\frac{d^2v}{dx^2} = -PL + Px \quad 1 \qquad v(0) = 0 \quad 2 \qquad \frac{dv}{dx}(0) = 0 \quad 3$$

Integrating Eq.1, we obtain,

$$EI\frac{dv}{dx} = -PLx + \frac{Px^2}{2} + C_1 \quad 4 \qquad EIv = \frac{-PLx^2}{2} + \frac{Px^3}{6} + C_1x + C_2 \quad 5$$

Substituting x = 0 in Eq.(4) and using Eq.(3), we obtain

$$EI\frac{dv}{dx}(0) = C_1 = 0 \quad 6$$

Substituting x = 0 in Eq.(5) and using Eq.(3), we obtain

$$EIv(0) = C_2 = 0 \quad 7$$

Substituting Eqs. (6) and (7) in Eq.(5) and using Eq.(3), we obtain the elastic curve below

$$v(x) = \frac{P}{EI}\left(\frac{-3Lx^2 + x^3}{6}\right) \text{ or } \quad \boxed{v(x) = \frac{Px^2}{6EI}(x - 3L)}$$

Substituting x = L in the elastic curve equation, we obtain

$$v(L) = \frac{PL^2}{6EI}(-2L) \text{ or } \quad \boxed{v(L) = -\frac{PL^3}{3EI}}$$

7.2 In terms of w, P, L, E, and I determine (a) equation of the elastic curve. (b) the deflection of the beam at point A

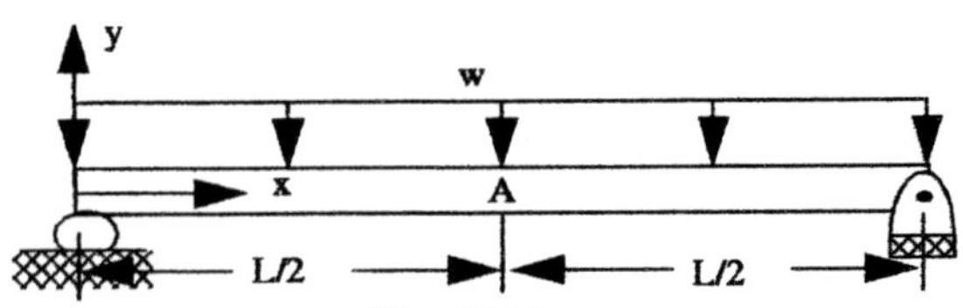

Fig. P7.2

Solution $v(x) = ?$ $v(L/2) = ?$

From symmetry, we know that the reaction at each support is half the total load. Fig(a) shows the free body diagram after an imaginary cut is made at some location x.Fig(b) shows the distributed force replaced by a equivalent force.

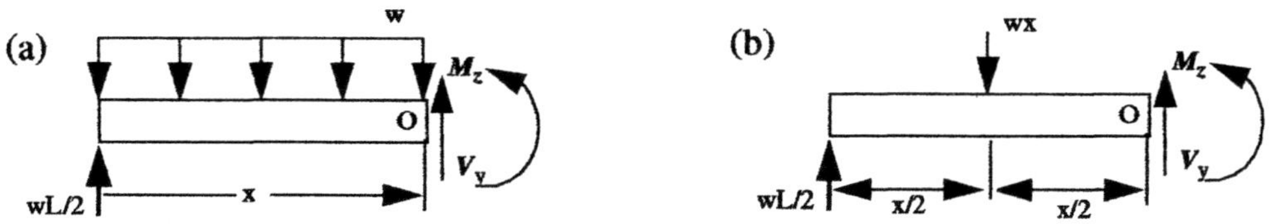

By equilibrium of moment about point O: $M_z + wx\left(\frac{x}{2}\right) - \frac{wL}{2}(x) = 0$ or $M_z = \frac{wLx}{2} - \frac{wx^2}{2}$

The boundary value statement is

$$EI\frac{d^2v}{dx^2} = \frac{wLx}{2} - \frac{wx^2}{2} \quad \mathbf{1} \qquad v(0) = 0 \quad \mathbf{2} \qquad v(L) = 0 \quad \mathbf{3}$$

Integrating Eq.1, we obtain,

$$EI\frac{dv}{dx} = \frac{wLx^2}{4} - \frac{wx^3}{6} + C_1 \quad \mathbf{4} \qquad EIv = \frac{wLx^3}{12} - \frac{wx^4}{24} + C_1x + C_2 \quad \mathbf{5}$$

Substituting x = 0 in Eq.(5) and using Eq.(2), we obtain

$$EIv(0) = C_2 = 0 \quad \mathbf{6}$$

Substituting x = 0 in Eq.(5) and using Eq.(3), we obtain

$$EIv(L) = \frac{wL^4}{12} - \frac{wL^4}{24} + C_1L = 0 \quad \text{or} \quad -\frac{wL^4}{24}(2-1) + C_1L = 0 \quad \text{or} \quad C_1 = -\frac{wL^3}{24} \quad \mathbf{7}$$

Substituting Eqs. (6) and (7) in Eq.(5) and using Eq.(3), we obtain the elastic curve below

$$v(x) = \frac{w}{EI}\left(\frac{2Lx^3 - x^4 - L^3x}{24}\right) \quad \text{or} \quad \boxed{v(x) = \frac{-wx}{24EI}(x^3 - 2Lx^2 + L^3)}$$

Substituting x = L/2 in the elastic curve equation, we obtain

$$v\left(\frac{L}{2}\right) = \frac{-wL}{48EI}\left(\frac{L^3}{8} - \frac{L^3}{2} + L^3\right) \quad \text{or} \quad \boxed{v\left(\frac{L}{2}\right) = -\left(\frac{5wL^4}{384EI}\right)}$$

7.3 In terms of w, P, L, E, and I determine (a) equation of the elastic curve. (b) the deflection of the beam at point A

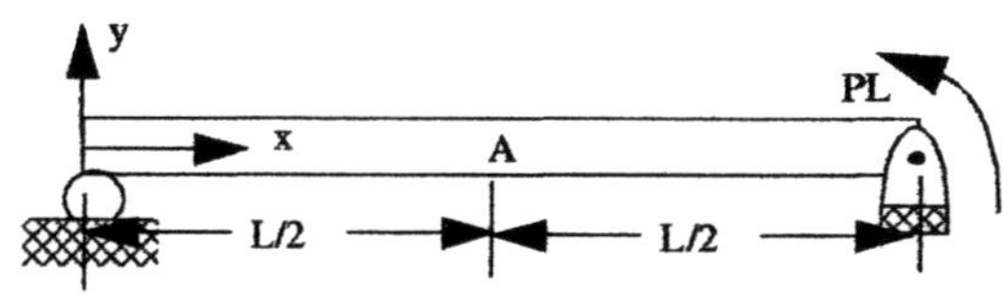

Fig. P7.3

Solution v(x) =? v(L/2) =?

Fig.(a) shows the free body diagram of the entire beam.By equilibrium of moment about point C, we obtain: $R_B(L) - PL = 0$ or $R_B = P$

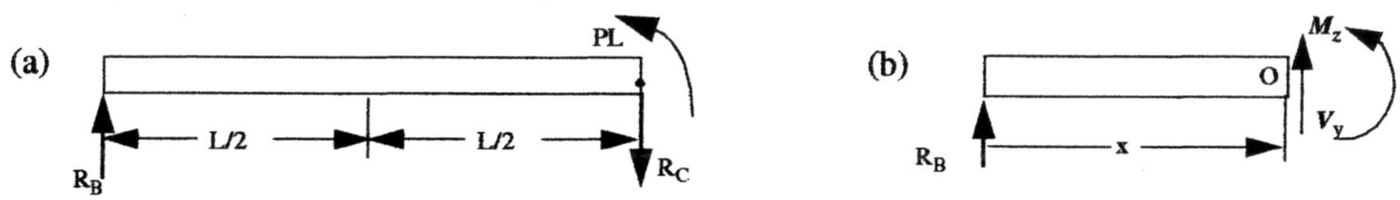

Fig.(b) shows the free body diagram after the imaginary cut is made at some location x.By equilibrium of moment about point O, we obtain: $M_z - R_Bx = 0$ or $M_z = Px$

The boundary value problem statement is,

$$EI\frac{d^2v}{dx^2} = Px \quad 1 \qquad v(0) = 0 \quad 2 \qquad v(L) = 0 \quad 3$$

Integrating Eq.1, we obtain,

$$EI\frac{dv}{dx} = \frac{Px^2}{2} + C_1 \quad 4 \qquad EIv = \frac{Px^3}{6} + C_1x + C_2 \quad 5$$

Substituting x = 0 in Eq.(5) and using Eq.(2), we obtain

$$EIv(0) = C_2 = 0 \quad 6$$

Substituting x = L in Eq.(5) and using Eq.(3), we obtain

$$EIv(L) = \frac{PL^3}{6} + C_1L = 0 \quad \text{or} \quad C_1 = -\frac{PL^2}{6} \quad 7$$

Substituting C_1 and C_2 in Eq.(5) and using Eq.(3), we obtain the elastic curve below

$$v(x) = \frac{P}{6EI}x^3 - \frac{P}{6EI}L^3x \text{ or } \qquad \boxed{v(x) = \frac{Px}{6EI}(x^2 - L^2)}$$

Substituting x = L/2 in the elastic curve equation, we obtain

$$v\left(\frac{L}{2}\right) = \frac{P}{6EI}\left(\frac{L}{2}\right)\left(\frac{L^2}{4} - L^2\right) \text{ or } \qquad \boxed{v\left(\frac{L}{2}\right) = -\left(\frac{PL^3}{16EI}\right)}$$

7.4 In terms of w, P, L, E, and I determine (a) equation of the elastic curve. (b) the deflection of the beam at point A

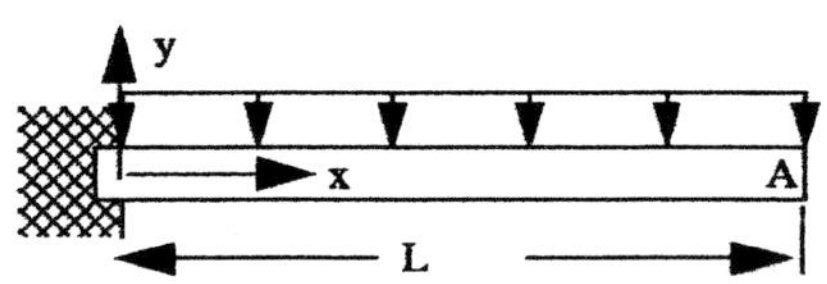

Fig. P7.4

Solution v(x) =? v(L) =?

Fig.(a) shows the free body diagram after an imaginary cut is made at some location x.Fig.(b) shows the distributed force replaced by a equivalent force.

By equilibrium of moment about point O, we obtain,

$$M_z + w(L-x)\left(\frac{L-x}{2}\right) = 0 \quad \text{or} \quad M_z = -\frac{w}{2}(L-x)^2 = -\frac{w}{2}(L^2 - 2Lx + x^2)$$

The boundary value statement is

$$EI\frac{d^2v}{dx^2} = -\frac{wL^2}{2} + wLx - \frac{wx^2}{2} \quad 1 \qquad v(0) = 0 \quad 2 \qquad \frac{dv}{dx}(0) = 0 \quad 3$$

Integrating Eq.1, we obtain,

$$EI\frac{dv}{dx} = -\frac{wL^2}{2}x + wL\frac{x^2}{2} - \frac{wx^3}{6} + C_1 \quad 4 \qquad EIv = -\frac{wL^2x^2}{4} + \frac{wLx^3}{6} - \frac{wx^4}{24} + C_1x + C_2 \quad 5$$

Substituting x = 0 in Eq.(4) and using Eq(3), we obtain

$$EI\frac{dv}{dx}(0) = C_1 = 0 \qquad \textbf{6}$$

Substituting x = 0 in Eq.(5) and using Eq.(2), we obtain

$$EIv(0) = C_2 = 0 \qquad \textbf{7}$$

Substituting $C_1$and C_2in Eq.(5), we obtain the elastic curve below

$$v(x) = \frac{w}{EI}\left(\frac{-6L^2x^2 + 4Lx^3 - x^4}{24}\right) \text{ or } \boxed{v(x) = \frac{-wx^2}{24EI}(x^2 - 4Lx + 6L^2)}$$

Substituting x = L in the elastic curve equation, we obtain

$$v(L) = \frac{-w(L^2)}{24EI}(L^2 - 4L^2 + 6L^2) \text{ or } \boxed{v(L) = -\left(\frac{wL^4}{8EI}\right)}$$

7.5 In terms of w, P, L, E, and I determine (a) equation of the elastic curve. (b) the deflection of the beam at point A.

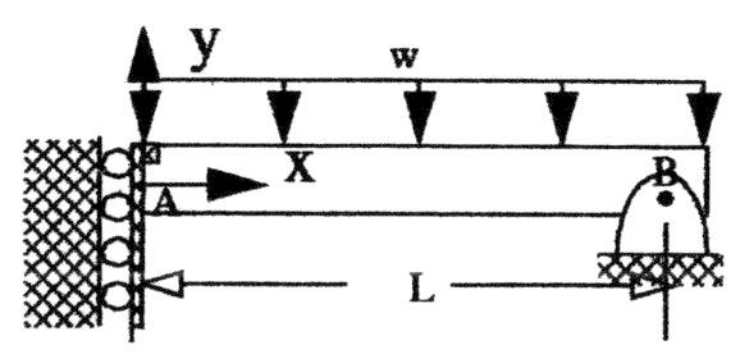

Fig. P7.5

Solution $v(x) = ?$ $v(0) = ?$

Fig.(a) shows the free body diagram of the entire beam with the distributed force replaced by a equivalent force.By equilibrium of moment about point B, we obtain

$$M_A - wL\left(\frac{L}{2}\right) = 0 \quad \text{or} \quad M_A = \frac{wL^2}{2}$$

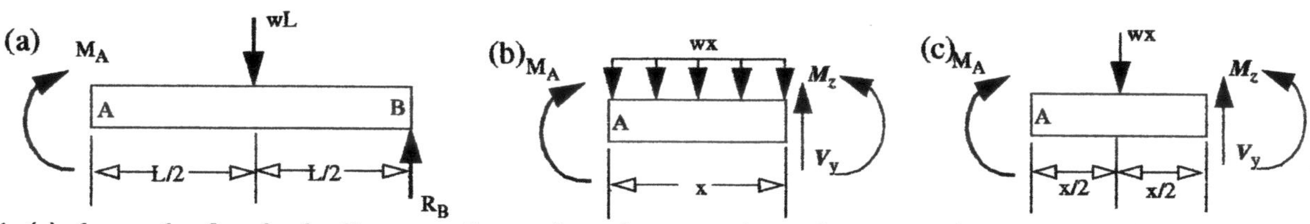

Fig(a) shows the free body diagram after an imaginary cut is made at some location x.Fig(b) shows the distributed force replaced by a equivalent force.By Equilibrium of moment about point O, we obtain,

$$M_z + wx\left(\frac{x}{2}\right) - M_A = 0 \quad \text{or} \quad M_z = \frac{wL^2}{2} - \frac{wx^2}{2}$$

The boundary value statement is

$$EI\frac{d^2v}{dx^2} = \frac{wL^2}{2} - \frac{wx^2}{2} \quad \textbf{1} \qquad \frac{dv}{dx}(0) = 0 \quad \textbf{2} \qquad v(L) = 0 \quad \textbf{3}$$

Integrating Eq.1, we obtain,

$$EI\frac{dv}{dx} = \frac{wL^2x}{2} - \frac{wx^3}{6} + C_1 \quad \textbf{4} \qquad EIv = \frac{wL^2x^2}{4} - \frac{wx^4}{24} + C_1x + C_2 \quad \textbf{5}$$

Substituting x = 0 in Eq.(4) and using Eq.(2), we obtain

$$EI\frac{dv}{dx}(0) = C_1 = 0 \qquad \textbf{6}$$

Substituting x = L in Eq.(5) and using Eq.(3), we obtain

$$EIv(L) = \frac{wL^4}{12} - \frac{wL^4}{24} + C_2 = 0 \qquad \text{or} \qquad C_2 = -\frac{5wL^4}{24} \qquad 7$$

Substituting C_1 and C_2 in Eq.(5), we obtain: $v = \frac{w}{EI}\left(\frac{6L^2x^2 - x^4 - 5L^4}{24}\right)$ or $\boxed{v(x) = \frac{w}{24EI}(6L^2x^2 - x^4 + 5L^4)}$

Substituting x = 0 in elastic curve, we obtain $\boxed{v(0) = -\left(\frac{5wL^4}{24}\right)}$

7.6 In terms of w, P, L, E, and I determine (a) equation of the elastic curve. (b) the deflection of the beam at point A

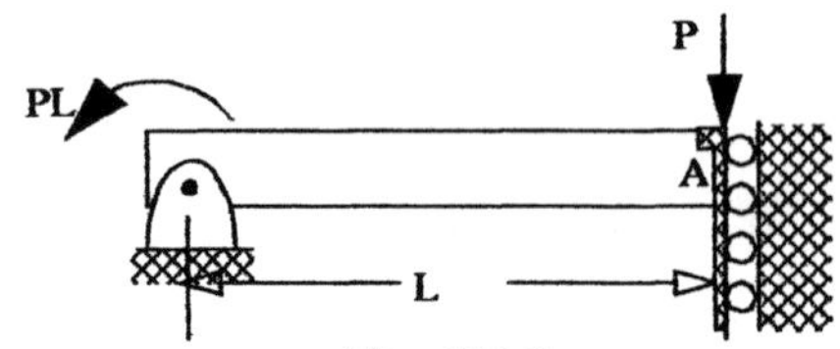

Fig. P7.6

Solution v(x) =? v(L) =?

Fig.(a) shows the free body diagram of the entire beam.By equilibrium of forces in the y-direction, we obtain $R_B = P$

(a)

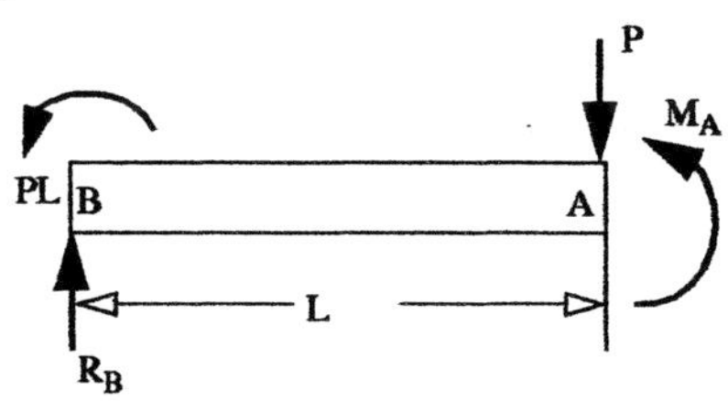

(b)

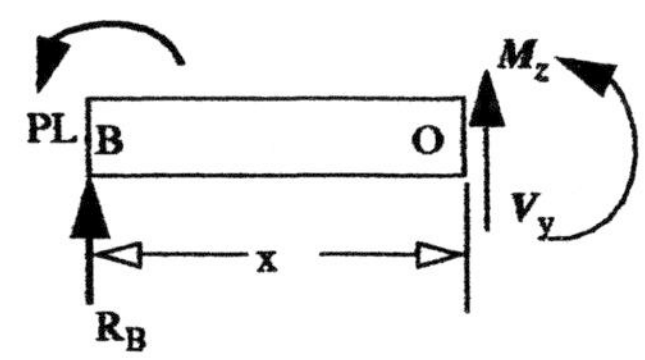

Fig(b) shows the free body diagram after the imaginary cut is made at some location x.By equilibrium of moment about point O, we obtain: $M_z + PL - R_Bx = 0$ or $M_z = Px - PL$

The boundary value problem statement is,

$$EI\frac{d^2v}{dx^2} = Px - PL \quad 1 \qquad v(0) = 0 \quad 2 \qquad \frac{dv}{dx}(L) = 0 \quad 3$$

Integrating Eq.1, we obtain,

$$EI\frac{dv}{dx} = \frac{Px^2}{2} - PLx + C_1 \quad 4 \qquad EIv = \frac{Px^3}{6} - \frac{PLx^2}{2} + C_1x + C_2 \quad 5$$

Substituting x = 0 in Eq.(5) and using Eq.(2), we obtain

$$EIv(0) = C_2 = 0 \qquad 6$$

Substituting x = L into Eq.(4) and using Eq.(3), we obtain

$$EI\frac{dv}{dx}(L) = \frac{PL^2}{2} - PL^2 + C_1 = 0 \qquad \text{or} \qquad C_1 = \frac{PL^2}{2} \qquad 7$$

Substituting C_1 and C_2 into Eq.(5) and using Eq.(3), we obtain the elastic curve below

$v(x) = \frac{P}{6EI}(x^3 - 3Lx^2 + 3L^2x)$ or $\boxed{v(x) = \frac{Px}{6EI}(x^2 - 3Lx + 3L^2)}$

Substituting x = L in the elastic curve equation, we obtain $\boxed{v(L) = \frac{PL^3}{6EI}}$

7.7 Determine the deflection at point A in terms of w, P, L, E, and I

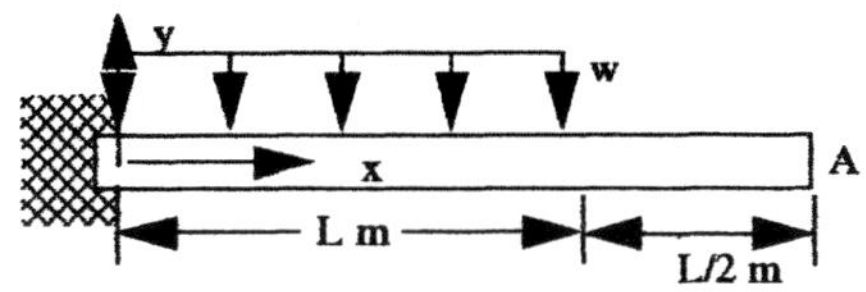

Fig. P7.7

Solution $v_A = ?$

Fig.(a) shows the elastic curve in the region BC is same as the problem 7-4.The elastic curve in the region CA is a straight line as the internal moment is zero.

(a)

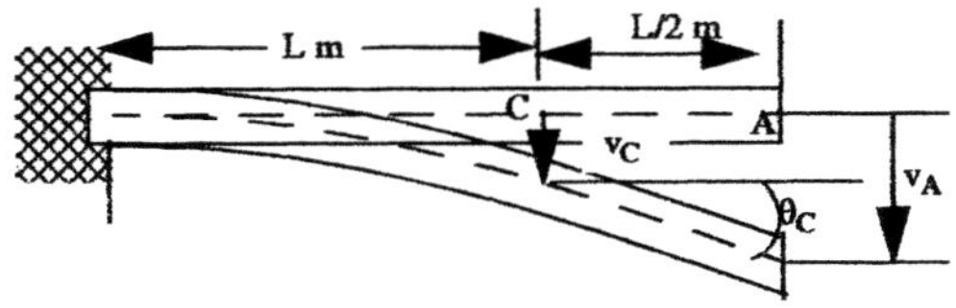

From Fig (a) we can write,

$$v_A = v_C + \frac{L}{2}\tan\theta_C \qquad 1$$

From problem 7-4, the elastic curve is

$$v(x) = \frac{-wx^2}{24EI}(x^2 - 4Lx + 6L^2) \qquad 2$$

Taking the first derivative of Eq.(2), we obtain,

$$\frac{dv}{dx} = \frac{-w}{24EI}(4x^3 - 12Lx^2 + 12L^2x) \qquad 3$$

Substituting x = L in Eq.(2) and (3)

$$v_C = v(L) = \frac{-wL^4}{8EI} \qquad 4$$

$$\tan\theta_C = \frac{dv}{dx}(L) = \frac{-wL^3}{24EI}(4 - 12 + 12) = \frac{-wL^3}{6EI} \qquad 5$$

Substituting Eq.(4) and (5) in Eq.(1)

$$v_A = \frac{-wL^4}{8EI} + \frac{L}{2}\left(\frac{-wL^3}{6EI}\right)$$

$$\boxed{v_A = -\left(\frac{5wL^4}{24EI}\right)}$$

7.8 Determine the deflection at point A in terms of w, P, L, E, and I

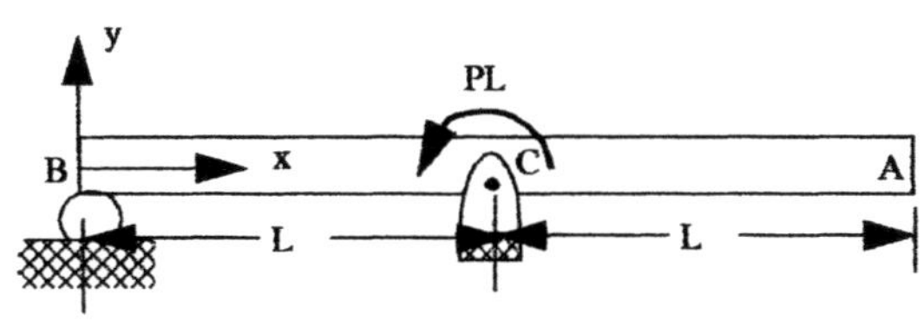

Fig. P7.8

Solution $v_A = ?$

Fig(a) shows the elastic curve in the region BC is same as that in problem 7-3.The elastic curve in the region CA is a straight line as the internal moment is zero.

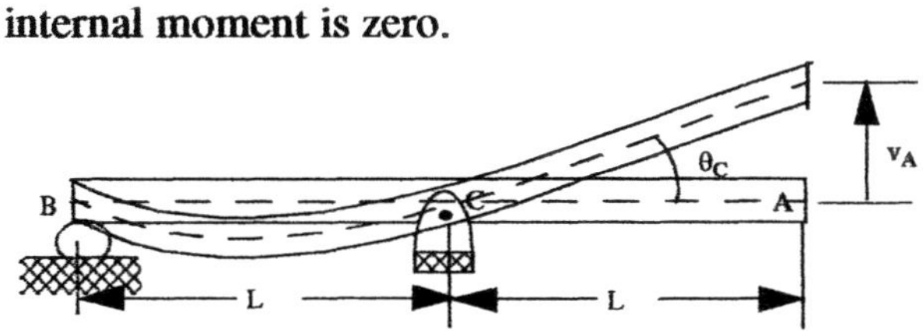

From Fig.(a) we can write,

$$v_A = L\tan\theta_C \qquad 1$$

From problem 7-3, the elastic curve is

$$v(x) = \frac{P}{6EI}(x^3 - L^2x) \qquad 2$$

Taking the first derivative of Eq.(2), we obtain,

$$\frac{dv}{dx} = \frac{P}{6EI}(3x^2 - L^2) \qquad 3$$

Substituting x = L in Eq.(2) and (3)

$$\tan\theta_C = \frac{dv}{dx}(L) = \frac{P}{6EI}(3L^2 - L^2) = \frac{PL^2}{3EI} \qquad 4$$

Substituting Eqs.(4) and (5) into Eq.(1) we obtain: $v_A = L\left(\frac{PL^2}{3EI}\right)$ or $\boxed{v_A = \frac{PL^3}{3EI}}$

7.9 In terms of w, L, E, and I, determine (a) the equation of the Elastic Curve. (b) the deflection at x = L.

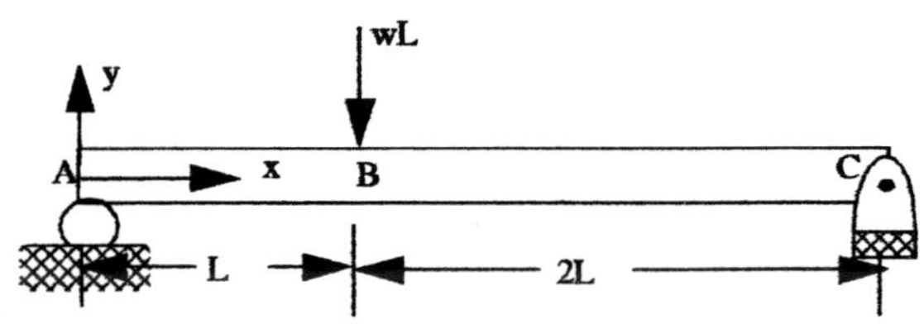

Fig. P7.9

Solution v(x) =? v(L) =?

Fig.(a) shows the free body diagram of the entire beam.By equilibrium of moment about point C, we obtain: $R_A(3L) - wL(2L) = 0$ or $R_A = \frac{2}{3}wL$

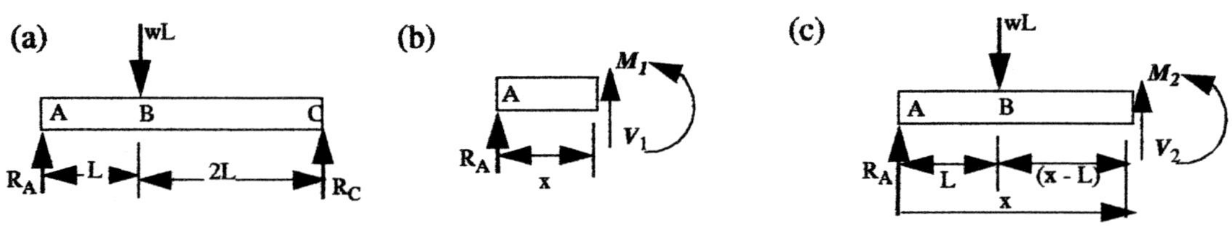

Fig.(b) shows the free body diagram after the imaginary cut between AB.By equilibrium of moment about point O_1, we obtain,

$$\boldsymbol{M_1} - R_Ax = 0 \quad \text{or} \quad \boldsymbol{M_1} = \frac{2}{3}wLx \qquad 1$$

Fig.(c) shows the free body diagram after the imaginary cut is made between AB.By equilibrium of moment about point O_2, we obtain,

$$\boldsymbol{M_2} - R_Ax + wL(x - L) = 0 \quad \text{or} \quad \boldsymbol{M_2} = \frac{2}{3}wLx - wL(x - L) \qquad 2$$

The boundary value problem statement is

$$EI\frac{d^2v_1}{dx^2} = \frac{2}{3}wLx \quad 0 \le x \le L \qquad 3 \qquad EI\frac{d^2v_2}{dx^2} = \frac{2}{3}wLx - wL(x - L) \quad L \le x \le 3L \qquad 4$$

$$v_1(0) = 0 \quad 5 \qquad v_2(3L) = 0 \quad 6 \qquad v_1(L) = v_2(L) \quad 7 \qquad \frac{dv_1}{dx}(L) = \frac{dv_2}{dx}(L) \quad 8$$

Integrating Eqs.(3) and (4), we obtain,

$$EI\frac{dv_1}{dx} = \frac{1}{3}wLx^2 + C_1 \quad 9$$

$$EI\frac{dv_2}{dx} = \frac{1}{3}wLx^2 - \frac{wL}{2}(x-L)^2 + C_2 \quad 10$$

Substituting x = L in Eqs.(9) and (10) and equating to satisfy Eq.(8) we obtain

$$\frac{1}{3}wL^3 + C_1 = \frac{1}{3}wL^3 + C_2 \qquad \text{or} \qquad C_1 = C_2 \quad 11$$

Integrating Eqs.(9) and (10) and using Eq.(11), we obtain,

$$EIv_1 = \frac{1}{9}wLx^3 + C_1x + C_3 \quad 12$$

$$EIv_2 = \frac{1}{9}wLx^3 - \frac{wL}{6}(x-L)^3 + C_1x + C_4 \quad 13$$

Substituting x = L in Eqs. (12) and (13) and equating to satisfy Eq.(7) we obtain

$$\frac{1}{9}wL^3 + C_1L + C_3 = \frac{1}{9}wL^3 + C_1L + C_4 \qquad \text{or} \qquad C_3 = C_4 \quad 14$$

Substituting x = 0 in Eqs. (12) and using Eq.(5) we obtain

$$EIv_1(0) = C_3 = 0 \quad 15$$

Substituting x = 3L in Eqs. (13) and using Eq.(6) we obtain

$$EIv_2(3L) = \frac{1}{9}wL(27L^3) - \frac{wL}{6}(8L^3) + C_1(3L) \text{ or}$$

$$C_1(3L) + \frac{wL^4(54-24)}{18} = 0 \qquad \text{or} \qquad C_1 = -\frac{5}{9}wL^3 \quad 16$$

The integration constant are $C_1 = C_2 = -\frac{5}{9}wL^3$ and $C_3 = C_4 = 0$

Substituting the above constants in Eqs.(12) and (13), we obtain

$$v_1 = \frac{w}{EI}\frac{(Lx^3 - 5L^3x)}{9} = \frac{wLx}{9EI}(x^2 - 5L^2) \quad 17$$

$$v_2 = \frac{w}{9EI}(Lx^3 - 5L^3x) - \frac{w}{6EI}(x-L)^3 \quad 18$$

The equation of the elastic curve is,

$$v(x) = \begin{cases} \frac{wLx}{9EI}(x^2 - 5L^2) & 0 \le x \le L \\ \frac{wLx}{9EI}(x^2 - 5L^2) - \frac{wL}{6EI}(x-L)^3 & L \le x \le 3L \end{cases}$$

Substitution in x= L in Eq.(17), we obtain

$$v(L) = \frac{wL^2}{9EI}(-4L^2) \text{ or} \qquad v(L) = -\left(\frac{4wL^4}{9EI}\right)$$

7.10 In terms of w, L, E, and I, determine (a) the equation of the Elastic Curve. (b) the deflection at x = L.

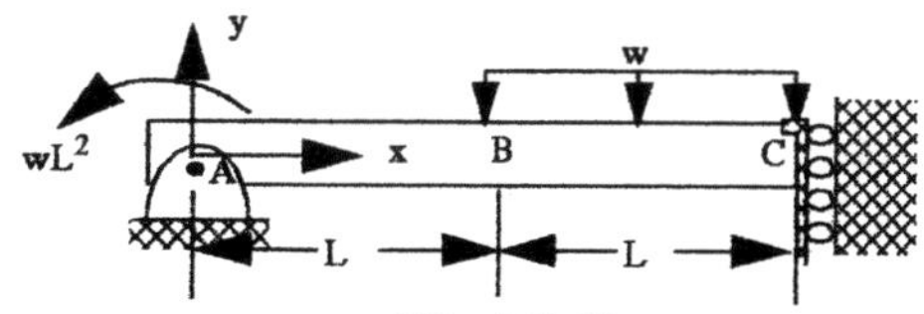

Fig. P7.10

Solution v(x) =? v(L) =?

Fig.(a) shows the free body diagram of the entire beam.By equilibrium of forces in the y-direction, we

obtain $R_A - wL = 0$ or $R_A = wL$

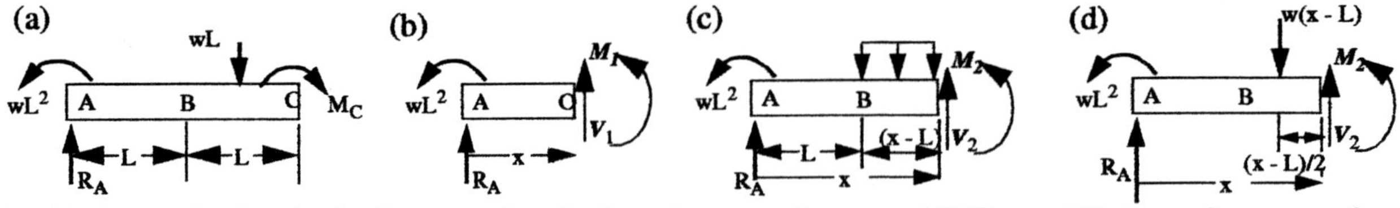

Fig.(b) shows the free body diagram after the imaginary cut between AB.By equilibrium of moment about point O_1, we obtain,

$$M_1 + wL^2 - R_A x = 0 \qquad \text{or} \qquad M_1 = wLx - wL^2 \qquad 1$$

Fig.(c) shows the free body diagram after the imaginary cut is a made between BC. By equilibrium of moment about point O_2, we obtain,

$$M_2 + wL^2 - R_A x + w(x-L)(x-L)/2 = 0 \qquad \text{or} \qquad M_2 = wLx - wL^2 - w(x-L)^2/2 \qquad 2$$

The boundary value problem statement is,

$$EI\frac{d^2v_1}{dx^2} = wLx - wL^2 \qquad 0 \le x \le L \qquad 3 \qquad EI\frac{d^2v_2}{dx^2} = wLx - wL^2 - w(x-L)^2/2 \qquad L \le x \le 2L \qquad 4$$

$$v_1(0) = 0 \qquad 5 \qquad \frac{dv_2}{dx}(2L) = 0 \qquad 6 \qquad v_1(L) = v_2(L) \qquad 7 \qquad \frac{dv_1}{dx}(L) = \frac{dv_2}{dx}(L) \qquad 8$$

Integrating Eq.(3) and (4), we obtain,

$$EI\frac{dv_1}{dx} = \frac{1}{2}wLx^2 - wL^2x + C_1 \qquad 9$$

$$EI\frac{dv_2}{dx} = \frac{1}{2}wLx^2 - wL^2x - \frac{w}{6}(x-L)^3 + C_2 \qquad 10$$

Substituting x = L in Eqs. (9) and (10) and equating to satisfy Eq.(8) we obtain

$$\frac{1}{2}wL^3 - wL^3 + C_1 = \frac{1}{2}wL^3 - wL^3 + C_2 \qquad \text{or} \qquad C_1 = C_2 \qquad 11$$

Integrating Eqs. (9) and (10), we obtain,

$$EIv_1 = \frac{wLx^3}{6} - \frac{wL^2x^2}{2} + C_1x + C_3 \qquad 12$$

$$EIv_2 = \frac{wLx^3}{6} - \frac{wL^2x^2}{2} - \frac{w(x-L)^4}{24} + C_1x + C_4 \qquad 13$$

Substituting x = L in Eqs.(12) and (13) and equating to satisfy Eq.(7) we obtain

$$\frac{1}{6}wL^4 - \frac{w}{2}L^4 + C_1L + C_3 = \frac{1}{6}wL^4 - \frac{w}{2}L^4 + C_1x + C_4 \qquad \text{or} \qquad C_3 = C_4 \qquad 14$$

Substituting x = 0 in Eqs. (12) and using Eq.(5) we obtain

$$EIv_1(0) = C_3 = 0 \qquad 15$$

Substituting x = 2Lin Eqs. (10) and using Eq.(6) we obtain

$$EI\frac{dv_2}{dx}(2L) = \frac{1}{2}wL(4L^2) - wL^2(2L) - \frac{1}{6}wL^3 + C_2 = 0 \text{ or}$$

$$C_2 + \frac{wL^3(12 - 12 - 1)}{6} = 0 \qquad \text{or} \qquad C_2 = \frac{1}{6}wL^3 \qquad 16$$

The value of integration constant are $C_1 = C_2 = \frac{1}{6}wL^3$ and $C_3 = C_4 = 0$

Substituting the above constants in Eqs. (12) and (13), we obtain

$$v_1 = \frac{w}{EI}\frac{(Lx^3 - 3L^2x^2 + L^3x)}{6} = \frac{wLx}{6EI}(x^2 - 3Lx + L^2) \qquad 17$$

$$v_2 = \frac{w}{EI}\frac{(Lx^3 - 3L^2x^2 + L^3x)}{6} - \frac{w}{24}(x-L)^4 \qquad 18$$

The equation of the elastic curve is,

$$\boxed{v(x) = \begin{cases} \dfrac{wLx}{6EI}(x^2 - 3Lx + L^2) & 0 \le x \le L \\ \dfrac{w}{EI}\dfrac{(Lx^3 - 3L^2x^2 + L^3x)}{6} - \dfrac{w}{24}(x-L)^4 & L \le x \le 2L \end{cases}}$$

Substitution in x= L in Eq(17), we obtain: $v(L) = \frac{wL^2}{6EI}(L^2 - 3L^2 + L^2)$ or $\boxed{v(L) = -\left(\frac{4wL^4}{6EI}\right)}$

7.11 In terms of w, L, E, and I, determine (a) the equation of the Elastic Curve. (b) the deflection at x = L.

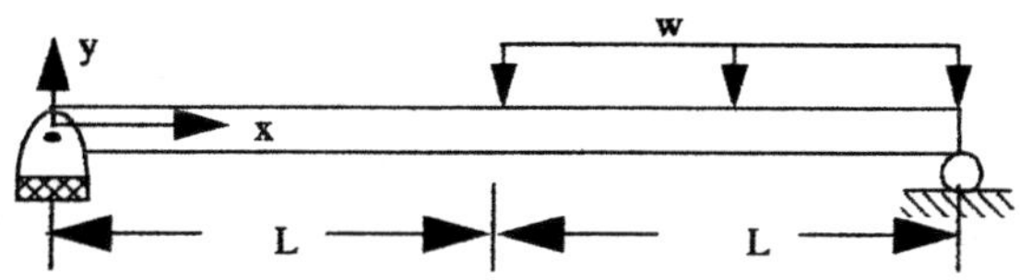

Fig. P7.11

Solution v(x) =? v(L) =?

From the Eqs. (7) and (8) in problem 6-55, we have

$$M_z = \frac{wLx}{4} \qquad 0 \le x \le L \qquad 1$$

$$M_z = \frac{wLx}{4} - \frac{w}{2}(x-L)^2 \qquad L \le x \le 2L \qquad 2$$

The boundary value problem statement is,

$$EI\frac{d^2v_1}{dx^2} = \frac{wLx}{4} \quad 0 \le x \le L \quad 3 \qquad EI\frac{d^2v_2}{dx^2} = \frac{wLx}{4} - \frac{w}{2}(x-L)^2 \quad L \le x \le 2L \quad 4$$

$$v_1(0) = 0 \quad 5 \qquad v_2(2L) = 0 \quad 6 \qquad v_1(L) = v_2(L) \quad 7 \qquad \frac{dv_1}{dx}(L) = \frac{dv_2}{dx}(L) \quad 8$$

Integrating Eq.(3) and (4), we obtain,

$$EI\frac{dv_1}{dx} = \frac{1}{8}wLx^2 + C_1 \qquad 9$$

$$EI\frac{dv_2}{dx} = \frac{1}{8}wLx^2 - \frac{w}{6}(x-L)^3 + C_2 \qquad 10$$

Substituting x = L in Eqs.(9) and (10) and equating to satisfy Eq.(8) we obtain

$$\frac{1}{8}wL^3 + C_1 = \frac{1}{8}wL^3 + C_2 \quad \text{or} \quad C_1 = C_2 \qquad 11$$

Integrating Eqs.(9) and (10), we obtain,

$$EIv_1 = \frac{wLx^3}{24} + C_1x + C_3 \qquad 12$$

$$EIv_2 = \frac{wLx^3}{24} - \frac{w(x-L)^4}{24} + C_1x + C_4 \qquad 13$$

Substituting x = L in Eqs. (12) and (13) and equating to satisfy Eq.(7) we obtain

$$\frac{wL^4}{24} + C_1L + C_3 = \frac{wL^4}{24} + C_1x + C_4 \qquad \text{or} \qquad C_3 = C_4 \qquad 14$$

Substituting x = 0 in Eqs.(12) and using Eq.(5) we obtain

$$EIv_1(0) = C_3 = 0 \qquad 15$$

Substituting x = 2Lin Eqs.(13) and using Eqs.(6),(14) and (15) we obtain

$$EIv_2(2L) = \frac{w}{24}L(8L^3) - \frac{wL^4}{24} + C_1(2L) = 0 \text{ or } C_1(2L) + \frac{wL^4(8-1)}{24} = 0 \qquad \text{or} \qquad C_1 = -\frac{7wL^3}{48} \qquad 16$$

The value of integration constants are $C_1 = C_2 = -\frac{7wL^3}{48}$ and $C_3 = C_4 = 0$

Substituting the above constants in Eqs.(12) and (13), we obtain

$$v_1 = \frac{w}{EI}\frac{(2Lx^3 - 7L^3x)}{48} = \frac{wLx}{48EI}(2x^2 - 7L^2) \qquad 17$$

$$v_2 = \frac{w}{EI}\frac{(2Lx^3 - 7L^3x)}{48} - \frac{w}{24EI}(x-L)^4 \qquad 18$$

The equation of the elastic curve is,

$$v(x) = \begin{cases} \frac{wLx}{48EI}(2x^2 - 7L^2) & 0 \le x \le L \\ \frac{w}{EI}\frac{(2Lx^3 - 7L^3x)}{48} - \frac{w}{24EI}(x-L)^4 & L \le x \le 2L \end{cases}$$

Substitution in x= L in Eq.(17), we obtain

$$v(L) = \frac{wL^2}{48EI}(2L^2 - 7L^2) \text{ or } \qquad v(L) = \frac{-5wL^4}{48EI}$$

7.12 In terms of w, L, E, and I, determine (a) the equation of the Elastic Curve. (b) the deflection at x = L.

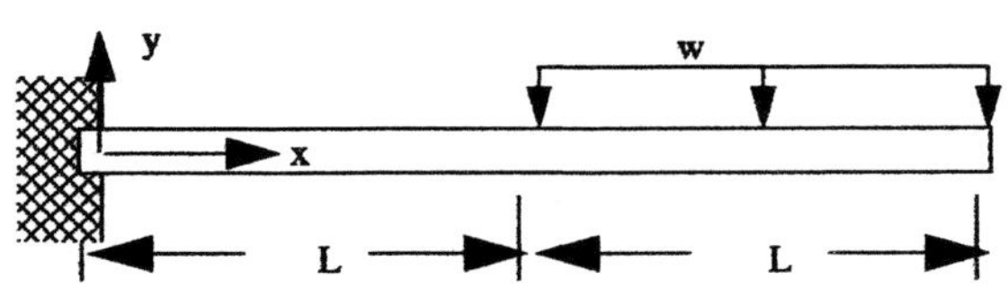

Fig. P7.12

Solution v(x) =? v(L) =?

From the Eqs.(7) and (8) in problem 6-56, we have

$$M_z = wLx - \frac{3}{2}wL^2 \qquad 0 \le x \le L \qquad 1$$

$$M_z = wLx - \frac{3}{2}wL^2 - \frac{w}{2}(x-L)^2 \qquad L \le x \le 2L \qquad 2$$

The boundary value problem statement is

$$EI\frac{d^2v_1}{dx^2} = wLx - \frac{3}{2}wL^2 \quad 0 \le x \le L \quad 3 \qquad EI\frac{d^2v_2}{dx^2} = wLx - \frac{3}{2}wL^2 - wL(x-L)^2 \quad L \le x \le 2L \quad 4$$

$$v_1(0) = 0 \quad 5 \qquad \frac{dv_1}{dx}(0) = 0 \quad 6 \qquad v_1(L) = v_2(L) \quad 7 \qquad \frac{dv_1}{dx}(L) = \frac{dv_2}{dx}(L) \quad 8$$

Integrating Eqs. (1) and (2), we obtain,

$$EI\frac{dv_1}{dx} = \frac{1}{2}wLx^2 - \frac{3}{2}wL^2x + C_1 \qquad 9$$

$$EI\frac{dv_2}{dx} = \frac{1}{2}wLx^2 - \frac{3}{2}wL^2x - \frac{w}{6}(x-L)^3 + C_2 \qquad 10$$

Substituting x = L in Eqs. (9) and (10) and equating to satisfy Eq. (8) we obtain

$$\frac{1}{2}wL^3 - \frac{3}{2}wL^3 + C_1 = \frac{1}{2}wL^3 - \frac{3}{2}wL^3 + C_2 \qquad \text{or} \qquad C_1 = C_2 \qquad 11$$

Integrating Eqs.(9) and (10) and using Eq.(11), we obtain,

$$EIv_1 = \frac{wLx^3}{6} - \frac{3wL^2x^2}{4} + C_1x + C_3 \qquad 12$$

$$EIv_2 = \frac{wLx^3}{6} - \frac{3wL^2x^2}{4} - \frac{w(x-L)^4}{24} + C_1x + C_4 \qquad 13$$

Substituting x = L in Eqs.(12) and (13) and equating to satisfy Eq.(7) we obtain

$$\frac{1}{6}wL^4 - \frac{3}{4}wL^4 + C_1L + C_3 = \frac{1}{6}wL^4 - \frac{3}{4}wL^4 + C_1x + C_4 \qquad \text{or} \qquad C_3 = C_4 \qquad 14$$

Substituting x = 0 in Eqs.(9) and using Eq.(6) we obtain

$$EI\frac{dv_1}{dx}(0) = C_1 = 0 \qquad 15$$

Substituting x = 0 in Eqs.(12) and using Eq.(5) we obtain

$$EIv_1(0) = C_3 = 0 \qquad 16$$

The value of integration constant are $C_1 = C_2 = 0$ and $C_3 = C_4 = 0$

Substituting the above constants in Eqs.(12) and (13), we obtain

$$v_1 = \frac{w}{EI}\frac{(2Lx^3 - 9L^2x^2)}{12} = \frac{wLx^2}{12EI}(2x - 9L) \qquad 17$$

$$v_2 = \frac{w}{EI}\frac{(2Lx^3 - 9L^2x^2)}{12} - \frac{w}{24EI}(x-L)^4 \qquad 18$$

The equation of the elastic curve is,

$$v(x) = \begin{cases} \dfrac{wLx^2}{12EI}(2x - 9L) & 0 \le x \le L \\ \dfrac{w}{EI}\dfrac{(2Lx^3 - 9L^2x^2)}{12} - \dfrac{w}{24EI}(x-L)^4 & L \le x \le 2L \end{cases}$$

Substitution in x= L in Eq.(17), we obtain

$$v(L) = \frac{wL^3}{12EI}(-7L) \quad \text{or} \qquad \boxed{v(L) = \frac{-7wL^4}{12EI}}$$

7.13 Write the boundary value problem for determining the deflection of the beam at any point x. Assume EI is constant. Do not integrate or solve

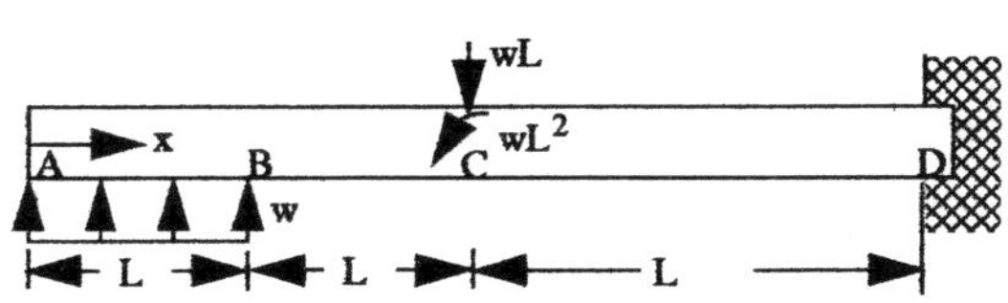

Fig. P7.13

Solution Boundary Value Problem =?

Fig.(a) shows the free body diagram after an imaginary cut is made in AB.In Fig.(b) the distributed force is replaced by an equivalent force.

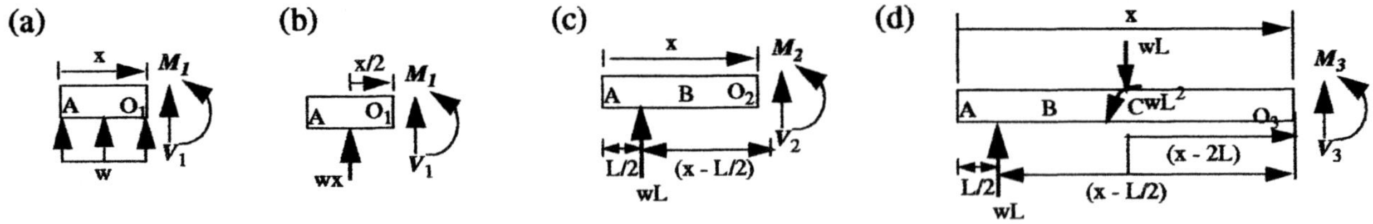

By equilibrium of moment about point O_1in Fig.(b), we obtain,

$$M_1 - wx\left(\frac{x}{2}\right) = 0 \qquad \text{or} \qquad M_1 = \frac{wx^2}{2} \qquad 1$$

Fig.(c) shows the free body diagram after an imaginary cut is made in BC and the distributed load is replaced by an equivalent force.By equilibrium of moment about point O_2, we obtain,

$$M_2 - wL\left(x - \frac{L}{2}\right) = 0 \qquad \text{or} \qquad M_2 = wL\left(x - \frac{L}{2}\right) \qquad 2$$

Fig.(d) shows the free body diagram after an imaginary cut is made in CD and the distributed load is replaced by an equivalent force. By equilibrium of moment about point O_3, we obtain,

$$M_3 - wL\left(x - \frac{L}{2}\right) + wL(x - 2L) + wL^2 = 0 \text{ or}$$

$$M_3 - wLx + \frac{wL^2}{2} + wLx - 2wL^2 + wL^2 = 0 \qquad \text{or} \qquad M_3 = \frac{wL^2}{2} \qquad 3$$

The boundary value problem statement is,

$$EI\frac{d^2v_1}{dx^2} = \frac{wx^2}{2} \quad 4 \qquad EI\frac{d^2v_2}{dx^2} = wL\left(x - \frac{L}{2}\right) \quad 5 \qquad EI\frac{d^2v_3}{dx^2} = \frac{wL^2}{2} \quad 6$$

$$v_3(3L) = 0 \quad 7 \qquad \frac{dv_3}{dx}(3L) = 0 \quad 8 \qquad v_1(L) = v_2(L) \quad 9$$

$$\frac{dv_1}{dx}(L) = \frac{dv_2}{dx}(L) \quad 10 \qquad v_2(2L) = v_3(2L) \quad 11 \qquad \frac{dv_1}{dx}(2L) = \frac{dv_2}{dx}(2L) \quad 12$$

7.14 In terms of w, L, E, and I, determine (a) the equation of the Elastic Curve. (b) the deflection at x = L.

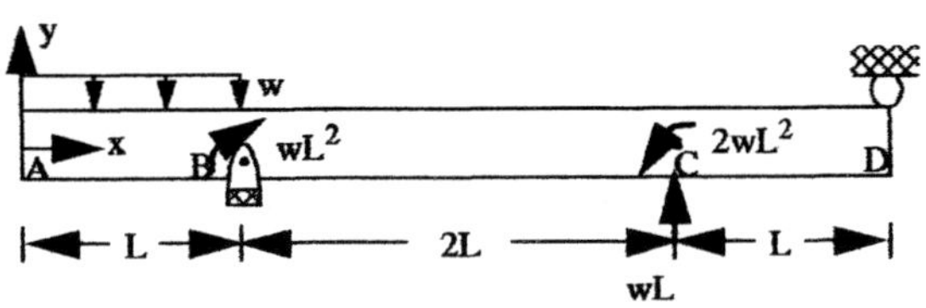

Fig. P7.14

Solution Boundary Value Problem =?

Fig.(a) shows the free body diagram of an entire beam with the distributed load is replaced by an equivalent force.By equilibrium of moment about point D, we obtain,

$$R_B(3L) - wL\left(3L - \frac{L}{2}\right) + wL^2 - 2wL^2 + wL(L) = 0 \qquad \text{or} \qquad R_B(3L) - \frac{7wL^2}{2} = 0$$

$$R_B = \frac{7wL}{6} \qquad 1$$

By equilibrium of forces in y-direction

$$R_B - R_D + wL - wL = 0 \qquad \text{or} \qquad R_B = R_D = \frac{7wL}{6} \qquad \mathbf{2}$$

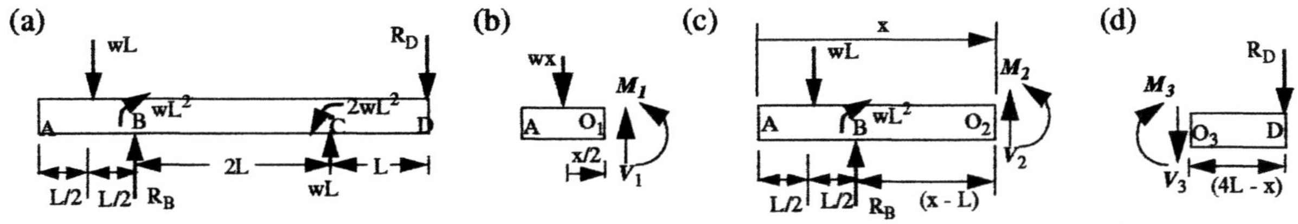

Fig.(b) shows the free body diagram after an imaginary cut is made in AB and the distributed force is replaced by an equivalent force.By equilibrium of moment about point O_1, we obtain,

$$M_1 + wx\left(\frac{x}{2}\right) = 0 \qquad \text{or} \qquad M_1 = -\frac{wx^2}{2} \qquad \mathbf{3}$$

Fig.(c) shows the free body diagram after an imaginary cut is made in BC and the distributed load is replaced by an equivalent force.By equilibrium of moment about point O_2, we obtain,

$$M_2 + wL\left(x - \frac{L}{2}\right) - wL^2 - R_B(x - L) = 0 \text{ or}$$

$$M_2 + wLx - \frac{wL^2}{2} - wL^2 - \frac{7w}{6}Lx + \frac{7}{6}wL^2 = 0 \qquad \text{or} \qquad M_2 = \frac{wLx}{6} + \frac{wL^2}{3} \qquad \mathbf{4}$$

Fig.(d) shows the free body diagram after an imaginary cut is made in CD.By equilibrium of moment about point O_3, we obtain,

$$M_3 + R_D(4L - x) = 0 \qquad \text{or} \qquad M_3 = \frac{7wLx}{6} - \frac{14wL^2}{3} \qquad \mathbf{5}$$

The boundary value problem statement is,

$$EI\frac{d^2v_1}{dx^2} = \frac{-wx^2}{2} \quad \mathbf{6} \qquad EI\frac{d^2v_2}{dx^2} = \frac{wLx}{6} + \frac{wL^2}{3} \quad \mathbf{7} \qquad EI\frac{d^2v_3}{dx^2} = \frac{7wLx}{6} - \frac{14wL^2}{3} \quad \mathbf{8}$$

$$v_1(L) = 0 \quad \mathbf{9} \qquad v_2(L) = 0 \quad \mathbf{10} \qquad v_3(4L) = 0 \quad \mathbf{11}$$

$$\frac{dv_1}{dx}(L) = \frac{dv_2}{dx}(L) \quad \mathbf{12} \qquad v_2(3L) = v_3(3L) \quad \mathbf{13} \qquad \frac{dv_2}{dx}(3L) = \frac{dv_3}{dx}(3L) \quad \mathbf{14}$$

7.15 A cantilever beam with variable depth h(x) and a constant width b is shown. The beam is to have a constant strength σ. In terms of b, L, E, x, σ, determine (a) the variation of h(x). (b) the maximum deflection.

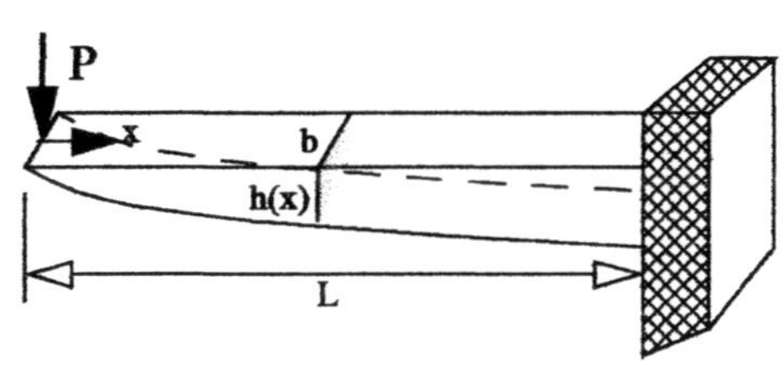

Fig. P7.15

Solution $h(x) = ?$ $v_{max} = ?$

Fig.(a) shows the free body diagram after an imaginary cut is made at any location x

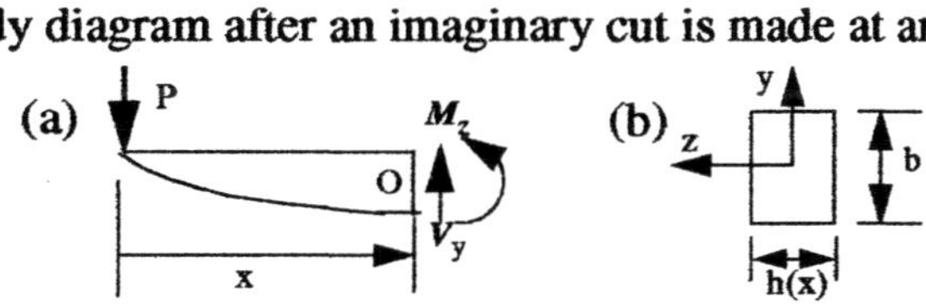

By equilibrium of moment about point O_1, we obtain,

$$M_z + Px = 0 \quad \text{or} \quad M_z = -Px \qquad 1$$

Fig(b) shows the cross section of the beam at any location x.The area moment of inertia is

$$I_{zz} = \frac{1}{12}bh^3(x) \qquad 2$$

The maximum bending normal stress will be at $y = \mp h(x)/2$. and its magnitude is

$$\sigma_{max} = \left|-\frac{M_z y}{I_{zz}}\right| = \frac{(Px)\left(\frac{h(x)}{2}\right)}{bh^3(x)/12} = \sigma \text{ or } \frac{6Px}{bh^2(x)} = \sigma \text{ or } \qquad \boxed{h(x) = \sqrt{\frac{6Px}{b\sigma}}}$$

Substituting h(x) into Eq.(2) we obtain

$$I_{zz} = \frac{1}{12}b\left(\frac{6Px}{b\sigma}\right)^{\frac{3}{2}} \qquad 3$$

The moment curvature equation can be written as

$$M_z = EI_{zz}\frac{d^2v_2}{dx^2} = -Px \text{ or } \frac{d^2v_2}{dx^2} = -\frac{Px}{E\sqrt{\frac{3P^3}{2b\sigma^3}}x^{3/2}} = -\sqrt{\frac{2b\sigma^3}{3PE^2}}\frac{1}{\sqrt{x}}.$$ We make the following substitution.

$$k = \sqrt{\frac{2b\sigma^3}{3PE^2}} \qquad 4$$

The boundary value problem statement is,

$$\frac{d^2v}{dx^2} = -\left(\frac{k}{\sqrt{x}}\right) \quad 5 \qquad v(L) = 0 \quad 6 \qquad \frac{dv}{dx}(L) = 0 \quad 7$$

Integrating Eq.(5)

$$\frac{dv}{dx} = -2kx^{1/2} + C_1 \quad 8 \qquad v = -\frac{4}{3}kx^{3/2} + C_1x + C_2 \quad 9$$

Substituting x = L in Eq.(8) and using Eq.(7) we obtain,

$$\frac{dv}{dx}(L) = -2kL^{1/2} + C_1 = 0 \quad \text{or} \quad C_1 = 2kL^{1/2} \qquad 10$$

Substituting x = L in Eq.(9) and using Eqs.(6) and (10) we obtain,

$$v = -k\frac{4}{3}L^{3/2} + 2k\,L^{1/2}x + C_2 \quad \text{or} \quad kL^{3/2}\left(\frac{-4+6}{3}\right) + C_2 = 0 \quad \text{or} \quad C_2 = -\left(\frac{2}{3}kL^{3/2}\right) \qquad 11$$

The maximum deflection will be at x = 0.Substituting x= 0 in Eq.(9) and using Eqs.(11) and (4), we obtain

$$v_{max} = C_2 = -\frac{2}{3}\sqrt{\frac{2b\sigma^3}{3PE^2}}L^{3/2} \text{ or } \qquad \boxed{v_{max} = -\sqrt{\frac{8b\sigma^3L^3}{27PE^2}}}$$

7.16 A cantilever tapered circular beam with variable radius R(x) is shown. The beam is to have a constant strength σ. In terms of L, E, x, σ, determine (a) the variation of R(x). (b) the maximum deflection.

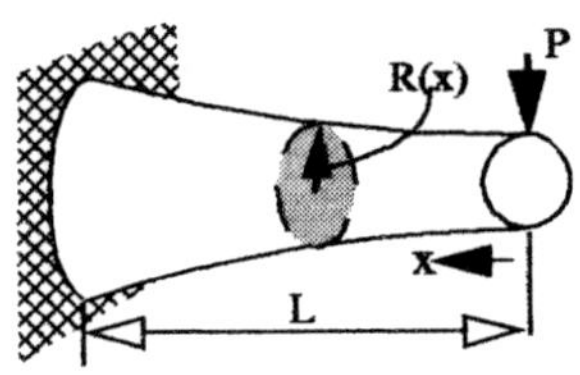

Fig. P7.16

Solution $R(x) = ?$ $v_{max} = ?$

Fig.(a) shows the free body diagram after an imaginary cut is made at any location x

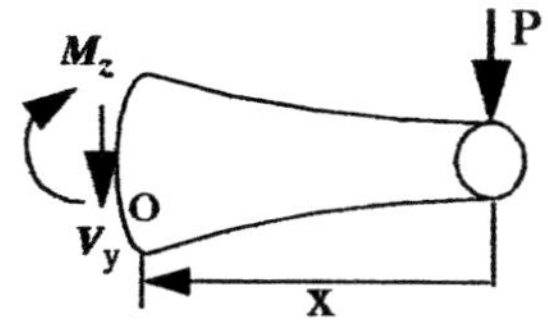

By equilibrium of moment about point O, we obtain,

$$M_z = -Px \quad 1$$

The area moment of inertia is,

$$I_{zz} = \frac{\pi}{4}R^4(x) \quad 2$$

The maximum bending normal stress will be at $y = \mp R(x)$. The magnitude of maximum bending stress is

$$\sigma_{max} = \left|-\frac{M_z y}{I_{zz}}\right| = \frac{(Px)R}{\frac{\pi}{4}R^4} = \sigma \text{ or } \qquad \boxed{R = \left(\frac{4Px}{\pi\sigma}\right)^{1/3}}$$

Substituting R(x) into Eq.(2) we obtain

$$I_{zz} = \frac{\pi}{4}\left(\frac{4Px}{\pi\sigma}\right)^{4/3} = \left(\frac{4P^4}{\pi\sigma^4}\right)^{1/3} x^{4/3} \quad 3$$

The moment curvature relationship can be written as

$$M_z = EI_{zz}\frac{d^2v}{dx^2} = -Px \text{ or } \frac{d^2v}{dx^2} = -\frac{Px}{E\left(\frac{4P^4}{3\sigma^4}\right)^{1/3} x^{4/3}} = -\left(\frac{3\sigma^4}{4PE^3}\right)^{1/3}\frac{1}{x^{1/3}}$$ We make the following substitution.

$$k = \left(\frac{3\sigma^4}{4PE^3}\right)^{1/3} \quad 4$$

The boundary value problem statement is,

$$\frac{d^2v}{dx^2} = -\frac{k}{x^{1/3}} \quad 5 \qquad v(L) = 0 \quad 6 \qquad \frac{dv}{dx}(L) = 0 \quad 7$$

Integrating Eq.(5)

$$\frac{dv}{dx} = -k\left(\frac{3}{2}x^{2/3}\right) + C_1 \quad 8 \qquad v = -k\left(\frac{9}{10}x^{5/3}\right) + C_1 x + C_2 \quad 9$$

Substituting x = L in Eq.(8) and using Eq.(6) we obtain,

$$-k\left(\frac{3}{2}L^{2/3}\right) + C_1 = 0 \quad \text{or} \quad C_1 = \frac{3}{2}kL^{2/3} \quad 10$$

Substituting x = L in Eq.(9) and using Eq.(6) and (10) we obtain,

$$v(L) = -\frac{9}{10}kL^{5/3} + \frac{3}{2}kL^{2/3}L + C_2 \quad \text{or} \quad kL^{5/3}\left(\frac{-9+15}{10}\right) + C_2 = 0 \quad \text{or} \quad C_2 = -\frac{3}{5}kL^{5/3} \quad 11$$

The maximum deflection will be at x = 0.Substituting x= 0 in Eq.(9) and using Eqs.(11) and (4), we obtain

$$v_{max} = C_2 = -\frac{3}{5}\left(\frac{3\sigma^4}{4PE^3}\right)^{1/3} L^{5/3} \text{ or } \qquad \boxed{v_{max} = -\left(\frac{81\sigma^4L^5}{500PE^3}\right)^{1/3}}$$

7.17 For the tapered beam shown, determine the maximum bending normal stress and maximum deflection in terms of E, w, b, h_o, and L.

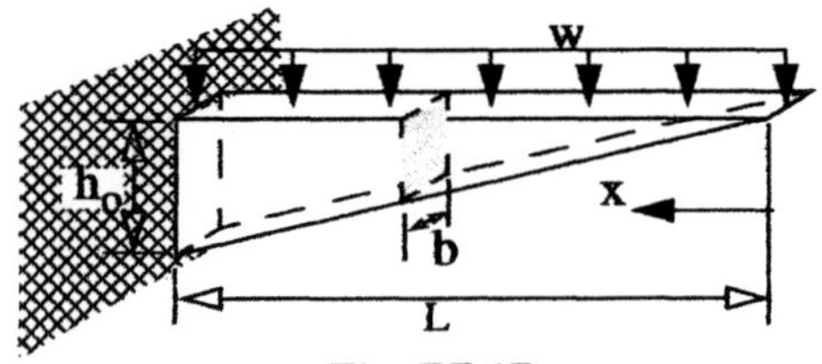

Fig. P7.17

Solution $\sigma_{max}=?$ $v_{max}=?$

Fig.(a) shows the free body diagram after an imaginary cut is made at any location x.

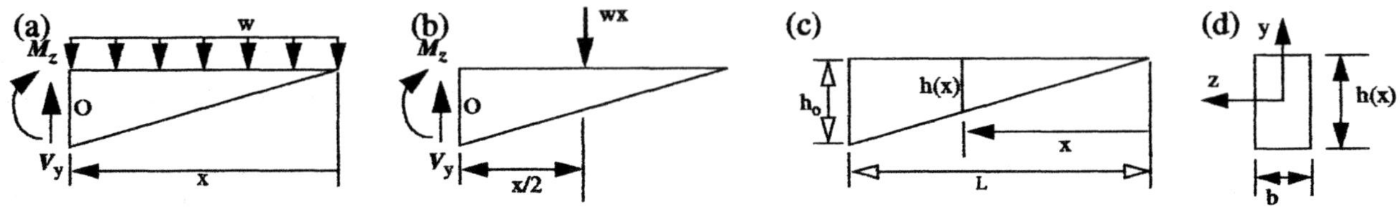

In Fig.(b) the distributed force is replaced by an equivalent force.By equilibrium of moment about point O, we obtain,

$$M_z = -\frac{wx^2}{2} \quad (1)$$

Using similar triangle in Fig.(c) we obtain

$$h = \frac{h_o}{L}x \quad (2)$$

Fig.(d) shows the cross-section of beam at x.The area moment of inertia is: $I_{zz} = \frac{1}{12}bh^3 = \frac{1}{12}b\frac{h_o^3}{L^3}x^3$

The maximum bending normal stress will be at $y = \mp\frac{h}{2}$.The magnitude of maximum bending stress is

$$\sigma_{max} = \left|-\frac{M_z y}{I_{zz}}\right| = \frac{\left(\frac{wx^2}{2}\right)\left(\frac{h}{2}\right)}{\frac{1}{12}bh^3} = \frac{3wx^2}{bh^2} = \frac{3wx^2}{b\frac{h_o^2}{L^2}x^2}$$

$$\boxed{\sigma_{max} = \frac{3wL^2}{bh_o^2}}$$

The moment curvature relationship can be written as

$$M_z = EI_{zz}\frac{d^2v}{dx^2} = -\frac{wx^2}{2} \quad \text{or} \quad E\left(\frac{3bh_o^3}{L^3}\right)x^3\frac{d^2v}{dx^2} = -\frac{wx^2}{2} \quad \text{or} \quad \frac{d^2v}{dx^2} = -\left(\frac{wL^3}{6Ebh_o^3}\right)\frac{1}{x} \quad (3)$$

We make the following substitution

$$k = \frac{wL^3}{6Ebh_o^3} \quad (4)$$

The boundary value problem statement is:

$$\frac{d^2v}{dx^2} = -\frac{k}{x} \quad (5) \qquad v(L) = 0 \quad (6) \qquad \frac{dv}{dx}(L) = 0 \quad (7)$$

Integrating Eq.(5)

$$\frac{dv}{dx} = -k\ln(x) + C_1 \quad (8) \qquad v = -k(x\ln(x) - x) + C_1x + C_2 \quad (9)$$

Substituting x = L in Eq.(8) and using Eq.(6) we obtain,

$$\frac{dv}{dx}(L) = -k\ln(L) + C_1 = 0 \qquad \text{or} \qquad C_1 = k\ln(L) \qquad \textbf{10}$$

Substituting x = L in Eq.(9) and using Eqs.(7) and (10) we obtain,

$$v(L) = -k(L\ln(L) - L) + (k\ln(L))L + C_2 \qquad \text{or} \qquad C_2 = -kL \qquad \textbf{11}$$

The maximum deflection will be at x = 0. Substituting x= 0 inEq.(9) and using Eqs.(11) and (4), we obtain

$v_{max} = C_2$ or

$$\boxed{v_{max} = -\left(\frac{wL^4}{6Ebh_o^3}\right)}$$

7.18 For the tapered circular beam shown, determine the maximum bending normal stress and maximum deflection in terms of E, P, d_o, and L.

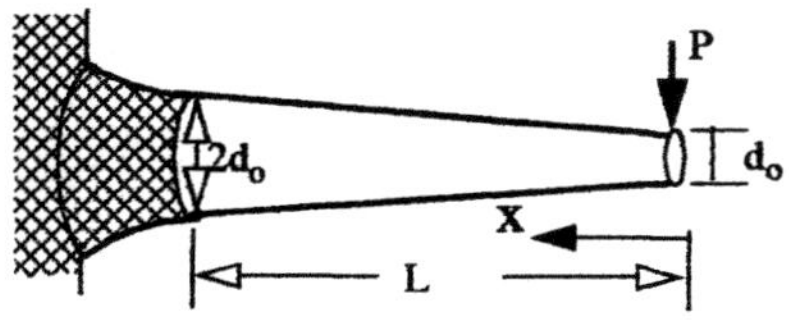

Fig. P7.18

Solution σ_{max}=? v_{max} =?

Fig.(a) shows the free body diagram after an imaginary cut is made at any location x. By equilibrium of moment about point O, we obtain,

$$M_z = -Px \qquad \textbf{1}$$

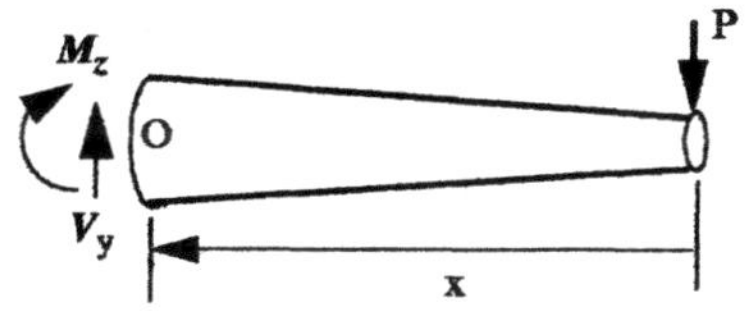

The diameter is a linear function of x and can be written as d = a+bx.At x = 0 d = d_o,hence a = d_o.At x - L, d = 2d_o,thus b = d_o/L.The variation of d can be written as

$$d = d_o + \frac{d_o}{L}x = \frac{d_o}{L}(x+L) \qquad \textbf{2}$$

The area moment of inertia of across-section at any x is,

$$I_{zz} = \frac{\pi}{64}d^4 = \frac{\pi}{64}\frac{d_o^4}{L^4}(x+L)^4 \qquad \textbf{3}$$

The maximum bending normal stress will be at $y = \pm\frac{d}{2}$.The magnitude of maximum bending stress is

$$\sigma = \left|-\frac{M_z y}{I_{zz}}\right| = \frac{(Px)\frac{d}{2}}{\pi d^4/64} = \frac{32}{\pi}P\frac{x}{d^3} \qquad \text{or} \qquad \sigma = \frac{32PL^3}{\pi d_o^3}\frac{x}{(x+L)^3} \qquad \textbf{4}$$

Let the maximum bending normal stress exist at x = x_1.Taking the first derivative of Eq.(4) with respect to x and setting it to zero at x_1, we obtain

$$\left.\frac{d\sigma}{dx}\right|_{x_1} = \left(\frac{32PL^3}{\pi d_o^3}\right)\left[\frac{1}{(x_1+L)^3} - \frac{3x_1}{(x_1+L)^4}\right] = 0 \qquad \text{or} \qquad (x_1+L) - 3x_1 = 0 \qquad \text{or} \qquad x_1 = \frac{L}{2} \qquad \textbf{5}$$

Substituting x= L/2 in Eq.(4) we obtain the maximum bending normal stress as

$$\sigma_{max} = \frac{32PL^3}{\pi d_o^3}\left[\frac{L/2}{(3L/2)^3}\right] \quad \text{or} \qquad \boxed{\sigma_{max} = \frac{128PL}{27\pi d_o^3}}$$

The moment curvature relationship can be written as

$$M_z = EI_{zz}\frac{d^2v}{dx^2} = -Px \text{ or } \frac{d^2v}{dx^2} = -\frac{Px}{E\left\{\frac{\pi}{64}\frac{d_o^4}{L^4}(x+L)^4\right\}} = -\frac{64PL^4}{E\pi d_o^4}\left[\frac{x}{(x+L)^4}\right]$$

. We substitute the following

$$k = \frac{64PL^4}{E\pi d_o^4} \qquad 6$$

We can write x = x+L-L in the numerator and state the boundary value problem as shown below,

$$\frac{d^2v}{dx^2} = -k\left[\frac{1}{(x+L)^3} - \frac{L}{(x+L)^4}\right] \quad 7 \qquad v(L) = 0 \quad 8 \qquad \frac{dv}{dx}(L) = 0 \quad 9$$

Integrating Eq.(5)

$$\frac{dv}{dx} = -k\left[-\frac{1}{2}\frac{1}{(x+L)^2} + \frac{1}{3}\frac{L}{(x+L)^3}\right] + C_1 \quad 10 \qquad v = -k\left[\frac{1}{2}\frac{1}{(x+L)} - \frac{1}{6}\frac{L}{(x+L)^2}\right] + C_1x + C_2 \quad 11$$

Substituting x = L in Eq.(10) and using Eq.(9) we obtain,

$$\frac{dv}{dx}(L) = -k\left[-\frac{1}{2}\left(\frac{1}{4L^2}\right) + \frac{1}{3}\left(\frac{L}{8L^3}\right)\right] + C_1 = 0 \text{ or } -k\left[\frac{-3+1}{24L^2}\right] + C_1 = 0 \quad \text{or} \quad C_1 = -\left(\frac{k}{12L^2}\right) \quad 12$$

Substituting x = L in Eq.(11) and using Eqs.(8) and (20) we obtain,

$$v(L) = -k\left[\frac{1}{2}\left(\frac{1}{2L}\right) - \frac{L}{6}\left(\frac{1}{4L^2}\right)\right] - \frac{kL}{12L^2} + C_2 \text{ or } -\frac{k}{24L}[6-1+2] + C_2 = 0 \quad \text{or} \quad C_2 = \frac{7k}{24L} \quad 13$$

The maximum deflection will be at x = 0.Substituting x= 0 in Eq.(11) and using Eqs.(13) and (6), we obtain:

$$v_{max} = -k\left[\frac{1}{2L} - \frac{1}{6L}\right] + \frac{7k}{24L} = \frac{k}{24L}[-12+4+7] = -\frac{k}{24L} \text{ or } \boxed{v_{max} = -\frac{8PL^3}{3E\pi d_o^4}}$$

7.19 A 2 in x 8 in wooden rectangular cross-section beam is braced at the support using 2 in x 1 in wooden pieces. The Modulus of Elasticity of wood is 2,000 ksi. Determine the maximum bending normal stress and maximum deflection.

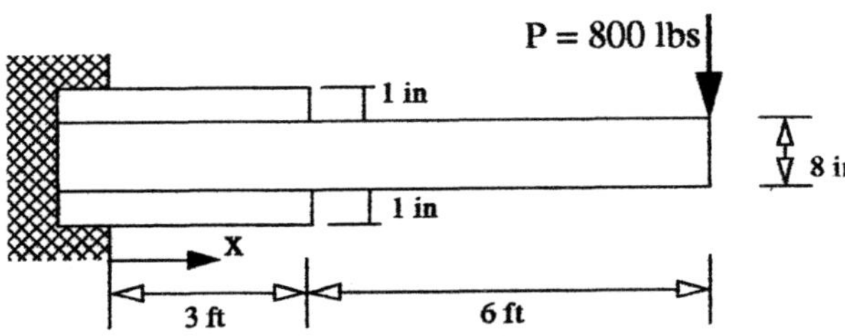

Fig. P7.19

Solution $E = 2000$ psi σ_{max}=? v_{max} =?

Fig.(a) shows the free body diagram after an imaginary cut is made at any location x.The length of the beam is 108 inch.By equilibrium of moment about point O, we obtain,

$$M_z = -800(108 - x) \qquad 1$$

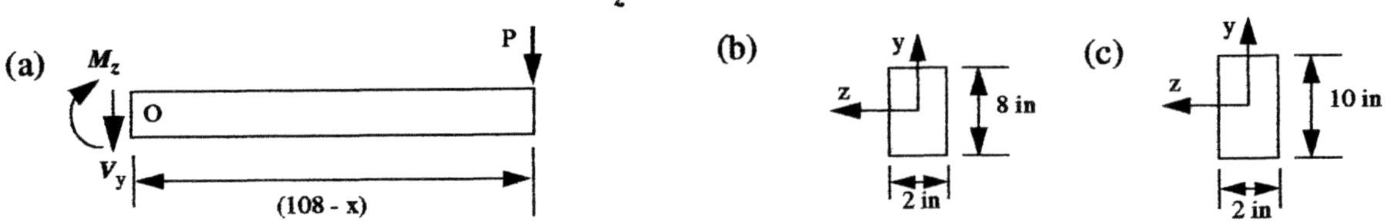

Fig.(b) and Fig.(c) shows the cross-section in segment BC and AB respectively.The area moment of inertia about the 2-axis in the segment AB and BC are

$$I_{AB} = \frac{1}{12}(2)(10)^3 = 166.66\ \text{in}^4 \qquad 0 \le x \le 36\ \text{in}$$ **2**

$$I_{BC} = \frac{1}{12}(2)(8)^3 = 85.33\ \text{in}^4 \qquad 36 \le x \le 108\ \text{in}$$ **3**

The maximum bending normal stress in BC will be on a section just right of Bi.e. at x = 36 inch and at $y = \pm 4$ in .The magnitude of the maximum bending stress in BC is

$$\sigma = \left|-\frac{M_z y}{I_{zz}}\right| = \frac{(800)(72)(4)}{85.33} = 2700\ \text{psi}$$ **4**

The maximum bending normal stress in AB will be at the wall i.e. at x = 0 inch and at $y = \pm 5$ in .The magnitude of the maximum bending stress in AB is

$$\sigma = \left|-\frac{M_z y}{I_{zz}}\right| = \frac{(800)(108)(5)}{166.66} = 2592\ \text{psi}$$ **5**

Comparing Eqs.(4) and (5) we note that the maximum bending stress will be in BC and its value is

$$\boxed{\sigma_{max} = 2.7\ \text{ksi}}$$

The bending rigidity in each segment is

$EI_{AB} = 2(10^6)(166.66) = 333.33(10^6)\ \text{lbs-in}^2$ and $EI_{BC} = 2(10^6)(85.33) = 170.67(10^6)\ \text{lbs-in}^2$

The boundary value problem is

$$\frac{d^2v_1}{dx^2} = \frac{M_z}{EI_{AB}} = -2.4(10^{-6})(108 - x) \qquad 0 \le x \le 36\ \text{in}$$ **6**

$$\frac{d^2v_2}{dx^2} = \frac{M_z}{EI_{BC}} = -4.6875(10^{-6})(108 - x) \qquad 36 \le x \le 108\ \text{in}$$ **7**

$v_1(0) = 0$ **8** $\quad \frac{dv_1}{dx}(0) = 0$ **9** $\quad v_1(36) = v_2(36)$ **10** $\quad \frac{dv_1}{dx}(36) = \frac{dv_2}{dx}(36)$ **11**

Integrating Eq.(6) and (7)

$$\frac{dv_1}{dx} = -2.4(10^{-6})\left(108x - \frac{x^2}{2}\right) + C_1$$ **12**

$$\frac{dv_2}{dx} = -4.6875(10^{-6})\left(108x - \frac{x^2}{2}\right) + C_2$$ **13**

Substituting x = 0 in Eq.(12) and using Eq.(8) we obtain,

$$\frac{dv_1}{dx}(0) = C_1 = 0$$ **14**

Substituting x = 36 in Eq.(12) and (13) and equating to satisfy Eq.(11) we obtain,

$$-2.4(10^{-6})\left(108(36) - \frac{(36)^2}{2}\right) = -4.6875(10^{-6})\left(108(36) - \frac{(36)^2}{2}\right) + C_2 \text{ or}$$

$$C_2 = -(2.4 - 4.6875)(10^{-6})(3240) = 7.406(10^{-3})$$ **15**

Integrating Eq.(12) and (13) and using Eqs. (14) and (15), we obtain,

$$v_1 = -2.4(10^{-6})\left(54x^2 - \frac{x^3}{6}\right) + C_3$$ **16**

$$v_2 = -4.6875(10^{-6})\left(54x^2 - \frac{x^3}{6}\right) + 7.406(10^{-3})x + C_4$$ **17**

Substituting x = 0 in Eq.(16) and using Eq.(8) we obtain,

$$v(0) = C_3 = 0 \qquad 18$$

Substituting x = 36 in Eq.(16) and (17) and equating to satisfy Eq.(10) we obtain,

$$-2.4(10^{-6})\left(54(36)^2 - \frac{(36)^3}{6}\right) = -4.6875(10^{-6})\left(54x^2 - \frac{x^3}{6}\right) + 7.406(10^{-3})x + C_4 \text{ or}$$

$$C_4 = -(2.4 - 4.6875)(10^{-6})(54432) - 266.62(10^{-3}) = -142.11(10^{-3}) \qquad 19$$

The maximum deflection will be at the free end i.e. x = 108 in. Substituting x= 108 into Eq.(17) and using Eqs.(19), we obtain

$$v_{max} = -4.6875(10^{-6})\left(54(108)^2 - \frac{(108)^3}{6}\right) + 7.406(10^{-3})(108) + -142.11(10^{-3}) = -1389.3(10^{-3}) \text{ in or}$$

$$\boxed{v_{max} = -1.4 \text{ in}}$$

Problem Set 7.2

7.20 The displacement in the y-direction in segment AB was found to be as $(x) = (20x^3 - 40x^2)(10^{-6})$ i. If the bending rigidity is 135 (10^6) lbs-in². Determine the reaction force and reaction moment at the wall at A.

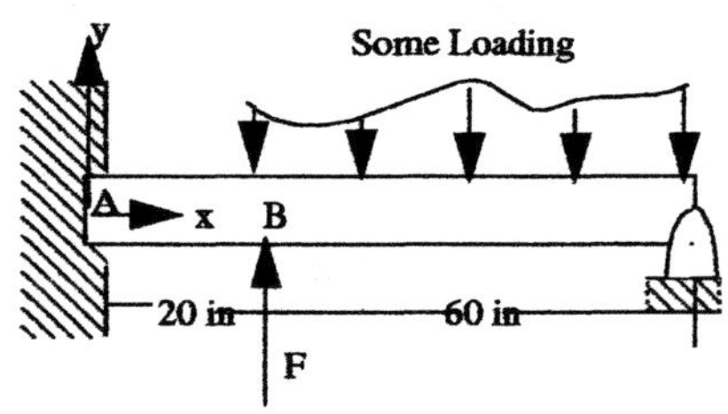

Fig. P7.20

Solution $v(x) = (20x^3 - 40x^2)(10)^{-6}$ in $EI = 135(10^6)$ lb $R_A = ?$ $M_A = ?$

Fig.(a) shows the free body diagram after an imaginary cut is made just right of A i.e. of an infinitesimal distance right of A.

(a)

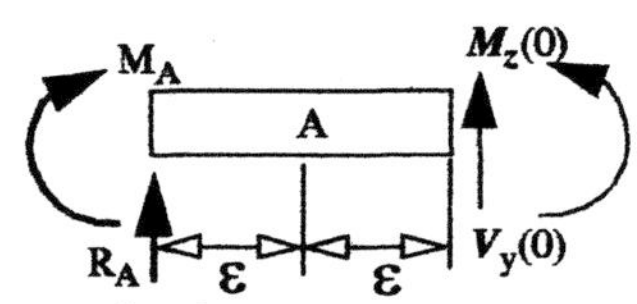

By equilibrium of forces and the moment, we obtain,

$$R_A = -V_y(0) \qquad 1$$

$$M_A = M_z(0) \qquad 2$$

The moment and shear force in AB can be found as shown

$$M_z(x) = EI\frac{d^2v}{dx^2} = EI(120x - 80)(10^{-6}) \qquad 3$$

$$V_y(x) = -\frac{dM_z}{dx} = -120EI(10^{-6}) \qquad 4$$

We can substitute x = 0 in Eqs.(3) and (4) and from Eqs.(1) and (2) determine the reactions as shown

$$R_A = -120EI(10^{-6}) = -(120)(135)(10^6)(10^{-6}) = -16200 \text{ lb} \qquad \boxed{R_A = 16.2 \text{ kips up}}$$

$$M_A = -80EI(10^{-6}) = -(80)(135)(10^6)(10^{-6}) = -10800\text{in lb} \qquad \boxed{M_A = 10.8 \text{ in-kips CCW}}$$

7.21 The displacement in the y direction in section AB is given by $_1(x) = -3(x^4 - 20x^3)(10^{-6})$ i and in BC by $_2(x) = -8(x^2 - 100x + 1600)(10^{-3})$ i. If the bending rigidity is 135 (10^6) lbs-in^2. Determine (a) the reaction force at B and applied moment M_B. (b) the reactions at the wall at A.

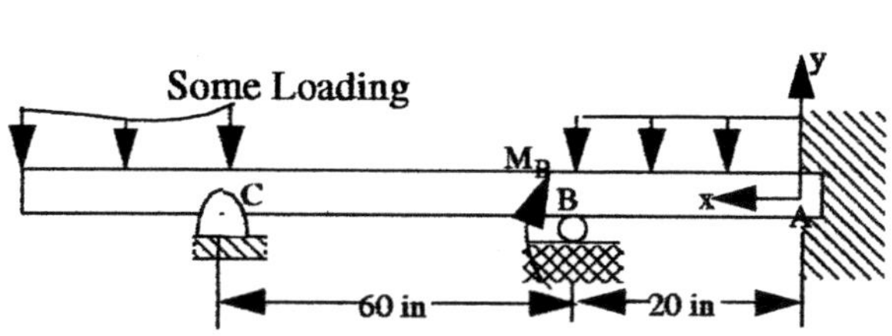

Fig. P7.21

Solution $_1(x) = -3(x^4 - 20x^3)(10^{-6})$ i $\qquad$ $_2(x) = -8(x^2 - 100x + 1600)(10^{-3})$ i

$EI = 135(10^6)$ lb-10^2 $\qquad$ $M_B = ?$ $\qquad$ $R_B = ?$

Fig.(a) shows the free body -diagram after imaginary cuts are made on either side of B.Shear force and moments are drawn as per our sign-convention

(a)

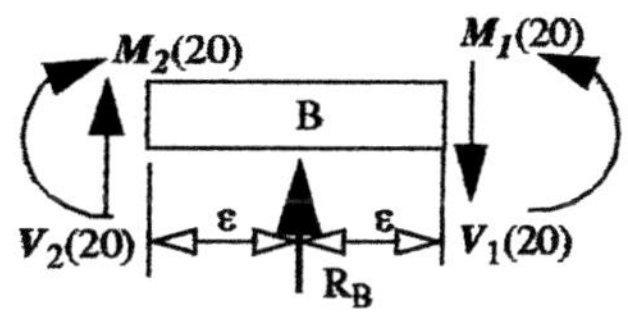

By equilibrium of forces and the moment, we obtain,

$$R_B + V_2(20) - V_1(20) = 0 \quad \text{or} \quad R_B = V_1(20) - V_2(20)$$ 1

$$M_B - M_1(20) + M_2(20) = 0 \quad \text{or} \quad M_B = M_1(20) - M_2(20)$$ 2

The moment and shear force in AB and BC can be found as shown

$$M_1(x) = EI\frac{d^2v_1}{dx^2} = EI[-3(12x^2 - 120x)(10^{-6})]$$ 3

$$V_1(x) = -\frac{d}{dx}(M_1) = EI[-3(24x - 120)(10^{-6})]$$ 4

$$M_2(x) = EI\frac{d^2v_2}{dx^2} = EI[-8(2)](10^{-3}) = -16EI(10^{-3})$$ 5

$$V_2(x) = -\frac{dM_2}{dx} = 0$$ 6

We can substitute x = 20in Eq.(3) through (6) and substituting the results into Eqs.(1) and (2) we obtain

$$R_B = EI[-3(24(20) - 120)] = -(1080(120))(135)(10^6)(10^{-6}) = 145.8(10^3)\text{ lb} \qquad \boxed{R_B = 145.8\text{ kips up}}$$

$$M_B = EI[-3(12(20)^2 - 120(20))(10^{-6})] - -16EI(10^{-3}) = (-7.2 + 16)EI(10^{-3}) = 1180(10^3)\text{ in lb}$$

$$\boxed{M_A = 1180\text{ in-kips}}$$

7.22 Determine the elastic curve and the reaction(s) at A in terms of E, I, P, w, and x.

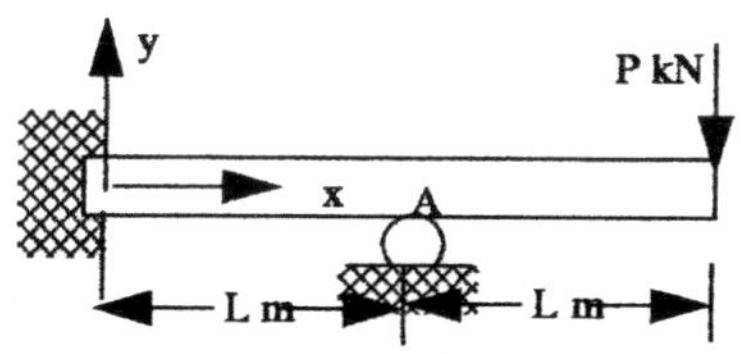

Fig. P7.22

Solution $v(x) = ?$ $\qquad$ $R_A = ?$

Fig.(a) and (b) shows the free body diagram after imaginary cuts are made in sections AB and BC respectively.

(a) 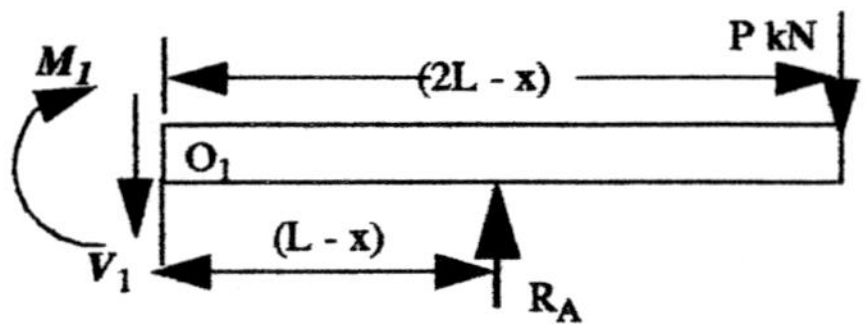

(b) 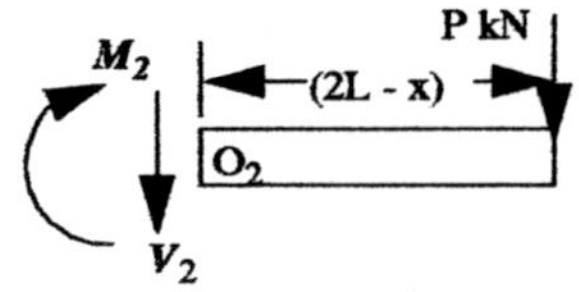

By equilibrium of moment about point O_1, we obtain

$$M_1 + P(2L-x) - R_A(L-x) = 0 \quad \text{or} \quad M_1 = P(x-2L) - R_A(x-L) \qquad 1$$

By equilibrium of moment about point O_2in Fig.(b), we obtain

$$M_2 + P(2L-x) = 0 \quad \text{or} \quad M_2 = P(x-2L) \qquad 2$$

The boundary value problem statement can be written as

$$EI\frac{d^2v_1}{dx^2} = P(x-2L) - R_A(x-L) \quad 3 \qquad EI\frac{d^2v_2}{dx^2} = P(x-2L) \quad 4$$

$$v_1(0) = 0 \quad 5 \qquad \frac{dv_1}{dx}(0) = 0 \quad 6 \qquad v_1(L) = 0 \quad 7 \qquad v_2(L) = 0 \quad 8 \qquad \frac{dv_1}{dx}(L) = \frac{dv_2}{dx}(L) \quad 9$$

Integrating Eq(3) and (4), we obtain,

$$EI\frac{dv_1}{dx} = \frac{P}{2}(x-2L)^2 - \frac{R_A}{2}(x-L)^2 + C_1 \qquad 10$$

$$EI\frac{dv_2}{dx} = \frac{P}{2}(x-2L)^2 + C_2 \qquad 11$$

Substituting x = 0 in Eqs. (10) and using Eq.(6) we obtain

$$EI\frac{dv_1}{dx}(0) = 2PL^2 - \frac{R_A}{2}L^2 + C_1 = 0 \quad \text{or} \quad C_1 = \frac{(R_A - 4P)}{2}L^2 \qquad 12$$

Substituting x = L in Eqs.(10) and (11) and equating to satisfy Eq(9) we obtain

$$\frac{PL^2}{2} + C_1 = \frac{PL^2}{2} + C_2 \quad \text{or} \quad C_1 = C_2 \qquad 13$$

Integrating Eqs.(10) and (11) and using Eq.(13), we obtain,

$$EIv_1 = \frac{P}{6}(x-2L)^3 - \frac{R_A}{6}(x-L)^3 + C_1x + C_3 \qquad 14$$

$$EIv_2 = \frac{P}{6}(x-2L)^3 + C_1x + C_4 \qquad 15$$

Substituting x = 0 in Eq.(14) and using Eq.(5) we obtain

$$EIv_1(0) = -\frac{8P}{6}L^3 - \frac{R_A}{6}L^3 + C_3 \quad \text{or} \quad C_3 = \frac{(8P - R_A)}{6}L^3 \qquad 16$$

Substituting x = L in Eqs.(14) and using Eq.(7) we obtain

$$EIv_1(L) = \frac{-P}{6}L^3 + C_1L + C_3 = 0 \qquad 17$$

Substituting Eqs.(12) and (16) into Eq.(17) we obtain

$$-\frac{P}{6}L^3 + \frac{(R_A - 4P)}{2}L^3 + \frac{(8P - R_A)}{6}L^3 = 0 \quad \text{or} \quad -P + 3R_A - 12P + 8P - R_A = 0 \text{ or } \boxed{R_A = \frac{5P}{2}}$$

Substituting R_A in Eqs.(12) and (16),we obtain

$$C_1 = C_2 = -\frac{3}{4}PL^2 \qquad C_3 = \frac{11}{12}PL^3 \qquad 18$$

Substituting x = L in Eqs(15) and using Eq(8) we obtain

$$EIv_2(L) = \frac{-P}{6}L^3 + C_1L + C_4 = 0 \quad \text{or} \quad \frac{-P}{6}L^3 + \left(\frac{-3P}{4}L^2\right)L + C_4 = 0 \qquad C_4 = \frac{11}{12}PL^3 \qquad 19$$

Substituting Eqs.(18) and (19)in Eqs(14) and (15), we obtain

$$v_1 = \frac{P}{12EI}[2(x-2L)^3 - 5(x-L)^3 - 9L^2x + 11L^3] \qquad 20$$

$$v_2 = \frac{P}{12EI}[2(x-2L)^3 - 9L^2x + 11L^3] \qquad 21$$

$$v(x) = \begin{cases} \frac{P}{12EI}[2(x-2L)^3 - 5(x-L)^3 - 9L^2x + 11L^3] & 0 \le x \le L \\ \frac{P}{12EI}[2(x-2L)^3 - 9L^2x + 11L^3] & L \le x \le 2L \end{cases}$$

7.23 Determine the elastic curve and the reaction(s) at A in terms of E, I, P, w, and x.

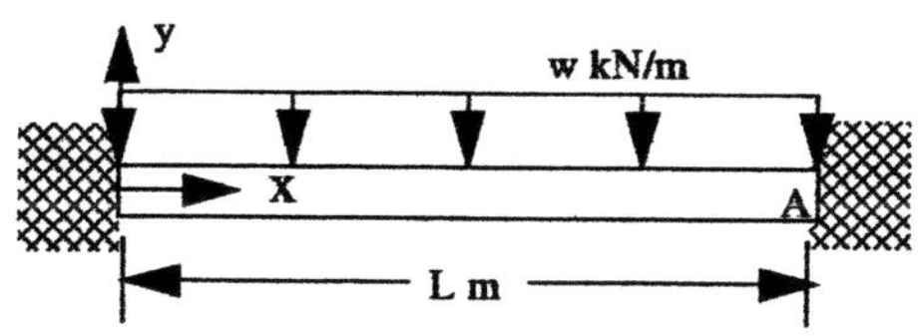

Fig. P7.23

Solution $v(x) = ?$ $R_A = ?$ $M_A = ?$

The boundary value problem is

$$\frac{d^2}{dx^2}\left(EI\frac{d^2v}{dx^2}\right) = -w \qquad 1$$

$$v(0) = 0 \quad 2 \qquad \frac{dv}{dx}(0) = 0 \quad 3 \qquad v(L) = 0 \quad 4 \qquad \frac{dv}{dx}(L) = 0 \quad 5$$

Integrating Eq.(1) four times, we obtain,

$$\frac{d}{dx}\left(EI\frac{d^2v}{dx^2}\right) = -wx + C_1 \qquad 6$$

$$EI\frac{d^2v}{dx^2} = -\frac{wx^2}{2} + C_1x + C_2 \qquad 7$$

$$EI\frac{dv}{dx} = -\frac{wx^3}{6} + \frac{C_1x^2}{2} + C_2x + C_3 \qquad 8$$

$$EIv = -\frac{wx^4}{24} + \frac{C_1x^3}{6} + \frac{C_2x^2}{2} + C_3x + C_4 \qquad 9$$

Substituting x = 0 in Eq.(8) and (9) and using Eq(2)and (3), we obtain

$$EIv(0) = C_3 = 0 \qquad 10$$

$$EIv(0) = C_4 = 0 \qquad 11$$

Substituting x = L in Eq.(8) and (9) and using Eq(4),(5),(10) and (11), we obtain

$$EI\frac{dv}{dx}(L) = -\frac{wL^3}{6} + \frac{C_1L^2}{2} + C_2L \quad \text{or} \quad \frac{C_1}{2}L + C_2 = \frac{wL^2}{6} \qquad 12$$

$$EIv(L) = -\frac{wL^4}{24} + \frac{C_1L^3}{6} + \frac{C_2L^2}{2} = 0 \quad \text{or} \quad \frac{C_1}{6}L + \frac{C_2}{2} = \frac{wL^2}{24} \qquad 13$$

Solving Eqs.(12) and (13), we obtain,

$$C_1 = \frac{wL}{2} \qquad C_2 = -\frac{wL^2}{12} \tag{14}$$

Substituting Eq.(8),(11),(13) and (14) into Eq(9), we obtain

$$v = \frac{w}{EI}\frac{(-x^4 + 2x^3L - L^2x^2)}{24} = -\frac{wx^2}{24EI}(x^2 - 2xL - L^2) \tag{15}$$

or

$$\boxed{v(x) = -\frac{wx^2}{24EI}(x - L)^2}$$

The moment and shear force can be found from Eqs.(15) as shown

$$\boldsymbol{M}_z = EI\frac{d^2v}{dx^2} = \frac{w}{24}(-12x^2 + 12xL - 2L^2) \tag{16}$$

$$\boldsymbol{V}_y = -\frac{d\boldsymbol{M}_z}{dx} = -\frac{w}{24}(-24x + 12L) \tag{17}$$

Substituting x = L in Eq.(16) and (17), we obtain

$$\boldsymbol{M}_z(L) = -\frac{wL^2}{12} \tag{18}$$

$$\boldsymbol{V}_y(L) = \frac{wL}{2} \tag{19}$$

Fig.(a) shows the free body diagram after an imaginary cut is made just left of A.The moment and shear force are drawn as per our sign convention.

(a)

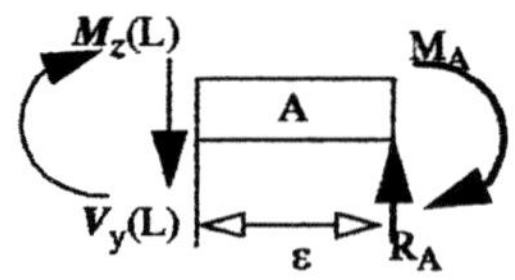

By equilibrium of forces and the moment and using Eqs.(18) and (19), we obtain,

$R_A = \boldsymbol{V}_y(L)$ or $\boxed{R_A = \frac{wL}{2} \text{ up}}$

$M_A = -\boldsymbol{M}_z(L)$ or $\boxed{M_A = \frac{wL^2}{12} \text{ CW}}$

7.24 Determine the slope at x =L and the reaction moment at the left wall in terms of E, I, w, and L.

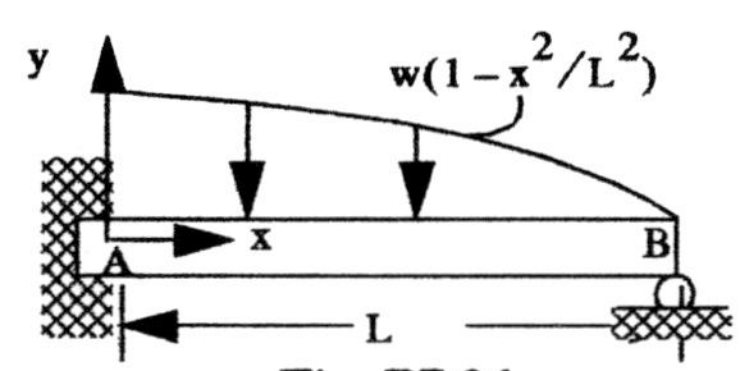

Fig. P7.24

Solution $\frac{dv}{dx}(L) = ?$ $R_A = ?$ $M_A = ?$

The boundary value problem can be written as:

$$\frac{d^2}{dx^2}\left(EI\frac{d^2v}{dx^2}\right) = -w\left(1 - \frac{x^2}{L^2}\right) \tag{1}$$

$v(0) = 0$ **2** $\quad \frac{dv}{dx}(0) = 0$ **3** $\quad v(L) = 0$ **4** $\quad EI\frac{d^2v}{dx^2}(L) = 0$ **5**

Integrating Eq.(1) four times, we obtain,

$$\frac{d}{dx}\left(EI\frac{d^2v}{dx^2}\right) = -w\left(x - \frac{x^3}{3L^2}\right) + C_1 \qquad 6$$

$$EI\frac{d^2v}{dx^2} = -w\left(\frac{x^2}{2} - \frac{x^4}{12L^2}\right) + C_1x + C_2 \qquad 7$$

$$EI\frac{dv}{dx} = -w\left(\frac{x^3}{6} - \frac{x^5}{60L^2}\right) + \frac{C_1x^2}{2} + C_2x + C_3 \qquad 8$$

$$EIv = -w\left(\frac{x^4}{24} - \frac{x^6}{360L^2}\right) + \frac{C_1x^3}{6} + \frac{C_2x^2}{2} + C_3x + C_4 \qquad 9$$

Substituting x = 0 in Eq.(8) and (9) and using Eq(2)and (3), we obtain

$$EI\frac{dv}{dx}(0) = C_3 = 0 \qquad 10$$

$$EIv(0) = C_4 = 0 \qquad 11$$

Substituting x = L in Eq.(7) and (9) and using Eq(4),(5),(10) and (11), we obtain

$$EI\frac{d^2v}{dx^2}(L) = -w\left(\frac{L^2}{2} - \frac{L^2}{12}\right) + C_1L + C_2 \quad \text{or} \quad C_1L + C_2 = \frac{5wL^2}{12} \qquad 12$$

$$EIv(L) = -w\left(\frac{L^4}{24} - \frac{L^4}{360}\right) + \frac{C_1L^3}{6} + \frac{C_2L^2}{2} = 0 \quad \text{or} \quad \frac{C_1}{6}L + \frac{C_2}{2} = \frac{7wL^2}{180} \qquad 13$$

Solving Eqs.(12) and (13), we obtain,

$$C_1 = \frac{61wL}{120} \quad \text{and} \quad C_2 = -\frac{11wL^2}{120} \qquad 14$$

Substituting x = L in Eq.(8) and using Eq(14),(10) and (11), we obtain

$$EI\frac{dv}{dx}(L) = -w\left(\frac{L^3}{6} - \frac{L^3}{60}\right) + \frac{61wL^3}{240} - \frac{11wL^3}{120} = \frac{wL^3}{80} \text{ or } \quad \boxed{\frac{dv}{dx}(L) = \frac{wL^3}{80EI}}$$

Fig.(a) shows the free body diagram after an imaginary cut is made just left of A.The bending moment and shear force are drawn as per our sign convention.

(a)

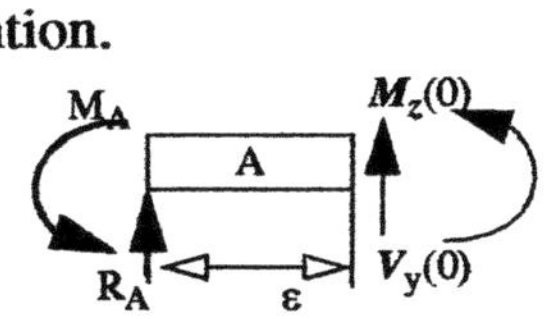

By equilibrium of forces and the moment, we obtain,

$$R_A = -V_y(0) \qquad 15$$

$$M_A = -M_z(0) \qquad 16$$

Substituting x = 0 in Eq.(6) and using the definition of shear and Eq. 15, we obtain

$$V_y(0) = -\frac{d}{dx}\left(EI\frac{d^2v}{dx^2}(L)\right)\bigg|_{x=0} = -C_1 = -\frac{61wL}{120} = -R_A \text{ or } \quad \boxed{R_A = \frac{61wL}{120} \text{ up}}$$

Substituting x = 0 in Eq.(7) and using the definition of shear and Eq. 16, we obtain

$$M_z(0) = EI\frac{d^2v}{dx^2}(L)\bigg|_{x=0} = C_2 = -\frac{11wL^2}{120} = -M_A \text{ or } \quad \boxed{M_A = \frac{11wL^2}{120} \text{ CW}}$$

7.25 Determine the deflection and moment reaction at x = L in terms of E, I, w, and L.

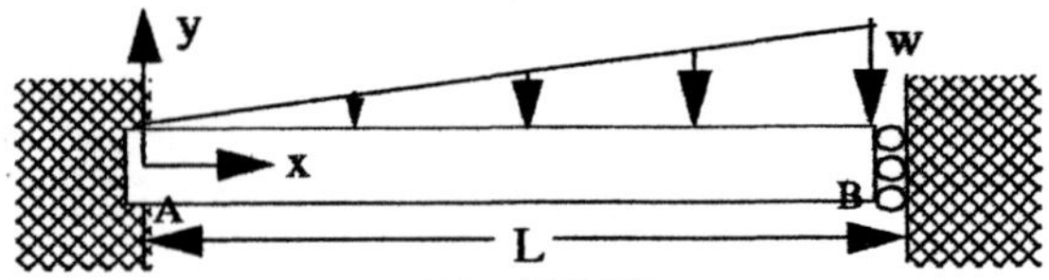

Fig. P7.25

Solution $v(L) = ?$ $M_B = ?$

The boundary value problem is,

$$\frac{d^2}{dx^2}\left(EI\frac{d^2v}{dx^2}\right) = -\frac{wx}{L} \quad 1$$

$$v(0) = 0 \quad 2 \qquad \frac{dv}{dx}(0) = 0 \quad 3 \qquad \frac{dv}{dx}(L) = 0 \quad 4 \qquad V_y(L) = -\frac{d}{dx}\left(EI\frac{d^2}{dx^2}v(L)\right)\bigg|_{x=L} = 0 \quad 5$$

Integrating Eq.(1), we obtain,

$$\frac{d}{dx}\left(EI\frac{d^2v}{dx^2}\right) = -\frac{wx^2}{2L} + C_1 \quad 6$$

$$EI\frac{d^2v}{dx^2} = -\frac{wx^3}{6L} + C_1x + C_2 \quad 7$$

$$EI\frac{dv}{dx} = -\frac{wx^4}{24L} + \frac{C_1x^2}{2} + C_2x + C_3 \quad 8$$

$$EIv = -\frac{wx^5}{120L} + \frac{C_1x^3}{6} + \frac{C_2x^2}{2} + C_3x + C_4 \quad 9$$

Substituting x = L in Eq.(6) and using Eq(5), we obtain

$$V_y(L) = -\left[-\frac{wL}{2} + C_1\right] = 0 \qquad \text{or} \qquad C_1 = \frac{wL}{2} \quad 10$$

Substituting x = 0 in Eq.(8) and using Eq. (3), we obtain

$$EI\frac{dv}{dx}(0) = C_3 = 0 \quad 11$$

Substituting x = L in Eq.(8) and using Eq(4),(10) and (11), we obtain

$$EI\frac{dv}{dx}(L) = -\frac{wL^3}{24} + \left(\frac{wL}{2}\right)\frac{L^2}{2} + C_2L = 0 \quad \text{or} \quad wL^3\frac{(-1+6)}{24} + C_2L = 0 \quad \text{or} \quad C_2 = -\frac{5}{24}wL^2 \quad 12$$

Substituting x = 0 in Eq. (9) and using Eq(2), we obtain

$$EIv(0) = C_4 = 0 \quad 13$$

Substituting Eqs.(10), (11), (12) and (13) into Eqs.(9), we obtain the equation elastic curve as

$$v = \frac{w}{240EIL}[-2x^5 + 20L^2x^3 - 25L^3x^2] \quad 14$$

Substituting x = L in Eq.(14), we obtain: $v(L) = \frac{w(L^5)}{240EIL}[-2+20-25]$ or $\boxed{v = \frac{-7wL^4}{240EI}}$

Fig.(a) shows the free body diagram after an imaginary cut is made just left of B

(a)
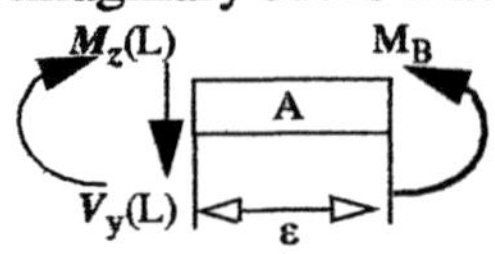

By equilibrium of moment we obtain: $M_B = -M_z(L)$. The moment of x = L can be found from Eq.(14) as

$$M_z(L) = EI\frac{d^2v}{dx^2}(L) = \frac{w}{240EIL}[-40L^3 + 120L^3 - 50L^3] = \frac{30wL^3}{240EIL} = -M_B \text{ or } \boxed{M_B = \frac{wL^2}{8EI} \text{ CW}}$$

7.26 Determine the deflection and slope at x = L in terms of E, I, w, and L.

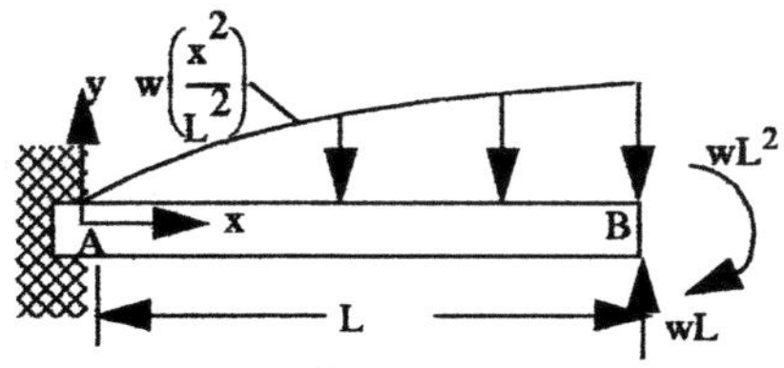

Fig. P7.26

Solution $v(L) = ?$ $\frac{dv}{dx}(L) = ?$

Fig.(a) shows the free body diagram after an imaginary cut is made just left of B

(a)

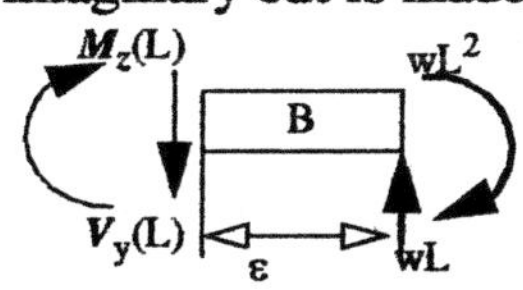

By equilibrium of forces and the moment, we obtain, $V_y(L) = wL$ and $M_z(L) = -wL^2$

The boundary value problem is,

$$\frac{d^2}{dx^2}\left(EI\frac{d^2v}{dx^2}\right) = -\frac{wx^2}{L^2} \quad 1$$

$$v(0) = 0 \quad 2 \qquad \frac{dv}{dx}(0) = 0 \quad 3$$

$$M_z(L) = EI\frac{d^2v}{dx^2}(L) = -wL^2 \quad 4 \qquad V_y(L) = -\frac{d}{dx}\left(EI\frac{d^2v}{dx^2}\right)\bigg|_{x=L} = wL \quad 5$$

Integrating Eq.(1), we obtain

$$\frac{d}{dx}\left(EI\frac{d^2v}{dx^2}\right) = -\frac{wx^3}{3L^2} + C_1 \quad 6$$

Substituting x = L in Eq.(6) and using Eq.(5), we obtain

$$V_y(L) = -\left[-\frac{wL}{3} + C_1\right] = wL \quad \text{or} \quad C_1 = \frac{-2wL}{3} \quad 7$$

Substituting Eq.(7) in Eq.(6) and integrating we obtain,

$$EI\frac{d^2v}{dx^2} = -\frac{wx^4}{12L^2} - \frac{2wL}{3}x + C_2 \quad 8$$

Substituting x = L in Eq.(8) and using Eq.(4), we obtain

$$M_z(L) = -\frac{wL^2}{12} - \frac{2wL^2}{3} + C_2 = -wL^2 \quad \text{or} \quad C_2 = \frac{-wL^2}{4} \quad 9$$

Substituting Eq.(9) in Eq.(8) and integrating we obtain

$$EI\frac{dv}{dx} = -\frac{wx^5}{60L^2} - \frac{wLx^2}{3} - \frac{wL^2}{4} + C_3 \quad 10$$

Substituting x = 0 in Eq.(10) and using Eq. (3), we obtain

$$EI\frac{dv}{dx}(0) = C_3 = 0 \qquad \textbf{11}$$

Substituting x = L in Eq.(10) and using Eq (11), we obtain

$$EI\frac{dv}{dx}(L) = -\frac{wL^3}{60}(-1-20+15) \qquad \text{or} \qquad \boxed{\frac{dv}{dx}(L) = -\frac{3wL^3}{5EI}}$$

Substituting Eq.(11) in Eq.(10) and integrating we obtain,

$$EIv = -\frac{wx^6}{360L^2} - \frac{wLx^3}{9} - \frac{wL^2x^2}{8} + C_4 \qquad \textbf{12}$$

Substituting x = 0 in Eq.(12) and using Eq. (2), we obtain

$$EIv(0) = C_4 = 0 \qquad \textbf{13}$$

Substituting x = L in Eq.(12) and using Eq (13), we obtain

$$EIv(L) = \frac{wL^4}{360}(-1-40-45) \qquad \text{or} \qquad \boxed{v(L) = -\frac{43wL^4}{180EI}}$$

7.27 A linear spring that has a spring constant K is attached to a beam at the end as shown. In terms of w, E, I, L and K. Write the boundary value problem but do not integrate or solve.

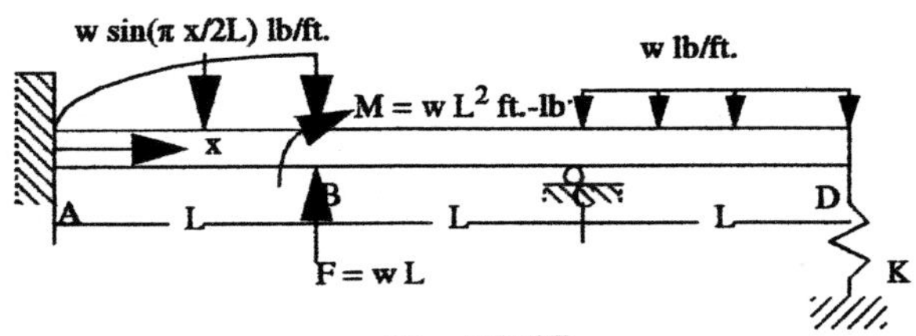

Fig. P7.27

Solution Boundary Value Problem

Fig.(a) shows the free body diagram after an imaginary cut is made just before and after of B.Fig.(b) shows the free body diagram after an imaginary cut is made just left of D.

(a)

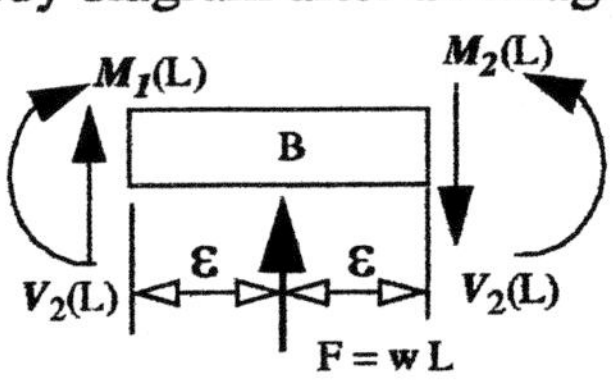

(b)

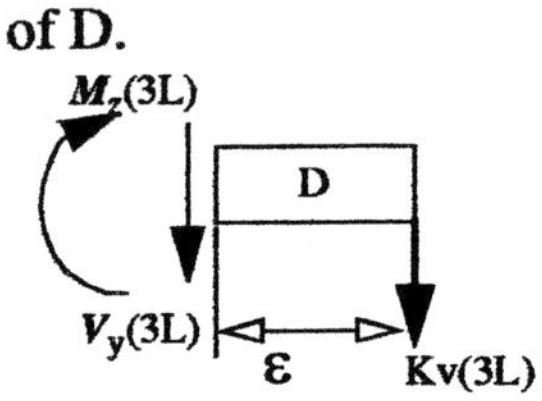

By equilibrium of forces and the moment in Fig.(a), we obtain,

$$V_2(L) - V_1(L) + wL = 0 \qquad \text{or} \qquad V_2(L) - V_1(L) = -wL \qquad \textbf{1}$$

$$M_2(L) - M_1(L) - wL^2 = 0 \qquad \text{or} \qquad M_2(L) - M_1(L) = wL^2 \qquad \textbf{2}$$

By equilibrium of forces and the moment in Fig.(b), we obtain,

$$V_3(3L) = -Kv_3(3L) \qquad \textbf{3}$$

$$M_3(3L) = 0 \qquad \textbf{4}$$

The boundary value problem is

$$\frac{d^2}{dx^2}\left(EI\frac{d^2v_1}{dx^2}\right) = -w\sin\frac{\pi x}{2L} \qquad 0 \le x < L \qquad \textbf{5}$$

$$\frac{d^2}{dx^2}\left(EI\frac{d^2v_2}{dx^2}\right) = 0 \qquad L < x < 2L \qquad \textbf{6}$$

$$\frac{d^2}{dx^2}\left(EI\frac{d^2v_3}{dx^2}\right) = -w \qquad 2L < x \le 3L \qquad 1$$

$$v_1(0) = 0 \quad 2 \qquad \frac{dv_1}{dx}(0) = 0 \quad 3 \qquad v_1(L) = v_2(L) \quad 4 \qquad \frac{dv_1}{dx}(L) = \frac{dv_2}{dx}(L) \quad 5$$

$$EI\frac{d^2v_2}{dx^2}(L) - EI\frac{d^2v_1}{dx^2}(L) = wL^2 \quad 6 \qquad -\frac{d}{dx}\left(EI\frac{d^2}{dx^2}v_2(L)\right)\bigg|_{x=L} + \frac{d}{dx}\left(EI\frac{d^2}{dx^2}v_1(L)\right)\bigg|_{x=L} = -wL \quad 7$$

$$v_2(2L) = 0 \quad 8 \qquad v_3(2L) = 0 \quad 9 \qquad \frac{dv_2}{dx}(2L) = \frac{dv_3}{dx}(2L) \quad 10 \qquad EI\frac{d^2v_2}{dx^2}(2L) = EI\frac{d^2v_3}{dx^2}(2L) \quad 11$$

$$EI\frac{d^2v_3}{dx^2}(3L) = 0 \quad 12 \qquad -\frac{d}{dx}\left(EI\frac{d^2v_3}{dx^2}\right)\bigg|_{x=3L} = -Kv_3(3L) \quad 13$$

7.28 The beam and loading shown was the first statically indeterminate beam for which solution was obtained by Navier. Show that Navier's solution for the reaction at A that is given below is correct.

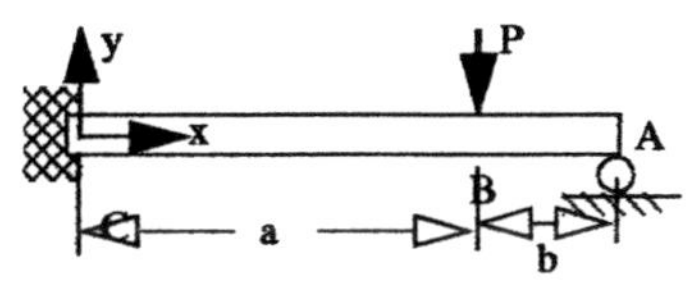

Fig. P7.28

$R_A = Pa^2(3L - a)/(2L^3)$ where, $L = (a + b)$.

Solution

Fig.(a)and (b) shows the free body diagram after imaginary cut are made in segment CB and BA.

(a) 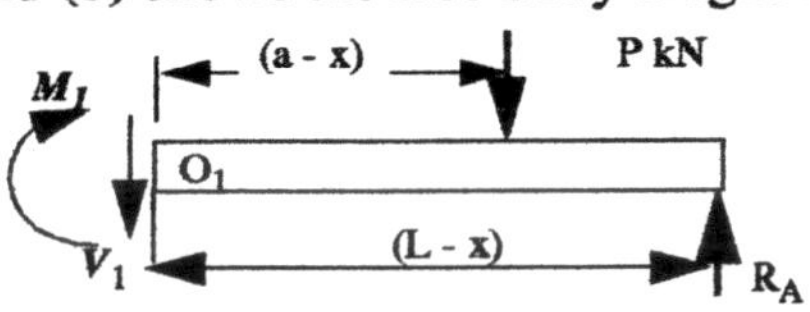

(b) 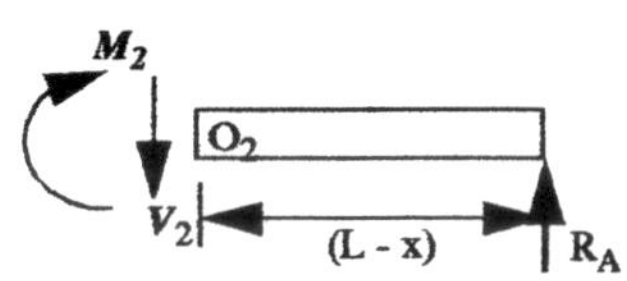

By equilibrium of moment about point O_1in Fig.(a), we obtain,

$M_1 + P(a - x) - R_A(L - x) = 0$ or $M_1 = P(x - a) - R_A(x - L)$

By equilibrium of moment about point O_2in Fig.(b), we obtain,

$M_2 - R_A(L - x) = 0$ or $M_2 = -R_A(x - L)$

The boundary value problem statement can be written as,

$$EI\frac{d^2v_1}{dx^2} = P(x - a) - R_A(x - L) \qquad 1$$

$$EI\frac{d^2v_2}{dx^2} = -R_A(x - L) \qquad 2$$

$$v_1(0) = 0 \quad 3 \qquad \frac{dv_1}{dx}(0) = 0 \quad 4 \qquad v_2(L) = 0 \quad 5$$

$$v_1(a) = v_2(a) \quad 6 \qquad \frac{dv_1}{dx}(a) = \frac{dv_2}{dx}(a) \quad 7$$

Integrating Eqs.(1) and (2), we obtain

$$EI\frac{dv_1}{dx} = \frac{P}{2}(x-a)^2 - \frac{R_A}{2}(x-L)^2 + C_1 \qquad 8$$

$$EI\frac{dv_2}{dx} = \frac{R_A}{2}(x-L)^2 + C_2 \qquad 9$$

Substituting x = a in Eq. (9) and equating to satisfy Eq(7) we obtain

$$-\frac{R_A}{2}(a-L)^2 + C_1 = -\frac{R_A}{2}(a-L)^2 + C_2 \quad \text{or} \quad C_1 = C_2 \qquad 10$$

Integrating Eqs.(8) and (9) and using Eq.(10), we obtain,

$$EIv_1 = \frac{P}{6}(x-a)^3 - \frac{R_A}{6}(x-L)^3 + C_1x + C_3 \qquad 11$$

$$EIv_2 = -\frac{R_A}{6}(x-L)^3 + C_1x + C_4 \qquad 12$$

Substituting x = a in Eqs. (11) and (12) and equating to satisfy Eq(6) we obtain

$$-\frac{R_A}{2}(a-L)^3 + C_1a + C_3 = -\frac{R_A}{2}(a-L)^3 + C_1a + C_4 \quad \text{or} \quad C_3 = C_4 \qquad 13$$

Substituting x = 0 in Eqs.(8) and (11) and using Eqs.(3) and (4), we obtain

$$EI\frac{dv_1}{dx}(0) = \frac{P}{2}a^2 - \frac{R_A}{2}L^2 + C_1 = 0 \quad \text{or} \quad C_1 = \frac{R_AL^2 - Pa^2}{2} \qquad 14$$

$$EIv_1(0) = -\frac{P}{6}a^3 + \frac{R_A}{6}L^3 + C_3 = 0 \quad \text{or} \quad C_3 = \frac{Pa^3 - R_AL^3}{6} \qquad 15$$

Substituting x = L in Eq. (12) and using Eqs. (5), (13), (14) and (15),we obtain

$$v_2(L) = C_1L + C_3 = 0 \quad \text{or} \quad \frac{R_AL^2 - Pa^2}{2}L + \frac{Pa^3 - R_AL^3}{6} = 0 \quad \text{or} \qquad 16$$

$$3R_AL^3 + R_AL^3 = 3Pa^3L - Pa^3 \quad \text{or} \quad \boxed{R_A = \frac{Pa^2}{2L^3}(3L-a)}$$

7.29 Jacob Bernoulli incorrectly assumed that the neutral axis was tangent to the concave side of the curve and obtained the following relationship:

$$\left(\frac{Ebh^3}{3}\right)\left(\frac{1}{R}\right) = Px$$

where, R is the radius of curvature of the beam at any location x. Bernoulli did assume plane sections remain plane during deformation. Derive the above equation based on Bernoulli's assumptions and show that it is incorrect by a factor of four.

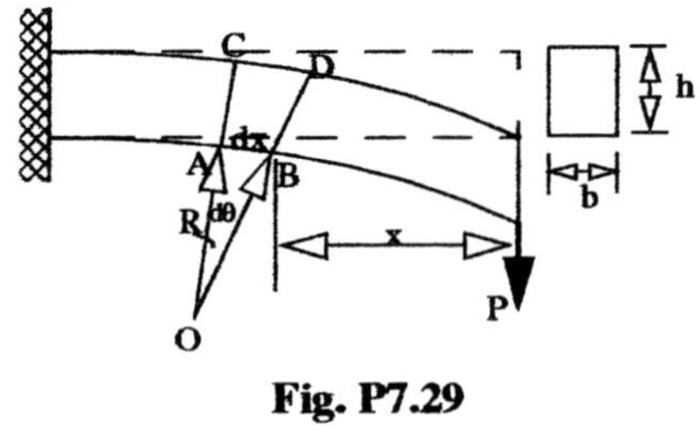

Fig. P7.29

Solution

Let R_c represents the correct radius of curvature. Noting that $\frac{d^2v_2}{dx^2}$ is positive concave upwards, we have the following relationship

$$R_c = -\frac{d^2v}{dx^2} \qquad 1$$

The moment of inertia of the cross section is

$$I_{zz} = \frac{1}{12}bh^3 \qquad 2$$

Fig.(a) shows the free body diagram after imaginary cut

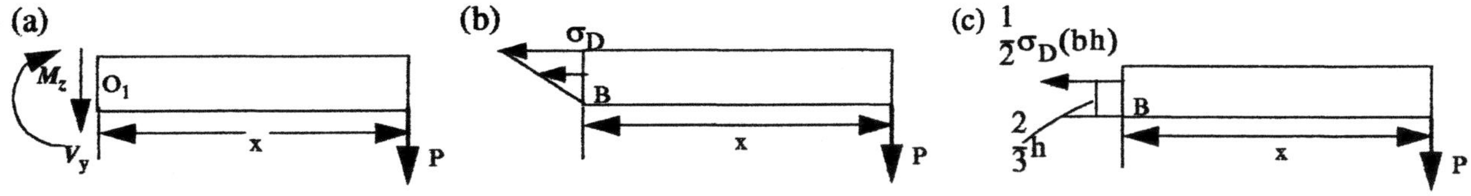

By equilibrium of moment about point O in Fig.(a), we obtain,

$$M_z + Px = 0 \qquad \text{or} \qquad M_z = -Px \qquad 3$$

Substituting Eqs(1), (2) and (3) in moment curvature equation, we obtain

$$EI_{zz}\frac{d^2v}{dx^2} = M_z \qquad \text{or} \qquad \frac{Ebh^3}{12}\frac{1}{R_c} = Px \qquad 4$$

To derive Bernoulli's equation we consider AB = R dθ and CD = (R+h)dθ.The strain in AB is zero, hence AB represents the original length that was equal to the original length of CD. Thus the strain in CD is,

$$\varepsilon_D = \frac{CD - AB}{AB} = \frac{(R+h)d\theta - Rd\theta}{Rd\theta} = \frac{h}{R} \qquad 5$$

The normal stress at D is

$$\sigma_D = E\varepsilon_D = \frac{Eh}{R} \qquad 6$$

As plane section remain plane, the bending normal strain and stress vary linearly as shown in Fig.(b). In Fig.(c) the stress is replaced by an equivalent force. By moment equilibrium about point B in Fig.(c), we have

$$\left(\frac{1}{2}\sigma_D bh\right)\left(2\frac{h}{3}\right) = Px \qquad 7$$

Substituting Eq.(6) in Eq.(7), we obtain

$$\frac{1}{3}bh^2\left(\frac{Eh}{R}\right) = Px \qquad \text{or} \qquad \boxed{\left(\frac{Ebh^3}{3R}\right)\left(\frac{1}{R}\right) = Px} \qquad 8$$

Comparing Eqs.(7) and (8) we note that Eq.(8) is incorrect by a factor of four.
Note:Fig.(c) shows that forces in the x- direction are not in equilibrium due to the incorrect location of neutral axis.

7.30 Clebsch considered a beam loaded by several concentrated forces P_i placed at a location x_i.He obtained the second order differential equation between the concentrated forces. By integration he obtained the slope and deflection as given below and concluded that all C_i's are equal and all D_i's are equal. Show that the conclusion of Clebsch is correct. 1

For $x_i < x_i < x_{i+1}$

$$EI\frac{d^2v}{dx^2} = Rx - \sum_{j=1}^{i} P_j(x - x_j)$$

$$EI\frac{dv}{dx} = R\frac{x^2}{2} - \sum_{j=1}^{i} P_i\frac{(x - x_i)^2}{2} + C_i$$

$$EIv = R\frac{x^3}{6} - \sum P_i\frac{(x - x_i)^3}{6} + C_i x + D_i$$

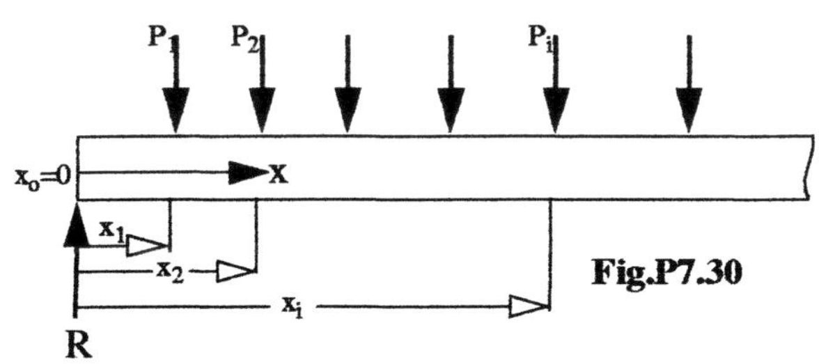

Fig.P7.30

Solution

Fig.(a) shows the free body diagram after imaginary cut is made in the interval between x_i and x_{i+1}.

(a)

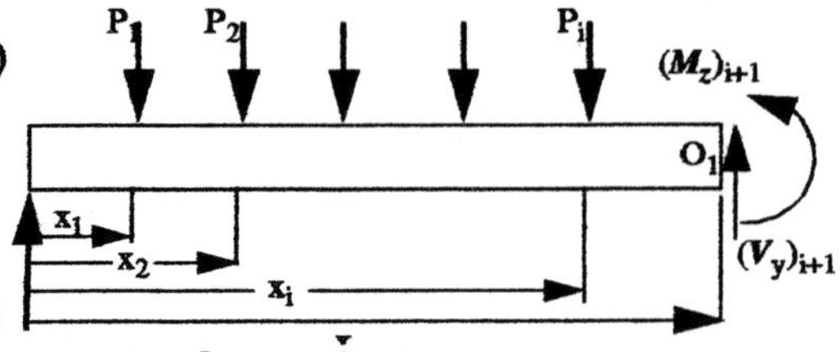

By equilibrium of f moment about point O, we obtain,

$$(M_z)_i = Rx - P_1(x - x_1) - P_2(x - x_2) \ldots\ldots - P_i(x - x_i) \qquad \text{or} \qquad (M_z)_i = Rx - \sum_{j=1}^{i} P_i(x - x_i)$$

The boundary value problem is,

$$EI\frac{d^2 v_i}{dx^2} = Rx - \sum_{j=1}^{i} P_j(x - x_i) \qquad i \geq 1 \qquad (1)$$

$$v_{i-1}(x_i) = v_i(x_i) \qquad i \geq 1 \qquad (2)$$

$$\frac{dv_{i-1}}{dx}(x_i) = \frac{dv_i}{dx}(x_i) \qquad i \geq 1 \qquad (3)$$

Integrating Eq(1), we obtain,

$$EI\frac{dv_i}{dx}(x) = \frac{Rx^2}{2} - \sum_{j=1}^{i} P_j\frac{(x - x_i)^2}{2} + C_i \qquad i \geq 1 \qquad (4)$$

Substituting $x = x_i$ in the $i^{th}$and i-1 slope representation of Eq.(4),we obtain

$$EI\frac{dv_{i-1}}{dx}(x_i) = \frac{Rx_i^2}{2} - \sum_{j=1}^{i-1} P_j\frac{(x_i - x_j)^2}{2} + C_{i-1} \qquad i \geq 1 \qquad (5)$$

$$EI\frac{dv_i}{dx}(x_i) = \frac{Rx_i^2}{2} - \sum_{j=1}^{i} P_j\frac{(x - x_i)^2}{2} + C_i \qquad i \geq 1 \qquad (6)$$

Substituting Eqs (5) and (6) into Eq.(3) we obtain

$$\boxed{C_{i-1} = C_i \qquad i \geq 1} \qquad (7)$$

Integrating Eq(4), we obtain,

$$EIv_i(x) = \frac{Rx_i^3}{6} - \sum_{j=1}^{i} P_j\frac{(x - x_i)^3}{6} + C_i x + D_i \qquad (8)$$

Substituting $x = x_i$ in the $i^{th}$and i-1 displacement representation of Eq.(4),we obtain

$$EIv_{i-1}(x_i) = \frac{Rx_i^3}{6} - \sum_{j=1}^{i-1} P_j\frac{(x_i - x_j)^3}{6} + C_{i-1}x_i + D_{i-1} \qquad (9)$$

$$EIv_i(x_i) = \frac{Rx_i^3}{6} - \sum_{j=1}^{i} P_j\frac{(x_i - x_j)^3}{6} + C_i x_i + D_i = \frac{Rx_i^3}{6} - \sum_{j=1}^{i-1} P_j\frac{(x_i - x_j)^3}{6} + C_i x_i + D_i \qquad (10)$$

Substituting Eqs (9) and (10) into Eq.(2) we obtain

$$C_{i-1}x_i + D_{i-1} = C_i x_i + D_i \qquad i \geq 1 \qquad 11$$

Substituting Eqs (7) into Eq.(11) we obtain

$$\boxed{D_{i-1} = D_i \qquad i \geq 1} \qquad 12$$

7.31 A beam resting on an elastic foundation has a distributed spring force that depends upon the deflections at a point acting as shown. Show the differential equation governing the deflection of the beam is as given below:

$$\frac{d^2}{dx^2}\left(EI_{zz}\frac{d^2v}{dx^2}\right) + kv = p \qquad 7.7$$

where k is the foundation modulus i.e., spring constant per unit length.

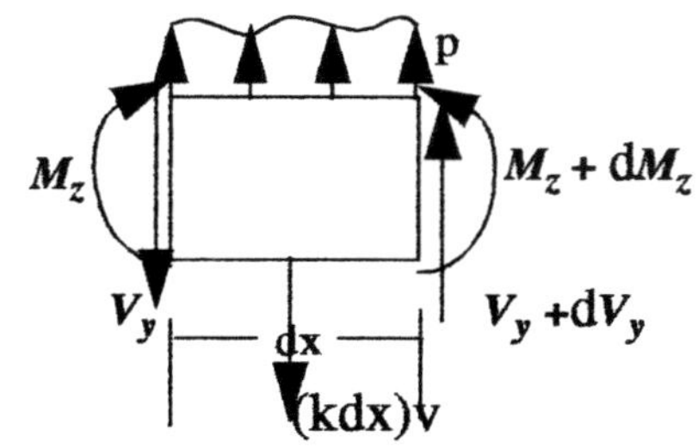

Fig. P7.31. Elastic Foundation Effect.

Solution

In the Fig.(a) the distributed force is replaced by an equivalent force

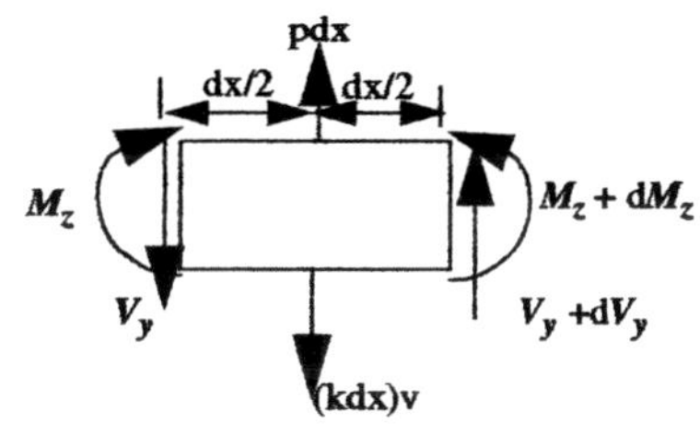

By the force equilibrium in the y-direction

$$dV_y + pdx - (kdx)v = 0 \qquad \text{or} \qquad \frac{dV_y}{dx} - kv = -p \qquad 1$$

By moment equilibrium about point O: $M_z + dM_z + (V_y + dV_y)\frac{dx}{2} + V_y\frac{dx}{2} - M_z = 0$

Neglecting the second order term ($dV_y dx$) we obtain

$$dM_z + V_y dx = 0 \qquad \text{or} \qquad V_y = -\frac{dM_z}{dx} \qquad 2$$

Substituting $M_z = EI_{zz}\frac{d^2v}{dx^2}$ and Eq.(2) into Eq.(1) we obtain

$$-\frac{d^2}{dx^2}\left(EI_{zz}\frac{d^2v}{dx^2}\right) - kv = -p \qquad \text{or} \qquad \boxed{\frac{d^2}{dx^2}\left(EI_{zz}\frac{d^2v}{dx^2}\right) + kv = p}$$

7.32 To account for shear, the assumption of planes remaining perpendicular to the axis of the beam (Assumption 3 in Section 6.2) is dropped, and it is assumed that plane rotates by the angle ψ from the vertical. This yields the following displacement equations: $u = -y\psi(x) \qquad v = v(x)$. The rest of the derivation is as before. Show the following equations apply:

$$\frac{d}{dx}\left[GA\left(\frac{dv}{dx} - \psi\right)\right] = -p$$

$$\frac{d}{dx}\left(EI_{zz}\frac{d\psi}{dx}\right) = -GA\left(\frac{dv}{dx} - \psi\right) \qquad 7.8$$

where, A is the area of cross-section and G is the Shear Modulus of Elasticity. Beams governed by above equations are called *Timoshenko Beams*.

Solution $u = -y\psi(x) \qquad v = v(x)$

The normal strain and shear strain can be written as

$$\varepsilon_{xx} = \frac{\partial u}{\partial x} = -y\frac{d}{dx}\psi(x) \qquad 3$$

$$\gamma_{xy} = \frac{\partial u}{\partial y} + \frac{\partial v}{\partial x} = \frac{dv}{dx} - \psi(x) \qquad 4$$

By Hooke's Law, we have

$$\sigma_{xx} = E\varepsilon_{xx} = -Ey\frac{d}{dx}\psi(x) \qquad 5$$

$$\tau_{xy} = G\gamma_{xy} = G\left(\frac{dv}{dx} - \psi(x)\right) \qquad 6$$

The equivalent internal shear force and moment can be written as shown below.we assume material is homogenous across the cross -section.we also have ψ and v are functions of x only and do not vary across the cross section.

$$V_y = \int_A \tau_{xy}\, dA = \int_A G\left(\frac{dv}{dx} - \psi(x)\right)dA = GA\left[\frac{dv}{dx} - \psi(x)\right] \qquad 7$$

$$M_z = -\int_A y\sigma_{xx} = \int_A Ey^2\frac{d\psi}{dx}dA = E\frac{d\psi}{dx}\int_A y^2 dA = EI_{zz}\frac{d\psi}{dx} \qquad 8$$

We can substitute Eq.(5) and (6) into the equilibrium equation to obtain the following

$$\frac{dV_y}{dx} = -p \quad \text{or} \quad \boxed{\frac{d}{dx}\left[GA\left(\frac{dv}{dx} - \psi(x)\right)\right] = -p}$$

$$\frac{dM_z}{dx} = -V_y \quad \text{or} \quad \boxed{\frac{d}{dx}\left[EI_{zz}\frac{d\psi}{dx}\right] = -GA\left[\frac{dv}{dx} - \psi(x)\right]}$$

7.33 Fig. P7.33shows a differential element of a beam that is free to vibrate, where ρ is the material density, A is the area of cross-section, and $\frac{\partial^2 v}{\partial t^2}$ is the linear acceleration. Show that the equation for dynamic equilibrium is given by:

$$\frac{\partial^2 v}{\partial t^2} + c^2\frac{\partial^4 v}{\partial x^4} = 0 \qquad \text{where } c = \sqrt{\frac{EI_{zz}}{\rho A}} \qquad 7.9$$

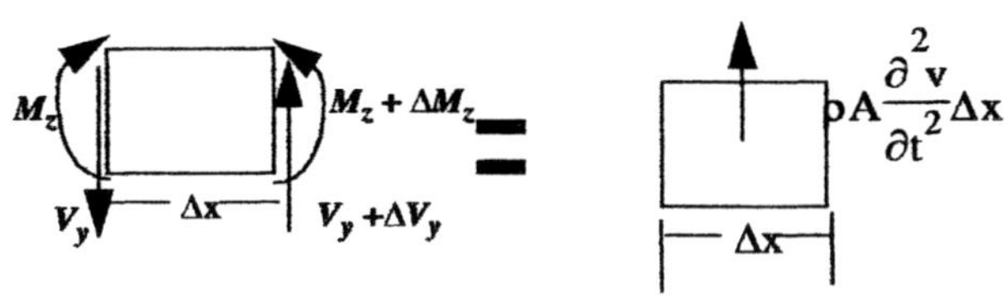

Fig. P7.33. Dynamic Equilibrium.

Solution

By the force equilibrium in Fig.7.33 we obtain: $V_y + \Delta V_y - V_y = \rho A\frac{\partial^2}{\partial t^2}(v)\Delta x$

Dividing by Δx and taking the limit$\Delta x \to 0$ we obtain,

$$\lim_{\Delta x \to 0}\left(\frac{\Delta V_y}{\Delta x}\right) = \rho A\frac{\partial^2}{\partial t^2}(v) \quad \text{or} \quad \frac{\partial V_y}{\partial x} = \rho A\frac{\partial^2}{\partial t^2}(v) \qquad 1$$

By the moment equilibrium about the center we obtain

$$\boldsymbol{M}_z + \Delta\boldsymbol{M}_z + (\boldsymbol{V}_y + \Delta\boldsymbol{V}_y)\frac{\Delta x}{2} + \boldsymbol{V}_y\frac{\Delta x}{2} - \boldsymbol{M}_z = 0$$

Neglecting $\Delta\boldsymbol{V}_y\Delta x$ as its tends to zero fast than other terms in the limiting operation: $\Delta\boldsymbol{M}_z + \boldsymbol{V}_y\Delta x = 0$

Dividing by Δx and taking the limit $\Delta x \to 0$ we obtain,

$$\lim_{\Delta x \to 0}\left(\frac{\Delta\boldsymbol{M}_z}{\Delta x}\right) = -\boldsymbol{V}_y \qquad \text{or} \qquad \frac{\partial\boldsymbol{M}_z}{\partial x} = -\boldsymbol{V}_y \qquad 2$$

Substituting $\boldsymbol{M}_z = EI_{zz}\frac{d^2v}{dx^2}$ and Eq.(2) in Eq.(1), we obtain

$$-\frac{\partial^2}{\partial x^2}\left(EI_{zz}\frac{\partial^2 v}{\partial x^2}\right) = \rho A\frac{\partial^2}{\partial t^2}(v) \qquad 3$$

Assuming $EI_{zz}$is a constant we can rewriteEq.(3) as

$$-\frac{\partial^4 v}{\partial x^4} = \frac{\rho A}{EI_{zz}}\frac{\partial^2}{\partial t^2}(v) = c^2\frac{\partial^2 v}{\partial t^2} \qquad \text{or} \qquad \boxed{\frac{\partial^2 v}{\partial t^2} + c^2\frac{\partial^4 v}{\partial x^4} = 0}$$

7.34 Show by substitution that the solution below satisfies Equation 7.9.

$$v(x, t) = G(x)H(t)$$

$$G(x) = A\cos\omega x + B\sin\omega x + C\cosh\omega x + D\sinh\omega x$$

$$H(t) = E\cos(c\omega^2)t + D\sin(c\omega^2)t$$

Solution

The x- derivatives of functions G are

$$\frac{\partial G}{\partial x} = -A\omega\cos\omega x + B\omega\sin\omega x + C\omega\cosh\omega x + D\omega\sinh\omega x$$

$$\frac{\partial^2 G}{\partial x^2} = -A\omega^2\cos\omega x - B\omega^2\sin\omega x + C\omega^2\cosh\omega x + D\omega^2\sinh\omega x$$

$$\frac{\partial^3 G}{\partial x^3} = A\omega^3\cos\omega x - B\omega^3\sin\omega x + \omega\cosh\omega x + D\omega^3\sinh\omega x$$

$$\frac{\partial^4 G}{\partial x^4} = A\omega^4\cos\omega x + B\omega^4\sin\omega x + C\omega^4\cosh\omega x + D\omega^4\sinh\omega x = \omega^4 G \qquad 1$$

Taking the time derivatives of the function H, we obtain

$$\frac{\partial H}{\partial t} = -E(c\omega^2)\cos(c\omega^2)t + D(c\omega^2)\sin(c\omega^2)t$$

$$\frac{\partial^2 H}{\partial t^2} = -E(c\omega^2)^2\cos(c\omega^2)t - D(c\omega^2)^2\sin(c\omega^2)t = -c^2\omega^4[E\cos(c\omega^2) + D\sin(c\omega^2)t] = -c^2\omega^4 H(t) \qquad 2$$

Substituting v(x,t) into Eq.(7.9) we obtain

$$G(x)\frac{\partial^2 H}{\partial t^2} + c^2 H(t)\frac{\partial^4 G}{\partial x^4} = 0 \qquad 3$$

Substituting Eq.(1) and (2) into Eq.(3) we obtain: $G(x)(-c^2\omega^4)H(t) + c^2\omega^4 H(t)G(x) = 0$

The terms on the left cancel, proving Eq.(7-10) is the solution of Eq.(7.9).

7.35 By substitution show that the following deflection solution satisfies the fourth order boundary value problem of the cantilever beam shown in Fig. P7.35.

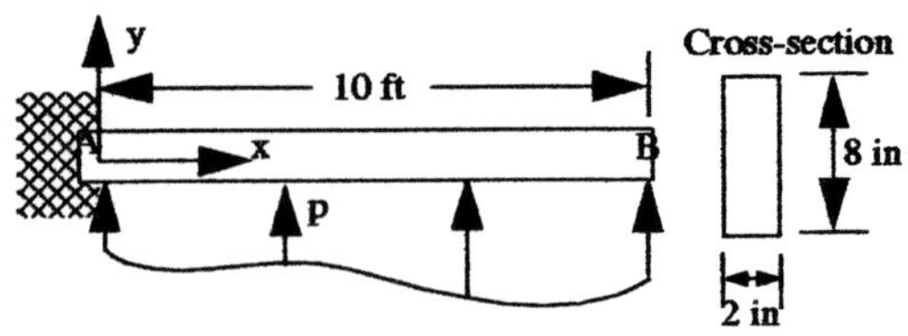

Fig. P7.35

$$v(x) = \frac{1}{6EI}\left[R_A x^3 + 3M_A x^2 + \int_0^x (x - x_1)^3 p(x_1)dx_1\right] \quad \mathbf{7.10}$$

where, $R_A = -\int_0^L p(x_1)dx_1$ and $M_A = \int_0^L x_1 p(x_1)dx_1$

Solution:

The boundary value problem for the cantilever beam shown in Fig. 7.35 is given below.

$$\frac{d^2}{dx^2}\left(EI\frac{d^2v}{dx^2}\right) = p \quad \mathbf{1}$$

$$v(0) = 0 \quad \mathbf{2} \qquad \frac{dv}{dx}(0) = 0 \quad \mathbf{3}$$

$$V_y(L) = -\frac{d}{dx}\left(EI\frac{d^2v}{dx^2}\right) = -EI\frac{d^3v}{dx^3}\bigg|_{x=L} = 0 \text{ or } \frac{d^3v}{dx^3}(L) = 0 \quad \mathbf{4}$$

$$M_z(L) = \left(EI\frac{d^2v}{dx^2}\right)\bigg|_{x=L} = 0 \text{ or } \frac{d^2v}{dx^2}(L) = 0 \quad \mathbf{5}$$

Substituting x=0 in Eq 7.10 we obtain the displacement is zero as in Eq. (2). We record the following formula for the derivative of the integral below.

$$\frac{d}{dx}\left[\int_0^x (x - x_1)^n p(x_1)dx_1\right] = (x - x_1)^n p(x_1)\Big|_{x_1 = x} + n\int_0^x (x - x_1)^{n-1} p(x_1)dx_1 = n\int_0^x (x - x_1)^{n-1} p(x_1)dx_1 \quad \mathbf{6}$$

Taking the x derivatives of Eq. 7.10 and using Eq. 6, we obtain:

$$\frac{dv}{dx} = \frac{1}{6EI}\left[3R_A x^2 + 6M_A x + 3\int_0^x (x - x_1)^2 p(x_1)dx_1\right] = \frac{1}{2EI}\left[R_A x^2 + 2M_A x + \int_0^x (x - x_1)^2 p(x_1)dx_1\right] \quad \mathbf{7}$$

$$\frac{d^2v}{dx^2} = \frac{1}{EI}\left[R_A x + 2M_A + \int_0^x (x - x_1)p(x_1)dx_1\right] \quad \mathbf{8}$$

$$\frac{d^3v}{dx^3} = \frac{1}{EI}\left[R_A + \int_0^x p(x_1)dx_1\right] \quad \mathbf{9}$$

$$\frac{d^4v}{dx^4} = \frac{1}{EI}p(x) \quad \mathbf{10}$$

We note that Eq. 10 is same as Eq. 1 for constant EI. Substituting x=0 in Eq. 7 we obtain the slope is zero as in Eq. (3). Substituting x = L in Eq. 9 and using Eq. 4, we obtain:

$$\frac{d^3v}{dx^3}(L) = \frac{1}{EI}\left[R_A + \int_0^L p(x_1)dx_1\right] = 0 \text{ or } R_A = -\int_0^L p(x_1)dx_1 \qquad \mathbf{11}$$

The value of R_A is same as that given in the problem.

Substituting $x = L$ in Eq. 8 and using Eq. 5, we obtain: $\frac{d^2v}{dx^2}(L) = \frac{1}{EI}\left[R_A L + M_A + \int_0^L (L - x_1)p(x_1)dx_1\right] = 0$

Substituting the value of R_A in Eq. 11 we obtain:

$$-L\int_0^L p(x_1)dx_1 + M_A + \int_0^L (L - x_1)p(x_1)dx_1 = 0 \text{ or } M_A + \int_0^L (L - x_1 - L)p(x_1)dx_1 = 0 \text{ or}$$

$$M_A = \int_0^L x_1 p(x_1)dx_1 \qquad \mathbf{12}$$

The value of M_A is same as given in the problem. Thus all equations of the boundary value problem are satisfied by the displacement solution given in Eq. 7.10.

7.36 Table 7.2 shows the value of distributed load at several point along the axis of a rectangular beam. Determine the slope and deflection at the free end using.Use Modulus of Elasticity as 2,000 ksi.

Table 7.2 Data in problem 7.36

x ft.	p (x) lbs/ft.	x ft.	p (x) lbs/ft.
0	275	6	377
1	348	7	316
2	398	8	233
3	426	9	128
4	432	10	0
5	416		

Solution: $E = 2000$ ksi $v_B = ?$ $\frac{dv_B}{dx} = ?$

The area moment of inertia is : $I_{zz} = \left(\frac{1}{12}\right)(2)(8^3) = 85.333\text{in}^4$

The bending rigidity is: $EI_{zz} = (2)(10^6)(85.333) = 170.667(10^6)\text{lb} - \text{in}^2$

Substituting the bending rigidity in Eq. 7.10 and Eq. 7 and evaluating at x = L = 10 ft = 120 inch we obtain:

$$v(120) = \frac{1}{1024(10^6)}\left[R_A(120)^3 + 3M_A(120)^2 + \int_0^{120}(120 - x_1)^3 p(x_1)dx_1\right] \qquad \mathbf{1}$$

$$\frac{dv}{dx} = \frac{1}{(341.33)(10^6)}\left[R_A(120)^2 + 2M_A(120) + \int_0^{120}(120 - x_1)^2 p(x_1)dx_1\right] \qquad \mathbf{2}$$

where

$$R_A = -\int_0^{120} p(x_1)dx_1 \qquad 3$$

$$M_A = \int_0^{120} x_1 p(x_1)dx_1 \qquad 4$$

We define $J_1 = \int_0^{120}(120-x_1)^3 p(x_1)dx_1$ and $J_2 = \int_0^{120}(120-x_1)^2 p(x_1)dx_1$. The integrals J_1, J_2 and the integrals in Eqs. 3 and 4 are evaluated using numerical integration on a spread sheet as described in Appendix B. In the table below the dimension x is in inches and the distributed load is converted from lb/ft to lb/in

x	p(x)	R_A	M_A	J_1	J_2
0	22.917	-3.115E+02	2.088E+03	4.568E+08	4.010E+06
12	29.000	-6.845E+02	8.952E+03	8.520E+08	7.873E+06
24	33.167	-1.097E+03	2.140E+04	1.154E+09	1.121E+07
36	35.500	-1.526E+03	3.943E+04	1.361E+09	1.383E+07
48	36.000	-1.950E+03	6.228E+04	1.487E+09	1.570E+07
60	34.667	-2.346E+03	8.833E+04	1.553E+09	1.688E+07
72	31.417	-2.693E+03	1.152E+05	1.581E+09	1.752E+07
84	26.333	-2.967E+03	1.396E+05	1.590E+09	1.780E+07
96	19.417	-3.148E+03	1.577E+05	1.591E+09	1.787E+07
108	10.667	-3.212E+03	1.646E+05	1.592E+09	1.788E+07
120	0.000				

From the table above we have the following:

$R_A = -3.212(10^3)$ $\quad M_A = 164.6(10^3)$ $\quad J_1 = 1.592(10^9)$ $\quad J_2 = 17.88(10^6)$

Substituting Eq. 5 in Eqs.(1) and (2) we obtain:

$$v(120) = \frac{1}{1024(10^6)}[(-3.212)(10^3)(120)^3 + 3(164.6)(10^3)(120)^2 + 1.592(10^9)] = 3.078\text{in} \text{ or}$$

$$\boxed{v(120) = 3.1\text{in}}$$

$$\frac{dv}{dx}(120) = \frac{1}{(341.33)(10^6)}[(-3.212)(10^3)(120)^2 + 2(164.6)(10^3)(120) + 17.88(10^6)] = 0.0326 \text{ or}$$

$$\boxed{\frac{dv}{dx}(120) = 0.033}$$

7.37 For the beam and loading given in problem 7.36, determine the slope and deflection at the free end in the following manner. First represent the distributed load by $p(x) = a + bx + cx^2$ and using the data in Table 7.2 determine the constant a, b, and c by the Least-Square Method. Then using fourth order differential equation solve the boundary value problem. Use Modulus of Elasticity as 2,000 ksi.

Solution: $E = 2000$ ksi, $a = ?$, $b = ?$, $c = ?$, $v_B = ?$ $\frac{dv_B}{dx} = ?$

Using the least square method described in Appendix B, we obtain the value of constants a, b, and c on a spread sheet as shown in the table below.

	x_i	$p(x_i)$	x_i^2	x_i^3	x_i^4	$x*p_i$	$x_i^2*p_i$
1	0	22.917	0.0	0.000E+00	0.000E+00	0.000E+00	0.000E+00
2	12	29.000	144.0	1.728E+03	2.074E+04	3.480E+02	4.176E+03
3	24	33.167	576.0	1.382E+04	3.318E+05	7.960E+02	1.910E+04
4	36	35.500	1296.0	4.666E+04	1.680E+06	1.278E+03	4.601E+04
5	48	36.000	2304.0	1.106E+05	5.308E+06	1.728E+03	8.294E+04
6	60	34.667	3600.0	2.160E+05	1.296E+07	2.080E+03	1.248E+05
7	72	31.417	5184.0	3.732E+05	2.687E+07	2.262E+03	1.629E+05
8	84	26.333	7056.0	5.927E+05	4.979E+07	2.212E+03	1.858E+05
9	96	19.417	9216.0	8.847E+05	8.493E+07	1.864E+03	1.789E+05
10	108	10.667	11664.0	1.260E+06	1.360E+08	1.152E+03	1.244E+05
11	120	0.000	14400.0	1.728E+06	2.074E+08	0.000E+00	0.000E+00
b_{ij} & r_i	660.0	279.1	55440.0	5227200.0	525305088.0	13720.0	929064.0
C_{ij}	1.799E+12	5.691E+10	3.764E+08	2.705E+09	2.091E+07	1.742E+05	
D	3.100E+12						
a_i	22.927	0.581	-0.006				

The value of the constants are

$$\boxed{a = 22.927}\ \boxed{b = 0.581}\ \boxed{c = -0.006} \tag{1}$$

The boundary value problem for the cantilever beam shown in Fig. p7.35 is given below.

$$EI\frac{d^4v}{dx^4} = a + bx + cx^2 \tag{2}$$

$$v(0) = 0 \tag{3}$$

$$\frac{dv}{dx}(0) = 0 \tag{4}$$

$$V_y(120) = -\frac{d}{dx}\left(EI\frac{d^2v}{dx^2}\right)\bigg|_{x=120} = -EI\frac{d^3v}{dx^3}\bigg|_{x=120} = 0 \text{ or } \frac{d^3v}{dx^3}(120) = 0 \tag{5}$$

$$M_z(120) = \left(EI\frac{d^2v}{dx^2}\right)\bigg|_{x=120} = 0 \text{ or } \frac{d^2v}{dx^2}(120) = 0 \tag{6}$$

Integrating Eq. 2 we obtain:

$$EI\frac{d^3v}{dx^3} = ax + \frac{bx^2}{2} + \frac{cx^3}{3} + c_1 \tag{7}$$

$$EI\frac{d^2v}{dx^2} = \frac{ax^2}{2} + \frac{bx^3}{6} + \frac{cx^4}{12} + c_1x + c_2 \tag{8}$$

$$EI\frac{dv}{dx} = \frac{ax^3}{6} + \frac{bx^4}{24} + \frac{cx^5}{60} + \frac{c_1x^2}{2} + c_2x + c_3 \tag{9}$$

$$EIv = \frac{ax^4}{24} + \frac{bx^5}{120} + \frac{cx^6}{360} + \frac{c_1x^3}{6} + \frac{c_2x^2}{2} + c_3x + c_4 \tag{10}$$

Substituting $x = 0$ in Eq. (10) and using Eq. (3) we obtain: $c_4 = 0$

Substituting $x = 0$ in Eq. (9) and using Eq. (4) we obtain: $c_3 = 0$

Substituting $x = 120$ in Eq. (7) and using Eq. (1) and (5) we obtain:

$$a(120) + \frac{b(120)^2}{2} + \frac{c(120)^3}{3} + c_1 = 0 \text{ or } c_1 = -3478.4 \qquad 11$$

Substituting x = 120 in Eq. (8) and using Eq. (1) and (6) we obtain:

$$\frac{a(120)^2}{2} + \frac{b(120)^3}{6} + \frac{c(120)^4}{12} + c_1(120) + c_2 \text{ or } c_2 = 188.69(10^3) \qquad 12$$

The bending rigidity was found in problem 7.36 as: $EI_{zz} = 170.667(10^6)\text{lb}-\text{in}^2$

Substituting x = 120 in Eq. 10 and all the constants we obtain:

$$EIv(120) = \frac{a(120)^4}{24} + \frac{b(120)^5}{120} + \frac{c(120)^6}{360} + \frac{c_1(120)^3}{6} + \frac{c_2(120)^2}{2} \text{ or } v(120) = 3.665\text{in}$$

$$\boxed{v(120) = 3.67\text{in}}$$

Substituting x = 120 in Eq. 9 and all the constants we obtain:

$$EI\frac{dv}{dx}(120) = \frac{a(120)^3}{6} + \frac{b(120)^4}{24} + \frac{c(120)^5}{60} + \frac{c_1(120)^2}{2} + c_2(120) \text{ or } \frac{dv}{dx}(120) = 0.03945$$

$$\boxed{\frac{dv}{dx}(120) = 0.039}$$

7.38 Table 7.3 shows the measured radii of a solid tapered circular beam at several point along the axis. The beam is made of aluminum (E = 28 GPa) and has a length of 1.5 meters. Determine the slope and deflection at point B

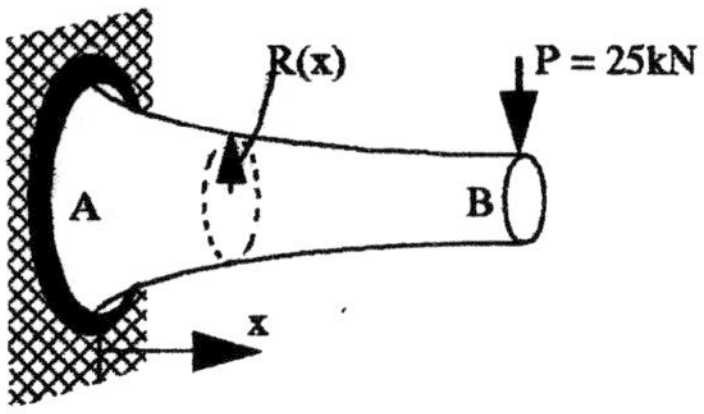

Fig. P7.38

Table 7.3. Data for Problem 7.38.

x (m)	R(x) (mm)	x (m)	R(x) (mm)
0.0	100.6	0.8	60.1
0.1	92.7	0.9	60.3
0.2	82.6	1.0	59.1
0.3	79.6	1.1	54.0
0.4	75.9	1.2	54.8
0.5	68.8	1.3	54.1
0.6	68.0	1.4	49.4
0.7	65.9	1.5	50.6

Solution: $E = 28\text{ GPa}$, $v_B = ?$ $\frac{dv_B}{dx} = ?$

The moment curvature equation can be written as: $\frac{d^2v}{dx^2} = \frac{M_z}{EI}$. Integrating from x = 0 and noting that the slope is zero at x = 0 we obtain:

$$\frac{dv}{dx}(x) = \int_0^x \frac{M_z}{EI}dx \qquad 1$$

Noting that deflection is zero at x = 0 we can write: $v = \int_0^x \frac{dv}{dx_1}dx_1$ Integrating by parts we obtain.

$$v(x) = x\frac{dv}{dx_1}\bigg|_0^x - \int_0^x x_1\frac{d^2v}{dx^2}dx_1 = x\int_0^x \frac{M_z}{EI}dx - \int_0^x x_1\frac{M_z}{EI}dx_1 = \int_0^x (x - x_1)\frac{M_z}{EI}dx_1 \qquad 2$$

Fig. (a) shows the free body diagram of the beam after an imaginary cut. By equilibrium of moment about point O we obtain:

$$M_z = -P(1.5 - x) \quad 3$$

The area moment of inertia is: $I = \frac{\pi R^4}{4} mm^4 = \frac{\pi R^4}{4}(10^{-12})m^4$, where R is in millimeter. Substituting $x = 1.5m$, $E = 28(10^9)\ N/m^2$. $P = 25\ (10^3)\ N$, I and Eq. 3 into Eqs. (1) and (2), we obtain the following.

$$v_B = v(1.5) = \int_0^{1.5} (1.5 - x)\frac{(-25)(10^3)(1.5 - x)}{(28)(10^9)\frac{\pi R^4}{4}(10^{-12})}dx = -1.1368(10^6)\int_0^{1.5}\frac{(1.5 - x)^2}{R^4}dx = \int_0^{1.5} F_1(x)dx \quad 4$$

$$\frac{dv_B}{dx} = \frac{dv}{dx}(1.5) = \int_0^{1.5} \frac{(-25)(10^3)(1.5 - x)}{(28)(10^9)\frac{\pi R^4}{4}(10^{-12})}dx = -1.1368(10^6)\int_0^{1.5}\frac{(1.5 - x)^2}{R^4}dx = \int_0^{1.5} F_2(x)dx \quad 5$$

The integrals in Eqs. (4) and (5) are evaluated using numerical integration on a spread sheet as described in Appendix B and the results are given in the table below.

x	R(x)	$F_1(x)$	v(x)	$F_2(x)$	$\frac{dv}{dx}(x)$
0.00	100.60	-2.497E-02	-2.757E-03	-1.665E-02	-1.910E-03
0.10	92.70	-3.017E-02	-6.330E-03	-2.155E-02	-4.575E-03
0.20	82.60	-4.127E-02	-1.043E-02	-3.175E-02	-7.861E-03
0.30	79.60	-4.078E-02	-1.454E-02	-3.398E-02	-1.144E-02
0.40	75.90	-4.145E-02	-1.915E-02	-3.768E-02	-1.587E-02
0.50	68.80	-5.074E-02	-2.384E-02	-5.074E-02	-2.079E-02
0.60	68.00	-4.307E-02	-2.792E-02	-4.785E-02	-2.560E-02
0.70	65.90	-3.858E-02	-3.199E-02	-4.822E-02	-3.106E-02
0.80	60.10	-4.270E-02	-3.567E-02	-6.099E-02	-3.669E-02
0.90	60.30	-3.095E-02	-3.838E-02	-5.159E-02	-4.160E-02
1.00	59.10	-2.330E-02	-4.062E-02	-4.659E-02	-4.660E-02
1.10	54.00	-2.139E-02	-4.225E-02	-5.348E-02	-5.117E-02
1.20	54.80	-1.134E-02	-4.309E-02	-3.782E-02	-5.438E-02
1.30	54.10	-5.308E-03	-4.345E-02	-2.654E-02	-5.666E-02
1.40	49.40	-1.909E-03	-4.354E-02	-1.909E-02	-5.762E-02
1.50	50.60	0.000E+00		0.000E+00	

From the table above we obtain the value of deflection and slope at point B as

$v_B = -43.54(10^{-3})m$ or $\quad$ $\boxed{v_B = -43.5mm}$

$\frac{dv_B}{dx} = -57.62(10^{-3})$ $\quad$ $\boxed{\frac{dv_B}{dx} = -0.058}$

7.39 Let the radius of the tapered beam in problem 7.38 be represented by the equation $R(x) = a + bx$. Using the data in Table 7.3 determine the constant a and b by the least-square method and then find the slope and deflection at point B by numerical integration.

Solution: $E = 28$ GPa, $a = ?$, $b = ?$, $v_B = ?$ $\frac{dv_B}{dx} = ?$

From results in problem 4-40, we obtain: $\boxed{a = 90.226mm}$ $\boxed{b = -30.593}$

Thus R(x) can be written as:

$$R(x) = 90.226 - 30.593x \qquad (6)$$

Eqs.(4) and (5) of problem 7.38 are:

$$v_B = -1.1368(10^6)\int_0^{1.5}\frac{(1.5-x)^2}{R^4}dx = \int_0^{1.5} F_1(x)dx \qquad (7)$$

$$\frac{dv_B}{dx} = -1.1368(10^6)\int_0^{1.5}\frac{(1.5-x)^2}{R^4}dx = \int_0^{1.5} F_2(x)dx \qquad (8)$$

The radius at any x can be found from Eq.(6). The integrals in Eqs. (7) and (8) are evaluated using numerical integration on a spreadsheet as described in Appendix B and the results are given in the table below.

x	R(x)	$F_1(x)$	v(x)	$F_2(x)$	$\frac{dv}{dx}(x)$
0.00	90.23	-3.860E-02	-3.860E-03	-2.573E-02	-2.665E-03
0.10	87.17	-3.860E-02	-7.709E-03	-2.757E-02	-5.520E-03
0.20	84.11	-3.839E-02	-1.153E-02	-2.953E-02	-8.577E-03
0.30	81.05	-3.794E-02	-1.528E-02	-3.162E-02	-1.185E-02
0.40	77.99	-3.718E-02	-1.894E-02	-3.380E-02	-1.534E-02
0.50	74.93	-3.606E-02	-2.247E-02	-3.606E-02	-1.906E-02
0.60	71.87	-3.451E-02	-2.582E-02	-3.835E-02	-2.301E-02
0.70	68.81	-3.245E-02	-2.893E-02	-4.056E-02	-2.716E-02
0.80	65.75	-2.980E-02	-3.175E-02	-4.258E-02	-3.150E-02
0.90	62.69	-2.649E-02	-3.420E-02	-4.415E-02	-3.596E-02
1.00	59.63	-2.247E-02	-3.621E-02	-4.495E-02	-4.042E-02
1.10	56.57	-1.776E-02	-3.772E-02	-4.439E-02	-4.472E-02
1.20	53.51	-1.248E-02	-3.869E-02	-4.158E-02	-4.856E-02
1.30	50.46	-7.017E-03	-3.916E-02	-3.508E-02	-5.144E-02
1.40	47.40	-2.253E-03	-3.927E-02	-2.253E-02	-5.256E-02
1.50	44.34	0.000E+00		0.000E+00	

From the table above we obtain the value of deflection and slope at point B as

$$v_B = -39.27(10^{-3})\text{m or} \qquad \boxed{v_B = -39.3\text{mm}}$$

$$\frac{dv_B}{dx} = -52.56(10^{-3}) \qquad \boxed{\frac{dv_B}{dx} = -0.053}$$

Problem Set 7.3

7.40 Determine the deflection at the free end of the beam shown in Fig. P7.12.

Solution: $v_A = v(2L) = ?$

The loading on the beam can be considered as the sum of two loading shown below:

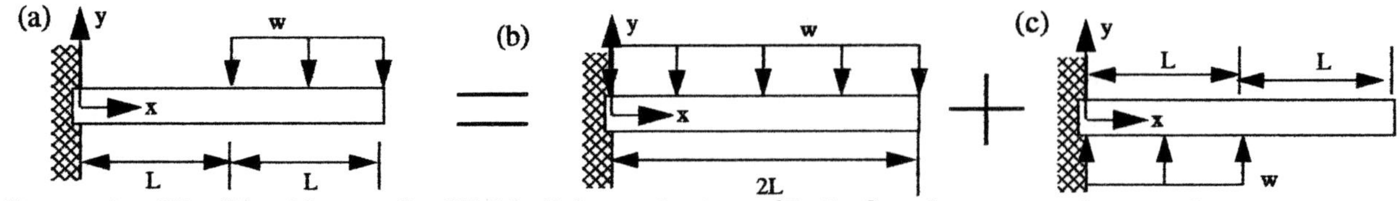

Comparing Fig.(b) with case 3 of Table 7.1 we obtain a=2L, b=0 and p_0=-w and can write

$$v_{A1} = \frac{-w(8L^3)(6L)}{24EI} = \frac{-2wL^4}{EI} \qquad (1)$$

Comparing Fig.(c) with case 4 of Table 7.1 we obtain a = L, b = L and p_0 = w and can write

$$v_{A2} = \frac{w(7L^3)(6L)}{24EI} = \frac{7wL^4}{24EI} \qquad 2$$

Superposing Eqs.(1) and (2) we obtain $v_A = v_{A1} + v_{A2} = \frac{wL^4}{24EI}(-48+7)$ or $\boxed{v_A = \frac{-41wL^4}{24EI}}$

7.41 Determine the reaction force at support A in Fig. P7.22.

Solution: $R_A = ?$

The loading on the beam can be considered as the sum of two loading shown below;

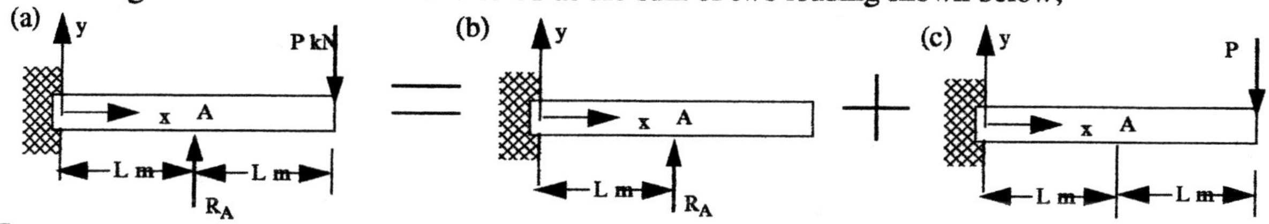

Comparing Fig.(b) with case 1 in Table 7.1 we obtain a = L, b = 0, P = R_A and can write

$$v_{A1} = \frac{R_A L^3}{3EI} \qquad 1$$

Comparing Fig.(b) with case 1 in Table 7.1 we obtain a = 2L, b = 0, P = -P. x = L and can write

$$v_{A2} = \frac{-PL^2}{6EI}(6L - L) = \frac{-5PL^3}{6EI} \qquad 2$$

The deflection of point A is zero and we obtain: $v = v_{A1} + v_{A2} = \frac{R_A L^3}{3EI} - \frac{5PL^3}{6EI} = 0$ or $\boxed{R_A = \frac{5}{2}P}$

7.42 Determine the deflection at point A on the beam shown in Fig. P7.42 in terms of w, L, E, and I.

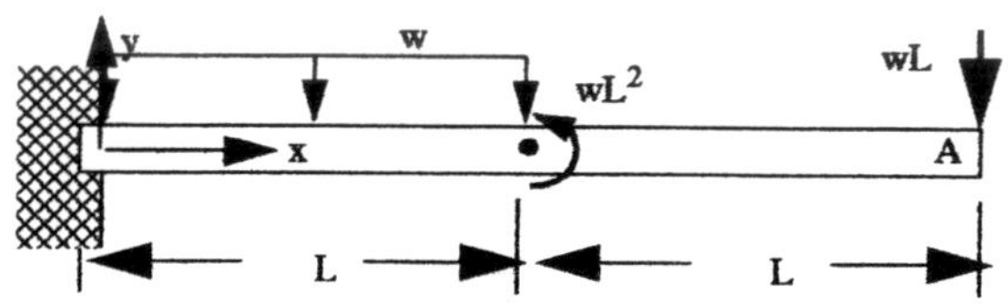

Fig. P7.42

Solution: $v_A = ?$

The beam and loading in Fig 7.42 is the sum of three loadings given below.

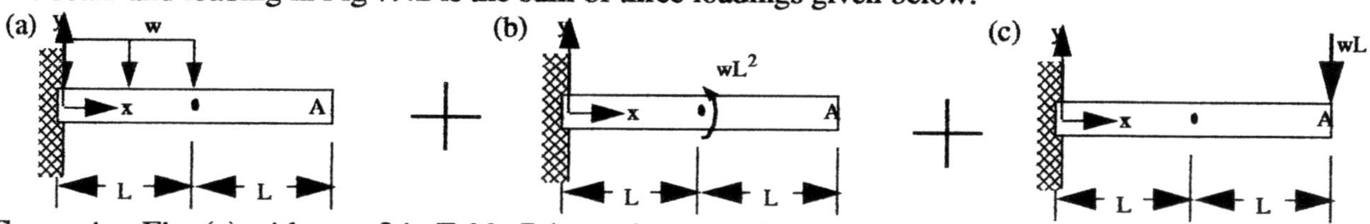

Comparing Fig. (a) with case 3 in Table 7.1 we obtain a = L , b = L and p_0 = -w and we can write

$$v_{A1} = \frac{-7wL^4}{24EI} \qquad 1$$

Comparing Fig.(b) with case 2 in Table 7.1 we obtain a = L, b = L and M = $wL^2$and we can write

$$v_{A2} = \frac{wL^2}{2EI}(3L) = \frac{3wL^4}{2EI} \qquad 2$$

Comparing Fig.(c) with case 1 in Table 7.1 we obtain a=2L, b=0 and p=-wL. we can write

$$v_{A3} = \frac{(-wL)}{6EI}(4L^2)(4L) = -\frac{8wL^3}{3EI}$$ 3

Superposing Eqs. (1), (2), (3) we obtain: $v = v_{A1} + v_{A2} + v_{A3} = \frac{wL^3}{EI}\left(\frac{-7+36-64}{6EI}\right)$ or $\boxed{v_A = \frac{-35wL^3}{24EI}}$

7.43 Determine the reaction force and the slope of the beam at A using superposition.

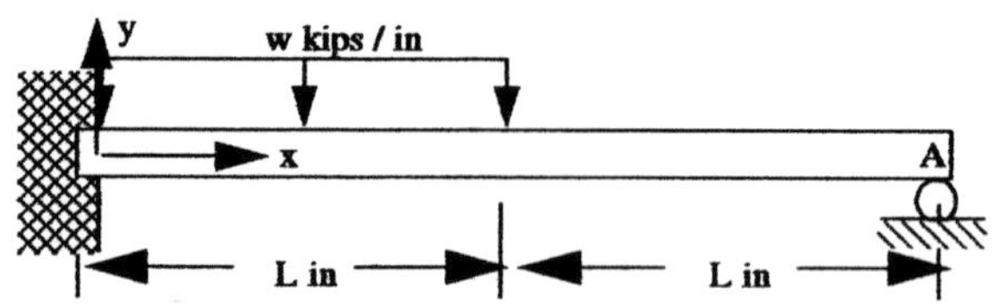

Fig. P7.43

Solution: $R_A = ?$ $\qquad \frac{dv_A}{dx} =$

The beam and loading shown in Fig. 7.43 can be considered as the sum of the two loadings shown:

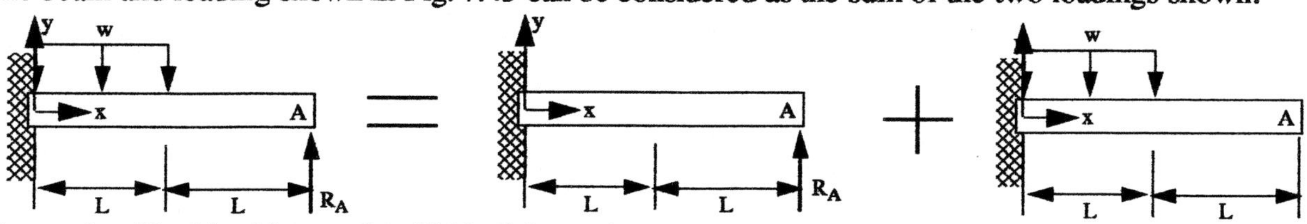

Comparing Fig.(a) with case 3 in Table 7.1 we obtain a = L , b = L and p_0 = -w and we can write

$$v_{A1} = \frac{(-wL^2)}{24EI}(7L) = -\frac{7wL^4}{24EI}$$ 1

$$\frac{dv_{A1}}{dx} = \left(\frac{wL^3}{6EI}\right)$$ 2

Comparing Fig.(b) with case 1 in Table 7.1 we obtain a = 2L, b = 0 and p = R_A, we can write

$$v_{A2} = (R_A)\frac{(4L^2)}{6EI}(4L) = \frac{8R_AL^3}{3EI}$$ 3

$$\frac{dv_{A2}}{dx} = \left(\frac{R_A(4L^2)}{2EI}\right) = \frac{2R_AL^2}{EI}$$ 4

The deflection at A is zero. So we can write: $v_A = v_{A1} + v_{A2} = -\frac{7wL^4}{24EI} + \frac{8R_AL^3}{3EI} = 0$ or $\boxed{R_A = \frac{7}{64}wL}$

Substituting R_A in Eq.(4) and superposing the results in Eq.(2) we obtain

$$\frac{dv_A}{dx} = \frac{dv_{A1}}{dx} + \frac{dv_{A2}}{dx} = -\frac{wL^3}{6EI} + \frac{7wL^3}{32EI}$$ or $\boxed{\frac{dv_A}{dx} = \frac{5wL^3}{96EI}}$

7.44 Two beams of length L and bending rigidity EI are simply supported at the ends and are in contact at the center. Determine the deflection at the center in terms of P, L, E, and I.

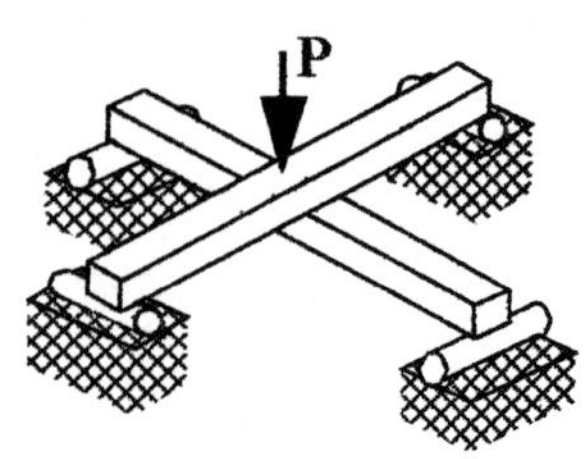

Fig. P7.44

Solution: v_A=?

The beams and loading in Fig 7.44 are the sum of beams and loading shown below:

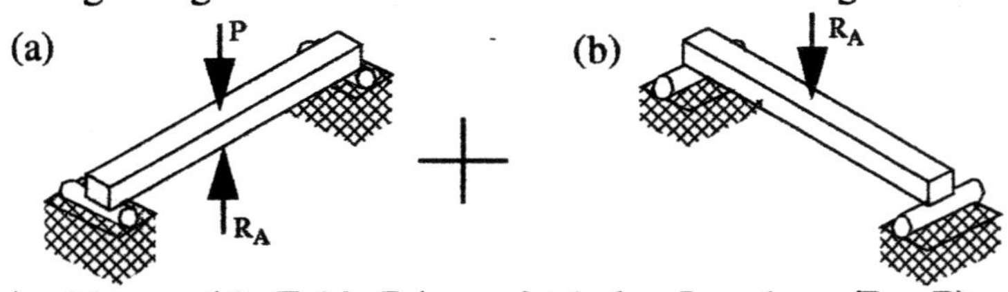

Comparing Fig.(a) with case 4 in Table 7.1 we obtain h = L and p = (R_A -P) and we can write

$$v_{A1} = \frac{(R_A - P)}{48EI}L^3 \qquad 1$$

Comparing Fig.(b) with case 4 in Table 7.1 we obtain h = L, P = -R_A and we can write

$$v_{A2} = -\frac{R_A L^3}{48EI} \qquad 2$$

The deflection at point A must be same for the two beams and we obtain $\frac{(R_A - P)}{48EI}L^3 = -\frac{R_A L^3}{48EI}$ or

$$\boxed{R_A = \frac{P}{2}} \qquad 3$$

Substituting Eq.(3) into Eq.(2) we obtain $\boxed{v_A = -\frac{PL^3}{96EI}}$

7.45 Two beams of length L and bending rigidity EI are simply supported at the ends and are in contact at the center. Determine the deflection at the center in terms of w, L, E, and I.

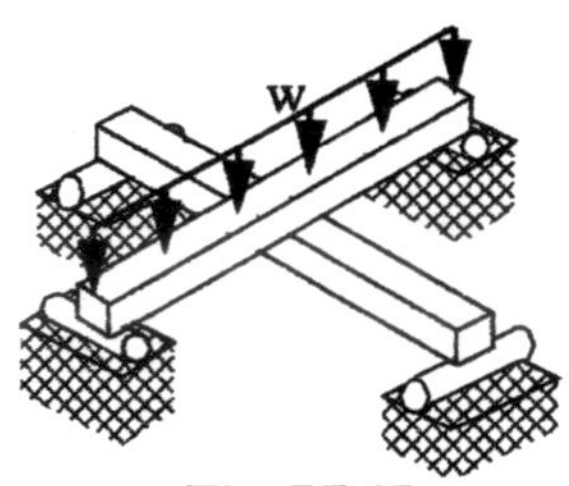

Fig. P7.45

Solution: v_A=?

The beams in Fig. 7.45 can be considered as the sum of three beams shown below:

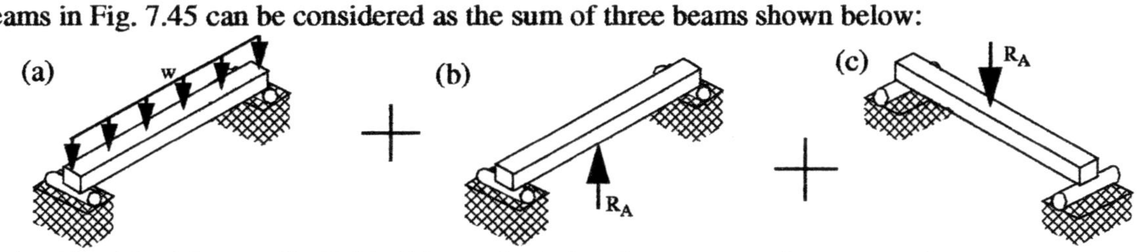

Comparing Fig.(a) with case 5in Table 7.1 we obtain h = L , p_0= -w and we can write

$$v_{A1} = \frac{-5wL^4}{384EI} \qquad 1$$

Comparing Fig.(b) with case 4 in Table 7.1 we obtain h = L, P = R_A and we can write

$$v_{A2} = \frac{R_A L^3}{48EI} \qquad 2$$

Comparing Fig.(c) with case 4 in Table 7.1 we obtain h = L, P = $-R_A$ and we can write

$$v_{A3} = -\frac{R_A L^3}{48EI} \qquad 3$$

The deflection of the top beam is the sum of deflection in Eq.(1) and Eq.(2) which must be equal to deflection of the bottom beam. We obtain

$$\frac{-5wL^4}{384EI} + \frac{R_A L^3}{48EI} = -\frac{R_A L^3}{48EI} \qquad 4$$

$$\frac{-5wL}{8} + 2R_A = 0 \text{ or } R_A = \frac{5wL}{16}$$

Substituting R_A in Eq.(3). We obtain $\boxed{v_A = \frac{-5wL^4}{758EI}}$

7.46 A cantilever beam's end rests on the middle of a simply supported beam as shown. Both beams have length L and bending rigidity EI. Determine the deflection at A and the reactions the wall at C in terms of P, L, E, and I.

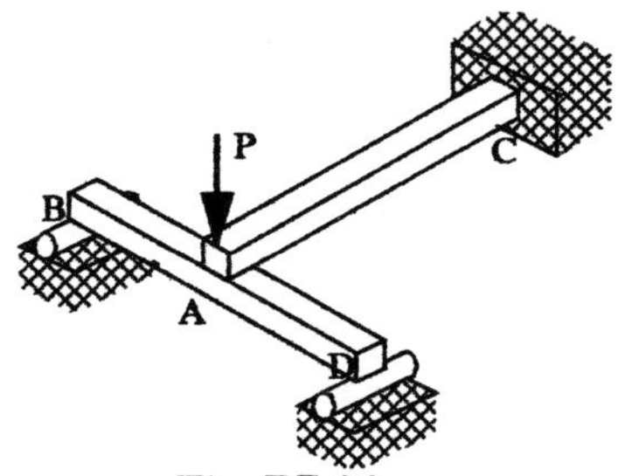

Fig. P7.46

Solution: v_A=?, R_C=?, M_C=?

The beam and loading shown in Fig. 7.46 can be considered as the sum of the two beams shown below

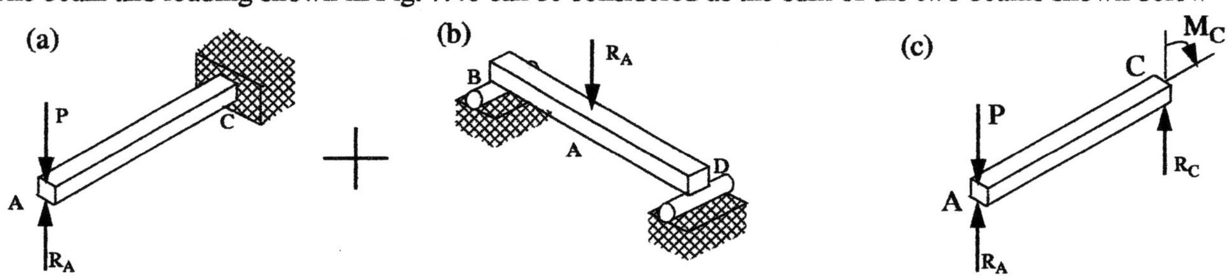

Comparing Fig.(a) with case 1 in Table 7.1 we obtain a = L, b = 0 and p = (R_A -P) and we can write

$$v_{A1} = \frac{(R_A - P)}{3EI}L^3 \qquad 1$$

Comparing Fig.(b) with case 4 in Table 7.1 we obtain h = L, P = $-R_A$ and we can write

$$v_{A2} = -\frac{R_A L^3}{48EI} \qquad 2$$

The deflection at point A must be same for the two beams and we obtain $\frac{(R_A - P)}{3EI}L^3 = -\frac{R_A L^3}{48EI}$ or

$$\boxed{R_A = \frac{16}{17}P} \qquad 3$$

Substituting R_A in Eq.(2) we obtain $\boxed{v_A = -\frac{PL^3}{51EI}}$

Fig. (c) shows the free body diagram of the beam shown in Fig(a).

By equilibrium of forces in Fig. (c) we obtain: $R_C + R_A - P = 0$ or $\boxed{R_C = \frac{P}{17}}$

By equilibrium of moment about point C in Fig.(c) we obtain: $M_C - (P - R_A)L = 0$ or $\boxed{M_c = \frac{PL}{17}}$

7.47 A cantilever beam's end rests on the middle of a simply supported beam as shown. Both beams have length L and bending rigidity EI. Determine the deflection at A and the reactions the wall at C in

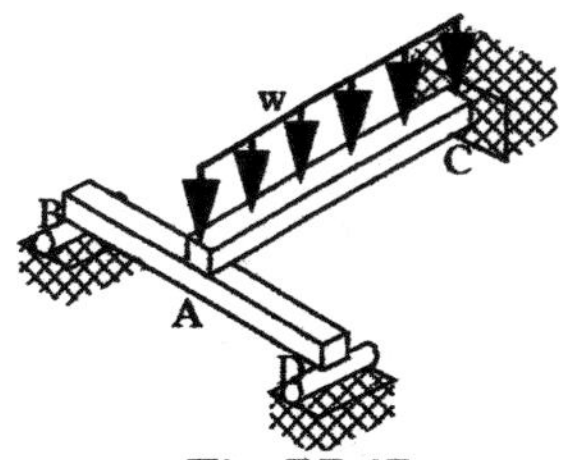

Fig. P7.47

Solution: $v_A = ?$, $R_C = ?$, $M_C = ?$

The beams in Fig. P7.47 can be considered as the sum of beams shown in figures (a), (b)

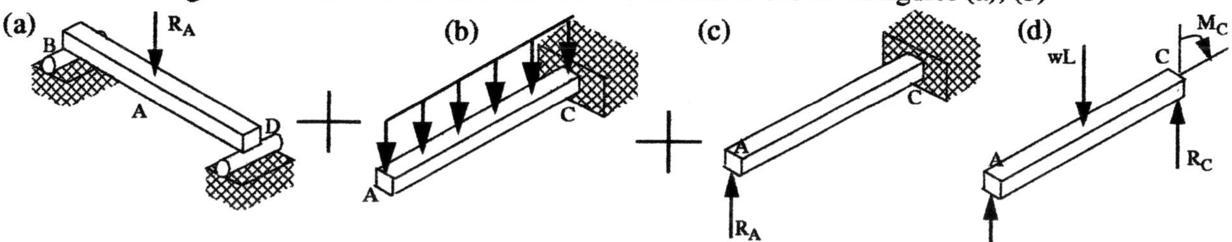

Comparing Fig.(a) with case 4 in Table 7.1 we obtain h = L, P = $-R_A$ and we can write

$$v_{A1} = -\frac{R_A L^3}{48EI} \quad 1$$

Comparing Fig. (b) with case 3 in Table 7.1 we obtain a = L, b=0, and $p_0 = -w$ and we can write

$$v_{A2} = \frac{-wL^4}{8EI} \quad 2$$

Comparing Fig. (c) with case 1 in Table 7.1 we obtain a = L, b = 0, P = R_A and we can write

$$v_{A3} = \frac{R_A L^3}{3EI} \quad 3$$

The deflection of the top beam is the sum of deflection in Eqs. (3) and (2) which must be equal to deflection of the bottom beam given by Eq.(1). We obtain

$$\frac{-wL^4}{8EI} + \frac{R_A L^3}{3EI} = -\frac{R_A L^3}{48EI} \quad 4$$

$$-6wL + 16R_A = -R_A \text{ or } R_A = \frac{6wL}{17} \quad 5$$

Substituting R_A in Eq.(1). We obtain $\boxed{v_A = \frac{-wL^4}{136EI}}$

Fig.(d) shows the free body diagram of the top beam with distributed force replaced by an equivalent force.

By equilibrium of forces in the y-direction, we obtain: $R_C + R_A - wL = 0$ or $\boxed{R_C = \frac{11wL}{17}}$

By equilibrium of moment about point C in Fig.(d),we obtain: $M_C + R_A L - wL(L/2) = 0$ or $\boxed{M_C = \frac{5wL^2}{34}}$

7.48 The end of one cantilever beam rests on the end of another cantilever beam as shown. Both beams have length L and bending rigidity EI. Determine the deflection at A and the reactions the wall at C in terms of w, L, E, and I.

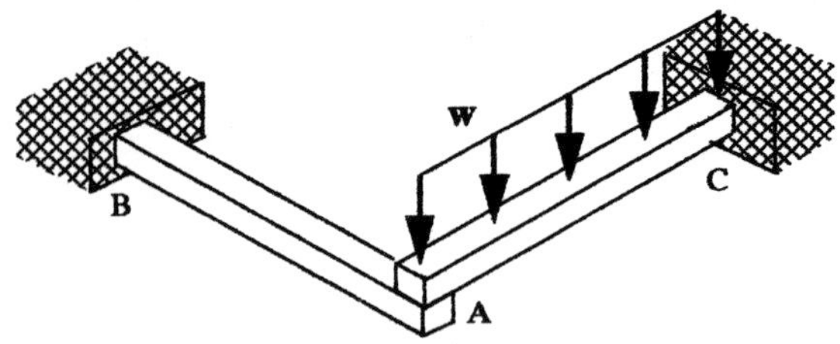

Fig. P7.48

Solution: $v_A=?$, $R_C = ?$, $M_C = ?$

The beams in Fig. P7.47 can be considered as the sum of beams shown in figures (a), (b)

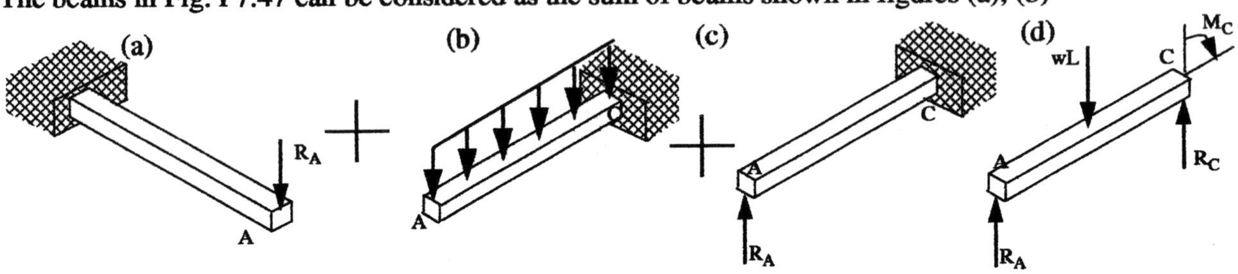

Comparing Fig.(a) with case 1 in Table 7.1 we obtain a = L, b = 0, P = $-R_A$ and we can write

$$v_{A1} = -\frac{R_A L^3}{3EI} \qquad (1)$$

Comparing Fig.(b) with case 3 in Table 7.1 we obtain a = L, b=0, and p_0 = -w and we can write

$$v_{A2} = \frac{-wL^4}{8EI} \qquad (2)$$

Comparing Fig.(c) with case 1 in Table 7.1 we obtain a = L, b = 0, P = R_A and we can write

$$v_{A3} = \frac{R_A L^3}{3EI} \qquad (3)$$

The deflection of the top beam is the sum of deflection in Eqs. (3) and (2) which must be equal to deflection of the bottom beam given by Eq.(1). We obtain

$$\frac{-wL^4}{8EI} + \frac{R_A L^3}{3EI} = -\frac{R_A L^3}{3EI} \qquad (4)$$

$$-3wL + 8R_A = -8R_A \text{ or } R_A = \frac{3wL}{16} \qquad (5)$$

Substituting R_A in Eq.(3) . We obtain $\boxed{v_A = \frac{-wL^4}{16EI}}$

Fig(d) shows the free body diagram of the top beam with distributed force replaced by an equivalent force.

By equilibrium of forces in the y-direction, we obtain $R_C + R_A - wL = 0$ or $\boxed{R_C = \frac{13wL}{16}}$

By equilibrium of moment about point C in Fig.(d). We obtain $M_C + R_A L - wL(L/2) = 0$ or $\boxed{M_C = \frac{5wL^2}{8}}$

7.49 Solve problem 7.9 using Discontinuity Functions.

Solution: $v(x) = ?$, $v(L) = ?$

Fig(a) shows the free body diagram of the beam. The reaction force R_A=2wL/3 was found in problem 7-9.

Using templates shown in Fig. 7-33 the moment equation can be written as

$$M_z = \frac{2}{3}wLx - wL\langle x-L\rangle^1 \qquad 1$$

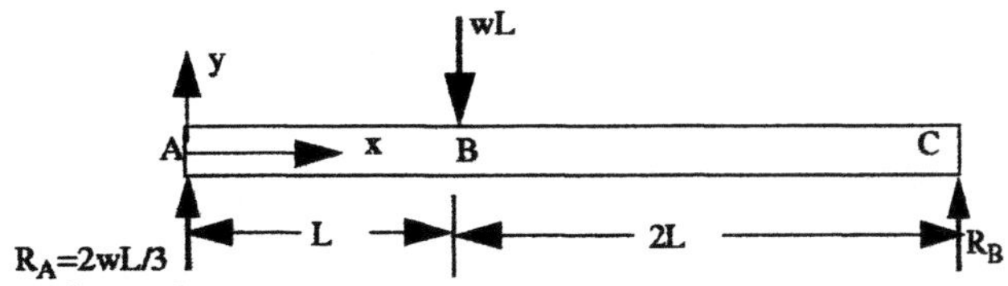

The boundary value problem can be written as

$$EI\frac{d^2v}{dx^2} = \frac{2}{3}wLx - wL\langle x-L\rangle^1 \qquad 2$$

$$v(0) = 0 \qquad 3$$

$$v(3L) = 0 \qquad 4$$

Integrating Eq.(2) we obtain

$$EI\frac{dv}{dx} = \frac{1}{3}wLx^2 - \frac{wL}{2}\langle x-L\rangle^2 + C_1 \qquad 5$$

$$EIv = \frac{1}{9}wLx^3 - \frac{wL}{6}\langle x-L\rangle^3 + C_1x + C_2 \qquad 6$$

Substituting x=0 in Eq.(6) and using Eq.(3) we obtain

$$EIv(0) = C_2 = 0 \qquad 7$$

Substituting x=3L in Eq.(6) and using Eq.(4) we obtain: $EIv(3L) = \frac{1}{9}wL(3L)^3 - \frac{wL}{6}(8L^3) + C_1(3L) = 0$ or

$$\frac{wL^4(54-24)}{18} + C_1(3L) = 0 \text{ or } C_1 = -\frac{5}{9}wL^3 \qquad 8$$

Substituting the constants in Eq.(6) we obtain the equation of the elastic- curve as

$$\boxed{v(x) = \frac{w}{18EI}[2Lx^3 - 3L\langle x-L\rangle^3 - 10L^3x]}$$

Substituting x = L in the elastic-curve, we obtain $v(L) = \frac{w}{18EI}[2L^4 - 10L^4]$ or $\boxed{v(L) = \frac{-4wL^4}{9EI}}$

7.50 Solve problem 7.10 using Discontinuity Functions.

Solution: $v(x) = ?$, $v(L) = ?$

Fig.(a) shows the free body diagram of the beam. The reaction force R_A was found to be wL in problem 7-10. Using templates shown in Fig. 7-33 the moment equation can be written in terms of discontinuity function as

$$M_z = wLx - wL^2 - \frac{w}{2}\langle x-L\rangle^2 \qquad 1$$

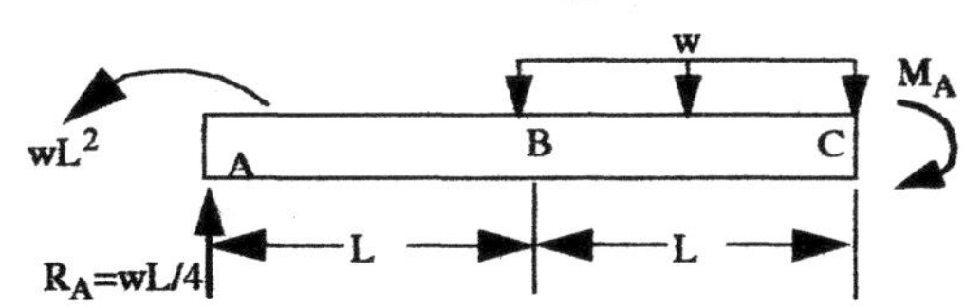

The boundary value problem can be written as

$$EI\frac{d^2v}{dx^2} = wLx - wL^2 - \frac{w}{2}\langle x-L\rangle^2 \qquad 2$$

$$v(0) = 0 \qquad 3$$

$$\frac{dv}{dx}(2L) = 0 \qquad 4$$

Integrating Eq.(2) we obtain

$$EI\frac{dv}{dx} = \frac{1}{2}wLx^2 - wL^2x - \frac{w}{6}\langle x-L\rangle^3 + C_1 \qquad 5$$

$$EIv = \frac{1}{6}wLx^3 - \frac{wL^2x^2}{2} - \frac{w}{24}\langle x-L\rangle^4 + C_1x + C_2 \qquad 6$$

Substituting x=0 in Eq.(6) and using Eq.(3) we obtain

$$EIv(0) = C_2 = 0 \qquad 7$$

Substituting x=2L in Eq. (6) and using Eq.(4) we obtain

$$EI\frac{dv}{dx}(2L) = \frac{w}{2}L(4L^2) - wL^2(2L) - \frac{w}{6}L^3 + C_1 = 0 \quad \text{or}$$

$$\frac{wL^3(12-12-1)}{6} + C_1 = 0 \text{ or } C_1 = \frac{wL^3}{6} \qquad 8$$

Substituting the above constants in Eq. (6) we obtain the equation of the elastic- curve as

$$\boxed{v(x) = \frac{w}{24EI}[4Lx^3 - 12L^2x^2 + 4L^3x - \langle x-L\rangle^4]}$$

Substituting x = L in the elastic-curve, we obtain: $v(L) = \frac{w}{24EI}[4L^4 - 12L^4 + 4L^4]$ or $\boxed{v(L) = \frac{-wL^4}{6EI}}$

7.51 Solve problem 7.11 using Discontinuity Functions.

Solution: v(x) = ?, v(L) = ?

Fig.(a) shows the free body diagram of the beam. The reaction force $R_A = wL/4$ was found in problem 6.55. Using templates shown in Fig. 7-33 the moment equation can be written in terms of discontinuity function as

$$M_z = \frac{wLx}{4} - \frac{w}{2}\langle x-L\rangle^2 \qquad 1$$

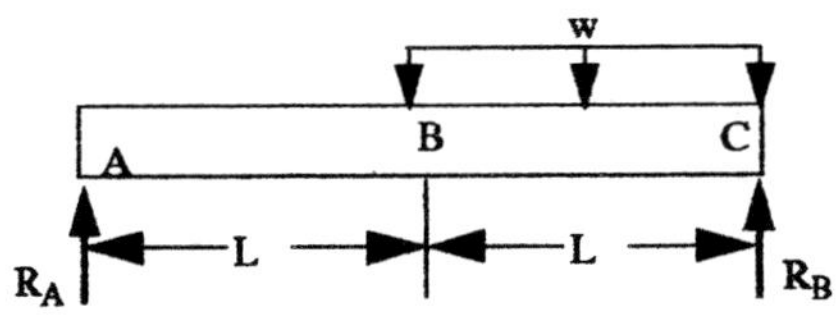

The boundary value problem can be written as

$$EI\frac{d^2v}{dx^2} = \frac{1}{4}wLx - \frac{w}{2}\langle x-L\rangle^2 \qquad 2$$

$$v(0) = 0 \qquad 3$$

$$v(2L) = 0 \qquad 4$$

Integrating Eq.(2) we obtain

$$EI\frac{dv}{dx} = \frac{1}{8}wLx^2 - \frac{w}{6}\langle x-L\rangle^3 + C_1 \qquad 5$$

$$EIv = \frac{1}{24}wLx^3 - \frac{w}{24}\langle x-L\rangle^4 + C_1x + C_2 \qquad 6$$

Substituting x=0 in Eq.(6) and using Eq.(3) we obtain

$$EIv(0) = C_2 = 0 \qquad 7$$

Substituting x=2L in Eq.(6) and using Eq.(4) we obtain

$$EIv(2L) = \frac{1}{24}wL(8L^3) - \frac{wL^4}{24} + C_1(2L) = 0 \text{ or } \frac{wL^4(8-1)}{24} + C_1(2L) = 0 \text{ or } C_1 = -\frac{7}{48}wL^3 \quad 8$$

Substituting the above constants in Eq.(6) we obtain the equation of the elastic- curve as

$$\boxed{v(x) = \frac{w}{48EI}[2Lx^3 - 7L^3x - 2\langle x-L\rangle^4]}$$

Substituting x=L in the elastic-curve, we obtain $v(L) = \frac{w}{48EI}[2L^4 - 7L^4]$ or $\boxed{v(L) = \frac{-5wL^4}{48EI}}$

7.52 Solve problem 7.12 using Discontinuity Functions.

Solution: $v(x) = ?$, $v(L) = ?$

Fig(a) shows the free body diagram of the beam. The reaction force $A_y = wL$ and $M_A = 3wL^2/2$ were found in problem 6.56. Using templates shown in Fig. 7-33 the moment equation can be written in terms of discontinuity function as

$$M_z = wLx - \frac{3}{2}wL^2 - \frac{w}{2}\langle x-L\rangle^2 \quad 1$$

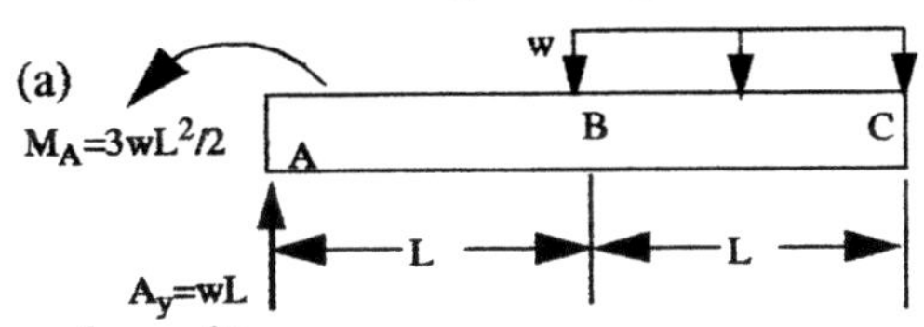

The boundary value problem can be written as

$$EI\frac{d^2v}{dx^2} = wLx - \frac{3}{2}wL^2 - \frac{w}{2}\langle x-L\rangle^2 \quad 2$$

$$v(0) = 0 \quad 3$$

$$\frac{dv}{dx}(0) = 0 \quad 4$$

Integrating Eq.(2) we obtain

$$EI\frac{dv}{dx} = \frac{1}{2}wLx^2 - \frac{3}{2}wL^2x - \frac{w}{6}\langle x-L\rangle^3 + C_1 \quad 5$$

$$EIv = \frac{1}{6}wLx^3 - \frac{3}{4}wL^2x^2 - \frac{w}{24}\langle x-L\rangle^4 + C_1x + C_2 \quad 6$$

Substituting x=0 in Eq.(5) and using Eq.(4) we obtain

$$EI\frac{dv}{dx}(0) = C_1 = 0 \quad 7$$

Substituting x=2L in Eq.(6) and using Eq.(3) we obtain

$$EIv(0) = C_2 = 0 \quad 8$$

Substituting the above constants in Eq.(6) we obtain the equation of the elastic- curve as

$$\boxed{v = \frac{w}{24EI}[4Lx^3 - 18L^2x^2 - \langle x-L\rangle^4]}$$

Substituting x = L in the equation of the elastic-curve, we obtain

$v(L) = \frac{w}{24EI}[4L^4 - 18L^4]$ or $\boxed{v(L) = \frac{-7wL^4}{12EI}}$

7.53 (a) Solve for the Elastic Curve for the beam and loading shown in Fig. P7.13. (b) Determine the slope and deflection at point C.

Solution: v(x) = ?, v(2L) = ?

Using templates shown in Fig. 7-33 the moment equation can be written in as

$$M_z = \frac{wx^2}{2} - \frac{w}{2}\langle x-L\rangle^2 - wL\langle x-2L\rangle^1 - wL^2\langle x-2L\rangle^0 \qquad 1$$

The boundary value problem can be written as

$$EI\frac{d^2v}{dx^2} = \frac{wx^2}{2} - \frac{w}{2}\langle x-L\rangle^2 - wL\langle x-2L\rangle^1 - wL^2\langle x-2L\rangle^0 \qquad 2$$

$$v(3L) = 0 \qquad 3$$

$$\frac{dv}{dx}(3L) = 0 \qquad 4$$

Integrating Eq.(2) we obtain

$$EI\frac{dv}{dx} = \frac{wx^3}{6} - \frac{w}{6}\langle x-L\rangle^3 - \frac{wL}{2}\langle x-2L\rangle^2 - wL^2\langle x-2L\rangle^1 + C_1 \qquad 5$$

$$EIv = \frac{wx^4}{24} - \frac{w}{24}\langle x-L\rangle^4 - \frac{wL}{6}\langle x-2L\rangle^3 - \frac{wL^2}{2}\langle x-2L\rangle^2 + C_1x + C_2 \qquad 6$$

Substituting x=3L in Eq.(5) and using Eq.(4) we obtain

$$EI\frac{dv}{dx}(3L) = \frac{w(27L^3)}{6} - \frac{w}{6}(8L^3) - \frac{wL^3}{2} - wL^3 + C_1 = 0 \text{ or}$$

$$\frac{wL^3(27-8-3-6)}{6} + C_1 = 0 \text{ or } C_1 = \frac{-5wL^3}{3} \qquad 7$$

Substituting x=3L in Eq.(6) and using Eq.(3) we obtain

$$EI\frac{dv}{dx}(3L) = \frac{w}{24}(81L^4) - \frac{w}{24}(16L^4) - \frac{wL^4}{6} - \frac{wL^4}{2} - \frac{5}{3}wL^3(3L) + C_2 = 0 \text{ or}$$

$$\frac{wL^4(81-16-4-12-120)}{24} + C_2 = 0 \text{ or } C_2 = \frac{71wL^4}{24} \qquad 8$$

Substituting the above constants in Eq.(6) we obtain the equation of the elastic- curve as

$$v(x) = \frac{w}{24EI}[x^4 - \langle x-L\rangle^4 - 4L\langle x-2L\rangle^3 - 12L^2\langle x-2L\rangle^2 - 40L^3x + 71L^4]$$

Substituting x = 2L in the equation of the elastic-curve, we obtain

$$v(L) = \frac{w}{24EI}[16L^4 - L^4 - 80L^4 + 71L^4] \text{ or} \qquad \boxed{v(2L) = \frac{wL^4}{4EI}}$$

7.54 Solve problem 7.22 using Discontinuity Functions.

Solution: v(x) = ? v(2L) = ?

Fig(a) shows the free body diagram of the beam.

(a)

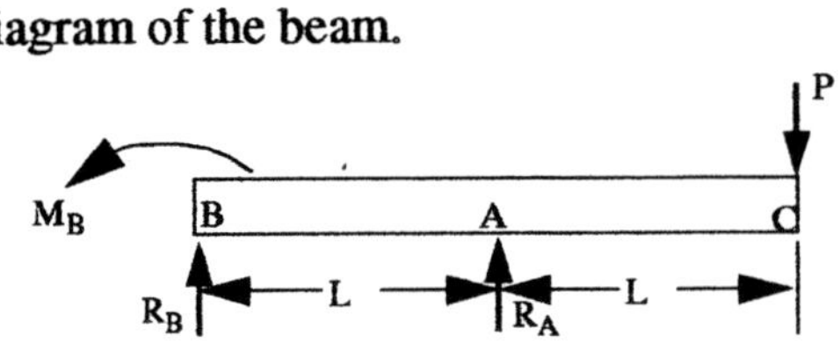

By equilibrium of forces in the y-direction, we obtain

$$R_B = P - R_A \qquad 1$$

By equilibrium of moment about point B, we obtain

$$M_B = 2PL - R_A L \qquad 2$$

Using templates shown in Fig. 7-33 the moment equation can be written as

$$M_z = -M_B + R_B x + R_A \langle x - L \rangle^1 \qquad 3$$

The boundary value problem can be written as

$$EI\frac{d^2v}{dx^2} = -M_B + R_B x + R_A \langle x - L \rangle^1 \qquad 4$$

$$\frac{dv}{dx}(0) = 0 \qquad 5$$

$$v(0) = 0 \qquad 6$$

$$v(L) = 0 \qquad 7$$

Integrating Eq.(4) we obtain

$$EI\frac{dv}{dx} = -M_B x + \frac{R_B x^2}{2} + \frac{R_A}{2}\langle x - L \rangle^2 + C_1 \qquad 8$$

$$EIv = -\frac{M_B}{2}x^2 + \frac{R_B x^3}{6} + \frac{R_A}{6}\langle x - L \rangle^3 + C_1 x + C_2 \qquad 9$$

Substituting x = 0 in Eqs.(2) and (9) and using Eqs.(5) and (6), we obtain

$$EI\frac{dv}{dx}(0) = C_1 = 0 \qquad 10$$

$$EIv(0) = C_2 = 0 \qquad 11$$

Substituting x = L in Eq.(9) and using Eq. (7) we obtain

$$EIv(L) = -\frac{M_B}{2}L^2 + \frac{R_B L^3}{6} = 0 \quad \text{or} \quad -3M_B + R_B L = 0 \qquad 12$$

Substituting Eqs.(1) and (2) into Eq.(12), we obtain

$$-3(2PL - R_A L) + (P - R_A)L = 0 \quad \text{and} \quad R_A = \frac{5}{2}P \qquad 13$$

Substituting R_A in Eqs.(1) and (2), we obtain

$$R_B = -\frac{3}{2}P \quad \text{and} \quad M_B = \frac{-PL}{2} \qquad 14$$

Substituting Eqs.(10), (11), (13) and (14) in Eq.(9), we obtain

$$v(x) = \frac{P}{EI}\left[\frac{x^2 L}{4} - \frac{x^3}{4} + \frac{5}{12}P\langle x - L \rangle^3\right] \quad \text{or} \quad \boxed{v(x) = \frac{P}{12EI}[3x^2 L - 3x^3 + 5\langle x - L \rangle^3]}$$

7.55 Using Area-Moment Method, determine the deflection in the middle for the beam shown in Fig. P7.2.

Solution $v_A = v(L/2) = ?$

The shear and moment diagrams are shown below. Also shown are the area under the moment curve and the location of the centroid. The deflection at C can be written as:

$$v(x_C) = v(x_B) + v'(x_B)(x_C - x_B) + \frac{1}{EI}[A_1(x_C - \bar{x}_1) + A_2(x_C - \bar{x}_2)] \qquad 1$$

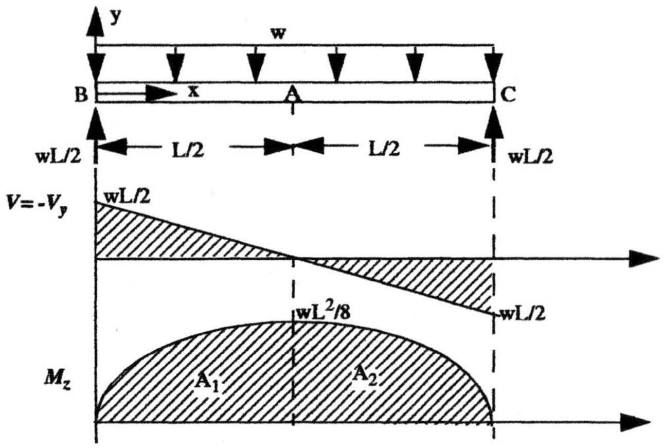

$$A_1 = A_2 = \frac{2}{3}\left(\frac{L}{2}\right)\left(\frac{wL^2}{8}\right) = \frac{wL^3}{24}$$

$$\bar{x}_1 = \frac{5}{8}\left(\frac{L}{2}\right) = \frac{5L}{16}$$

$$\bar{x}_2 = \frac{L}{2} + \frac{3}{8}\left(\frac{L}{2}\right) = \frac{11L}{16}$$

The deflections at the support are zero. Substituting $v(x_C) = 0$ and $v(x_B) = 0$ and the values of the areas and centroids in Eq. 1, the slope at B can be found as shown below.

$$v'(x_B)(L) + \frac{1}{EI}\left[\frac{wL^3}{24}\left(L - \frac{5L}{16}\right) + \frac{wL^3}{24}\left(L - \frac{11L}{16}\right)\right] = 0 \qquad \text{or} \qquad v'(x_B) = -\left(\frac{wL^3}{24EI}\right) \qquad 2$$

The deflection at A can be written as:

$$v(x_A) = v(x_B) + v'(x_B)(x_A - x_B) + \frac{1}{EI}[A_1(x_A - \bar{x}_1)]$$

Substituting the calculated values in the above equation, the deflection at A can be found as shown below:

$$v(x_A) = -\left(\frac{wL^3}{24EI}\right)\left(\frac{L}{2}\right) + \frac{1}{EI}\left[\frac{wL^3}{24}\left(\frac{L}{2} - \frac{5L}{16}\right)\right] \quad \text{or} \quad \boxed{v(x_A) = -\left(\frac{5wL^4}{384EI}\right)}$$

7.56 Using Area-Moment method, determine the deflection in the middle of the beam shown in Fig. P7.3.

Solution $v(L/2) = ?$

Fig.(a) shows the free body diagram of the beam. By equilibrium of moment about point C, we obtain $R_B = P$. By equilibrium of forces in the y-direction, we obtain $R_C = P$. The shear and moment diagrams are shown in Fig. (b) and (c). Also shown are the area under the moment curve and the location of the centroid.

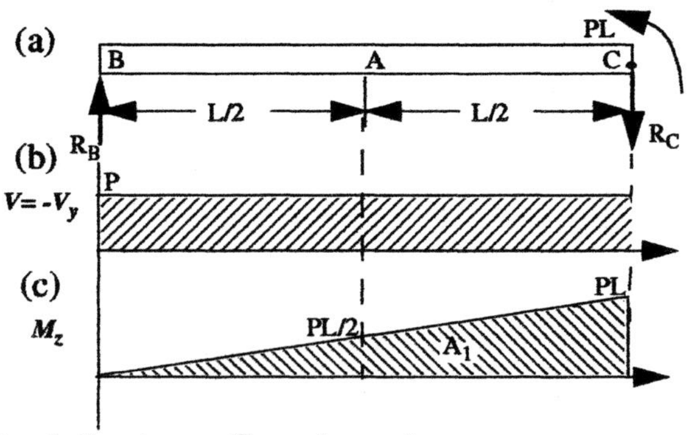

$$A_1 = \frac{1}{2}(PL)(L) = \frac{PL^2}{2}$$

$$\bar{x}_1 = \frac{2L}{3}$$

The deflection at C can be written as:

$$v(x_C) = v(x_B) + v'(x_B)(x_C - x_B) + \frac{1}{EI}[A_1(x_C - \bar{x}_1)] \qquad 1$$

The deflections at the support are zero. Substituting $v(x_C) = 0$ and $v(x_B) = 0$ and the values of the area and centroids in Eq. 1, the slope at B can be found as shown below.

$$v'(x_B)(L) + \frac{1}{EI}\left[\frac{PL^2}{2}\left(L - \frac{2L}{3}\right)\right] = 0 \qquad \text{or} \qquad v'(x_B) = -\left(\frac{PL^2}{6EI}\right) \qquad \textbf{2}$$

The area under the moment curve up to the middle of the beam is $A_2 = \frac{1}{2}\left(\frac{PL}{2}\right)\left(\frac{L}{2}\right) = \frac{PL^2}{8}$ and its centroid is at $\bar{x}_2 = \left(\frac{2}{3}\right)\left(\frac{L}{2}\right) = \frac{L}{3}$. The deflection at A can be written as:

$$v(x_A) = v(x_B) + v'(x_B)(x_A - x_B) + \frac{1}{EI}[A_2(x_A - \bar{x}_2)]$$

Substituting the calculated values in the above equation, the deflection at A can be found as shown below:

$$v(x_A) = -\left(\frac{PL^2}{6EI}\right)\left(\frac{L}{2}\right) + \frac{1}{EI}\left[\frac{PL^2}{8}\left(\frac{L}{2} - \frac{L}{3}\right)\right] \quad \text{or} \quad \boxed{v(x_A) = -\left(\frac{PL^3}{16EI}\right)}$$

7.57 Using Area-Moment Method, determine the deflection and slope at the free end of the beam shown in Fig. P7.4.

Solution $v_A = ?$, $\quad v'(x_A) = \frac{dv_A}{dx} = ?$

The shear-moment diagrams are shown below. Also shown are the area under the moment curve and the location of the centroid.

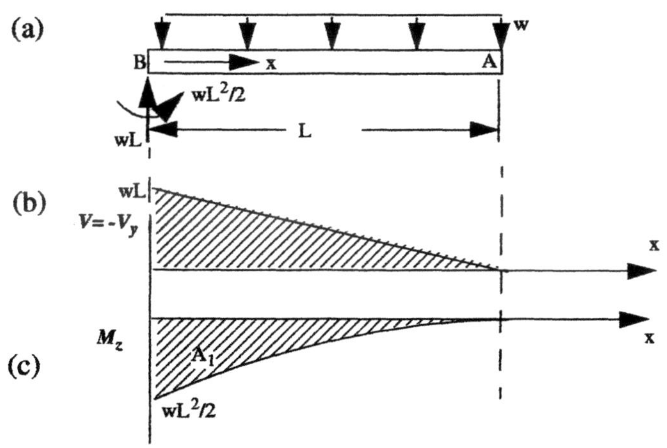

$$A_1 = -\frac{1}{3}\left(\frac{wL^2}{2}\right)(L) = -\left(\frac{wL^3}{6}\right)$$

$$\bar{x}_1 = \frac{L}{4}$$

Noting that the slope at B is zero, the slope at A can be found as shown below.

$$v'(x_A) = v'(x_B) + \frac{A_1}{EI} \quad \text{or} \quad \boxed{v'(x_A) = -\left(\frac{wL^3}{6EI}\right)}$$

Noting that the deflection and slope at B is zero, the deflection at A can be found as shown below.

$$v(x_A) = v(x_B) + v'(x_B)(x_A - x_B) + \frac{1}{EI}[A_1(x_A - \bar{x}_1)] = \frac{1}{EI}\left[\frac{wL^3}{6}\left(L - \frac{L}{4}\right)\right] \quad \text{or} \quad \boxed{v(x_A) = -\left(\frac{wL^4}{8EI}\right)}$$

7.58 Using Area-Moment Method, determine the slope at x = 0 and deflection at point x = L of the beam shown in Fig. P7.6.

Solution $v'(x_B) = \frac{dv_B}{dx}(0) = ?$ $\quad v(L) = v_A = ?$

Fig. (a) shows the free body diagram of the beam. By equilibrium of forces in the y-direction we obtain $R_B = P$. By equilibrium of moment about point A we obtain $M_A = 0$. Fig. (b) and (c) shows the bending-moment diagram. Also shown are the area under the moment curve and the location of the centroid.

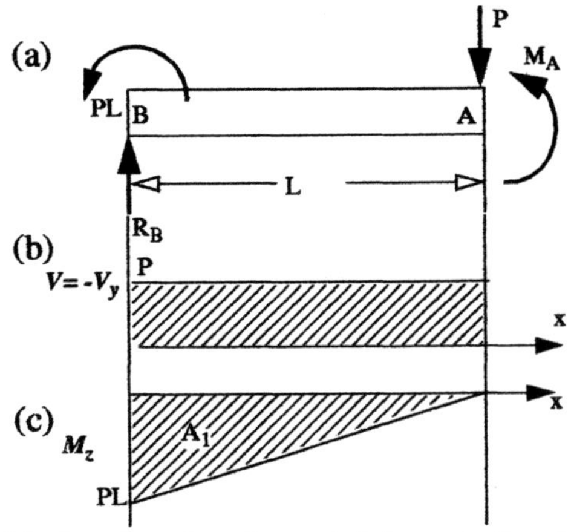

$$A_1 = -\left(\frac{1}{2}\right)(PL)(L) = -\left(\frac{PL^2}{2}\right)$$

$$\bar{x}_1 = \frac{L}{3}$$

Noting that the slope at A is zero the slope at B can be found as:

$$v'(x_A) = v'(x_B) + \frac{A_1}{EI} \quad \text{or} \quad v'(x_B) = -\left(\frac{A_1}{EI}\right) \quad \text{or} \qquad \boxed{v'(x_B) = \frac{PL^2}{2EI}}$$

Noting that the deflection B is zero, the deflection at A can be can be found as shown below.

$$v(x_A) = v(x_B) + v'(x_B)(x_A - x_B) + \frac{1}{EI}[A_1(x_A - \bar{x}_1)] = 0 + \left(\frac{PL^2}{2EI}\right)(L) + \frac{1}{EI}\left(-\frac{PL^2}{2}\right)\left(L - \frac{L}{3}\right) \quad \text{or}$$

$$\boxed{v(x_A) = \left(\frac{PL^3}{6EI}\right)}$$

7.59 Using Area-Moment Method, determine the slope at $x = 0$ and deflection at point $x = L$ of the beam shown in Fig. P7.9.

Solution $\quad v'(x_A) = \frac{dv_A}{dx}(0) = ? \quad v(L) = v_B = ?$

Fig. (a) shows the free body diagram of the beam. By equilibrium of moment about point C we obtain $R_B = 2wL/3$. By equilibrium of forces in the y-direction we obtain $R_C = wL/3$. Figs. (b) and (c) show the shear and moment diagram.Also shown are the areas under the moment curve and the location of the centroids

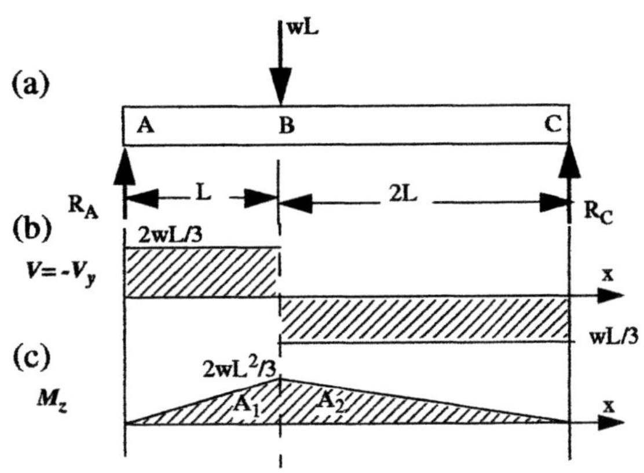

$$A_1 = \left(\frac{1}{2}\right)\left(\frac{2wL^2}{3}\right)(L) = \left(\frac{wL^3}{3}\right)$$

$$A_2 = \left(\frac{1}{2}\right)\left(\frac{2wL^2}{3}\right)(2L) = \left(\frac{2wL^3}{3}\right)$$

$$\bar{x}_1 = \frac{2L}{3}$$

$$\bar{x}_2 = L + \frac{2L}{3} = \frac{5L}{3}$$

The deflection at C can be written as:

$$v(x_C) = v(x_A) + v'(x_A)(x_C - x_A) + \frac{1}{EI}[A_1(x_C - \bar{x}_1) + A_2(x_C - \bar{x}_2)] \quad 1$$

The deflections at the support are zero. Substituting $v(x_C) = 0$ and $v(x_A) = 0$ and the values of the area and centroids in Eq. 1, the slope at A can be found as shown below.

$$v'(x_A)(3L) + \frac{1}{EI}\left[\left(\frac{wL^3}{3}\right)\left(3L - \frac{2L}{3}\right) + \frac{2wL^3}{3}\left(3L - \frac{5L}{3}\right)\right] = 0 \text{ or } \boxed{v'(x_A) = -\left(\frac{5wL^3}{9EI}\right)}$$

The deflection at B can be written as:

$$v(x_B) = v(x_A) + v'(x_A)(x_B - x_A) + \frac{1}{EI}[A_1(x_B - \bar{x}_1)]$$

Substituting the calculated values in the above equation, the deflection at B can be found as shown below:

$$v(x_B) = -\left(\frac{5wL^3}{9EI}\right)(L) + \frac{1}{EI}\left[\frac{wL^3}{3}\left(L - \frac{2L}{3}\right)\right] \text{ or } \boxed{v(x_B) = -\left(\frac{4wL^4}{9EI}\right)}$$

7.60 Using Area-Moment Method, determine the slope at x = 0 and deflection at point x = L of the beam shown in Fig. P7.10.

Solution $v'(x_A) = \frac{dv_A}{dx}(0) = ?$ $\quad v(L) = v(x_B) = ?$

Fig. (a) shows the free body diagram of the beam. By equilibrium of forces in the y-direction we obtain $R_A = wL$. By equilibrium of moment about point C we obtain $M_C = wL^2/2$.Figs. (b) and (c) show the shear and moment diagram.Also shown are the areas under the moment curve and the location of the centroids

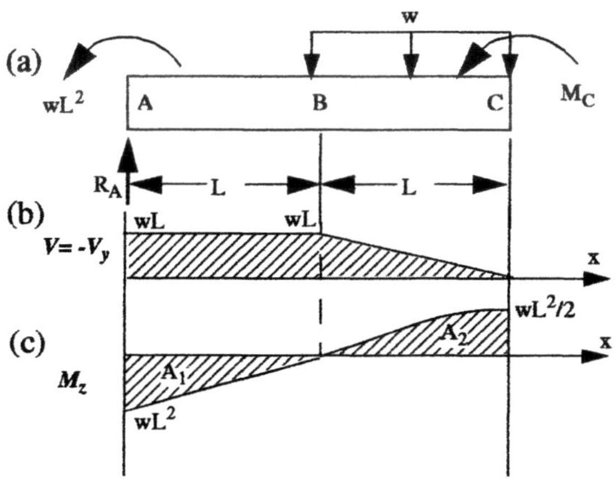

$$A_1 = -\left(\frac{1}{2}\right)(wL^2)(L) = -\left(\frac{wL^3}{2}\right)$$

$$A_2 = \left(\frac{2}{3}\right)\left(\frac{wL^2}{2}\right)(L) = \left(\frac{wL^3}{3}\right)$$

$$\bar{x}_1 = \frac{L}{3}$$

$$\bar{x}_2 = L + \frac{L}{4} = \frac{5L}{4}$$

Noting that the slope at C is zero, the slope at A can be found as shown below.

$$v'(x_C) = v'(x_A) + \frac{1}{EI}(A_1 + A_2) \quad \text{or} \quad v'(x_A) = -\frac{1}{EI}\left(-\frac{wL^3}{2} + \frac{wL^3}{3}\right) \text{ or } \boxed{v'(x_A) = \left(\frac{wL^3}{6EI}\right)}$$

The deflection at B can be written as:

$$v(x_B) = v(x_A) + v'(x_A)(x_B - x_A) + \frac{1}{EI}[A_1(x_B - \bar{x}_1)]$$

Substituting the calculated values in the above equation, the deflection at B can be found as shown below:

$$v(x_B) = \left(\frac{wL^3}{6EI}\right)(L) + \frac{1}{EI}\left[-\left(\frac{wL^3}{2}\right)\left(L - \frac{L}{3}\right)\right] \text{ or } \boxed{v(x_B) = -\left(\frac{wL^4}{6EI}\right)}$$

7.61 Using Area-Moment Method, determine the deflection and slope at the free end of the beam shown in Fig. P7.12.

Solution $v'(x_C) = \frac{dv_C}{dx}(2L) = ?$ $\quad v(2L) = v(x_C) = ?$

Fig. (a) shows the free body diagram of the beam. By equilibrium of forces in the y-direction we obtain $R_A = wL$. By equilibrium of moment about point A we obtain $M_A = 3wL^2/4$. Figs. (b) and (c) show the shear and moment diagram. Also shown are the areas under the moment curve and the location of the centroids

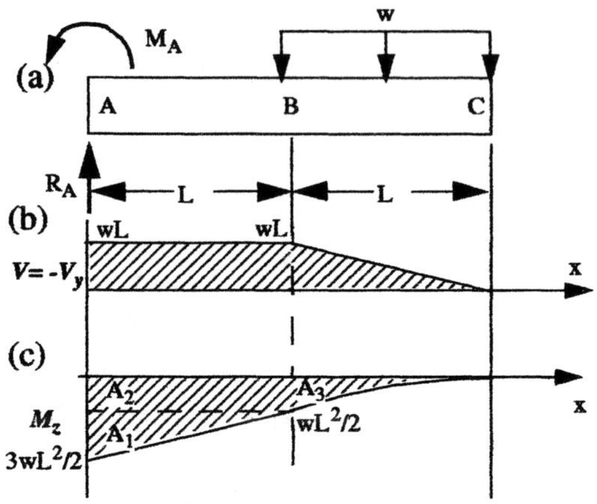

$$A_1 = -\left(\frac{1}{2}\right)(wL^2)(L) = -\left(\frac{wL^3}{2}\right) \qquad \bar{x}_1 = \frac{L}{3}$$

$$A_2 = -\left(\frac{wL^2}{2}\right)(L) = -\left(\frac{wL^3}{2}\right) \qquad \bar{x}_2 = \frac{L}{2}$$

$$A_3 = -\frac{1}{3}\left(\frac{wL^2}{2}\right)(L) = -\left(\frac{wL^3}{6}\right) \qquad \bar{x}_3 = L + \frac{1}{4}(L) = \frac{5L}{4}$$

The slope at point C can be written as:

$$v'(x_C) = v'(x_A) + \frac{1}{EI}[A_1 + A_2 + A_3] \qquad 1$$

The slope at point A is zero. Substituting $v'(x_A) = 0$ and the values of the area in Eq. 1, the slope at A can be found as shown below.

$$v'(x_C) = \frac{1}{EI}\left[-\frac{wL^3}{2} - \frac{wL^3}{2} - \frac{wL^3}{6}\right] \quad \text{or} \qquad \boxed{v'(x_C) = -\left(\frac{7wL^3}{6EI}\right)}$$

The deflection at C can be written as:

$$v(x_C) = v(x_A) + v'(x_A)(x_C - x_A) + \frac{1}{EI}[A_1(x_C - \bar{x}_1) + A_2(x_C - \bar{x}_2) + A_3(x_C - \bar{x}_3)] \qquad 2$$

The deflections and slope at A are zero. Substituting $v(x_A) = 0$ and $v'(x_A) = 0$ and the values of the area and centroids in Eq. 2, the deflection at C can be found as shown below.

$$v(x_C) = \frac{1}{EI}\left[\left(-\frac{wL^3}{2}\right)\left(2L - \frac{L}{3}\right) + \left(-\frac{wL^3}{2}\right)\left(2L - \frac{L}{2}\right) + \left(-\frac{wL^3}{6}\right)\left(2L - \frac{5L}{4}\right)\right] \quad \text{or}$$

$$v(x_C) = \frac{1}{EI}\left[\left(-\frac{5wL^4}{6}\right) + \left(-\frac{3wL^4}{4}\right) + \left(-\frac{wL^4}{8}\right)\right] \quad \text{or} \qquad \boxed{v(x_C) = -\left(\frac{41wL^4}{24EI}\right)}$$

Problem Set 8.1

8.1—8.9 In the following problems one could say that the normal stress on the incline AA is in tension, compression or can't be determined by inspection. Similarly we could say that the shear stress on the incline AA is positive, negative or can't be determined by inspection. Choose the correct answers for normal and shear stress on the incline AA by inspection. Assume coordinate z is perpendicular to this page and towards you.

8.1

Solution

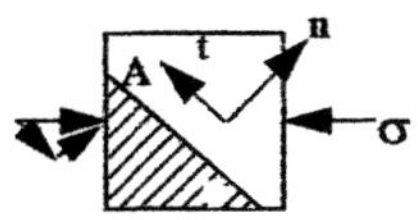

$\sigma_{nn} \Rightarrow$ (C)

$\tau_{nt} \Rightarrow$ positive

8.2

Solution

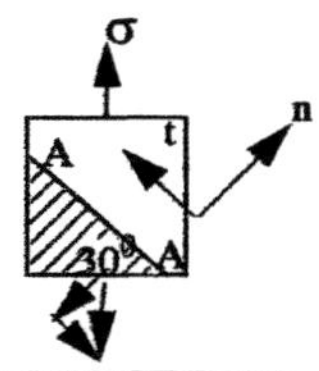

$\sigma_{nn} \Rightarrow$ (T)

$\tau_{nt} \Rightarrow$ positive

8.3

Solution

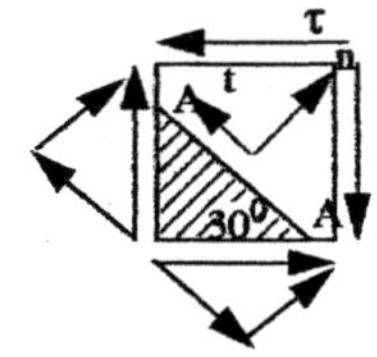

$\sigma_{nn} \Rightarrow$ (C)

$\tau_{nt} \Rightarrow$ Can't Say

8.4

Solution

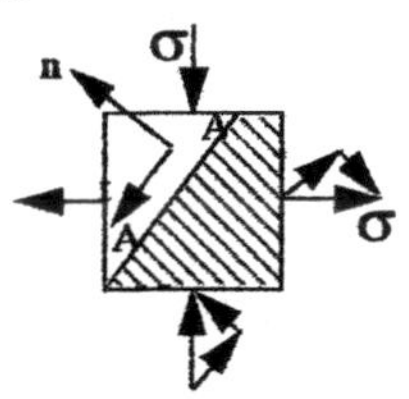

$\sigma_{nn} \Rightarrow$ Can't Say

$\tau_{nt} \Rightarrow$ positive

8.5

Solution

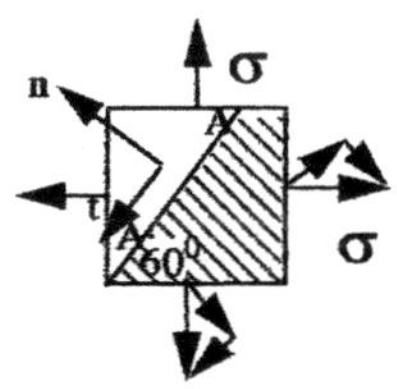

$\sigma_{nn} \Rightarrow$ (T)

$\tau_{nt} \Rightarrow$ Can't Say

8.6

Solution

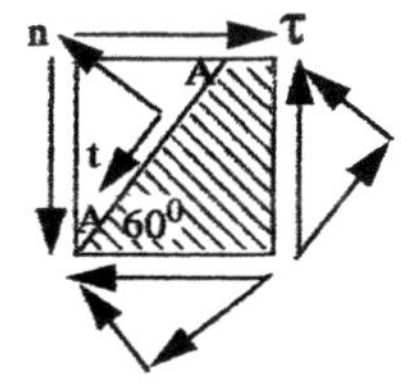

$\sigma_{nn} \Rightarrow$ (C)

$\tau_{nt} \Rightarrow$ Can't Say

8.7

Solution

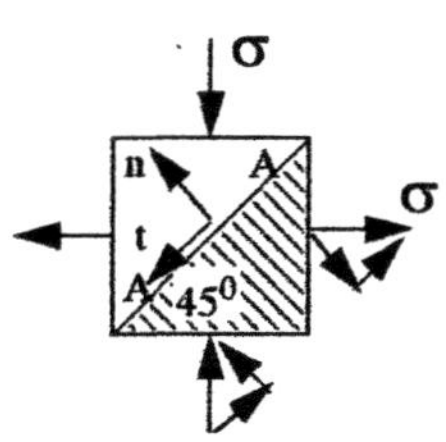

$\sigma_{nn} \Rightarrow$ zero or Can't Say

$\tau_{nt} \Rightarrow$ positive

8.8

Solution

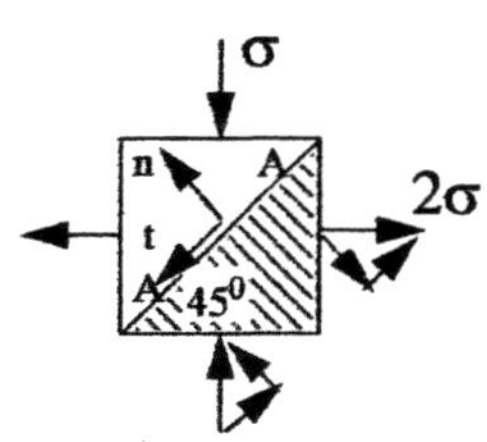

$\sigma_{nn} \Rightarrow$ (T) or Can't Say

$\tau_{nt} \Rightarrow$ positive

8.9

Solution

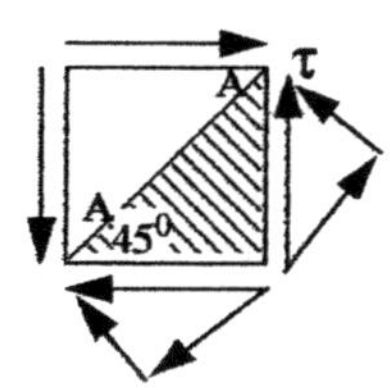

$\sigma_{nn} \Rightarrow$ (C)

$\tau_{nt} \Rightarrow$ zero or Can't Say

8.10 Determine the normal and shear stress on plane AA in problem 8.1 for $\sigma = 10$ ksi.

Solution $\sigma = 10$ ksi $\quad \sigma_{nn} = ?$ $\quad \tau_{nt} = ?$

Step 1 The stress cube is as shown in 8.1.

Step 2 The stress wedge and the local coordinate system is shown in Fig. (a).

Step 3 The force wedge is shown in Fig. (b). The dotted lines show the components of the force in n and t direction.

(a) Stress Wedge

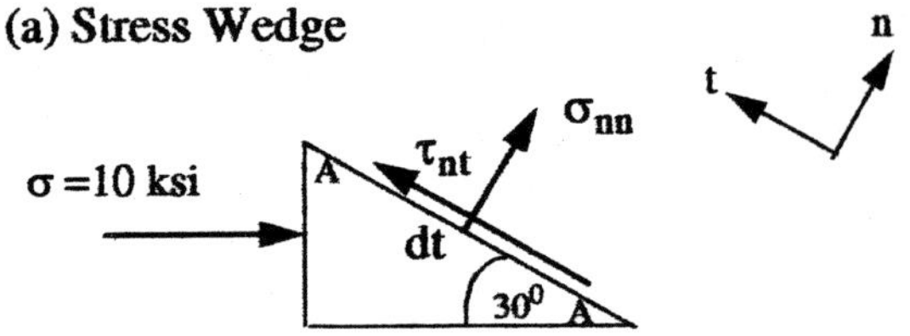

(b) Force Wedge

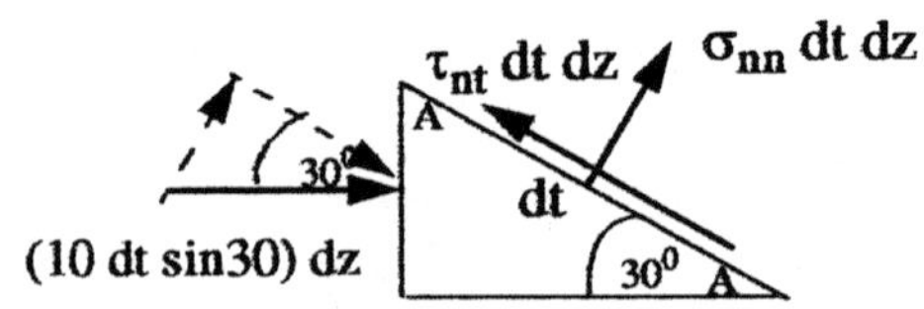

Step 4 By equilibrium of forces in the n-direction we obtain

$(10\ \ dt\ \ \sin 30)dz\ \ \sin 30 + \sigma_{nn}\ \ dt\ \ dz\ =\ 0$ or $\sigma_{nn}\ =\ -10\sin^2 30$ or $\boxed{\sigma_{nn}\ =\ 2.5\ \ ksi\ \ (C)}$

By equilibrium of forces in the t-direction we obtain

$-(10\ \ dt\ \ \sin 30)dz\ \ \cos 30 + \tau_{nt}\ \ dt\ \ dz\ =\ 0$ or $\tau_{nt}\ =\ 10\sin 30\cos 30$ or $\boxed{\tau_{nt}\ =\ 4.33\ \ ksi}$

8.11 Determine the normal and shear stress on plane AA in problem 8.4 for $\sigma = 10$ ksi

Solution $\sigma = 10$ ksi $\sigma_{nn} = ?$ $\tau_{nt} = ?$

Step 1 The stress cube is as shown in 8.4.

Step 2 The stress wedge and the local coordinate system is shown in Fig. (a).

Step 3 The force wedge is shown in Fig. (b). The dotted lines show the components of the force in n and t direction.

Step 4 By equilibrium of forces in the n-direction we obtain

$-(10\ \ dt\ \ \sin 60)dz\ \ \sin 60 + (10\ \ dt\ \ \cos 60)dz\ \ \cos 60 + \sigma_{nn}\ \ dt\ \ dz\ =\ 0$

$\sigma_{nn}\ =\ 10\sin^2 60 - 10\cos^2 60$ or $\boxed{\sigma_{nn}\ =\ 5.0\ \ ksi\ \ (T)}$

(a) Stress Wedge

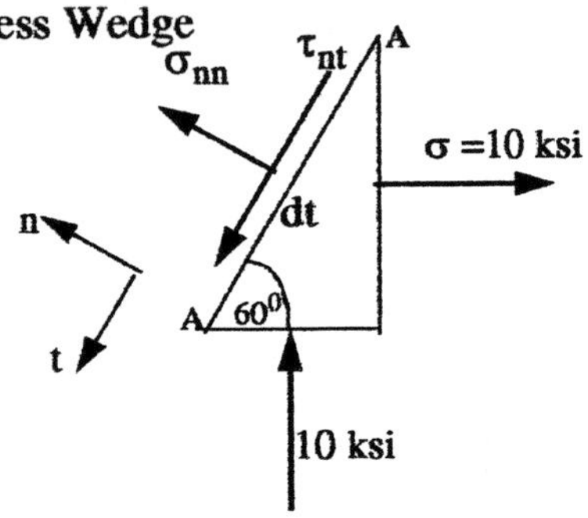

(b) Force Wedge

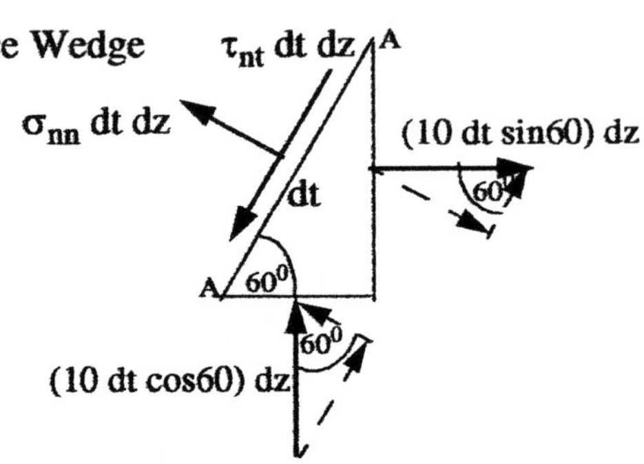

By equilibrium of forces in the t-direction we obtain

$-(10\ \ dt\ \ \sin 60)dz\ \ \cos 60 - (10\ \ dt\ \ \cos 60)dz\ \ \sin 60 + \tau_{nt}\ \ dt\ \ dz\ =\ 0$

$\tau_{nt}\ =\ 10\sin 60\cos 60 + 10\cos 60\sin 60$ or $\boxed{\tau_{nt}\ =\ 8.66\ \ ksi}$

8.12 Determine the normal and shear stress on plane AA in problem 8.6 for $\tau = 10$ ksi.

Solution $\sigma = 10$ ksi $\sigma_{nn} = ?$ $\tau_{nt} = ?$

Step 1 The stress cube is as shown in 8.6.

Step 2 The stress wedge and the local coordinate system is shown in Fig. (a).

Step 3 The force wedge is shown in Fig. (b). The dotted lines show the components of the force in n and t direction.

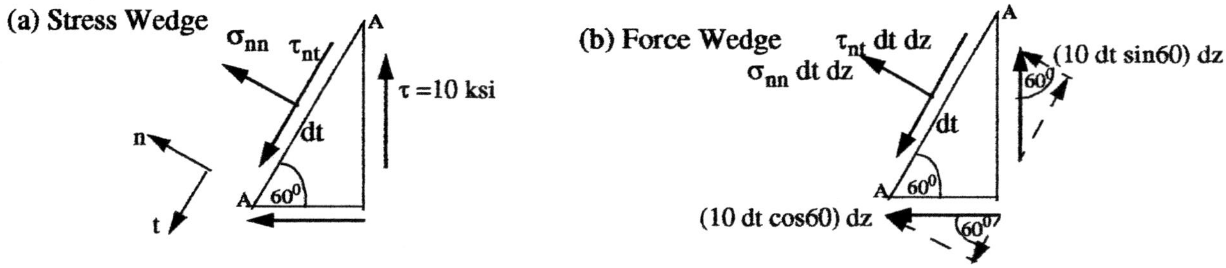

Step 4 By equilibrium of forces in the n-direction we obtain

$(10\ dt\ \sin 60)dz\ \cos 60 + (10\ dt\ \cos 60)dz\ \sin 60 + \sigma_{nn}\ dt\ dz = 0$

$\sigma_{nn} = -10\sin 60\cos 60 - 10\cos 60\sin 60$ or $\boxed{\sigma_{nn} = 8.66\ \text{ksi (C)}}$

By equilibrium of forces in the t-direction we obtain

$-(10\ dt\ \sin 60)dz\ \sin 60 + (10\ dt\ \cos 60)dz\ \cos 60 + \tau_{nt}\ dt\ dz = 0$

$\tau_{nt} = 10\sin^2 60 - 10\cos^2 60$ or $\boxed{\tau_{nt} = 5.0\ \text{ksi}}$

8.13 Determine the normal and shear stress on plane AA in problem 8.7 for $\sigma = 60$ MPa.

Solution $\sigma = 60$ MPa $\sigma_{nn} = ?$ $\tau_{nt} = ?$

Step 1 The stress cube is as shown in 8.7.
Step 2 The stress wedge and the local coordinate system is shown in Fig. (a).
Step 3 The force wedge is shown in Fig. (b). The dotted lines show the components of the force in n and t direction.

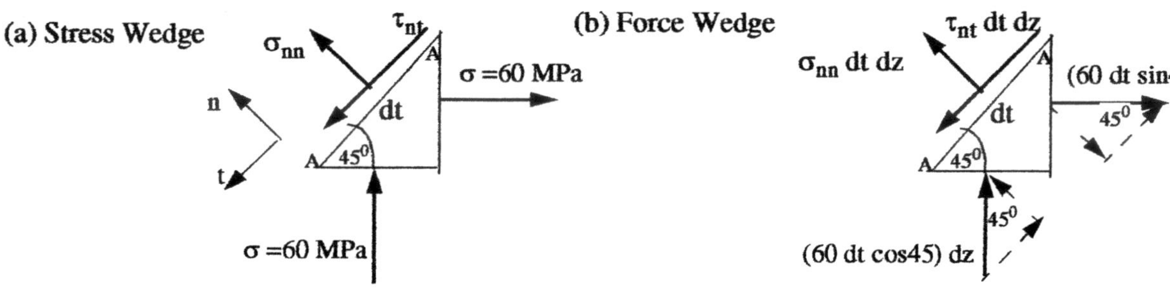

Step 4 By equilibrium of forces in the n-direction we obtain

$-(60\ dt\ \sin 45)dz\ \sin 45 + (60\ dt\ \cos 45)dz\ \cos 45 + \sigma_{nn}\ dt\ dz = 0$

$\sigma_{nn} = 60\sin^2 45 - 60\cos^2 45$ or $\boxed{\sigma_{nn} = 0}$

By equilibrium of forces in the t-direction we obtain

$-(60\ dt\ \sin 45)dz\ \cos 45 - (60\ dt\ \cos 45)dz\ \sin 45 + \tau_{nt}\ dt\ dz = 0$

$\tau_{nt} = 60\sin 45\cos 45 + 60\cos 45\sin 45$ or $\boxed{\tau_{nt} = 60\ \text{MPa}}$

8.14 Determine the normal and shear stress on plane AA in problem 8.9 for $\tau = 60$ MPa.

Solution $\sigma = 60$ MPa $\sigma_{nn} = ?$ $\tau_{nt} = ?$

Step 1 The stress cube is as shown in 8.7.
Step 2 The stress wedge and the local coordinate system is shown in Fig. (a).
Step 3 The force wedge is shown in Fig. (b). The dotted lines show the components of the force in n and t direction.

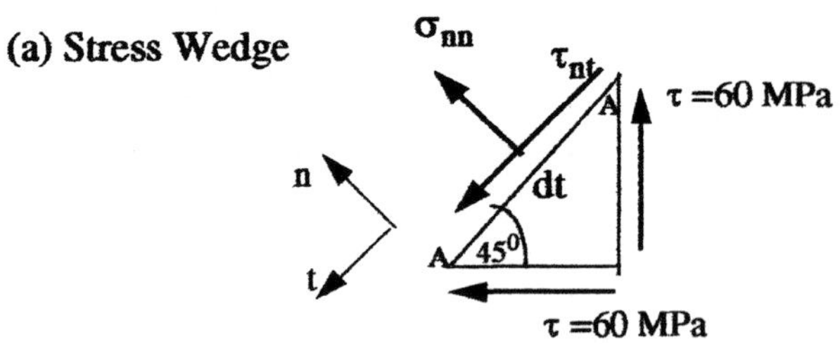

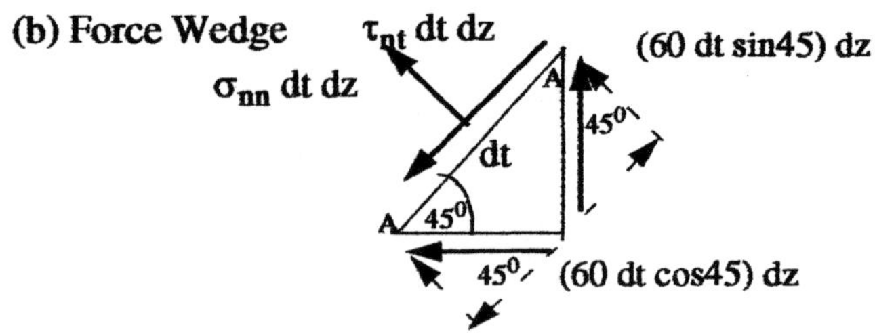

Step 4 By equilibrium of forces in the n-direction we obtain

$$(60\ dt\ \sin 45)dz\ \cos 45 + (60\ dt\ \cos 45)dz\ \sin 45 + \sigma_{nn}\ dt\ dz = 0$$

$\sigma_{nn} = -60\sin 45\cos 45 - 60\cos 45\sin 45$ or $\boxed{\sigma_{nn} = 60\ \text{MPa(C)}}$

By equilibrium of forces in the t-direction we obtain

$$-(60\ dt\ \sin 45)dz\ \sin 45 + (60\ dt\ \cos 45)dz\ \cos 45 + \tau_{nt}\ dt\ dz = 0$$

$\tau_{nt} = 60\sin^2 45 - 60\cos^2 45$ or $\boxed{\tau_{nt} = 0}$

In Problems 8.15 through 8.18 a shaft is adhesively bonded along the seam shown. By inspection determine if the adhesive will be in tension or compression.

8.15

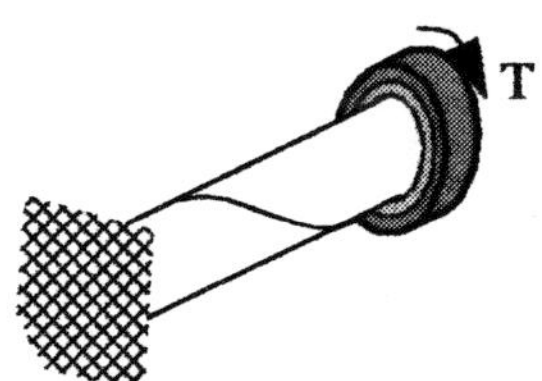

Fig. P8.15

Solution
The upper surface will be pressed into the lower surface. The normal stress in adhesive is: $\boxed{\text{Compression}}$

8.16

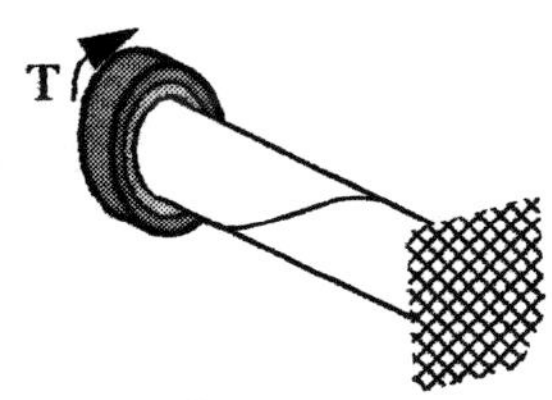

Fig. P8.16

Solution
The upper surface will be move away from the lower surface. The normal stress in adhesive is: $\boxed{\text{Tension}}$

8.17

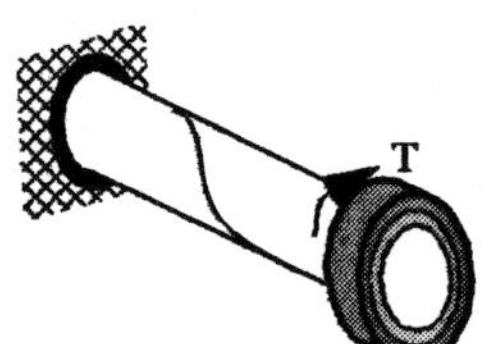

Fig. P8.17

Solution
The upper surface will be move away from the lower surface. The normal stress in adhesive is: $\boxed{\text{Tension}}$

8.18

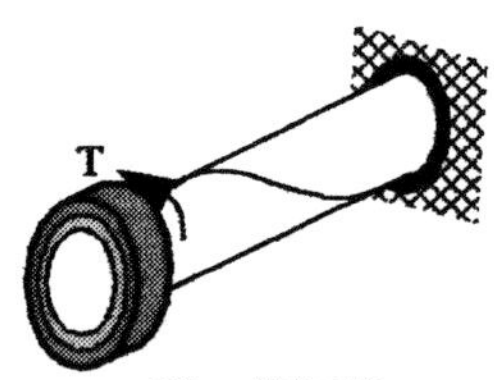

Fig. P8.18

Solution
The lower surface will be pressed into the upper surface. The normal stress in adhesive is: $\boxed{\text{Compression}}$

8.19 Determine the normal and shear stress on plane AA

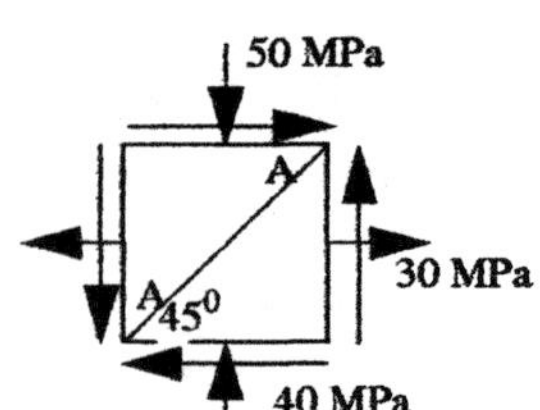

Fig. P8.19

Solution $\sigma_{nn} = ?$ $\tau_{nt} = ?$

Step 1 The stress cube is as shown in Fig. P8.19.

Step 2 The stress wedge and the local coordinate system is shown in Fig. (a).

Step 3 The force wedge is shown in Fig. (b).

(a) Stress Wedge (b) Force Wedge (c) Force Resolution

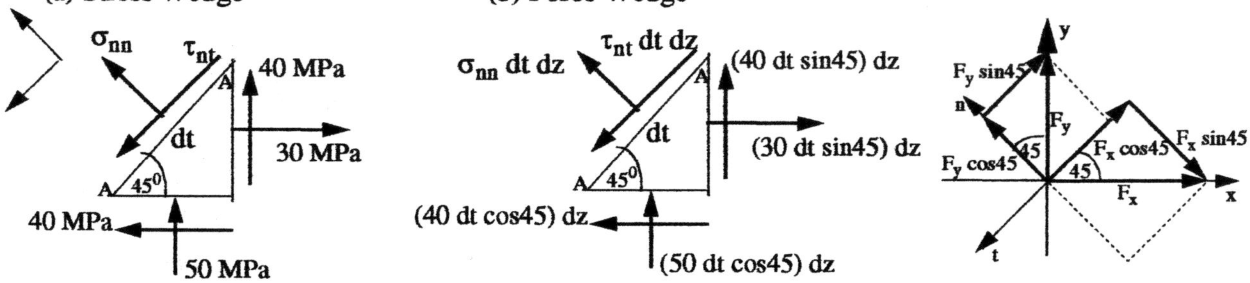

Step 4 Fig.(c) shows the resolution of forces from cartesian coordinates to normal and tangential coordinate system. From Fig.(b) we obtain:

$$F_x = [(30 dt \sin 45)dz - (40 dt \cos 45)dz] \quad \text{and} \quad F_y = [(40 dt \sin 45)dz + (50 dt \cos 45)dz] \qquad 1$$

By equilibrium of forces in the n-direction we obtain: $\sigma_{nn}\ dt\ dz - F_x \sin 45 + F_y \cos 45 = 0$ or

$$\sigma_{nn}\ dt\ dz - [(30 dt \sin 45)dz - (40 dt \cos 45)dz]\sin 45 + [(40 dt \sin 45)dz + (50 dt \cos 45)dz]\cos 45 = 0 \text{ or}$$

$$\sigma_{nn} = 30\sin^2 45 - 50\cos^2 45 - 2(40)\sin 45\cos 45 \text{ or} \qquad \boxed{\sigma_{nn} = 50\ \text{MPa(C)}}$$

By equilibrium of forces in the t-direction we obtain: $\tau_{nt}\ dt\ dz - F_x \cos 45 - F_y \sin 45 = 0$ or

$$\tau_{nt}\ dt\ dz - [(30 dt \sin 45)dz - (40 dt \cos 45)dz]\cos 45 - [(40 dt \sin 45)dz + (50 dt \cos 45)dz]\sin 45 = 0 \text{ or}$$

$$\tau_{nt} = 30\sin 45\cos 45 - 40\cos^2 45 + 40\sin^2 45 + 50\cos 45\sin 45 \text{ or} \qquad \boxed{\tau_{nt} = 40\ \text{MPa}}$$

8.20 Determine the normal and shear stress on plane AA

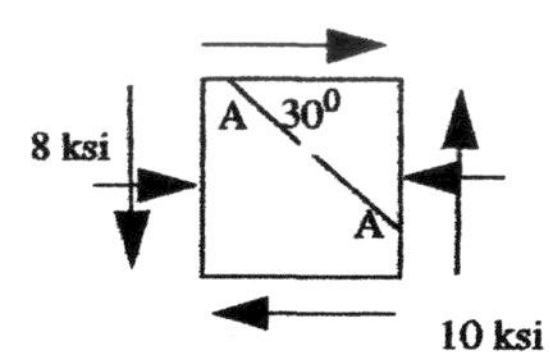

Fig. P8.20

Solution $\sigma_{nn} = ?$ $\tau_{nt} = ?$

Step 1 The stress cube is as shown in Fig. P8.20.

Step 2 The stress wedge and the local coordinate system is shown in Fig. (a).

Step 3 The force wedge is shown in Fig. (b).

(a) Stress Wedge (b) Force Wedge (c) Force Resolution

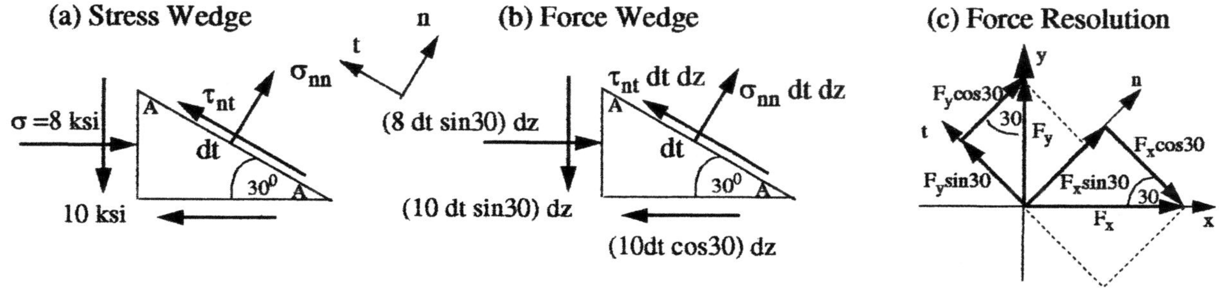

Step 4 Fig.(c) shows the resolution of forces from cartesian coordinates to normal and tangential coordinate system. From Fig.(b) we obtain:

$$F_x = [(8dt\sin 30)dz - (10dt\cos 30)dz] \quad \text{and} \quad F_y = -(10dt\sin 30)dz \qquad 1$$

By equilibrium of forces in the n-direction we obtain: $\sigma_{nn}\ dt\ dz + F_x\sin 30 + F_y\cos 30 = 0$ or

$$\sigma_{nn}\ dt\ dz + [(8dt\sin 30)dz - (10dt\cos 30)dz]\sin 30 + [-(10dt\sin 30)dz]\cos 30 = 0 \text{ or}$$

$$\sigma_{nn} = -8\sin^2 30 + 2(10)\sin 30\cos 30 \text{ or} \qquad \boxed{\sigma_{nn} = 6.66\ \text{ksi(T)}}$$

By equilibrium of forces in the t-direction we obtain: $\tau_{nt}\ dt\ dz - F_x\cos 30 + F_y\sin 30 = 0$ or

$$\tau_{nt}\ dt\ dz - [(8dt\sin 30)dz - (10dt\cos 30)dz]\cos 30 + [-(10dt\sin 30)dz]\sin 30 = 0 \text{ or}$$

$$\tau_{nt} = 8\sin 30\cos 30 - 10\cos^2 30 + 10\sin^2 30 \text{ or} \qquad \boxed{\tau_{nt} = -1.53\ \text{ksi}}$$

8.21 Determine the normal and shear stress on plane AA

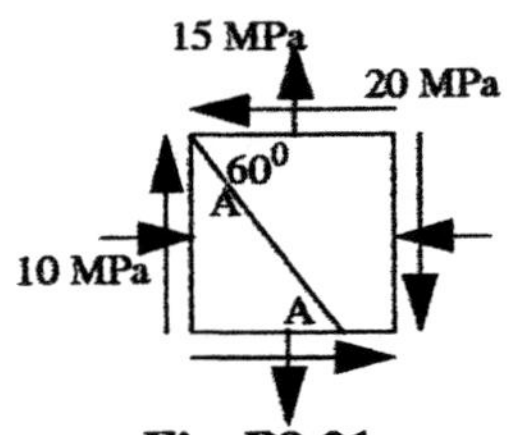

Fig. P8.21

Solution $\sigma_{nn} = ?$ $\tau_{nt} = ?$

Step 1 The stress cube is as shown in Fig. P8.21.
Step 2 The stress wedge and the local coordinate system is shown in Fig. (a).
Step 3 The force wedge is shown in Fig. (b).

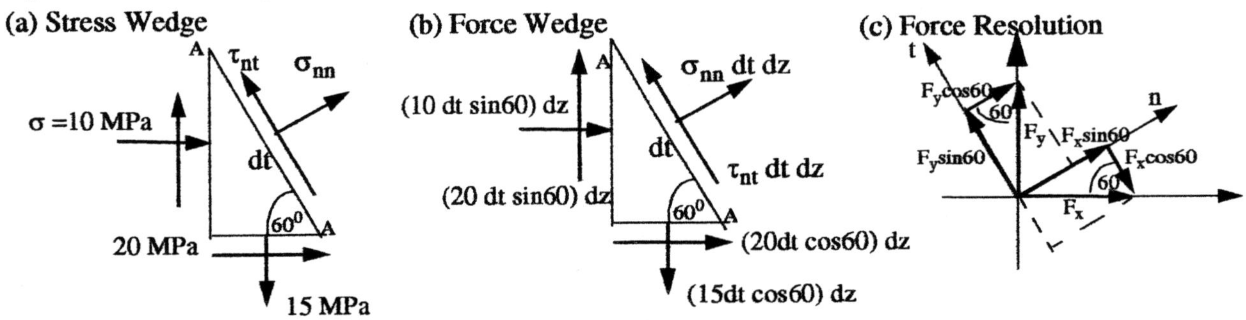

Step 4 Fig.(c) shows the resolution of forces from cartesian coordinates to normal and tangential coordinate system. From Fig.(b) we obtain:

$$F_x = [(10dt\sin 60)dz + (20dt\cos 60)dz] \quad \text{and} \quad F_y = [(20dt\sin 60)dz - (15dt\cos 60)dz] \qquad 1$$

By equilibrium of forces in the n-direction we obtain: $\sigma_{nn}\ dt\ dz + F_x\sin 60 + F_y\cos 60 = 0$ or

$$\sigma_{nn}\ dt\ dz + [(10dt\sin 60)dz + (20dt\cos 60)dz]\sin 60 + [(20dt\sin 60)dz - (15dt\cos 60)dz]\cos 60 = 0 \text{ or}$$

$$\sigma_{nn} = -10\sin^2 60 + 15\cos^2 60 - 2(20)\sin 60\cos 60 \text{ or} \qquad \boxed{\sigma_{nn} = 21.07\ \text{MPa(C)}}$$

By equilibrium of forces in the t-direction we obtain: $\tau_{nt}\ dt\ dz - F_x\cos 60 + F_y\sin 60 = 0$ or

$$\tau_{nt}\ dt\ dz - [(10dt\sin 60)dz + (20dt\cos 60)dz]\cos 60 + [(20dt\sin 60)dz - (15dt\cos 60)dz]\sin 60 = 0 \text{ or}$$

$$\tau_{nt} = 10\sin 60\cos 60 + 20\cos^2 60 - 20\sin^2 60 + 15\sin 60\cos 60 \text{ or} \qquad \boxed{\tau_{nt} = 0.825\ \text{MPa}}$$

8.22 The stresses at a point in plane stress are: $\sigma_{xx} = 45$ MPa (T), $\sigma_{yy} = 15$ MPa (T), and

$\tau_{xy} = -20$ MPa. Determine the normal and shear stresses on a plane passing through the point at 28° counterclockwise to the x axis.

Solution $\sigma_{nn} = ?$ $\tau_{nt} = ?$

Step 1 The stress cube is as shown in Fig. (a).
Step 2 The stress wedge and the local coordinate system is shown in Fig. (b).
Step 3 The force wedge is shown in Fig. (c).
Step 4 Fig.(d) shows the resolution of forces from cartesian coordinates to normal and tangential coordinate system. From Fig.(c) we obtain:

$$F_x = [(45dt\sin 28)dz + (20dt\cos 28)dz] \quad \text{and} \quad F_y = [-(20dt\sin 28)dz - (15dt\cos 28)dz] \tag{1}$$

By equilibrium of forces in the n-direction we obtain: $\sigma_{nn}\ dt\ dz - F_x\sin 28 + F_y\cos 28 = 0$ or

$\sigma_{nn}\ dt\ dz - [(45dt\sin 28)dz + (20dt\cos 28)dz]\sin 28 + [-(20dt\sin 28)dz - (15dt\cos 28)dz]\cos 28 = 0$ or

$\sigma_{nn} = 45\sin^2 28 + 15\cos^2 28 + 2(20)\sin 28\cos 28$ or $\boxed{\sigma_{nn} = 38.19\ \text{MPa(T)}}$

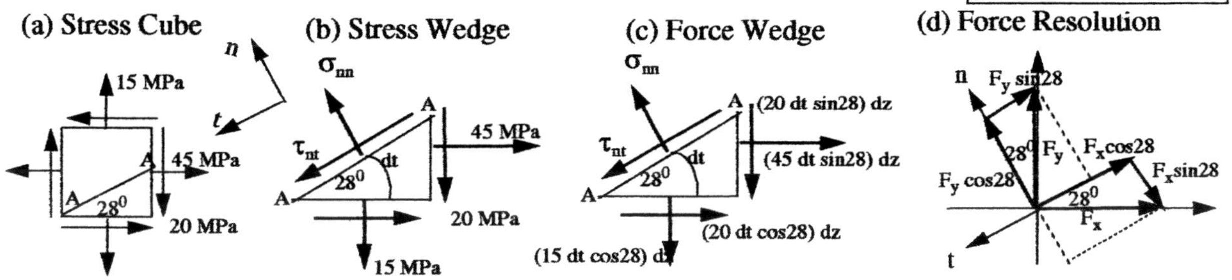

By equilibrium of forces in the t-direction we obtain: $\tau_{nt}\ dt\ dz - F_x\cos 28 - F_y\sin 28 = 0$ or

$\tau_{nt}\ dt\ dz - [(45dt\sin 28)dz + (20dt\cos 28)dz]\cos 28 - [-(20dt\sin 28)dz - (15dt\cos 28)dz]\sin 28 = 0$ or

$\tau_{nt} = 45\sin 28\cos 28 + 20\cos^2 28 - 20\sin^2 28 - 15\cos 28\sin 28$ or $\boxed{\tau_{nt} = 23.62\ \text{MPa}}$

8.23 The stresses at a point in plane stress are: $\sigma_{xx} = 45$ MPa (T), $\sigma_{yy} = 15$ MPa (C), and $\tau_{xy} = -20$ MPa. Determine the normal and shear stresses on a plane passing through the point at 38°clockwise to the x-axis.

Solution $\sigma_{nn} = ?$ $\tau_{nt} = ?$

Step 1 The stress cube is as shown in Fig. (a).
Step 2 The stress wedge and the local coordinate system is shown in Fig. (b).
Step 3 The force wedge is shown in Fig. (c).

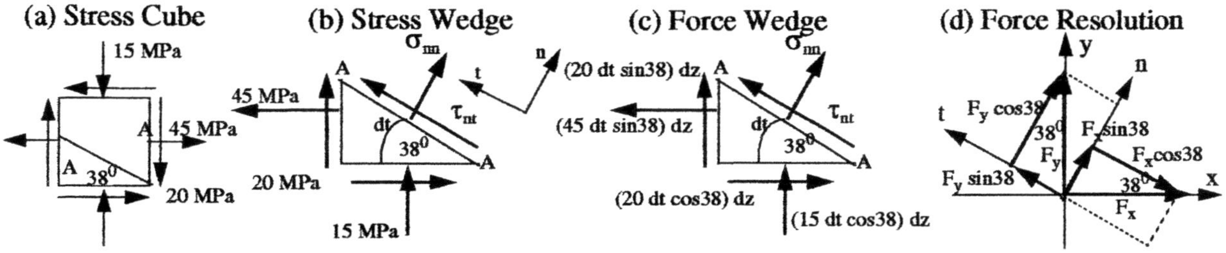

Step 4 Fig.(d) shows the resolution of forces from cartesian coordinates to normal and tangential coordinate system. From Fig.(c) we obtain:

$F_x = [(-45dt\sin 38)dz + (20dt\cos 38)dz]$ and $F_y = [(20dt\sin 38)dz + (15dt\cos 38)dz]$ 1

By equilibrium of forces in the n-direction we obtain: $\sigma_{nn}\ dt\ dz + F_x\sin 38 + F_y\cos 38 = 0$ or

$\sigma_{nn}\ dt\ dz + [(-45dt\sin 38)dz + (20dt\cos 38)dz]\sin 38 + [(20dt\sin 38)dz + (15dt\cos 38)dz]\cos 38 = 0$ or

$\sigma_{nn} = 45\sin^2 38 - 15\cos^2 38 - 2(20)\sin 38\cos 38$ or $\boxed{\sigma_{nn} = 11.66\ \text{MPa(C)}}$

By equilibrium of forces in the t-direction we obtain: $\tau_{nt}\ dt\ dz - F_x\cos 38 + F_y\sin 38 = 0$ or

$\tau_{nt}\ dt\ dz - [(-45dt\sin 38)dz + (20dt\cos 38)dz]\cos 38 + [(20dt\sin 38)dz + (15dt\cos 38)dz]\sin 38 = 0$ or

$\tau_{nt} = -45\sin 38\cos 38 + 20\cos^2 38 - 20\sin^2 38 - 15\cos 38\sin 38$ or $\boxed{\tau_{nt} = -24.27\ \text{MPa}}$

8.24 The stresses at a point in plane stress are: $\sigma_{xx} = 10\ \text{ksi (C)}$, $\sigma_{yy} = 20\ \text{ksi (C)}$, and $\tau_{xy} = 30\ \text{ksi}$. Determine the normal and shear stresses on a plane passing through the point that is 42^0 counter-clockwise to the x-axis.

Solution $\sigma_{nn} = ?$ $\tau_{nt} = ?$

Step 1 The stress cube is as shown in Fig. (a).
Step 2 The stress wedge and the local coordinate system is shown in Fig. (b).
Step 3 The force wedge is shown in Fig. (c).

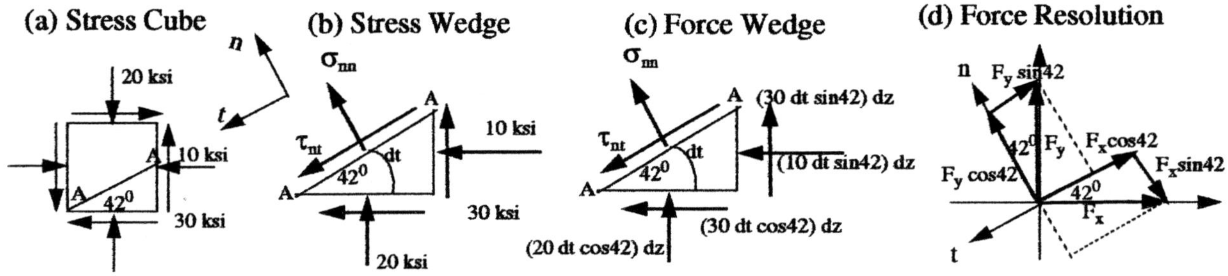

Step 4 Fig.(d) shows the resolution of forces from cartesian coordinates to normal and tangential coordinate system. From Fig.(c) we obtain:

$F_x = [-(10dt\sin 42)dz - (30dt\cos 42)dz]$ and $F_y = [(30dt\sin 42)dz + (20dt\cos 42)dz]$ 1

By equilibrium of forces in the n-direction we obtain: $\sigma_{nn}\ dt\ dz - F_x\sin 42 + F_y\cos 42 = 0$ or

$\sigma_{nn}\ dt\ dz - [-(10dt\sin 42)dz - (30dt\cos 42)dz]\sin 42 + [(30dt\sin 42)dz + (20dt\cos 42)dz]\cos 42 = 0$ or

$\sigma_{nn} = -10\sin^2 42 - 20\cos^2 42 - 2(30)\sin 42\cos 42$ or $\boxed{\sigma_{nn} = 45.36\ \text{ksi(C)}}$

By equilibrium of forces in the t-direction we obtain: $\tau_{nt}\ dt\ dz - F_x\cos 42 - F_y\sin 42 = 0$ or

$\tau_{nt}\ dt\ dz - [-(10dt\sin 42)dz - (30dt\cos 42)dz]\cos 42 - [(30dt\sin 42)dz + (20dt\cos 42)dz]\sin 42 = 0$ or

$\tau_{nt} = -10\sin 42\cos 42 - 30\cos^2 42 + 30\sin^2 42 + 20\cos 42\sin 42$ or $\boxed{\tau_{nt} = 1.84\ \text{ksi}}$

8.25 A cast iron shaft of diameter 25 mm fractured along a surface that is 45° to the axis of the shaft. The shear stress τ due to torsion is as shown. If the ultimate normal stress for the brittle cast iron material is 330 MPa (T), determine the torque that caused the fracture.

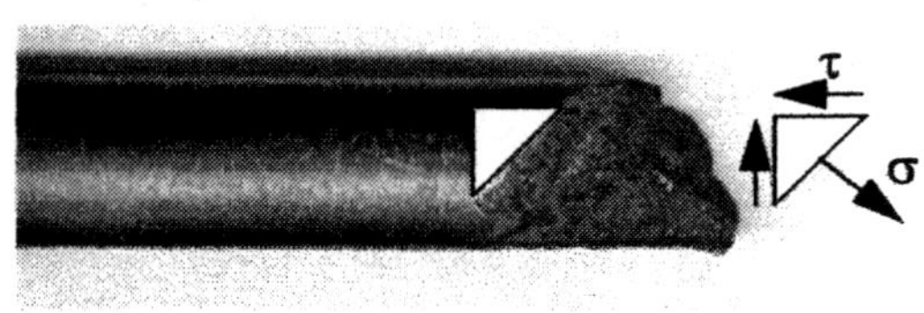

Fig. P8.25

Solution $d = 25$ mm $\sigma = 330$ MPa $T = ?$

The stress wedge and the local coordinate system is shown in Fig. (a). The force wedge is shown in Fig. (b). The dotted lines show the components of the force in n and t direction.

(a) Stress Wedge

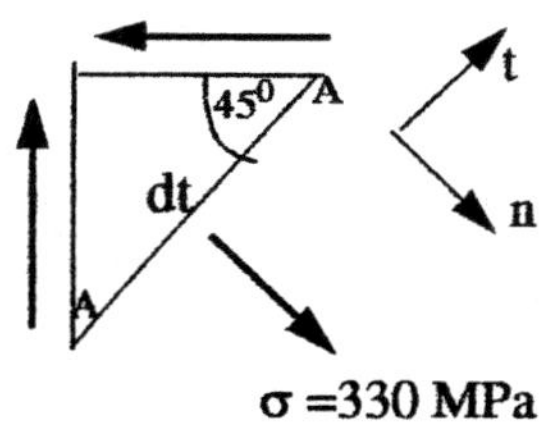

(b) Force Wedge

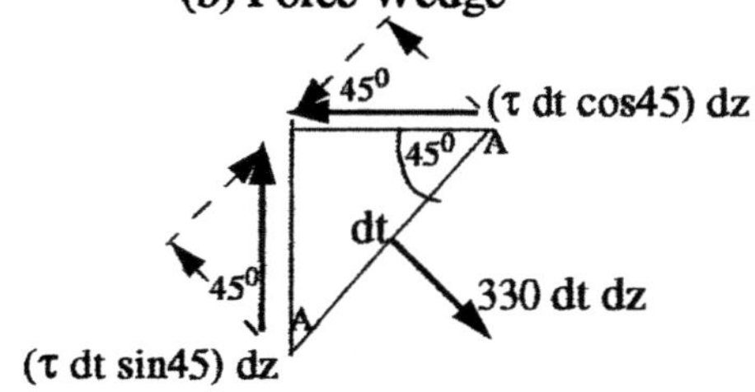

By equilibrium of forces in the n-direction we obtain

$330\ dt\ dz - [(\tau dt \cos 45)dz]\sin 45 - [(\tau dt \sin 45)dz]\cos 45 = 0$ or $2\tau \cos 45 \sin 45 = 330$ or $\tau = 330$ MPa

The polar moment of inertia is: $J = \frac{\pi}{32}(0.025^4) = 38.35(10^{-9})\text{m}^4$

The maximum torsional shear stress will be at the surface of the shaft i.e., $\rho = 0.0125$ m. From torsional shear stress formula we obtain:

$$\tau = \frac{T\rho}{J} = \frac{T(0.0125)}{38.35(10^{-9})} = 330(10^6) \quad \text{or} \qquad \boxed{T = 1012\ \text{N-m}}$$

8.26 In a wooden structure, a member was adhesively bonded along a plane 40° to the horizontal plane as shown. Due to a load P on the structure, the stresses at a point on the bonded plane were estimated as shown, where P is in lbs. If the adhesive strength in tension is 500 psi and its strength in shear is 200 psi, determine the maximum permissible load the structure can support without breaking the adhesive joint.

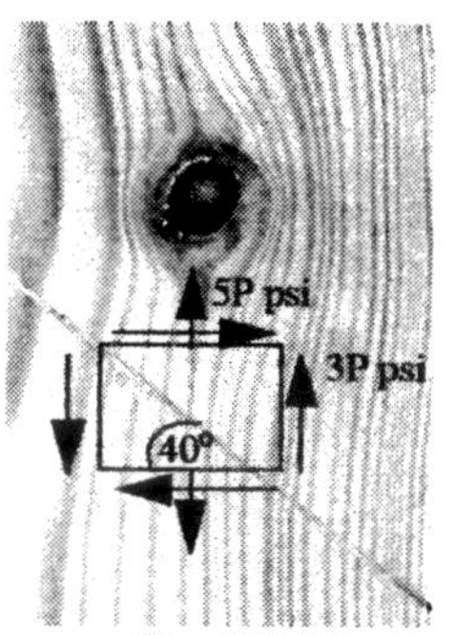

Fig. P8.26

Solution $\sigma_a \le 500\text{psi}$ $\tau_a \le 200\text{psi}$ $P = ?$

Fig. P8.26 shows the stress cube. Fig. (a) shows the stress wedge. Fig. (b) shows the force wedge.

(a) Stress Wedge

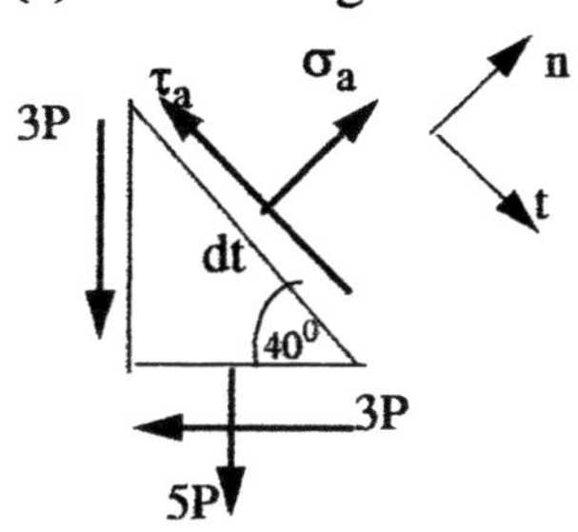

(b) Force Wedge

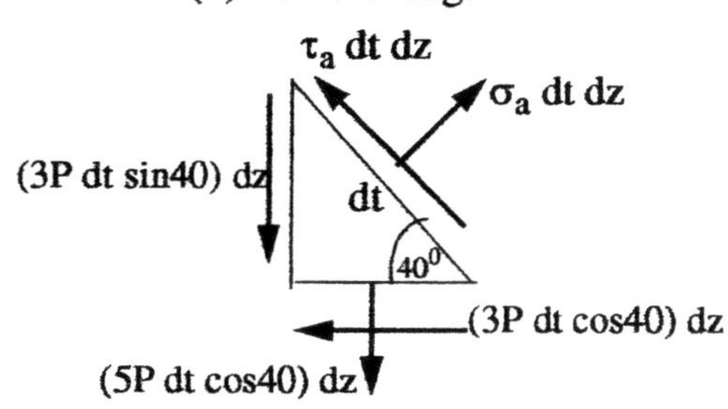

(c) Force Resolution

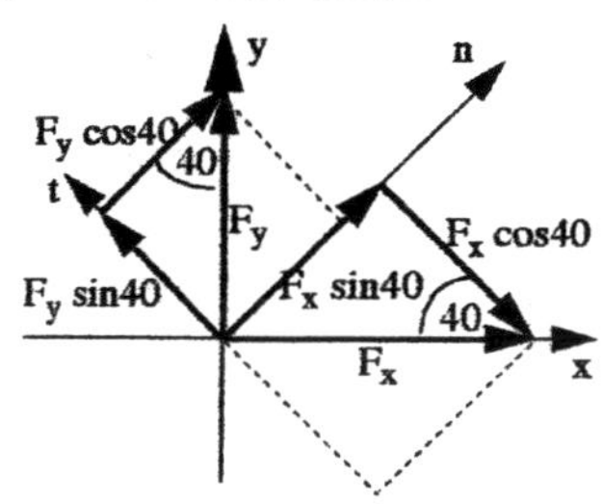

Fig.(c) shows the resolution of forces from cartesian coordinates to normal and tangential coordinate system. From Fig.(b) we obtain:

$$F_x = -(3Pdt\cos 40)dz \quad \text{and} \quad F_y = [(-3Pdt\sin 40)dz - (5Pdt\cos 40)dz] \qquad 1$$

By equilibrium of forces in the n-direction we obtain: $\sigma_a\ dt\ dz + F_x \sin 40 + F_y \cos 40 = 0$ or

$\sigma_a\ dt\ dz + [-(3Pdt\cos 40)dz]\sin 40 + [(-3Pdt\sin 40)dz - (5Pdt\cos 40)dz]\cos 40 = 0$ or

$$\sigma_a = 6P\sin 40\cos 40 + 5P\cos^2 40 = 5.888P \le 500 \quad \text{or} \quad P \le 84.9\ \text{lb} \qquad 1$$

By equilibrium of forces in the t-direction we obtain: $\tau_a\ dt\ dz - F_x\cos 40 + F_y\sin 40 = 0$ or

$\tau_a\ dt\ dz - [-(3Pdt\cos 40)dz]\cos 40 + [(-3Pdt\sin 40)dz - (5Pdt\cos 40)dz]\sin 40 = 0$ or

$$\tau_a = -3P\cos^2 40 + 3P\sin^2 40 + 5P\cos 40\sin 40 = 1.941P \le 200 \quad \text{or} \quad P \le 103\ \text{lb} \qquad 2$$

The maximum value of P that satisfy Eqs. (1) and (2) is $\boxed{P_{max} = 84.9\ \text{lb}}$

8.27 The stress at a point are $\sigma_{xx} = 8\ \text{ksi(T)}$ $\sigma_{yy} = 12\ \text{ksi(T)}$ $\sigma_{zz} = 8\ \text{ksi(C)}$. Determine the normal stress on a plane that has an outward normal at 60°, -60°, and 45°, to x, y, and z direction respectively.

Solution $\theta_x = 60°$ $\theta_y = -60°$ $\theta_z = 45°$ $\sigma_{nn} = ?$

The unit normal vector to the inclined plane is:

$$\bar{i}_n = \cos\theta_x\bar{i} + \cos\theta_y\bar{j} + \cos\theta_z\bar{k} = 0.5\bar{i} + 0.5\bar{j} + 0.7071\bar{k} \qquad 1$$

The projected areas in the coordinate directions are:

$$A_x = \cos\theta_x A = 0.5A \qquad A_y = \cos\theta_y A = 0.5A \qquad A_z = \cos\theta_z A = 0.7071A \qquad 2$$

Fig. (a) shows the stress wedge. Fig.(b) shows the force wedge.

(a) Stress Wedge

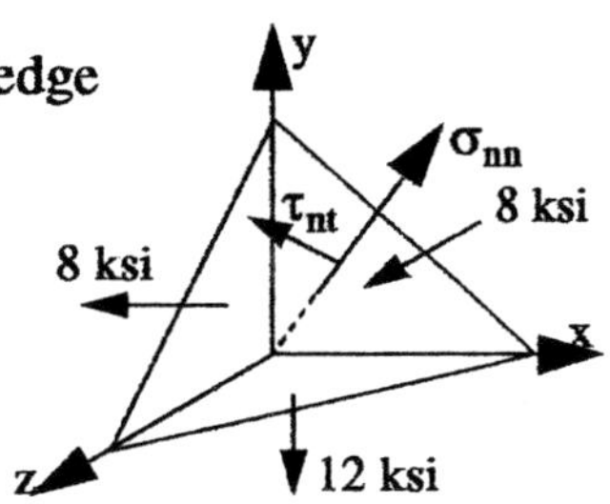

(a) Force Wedge

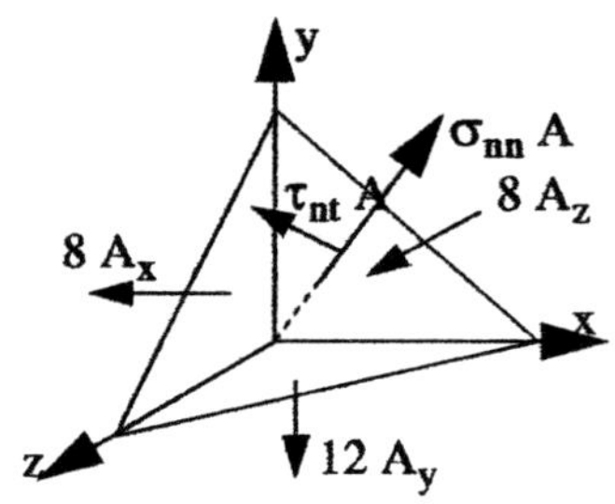

The forces on the coordinate planes can be written as a force vector shown below.

$$\bar{F} = -8A_x\bar{i} - 12A_y\bar{j} + 8A_z\bar{k} = (-4\bar{i} - 6\bar{j} + 5.657\bar{k})A \qquad 3$$

By equilibrium of forces in the n direction we obtain the following.

$\sigma_{nn}A + \bar{F} \bullet \bar{i}_n = 0$ or $\sigma_{nn}A + (-4\bar{i} - 6\bar{j} + 5.657\bar{k})A \bullet 0.5\bar{i} + 0.5\bar{j} + 0.7071\bar{k} = 0$ or

$\sigma_{nn} - (4)(0.5) - (6)(0.5) + (5.657)(0.7071) = 0$ or $\boxed{\sigma_{nn} = 1.0\ \text{ksi(T)}}$

8.28 The stress at a point are $\tau_{xy} = 125\ \text{MPa}$, $\tau_{xz} = -150\ \text{MPa}$. Determine the normal stress on a plane that has an outward normal at 72.54°, 120°, and 35.67°, to x, y, and z direction respectively.

Solution $\theta_x = 72.54°$ $\theta_y = 120°$ $\theta_z = 35.67°$ $\sigma_{nn} = ?$

The unit normal vector to the inclined plane is:

$$\bar{i}_n = \cos\theta_x\bar{i} + \cos\theta_y\bar{j} + \cos\theta_z\bar{k} = 0.3\bar{i} - 0.5\bar{j} + 0.8123\bar{k} \qquad 1$$

The projected areas in the coordinate directions are:

$$A_x = \cos\theta_x A = 0.3A \qquad A_y = \cos\theta_y A = -0.5A \qquad A_z = \cos\theta_z A = 0.8123A \qquad 2$$

Fig. (a) shows the stress wedge. Fig.(b) shows the force wedge.

(a) Stress Wedge

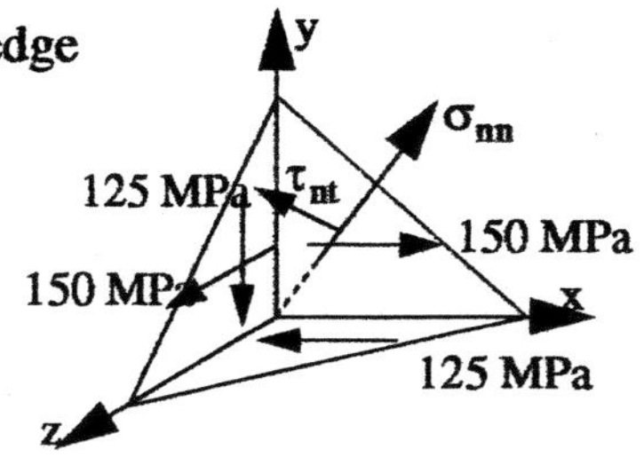

(a) Force Wedge

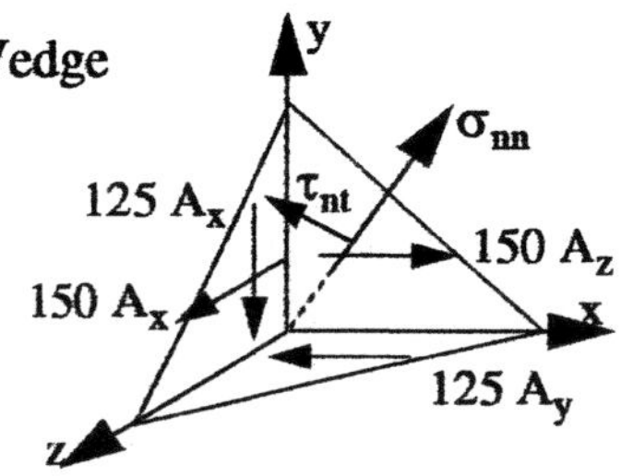

The forces on the coordinate planes can be written as a force vector shown below.

$$\bar{F} = (150A_z - 125A_y)\bar{i} - 125A_x\bar{j} + 150A_x\bar{k} = (184.345\bar{i} - 37.5\bar{j} + 45\bar{k})A \qquad 3$$

By equilibrium of forces in the n direction we obtain the following.

$$\sigma_{nn}A + \bar{F} \bullet \bar{i}_n = 0 \quad \text{or} \quad \sigma_{nn}A + (184.345\bar{i} - 37.5\bar{j} + 45\bar{k})A \bullet (0.3\bar{i} - 0.5\bar{j} + 0.8123\bar{k}) = 0 \text{ or}$$

$$\sigma_{nn} + (184.345)(0.3) - (37.5)(0.5) + (45)(0.8123) = 0 \text{ or} \qquad \boxed{\sigma_{nn} = 73.1 \text{ ksi(C)}}$$

Problem Set 8.2

8.29 Show that Equations 8.6 and 8.7 are correct by substituting the values Sines and Cosines given in Figure 8.14 into Equation 8.3 and 8.4.

Solution

From Fig. 8.14 we have:

$$R = \sqrt{\left(\frac{\sigma_{xx} - \sigma_{yy}}{2}\right)^2 + \tau_{xy}^2} \qquad \sin 2\theta_{1,2} = \pm\tau_{xy}/R \qquad \cos 2\theta_{1,2} = \pm\left(\frac{\sigma_{xx} - \sigma_{yy}}{2}\right)/R \qquad 1$$

Substituting Eq. 1 into Eq. 8.3 we obtain:

$$\sigma_{1,2} = \frac{(\sigma_{xx} + \sigma_{yy})}{2} + \frac{(\sigma_{xx} - \sigma_{yy})}{2}\left[\pm\left(\frac{\sigma_{xx} - \sigma_{yy}}{2}\right)/R\right] + \tau_{xy}[\pm\tau_{xy}/R] = \frac{(\sigma_{xx} + \sigma_{yy})}{2} \pm \frac{1}{R}\left[\left(\frac{\sigma_{xx} - \sigma_{yy}}{2}\right)^2 + \tau_{xy}^2\right] \text{ or}$$

$$\sigma_{1,2} = \frac{(\sigma_{xx} + \sigma_{yy})}{2} \pm \frac{1}{R}[R^2] = \frac{(\sigma_{xx} + \sigma_{yy})}{2} + R \qquad 2$$

Eq. 2 is the same as Eq. 8.6

Substituting Eq. 1 into Eq. 8.4 we obtain:

$$\tau_{1,2} = -\frac{(\sigma_{xx} - \sigma_{yy})}{2}[\pm\tau_{xy}/R] + \tau_{xy}\left[\pm\left(\frac{\sigma_{xx} - \sigma_{yy}}{2}\right)/R\right] = 0 \qquad 3$$

Eq. 3 is the same as Eq. 8.7.

8.30 Show that Equations 8.12 and 8.13 are correct by substituting the appropriate Sines and Cosines given in Figure 8.15 into Equation 8.3 and 8.4.

Solution

From Fig. 8.15 we have:

$$R = \sqrt{\left(\frac{\sigma_{xx} - \sigma_{yy}}{2}\right)^2 + \tau_{xy}^2} \qquad \sin 2\bar{\theta}_{1,2} = \mp\left(\frac{\sigma_{xx} - \sigma_{yy}}{2}\right)/R \qquad \cos 2\bar{\theta}_{1,2} = \pm\tau_{xy}/R \qquad 1$$

Substituting Eq. 1 into Eq. 8.3 we obtain:

$$\sigma_{nn} = \frac{(\sigma_{xx} + \sigma_{yy})}{2} + \frac{(\sigma_{xx} - \sigma_{yy})}{2}[\pm\tau_{xy}/R] + \tau_{xy}\left[\mp\left(\frac{\sigma_{xx} - \sigma_{yy}}{2}\right)/R\right] = \frac{(\sigma_{xx} + \sigma_{yy})}{2} \qquad 2$$

Eq. 2 is the same as Eq. 8.12.
Substituting Eq. 1 into Eq. 8.4 we obtain:

$$\tau_{nt} = -\frac{(\sigma_{xx}-\sigma_{yy})}{2}\left[\mp\left(\frac{\sigma_{xx}-\sigma_{yy}}{2}\right)/R\right] + \tau_{xy}[\pm\tau_{xy}/R] = \pm\frac{1}{R}\left[\left(\frac{\sigma_{xx}-\sigma_{yy}}{2}\right)^2 + \tau_{xy}^2\right] = \pm R = \pm[\sigma_1 - \sigma_2] \qquad 3$$

Eq. 3 is equivalent to Eq. 8.13

8.31 Derive Equation 8.6 by starting from Equation 8.8.

Solution

From Eq. 8.8 we have:

$$\sigma_{tt} = \sigma_{xx}\sin^2\theta + \sigma_{yy}\cos^2\theta - 2\tau_{xy}\cos\theta\sin\theta = \frac{(\sigma_{xx}+\sigma_{yy})}{2} + \frac{(\sigma_{yy}-\sigma_{xx})}{2}\cos 2\theta - \tau_{xy}\sin 2\theta \qquad 1$$

By differentiating Eq. 1 with respect to θ and letting its value be zero at $\theta = \theta_t$, we obtain

$$\left.\frac{d\sigma_{tt}}{d\theta}\right|_{\theta=\theta_t} = \left.\left(-2\frac{(\sigma_{yy}-\sigma_{xx})\sin 2\theta}{2} - 2\tau_{xy}\cos 2\theta\right)\right|_{\theta=\theta_t} = 0 \text{ or}$$

$$\tan 2\theta_t = \frac{-2\tau_{xy}}{(\sigma_{yy}-\sigma_{xx})} = \frac{2\tau_{xy}}{(\sigma_{xx}-\sigma_{yy})} = \tan 2\theta_p \qquad 2$$

As θ_t and θ_p are equivalent, we obtain from Fig. 8.14:

$$R = \sqrt{\left(\frac{\sigma_{xx}-\sigma_{yy}}{2}\right)^2 + \tau_{xy}^2} \qquad \sin 2\theta_{1,2} = \pm\tau_{xy}/R \qquad \cos 2\theta_{1,2} = \pm\left(\frac{\sigma_{xx}-\sigma_{yy}}{2}\right)/R \qquad 3$$

Substituting Eq. 3 into Eq. 1 we obtain:

$$\sigma_{1,2} = \frac{(\sigma_{xx}+\sigma_{yy})}{2} + \frac{(\sigma_{yy}-\sigma_{xx})}{2}\left[\pm\left(\frac{\sigma_{xx}-\sigma_{yy}}{2}\right)/R\right] - \tau_{xy}[\pm\tau_{xy}/R] = \frac{(\sigma_{xx}+\sigma_{yy})}{2} - \left[\pm\frac{1}{R}\left[\left(\frac{\sigma_{xx}-\sigma_{yy}}{2}\right)^2 + \tau_{xy}^2\right]\right]$$

$$\text{or } \sigma_{1,2} = \frac{(\sigma_{xx}+\sigma_{yy})}{2} \mp \frac{1}{R}[R^2] = \frac{(\sigma_{xx}+\sigma_{yy})}{2} \mp R = \frac{(\sigma_{xx}+\sigma_{yy})}{2} \mp \sqrt{\left(\frac{\sigma_{xx}-\sigma_{yy}}{2}\right)^2 + \tau_{xy}^2} \qquad 4$$

Eq. 4 is similar to Eq. 8.8. The difference is that in Eq. 4 the minus sign is associated with σ_1 and plus is with σ_2, which implies that $\theta_{1,2}$ in Eq. 3 are ninety degrees from those in Fig. 8.14.

8.32 Draw the Mohr's circle and determine the normal and shear stress on plane A.

Fig. P8.32

Solution $\sigma_A = ?$ $\tau_A = ?$

Figs. (a) and (c) below are the Mohr's circle associated with the stress cubes in Figs. P8.32(a) and (b). Figs. (b) and (d) show the direction of shear stress such that it causes rotation of plane A as per the location on the Mohr's circle. Local coordinate system is also shown.

(a) From Figs. (a) and (b) below we obtain: $\boxed{\sigma_A = (\sigma/2)\ (C)}$ and $\boxed{\tau_A = -(\sigma/2)}$

(b) From Figs. (c) and (d) below we obtain: $\boxed{\sigma_A = (\sigma/2)\ (T)}$ and $\boxed{\tau_A = (\sigma/2)}$

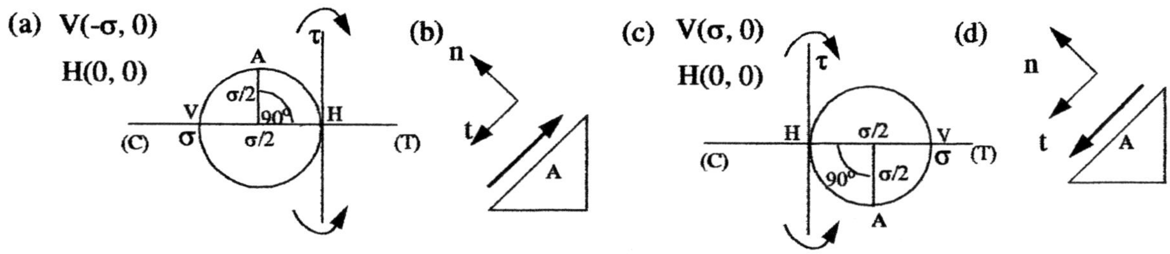

8.33 Draw the Mohr's circle and determine the normal and shear stress on plane A.

Fig. P8.33

Solution $\sigma_A = ?$ $\tau_A = ?$

(a) Fig.(a) shows that the Mohr's circle associated with Fig.P8.33(a) is a point. Thus all planes have the same normal stress s and zero shear stress. Thus: $\boxed{\sigma_A = \sigma \text{ (T)}}$ and $\boxed{\tau_A = 0}$

(b) Fig. (b) below is the Mohr's circle associated with the stress cubes in Fig. P8.33(b). Fig.(d) shows the direction of shear stress such that it causes rotation of plane A as per the location on the Mohr's circle. Local coordinate system is also shown.

(b) From Figs. (b) and (c) we obtain: $\boxed{\sigma_A = 0}$ and $\boxed{\tau_A = -\sigma}$

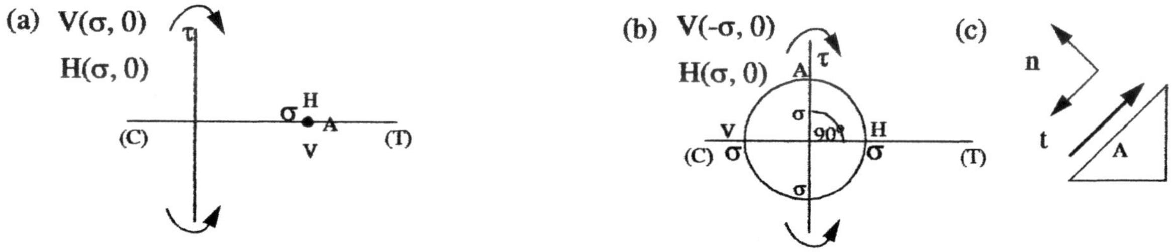

8.34 Draw the Mohr's circle and determine the normal and shear stress on plane A.

Fig. P8.34

Solution $\sigma_A = ?$ $\tau_A = ?$

Figs. (a) and (c) below are the Mohr's circle associated with the stress cubes in Figs. P8.32(a) and (b). Figs. (b) and (d) show the direction of shear stress such that it causes rotation of plane A as per the location on the Mohr's circle. Local coordinate system is also shown.

(a) From Figs. (a) and (b) below we obtain: $\boxed{\sigma_A = \tau \text{ (T)}}$ and $\boxed{\tau_A = 0}$

(b) From Figs. (c) and (d) below we obtain: $\boxed{\sigma_A = \tau \text{ (C)}}$ and $\boxed{\tau_A = 0}$

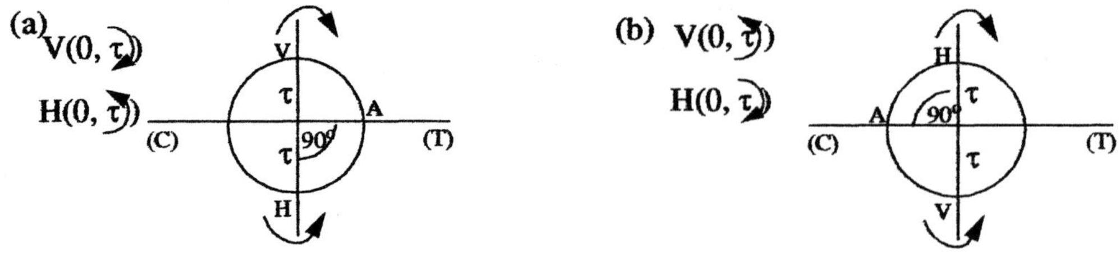

8.35 Explain the failure surfaces due to torsion that are shown in Figure 8.1.
Solution

Fig. (a) shows the stress cube with the torsional shear stress as shown in Fig. 8.1. Fig. (b) shows the Mohr's circle associated with the stress cube Fig. (c) shows the principal element.

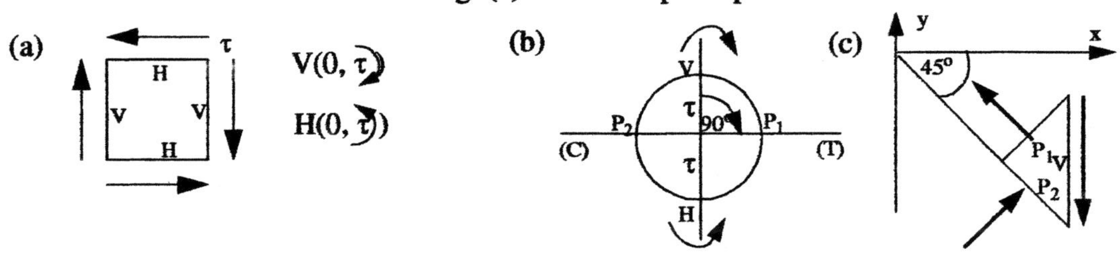

Cast iron being a brittle material will fail from maximum tensile stress that is from principle stress one. From the principal element in Fig. (c) we see that the orientation of the principal plane one (P_1) is as shown in Fig. 8.1 for cast iron under torsion. Aluminum being a ductile material will fail from maximum shear stress which is on the vertical (V) plane of the principal element. The orientation is the same as shown in Fig. 8.1 for aluminum bar under torsion.

8.36 Solve problem 8.19 by the Method of Equations.
Solution $\sigma_A = ?$ $\tau_A = ?$

Fig. P8.19 is shown below as Fig.(a). Fig.(b) shows the orientation of the outward normal to the plane AA.

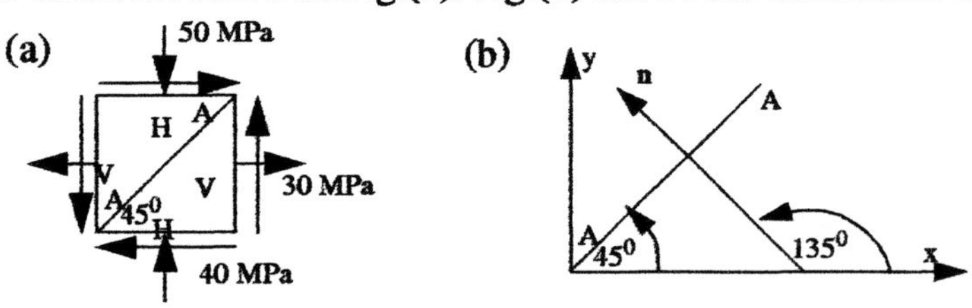

From Figs.(a) and (b) we obtain:

$$\sigma_{xx} = 30\text{MPa} \qquad \sigma_{yy} = -50\text{MPa} \qquad \tau_{xy} = 40\text{MPa} \qquad \theta = 135^{\circ} \qquad (1)$$

Substituting Eq. 1 into Eqs. 8.1 and 8.2 we obtain:

$$\sigma_A = (30)\cos^2 135 + (-50)\sin^2 135 + 2(40)\sin 135 \cos 135 \text{ or} \qquad \boxed{\sigma_A = 50\text{MPa(C)}}$$

$$\tau_A = -(30)\cos 135 \sin 135 + (-50)\sin 135 \cos 135 + (40)(\cos^2 135 - \sin^2 135) \qquad \boxed{\tau_A = 40\text{MPa}}$$

8.37 Solve problem 8.19 by the Mohr's Circle.
Solution $\sigma_A = ?$ $\tau_A = ?$

Using Fig. (a) in problem 8.37, we can write the coordinates of V and H plane as: V(30, 40↻) H(-50, 40↺).

Fig. (a) below shows the Mohr's circle. Fig. (b) shows the direction of the shear stress on plane A and the local coordinate system.

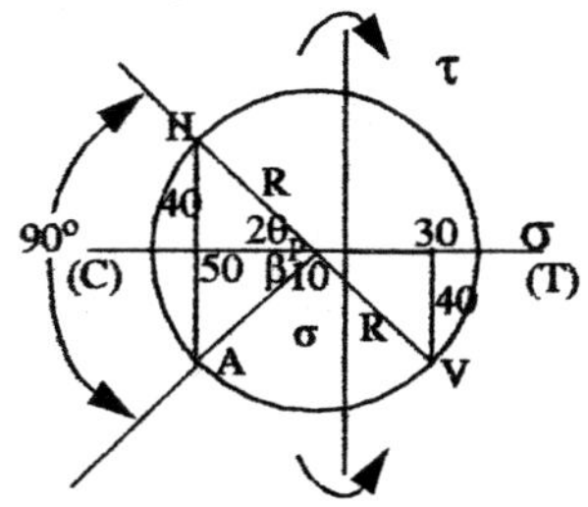

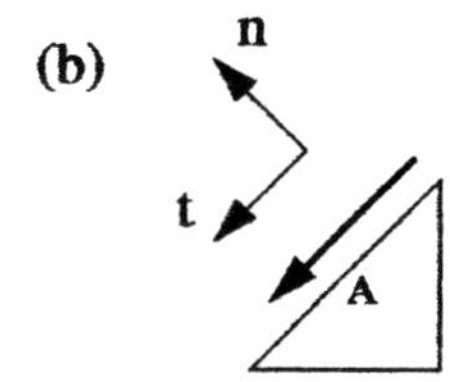

The radius of the circle is: $R = \sqrt{40^2 + 40^2} = 56.57\text{MPa}$

The angle $2\theta_p$ is: $2\theta_p = \text{atan}(40/40) = 45^\circ$. The angle β is: $\beta = 90^\circ - 45^\circ = 45^\circ$

The normal stress on plane A is: $\sigma_A = -10 - R\cos\beta = -50$ or $\boxed{\sigma_A = 50\text{MPa(C)}}$

The magnitude of shear stress on plane A is: $|\tau_A| = R\sin\beta = 40$. From Fig. (b) we see that shear stress on plane a is positive. $\boxed{\tau_A = 40\text{MPa}}$

8.38 Solve problem 8.20 the Method of Equations.

Solution $\sigma_A = ?$ $\tau_A = ?$

Fig. P8.20 is shown below as Fig.(a). Fig.(b) shows the orientation of the outward normal to the plane AA.

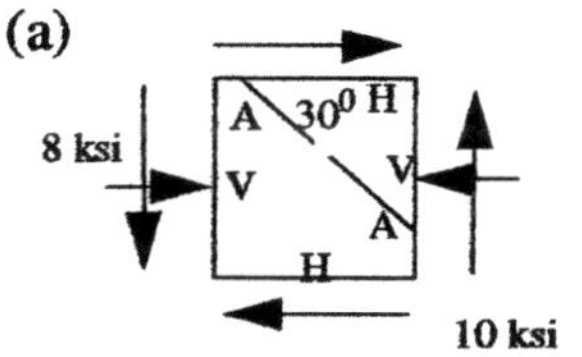

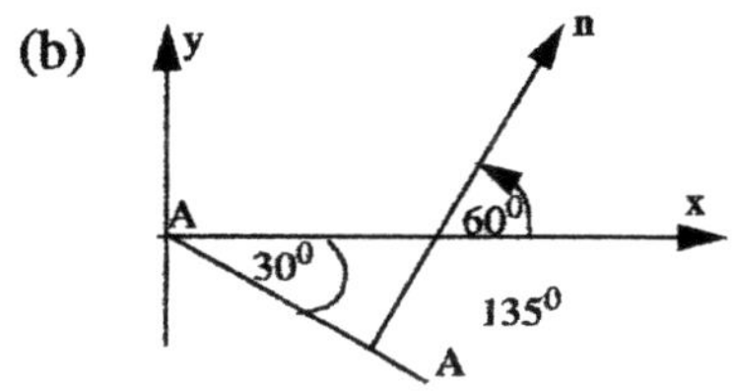

From Figs.(a) and (b) we obtain:

$$\sigma_{xx} = -8\text{ksi} \qquad \sigma_{yy} = 0 \qquad \tau_{xy} = 10\text{ksi} \qquad \theta = 60^\circ \tag{1}$$

Substituting Eq. 1 into Eqs. 8.1 and 8.2 we obtain:

$\sigma_A = (-8)\cos^2 60 + (0)\sin^2 60 + 2(10)\sin 60\cos 60$ or $\boxed{\sigma_A = 6.66\text{ksi(T)}}$

$\tau_A = -(30)\cos 60\sin 60 + (-50)\sin 60\cos 60 + (40)(\cos^2 60 - \sin^2 60)$ $\boxed{\tau_A = -1.54\text{ksi}}$

8.39 Solve problem 8.20 by the Mohr's Circle.

Solution $\sigma_A = ?$ $\tau_A = ?$

Using Fig. (a) in problem 8.39, we can write the coordinates of V and H plane as: V(-8, 10↻) H(0, 10↺). Fig. (a) below shows the Mohr's circle. Fig. (b) shows the direction of the shear stress on plane A and the local coordinate system.

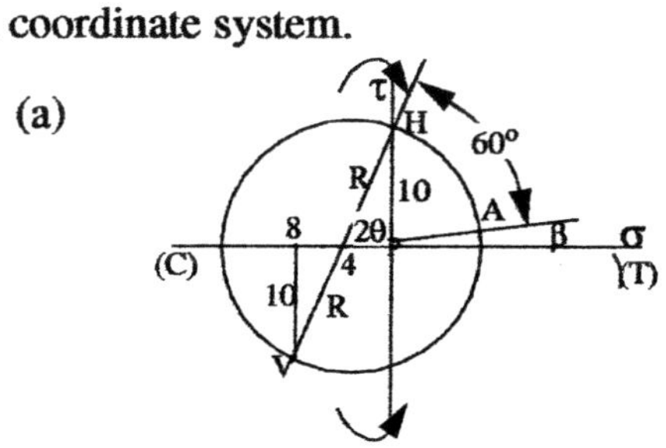

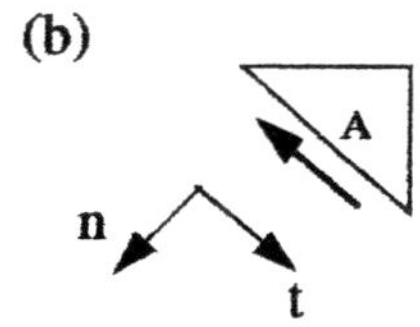

The radius of the circle is: $R = \sqrt{4^2 + 10^2} = 10.77\text{ksi}$

The angle $2\theta_p$ is: $2\theta_p = \text{atan}(10/4) = 68.2^\circ$. The angle β is: $\beta = 68.2^\circ - 60^\circ = 8.2^\circ$

The normal stress on plane A is: $\sigma_A = -4 + R\cos\beta = 6.66$ or $\boxed{\sigma_A = 6.66\text{ksi(T)}}$

The magnitude of shear stress on plane A is: $|\tau_A| = R\sin\beta = 1.54$. From Fig. (b) we see that shear stress on plane A is negative. $\boxed{\tau_A = -1.54\text{ksi}}$

8.40 Solve problem 8.21 the Method of Equations.

Solution $\sigma_A = ?$ $\tau_A = ?$

Fig. P8.20 is shown below as Fig.(a). Fig.(b) shows the orientation of the outward normal to the plane AA.

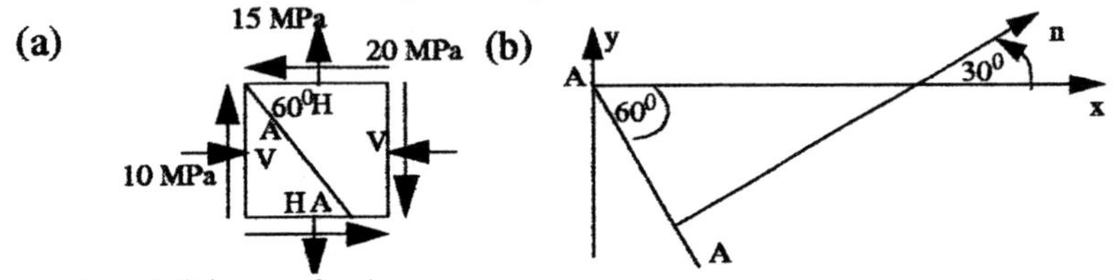

From Figs.(a) and (b) we obtain:

$$\sigma_{xx} = -10\text{MPa} \qquad \sigma_{yy} = 15\text{MPa} \qquad \tau_{xy} = -20\text{MPa} \qquad \theta = 30^\circ \qquad (1)$$

Substituting Eq. 1 into Eqs. 8.1 and 8.2 we obtain:

$\sigma_A = (-10)\cos^2 30 + (15)\sin^2 30 + 2(-20)\sin 30\cos 30$ or $\boxed{\sigma_A = 21.1\text{MPa(C)}}$

$\tau_A = -(-10)\cos 30\sin 30 + (15)\sin 30\cos 30 + (-20)(\cos^2 30 - \sin^2 30)$ $\boxed{\tau_A = 0.83\text{MPa}}$

8.41 Solve problem 8.21 by the Mohr's Circle.

Solution $\sigma_A = ?$ $\tau_A = ?$

Using Fig. (a) in problem 8.40, we can write the coordinates of V and H plane as: V(-10, 20↻) H(15, 20↺).
Fig. (a) below shows the Mohr's circle. Fig. (b) shows the direction of the shear stress on plane A and the local coordinate system.

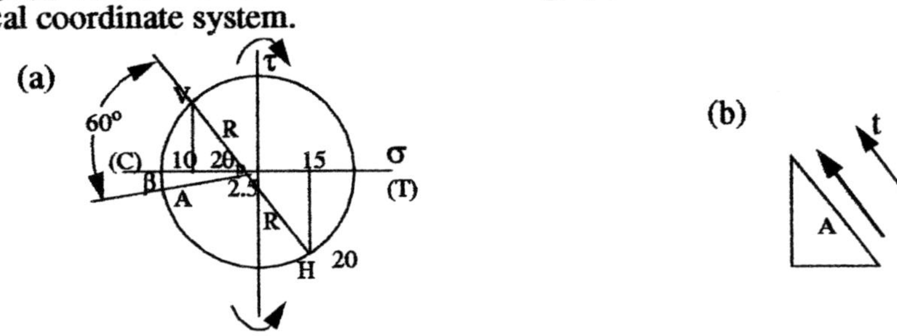

The radius of the circle is: $R = \sqrt{17.5^2 + 20^2} = 23.58\text{MPa}$

The angle $2\theta_p$ is: $2\theta_p = \text{atan}(20/17.5) = 57.995^\circ$. The angle β is: $\beta = 60^\circ - 57.995^\circ = 2.005^\circ$

The normal stress on plane A is: $\sigma_A = -2.5 - R\cos\beta = -21.07$ or $\boxed{\sigma_A = 21.1\text{MPa(C)}}$

The magnitude of shear stress on plane A is: $|\tau_A| = R\sin\beta = 0.825$. From Fig. (b) we see that shear stress on plane a is positive. $\boxed{\tau_A = 0.83\text{MPa}}$

8.42 In a thin body (*plane stress*) the stresses in the x-y plane are as shown on each stress element. (a) Determine the normal and shear stresses on plane A. (b) Determine the principal stresses at the point. (c) Determine the maximum shear stress at the point. (d) Draw the principal element.

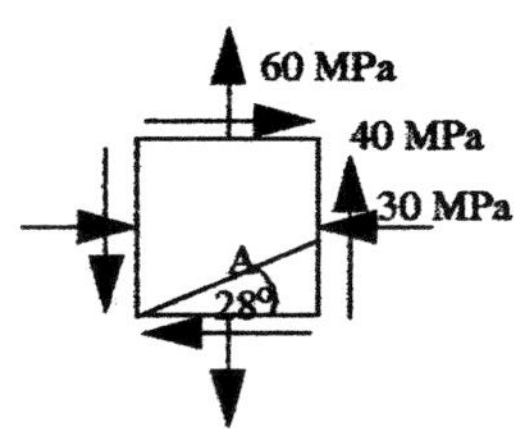

Fig. P8.42

Solution Plane Stress $\sigma_A = ?$ $\tau_A = ?$ $\sigma_1 = ?$ $\sigma_2 = ?$ $\sigma_3 = ?$ $\theta_1 = ?$ $\tau_{max} = ?$

Fig. (a) shows the orientation of plane A and the angle of the outward normal to the plane. From Fig. P8.42 and Fig. (a) we obtain:

$$\sigma_{xx} = -30\text{MPa} \qquad \sigma_{yy} = 60\text{MPa} \qquad \tau_{xy} = 40\text{MPa} \qquad \theta = 118^\circ \tag{1}$$

(a)

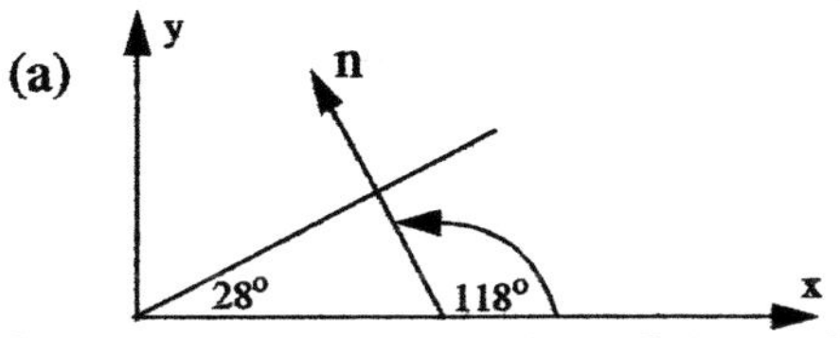

(b)

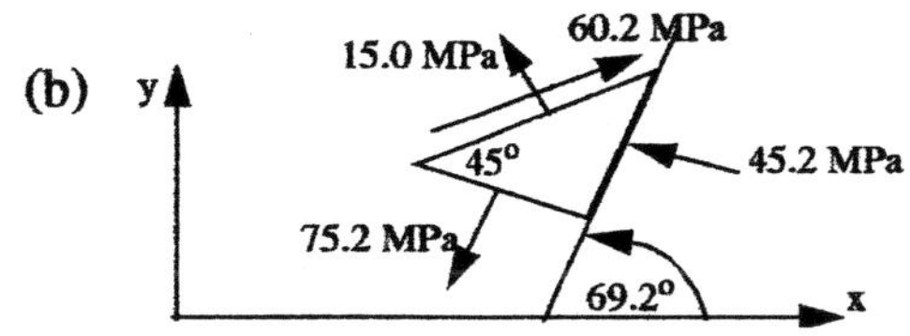

(a) Substituting Eq. 1 into Eqs. 8.1 and 8.2 we obtain:

$\sigma_A = (-30)\cos^2 118 + (60)\sin^2 118 + 2(40)\sin 118 \cos 118$ or $\boxed{\sigma_A = 7.0\text{MPa(C)}}$

$\tau_A = -(-30)\cos 118 \sin 118 + (60)\sin 118 \cos 118 + (40)(\cos^2 118 - \sin^2 118)$ $\boxed{\tau_A = -59.67\text{MPa}}$

(b) The principal angle θ_p is: $\theta_p = 0.5 \ \operatorname{atan}\left(\frac{(2)(40)}{(-30-60)}\right) = -20.817^\circ$

Substituting θ_p and the stresses from Eq. 1 into Eq. 8.1, we obtain one principal stress as:

$$\sigma_p = (-30)\cos^2(-20.817) + (60)\sin^2(-20.817) + 2(40)\sin(-20.817)\cos(-20.817) = -45.2\text{MPa} \tag{2}$$

From Eq. 8.9 we can find the other principal stress as: $-30 + 60 - (45.2) = 75.2\text{MPa}$. As this stress is greater than the stress in Eq. 2, therefore this is principal stress one. Thus principal angle 1 is $\theta_1 = \theta_p \pm 90$

For plane stress we have the following answers:

$$\boxed{\sigma_1 = 75.2\text{MPa(T)} \qquad \sigma_2 = 45.2\text{MPa(C)} \qquad \sigma_3 = 0 \qquad \theta_1 = 69.2^\circ \text{ or } -110.8^\circ}$$

Check: From Eq. 8.6 we have: $\sigma_{1,2} = \left(\frac{-30+60}{2}\right) \pm \sqrt{\left(\frac{-30-60}{2}\right)^2 + 40^2} = 15 \pm 60.2$ or $\sigma_1 = 75.2\text{MPa}$

and $\sigma_2 = -45.2\text{MPa}$,which are same as above.

(c) The difference between σ_1 and σ_2 is the largest difference between principal stresses. Thus, the magnitude of the maximum shear stress is same as maximum inplane shear stress and is given by:

$\tau_{max} = |(\sigma_1 - \sigma_2)/2| = |(75.2 + 45.2)/2|$ or $\boxed{\tau_{max} = 60.2\text{MPa}}$

(d) The average normal stress is: $\sigma_{avg} = (\sigma_{xx} + \sigma_{yy})/2 = 15\text{MPa(T)}$. The principal element is shown in Fig.(b). The direction of maximum shear stress on the plane is determine by inspection to oppose the possible motion of the inclined plane due to principal stresses.

8.43 In a thin body (*plane stress*) the stresses in the x-y plane are as shown on each stress element. (a) Determine the normal and shear stresses on plane A. (b) Determine the principal stresses at the point. (c) Determine the maximum shear stress at the point. (d) Draw the principal element.

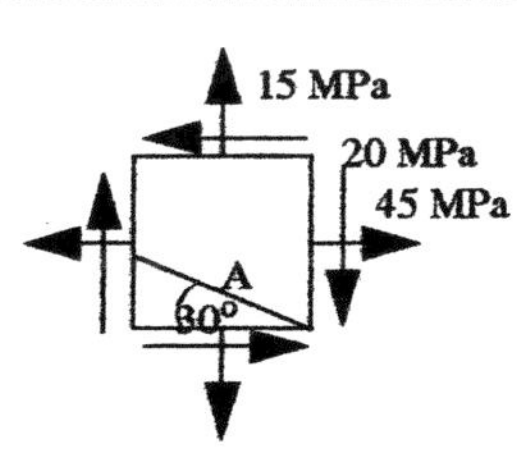

Fig. P8.43

Solution Plane Stress $\sigma_A = ?$ $\tau_A = ?$ $\sigma_1 = ?$ $\sigma_2 = ?$ $\sigma_3 = ?$ $\theta_1 = ?$ $\tau_{max} = ?$

Fig. (a) shows the orientation of plane A and the angle of the outward normal to the plane. From Fig. P8.43 and Fig. (a) we obtain:

$$\sigma_{xx} = 45\text{MPa} \qquad \sigma_{yy} = 15\ \text{MPa} \qquad \tau_{xy} = -20\text{MPa} \qquad \theta = 60^\circ \tag{1}$$

(a)

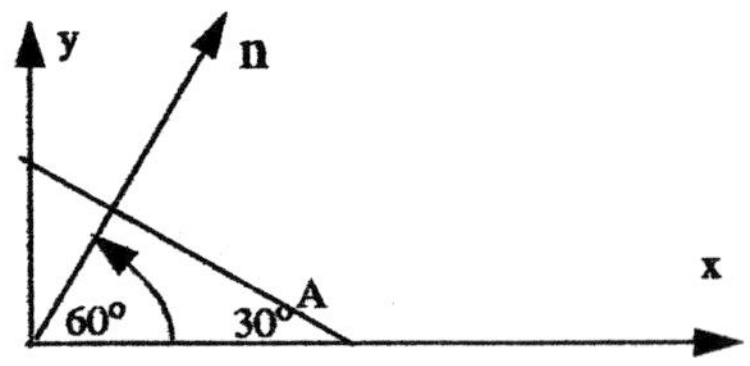

(b)

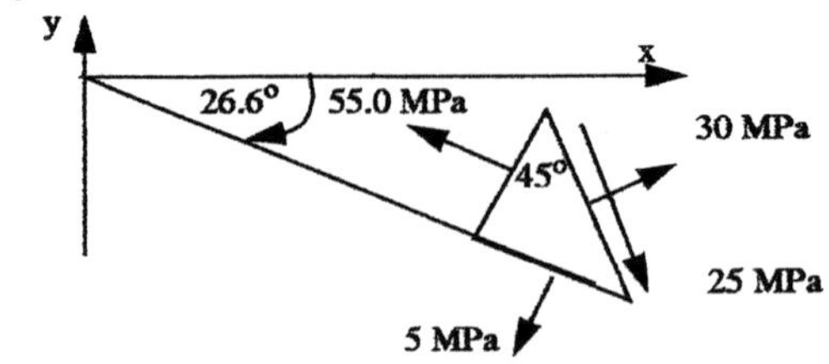

(a) Substituting Eq. 1 into Eqs. 8.1 and 8.2 we obtain:

$\sigma_A = (45)\cos^2 60 + (15)\sin^2 60 + 2(-20)\sin 60\cos 60$ or $\boxed{\sigma_A = 5.2\text{MPa(T)}}$

$\tau_A = -(45)\cos 60\sin 60 + (15)\sin 60\cos 60 + (-20)(\cos^2 60 - \sin^2 60)$ $\boxed{\tau_A = -3.0\text{MPa}}$

(b) The principal angle θ_p is: $\theta_p = 0.5\ \text{atan}\left(\frac{(2)(-20)}{(45-15)}\right) = -26.565^\circ$

Substituting θ_p and the stresses from Eq. 1 into Eq. 8.1, we obtain one principal stress as:

$$\sigma_p = (45)\cos^2(-26.565) + (15)\sin^2(-26.565) + 2(-20)\sin(-26.565)\cos(-26.565) = 55\text{MPa} \tag{2}$$

From Eq. 8.9 we can find the other principal stress as: $45 + 15 - (55) = 5.0\text{MPa}$. As this stress is smaller than the stress in Eq. 2, therefore the stress in Eq.2 is principal stress one. Thus principal angle 1 is θ_p. For plane stress we have the following answers:

$$\boxed{\sigma_1 = 55\text{MPa(T)} \qquad \sigma_2 = 5\text{MPa(T)} \qquad \sigma_3 = 0 \qquad \theta_1 = -26.6^\circ \text{ or } 153.4^\circ}$$

Check: From Eq. 8.6 we have: $\sigma_{1,2} = \left(\frac{45+15}{2}\right) \pm \sqrt{\left(\frac{45-15}{2}\right)^2 + 20^2} = 30 \pm 25$ or $\sigma_1 = 55\text{MPa}$ and $\sigma_2 = 5\text{MPa}$, which are same as above.

(c) The difference between σ_1 and σ_3 is the largest difference between principal stresses. Thus, the magnitude of the maximum shear stress is same as maximum inplane shear stress and is given by:

$\tau_{max} = |(\sigma_1 - \sigma_3)/2| = |(55-0)/2|$ or $\boxed{\tau_{max} = 27.5\text{MPa}}$

(d) The average in-plane normal stress is: $\sigma_{avg} = (\sigma_{xx} + \sigma_{yy})/2 = 15\text{MPa(T)}$. The inplane maximum shear stress is: $\tau_p = |(\sigma_1 - \sigma_2)/2| = |(55-5)/2| = 25$. The inplane principal element is shown in Fig.(b). The direction of maximum shear stress on the plane is determined by inspection to oppose the possible motion of the inclined plane due to principal stresses.

8.44 In a thin body (*plane stress*) the stresses in the x-y plane are as shown on each stress element. (a) Determine the normal and shear stresses on plane A. (b) Determine the principal stresses at the point. (c) Determine the maximum shear stress at the point. (d) Draw the principal element.

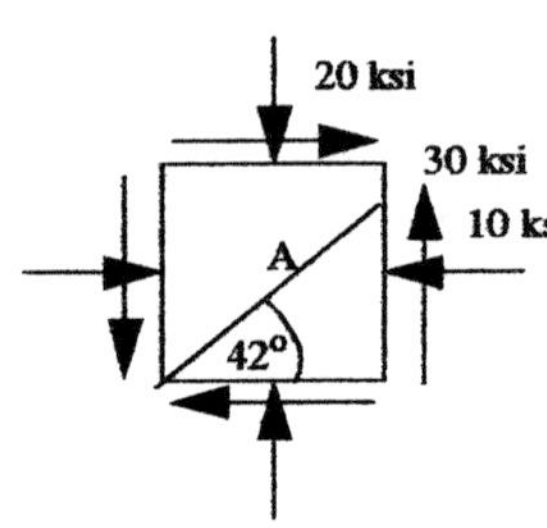

Fig. P8.44

Solution Plane Stress $\sigma_A = ?$ $\tau_A = ?$ $\sigma_1 = ?$ $\sigma_2 = ?$ $\sigma_3 = ?$ $\theta_1 = ?$ $\tau_{max} = ?$

Fig. (a) shows the orientation of plane A and the angle of the outward normal to the plane. From Fig. P8.44 and Fig. (a) we obtain:

$$\sigma_{xx} = -10\text{ksi} \qquad \sigma_{yy} = -20\text{ksi} \qquad \tau_{xy} = 30\text{ksi} \qquad \theta = 132^{\circ} \qquad (1)$$

(a)

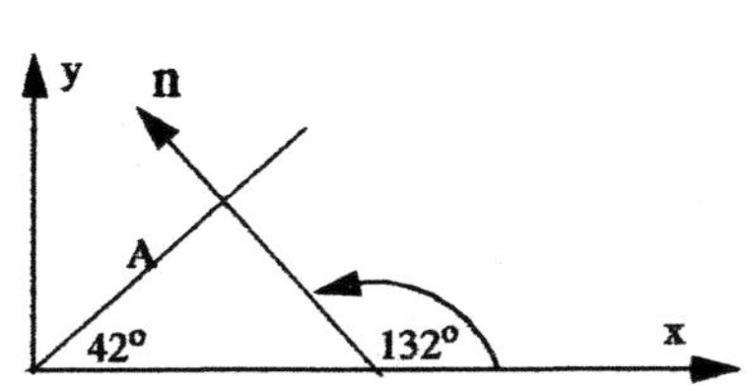

(b)

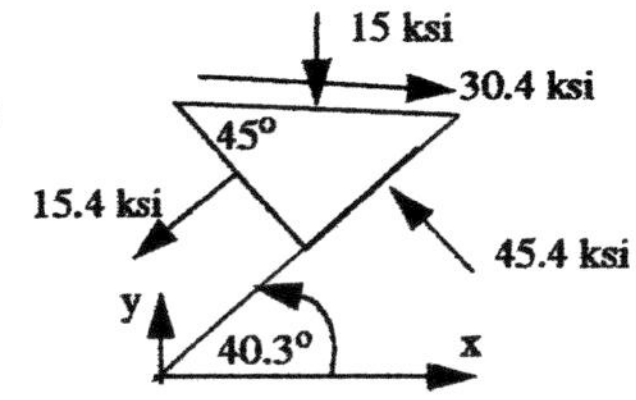

(a) Substituting Eq. 1 into Eqs. 8.1 and 8.2 we obtain:

$\sigma_A = (-10)\cos^2 132 + (-20)\sin^2 132 + 2(30)\sin 132 \cos 132$ or $\boxed{\sigma_A = 45.36\text{ksi}(C)}$

$\tau_A = -(-10)\cos 132 \sin 132 + (-20)\sin 132 \cos 132 + (30)(\cos^2 132 - \sin^2 132)$ $\boxed{\tau_A = 1.84\text{ksi}}$

(b) The principal angle θ_p is: $\theta_p = 0.5 \ \operatorname{atan}\left(\frac{(2)(30)}{(-10-(-20))}\right) = 40.269^{\circ}$

Substituting θ_p and the stresses from Eq. 1 into Eq. 8.1, we obtain one principal stress as:

$$\sigma_p = (-10)\cos^2(40.269) + (-20)\sin^2(40.269) + 2(30)\sin(40.269)\cos(40.269) = 15.414 \qquad (2)$$

From Eq. 8.9 we can find the other principal stress as: $-10-20-(15.414) = -45.414\text{ksi}$. As this stress is smaller than the stress in Eq. 2, therefore the stress in Eq.2 is principal stress one. Thus principal angle 1 is θ_p. For plane stress we have the following answers:

$$\boxed{\sigma_1 = 15.4\text{ksi}(T) \qquad \sigma_2 = 45.4\text{ksi}(C) \qquad \sigma_3 = 0 \qquad \theta_1 = 40.3^{\circ} \text{ or } -139.7^{\circ}}$$

Check: From Eq. 8.6 we have: $\sigma_{1,2} = \left(\frac{-10+(-20)}{2}\right) \pm \sqrt{\left(\frac{-10-(-20)}{2}\right)^2 + 30^2} = -15 \pm 30.4$ or

$\sigma_1 = 55MPa$ and $\sigma_2 = 5MPa$,which are same as above.

(c) The difference between σ_1 and σ_2 is the largest difference between principal stresses. Thus, the magnitude of the maximum shear stress is same as maximum inplane shear stress and is given by:

$\tau_{max} = |(\sigma_1 - \sigma_2)/2| = |(15.4-(-45.4))/2|$ or $\boxed{\tau_{max} = 30.4\text{ksi}}$

(d) The average in-plane normal stress is: $\sigma_{avg} = (\sigma_{xx} + \sigma_{yy})/2 = 15\text{ksi}(C)$. The principal stress element is shown in Fig.(b). The direction of maximum shear stress on the plane is determine by inspection to oppose the possible motion of the inclined plane due to principal stresses.

8.45 In a thick body (*plane strain*) the stresses in the x-y plane are as shown on each stress element. The Poisson's ratio of the material is 0.3. (a) Determine the normal and shear stresses on plane A. (b) Determine the principal stresses at the point. (c) Determine the maximum shear stress at the point. (d) Draw the principal element.

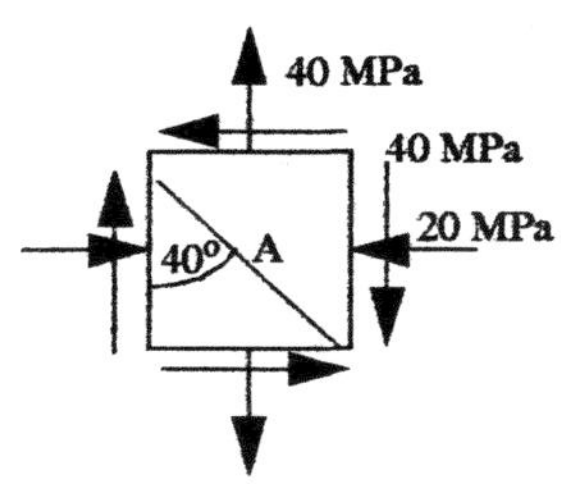

Fig. P8.45

Solution Plane Strain $\nu = 0.3$ $\quad \sigma_A = ? \ \tau_A = ? \ \sigma_1 = ? \ \sigma_2 = ? \ \sigma_3 = ? \ \theta_1 = ? \ \tau_{max} = ?$

Fig. (a) shows the orientation of plane A and the angle of the outward normal to the plane. From Fig. P8.45 and Fig. (a) we obtain:

$$\sigma_{xx} = -20MPa \qquad \sigma_{yy} = 40MPa \qquad \tau_{xy} = -40MPa \qquad \theta = 40^{\circ} \qquad (1)$$

(a)

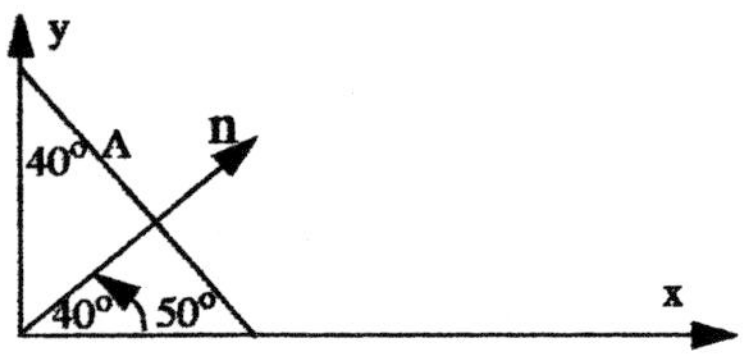

(b)

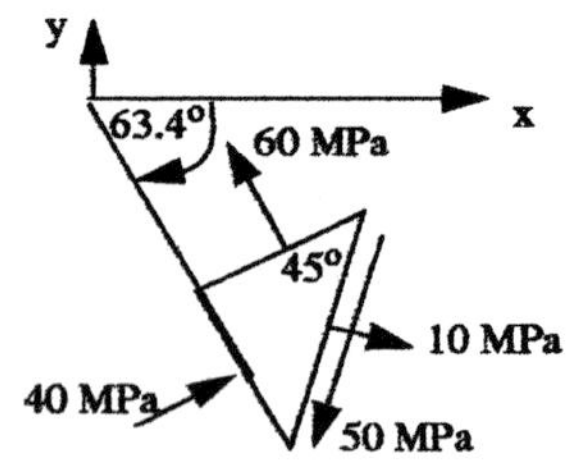

(a) Substituting Eq. 1 into Eqs. 8.1 and 8.2 we obtain:

$\sigma_A = (-20)\cos^2 40 + (40)\sin^2 40 + 2(-40)\sin 40 \cos 40$ or $\boxed{\sigma_A = 34.6\text{MPa(C)}}$

$\tau_A = -(-20)\cos 40 \sin 40 + (40)\sin 40 \cos 40 + (-40)(\cos^2 40 - \sin^2 40)$ $\boxed{\tau_A = 22.6\text{MPa}}$

(b) The principal angle θ_p is: $\theta_p = 0.5 \ \text{atan}\left(\frac{(2)(-40)}{(-20-(40))}\right) = 26.565^\circ$

Substituting θ_p and the stresses from Eq. 1 into Eq. 8.1, we obtain one principal stress as:

$$\sigma_p = (-20)\cos^2(26.565) + (40)\sin^2(26.565) + 2(-40)\sin(26.565)\cos(26.565) = -40 \qquad 2$$

From Eq. 8.9 we can find the other principal stress as: $-20 + 40 - (-40) = 60\text{MPa}$. As this stress is greater than the stress in Eq. 2, therefore this stress is principal stress one. Thus principal angle 1 is $\theta_1 = \theta_p \pm 90$.

For plane strain $\sigma_3 = \sigma_{zz} = \nu(\sigma_{xx} + \sigma_{yy}) = 0.3(-20+40) = 6$. We have the following answers:

$$\boxed{\sigma_1 = 60\text{MPa(T)} \qquad \sigma_2 = 40\text{MPa(C)} \qquad \sigma_3 = 6\text{MPa(T)} \qquad \theta_1 = 116.6^\circ \text{ or } -63.4^\circ}$$

Check: From Eq. 8.6 we have: $\sigma_{1,2} = \left(\frac{-20+(40)}{2}\right) \pm \sqrt{\left(\frac{-20-(40)}{2}\right)^2 + (-40)^2} = 10 \pm 50$ or $\sigma_1 = 60\text{MPa}$ and $\sigma_2 = -40\text{MPa}$, which are same as above.

(c) The difference between σ_1 and σ_2 is the largest difference between principal stresses. Thus, the magnitude of the maximum shear stress is same as maximum inplane shear stress and is given by:

$\tau_{max} = |(\sigma_1 - \sigma_2)/2| = |(60-(-40))/2|$ or $\boxed{\tau_{max} = 50\text{MPa}}$

(d) The average in-plane normal stress is: $\sigma_{avg} = (\sigma_{xx} + \sigma_{yy})/2 = 10\text{MPa(T)}$. The principal stress element is shown in Fig.(b). The direction of maximum shear stress on the plane is determine by inspection to oppose the possible motion of the inclined plane due to principal stresses.

8.46 In a thick body (*plane strain*) the stresses in the x-y plane are as shown on each stress element. The Poisson's ratio of the material is 0.3. (a) Determine the normal and shear stresses on plane A. (b) Determine the principal stresses at the point. (c) Determine the maximum shear stress at the point. (d) Draw the principal element.

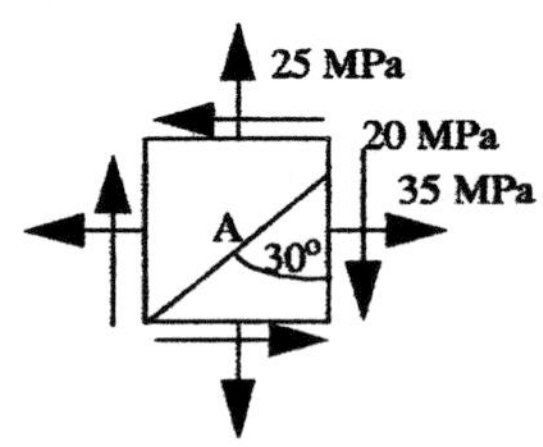

Fig. P8.46

Solution Plane Strain $\nu = 0.3$ $\quad \sigma_A = ?$ $\tau_A = ?$ $\sigma_1 = ?$ $\sigma_2 = ?$ $\sigma_3 = ?$ $\theta_1 = ?$ $\tau_{max} = ?$

Fig. (a) shows the orientation of plane A and the angle of the outward normal to the plane. From Fig. P8.45 and Fig. (a) we obtain:

$$\sigma_{xx} = 35\text{MPa} \qquad \sigma_{yy} = 25\text{MPa} \qquad \tau_{xy} = -20\text{MPa} \qquad \theta = 150^\circ \qquad 1$$

(a) Substituting Eq. 1 into Eqs. 8.1 and 8.2 we obtain:

$\sigma_A = (35)\cos^2 150 + (25)\sin^2 150 + 2(-20)\sin 150 \cos 150$ or $\boxed{\sigma_A = 49.8\text{MPa(T)}}$

$\tau_A = -(35)\cos 150 \sin 150 + (25)\sin 150 \cos 150 + (-20)(\cos^2 150 - \sin^2 150)$ $\boxed{\tau_A = -5.67\text{MPa}}$

(a)

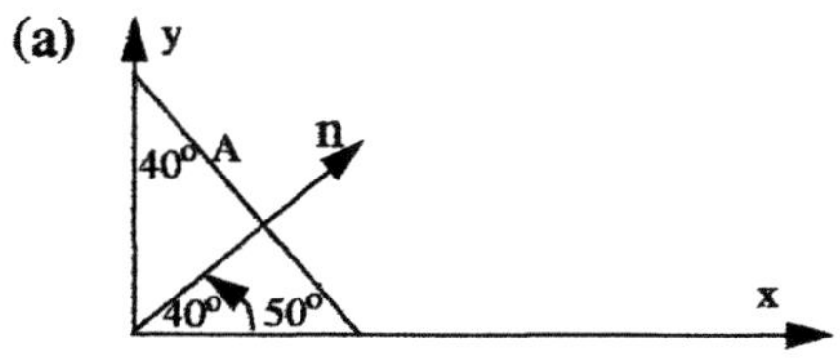

(b)

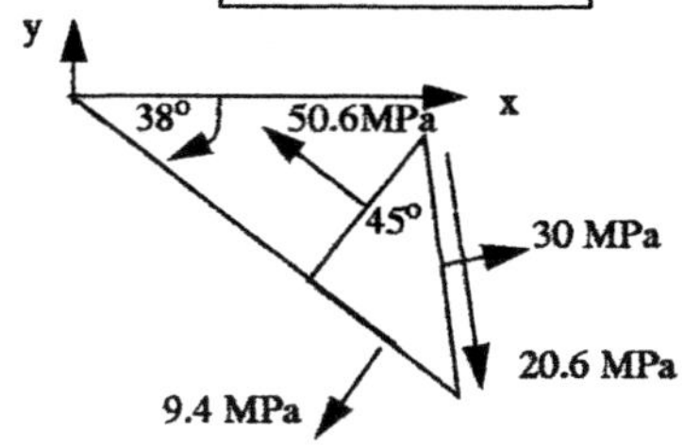

(b) The principal angle θ_p is: $\theta_p = 0.5 \ \text{atan}\left(\frac{(2)(-20)}{(35-(25))}\right) = -37.982^\circ$

Substituting θ_p and the stresses from Eq. 1 into Eq. 8.1, we obtain one principal stress as:

$$\sigma_p = (35)\cos^2(26.565) + (25)\sin^2(26.565) + 2(-20)\sin(26.565)\cos(26.565) = 50.616 \qquad \textbf{2}$$

From Eq. 8.9 we can find the other principal stress as: $35 + 25 - (50.616) = 9.384\text{MPa}$. As this stress is smaller than the stress in Eq. 2, therefore the stress in Eq. 2 is principal stress one. Thus principal angle 1 is θ_p. For plane strain $\sigma_3 = \sigma_{zz} = \nu(\sigma_{xx} + \sigma_{yy}) = 0.3(35 + 25) = 18$. We have the following answers:

$$\boxed{\sigma_1 = 50.6\text{MPa(T)} \qquad \sigma_2 = 9.4\text{MPa(T)} \qquad \sigma_3 = 18\text{MPa(T)} \qquad \theta_1 = -38^\circ \text{ or } 142^\circ}$$

Check: From Eq. 8.6 we have: $\sigma_{1,2} = \left(\frac{35+25}{2}\right) \pm \sqrt{\left(\frac{35-25}{2}\right)^2 + (-20)^2} = 30 \pm 20.6$ or $\sigma_1 = 50.6\text{MPa}$ and $\sigma_2 = -9.4\text{MPa}$,which are same as above.

(c) The difference between σ_1 and σ_2 is the largest difference between principal stresses. Thus, the magnitude of the maximum shear stress is same as maximum inplane shear stress and is given by:

$\tau_{max} = |(\sigma_1 - \sigma_2)/2| = |(50.6 - 9.4)/2|$ or $\boxed{\tau_{max} = 20.6\text{MPa}}$

(d) The average in-plane normal stress is: $\sigma_{avg} = (\sigma_{xx} + \sigma_{yy})/2 = 30\text{MPa(T)}$. The principal stress element is shown in Fig.(b). The direction of maximum shear stress on the plane is determine by inspection to oppose the possible motion of the inclined plane due to principal stresses.

8.47 In a thick body (*plane strain*) the stresses in the x-y plane are as shown on each stress element. The Poisson's ratio of the material is 0.3. (a) Determine the normal and shear stresses on plane A. (b) Determine the principal stresses at the point. (c) Determine the maximum shear stress at the point. (d) Draw the principal element.

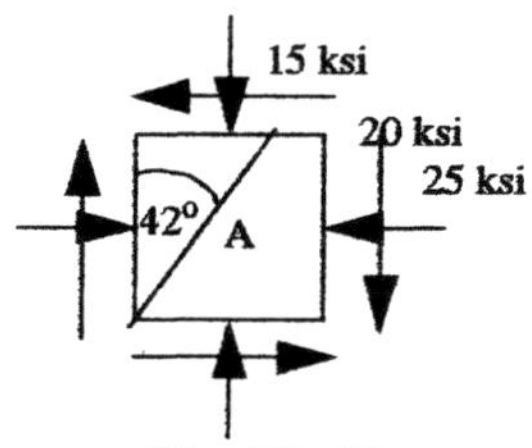

Fig. P8.47

Solution Plane Strain $\nu = 0.3$ $\sigma_A = ?$ $\tau_A = ?$ $\sigma_1 = ?$ $\sigma_2 = ?$ $\sigma_3 = ?$ $\theta_1 = ?$ $\tau_{max} = ?$

Fig. (a) shows the orientation of plane A and the angle of the outward normal to the plane. From Fig. P8.45 and Fig. (a) we obtain:

$$\sigma_{xx} = -25\text{ksi} \qquad \sigma_{yy} = -15\text{ksi} \qquad \tau_{xy} = -20\text{ksi} \qquad \theta = 138^\circ \qquad \textbf{1}$$

(a)

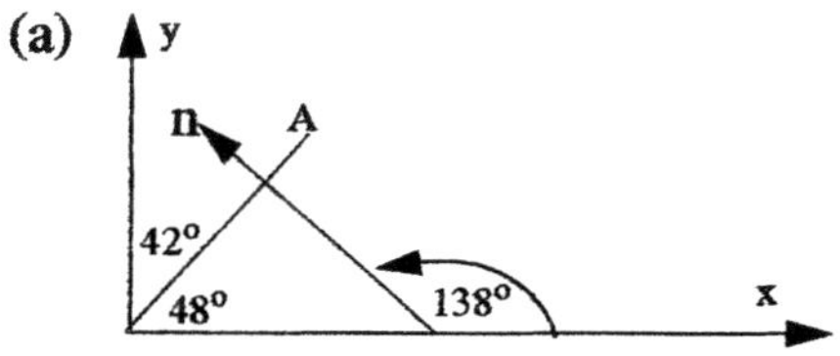

(b)

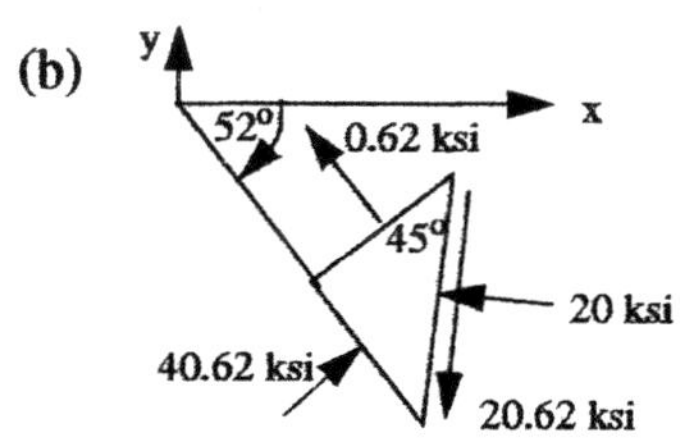

(a) Substituting Eq. 1 into Eqs. 8.1 and 8.2 we obtain:

$\sigma_A = (-25)\cos^2 138 + (-15)\sin^2 138 + 2(-20)\sin 138 \cos 138$ or $\boxed{\sigma_A = 0.63\text{ksi(C)}}$

$\tau_A = -(-25)\cos 138 \sin 138 + (-15)\sin 138 \cos 138 + (-20)(\cos^2 138 - \sin^2 138)$ $\boxed{\tau_A = -7.06\text{ksi}}$

(b) The principal angle θ_p is: $\theta_p = 0.5 \ \text{atan}\left(\frac{(2)(-20)}{(-25-(-15))}\right) = 37.982°$

Substituting θ_p and the stresses from Eq. 1 into Eq. 8.1, we obtain one principal stress as:

$$\sigma_p = (-25)\cos^2(37.982) + (-15)\sin^2(37.982) + 2(-20)\sin(37.982)\cos(37.982) = -40.616 \quad 2$$

From Eq. 8.9 we can find the other principal stress as: $-25 - 15 - (-40.616) = 0.616\text{ksi}$. As this stress is greater than the stress in Eq. 2, therefore this stress is principal stress one. Thus principal angle 1 is $\theta_1 = \theta_p \pm 90$. For plane strain $\sigma_3 = \sigma_{zz} = \nu(\sigma_{xx} + \sigma_{yy}) = 0.3(-25 - 15) = -12$. We have the following answers:

$\boxed{\sigma_1 = 0.62\text{ksi(T)} \qquad \sigma_2 = 40.62\text{ksi(C)} \qquad \sigma_3 = 12\text{ksi(C)} \qquad \theta_1 = 128° \text{ or } -52°}$

Check: From Eq. 8.6 we have: $\sigma_{1,2} = \left(\frac{-25+(-15)}{2}\right) \pm \sqrt{\left(\frac{-25-(-15)}{2}\right)^2 + (-20)^2} = -20 \pm 20.62$ or

$\sigma_1 = 0.62\text{ksi}$ and $\sigma_2 = -40.62\text{ksi}$,which are same as above.

(c) The difference between σ_1 and σ_2 is the largest difference between principal stresses. Thus, the magnitude of the maximum shear stress is same as maximum inplane shear stress and is given by:

$\tau_{max} = |(\sigma_1 - \sigma_2)/2| = |(0.62 - (-40.62))/2|$ or $\boxed{\tau_{max} = 20.62\text{ksi}}$

(d) The average in-plane normal stress is: $\sigma_{avg} = (\sigma_{xx} + \sigma_{yy})/2 = 20\text{MPa(C)}$. The principal stress element is shown in Fig.(b). The direction of maximum shear stress on the plane is determine by inspection to oppose the possible motion of the inclined plane due to principal stresses.

8.48 The strains at a point and the material properties are as given in each problem. Assuming plane stress, determine the principal stresses, principal angle one, and the maximum shear stress at the point.

$\varepsilon_{xx} = 500\ \mu \qquad \varepsilon_{yy} = 400\ \mu \qquad \gamma_{xy} = -300\ \mu \qquad E = 200\ \text{GPa} \qquad \nu = 0.32$

Solution Plane stress $\sigma_1 = ?$ $\sigma_2 = ?$ $\sigma_3 = ?$ $\theta_1 = ?$ $\tau_{max} = ?$

From problem 3.70 we have

$$\sigma_{xx} = 139.93\text{MPa} \qquad \sigma_{yy} = 124.78\text{MPa} \qquad \tau_{xy} = -22.7\text{MPa} \quad 1$$

The principal angle θ_p is: $\theta_p = 0.5 \ \text{atan}\left(\frac{(2)(-22.7)}{(139.93-124.78)}\right) = -35.78°$

Substituting θ_p and the stresses from Eq. 1 into Eq. 8.1, we obtain one principal stress as:

$$\sigma_p = (139.93)\cos^2(-35.78) + (124.78)\sin^2(-35.78) + 2(-22.7)\sin(-35.78)\cos(-35.78) = 156.31\text{MPa} \quad 2$$

From Eq. 8.9 we can find the other principal stress as: $139.93 + 124.78 - 156.31 = 108.40\text{MPa}$. As this stress is smaller than the stress in Eq. 2, therefore the stress in Eq.2 is principal stress one. Thus principal angle 1 is θ_p. For plane stress we have the following answers:

$\boxed{\sigma_1 = 156.3\text{MPa(T)} \qquad \sigma_2 = 108.4\text{MPa(T)} \qquad \sigma_3 = 0 \qquad \theta_1 = -35.78° \text{ or } 144.22°}$

Check: From Eq. 8.6 we have:

$\sigma_{1,2} = \left(\frac{139.93+124.78}{2}\right) \pm \sqrt{\left(\frac{139.93-124.78}{2}\right)^2 + (-22.7)^2} = 132.35 \pm 23.96$ or $\sigma_1 = 156.3\text{MPa}$ and

$\sigma_2 = 108.4\text{MPa}$, which are same as above.

The difference between σ_1 and σ_3 is the largest difference between principal stresses. Thus, the magnitude of the maximum shear stress is same as maximum inplane shear stress and is given by:

$\tau_{max} = |(\sigma_1 - \sigma_3)/2| = |(156.3 - 0)/2|$ or $\boxed{\tau_{max} = 78.2\text{MPa}}$

8.49 The strains at a point and the material properties are as given in each problem. Assuming plane stress, determine the principal stresses, principal angle one, and the maximum shear stress at the point.

$\varepsilon_{xx} = -3000\ \mu \qquad \varepsilon_{yy} = 1500\ \mu \qquad \gamma_{xy} = 2000\ \mu \qquad E = 70\ \text{GPa} \qquad G = 28\ \text{GPa}$

Solution Plane stress $\sigma_1 = ?$ $\sigma_2 = ?$ $\sigma_3 = ?$ $\theta_1 = ?$ $\tau_{max} = ?$

From problem 3.71 we have

$$\sigma_{xx} = -196\text{MPa} \qquad \sigma_{yy} = 56\text{MPa} \qquad \tau_{xy} = 56\text{MPa} \qquad (1)$$

The principal angle θ_p is: $\theta_p = 0.5\ \text{atan}\left(\frac{(2)(56)}{((-196) - 124.78)}\right) = -11.98°$

Substituting θ_p and the stresses from Eq. 1 into Eq. 8.1, we obtain one principal stress as:

$$\sigma_p = (-196)\cos^2(-11.98) + (56)\sin^2(-11.98) + 2(56)\sin(-11.98)\cos(-11.98) = -207.88\text{MPa} \qquad (2)$$

From Eq. 8.9 we can find the other principal stress as: $(-196) + 56 - (-207.88) = 67.88\text{MPa}$. As this stress is greater than the stress in Eq. 2, therefore this stress is principal stress one. Thus principal angle 1 is $\theta_1 = \theta_p \pm 90$. For plane stress we have the following answers:

$$\boxed{\sigma_1 = 67.9\text{MPa(T)} \qquad \sigma_2 = 207.9\text{MPa(C)} \qquad \sigma_3 = 0 \qquad \theta_1 = 78.02° \text{ or } -101.98°}$$

Check: From Eq. 8.6 we have: $\sigma_{1,2} = \left(\frac{(-196) + 56}{2}\right) \pm \sqrt{\left(\frac{(-196) - 56}{2}\right)^2 + (56)^2} = -70 \pm 137.88$ or

$\sigma_1 = 67.88\text{MPa}$ and $\sigma_2 = -207.88\text{MPa}$, which are same as above.

The difference between σ_1 and σ_2 is the largest difference between principal stresses. Thus, the magnitude of the maximum shear stress is same as maximum inplane shear stress and is given by:

$\tau_{max} = |(\sigma_1 - \sigma_2)/2| = |(67.88 - (-207.88))/2|$ or $\boxed{\tau_{max} = 137.88\text{MPa}}$

8.50 The strains at a point and the material properties are as given in each problem. Assuming plane stress, determine the principal stresses, principal angle one, and the maximum shear stress at the point.

$\varepsilon_{xx} = -800\ \mu \qquad \varepsilon_{yy} = -1000\ \mu \qquad \gamma_{xy} = -500\ \mu \qquad E = 30,000\ \text{ksi} \qquad \nu = 0.3$

Solution Plane stress $\sigma_1 = ?$ $\sigma_2 = ?$ $\sigma_3 = ?$ $\theta_1 = ?$ $\tau_{max} = ?$

From problem 3.72 we have

$$\sigma_{xx} = -36.26\text{ksi} \qquad \sigma_{yy} = -40.90\text{ksi} \qquad \tau_{xy} = -5.77\text{ksi} \qquad (1)$$

The principal angle θ_p is: $\theta_p = 0.5\ \text{atan}\left(\frac{(2)(-5.77)}{(-36.26) - (-40.9)}\right) = -34.10°$

Substituting θ_p and the stresses from Eq. 1 into Eq. 8.1, we obtain one principal stress as:

$$\sigma_p = (-36.26)\cos^2(-34.10) + (-40.9)\sin^2(-34.10) + 2(-5.77)\sin(-34.10)\cos(-34.10) = -32.36\text{ksi} \qquad (2)$$

From Eq. 8.9 we can find the other principal stress as: $(-36.26) + (-40.9) - (-32.36) = -44.79\text{ksi}$. As this stress is smaller than the stress in Eq. 2, therefore the stress in Eq.2 is principal stress one. Thus principal angle 1 is θ_p. For plane stress we have the following answers:

$$\boxed{\sigma_1 = 32.4\text{ksi(C)} \qquad \sigma_2 = 44.8\text{ksi(C)} \qquad \sigma_3 = 0 \qquad \theta_1 = -34.10^\circ \text{ or } 145.9^\circ}$$

Check: From Eq. 8.6 we have:

$$\sigma_{1,2} = \left(\frac{-36.26+(-40.9)}{2}\right) \pm \sqrt{\left(\frac{-36.26-(-40.9)}{2}\right)^2 + (-5.77)^2} = -38.57 \pm 6.21 \text{ or } \sigma_1 = -32.4\text{ksi and}$$

$\sigma_2 = -44.8\text{MPa}$,which are same as above.

The difference between σ_2 and σ_3 is the largest difference between principal stresses. Thus, the magnitude of the maximum shear stress is same as maximum inplane shear stress and is given by:

$\tau_{max} = |(\sigma_2 - \sigma_3)/2| = |(-44.8 - 0)/2|$ or $\boxed{\tau_{max} = 22.4\text{ksi}}$

8.51 The difference in principal stresses $\sigma_1 - \sigma_2$ and the principal direction one θ_1 from the x-axis were measured by photoelasticity at several points and are shown below. The sum of the principal stresses $\sigma_1 + \sigma_2$ were found from elasticity and are also shown below. Assuming plane stress, determine the stresses σ_{xx}, σ_{yy}, and τ_{xy} at the point. $\sigma_1 - \sigma_2 = 10\text{ksi}$ $\quad \theta_1 = -15^\circ$ $\quad \sigma_1 + \sigma_2 = 6\text{ksi}$

Solution plane stress $\sigma_{xx} = ?$ $\quad \sigma_{yy} = ?$ $\quad \tau_{xy} = ?$

Solving the two equations $\sigma_1 - \sigma_2 = 10\text{ksi}$ and $\sigma_1 + \sigma_2 = 6\text{ksi}$ we obtain: $\sigma_1 = 8\text{ksi}$ and $\sigma_2 = -2\text{ksi}$. Fig. (a) shows the Mohr's circle for the given state of stress. The center is at 3 ksi and the radius is R = 5 ksi. Fig. (b) shows the orientation of principal planes P_1 and P_2 at 15° clockwise from the x axis as per the value of θ_1. To get to the vertical plane V and horizontal plane H, we must rotate counter-clockwise from the principal planes. Starting from P_1 and P_2 on the Mohr's circle we rotate 30° counter-clockwise to get planes V and H as shown in Fig. (a).

(a)

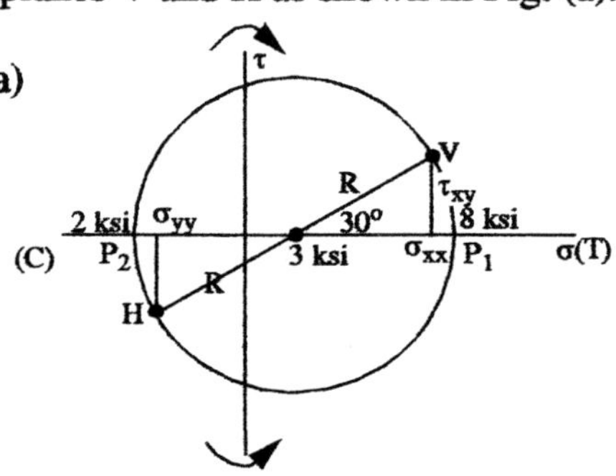

(b)

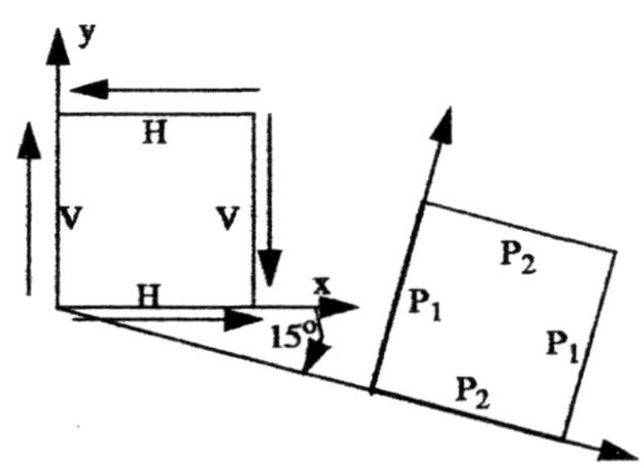

The normal stresses in the x and y coordinate system are:

$\sigma_{xx} = 3 + R\cos 30 = 7.33$ or $\boxed{\sigma_{xx} = 7.33\text{ksi(T)}}$

$\sigma_{yy} = 3 - R\cos 30 = -1.33$ $\boxed{\sigma_{yy} = 1.33\text{ksi(C)}}$

The magnitude of shear stress is: $|\tau_{xy}| = R\sin 30 = 2.5\text{ksi}$. From Fig. (a) we note that the shear stress on V plane rotates the plane clockwise and on H plane rotates it counter-clockwise. We draw the shear stress on the cube in Fig. (b) to rotate the planes in the direction deduced from Fig. (a). We note from Fig. (b) that shear stress τ_{xy} is negative. $\boxed{\tau_{xy} = -2.5\text{ksi}}$

8.52 The difference in principal stresses $\sigma_1 - \sigma_2$ and the principal direction one θ_1 from the x-axis were measured by photoelasticity at several points and are shown below. The sum of the principal stresses

$\sigma_1 + \sigma_2$ were found from elasticity and are also shown below. Assuming plane stress, determine the stresses σ_{xx}, σ_{yy}, and τ_{xy} at the point. $\sigma_1 - \sigma_2 = 3\text{ksi}$ $\theta_1 = 25^{\circ}$ $\sigma_1 + \sigma_2 = -17\text{ksi}$

Solution plane stress $\sigma_{xx} = ?$ $\sigma_{yy} = ?$ $\tau_{xy} = ?$

Solving the two equations $\sigma_1 - \sigma_2 = 3\text{ksi}$ and $\sigma_1 + \sigma_2 = -17\text{ksi}$ we obtain: $\sigma_1 = -7\text{ksi}$ and $\sigma_2 = -10\text{ksi}$. Fig. (a) shows the Mohr's circle for the given state of stress. The center is at -8.5 ksi and the radius is R = 1.5 ksi. Fig. (b) shows the orientation of principal planes P_1 and P_2 at 25° counter-clockwise from the x axis as per the value of θ_1. To get to the vertical plane V and horizontal plane H, we must rotate clockwise from the principal planes. Starting from P_1 and P_2 on the Mohr's circle we rotate 50° clockwise to get planes V and H as shown in Fig. (a).

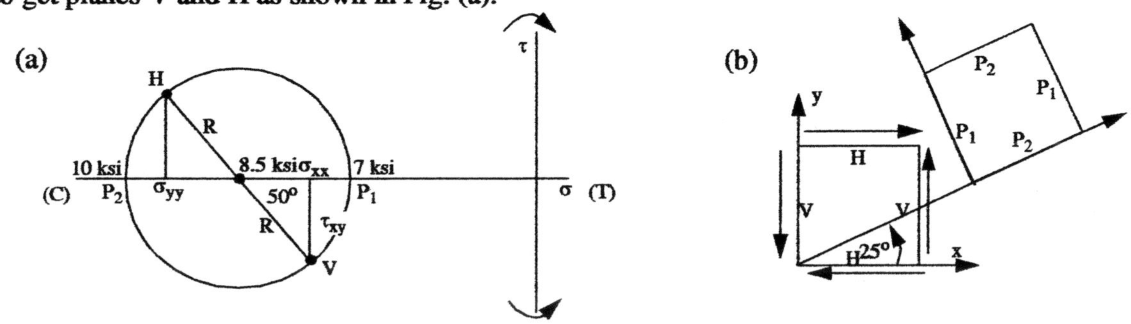

The normal stresses in the x and y coordinate system are:

$\sigma_{xx} = -8.5 + R\cos 50 = -7.54$ or $\boxed{\sigma_{xx} = 7.54\text{ksi(C)}}$

$\sigma_{yy} = -8.5 - R\cos 50 = -9.46$ $\boxed{\sigma_{yy} = 9.46\text{ksi(C)}}$

The magnitude of shear stress is: $|\tau_{xy}| = R\sin 50 = 1.15\text{ksi}$. From Fig. (a) we note that the shear stress on V plane rotates the plane counter-clockwise and on H plane rotates it clockwise. We draw the shear stress on the cube in Fig. (b) to rotate the planes in the direction deduced from Fig. (a). We note from Fig. (b) that shear stress τ_{xy} is positive. $\boxed{\tau_{xy} = 1.15\text{ksi}}$

8.53 The difference in principal stresses $\sigma_1 - \sigma_2$ and the principal direction one θ_1 from the x-axis were measured by photoelasticity at several points and are shown below. The sum of the principal stresses $\sigma_1 + \sigma_2$ were found from elasticity and are also shown below. Assuming plane stress, determine the stresses σ_{xx}, σ_{yy}, and τ_{xy} at the point. $\sigma_1 - \sigma_2 = 5\text{ksi}$ $\theta_1 = 35^{\circ}$ $\sigma_1 + \sigma_2 = 5\text{ksi}$

Solution plane stress $\sigma_{xx} = ?$ $\sigma_{yy} = ?$ $\tau_{xy} = ?$

Solving the two equations $\sigma_1 - \sigma_2 = 5\text{ksi}$ and $\sigma_1 + \sigma_2 = 5\text{ksi}$ we obtain: $\sigma_1 = 5\text{ksi}$ and $\sigma_2 = 0$. Fig. (a) shows the Mohr's circle for the given state of stress. The center is at 2.5 ksi and the radius is R = 2.5 ksi. Fig. (b) shows the orientation of principal planes P_1 and P_2 at 35° counter-clockwise from the x axis as per the value of θ_1. To get to the vertical plane V and horizontal plane H, we must rotate clockwise from the principal planes. Starting from P_1 and P_2 on the Mohr's circle we rotate 50° clockwise to get planes V and H as shown in Fig. (a).

The normal stresses in the x and y coordinate system are:

$\sigma_{xx} = 2.5 + R\cos 70 = 3.36$ or $\boxed{\sigma_{xx} = 3.36\text{ksi(T)}}$

$\sigma_{yy} = 2.5 - R\cos 70 = 1.64$ $\boxed{\sigma_{yy} = 1.64\text{ksi(C)}}$

(a)

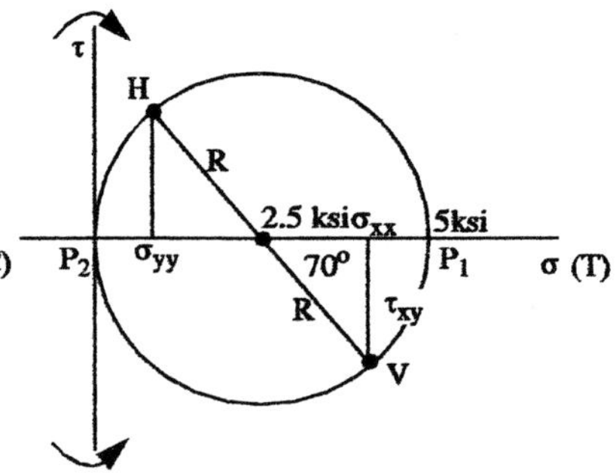

(b)

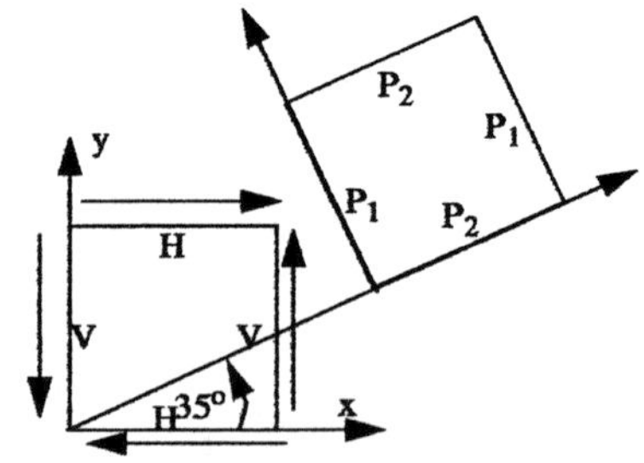

The magnitude of shear stress is: $|\tau_{xy}| = R\sin 70 = 2.35\text{ksi}$. From Fig. (a) we note that the shear stress on V plane rotates the plane counter-clockwise and on H plane rotates it clockwise. We draw the shear stress on the cube in Fig. (b) to rotate the planes in the direction deduced from Fig. (a). We note from Fig. (b) that shear stress τ_{xy} is positive. $\boxed{\tau_{xy} = 2.35\text{ksi}}$

8.54 A broken 2 in x 6 in wooden bar was glued together as shown. Determine the normal and shear stress in the glue.

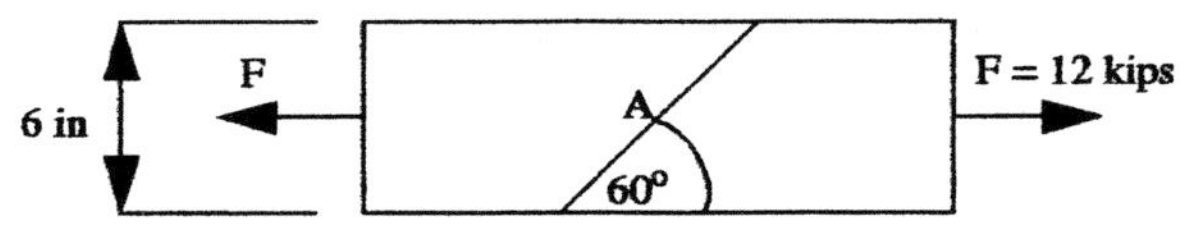

Fig. P8.54

Solution $\sigma_A = ?$ $\tau_A = ?$

The axial stress is: $\sigma_{xx} = \frac{12}{(2)(6)} = 1\text{ksi(T)}$ All other stress components in the x-y coordinate system are zero. Fig. (a) shows the stress cube. Fig. (b) shows the orientation of the glue plane and the angle of the outward normal.

(a)

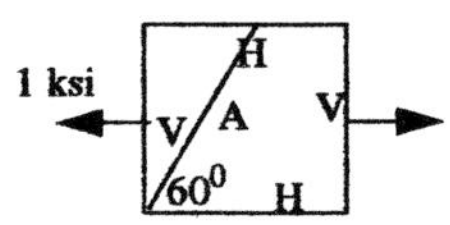

(b)

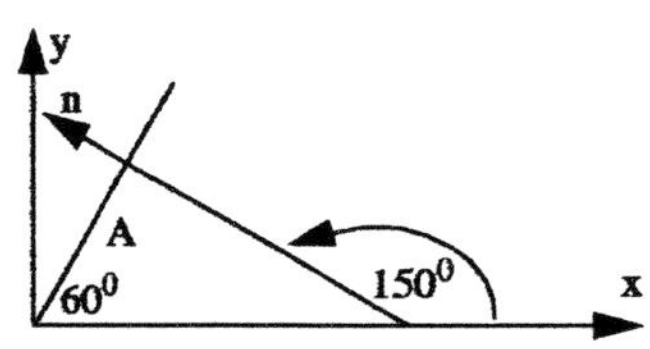

From Figs.(a) and (b) we obtain:

$$\sigma_{xx} = 1\text{ksi} \qquad \sigma_{yy} = 0 \qquad \tau_{xy} = 0 \qquad \theta = 150^\circ \tag{1}$$

Substituting Eq. 1 into Eqs. 8.1 and 8.2 we obtain:

$\sigma_A = (1)\cos^2 150 + (0)\sin^2 150 + 2(0)\sin 150\cos 150 = 0.75\text{ksi}$ or $\boxed{\sigma_A = 750\text{psi(T)}}$

$\tau_A = -(1)\cos 150\sin 150 + (0)\sin 150\cos 150 + (0)(\cos^2 150 - \sin^2 150) = 0.433\text{ksi}$ or $\boxed{\tau_A = 433\text{psi}}$

8.55 Determine the normal and shear stress in the seam of the shaft passing through point A an angle of $\theta = 60^\circ$ to the axis a solid shaft of 2 inch diameter.

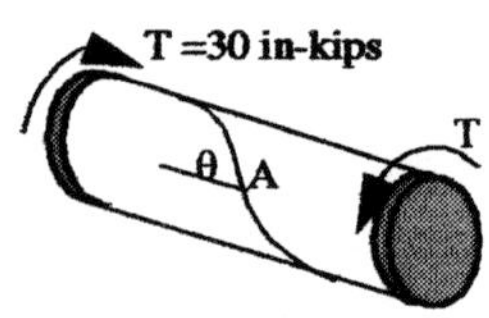

Fig. P8.55

Solution $\sigma_A = ?$ $\tau_A = ?$

The magnitude of torsional shear stress is at outer surface is: $|\tau_{x\theta}| = \dfrac{30(1)}{\pi 2^4/32} = 19.1\,\text{ksi}$. Fig. (a) shows the direction of torsional shear stress determined by inspection. Fig. (b) shows the stress cube. Fig. (c) shows the orientation of the plane of the seam and the angle of the outward normal to it.

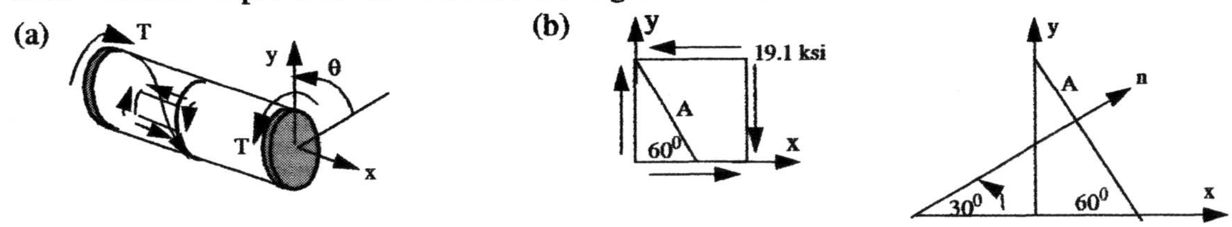

From Figs.(b) and (c) we obtain:

$$\sigma_{xx} = 0 \qquad \sigma_{yy} = 0 \qquad \tau_{xy} = -19.1\,\text{ksi} \qquad \theta = 30^\circ \qquad (1)$$

Substituting Eq. 1 into Eqs. 8.1 and 8.2 we obtain:

$\sigma_A = (0)\cos^2 30 + (0)\sin^2 30 + 2(-19.1)\sin 30\cos 30$ or $\boxed{\sigma_A = 16.5\,\text{ksi(C)}}$

$\tau_A = -(0)\cos 30\sin 30 + (0)\sin 30\cos 30 + (-19.1)(\cos^2 30 - \sin^2 30)$ or $\boxed{\tau_A = -9.55\,\text{ksi}}$

8.56 If the applied force P = 1.8 kN, determine the principal stresses and maximum shear stress at points A, B, and C which are on the surface of the beam.

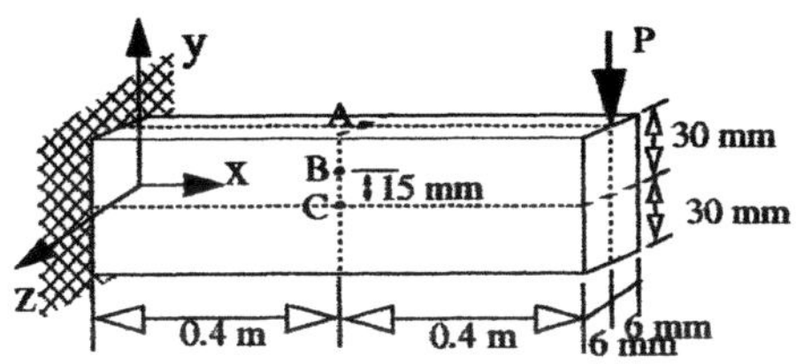

Fig. P8.56

Solution P = 1.8 kN $\sigma_1 = ?$ $\sigma_2 = ?$ $\sigma_3 = ?$ $\tau_{max} = ?$ @A, B, and C

Fig. (a) shows the free body diagram after an imaginary cut is made through the section containing points A, B, and C. By equilibrium of forces and moment we obtain:

$$V_y = -1.8\,\text{kN} \qquad M_z = -(1.8)(0.4) = -0.72\,\text{kN-m} \qquad (1)$$

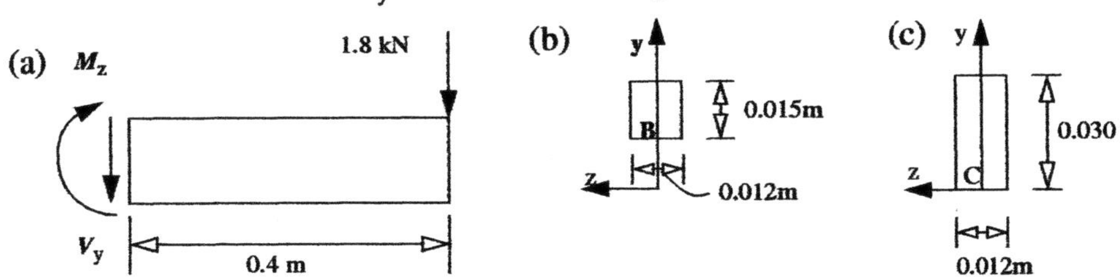

The area moment of inertia about the z-axis is: $I_{zz} = (0.012)(0.060)^3/12 = 216(10^{-9})\ \text{m}^4$

The bending normal stress at point B will be half that at point A as $y_B = y_A/2$ and will be zero at point C.

$$(\sigma_{xx})_A = -\frac{M_z y_A}{I_{zz}} = -\frac{(-0.72)(10^3)(0.03)}{216(10^{-9})} = 100\,\text{MPa(T)} \qquad (\sigma_{xx})_B = 50\,\text{MPa(T)} \qquad (\sigma_{xx})_C = 0 \qquad (2)$$

Figs. (b) and (c) show the area A_s used in calculation of bending shear stress. The first moment of the areas are:

$(Q_z)_B = (0.012)(0.015)(0.015 + 0.0075) = 4.05(10^{-6})\text{m}^3$ $\quad (Q_z)_C = (0.012)(0.030)(0.015) = 5.40(10^{-6})\text{m}^3$

The bending shear stress at A is zero. The bending shear stress τ_{xy} at points B and C will be negative as the shear force V_y is negative.The bending shear stresses at these points are:

$$(\tau_{xy})_A = 0 \qquad (\tau_{xy})_B = -\left|\frac{V_y(Q_z)_B}{I_{zz}t}\right| = -\left|\frac{1.8(10^3)(4.05)(10^{-6})}{216(10^{-9})(0.012)}\right| = -2.8125\text{MPa} \quad 3$$

$$(\tau_{xy})_C = -\left|\frac{V_y(Q_z)_B}{I_{zz}t}\right| = -\left|\frac{1.8(10^3)(5.40)(10^{-6})}{216(10^{-9})(0.012)}\right| = -3.75\text{MPa} \quad 4$$

The third principal stress at the three points is zero as these points are on free surfaces. At point A the bending normal stress is the principal stress 1 as all other stress components in the x, y, z coordinate system are zero. The maximum shear stress is half the principal stress 1 as other principal stresses are zero.

$$\boxed{(\sigma_1)_A = 100\text{MPa(T)} \qquad (\sigma_2)_A = 0 \qquad (\sigma_3)_A = 0 \qquad (\tau_{max})_A = 50\text{MPa}}$$

Noting σ_{yy} is zero, we obtain the principal stress 1 and 2 at point B as:

$$(\sigma_{1,2})_B = \left(\frac{50+0}{2}\right) \pm \sqrt{\left(\frac{50-0}{2}\right)^2 + (2.8125)^2} = 25 \pm 25.16 \text{ or}$$

$$\boxed{(\sigma_1)_B = 50.16\text{MPa(T)} \qquad (\sigma_2)_B = 0.16\text{MPa(C)} \qquad (\sigma_3)_B = 0 \qquad (\tau_{max})_B = 25.16\text{MPa}}$$

Noting that σ_{xx} and σ_{yy} are zero, we obtain the principal stress 1 and 2 at point C as:

$$(\sigma_{1,2})_C = 0 \pm \sqrt{0 + (3.75)^2} = 0 \pm 3.75 \text{ or}$$

$$\boxed{(\sigma_1)_C = 3.75\text{MPa(T)} \qquad (\sigma_2)_C = 3.75\text{MPa(C)} \qquad (\sigma_3)_C = 0 \qquad (\tau_{max})_C = 3.75\text{MPa}}$$

8.57 If the applied force P = 1.8 kN, determine the principal stresses and maximum shear stress at points A, B, and C which are on the surface of the beam.

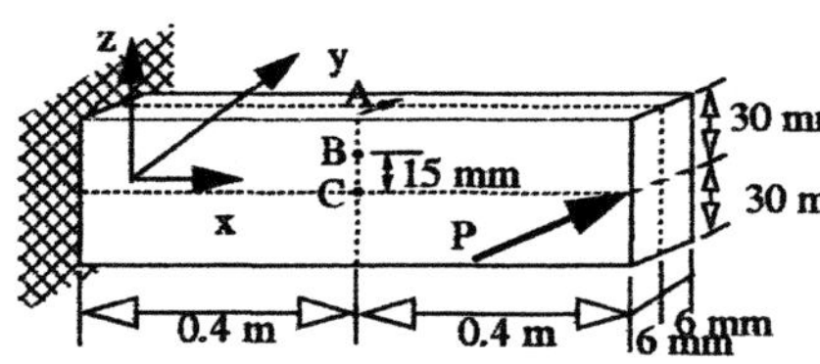

Fig. P8.57

Solution P = 1.8 kN $\sigma_1 = ?$ $\sigma_2 = ?$ $\sigma_3 = ?$ $\tau_{max} = ?$ @ A, B, and C

Fig. (a) shows the free body diagram after an imaginary cut is made through the section containing points A, B, and C. By equilibrium of forces and moment we obtain:

$$V_y = 1.8\text{kN} \qquad M_z = (1.8)(0.4) = 0.72\text{kN-m} \quad 1$$

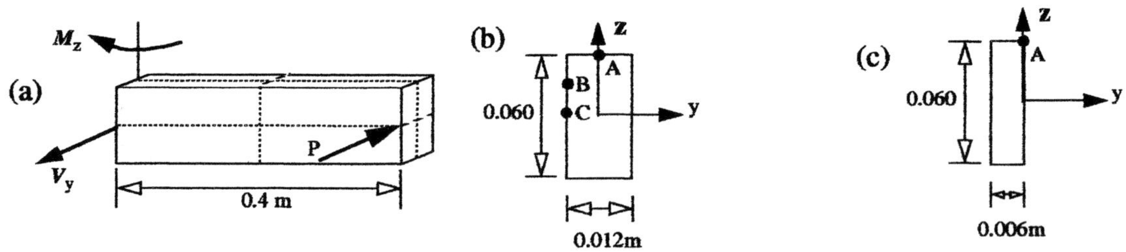

The area moment of inertia about the z-axis is: $I_{zz} = (0.060)(0.012)^3/12 = 8.64(10^{-9})\ \text{m}^4$

The bending normal stress at point B and C will be same as $y_B = y_C = -0.006$ m and will be zero at point A.

$$(\sigma_{xx})_A = 0 \qquad (\sigma_{xx})_B = (\sigma_{xx})_C = -\frac{M_z y_{B,C}}{I_{zz}} = -\frac{(0.72)(10^3)(-0.006)}{8.64(10^{-9})} = 500\text{MPa(T)} \quad 2$$

Fig. (c) show the area A_s used in calculation of bending shear stress. The first moment of the area is:

$$(Q_z)_A = (0.060)(0.006)(0.003) = 1.08(10^{-6})\text{m}^3$$

The bending shear stress τ_{xy} at points B and C will be zero as these points are on the free surface. The bending shear stress at A is positive as the shear force V_y is positive.The bending shear stress at A is:

$$(\tau_{xy})_A = \left|\frac{V_y(Q_z)_A}{I_{zz}t}\right| = \left|\frac{1.8(10^3)(1.08)(10^{-9})}{8.64(10^{-9})(0.06)}\right| = 3.75\text{MPa} \qquad (\tau_{xy})_B = (\tau_{xy})_C = 0 \qquad 3$$

The third principal stress at the three points is zero as these points are on free surfaces. Noting that σ_{xx} and σ_{yy} are zero at point A the principal stress 1 and 2 are:

$$(\sigma_{1,2})_A = 0 \pm \sqrt{0+(3.75)^2} = 0 \pm 3.75 \quad \text{or}$$

$$\boxed{(\sigma_1)_A = 3.75\text{MPa(T)} \qquad (\sigma_2)_A = 3.75\text{MPa(C)} \qquad (\sigma_3)_A = 0 \qquad (\tau_{max})_A = 3.75\text{MPa}}$$

At points B and C the bending normal stress is the principal stress 1 as all other stress components in the x, y, z coordinate system are zero. The maximum shear stress is half the principal stress 1 as other principal stresses are zero.

$$\boxed{(\sigma_1)_B = 500\text{MPa(T)} \qquad (\sigma_2)_B = 0 \qquad (\sigma_3)_B = 0 \qquad (\tau_{max})_B = 250\text{MPa}}$$

$$\boxed{(\sigma_1)_C = 500\text{MPa(T)} \qquad (\sigma_2)_C = 0 \qquad (\sigma_3)_C = 0 \qquad (\tau_{max})_C = 250\text{MPa}}$$

8.58 A thin tube of 1/8 in thickness has a mean diameter of 6 in. What is the maximum torque the tube can transmit if the allowable normal stress in compression is 10 ksi?

Solution t= 1/8 in $\quad d_{mean}$ = 6 in $\quad \sigma_2 \le 10\text{ksi(C)}$ $\quad T_{max}$ = ?

The polar moment of inertia is: $J = \pi(6.125^4 - 5.875^4)/32 = 21.21\text{in}^4$

The outer radius is $\rho_{max} = (d_{mean}/2 + t/2) = 3.0625$ in

The magnitude of maximum torsional shear stress is: $\tau = \dfrac{T\rho_{max}}{J} = \dfrac{T(3.0625)}{21.21}$

As the normal stresses on the cross-section plane are zero, the principal stresses are equal to the shear stress. Thus: $\sigma_2 = \tau = \dfrac{T(3.0625)}{21.21} \le 10$ $\quad$ or $\quad T \le 69.27$ in – kip or $\boxed{T_{max} = 69.2 \text{ in} - \text{kip}}$

8.59 Re-solve Example 8.6 without the thin tube approximation.

Solution L = 30 in $\quad$ T= 25π in-kips $\quad$ t= 1/16 in $\quad$ G=12000ksi $\quad \sigma_S \le 10\text{ksi(C)}$

$\tau_W \le 12\text{ksi}$ $\quad \sigma_W \le 20\text{ksi(T)}$ $\quad \Delta\phi \le 3^\circ$ $\quad$ minimum R_o = ? to nearest 1/16 in

---

Fig. (a) shows the hollow shaft with the applied torque. By inspection we see that weld would be subjected to compressive stress, hence we do not have to account for the limitation on tensile stress in the weld. The maximum compressive stress in steel is equal to the torsional shear stress τ_{tor}. Thus, we have:

$$\sigma_2 = \tau_{tor} \le 10\text{ksi} \qquad 1$$

(a)

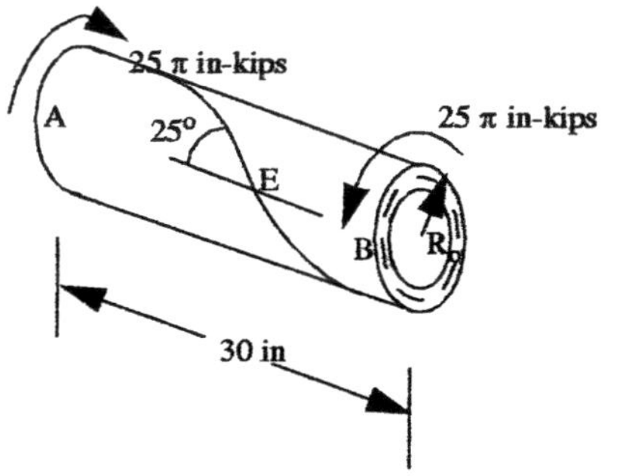

(b)

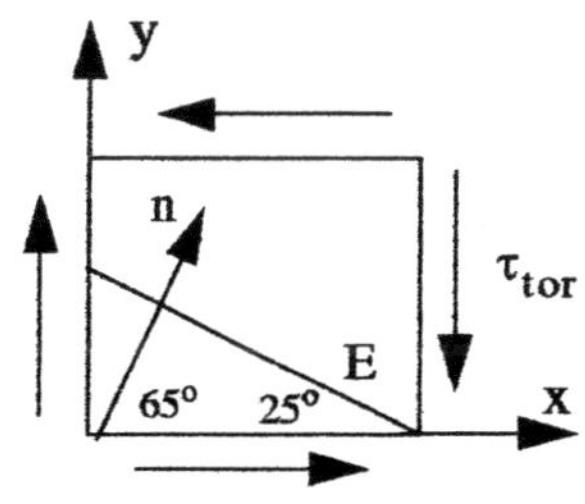

Fig. (b) shows the stress cube with the plane containing the weld and the angle of the outward normal. from Fig. (b) we have:

$$\sigma_{xx} = 0 \qquad \sigma_{yy} = 0 \qquad \tau_{xy} = -\tau_{tor} \qquad \theta = 65^\circ \tag{2}$$

Substituting Eq. 1 into Eq. 8.2 we obtain:

$$\tau_W = -(0)\cos 65 \sin 65 + (0)\sin 65 \cos 65 + (-\tau_{tor})(\cos^2 65 - \sin^2 65) = 0.6428\tau_{tor} \le 12 \text{ or } \tau_{tor} \le 18.7\text{ksi} \tag{3}$$

If torsional shear stress is less than or equal to 10 ksi then it will satisfy Eq. 1 and 3. The magnitude of maximum torsional shear stress is:

$$\tau_{tor} = \frac{T\rho_{max}}{J} = \frac{(25\pi)R_o}{\pi[R_o^4 - (R_o - 0.0625)^4]/2} \le 10 \quad \text{or} \quad R_o^4 - (R_o - 0.0625)^4 - 5R_o \ge 0 \tag{4}$$

The relative rotation is

$$\Delta\phi = \frac{TL}{GJ} = \frac{(25\pi)(30)}{(12000)\pi[R_o^4 - (R_o - 0.0625)^4]/2} \le \frac{3\pi}{180} \quad \text{or} \quad R_o^4 - (R_o - 0.0625)^4 - 2.3873 \ge 0 \tag{5}$$

Let $f_1(R_o) = R_o^4 - (R_o - 0.0625)^4 - 5R_o \qquad f_2(R_o) = R_o^4 - (R_o - 0.0625)^4 - 2.3873$

We consider the values of the two functions for R_o starting at 4 in. in steps of 1/16th in. The calculations are done on a spreadsheet and results reported below.

R_o	$f_1(R_o)$	$f_2(R_o)$
4.0000	-4.37	13.34
4.0625	-3.93	14.09
4.1250	-3.47	14.87
4.1875	-2.99	15.66
4.2500	-2.48	16.48
4.3125	-1.94	17.33
4.3750	-1.38	18.20
4.4375	-0.80	19.10
4.5000	-0.19	20.02
4.5625	0.45	20.97
4.6250	1.11	21.95
4.6875	1.80	22.95
4.7500	2.52	23.98

In the range considered, $f_2(R_o)$ is always positive, hence condition in Eq. 5 is met. The smallest value of R_o for which $f_1(R_o)$ becomes positive is R_o= 4.5625 in. The minimum value of R_o to the nearest 1/16th inch is:

$$\boxed{R_o = 4\tfrac{9}{16}\text{in}}$$

8.60 An aluminum rod (E_{al} = 70 GPa) and a steel rod (E_s = 210 GPa) are securely fastened to a rigid plate that does not rotate during the application of the load P. The diameter of aluminum and steel rods are 20 mm and 10 mm, respectively. The allowable shear stress in aluminum and steel are 120 MPa and 150 MPa. Determine the maximum force P that can be applied to the rigid plate.

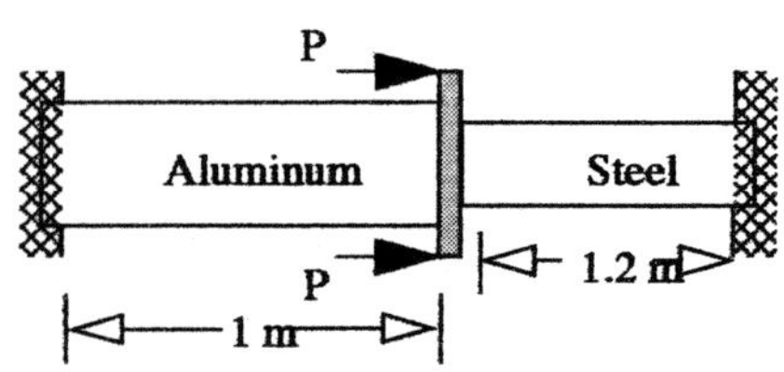

Fig. P8.60

Solution E_{al} = 70 GPa $\quad E_{steel}$ = 210 GPa $\quad d_{al}$ = 20 mm $\quad d_s$ = 10 mm

$\tau_{al} \le 120\text{MPa} \quad \tau_s \le 150\text{MPa} \quad P_{max} = ?$

Fig. (a) is the free body diagrams after an imaginary cut is made through aluminum and steel. By equilibrium of forces we obtain

$$N_S + N_{al} = 2P \qquad 1$$

(a)

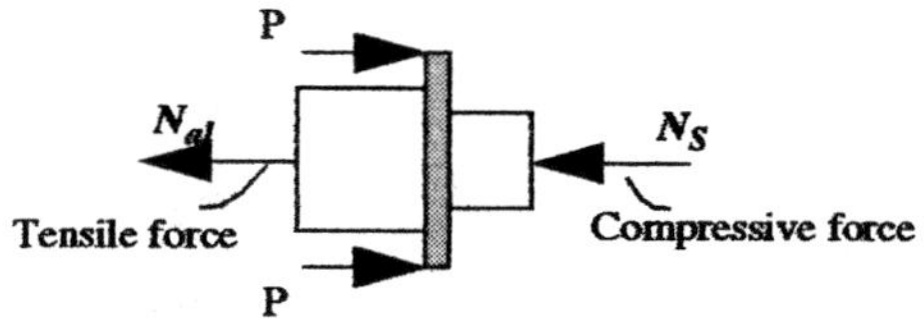

(b)

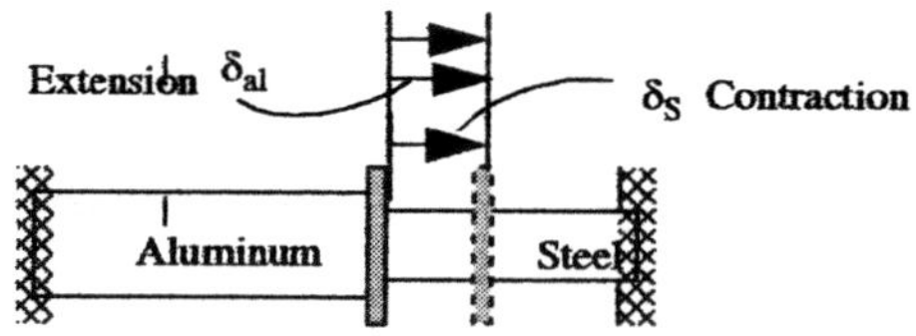

Fig.(b) shows draw the exaggerated deformed shape. The deformation relationship is:

$$\delta_{al} = \delta_S \qquad 2$$

The cross-sectional areas of the two bars are:

$$A_s = \pi(0.01)^2/4 = 78.54(10^{-6})m^2 \qquad A_{al} = \pi(0.02)^2/4 = 314.6(10^{-6})m^2 \qquad 3$$

The deformations of the bars can be written as

$$\delta_S = \frac{N_S L_S}{E_S A_S} = \frac{N_S(1.2)}{210(10^9)78.54(10^{-6})} = 0.07277N_S(10^{-6}) \qquad 4$$

$$\delta_{al} = \frac{N_{al} L_{al}}{E_{al} A_{al}} = \frac{N_{al}(1)}{70(10^9)314.6(10^{-6})} = 0.04547N_{al}(10^{-6}) \qquad 5$$

Substituting Eqs. 3 and 4 into Eq. 2 we obtain

$$0.04547N_{al}(10^{-6}) = 0.07277N_S(10^{-6}) \qquad \text{or} \qquad N_{al} = 1.6N_S \qquad 6$$

Solving Eq. 1 and 5, we obtain: $N_S = 0.7691P$ and $N_{al} = 1.2306P$

The axial stresses in steel and aluminum are:

$$s = \frac{0.7691P}{78.54(10^{-6})} = 9792.4P \qquad \text{and} \qquad \sigma_{al} = \frac{1.2306P}{314.6(10^{-6})} = 3911.64 \qquad 7$$

The maximum shear stress is half the axial stress. We obtain:

$$\tau_s = \sigma_S/2 = 4896.2P \le 150(10^6) \qquad \text{or} \qquad P \le 30.64(10^3)N \qquad 8$$

$$\tau_{al} = \sigma_{al}/2 = 1955.82P \le 120(10^6) \qquad \text{or} \qquad P \le 61.35(10^3)N \qquad 9$$

The maximum value of P that satisfies Eqs. 8 and 9 is: $\boxed{P_{max} = 30.6kN}$

8.61 Two pieces of solid shafts of diameter 75 mm are securely connected to a rigid wheel. The shaft material has a Modulus of Shear Rigidity G = 80 GPa and an allowable normal stress in tension or compression of 90 MPa. Determine the maximum torque T that can act on the wheel

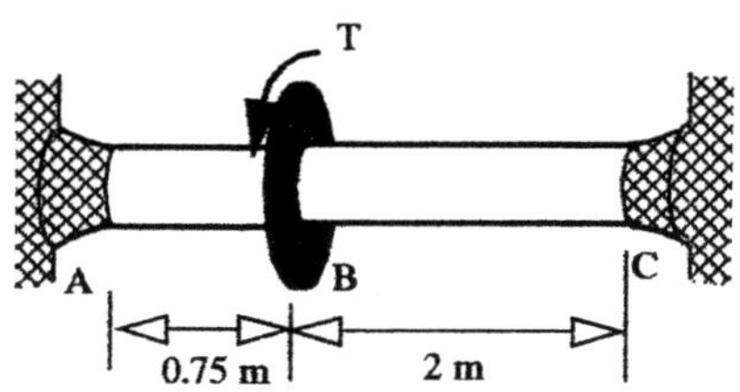

Fig. P8.61

Solution G = 80 GPa d = 75 mm $\sigma_{max} \le 90MPa$ $T_{max} = ?$

The polar moment of inertia is: $J = \pi(0.075)^4/32 = 3.106(10^{-6})m^4$

Fig.(a) shows the free body diagram of the shaft after imaginary cuts are made in section AB and BC. By moment equilibrium we obtain:

$$T_{AB} = T_A \qquad T_{BC} = T_A - T \qquad 1$$

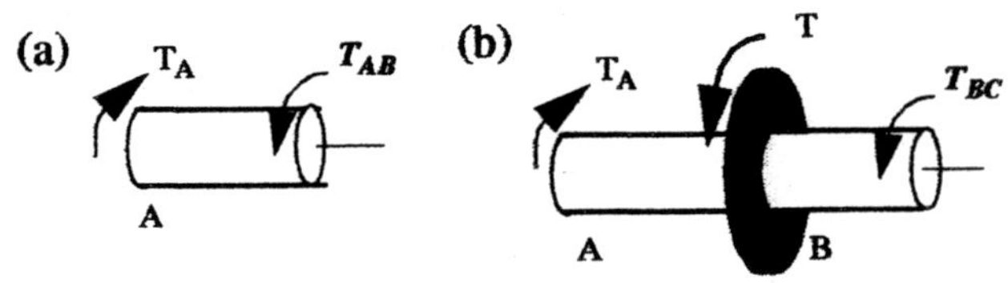

The relative rotation of the ends can be written as:

$$\phi_B - \phi_A = \frac{T_{AB}(x_B - x_A)}{GJ} = \frac{(T_A)(0.75)}{GJ} \qquad 2$$

$$\phi_C - \phi_B = \frac{T_{BC}(x_C - x_B)}{GJ} = \frac{(T_A - T)(2)}{GJ} \qquad 3$$

Adding Eqs. 2 and 3 and noting that $\phi_C = 0$ and $\phi_A = 0$ we obtain:

$$\phi_C - \phi_A = \frac{(T_A)(0.75)}{GJ} + \frac{(T_A - T)(2)}{GJ} = 0 \quad \text{or} \quad T_A = \frac{2T}{2.75} = 0.7273T \qquad 4$$

From Eq. 1 we obtain: $T_{AB} = 0.7273T$ $\quad T_{BC} = -0.2727T$. The maximum torsional shear stress will be in segment AB and its value is: $\tau_{AB} = \frac{T_{AB}\rho_{max}}{J} = \frac{(0.7273T)(0.0375)}{3.106(10^{-6})} = 8.7807T(10^3)$. The principal stresses are equal to the torsional shear stress. Thus the maximum normal stress in AB is:

$\sigma_{AB} = \tau_{AB} = 8.7807T(10^3) \le 90(10^6)$ $\quad$ or $\quad$ $T \le 10.25(10^3)N-m$

The maximum torque that can be applied is: $\boxed{T_{max} = 10.2 \text{ kN}-\text{m}}$

8.62 A cantilever beam is constructed by gluing three pieces of timber as shown. The allowable shear in the adhesive is 300 psi and the allowable tensile stress is 200 psi. The allowable tensile or compressive stress in wood is 2000 psi. Determine the maximum intensity of the distributed load w.

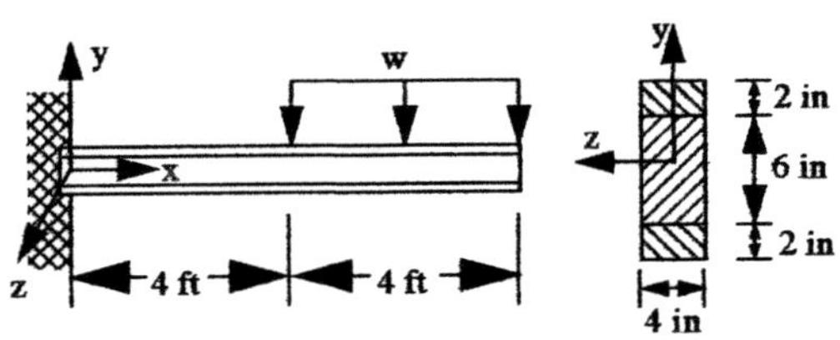

Fig. P8.62

Solution $\quad \tau_G \le 300\text{psi} \quad \sigma_G \le 200\text{psi(T)} \quad \sigma_W \le 2000\text{psi} \quad w_{max} = ?$

The area moment of inertia is: $I_{zz} = (4)(10)^3/12 = 333.3 \text{ in}^4$. Fig.(a) shows the free body diagram of the beam. By equilibrium of forces and moments we obtain: $R = 4w$ $\quad M = 24w$. Fig.(b) and (c) are the shear force and bending moment diagrams.

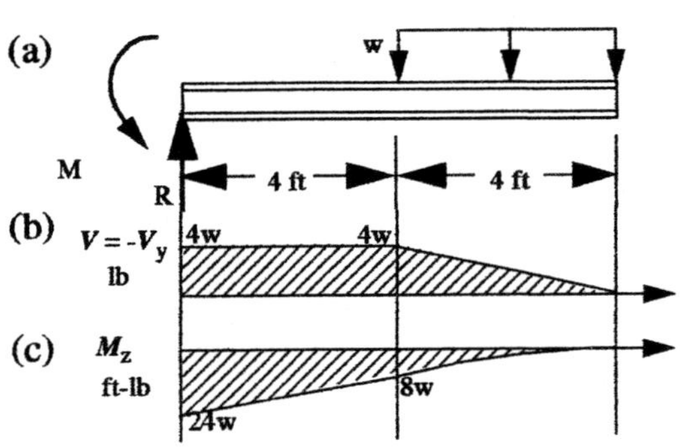

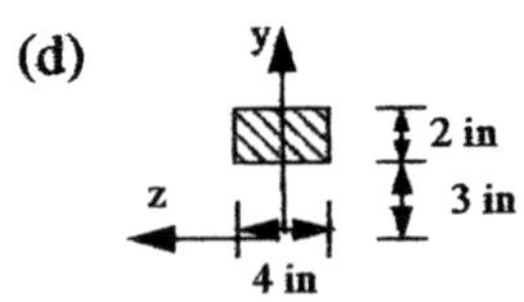

The maximum bending moment and shear force occur in the same section at the wall. Thus the maximum bending normal and shear stress will occur at the wall. Fig.(d) shows the area A_s used in calculating the shear stress in the glue. The first moment of the area is: $Q_z = (4)(2)(4) = 32in^3$. By inspection the glue on the top junction will be in tensile. As $V_{max} = -V_y = 4w$, the shear stress τ_{xy} will be negative. The normal and shear stress in the glue are:

$$(\sigma_{xx})_G = -\frac{M_z y_G}{I_{zz}} = -\frac{(-24w)(12)(3)}{333.3} = 2.592w \text{ and } (\tau_{xy})_G = -\left|\frac{V_y Q_z}{I_{zz}t}\right| = -\left|\frac{(4w)(32)}{(333.3)(4)}\right| = 0.096w$$

Noting that $(\sigma_{yy})_G = 0$, the principal stresses in the glue are:

$$(\sigma_{1,2})_G = \left(\frac{2.592w + 0}{2}\right) \pm \sqrt{\left(\frac{2.592w - 0}{2}\right)^2 + (0.096w)^2} = 1.296w \pm 1.300w \text{ or}$$

$(\sigma_1)_G = 2.596w$ and $(\sigma_2)_G = -0.004w$.

Noting that the tensile stress in glue should be less than 200psi we obtain:

$$(\sigma_1)_G = 2.596w \le 200 \qquad \text{or} \qquad w \le 77.04 lb/in \qquad 5$$

The maximum shear stress in the glue is:

$$\tau_{max} = \left|\frac{(\sigma_1)_G - (\sigma_2)_G}{2}\right| = 1.300w \le 300 \qquad \text{or} \qquad w \le 230.7 lb/in \qquad 6$$

The maximum normal stress in wood is the maximum bending normal stress. We obtain:

$$\sigma_W = -\frac{(-24w)(12)(5)}{333.3} = 4.32w \le 2000 \qquad \text{or} \qquad w \le 462.9 lb/in \qquad 7$$

The maximum value of w that satisfies Eqs. 5, 6, and 7 is: $\boxed{w_{max} = 77 lb/in}$

8.63 Determine the thickness of a steel plate required for a thin-cylindrical boiler with a center-line diameter of 2.5 m, if the maximum tensile stress is not to exceed 100 MPa and maximum shear stress is not to exceed 60 MPa, when the pressure in the boiler is 1800 kPa.

Solution $d = 2.5$ m $\quad \sigma_{max} \le 100MPa(T) \quad \tau_{max} \le 60MPa \quad p = 1800$ kPa $\quad t = ?$

The axial stress is: $\sigma_{xx} = \frac{pr}{2t} = \frac{(1800)(10^3)(1.25)}{2t} = \frac{1.125(10^6)}{t}$

The hoop stress is: $\sigma_{yy} = \frac{pr}{t} = \frac{(1800)(10^3)(1.25)}{t} = \frac{2.25(10^6)}{t}$

The shear stress $\tau_{xy} = 0$. Thus, hoop stress is principal stress 1 and axial stress is principal stress 2. The maximum normal and shear stress can be written as:

$$\sigma_{max} = \frac{2.25(10^6)}{t} \le 100(10^6) \qquad \text{or} \qquad t \ge 0.0225m \qquad 1$$

$$\tau_{max} = \left|\frac{2.25(10^6)}{t} - \frac{1.125(10^6)}{t}\right|/2 = \frac{0.5625(10^6)}{t} \le 60(10^6) \qquad \text{or} \qquad t \ge 0.0094m \qquad 2$$

The minimum value of t that satisfies Eqs. 1 and 2 is: $\boxed{t_{min} = 22.5mm}$

8.64 A thin-cylindrical tank is fabricated by butt welding a 1/2 in. thick plate. The center-line diameter of the tank is 4 ft.The maximum tensile stress of the plate cannot exceed 30 ksi. The normal and shear stress in the weld is limited to 25 ksi and 18 ksi, respectively. What is the maximum pressure the tank can hold?

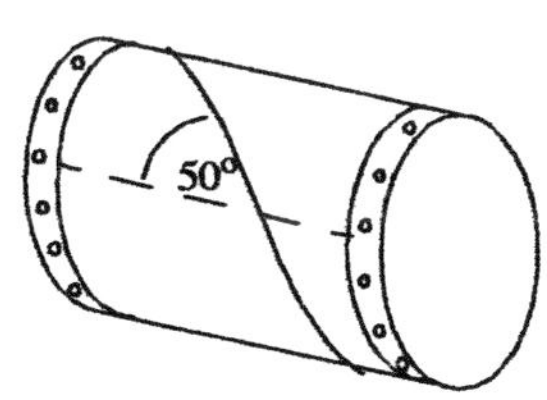

Fig. P8.64

Solution $d = 4\text{ ft} = 48\text{ in}$ $t = 1/2\text{ in.}$ $\sigma_{max} \le 30\text{ksi(T)}$ $\sigma_W \le 25\text{ksi}$ $\tau_W \le 18\text{ksi}$ $p_{max} = ?$

The axial stress is: $\sigma_{xx} = \frac{pr}{2t} = \frac{p(24)}{2(0.5)} = 24p$

The hoop stress is: $\sigma_{yy} = \frac{pr}{t} = \frac{p(24)}{0.5} = 48p$

The shear stress $\tau_{xy} = 0$. Thus, hoop stress is principal stress 1 and axial stress is principal stress 2.

$$\sigma_{max} = 48p \le 30 \qquad \text{or} \qquad p \le 0.625\text{ksi} \qquad (1)$$

Fig. (a) shows the weld and the angle of the outward normal to the weld.

(a)

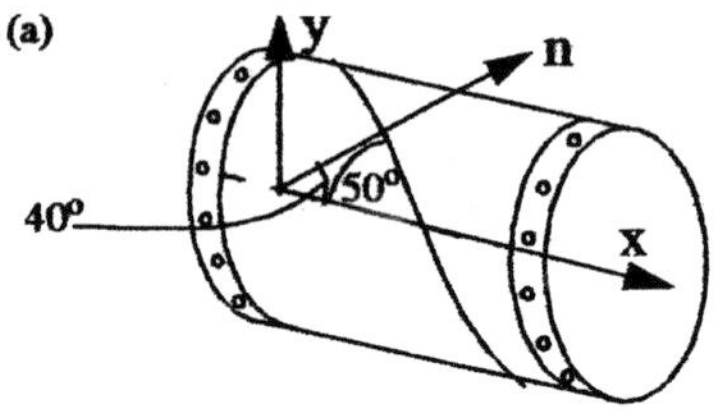

Substituting the stresses and angle normal to the weld as $\theta = 40°$ into Eq. 8.1 and 8.2 we obtain:

$$\sigma_W = (24p)\cos^2 40 + (48p)\sin^2 40 = 33.9162p \le 25 \qquad \text{or} \qquad p \le 0.737\text{ksi} \qquad (2)$$

$$\tau_W = -(24p)\cos 40 \sin 40 + (48p)\sin 40 \cos 40 = 11.818p \le 18 \qquad \text{or} \qquad p \le 1.523\text{ksi} \qquad (3)$$

The maximum value of p that meets the requirements of Eqs. 1, 2, and 3 is: $\boxed{p_{max} = 625\text{psi}}$

8.65 By multiplying the matrices show that the matrix equations below are same as Equations 8.1, 8.2, and 8.8.

$$[\sigma]_{nt} = [T]^T[\sigma][T] \qquad \textbf{8.16}$$

where, $[\sigma]_{nt} = \begin{bmatrix} \sigma_{nn} & \tau_{nt} \\ \tau_{tn} & \sigma_{tt} \end{bmatrix}$ $\quad [T] = \begin{bmatrix} \cos\theta & -\sin\theta \\ \sin\theta & \cos\theta \end{bmatrix}$ $\quad [\sigma] = \begin{bmatrix} \sigma_{xx} & \tau_{xy} \\ \tau_{yx} & \sigma_{yy} \end{bmatrix}$

and $[T]^T$ represents the transpose of the matrix [T]. The matrix [T] is the transformation matrix that relates the x and the y coordinates to n and t coordinates.

Solution

We can write the following.

$$[\sigma]_{nt} = [T]^T[\sigma][T] = [T]^T\begin{bmatrix} \sigma_{xx} & \tau_{xy} \\ \tau_{yx} & \sigma_{yy} \end{bmatrix}\begin{bmatrix} \cos\theta & -\sin\theta \\ \sin\theta & \cos\theta \end{bmatrix} = [T]^T\begin{bmatrix} \sigma_{xx}\cos\theta + \tau_{xy}\sin\theta & -\sigma_{xx}\sin\theta + \tau_{xy}\cos\theta \\ \tau_{yx}\cos\theta + \sigma_{yy}\sin\theta & -\tau_{yx}\sin\theta + \sigma_{yy}\cos\theta \end{bmatrix}$$

Substituting $[T]^T$ and multiplying we obtain:

$$[\sigma]_{nt} = \begin{bmatrix} \cos\theta & \sin\theta \\ -\sin\theta & \cos\theta \end{bmatrix}\begin{bmatrix} \sigma_{xx}\cos\theta + \tau_{xy}\sin\theta & -\sigma_{xx}\sin\theta + \tau_{xy}\cos\theta \\ \tau_{yx}\cos\theta + \sigma_{yy}\sin\theta & -\tau_{yx}\sin\theta + \sigma_{yy}\cos\theta \end{bmatrix} = \begin{bmatrix} \sigma_{nn} & \tau_{nt} \\ \tau_{tn} & \sigma_{tt} \end{bmatrix}$$

We evaluate each term and use the identity $\tau_{xy} = \tau_{yx}$ to obtain the following:

$$\sigma_{nn} = (\sigma_{xx}\cos\theta + \tau_{xy}\sin\theta)\cos\theta + (\tau_{yx}\cos\theta + \sigma_{yy}\sin\theta)\sin\theta = \sigma_{xx}\cos^2\theta + \sigma_{yy}\sin^2\theta + 2\tau_{xy}\sin\theta\cos\theta \qquad (1)$$

$$\tau_{nt} = (-\sigma_{xx}\sin\theta + \tau_{xy}\cos\theta)\cos\theta + (-\tau_{yx}\sin\theta + \sigma_{yy}\cos\theta)\sin\theta = (-\sigma_{xx} + \sigma_{yy})\cos\theta\sin\theta + \tau_{xy}(\cos^2\theta - \sin^2\theta) \qquad (2)$$

$$\tau_{tn} = -(\sigma_{xx}\cos\theta + \tau_{xy}\sin\theta)\sin\theta + (\tau_{yx}\cos\theta + \sigma_{yy}\sin\theta)\cos\theta = (-\sigma_{xx} + \sigma_{yy})\cos\theta\sin\theta + \tau_{xy}(\cos^2\theta - \sin^2\theta) \qquad (3)$$

$$\sigma_{tt} = -(-\sigma_{xx}\sin\theta + \tau_{xy}\cos\theta)\sin\theta + (-\tau_{yx}\sin\theta + \sigma_{yy}\cos\theta)\cos\theta = \sigma_{xx}\sin^2\theta + \sigma_{yy}\cos^2\theta - 2\tau_{xy}\cos\theta\sin\theta \qquad (4)$$

Eqs. 1 and 4 are identical to Eqs. 8.1 and 8.8, respectively. Eqs. 2 and 3 imply $\tau_{nt} = \tau_{tn}$ and Eq. 2 is identical to Eq. 8.2.

8.66 Show that the eigenvalues of the matrix $[\sigma]$ are the principal stresses given by Equation 8.8.

Solution

Let λ be the eigenvalue of the matrix $[\sigma]$. Thus the following determinant must be zero.

$$\begin{vmatrix} \sigma_{xx}-\lambda & \tau_{xy} \\ \tau_{yx} & \sigma_{yy}-\lambda \end{vmatrix} = 0 \quad \text{or} \quad (\sigma_{xx}-\lambda)(\sigma_{yy}-\lambda)-\tau_{xy}\tau_{yx} = 0 \qquad 1$$

Noting that $\tau_{xy} = \tau_{yx}$, we can write Eq. 1 as

$$\sigma_{xx}\sigma_{yy}-\lambda(\sigma_{xx}+\sigma_{yy})+\lambda^2-\tau_{xy}^2 = 0 \qquad 2$$

The roots of the Eq. 2 are:

$$\lambda_{1,2} = \frac{(\sigma_{xx}+\sigma_{yy})\pm\sqrt{(\sigma_{xx}+\sigma_{yy})^2-4(\sigma_{xx}\sigma_{yy}-\tau_{xy}^2)}}{2} = \frac{(\sigma_{xx}+\sigma_{yy})}{2}\pm\sqrt{\frac{\sigma_{xx}^2+\sigma_{yy}^2-2\sigma_{xx}\sigma_{yy}+4\tau_{xy}^2}{4}} \quad \text{or}$$

$$\lambda_{1,2} = \frac{(\sigma_{xx}+\sigma_{yy})}{2}\pm\sqrt{\frac{(\sigma_{xx}-\sigma_{yy})^2}{4}+\tau_{xy}^2} = \frac{(\sigma_{xx}+\sigma_{yy})}{2}\pm\sqrt{\left(\frac{\sigma_{xx}-\sigma_{yy}}{2}\right)^2+\tau_{xy}^2} \qquad 3$$

Eq. 3 is same Eq.8.8 with λ_1 and λ_2 representing principal stresses.

8.67 Using the wedge shown in Fig. P8.27 show that the normal stress on an inclined plane is related to the stresses in cartesian coordinates by the following equation.

$$\sigma_{nn} = \sigma_{xx}n_x^2+\sigma_{yy}n_y^2+\sigma_{zz}n_z^2+2\tau_{xy}n_xn_y+2\tau_{yz}n_yn_z+2\tau_{zx}n_zn_x \qquad \textbf{8.17}$$

Solution

From Fig. P8.27 we have the following relation between the area A of the inclined plane and the area of the cartesian plane on the wedge.

$$A_x = n_xA \qquad A_y = n_yA \qquad A_z = n_zA \qquad 1$$

(a) Force Wedge

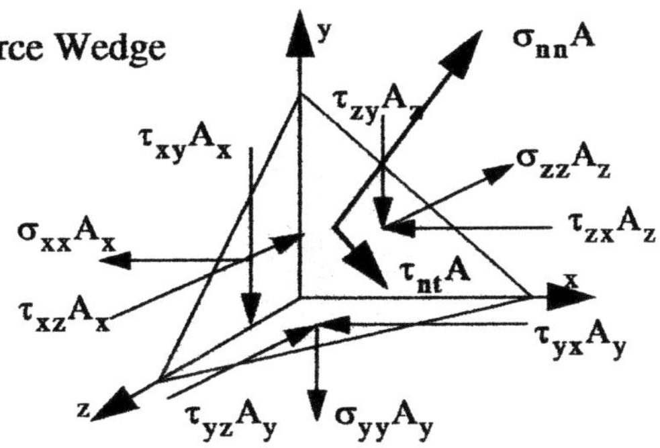

Fig. (a) shows the force wedge obtained by multiplying the stress components with the area of the plane on which they act. We can find the components of the force vectors in the n-direction by taking the dot product of the force vector and the unit vector in the n-direction. By force equilibrium in the n-direction we obtain the following.

$$\sigma_{nn}A+(n_x\bar{i}+n_y\bar{j}+n_z\bar{k})\bullet[(-\sigma_{xx}A_x-\tau_{yx}A_y-\tau_{zx}A_z)\bar{i}+(-\tau_{xy}A_x-\sigma_{yy}A_y-\tau_{zy}A_z)\bar{j}$$
$$(-\tau_{xz}A_x-\tau_{yz}A_y-\sigma_{zz}A_z)\bar{k}] = 0 \qquad 2$$

Substituting Eq. 1 into Eq. 2 and performing the dot product we obtain

$$\sigma_{nn}A + [(-\sigma_{xx}n_xA - \tau_{yx}n_yA - \tau_{zx}n_zA)n_x + (-\tau_{xy}n_xA - \sigma_{yy}n_yA - \tau_{zy}n_zA)n_y$$
$$(-\tau_{xz}n_xA - \tau_{yz}n_yA - \sigma_{zz}n_zA)n_z] = 0 \qquad 3$$

Noting that the area A is a common factor in all terms we can write:

$$\sigma_{nn} = \sigma_{xx}n_x^2 + \sigma_{yy}n_y^2 + \sigma_{zz}n_z^2 + (\tau_{xy} + \tau_{yx})n_xn_y + (\tau_{yz} + \tau_{zy})n_yn_z + (\tau_{zx} + \tau_{zx})n_zn_x \qquad 4$$

By making use of the symmetry of shear stresses in Eq. 4, we obtain Eq. 8.17.

8.68 Fig. P8.68 show eight (octal) planes that make equal angles with the principal planes. These planes are called the octahedral planes. Though the sign of direction cosines change with each plane, the magnitude of direction cosines is the same for all eight planes i.e., $|n_x| = 1/\sqrt{3}$ $\quad |n_y| = 1/\sqrt{3}$ $\quad |n_z| = 1/\sqrt{3}$, The normal stress and shear stress on the octahedral planes σ_{oct} and τ_{oct} are given by the Equations 8.18 and 8.19. Using Equation 8.17 obtain Equation 8.18.

$$\sigma_{oct} = (\sigma_1 + \sigma_2 + \sigma_3)/3 \qquad \textbf{8.18}$$

$$\tau_{oct} = \frac{1}{3}\sqrt{(\sigma_1 - \sigma_2)^2 + (\sigma_2 - \sigma_3)^2 + (\sigma_3 - \sigma_1)^2} \qquad \textbf{8.19}$$

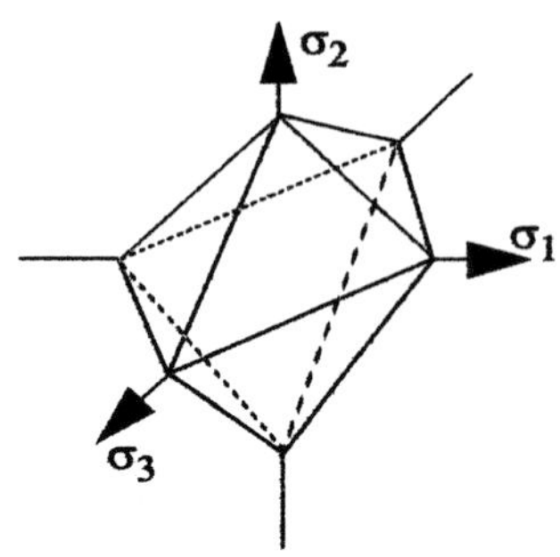

Fig. P8.68The eight octahedral planes

Solution

In terms of principal stresses, Eq. 8.17 can be written as:

$$\sigma_{nn} = \sigma_1n_x^2 + \sigma_2n_y^2 + \sigma_3n_z^2 \qquad 1$$

where n_x, n_y, and n_z, are now measured from the principal planes. Substituting $|n_x| = 1/\sqrt{3}$ $\quad |n_y| = 1/\sqrt{3}$ $\quad |n_z| = 1/\sqrt{3}$ into Eq. 1, we obtain:

$$\sigma_{nn} = \sigma_1\left(\frac{1}{\sqrt{3}}\right)^2 + \sigma_2\left(\frac{1}{\sqrt{3}}\right)^2 + \sigma_3\left(\frac{1}{\sqrt{3}}\right)^2 = \frac{(\sigma_1 + \sigma_2 + \sigma_3)}{3} \qquad 2$$

Eq. 2 is the same as Eq. 8.18.

8.69 On a machine component made of steel (E= 30,000 ksi, G = 11,600 ksi) the following strains were found $\varepsilon_{xx} = [100(2x+y)+50]\ \mu$, $\varepsilon_{yy} = -100(2x+y)\ \mu$, and $\gamma_{xy} = 200(x-2y)\ \mu$. Assuming plane stress, determine the principal stresses, the principal angle one, and the maximum shear stress every 30° on a circle of radius one around the origin. Use a spread sheet or write a computer program for the calculation.

Solution Plane stress $\quad \sigma_1 = ?\ \ \sigma_2 = ?\ \ \sigma_3 = ?\ \ \theta_1 = ?\ \ \tau_{max} = ?$ every 30° with r =1in

---

The Poisson's ratio is: $\nu = \frac{E}{2G} - 1 = \frac{30000}{2(11600)} - 1 = 0.2931$

The given strains are:

$$\varepsilon_{xx} = [100(2x+y)+50]\ \mu \qquad \varepsilon_{yy} = -100(2x+y)\ \mu \qquad \gamma_{xy} = 200(x-2y)\ \mu \qquad 1$$

For plane stress, using Eq. 3.15 we can write the following.

$$\sigma_{xx} = [\varepsilon_{xx} + \nu\varepsilon_{yy}]\frac{E}{(1-\nu^2)} \qquad \sigma_{yy} = [\varepsilon_{yy} + \nu\varepsilon_{xx}]\frac{E}{(1-\nu^2)} \qquad \tau_{xy} = G\gamma_{xy} \qquad 2$$

The x and y coordinates can be written as:

$$x = r\cos\theta = \cos\theta \qquad y = r\sin\theta = \sin\theta \qquad 3$$

Starting with $\theta = 0$ in steps of 30°, we can find the x and y coordinates on a spreadsheet. The strains can be found by substituting the coordinates into Eq. 1. The stresses can be found by substituting the strains into Eq. 2. Principal stresses can be found using Eq. 8.6. Principal angle θ_p can be found using Eq. 8.5. Substituting θ_p into Eq. 8.1 the principal stress corresponding to θ_p can be found, if this stress is principal stress then θ_p is reported as θ_1, otherwise 90° are added and reported as θ_1. The maximum shear stress in this case is equal to inplane maximum shear stress as the principal stresses have opposite signs. All these calculations can be done on a spreadsheet. The results are reported in the table below.

θ	x	y	ε_{xx} (μ)	ε_{yy} (μ)	γ_{xy} (μ)	σ_{xx} (psi)	σ_{yy} (psi)	τ_{xy} (psi)	σ_1 (psi)	σ_2 (psi)	θ_p	σ_p (psi)	θ_1	τ_{max} (psi)
0	1.000	0.000	250	-200	200	6281	-4159	2320	6773	-4651	12.0	6773	12.0	5712
30	0.866	0.500	273	-223	-27	6819	-4697	-311	6828	-4706	-1.5	6828	-1.5	5767
60	0.500	0.866	237	-187	-246	5970	-3848	-2858	6742	-4620	-15.1	6742	-15.1	5681
90	0.000	1.000	150	-100	-400	3961	-1839	-4640	6533	-4411	-29.0	6533	-29.0	5472
120	-0.500	0.866	37	13	-446	1330	792	-5178	6246	-4124	-43.5	6246	-43.5	5185
150	-0.866	0.500	-73	123	-373	-1217	3339	-4329	5953	-3831	31.1	-3831	121.1	4892
180	-1.000	0.000	-150	200	-200	-2999	5121	-2320	5737	-3615	14.9	-3615	104.9	4676
210	-0.866	-0.500	-173	223	27	-3537	5659	311	5670	-3548	-1.9	-3548	88.1	4609
240	-0.500	-0.866	-137	187	246	-2688	4810	2858	5775	-3654	-18.7	-3654	71.3	4715
270	-0.000	-1.000	-50	100	400	-679	2801	4640	6016	-3895	-34.7	-3895	55.3	4956
300	0.500	-0.866	63	-13	446	1952	170	5178	6315	-4193	40.1	6315	40.1	5254
330	0.866	-0.500	173	-123	373	4499	-2377	4329	6589	-4468	25.8	6589	25.8	5528

8.70 The stresses around a hole in a very large plate subject to a uniform stress σ are in polar coordinates.

$$\sigma_{rr} = \frac{\sigma}{2}\left(1 - \frac{a^2}{r^2}\right) - \frac{\sigma}{2}\left(1 - \frac{4a^2}{r^2} + \frac{3a^4}{r^4}\right)\cos 2\theta$$

$$\sigma_{\theta\theta} = \frac{\sigma}{2}\left(1 + \frac{a^2}{r^2}\right) + \frac{\sigma}{2}\left(1 + \frac{3a^4}{r^4}\right)\cos 2\theta \qquad \textbf{8.20}$$

$$\tau_{r\theta} = \frac{\sigma}{2}\left(1 + \frac{2a^2}{r^2} - \frac{3a^4}{r^4}\right)\sin 2\theta$$

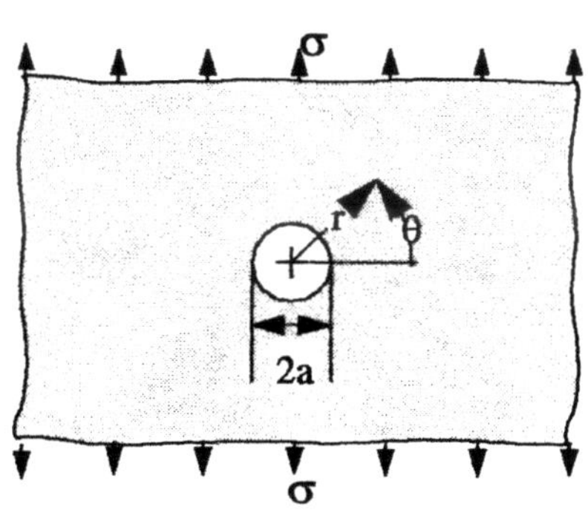

Fig. P8.70

On a ship deck with a manhole that had a diameter of 2 feet, it was estimated that $\sigma = 10$ ksi. Calculate the principal stresses every 15° at a radius of 18 inches. Use a spread sheet or write a computer program for the calculations.

Solution a= 12 in $\sigma_1 = ?$ $\sigma_2 = ?$ every 15° with r =18in

---

In Eq. 8.20, a = 12, r = 18, and θ starting from zero in steps of 15° can be substituted in a spreadsheet to obtain the stresses in polar coordinates. Principal stresses can be found using Eq. 8.6. The calculations done on a spreadsheet are reported in the table below.

θ	σ_{rr} (ksi)	$\sigma_{\theta\theta}$ (ksi)	$\tau_{r\theta}$ (ksi)	σ_1 (ksi)	σ_2 (ksi)
0	3.70	15.19	0.00	15.19	3.70
15	3.58	14.12	3.24	15.04	2.66
30	3.24	11.20	5.61	14.10	0.34
45	2.78	7.22	6.48	11.85	-1.85
60	2.31	3.24	5.61	8.41	-2.85
75	1.98	0.33	3.24	4.50	-2.19
90	1.85	-0.74	0.00	1.85	-0.74
105	1.98	0.33	-3.24	4.50	-2.19
120	2.31	3.24	-5.61	8.41	-2.85
135	2.78	7.22	-6.48	11.85	-1.85
150	3.24	11.20	-5.61	14.10	0.34
165	3.58	14.12	-3.24	15.04	2.66
180	3.70	15.19	-0.00	15.19	3.70
195	3.58	14.12	3.24	15.04	2.66
210	3.24	11.20	5.61	14.10	0.34
225	2.78	7.22	6.48	11.85	-1.85
240	2.31	3.24	5.61	8.41	-2.85
255	1.98	0.33	3.24	4.50	-2.19
270	1.85	-0.74	0.00	1.85	-0.74
285	1.98	0.33	-3.24	4.50	-2.19
300	2.31	3.24	-5.61	8.41	-2.85
315	2.78	7.22	-6.48	11.85	-1.85
330	3.24	11.20	-5.61	14.10	0.34
345	3.58	14.12	-3.24	15.04	2.66

Problem Set 9.1

9.1 Determine the rotation of line OP and the normal strain in the direction of OP due to the given strain component below.

$$\varepsilon_{xx} = 500\mu$$

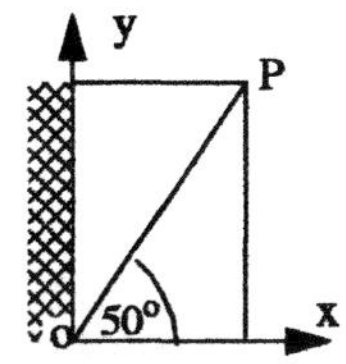

Fig. P9.1

Solution $\varepsilon_{OP} = ?$ $\phi = ?$

Fig.(a) shows the geometry of the rectangle and its deformed shape. From geometry we can write:

$$\Delta x = \Delta n \cos 50 \qquad PP_1 = \varepsilon_{xx}\Delta x = 500(10^{-6})\Delta n \ \cos 50 = 321.39(\Delta n)(10^{-6}) \qquad 1$$

(a)

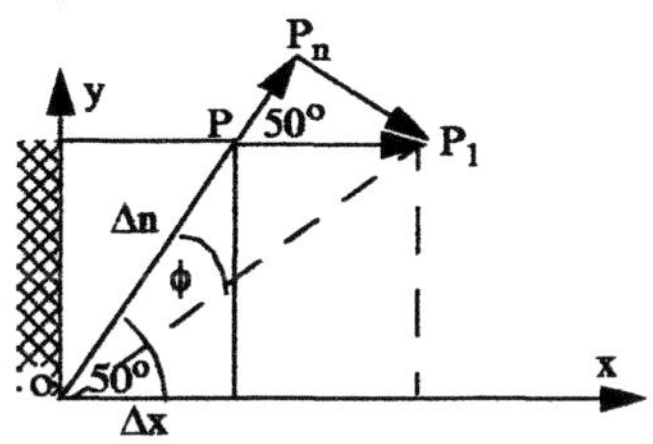

From triangle P_nPP_1 and Eq. 1 we obtain:

$$PP_n = PP_1 \cos 50 = 206.59(\Delta n)(10^{-6}) \qquad P_nP_1 = PP_1 \sin 50 = 246.20(\Delta n)(10^{-6}) \qquad 2$$

The normal strain in direction of OP is: $\varepsilon_{OP} = \dfrac{PP_n}{OP} = \dfrac{206.59(\Delta n)(10^{-6})}{\Delta n}$ or $\boxed{\varepsilon_{OP} = 206.59\mu}$

The length OP_n can be approximated as: $OP_n = OP + PP_n = (\Delta n)[1 + 206.59(10^{-6})] \approx (\Delta n)$ and for small angles in small strains we can write $\tan\phi \approx \phi$. From triangle P_nOP_1 we obtain the rotation of the line OP as shown below.

$$\tan\phi \approx \phi = \frac{P_nP_1}{OP_n} \approx \frac{246.20(\Delta n)(10^{-6})}{\Delta n} \quad \text{or} \qquad \boxed{\phi = 246.20\mu\text{rads CW}}$$

9.2 Determine the rotation of line OP and the normal strain in the direction of OP due to the given strain component below.

$$\varepsilon_{yy} = -400\mu$$

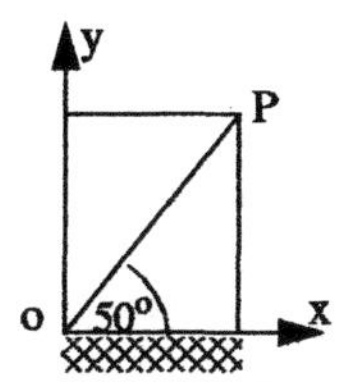

Fig. P9.2

Solution $\varepsilon_{OP} = ?$ $\phi = ?$

Fig.(a) shows the geometry of the rectangle and its deformed shape. From geometry we can write:

$$\Delta y = \Delta n \sin 50 \qquad PP_1 = |\varepsilon_{yy}\Delta x| = 400(10^{-6})\Delta n \ \sin 50 = 306.42(\Delta n)(10^{-6}) \qquad 1$$

(a)

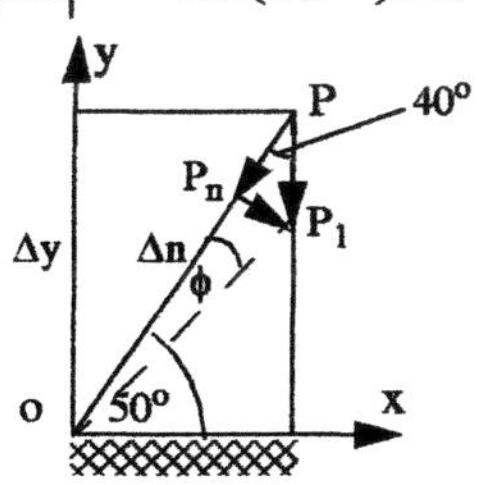

From triangle P_nPP_1 and Eq. 1 we obtain:

$$PP_n = PP_1\cos 40 = 234.72(\Delta n)(10^{-6}) \qquad P_nP_1 = PP_1\sin 40 = 196.96(\Delta n)(10^{-6}) \qquad \textbf{2}$$

The normal strain in direction of OP: $\varepsilon_{OP} = \frac{-PP_n}{OP} = \frac{-234.72(\Delta n)(10^{-6})}{\Delta n}$ or $\boxed{\varepsilon_{OP} = -234.72\mu}$

The length OP_n can be approximated as: $OP_n = OP - PP_n = (\Delta n)[1 - 234.72(10^{-6})] \approx (\Delta n)$ and for small angles in small strains we can write $\tan\phi \approx \phi$. From triangle P_nOP_1 we obtain the rotation of the line OP as shown below.

$\tan\phi \approx \phi = \frac{P_nP_1}{OP_n} \approx \frac{196.96(\Delta n)(10^{-6})}{\Delta n}$ or $\boxed{\phi = 196.96\mu\text{rads CW}}$

9.3 Determine the rotation of line OP and the normal strain in the direction of OP due to the given strain component below.

$$\gamma_{xy} = 300\mu$$

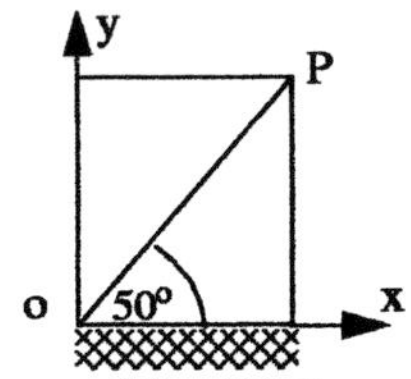

Fig. P9.3

Solution $\varepsilon_{OP} = ?$ $\phi = ?$

Fig.(a) shows the geometry of the rectangle and its deformed shape. From geometry we can write:

$$\Delta y = \Delta n\sin 50 \qquad PP_1 = \gamma_{xy}\Delta y = 300(10^{-6})\Delta n\ \sin 50 = 229.81(\Delta n)(10^{-6}) \qquad \textbf{1}$$

(a)

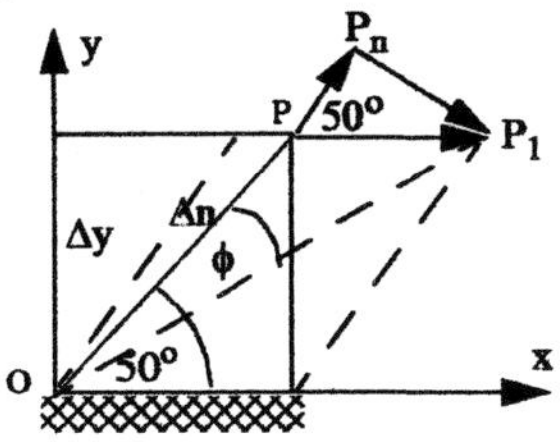

From triangle P_nPP_1 and Eq. 1 we obtain:

$$PP_n = PP_1\cos 50 = 147.72(\Delta n)(10^{-6}) \qquad P_nP_1 = PP_1\sin 50 = 176.05(\Delta n)(10^{-6}) \qquad \textbf{2}$$

The normal strain in direction of OP is: $\varepsilon_{OP} = \frac{PP_n}{OP} = \frac{147.72(\Delta n)(10^{-6})}{\Delta n}$ or $\boxed{\varepsilon_{OP} = 147.72\mu}$

The length OP_n can be approximated as: $OP_n = OP + PP_n = (\Delta n)[1 + 147.72(10^{-6})] \approx (\Delta n)$ and for small angles in small strains we can write $\tan\phi \approx \phi$. From triangle P_nOP_1 we obtain the rotation of the line OP as shown below.

$\tan\phi \approx \phi = \frac{P_nP_1}{OP_n} \approx \frac{176.05(\Delta n)(10^{-6})}{\Delta n}$ or $\boxed{\phi = 176.05\mu\text{rads CW}}$

9.4 Determine the rotation of line OP and the normal strain in the direction of OP due to the given strain component below.

$$\gamma_{yx} = 300\mu$$

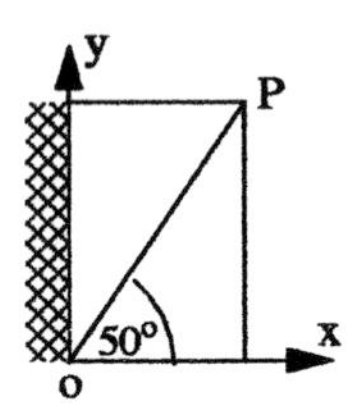

Fig. P9.4

Solution $\varepsilon_{OP} = ?$ $\phi = ?$

Fig.(a) shows the geometry of the rectangle and its deformed shape.From geometry we can write:

$$\Delta x = \Delta n \cos 50 \qquad PP_1 = AP\tan\gamma_{xy} \approx \Delta x \gamma_{yx} = 300(10^{-6})\Delta n \ \cos 50 = 192.84(\Delta n)(10^{-6}) \qquad 1$$

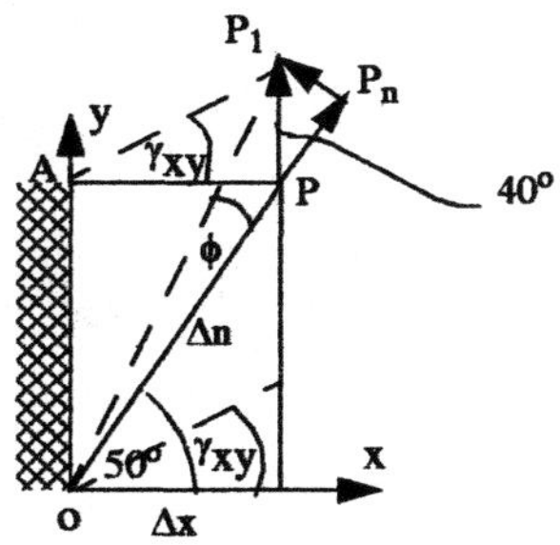

From triangle P_nPP_1 and Eq. 1 we obtain:

$$PP_n = PP_1\cos 40 = 147.72(\Delta n)(10^{-6}) \qquad P_nP_1 = PP_1\sin 40 = 123.95(\Delta n)(10^{-6}) \qquad 2$$

The normal strain in direction of OP: $\varepsilon_{OP} = \dfrac{PP_n}{OP} = \dfrac{147.72(\Delta n)(10^{-6})}{\Delta n}$ or $\boxed{\varepsilon_{OP} = 147.72\mu}$

The length OP_n can be approximated as: $OP_n = OP + PP_n = (\Delta n)[1 + 147.72(10^{-6})] \approx (\Delta n)$ and for small angles in small strains we can write $\tan\phi \approx \phi$. From triangle P_nOP_1 we obtain the rotation of the line OP as shown below.

$\tan\phi \approx \phi = \dfrac{P_nP_1}{OP_n} \approx \dfrac{123.95(\Delta n)(10^{-6})}{\Delta n}$ or $\boxed{\phi = 123.95\mu\text{rads CCW}}$

9.5 At a point, the only non-zero strain component is as given below. Determine the strain components in 'n' and 't' coordinate system shown in Fig. P9.5.

$$\varepsilon_{xx} = -400\ \mu.$$

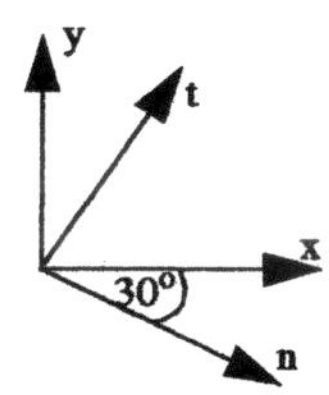

Fig. P9.5

Solution $\varepsilon_{nn} = ?$ $\varepsilon_{tt} = ?$ $\gamma_{nt} = ?$

Fig.(a) shows the rotation of the n and t lines, while Figs. (b) and (c) show rectangles with diagonals in the n and t directions deforming due to ε_{xx}.

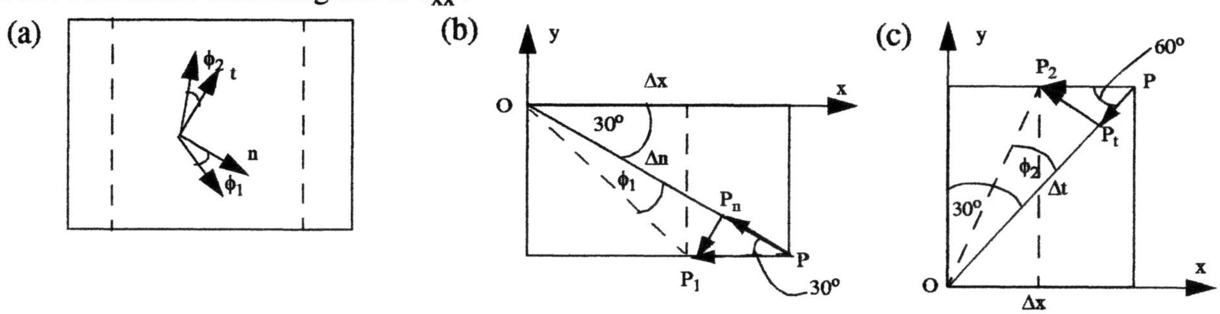

From Fig.(b) we obtain:

$$\Delta x = \Delta n \cos 30 \qquad PP_1 = |\varepsilon_{xx}|\Delta x = 400(10^{-6})\Delta n \ \cos 30 = 346.41(\Delta n)(10^{-6}) \qquad 1$$

From triangle P_nPP_1 and Eq. 1 we obtain:

$$PP_n = PP_1\cos 30 = 300(\Delta n)(10^{-6}) \qquad P_nP_1 = PP_1\sin 30 = 173.2(\Delta n)(10^{-6}) \qquad (2)$$

The normal strain in the n-direction is: $\varepsilon_{nn} = \dfrac{-PP_n}{OP} = \dfrac{(-300)(\Delta n)(10^{-6})}{\Delta n}$ or $\boxed{\varepsilon_{nn} = -300\mu}$

The length OP_n can be approximated as: $OP_n = OP - PP_n = (\Delta n)[1 - 300(10^{-6})] \approx (\Delta n)$ and for small angles in small strains we can write $\tan\phi_1 \approx \phi_1$. From triangle P_nOP_1 we obtain the rotation of the line OP as shown below.

$$\tan\phi_1 \approx \phi_1 = \frac{P_nP_1}{OP_n} \approx \frac{173.2(\Delta n)(10^{-6})}{\Delta n} \quad \text{or} \quad \phi_1 = 173.2\mu\text{rads} \qquad (3)$$

From Fig. (c) we obtain

$$\Delta x = \Delta t\sin 30 \qquad PP_2 = |\varepsilon_{xx}|\Delta x = 400(10^{-6})\Delta t\ \sin 30 = 200(\Delta t)(10^{-6}) \qquad (4)$$

From triangle P_tPP_2 and Eq. 4 we obtain:

$$PP_t = PP_2\cos 60 = 100(\Delta t)(10^{-6}) \qquad P_tP_2 = PP_2\sin 60 = 173.2(\Delta t)(10^{-6}) \qquad (5)$$

The normal strain in the t-direction is: $\varepsilon_{tt} = \dfrac{-PP_t}{OP} = \dfrac{(-100)(\Delta t)(10^{-6})}{\Delta t}$ or $\boxed{\varepsilon_{tt} = -100\mu}$

The length OP_t can be approximated as: $OP_t = OP - PP_t = (\Delta t)[1 - 100(10^{-6})] \approx (\Delta t)$ and for small angles in small strains we can write $\tan\phi_2 \approx \phi_2$. From triangle P_tOP_2 we obtain the rotation of the line OP as shown below.

$$\tan\phi_2 \approx \phi_2 = \frac{P_tP_2}{OP_t} \approx \frac{173.2(\Delta t)(10^{-6})}{\Delta t} \quad \text{or} \quad \phi_2 = 173.2\mu\text{rads} \qquad (6)$$

From Fig. (a) we see that the angle between n and t increases by ϕ_1 and ϕ_2. Using Eqs. 3 and 6 we obtain:

$\gamma_{nt} = -(\phi_1 + \phi_2) = -(173.2 + 173.2)$ or $\boxed{\gamma_{nt} = -346.4\mu\text{rad}}$

9.6 At a point, the only non-zero strain component is as given below. Determine the strain components in 'n' and 't' coordinate system shown in Fig. P9.6.

$$\varepsilon_{yy} = 600\ \mu.$$

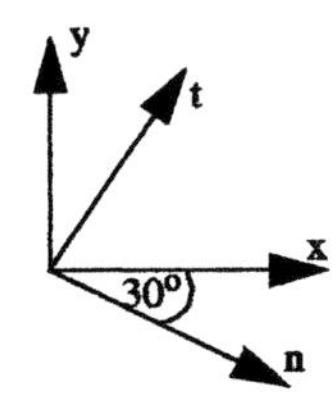

Fig. P9.6

Solution $\varepsilon_{nn} = ?$ $\varepsilon_{tt} = ?$ $\gamma_{nt} = ?$

Fig.(a) shows the rotation of the n and t lines, while Figs. (b) and (c) show rectangles with diagonals in the n and t directions deforming due to ε_{yy}.

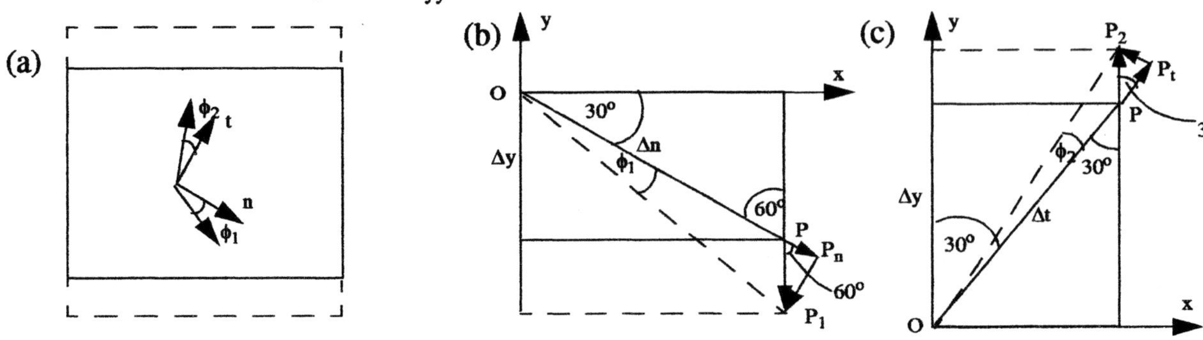

From Fig.(b) we obtain:

$$\Delta y = \Delta n \sin 30 \qquad PP_1 = |\varepsilon_{yy}|\Delta y = 600(10^{-6})\Delta n \ \sin 30 = 300(\Delta n)(10^{-6}) \qquad 1$$

From triangle P_nPP_1 and Eq. 1 we obtain:

$$PP_n = PP_1\cos 60 = 150(\Delta n)(10^{-6}) \qquad P_nP_1 = PP_1\sin 60 = 259.8(\Delta n)(10^{-6}) \qquad 2$$

The normal strain in the n-direction is: $\varepsilon_{nn} = \frac{PP_n}{OP} = \frac{(150)(\Delta n)(10^{-6})}{\Delta n}$ or $\boxed{\varepsilon_{nn} = 150\mu}$

The length OP_n can be approximated as: $OP_n = OP + PP_n = (\Delta n)[1 + 300(10^{-6})] \approx (\Delta n)$ and for small angles in small strains we can write $\tan\phi_1 \approx \phi_1$. From triangle P_nOP_1 we obtain the rotation of the line OP as shown below.

$$\tan\phi_1 \approx \phi_1 = \frac{P_nP_1}{OP_n} \approx \frac{259.8(\Delta n)(10^{-6})}{\Delta n} \quad \text{or} \quad \phi_1 = 259.8\mu\text{rads} \qquad 3$$

From Fig. (c) we obtain

$$\Delta y = \Delta t \cos 30 \qquad PP_2 = |\varepsilon_{yy}|\Delta y = 600(10^{-6})\Delta t \ \cos 30 = 519.62(\Delta t)(10^{-6}) \qquad 4$$

From triangle P_tPP_2 and Eq. 4 we obtain:

$$PP_t = PP_2\cos 30 = 450(\Delta t)(10^{-6}) \qquad P_tP_2 = PP_2\sin 30 = 259.8(\Delta t)(10^{-6}) \qquad 5$$

The normal strain in the t-direction is: $\varepsilon_{tt} = \frac{PP_t}{OP} = \frac{(450)(\Delta t)(10^{-6})}{\Delta t}$ or $\boxed{\varepsilon_{tt} = 450\mu}$

The length OP_t can be approximated as: $OP_t = OP + PP_t = (\Delta t)[1 + 450(10^{-6})] \approx (\Delta t)$ and for small angles in small strains we can write $\tan\phi_2 \approx \phi_2$. From triangle $P_tOP_2$we obtain the rotation of the line OP as shown below.

$$\tan\phi_2 \approx \phi_2 = \frac{P_tP_2}{OP_t} \approx \frac{259.8(\Delta t)(10^{-6})}{\Delta t} \quad \text{or} \quad \phi_2 = 259.8\mu\text{rads} \qquad 6$$

From Fig. (a) we see that the angle between n and t increases by $\phi_1$and ϕ_2. Using Eqs. 3 and 6 we obtain:

$\gamma_{nt} = -(\phi_1 + \phi_2) = -(259.8 + 259.8)$ or $\boxed{\gamma_{nt} = -519.6\mu\text{rad}}$

9.7 At a point, the only non-zero strain component is as given below. Determine the strain components in 'n' and 't' coordinate system shown in Fig. P9.7.

$$\gamma_{xy} = -500\ \mu.$$

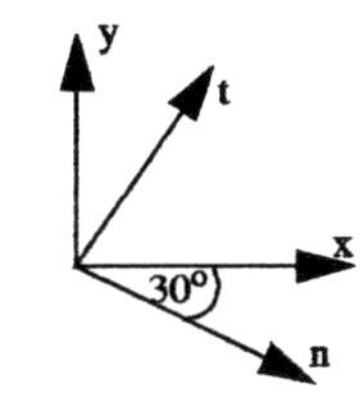

Fig. P9.7

Solution $\varepsilon_{nn} = ?$ $\varepsilon_{tt} = ?$ $\gamma_{nt} = ?$

Fig.(a) shows the rotation of the n and t lines as the angle between x and y coordinate increases by 500μ. Note we could have rotated the vertical lines (counter-clockwise) rather then the rotation of horizontal lines clockwise as in Fig. (a). Figs. (b) and (c) show rectangles with diagonals in the n and t directions deforming due to γ_{xy}.

From Fig.(b) we obtain:

$$\Delta x = \Delta n \cos 30 \qquad PP_1 = AP\tan|\gamma_{xy}| \approx \Delta x|\gamma_{xy}| = 500(10^{-6})\Delta n \ \cos 30 = 433.01(\Delta n)(10^{-6}) \qquad 1$$

From triangle P_nPP_1 and Eq. 1 we obtain:

$$PP_n = PP_1\cos 60 = 216.5(\Delta n)(10^{-6}) \qquad P_nP_1 = PP_1\sin 60 = 375(\Delta n)(10^{-6}) \qquad 2$$

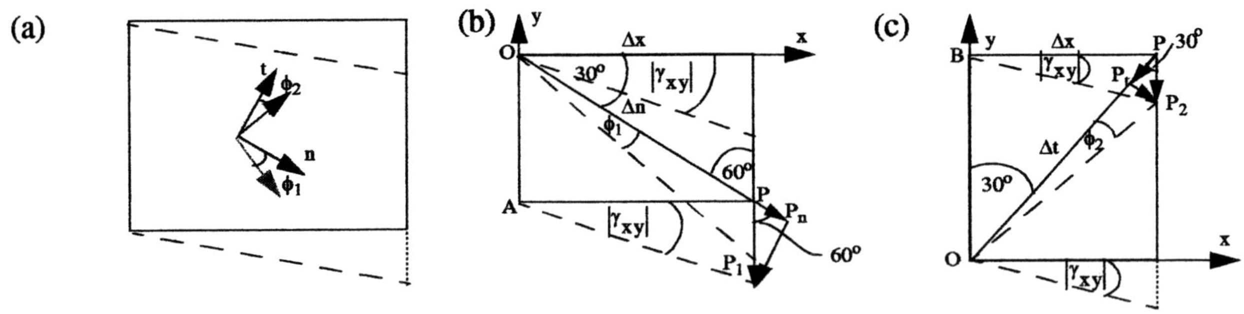

The normal strain in the n-direction is: $\varepsilon_{nn} = \dfrac{PP_n}{OP} = \dfrac{(216.5)(\Delta n)(10^{-6})}{\Delta n}$ or $\boxed{\varepsilon_{nn} = 216.5\mu}$

The length OP_n can be approximated as: $OP_n = OP + PP_n = (\Delta n)[1 + 216.5(10^{-6})] \approx (\Delta n)$ and for small angles in small strains we can write $\tan\phi_1 \approx \phi_1$. From triangle P_nOP_1 we obtain the rotation of the line OP as shown below.

$$\tan\phi_1 \approx \phi_1 = \frac{P_nP_1}{OP_n} \approx \frac{375(\Delta n)(10^{-6})}{\Delta n} \qquad \text{or} \qquad \phi_1 = 375\mu\text{rads} \qquad 3$$

From Fig. (c) we obtain

$$\Delta x = \Delta t \sin 30 \qquad PP_2 = BP\tan|\gamma_{xy}| \approx \Delta x|\gamma_{xy}| = 500(10^{-6})\Delta t \;\; \sin 30 = 250(\Delta t)(10^{-6}) \qquad 4$$

From triangle P_tPP_2 and Eq. 4 we obtain:

$$PP_t = PP_2\cos 30 = 216.5(\Delta t)(10^{-6}) \qquad P_tP_2 = PP_2\sin 30 = 125(\Delta t)(10^{-6}) \qquad 5$$

The normal strain in the t-direction is: $\varepsilon_{tt} = \dfrac{-PP_t}{OP} = \dfrac{(-216.5)(\Delta t)(10^{-6})}{\Delta t}$ or $\boxed{\varepsilon_{tt} = -216.5\mu}$

The length OP_t can be approximated as: $OP_t = OP - PP_t = (\Delta t)[1 - 216.5(10^{-6})] \approx (\Delta t)$ and for small angles in small strains we can write $\tan\phi_2 \approx \phi_2$. From triangle P_tOP_2 we obtain the rotation of the line OP as shown below.

$$\tan\phi_2 \approx \phi_2 = \frac{P_tP_2}{OP_t} \approx \frac{125(\Delta t)(10^{-6})}{\Delta t} \qquad \text{or} \qquad \phi_2 = 125\mu\text{rads} \qquad 6$$

From Fig. (a) we see that the angle between n and t increases by (ϕ_1- ϕ_2). Using Eqs. 3 and 6 we obtain:

$\gamma_{nt} = -(\phi_1 - \phi_2) = -(375 - 125)$ or $\boxed{\gamma_{nt} = -250\mu\text{rad}}$

9.8 At a point in plane strain the strain components are

$\varepsilon_{xx} = -400\ \mu \qquad \varepsilon_{yy} = 600\ \mu. \qquad \gamma_{xy} = -500\ \mu.$

Using the results obtained in problems 9.5, 9.6, and 9.7, determine the strain components in 'n' and 't' coordinate system shown in Fig. P9.8

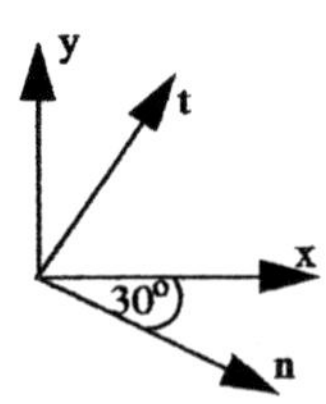

Fig. P9.8

Solution $\qquad \varepsilon_{nn} = ? \qquad \varepsilon_{tt} = ? \qquad \gamma_{nt} = ?$

From problem 9.5 we have: $\varepsilon_{nn}^{(1)} = -300\mu \qquad \varepsilon_{tt}^{(1)} = -100\mu \qquad \gamma_{nt}^{(1)} = -346.4\mu\text{rad}$

From problem 9.6 we have: $\varepsilon_{nn}^{(2)} = 150\mu \qquad \varepsilon_{tt}^{(2)} = 450\mu \qquad \gamma_{nt}^{(2)} = -519.62\mu\text{rad}$

From problem 9.7 we have: $\varepsilon_{nn}^{(3)} = 216.5\mu \qquad \varepsilon_{tt}^{(3)} = -216.5\mu \qquad \gamma_{nt}^{(3)} = -250\mu\text{rad}$

The total strain in is superposition of the above strains for each component. We obtain:

$\varepsilon_{nn} = \varepsilon_{nn}^{(1)} + \varepsilon_{nn}^{(2)} + \varepsilon_{nn}^{(3)} = (-300 + 150 + 216.5)$ or $\boxed{\varepsilon_{nn} = 66.5\mu}$

$\varepsilon_{tt} = \varepsilon_{tt}^{(1)} + \varepsilon_{tt}^{(2)} + \varepsilon_{tt}^{(3)} = (-100 + 450 - 216.5)$ or $\boxed{\varepsilon_{tt} = 133.5\mu}$

$\gamma_{nt} = \gamma_{nt}^{(1)} + \gamma_{nt}^{(2)} + \gamma_{nt}^{(3)} = (-346.4 - 519.62 - 250)$ or $\boxed{\gamma_{nt} = -1116\mu}$

9.9 At a point, the only non-zero strain component is as given below. Determine the strain components in 'n' and 't' coordinate system shown in Fig. P9.9.

$$\varepsilon_{xx} = -600\ \mu.$$

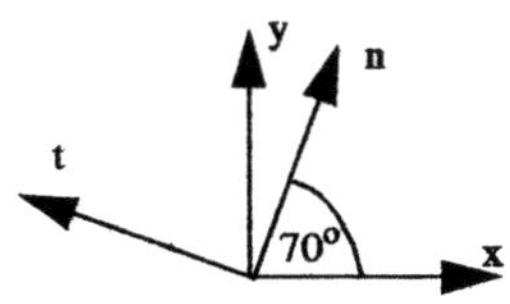

Fig. P9.9

Solution $\varepsilon_{nn} = ?$ $\varepsilon_{tt} = ?$ $\gamma_{nt} = ?$

Fig.(a) shows the rotation of the n and t lines, while Figs. (b) and (c) show rectangles with diagonals in the n and t directions deforming due to ε_{xx}.

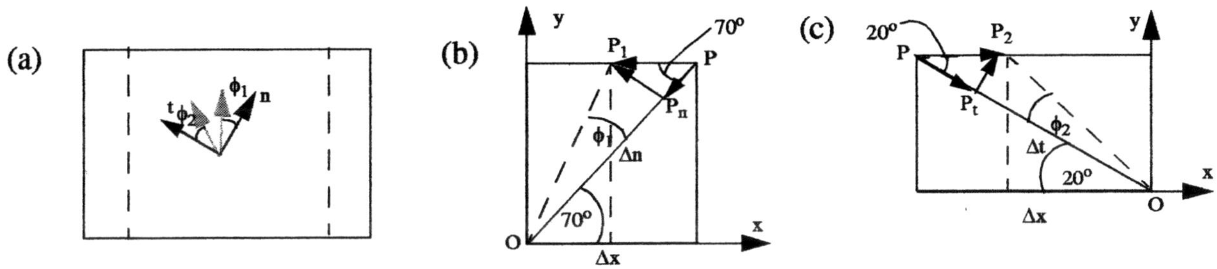

From Fig.(b) we obtain:

$$\Delta x = \Delta n \cos 70 \qquad PP_1 = |\varepsilon_{xx}|\Delta x = 600(10^{-6})\Delta n \ \cos 70 = 205.21(\Delta n)(10^{-6}) \qquad 1$$

From triangle P_nPP_1 and Eq. 1 we obtain:

$$PP_n = PP_1\cos 70 = 70.19(\Delta n)(10^{-6}) \qquad P_nP_1 = PP_1\sin 70 = 192.84(\Delta n)(10^{-6}) \qquad 2$$

The normal strain in the n-direction is: $\varepsilon_{nn} = \dfrac{-PP_n}{OP} = \dfrac{(-70.19)(\Delta n)(10^{-6})}{\Delta n}$ or $\boxed{\varepsilon_{nn} = -70.19\mu}$

The length OP_n can be approximated as: $OP_n = OP - PP_n = (\Delta n)[1 - 70.19(10^{-6})] \approx (\Delta n)$ and for small angles in small strains we can write $\tan\phi_1 \approx \phi_1$. From triangle P_nOP_1 we obtain the rotation of the line OP as shown below.

$$\tan\phi_1 \approx \phi_1 = \frac{P_nP_1}{OP_n} \approx \frac{192.84(\Delta n)(10^{-6})}{\Delta n} \qquad \text{or} \qquad \phi_1 = 192.84\mu\text{rads} \qquad 3$$

From Fig. (c) we obtain

$$\Delta x = \Delta t\cos 20 \qquad PP_2 = |\varepsilon_{xx}|\Delta x = 600(10^{-6})\Delta t \ \cos 20 = 563.82(\Delta t)(10^{-6}) \qquad 4$$

From triangle P_tPP_2 and Eq. 4 we obtain:

$$PP_t = PP_2\cos 20 = 529.81(\Delta t)(10^{-6}) \qquad P_tP_2 = PP_2\sin 20 = 192.84(\Delta t)(10^{-6}) \qquad 5$$

The normal strain in the t-direction is: $\varepsilon_{tt} = \dfrac{-PP_t}{OP} = \dfrac{(-529.81)(\Delta t)(10^{-6})}{\Delta t}$ or $\boxed{\varepsilon_{tt} = -529.81\mu}$

The length OP_t can be approximated as: $OP_t = OP - PP_t = (\Delta t)[1 - 529.81(10^{-6})] \approx (\Delta t)$ and for small angles in small strains we can write $\tan\phi_2 \approx \phi_2$. From triangle $P_tOP_2$we obtain the rotation of the line OP as

shown below.

$$\tan\phi_2 \approx \phi_2 = \frac{P_tP_2}{OP_t} \approx \frac{192.84(\Delta t)(10^{-6})}{\Delta t} \quad \text{or} \quad \phi_2 = 192.84\mu\text{rads} \qquad \mathbf{6}$$

From Fig. (a) we see that the angle between n and t decreases by ϕ_1 and ϕ_2. Using Eqs. 3 and 6 we obtain:

$\gamma_{nt} = (\phi_1 + \phi_2) = (192.84 + 192.84)$ or $\boxed{\gamma_{nt} = 385.7\mu\text{rad}}$

9.10 At a point, the only non-zero strain component is as given below. Determine the strain components in 'n' and 't' coordinate system shown in Fig. P9.10.

$$\varepsilon_{yy} = -800\ \mu.$$

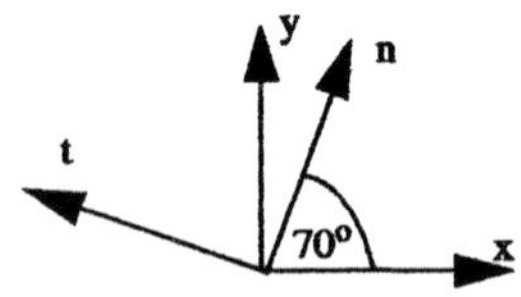

Fig. P9.10

Solution $\varepsilon_{nn} = ?$ $\varepsilon_{tt} = ?$ $\gamma_{nt} = ?$

Fig.(a) shows the rotation of the n and t lines, while Figs. (b) and (c) show rectangles with diagonals in the n and t directions deforming due to ε_{yy}.

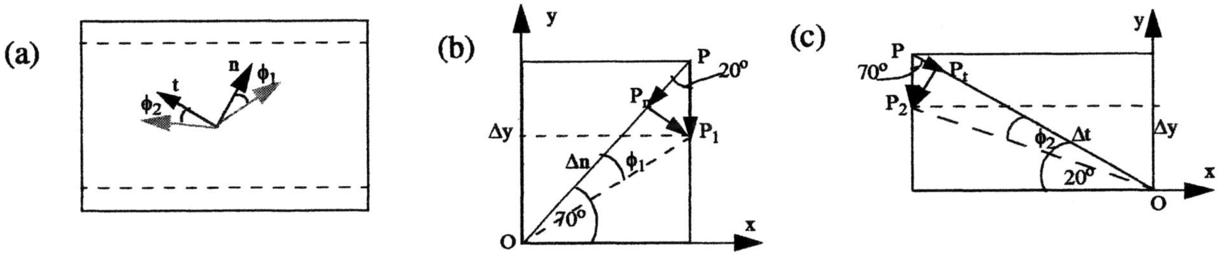

From Fig.(b) we obtain:

$$\Delta y = \Delta n \sin 70 \qquad PP_1 = |\varepsilon_{yy}|\Delta y = 800(10^{-6})\Delta n\ \sin 70 = 751.75(\Delta n)(10^{-6}) \qquad \mathbf{1}$$

From triangle P_nPP_1 and Eq. 1 we obtain:

$$PP_n = PP_1\cos 20 = 706.42(\Delta n)(10^{-6}) \qquad P_nP_1 = PP_1\sin 20 = 257.12(\Delta n)(10^{-6}) \qquad \mathbf{2}$$

The normal strain in the n-direction is: $\varepsilon_{nn} = \dfrac{-PP_n}{OP} = \dfrac{-706.42(\Delta n)(10^{-6})}{\Delta n}$ or $\boxed{\varepsilon_{nn} = -706.4\mu}$

The length OP_n can be approximated as: $OP_n = OP - PP_n = (\Delta n)[1 - 706.42(10^{-6})] \approx (\Delta n)$ and for small angles in small strains we can write $\tan\phi_1 \approx \phi_1$. From triangle P_nOP_1 we obtain the rotation of the line OP as shown below.

$$\tan\phi_1 \approx \phi_1 = \frac{P_nP_1}{OP_n} \approx \frac{257.12(\Delta n)(10^{-6})}{\Delta n} \quad \text{or} \quad \phi_1 = 257.12\mu\text{rads} \qquad \mathbf{3}$$

From Fig. (c) we obtain

$$\Delta y = \Delta t \sin 20 \qquad PP_2 = |\varepsilon_{yy}|\Delta y = 800(10^{-6})\Delta t\ \sin 20 = 273.62(\Delta t)(10^{-6}) \qquad \mathbf{4}$$

From triangle P_tPP_2 and Eq. 4 we obtain:

$$PP_t = PP_2\cos 70 = 93.58(\Delta t)(10^{-6}) \qquad P_tP_2 = PP_2\sin 70 = 257.12(\Delta t)(10^{-6}) \qquad \mathbf{5}$$

The normal strain in the t-direction is: $\varepsilon_{tt} = \dfrac{-PP_t}{OP} = \dfrac{-93.58(\Delta t)(10^{-6})}{\Delta t}$ or $\boxed{\varepsilon_{tt} = -93.6\mu}$

The length OP_t can be approximated as: $OP_t = OP - PP_t = (\Delta t)[1 - 93.58(10^{-6})] \approx (\Delta t)$ and for small angles in small strains we can write $\tan\phi_2 \approx \phi_2$. From triangle P_tOP_2 we obtain the rotation of the line OP as

shown below.

$$\tan\phi_2 \approx \phi_2 = \frac{P_tP_2}{OP_t} \approx \frac{257.12(\Delta t)(10^{-6})}{\Delta t} \quad \text{or} \quad \phi_2 = 257.12\mu\text{rads} \qquad 6$$

From Fig. (a) we see that the angle between n and t increases by ϕ_1 and ϕ_2. Using Eqs. 3 and 6 we obtain:

$\gamma_{nt} = -(\phi_1+\phi_2) = -(257.12+257.12)$ or $\boxed{\gamma_{nt} = -514.2\mu\text{rad}}$

9.11 At a point, the only non-zero strain component is as given below. Determine the strain components in 'n' and 't' coordinate system shown in Fig. P9.11.

$$\gamma_{xy} = \gamma_{xy} = 500\ \mu.$$

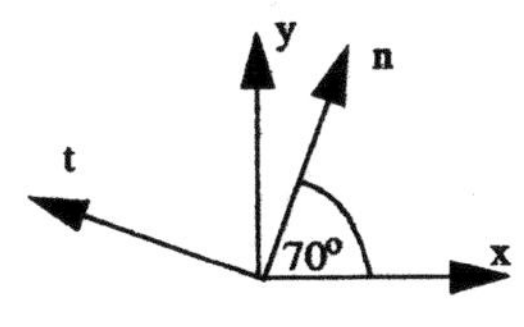

Fig. P9.11

Solution $\varepsilon_{nn} = ?$ $\varepsilon_{tt} = ?$ $\gamma_{nt} = ?$

Fig.(a) shows the rotation of the n and t lines as the angle between x and y coordinate increases by 500μ. Note we could have rotated the vertical lines (counter-clockwise) rather then the rotation of horizontal lines clockwise as in Fig. (a). Figs. (b) and (c) show rectangles with diagonals in the n and t directions deforming due to γ_{xy}.

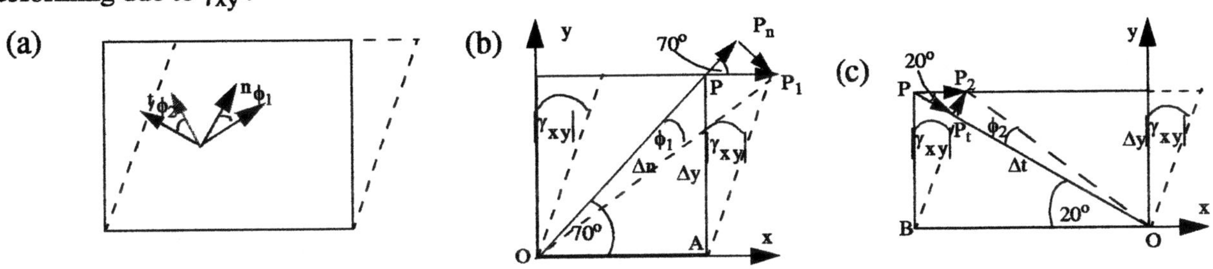

From Fig.(b) we obtain:

$$\Delta y = \Delta n \sin 70 \qquad PP_1 = AP\tan|\gamma_{xy}| \approx \Delta y|\gamma_{xy}| = 500(10^{-6})\Delta n\ \sin 70 = 469.85(\Delta n)(10^{-6}) \qquad 1$$

From triangle P_nPP_1 and Eq. 1 we obtain:

$$PP_n = PP_1\cos 70 = 160.7(\Delta n)(10^{-6}) \qquad P_nP_1 = PP_1\sin 70 = 441.51(\Delta n)(10^{-6}) \qquad 2$$

The normal strain in the n-direction is: $\varepsilon_{nn} = \dfrac{PP_n}{OP} = \dfrac{(160.7)(\Delta n)(10^{-6})}{\Delta n}$ or $\boxed{\varepsilon_{nn} = 160.7\mu}$

The length OP_n can be approximated as: $OP_n = OP + PP_n = (\Delta n)[1+160.7(10^{-6})] \approx (\Delta n)$ and for small angles in small strains we can write $\tan\phi_1 \approx \phi_1$. From triangle P_nOP_1 we obtain the rotation of the line OP as shown below.

$$\tan\phi_1 \approx \phi_1 = \frac{P_nP_1}{OP_n} \approx \frac{441.51(\Delta n)(10^{-6})}{\Delta n} \quad \text{or} \quad \phi_1 = 441.51\mu\text{rads} \qquad 3$$

From Fig. (c) we obtain

$$\Delta y = \Delta t \sin 20 \qquad PP_2 = BP\tan|\gamma_{xy}| \approx \Delta y|\gamma_{xy}| = 500(10^{-6})\Delta t\ \sin 20 = 171.01(\Delta t)(10^{-6}) \qquad 4$$

From triangle P_tPP_2 and Eq. 4 we obtain:

$$PP_t = PP_2\cos 20 = 160.7(\Delta t)(10^{-6}) \qquad P_tP_2 = PP_2\sin 20 = 58.49(\Delta t)(10^{-6}) \qquad 5$$

The normal strain in the t-direction is: $\varepsilon_{tt} = \dfrac{-PP_t}{OP} = \dfrac{(-160.7)(\Delta t)(10^{-6})}{\Delta t}$ or $\boxed{\varepsilon_{tt} = -160.7\mu}$

The length OP_t can be approximated as: $OP_t = OP - PP_t = (\Delta t)[1-160.7(10^{-6})] \approx (\Delta t)$ and for small angles in small strains we can write $\tan\phi_2 \approx \phi_2$. From triangle P_tOP_2 we obtain the rotation of the line OP as shown below.

$$\tan\phi_2 \approx \phi_2 = \frac{P_tP_2}{OP_t} \approx \frac{58.49(\Delta t)(10^{-6})}{\Delta t} \quad \text{or} \quad \phi_2 = 58.49\mu\text{rads} \qquad 6$$

From Fig. (a) we see that the angle between n and t increases by $(\phi_1 - \phi_2)$. Using Eqs. 3 and 6 we obtain:

$\gamma_{nt} = -(\phi_1 - \phi_2) = -(441.51 - 58.49)$ or $\boxed{\gamma_{nt} = -383\mu\text{rad}}$

9.12 At a point in plane strain the strain components are

$\varepsilon_{xx} = -600\ \mu \quad \varepsilon_{yy} = -800\ \mu. \quad \gamma_{xy} = 500\ \mu.$

Using the results obtained in problems 9.9, 9.10, and 9.12, determine the strain components in 'n' and 't' coordinate system shown in Fig. P9.12.

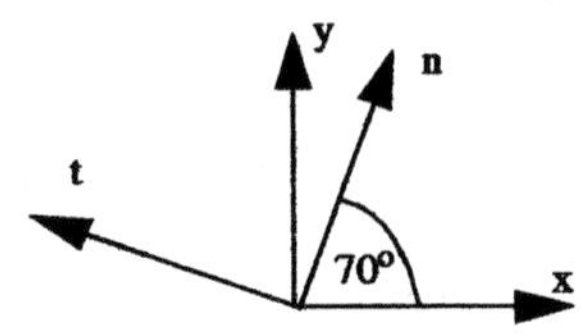

Fig. P9.12

Solution $\varepsilon_{nn} = ?$ $\varepsilon_{tt} = ?$ $\gamma_{nt} = ?$

From problem 9.9 we have: $\varepsilon_{nn}^{(1)} = -70.2\mu$ $\varepsilon_{tt}^{(1)} = -529.8\mu$ $\gamma_{nt}^{(1)} = 385.7\mu\text{rad}$

From problem 9.10 we have: $\varepsilon_{nn}^{(2)} = -706.4\mu$ $\varepsilon_{tt}^{(2)} = -93.6\mu$ $\gamma_{nt}^{(2)} = -514.2\mu\text{rad}$

From problem 9.12 we have: $\varepsilon_{nn}^{(3)} = 160.7\mu$ $\varepsilon_{tt}^{(3)} = -160.7\mu$ $\gamma_{nt}^{(3)} = -383.0\mu\text{rad}$

The total strain in is superposition of the above strains for each component. We obtain:

$\varepsilon_{nn} = \varepsilon_{nn}^{(1)} + \varepsilon_{nn}^{(2)} + \varepsilon_{nn}^{(3)} = (-70.2 - 706.4 + 160.7)$ or $\boxed{\varepsilon_{nn} = -615.9\mu}$

$\varepsilon_{tt} = \varepsilon_{tt}^{(1)} + \varepsilon_{tt}^{(2)} + \varepsilon_{tt}^{(3)} = (-529.8 - 93.6 - 160.7)$ or $\boxed{\varepsilon_{tt} = -784.1\mu}$

$\gamma_{nt} = \gamma_{nt}^{(1)} + \gamma_{nt}^{(2)} + \gamma_{nt}^{(3)} = (385.7 - 514.2 - 383)$ or $\boxed{\gamma_{nt} = -511.6\mu}$

9.13 At a point, the only non-zero strain component is as given below. Determine the strain components in 'n' and 't' coordinate system shown in Fig. P9.13.

$$\varepsilon_{xx} = 600\ \mu$$

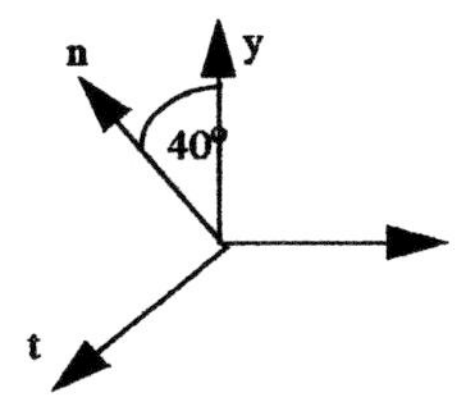

Fig. P9.13

Solution $\varepsilon_{nn} = ?$ $\varepsilon_{tt} = ?$ $\gamma_{nt} = ?$

Fig.(a) shows the rotation of the n and t lines, while Figs. (b) and (c) show rectangles with diagonals in the n and t directions deforming due to ε_{xx}.

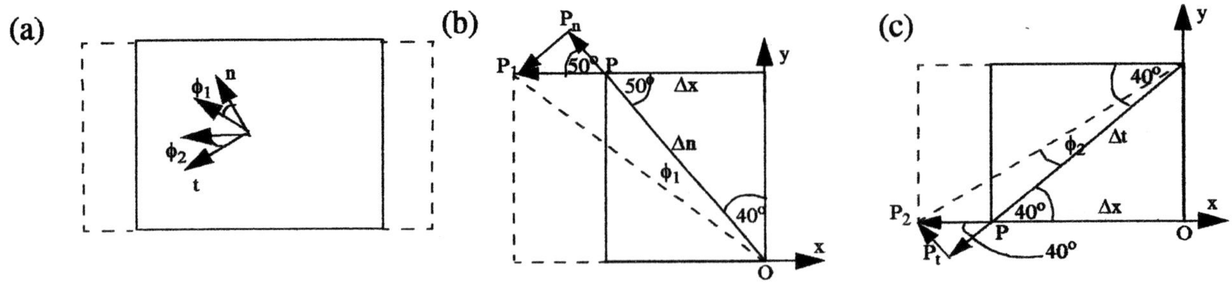

From Fig.(b) we obtain:

$$\Delta x = \Delta n \sin 40 \qquad PP_1 = |\varepsilon_{xx}|\Delta x = 600(10^{-6})\Delta n \ \sin 40 = 385.67(\Delta n)(10^{-6}) \qquad 1$$

From triangle P_nPP_1 and Eq. 1 we obtain:

$$PP_n = PP_1 \cos 50 = 247.91(\Delta n)(10^{-6}) \qquad P_nP_1 = PP_1 \sin 50 = 295.44(\Delta n)(10^{-6}) \qquad 2$$

The normal strain in the n-direction is: $\varepsilon_{nn} = \dfrac{PP_n}{OP} = \dfrac{(247.91)(\Delta n)(10^{-6})}{\Delta n}$ or $\boxed{\varepsilon_{nn} = 247.9\mu}$

The length OP_n can be approximated as: $OP_n = OP + PP_n = (\Delta n)[1 + 247.91(10^{-6})] \approx (\Delta n)$ and for small angles in small strains we can write $\tan\phi_1 \approx \phi_1$. From triangle P_nOP_1 we obtain the rotation of the line OP as shown below.

$$\tan\phi_1 \approx \phi_1 = \frac{P_nP_1}{OP_n} \approx \frac{295.44(\Delta n)(10^{-6})}{\Delta n} \qquad \text{or} \qquad \phi_1 = 295.44\mu\text{rads} \qquad 3$$

From Fig. (c) we obtain

$$\Delta x = \Delta t \cos 40 \qquad PP_2 = |\varepsilon_{xx}|\Delta x = 600(10^{-6})\Delta t \ \cos 40 = 459.63(\Delta t)(10^{-6}) \qquad 4$$

From triangle P_tPP_2 and Eq. 4 we obtain:

$$PP_t = PP_2 \cos 40 = 352.09(\Delta t)(10^{-6}) \qquad P_tP_2 = PP_2 \sin 40 = 295.44(\Delta t)(10^{-6}) \qquad 5$$

The normal strain in the t-direction is: $\varepsilon_{tt} = \dfrac{PP_t}{OP} = \dfrac{(352.09)(\Delta t)(10^{-6})}{\Delta t}$ or $\boxed{\varepsilon_{tt} = 352.1\mu}$

The length OP_t can be approximated as: $OP_t = OP - PP_t = (\Delta t)[1 - 100(10^{-6})] \approx (\Delta t)$ and for small angles in small strains we can write $\tan\phi_2 \approx \phi_2$. From triangle $P_tOP_2$we obtain the rotation of the line OP as shown below.

$$\tan\phi_2 \approx \phi_2 = \frac{P_tP_2}{OP_t} \approx \frac{295.44(\Delta t)(10^{-6})}{\Delta t} \qquad \text{or} \qquad \phi_2 = 295.44\mu\text{rads} \qquad 6$$

From Fig. (a) we see that the angle between n and t decreases by $\phi_1$and ϕ_2. Using Eqs. 3 and 6 we obtain:

$\gamma_{nt} = (\phi_1 + \phi_2) = (295.44 + 295.44)$ or $\boxed{\gamma_{nt} = 590.9\mu\text{rad}}$

9.14 At a point, the only non-zero strain component is as given below. Determine the strain components in 'n' and 't' coordinate system shown in Fig. P9.14.

$$\varepsilon_{yy} = 600\ \mu.$$

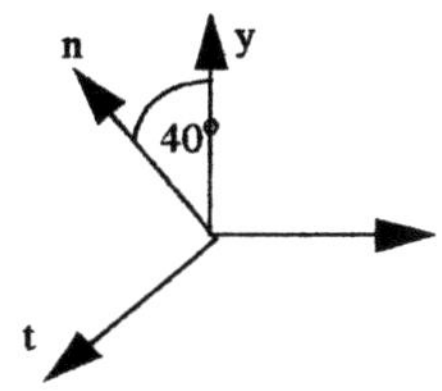

Fig. P9.14

Fig.(a) shows the rotation of the n and t lines, while Figs. (b) and (c) show rectangles with diagonals in the n and t directions deforming due to ε_{yy}.

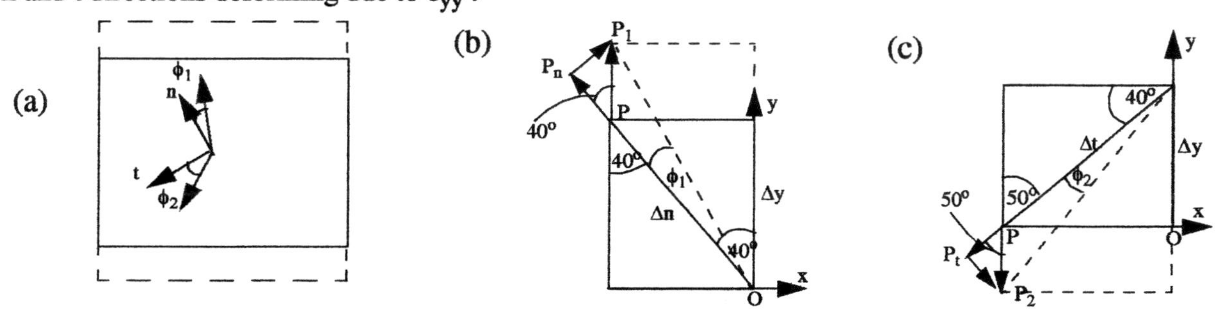

From Fig.(b) we obtain:

$$\Delta y = \Delta n \cos 40 \qquad PP_1 = |\varepsilon_{yy}|\Delta y = 600(10^{-6})\Delta n \ \cos 40 = 459.63(\Delta n)(10^{-6}) \qquad 1$$

From triangle P_nPP_1 and Eq. 1 we obtain:

$$PP_n = PP_1 \cos 40 = 352.09(\Delta n)(10^{-6}) \qquad P_nP_1 = PP_1 \sin 40 = 295.44(\Delta n)(10^{-6}) \qquad 2$$

The normal strain in the n-direction is: $\varepsilon_{nn} = \dfrac{PP_n}{OP} = \dfrac{352.09(\Delta n)(10^{-6})}{\Delta n}$ or $\boxed{\varepsilon_{nn} = 352.1\mu}$

The length OP_n can be approximated as: $OP_n = OP + PP_n = (\Delta n)[1 + 352.09(10^{-6})] \approx (\Delta n)$ and for small angles in small strains we can write $\tan\phi_1 \approx \phi_1$. From triangle P_nOP_1 we obtain the rotation of the line OP as shown below.

$$\tan\phi_1 \approx \phi_1 = \frac{P_nP_1}{OP_n} \approx \frac{295.44(\Delta n)(10^{-6})}{\Delta n} \quad \text{or} \quad \phi_1 = 295.44\mu\text{rads} \qquad 3$$

From Fig. (c) we obtain

$$\Delta y = \Delta t \sin 40 \qquad PP_2 = |\varepsilon_{yy}|\Delta y = 600(10^{-6})\Delta t \ \sin 40 = 385.67(\Delta t)(10^{-6}) \qquad 4$$

From triangle P_tPP_2 and Eq. 4 we obtain:

$$PP_t = PP_2 \cos 50 = 247.91(\Delta t)(10^{-6}) \qquad P_tP_2 = PP_2 \sin 50 = 295.44(\Delta t)(10^{-6}) \qquad 5$$

The normal strain in the t-direction is: $\varepsilon_{tt} = \dfrac{PP_t}{OP} = \dfrac{(247.91)(\Delta t)(10^{-6})}{\Delta t}$ or $\boxed{\varepsilon_{tt} = 247.9\mu}$

The length OP_t can be approximated as: $OP_t = OP + PP_t = (\Delta t)[1 + 247.91(10^{-6})] \approx (\Delta t)$ and for small angles in small strains we can write $\tan\phi_2 \approx \phi_2$. From triangle $P_tOP_2$we obtain the rotation of the line OP as shown below.

$$\tan\phi_2 \approx \phi_2 = \frac{P_tP_2}{OP_t} \approx \frac{295.44(\Delta t)(10^{-6})}{\Delta t} \quad \text{or} \quad \phi_2 = 295.44\mu\text{rads} \qquad 6$$

From Fig. (a) we see that the angle between n and t increases by $\phi_1$and ϕ_2. Using Eqs. 3 and 6 we obtain:

$_{nt} = -(\phi_1 + \phi_2) = -(295.44 + 295.44$ or $\boxed{\gamma_{nt} = -590.9\mu\text{rad}}$

9.15 At a point, the only non-zero strain component is as given below. Determine the strain components in 'n' and 't' coordinate system shown in Fig. P9.15.

$$\gamma_{xy} = \gamma_{xy} = 600\ \mu.$$

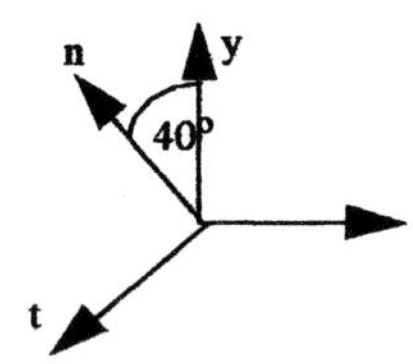

Fig. P9.15

Solution $\varepsilon_{nn} = ?$ $\varepsilon_{tt} = ?$ $\gamma_{nt} = ?$

Fig.(a) shows the rotation of the n and t lines as the angle between x and y coordinate increases by 500μ. Note we could have rotated the vertical lines (counter-clockwise) rather then the rotation of horizontal lines clockwise as in Fig. (a). Figs. (b) and (c) show rectangles with diagonals in the n and t directions deforming due to γ_{xy}.

From Fig.(b) we obtain:

$$\Delta x = \Delta n \sin 40 \qquad PP_1 = AP\tan|\gamma_{xy}| \approx \Delta x|\gamma_{xy}| = 600(10^{-6})\Delta n \ \sin 40 = 385.67(\Delta n)(10^{-6}) \qquad 1$$

From triangle P_nPP_1 and Eq. 1 we obtain:

$$PP_n = PP_1 \cos 40 = 295.44(\Delta n)(10^{-6}) \qquad P_nP_1 = PP_1 \sin 40 = 247.91(\Delta n)(10^{-6}) \qquad 2$$

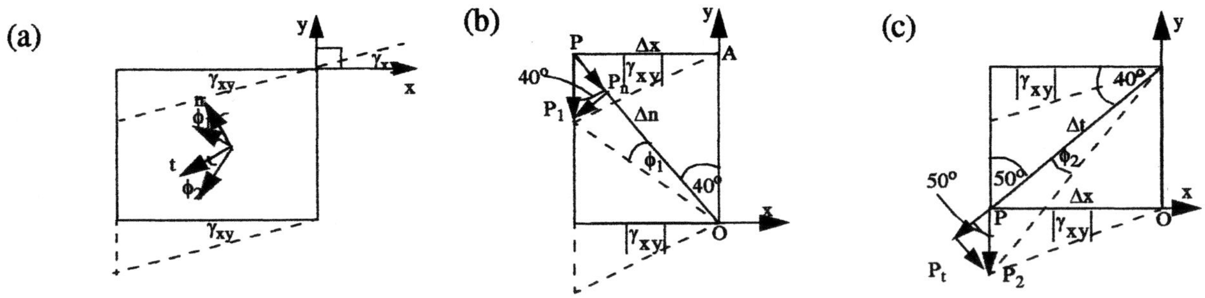

The normal strain in the n-direction is: $\varepsilon_{nn} = \dfrac{-PP_n}{OP} = \dfrac{(-295.44)(\Delta n)(10^{-6})}{\Delta n}$ or $\boxed{\varepsilon_{nn} = -295.4\mu}$

The length OP_n can be approximated as: $OP_n = OP - PP_n = (\Delta n)[1 - 295.44(10^{-6})] \approx (\Delta n)$ and for small angles in small strains we can write $\tan\phi_1 \approx \phi_1$. From triangle P_nOP_1 we obtain the rotation of the line OP as shown below.

$$\tan\phi_1 \approx \phi_1 = \frac{P_nP_1}{OP_n} \approx \frac{247.91(\Delta n)(10^{-6})}{\Delta n} \quad \text{or} \quad \phi_1 = 247.91\mu\text{rads} \qquad 3$$

From Fig. (c) we obtain

$$\Delta x = \Delta t\cos 40 \qquad PP_2 = OP\tan|\gamma_{xy}| \approx \Delta x|\gamma_{xy}| = 600(10^{-6})\Delta t\ \cos 40 = 459.63(\Delta t)(10^{-6}) \qquad 4$$

From triangle P_tPP_2 and Eq. 4 we obtain:

$$PP_t = PP_2\cos 50 = 295.44(\Delta t)(10^{-6}) \qquad P_tP_2 = PP_2\sin 50 = 352.09(\Delta t)(10^{-6}) \qquad 5$$

The normal strain in the t-direction is: $\varepsilon_{tt} = \dfrac{PP_t}{OP} = \dfrac{(295.44)(\Delta t)(10^{-6})}{\Delta t}$ or $\boxed{\varepsilon_{tt} = 295.4\mu}$

The length OP_t can be approximated as: $OP_t = OP + PP_t = (\Delta t)[1 + 295.44(10^{-6})] \approx (\Delta t)$ and for small angles in small strains we can write $\tan\phi_2 \approx \phi_2$. From triangle P_tOP_2 we obtain the rotation of the line OP as shown below.

$$\tan\phi_2 \approx \phi_2 = \frac{P_tP_2}{OP_t} \approx \frac{352.09(\Delta t)(10^{-6})}{\Delta t} \quad \text{or} \quad \phi_2 = 352.09\mu\text{rads} \qquad 6$$

From Fig. (a) we see that the angle between n and t increases by $(\phi_2 - \phi_1)$. Using Eqs. 3 and 6 we obtain:

$\gamma_{nt} = -(\phi_2 - \phi_1) = -(352.09 - 247.91)$ or $\boxed{\gamma_{nt} = -104.2\mu\text{rad}}$

9.16 At a point in plane strain the strain components are

$\varepsilon_{xx} = 600\ \mu \qquad \varepsilon_{yy} = 600\ \mu. \qquad \gamma_{xy} = 600\ \mu.$

Using the results obtained in problems 9.13, 9.14, and 9.15, determine the strain components in 'n' and 't' coordinate system shown in Fig. P9.16.

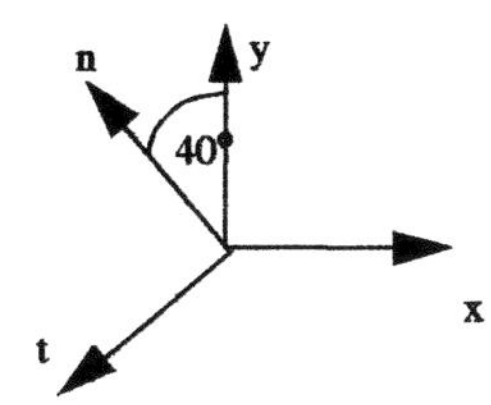

Fig. P9.16

Solution $\quad \varepsilon_{nn} = ? \qquad \varepsilon_{tt} = ? \qquad \gamma_{nt} = ?$

From problem 9.13 we have: $\varepsilon_{nn}^{(1)} = 247.9\mu \qquad \varepsilon_{tt}^{(1)} = 352.1\mu \qquad \gamma_{nt}^{(1)} = 590.9\mu\text{rad}$

From problem 9.14 we have: $\varepsilon_{nn}^{(2)} = 352.1\mu \qquad \varepsilon_{tt}^{(2)} = 247.9\mu \qquad \gamma_{nt}^{(2)} = -590.9\mu\text{rad}$

From problem 9.15 we have: $\varepsilon_{nn}^{(3)} = -295.4\mu \qquad \varepsilon_{tt}^{(3)} = 895.4\mu \qquad \gamma_{nt}^{(3)} = -104.2\mu\text{rad}$

The total strain in is superposition of the above strains for each component. We obtain:

$\varepsilon_{nn} = \varepsilon_{nn}^{(1)} + \varepsilon_{nn}^{(2)} + \varepsilon_{nn}^{(3)} = (247.9 + 352.1 - 295.4)$ or $\boxed{\varepsilon_{nn} = 304.6\mu}$

$\varepsilon_{tt} = \varepsilon_{tt}^{(1)} + \varepsilon_{tt}^{(2)} + \varepsilon_{tt}^{(3)} = (352.1 + 247.9 + 295.4)$ or $\boxed{\varepsilon_{tt} = 895.4\mu}$

$\gamma_{nt} = \gamma_{nt}^{(1)} + \gamma_{nt}^{(2)} + \gamma_{nt}^{(3)} = (590.9 - 590.9 - 104.2)$ or $\boxed{\gamma_{nt} = -104.2\mu}$

9.17 The state of strain at a point in plane strain is as given below. Estimate the orientation of the principal directions and report your results using sectors shown in Figure 9.6.

$$\varepsilon_{xx} = -400\ \mu \qquad \varepsilon_{yy} = 600\ \mu \qquad \gamma_{xy} = -500\ \mu$$

Solution principal directions?

Principal direction one will be in sector 3 and principal direction two will be in sector 5

Principal direction one will be in sector 7 and principal direction two will be in sector 1

9.18 The state of strain at a point in plane strain is as given below. Estimate the orientation of the principal directions and report your results using sectors shown in Figure 9.6

$$\varepsilon_{xx} = -600\ \mu \qquad \varepsilon_{yy} = -800\ \mu \qquad \gamma_{xy} = 500\ \mu$$

Solution principal directions?

Principal direction one will be in sector 1 and principal direction two will be in sector 3

Principal direction one will be in sector 5 and principal direction two will be in sector 7

9.19 The state of strain at a point in plane strain is as given below. Estimate the orientation of the principal directions and report your results using sectors shown in Figure 9.6.

$$\varepsilon_{xx} = 800\ \mu \qquad \varepsilon_{yy} = 600\ \mu \qquad \gamma_{xy} = -1000\ \mu$$

Solution principal directions?

Principal direction one will be in sector 4 and principal direction two will be in sector 6

Principal direction one will be in sector 8 and principal direction two will be in sector 2

9.20 The state of strain at a point in plane strain is as given below. Estimate the orientation of the principal directions and report your results using sectors shown in Figure 9.6.

$$\varepsilon_{xx} = 0 \qquad \varepsilon_{yy} = 600\ \mu \qquad \gamma_{xy} = -500\ \mu$$

Solution principal directions?

Principal direction one will be in sector 3 and principal direction two will be in sector 5

Principal direction one will be in sector 7 and principal direction two will be in sector 1

9.21 The state of strain at a point in plane strain is as given below. Estimate the orientation of the principal directions and report your results using sectors shown in Figure 9.6.

$$\varepsilon_{xx} = -1000\ \mu \qquad \varepsilon_{yy} = -500\ \mu \qquad \gamma_{xy} = 700\ \mu$$

Solution principal directions?

Principal direction one will be in sector 2 and principal direction two will be in sector 4
Principal direction one will be in sector 6 and principal direction two will be in sector 8

Problem Set 9.2

9.22 Starting from Equation 9.10, show that maximum or minimum normal strain will exist in the direction of θ_p as given by Equation 9.13. ((Hint: See similar derivation in stress transformation).

Solution

Eqs. 9.10 can be re-written in terms of double angle as below.

$$\varepsilon_{nn} = \frac{(\varepsilon_{xx}+\varepsilon_{yy})}{2} + \frac{(\varepsilon_{xx}-\varepsilon_{yy})}{2}\cos 2\theta + \frac{\gamma_{xy}}{2}\sin 2\theta \qquad 1$$

Differentiating ε_{nn} with respect θ and assuming the slope is zero at $\theta = \theta_p$ we obtain:

$$\frac{d\varepsilon_{nn}}{d\theta} = \frac{(\varepsilon_{xx}-\varepsilon_{yy})}{2}(-2\sin 2\theta) + \frac{\gamma_{xy}}{2}(2\cos 2\theta)\bigg|_{\theta=\theta_p} \quad \text{or} \quad \tan 2\theta_p = \frac{\gamma_{xy}}{\varepsilon_{xx}-\varepsilon_{yy}} \qquad 2$$

Eq.2 is same as Eq. 9.13

9.23 Show that the values of the maximum and minimum normal strains are given by Equation 9.14. (Hint: See similar derivation in stress transformation).

Solution

From Eq. 2 in problem 9.22 we have the angle at which the normal strain is maximum or minimum.We can write the following:

$$\sin 2\theta_p = \pm\frac{\gamma_{xy}}{R} \qquad \cos 2\theta_p = \pm\frac{(\varepsilon_{xx}-\varepsilon_{yy})}{R} \qquad R = \sqrt{(\varepsilon_{xx}-\varepsilon_{yy})^2 + \gamma_{xy}^2} \qquad 1$$

Substituting Eq. 1 into Eq. 1 of problem 9.22, we obtain:

$$\varepsilon_{nn} = \frac{(\varepsilon_{xx}+\varepsilon_{yy})}{2} + \frac{(\varepsilon_{xx}-\varepsilon_{yy})}{2}\left[\pm\frac{(\varepsilon_{xx}-\varepsilon_{yy})}{R}\right] + \frac{\gamma_{xy}}{2}\left[\pm\frac{\gamma_{xy}}{R}\right] = \frac{(\varepsilon_{xx}+\varepsilon_{yy})}{2} \pm \frac{(\varepsilon_{xx}-\varepsilon_{yy})^2 + \gamma_{xy}^2}{2R} \quad \text{or}$$

$$\varepsilon_{nn} = \frac{(\varepsilon_{xx}+\varepsilon_{yy})}{2} \pm \frac{R^2}{2R} = \frac{(\varepsilon_{xx}+\varepsilon_{yy})}{2} \pm \frac{\sqrt{(\varepsilon_{xx}-\varepsilon_{yy})^2+\gamma_{xy}^2}}{2} = \frac{(\varepsilon_{xx}+\varepsilon_{yy})}{2} \pm \sqrt{\left(\frac{\varepsilon_{xx}-\varepsilon_{yy}}{2}\right)^2 + \frac{\gamma_{xy}}{2}^2} \qquad 2$$

Eq. 2 is same as Eq. 9.14

9.24 Show the angle θ_p as given by Equation 9.13 is the principal angle, i.e., shear strain is zero in a coordinate system that is at an angle of θ_p to the cartesian coordinate system. (Hint: See similar derivation in stress transformation).

Solution

The shear stress in Eq. 9.12 can be re-written in double angles as:

$$\gamma_{nt} = -(\varepsilon_{xx}-\varepsilon_{yy})\sin 2\theta + \gamma_{xy}\cos 2\theta \qquad 1$$

Substituting Eq. 1 of problem 9.23 into the above equation we obtain:

$$\gamma_{nt} = -(\varepsilon_{xx}-\varepsilon_{yy})\left[\pm\frac{\gamma_{xy}}{R}\right]+\gamma_{xy}\left[\pm\frac{(\varepsilon_{xx}-\varepsilon_{yy})}{R}\right] = 0 \qquad 2$$

Eq. 2 shows that the shear stress is zero on the plane with outward normal at an angle of θ_p, which make these planes the principal planes.

9.25 Show that the coordinate system of maximum in-plane shear strain is 45° to the principal coordinate system. (Hint: See similar derivation in stress transformation).

Solution

The principal angle is given by $\tan 2\theta_p = \frac{\gamma_{xy}}{\varepsilon_{xx}-\varepsilon_{yy}}$

Differentiating γ_{nt} in Eq. 1 of problem 9.24 with respect θ and assuming the slope is zero at $\theta = \theta_s$ we obtain:

$$\frac{d\gamma_{nt}}{d\theta} = -(\varepsilon_{xx}-\varepsilon_{yy})(2\cos 2\theta)+\gamma_{xy}(-2\sin 2\theta)\Big|_{\theta=\theta_s} \quad \text{or} \quad \tan 2\theta_s = -\frac{(\varepsilon_{xx}-\varepsilon_{yy})}{\gamma_{xy}} \qquad 1$$

Thus we have:

$$\tan 2\theta_s \tan 2\theta_p = -1 \quad \text{or} \quad \tan 2\theta_s = -\cot 2\theta_p = \tan\left(2\theta_p \pm \frac{\pi}{2}\right) \quad \text{or} \quad \theta_s = \theta_p \pm \frac{\pi}{4} \qquad 2$$

Eq. 2 shows that the plane of maximum shear stress is 45° to the principal planes.

9.26 Show that the maximum in-plane shear strain is given by Equation 9.18. (Hint: See similar derivation in stress transformation).

Solution

From Eq. 1 in problem 9.25 we have: $\tan 2\theta_s = -(\varepsilon_{xx}-\varepsilon_{yy})/\gamma_{xy}$. We can write the following:

$$\cos 2\theta_s = \pm\frac{\gamma_{xy}}{R} \qquad \sin 2\theta_s = \mp\frac{(\varepsilon_{xx}-\varepsilon_{yy})}{R} \qquad R = \sqrt{\left(\frac{\varepsilon_{xx}-\varepsilon_{yy}}{2}\right)^2+\frac{\gamma_{xy}}{2}^2} \qquad 1$$

Substituting Eq.1 into Eq. 1 of problem 9.24, we obtain:

$$\gamma_{nt} = -(\varepsilon_{xx}-\varepsilon_{yy})\left[\mp\frac{(\varepsilon_{xx}-\varepsilon_{yy})}{R}\right]+\gamma_{xy}\left[\pm\frac{\gamma_{xy}}{R}\right] = \pm\frac{\left(\frac{\varepsilon_{xx}-\varepsilon_{yy}}{2}\right)^2+\frac{\gamma_{xy}}{2}^2}{R} = R = |\varepsilon_1-\varepsilon_2| \qquad 2$$

Eq. 2 is same as Eq. 9.18 for maximum inplane shear stress.

9.27 Starting from Equations 9.10 and 9.12, obtain the equation of Mohr's circle given by Equation 9.19. (Hint: See similar derivation in stress transformation).

Solution

Eqs. 9.10 and 9.12 can be re-written in terms of double angle as below.

$$\varepsilon_{nn} = \frac{(\varepsilon_{xx}+\varepsilon_{yy})}{2}+\frac{(\varepsilon_{xx}-\varepsilon_{yy})}{2}\cos 2\theta+\frac{\gamma_{xy}}{2}\sin 2\theta \text{ or}$$

$$\varepsilon_{nn}-\frac{\varepsilon_{xx}+\varepsilon_{yy}}{2} = \frac{(\varepsilon_{xx}-\varepsilon_{yy})}{2}\cos 2\theta+\frac{\gamma_{xy}}{2}\sin 2\theta \qquad 1$$

$$\gamma_{nt} = -(\varepsilon_{xx}-\varepsilon_{yy})\sin 2\theta+\gamma_{xy}\cos 2\theta \text{ or}$$

$$\frac{\gamma_{nt}}{2} = -\frac{(\varepsilon_{xx}-\varepsilon_{yy})}{2}\sin 2\theta + \frac{\gamma_{xy}}{2}\cos 2\theta \qquad \mathbf{2}$$

Squaring Eqs. 1 and 2 and adding we obtain:

$$\left(\varepsilon_{nn}-\frac{\varepsilon_{xx}+\varepsilon_{yy}}{2}\right)^2+\left(\frac{\gamma_{nt}}{2}\right)^2 = \left(\frac{\varepsilon_{xx}-\varepsilon_{yy}}{2}\right)^2[\cos^2 2\theta+\sin^2 2\theta]+\left(\frac{\gamma_{xy}}{2}\right)^2[\sin^2 2\theta+\cos^2 2\theta] \quad \text{or}$$

$$\left(\varepsilon_{nn}-\frac{\varepsilon_{xx}+\varepsilon_{yy}}{2}\right)^2+\left(\frac{\gamma_{nt}}{2}\right)^2 = \left(\frac{\varepsilon_{xx}-\varepsilon_{yy}}{2}\right)^2+\left(\frac{\gamma_{xy}}{2}\right)^2 \qquad \mathbf{3}$$

Eq. 3 is equation for the Mohr's circle for strain given by Eq. 9.19.

9.28 Solve problem 9.5 by Method of Equations.

Solution $\varepsilon_{xx} = -400\ \mu.$ $\varepsilon_{nn} = ?$ $\varepsilon_{tt} = ?$ $\gamma_{nt} = ?$ $\theta = -30^o$

Substituting $\varepsilon_{xx} = -400\ \mu.$, $\varepsilon_{yy} = 0$, $\gamma_{xy} = 0$ and $\theta = -30^o$ into Eqs. 9.10, 9.11 and 9.12 we obtain:

$\varepsilon_{nn} = -400\cos^2(-30)$ or $\boxed{\varepsilon_{nn} = -300\mu}$

$\varepsilon_{tt} = -400\sin^2\theta$ or $\boxed{\varepsilon_{tt} = -100\mu}$

$\gamma_{nt} = -2(-400)\sin(-30)\cos(-30)$ or $\boxed{\gamma_{nt} = -346.4\mu}$

Check: $\varepsilon_{nn}+\varepsilon_{tt} = -400$ and $\varepsilon_{xx}+\varepsilon_{yy} = -400$ which is consistent with Eq. 9.14.

9.29 Solve problem 9.5 by Mohr's Circle.

Solution $\varepsilon_{xx} = -400\ \mu.$ $\varepsilon_{nn} = ?$ $\varepsilon_{tt} = ?$ $\gamma_{nt} = ?$ $\theta = -30^o$

Fig. (a) shows the un-deformed squares in x-y and n-t coordinate systems, with V,H, N, and T points identified.Fig.(b) shows the Mohr's circle with V, H, N, and T points identified

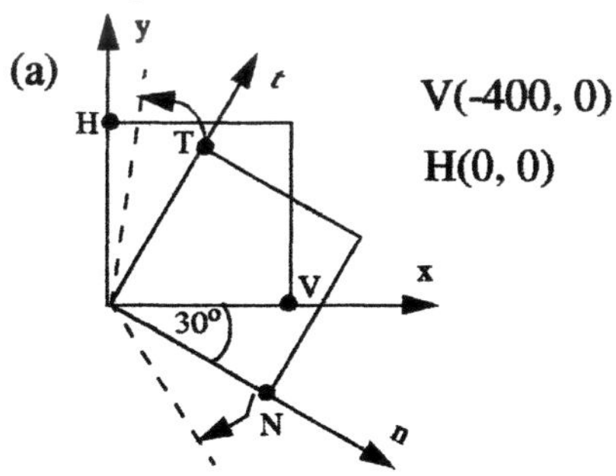

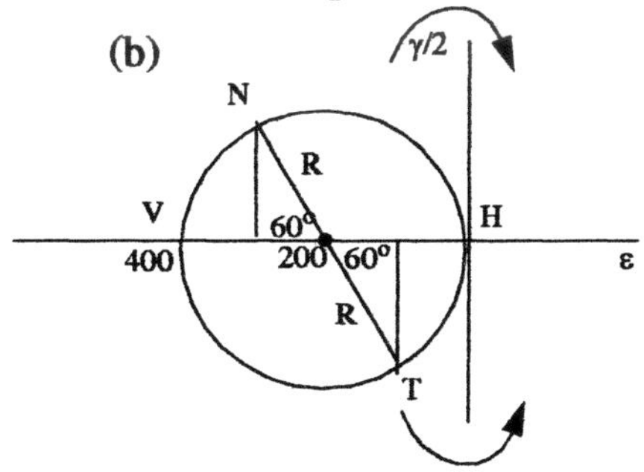

The center of the Mohr's circle is located at -200 μ, and the radius is R = 200μ. From Fig.(b) we obtain the following:

$\varepsilon_{nn} = -200 - R\cos 60 = -300$ or $\boxed{\varepsilon_{nn} = -300\mu}$

$\varepsilon_{tt} = -200 + R\cos 60 = -100$ or $\boxed{\varepsilon_{tt} = -100\mu}$

$|\gamma_{nt}/2| = R\sin 60$ or $|\gamma_{nt}| = 346.4\mu$. Noting the rotation direction of N and T on the Mohr's circle, we draw the deformed (dashed) lines in Fig. (a) and note that the angle increase. Thus, $\boxed{\gamma_{nt} = -346.4\mu}$

9.30 Solve problem 9.6 by Method of Equations.

Solution $\varepsilon_{yy} = 600\mu.$ $\varepsilon_{nn} = ?$ $\varepsilon_{tt} = ?$ $\gamma_{nt} = ?$ $\theta = -30^o$

Substituting $\varepsilon_{xx} = 0$, $\varepsilon_{yy} = 600\mu$, $\gamma_{xy} = 0$ and $\theta = -30^{o}$ into Eqs. 9.10,9.11 and 9.12 we obtain the following

$\varepsilon_{nn} = 600\sin^2(-30)$ or $\boxed{\varepsilon_{nn} = 150\mu}$

$\varepsilon_{tt} = 600\cos^2(-30)$ or $\boxed{\varepsilon_{tt} = 450\mu}$

$\gamma_{nt} = 2(600)\sin(-30)\cos(-30)$ or $\boxed{\gamma_{nt} = -519.6\mu}$

Check: $\varepsilon_{nn} + \varepsilon_{tt} = 600$ and $\varepsilon_{xx} + \varepsilon_{yy} = 600$ which is consistent with Eq. 9.14.

9.31 Solve problem 9.6 by Mohr's Circle.

Solution $\varepsilon_{yy} = 600\mu$. $\varepsilon_{nn} = ?$ $\varepsilon_{tt} = ?$ $\gamma_{nt} = ?$ $\theta = -30^{o}$

Fig. (a) shows the un-deformed squares in x-y and n-t coordinate systems, with V,H, N, and T points identified.Fig.(b) shows the Mohr's circle with V, H, N, and T points identified

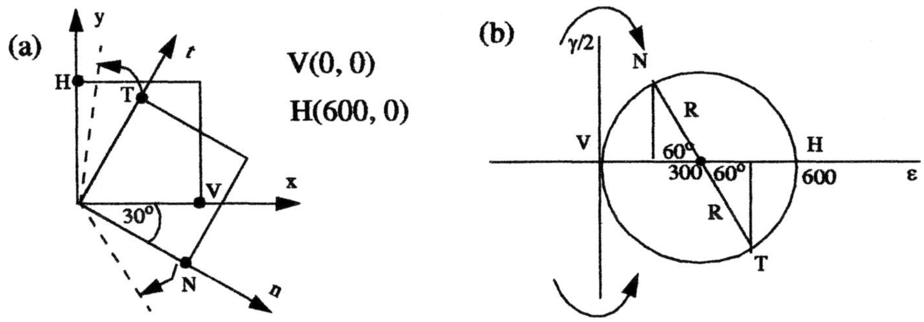

The center of the Mohr's circle is located at 300 μ, and the radius is R = 300μ. From Fig.(b) we obtain the following:

$\varepsilon_{nn} = 300 - R\cos 60 = 150$ or $\boxed{\varepsilon_{nn} = 150\mu}$

$\varepsilon_{tt} = 300 + R\cos 60 = 450$ or $\boxed{\varepsilon_{tt} = 450\mu}$

$|\gamma_{nt}/2| = R\sin 60$ or $|\gamma_{nt}| = 519.62\mu$. Noting the rotation direction of N and T on the Mohr's circle, we draw the deformed (dashed) lines in Fig. (a) and note that the angle increase. Thus, $\boxed{\gamma_{nt} = -519.6\mu}$

9.32 Solve problem 9.7 by Method of Equations.

Solution $\gamma_{xy} = -500\mu$. $\varepsilon_{nn} = ?$ $\varepsilon_{tt} = ?$ $\gamma_{nt} = ?$ $\theta = -30^{o}$

Substituting $\varepsilon_{xx} = 0$, $\varepsilon_{yy} = $, $\gamma_{xy} = -500\mu$ and $\theta = -30^{o}$ into Eqs. 9.10,9.11 and 9.12 we obtain the following

$\varepsilon_{nn} = -500\sin(-30)\cos(-30)$ or $\boxed{\varepsilon_{nn} = 216.5\mu}$

$\varepsilon_{tt} = -(-500)\sin(-30)\cos(-30)$ or $\boxed{\varepsilon_{tt} = -216.5\mu}$

$\gamma_{nt} = (-500)(\cos^2(-30) - \sin^2(-30))$ or $\boxed{\gamma_{nt} = -250\mu}$

Check: $\varepsilon_{nn} + \varepsilon_{tt} = 0$ and $\varepsilon_{xx} + \varepsilon_{yy} = 0$ which is consistent with Eq. 9.14.

9.33 Solve problem 9.7 by Mohr's Circle.

Solution $\gamma_{xy} = -500\mu$. $\varepsilon_{nn} = ?$ $\varepsilon_{tt} = ?$ $\gamma_{nt} = ?$ $\theta = -30^{o}$

Fig. (a) shows the un-deformed squares in x-y and n-t coordinate systems, with V,H, N, and T points iden-

tified. As shear strain is negative, the angle in the x-y coordinate increases as shown by the dashed lines. Fig.(b) shows the Mohr's circle with V, H, N, and T points identified. The center of the Mohr's circle is located at the origin, and the radius is R = 250μ.

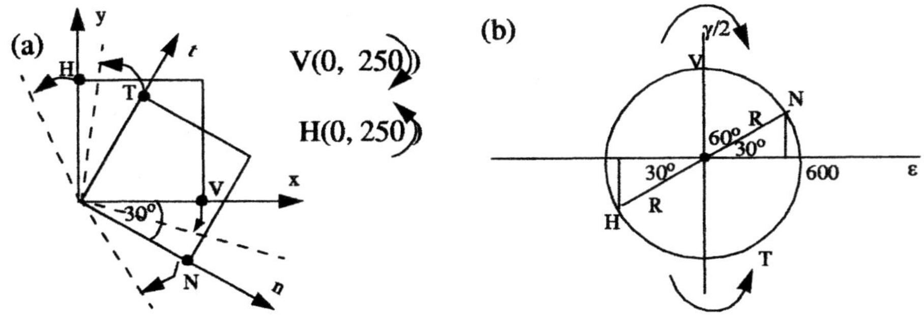

From Fig.(b) we obtain the following:

$\varepsilon_{nn} = R\cos 30 = 216.51$ or $\boxed{\varepsilon_{nn} = 216.5\mu}$

$\varepsilon_{tt} - R\cos 30 = -216.51$ or $\boxed{\varepsilon_{tt} = -216.5\mu}$

$|\gamma_{nt}/2| = R\sin 30$ or $|\gamma_{nt}| = 250$. Noting the rotation direction of N and T on the Mohr's circle, we draw the deformed (dashed) lines in Fig. (a) and note that the angle increase. Thus, $\boxed{\gamma_{nt} = -250\mu}$

9.34 At a point in *plane strain*, the strain components in the x-y coordinate system are as given below. Determine

(a) the principal strains and principal angle one.
(b) the maximum shear strain.
(c) the strain components in the n-t coordinate system shown.

$\varepsilon_{xx} = -400\ \mu \qquad \varepsilon_{yy} = 600\ \mu \qquad \gamma_{xy} = -500\ \mu$

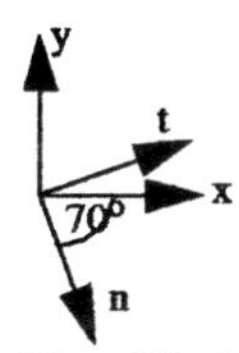

Fig. P9.34

Solution Plane Strain $\varepsilon_1 = ?\ \ \varepsilon_2 = ?\ \ \varepsilon_3 = ?\ \ \theta_1 = ?\ \ \gamma_{max} = ?\ \varepsilon_{nn} = ?\ \varepsilon_{tt} = ?\ \ \gamma_{nt} = ?$

(a) Substituting the given strains into Eq. 9.13 we obtain the principal angle θ_p as:

$$\theta_p = 0.5\ \text{atan}\left(\frac{(-500)}{(-400-600)}\right) = 13.28^{\circ} \qquad 1$$

Substituting θ_p and the given strains into Eq. 9.10, we obtain one principal strain as:

$$\varepsilon_p = (-400)\cos^2(13.28) + (600)\sin^2(13.28) + (-500)\sin(13.28)\cos(13.28) = -459\mu \qquad 2$$

From Eq. 9.15 we can find the other principal strain as: $-400 + 600 - (-459) = 659\mu$. As this strain is greater than the strain in Eq. 2, therefore this is principal strain one. Thus principal angle 1 is $\theta_1 = \theta_p \pm 90$

For plane strain we have the following answers:

$$\boxed{\varepsilon_1 = 659\mu \qquad \varepsilon_2 = -459\mu \qquad \varepsilon_3 = 0 \qquad \theta_1 = 103.28^{\circ} \text{ or } -76.72^{\circ}}$$

Check: From Eq. 9.14 we have: $\varepsilon_{1,2} = \left(\frac{-400+600}{2}\right) \pm \sqrt{\left(\frac{-400-600}{2}\right)^2 + (-250)^2} = 100 \pm 559$ or

$\varepsilon_1 = 659\mu$ and $\varepsilon_2 = -459\mu$, which are same as above.

(b) The difference between ε_1 and ε_2 is the largest difference between principal strains. Thus, the magnitude of the maximum shear strain is same as maximum inplane shear strain and is given by:

$\gamma_{max}/2 = |(\varepsilon_1 - \varepsilon_2)/2| = |(659 + 459)/2|$ or $\boxed{\gamma_{max} = 1118\mu}$

(c) Substituting the given strains and θ = -70° into Eqs. 9.10, 9.11 and 9.12 we obtain:

$\varepsilon_{nn} = (-400)\cos^2(-70) + (600)\sin^2(-70) + (-500)\sin(-70)\cos(-70)$ or $\boxed{\varepsilon_{nn} = 643.7\mu}$

$\varepsilon_{tt} = (-400)\cos^2(-70) + (600)\sin^2(-70) - (-500)\sin(-70)\cos(-70)$ or $\boxed{\varepsilon_{tt} = -443.7\mu}$

$\gamma_{nt} = -2(-400 - 600)\cos(-70)\sin(-70) + (-500)(\cos^2(-70) - \sin^2(-70))$ or $\boxed{\gamma_{nt} = -259.8\mu}$

Check: $\varepsilon_{nn} + \varepsilon_{tt} = 200$ and $\varepsilon_{xx} + \varepsilon_{yy} = 200$ which is consistent with Eq. 9.14.

9.35 At a point in *plane strain*, the strain components in the x-y coordinate system are as given below. Determine

(a) the principal strains and principal angle one.
(b) the maximum shear strain.
(c) the strain components in the n-t coordinate system shown.

$\varepsilon_{xx} = -600\ \mu \qquad \varepsilon_{yy} = -800\ \mu \qquad \gamma_{xy} = 500\ \mu$

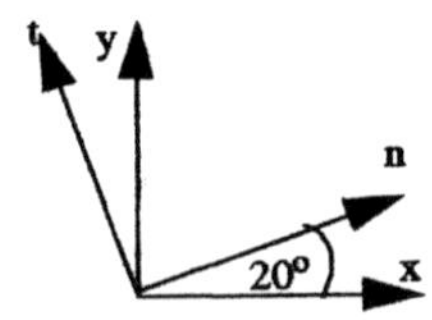

Fig. P9.35

Solution Plane Strain $\varepsilon_1 = ?\ \ \varepsilon_2 = ?\ \ \varepsilon_3 = ?\ \ \theta_1 = ?\ \ \gamma_{max} = ?\ \varepsilon_{nn} = ?\ \varepsilon_{tt} = ?\ \ \gamma_{nt} = ?$

(a) Substituting the given strains into Eq. 9.13 we obtain the principal angle θ_p as:

$$\theta_p = 0.5\ \operatorname{atan}\left(\frac{(500)}{(-600+800)}\right) = 34.1^\circ \tag{1}$$

Substituting θ_p and the given strains into Eq. 9.10, we obtain one principal strain as:

$$\varepsilon_p = (-600)\cos^2(34.1) + (-800)\sin^2(34.1) + (500)\sin(34.1)\cos(34.1) = -430.7\mu \tag{2}$$

From Eq. 9.15 we can find the other principal strain as: $-600 - 800 - (-430.7) = -969.3\mu$. As this strain is smaller than the strain in Eq. 2, therefore the strain in Eq. 2 is principal strain one. Thus principal angle 1 is $\theta_1 = \theta_p$ For plane strain we have the following answers:

$$\boxed{\varepsilon_1 = -430.7\mu \qquad \varepsilon_2 = -969.3\mu \qquad \varepsilon_3 = 0 \qquad \theta_1 = 34.1^\circ \text{ or } -145.9^\circ}$$

Check: From Eq. 9.14 we have: $\varepsilon_{1,2} = \left(\frac{-600-800}{2}\right) \pm \sqrt{\left(\frac{-600+800}{2}\right)^2 + (250)^2} = -700 \pm 269.3$ or

$\varepsilon_1 = -430.7\mu\mu$ and $\varepsilon_2 = -969.3\mu$,which are same as above.

(b) The difference between ε_2 and ε_3 is the largest difference between principal strains. Thus, the magnitude of the maximum shear strain is given by:

$\gamma_{max}/2 = |(\varepsilon_3 - \varepsilon_2)/2| = |(0 + 969.3)/2|$ or $\boxed{\gamma_{max} = 969.3\mu}$

(c) Substituting the given strains and $\theta = 20^\circ$ into Eqs. 9.10, 9.11 and 9.12 we obtain:

$\varepsilon_{nn} = (-600)\cos^2(20) + (-800)\sin^2(20) + (500)\sin(20)\cos(20)$ or $\boxed{\varepsilon_{nn} = -462.7\mu}$

$\varepsilon_{tt} = (-600)\cos^2(20) + (-800)\sin^2(20) - (500)\sin(20)\cos(20)$ or $\boxed{\varepsilon_{tt} = -937.3\mu}$

$\gamma_{nt} = -2(-600 - (-800))\cos(20)\sin(20) + (500)(\cos^2(20) - \sin^2(20))$ or $\boxed{\gamma_{nt} = 254.5\mu}$

Check: $\varepsilon_{nn} + \varepsilon_{tt} = -1400$ and $\varepsilon_{xx} + \varepsilon_{yy} = -1400$ which is consistent with Eq. 9.14.

9.36 At a point in *plane strain*, the strain components in the x-y coordinate system are as given below. Determine

(a) the principal strains and principal angle one.
(b) the maximum shear strain.
(c) the strain components in the n-t coordinate system shown.

$\varepsilon_{xx} = 250\ \mu \qquad \varepsilon_{yy} = 850\ \mu \qquad \gamma_{xy} = 1600\ \mu$

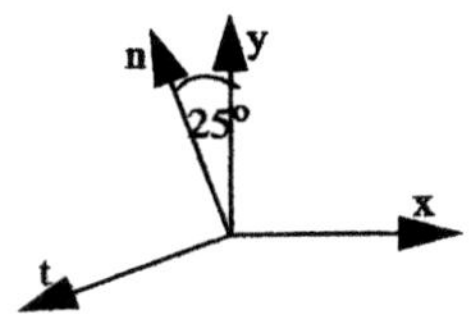

Fig. P9.36

Solution Plane Strain $\varepsilon_1 = ?\ \ \varepsilon_2 = ?\ \ \varepsilon_3 = ?\ \ \theta_1 = ?\ \ \gamma_{max} = ?\ \varepsilon_{nn} = ?\ \varepsilon_{tt} = ?\ \ \gamma_{nt} = ?$

(a) Substituting the given strains into Eq. 9.13 we obtain the principal angle θ_p as:

$$\theta_p = 0.5 \ \text{atan}\left(\frac{1600}{250-850}\right) = -34.7^o \qquad 1$$

Substituting θ_p and the given strains into Eq. 9.10, we obtain one principal strain as:

$$\varepsilon_p = (250)\cos^2(13.28) + (850)\sin^2(13.28) + (1600)\sin(13.28)\cos(13.28) = -304.4\mu \qquad 2$$

From Eq. 9.15 we can find the other principal strain as: $250 + 850 - (-304.4) = 1404.4\mu$. As this strain is greater than the strain in Eq. 2, therefore this is principal strain one. Thus principal angle 1 is $\theta_1 = \theta_p \pm 90$ For plane strain we have the following answers:

$$\boxed{\varepsilon_1 = 1404.4\mu \qquad \varepsilon_2 = -304.4\mu \qquad \varepsilon_3 = 0 \qquad \theta_1 = 55.3^o \ \text{or} \ -124.7^o}$$

Check: From Eq. 9.14 we have: $\varepsilon_{1,2} = \left(\frac{250+850}{2}\right) \pm \sqrt{\left(\frac{250-850}{2}\right)^2 + (800)^2} = 550 \pm 854.4$ or

$\varepsilon_1 = 1404.4\mu$ and $\varepsilon_2 = -304.4\mu$,which are same as above.

(b) The difference between ε_1 and ε_2 is the largest difference between principal strains. Thus, the magnitude of the maximum shear strain is same as maximum inplane shear strain and is given by:

$\gamma_{max}/2 = |(\varepsilon_1 - \varepsilon_2)/2| = |(1404.4 + 304.4)/2|$ or $\boxed{\gamma_{max} = 1708.8\mu}$

(c) Substituting the given strains and $\theta = 115^o$ into Eqs. 9.10, 9.11 and 9.12 we obtain:

$\varepsilon_{nn} = (250)\cos^2(115) + (850)\sin^2(115) + (1600)\sin(115)\cos(115)$ or $\boxed{\varepsilon_{nn} = 130\mu}$

$\varepsilon_{tt} = (250)\cos^2(115) + (850)\sin^2(115) - (1600)\sin(115)\cos(115)$ or $\boxed{\varepsilon_{tt} = 970\mu}$

$\gamma_{nt} = -2(250 - 850)\cos(115)\sin(115) + (1600)(\cos^2(115) - \sin^2(115))$ or $\boxed{\gamma_{nt} = -1488\mu}$

Check: $\varepsilon_{nn} + \varepsilon_{tt} = 1100$ and $\varepsilon_{xx} + \varepsilon_{yy} = 1100$ which is consistent with Eq. 9.14.

9.37 At a point in *plane strain*, the strain components in the x-y coordinate system are as given below. Determine

(a) the principal strains and principal angle one.
(b) the maximum shear strain.
(c) the strain components in the n-t coordinate system shown.

$\varepsilon_{xx} = -1200 \ \mu \qquad \varepsilon_{yy} = -2400 \ \mu \qquad \gamma_{xy} = -1000 \ \mu$

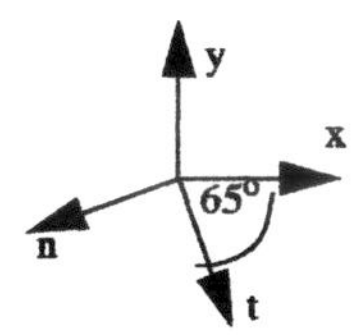

Fig. P9.37

Solution Plane Strain $\varepsilon_1 = ?$ $\varepsilon_2 = ?$ $\varepsilon_3 = ?$ $\theta_1 = ?$ $\gamma_{max} = ?$ $\varepsilon_{nn} = ?$ $\varepsilon_{tt} = ?$ $\gamma_{nt} = ?$

(a) Substituting the given strains into Eq. 9.13 we obtain the principal angle θ_p as:

$$\theta_p = 0.5 \ \text{atan}\left(\frac{(-1000)}{(-1200+2400)}\right) = -19.9^o \qquad 1$$

Substituting θ_p and the given strains into Eq. 9.10, we obtain one principal strain as:

$$\varepsilon_p = (-1200)\cos^2(-19.9) + (-2400)\sin^2(-19.9) + (-1000)\sin(-19.9)\cos(-19.9) = -1019\mu \qquad 2$$

From Eq. 9.15 we can find the other principal strain as: $-1200 - 2400 - (-1019) = -2581\mu$. As this strain is smaller than the strain in Eq. 2, therefore the strain in Eq. 2 is principal strain one. Thus principal angle 1 is $\theta_1 = \theta_p \pm 90$ For plane strain we have the following answers:

$$\boxed{\varepsilon_1 = -1019\mu \qquad \varepsilon_2 = -2581\mu \qquad \varepsilon_3 = 0 \qquad \theta_1 = -19.9^o \ \text{or} \ 160.1^o}$$

Check: From Eq. 9.14 we have: $\varepsilon_{1,2} = \left(\frac{-1200-2400}{2}\right) \pm \sqrt{\left(\frac{-1200+2400}{2}\right)^2 + (-500)^2} = -1800 \pm 781$ or

$\varepsilon_1 = -1019\mu$ and $\varepsilon_2 = -2581\mu$,which are same as above.

(b) The difference between ε_2 and ε_3 is the largest difference between principal strains. Thus, the magnitude of the maximum shear strain is given by:

$\gamma_{max}/2 = |(\varepsilon_3 - \varepsilon_2)/2| = |(0+2581)/2|$ or $\boxed{\gamma_{max} = 2581\mu}$

(c) Substituting the given strains and θ = - 65° into Eqs. 9.10, 9.11 and 9.12 we obtain:

$\varepsilon_{nn} = (-1200)\cos^2(-65) + (-2400)\sin^2(-65) + (-1000)\sin(-65)\cos(-65)$ or $\boxed{\varepsilon_{nn} = -1803\mu}$

$\varepsilon_{tt} = (-1200)\cos^2(-65) + (-2400)\sin^2(-65) - (-1000)\sin(-65)\cos(-65)$ or $\boxed{\varepsilon_{tt} = -1797\mu}$

$\gamma_{nt} = -2(-1200-(-2400))\cos(-65)\sin(-65) + (-1000)(\cos^2(-65) - \sin^2(-65))$ or $\boxed{\gamma_{nt} = 1562\mu}$

Check: $\varepsilon_{nn} + \varepsilon_{tt} = -3600$ and $\varepsilon_{xx} + \varepsilon_{yy} = -3600$ which is consistent with Eq. 9.14.

9.38 At a point in *plane strain*, the strain components in the n-t coordinate system are as given in each problem. Determine

(a) the principal strains.
(b) the maximum shear strain.
(c) the strain components in the x-y coordinate system.

$\varepsilon_{nn} = 2000\ \mu \qquad \varepsilon_{tt} = -800\ \mu \qquad \gamma_{nt} = 750\ \mu$

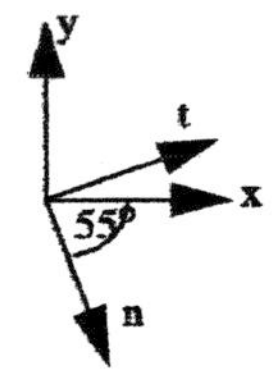

Fig. P9.38

Solution Plane Strain $\varepsilon_1 = ?\ \varepsilon_2 = ?\ \varepsilon_3 = ?\ \theta_1 = ?\ \gamma_{max} = ?\ \varepsilon_{xx} = ?\ \varepsilon_{yy} = ?\ \gamma_{xy} = ?$

(a) We can use n and t as the reference frame as principal strains, and maximum shear strains are unique at a point. We also note that we do not need to calculate the principal angle one, thus we can find principal strains from Eq. 9.14.

The principal strains are: $\varepsilon_{1,2} = \left(\frac{2000-800}{2}\right) \pm \sqrt{\left(\frac{2000-(-800)}{2}\right)^2 + (375)^2} = 600 \pm 1449.4$. For plane strain the principal strains are: $\boxed{\varepsilon_1 = 2049\mu \qquad \varepsilon_2 = -849\mu \qquad \varepsilon_3 = 0}$

Check: $\varepsilon_{nn} + \varepsilon_{tt} = 1200$ and $\varepsilon_1 + \varepsilon_2 = 1200$, which is consistent with Eq. 9.14.

(b) The difference between ε_1 and ε_2 is the largest difference between principal strains. Thus, the magnitude of the maximum shear strain is given by:

$\gamma_{max}/2 = |(\varepsilon_1 - \varepsilon_2)/2| = |(2049+849)/2|$ or $\boxed{\gamma_{max} = 2898\mu}$

(c) Using n and t as the reference coordinate, we note that the x-direction is at θ = 55° from the n-direction. Substituting the given strains in place of ε_{xx}, ε_{yy}, and γ_{xy} and θ = 55° in Eqs. 9.10, 9.11 and 9.12 we obtain:

$\varepsilon_{xx} = (2000)\cos^2(55) + (-800)\sin^2(55) + (750)\sin(55)\cos(55)$ or $\boxed{\varepsilon_{xx} = 473.6\mu}$

$\varepsilon_{yy} = (2000)\cos^2(55) + (-800)\sin^2(55) - (750)\sin(55)\cos(55)$ or $\boxed{\varepsilon_{yy} = 726.4\mu}$

$\gamma_{xy} = -2(2000-(-800))\cos(55)\sin(55) + (750)(\cos^2(55) - \sin^2(55))$ or $\boxed{\gamma_{nt} = -2888\mu}$

Check: $\varepsilon_{nn} + \varepsilon_{tt} = 1200$ and $\varepsilon_{xx} + \varepsilon_{yy} = 1200$ which is consistent with Eq. 9.14.

9.39 At a point in *plane strain*, the strain components in the n-t coordinate system are as given in each problem. Determine

(a) the principal strains.
(b) the maximum shear strain.
(c) the strain components in the x-y coordinate system.

$\varepsilon_{nn} = -2000\ \mu \qquad \varepsilon_{tt} = -800\ \mu \qquad \gamma_{nt} = -600\ \mu$

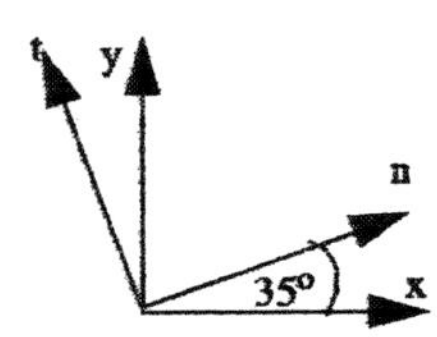

Fig. P9.39

Solution **Plane Strain** $\varepsilon_1 = ?$ $\varepsilon_2 = ?$ $\varepsilon_3 = ?$ $\theta_1 = ?$ $\gamma_{max} = ?$ $\varepsilon_{xx} = ?$ $\varepsilon_{yy} = ?$ $\gamma_{xy} = ?$

(a) We can use n and t as the reference frame as principal strains, and maximum shear strains are unique at a point. We also note that we do not need to calculate the principal angle one, thus we can find principal strains from Eq. 9.14.

The principal strains are: $\varepsilon_{1,2} = \left(\frac{-2000-800}{2}\right) \pm \sqrt{\left(\frac{-2000-(-800)}{2}\right)^2 + (-300)^2} = -1400 \pm 670.8$. For plane strain the principal strains are: $\boxed{\varepsilon_1 = -729\mu \quad \varepsilon_2 = -2071\mu \quad \varepsilon_3 = 0}$

Check: $\varepsilon_{nn} + \varepsilon_{tt} = 1200$ and $\varepsilon_1 + \varepsilon_2 = 1200$, which is consistent with Eq. 9.14.

(b) The difference between ε_3 and ε_2 is the largest difference between principal strains. Thus, the magnitude of the maximum shear strain is given by:

$\gamma_{max}/2 = |(\varepsilon_3 - \varepsilon_2)/2| = |(0 + 2070.8)/2|$ or $\boxed{\gamma_{max} = 2071\mu}$

(c) Using n and t as the reference coordinate, we note that the x-direction is at $\theta = 55^\circ$ from the n-direction. Substituting the given strains in place of ε_{xx}, ε_{yy}, and γ_{xy} and $\theta = -35^\circ$ in Eqs. 9.10, 9.11 and 9.12 we obtain:

$\varepsilon_{xx} = (-2000)\cos^2(-35) + (-800)\sin^2(-35) + (-600)\sin(-35)\cos(-35)$ or $\boxed{\varepsilon_{xx} = -1323\mu}$

$\varepsilon_{yy} = (-2000)\cos^2(-35) + (-800)\sin^2(-35) - (-600)\sin(-35)\cos(-35)$ or $\boxed{\varepsilon_{yy} = -1477\mu}$

$\gamma_{xy} = -2(-2000-(-800))\cos(-35)\sin(-35) + (-600)(\cos^2(-35) - \sin^2(-35))$ or $\boxed{\gamma_{nt} = -1333\mu}$

Check: $\varepsilon_{nn} + \varepsilon_{tt} = -2800$ and $\varepsilon_{xx} + \varepsilon_{yy} = -2800$ which is consistent with Eq. 9.14.

9.40 At a point in *plane strain*, the strain components in the n-t coordinate system are as given in each problem. Determine

(a) the principal strains.
(b) the maximum shear strain.
(c) the strain components in the x-y coordinate system.

$\varepsilon_{nn} = 350\ \mu \qquad \varepsilon_{tt} = 700\ \mu \qquad \gamma_{nt} = 1400\ \mu$

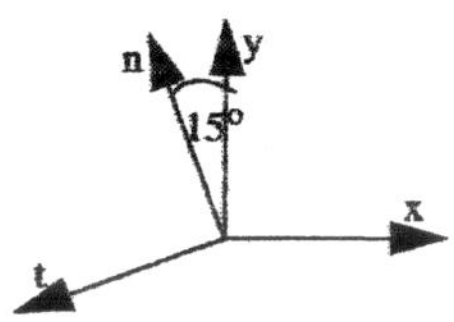

Fig. P9.40

Solution **Plane Strain** $\varepsilon_1 = ?$ $\varepsilon_2 = ?$ $\varepsilon_3 = ?$ $\theta_1 = ?$ $\gamma_{max} = ?$ $\varepsilon_{xx} = ?$ $\varepsilon_{yy} = ?$ $\gamma_{xy} = ?$

(a) We can use n and t as the reference frame as principal strains, and maximum shear strains are unique at a point. We also note that we do not need to calculate the principal angle one, thus we can find principal strains from Eq. 9.14.

The principal strains are: $\varepsilon_{1,2} = \left(\frac{350+700}{2}\right) \pm \sqrt{\left(\frac{350-700}{2}\right)^2 + (700)^2} = 525 \pm 721.5$. For plane strain the principal strains are: $\boxed{\varepsilon_1 = 1246.5\mu \quad \varepsilon_2 = -196.5\mu \quad \varepsilon_3 = 0}$

Check: $\varepsilon_{nn} + \varepsilon_{tt} = 1050$ and $\varepsilon_1 + \varepsilon_2 = 1050$, which is consistent with Eq. 9.14.

(b) The difference between ε_1 and ε_2 is the largest difference between principal strains. Thus, the magnitude of the maximum shear strain is given by:

$\gamma_{max}/2 = |(\varepsilon_1 - \varepsilon_2)/2| = |(1246.5 + 196.5)/2|$ or $\boxed{\gamma_{max} = 1443\mu}$

(c) Using n and t as the reference coordinate, we note that the x-direction is at $\theta = 55^\circ$ from the n-direction. Substituting the given strains in place of ε_{xx}, ε_{yy}, and γ_{xy} and $\theta = -105^\circ$ in Eqs. 9.10, 9.11 and 9.12 we obtain:

$\varepsilon_{xx} = 350\cos^2(-105) + 700\sin^2(-105) + 1400\sin(-105)\cos(-105)$ or $\boxed{\varepsilon_{xx} = 1027\mu}$

$\varepsilon_{yy} = 350\cos^2(-105) + 700\sin^2(-105) - 1400\sin(-105)\cos(-105)$ or $\boxed{\varepsilon_{yy} = 23\mu}$

$\gamma_{xy} = -2(350 - 700)\cos(-105)\sin(-105) + 1400(\cos^2(-105) - \sin^2(-105))$ or $\boxed{\gamma_{xy} = -1037\mu}$

Check: $\varepsilon_{nn} + \varepsilon_{tt} = 1050$ and $\varepsilon_{xx} + \varepsilon_{yy} = 1050$ which is consistent with Eq. 9.14.

9.41 At a point in *plane strain*, the strain components in the n-t coordinate system are as given in each problem. Determine

(a) the principal strains.
(b) the maximum shear strain.
(c) the strain components in the x-y coordinate system.

$\varepsilon_{nn} = -3600\ \mu \qquad \varepsilon_{tt} = 2500\ \mu \qquad \gamma_{nt} = -1000\mu$

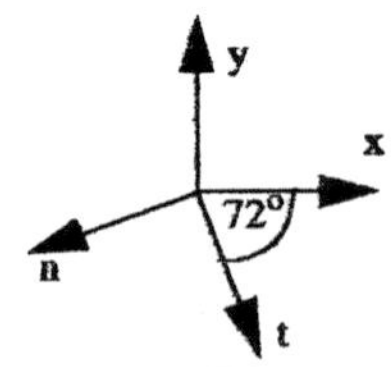

Fig. P9.41

Solution Plane Strain $\varepsilon_1 = ?\quad \varepsilon_2 = ?\quad \varepsilon_3 = ?\quad \theta_1 = ?\quad \gamma_{max} = ?\ \varepsilon_{xx} = ?\ \varepsilon_{yy} = ?\quad \gamma_{xy} = ?$

(a) We can use n and t as the reference frame as principal strains, and maximum shear strains are unique at a point. We also note that we do not need to calculate the principal angle one, thus we can find principal strains from Eq. 9.14.

The principal strains are: $\varepsilon_{1,2} = \left(\frac{-3600 + 2500}{2}\right) \pm \sqrt{\left(\frac{-3600 - 2500}{2}\right)^2 + (-500)^2} = -550 \pm 3090.7$. For plane strain the principal strains are: $\boxed{\varepsilon_1 = 2541\mu \qquad \varepsilon_2 = -3641\mu \qquad \varepsilon_3 = 0}$

Check: $\varepsilon_{nn} + \varepsilon_{tt} = -1100$ and $\varepsilon_1 + \varepsilon_2 = -1100$, which is consistent with Eq. 9.14.

(b) The difference between ε_1 and ε_2 is the largest difference between principal strains. Thus, the magnitude of the maximum shear strain is given by:

$\gamma_{max}/2 = |(\varepsilon_1 - \varepsilon_2)/2| = |(2540.7 + 3640.7)/2|$ or $\boxed{\gamma_{max} = 6181\mu}$

(c) Using n and t as the reference coordinate, we note that the x-direction is at $\theta = 55^\circ$ from the n-direction. Substituting the given strains in place of ε_{xx}, ε_{yy}, and γ_{xy} and $\theta = 162^\circ$ in Eqs. 9.10, 9.11 and 9.12 we obtain:

$\varepsilon_{xx} = (-3600)\cos^2(162) + 2500\sin^2(162) + (-1000)\sin(162)\cos(162)$ or $\boxed{\varepsilon_{xx} = -2724\mu}$

$\varepsilon_{yy} = (-3600)\cos^2(162) + 2500\sin^2(162) - (-1000)\sin(162)\cos(162)$ or $\boxed{\varepsilon_{yy} = 1624\mu}$

$\gamma_{xy} = -2(-3600 - 2500)\cos(162)\sin(162) + (-1000)(\cos^2(162) - \sin^2(162))$ or $\boxed{\gamma_{nt} = -4395\mu}$

Check: $\varepsilon_{nn} + \varepsilon_{tt} = -1100$ and $\varepsilon_{xx} + \varepsilon_{yy} = -1100$ which is consistent with Eq. 9.14.

9.42 The principal strains (ε_1 and ε_2) and the direction of principal direction one (θ_1) from the x-axis are given as given below. Determine the strains ε_{xx}, ε_{yy}, and γ_{xy} at the point.

$\varepsilon_1 = 1200\ \mu \qquad \varepsilon_2 = 300\ \mu \qquad \theta_1 = 27.5^0$

Solution $\varepsilon_{xx} = ?\qquad \varepsilon_{yy} = ?\qquad \gamma_{xy} = ?$

Fig. (a) shows the Mohr's circle for the given state of strain. The center is at 750 μ and the radius is R = 450 μ. Fig. (b) shows the orientation of principal directions P_1 and P_2 at 27.5° counter-clockwise from the x axis as per the value of θ_1. To get to points V and H, we must rotate clockwise from the principal direction points P_1 and P_2. Starting from P_1 and P_2 on the Mohr's circle we rotate 55° clockwise to get points V and H as shown in Fig. (a).

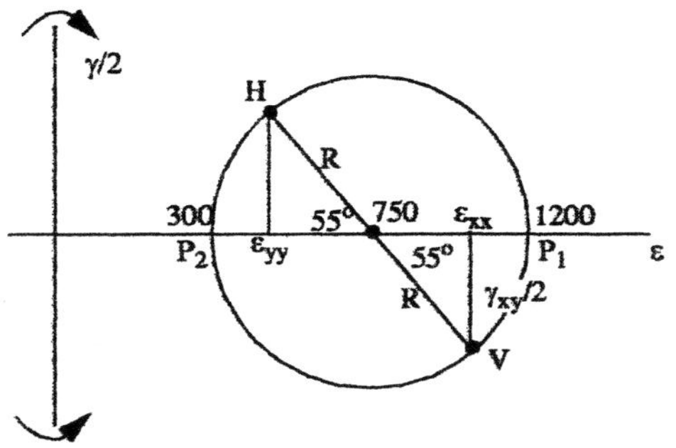

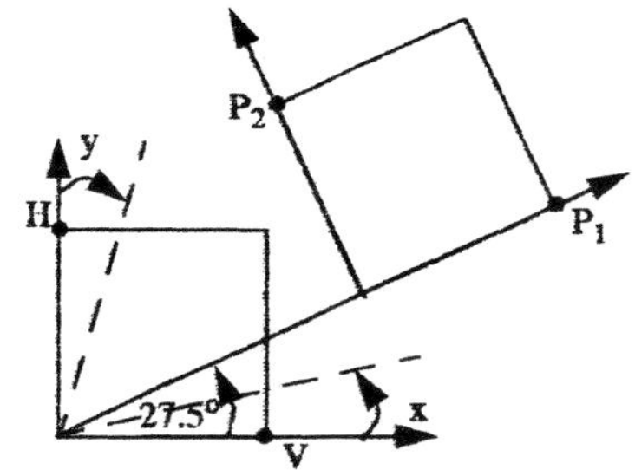

The normal strains in the x and y coordinate system are:

$\varepsilon_{xx} = 750 + R\cos 55 = 1008.11$ or $\boxed{\varepsilon_{xx} = 1008.1\mu}$

$\varepsilon_{yy} = 750 - R\cos 55 = 491.89$ $\boxed{\varepsilon_{yy} = 491.9\mu}$

The magnitude of shear strain is: $|\gamma_{xy}/2| = R\sin 55 = 368.62$. From Fig. (a) we note that at point V the rotation is counter-clockwise and H the rotation is clockwise. We draw dashed lines in Fig. (b) such as to cause the lines at V and H to rotate in the direction deduced from Fig. (a). We note from Fig. (b) that angle decreases, hence shear strain is positive. $\boxed{\gamma_{xy} = 737.2\mu}$

9.43 The principal strains (ε_1 and ε_2) and the direction of principal direction one (θ_1) from the x-axis are given as given below. Determine the strains ε_{xx}, ε_{yy}, and γ_{xy} at the point.

$$\varepsilon_1 = 900\ \mu \qquad \varepsilon_2 = -600\ \mu \qquad \theta_1 = -20^0$$

Solution $\varepsilon_{xx} = ?$ $\varepsilon_{yy} = ?$ $\gamma_{xy} = ?$

Fig. (a) shows the Mohr's circle for the given state of strain. The center is at 150 μ and the radius is R = 750 μ. Fig. (b) shows the orientation of principal directions P_1 and P_2 at 27.5° counter-clockwise from the x axis as per the value of θ_1. To get to points V and H, we must rotate clockwise from the principal direction points P_1 and P_2. Starting from P_1 and P_2 on the Mohr's circle we rotate 55° clockwise to get points V and H as shown in Fig. (a).

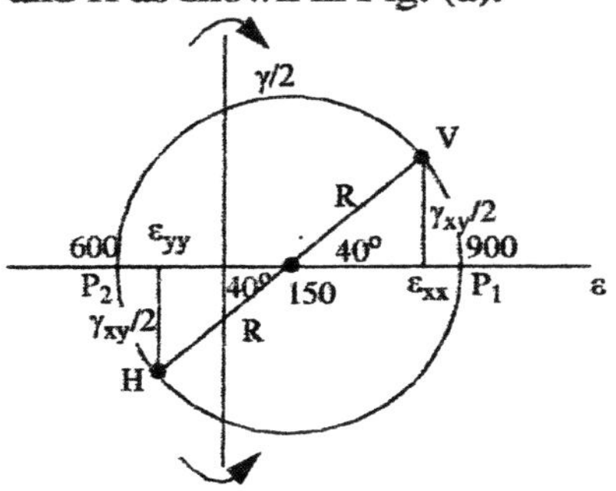

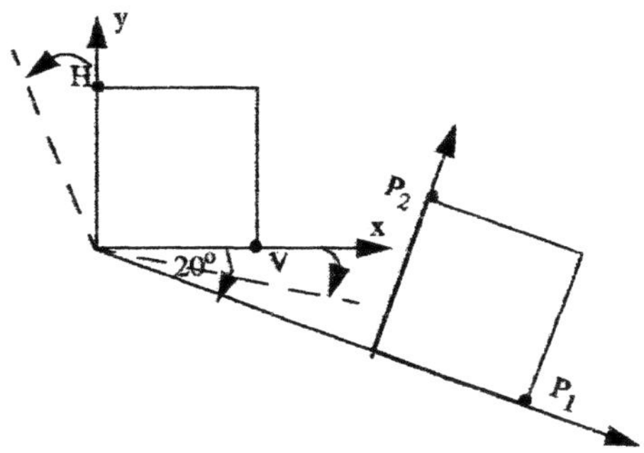

The normal strains in the x and y coordinate system are:

$\varepsilon_{xx} = 150 + R\cos 40 = 724.5$ or $\boxed{\varepsilon_{xx} = 724.5\mu}$

$\varepsilon_{yy} = 150 - R\cos 40 = -424.5$ $\boxed{\varepsilon_{yy} = -424.5\mu}$

The magnitude of shear strain is: $|\gamma_{xy}/2| = R\sin 40 = 482.09$. From Fig. (a) we note that at point V the rotation is clockwise and H the rotation is counter-clockwise. We draw dashed lines in Fig. (b) such as to cause the lines at V and H to rotate in the direction deduced from Fig. (a). We note from Fig. (b) that angle increases, hence shear strain is negative. $\boxed{\gamma_{xy} = -964.2\mu}$

9.44 The principal strains (ε_1 and ε_2) and the direction of principal direction one (θ_1) from the x-axis are given as given below. Determine the strains ε_{xx}, ε_{yy}, and γ_{xy} at the point.

$$\varepsilon_1 = -200\ \mu \qquad \varepsilon_2 = -2000\ \mu \qquad \theta_1 = 105^0$$

Solution $\varepsilon_{xx} = ?$ $\varepsilon_{yy} = ?$ $\gamma_{xy} = ?$

Fig. (a) shows the Mohr's circle for the given state of strain. The center is at -1100 μ and the radius is R = 900 μ. Fig. (b) shows the orientation of principal directions P_1 and P_2 at 105° counter-clockwise from the x axis as per the value of θ_1. To get to points V and H, we must rotate clockwise from the principal direction points P_1 and P_2. Starting from P_1 and P_2 on the Mohr's circle we rotate 210° clockwise to get points V and H as shown in Fig. (a).

(a)

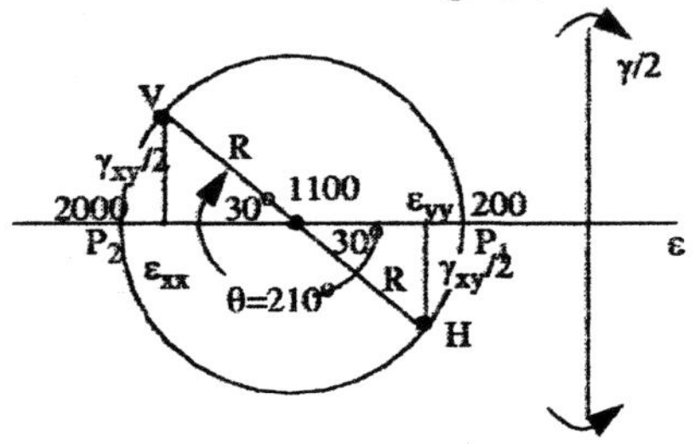

(b)

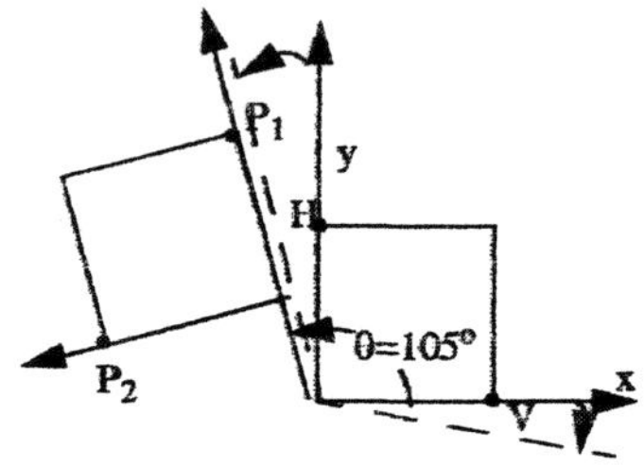

The normal strains in the x and y coordinate system are:

$\varepsilon_{xx} = -1100 - R\cos 30 = -1879.42$ or $\boxed{\varepsilon_{xx} = -1879\mu}$

$\varepsilon_{yy} = -100 + R\cos 30 = -320.6$ or $\boxed{\varepsilon_{yy} = -320.6\mu}$

The magnitude of shear strain is: $|\gamma_{xy}/2| = R\sin 30 = 450$. From Fig. (a) we note that at point V the rotation is clockwise and H the rotation is counter-clockwise. We draw dashed lines in Fig. (b) such as to cause the lines at V and H to rotate in the direction deduced from Fig. (a). We note from Fig. (b) that angle increases, hence shear strain is negative. $\boxed{\gamma_{xy} = -900\mu}$

9.45 The principal strains (ε_1 and ε_2) and the direction of principal direction one (θ_1) from the x-axis are given as given below. Determine the strains ε_{xx}, ε_{yy}, and γ_{xy} at the point.

$$\varepsilon_1 = 1400\ \mu \qquad \varepsilon_2 = -600\ \mu \qquad \theta_1 = -75^0$$

Solution $\varepsilon_{xx} = ?$ $\varepsilon_{yy} = ?$ $\gamma_{xy} = ?$

Fig. (a) shows the Mohr's circle for the given state of strain. The center is at 400 μ and the radius is R = 1000 μ. Fig. (b) shows the orientation of principal directions P_1 and P_2 at 75° clockwise from the x axis as per the value of θ_1. To get to points V and H, we must rotate counter-clockwise from the principal direction points P_1 and P_2. Starting from P_1 and P_2 on the Mohr's circle we rotate 150° counter-clockwise to get points V and H as shown in Fig. (a).

(a)

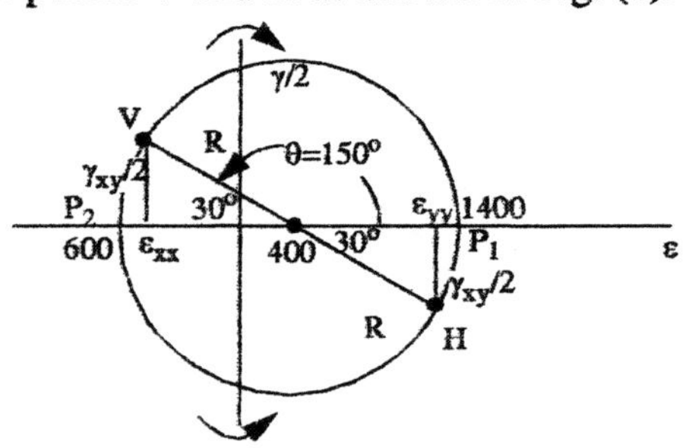

(b)

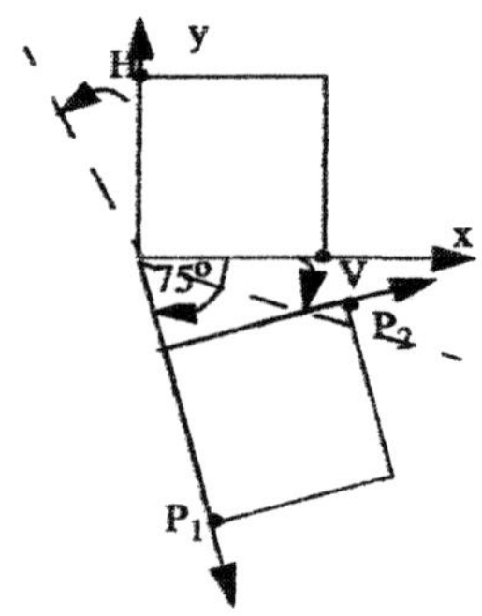

The normal strains in the x and y coordinate system are:

$\varepsilon_{xx} = 400 - R\cos 30 = -466.03$ or $\boxed{\varepsilon_{xx} = -466\mu}$

$\varepsilon_{yy} = 400 + R\cos 30 = 1266.03$ or $\boxed{\varepsilon_{yy} = 1266\mu}$

The magnitude of shear strain is: $|\gamma_{xy}/2| = R\sin 30 = 500$. From Fig. (a) we note that at point V the rotation is clockwise and H the rotation is counter-clockwise. We draw dashed lines in Fig. (b) such as to cause the lines at V and H to rotate in the direction deduced from Fig. (a). We note from Fig. (b) that angle increases, hence shear strain is negative. $\boxed{\gamma_{xy} = -1000\mu}$

9.46 In a thin body (*plane stress*) the stresses in the x-y plane are as shown on each stress element. The material has a modulus of elasticity E = 70 GPa and Poisson's ratio ν = 0.25. Determine

(a) the principal strains and the principal angle one at the point.
(b) the maximum shear strain at the point.

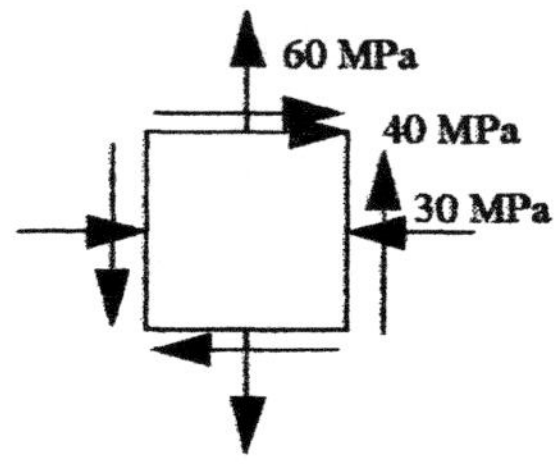

Fig. P9.46

Solution Plane Stress $\varepsilon_1 = ?$ $\varepsilon_2 = ?$ $\varepsilon_3 = ?$ $\theta_1 = ?$ $\gamma_{max} = ?$

From Fig. 9.46 we obtain:

$$\sigma_{xx} = -30\text{MPa} \qquad \sigma_{yy} = 60\text{MPa} \qquad \tau_{xy} = 40\text{MPa} \qquad (1)$$

The principal angle θ_p is: $\theta_p = 0.5\ \text{atan}\left(\frac{(2)(40)}{((-30)-60)}\right) = -20.82^\circ$

Substituting θ_p and the stresses from Eq. 1 into Eq. 8.1, we obtain one principal stress as:

$$\sigma_p = (-30)\cos^2(-20.82) + (60)\sin^2(-20.82) + 2(40)\sin(-20.82)\cos(-20.82) = -45.2\text{MPa} \qquad (2)$$

From Eq. 8.9 we can find the other principal stress as: $-30 + 60 - (-45.2) = 75.2\text{MPa}$. As this stress is greater than the stress in Eq. 2, therefore this is principal stress one. Thus principal angle 1 is $\theta_1 = \theta_p \pm 90$. We have the following results:

$$\sigma_1 = 75.2\text{MPa} \qquad \sigma_2 = -45.2\text{MPa} \qquad \sigma_3 = 0 \qquad (3)$$

The principal angle 1 is: $\boxed{\theta_1 = -69.2^\circ \text{ or } 110.8^\circ}$

Check: From Eq. 8.6 we have: $\sigma_{1,2} = \left(\frac{-30+60}{2}\right) \pm \sqrt{\left(\frac{-30-60}{2}\right)^2 + (40)^2} = 15 \pm 60.2$ or $\sigma_1 = 75.2\text{MPa}$ and $\sigma_2 = -45.2\text{MPa}$, which are same as above.

From Eqs. 9.21a, 9.21b, and 9.21c we obtain:

$$\varepsilon_1 = \frac{[\sigma_1 - \nu(\sigma_2 + \sigma_3)]}{E} = \frac{[75.2 - 0.25(-45.2)](10^6)}{70(10^9)} \text{ or } \boxed{\varepsilon_1 = 1236\mu}$$

$$\varepsilon_2 = \frac{[\sigma_2 - \nu(\sigma_3 + \sigma_1)]}{E} = \frac{[(-45.2) - 0.25(75.2)](10^6)}{70(10^9)} \qquad \boxed{\varepsilon_2 = -914.4\mu}$$

$$\varepsilon_3 = \frac{[\sigma_3 - \nu(\sigma_1 + \sigma_2)]}{E} = \frac{[0 - 0.25(75.2 - 45.2)](10^6)}{70(10^9)} \text{ or } \boxed{\varepsilon_3 = -107.1\mu}$$

From the above results we see that the difference between ε_1 and ε_2 is the largest difference between principal strains. Thus, the magnitude of the maximum shear strain is:

$\gamma_{max}/2 = |(\varepsilon_1 - \varepsilon_2)/2| = |(1236 + 914.4)/2|$ or $\boxed{\gamma_{max} = 2150\mu}$

9.47 In a thin body (*plane stress*) the stresses in the x-y plane are as shown on each stress element. The material has a modulus of elasticity E = 70 GPa and Poisson's ratio $\nu = 0.25$. Determine

(a) the principal strains and the principal angle one at the point.
(b) the maximum shear strain at the point.

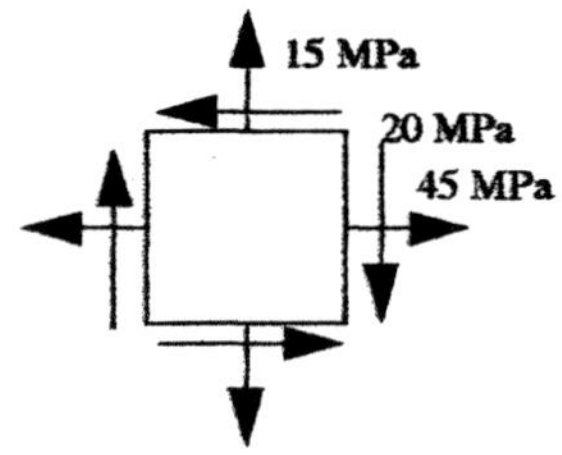

Fig. P9.47

Solution Plane Stress $\varepsilon_1 = ?$ $\varepsilon_2 = ?$ $\varepsilon_3 = ?$ $\theta_1 = ?$ $\gamma_{max} = ?$

From Fig. 9.47 we obtain:

$$\sigma_{xx} = 45\text{MPa} \qquad \sigma_{yy} = 15\text{MPa} \qquad \tau_{xy} = -20\text{MPa} \tag{1}$$

The principal angle θ_p is: $\theta_p = 0.5\ \text{atan}\left(\frac{(2)(-20)}{(45-15)}\right) = -26.57^\circ$

Substituting θ_p and the stresses from Eq. 1 into Eq. 8.1, we obtain one principal stress as:

$$\sigma_p = (45)\cos^2(-26.57) + (15)\sin^2(-26.57) + 2(-20)\sin(-26.57)\cos(-26.57) = 55\text{MPa} \tag{2}$$

From Eq. 8.9 we can find the other principal stress as: $45 + 15 - 55 = 5\text{MPa}$. As this stress is smaller than the stress in Eq. 2, therefore the stress in Eq. 2 is principal stress one. Thus principal angle 1 is $\theta_1 = \theta_p$. We have the following results:

$$\sigma_1 = 55\text{MPa} \qquad \sigma_2 = 5\text{MPa} \qquad \sigma_3 = 0 \tag{3}$$

The principal angle 1 is: $\boxed{\theta_1 = -26.57^\circ \text{ or } 153.4^\circ}$

Check: From Eq. 8.6 we have: $\sigma_{1,2} = \left(\frac{45+15}{2}\right) \pm \sqrt{\left(\frac{45-15}{2}\right)^2 + (-20)^2} = 30 \pm 25$ or $\sigma_1 = 55\text{MPa}$ and $\sigma_2 = 5\text{MPa}$, which are same as above.

From Eqs. 9.21a, 9.21b, and 9.21c we obtain:

$$\varepsilon_1 = \frac{[\sigma_1 - \nu(\sigma_2+\sigma_3)]}{E} = \frac{[55 - 0.25(5)](10^6)}{70(10^9)} \text{ or } \qquad \boxed{\varepsilon_1 = 767.9\mu}$$

$$\varepsilon_2 = \frac{[\sigma_2 - \nu(\sigma_3+\sigma_1)]}{E} = \frac{[5 - 0.25(55)](10^6)}{70(10^9)} \qquad \boxed{\varepsilon_2 = -125.0\mu}$$

$$\varepsilon_3 = \frac{[\sigma_3 - \nu(\sigma_1+\sigma_2)]}{E} = \frac{[0 - 0.25(5+55)](10^6)}{70(10^9)} \text{ or } \qquad \boxed{\varepsilon_3 = -214.3\mu}$$

From the above results we see that the difference between ε_1 and ε_3 is the largest difference between principal strains. Thus, the magnitude of the maximum shear strain is:

$$\gamma_{max}/2 = |(\varepsilon_1 - \varepsilon_3)/2| = |(767.9 + 214.3)/2| \text{ or } \qquad \boxed{\gamma_{max} = 982.1\mu}$$

9.48 In a thin body (*plane stress*) the stresses in the x-y plane are as shown on each stress element. The material has a modulus of elasticity E = 30,000 ksi and Poisson's ratio $\nu = 0.28$. Determine

(a) the principal strains and the principal angle one at the point.
(b) the maximum shear strain at the point.

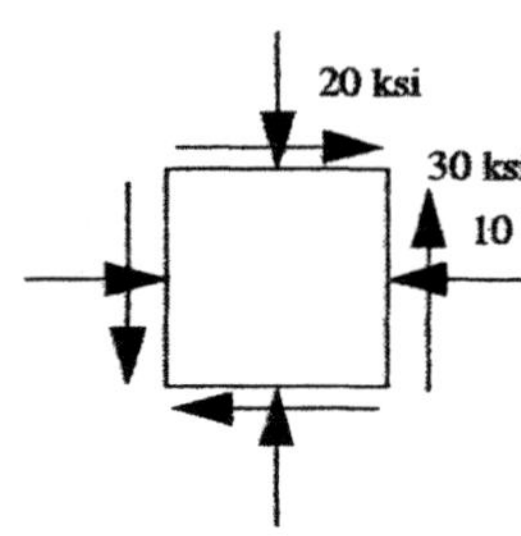

Fig. P9.48

Solution Plane Stress $\varepsilon_1 = ?$ $\varepsilon_2 = ?$ $\varepsilon_3 = ?$ $\theta_1 = ?$ $\gamma_{max} = ?$

From Fig. 9.48 we obtain:

$$\sigma_{xx} = -10\text{ksi} \qquad \sigma_{yy} = -20\text{ksi} \qquad \tau_{xy} = 30\text{ksi} \tag{1}$$

The principal angle θ_p is: $\theta_p = 0.5\ \text{atan}\left(\frac{(2)(-20)}{(-10-(-20))}\right) = -40.27^\circ$

Substituting θ_p and the stresses from Eq. 1 into Eq. 8.1, we obtain one principal stress as:

$$\sigma_p = (-10)\cos^2(-40.27) + (-20)\sin^2(-40.27) + 2(30)\sin(-40.27)\cos(-40.27) = 15.4\text{ksi} \tag{2}$$

From Eq. 8.9 we can find the other principal stress as: $-10 - 20 - 15.4 = -45.4\text{ksi}$. As this stress is smaller than the stress in Eq. 2, therefore the stress in Eq. 2 is principal stress one. Thus principal angle 1 is $\theta_1 = \theta_p$. We have the following results:

$$\sigma_1 = 15.4\text{ksi} \qquad \sigma_2 = -45.4\text{ksi} \qquad \sigma_3 = 0 \tag{3}$$

The principal angle 1 is: $\boxed{\theta_1 = -40.27^\circ \text{ or } 139.73^\circ}$

Check: From Eq. 8.6 we have: $\sigma_{1,2} = \left(\frac{-10+(-20)}{2}\right) \pm \sqrt{\left(\frac{45-(-20)}{2}\right)^2 + (30)^2} = -15 \pm 30.4$ or

$\sigma_1 = 15.4\text{ksi}$ and $\sigma_2 = -45.4\text{ksi}$, which are same as above.

From Eqs. 9.21a, 9.21b, and 9.21c we obtain:

$$\varepsilon_1 = \frac{[\sigma_1 - \nu(\sigma_2 + \sigma_3)]}{E} = \frac{[15.4 - 0.28(-45.4)]}{30000} \text{ or } \qquad \boxed{\varepsilon_1 = 938\mu}$$

$$\varepsilon_2 = \frac{[\sigma_2 - \nu(\sigma_3 + \sigma_1)]}{E} = \frac{[(-45.4) - 0.28(15.4)]}{30000} \qquad \boxed{\varepsilon_2 = -1658\mu}$$

$$\varepsilon_3 = \frac{[\sigma_3 - \nu(\sigma_1 + \sigma_2)]}{E} = \frac{[0 - 0.28(15.4 - 45.4)]}{30000} \text{ or } \qquad \boxed{\varepsilon_3 = 280\mu}$$

From the above results we see that the difference between ε_1 and ε_2 is the largest difference between principal strains. Thus, the magnitude of the maximum shear strain is:

$$\gamma_{max}/2 = |(\varepsilon_1 - \varepsilon_2)/2| = |(937.7 + 1657.7)/2| \text{ or } \qquad \boxed{\gamma_{max} = 2595\mu}$$

9.49 In a thick body (*plane strain*) the stresses in the x-y plane are as shown on the stress element. The material has a modulus of elasticity E = 70 GPa and Poisson's ratio $\nu = 0.25$. Determine

(a) the principal strains and the principal angle one at the point.
(b) the maximum shear strain at the point.

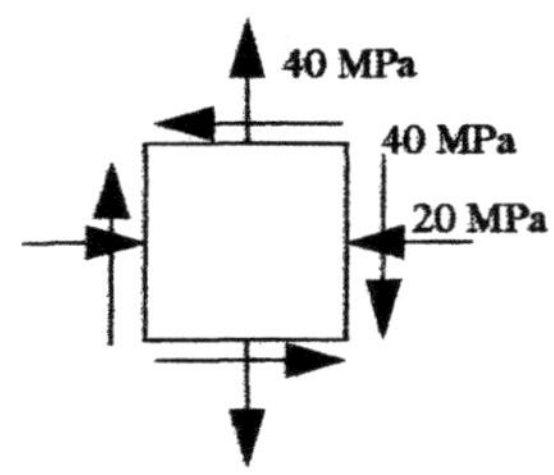

Fig. P9.49

Solution Plane Strain $\varepsilon_1 = ?$ $\varepsilon_2 = ?$ $\varepsilon_3 = ?$ $\theta_1 = ?$ $\gamma_{max} = ?$

From Fig. 9.49 we obtain:

$$\sigma_{xx} = -20\text{MPa} \qquad \sigma_{yy} = 40\text{MPa} \qquad \tau_{xy} = -40\text{MPa} \tag{1}$$

The principal angle θ_p is: $\theta_p = 0.5\ \text{atan}\left(\frac{(2)(-40)}{(-20-40)}\right) = 26.57^\circ$

Substituting θ_p and the stresses from Eq. 1 into Eq. 8.1, we obtain one principal stress as:

$$\sigma_p = (-20)\cos^2(26.57) + (40)\sin^2(26.57) + 2(-40)\sin(26.57)\cos(26.57) = -40\text{MPa} \tag{2}$$

From Eq. 8.9 we can find the other principal stress as: $-20+40-(-40) = 60\text{MPa}$. As this stress is greater than the stress in Eq. 2, therefore this stress is principal stress one. Thus principal angle 1 is $\theta_1 = \theta_p \pm 90$. We have the following results:

$$\sigma_1 = 60\text{MPa} \qquad \sigma_2 = -40\text{MPa} \qquad (3)$$

The principal angle 1 is: $\boxed{\theta_1 = 26.57^\circ \text{ or } -153.4^\circ}$

Check: From Eq. 8.6 we have: $\sigma_{1,2} = \left(\frac{-20+40}{2}\right) \pm \sqrt{\left(\frac{-20-40}{2}\right)^2 + (-40)^2} = 10 \pm 50$ or $\sigma_1 = 60\text{MPa}$ and $\sigma_2 = -40\text{MPa}$, which are same as above.

For plane strain we have: $\sigma_3 = \sigma_{zz} = \nu(\sigma_{xx}+\sigma_{yy}) = 0.25(-20+40) = 5\text{MPa}$ and $\boxed{\varepsilon_3 = 0}$

From Eqs. 9.21a and 9.21b, we obtain:

$$\varepsilon_1 = \frac{[\sigma_1 - \nu(\sigma_2+\sigma_3)]}{E} = \frac{[60-0.25(-40+5)](10^6)}{70(10^9)} \text{ or } \boxed{\varepsilon_1 = 982.1\mu}$$

$$\varepsilon_2 = \frac{[\sigma_2 - \nu(\sigma_3+\sigma_1)]}{E} = \frac{[-40-0.25(60+5)](10^6)}{70(10^9)} \text{ or } \boxed{\varepsilon_2 = -803.6\mu}$$

From the above results we see that the difference between ε_1 and ε_2 is the largest difference between principal strains. Thus, the magnitude of the maximum shear strain is:

$$\gamma_{max}/2 = |(\varepsilon_1 - \varepsilon_3)/2| = |(982.1+803.6)/2| \text{ or } \boxed{\gamma_{max} = 1786\mu}$$

9.50 In a thick body (*plane strain*) the stresses in the x-y plane are as shown on the stress element. The material has a modulus of elasticity E = 70 GPa and Poisson's ratio ν = 0.25. Determine

(a) the principal strains and the principal angle one at the point.
(b) the maximum shear strain at the point.

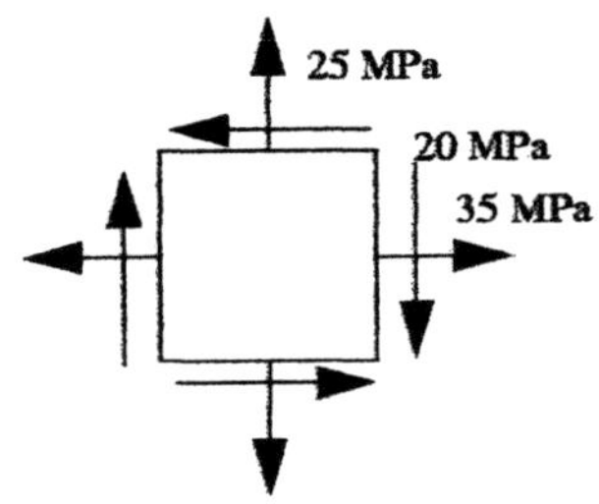

Fig. P9.50

Solution Plane Strain $\varepsilon_1 = ?$ $\varepsilon_2 = ?$ $\varepsilon_3 = ?$ $\theta_1 = ?$ $\gamma_{max} = ?$

From Fig. 9.49 we obtain:

$$\sigma_{xx} = 35\text{MPa} \qquad \sigma_{yy} = 25\text{MPa} \qquad \tau_{xy} = -20\text{MPa} \qquad (1)$$

The principal angle θ_p is: $\theta_p = 0.5 \operatorname{atan}\left(\frac{(2)(-20)}{(35-25)}\right) = -37.98^\circ$

Substituting θ_p and the stresses from Eq. 1 into Eq. 8.1, we obtain one principal stress as:

$$\sigma_p = (35)\cos^2(-37.98) + (25)\sin^2(-37.98) + 2(-20)\sin(-37.98)\cos(-37.98) = 50.6\text{MPa} \qquad (2)$$

From Eq. 8.9 we can find the other principal stress as: $35+25-50.6 = 9.4\text{MPa}$. As this stress is smaller than the stress in Eq. 2, therefore the stress in Eq. 2 is principal stress one. Thus principal angle 1 is $\theta_1 = \theta_p$. We have the following results:

$$\sigma_1 = 50.6\text{MPa} \qquad \sigma_2 = 9.4\text{MPa} \qquad (3)$$

The principal angle 1 is: $\boxed{\theta_1 = -37.98^\circ \text{ or } 142.2^\circ}$

Check: From Eq. 8.6 we have: $\sigma_{1,2} = \left(\frac{35+25}{2}\right) \pm \sqrt{\left(\frac{35-25}{2}\right)^2 + (-20)^2} = 30 \pm 20.6$ or $\sigma_1 = 50.6\text{MPa}$ and

$\sigma_2 = 9.4\text{MPa}$, which are same as above.

For plane strain we have: $\sigma_3 = \sigma_{zz} = \nu(\sigma_{xx} + \sigma_{yy}) = 0.25(35 + 25) = 15\text{MPa}$ and $\boxed{\varepsilon_3 = 0}$

From Eqs. 9.21a and 9.21b, we obtain:

$$\varepsilon_1 = \frac{[\sigma_1 - \nu(\sigma_2 + \sigma_3)]}{E} = \frac{[50.6 - 0.25(9.4 + 15)](10^6)}{70(10^9)} \text{ or } \boxed{\varepsilon_1 = 636\mu}$$

$$\varepsilon_2 = \frac{[\sigma_2 - \nu(\sigma_3 + \sigma_1)]}{E} = \frac{[9.4 - 0.25(50.6 + 15)](10^6)}{70(10^9)} \text{ or } \boxed{\varepsilon_2 = -100.3\mu}$$

From the above results we see that the difference between ε_1 and ε_2 is the largest difference between principal strains. Thus, the magnitude of the maximum shear strain is:

$$\gamma_{max}/2 = |(\varepsilon_1 - \varepsilon_3)/2| = |(636 + 100.3)/2| \text{ or } \boxed{\gamma_{max} = 736.3\mu}$$

9.51 In a thick body (*plane strain*) the stresses in the x-y plane are as shown on the stress element. The material has a modulus of elasticity E = 30,000 ksi and Poisson's ratio ν = 0.28. Determine

(a) the principal strains and the principal angle one at the point.
(b) the maximum shear strain at the point.

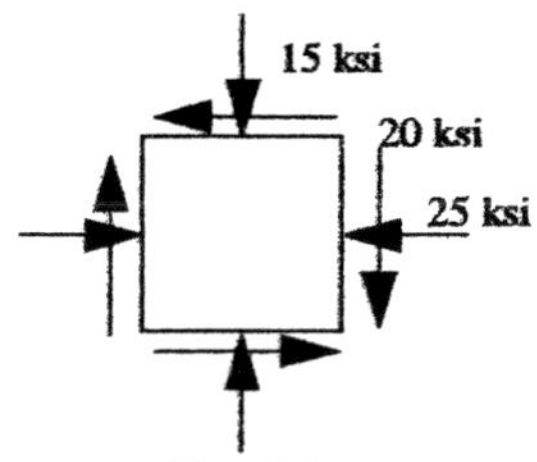

Fig. P9.51

Solution Plane Strain $\varepsilon_1 = ?$ $\varepsilon_2 = ?$ $\varepsilon_3 = ?$ $\theta_1 = ?$ $\gamma_{max} = ?$

From Fig. 9.49 we obtain:

$$\sigma_{xx} = -25\text{ksi} \qquad \sigma_{yy} = -15\text{ksi} \qquad \tau_{xy} = -20\text{ksi} \qquad (1)$$

The principal angle θ_p is: $\theta_p = 0.5 \ \text{atan}\left(\frac{(2)(-20)}{(-25-(-15))}\right) = 37.98°$

Substituting θ_p and the stresses from Eq. 1 into Eq. 8.1, we obtain one principal stress as:

$$\sigma_p = (-25)\cos^2(37.98) + (-15)\sin^2(37.98) + 2(-20)\sin(37.98)\cos(37.98) = -40.62\text{ksi} \qquad (2)$$

From Eq. 8.9 we can find the other principal stress as: $-25 + (-15) - (-40.62) = 0.62\text{ksi}$. As this stress is greater than the stress in Eq. 2, therefore this stress is principal stress one. Thus principal angle 1 is $\theta_1 = \theta_p \pm 90$. We have the following results:

$$\sigma_1 = 0.62\text{ksi} \qquad \sigma_2 = -40.62\text{ksi} \qquad (3)$$

The principal angle 1 is: $\boxed{\theta_1 = 37.98° \text{ or } -142.2°}$

Check: From Eq. 8.6 we have: $\sigma_{1,2} = \left(\frac{-25-15}{2}\right) \pm \sqrt{\left(\frac{-25-(-15)}{2}\right)^2 + (-20)^2} = -20 \pm 20.62$ or

$\sigma_1 = 0.62\text{ksi}$ and $\sigma_2 = -40.62\text{ksi}$, which are same as above.

For plane strain we have: $\sigma_3 = \sigma_{zz} = \nu(\sigma_{xx} + \sigma_{yy}) = 0.28(-25 - 15) = -11.2\text{ksi}$ and $\boxed{\varepsilon_3 = 0}$

From Eqs. 9.21a and 9.21b, we obtain:

$$\varepsilon_1 = \frac{[\sigma_1 - \nu(\sigma_2 + \sigma_3)]}{E} = \frac{[0.62 - 0.28(-40.62 - 11.2)]}{30000} \text{ or } \boxed{\varepsilon_1 = 504\mu}$$

$$\varepsilon_2 = \frac{[\sigma_2 - \nu(\sigma_3 + \sigma_1)]}{E} = \frac{[-40.62 - 0.28(0.62 - 11.2)]}{30000} \text{ or } \boxed{\varepsilon_2 = -1255\mu}$$

From the above results we see that the difference between ε_1 and ε_2 is the largest difference between prin-

cipal strains. Thus, the magnitude of the maximum shear strain is:

$\gamma_{max}/2 = |(\varepsilon_1 - \varepsilon_3)/2| = |(504 + 1255)/2|$ or $\boxed{\gamma_{max} = 1759\mu}$

9.52 An orthotropic material has the following properties E_x=7,500ksi, E_y= 2,500 ksi, G_{xy} = 1,250 ksi and ν_{xy}= 0.25. Using Equation 9.22, determine the principal directions for stresses and strains at a point on a free surface where the following strains were measured.

$$\varepsilon_{xx} = -400\ \mu \qquad \varepsilon_{yy} = 600\ \mu \qquad \gamma_{xy} = -500\ \mu$$

Solution E_x=7,500ksi E_y= 2,500 ksi G_{xy} = 1,250 ksi ν_{xy}= 0.25.
$\theta_1 = ?$ for strains $\bar{\theta}_1 = ?$ for stresses

Substituting the given strains into Eq. 9.13 we obtain the principal angle θ_p for strain as:

$$\theta_p = 0.5\ \text{atan}\left(\frac{(-500)}{(-400-600)}\right) = 13.3^\circ \qquad 1$$

Substituting θ_p and the given strains into Eq. 9.10, we obtain one principal strain as:

$$\varepsilon_p = (-400)\cos^2(13.3) + (600)\sin^2(13.3) + (-500)\sin(13.3)\cos(13.3) = -459\mu \qquad 2$$

From Eq. 9.15 we can find the other principal strain as: $-400 + 600 - (-459) = 659\mu$. As this strain is greater than the strain in Eq. 2, therefore this strain is principal strain one. Thus principal angle 1 is $\theta_1 = \theta_p \pm 90$ For strain, the principal angle 1 is: $\boxed{\theta_1 = 103.3^\circ \text{ or } -76.7^\circ}$

From the given material constants and Eq. 9.22 we have: $\nu_{yx} = \dfrac{E_y \nu_{xy}}{E_x} = \left(\dfrac{2,500}{7,500}\right)(0.25) = 0.0833$

Substituting the given strain values and the constants above in Eqs. 9.22 we obtain the following

$$\varepsilon_{xx} = \frac{\sigma_{xx}}{E_x} - \frac{\nu_{yx}}{E_y}\sigma_{yy} = \frac{\sigma_{xx}}{7,500} - \frac{0.0833}{2,500}\sigma_{yy} = (-400)(10^{-6}) \quad \text{or} \quad \sigma_{xx} - 0.25\sigma_{yy} = -3\text{ksi} \qquad 3$$

$$\varepsilon_{yy} = \frac{\sigma_{yy}}{E_y} - \frac{\nu_{xy}}{E_x}\sigma_{xx} = \frac{\sigma_{yy}}{2,500} - \frac{(0.25)}{7500}(\sigma_{xx}) = 600(10^{-6}) \quad \text{or} \quad 3\sigma_{yy} - 0.25\sigma_{xx} = 4.5\text{ksi} \qquad 4$$

$$\tau_{xy} = G_{xy}\gamma_{xy} = (1250)(-500)(10^{-6}) = -0.625\text{ksi} \qquad 5$$

Solving Eqs. (3) and (4), we obtain

$$\sigma_{xx} = -2.681\text{ksi} \qquad \sigma_{yy} = 1.277\text{ksi} \qquad 6$$

Using Eq. 5 and 6, the principal angle $\bar{\theta}_p$ for stress is: $\bar{\theta}_p = 0.5\ \text{atan}\left(\dfrac{(2)(-0.625)}{(-2.681-1.277)}\right) = 8.76^\circ$

Substituting $\bar{\theta}_p$ and the stresses from Eqs. 5 and 6 into Eq. 8.1, we obtain one principal stress as:

$$\sigma_p = (-2.681)\cos^2(8.76) + (1.277)\sin^2(8.76) + 2(-0.625)\sin(8.76)\cos(8.76) = -2.78\text{ksi} \qquad 7$$

From Eq. 8.9 we can find the other principal stress as: $-2.68 + 1.28 - (-2.78) = 1.38\text{ksi}$. As this stress is greater than the stress in Eq. 7, therefore this stress is principal stress one. Thus principal angle 1 is $\bar{\theta}_1 = \bar{\theta}_p \pm 90$. The principal angle 1 for stress is; $\boxed{\bar{\theta}_1 = 98.76^\circ \text{ or } -81.24^\circ}$

9.53 An orthotropic material has the following properties E_x=7,500ksi, E_y= 2,500 ksi, G_{xy} = 1,250 ksi and ν_{xy}= 0.25. Using Equation 9.22, determine the principal directions for stresses and strain at a point on a free surface where the following stresses were computed.

$$\sigma_{xx} = 10\ \text{ksi(T)} \qquad \sigma_{yy} = 7\ \text{ksi(C)} \qquad \tau_{xy} = 5\ \text{ksi}.$$

Solution E_x=7,500ksi E_y= 2,500 ksi G_{xy} = 1,250 ksi ν_{xy}= 0.25.
θ_1 = ? for strains $\bar{\theta}_1$ = ? for stresses

Substituting the given stresses into Eq. 8.5, we obtain the principal angle θ_p for stress as:

$$\bar{\theta}_p = 0.5 \ \operatorname{atan}\left(\frac{(2)(5)}{(10-(-7))}\right) = 15.23^\circ \qquad 1$$

Substituting $\bar{\theta}_p$ and the stresses into Eq. 8.1, we obtain one principal stress as:

$$\sigma_p = (10)\cos^2(15.23) + (-7)\sin^2(15.23) + 2(5)\sin(15.23)\cos(15.23) = 11.36\text{ksi} \qquad 2$$

From Eq. 8.9 we can find the other principal stress as: $10 - 7 - 11.36 = -8.36\text{ksi}$. As this stress is smaller than the stress in Eq. 2, therefore the stress in Eq. 2 is principal stress one. Thus principal angle 1 is $\bar{\theta}_1 = \bar{\theta}_p$. The principal angle 1 for stress is; $\boxed{\bar{\theta}_1 = 15.23^\circ \text{ or } -164.77^\circ}$

From the given material constants and Eq. 9.22 we have: $\nu_{yx} = \dfrac{E_y \nu_{xy}}{E_x} = \left(\dfrac{2,500}{7,500}\right)(0.25) = 0.0833$

Substituting the given stress values and the constants above in Eqs. 9.22 we obtain the following

$$\varepsilon_{xx} = \frac{\sigma_{xx}}{E_x} - \frac{\nu_{yx}}{E_y}\sigma_{yy} = \frac{10}{7,500} - \frac{0.0833}{2,500}(-7) = 1567(10^{-6}) = 1567\mu \qquad 3$$

$$\varepsilon_{yy} = \frac{\sigma_{yy}}{E_y} - \frac{\nu_{xy}}{E_x}\sigma_{xx} = \frac{(-7)}{2,500} - \frac{(0.25)}{7500}(10) = -3133(10^{-6}) = -3133\mu \qquad 4$$

$$\gamma_{xy} = \frac{\tau_{xy}}{G_{xy}} = \frac{5}{1250} = 4000(10^{-6}) = 4000\mu \qquad 5$$

Substituting the strains from Eqs. 3, 4, and 5 into Eq. 9.13 we obtain the principal angle θ_p for strain as:

$$\theta_p = 0.5 \ \operatorname{atan}\left(\frac{(4000)}{(1567-(-3133))}\right) = 20.2^\circ \qquad 6$$

Substituting θ_p and the strains from Eqs. 3, 4, and 5 into Eq. 9.10, we obtain one principal strain as:

$$\varepsilon_p = (1567)\cos^2(20.2) + (-3133)\sin^2(20.2) + (4000)\sin(20.2)\cos(20.2) = 2302\mu \qquad 7$$

From Eq. 9.15 we can find the other principal strain as: $1567 - 3133 - 2302 = -3868\mu$. As this strain is smaller than the strain in Eq. 7, therefore the strain in Eq. 7 is principal strain one. Thus principal angle 1 is $\theta_1 = \theta_p$ For strain, the principal angle 1 is: $\boxed{\theta_1 = 20.2^\circ \text{ or } -159.8^\circ}$

9.54 An orthotropic material has the following properties E_x=50 MPa, E_y= 18 MPa, G_{xy} = 9 MPa and ν_{xy}= 0.25. Using Equation 9.22, determine the principal directions for stresses and strains at a point on a free surface where the following strains were measured.

$$\varepsilon_{xx} = 800\ \mu \qquad \varepsilon_{yy} = 200\ \mu \qquad \gamma_{xy} = 300\ \mu.$$

Solution E_x=50 MPa E_y= 18 MPa G_{xy} = 9 MPa ν_{xy}= 0.25.
θ_1 = ? for strains $\bar{\theta}_1$ = ? for stresses

Substituting the given strains into Eq. 9.13 we obtain the principal angle θ_p for strain as:

$$\theta_p = 0.5 \ \operatorname{atan}\left(\frac{300}{(800-200)}\right) = 13.3^\circ \qquad 1$$

Substituting θ_p and the given strains into Eq. 9.10, we obtain one principal strain as:

$$\varepsilon_p = 800\cos^2(13.3) + 200\sin^2(13.3) + 300\sin(13.3)\cos(13.3) = 835.4\mu \qquad 2$$

From Eq. 9.15 we can find the other principal strain as: $800 + 200 - 835.4 = 164.6\mu$. As this strain is smaller than the strain in Eq. 2, therefore the strain in Eq. 2 is principal strain one. Thus principal angle 1 is $\theta_1 = \theta_p$ For strain, the principal angle 1 is: $\boxed{\theta_1 = 13.3^o \text{ or } -166.7^o}$

From the given material constants and Eq. 9.22 we have: $\nu_{yx} = \dfrac{E_y \nu_{xy}}{E_x} = \left(\dfrac{18}{50}\right)(0.25) = 0.09$

Substituting the given strain values and the constants above in Eqs. 9.22 we obtain the following

$$\varepsilon_{xx} = \frac{\sigma_{xx}}{E_x} - \frac{\nu_{yx}}{E_y}\sigma_{yy} = \frac{\sigma_{xx}}{50(10^9)} - \frac{0.09}{18(10^9)}\sigma_{yy} = 800(10^{-6}) \quad \text{or} \quad \sigma_{xx} - 0.25\sigma_{yy} = 40\text{MPa} \qquad 3$$

$$\varepsilon_{yy} = \frac{\sigma_{yy}}{E_y} - \frac{\nu_{xy}}{E_x}\sigma_{xx} = \frac{\sigma_{yy}}{18(10^9)} - \frac{(0.25)}{50(10^9)}(\sigma_{xx}) = 200(10^{-6}) \quad \text{or} \quad 2.7778\sigma_{yy} - 0.25\sigma_{xx} = 10\text{MPa} \qquad 4$$

$$\tau_{xy} = G_{xy}\gamma_{xy} = (9)(10^9)(300)(10^{-6}) = 2.70\text{MPa} \qquad 5$$

Solving Eqs. (3) and (4), we obtain

$$\sigma_{xx} = 41.84\text{MPa} \qquad \sigma_{yy} = 7.37\text{MPa} \qquad 6$$

Using Eq. 5 and 6, the principal angle $\bar{\theta}_p$ for stress is: $\bar{\theta}_p = 0.5 \ \text{atan}\left(\dfrac{(2)(2.70)}{(41.84 - 7.37)}\right) = 4.45^o$

Substituting $\bar{\theta}_p$ and the stresses from Eqs. 5 and 6 into Eq. 8.1, we obtain one principal stress as:

$$\sigma_p = (41.84)\cos^2(4.45) + (7.37)\sin^2(4.45) + 2(2.70)\sin(4.45)\cos(4.45) = 42.05\text{MPa} \qquad 7$$

From Eq. 8.9 we can find the other principal stress as: $41.84 + 7.37 - 42.05 = 7.16\text{MPa}$. As this stress is smaller than the stress in Eq. 7, therefore the stress in Eq. 7 is principal stress one. Thus principal angle 1 is $\bar{\theta}_1 = \bar{\theta}_p$. The principal angle 1 for stress is; $\boxed{\bar{\theta}_1 = 4.45^o \text{ or } -175.55^o}$

9.55 An orthotropic material has the following properties E_x=50 MPa, E_y= 18 MPa, G_{xy} = 9 MPa and ν_{xy}= 0.25. Using Equation 9.22, determine the principal directions for stresses and strain at a point on a free surface where the following stresses were computed.

$$\sigma_{xx} = 70 \ \text{MPa(C)} \qquad \sigma_{yy} = 49 \ \text{MPa(C)} \qquad \tau_{xy} = -30 \ \text{MPa}.$$

Solution E_x=50 MPa $\quad E_y$= 18 MPa $\quad G_{xy}$ = 9 MPa $\quad \nu_{xy}$= 0.25.

$\theta_1 = ?$ for strains $\qquad \bar{\theta}_1 = ?$ for stresses

---

Substituting the given stresses into Eq. 8.5, we obtain the principal angle θ_p for stress as:

$$\bar{\theta}_p = 0.5 \ \text{atan}\left(\frac{(2)(-30)}{(-70 - (-49))}\right) = 35.35^o \qquad 1$$

Substituting $\bar{\theta}_p$ and the stresses into Eq. 8.1, we obtain one principal stress as:

$$\sigma_p = (-70)\cos^2(35.35) + (-49)\sin^2(35.35) + 2(-30)\sin(35.35)\cos(35.35) = -91.3\text{MPa} \qquad 2$$

From Eq. 8.9 we can find the other principal stress as: $-70 - 49 - (-91.3) = -27.7\text{MPa}$. As this stress is greater than the stress in Eq. 2, therefore this stress is principal stress one. Thus principal angle 1 is $\bar{\theta}_1 = \bar{\theta}_p \pm 90^o$. The principal angle 1 for stress is; $\boxed{\bar{\theta}_1 = 125.35^o \text{ or } -54.65^o}$

From the given material constants and Eq. 9.22 we have: $\nu_{yx} = \dfrac{E_y \nu_{xy}}{E_x} = \left(\dfrac{18}{50}\right)(0.25) = 0.09$

Substituting the given stress values and the constants above in Eqs. 9.22 we obtain the following

$$\varepsilon_{xx} = \frac{\sigma_{xx}}{E_x} - \frac{\nu_{yx}}{E_y}\sigma_{yy} = \frac{(-70)(10^6)}{50(10^9)} - \frac{0.09}{18(10^9)}(-49)(10^6) = -1155(10^{-6}) = -1155\mu \quad 3$$

$$\varepsilon_{yy} = \frac{\sigma_{yy}}{E_y} - \frac{\nu_{xy}}{E_x}\sigma_{xx} = \frac{(-49)(10^6)}{18(10^9)} - \frac{0.25}{50(10^9)}(-70) = -2372(10^{-6}) = -2372\mu(10^6) \quad 4$$

$$\gamma_{xy} = \frac{\tau_{xy}}{G_{xy}} = \frac{(-30)}{9(10^9)} = -3333(10^{-6}) = -3333\mu \quad 5$$

Substituting the strains from Eqs. 3, 4, and 5 into Eq. 9.13 we obtain the principal angle θ_p for strain as:

$$\theta_p = 0.5 \ \text{atan}\left(\frac{(-3333)}{(-1155-(-2372))}\right) = -34.97^\circ \quad 6$$

Substituting θ_p and the strains from Eqs. 3, 4, and 5 into Eq. 9.10, we obtain one principal strain as:

$$\varepsilon_p = (-1155)\cos^2(-34.97) + (-2372)\sin^2(-34.97) + (-3333)\sin(-34.97)\cos(-34.97) = 10.7\mu \quad 7$$

From Eq. 9.15 we can find the other principal strain as: $-1155 - 2372 - 10.7 = -3537.7\mu$. As this strain is smaller than the strain in Eq. 7, therefore the strain in Eq. 7 is principal strain one. Thus principal angle 1 is $\theta_1 = \theta_p$ For strain, the principal angle 1 is: $\boxed{\theta_1 = -34.97^\circ \text{ or } 145.03^\circ}$

Problem Set 9.3

9.56 Show that by substituting $(\theta \pm 180^0)$ in place of θ, the strain transformation equation 9.10 is unchanged.

Solution

Substituting $(\theta \pm 180^0)$ in place of θ in Eq. 9.10 and noting that $\sin(\theta \pm 180^0) = -\sin\theta$ and $\cos(\theta \pm 180^0) = -\cos\theta$ we obtain:

$$\varepsilon_{nn} = \varepsilon_{xx}\cos^2(\theta \pm 180^0) + \varepsilon_{yy}\sin^2(\theta \pm 180^0) + \gamma_{xy}\sin(\theta \pm 180^0)\cos(\theta \pm 180^0) \text{ or}$$

$$\varepsilon_{nn} = \varepsilon_{xx}[-\cos\theta]^2 + \varepsilon_{yy}[-\sin\theta]^2 + \gamma_{xy}[-\sin\theta][-\cos\theta] = \varepsilon_{nn} = \varepsilon_{xx}\cos^2\theta + \varepsilon_{yy}\sin^2\theta + \gamma_{xy}\sin\theta\cos\theta \quad 1$$

Eq. 1 is same as Eq. 9.10.

9.57 At a point on a free surface the strain components in x-y coordinates are calculated as

$\varepsilon_{xx} = 400\mu$ in/in $\qquad \varepsilon_{yy} = -200\mu$ in/in $\qquad \gamma_{xy} = 500\mu$ rad

Predict the strains that the strain gages shown in Fig. P9.57 would record.

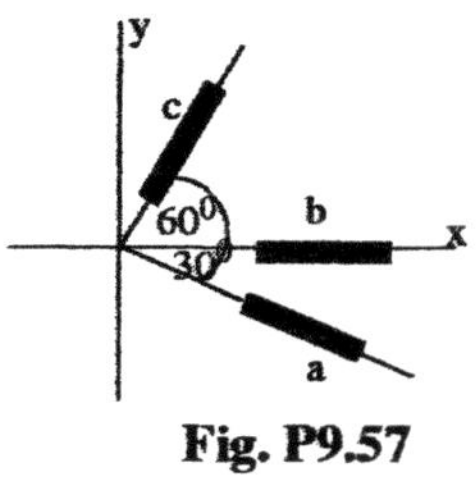

Fig. P9.57

Solution $\qquad \varepsilon_a = ? \qquad \varepsilon_b = ? \qquad \varepsilon_c = ?$

Substituting the given strains and $\theta_a = -30^\circ$ in Eq. 9.10, we obtain

$$\varepsilon_a = (400)\cos^2(-30) + (-200)\sin^2(-30) + 500\sin(-30)\cos(-30)$$ $\boxed{\varepsilon_a = 33.5\mu \text{ in/in}}$

By inspection of Fig. (b), we obtain: $\boxed{\varepsilon_b = 400\mu \text{ in/in}}$

Substituting the given strains and $\theta_c = 60^o$ in Eq. 9.10, we obtain

$\varepsilon_a = (400)\cos^2(60) + (-200)\sin^2(60) + 500\sin(60)\cos(60)$ $\boxed{\varepsilon_c = 166.5\mu \text{ in/in}}$

9.58 At a point on a free surface the strains recorded by the three strain gages shown in Fig. P9.58 are

$\varepsilon_a = 200\mu \text{ in/in}$ $\varepsilon_b = 100\mu \text{ in/in}$ $\varepsilon_c = -400\mu \text{ in/in}$

Determine the strains ε_{xx}, ε_{yy}, and γ_{xy}.

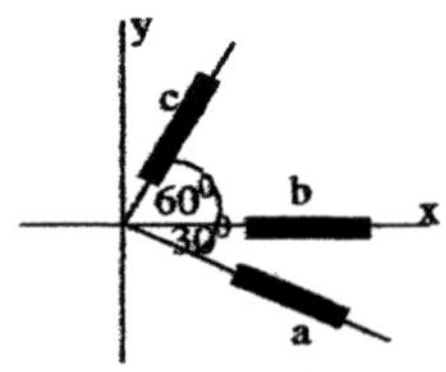

Fig. P9.58

Solution $\varepsilon_{xx} = ?$ $\varepsilon_{yy} = ?$ $\gamma_{xy} = ?$

By inspection of Fig. P9.58 we obtain: $\boxed{\varepsilon_{xx} = 100\mu \text{ in/in}}$

Substituting $\theta_a = -30^o$ and $\theta_c = 60^o$ and the given strain values into Eq. 9.10 we obtain:

$$\varepsilon_a = 100\cos^2(-30) + \varepsilon_{yy}\sin^2(-30) + \gamma_{xy}\sin(-30)\cos(-30) = 200 \quad \text{or} \quad 0.25\varepsilon_{yy} - 0.433\gamma_{xy} = 125 \qquad 1$$

$$\varepsilon_c = 100\cos^2(60) + \varepsilon_{yy}\sin^2(60) + \gamma_{xy}\sin(60)\cos(60) = -400 \quad \text{or} \quad 0.75\varepsilon_{yy} + 0.433\gamma_{xy} = -425 \qquad 2$$

Solving Eq. 1 and 2 we obtain: $\boxed{\varepsilon_{yy} = -300\mu \text{ in/in}}$ and $\boxed{\gamma_{xy} = -461.9\mu \text{ rad}}$

9.59 At a point on a free surface of aluminum (E = 10,000 ksi and G =4,000 ksi) machine component the stress components in x-y coordinates were calculated by Finite Element Method as

$\sigma_{xx} = 22 \text{ ksi(T)}$ $\sigma_{yy} = 15 \text{ ksi(C)}$ $\tau_{xy} = -10 \text{ ksi}$

Predict the strains that the strain gages shown in Fig. P9.59 would show.

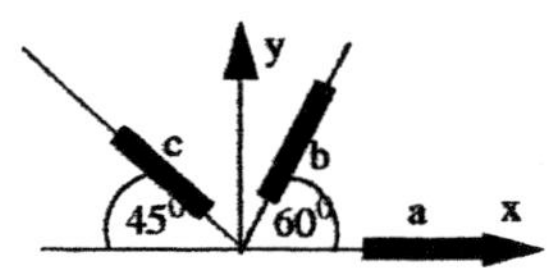

Fig. P9.59

Solution $\varepsilon_a = ?$ $\varepsilon_b = ?$ $\varepsilon_c = ?$ Plane stress

---

The Poisson's ratio is: $\nu = \frac{E}{2G} - 1 = \frac{10000}{2(4000)} - 1 = 0.25$ .The shear strain is:

$$\gamma_{xy} = \tau_{xy}/G = (-10)/4000 = -2500\mu \qquad 1$$

From generalized Hooke's law, for plane stress ($\sigma_{zz} = 0$) we obtain:

$$\varepsilon_{xx} = \frac{\sigma_{xx}}{E} - \frac{\nu}{E}(\sigma_{yy} + \sigma_{zz}) = \frac{22}{10000} - \frac{0.25}{10000}(-15) = 2575\mu \qquad 2$$

$$\varepsilon_{yy} = \frac{\sigma_{yy}}{E} - \frac{\nu}{E}(\sigma_{xx} + \sigma_{zz}) = \frac{(-15)}{10000} - \frac{0.25}{10000}(22) = -2050\mu \qquad 3$$

By inspection of Fig. P9.59, we obtain: $\boxed{\varepsilon_a = 2575\mu}$

Substituting $\theta_b = 60^o$, $\theta_c = 135^o$ and Eqs. 1, 2, and 3 we obtain:

$\varepsilon_b = 2575\cos^2(60) + (-2050)\sin^2(60) + (-2500)\sin(60)\cos(60) = -1976.3$ $\boxed{\varepsilon_b = -1976\mu}$

$\varepsilon_c = 2575\cos^2(135) + (-2050)\sin^2(135) + (-2500)\sin(135)\cos(135) = 1512.5$ $\boxed{\varepsilon_c = 1513\mu}$

9.60 At a point on a free surface of aluminum ($E = 10{,}000$ ksi and $G = 4{,}000$ ksi) the strains recorded by the three strain gages shown in Fig. P9.59 are

$\varepsilon_a = -600\ \mu\ \text{in/in}$ $\quad\varepsilon_b = 500\ \mu\ \text{in/in}$ $\quad\varepsilon_c = 400\ \mu\ \text{in/in}$

Determine the stresses σ_{xx}, σ_{yy}, and τ_{xy}.

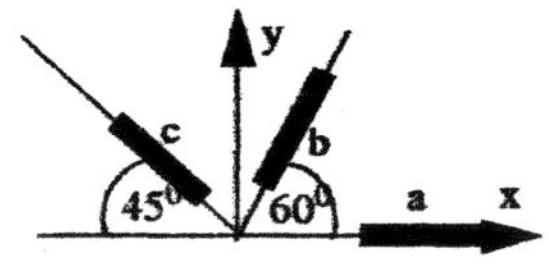

Fig. P9.60

Solution $E = 10{,}000$ ksi $G = 4{,}000$ ksi $\quad\sigma_{xx} = ?\quad \sigma_{yy} = ?\quad \tau_{xy} = ?$

By inspection of Fig. P9.60 we obtain: $\varepsilon_{xx} = -600\mu\ \text{in/in}$

Substituting $\theta_b = 60°$ and $\theta_c = 135°$ and the given strain values into Eq. 9.10 we obtain:

$$\varepsilon_b = -600\cos^2(60) + \varepsilon_{yy}\sin^2(60) + \gamma_{xy}\sin(60)\cos(60) = 500 \quad \text{or} \quad 0.75\varepsilon_{yy} + 0.433\gamma_{xy} = 650 \qquad 1$$

$$\varepsilon_c = -600\cos^2(135) + \varepsilon_{yy}\sin^2(135) + \gamma_{xy}\sin(135)\cos(135) = 400 \quad \text{or} \quad 0.5\varepsilon_{yy} - 0.5\gamma_{xy} = 700 \qquad 2$$

Solving Eq. 1 and 2 we obtain:

$$\varepsilon_{yy} = 1061.9\mu\ \text{in/in and}\ \gamma_{xy} = -338.1\mu\ \text{rad} \qquad 3$$

The Poisson's ratio is: $\nu = \dfrac{E}{2G} - 1 = \dfrac{10000}{2(4000)} - 1 = 0.25$.

The shear stress is: $\tau_{xy} = G\gamma_{xy} = (4000)(-338.1)(10^{-6}) = -1.352$ or $\boxed{\tau_{xy} = -1.35\text{ksi}}$

From generalized Hooke's law, for plane stress ($\sigma_{zz} = 0$) we obtain:

$$\varepsilon_{xx} = \frac{\sigma_{xx}}{E} - \frac{\nu}{E}(\sigma_{yy} + \sigma_{zz}) \quad \text{or} \quad \sigma_{xx} - 0.25\sigma_{yy} = E\varepsilon_{xx} = (10000)(-600)(10^{-6}) = -6\text{ksi} \qquad 4$$

$$\varepsilon_{yy} = \frac{\sigma_{yy}}{E} - \frac{\nu}{E}(\sigma_{xx} + \sigma_{zz}) \quad \text{or} \quad \sigma_{yy} - 0.25\sigma_{xx} = E\varepsilon_{yy} = (10000)(1061.9)(10^{-6}) = 10.62\text{ksi} \qquad 5$$

Solving Eqs. 4 and 5 we obtain: $\boxed{\sigma_{xx} = 3.57\text{ksi(C)} \quad \text{and} \quad \sigma_{yy} = 9.73\text{ksi(T)}}$

9.61 At a point on a free surface of machine component ($E = 80$ GPa and $G = 32$ GPa) the stress components in x-y coordinates were calculated by Finite Element Method as

$\sigma_{xx} = 50\ \text{MPa(T)} \quad \sigma_{yy} = 20\ \text{MPa(C)} \quad \tau_{xy} = 96\ \text{MPa}$

Predict the strains that the strain gages shown in Fig. P9.61 would show.

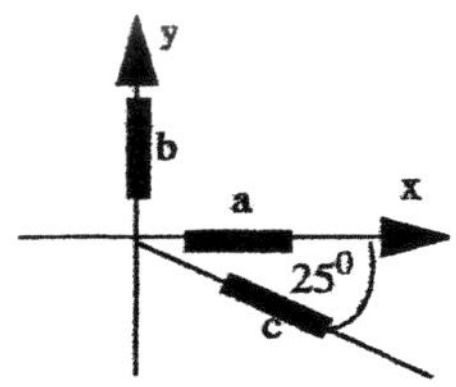

Fig. P9.61

Solution $\varepsilon_a = ?\quad \varepsilon_b = ?\quad \varepsilon_c = ?$ Plane stress

---

The Poisson's ratio is: $\nu = \dfrac{E}{2G} - 1 = \dfrac{80}{2(32)} - 1 = 0.25$. The shear strain is:

$$\gamma_{xy} = \frac{\tau_{xy}}{G} = \frac{(96)(10^6)}{32(10^9)} = 3000\mu \qquad 1$$

From generalized Hooke's law, for plane stress ($\sigma_{zz} = 0$) we obtain:

$$\varepsilon_{xx} = \frac{\sigma_{xx}}{E} - \frac{\nu}{E}(\sigma_{yy} + \sigma_{zz}) = \frac{(50)}{(80)(10^9)} - \frac{0.25}{(80)(10^9)}(-20) = 687.5\mu \qquad 2$$

$$\varepsilon_{yy} = \frac{\sigma_{yy}}{E} - \frac{\nu}{E}(\sigma_{xx} + \sigma_{zz}) = \frac{(-20)}{(80)(10^9)} - \frac{0.25}{(80)(10^9)}(50) = -406.3\mu \qquad 3$$

By inspection of Fig. P9.61, we obtain: $\boxed{\varepsilon_a = 687.5\mu}$ and $\boxed{\varepsilon_b = -406.3\mu}$

Substituting $\theta_c = -25^o$ and Eqs. 1, 2, and 3 into Eq. 9.10 we obtain:

$$\varepsilon_c = 687.5\cos^2(-25) + (-406.3)\sin^2(-25) + (3000)\sin(-25)\cos(-25) = -656.9 \qquad \boxed{\varepsilon_c = -656.9\mu}$$

9.62 At a point on a free surface of aluminum (E = 80 GPa and G = 32 GPa) the strains recorded by the three strain gages shown in Fig. P9.61 are

$\varepsilon_a = 1000\ \mu\ m/m \qquad \varepsilon_b = 1500\mu\ m/m \qquad \varepsilon_c = -450\ \mu\ m/m$

Determine the stresses σ_{xx}, σ_{yy}, and τ_{xy}.

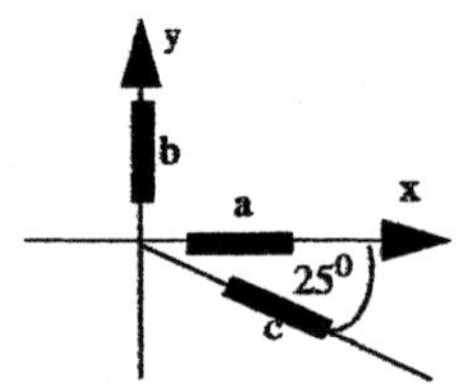

Fig. P9.62

Solution $E = 80GPa \qquad G = 32GPa \qquad \sigma_{xx} = ? \qquad \sigma_{yy} = ? \qquad \tau_{xy} = ?$

By inspection of Fig. P9.62 we obtain: $\varepsilon_{xx} = 1000\mu\ m/m$ and $\varepsilon_{yy} = 1500\mu\ m/m$

Substituting $\theta_c = -25^o$ the given strain values into Eq. 9.10 we obtain:

$$\varepsilon_c = 1000\cos^2(-25) + 1500\sin^2(-25) + \gamma_{xy}\sin(-25)\cos(-25) = -450 \quad \text{or} \quad \gamma_{xy} = 4018.8\mu\ rad \qquad 1$$

The Poisson's ratio is: $\nu = \frac{E}{2G} - 1 = \frac{80}{2(32)} - 1 = 0.25$.

The shear stress is: $\tau_{xy} = G\gamma_{xy} = (32)(10^9)(4018.8)(10^{-6})$ or $\boxed{\tau_{xy} = 128.6MPa}$

From generalized Hooke's law, for plane stress ($\sigma_{zz} = 0$) we obtain:

$$\varepsilon_{xx} = \frac{\sigma_{xx}}{E} - \frac{\nu}{E}(\sigma_{yy} + \sigma_{zz}) \quad \text{or} \quad \sigma_{xx} - 0.25\sigma_{yy} = E\varepsilon_{xx} = (80)(10^9)(1000)(10^{-6}) = 80MPa \qquad 2$$

$$\varepsilon_{yy} = \frac{\sigma_{yy}}{E} - \frac{\nu}{E}(\sigma_{xx} + \sigma_{zz}) \quad \text{or} \quad \sigma_{yy} - 0.25\sigma_{xx} = E\varepsilon_{yy} = (80)(10^9)(1500)(10^{-6}) = 120MPa \qquad 3$$

Solving Eqs. 2 and 3 we obtain: $\boxed{\sigma_{xx} = 117.3MPa \quad \text{and} \quad \sigma_{yy} = 149.3MPa}$

9.63 On a free surface of steel (E = 210 GPa and ν = 0.28) the strains recorded by the three strain gages shown in Fig. P9.63 are

$\varepsilon_a = -800\ \mu\ m/m \qquad \varepsilon_b = -300\ \mu\ m/m \qquad \varepsilon_c = -700\ \mu\ m/m$

Determine the principal *strains*, principal angle one and the maximum shear *strain*.

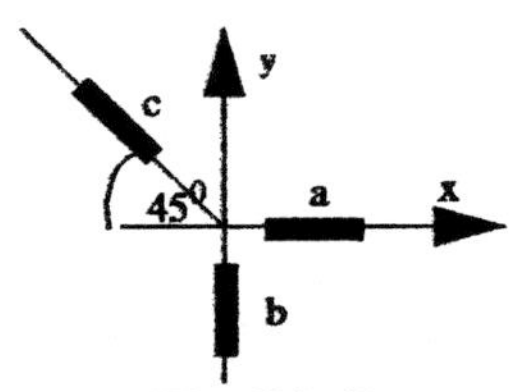

Fig. P9.63

Solution Plane Stress $\varepsilon_1 = ? \qquad \varepsilon_2 = ? \qquad \varepsilon_3 = ? \quad \theta_1 = ? \qquad \gamma_{max} = ?$

By inspection of Fig. P9.63 we obtain: $\varepsilon_{xx} = -800\mu\ m/m$ and $\varepsilon_{yy} = -300\mu\ m/m$

Substituting $\theta_c = 135^o$ the given strain values into Eq. 9.10 we obtain:

$$\varepsilon_c = (-800)\cos^2(135) + (-300)\sin^2(135) + \gamma_{xy}\sin(135)\cos(135) = -700 \quad \text{or} \quad \gamma_{xy} = 300\mu\ rad \qquad 1$$

From Eq. 9.13 we obtain the principal angle θ_p as:

$$\theta_p = 0.5\ \text{atan}\left(\frac{300}{-800+300}\right) = -15.48^o \qquad 2$$

Substituting θ_p and the given strains into Eq. 9.10, we obtain one principal strain as:

$$\varepsilon_p = (-800)\cos^2(-15.48) + (-300)\sin^2(-15.48) + (300)\sin(-15.48)\cos(-15.48) = -841.5\mu \qquad 3$$

From Eq. 9.15 we can find the other principal strain as: $-800-300-(-841.5) = -258.5\mu$. As this strain is greater than the strain in Eq. 3, therefore this strain is principal strain one. Thus principal angle 1 is $\theta_1 = \theta_p \pm 90$. We have the following answers:

$$\boxed{\varepsilon_1 = -258.5\mu \qquad \varepsilon_2 = -841.5\mu \qquad \theta_1 = 74.52^\circ \text{ or } -105.48^\circ}$$

Check: From Eq. 9.14 we have: $\varepsilon_{1,2} = \left(\frac{-800-300}{2}\right) \pm \sqrt{\left(\frac{-800+300}{2}\right)^2 + (150)^2} = -550 \pm 291.5$ or

$\varepsilon_1 = -258.5\mu$ and $\varepsilon_2 = -841.5\mu$, which are same as above.

For plane stress: $\varepsilon_3 = \varepsilon_z = -\frac{\nu}{(1-\nu)}(\varepsilon_{xx}+\varepsilon_{yy}) = -\frac{0.28}{(1-0.28)}(-800-300)$ $\boxed{\varepsilon_3 = 427.8\mu}$

The difference between ε_2 and ε_3 is the largest difference between principal strains. Thus, the magnitude of the maximum shear strain is given by:

$\gamma_{max}/2 = |(\varepsilon_3-\varepsilon_2)/2| = |(427.8+841.5)/2|$ or $\boxed{\gamma_{max} = 1269\mu}$

9.64 On a free surface of steel (E = 210 GPa and ν = 0.28) the strains recorded by the three strain gages shown in Fig. P9.63 are

$$\varepsilon_a = 200\ \mu\ m/m \qquad \varepsilon_b = 100\ \mu\ m/m \qquad \varepsilon_c = 0$$

Determine the principal *stresses*, principal angle one and the maximum shear *stress*.

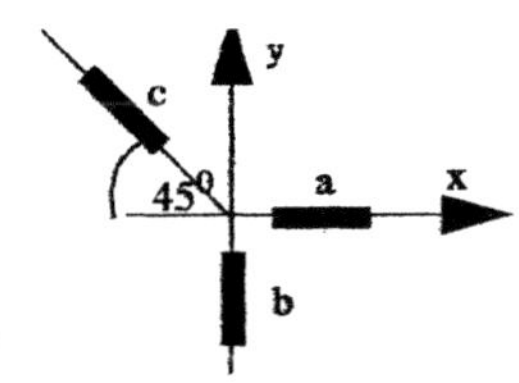

Fig. P9.64

Solution Plane Stress $\sigma_1 = ?$ $\sigma_2 = ?$ $\sigma_3 = ?$ $\theta_1 = ?$ $\tau_{max} = ?$

By inspection of Fig. P9.64 we obtain: $\varepsilon_{xx} = 200\mu\ m/m$ and $\varepsilon_{yy} = 100\mu\ m/m$

Substituting $\theta_c = 135^\circ$ the given strain values into Eq. 9.10 we obtain:

$$\varepsilon_c = (200)\cos^2(135) + (100)\sin^2(135) + \gamma_{xy}\sin(135)\cos(135) = 0 \quad \text{or} \quad \gamma_{xy} = 300\mu \text{ rad} \qquad 1$$

The shear modulus of elasticity is: $G = \frac{E}{2(1+\nu)} = \frac{210}{2(1+0.28)} = 82.03\text{GPa}$.

The shear stress is: $\tau_{xy} = G\gamma_{xy} = (82.03)(10^9)(300)(10^{-6}) = 24.61\text{MPa}$

From generalized Hooke's law, for plane stress ($\sigma_{zz} = 0$) we obtain:

$$\varepsilon_{xx} = \frac{\sigma_{xx}}{E} - \frac{\nu}{E}(\sigma_{yy}+\sigma_{zz}) \quad \text{or} \quad \sigma_{xx} - 0.28\sigma_{yy} = E\varepsilon_{xx} = (210)(10^9)(200)(10^{-6}) = 42\text{MPa} \qquad 2$$

$$\varepsilon_{yy} = \frac{\sigma_{yy}}{E} - \frac{\nu}{E}(\sigma_{xx}+\sigma_{zz}) \quad \text{or} \quad \sigma_{yy} - 0.28\sigma_{xx} = E\varepsilon_{yy} = (210)(10^9)(100)(10^{-6}) = 21\text{MPa} \qquad 3$$

Solving Eqs. 2 and 3 we obtain: $\sigma_{xx} = 51.95\text{MPa}$ and $\sigma_{yy} = 35.55\text{MPa}$

The principal angle θ_p is: $\theta_p = 0.5\ \text{atan}\left(\frac{(2)(24.61)}{51.95-35.55}\right) = 35.79^\circ$

Substituting θ_p and the stresses from Eq. 1 into Eq. 8.1, we obtain one principal stress as:

$$\sigma_p = (51.95)\cos^2(35.79) + (35.55)\sin^2(35.79) + 2(24.61)\sin(35.79)\cos(35.79) = 69.69\text{MPa} \qquad 4$$

From Eq. 8.9 we can find the other principal stress as: $51.95 + 35.55 - 69.69 = 17.81\text{MPa}$. As this stress is smaller than the stress in Eq. 4, therefore the stress in Eq. 4 is principal stress one. Thus principal angle 1 is $\theta_1 = \theta_p$. We have the following results:

$$\boxed{\sigma_1 = 69.7\text{MPa(T)} \qquad \sigma_2 = 17.8\text{MPa(T)} \qquad \sigma_3 = 0 \qquad \theta_1 = 35.79^\circ \text{ or } -144.2^\circ}$$

Check: From Eq. 8.6 we have: $\sigma_{1,2} = \left(\frac{51.95 + 35.55}{2}\right) \pm \sqrt{\left(\frac{51.95 - 35.55}{2}\right)^2 + (24.61)^2} = 43.75 \pm 25.94$ or

$\sigma_1 = 69.7\text{MPa}$ and $\sigma_2 = 17.8\text{MPa}$, which are same as above.

The difference between σ_1 and σ_3 is the largest difference between principal stresses. Thus, the magnitude of the maximum shear stress is given by:

$\tau_{max} = |(\sigma_1 - \sigma_3)/2| = |69.69/2|$ or $\boxed{\tau_{max} = 34.8\text{MPa}}$

9.65 On a free surface of an aluminum machine component (E = 10,000 ksi and ν = 0.25) the strains recorded by the three strain gages shown in Fig. P9.65 are

$\varepsilon_a = -100\ \mu\ \text{in/in} \qquad \varepsilon_b = 200\ \mu\ \text{in/in} \qquad \varepsilon_c = 300\ \mu\ \text{in/in}$

Determine the principal *strains*, principal angle one and the maximum shear *strain*.

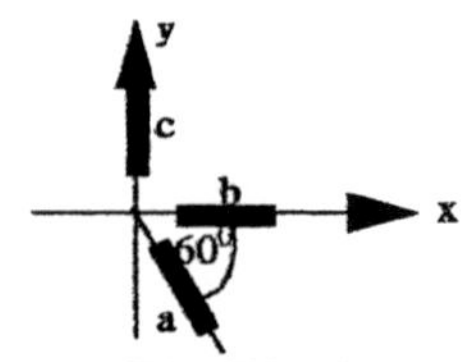

Fig. P9.65

Solution Plane Stress $\varepsilon_1 = ?$ $\varepsilon_2 = ?$ $\varepsilon_3 = ?$ $\theta_1 = ?$ $\gamma_{max} = ?$

By inspection of Fig. P9.65 we obtain: $\varepsilon_{xx} = 200\mu\ \text{m/m}$ and $\varepsilon_{yy} = 300\mu\ \text{m/m}$

Substituting $\theta_a = -60°$ the given strain values into Eq. 9.10 we obtain:

$$\varepsilon_a = (200)\cos^2(-60) + (300)\sin^2(-60) + \gamma_{xy}\sin(-60)\cos(-60) = -100 \quad \text{or} \quad \gamma_{xy} = 866\mu\ \text{rad} \qquad 1$$

From Eq. 9.13 we obtain the principal angle θ_p as:

$$\theta_p = 0.5\ \text{atan}\left(\frac{866}{200 - 300}\right) = -41.71^\circ \qquad 2$$

Substituting θ_p and the given strains into Eq. 9.10, we obtain one principal strain as:

$$\varepsilon_p = (200)\cos^2(-41.71) + (300)\sin^2(-41.71) + (866)\sin(-41.71)\cos(-41.71) = -185.9\mu \qquad 3$$

From Eq. 9.15 we can find the other principal strain as: $200 + 300 - (-185.9) = 685.9\mu$. As this strain is greater than the strain in Eq. 3, therefore this strain is principal strain one. Thus principal angle 1 is $\theta_1 = \theta_p \pm 90$ For we have the following answers:

$$\boxed{\varepsilon_1 = 685.9\mu \qquad \varepsilon_2 = -185.9\mu \qquad \theta_1 = 48.29^\circ \text{ or } -131.7^\circ}$$

Check: From Eq. 9.14 we have: $\varepsilon_{1,2} = \left(\frac{200 + 300}{2}\right) \pm \sqrt{\left(\frac{200 - 300}{2}\right)^2 + (433)^2} = 250 \pm 435.9$ or

$\varepsilon_1 = 685.9\mu$ and $\varepsilon_2 = -185.9\mu$, which are same as above.

For plane stress: $\varepsilon_3 = \varepsilon_z = -\frac{\nu}{(1-\nu)}(\varepsilon_{xx} + \varepsilon_{yy}) = -\frac{0.25}{(1-0.25)}(200 + 300)$ $\boxed{\varepsilon_3 = -167.7\mu}$

The difference between ε_1 and ε_2 is the largest difference between principal strains. Thus, the magnitude of the maximum shear strain is given by:

$\gamma_{max}/2 = |(\varepsilon_1 - \varepsilon_2)/2| = |(685.9 + 185.9)/2|$ or $\boxed{\gamma_{max} = 871.8\mu}$

9.66 On a free surface of an aluminum machine component (E = 10,000 ksi and ν = 0.25) the strains recorded by the three strain gages shown in Fig. P9.65 are

$\varepsilon_a = 500\ \mu\ \text{in/in} \qquad \varepsilon_b = 500\ \mu\ \text{in/in} \qquad \varepsilon_c = 500\ \mu\ \text{in/in}$

Determine the principal *stresses*, principal angle one and the maximum shear *stress*.

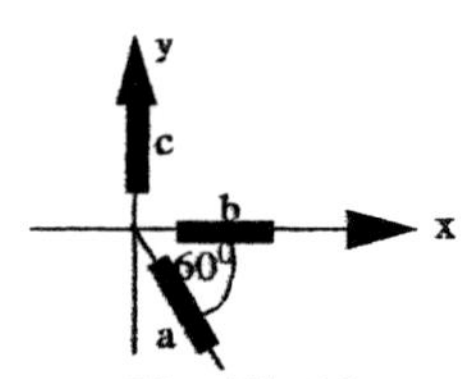

Fig. P9.66

Solution Plane Stress $\sigma_1 = ?$ $\sigma_2 = ?$ $\sigma_3 = ?$ $\theta_1 = ?$ $\tau_{max} = ?$

By inspection of Fig. P9.65 we obtain: $\varepsilon_{xx} = 500\mu \text{ m/m}$ and $\varepsilon_{yy} = 500\mu \text{ m/m}$

Substituting $\theta_a = -60°$ the given strain values into Eq. 9.10 we obtain:

$$\varepsilon_a = (500)\cos^2(-60) + (500)\sin^2(-60) + \gamma_{xy}\sin(-60)\cos(-60) = 500 \quad \text{or} \quad \gamma_{xy} = 0 \qquad 1$$

The shear modulus of elasticity is: $G = \frac{E}{2(1+\nu)} = \frac{10000}{2(1+0.25)} = 4000\text{ksi}$.

The shear stress is: $\tau_{xy} = G\gamma_{xy} = (4000)(0) = 0$

From generalized Hooke's law, for plane stress ($\sigma_{zz} = 0$) we obtain:

$$\varepsilon_{xx} = \frac{\sigma_{xx}}{E} - \frac{\nu}{E}(\sigma_{yy} + \sigma_{zz}) \quad \text{or} \quad \sigma_{xx} - 0.25\sigma_{yy} = E\varepsilon_{xx} = (10000)(500)(10^{-6}) = 5\text{ksi} \qquad 2$$

$$\varepsilon_{yy} = \frac{\sigma_{yy}}{E} - \frac{\nu}{E}(\sigma_{xx} + \sigma_{zz}) \quad \text{or} \quad \sigma_{yy} - 0.25\sigma_{xx} = E\varepsilon_{yy} = (10000)(500)(10^{-6}) = 5\text{ksi} \qquad 3$$

Solving Eqs. 2 and 3 we obtain: $\sigma_{xx} = 6.67\text{ksi}$ and $\sigma_{yy} = 6.67\text{ksi}$

As shear stress is zero, σ_{xx} and σ_{yy} are the principal stress. As these two stress components are the same, all planes have the same normal stress.We have the following results:

$$\boxed{\sigma_1 = 6.67\text{ksi(T)} \quad \sigma_2 = 6.67\text{ksi(T)} \quad \sigma_3 = 0 \quad \theta_1 = \text{all directions}}$$

The difference between σ_1 (or σ_2) and σ_3 is the largest difference between principal stresses. Thus, the magnitude of the maximum shear stress is given by:

$\tau_{max} = |(\sigma_1 - \sigma_3)/2| = |6.67/2|$ or $\boxed{\tau_{max} = 3.34\text{ksi}}$

9.67 An aluminum 50 mm x 50 mm square bar is axially loaded with a force of F = 100 kN. The Modulus of Elasticity is E = 70 GPa and the Shear Modulus G = 28 GPa. Determine the strain that will be recorded by a strain gage shown in Fig. P9.67.

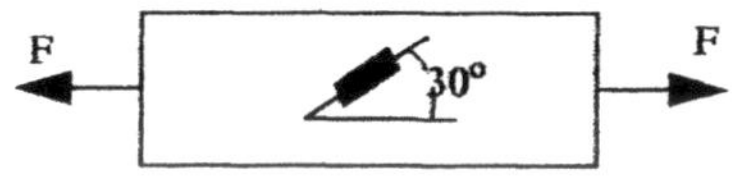

Fig. P9.67

Solution 50 mm x 50 mm F = 100 kN E = 70 GPa G = 28 GPa $\varepsilon_{30} = ?$

The axial stress is: $\sigma_{xx} = \frac{100(10^3)}{(0.05)(0.05)} = 40(10^6)\text{N/m}^2$

The axial strain is: $\varepsilon_{xx} = \frac{\sigma_{xx}}{E} = \frac{40(10^6)}{70(10^9)} = 571.4\mu$

The Poisson's ratio is: $\nu = \frac{E}{2G} - 1 = \frac{70}{2(28)} - 1 = 0.25$

The transverse strain is: $\varepsilon_{yy} = -\nu\varepsilon_{xx} = -0.25(571.4) = -142.85\mu$

The shear strain is $\gamma_{xy} = 0$. Substituting the strains and $\theta = 30°$ into Eq. 9.10, we obtain

$\varepsilon_{30} = (571.4)\cos^2(30) + (-142.85)\sin^2(30)$ or $\boxed{\varepsilon_{30} = 392.9\mu}$

9.68 A circular steel bar is axially loaded. The diameter of the steel bar is 2 inches. and has Modulus of Elasticity E = 30,000 ksi and a Poisson's ratio of ν=0.3. Determine the applied axial force F when the strain gage shown in Fig. P9.68 records a reading of 1000 μ in/in.

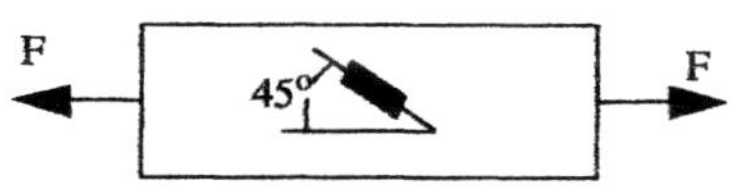

Fig. P9.68

Solution $d = 2$ in $E = 30000$ ksi $\nu = 0.3$ $\varepsilon_{45} = 1000\ \mu$ $F = ?$

The axial stress is: $\sigma_{xx} = \dfrac{F}{(\pi 2^2)/4} = 0.3183F$ ksi

The axial strain is: $\varepsilon_{xx} = \dfrac{\sigma_{xx}}{E} = \dfrac{0.3183F}{30000} = 10.61F\mu$

The transverse strain is: $\varepsilon_{yy} = -\nu\varepsilon_{xx} = -0.3(10.61F) = -3.183F\mu$

The shear strain is $\gamma_{xy} = 0$. Substituting the strains and $\theta = -45°$ into Eq. 9.10, we obtain

$\varepsilon_{45} = [(10.61F)\cos^2(-45) + (-3.183F)\sin^2(-45)](10^{-6}) = 3.716F(10^{-6}) = 1000(10^{-6})$ or $\boxed{F = 269\text{kips}}$

9.69 A circular shaft of 2 inch diameter has a torque applied to it as shown in Fig. P9.69. The shaft material has a Modulus of Elasticity of 30,000 ksi and a Poisson's ratio of 0.3. Determine the strain that will be recorded by a strain gage.

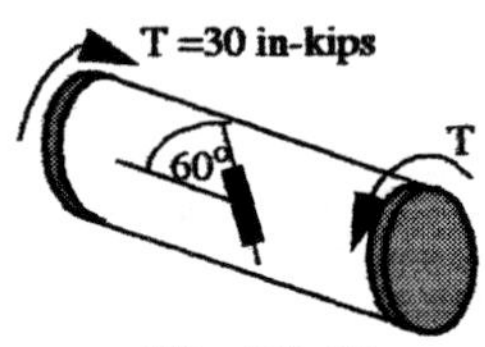

Fig. P9.69

Solution $d = 2$ in $E = 30000$ ksi $\nu = 0.3$ $\varepsilon_{60} = ?$

The magnitude of torsional shear stress is: $\tau = \dfrac{30(1)}{\pi(2^4)/32} = 19.1\text{ksi}$

Fig. (a) shows the direction of shear stress determined by inspection. In the x-y coordinate we have

$$\tau_{xy} = -19.1\text{ksi}$$ 1

(a)

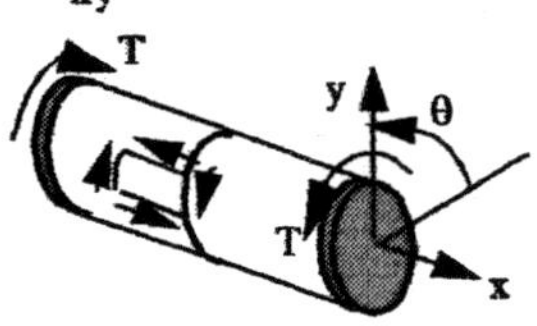

The shear modulus of elasticity is $G = \dfrac{E}{2(1+\nu)} = \dfrac{30000}{2(1.3)} = 11538.5\text{ksi}$

The shear strain is: $\gamma_{xy} = \dfrac{\tau_{xy}}{G} = -\dfrac{19.1}{11538.5} = -1655.3\mu$

The normal strains are: $\varepsilon_{xx} = 0$ $\varepsilon_{yy} = 0$. Substituting the strains and $\theta = -60°$, we obtain

$\varepsilon_{60} = (-1655.3)\sin(-60)\cos(-60) = 716.77$ or $\boxed{\varepsilon_{60} = 716.8\mu}$

9.70 A circular shaft of 50 mm diameter has a torque applied to it as shown in Fig. P9.70. The shaft material has a Modulus of Elasticity of E = 70 GPa and the Shear Modulus of G = 28 GPa. If the strain gage shows a reading of -600 μ, determine the applied torque T.

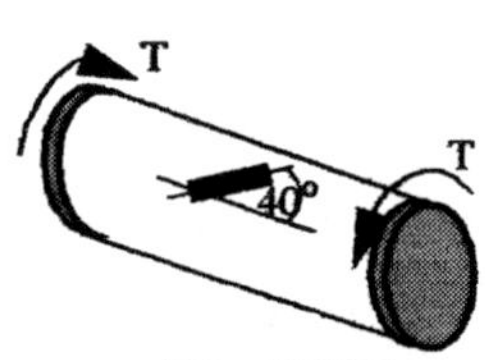

Fig. P9.70

Solution $d = 50$ mm $E = 70$ GPa $G = 28$ GPa $\varepsilon_{40} = -600\ \mu$ $T = ?$

The magnitude of torsional shear stress is: $\tau = \dfrac{T(0.025)}{\pi(0.05^4)/32} = 40.743T(10^3)\text{N/m}^2$

Fig. (a) shows the direction of shear stress determined by inspection. In the x-y coordinate we have

$$\tau_{xy} = -40.743T(10^3)N/m^2 \qquad 1$$

(a)

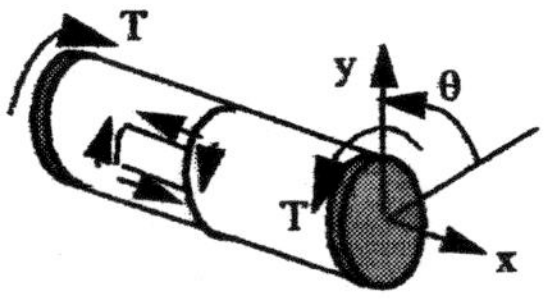

The shear strain is: $\gamma_{xy} = \frac{\tau_{xy}}{G} = -\frac{40.743T(10^3)}{28(10^9)} = -1.4551T\mu$

The normal strains are: $\varepsilon_{xx} = 0 \qquad \varepsilon_{yy} = 0$. Substituting the strains and $\theta = 40^o$, we obtain

$\varepsilon_{40} = (-1.4551T)\sin(40)\cos(40) = -600$ or $\boxed{T = 837.4\ N-m}$

9.71 A steel (E = 210GPa and ν= 0.28) cylindrical pressure vessel has a mean diameter of 1000 mm. The wall of the cylinder is 10 mm thick and the gas pressure is 200 kPa. Determine the strain recorded by the two strain gages attached on the surface of the cylinder.

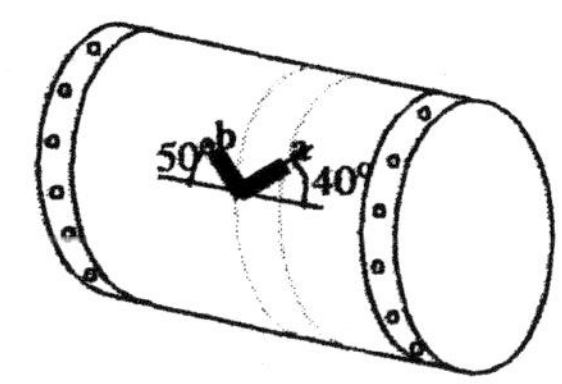

Fig. P9.71

Solution d = 1000 mm t = 10 mm p = 200 kPa E = 210GPa ν= 0.28 $\varepsilon_a = ?\quad \varepsilon_b = ?$

The axial stress is: $\sigma_{xx} = \frac{pr}{2t} = \frac{(200)(10^3)(0.5)}{2(0.01)} = 5MPa$

The hoop stress is: $\sigma_{yy} = \frac{pr}{t} = \frac{(200)(10^3)(0.5)}{(0.01)} = 10MPa$

The normal strains are:

$$\varepsilon_{xx} = \frac{\sigma_{xx}}{E} - \frac{\nu}{E}(\sigma_{yy} + \sigma_{zz}) = \frac{5(10^6)}{210(10^9)} - \frac{0.28(10)(10^6)}{210(10^9)} = 10.476\mu \qquad 2$$

$$\varepsilon_{yy} = \frac{\sigma_{yy}}{E} - \frac{\nu}{E}(\sigma_{xx} + \sigma_{zz}) = \frac{(10)(10^6)}{210(10^9)} - \frac{0.28(5)(10^6)}{210(10^9)} = 40.952\mu \qquad 3$$

The shear stress $\tau_{xy} = 0$, hence $\gamma_{xy} = 0$.

Substituting the strains and $\theta_a = 40^o$ and $\theta_b = -50^o$ into Eq. 9.10 we obtain the following.

$\varepsilon_a = (10.476)\cos^2 40 + (40.952)\sin^2 40$ or $\boxed{\varepsilon_a = 23.1\mu}$

$\varepsilon_b = (10.476)\cos^2(-50) + (40.952)\sin^2(-50)$ or $\boxed{\varepsilon_b = 28.4\mu}$

9.72—9.73 An aluminum (E = 70 GPa, and ν=0.25) beam is loaded by a force P and moment M at the free end as shown in Fig. P9.73. The two strain gages shown are at an angle of 25° to the longitudinal axis. Use this information to solve the two problems below.

9.72 If P= 10kN and M = 5kN-m, determine the strains in the gages

9.73 Determine the applied force P and applied moment M if the strain gages recorded the following strains:

$$\varepsilon_a = -386\ \mu\ m/m \qquad \varepsilon_b = 4092\ \mu m/m$$

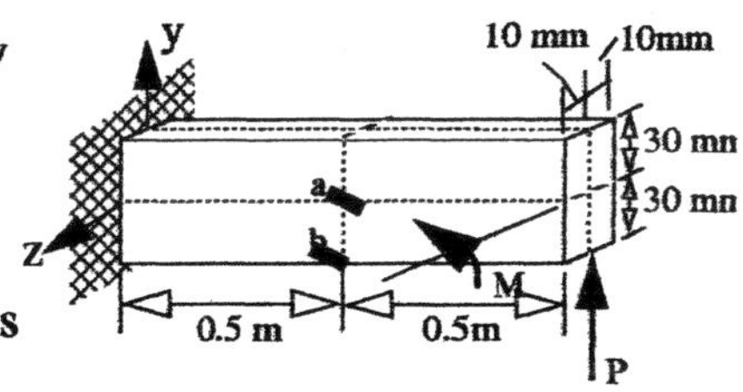

Fig. P9.73

Solution E = 70 GPa ν=0.25

The area moment of inertia is: $I_{zz} = \frac{(0.02)(0.06^3)}{12} = 0.36(10^{-6})m^4$

Fig. (a) shows the free body diagram after an imaginary cut through the section containing the gages.

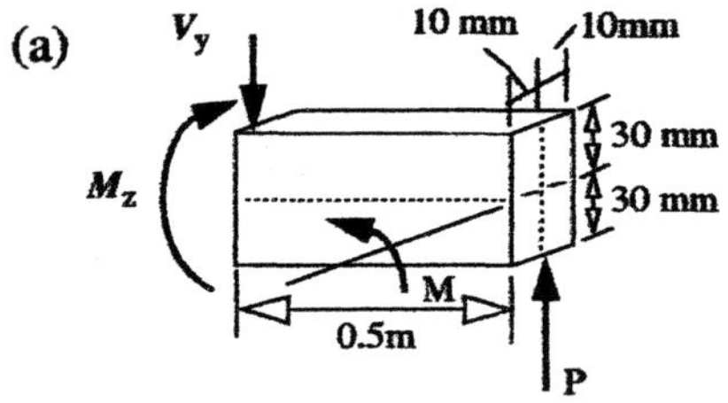

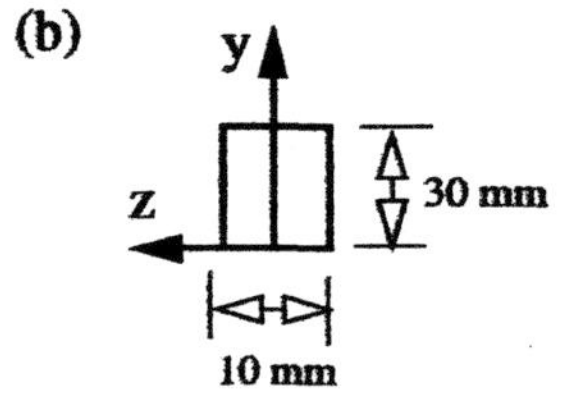

By equilibrium of forces and moment in Fig. (a) we obtain:

$$V_y = P \qquad M_z = M + 0.5P \tag{1}$$

The bending normal stresses at the two gage locations are:

$$(\sigma_{xx})_a = 0 \qquad (\sigma_{xx})_b = \frac{-M_z y_b}{I_{zz}} = \frac{-(M+0.5P)(-0.03)}{0.36(10^{-6})} = 83.33(M+0.5P)(10^3) \tag{2}$$

Fig. (b) shows the area As used in calculating the shear stress at the neutral axis. The first moment of the area is: $(Q_z)_a = (0.01)(0.03)(0.015) = 4.5(10^{-6})m^3$

As the shear force V_y is positive, the bending shear stress τ_{xy} will be positive at the neutral axis. The bending shear stresses at the two gage locations is:

$$(\tau_{xy})_a = \left|\frac{V_y(Q_z)_a}{I_{zz}t}\right| = \frac{(P)(4.5)(10^{-6})}{0.36(10^{-6})(0.01)} = 1250P \qquad (\tau_{xy})_b = 0 \tag{3}$$

The shear modulus of elasticity is: $G = E/[2(1+\nu)] = 70/[2(1+0.25)] = 28GPa$

The bending normal and shear strains at the gage location are:

$$(\varepsilon_{xx})_a = 0 \qquad (\varepsilon_{yy})_a = 0 \qquad (\gamma_{xy})_a = \frac{(\tau_{xy})_a}{G} = \frac{1250P}{28(10^9)} = 44.643P(10^{-9}) \tag{4}$$

$$(\varepsilon_{xx})_b = \frac{(\sigma_{xx})_b}{E} = \frac{83.33(M+0.5P)(10^3)}{70(10^9)} = 1.1904(M+0.5P)(10^{-6}) \tag{5}$$

$$(\varepsilon_{yy})_b = -\nu(\varepsilon_{xx})_b = -0.25(\varepsilon_{xx})_b \qquad (\gamma_{xy})_a = 0 \tag{6}$$

Problem 9-72 Substituting $\theta = -25°$ and the above strains in Eq. 9.10 we obtain the following

$$\varepsilon_a = 44.643P(10^{-9})\sin(-25)\cos(-25) = -17.0993P(10^{-9}) \tag{7}$$

$\varepsilon_b = (\varepsilon_{xx})_b\cos^2(-25) + [-0.25(\varepsilon_{xx})_b]\sin^2(-25) = 0.7767(\varepsilon_{xx})_b = 0.7767[1.1904(M+0.5P)(10^{-6})]$ or

$$\varepsilon_b = 0.9246(M+0.5P)(10^{-6}) \tag{8}$$

Substituting P= 10kN and M = 5kN-m in Eq. 7, we obtain:

$\varepsilon_a = -17.0993(10)(10^3)(10^{-9}) = -170.99(10^{-6})$ $\boxed{\varepsilon_a = -171\mu}$

$\varepsilon_b = 0.9246[5+0.5(10)](10^3)(10^{-6}) = 9.246(10^{-3})$ $\boxed{\varepsilon_b = 9246\mu}$

Problem 9-73 We are given: $\varepsilon_a = -386\ \mu\ m/m$ $\varepsilon_b = 4092\ \mu m/m$. Substituting $\theta = -30°$ and the strains in Eqs. 4,5, and 6 into Eq. 9.10 and equating to the given strain gage values, we obtain:

$\varepsilon_a = 44.643P(10^{-9})\sin(-30)\cos(-30) = -19.331P(10^{-9}) = -386(10^{-6})$ or $\boxed{P = 19.97kN}$

$\varepsilon_b = (\varepsilon_{xx})_b\cos^2(-30) + [-0.25(\varepsilon_{xx})_b]\sin^2(-30) = 0.6875(\varepsilon_{xx})_b = 0.6875[1.1904(M + 0.5P)(10^{-6})]$ or

$\varepsilon_b = 0.8184(M + 0.5P)(10^{-6}) = 4092(10^{-6})$ or $M + 0.5P = 5000$

Substituting P=19.97(10^3) we obtain: $M = -4985N - m$ or $\boxed{M = -4.985kN - m}$

9.74 A steel (E = 210 GPa and ν=0.28) rod of diameter 50 mm is loaded by axial forces P = 100 kN as shown in Fig. P9.75. Determine the strain that will be recorded by the strain gage.

9.75 The strain gage mounted on the surface of a solid axial steel (E = 210 GPa and ν=0.28) rod in Fig. P9.75 showed a strain of -214 μ m/m. If the diameter of the shaft is 50 mm, determine the applied axial force P.

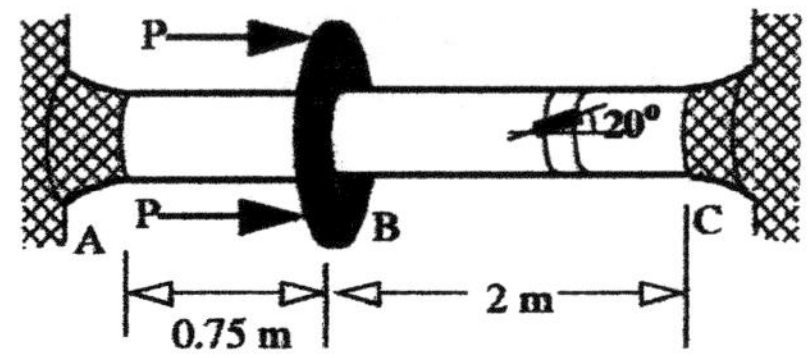

Fig. P9.75

Solution E = 210 GPa ν=0.28 d = 50 mm

Fig.(a) shows the free body diagram of the rigid plate.

(a)

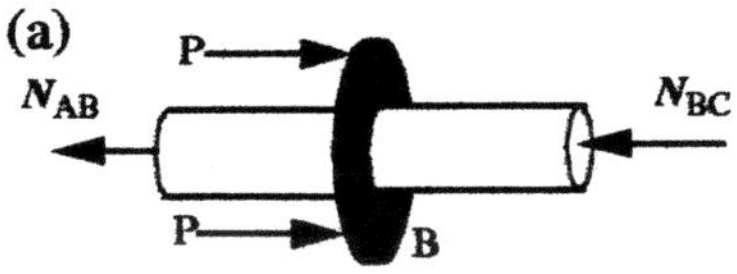

By equilibrium of forces we obtain:

$$N_{AB} + N_{BC} = 2P \quad (1)$$

The deformation of the two bars is the same as shown below.

$$\delta_{AB} = \delta_{BC} \quad (2)$$

The cross-sectional area of the rods is: $A = \pi(0.05)^2/4 = 1.963(10^{-3})$

The deformation of the bars can be written as:

$$\delta_{AB} = \frac{N_{AB}L_{AB}}{EA} = \frac{N_{AB}(0.75)}{210(10^9)(1.963)(10^{-3})} = 1.8189N_{AB}(10^{-9}) \quad (3)$$

$$\delta_{BC} = \frac{N_{BC}L_{BC}}{EA} = \frac{N_{BC}(2)}{210(10^9)(1.963)(10^{-3})} = 4.8504N_{BC}(10^{-9}) \quad (4)$$

Substituting Eqs. 3 and 4 into Eq. 2, we obtain:

$$1.8189N_{AB}(10^{-9}) = 4.8504N_{BC}(10^{-9}) \quad \text{or} \quad N_{BC} = 0.375N_{AB} \quad (5)$$

Solving Eqs. 1 and 5 we obtain:

$$N_{AB} = 1.4545P \quad \text{and} \quad N_{BC} = 0.5455P \quad (6)$$

The axial stress in segment BC is: $\sigma_{xx} = N_{BC}/A = 0.5455P/[1.963(10^{-3})] = 0.2779P(10^3)\ \frac{N}{m^2}(C)$

The axial strain in segment BC is: $\varepsilon_{xx} = \sigma_{xx}/E = -0.2779P(10^3)/[210(10^9)] = -1.3232P(10^{-9})$

The transverse strain in segment BC is: $\varepsilon_{yy} = -\nu\varepsilon_{xx} = -0.28[-1.3232P(10^{-9})] = 0.3705P(10^{-9})$

The shear strain $\gamma_{xy} = 0$. Substituting the strains in BC and θ = 20° into Eq. 9.10, we obtain the strain in the gage as:

$$\varepsilon_{20} = [-1.3232P(10^{-9})]\cos^2(20) + [0.3705P]\sin^2(20) = -1.1251P(10^{-9}) \quad (7)$$

Problem 9.74: Substituting $P = 100(10^3)$ N into Eq. 7, we obtain:

$\varepsilon_{20} = -1.1251(100)(10^3)(10^{-9})$ or $\boxed{\varepsilon_{20} = -112.5\mu}$

Problem 9.75: Substituting $\varepsilon_{20} = -214(10^{-6})$ into Eq. 7, we obtain:

$\varepsilon_{20} = -1.1251P(10^{-9}) = -214(10^{-6})$ or $\boxed{P = 190.2\text{kN}}$

9.76 A steel (E = 210 GPa and ν=0.28) shaft of diameter 50 mm is loaded by a torque of T = 100 kN-m as shown in Fig. P9.77. Determine the strain that will be recorded by the strain gage.

9.77 The strain gage mounted on the surface of a solid steel (E = 210 GPa and ν=0.28) shaft shown in Fig. P9.77 recorded a strain of 1088 μ. If the diameter of the shaft is 50 mm, determine

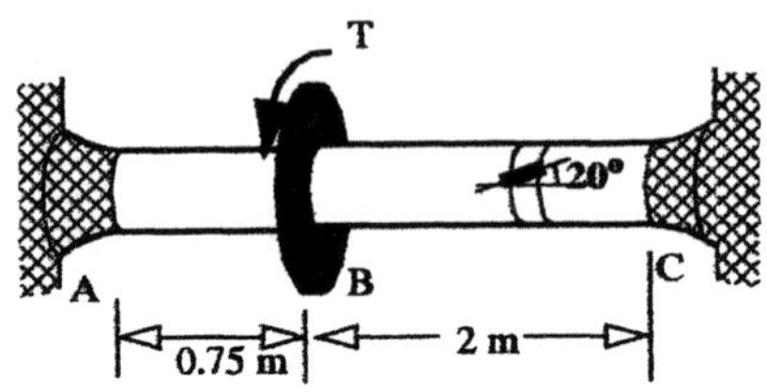

Solution E = 210 GPa ν=0.28 d = 50 mm

The polar moment of inertia is: $J = \pi(0.05)^4/32 = 0.6136(10^{-6})\text{m}^4$

Fig.(a) shows the free body diagram of the shaft after imaginary cuts are made in section AB and BC. By moment equilibrium we obtain:

$$T_{AB} = T_A \qquad T_{BC} = T_A - T \tag{1}$$

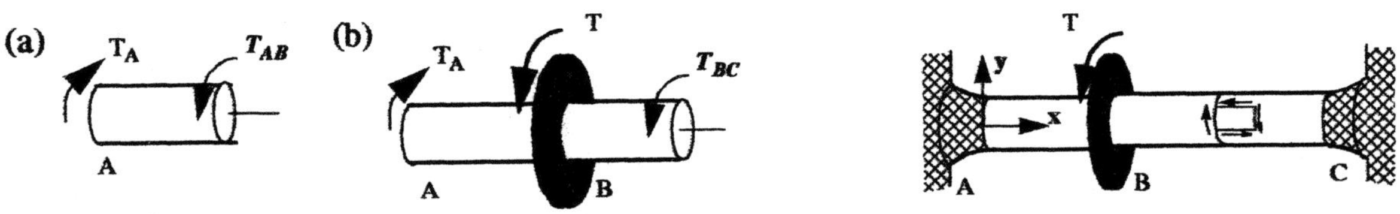

The relative rotation of the ends can be written as:

$$\phi_B - \phi_A = \frac{T_{AB}(x_B - x_A)}{GJ} = \frac{(T_A)(0.75)}{GJ} \tag{2}$$

$$\phi_C - \phi_B = \frac{T_{BC}(x_C - x_B)}{GJ} = \frac{(T_A - T)(2)}{GJ} \tag{3}$$

Adding Eqs. 2 and 3 and noting that $\phi_C = 0$ and $\phi_A = 0$ we obtain:

$$\phi_C - \phi_A = \frac{(T_A)(0.75)}{GJ} + \frac{(T_A - T)(2)}{GJ} = 0 \quad \text{or} \quad T_A = \frac{2T}{2.75} = 0.7273T \tag{4}$$

From Eq. 1 we obtain: $T_{AB} = 0.7273T$ $\quad T_{BC} = -0.2727T$.

The magnitude of torsional shear stress on the surface of segment BC is:

$$\tau = \left|\frac{T_{BC}\rho_{max}}{J}\right| = \frac{(0.2727T)(0.025)}{0.6136(10^{-6})} = 11.11T(10^3)\frac{N}{m^2}$$

Fig. (c) shows direction of torsional shear stress that is determined by inspection. In the x, y coordinates the shear stress is negative. We obtain: $\tau_{xy} = -11.11T(10^3)\frac{N}{m^2}$.

The shear modulus of elasticity is: $G = E/[2(1+\nu)] = 210/[2(1+0.28)] = 82.03\text{GPa}$

The shear strain is: $\gamma_{xy} = \tau_{xy}/G = -11.11T(10^3)/[82.03(10^9)] = -0.1354T(10^{-6}) = -0.1354T\mu$

The normal strains in the x, y coordinate are zero. Substituting the strains and θ = 20° into Eq. 9.10, we obtain the strain in the gage as:

$$\varepsilon_{20} = -0.1354T\cos(20)\sin(20) = -0.04352T\mu \qquad 5$$

Problem 9.76: Substituting T = 100(10^3) N-m into Eq. 5, we obtain:

$\varepsilon_{20} = -0.04352(100)(10^3)\mu$ or $\boxed{\varepsilon_{20} = -4352\mu}$

Problem 9.76: Substituting $\varepsilon_{20} = 1088\ \mu$ into Eq. 5, we obtain

$\varepsilon_{20} = 0.04352T\mu = 1088\mu$ or $\boxed{T = -250\text{kN}-\text{m}}$

9.78 Starting with $\varepsilon_{nn} = \dfrac{\partial u_n}{\partial n}$, using Equations 9.24a through 9.24d and chain rule for differentiation, derive Equation 9.10.

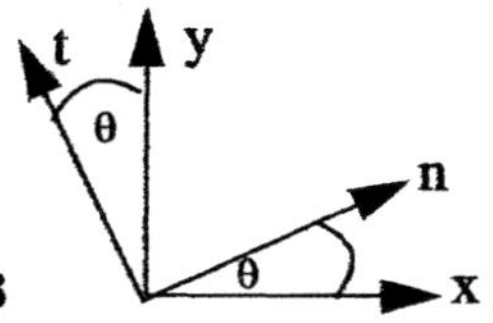

Fig. P9.78

$n = x\cos\theta + y\sin\theta$ **9.24a** $\qquad u_n = u\cos\theta + v\sin\theta$ **9.24c**

$t = -x\sin\theta + y\cos\theta$ **9.24b** $\qquad u_t = -u\sin\theta + v\cos\theta$ **9.24d**

Solution

By chain rule we have

$$\varepsilon_{nn} = \frac{\partial u_n}{\partial n} = \left(\frac{\partial u_n}{\partial x}\right)\frac{\partial x}{\partial n} + \left(\frac{\partial u_n}{\partial y}\right)\frac{\partial y}{\partial n} \qquad 1$$

Eqs. 9.24a and b can be solved for x and y and written as:

$$x = n\cos\theta - t\sin\theta \qquad y = n\sin\theta + t\cos\theta \qquad 2$$

From Eq. 9.24c we obtain the following:

$$\frac{\partial u_n}{\partial x} = \frac{\partial u}{\partial x}\cos\theta + \frac{\partial v}{\partial x}\sin\theta \qquad \frac{\partial u_n}{\partial y} = \frac{\partial u}{\partial y}\cos\theta + \frac{\partial v}{\partial y}\sin\theta \qquad 3$$

From Eq. 2 we obtain the following:

$$\frac{\partial x}{\partial n} = \cos\theta \qquad \frac{\partial y}{\partial n} = \sin\theta \qquad 4$$

Substituting Eq. 3 and 4 into Eq. 1

$$\varepsilon_{nn} = \left(\frac{\partial u}{\partial x}\cos\theta + \frac{\partial v}{\partial x}\sin\theta\right)(\cos\theta) + \left(\frac{\partial u}{\partial y}\cos\theta + \frac{\partial v}{\partial y}\sin\theta\right)(\sin\theta) = \frac{\partial u}{\partial x}\cos^2\theta + \frac{\partial v}{\partial y}\sin^2\theta + \left(\frac{\partial u}{\partial y} + \frac{\partial v}{\partial x}\right)\cos\theta\sin\theta \qquad 5$$

Substituting Eqs. 9.24a through 9.24d into Eq.5 we obtain Eq. 9.10.

9.79 Starting with $\varepsilon_{tt} = \dfrac{\partial u_t}{\partial t}$, using Equations 9.24a through 9.24d and chain rule for differentiation, derive Equation 9.11.

Solution

By chain rule we have

$$\varepsilon_{tt} = \frac{\partial u_t}{\partial t} = \left(\frac{\partial u_t}{\partial x}\right)\frac{\partial x}{\partial t} + \left(\frac{\partial u_t}{\partial y}\right)\frac{\partial y}{\partial t} \qquad 1$$

Eqs. 9.24a and b can be solved for x and y and written as:

$$x = n\cos\theta - t\sin\theta \qquad y = n\sin\theta + t\cos\theta \qquad 2$$

From Eqs. 9.24d we obtain the following:

$$\frac{\partial u_t}{\partial x} = -\frac{\partial u}{\partial x}\sin\theta + \frac{\partial v}{\partial x}\cos\theta \qquad \frac{\partial u_t}{\partial y} = -\frac{\partial u}{\partial y}\sin\theta + \frac{\partial v}{\partial y}\cos\theta \qquad 3$$

From Eq. 2 we obtain the following:

$$\frac{\partial x}{\partial t} = -\sin\theta \qquad \frac{\partial y}{\partial t} = \cos\theta \qquad 4$$

Substituting Eq. 3 and 4 into Eq. 1

$$\varepsilon_{tt} = \left(-\frac{\partial u}{\partial x}\sin\theta + \frac{\partial v}{\partial x}\cos\theta\right)(-\sin\theta) + \left(-\frac{\partial u}{\partial y}\sin\theta + \frac{\partial v}{\partial y}\cos\theta\right)(\cos\theta) = \frac{\partial u}{\partial x}\sin^2\theta + \frac{\partial v}{\partial y}\cos^2\theta - \left(\frac{\partial u}{\partial y} + \frac{\partial v}{\partial x}\right)\cos\theta\sin\theta \qquad 5$$

Substituting Eqs. 9.24a through 9.24d into Eq.5 we obtain Eq. 9.11.

9.80 Starting with $\gamma_{nt} = \frac{\partial u_t}{\partial n} + \frac{\partial u_n}{\partial t}$, using Equations 9.24a through 9.24d and chain rule for differentiation, derive Equation 9.12.

Solution

By chain rule we have

$$\frac{\partial u_t}{\partial n} = \left(\frac{\partial u_t}{\partial x}\right)\frac{\partial x}{\partial n} + \left(\frac{\partial u_t}{\partial y}\right)\frac{\partial y}{\partial n} \qquad \frac{\partial u_n}{\partial t} = \left(\frac{\partial u_n}{\partial x}\right)\frac{\partial x}{\partial t} + \left(\frac{\partial u_n}{\partial y}\right)\frac{\partial y}{\partial t} \qquad 1$$

Eqs. 9.24a and b can be solved for x and y and written as:

$$x = n\cos\theta - t\sin\theta \qquad y = n\sin\theta + t\cos\theta \qquad 2$$

From Eqs. 9.24d we obtain the following:

$$\frac{\partial u_t}{\partial x} = -\frac{\partial u}{\partial x}\sin\theta + \frac{\partial v}{\partial x}\cos\theta \qquad \frac{\partial u_t}{\partial y} = -\frac{\partial u}{\partial y}\sin\theta + \frac{\partial v}{\partial y}\cos\theta \qquad 3$$

From Eq. 2 we obtain the following:

$$\frac{\partial x}{\partial n} = \cos\theta \qquad \frac{\partial y}{\partial n} = \sin\theta \qquad 4$$

Substituting Eqs. 3 and 4 into Eq. 1, we obtain:

$$\frac{\partial u_t}{\partial n} = \left(-\frac{\partial u}{\partial x}\sin\theta + \frac{\partial v}{\partial x}\cos\theta\right)\cos\theta + \left(-\frac{\partial u}{\partial y}\sin\theta + \frac{\partial v}{\partial y}\cos\theta\right)\sin\theta = \left(-\frac{\partial u}{\partial x} + \frac{\partial v}{\partial y}\right)\sin\theta\cos\theta + \frac{\partial v}{\partial x}\cos^2\theta - \frac{\partial u}{\partial y}\sin^2\theta \qquad 5$$

From Eq. 9.24c we obtain the following:

$$\frac{\partial u_n}{\partial x} = \frac{\partial u}{\partial x}\cos\theta + \frac{\partial v}{\partial x}\sin\theta \qquad \frac{\partial u_n}{\partial y} = \frac{\partial u}{\partial y}\cos\theta + \frac{\partial v}{\partial y}\sin\theta \qquad 6$$

From Eq. 2 we obtain the following:

$$\frac{\partial x}{\partial t} = -\sin\theta \qquad \frac{\partial y}{\partial t} = \cos\theta \qquad 7$$

Substituting Eqs. 6 and 7 into Eq. 1, we obtain

$$\frac{\partial u_n}{\partial t} = \left(\frac{\partial u}{\partial x}\cos\theta + \frac{\partial v}{\partial x}\sin\theta\right)(-\sin\theta) + \left(\frac{\partial u}{\partial y}\cos\theta + \frac{\partial v}{\partial y}\sin\theta\right)(\cos\theta) = \left(-\frac{\partial u}{\partial x} + \frac{\partial v}{\partial y}\right)\sin\theta\cos\theta + \frac{\partial u}{\partial y}\cos^2\theta - \frac{\partial v}{\partial x}\sin^2\theta \qquad 8$$

Adding Eqs. 5 and 8, we obtain:

$$\gamma_{nt} = \frac{\partial u_t}{\partial n} + \frac{\partial u_n}{\partial t} = 2\left(-\frac{\partial u}{\partial x} + \frac{\partial v}{\partial y}\right)\sin\theta\cos\theta + \left(\frac{\partial u}{\partial y} + \frac{\partial v}{\partial x}\right)(\cos^2\theta - \sin^2\theta) \qquad 9$$

Substituting Eqs. 9.24a through 9.24d into Eq.9 we obtain Eq. 9.12.

9.81 Starting from Equation 9.22, show that for isotropic materials $E_x = E_y$ and $G_{xy} = E_x/[2(1+\nu)]$.

Solution

Eq. 9.22 is written below:

$$\varepsilon_{xx} = \frac{\sigma_{xx}}{E_x} - \frac{\nu_{yx}}{E_y}\sigma_{yy} \qquad \varepsilon_{yy} = \frac{\sigma_{yy}}{E_y} - \frac{\nu_{xy}}{E_x}\sigma_{xx} \qquad \gamma_{xy} = \frac{\tau_{xy}}{G_{xy}} \qquad \frac{\nu_{yx}}{E_y} = \frac{\nu_{xy}}{E_x} \tag{1}$$

Consider the following state of stress:

$$\sigma_{xx} = \sigma \qquad \sigma_{yy} = -\sigma \qquad \tau_{xy} = 0 \tag{2}$$

Substituting Eq. 2 into Eqs. 8.1 and 8.8, we obtain:

$$\sigma_{nn} = \sigma\cos^2\theta + (-\sigma)\sin^2\theta = \sigma\cos 2\theta \tag{3}$$

$$\sigma_{tt} = \sigma\sin^2\theta + (-\sigma)\cos^2\theta = -\sigma\cos 2\theta \tag{4}$$

Substituting Eq. 2 into Eq. 1, we obtain:

$$\varepsilon_{xx} = \left(\frac{1}{E_x} + \frac{\nu_{yx}}{E_y}\right)\sigma \qquad \varepsilon_{yy} = -\left(\frac{1}{E_y} + \frac{\nu_{xy}}{E_x}\right)\sigma \qquad \gamma_{xy} = 0 \tag{5}$$

Substituting Eq. 5 into Eqs. 9.10 we obtain:

$$\varepsilon_{nn} = \left[\left(\frac{1}{E_x} + \frac{\nu_{yx}}{E_y}\right)\sigma\right]\cos^2\theta + \left[-\left(\frac{1}{E_y} + \frac{\nu_{xy}}{E_x}\right)\sigma\right]\sin^2\theta \text{ or}$$

$$\varepsilon_{nn} = \left[\left(\frac{1}{E_x} + \frac{\nu_{yx}}{E_y}\right)\sigma\right]\frac{(1+\cos 2\theta)}{2} + \left[-\left(\frac{1}{E_y} + \frac{\nu_{xy}}{E_x}\right)\sigma\right]\frac{(1-\cos 2\theta)}{2} \text{ or}$$

$$\varepsilon_{nn} = \left(\frac{1}{E_x} - \frac{1}{E_y} + \frac{\nu_{yx}}{E_y} - \frac{\nu_{xy}}{E_x}\right)\frac{\sigma}{2} + \left[\frac{1}{E_x} + \frac{1}{E_y} + \frac{\nu_{yx}}{E_y} + \frac{\nu_{xy}}{E_x}\right]\frac{\sigma\cos 2\theta}{2} \tag{6}$$

Substituting Eqs. 3, 4, and 6 into Generalized Hooke's law for isotropic materials in plane stress we obtain:

$$\varepsilon_{nn} = \frac{\sigma_{nn}}{E} - \frac{\nu\sigma_{tt}}{E} = \frac{(1+\nu)}{E}\sigma\cos 2\theta = \left(\frac{1}{E_x} - \frac{1}{E_y} + \frac{\nu_{yx}}{E_y} - \frac{\nu_{xy}}{E_x}\right)\frac{\sigma}{2} + \left[\frac{1}{E_x} + \frac{1}{E_y} + \frac{\nu_{yx}}{E_y} + \frac{\nu_{xy}}{E_x}\right]\frac{\sigma\cos 2\theta}{2} \tag{7}$$

If Eq. 7 is to be valid for any angle θ, then the following conditions must be satisfied.

$$\left(\frac{1}{E_x} - \frac{1}{E_y} + \frac{\nu_{yx}}{E_y} - \frac{\nu_{xy}}{E_x}\right) = 0 \tag{8}$$

$$\frac{1}{2}\left[\frac{1}{E_x} + \frac{1}{E_y} + \frac{\nu_{yx}}{E_y} + \frac{\nu_{xy}}{E_x}\right] = \frac{(1+\nu)}{E} \tag{9}$$

Noting that $\frac{\nu_{yx}}{E_y} = \frac{\nu_{xy}}{E_x}$, Eq. 9 implies $E_x = E_y$, which in turn implies $\nu_{yx} = \nu_{xy}$. Substituting these identities into Eq. 10, we obtain $\frac{2}{2}\left[\frac{1}{E_x} + \frac{\nu_{xy}}{E_x}\right] = \frac{(1+\nu)}{E}$, which implies

$$\frac{E}{(1+\nu)} = \frac{E_x}{(1+\nu_{xy})} \tag{10}$$

We now consider another state of stress

$$\sigma_{xx} = 0 \qquad \sigma_{yy} = 0 \qquad \tau_{xy} = \tau \tag{11}$$

From Eq. 8.2, we obtain:

$$\tau_{nt} = \tau(\cos^2\theta - \sin^2\theta) = \tau\cos 2\theta \tag{12}$$

Substituting Eq. 11 into Eq. 1, we obtain

$$\varepsilon_{xx} = 0 \qquad \varepsilon_{yy} = 0 \qquad \gamma_{xy} = \tau/G_{xy} \tag{13}$$

Substituting Eq. 13 into Eq. 9.12, we obtain:

$$\gamma_{nt} = \tau(\cos^2\theta - \sin^2\theta)/G_{xy} = \tau\cos 2\theta/G_{xy} \tag{14}$$

Substituting Eqs. 12 and 14 into the generalized Hooke's law for isotropic material, we obtain:

$$\tau_{nt} = G\gamma_{nt} = \frac{E}{2(1+\nu)}\gamma_{nt} \quad \text{or} \quad \tau\cos 2\theta = \frac{E}{2(1+\nu)}\left[\frac{\tau}{G_{xy}}\cos 2\theta\right] \quad \text{or} \quad G_{xy} = \frac{E}{2(1+\nu)} \tag{15}$$

Substituting Eq. 10 into Eq. 15 we obtain: $G_{xy} = E_x/(1+\nu_{xy})$

9.82 The displacements u and v in the x and y-direction are given by the following equations

$$u = [0.5(x^2 - y^2) + 0.5xy + 0.25x](10^{-3})\text{mm} \qquad v = [0.25(x^2 - y^2) - xy](10^{-3})\text{mm}$$

Assuming plane strain, determine the principal strains, the principal angle one, and the maximum shear strain every 30° on a circle of radius one around the origin. Use a spread sheet or write a computer program for the calculation.

Solution Plane Strain $\varepsilon_1 = ?$ $\varepsilon_2 = ?$ $\varepsilon_3 = ?$ $\theta_1 = ?$ $\gamma_{max} = ?$

For plane strain the third principal strain is: $\boxed{\varepsilon_3 = 0}$

From Eqs. 2.10a, 2.10b, and 2.10d we obtain the following strains from the given displacements.

$$\varepsilon_{xx} = \frac{\partial u}{\partial x} = [0.5(2x) + 0.5y + 0.25](10^{-3}) = [x + 0.5y + 0.25](10^{-3}) \tag{1}$$

$$\varepsilon_{yy} = \frac{\partial v}{\partial y} = [0.25(-2y) - x](10^{-3}) = -[x + 0.5y](10^{-3}) \tag{2}$$

$$\gamma_{xy} = \frac{\partial u}{\partial y} + \frac{\partial v}{\partial x} = [0.5(-2y) + 0.5x + 0.25(2x) - y](10^{-3}) = [x - 2y](10^{-3}) \tag{3}$$

The coordinates of a point are: $x = \cos\theta$ $y = \sin\theta$. These coordinates can be found at 30° interval values of θ. Using Eqs. 1,2, and 3, the strains ε_{xx}, ε_{yy}, and γ_{xy} can be found. The principal angle (θ_p) can be found from Eq. 9.13 and a principal strain (ε_p) found using Eq. 9.10. The other principal strain (ε_{p2}) can be found from Eq. 9.15. Using the results decision can be made on principal strain one and two and principal angle one. These calculations can be done on a spreadsheet and the results reported in the table below.

θ (°)	x (mm)	y (mm)	ε_{xx} μ	ε_{yy} μ	γ_{xy} μ	θ_p (rads)	ε_p μ	ε_{p2} μ	ε_1 μ	ε_2 μ	θ_1 (°)
0	1.000	0.000	1250	-1000	1000	0.209	1356	-1106	1356	-1106	11.98
30	0.866	0.500	1366	-1116	-134	-0.027	1368	-1118	1368	-1118	-1.54
60	0.500	0.866	1183	-933	-1232	-0.264	1349	-1099	1349	-1099	-15.11
90	0.000	1.000	750	-500	-2000	-0.506	1304	-1054	1304	-1054	-29.00
120	-0.500	0.866	183	67	-2232	-0.759	1243	-993	1243	-993	-43.51
150	-0.866	0.500	-366	616	-1866	0.543	-929	1179	1179	-929	121.12
180	-1.000	0.000	-750	1000	-1000	0.260	-883	1133	1133	-883	104.87
210	-0.866	-0.500	-866	1116	134	-0.034	-868	1118	1118	-868	88.07
240	-0.500	-0.866	-683	933	1232	-0.326	-891	1141	1141	-891	71.34
270	-0.000	-1.000	-250	500	2000	-0.606	-943	1193	1193	-943	55.28
300	0.500	-0.866	317	-67	2232	0.700	1257	-1007	1257	-1007	40.12
330	0.866	-0.500	866	-616	1866	0.450	1316	-1066	1316	-1066	25.77

9.83 On a aluminum (E = 70 GPa and ν = 0.25) beam two strain gages were attached to monitor loads P and w that vary slowly over time. The measured values of strain gage readings are given in Table 9.1. Determine the values of P and w at the times the strain were measured.

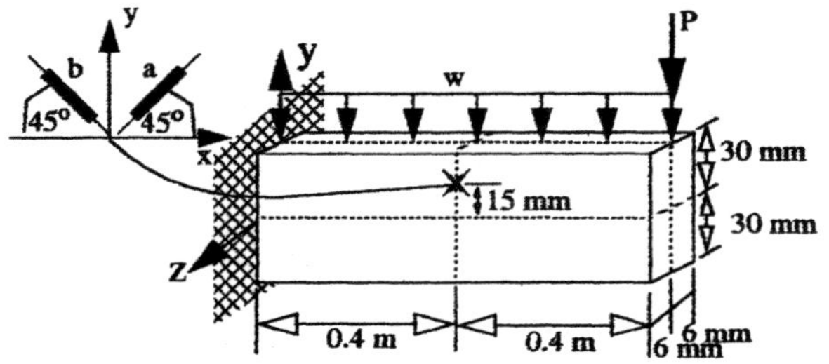

Fig. P9.83

Table 9.1 Strain Values

	ε_a μ	ε_b μ
1	1501	2368
2	1433	2276
3	1385	2193
4	1483	2336
5	1470	2331

	ε_a μ	ε_b μ
6	1380	2191
7	1448	2282
8	1496	2366
9	1398	2223
10	1411	2228

Solution E = 70 GPa ν = 0.25 P= ? w = ?

The area moment of inertia is: $I_{zz} = (0.012)(0.06^3)/12 = 0.216(10^{-6})$

Fig. (a) shows the free body diagram after an imaginary cut through the section containing the gages.

(a)

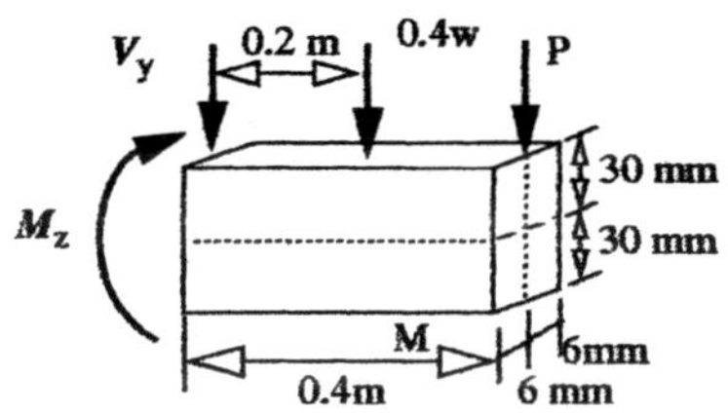

(b)

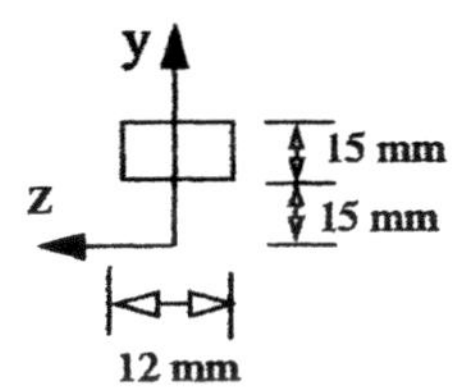

By equilibrium of forces and moment in Fig. (a) we obtain:

$$V_y = -(P+0.4w) \qquad M_z = -[0.4P+(0.4w)(0.2)] = -0.4(P+0.2w) \quad (1)$$

The bending normal stress at the two gage locations is:

$$\sigma_{xx} = \frac{-M_z y}{I_{zz}} = \frac{[-0.4(P+0.2w)](0.015)}{(0.216)(10^{-6})} = 27.777(P+0.2w)(10^3) \quad (2)$$

The bending normal strain at the gage location is:

$$\varepsilon_{xx} = \frac{\sigma_{xx}}{E} = \frac{27.777(P+0.2w)(10^3)(10^3)}{70(10^9)} = 0.3968(P+0.2w)(10^{-6}) \quad (3)$$

Fig. (b) shows the area A_s used in calculating the shear stress at the neutral axis. The first moment of the area is: $Q_z = (0.012)(0.015)(0.0225) = 4.05(10^{-6})m^3$

As the shear force V_y is negative, the bending shear stress τ_{xy} will be negative. The bending shear stresses at the two gage locations is:

$$\tau_{xy} = -\left|\frac{V_y(Q_z)_a}{I_{zz}t}\right| = -\frac{(P+0.4w)(4.05)(10^{-6})}{(0.216)(10^{-6})(0.012)} = -1562.5(P+0.4w) \quad (4)$$

The shear modulus of elasticity is: $G = E/[2(1+\nu)] = 70/[2(1+0.25)] = 28\text{GPa}$

The bending shear strain at the gage location is:

$$\gamma_{xy} = \frac{\tau_{xy}}{G} = \frac{-1562.5(P+0.4w)}{28(10^9)} = -0.0558(P+0.4w)(10^{-6}) \quad (5)$$

Substituting $\varepsilon_{yy} = -\nu\varepsilon_{xx}$ and $\theta_a = 45°$ and $\theta_b = -45°$ into Eq. 9.10, we obtain:

$$\varepsilon_a = \varepsilon_{xx}\cos^2(45) + (-0.28\varepsilon_{xx})\sin^2(45) + \gamma_{xy}\sin(45)\cos(45) = \frac{(1-\nu)\varepsilon_{xx}}{2} + \frac{\gamma_{xy}}{2} \quad \textbf{6}$$

$$\varepsilon_b = \varepsilon_{xx}\cos^2(-45) + (-\nu\varepsilon_{xx})\sin^2(-45) + \gamma_{xy}\sin(-45)\cos-(45) = \frac{(1-\nu)\varepsilon_{xx}}{2} - \frac{\gamma_{xy}}{2} \quad \textbf{7}$$

Solving Eqs. 6 and 7, we obtain: $\varepsilon_{xx} = \frac{1}{(1-\nu)}(\varepsilon_a + \varepsilon_b)$ $\gamma_{xy} = (\varepsilon_a - \varepsilon_b)$. **Substituting Eqs. 3 and 5 we obtain:**

$$0.3968(P + 0.2w)(10^{-6}) = \frac{1}{(1-\nu)}(\varepsilon_a + \varepsilon_b) \quad \text{or} \quad P + 0.2w = 3.36(\varepsilon_a + \varepsilon_b)(10^6) \quad \textbf{8}$$

$$-0.0558(P + 0.4w)(10^{-6}) = (\varepsilon_a - \varepsilon_b) \quad \text{or} \quad P + 0.4w = -17.92(\varepsilon_a - \varepsilon_b)(10^6) \quad \textbf{9}$$

Solving Eqs. 8 and 9 we obtain:

$$P = [6.72(\varepsilon_a + \varepsilon_b) + 17.92(\varepsilon_a - \varepsilon_b)](10^6) \quad \textbf{10}$$

$$w = -[16.8(\varepsilon_a + \varepsilon_b) + 89.6(\varepsilon_a - \varepsilon_b)] \quad \textbf{11}$$

Using Eqs. 10 and 11 the various values of P and w can be found on a spreadsheet from the reading of ε_a and ε_b as shown in the table below.

ε_a (μ)	ε_b (μ)	P N	w N/m
1501	2368	10463	12297
1433	2276	9818	12851
1385	2193	9565	11929
1483	2336	10378	11888
1470	2331	10114	12909
1380	2191	9464	12316
1448	2282	10120	11689
1496	2366	10362	12684
1398	2223	9549	12725
1411	2228	9813	11704

Problem Set 10.1

10.1 Determine the normal and shear stress in the seam of the shaft passing through point A. The seam is at an angle of 60° to the axis of a solid shaft of 2 inch diameter.

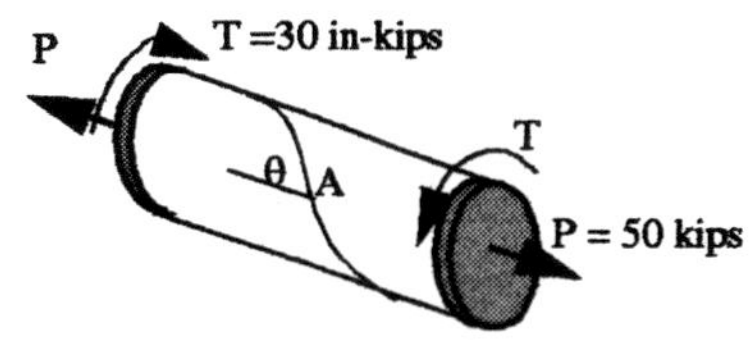

Fig. P10.1

Solution: $d = 2\text{in}$ $\theta = 60^{\circ}$ $\sigma_{nn} = ?$ $\tau_{nt} = ?$

Fig(a) shows the free body diagram with the internal force and torque drawn as per our sign convention.

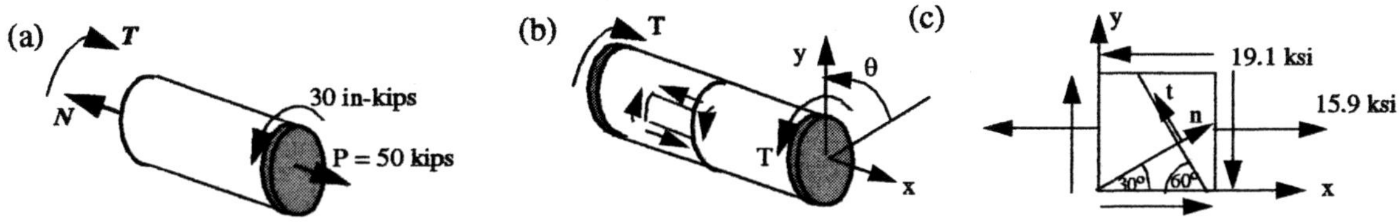

By equilibrium of forces and moment in Fig (a) we obtain: $N = 50$ kips and $T = 30$ in - kips
The area of cross section and polar moment is.

$$A = \frac{\pi}{4}(2)^2 = 3.1419\text{in}^2 \qquad J = \frac{\pi}{32}(2)^4 = 1.5708\text{in}^4$$

The axial stress is $\sigma_{x\theta} = \frac{N}{A} = \frac{50}{3.1419} = 15.92\text{ksi}$

The torsional shear stress is $\tau_{x\theta} = \frac{T\rho}{J} = \frac{(30)(1)}{1.5708} = 19.1\text{ksi}$

Fig (b) shows the torsional shear stress direction. Fig (c) shows the stress cube at point A. We note $\sigma_{xx} = 15.92\text{ksi}$, $\sigma_{yy} = 0\text{ksi}$, $\sigma_{xy} = (-19.1)\text{ksi}$. The angle the outward normal to plane is 30°
From Eqs. 8.1 and 8.2 we obtain

$\sigma_{nn} = 15.19(\cos 30)^2 + (2)(-19.1)\cos 30 \sin 30$ or $\boxed{\sigma_{nn} = 5.15\text{ksi}(C))}$

$\tau_{nt} = (-15.19)(\cos 30 \sin 30) + (-19.1)[(\cos 30)^2 - (\sin 30)^2]$ or $\boxed{\tau_{nt} = -16.13\text{ksi}}$

10.2 A 4 inch diameter solid circular steel shaft is loaded as shown. Determine the shear stress and normal stress on a plane passing through point E. Point E is on the surface of the shaft.

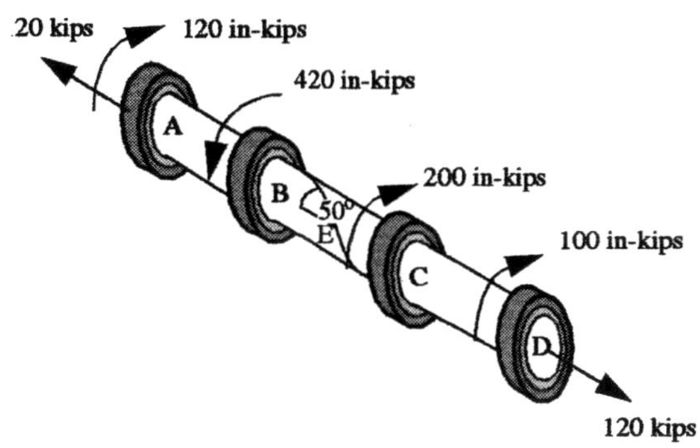

Fig. P10.2

Solution: $d = 4$ in $\sigma_{nn} = ?$ $\tau_{nt} = ?$

Fig(a) shows the free body diagram with the internal force and torque drawn as per our sign convention.

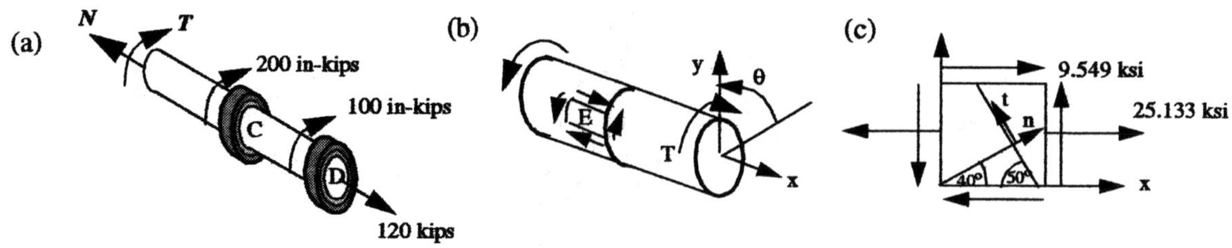

By equilibrium of forces and moments in Fig (a) we obtain: $N = 120$ kips and $T = -300$ in - kips
The area of cross section and area moment of inertia is.

$A = \frac{\pi}{4}(4)^2 = 12.566\text{in}^2 \qquad J = \frac{\pi}{32}(4)^4 = 25.133\text{in}^4$

The axial stress is $\sigma_{xx} = \frac{N}{A} = \frac{120}{12.566} = 9.549\text{ksi}$

The torsional shear stress is $\tau_{x\theta} = \frac{T\rho}{J} = \frac{(-300)(2)}{25.133} = -23.872\text{ksi}$

Fig (b) shows the torsional shear stress direction. Fig (c) shows the stress cube at point E. We note $\sigma_{xx} = 9.549\text{ksi}$, $\sigma_{yy} = 0\text{ksi}$, $\sigma_{xy} = 23.872\text{ksi}$. The angle the outward normal to plane is 40°
From Eq. 8.1 and 8.2 we obtain

$\sigma_{nn} = 9.549(\cos 40)^2 + (2)(23.872)\cos 40 \sin 40$ or $\boxed{\sigma_{nn} = 29.11\text{ksi(T)}}$

$\tau_{nt} = (-9.549)(\cos 40 \sin 40) + (23.872)[(\cos 40)^2 - (\sin 40)^2]$ or $\boxed{\tau_{nt} = -0.557\text{ksi}}$

10.3—10.5 A solid shaft of 75 mm diameter is loaded as shown in Fig. P10.5. The strain gage is 20° to the axis of the shaft and the shaft material has a Modulus of Elasticity E = 250 GPa and Poisson's ratio of ν = 0.3. Use this data in the problems below.

10.3 If T = 20 kN-m and P = 50 kN what strain will the strain gage show.

10.4 If the strain gage shows a reading of - 450 μ m/m, and T = 10 kN, determine the axial load P.

10.5 If the strain gage shows a reading of -300 μ m/m and P = 55 kN, determine the applied torque T.

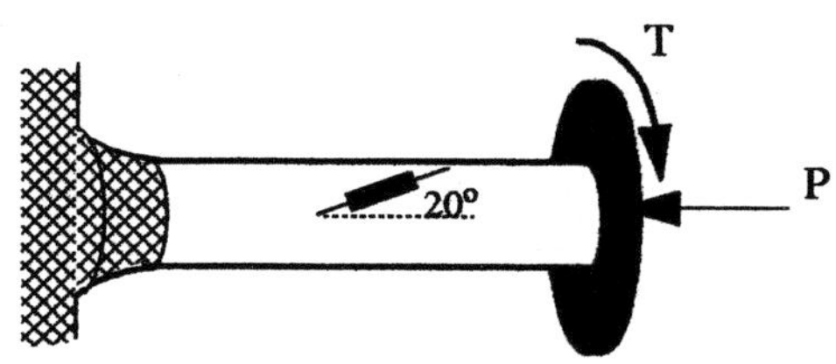

Fig. P10.5

Solution: d = 75mm θ = 20° E = 250 GPa ν = 0.3

Fig(a) shows the free body diagram with the internal force and torque drawn as per our sign convention.

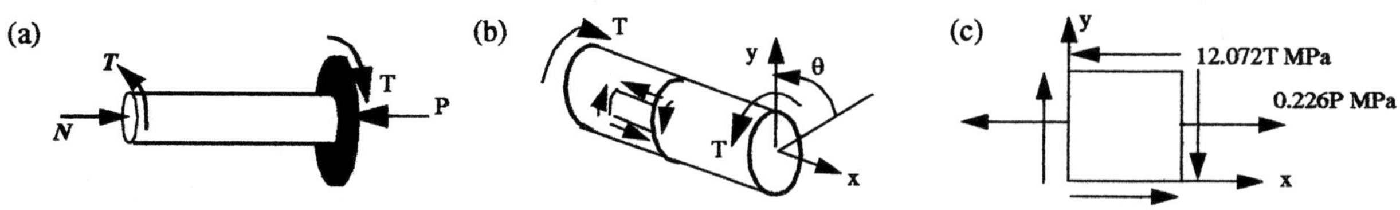

By equilibrium of forces and moments in Fig (a) we obtain: T = T kN-m and N = -P kN
The area of cross section and area moment of inertia is.

$A = \frac{\pi}{4}(0.075)^2 = 4.418(10^{-3})\text{m}^2 \qquad J = \frac{\pi}{32}(0.075)^4 = 3.106(10^{-6})\text{m}^4$

The axial stress is: $\sigma_{xx} = \dfrac{N}{A} = \dfrac{-P(10^3)}{4.418(10^{-3})} = -0.226P(10^6)\dfrac{N}{m^2}$

The torsional shear stress is: $\tau_{x\theta} = \dfrac{T\rho}{J} = \dfrac{T(10^3)(37.5)(10^{-3})}{3.106(10^{-6})} = 12.072T(10^6)\dfrac{N}{m^2}$

The normal strains are $\varepsilon_{xx} = \dfrac{\sigma_{xx}}{E} = -\dfrac{0.226P(10^6)}{250(10^9)} = -0.9054P\mu$ and $\varepsilon_{yy} = -\nu\varepsilon_{xx} = 0.2716P\mu$

Fig. (b) shows the torsional shear stress. Fig. (c) shows the stress cube from which we note $\tau_{xy} = -12.072T(10^6)\dfrac{N}{m^2}$. The shear modulus is $G = \dfrac{E}{2(1+\nu)} = 96.15GPa$

The shear strain is $\gamma_{xy} = \dfrac{\tau_{xy}}{G} = \dfrac{-12.072T(10^6)}{96.15(10^9)} = -125.5T\mu$

From Eq. 9.1 we obtain

$\varepsilon_{20} = (-0.9054P)(\cos 20)^2 + (0.2716P)(\sin 20)^2 + (-125.5T)\cos 20 \sin 20$ or

$$\varepsilon_{20} = (-0.7677P - 40.335T)\mu \qquad 1$$

Problem 10-3: We are given P = 50 kN and T = 20 kN-m. Substituting these values in Eq. 1 we obtain

$\varepsilon_{20} = -(0.7677)(50) - (40.335)(20)$ or $\boxed{\varepsilon_{20} = -845(\mu m)/m}$

Problem 10-4: We are given $\varepsilon_{20} = -450\mu$ and T = 10 kN-m. Substituting these values in Eq. 1 we obtain

$\varepsilon_{20} = -(0.7677)(P) - (40.335)(10)$ or $\boxed{P = 60.76kN}$

Problem 10-5: We are given $\varepsilon_{20} = -300\mu$ and P = 55kN. Substituting these values in Eq. 1 we obtain

$\varepsilon_{20} = -(0.7677)(55) - (40.335)T = -300$ or $\boxed{T = 6.39kN-m}$

10.6—10.7 A solid shaft of 2 inch diameter is loaded as shown in Fig. P10.7. The shaft material has a Modulus of Elasticity of E = 30,000 ksi and a Poisson's ratio of ν = 0.3. Use this data in the problems below.

10.6 Determine the strain the gage would show if P = 70 kips and T = 50 in-kips.

10.7 The strain gages mounted on the surface of the shaft recorded the following strain values: $\varepsilon_a = 2078\ \mu$ $\varepsilon_b = -1410\mu$. Determine the axial force P and the torque T.

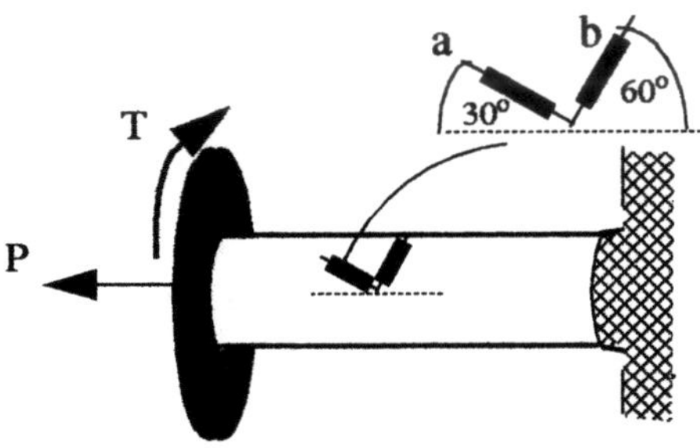

Fig. P10.7

Solution: d=2 E=30,000 ν = 0.3

By equilibrium of forces and moments in Fig (a) we obtain: ***T*** = T and ***N*** = P kips

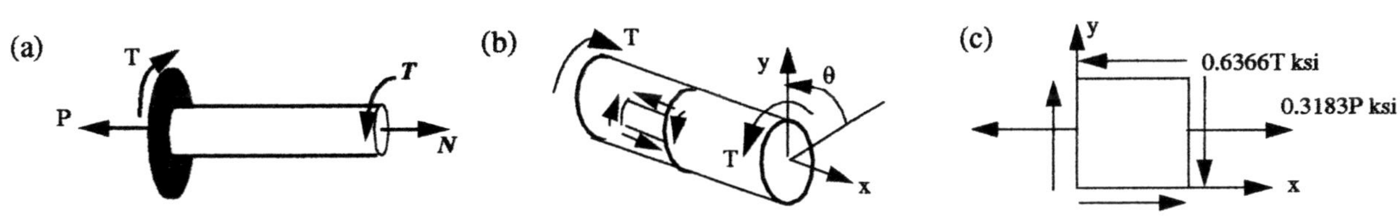

The area of cross section and area moment of inertia is.

$A = \frac{\pi}{4}(2)^2 = 3.1416\text{in}^2$ $\qquad J = 1.5708\text{in}^4$

The axial stress is $\sigma_{xx} = \frac{N}{A} = \frac{P}{3.1416} = 0.3183P\text{ksi}$

The torsional shear stress is $\tau_{x\theta} = \frac{T\rho}{J} = \frac{T(10^3)(1)}{1.5708} = 0.6366T\text{ksi}$

The normal strains are: $\varepsilon_{xx} = \frac{\sigma_{xx}}{E} = \frac{0.3183P}{30000} = 10.61P\mu$ and $\varepsilon_{yy} = -\nu\varepsilon_{xx} = (-3.183)P\mu$

Fig. (b) shows the torsional shear stress. Fig (c) shows the stress cube. we note $\tau_{xy} = -0.6366T\text{ksi}$. The shear modulus is $G = \frac{E}{2(1+\nu)} = 11538\text{ksi}$

The shear strain is $\gamma_{xy} = \frac{\tau_{xy}}{G} = \frac{-0.6366T}{11538} = -55.174T\mu$

Substituting $\theta_a = 150°$ and the value of strains in Eq. 9.1 we obtain

$\varepsilon_a = (10.61P)(\cos 150)^2 + (-3.183P)(\sin 150)^2 + (-55.174T)\cos 150 \sin 150$ or

$$\varepsilon_a = (7.1617P + 23.891T)\mu \qquad 1$$

Substituting $\theta_b = 60°$ and the value of strains in Eq. 9.1 we obtain

$\varepsilon_b = (10.61P)(\cos 60)^2 + (-3.183P)(\sin 60)^2 + (-55.174T)\cos 60 \sin 60$ or

$$\varepsilon_b = (0.26525P - 23.891T)\mu \qquad 2$$

Problem 10-6: We are given P = 70 kips and T = 50 in-kips. Substituting these values in Eq. (1) and (2)we obtain

$\varepsilon_a = (7.1617)(70) + (23.891)(50)$ or $\qquad \boxed{\varepsilon_a = 1696\mu}$

$\varepsilon_b = (0.26525)(70) - (23.891)(50)$ or $\qquad \boxed{\varepsilon_b = -1176\mu}$

Problem 10-7: We are given $\varepsilon_a = 2078\mu$ and $\varepsilon_b = -1410\mu$. Substituting these values in Eq. (1) and (2)we obtain

$$(7.1617)P + (23.891)T = 2078 \qquad 3$$

$$(0.26525)P - (23.891)T = -1410 \qquad 4$$

Solving Eq. (3) and (4) we obtain $\qquad \boxed{P = 89.9\text{kips}} \quad \boxed{T = 60.0\text{in-kips}}$

10.8 Two solid circular steel (E_s= 200GPa, G_s= 80 GPa) shafts and a solid circular bronze (E_{br} = 100 GPa, G_{br} = 40 GPa) shaft are securely connected and loaded as shown Fig. P10.8. Determine the maximum normal and shear stress in the shaft.

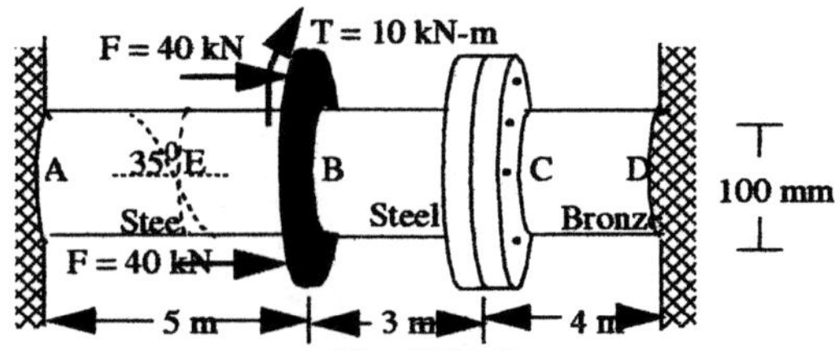

Fig. P10.8

Solution: E_s = 200 GPa $\qquad G_s$ = 80 GPa $\qquad E_{br}$ = 100 GPa $\qquad G_{br}$ = 40 GPa

σ_{max} =? $\qquad \tau_{max}$ =?

Figs. (a) and (b) show the free body diagram after an imaginary cut is made in AB and BC respectively.

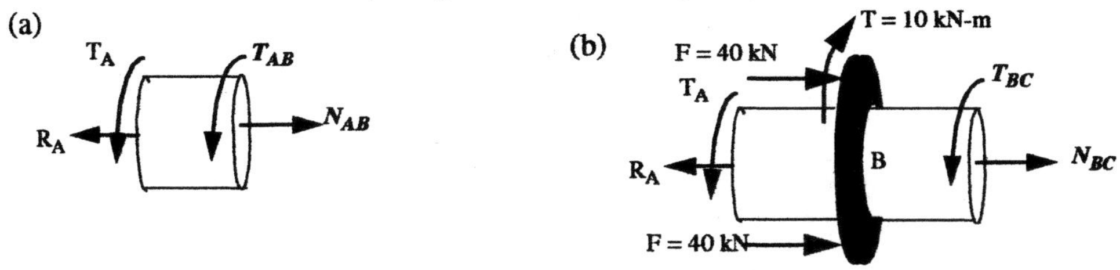

By equilibrium of forces and moment in Figs (a) and (b) we obtain.

$$N_{AB} = R_A \text{ and } T_{AB} = -T_A \quad (1)$$

$$N_{BC} = R_A - 80 \text{ and } T_{BC} = -T_A + 10 \quad (2)$$

Noting that there are no external forces or moments at C we obtain

$$N_{CD} = N_{BC} = R_A - 80 \text{ and } T_{CD} = T_{BC} = -T_A + 10 \quad (3)$$

The area of cross section and polar moment J are

$$A = \frac{\pi}{4}(0.1^2) = 7.854(0.1^{-3})\text{m}^2 \quad J = \frac{\pi}{32}(0.1^4) = 9.817(0.1^{-6})\text{m}^4 \quad (4)$$

The relative displacement of ends of each segment can be written as shown below.

$$u_B - u_A = \frac{N_{AB}(x_B - x_A)}{E_{AB}A_{AB}} = \frac{R_A(5)}{(200)(10^9)(7.854)(10^{-3})} = 3.183R_A(10^{-9}) \quad (5)$$

$$u_C - u_B = \frac{N_{BC}(x_C - x_B)}{E_{BC}A_{BC}} = \frac{(R_A - 80)(3)}{(200)(10^9)(7.854)(10^{-3})} = 1.9099(R_A - 80)(10^{-9}) \quad (6)$$

$$u_D - u_C = \frac{N_{CD}(x_D - x_C)}{E_{CD}A_{CD}} = \frac{(R_A - 80)(4)}{(100)(10^9)(7.854)(10^{-3})} = 5.0929(R_A - 80)(10^{-9}) \quad (7)$$

Adding Eqs. (5) (6) and (7) and noting $u_D = 0$ and $u_A = 0$ we obtain

$$u_D - u_A = [3.183R_A + 1.9099(R_A - 80) + 5.0929(R_A - 80)](10^{-9}) = 0 \text{ or}$$

$$R_A = \frac{[1.9099 + 5.0929](80)}{(3.183 + 1.9099 + 5.0929)} = 55\text{kN} \quad (8)$$

The relative rotation of ends of each segment can be written as

$$\phi_B - \phi_A = \frac{T_{AB}(x_B - x_A)}{G_{AB}J_{AB}} = \frac{(-T_A)(5)}{(80)(10^9)(9.817)(10^{-6})} = -6.3665T_A(10^{-6}) \quad (9)$$

$$\phi_C - \phi_B = \frac{T_{BC}(x_C - x_B)}{G_{BC}J_{BC}} = \frac{(-T_A + 10)(3)}{(80)(10^9)(9.817)(10^{-6})} = 3.8199(-T_A + 10)(10^{-6}) \quad (10)$$

$$\phi_D - \phi_C = \frac{T_{CD}(x_D - x_C)}{G_{DC}J_{DC}} = \frac{(-T_A + 10)(4)}{(40)(10^9)(9.817)(10^{-6})} = 10.186(-T_A + 10)(10^{-6}) \quad (11)$$

Adding Eqs. (9) (10) and (11) and noting $\phi_A = 0$ and $\phi_D = 0$, we obtain

$$\phi_D - \phi_A = [-6.3665T_A + 3.8199(-T_A + 10) + 10.186(-T_A + 10)](10^{-6}) = 0 \text{ or}$$

$$T_A = \frac{[3.8199 + 10.186](10)}{(6.3665 + 3.8199 + 10.186)} = 6.875\text{kN} - \text{m} \quad (12)$$

Substituting Eqs. (8) and (12) into Eqs. (1), (2), (3) we obtain

$N_{AB} = 55$ kN $T_{AB} = -6.875$ kN-m

$N_{BC} = -25$kN $T_{AB} = 3.125$kN-m

$N_{CD} = -25$kN $T_{AB} = 3.125$kN-m

As diameter is the same in all segments the maximum axial and torsional shear stress will exist in segments AB, The value of stresses are

$$\sigma_{xx} = \frac{N}{A} = \frac{55(10^3)}{7.854(10^{-3})} = 7.0(10^6)\frac{N}{m^2} = 7.0\text{MPa(T)} \quad 13$$

$$\tau_{x\theta} = \frac{T_{AB}\rho}{J} = \frac{(-6.875)(10^3)(0.05)}{9.817(10^{-6})} = 35.0(10^6)\frac{N}{m^2} = -35.0\text{MPa(T)} \quad 14$$

Noting that $\sigma_{yy} = 0$, the principal stresses are

$$\sigma_{1,2} = \frac{7+0}{2} \pm \sqrt{\left(\frac{7}{2}\right)^2 + (35)^2} = 3.5 \pm 35.17 \text{ or } \sigma_1 = 38.67\text{MPa and } \sigma_2 = -31.67\text{MPa}$$

The maximum normal stress is $\boxed{\sigma_{max} = 38.7\text{MPa}}$

As $\sigma_3 = 0$, the maximum shear stress is $\tau_{max} = \frac{\sigma_1 - \sigma_2}{2} = 35.17$ or $\boxed{\tau_{max} = 35.2\text{MPa}}$

10.9 Determine the normal and shear stress on a plane 35° to the axis of the shaft at point E in Fig. P10.8. Point E is on the surface of the shaft.

Solution: $E_s = 200$ GPa $G_s = 80$ GPa $E_{br} = 100$ GPa $G_{br} = 40$ GPa

$\sigma_{nn} = ?$ $\tau_{nt} = ?$

Solution

From Eqs. (14) and (15) of problem 10.8 we have $\sigma_{xx} = 7\text{MPa(T)}$, $\tau_{x\theta} = -35\text{MPa}$.

Fig(a) shows the torsional shear stress. Fig (b) shows the stress cube.

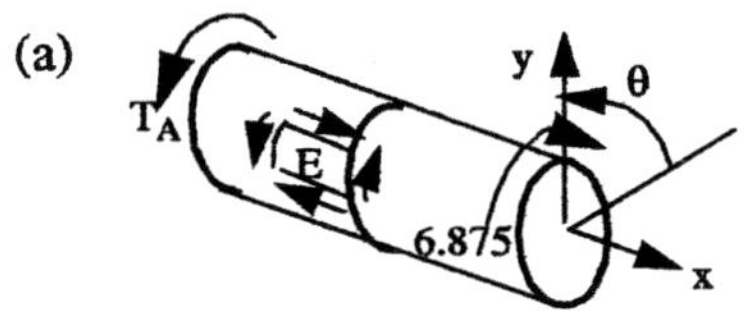

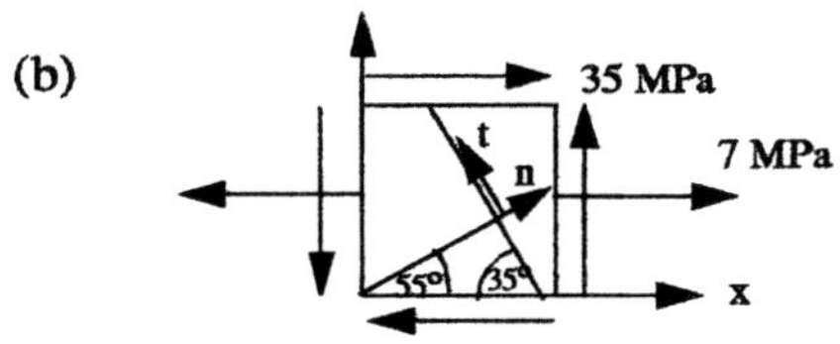

From Fig (b) we have the following stresses $\sigma_{xx} = 7\text{MPa}, \sigma_{yy} = 0$, $\tau_{xy} = 35\text{MPa}$. The outward normal to the inclined plane is at an angle of 55° from the X-axis. Substituting theses values in Eqs. (8.1) and (8.21) we obtain.

$$\sigma_{nn} = (7)(\cos 55)^2 + (2)(35)\cos 55 \sin 55 = 35.19 \text{ or } \boxed{\sigma_{nn} = 35.2\text{MPa(T)}}$$

$$\tau_{nt} = (-7)\cos 55 \sin 55 + (35)[(\cos 55)^2 - (\sin 55)^2] = -15.26 \text{ or } \boxed{\tau_{nt} = -15.3\text{MPa}}$$

10.10 A rectangular hollow member is constructed from a 1/2 inch thick sheet metal and loaded as shown Fig. P10.10. Determine the normal and shear stresses at points A and B and show it on the stress cubes for $P_1 = 72$ kips, $P_2 = 0$, and $P_3 = 6$ kips.

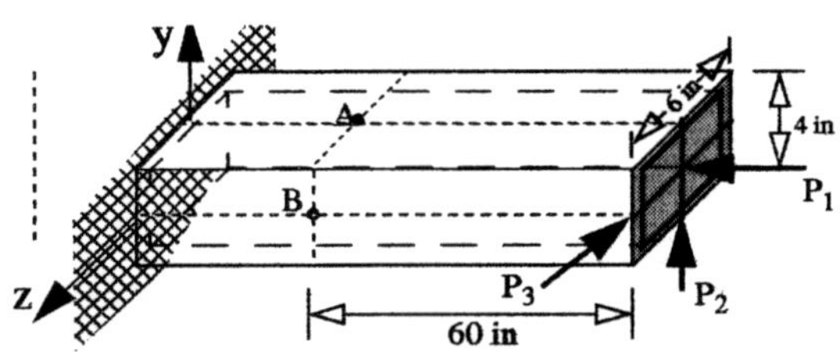

Fig. P10.10

Solution: $t = \frac{1}{2}\text{in}$, $P_1 = 72$ kips, $P_2 = 0$, $P_3 = 6$ kips, $\sigma_A = ?$, $\tau_A = ?$, $\sigma_B = ?$, $\tau_B = ?$

Fig(a) shows the free body diagram with the internal force and moments drawn as per our sign convention.

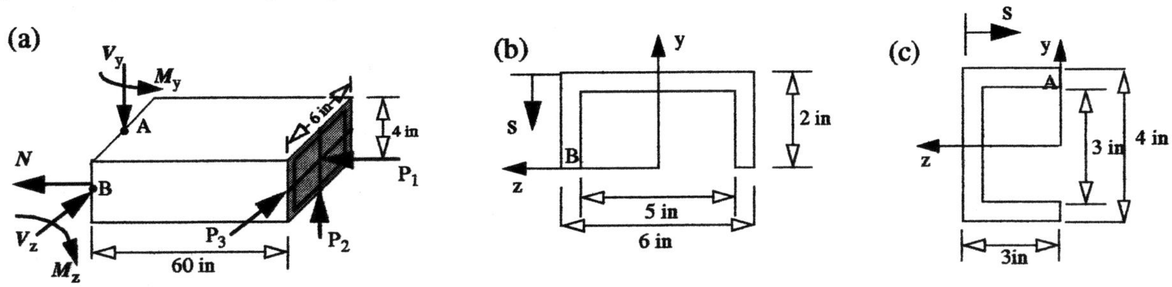

By equilibrium of forces and moments we obtain

$$N = -P_1 \quad (1)$$

$$V_y = P_2 \quad \text{and} \quad M_z = 60\,P_2 \quad (2)$$

$$V_z = -P_3 \quad \text{and} \quad M_y = -60\,P_3 \quad (3)$$

with $P_2 = 0$, we have axial stress, and stresses due to bending about y-axis. The cross-sectional area and second area moment of inertia are

$$A = (6)(4)-(5)(3) = 9\text{in}^2 \quad I_{yy} = \frac{1}{12}(4)(6^3)-\frac{1}{12}(3)(5^3) = 40.75\text{in}^4 \quad (4)$$

Figs (b) and (c) shows the area As for calculation of shear stress at points B and A. From Fig. (b) we see $(Q_y)_B = 0$, thus $(\tau_{xy})_B = 0$. From Fig (c) we can find the first moment of inertia as

$$(Q_y)_A = (2)(3)(0.5)(1.5)+(3)(0.5)(2.75) = 8.625\text{in}^3 \quad (5)$$

The axial stress at A and B is

$$(\sigma_{xx})_{A,B} = \frac{N}{A} = \frac{-72}{9} = -8\text{ksi} \quad (6)$$

The normal stresses due to bending about y-axis are

$$\sigma_A = 0 \qquad \sigma_B = \frac{-M_y z_B}{I_{yy}} = \frac{-(-60)(6)(3)}{40.75} = 26.5\text{ksi} \quad (7)$$

The shear stress of point A is

$$(\tau_{xs})_A = \frac{-V_z(Q_y)_A}{I_{yy}t} = \frac{-(-6)(8.625)}{(40.75)(1)} = 1.27\text{ksi}$$

Noting that s in Fig. (b) is in opposite direction to z we obtain

$$(\tau_{xz})_A = -(\tau_{xs})_A = -1.27\text{ksi} \quad (8)$$

The total stresses at point A and B are

$(\sigma_{xx})_A = -8\text{ksi}$ or $\boxed{(\sigma_{xx})_A = 8\text{ksi(C)} \quad (\tau_{xy})_A = 0 \quad (\tau_{xz})_A = (-1.27)\text{ksi}}$

$(\sigma_{xx})_B = -8+26.5 = 18.5\text{ksi}$ or $\boxed{(\sigma_{xx})_B = 18.5\text{ksi} \quad (\tau_{xy})_B = 0 \quad (\tau_{xz})_B = 0}$

Figs. (d) and (e) show the stress cube at points A and B.

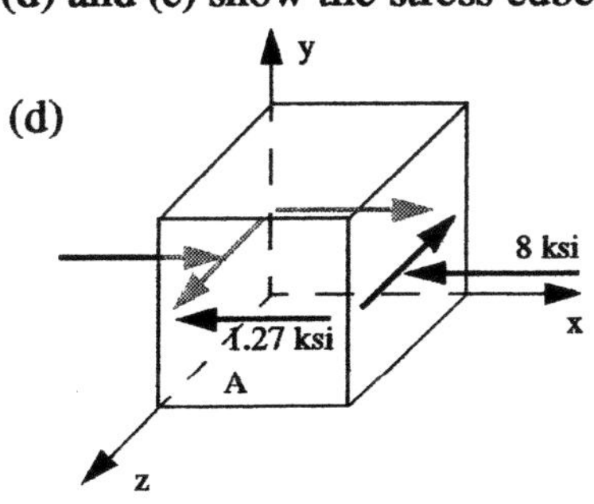

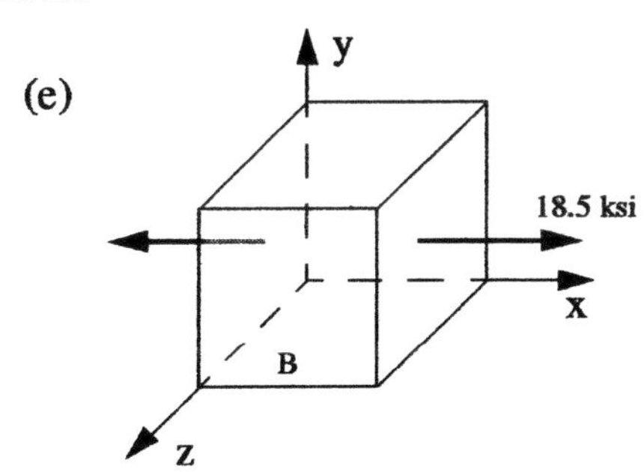

10.11 Determine the principal stresses and maximum shear stress at points A and B in problem Fig. P10.10 for $P_1 = 72$ kips, $P_2 = 3$ kips, and $P_3 = 0$.

Solution: $P_1 = 72$ kips, $P_2 = 3$ kips, $P_3 = 0$, $(\sigma_1)_{A,B} = ?$, $(\sigma_2)_{A,B} = ?$, $(\tau_{max})_{A,B} = ?$

With $P_3 = 0$ we have axial and stresses due to bending about z -axis. The second area moment of inertia.

$$I_{zz} = \frac{1}{12}(6)(4^3) - \frac{1}{12}(5)(3^3) = 20.75\text{in}^4 \quad (1)$$

From Fig (c) in problem 10.10 we see $(Q_z)_A=0$, thus $(\tau_{xz})_A = 0$. From Fig. (b) in problem 10.10 we can find the first moment of inertia as.

$$(Q_z)_B = (2)(2)(0.5)(1) + (5)(0.5)(1.75) = 6.375\text{in}^4 \quad (2)$$

From Eqs. (1) and (2) in problem 10.10, we have

$$N = -P_1 = -72\text{kips} \quad (3)$$

$$V_y = P_2 = 3\text{kips} \qquad M_z = 60P_2 = 180\text{in–kips} \quad (4)$$

The axial stresses at points A and B are

$$(\sigma_{xx})_{A,B} = \frac{N}{A} = \frac{-72}{9} = -8\text{ksi} \quad (5)$$

The normal stresses due to bending about z-axis are

$$(\sigma_{xx})_A = \frac{-M_z y_B}{I_{zz}} = \frac{-(180)(2)}{20.75} = -17.35\text{ksi} \quad \text{and } (\sigma_{xx})_B = 0 \quad (6)$$

The shear stress of point B is $(\tau_{xs})_B = \dfrac{-V_y(Q_z)_B}{I_{zz}t} = \dfrac{-(3)(6.375)}{(20.75)(1)} = -0.922\text{ksi}$

From Fig. (b) in problem 10.10 we note that the direction of coordinate s is opposite to the direction of y. Hence

$$(\tau_{xy})_B = -(\tau_{xs})_B = 0.922\text{ksi} \quad (7)$$

The total stresses at points A and B are

$$(\sigma_{xx})_A = -8 - 17.35 = -25.35\text{ksi and } (\tau_{xy})_A = (\tau_{xz})_A = 0 \quad (8)$$

$$(\sigma_{xx})_B = -8\text{ksi and } (\tau_{xy})_B = 0.922\text{ksi and } (\tau_{xz})_B = 0 \quad (9)$$

From Eq. (8) we note that shear stresses are zero. Thus the principal stresses are

$$\boxed{(\sigma_1)_A = 0 \quad \text{and} \quad (\sigma_2)_A = 25.35\text{ksi(C)}}$$

The maximum shear stress is:

$$(\tau_{max})_A = \left|\frac{\sigma_1 - \sigma_2}{2}\right| = \frac{25.35}{2} \text{ or } \qquad \boxed{(\tau_{max})_A = 12.7\text{ksi}}$$

From Eq. (9) we obtain the principal stresses as

$$(\sigma_{1,2})_B = \left(\frac{-8}{2}\right) \pm \sqrt{\left(\frac{-8}{2}\right)^2 + (0.922)^2} = -4 \pm 4.1 \text{ or } \boxed{(\sigma_1)_B = 0.1\text{ksi(T)} \quad \text{and} \quad (\sigma_2)_B = 8.1\text{ksi(C)}}$$

The maximum shear stress at point B is:

$$(\tau_{max})_B = \left|\frac{\sigma_1 - \sigma_2}{2}\right| = \left|\frac{0.1 + 8.1}{2}\right| \text{ or } \qquad \boxed{(\tau_{max})_B = 4.1\text{ksi}}$$

10.12 Determine the strain shown by the strain gages in Fig. P10.13, if $P_1 = 3$ kN, $P_2 = 40$ kN and the Modulus of Elasticity is 200 GPa and the Poisson's ratio is 0.3. The strain gages are parallel to the axis of the beam.

10.13 The strain gages shown in Fig. P10.13 recorded the following strain values: $\varepsilon_a = 1000\ \mu$; $\varepsilon_b = -750\ \mu$. Determine the loads P_1 and P_2. The Modulus of Elasticity is 200 GPa and the Poisson's ratio is 0.3.

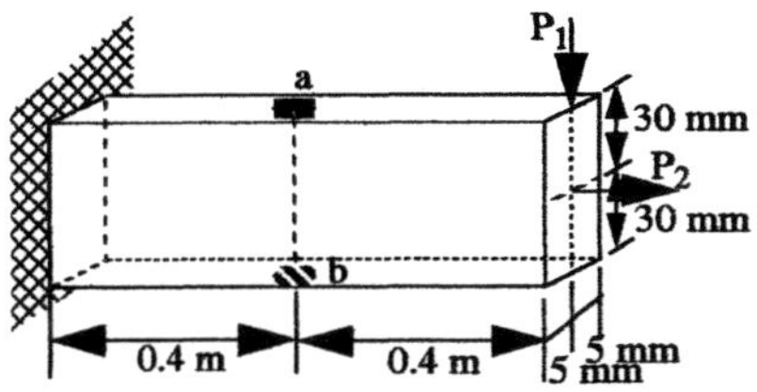

Fig. P10.13

Solution: **Problem 10.12:** E = 200 GPa ν = 0.3 $P_1 = 3$ kN; $P_2 = 40$ kN $\varepsilon_a = ?$ $\varepsilon_b = ?$

Problem 10.13: E = 200 GPa ν = 0.3 $\varepsilon_a = 1000\ \mu$; $\varepsilon_b = -750\ \mu$ $P_1 = ?$ kN, $P_2 = ?$

Fig (a) shows the free body diagram after an imaginary cut.

(a)

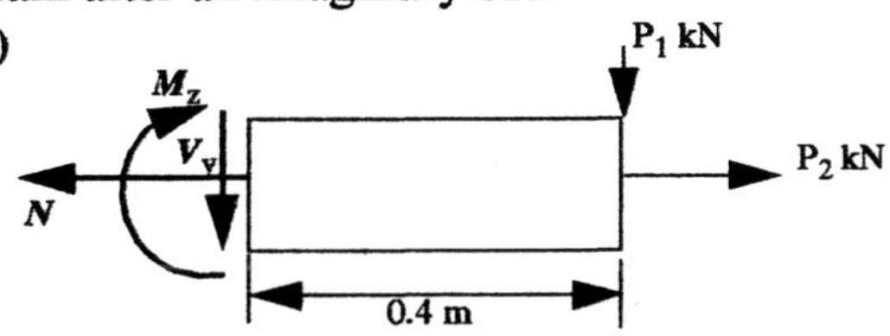

By equilibrium of forces and moments we obtain

$$V_y = -P_1 \text{ kN} \qquad N = P_2 \text{ kN} \qquad M_z = -0.4P_1 \text{ kN-m} \tag{1}$$

The area of cross section and second area moment of inertia is.

$$A = (0.01)(0.06) = 0.6(10^{-3})\text{m}^2 \quad \text{and} \quad I_{zz} = \frac{1}{12}(0.01)(0.06^3) = 0.18(10^{-6})\text{m}^4 \tag{2}$$

The axial stresses at strain gage points a and b are

$$(\sigma_{xx})_{a,\,b} = \frac{N}{A} = \frac{P_2(10^3)}{(0.6)(10^{-3})} = 1.667P_2(10^6)\frac{\text{N}}{\text{m}^2} \tag{3}$$

The bending normal stresses at points A and B are

$$(\sigma_{xx})_{a,\,b} = \frac{-M_z y_{a,\,b}}{I_{zz}} = \frac{-(-0.4P_1)(10^3)(\pm 30)(10^{-3})}{0.18(10^{-6})} = \pm 66.67P_1(10^6)\frac{\text{N}}{\text{m}^2} \tag{4}$$

There are no shear stresses at a and b. The total normal stresses at a and b is are:

$$(\sigma_{xx})_a = (1.667P_2 + 66.67P_1)(10^6)\frac{\text{N}}{\text{m}^2} \text{ and } (\sigma_{xx})_b = (1.667P_2 - 66.67P_1)(10^6)\frac{\text{N}}{\text{m}^2}$$

The normal strain in the x -direction is

$$\varepsilon_a = (\varepsilon_{xx})_a = \frac{(\sigma_{xx})_a}{E} = (8.333P_2 + 333.3P_1)\mu \tag{5}$$

$$\varepsilon_b = (\varepsilon_{xx})_b = \frac{(\sigma_{xx})_b}{E} = (8.333P_2 - 333.3P_1)\mu \tag{6}$$

Problem 10.12: Substituting $P_1 = 3$ kN and $P_2 = 40$ kN. into Eqs. (5) and (6) we obtain

$\varepsilon_a = 8.333(40) + 333.3(3) = 1333.33$ or $\boxed{\varepsilon_a = 1333\mu}$

$\varepsilon_b = 8.333(40) - (333.3)(3) = -666.66$ or $\boxed{\varepsilon_b = -666.66\mu}$

Problem 10.13: Substituting $\varepsilon_a = 1000\mu$ and $\varepsilon_b = -750\mu$. into Eqs. (5) and (6) we obtain

$$8.333P_2 + 333.3P_1 = 1000 \tag{7}$$

$$8.333P_2 - 333.3P_1 = -750 \tag{8}$$

Solving Eqs. (8) and (9) we obtain: $\boxed{P_1 = 2.625\text{kN}}$ and $\boxed{P_2 = 15\text{kN}}$

10.14 Determine the strain shown by the strain gages in Fig. P10.15, if P_1= 3 kN, P_2= 40 kN and the Modulus of Elasticity is 200 GPa and the Poisson's ratio is 0.3.

10.15 The strain gages shown in Fig. P10.15 recorded the following strain values: $\varepsilon_a = 133\ \mu$ $\varepsilon_b = 159\ \mu$. Determine the loads P_1 and P_2. The Modulus of Elasticity is 200 GPa and the Poisson's ratio is 0.3.

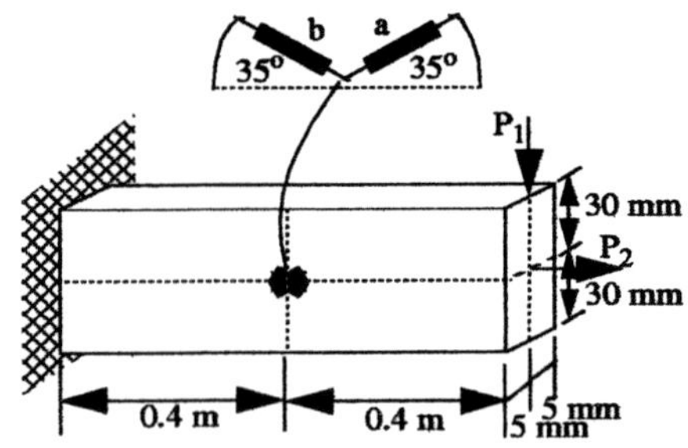

Fig. P10.15

Solution: **Problem 10.14:** E = 200 GPa ν = 0.3 P_1= 3 kN; P_2= 40 kN $\varepsilon_a = ?$ $\varepsilon_b = ?$

Problem 10.15: E = 200 GPa ν = 0.3 $\varepsilon_a = 133\ \mu$; $\varepsilon_b = -159\ \mu$ P_1= ? kN, P_2= ?

From Eq. (1) in problem 10.12 we have

$$N = P_2 \text{ kN} \qquad V_y = -P_1 \text{ kN} \qquad M_z = -0.4P_1 \text{ kN-m}$$ 1

The axial stress at strain gauge a and b is given by Eq.(3) in problem 10-12 as

$$(\sigma_{xx})_{a,\,b} = 1.667P_2(10^6)\frac{N}{m^2}$$ 2

The normal stress due to bending at strain gage location is zero. Fig (a) shows the area A_s used in calculation of the shear stress.

(a)

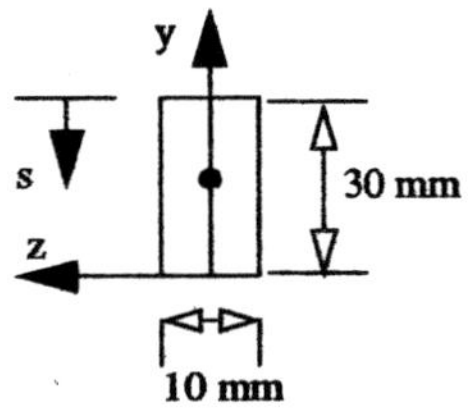

The first moment of the area A_s is: $Q_t = (0.01)(0.03)(0.015) = 4.5(10^{-6})m^3$

The bending shear stress is: $\tau_{xs} = \dfrac{-V_yQ_z}{I_zt} = \dfrac{-(-P_1)(10^3)(4.5)(10^{-6})}{(0.18)(10^{-6})(0.01)} = 2.5P_1(10^6)\dfrac{N}{m^2}$

Noting that s and y are in opposite direction in Fig. (a) we obtain: $\tau_{xy} = (-2.5)P_1(10^6)\dfrac{N}{m^2}$

The normal strains are

$$\varepsilon_{xx} = \frac{\sigma_{xx}}{E} = \frac{1.667P_2(10^6)}{200(10^9)} = 8.333P_2\mu$$ 3

$$\varepsilon_{yy} = -\nu\sigma_{xx} = -2.5P_2\mu$$ 4

The shear modulus is $G = \dfrac{E}{2(1+\nu)} = 76.923\text{GPa}$. The shear strain is

$$\gamma_{xy} = \frac{\tau_{xy}}{G} = -32.5P_1\mu$$ 5

Substituting $\theta_a = 35°$ and the above strains in Eq. 9.1 we obtain

$\varepsilon_a = 8.333P_2(\cos 35)^2 - 2.5P_2(\sin 35)^2 - 32.5P_1\cos 35\sin 35$ or

$$\varepsilon_a = (4.769P_2 - 15.27P_1)\mu$$ 6

Substituting $\theta_b = -35°$ and the above strains in Eq. 9.1 we obtain

$$\varepsilon_b = 8.333P_2(\cos(-35))^2 - 2.5P_2(\sin(-35))^2 - 32.5P_1\cos(-35)\sin(-35) \text{ or}$$

$$\varepsilon_b = (4.769P_2 + 15.27P_1)\mu \quad 7$$

Problem 10.14: Substituting $P_1 = 3$ kN and $P_2 = 40$ kN into Eqs. (6) and (7) we obtain.

$\varepsilon_a = 4.769(40) - 15.27(3)$ or $\boxed{\varepsilon_a = 145\mu}$

$\varepsilon_b = 4.769(40) + 15.27(3)$ or $\boxed{\varepsilon_a = 236.6\mu}$

Problem 10.15: Substituting $\varepsilon_a = 133\ \mu$ and $\varepsilon_b = 159\mu$ into Eqs. (6) and (7) we obtain.

$$4.769(P_2) - 15.27(P_1) = 133 \quad 8$$

$$4.769(P_2) + 15.27(P_1) = 159 \quad 9$$

Solving Eqs. (8) and (9) we obtain. $\boxed{P_1 = 0.85\text{kN}}$ $\boxed{P_2 = 30.61\,\text{kN}}$

10.16 Determine the strain recorded by the gages at point A and B. Both gages are at 30° to the axis of the beam. The Modulus of Elasticity is E = 30,000 ksi and ν = 0.3

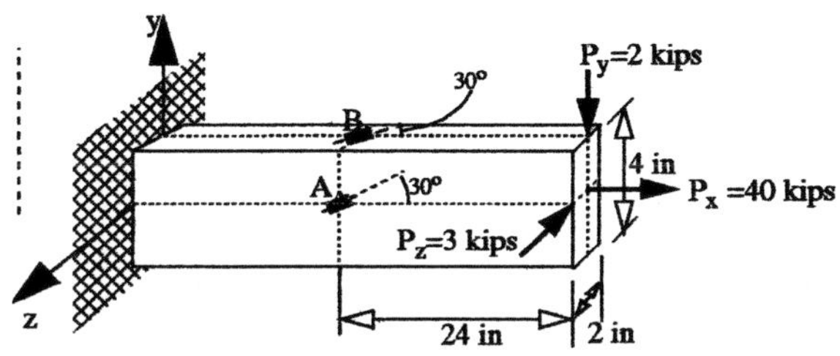

Fig. P10.16

Solution: E = 30,000 ν = 0.3 ε_A =? ε_B =?

Figs. (a) and (b) show the free body diagram after an imaginary cut is made through section with the strain gages.

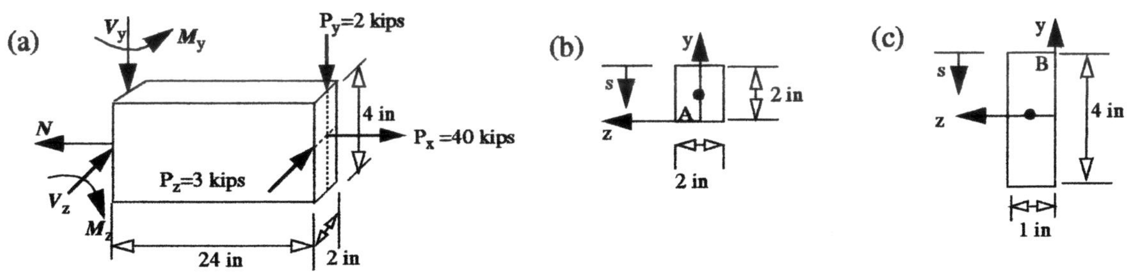

By equilibrium of forces and moments we obtain

$$N = 40 \text{ kips} \quad 1$$

$$V_y = -2 \text{ kips} \qquad M_z = -48 \text{ in - kips} \quad 2$$

$$V_z = -3 \text{ kips} \qquad M_y = -72 \text{ in - kips} \quad 3$$

The cross sectional area and area moment of inertia are

$$A = (2)(4)^3 = 8\text{in}^2 \qquad I_{yy} = \frac{1}{12}(4)(2^3) = 2.667\text{in}^4 \qquad I_{zz} = \frac{1}{12}(2)(4^3) = 10.667\text{in}^4 \quad 4$$

Fig (b) and (c) show the areas A_s used for finding shear stresses at A and B. The first moment of the areas can be found as

$$(Q_z)_A = (2)(2)(1) = 4\text{in}^2 \qquad (Q_y)_B = (4)(1)(0.5) = 2\text{in}^3 \quad 5$$

We note that the x and y direction in Fig. (b) are opposite. The shear stress at A can be written as

$$-(\tau_{xy})_A = (\tau_{xs})_A = \frac{-V_y(Q_z)_A}{I_{zz}t} = \frac{-(-2)(4)}{(10.667)(2)} = 0.375\text{ksi} \quad \text{or} \quad (\tau_{xy})_A = (-0.375)\text{ksi} \quad 6$$

We note that the x and z direction are opposite in Fig. (c), we obtain

$$-(\tau_{xz})_B = (\tau_{xs})_B = \frac{-V_z(Q_y)_B}{I_{yy}t} = \frac{-(-3)(2)}{(2.667)(4)} = 0.5625\text{ksi} \quad \text{or} \quad (\tau_{xz})_B = (-0.5625)\text{ksi} \qquad \mathbf{7}$$

The axial stresses at A and B are: $(\sigma_{xx})_{A,B} = N/A = 40/8 = 5\text{ksi}$

The normal stresses due to bending about z -axis are: $(\sigma_{xx})_A = 0 \quad (\sigma_{xx})_B = \frac{-M_z y_B}{I_{zz}} = \frac{-(-48)(2)}{10.667} = 9\text{ksi}$

The normal stresses due to bending about y-axis are $(\sigma_{xx})_A = \frac{-M_y z_A}{I_{yy}} = \frac{-(-72)(1)}{2.667} = 27\text{ksi} \quad (\sigma_{xx})_B = 0$

The total normal stresses at points A and B are

$(\sigma_{xx})_A = 5+27 = 32\text{ksi}$ and $(\sigma_{xx})_B = 5+9 = 14\text{ksi}$

The normal strains at points A and B are

$$(\varepsilon_{xx})_A = \frac{(\sigma_{xx})_A}{E} = \frac{32}{30000} = 1067\mu \text{ and } (\varepsilon_{yy})_A = -\nu(\varepsilon_{xx})_A = (-320)\mu$$

$$(\varepsilon_{xx})_B = \frac{(\sigma_{xx})_B}{E} = \frac{14}{30000} = 467\mu \text{ and } (\varepsilon_{zz})_B = -\nu(\varepsilon_{xx})_B = (-140)\mu$$

The shear modulus is: $G = \frac{E}{2(1+\nu)} = 11538$

The shear strains at A and B are: $(\gamma_{xy})_A = \frac{(\tau_{xy})_A}{G} = -32.5\mu \quad (\gamma_{xz})_B = \frac{(\tau_{xz})_B}{G} = -48.75\mu$

Substituting $\theta_A = 30^\circ$ and $\theta_B = 30^\circ$ and the corresponding strains at A and B in Eq. (9.1) we obtain

$\varepsilon_a = 1067(\cos 30)^2 - 320(\sin 30)^2 - 32.5\cos 30 \sin 30$ or $\boxed{\varepsilon_a = 706.2\mu}$

$\varepsilon_b = 467(\cos 30)^2 - 140(\sin 30)^2 - 48.75\cos 30 \sin 30$ or $\boxed{\varepsilon_b = 294.1\mu}$

10.17 A thin cylinder with an outer diameter of 100 mm and thickness of 10 mm is loaded as shown in Fig. P10.18. Points A and B are on the surface of the shaft. Determine the normal and shear stresses at points A and B in the x, y, and z coordinate system and show your results on stress cubes.

10.18 Determine the principal stresses and the maximum shear stress at point B on the shaft shown in Fig. P10.18.

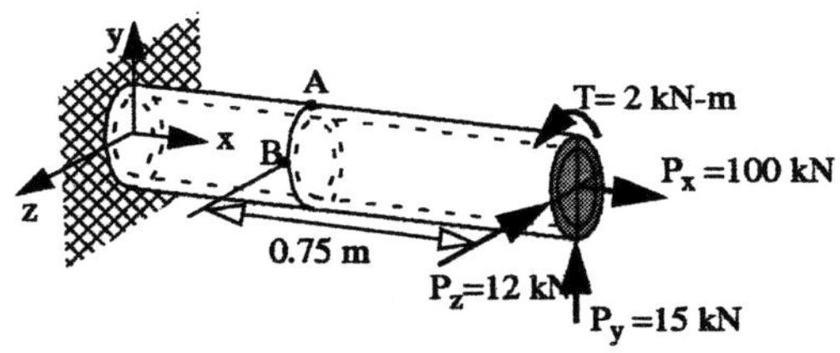

Fig. P10.18

Solution: $(\sigma_{xx})_A = ?$ $(\tau_{xy})_A = ?$ $(\tau_{xz})_A = ?$ $(\sigma_{xx})_B = ?$ $(\tau_{xy})_B = ?$ $(\tau_{xz})_B = ?$

Fig. (a) shows the free body diagram after an imaginary cut is made through section containing points A and B. By equilibrium of forces and moments we obtain

$N = 100$ kN $\quad V_y = 15$kN $\quad V_z = -12$ kN

$T = 2$kN-m $\quad M_y = (-12)(0.75) = -9$ kN-m $\quad M_z = (15)(0.75) = 11.25$ kN-m

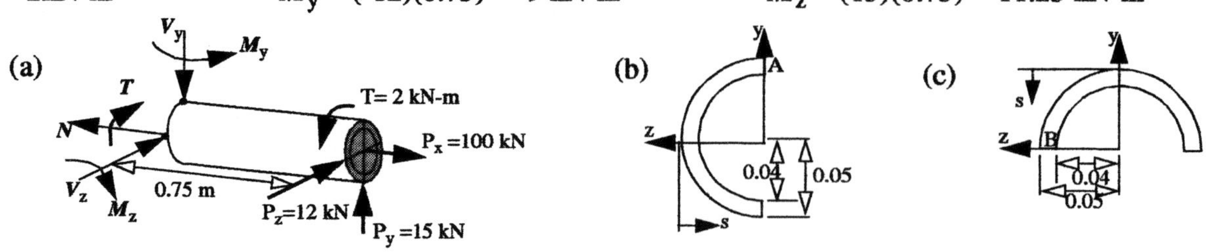

The cross sectional area and area moment of inertia are

$$A = \frac{\pi}{4}[(0.1)^2 - (0.08)^2] = 2.827(10^{-3})\text{m}^2 \quad J = \frac{\pi}{32}[(0.1)^4 - (0.08)^4] = 5.796(10^{-6})\text{m}^4$$

$$I_{yy} = I_{zz} = \frac{J}{2} = 2.898(10^{-6})\text{m}^4$$

Fig. (b) and (c) show the area A_s that is used in calculating shear stresses at points A and B. The first moment of inertia are.

$$(Q_y)_A = \left[\left(\frac{4R_o}{3\pi}\right)\left(\frac{\pi R_o^2}{2}\right) - \left(\frac{4R_i}{3\pi}\right)\left(\frac{\pi R_i^2}{2}\right)\right] = \frac{2}{3}(R_o^3 - R_i^3) = \frac{2}{3}(0.05^3 - 0.04^3) = 40.67(10^{-6})\text{m}^3$$

In magnitude $(Q_z)_B$ is same as $(Q_y)_A$. Thus $(Q_z)_B = 40.67(10^{-6})\text{m}^3$

The axial stresses at point A and B are: $(\sigma_{xx})_{A,B} = \dfrac{N}{A} = \dfrac{100(10^3)}{2.827(10^{-3})} = 35.37(10^6)\text{N/m}^2$

The normal stresses due to bending about y-axis are:

$$(\sigma_{xx})_A = 0 \quad (\sigma_{xx})_B = \frac{-M_y z_B}{I_{yy}} = \frac{-(-9)(10^3)(0.05)}{2.898(10^{-6})} = 155.3(10^6)\text{N/m}^2$$

The normal stress due to bending about z-axis is

$$(\sigma_{xx})_A = \frac{-M_z y_A}{I_{zz}} = \frac{-(11.25)(10^3)(0.05)}{2.898(10^{-6})} = -194.1(10^6)\text{N/m}^2 \quad (\sigma_{xx})_B = 0$$

The total stress at point A and B is

$(\sigma_{xx})_A = (35.37 - 194.1)(10^6)\text{N/m}^2$ or $\boxed{(\sigma_{xx})_A = 158.7\text{MPa(C)}}$

$(\sigma_{xx})_B = (35.37 + 155.3)(10^6)\text{N/m}^2$ or $\boxed{(\sigma_{xx})_B = 190.7\text{MPa(T)}}$

The shear stress due to bending about y-axis can be found out as shown below. Note s and z direction in Fig. (b) are opposite. Hence

$$-(\tau_{xz})_A = (\tau_{xs})_A = \frac{-V_z(Q_y)_A}{I_{yy}t} = \frac{-(-12)(10^3)(40.67)(10^{-6})}{(2.898)(10^{-6})(0.02)} \text{ or } (\tau_{xz})_A = -8.42(10^6)\frac{\text{N}}{\text{m}^2} \text{ and } (\tau_{xz})_B = 0$$

The shear stress due to bending about z -axis can be found out as shown below. Note sand y direction in Fig. (c) are opposite, hence

$$-(\tau_{xy})_B = (\tau_{xx})_B = \frac{-V_y Q_z}{I_{zz}t} = \frac{-(15)(10^3)(40.67)(10^{-2})}{(2.898)(0.02)} \text{ or } (\tau_{xy})_B = 10.52(10^6)\text{N/m}^2$$

The shear stress due to torsion is $\tau_{x\theta} = \dfrac{T\rho}{J} = \dfrac{(2)(10^3)(0.05)}{5.796(10^{-6})} = 17.25(10^6)\text{N/m}^2$

Fig. (d) shows the direction of torsional shear stress, we obtain.

$(\tau_{xz})_A = 17.25(10^6)\text{N/m}^2$ and $(\tau_{xy})_B = -17.25(10^6)\text{N/m}^2$

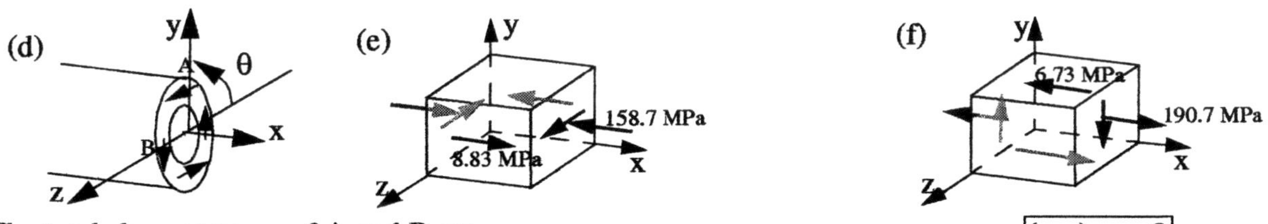

The total shear stresses of A and B are $\boxed{(\tau_{xy})_A = 0}$

$(\tau_{xz})_A = (-8.42 + 17.25)(10^6)\text{N/m}^2$ or $\boxed{(\tau_{xz})_A = 8.83\text{MPa}}$

$(\tau_{xy})_B = (10.52-17.25)(10^6)\text{N/m}^2$ or $\boxed{(\tau_{xy})_B = -6.73\text{MPa}}$

Figs. (e) and (f) shows the stress cubes at points A and B.

10.19 By inspection determine and show the total stresses at points A and B on stress cubes using the following notation for the *magnitude* of stress components:

σ_{axial} —axial normal stress; τ_{tor}—torsional shear stress;
$\sigma_{bend\text{-}y}$ —normal stress due to bending about y-axis;
$\tau_{bend\text{-}y}$ —shear stress due to bending about y-axis;
$\sigma_{bend\text{-}z}$ —normal stress due to bending about z-axis;
$\tau_{bend\text{-}z}$ —shear stress due to bending about z-axis.

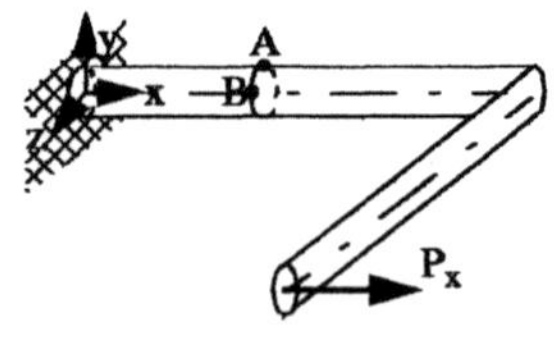

Fig. P10.19

Solution Stresses at points A and B.

The normal stresses are: $(\sigma_{xx})_A = \sigma_{axial}$ and $(\sigma_{xx})_B = \sigma_{axial} + \sigma_{bend-y}$

All shear stresses are zero.

Figs. (a) and (b) show the stress cubes at points A and B, respectively.

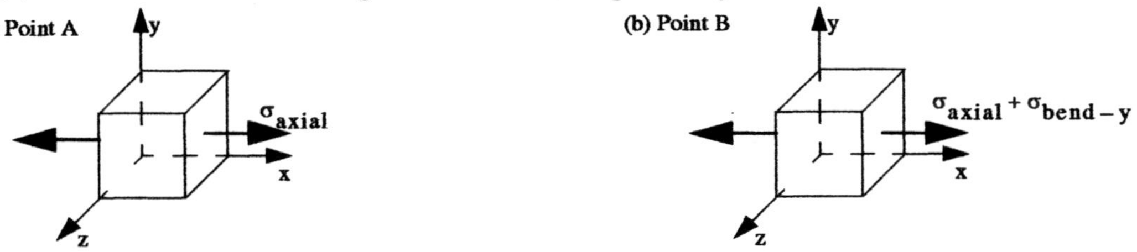

10.20 By inspection determine and show the total stresses at points A and B on stress cubes using the following notation for the *magnitude* of stress components:

σ_{axial} —axial normal stress; τ_{tor}—torsional shear stress;
$\sigma_{bend\text{-}y}$ —normal stress due to bending about y-axis;
$\tau_{bend\text{-}y}$ —shear stress due to bending about y-axis;
$\sigma_{bend\text{-}z}$ —normal stress due to bending about z-axis;
$\tau_{bend\text{-}z}$ —shear stress due to bending about z-axis.

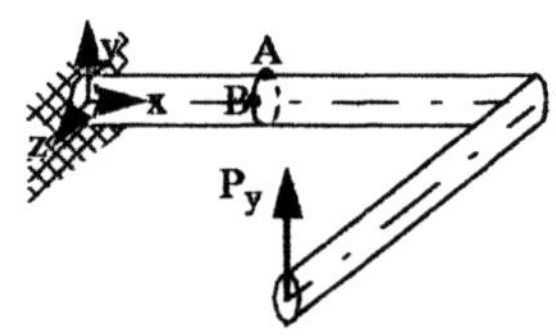

Fig. P10.20

Solution Stresses at points A and B.

The normal stresses are: $(\sigma_{xx})_A = -\sigma_{bend-z}$ and $(\sigma_{xx})_B = 0$

Fig.(a) shows the shear stress direction, determined by inspection, due to the torsional moment caused by P_y. Fig.(b) shows the direction of shear stress, determined by inspection, due to a positive shear force $V_y = P_y$.

The shear stresses at points A and B are: $(\tau_{xz})_A = -\tau_{tor}$ and $(\tau_{xy})_B = \tau_{tor} + \tau_{bend-z}$

Figs. (c) and (d) show the stress cubes at points A and B, respectively.

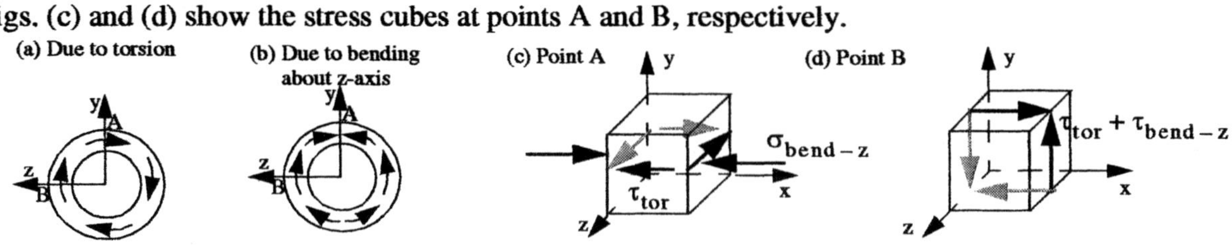

10.21 By inspection determine and show the total stresses at points A and B on stress cubes using the following notation for the *magnitude* of stress components:

σ_{axial} —axial normal stress; τ_{tor}—torsional shear stress;
$\sigma_{bend\text{-}y}$ —normal stress due to bending about y-axis;
$\tau_{bend\text{-}y}$ —shear stress due to bending about y-axis;
$\sigma_{bend\text{-}z}$ —normal stress due to bending about z-axis;
$\tau_{bend\text{-}z}$ —shear stress due to bending about z-axis.

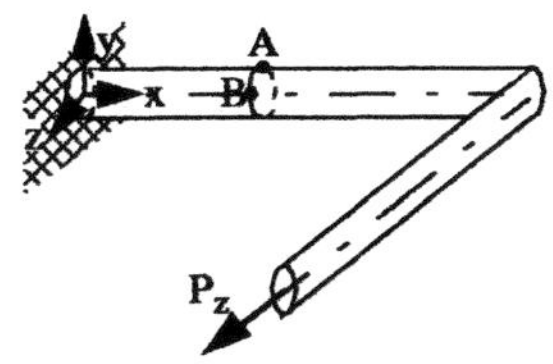

Fig. P10.21

Solution Stresses at points A and B.

The normal stresses are: $(\sigma_{xx})_A = 0$ and $(\sigma_{xx})_B = -\sigma_{bend-y}$

Fig.(a) shows the direction of shear stress, determined by inspection, due to a positive shear force $V_z = P_z$.

The shear stresses at points A and B are: $(\tau_{xz})_A = \tau_{bend-y}$ and $(\tau_{xy})_B = 0$

Figs. (b) and (c) show the stress cubes at points A and B, respectively.

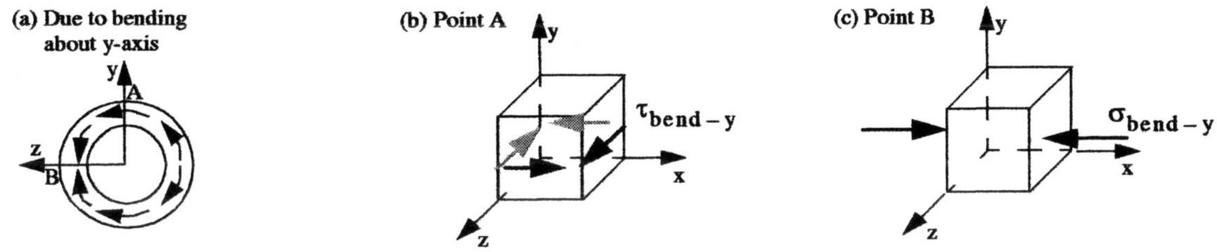

10.22 By inspection determine and show the total stresses at points A and B on stress cubes using the following notation for the *magnitude* of stress components:

σ_{axial} —axial normal stress; τ_{tor}—torsional shear stress;
$\sigma_{bend\text{-}y}$ —normal stress due to bending about y-axis;
$\tau_{bend\text{-}y}$ —shear stress due to bending about y-axis;
$\sigma_{bend\text{-}z}$ —normal stress due to bending about z-axis;
$\tau_{bend\text{-}z}$ —shear stress due to bending about z-axis.

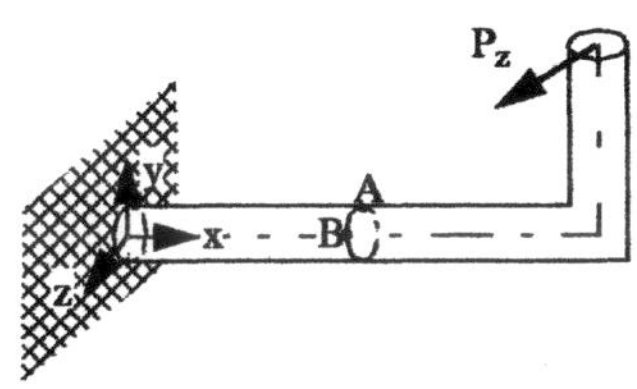

Fig. P10.22

Solution Stresses at points A and B.

The normal stresses are: $(\sigma_{xx})_A = 0$ and $(\sigma_{xx})_B = -\sigma_{bend-y}$

Fig.(a) shows the shear stress direction, determined by inspection, due to the torsional moment caused by P_z. Fig.(b) shows the direction of shear stress, determined by inspection, due to a positive shear force $V_z = P_z$.

The shear stresses at points A and B are: $(\tau_{xz})_A = \tau_{tor} + \tau_{bend-y}$ and $(\tau_{xy})_B = -\tau_{tor}$

Figs. (c) and (d) show the stress cubes at points A and B, respectively.

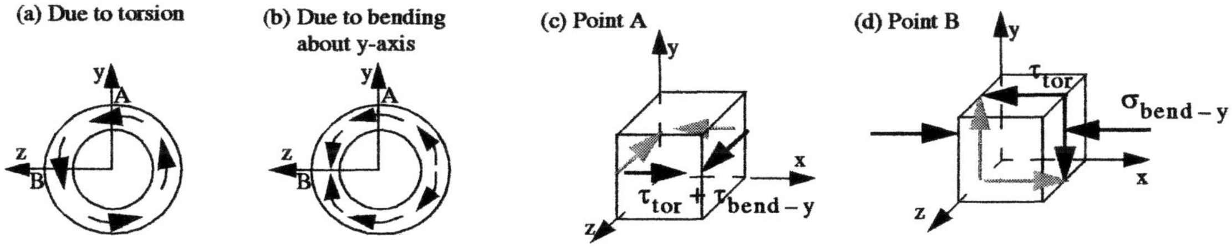

10.23 By inspection determine and show the total stresses at points A and B on stress cubes using the following notation for the *magnitude* of stress components:

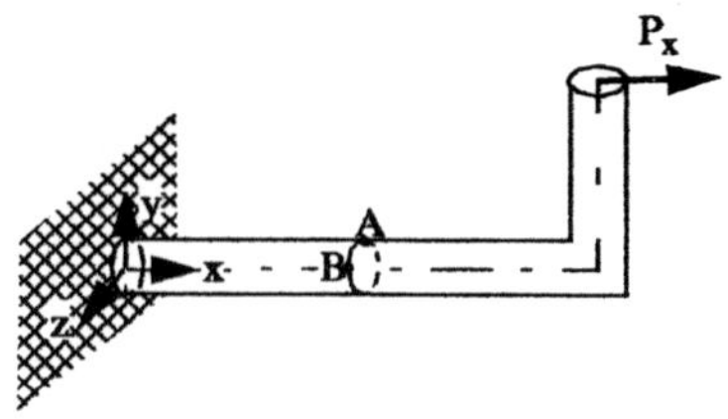

Fig. P10.23

σ_{axial} —axial normal stress; τ_{tor}—torsional shear stress;
$\sigma_{bend\text{-}y}$ —normal stress due to bending about y-axis;
$\tau_{bend\text{-}y}$ —shear stress due to bending about y-axis;
$\sigma_{bend\text{-}z}$ —normal stress due to bending about z-axis;
$\tau_{bend\text{-}z}$ —shear stress due to bending about z-axis.

Solution Stresses at points A and B.

The normal stresses are: $(\sigma_{xx})_A = \sigma_{axial} + \sigma_{bend-z}$ and $(\sigma_{xx})_B = \sigma_{axial}$

All shear stresses are zero.

Figs. (a) and (b) show the stress cubes at points A and B, respectively.

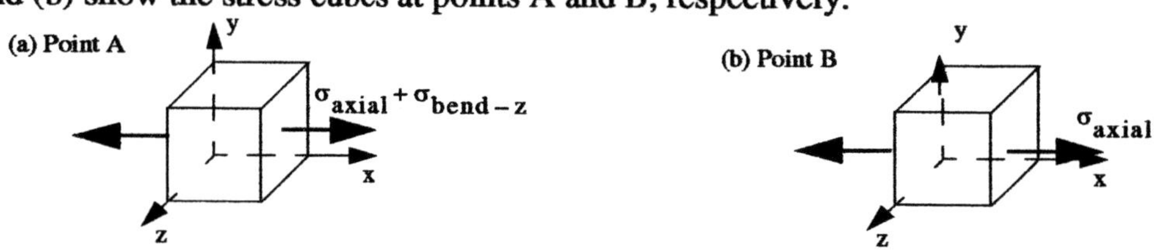

10.24 By inspection determine and show the total stresses at points A and B on stress cubes using the following notation for the *magnitude* of stress components:

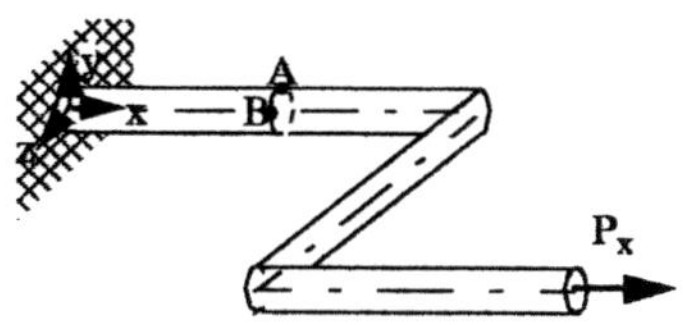

Fig. P10.24

σ_{axial} —axial normal stress; τ_{tor}—torsional shear stress;
$\sigma_{bend\text{-}y}$ —normal stress due to bending about y-axis;
$\tau_{bend\text{-}y}$ —shear stress due to bending about y-axis;
$\sigma_{bend\text{-}z}$ —normal stress due to bending about z-axis;
$\tau_{bend\text{-}z}$ —shear stress due to bending about z-axis.

Solution Stresses at points A and B.

The normal stresses are: $(\sigma_{xx})_A = \sigma_{axial}$ and $(\sigma_{xx})_B = \sigma_{axial} + \sigma_{bend-y}$

All shear stresses are zero.

Figs. (a) and (b) show the stress cubes at points A and B, respectively.

10.25 By inspection determine and show the total stresses at points A and B on stress cubes using the following notation for the *magnitude* of stress components:

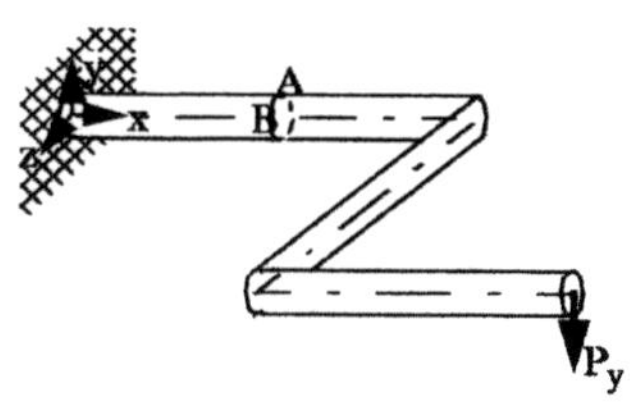

Fig. P10.25

σ_{axial} —axial normal stress; τ_{tor}—torsional shear stress;
$\sigma_{bend\text{-}y}$ —normal stress due to bending about y-axis;
$\tau_{bend\text{-}y}$ —shear stress due to bending about y-axis;
$\sigma_{bend\text{-}z}$ —normal stress due to bending about z-axis;
$\tau_{bend\text{-}z}$ —shear stress due to bending about z-axis.

Solution Stresses at points A and B.

The normal stresses are: $(\sigma_{xx})_A = \sigma_{bend-z}$ and $(\sigma_{xx})_B = 0$

Fig.(a) shows the shear stress direction, determined by inspection, due to the torsional moment caused by P_y. Fig.(b) shows the direction of shear stress, determined by inspection, due to a negative shear force $V_y = -P_y$.

The shear stresses at points A and B are: $(\tau_{xz})_A = \tau_{tor}$ and $(\tau_{xy})_B = -\tau_{tor} - \tau_{bend-z}$

Figs. (c) and (d) show the stress cubes at points A and B, respectively.

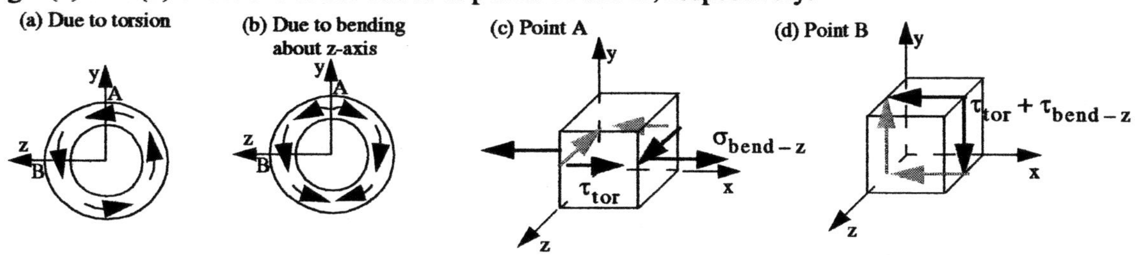

10.26 By inspection determine and show the total stresses at points A and B on stress cubes using the following notation for the *magnitude* of stress components:

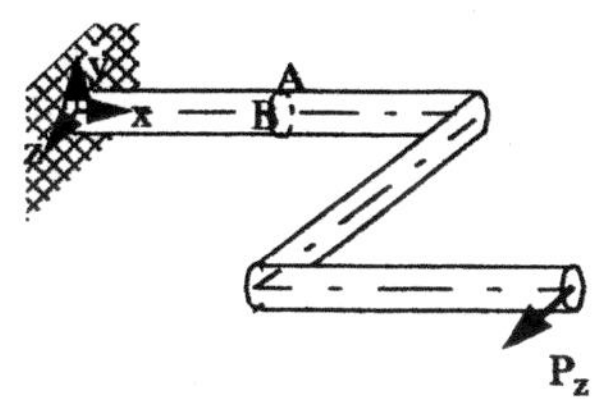

Fig. P10.26

σ_{axial} —axial normal stress; τ_{tor}—torsional shear stress;
$\sigma_{bend\text{-}y}$ —normal stress due to bending about y-axis;
$\tau_{bend\text{-}y}$ —shear stress due to bending about y-axis;
$\sigma_{bend\text{-}z}$ —normal stress due to bending about z-axis;
$\tau_{bend\text{-}z}$ —shear stress due to bending about z-axis.

Solution Stresses at points A and B.

The normal stresses are: $(\sigma_{xx})_A = 0$ and $(\sigma_{xx})_B = -\sigma_{bend-y}$

Fig.(a) shows the direction of shear stress, determined by inspection, due to a positive shear force $V_z = P_z$.

The shear stresses at points A and B are: $(\tau_{xz})_A = \tau_{bend-y}$ and $(\tau_{xy})_B = 0$

Figs. (b) and (c) show the stress cubes at points A and B, respectively.

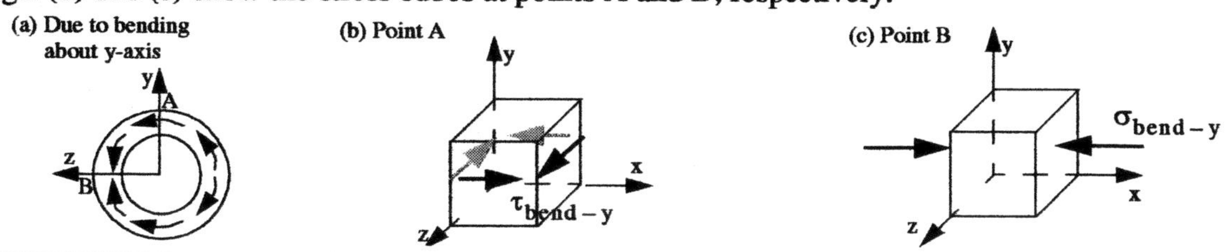

10.27 By inspection determine and show the total stresses at points A and B on stress cubes using the following notation for the *magnitude* of stress components:

σ_{axial} —axial normal stress; τ_{tor}—torsional shear stress;
$\sigma_{bend\text{-}y}$ —normal stress due to bending about y-axis;
$\tau_{bend\text{-}y}$ —shear stress due to bending about y-axis;
$\sigma_{bend\text{-}z}$ —normal stress due to bending about z-axis;
$\tau_{bend\text{-}z}$ —shear stress due to bending about z-axis.

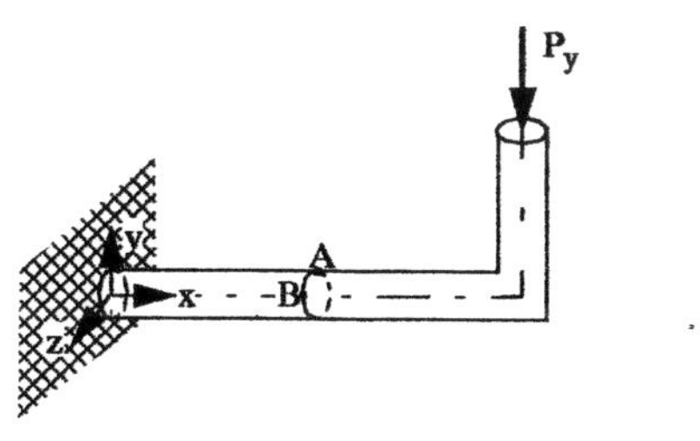

Fig. P10.27

Solution Stresses at points A and B.

The normal stresses are: $(\sigma_{xx})_A = \sigma_{bend-z}$ and $(\sigma_{xx})_B = 0$

Fig.(a) shows the direction of shear stress, determined by inspection, due to a negative shear force $V_y = -P_y$.

The shear stresses at points A and B are: $(\tau_{xz})_A = 0$ and $(\tau_{xy})_B = -\tau_{bend-z}$

Figs. (b) and (c) show the stress cubes at points A and B, respectively.

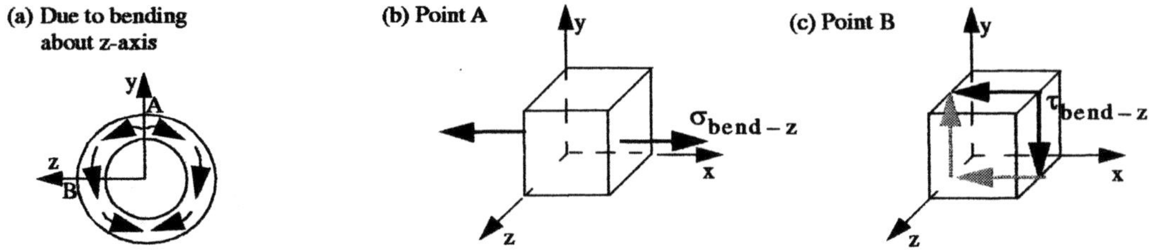

10.28 By inspection determine and show the total stresses at points A and B on stress cubes using the following notation for the *magnitude* of stress components:

σ_{axial} —axial normal stress; τ_{tor}—torsional shear stress;
σ_{bend-y} —normal stress due to bending about y-axis;
τ_{bend-y} —shear stress due to bending about y-axis;
σ_{bend-z} —normal stress due to bending about z-axis;
τ_{bend-z} —shear stress due to bending about z-axis.

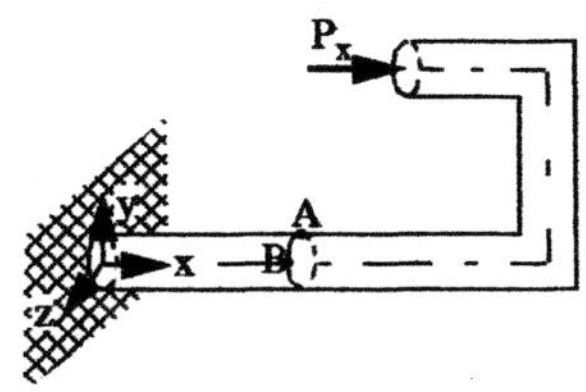

Fig. P10.28

Solution Stresses at points A and B.

The normal stresses are: $(\sigma_{xx})_A = \sigma_{axial} + \sigma_{bend-z}$ and $(\sigma_{xx})_B = \sigma_{axial}$

All shear stresses are zero.

Figs. (a) and (b) show the stress cubes at points A and B, respectively.

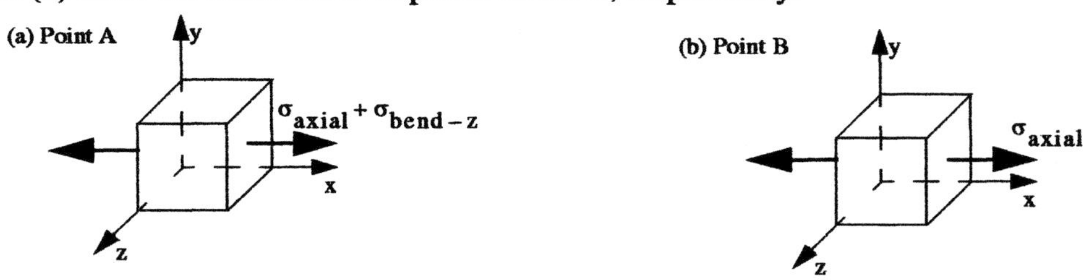

10.29 By inspection determine and show the total stresses at points A and B on stress cubes using the following notation for the *magnitude* of stress components:

σ_{axial} —axial normal stress; τ_{tor}—torsional shear stress;
σ_{bend-y} —normal stress due to bending about y-axis;
τ_{bend-y} —shear stress due to bending about y-axis;
σ_{bend-z} —normal stress due to bending about z-axis;
τ_{bend-z} —shear stress due to bending about z-axis.

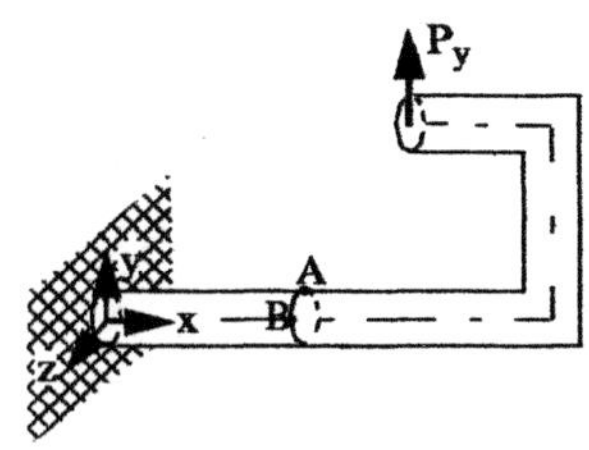

Fig. P10.29

Solution Stresses at points A and B.

The normal stresses are: $(\sigma_{xx})_A = -\sigma_{bend-z}$ and $(\sigma_{xx})_B = 0$

Fig.(a) shows the direction of shear stress, determined by inspection, due to a positive shear force $V_y = P_y$.

The shear stresses at points A and B are: $(\tau_{xz})_A = 0$ and $(\tau_{xy})_B = \tau_{bend-z}$

Figs. (b) and (c) show the stress cubes at points A and B, respectively.

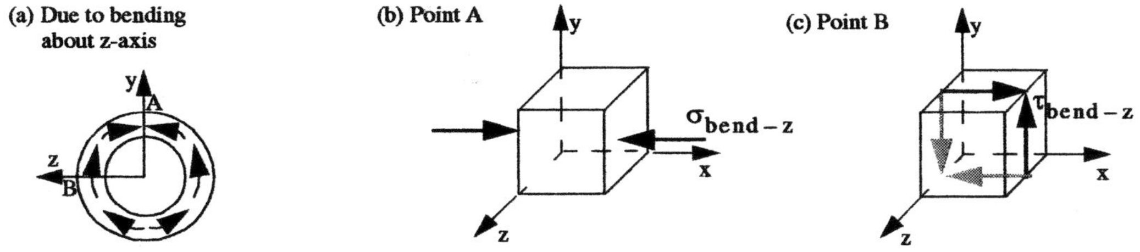

10.30 By inspection determine and show the total stresses at points A and B on stress cubes using the following notation for the *magnitude* of stress components:

σ_{axial} —axial normal stress; τ_{tor}—torsional shear stress;
σ_{bend-y} —normal stress due to bending about y-axis;
τ_{bend-y} —shear stress due to bending about y-axis;
σ_{bend-z} —normal stress due to bending about z-axis;
τ_{bend-z} —shear stress due to bending about z-axis.

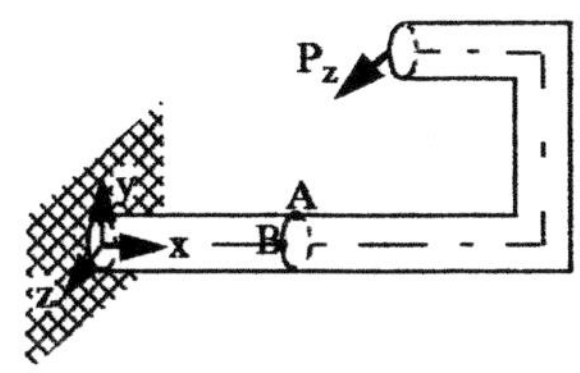

Fig. P10.30

Solution Stresses at points A and B.

The normal stresses are: $(\sigma_{xx})_A = 0$ and $(\sigma_{xx})_B = -\sigma_{bend-y}$

Fig.(a) shows the shear stress direction, determined by inspection, due to the torsional moment caused by P_z. Fig.(b) shows the direction of shear stress, determined by inspection, due to a positive shear force $V_z = P_z$.

The shear stresses at points A and B are: $(\tau_{xz})_A = \tau_{tor} + \tau_{bend-y}$ and $(\tau_{xy})_B = -\tau_{tor}$

Figs. (c) and (d) show the stress cubes at points A and B, respectively.

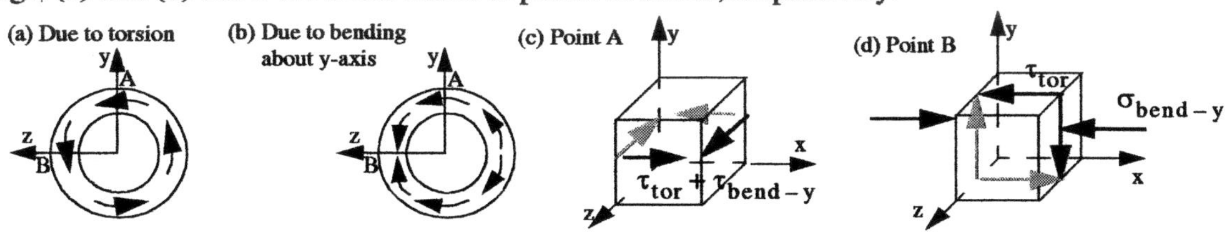

10.31 A pipe with an outside diameter of 2.0 inch and wall thickness of 1/4 inch is loaded as shown in Fig. P10.31. Determine the normal and shear stresses at point A and B in the x, y, and z coordinate system and show it on a stress cube. Points A and B are on the surface of the pipe.

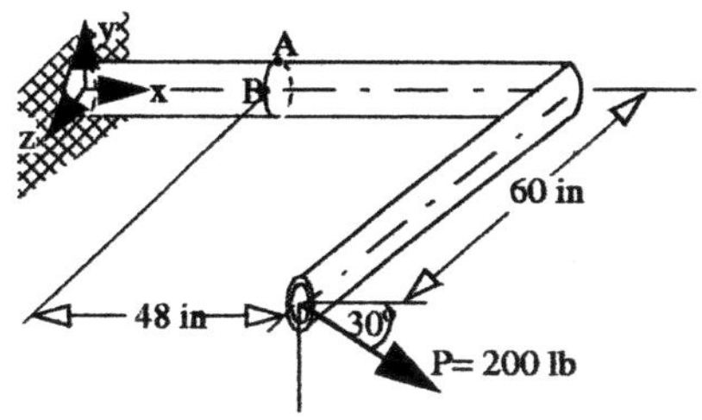

Fig. P10.31

Solution $d_0 = 2$ in, $t = \frac{1}{4}$in, Stresses at points A and B =?

Fig. (a) shows the free body diagram after an imaginary cut is made through section containing points A and B. By equilibrium of forces and moments we obtain.

$N = 200\cos(30) = 173.2\text{lb}$ $\quad T = [200\sin(30)](60) = 6000\text{in}-\text{lb}$ $\quad V_y = (-200)\sin(30) = 100\text{lb}$

$M_y = [(-200)\cos(30)](60) = -10392.3\text{in}-\text{lb}$ $M_z = [(-200)\sin(30)](48) = -4800\text{in}-\text{lb}$

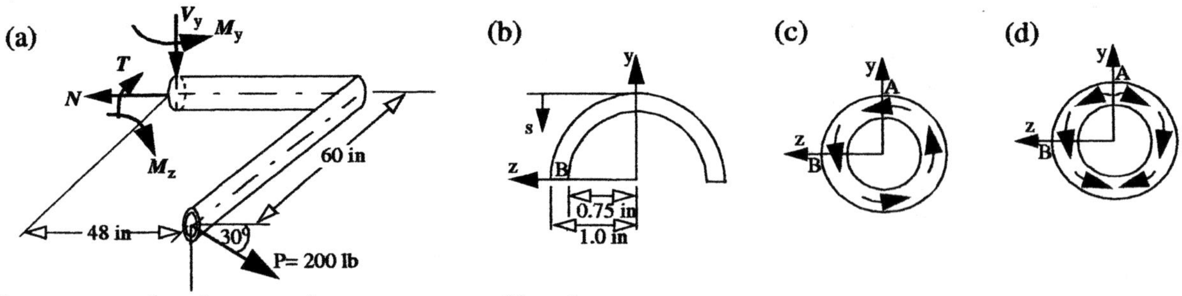

The cross sectional area and area moment of inertias are

$$A = \frac{\pi}{4}[(2)^2-(1.5)^2] = 1.374\text{in}^2 \quad J = \frac{\pi}{32}[(2)^4-(1.5)^4] = 1.0738\text{in}^4 \quad I_{yy} = I_{zz} = \frac{J}{2} = 0.5369\text{in}^4$$

The axial stress at points A and B will be in tension and the value is $(\sigma_{xx})_{A,B} = \frac{N}{A} = \frac{173.2}{1.374} = 126\text{psi(T)}$

The normal stress due to bending about z -axis will be tension at A and zero at B. Thus

$$(\sigma_{xx})_A = \left|\frac{M_z y_A}{I_{zz}}\right| = \frac{(4800)(1)}{0.5369} = 8940\text{psi(T)} \quad (\sigma_{xx})_B = 0$$

The normal stress due to bending about y-axis will be tension at B and zero at A. Thus

$$(\sigma_{xx})_A = 0 \quad (\sigma_{xx})_B = \left|\frac{M_y z_A}{I_{yy}}\right| = \frac{(10392.3)(1)}{0.5369} = 19356\text{psi(T)}$$

The total normal stress at points A and B is

$$\boxed{(\sigma_{xx})_A = 126+8940 = 9066\text{psi(T)}}$$ 1

$$\boxed{(\sigma_{xx})_B = 126+19356 = 19482\text{psi(T)}}$$ 2

Fig. (b) shows the area A_s used in calculation of shear stress at B: $(Q_z)_B = \frac{2}{3}[(1)^3-(0.75)^3] = 0.3854\text{in}^3$

Fig.(c) shows the direction of shear stress due to V_y. We note that the shear stress at A is zero and at B is downwards. Thus: $(\tau_{xz})_A = 0$ and $(\tau_{xy})_B = -\left|\frac{V_y(Q_z)_B}{I_{zz}t}\right| = \frac{-(100)(0.3854)}{(0.5369)(0.5)} = -143.6\text{psi}$

The magnitude of torsional shear stress is: $\tau_{tor} = \left|\frac{T\rho_{max}}{J}\right| = \frac{(6000)(1)}{1.0738} = 5587.6\text{psi}$

Fig. (d) shows the direction of torsional shear stress. At point A it is in the positive z - direction and at point B it is in the negative y-direction.

$(\tau_{xz})_A = 5587.6\text{psi}$ and $(\tau_{xy})_B = -5587.6\text{psi}$

The total shear stresses of A and B are

$$\boxed{(\tau_{xz})_A = 5588\text{psi}}$$ 3

$$\boxed{(\tau_{xy})_B = (-143.6-5587.6) = -5731.2\text{psi}}$$ 4

Fig. (e) and (f) show the stress cubes of points A and B.

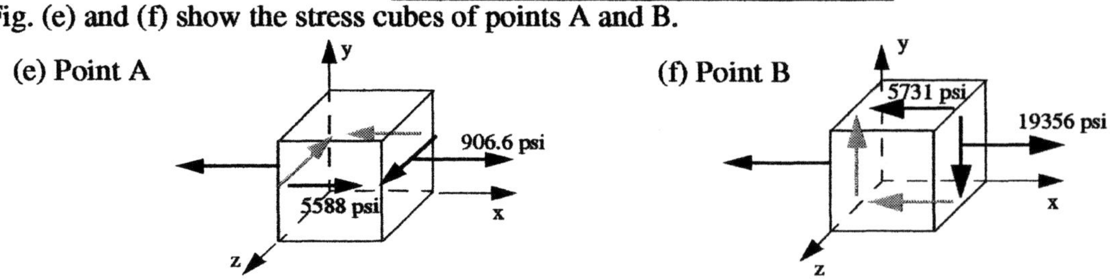

10.32 Determine the maximum normal stress and maximum shear stress at point B on the pipe shown in Fig. P10.31.

Solution: $\sigma_{max} = ?$ $\tau_{max} = ?$

From Eqs. (2) and (4) of problem 10.31 we have $(\sigma_{xx})_B = 19482\text{psi(T)}$ and $(\tau_{xy})_B = -5731\text{psi}$

The principal stresses are

$$\sigma_{1,2} = \left(\frac{19482}{2}\right) \pm \sqrt{\left(\frac{19482}{2}\right)^2 + (5731)^2} \text{ or } \sigma_1 = 9741 + 11302 = 21043\text{psi} \quad \sigma_2 = 9741 - 11302 = -1561\text{psi}$$

The maximum normal stress is $\boxed{\sigma_{max} = 21043\text{psi}}$

The magnitude of maximum shear stress is $\tau_{max} = \left|\frac{\sigma_1 - \sigma_2}{2}\right|$ or $\boxed{\tau_{max} = 11302\text{psi}}$

10.33 A pipe with an outside diameter of 40 mm and wall thickness of 10 mm is loaded as shown in Fig. P10.33. Determine the normal and shear stresses at point A and B in the x, y, and z coordinate system and show it on a stress cube. Points A and B are on the surface of the pipe.

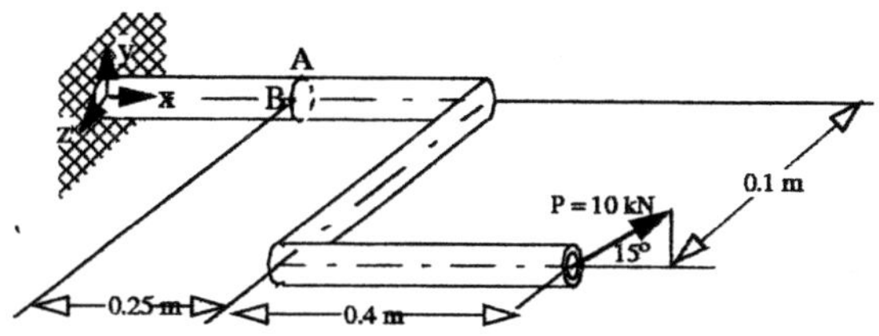

Fig. P10.33

Solution: $d_0 = 40$ mm, $t = 10$ mm, stresses at A and B =?

Fig (a) shows the free body diagram after an imaginary cut is made through section containing points A and B. By equilibrium of forces and moments we obtain

$N = 10\cos(15) = 9.659\text{kN}$ $T = (-10\sin(15))(0.1) = (-0.2588)\text{kN}-\text{m}$ $V_y = 10\sin(15) = 2.588\text{kN}$

$M_y = (-10\cos(15))(0.1) = -0.9659\text{kN}-\text{m}$ $M_z = (10)\sin(15)(0.65) = 1.6823\text{kN}-\text{m}$

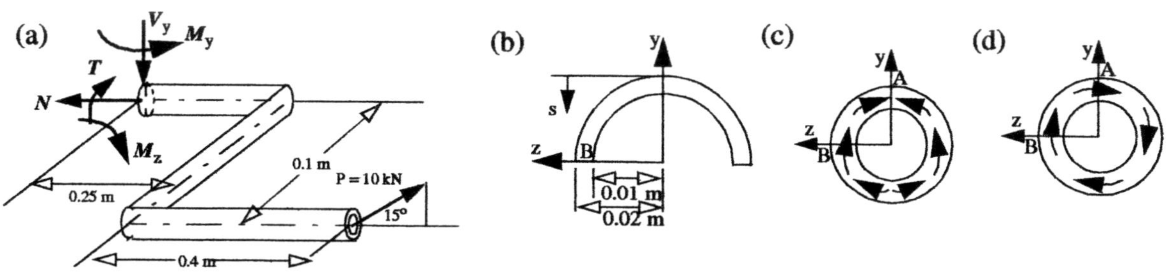

The cross sectional area and area moment of inertia are

$$A = \frac{\pi}{4}[(0.04)^2 - (0.02)^2] = 0.9425(10^{-3})\text{m}^2$$

$$J = \frac{\pi}{32}[(0.04)^4 - (0.02)^4] = 0.2356(10^{-6})\text{m}^4 \quad I_{yy} = I_{zz} = \frac{J}{2} = 0.1178(10^{-6})\text{m}^4$$

The axial stress at points A and B will be in tension

$$(\sigma_{xx})_{A,B} = \left|\frac{N}{A}\right| = \frac{9.659(10^3)}{0.9425(10^{-3})} = 10.25(10^6)\frac{\text{N}}{\text{m}^2} = 10.25\text{MPa(T)}$$

The normal stress due to bending about the z-axis will be compression at point A and zero at point B

$$(\sigma_{xx})_A = \left|\frac{M_z y_A}{I_{zz}}\right| = \frac{(1.6823)(10^3)(0.02)}{(0.1178)(10^{-6})} = 285.6(10^6)\frac{\text{N}}{\text{m}^2} = 285.6\text{MPa(C)} \quad (\sigma_{xx})_B = 0$$

The normal stress due to bending about y-axis will be tension at point B and zero at point A

$$(\sigma_{xx})_A = 0 \quad (\sigma_{xx})_B = \left|\frac{M_y z_A}{I_{yy}}\right| = \frac{(0.9659)(10^3)(0.02)}{(0.1178)(10^{-6})} = 163.98(10^6)\frac{N}{m^2} = 164\text{MPa(T)}$$

The total stress at point A and B is

$$\boxed{(\sigma_{xx})_A = 10.25 - 285.6 = -275.35\text{MPa}}$$ 1

$$\boxed{(\sigma_{xx})_B = 10.25 + 164 = 174.25\text{MPa}}$$ 2

Fig. (b) shows the area A_s used in calculation of shear stress at B.

$$(Q_z)_B = \frac{2}{3}(0.02^3 - 0.01^3) = 4.667(10^{-6})\text{m}^3$$

Fig (c) shows the direction of shear stress due to V_y. We note that the shear stress at A is zero and at B is upwards. Thus

$$(\tau_{xz})_A = 0 \text{ and } (\tau_{xy})_B = \left|\frac{V_y(Q_t)_B}{I_{zz}t}\right| = \frac{(2.588)(10^3)(4.667)(10^{-6})}{(0.1178)(10^{-6})(0.02)} = 5.13(10^6)\frac{N}{m^2} = 5.13\text{MPa}$$

The magnitude of torsional shear stress is:

$$\tau_{tor} = \left|\frac{T\rho_{max}}{J}\right| = \frac{(0.2588)(10^3)(0.02)}{0.2356(10^{-6})} = 21.97(10^6)\frac{N}{m^2} = 21.97\text{MPa}$$

Fig. (d) shows the direction of torsional shear stress. At point A it is in the negative z - direction and at point B it is in the positive y-direction. Thus

$(\tau_{xz})_A = -21.97\text{MPa}$ and $(\tau_{xy})_B = 21.97\text{MPa}$

The total shear stresses of points A and B are

$$\boxed{(\tau_{xz})_A = -22\text{MPa}}$$ 3

$$\boxed{(\tau_{xy})_B = (5.13 + 21.97) = 27.1\text{MPa}}$$ 4

Fig. (e) and (f) show the stress cubes of points A and B.

(c) Point A

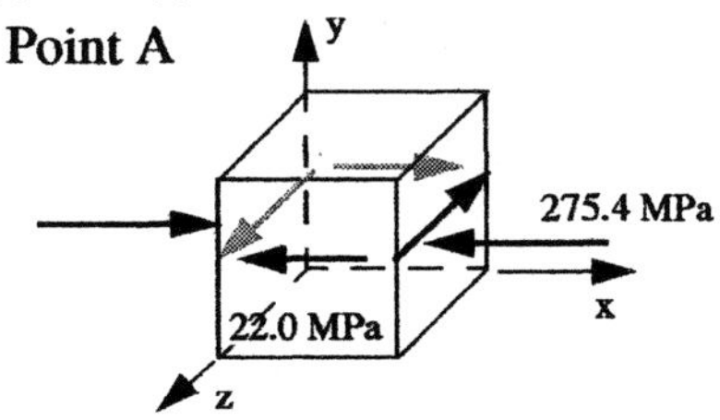

(d) Point B

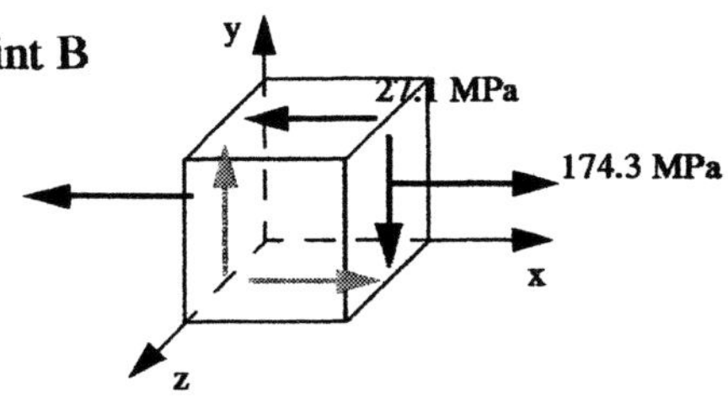

10.34 Determine the maximum normal stress and maximum shear stress at point B on the pipe shown in Fig. P10.33.

Solution: $\sigma_{max} = ?$ $\tau_{max} = ?$

From Eqs. (2) and (4) of problem 10.33 we have $(\sigma_{xx})_B = 174.3\text{MPa}$ and $(\tau_{xy})_B = 27.1\text{MPa}$

The principal stresses are: $\sigma_{1,2} = \left(\frac{174.3}{2}\right) \pm \sqrt{\left(\frac{174.3}{2}\right)^2 + (27.1)^2}$ or

$\sigma_1 = 87.15 + 91.27 = 178.42\text{MPa}$ and $\sigma_2 = 87.15 - 91.27 = -4.12\text{MPa}$

The maximum normal stress is $\boxed{\sigma_{max} = 178.4\text{MPa}}$

The magnitude of maximum shear stress is $\tau_{max} = \left|\frac{\sigma_1 - \sigma_2}{2}\right|$ or $\boxed{\tau_{max} = 91.3\text{MPa}}$

10.35 A bent pipe of 2 inch outside diameter and a wall thickness of 1/4 inch is loaded as shown in Fig. P10.35. Determine the stress components at point A which is on the surface of the shaft. Show your answer on a stress cube.

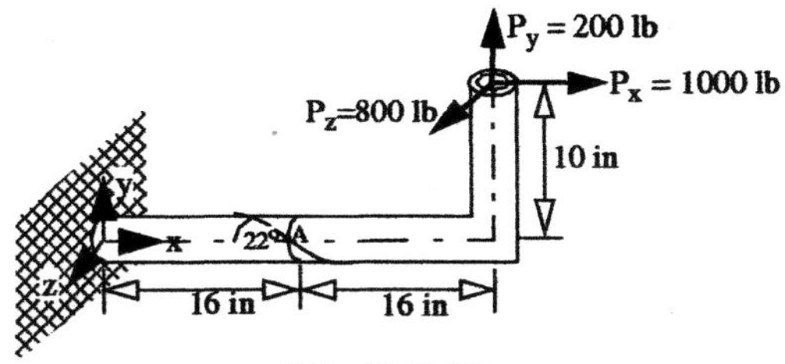

Fig. P10.35

Solution $d_0 = 2$ in, $t = \frac{1}{4}$in, Stresses at points A =?

Figs. (a) shows the free body diagram after an imaginary cut is made through section containing points A.

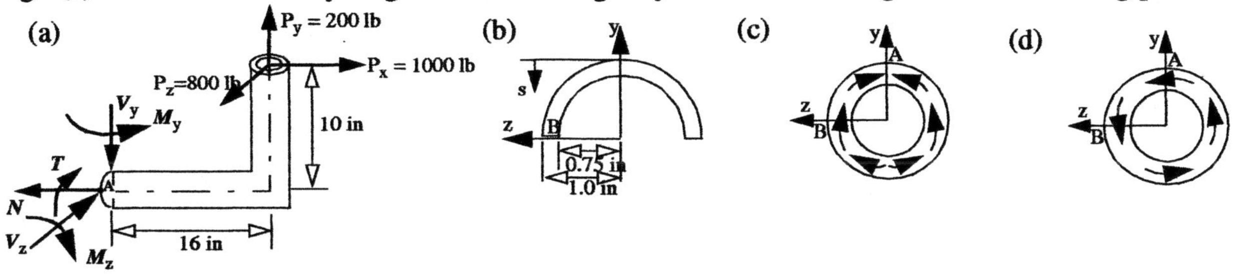

By equilibrium of forces and moments we obtain

$N = 1000\text{lb}$ $T = (800)(10) = 8000\text{in-lb}$ $V_y = 200\text{in-lb}$ $V_z = 800\text{in-lb}$

$M_y = (800)(16) = 12800\text{in-lb}$ $M_z = (200)(16) = 3200\text{in-lb}$

The cross sectional area and area moment of inertia are

$A = \frac{\pi}{4}[(2)^2 - (1.5)^2] = 1.3745\text{in}^2$ $J = \frac{\pi}{32}[(2)^4 - (1.5)^4] = 1.0738\text{in}^4$ $I_{yy} = I_{zz} = 0.5369\text{in}^4$

The axial stress at points A is in tension: $(\sigma_{xx})_A = \left|\frac{N}{A}\right| = \frac{1000}{1.3745} = 728\text{psi}$

The normal stress due to bending about z-axis is zero at point A. The normal stress due to bending about y axis is compressive at point A

$$(\sigma_{xx})_A = \left|\frac{M_y z_A}{I_{yy}}\right| = \frac{12800(1)}{0.5369} = 23841\text{psi(C)}$$

The total normal stress at point A is:

$$\boxed{(\sigma_{xx})_A = -23841 + 728 = -23113\text{psi}}$$ 1

Fig. (b) shows the area A_s used in calculation of shear stress at point A. $(Q_z)_A = \frac{2}{3}(1^3 - 0.75^3) = 0.3854\text{in}^3$

Fig (c) shows the direction of shear stress due to V_y. We note that the shear stress at A is in the positive direction.

$$(\tau_{xy})_A = \left|\frac{V_y (Q_z)_A}{I_{zz} t}\right| = \frac{(200)(0.3854)}{(0.5369)(0.5)} = 287\text{psi}$$

Fig. (a) shows the direction of torsional shear stress. At point A the direction of shear stress is in the negative y- direction.

$$(\tau_{xy})_A = -\left|\frac{T\rho}{J}\right| = \frac{-(8000)(1)}{1.0738} = -7450\text{psi}$$

The total shear stresses of points A is:

$$(\tau_{xy})_A = 287 - 7450 = -7163\text{psi}$$ **2**

Fig. (e) shows the stress cube at point A.

(e)

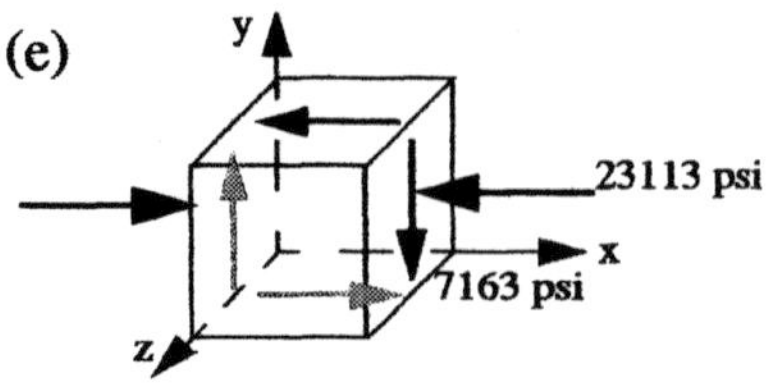

10.36 Determine the normal and shear stress on a seam through point A that is 22° to the axis of the pipe shown in Fig. P10.35.

Solution: σ_{nn} =? τ_{nn} =?

From Eqs. (1) and (2) in problem 10.35 we have: $(\sigma_{xx})_A = -23113\text{psi}$ and $(\tau_{xy})_A = -7163\text{psi}$

Fig (a) shows the stress cube with inclined plane at point A

(a)

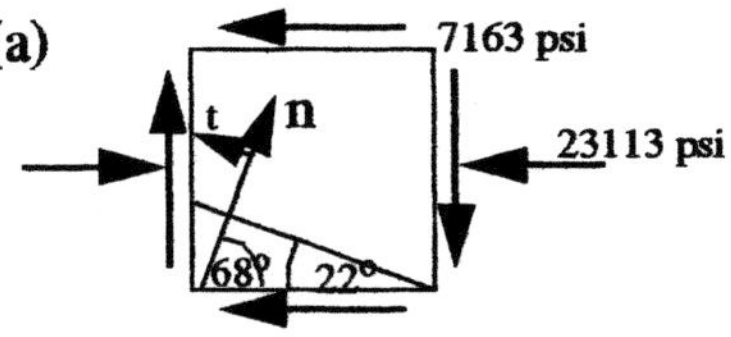

Substituting the stresses of Eq. (1) and the angle of the normal to the plane as 68o in Eqs. 8.1 and 8.2 we obtain.

$\sigma_{nn} = (-23113)(\cos 68)^2 + 2(-7163)\cos 68 \sin 68$ or $\boxed{\sigma_{nn} = 8219\text{psi}(C)}$

$\tau_{nt} = -(-23113)\cos 68 \sin 68 + (-7163)[(\cos 68)^2 - (\sin 68)^2]$ or $\boxed{\tau_{nt} = 3180\text{psi}}$

10.37 A hollow steel shaft has an outside diameter of 4 inches and an inside diameter of 3 inches. Two pulleys of 24 inch diameter carry belts that have the given tensions. The shaft is supported at the walls using flexible bearings permitting rotation in all directions. Determine the maximum normal and shear stress in the shaft.

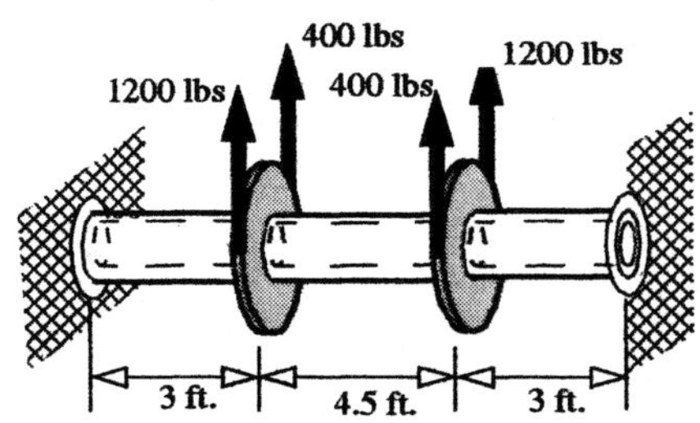

Fig. P10.37

Solution: $d_o = 4$ in, $d_i = 3$ in, σ_{max} =?, τ_{max} =?

By symmetry the reaction force at each wall is half the total force i.e. $R_A = R_B = 1600\text{lb}$. The magnitude of the torque on each pulley is: $T_{ext} = (1200)(12) - (400)(12) = 9600\text{in-lb}$. Fig (a) shows the shaft with equivalent loads. Fig (b) shows the shear force and bending moment diagram.

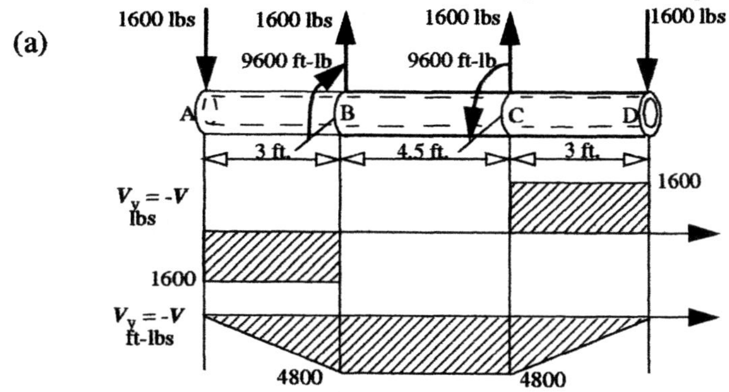

The area moment of inertias are: $J = \frac{\pi}{32}[(4)^4 - (3)^4] = 17.18\text{in}^4$ and $I_{zz} = \frac{J}{2} = 8.59\text{in}^4$

In segment AB and CD we have bending stresses only. The normal stress due to bending will be at top and bottom surface but the shear stress is zero at top and bottom. In segment BC in addition to the bending stress at top and bottom we also have torsional shear stress. Thus the maximum normal stress (principal stress and maximum shear stress will be in segment BC.

The magnitude of maximum bending normal stress is: $\sigma_{bend} = \left|\frac{M_z y_{max}}{I_{zz}}\right| = \frac{(4800)(12)(2)}{8.59} = 13411\text{psi}$

The magnitude of maximum torsional shear stress is: $\tau_{tor} = \left|\frac{T\rho_{max}}{J}\right| = \frac{(9600)(2)}{17.18} = 1118\text{psi}$

The magnitude of maximum normal stress is: $\sigma_{max} = \left(\frac{13411}{2}\right) \pm \sqrt{\left(\frac{13411}{2}\right)^2 + (1118)^2}$ or $\boxed{\sigma_{max} = 13504\text{psi}}$

The magnitude of the maximum shear stress is: $\tau_{max} = \sqrt{\left(\frac{13411}{2}\right)^2 + (1118)^2}$ or $\boxed{\tau_{max} = 6798\text{psi}}$

10.38 A thin cylinder is subjected to a uniform pressure of 300 psi and torques as shown. The outer radius of the cylinder is 10 inches and has a wall thickness of 0.25 inch. Determine the normal and shear stress at point A and show it on a stress element in cylindrical coordinates.

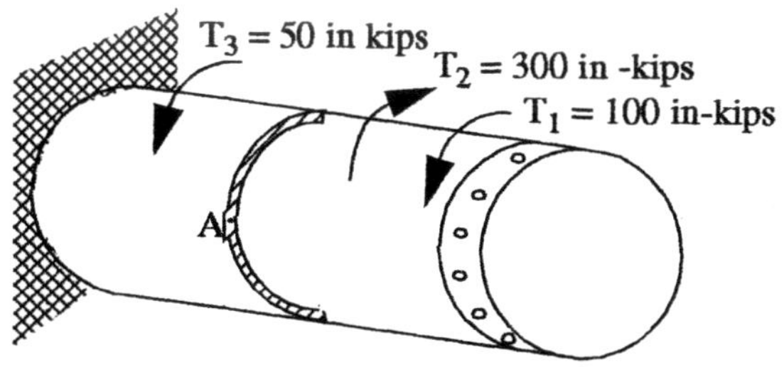

Fig. P10.38

Solution p = 300 psi, $d_0 = 10$ in, $t = 0.25$, Stresses at A =?

Figs. (a) shows the free body diagram for purpose of calculating the internal torque.

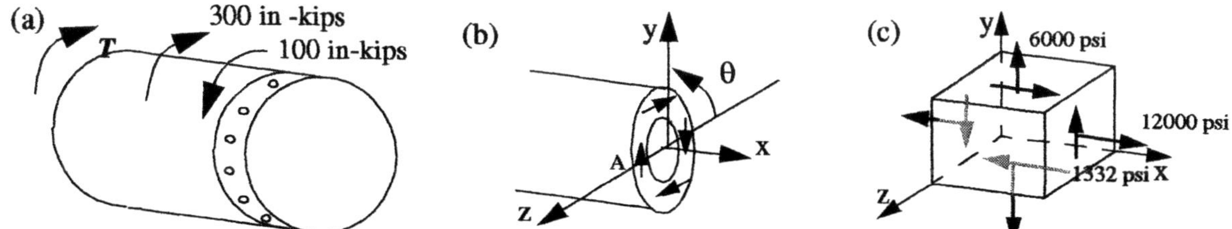

By equilibrium of moment $T = -200$ in-kips.

The polar moment of inertia is: $J = \frac{\pi}{2}[(10)^4 - (9.75)^4] = 1512.9\text{in}^4$

The torsional shear stress is: $\tau_{x\theta} = \left|\frac{T\rho}{J}\right| = \frac{(-200)(10)}{1512.9} = -1.322\text{ksi} = -1322\text{psi}$

Fig (b) shows the direction of torsional shear stress. At point A it is in the positive y - direction. Hence $(\tau_{xy})_A = 1322\text{psi}$

The hoop and axial stress due to pressure are

$\sigma_{\theta\theta} = \frac{pr}{t} = \frac{(300)(10)}{0.25} = 12000\text{psi}$ and $\sigma_{xx} = \frac{pr}{2t} = 6000\text{psi}$

Fig. (c) shows the stress cube at point A

10.39 A hollow shaft that has an outside diameter of 100 mm and an inside diameter of 50 mm is loaded as shown. The normal stress and shear stress in the shaft must be limited to 200 MPa and 115 MPa, respectively. (a) Determine the maximum value of the torque T that can be applied to the shaft. (b) Using the result of part (a), determine the strain that will be shown by the strain gage that is mounted on the surface at an angle of 35^0 to the axis of the shaft. Use E = 200 GPa, G = 80 GPa, ν = 0.25

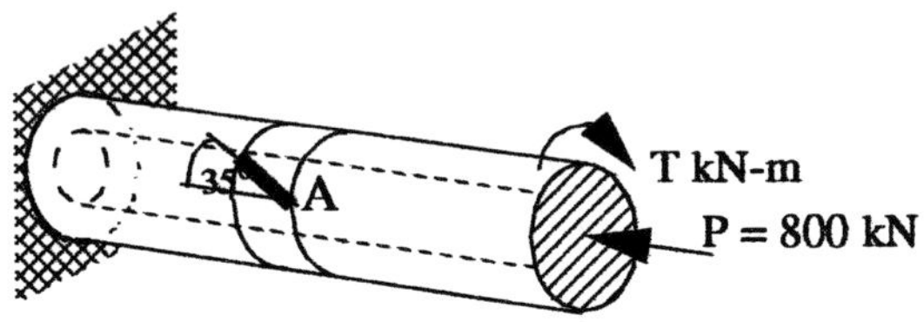

Fig. P10.39

Solution: $d_o = 100$ mm, $d_i = 50$ mm, $\sigma_{max} \le 200\text{MPa}$, $\tau_{max} \le 115$ MPa, E = 200 GPa G = 80 GPa, ν = 0.25 T_{max} =?, ε_A =?,

---

The cross sectional area and polar moment of inertia are

$A = \frac{\pi}{4}[(0.1)^2 - (0.05)^2] = 5.89(10^{-3})\text{m}^2$ and $J = \frac{\pi}{32}[(0.1)^4 - (0.05)^4] = 9.204(10^{-6})\text{m}^4$

The axial stress will be compressive and its value is

$$\sigma_{xx} = \frac{800(10^3)}{5.89(10^{-3})} = 135.8(10^6)\frac{N}{m^2} = 135.8\text{MPa(C)} \quad 1$$

The magnitude of torsional shear stress is: $\tau_{tor} = \left|\frac{T\rho}{J}\right| = \frac{T(10^3)(0.05)}{9.204(10^{-6})} = 5.432(T)\text{MPa}$

Fig (a) shows the direction of shear stress

(a)

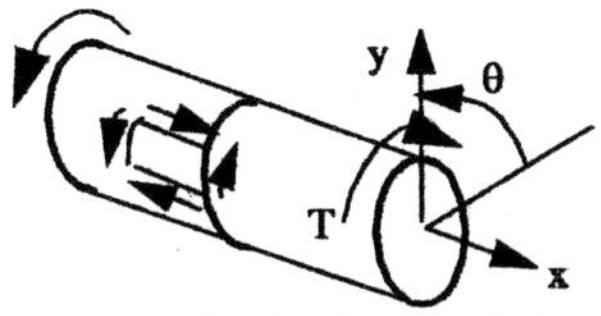

From Fig (a) we note that the torsional shear stress is in the positive y-direction at A. Thus,

$$(\tau_{xy})_A = 5.432(T)\text{MPa} \quad 2$$

The maximum normal stress is: $\sigma_{max} = \left|\left(\frac{-135.8}{2}\right) - \sqrt{\left(\frac{-135.8}{2}\right)^2 + (5.432T)^2}\right| \le 200$ or

$$\sqrt{\left(\frac{135.8}{2}\right)^2 + (5.432T)^2} \le 132.1 \quad 3$$

The maximum shear stress is

$$\tau_{max} = \sqrt{\left(\frac{135.8}{2}\right)^2 + (5.432T)^2} \le 115 \qquad \textbf{4}$$

If T is found from Eq. (4) than the condition of Eq. (3) will be met.

$\left(\frac{135.8}{2}\right)^2 + (5.432T)^2 \le 13225$ or $T \le 17.08 \text{in-kips}$ or $\boxed{T_{max} = 17.0\text{in-kips}}$

The strains are:

$$\varepsilon_{xx} = \frac{(-135.8)(10^6)}{200(10^9)} = -679\mu \text{ and } \varepsilon_{yy} = -\nu\varepsilon_{xx} = 169.75\mu \quad \gamma_{xy} = \frac{\tau_{xy}}{G} = \frac{(5.432)(17)(10^6)}{(80)(10^9)} = 1154\mu$$

Substituting the above strains and $\theta = -35^\circ$ in Eq. 9.1 we obtain

$$\varepsilon_A = -(679)(\cos(-35))^2 + (169.75)(\sin(-35))^2 + 1154(\cos(-35)\sin(-35)) \qquad \boxed{\varepsilon_A = -942\mu}$$

10.40 On the C-clamp, determine the maximum clamping force P, if the allowable normal stress is 160 MPa in tension and 120 MPa in compression.

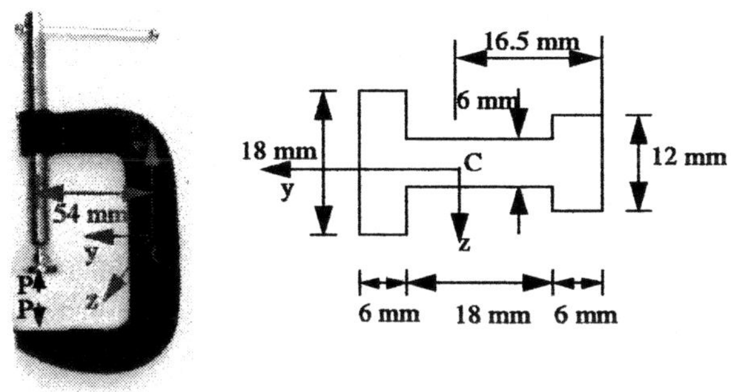

Fig. P10.40

Solution: $\sigma_T \le 160$ MPa (T) $\qquad \sigma_C \le 120$ MPa (C)

The cross sectional area is $A = (12)(6) + (18)(6) = 288\text{mm}^2$. The area moment of inertia about z -axis is.

$$I_{zz} = \frac{1}{12}(12)(6^3) + (12)(6)(16.5-3)^2 + \frac{1}{12}(6)(18^3) + (6)(18)(15-16.5)^2 + \frac{1}{12}(18)(6^3) + (18)(6)(27-16.5)^2$$

$$I_{zz} = 28.728(10^3)\text{mm}^4$$

Fig (a) shows the free body diagram after an imaginary cut is made.

(a)

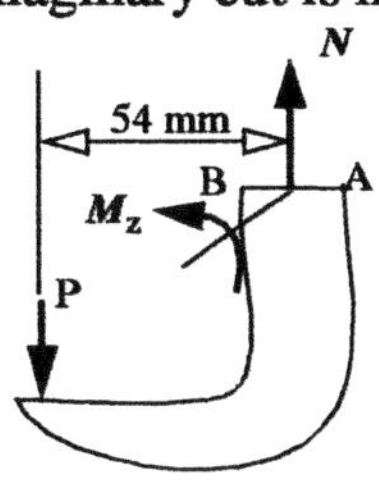

By equilibrium of forces and moments $N = P$ and $M_z = -0.054P$

The axial stress will be in tension. Its value is: $(\sigma_{xx})_{A,B} = \frac{P}{288(10^{-6})} = 3.472P(10^3)\frac{N}{m^2}(T)$

Due to bending about z -axis, point B will be in tension and point A will be in compression.

$$(\sigma_{xx})_B = \left|\frac{M_z y_B}{I_{zz}}\right| = \frac{(0.054P)(30-16.5)(10^{-3})}{28.728(10^{-9})} = 25.376P(10^3)\frac{N}{m^2}(T)$$

$$(\sigma_{xx})_A = \left|\frac{M_z y_A}{I_{zz}}\right| = \frac{(0.054P)(16.5)(10^{-3})}{28.728(10^{-9})} = 31.01P(10^3)\frac{N}{m^2}(C)$$

The total normal stresses at points A and B can be found and compared with the limiting values as shown below.

$$(\sigma_{xx})_A = (31.01P - 3.472P)(10^3) = 27.538P(10^3) \le 120(10^6) \quad \text{or} \quad P \le 4.357(10^3)N \qquad 5$$

$$(\sigma_{xx})_B = (25.376P + 3.472P)(10^3) = 28.848P(10^3) \le 160(10^6) \quad \text{or} \quad P \le 5.546(10^3)N \qquad 6$$

The maximum value of P that meets the requirements in Eqs. (1) and (2) is $\boxed{P_{max} = 4.3\text{kN}}$

10.41 The T cross-section of the beam was constructed by gluing two rectangular pieces together. A small crack was detected in the glue joint at section AA. Determine the maximum value of the applied load P if the normal stress in the glue at section AA is to be limited to 20MPa in tension and 12 MPa in shear. The load P acts at the centroid of the cross-section at C as shown.

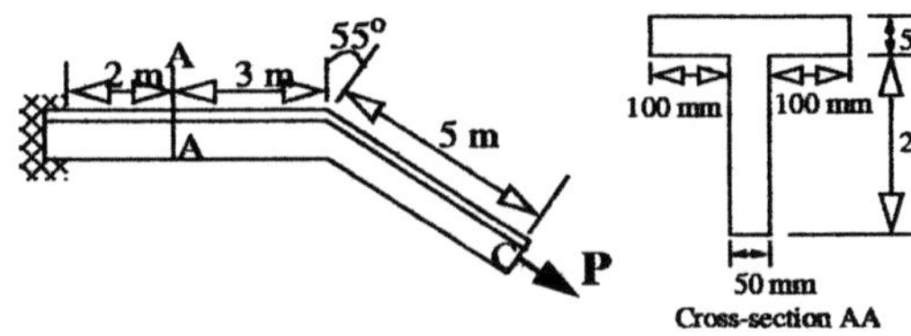

Fig. P10.41

Solution: $\sigma_G \le 20\text{MPa}$, $\tau_G \le 12\text{ MPa}$, $P_{max} = ?$

Fig (a) shows the cross section with centroid at C. The location of the centroid can be found as

$$\eta_c = \frac{(50)(250)(125) + (50)(250)(275)}{(50)(250) + (50)(250)} = 200\text{mm}$$

The cross sectional area and area moment of inertia are.

$$A = (50)(250) + (50)(250) = 25(10^3)\text{mm}^2$$

$$I_{zz} = \frac{1}{12}(50)(250^3) + (50)(250)(200 - 125)^2 + \frac{1}{12}(250)(50^3) + (50)(250)(275 - 200)^2$$

$$I_{zz} = 208.3(10^6)\text{mm}^4$$

Fig. (a) shows the free body diagram after an imaginary cut is made at section AA.

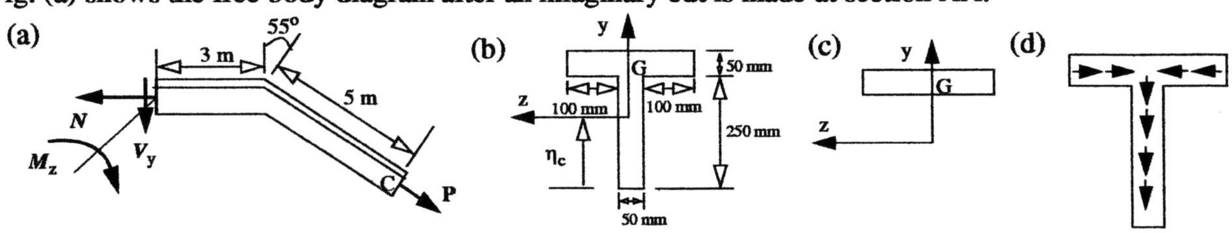

By equilibrium of forces and moment we obtain:
$N = P\cos55 = 0.5736P$ and $V_y = -P\sin55 = -0.8191P$ and $M_z = (-P\sin55)(3) = -2.457P$

The axial stress is tensile and its value is: $(\sigma_{xx})_G = \dfrac{0.5736P}{25(10^{-3})} = 22.943P$

The bending normal stress on upper side of the neutral axis will be in tension. Its value is

$$(\sigma_{xx})_G = \left|\frac{M_z y_G}{I_{zz}}\right| = \frac{(2.457P)(250 - 200)(10^{-3})}{208.3(10^{-6})} = 589.77P$$

The total normal stress in glue is: $(\sigma_{xx})_G = (22.943 + 589.77)P = 612.72P$

Fig.(c) shows the area As used in calculating shear stress in the glue. The first moment of area is.

$$(Q_z)_G = (250)(50)(75) = 937.5(10^3)\text{mm}^3$$

The magnitude of shear stress due to bending is:

$$(\tau_{xy})_G = \left|\frac{V_y(Q_t)_G}{I_{zz}t}\right| = \frac{(0.8191P)(937.5)(10^{-6})}{(208.3)(10^{-6})(50)(10^{-3})} = 73.73P$$

The magnitude of maximum normal stress in glue is

$$\sigma_{max} = \left(\frac{612.72P}{2}\right) + \sqrt{\left(\frac{612.72P}{2}\right)^2 + (73.73P)^2} = (306.4 + 315.1)P = 621.5P \le 20(10^6) \text{ or } P \le 32.18(10^3)N \tag{1}$$

The magnitude of maximum shear stress in glue is

$$\tau_{max} = \sqrt{\left(\frac{612.72P}{2}\right)^2 + (73.73P)^2} = 315.1P = 12(10^6) \text{ or } P \le 38.08(10^3)N \tag{2}$$

The maximum value of P that meets the requirements of Eq.(1) and (2) is $\boxed{P_{max} = 32.1kN}$

10.42 The bars in the pin connected structure shown in Fig. P10.42 are circular bars of diameters that are available in increments of 5 mm. The allowable shear stress in the bars is 90 MPa. Determine the diameters of the bars for designing the lightest structure to support a force of P = 40 kN.

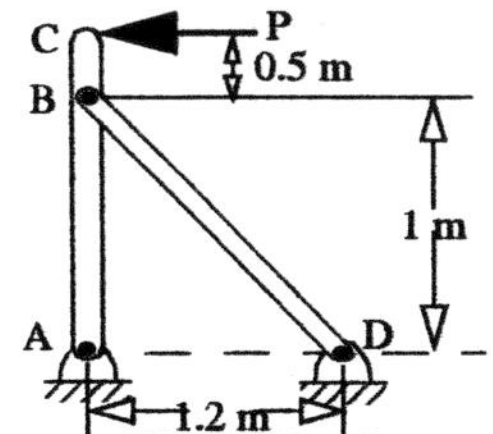

Fig. P10.42

Solution: $\tau_{max} < 90$ MPa, P = 40 kN, d_{BD} =?, d_{AB} =? to nearest 5 mm

---

Fig (a) shows the free body diagram of the structure.

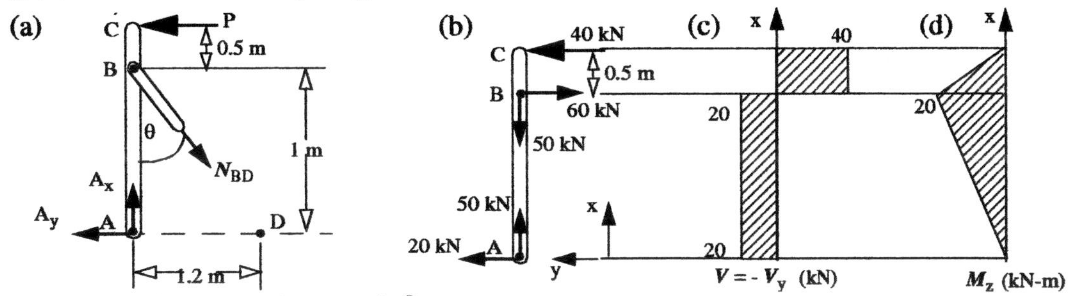

In Fig. (a) $\tan\theta = 1.2$ or $\theta = 50.194°$. By equilibrium of moment about point A, we obtain.

$$(N_{BD}\sin\theta)(1) - P(1.5) = 0 \text{ or } N_{BD} = 78.10kN \tag{1}$$

By equilibrium of forces we obtain.

$$A_y + P - (N_{BD}\sin\theta) = 0 \text{ or } A_y = 20kN \tag{2}$$

$$A_x - (N_{BD}\cos\theta) = 0 \text{ or } A_x = 50kN \tag{3}$$

Maximum shear stress in axial member BD is half the axial stress. We obtain.

$$\tau_{max} = \frac{\sigma_{BD}}{2} = \frac{1}{2}\frac{N_{BD}}{A_{BD}} = \frac{1}{2}\frac{(78.10)(10^3)}{\pi d_{BD}^2/4} \le 90(10^6) \qquad \text{or} \qquad d_{BD} \ge 0.0235m$$

The minimum diameter to the nearest 5 mm is $\boxed{d_{BD} = 25mm}$

Fig (b) shows the forces in component form that are acting on member ABC. Figs.(b) and (c) show the shear force and bending moment diagrams. Noting that the axial stress in segment AB is compressive, the maximum normal stress is σ_{xx} is sum of maximum bending normal stress and axial stress.

$$\sigma_{xx} = \frac{(20)(10^3)(d_{AB}/2)}{\pi d_{AB}^4/64} + \frac{(50)(10^3)}{\pi d_{AB}^2/4} = \frac{203.72(10^3)}{d_{AB}^3} + \frac{63.66(10^3)}{d_{AB}^2} \tag{4}$$

The normal stress in Eq. (4) is on the left surface of segment AB, where the bending shear stress is zero. The maximum shear stress is thus

$$\tau_{max} = \frac{\sigma_{xx}}{2} = \frac{101.86(10^3)}{d_{AB}^3} + \frac{31.83(10^3)}{d_{AB}^2} = 90(10^6) \tag{5}$$

For an initial guess we neglect the second term and obtain: $d_{AB}^3 \geq \left(\frac{101.86}{90}\right)(10^{-3})$ or $d_{AB} \geq 0.104\text{m}$

We consider values of $d_{AB} = 0.105$ m and obtain $\tau_{max} = 90.93\ (10^6)$ which exceeds the allowable stress.
We next consider $d_{AB} = 0.11$ and obtain $\tau_{max} = 79.2.93\ (10^6)$ which is below the allowable stress value.
The minimum diameter of member ABC is: $\boxed{d_{AB} = 110\text{mm}}$

10.43 Member AB has a circular cross-section with a diameter of 0.75 inch. Member BC has a square cross-section of 2 in x 2 in.Determine the maximum normal stress in members AB and BC.

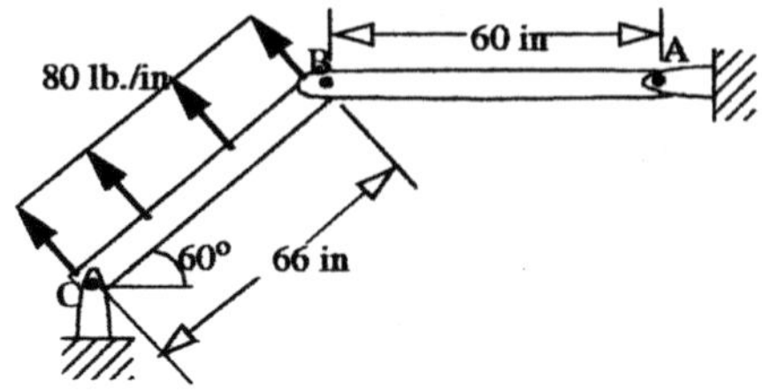

Fig. P10.43

Solution: $d_{AB} = 0.75\text{in}$, $A_{BC} = 2$ in x 2 in, $\sigma_{AB} = ?$, $\sigma_{BC} = ?$,

---

Fig (a) shows the free body diagram with distributed force replaced by an equivalent force.

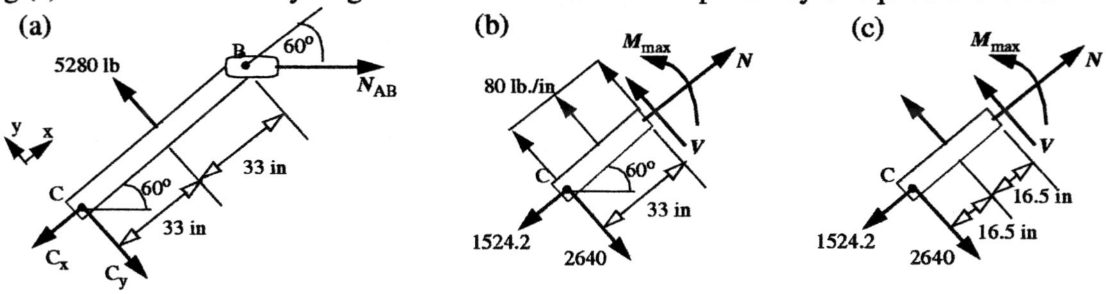

By equilibrium of moment about point C, we obtain: $N_{AB}(66\sin 60) - (5280)(33) = 0$ or $N_{AB} = 3048.4\text{lbs}$

By equilibrium of forces we obtain

$C_y - 5280 + N_{AB}\sin 60 = 0$ or $C_y = 2640\text{lb}$ and $C_x - N_{AB}\cos 60 = 0$ or $C_x = 1524.2\text{lb}$

The axial stress in AB is: $\sigma_{AB} = \frac{N_{AB}}{A_{AB}} = \frac{3048.4}{\pi(0.75)^2/4} = \frac{3048.4}{0.4418}\text{psi}$ $\boxed{\sigma_{AB} = 6.9\text{ksi(T)}}$

The maximum bending moment will occur at the center of CB. Fig (b) shows the free body diagram after an imaginary cut is made through the center of CB. Fig (C) shows the statically equivalent diagram.
By equilibrium of force and moment about point O we obtain.

$N_{BC} = 1524.2$ and $M_{max} + (2640)(33) - (2640)(16.5) = 0$ or $M_{max} = 43560$

The maximum normal stress in BC is

$$\sigma_{CD} = \frac{1524.2}{(2)(2)} + \frac{(43560)(1)}{(2)(2^3)/12} = (381 + 32670)\text{psi or}$$ $\boxed{\sigma_{CD} = 33.05\text{ksi}}$

10.44 The members of the structure have rectangular cross-sections and are pin connected. Cross-section dimensions for members are: 100 mm x 150 mm for ABC, 100 mm x 200 mm for CDE, and 100 mm x 50 mm for BD. The allowable normal stress in the members is 20 MPa. Determine the maximum intensity of the distributed load w.

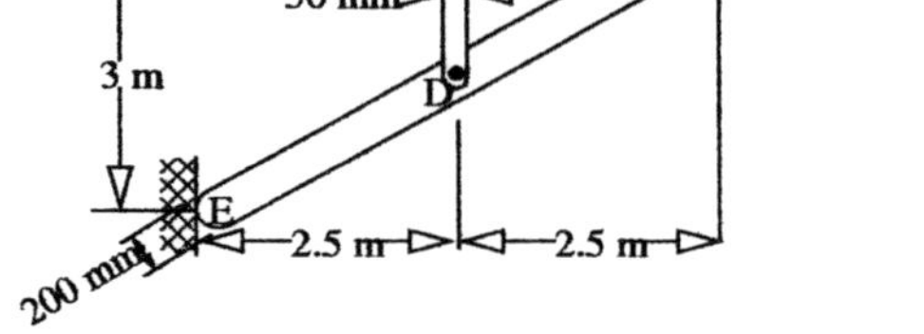

Fig. P10.44

Solution: $A_{ABC} = 100\text{ mm x }150\text{ mm}$, $A_{CDE} = 100\text{ mm x }200\text{ mm}$, $A_{BD} = 100\text{ mm x }50\text{ mm}$, $|\sigma_{max}| \le 20\text{MPa}$, $w_{max} = ?$

Fig (a) shows the free body diagram of the entire structure. By equilibrium of moment about point A we obtain. $E_x(3) - (2.5)(5w) = 0 \quad E_x = 4.1667w$

By equilibrium of forces we obtain: $A_x = E_x = 4.1667w$ and $A_y = 5w$

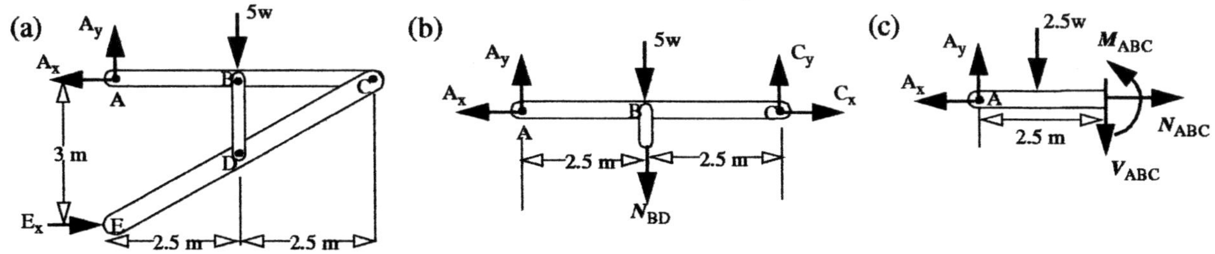

Fig. (b) shows the free body diagram of the member ABC. By equilibrium of moment about (C) we obtain. $A_y(5) - 5w(2.5) - N_{BD}(2.5) = 0$ or $N_{BD} = 5w$

By equilibrium of forces we obtain: $C_x = A_x = 4.1667w$ and $C_y = 5w + N_{BD} - A_y = 5w$

The axial stress in member BD is $\sigma_{BD} = \dfrac{N_{BD}}{A_{BD}} = \dfrac{5w}{(0.1)(0.05)} = 1000w \le 20(10^6)$ or

$$w \le 20(10^3)\frac{N}{m} \qquad 1$$

The maximum moment in member ABC will be in the center. Fig (c) shows the free body diagram after an imaginary cut is made through middle of ABC. By equilibrium of force and moment about O we obtain.
$N_{ABC} = A_x = 4.1667w$ and $M_{ABC} = A_y(2.5) - (2.5w)(1.25) = 9.375w$

The maximum normal stress in member ABC is

$$\sigma_{ABC} = \frac{4.1667w}{(0.1)(0.15)} + \frac{(9.375w)(0.075)}{(0.1)(0.15^3)/12} = 277.78w + 25000w = 25278w \le 20(10^6) \text{ or } w \le 791.2\frac{N}{m} \qquad 2$$

Fig (d) shows member CDE with all calculated forces acting on it. Fig (e) shows the equivalent diagram with forces resolved in axial and transverse direction.

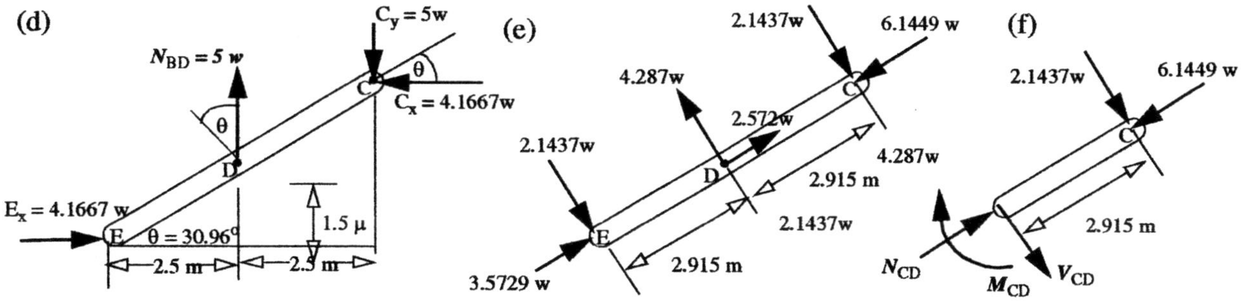

The maximum bending moment will be in the middle of member CDE. The maximum axial stress will be in segment CD. Fig (f) shows the free body diagram after an imaginary cut is made just right of D. By equilibrium of forces and moment about point O_2, we obtain.

$N_{CD} = 6.1449w$ and $M_{CD} = (-2.1437w)(2.915) = -6.25w$

The maximum normal stress in segment CD will be compressive and its value is

$$\sigma_{CD} = \frac{6.1449w}{(0.1)(0.2)} + \frac{(6.25w)(0.1)}{\frac{1}{12}(0.1)(0.2^3)} = 9682.1w \le 20(10^6) \quad \text{or} \quad w \le 2.06(10^3)\frac{N}{m} \qquad 3$$

The maximum value of w that satisfies Eqs. (1), (2) and (3) is $\boxed{w = 791.2\frac{N}{m}}$

10.45 A hoist is to be designed for lifting a maximum weight of W = 300 lbs. The hoist will be installed at a certain height above ground and will be constructed using lumber and assembled using steel bolts. The lumber rectangular cross-section dimensions are listed in Table 10.2. The bolt joints will be modeled as pins in single shear. Same size bolts will be used in all joints. The allowable normal stress in the wood is 1.2 ksi and the allowable shear stress in bolts is 6 ksi. Design the lightest hoist by choosing the lumber from Table 10.2 and bolt size to the nearest 1/8 inch diameter.

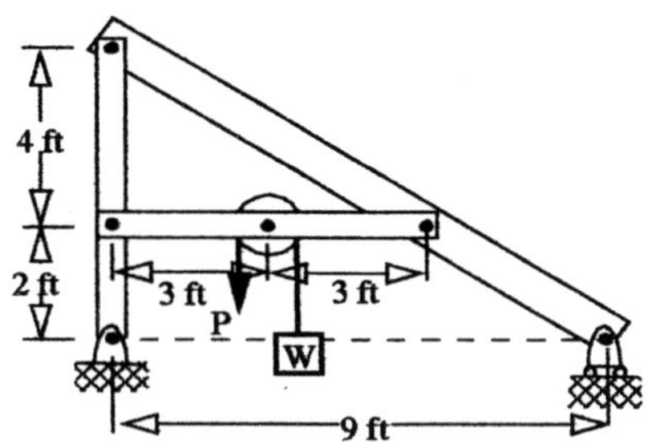

Fig. P10.45

Solution: W = 300lb, $\sigma_w \le 1.2$ psi, $\tau_w \le 6$ ksi, Lumber =?, d_b =? nearest 118

Fig (a) shows the free body diagram of the entire structure. By equilibrium of moment about point (A) and equilibrium of forces we obtain.

$F_y(9) = (600)(3)$ or $F_y = 200\text{lb}$ and $A_y + F_y = 600$ or $A_y = 400\text{lb}$ and $A_x = 0$

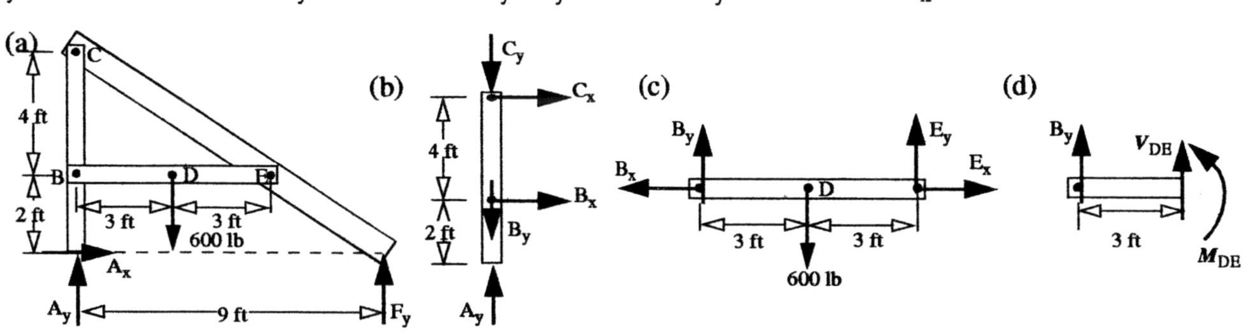

Figs. (b), (c) and (d) are the free body diagram of members ABC, BDE and CEF. By equilibrium of moment about point B in Fig (b) and equilibrium of force in the x - direction we obtain.

$C_x(4) = 0$ or $C_x = 0$ and $B_x = -C_x = 0$

By equilibrium of moment about point (E) and force equilibrium we obtain.

$B_y(6)-(600)(3) = 0$ or $B_y = 300\text{lb}$ and $B_y + E_y = 600$ or $E_y = 300\text{lb}$ and $E_x = B_x = 0$

By equilibrium of forces in y-direction in Fig.(b) we obtain: $C_y + B_y - A_y = 0$ or $C_y = 100\text{lb}$

The largest force acts on pin A.The shear stress in pin A is. $\tau_A = \dfrac{400}{\frac{\pi}{4}d_b^2} \le 6(10^3)$ or $d_b \ge 0.291\text{in}$

The diameter of the bolt to the nearest 1/8 is $\boxed{d_b = \frac{3}{8}\text{in}}$

Segment AB of member ABC will have the largest compressive axial stress.

$$\sigma_{AB} = \frac{400}{A_{AB}} \le 1200\text{psi} \text{ or } A_{AB} \ge 0.333\text{in}^2$$

For member ABC, use lumber with cross section dimension of 2 in x 4 in

The maximum bending moment in member BDE will be at the center, Fig (d) shows the free body diagram of member BDE after an imaginary cut is made in the middle.

By equilibrium of moment: $M_{DE} = (3)(B_y) = 900\text{ft}-\text{lb} = 10800\text{in}-\text{lb}$

The maximum normal stress in BDE can be written as $\sigma_{BDE} = \dfrac{10800}{S_{BDE}} \le 1200$ or $S_{DE} \ge 9\text{in}^3$

The lightest lumber for which the section modulus is greater than 9 in^3 in Table 10.3 is 2 in x 6 in

For member BDE, use lumber with cross section dimension of 2 in x 6in

Fig (e) shows the forces acting on member CEF. Fig (f) shows the equivalent diagram with forces resolved in axial and transverse direction.

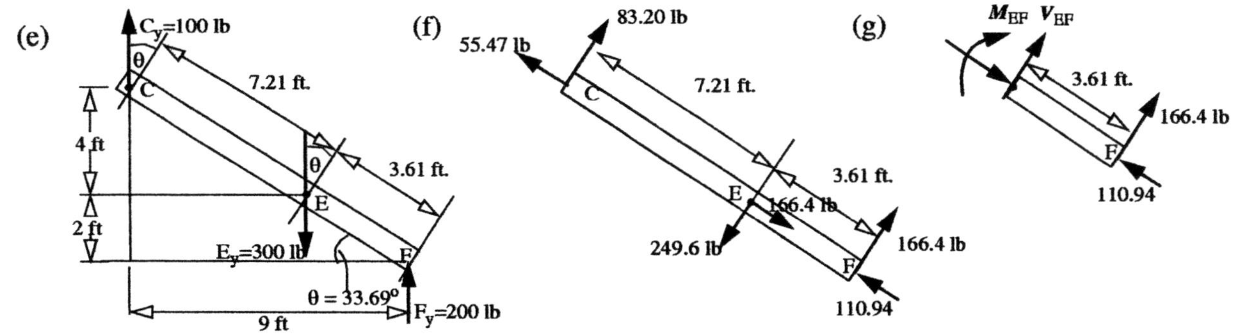

The maximum axial stress will be in segment EF. The maximum bending moment will be at E in member CEF. Fig (g) shows the free body diagram of segment EF. By equilibrium of forces and moment we obtain.

$N_{EF} = 110.94\text{lb}$

$M_{EF} = (166.4)(3.61) = 600\text{ft}-\text{lb} = 7200\text{in}-\text{lb}$

The maximum compressive stress in member CEF is

$$\sigma_{CEF} = \frac{110.94}{A_{CEF}} + \frac{7200}{S_{CEF}} \leq 1200 \qquad 4$$

From Table 10.3, the 2 in. x 4 in. cross-section has as area $A_{CEF} = 8$ and section modulus $S_{CEF} = 5.3$. Substituting these values in Eq. (1) we obtain, $\sigma_{CEF} = 1372$ psi, which exceeds 1200 psi. We consider 2 in. x 6 in. cross-section for which $A_{CEF} = 12$ and $S_{EF} = 12$. Substituting these values in Eq. (1), we obtain $\sigma_{CEF} =$ 609 psi which is less than 1200 psi.

For member CEF, use lumber with cross section dimension of 2 in x 6in

---

10.46 A rectangular wooden beam of 4 in x 8 in cross-section is supported at the right end by an aluminum circular rod of 1/2 in diameter. The allowable normal stress in the wood is 1.5 ksi and the allowable shear stress in aluminum 8 ksi. The Modulus of Elasticity for wood and aluminum are $E_W = 1{,}800$ ksi and $E_{al} = 10{,}000$ ksi. Determine the maximum force P that the structure can support.

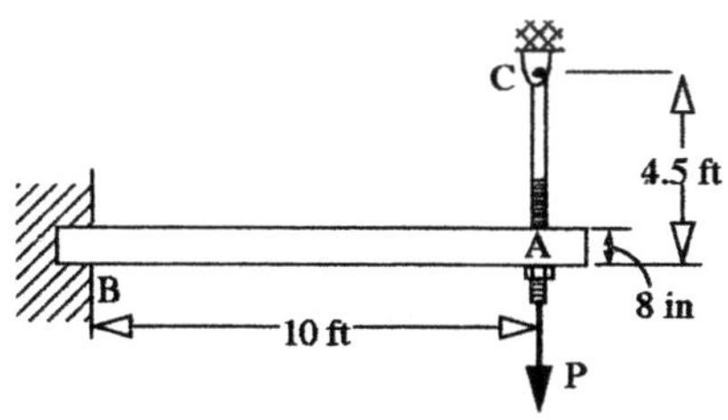

Fig. P10.46

Solution: $A_{AB} = 4\text{ in. x } 8\text{ in.}$, $d_{al} = \frac{1}{2}\text{in}$, $\sigma_w \leq 1.5$ ksi, $\tau_{al} \leq 8$ ksi,

$E_w = 1800$ ksi, $E_{al} = 10{,}000$ $P_{max} = ?$

Fig (a) and (b) shows the cantilever beam and the axial rod.

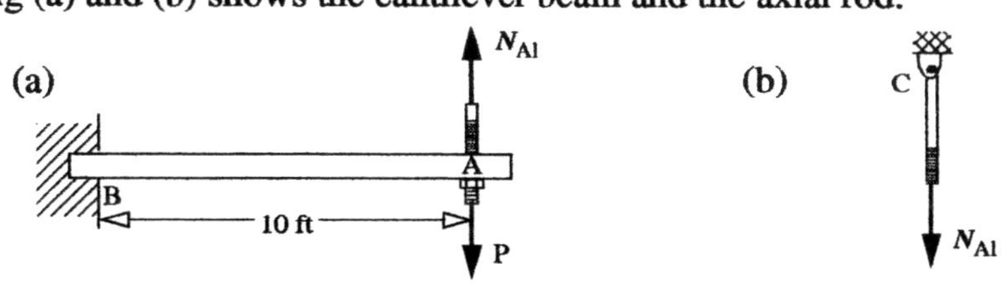

Comparing Fig (a) to case 1 in Table 7.1 we obtain. a = 120 in., b = 0, P = (P - N_{al}), E = 1800 and

$I = \frac{1}{12}(4)(8^3) = 170.67 in^4$ we obtain

$$v_A = \frac{(P - N_{al})(120^3)}{(3)(1800)(170.67)} = 1.875(P - N_{al}) \quad 1$$

The extension of aluminium rod can be found using Eq. (4.27)

$$\delta_{al} = \frac{N_{al}(54)}{(10000)(\pi)(1/2)^2/4} = 0.0275 N_{al} \quad 2$$

Equating Eq. (1) and (2) we obtain: $1.875P - 1.875N_{al} = 0.0275N_{al}$ or

$$N_{al} = \frac{1.875}{1.9025}P = 0.9855P \quad 3$$

The axial stress in aluminium rod is: $\sigma_{al} = \frac{0.9855P}{\pi(1/2)^2/4} = 5.0193P$

The maximum shear stress in aluminium is: $\tau_{al} = \sigma_{al}/2 = 2.5097P \le 8$ or

$$P \le 3.19 kips \quad 4$$

The maximum bending moment in the beam will be at the left wall. Its value is $M_{max} = (P - N_{al})(120) = 1.735P$. The maximum normal stress in the beam is

$$\sigma_w = \frac{(1.735P)(4)}{(170.67)} \le 1.5 \text{ or } P \le 36.9 kips \quad 5$$

The maximum value of P that meets the requirements in Eqs. (4) and (5) is $\boxed{P_{max} = 3.1 kips}$

10.47 A steel pipe with an outside diameter of 1.5 inch and wall thickness of 1/4 inch is simply supported at D. A torque of 120 in-kips is applied as shown in Fig. P10.47. Determine the normal and shear stresses at point A and B in the x, y, and z coordinate system and show it on a stress cube. Points A and B are on the surface of the pipe. The Modulus of Elasticity is E = 30,000 ksi and the Poisson's ratio ν = 0.28.

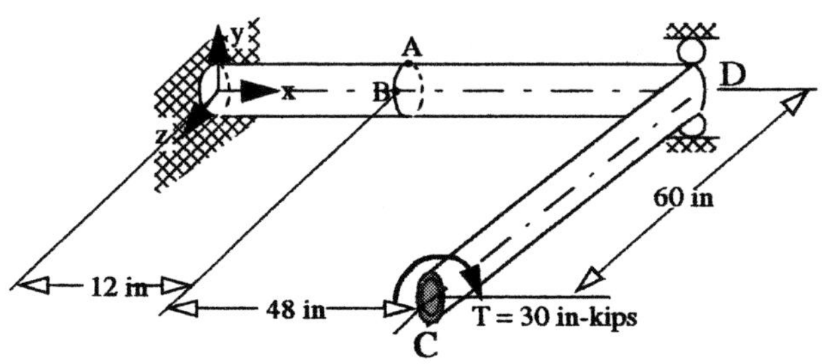

Fig. P10.47

Solution: $d_o = 1.5$ in. $t = \frac{1}{4}$ in $E = 30,000$ ksi, $\nu = 0.28$ Stresses at A and B.

In Fig (a) support at D is replaced by a reaction force.

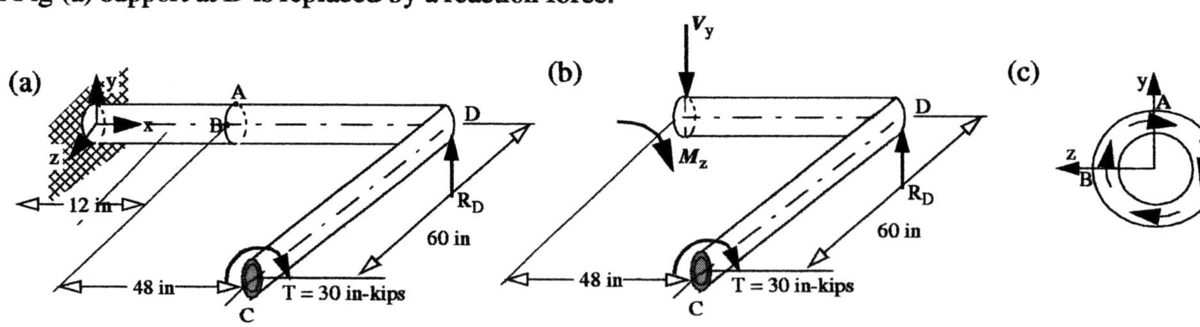

The beam ED is subjected to a couple of 30 in-kips and a force R_D comparing beam in Fig. (a) to cases (1) and (2) in Table 1. We have M = -30 in-kips, $P = R_D$, $L = 60$, $E = 30,000$ ksi.

$I = \frac{\pi}{64}[1.5^4 - 1^4] = 0.1994\text{in}^4$. The deflection of D is the superposition of deflection in cases 1 and 2 and should equal to zero.

$$v_D = \frac{(-30)(60^2)}{(2)(30000)(0.1994)} + \frac{(R_o)(60^3)}{(2)(30000)(0.1994)} = 0 \text{ or } R_D = \frac{3}{2}\frac{(30)}{60} = 0.75\text{kip}$$

Fig (b) shows the free body diagram after an imaginary cut is made through section containing points A and B. By equilibrium of forces and moment.

$V_y = R_D = 0.75\text{kip}$ and $M_z + 30 - R_D(48) = 0$ or $M_z = 6\text{in}-\text{kips}$

The normal stress at point A is compressive and at B it is zero.

$$(\sigma_{xx})_A = \frac{(6)(0.75)}{(0.1994)} \text{ or } (\sigma_{xx})_A = 22.6\text{ksi(C)} \text{ and } (\sigma_{xx})_B = 0$$

Fig.(c) shows the direction of shear stress. At point A the shear stress is zero and at point B it is in the positive y - direction. The first moment of area A_s shown in Fig. (d) is.

$$(Q_z)_B = \frac{2}{3}\left[\left(\frac{1.5}{2}\right)^3 - \left(\frac{1}{2}\right)^3\right] = 0.1979\text{in}^3$$

The shear stress at B is $(\tau_{xy})_B = \frac{(0.75)(0.1979)}{(0.1994)(0.5)}$ or $(\tau_{xy})_B = 1.49\text{ksi}$ and $(\tau_{xy})_A = 0$

Fig. (e) and (f) show the stress cube at points A and B

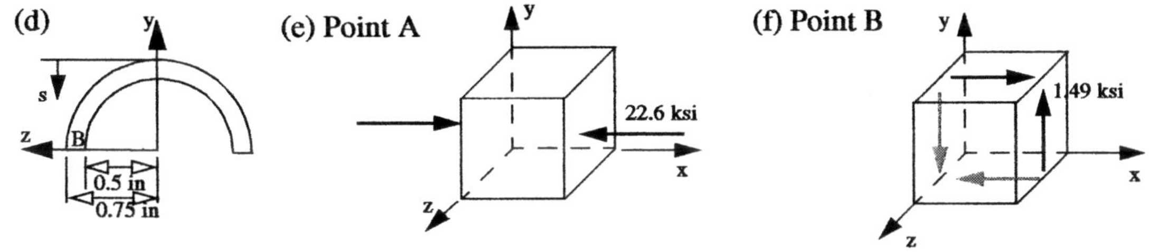

10.48 A composite beam is constructed by attaching steel strips on top and bottom of a wooden beam. The beam is supported at the right end by an aluminum circular rod of 8 mm diameter. The allowable normal stresses in the wood and steel are 14 MPa and 140 MPa, respectively. The allowable shear stress in aluminum is 60 MPa. The Modulus of Elasticity for wood, steel, and aluminum are $E_W = 12.6$ GPa, $E_S = 200$ GPa and $E_{al} = 70$ GPa, respectively. Determine the maximum intensity w of the distributed load that the structure can support.

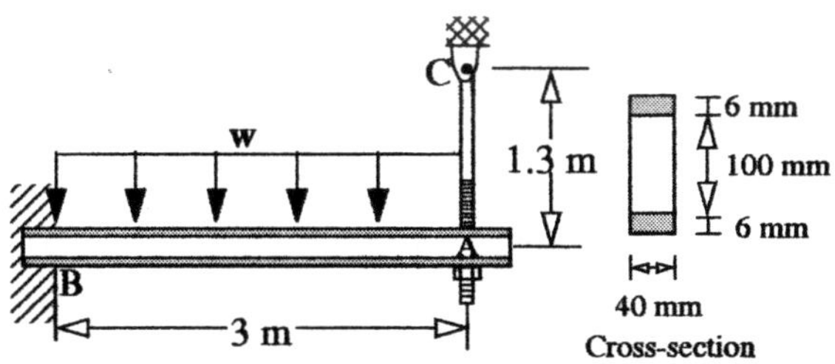

Fig. P10.48

Solution: $d_{al} = 8$ mm, $\sigma_W \le 14$ MPa, $\sigma_S \le 140$ MPa, $\tau_{al} \le 60$ MPa, $E_W = 12.6$ GPa, $E_S = 200$ GPa $E_{al} = 70$ GPa, w_{max} =?

The area moment of inertia for each material cross section is

$$(I_{zz})_w = \frac{1}{12}(40)(100)^3 = 3.333(10^6)\text{mm}^4 \text{ and } (I_{zz})_s = \frac{1}{12}(40)(6)^3 + (40)(6)(53^2) = 0.6749(10^6)\text{mm}^4$$

The bending rigidity of the cross section is

$$\sum E_j(I_{zz})_j = (12.6)(10^9)(3.333)(10^{-6}) + (2)(200)(10^9)(0.6749)(10^{-6}) = 311.95(10^3)\frac{N}{m^2} \quad 1$$

The cross sectional area of aluminium rod is: $A_{al} = \frac{\pi}{4}(8)^2 = 50.26\text{mm}^2$

Fig (a) and (b) show the cantilever beam and axial rod. Comparing Fig (a) to case 1 and 3 in Table 7.1, we obtain a = 3m, b = 0, $P = -N_{al}$, $p_o = w$. We note that the bending rigidity (EI) is replaced by value given in Eq. (1). We obtain

$$v_A = \frac{-N_{al}(3^3)}{(3)(311.95)(10^3)} + \frac{w(3^4)}{(8)(311.95)(10^3)} = [-28.95N_{al} + 32.46w](10^{-6}) \qquad 2$$

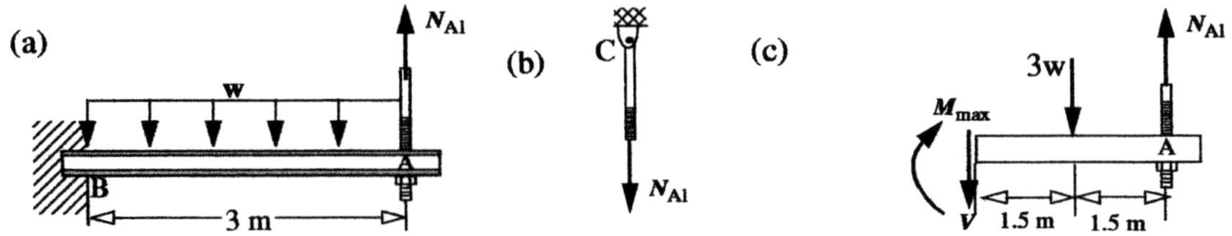

The extension of aluminium rod can be found using Eq. (4.27)

$$\delta_{al} = \frac{N_{al}(1.3)}{(70)(10^9)(50.26)(10^{-6})} = 0.3695N_{al}(10^{-6}) \qquad 3$$

Equating Eq. (2) and (3) we obtain: $-28.95N_{al} + 32.46w = 0.3695N_{al}$ or

$$N_{al} = 1.107W \qquad 4$$

The axial stress in aluminium rod is: $\sigma_{al} = \frac{1.107w}{(50.26)(10^{-6})} = 22.02w(10^3)$

The maximum shear stress in aluminium rod is

$$\tau_{max} = \frac{\sigma_{al}}{2} = 11.01w(10^3) \le 60(10^6) \text{ or } w \le 5.45(10^3)\frac{N}{m} \qquad 5$$

As in Example 10.7, the maximum bending moment will be at the wall. Its value can be found from the free body diagram in Fig. (c).

$M_{max} + 3w(1.5) - N_{al}(3) = 0$ or $M_{max} = -1.179w$

The maximum normal stress in wood will be at $y_w = \pm 0.05\text{m}$. Its value is

$$\sigma_w = \left|-\frac{E_w M_z y_w}{\sum E_j (I_{zz})_j}\right| = \left|\frac{(12.6)(10^9)(1.179w)(0.05)}{(311.95)(10^3)}\right| = 2.381w(10^3) \le 14(10^6) \text{ or } w \le 5.88(10^3)\frac{N}{m} \qquad 6$$

The maximum normal stress in steel will be at $y_s = \pm 0.056\text{m}$. Its value is

$$\sigma_s = \left|-\frac{E_s M_z y_s}{\sum E_j (I_{zz})_j}\right| = \left|\frac{(200)(10^9)(1.179w)(0.05)}{(311.95)(10^3)}\right| = 42.33w(10^3) \le 140(10^6) \text{ or } w \le 3.307(10^3)\frac{N}{m} \qquad 7$$

The maximum value of w that meets the requirements of Eq. (5),(6) and (7) is $\boxed{w_{max} = 3.3\text{kN/m}}$

10.49 A park structure is modeled with pin joints at the points shown. Members BD and CE have a cross-sectional dimensions of 6 in x 6 in, while the members AB, AC, and BC have a cross-sectional dimensions of 2in x 8 in. Determine the maximum normal and shear stress in each of the members due to the estimated snow load shown on the structure.

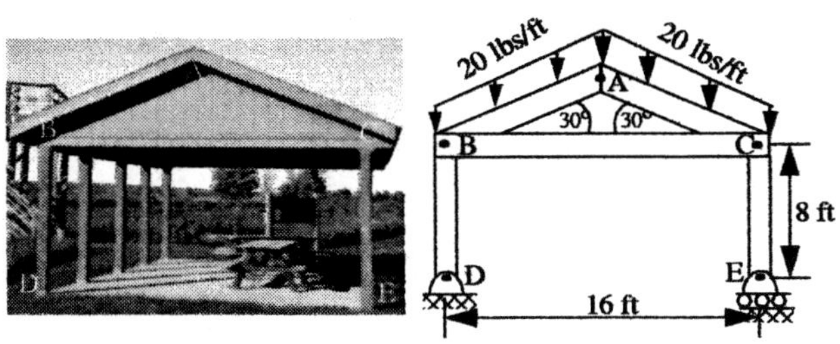

Fig. P10.49

Solution: $A_{BD} = A_{CE} = 6$ in x 6in, $A_{AB} = A_{AC} = A_{BC} = 2$ in x 8 in, $\sigma_{max} =?$ $\tau_{max} =?$ in each member

---

Figs. (a) through (d) shows free body diagrams of various members, with distributed force replaced by equivalent load

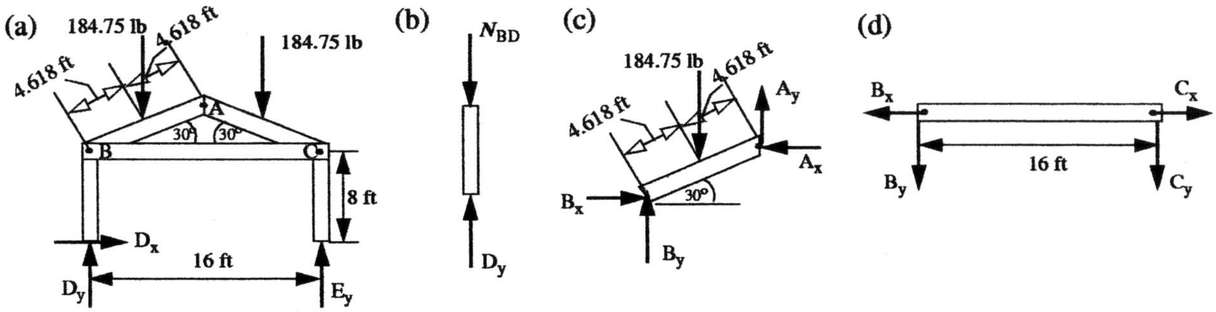

From equilibrium in Fig. (a)

$$D_x = 0 \quad D_y = E_y = 184.75 \qquad (1)$$

From Fig. (b) $N_{BD} = D_y = 184.75$. Thus the axial stress in BD and CE is

$$\sigma_{BD} = \sigma_{CE} = \frac{184.75}{(6)(6)} \text{ or } \boxed{\sigma_{BD} = \sigma_{CE} = 5.13\text{psi(C)}}$$

By equilibrium of moment about point D in Fig (c) we obtain

$$(184.75 - B_y) = 0 \text{ or } B_y = 184.75 \qquad (2)$$

By equilibrium of moment about point A in Fig.(c) we obtain

$$B_x(9.236\sin 30) - B_y(9.236\cos 30) + 184.75(4.68\cos 30) = 0 \text{ or } B_x = 159.78\text{lb} \qquad (3)$$

By equilibrium of forces we obtain

$$A_x = B_x = 159.78 \qquad (4)$$

$$(A_y + B_y) = 184.75 \text{ or } A_y = 24.97\text{lb} \qquad (5)$$

The axial stress in member BC is tensile its value is: $\sigma_{BC} = \frac{159.78}{(2)(8)} = 9.986$ or $\boxed{\sigma_{BC} = 10\text{psi(T)}}$

Fig. (e) shows the coordinates x, y and coordinates $x^{|}$, $y^{|}$ that are in axial and in transverse direction to member ***M***.

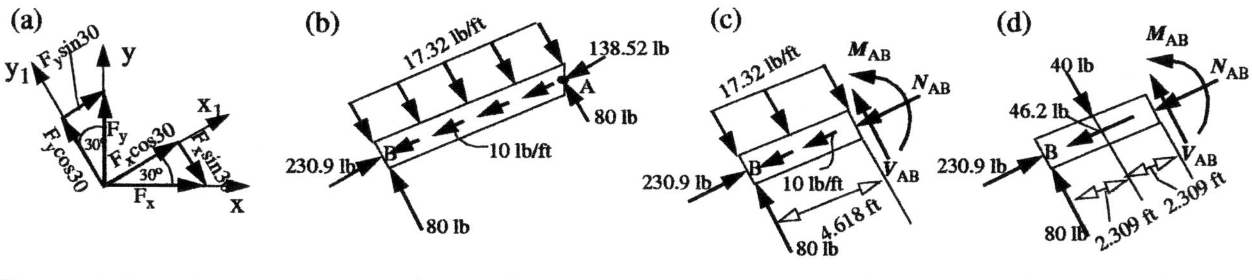

The resolution of forces can be written as: $F_{x1} = F_x\cos 30 + F_y\sin 30$ and $F_{y1} = -F_x\sin 30 + F_y\cos 30$

Using the above transformation equations the forces in Fig. (e) are transformed to those shown in Fig. (f). The bending moment will be maximum at the center of member AB. The axial force will be maximum at the center of member AB. The axial force will be maximum at a point other than center. Noting that bending stresses are much greater than axial stresses for same magnitude of external force, we expect the maximum normal stress to reach a maximum at the center. Fig. (g.) shows the free body diagram after an imaginary cut is made in the center. Fig. (b) is the statically equivalent diagram. By equilibrium of forces and moments we obtain.

$N_{AB} = 230.9 - 46.2 = 184.7\text{lb}$ and $M_{AB} = (80)(4.618) - (40)(2.309) = 277.1\text{lb}-\text{ft}$

The cross sectional area and area moment of inertia for member AB is

$A_{AB} = (2)(8) = 16\text{in}^3$ $\qquad I_{AB} = \frac{1}{12}(2)(8)^3 = 85.33\text{in}^4$

The maximum compressive stress in AB is

$$\sigma_{AB} = \frac{184.7}{16} + \frac{(277.1)(12)(4)}{85.33} = 11.5 + 155.9 \qquad \boxed{\sigma_{AB} = 167.4\text{psi(C)}}$$

10.50 A highway sign uses a 16 inch hollow pipe as a vertical post and 12 inch hollow pipe for horizontal arms. The pipes are one inch thick. Assume a uniform wind pressure of 20 lbs/ft^2 acts on the sign boards and the pipes. Note the pressure on the pipes acts on the projected area of L d, where L is the length of pipe and d is the diameter of the pipe. Neglecting the weight of the pipe, determine the normal and shear stresses at points A and B and show these stresses on stress cubes.

Fig. P10.50

Solution: $(d_{CD})_O = 16$ in, $\quad (d_{EF})_O = (d_{PG})_O = 12$ in, $\quad t = 1$,

$p = 20\ \text{lb/ft}^2$ $\quad$ Stresses at A and B=?

Fig (a) shows the free body diagram after an imaginary cut is made through section containing points A and B. The pressure on the pipes and signboards are replaced by statically equivalent forces.

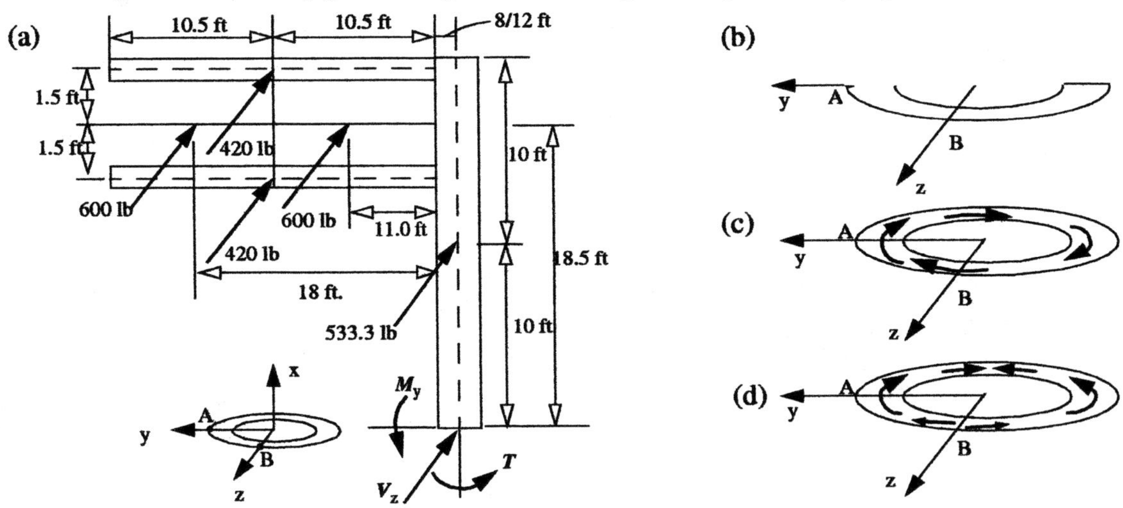

By equilibrium of forces and moment we obtain

$$V_z = -[(2)(600) + (2)(420) + 533.3] = -2573.3\text{lb} \qquad 1$$

$$T = -\left[(2)(420)\left(10.5 + \frac{8}{12}\right) + 600\left(18 + \frac{8}{12}\right) + 600\left(11 + \frac{8}{12}\right)\right] = -27580\text{ft}-\text{lb} = -330960\text{in}-\text{lb} \qquad 2$$

$M_y = 420(20) + 420(17) + 2(600)(18.5) + 533.3(10)$ or

$$M_y = 43073\text{ft}-\text{lb} = 516876\text{in}-\text{lb} \qquad 3$$

The polar moment of inertia for pipe CDE is $J = \frac{\pi}{32}[(16)^4 - (14)^4] = 2662.5\text{in}^4$ and $I = \frac{J}{2} = 1331.3\text{in}^4$

The normal stress at point A is zero and at point B is tensile.

$(\sigma_{xx})_A = 0 \quad (\sigma_{xx})_B = \frac{(516876)(8)}{1331.3}$ or $\boxed{(\sigma_{xx})_B = 3106\text{psi(T)}}$

Fig (b) shows the area A_s used in calculating the shear stress of A. The first moment of area A_s is

$Q_y = \frac{2}{3}[8^3 - 7^3] = 112.67\ \text{in}^3$

The torsional shear stress is on the outer radius of pipe CDE is: $\tau_{x\theta} = \frac{(330960)(8)}{2662.5} = -994.4\text{psi}$

Fig. (c) shows the direction of torsional shear stress. From Fig. (c) we obtain

$(\tau_{xz})_A = -994.4\text{psi}$ and $(\tau_{xy})_B = 994.4\text{psi}$

Fig. (d) shows the direction of shear stress due to bending about y - axis. At point B the shear stress is zero and at point A it is in negative z - direction.

$(\tau_{xz})_A = -\frac{(2573.3)(112.67)}{(1331.3)(2)} = -108.9\text{psi}$ and $(\tau_{xy})_B = 0$

The total shear stress at points A and B are

$(\tau_{xz})_A = (-994.4 - 108.9)\text{psi}$ or $\boxed{(\tau_{xz})_A = -1103.3\text{psi}}$ and $\boxed{(\tau_{xy})_B = 994.4\text{psi}}$

10.51 A bicycle rack is made from thin aluminum tubes of thickness 1/16th inch, and outer diameter 1 inch. Weight of the bicycles is supported by the belts from C to D and members between C and B. Members AC carries negligible force and is neglected in the stress analysis as shown on the model to the right. If the allowable normal stress in the steel tubes is 12 ksi and the allowable shear stress is 8 ksi, determine the maximum weight W to the nearest lb of each bicycle that can be put on the rack.

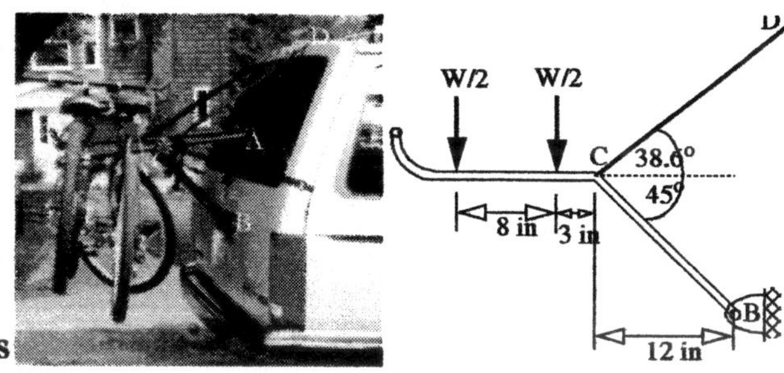

Fig. P10.51

Solution: $d_o = 1$ in., $t = \frac{1}{16}$in, $\sigma_{max} \le 12$ksi, $\tau_{max} \le 8$ ksi, W_{max} =?, to nearest lb.

Fig. (a) shows the free body diagram of the bicycle rack.

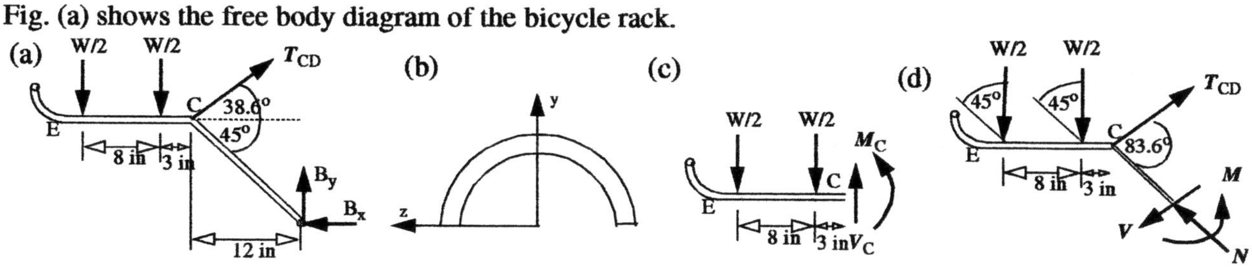

By equilibrium of moment about B, we obtain:

$$\frac{W}{2}(23) + \frac{W}{2}(15) - (T_{CD}\sin 38.6)(12) - (T_{CD}\cos 38.6)(12) = 0 \text{ or } T_{CD} = 1.126W$$

By equilibrium of forces we obtain: $B_x = T_{CD}\cos 38.6 = 0.782W$ and $B_y = W - T_{CD}\sin 38.6 = 0.299W$

In EC, the maximum moment and shear force will be just before C as shown in Fig. (c).

$M_c = -(W/2)(11) + (W/2)(3) = 7W$ and $V_c = W$

The cross sectional area and area moment of inertia can be found as

$A = \pi(1^2 - 0.875^2)/4 = 0.184\text{in}^2$ $\qquad$ $I = \pi(1^4 - 0.875^4)/64 = 20.31(10^{-3})\text{in}^4$

Fig. (b) shows the area As used in calculating shear stress. The first moment of the area is

$Q_z = \frac{2}{3}[1^3 - 0.875^3] = 0.220\text{in}^3$

The magnitude of maximum bending normal and shear stress can be found as

$$(\sigma_{xx})_{max} = \frac{(7W)(0.5)}{20.31(10^{-3})} = 172.31W \le 12(10^5) \quad \text{or} \quad W \le 69.6\text{lb} \qquad 1$$

$$(\tau_{xy})_{max} = \frac{(W)(0.22)}{20.31(10^{-3})(0.125)} = 86.7W \le 8(10^3) \quad \text{or} \quad W \le 92.3\text{lb} \qquad 2$$

The maximum shear stress on the top and bottom surface on a plane 45o to the axis of EC is

$$\tau_{max} = \frac{(\sigma_{xx})_{max}}{2} = 86.15W \le 8(10^3) \quad \text{or} \quad W \le 92.86\text{lb} \qquad 3$$

In BC, the maximum moment will be next to C. Fig (d). shows the free body diagram after an imaginary cut is made just next to C in member BC. The moment value is same as before. By equilibrium of forces we obtain.

$$N = T_{CD}(\cos 83.6) + 2\left(\frac{W}{2}\right)(\cos 45) = 0.833W \text{ and } V = T_{CD}(\sin 83.6) - 2\left(\frac{W}{2}\right)(\sin 45) = 0.412W$$

The magnitude of maximum normal stress is

$$\sigma = \frac{0.833W}{0.184} + \frac{(7W)(0.5)}{(20.31)(10^{-3})} = 176.86W \le 12(10^5) \quad \text{or} \quad W \le 67.86\text{lb} \qquad 4$$

Eqs. (1), (2) and (3) show that the limiting design constraint is the normal stress in member EC. This will be also true in member BC. The maximum value of W that meets requirements of Eqs. (1) through (4) to the nearest pound is $\boxed{W_{max} = 67 \text{ lb}}$

10.52 The hoist shown was used to lift heavy loads in a mining operation. Member EF supported load only if the load being lifted was un-symmetric with respect to the pulley, otherwise it carried no load and can be neglected in the stress analysis. If the allowable normal stress in steel is 18 ksi and the allowable shear stress is 10 ksi, determine the maximum load W that could be lifted using the hoist.

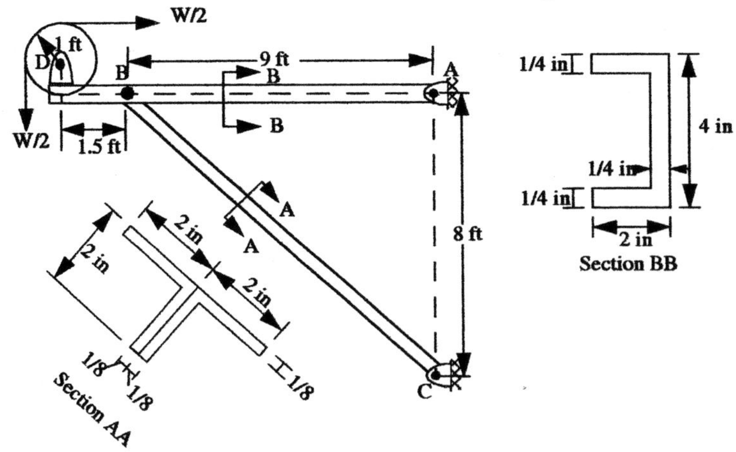

Solution: $\sigma_{max} \le 18\text{ksi}$, $W_{max} = ?$

Fig. (a) shows the free body diagram of member ABD

(a)

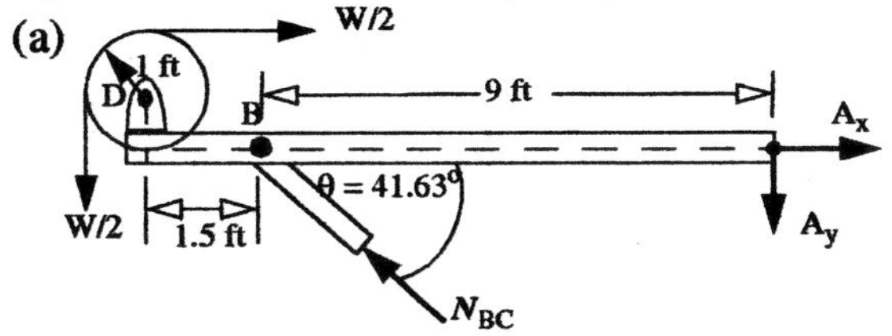

(b)

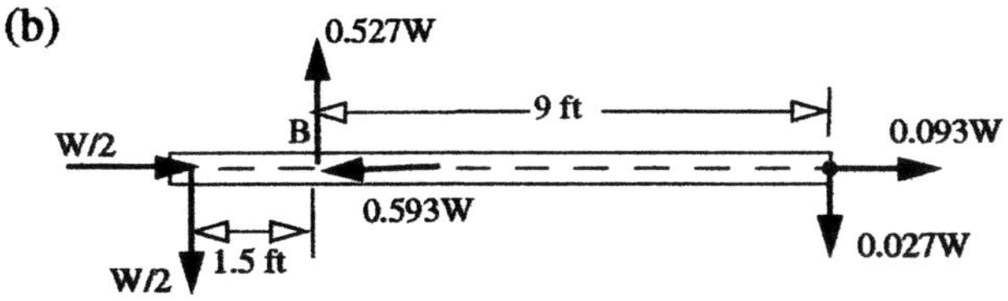

By equilibrium of moment about point A, we obtain.

$$(N_{BC}\sin 41.63^{o})(9) + \frac{W}{2}(2) - \frac{W}{2}(11.5) = 0 \text{ or } N_{BC} = 0.794W$$

By equilibrium of forces we obtain:

$$A_y = N_{BC}\sin 41.63^{o} - \frac{W}{2} = 0.027W \text{ and } A_x = N_{BC}\cos 41.63^{o} - \frac{W}{2} = 0.093W$$

The area of cross section of member BC is

$$A_{AA} = 2\left[(2)\left(\frac{1}{8}\right) + \left(2 - \frac{1}{8}\right)\left(\frac{1}{8}\right)\right] = 0.969\text{in}^2$$

The axial stress in BC is

$$\sigma_{BC} = \frac{0.794W}{0.969} = 0.8196W \le 18\text{ksi} \text{ or } W \le 21.96\text{kip} \qquad 1$$

Fig. (b) shows the forces acting in axial and transverse direction on member ABD. The maximum moment and axial force will be just left of B. The magnitudes are

$N = 0.5W$ and $M = (0.5W)(1.5) = 0.75W\text{ft}-\text{kips} = 9W\text{in}-\text{kips}$

The cross sectional area and area moment of inertia for cross section BB can be found as

$$A_{BB} = [(4)(0.25) + (2)(1.75)(0.25)] = 1.875\text{in}^2$$

$$I_{zz} = \frac{1}{12}(0.25)(4^3) + 2\left[\frac{1}{12}(1.75)(0.25^3) + (1.75)(0.25)(1.875^2)\right] = 4.414\text{in}^4$$

The maximum compressive stress in segment BD is

$$(\sigma_{BD})_{max} = \frac{(9W)(2)}{4.414} + \frac{(0.5)W}{1.875} = 4.375W \le 18\text{ksi} \text{ or } W \le 4.14\text{kip} \qquad 2$$

The maximum weight that satisfies the requirements in Eqs. (1) and (2) is $\boxed{W = 4.1\text{kip}}$

Member BC should be checked for buckling.

10.53 A solid shaft of 50 mm diameter is made from a brittle material that has an allowable tensile stress of 100 MPa. Draw a failure envelope representing the maximum permissible positive values of T and P.

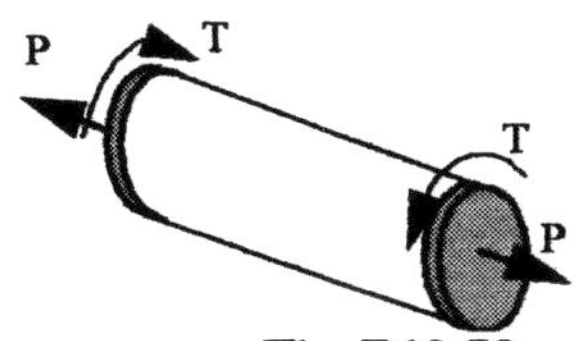

Fig. P10.53

Solution: d = 50 mm, $\sigma_{max} \le$ 100 MPa, Failure Envelope =?

The area of cross section and the polar moment of inertia is.

$$A = \frac{\pi}{4}(0.05)^2 = 1.963(10^{-3})\text{mm}^2 \qquad 1$$

$$J = \frac{\pi}{32}(0.05)^4 = 0.613(10^{-6})\text{mm}^4 \qquad 2$$

The axial stress can be written as

$$\sigma_{xx} = \frac{P(10^3)}{1.963(10^{-3})} = 0.5094P(10^6)\frac{N}{m^2} = 0.5094P \text{ MPa} \qquad 3$$

where P is in kN. The torsional shear stress is

$$\tau_{x\theta} = \frac{T(0.025)}{0.613(10^{-6})} = 0.0408T(10^6)\frac{N}{m^2} = 0.0408T\text{MPa} \qquad 4$$

where T is in N-m. The principal stress 1 is $\sigma_1 = \left(\frac{0.5094P}{2}\right) + \sqrt{\left(\frac{0.5094P}{2}\right)^2 + (0.0408T)^2}$ or

$$\sigma_1 = 0.2547P + \sqrt{(0.2547P)^2 + (0.0408T)^2} \le 100 \qquad 5$$

when $T = 0$ the value of P = 196.3 kN. The table below show the values of P and T that satisfy the equality in Eq. (1). Failure envelope is shown in figure below.

P (kN)	T (N-m)
0.00	2450.98
15.00	2355.48
30.00	2255.94
45.00	2151.80
60.00	2042.36
75.00	1926.71
90.00	1803.66
105.00	1671.58

P (kN)	T (N-m)
120.00	1528.12
135.00	1369.72
150.00	1190.43
165.00	978.83
180.00	706.46
195.00	200.17
196.31	0.00

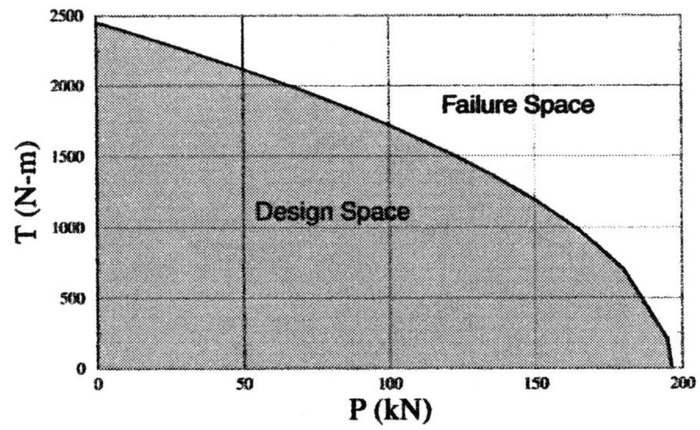

10.54 The shaft shown in Fig. P10.53 is made from a ductile material and has an allowable shear stress of 75 MPa. Draw a failure envelope representing the maximum permissible positive values of T and P.

Solution: $\tau_{max} \le 75$ MPa, Failure Envelope =?

From Eq. (5) in problem 10.53 we have: $\sigma_1 = 0.2547P + \sqrt{(0.2547P)^2 + (0.0408T)^2}$

The principal stresses two can be written as: $\sigma_2 = 0.2547P - \sqrt{(0.2547P)^2 + (0.0408T)^2}$

The maximum shear stress is : $\tau_{max} = \left|\frac{\sigma_1 - \sigma_2}{2}\right| = \sqrt{(0.2547P)^2 + (0.0408T)^2} \le 75$

when $T = 0$, the value of P = 294.46 kN. The table below shows the values of P and T that satisfy the equality in Eq. (1). Failure envelope is shown in figure below.

P (kN)	T (N-m)
0.00	1840.94
20.00	1836.69
40.00	1823.89
60.00	1802.34
80.00	1771.73
100.00	1731.58
120.00	1681.21
140.00	1619.66

P (kN)	T (N-m)
160.00	1545.60
180.00	1457.13
200.00	1351.40
220.00	1223.97
240.00	1067.08
260.00	864.85
280.00	570.98
294.52	0.07

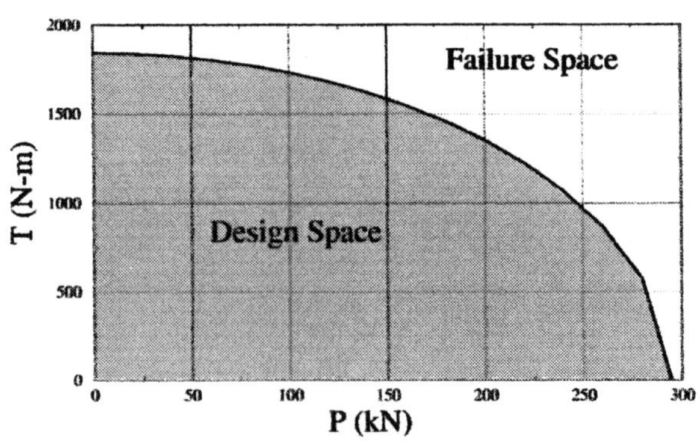

10.55 The shaft in Problem 10.53 is 1.5 m long and has a Modulus of Elasticity E = 200 GPa and a Modulus of Rigidity G = 80 GPa. Modify the failure envelope of Problem 10.53 to incorporate the limitation that the elongation cannot exceed 0.5 mm and the relative rotation of the right end with respect to the left end cannot exceed 3°.

Solution: $L = 1.5$ m $E = 200$ GPa $G = 80$ GPa $\delta \leq 0.5$ mm
$\Delta\phi \leq 3°$ $\sigma_{max} \leq 100$ MPa, Failure Envelope =?

From Eqs. (1), (2) and (5) in problem 10.53, we have: $A = 1.963(10^{-3})\text{mm}^2$, $J = 0.613(10^{-6})\text{mm}^4$ and

$$\sigma_1 = 0.2547P + \sqrt{(0.2547P)^2 + (0.0408T)^2} \leq 100 \quad (1)$$

The deformation can be written as: $\delta = \dfrac{P(10^3)1.5}{(200)(10^9)(1.963)(10^{-3})} = 3.820P(10^{-6}) \leq 0.5(10^{-3})$ or

$$P \leq 130.9\text{kN} \quad (2)$$

The relative rotation in radians is: $\Delta\phi = (3)(\pi/180) = 52.36(10^{-3})\text{rads}$ We can write

$$\Delta\phi = \frac{T(1.5)}{(80)(10^9)(0.613)(10^{-6})} = 30.59T(10^{-6}) \leq 52.36(10^{-3}) \quad \text{or} \quad T \leq 1.712(10^3) \quad (3)$$

Table in problem 10.59 shows the values of T and P corresponding to the equality sign in Eq. (1). Figure below shows the failure envelope corresponding to the equality sign in Eqs. (1), (2) and (3).

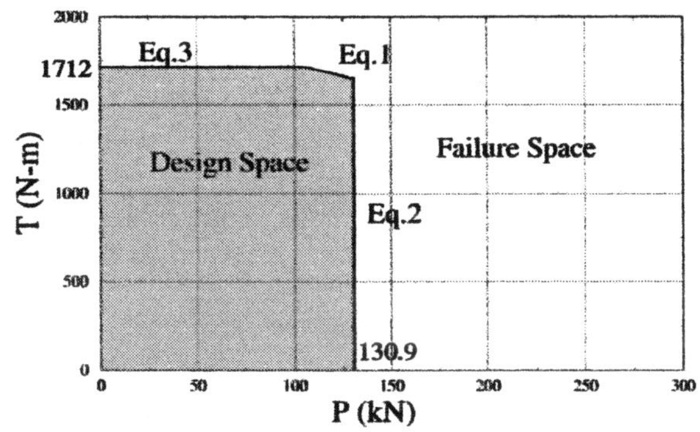

10.56 A pipe with an outside diameter of 40 mm and wall thickness of 10 mm is loaded as shown in Fig. P10.56. At section AA, the allowable shear stress 60 MPa. Draw the failure envelope for the applied loads P_1 and P_2.

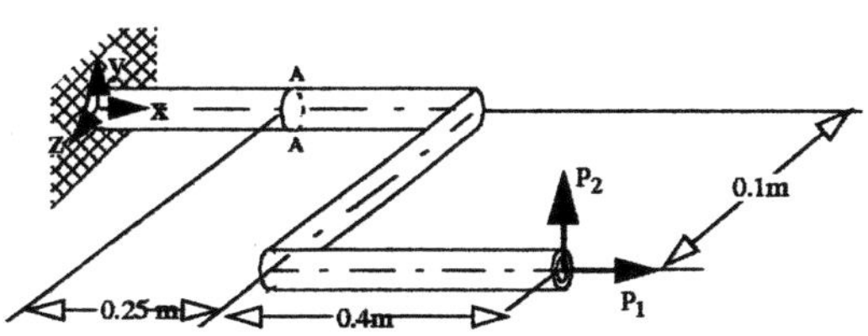

Fig. P10.56

Solution: $d_o = 40$ mm. $t = 10\text{mm}$, $\tau_{max} \leq 60\text{MPa}$, Failure Envelope =?

Figure (a) shows the free body diagram after an imaginary cut is made through section.

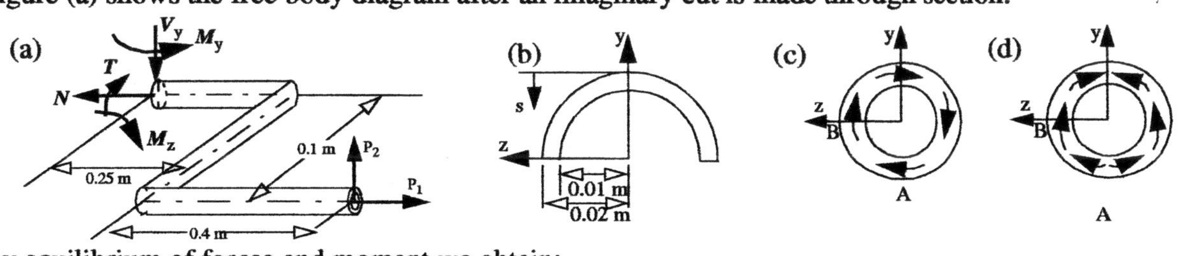

By equilibrium of forces and moment we obtain:

$N = P_1 \quad V_y = P_2 \quad M_z = 0.65P_2 \quad T = (-0.1)P_1 \quad M_y = (-0.1)P_1$

The area of cross section and the area moment of inertia are as follows.

$A = \pi[(0.04)^2 - (0.02)^2]/4 = 0.9425(10^{-3})m^2$

$J = \pi[(0.04)^4 - (0.02)^4]/32 = 0.2356(10^{-6})m^4$ and $I = J/2 = 0.1178(10^{-6})m^4$

We assume that P_1 and P_2 are in kN. The normal stresses at points A and B are

$$(\sigma_{xx})_A = \frac{P_1(10^3)}{0.9425(10^{-3})} + \frac{(0.65P_2)(10^3)(0.02)}{0.1178(10^{-6})} = (1.061P_1 + 110.36P_2)(10^6) = (1.061P_1 + 110.36P_2)MPa \quad 1$$

$$(\sigma_{xx})_B = \frac{P_1(10^3)}{0.9425(10^{-3})} + \frac{(0.1P_1)(10^3)(0.02)}{0.1178(10^{-6})} = 18.039P_1(10^6) = 18.039P_1 MPa \quad 2$$

Fig. (b) shows the area A_s used for calculating the shear stress at the neutral axis. The first moment of the area is: $Q_z = \frac{2}{3}[(0.02)^3 - (0.01)^3] = 4.67(10^{-6})m^3$

Fig. (c) shows the direction of torsional shear stress. Fig. (d) shows direction of shear stress due bending about z - axis. The shear stress at points A and B are

$$(\tau_{xz})_A = \frac{(0.1)P_2(10^3)(0.02)}{0.2356(10^{-6})} = 8.489P_2(10^6) = 8.489P_2 MPa \quad 3$$

$$(\tau_{xz})_B = \frac{(0.1)P_2(10^3)(0.02)}{0.2356(10^{-6})} + \frac{P_2(10^3)(4.67)(10^{-6})}{0.1178(10^{-6})(0.02)} = 10.471P_2(10^6) = 10.471P_2 MPa \quad 4$$

The maximum shear stress at points A and B are

$$(\tau_{max})_A = \sqrt{\left(\frac{1.061P_1 + 110.36P_2}{2}\right)^2 + (8.489P_2)^2} \le 60 \quad 5$$

$$(\tau_{max})_B = \sqrt{\left(\frac{18.039P_1}{2}\right)^2 + (10.471P_2)^2} \le 60 \quad 6$$

Using the equality sign, Eqs. (5) and (6) can be rewritten as

$$P_1 = \left[\sqrt{(60)^2 - (8.489P_2)^2} - 55.18P_2\right]/0.5305 \quad 7$$

$$P_1 = \left[\sqrt{(60)^2 - (10.471P_2)^2}\right]/9.0195 \quad 8$$

Table below can be constructed on a spread sheet and failure envelope constructed as shown below

P_2 Eq.(7)	P_1 Eq. (7)	P_2 Eq. (8)	P_1 Eq. (8)
0.000	113.10	0.00	6.65
0.100	102.69	0.50	6.63
0.200	92.25	1.00	6.55
0.300	81.79	1.50	6.42
0.400	71.31	2.00	6.23
0.500	60.81	2.50	5.99
0.600	50.28	3.00	5.67
0.700	39.73	3.50	5.27
0.800	29.16	4.00	4.76
0.900	18.57	4.50	4.12
1.000	7.95	5.00	3.25
1.075	0.00	5.73	0.00

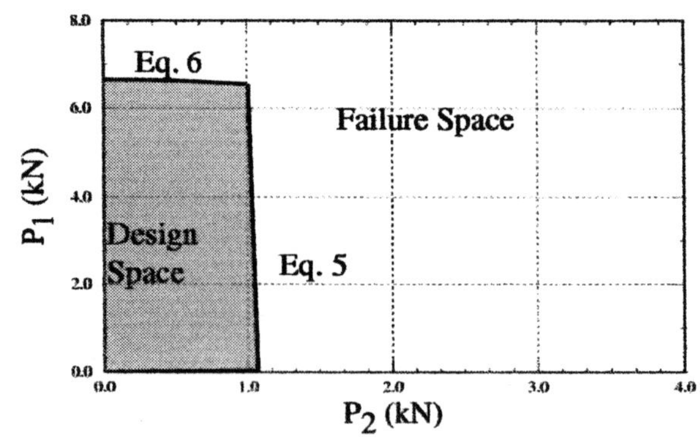

10.57 A bent pipe of 2 inch outside diameter and a wall thickness of 1/4 inch is loaded as shown. The maximum shear stress the pipe material can support is 24 ksi. Draw the failure envelope for the applied loads P_1 and P_2.

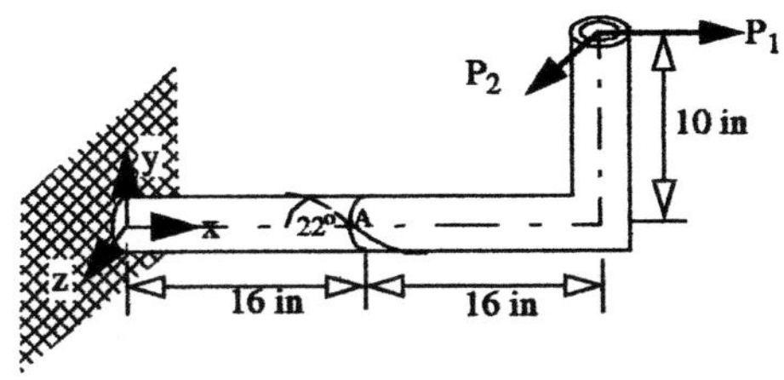

Fig. P10.57

Solution: $d_o = 2$ in., $t = 1/4$in, $\tau_{max} \le 24$ ksi, Failure Envelope =?

Figure (a) shows the free body diagram after an imaginary cut is made through section containing point A.By equilibrium of forces and moment we obtain.

$N = P_1$ $V_z = P_2$ $T = 10P_2$ $M_y = 16P_2$ $M_z = (-10)P_1$

The area of cross section and the area moment of inertia are as follows.

$A = \pi[(2)^2-(1.5)^2]/4 = 1.374\text{in}^2$ $J = \pi[(2)^4-(1.5)^4]/32 = 1.074\text{in}^4$ $I = J/2 = 0.5369\text{in}^4$

The normal stresses at points A and B are

$$(\sigma_{xx})_A = \frac{P_1}{1.374} + \frac{(10P_1)(1)}{0.5369} = 19.353P_1 \qquad 1$$

$$(\sigma_{xx})_B = \frac{P_1}{1.374} + \frac{(16P_2)(1)}{0.5369} = 0.7278P_1 + 29.801P_2 \qquad 2$$

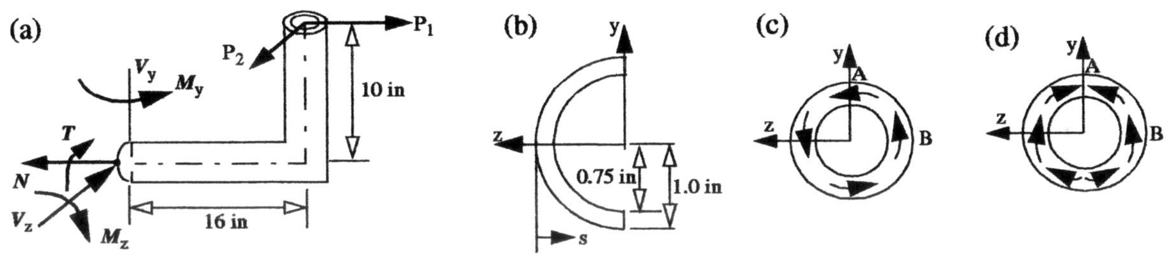

Fig. (b) shows the area A_s used for calculating the shear stress due to bending:

$$Q_y = \frac{2}{3}[(1)^3-(0.75)^3] = 0.3854\text{in}^3$$

Fig. (c) shows the direction of torsional shear stress. Fig. (d) shows direction of shear stress due bending about z - axis. The shear stress at points A and B are

$$(\tau_{xz})_A = \frac{(10)P_2(1)}{1.074} + \frac{P_2(0.3854)}{(0.5369)(0.5)} = 10.746P_2 \qquad 3$$

$$(\tau_{xz})_B = \frac{(10)P_2(1)}{1.074} = 9.311P_2 \qquad 4$$

The maximum shear stress at points A and B are

$$(\tau_{max})_A = \sqrt{\left(\frac{19.353P_1}{2}\right)^2 + (10.746P_2)^2} \le 24 \qquad 5$$

$$(\tau_{max})_B = \sqrt{\left(\frac{0.7278P_1 + 29.801P_2}{2}\right)^2 + (9.311P_2)^2} \le 24 \qquad 6$$

Eqs. (5) and (6) can be rewritten as

$$P_1 = \frac{1}{9.676}\left[\sqrt{(24)^2-(10.746P_2)^2}\right] \qquad 7$$

$$P_1 = \frac{1}{0.3637}\left[\sqrt{(24)^2-(9.311P_2)^2} - 14.90P_2\right] \qquad 8$$

Table below can be constructed on a spread sheet and failure envelope constructed as shown below

P_2 Eq.(7)	P_1 Eq. (7)	P_2 Eq. (8)	P_1 Eq. (8)
0.000	2.480	0.000	65.952
0.200	2.470	0.150	59.699
0.400	2.440	0.300	53.220
0.600	2.389	0.450	46.514
0.800	2.316	0.600	39.573
1.000	2.218	0.750	32.390
1.200	2.092	0.900	24.951
1.400	1.933	1.050	17.240
1.600	1.731	1.200	9.235
1.800	1.468	1.350	0.907
2.000	1.104	1.360	0.339
2.200	0.427	1.365	0.055
2.233	0.000	1.366	0.000

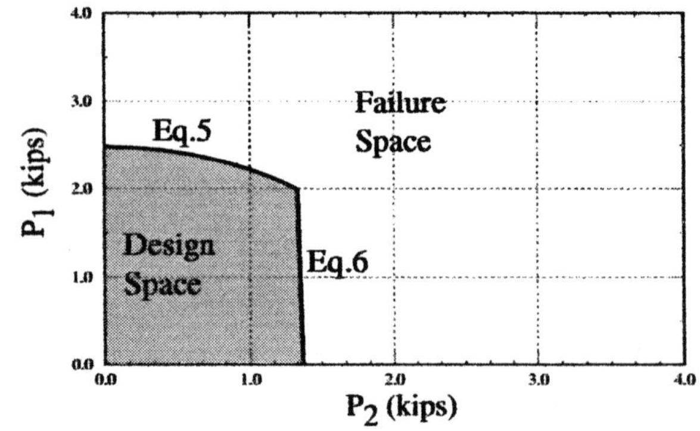

10.58 A hollow aluminum shaft of 5 feet length is to carry a torque of 200 in-kips and an axial force of 100 kips. The inner radius of the shaft is 1 inch. If the allowable shear stress in the shaft is 10 ksi, determine the outer radius of the lightest shaft.

Solution: $L = 5$ ft. $T = 200$ in-kips, $P = 100$ kips, $R_i = 1$, $\tau_{max} \le 10$ ksi R_o =? lightest

Fig. (a) shows the shaft with loading

(a)

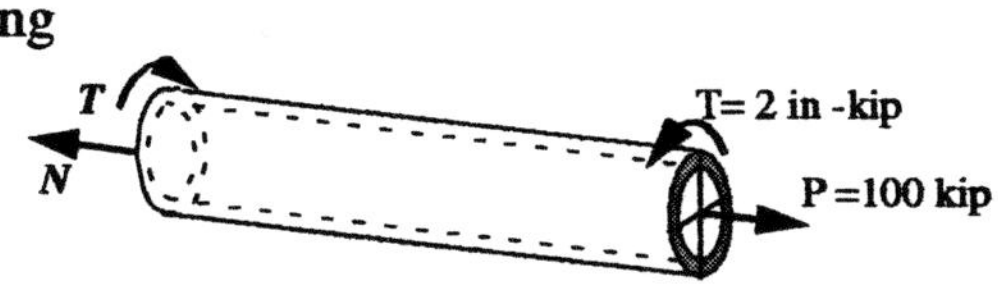

The axial stress and maximum torsional shear stress can be written as

$$\sigma_{xx} = \frac{100}{\pi(R_o^2 - 1)} = \frac{31.831}{(R_o^2 - 1)} \quad \text{and} \quad \tau_{x\theta} = \frac{200(R_o)}{\pi(R_o^4 - 1)} = \frac{127.323(R_o)}{(R_o^4 - 1)}$$

The maximum shear stress is

$$\tau_{max} = \sqrt{\left(\frac{15.915}{(R_o^2 - 1)}\right)^2 + \left(\frac{127.323(R_o)}{(R_o^4 - 1)}\right)^2} \le 10 \qquad (1)$$

We let $f(R_o) = \sqrt{\left(\frac{15.915}{(R_o^2 - 1)}\right)^2 + \left(\frac{127.323(R_o)}{(R_o^4 - 1)}\right)^2} - 10$

The root of $f(R_o)$ corresponding to the equality sign can be found on a spreadsheet as shown in table below.

R_o	$f(R_o)$	R_o	$f(R_o)$
2.000	7.786	2.4000	0.068
2.100	5.227	2.4025	0.036
2.200	3.160	**2.4050**	0.004
2.300	1.469	2.4075	-0.027
2.400	0.068	2.4100	-0.059
2.500	-1.105	2.4125	-0.090
2.600	-2.095	2.4150	-0.121
2.700	-2.938	2.4175	-0.152

The smallest value of R_o that satisfies the Eq. (1) is $\boxed{R_o = 2.405\text{in}}$

10.59 A hollow cylinder is fabricated from a sheet metal of thickness 15 mm. Determine the minimum outer radius to the nearest millimeter if the allowable normal stress is 150 MPa in tension or compression.

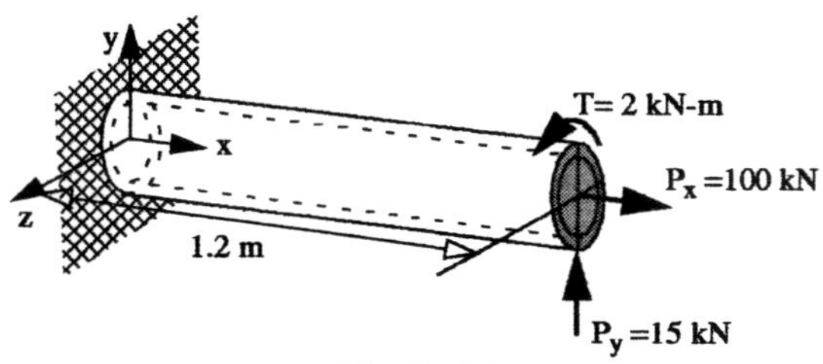

Fig. P10.59

Solution $t = 15$ mm $\sigma_{max} \le 150$ MPa $R_o = ?$ nearest millimeter

---

Fig.(a) shows the free body diagram after an imaginary cut is made next to the wall.

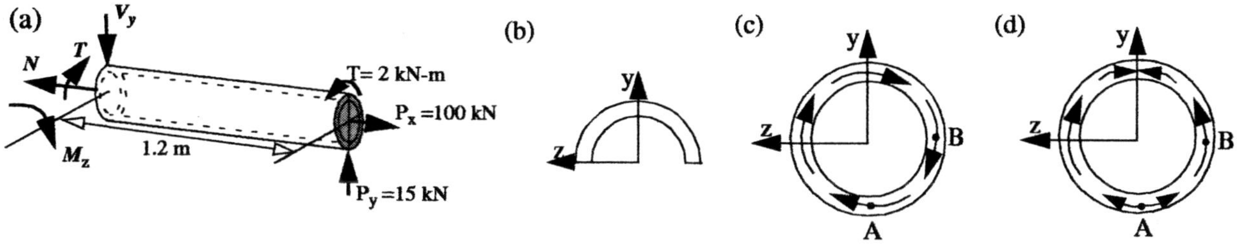

By force and moment equilibrium we obtain the following.

$N = 100 \text{ kN} \qquad V_y = 15 \text{ kN} \qquad T = 2 \text{ kN-m} \qquad M_z = (15)(1.2) = 18 \text{ kN-m}.$

The area of cross-section and area moment of inertias can be written as shown below.

$$A = \pi[R_o^2 - (R_o - 0.015)^2] \text{ m}^2 \qquad I = \frac{\pi}{4}[R_o^4 - (R_o - 0.015)^4] \text{ m}^4 \qquad J = 2I$$

Fig.(b) shows the area A_s used in the calculation of shear stress. The first moment of the area A_s can be written as

$$Q_z = \frac{2}{3}[R_o^3 - (R_o - 0.015)^3] \text{ m}^3$$

The normal stress at points A and B can be written as

$$(\sigma_{xx})_A = \frac{100(10^3)}{A} + \frac{(18)(10^3)(R_o)}{I} \quad \text{and} \quad (\sigma_{xx})_B = \frac{100(10^3)}{A}$$

Fig.(c) shows the direction of torsional shear stress and Fig. (d) shows the direction of shear stress due to bending about the z-axis. The shear stresses at A and B can be written as

$$(\tau_{xz})_A = \frac{(2)(10^3)(R_o)}{J} + \frac{(15)(10^3)(Q_z)}{I(0.015)} \quad \text{and} \quad (\tau_{xy})_B = \frac{(2)(10^3)(R_o)}{J}.$$

Principal stress one is the maximum normal stress at points A and B. The principal stress should be less than or equal to 150 MPa. The value can be written as:

$$(\sigma_{max})_A = \frac{(\sigma_{xx})_A}{2} + \sqrt{\left[\frac{(\sigma_{xx})_A}{2}\right]^2 + (\tau_{xz})_A^2} \le 150(10^6) \text{ or}$$

$$f_1(R_o) = \frac{(\sigma_{xx})_A}{2} + \sqrt{\left[\frac{(\sigma_{xx})_A}{2}\right]^2 + (\tau_{xz})_A^2} - 150(10^6) \le 0$$

$$(\sigma_{max})_B = \frac{(\sigma_{xx})_B}{2} + \sqrt{\left[\frac{(\sigma_{xx})_B}{2}\right]^2 + (\tau_{xy})_B^2} \le 150(10^6) \text{ or}$$

$$f_2(R_o) = \frac{(\sigma_{xx})_B}{2} + \sqrt{\left[\frac{(\sigma_{xx})_B}{2}\right]^2 + (\tau_{xy})_B^2} - 150(10^6) \le 0$$

The calculation of the various quantities as a function of R_o can be done on a spread sheet. The results are shown in the table below. $f_2(R_o)$ is always less than zero for values of R_o shown in column one. $f_1(R_o)$ changes sign at the value of $R_o = 0.064$ m. Thus the minimum value of R_o is $R_o = 64$ mm

R_o m	A m^2	I m^4	Q_z m^3	$(\sigma_{xx})_A$ MPa	$(\tau_{xz})_A$ MPa	$f_1(R_o)$ MPa	$(\sigma_{xx})_B$ MPa	$(\tau_{xy})_B$ MPa	$f_2(R_o)$ MPa
0.055	4.477E-03	5.176E-06	6.825E-05	213.6	10.6	64.1	22.3	6.6	-125.9
0.056	4.571E-03	5.505E-06	7.113E-05	205.0	10.2	55.5	21.9	6.5	-126.4
0.057	4.665E-03	5.847E-06	7.407E-05	196.9	9.7	47.4	21.4	6.3	-126.8
0.058	4.760E-03	6.203E-06	7.707E-05	189.3	9.4	39.8	21.0	6.2	-127.3
0.059	4.854E-03	6.573E-06	8.013E-05	182.2	9.0	32.6	20.6	6.1	-127.7
0.060	4.948E-03	6.958E-06	8.325E-05	175.4	8.6	25.8	20.2	6.0	-128.2
0.061	5.042E-03	7.358E-06	8.643E-05	169.1	8.3	19.5	19.8	5.9	-128.6
0.062	5.137E-03	7.773E-06	8.967E-05	163.0	8.0	13.4	19.5	5.8	-129.0
0.063	5.231E-03	8.203E-06	9.297E-05	157.4	7.7	7.7	19.1	5.7	-129.3
0.064	5.325E-03	8.649E-06	9.633E-05	152.0	7.4	2.3	18.8	5.6	-129.7
0.065	5.419E-03	9.111E-06	9.975E-05	146.9	7.1	-2.8	18.5	5.5	-130.0
0.066	5.513E-03	9.589E-06	1.032E-04	142.0	6.9	-7.6	18.1	5.4	-130.4
0.067	5.608E-03	1.008E-05	1.068E-04	137.4	6.6	-12.3	17.8	5.3	-130.7

10.60 Table 10.1 shows the measured radii of a solid tapered member at several point along the axis of the shaft. The member is subjected to a torque T = 30 kN-m and an axial force P = 100 kN. Plot the maximum normal and shear stress as a function of x.

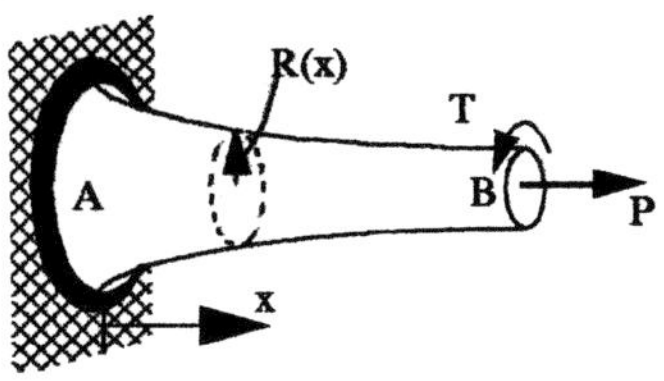

Fig. P10.60

Table 10.1. Data for Problem 10.60.

x (m)	R(x) (mm)	x (m)	R(x) (mm)
0.0	100.6	0.8	60.1
0.1	92.7	0.9	60.3
0.2	82.6	1.0	59.1
0.3	79.6	1.1	54.0
0.4	75.9	1.2	54.8
0.5	68.8	1.3	54.1
0.6	68.0	1.4	49.4
0.7	65.9	1.5	50.6

Solution T = 30 kN-m P = 100 kN σ_{max} and τ_{max} vs. x

The axial stress is: $\sigma_{xx} = \dfrac{100(10^3)}{\pi R^2} = \dfrac{31.83(10^3)}{R^2}$

The maximum torsional shear stress is: $\tau_{x\theta} = \dfrac{(30)(10^3)(R)}{(\pi R^4/2)} = \dfrac{19.098(10^3)}{R^3}$

The maximum normal stress is: $\sigma_{max} = \dfrac{\sigma_{xx}}{2} + \sqrt{\left(\dfrac{\sigma_{xx}}{2}\right)^2 + \tau_{x\theta}^2}$

The maximum shear stress is: $\tau_{max} = \sqrt{\left(\dfrac{\sigma_{xx}}{2}\right)^2 + \tau_{x\theta}^2}$

The table below shows the calculation of the stresses and the figure shows the plot of σ_{max} and τ_{max} vs. x.

x (m	R (mm)	σ_{xx} (MPa)	$\tau_{x\theta}$ (MPa)	σ_{max} (MPa)	τ_{max} (MPa)
0.00	100.60	3.15	18.76	20.40	18.82
0.10	92.70	3.70	23.98	25.90	24.05
0.20	82.60	4.67	33.89	36.30	33.97
0.30	79.60	5.02	37.87	40.46	37.95
0.40	75.90	5.53	43.68	46.53	43.77
0.50	68.80	6.72	58.65	62.10	58.74
0.60	68.00	6.88	60.74	64.28	60.84
0.70	65.90	7.33	66.73	70.50	66.83
0.80	60.10	8.81	87.98	92.50	88.09
0.90	60.30	8.75	87.11	91.59	87.22
1.00	59.10	9.11	92.52	97.19	92.63
1.10	54.00	10.92	121.29	126.87	121.41
1.20	54.80	10.60	116.05	121.47	116.17
1.30	54.10	10.88	120.62	126.18	120.74
1.40	49.40	13.04	158.42	165.08	158.56
1.50	50.60	12.43	147.42	153.76	147.55

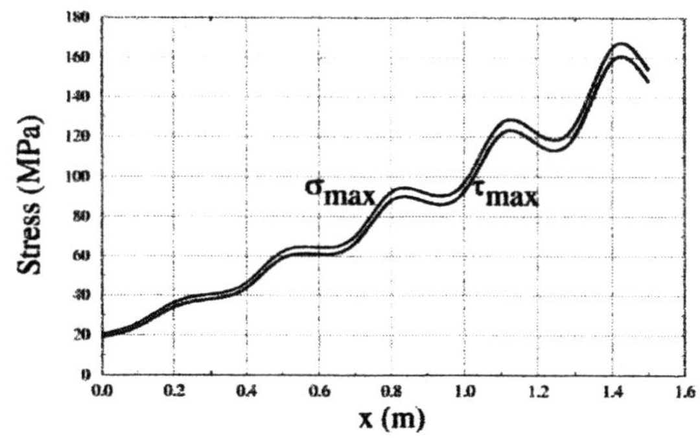

Problem Set 10.2

10.61 An axial bar was welded together along the weld-line shown in Fig. P10.61. The bar material has a critical stress intensity factor of $20\text{ksi}\sqrt{\text{in}}$. The applied load P = 12 kips. What is the critical length of a crack at which the axial member should be taken out of service?

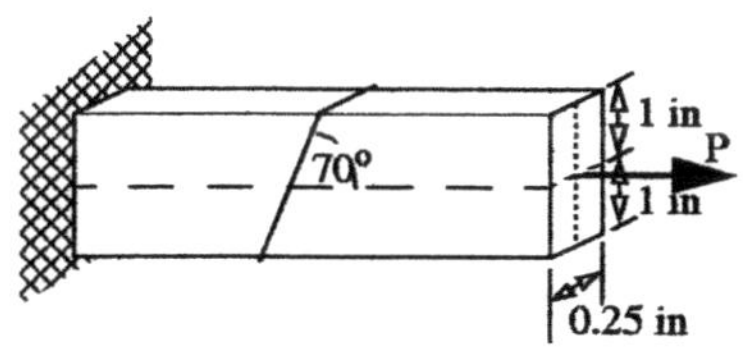

Fig. P10.61

Solution $k_{crit} = 20\text{ksi}\sqrt{\text{in}}$ P = 12 kips $2a_{crit} = ?$

The axial stress is: $\sigma_{xx} = \dfrac{12}{(0.25)(2)} = 24\text{ksi}$

From Eq.(10-5a) we have : $k_{crit} = \sigma_{xx}\sqrt{\pi a_{crit}}$ or $20 = 24\sqrt{\pi a_{crit}}$ or $a_{crit} = 0.221$ or

$$\boxed{2a_{crit} = 0.442}$$

10.62 An axial bar was welded together along the weld-line shown in Fig. P10.62. The bar material has a critical stress intensity factor of $20\text{ksi}\sqrt{\text{in}}$. Before the axial load of P = 10 kips was applied, a crack was observed along the weld line shown in Fig. P10.61. (a) What is the critical length of a crack length along the weld line? (b) What is the critical length of a crack at which the axial member should be taken out of service?

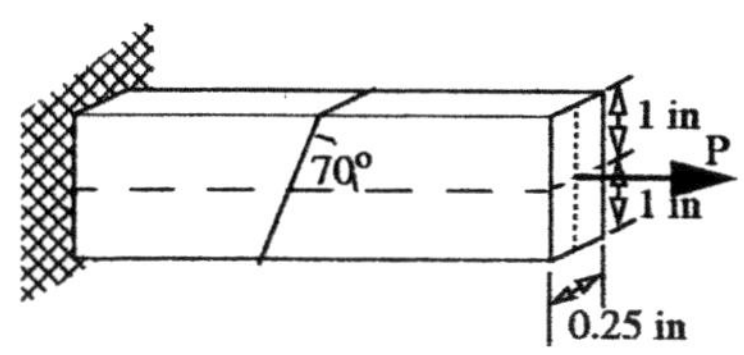

Fig. P10.62

Solution $k_{crit} = 20\text{ksi}\sqrt{\text{in}}$ P = 10 kips $2a_{crit} = ?$

The axial stress is: $\sigma_{xx} = \dfrac{10}{(0.25)(2)} = 20\text{ksi}$

The outward normal of the plane makes an angle $\theta = 70° + 90° = 160°$ with the x-axis.Substituting $\sigma_{xx} = 20\text{ksi}$, $\sigma_{yy} = 0$, $\tau_{xy} = 0$ and $\theta = 160°$ in Eqs.(8.1) and (8.2) we obtain the normal and shear on the plane as

$\sigma_{xx} = 20\cos 160^2 = 17.66\text{ksi}$ and $\tau_{nt} = -\sigma\sin 160\cos 160 = 6.428\text{ksi}$

From Eq.(10-5a) and (10-5b) we have:

$k_I = 17.66\sqrt{\pi a_{crit}} = 31.302\sqrt{a_{crit}}$ and $k_{II} = 6.428\sqrt{\pi a_{crit}} = 11.393\sqrt{a_{crit}}$

The equivalent stress intensity factor is: $k_{eq} = \sqrt{k_I^2 + k_{II}^2} = \sqrt{(31.302)^2 + (11.393)^2}(\sqrt{a_{crit}})$ or $a_{crit} = 0.36$

$\boxed{2a_{crit} = 0.72\text{in}}$

10.63 An axial bar was welded together along the weld-line shown in Fig. P10.63. The bar material has a critical stress intensity factor of $20\text{ksi}\sqrt{\text{in}}$. The smallest cracks length of 0.06 inches represents the limit of the flaw detection system. What is the maximum axial force that should be permitted to act on the member, assuming there is no pre-existing crack?

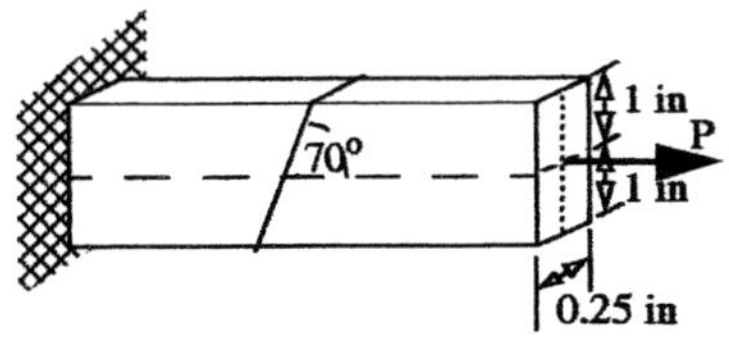

Fig. P10.63

Solution $k_{crit} = 20\text{ksi}\sqrt{\text{in}}$ $2a_{crit} = 0.06$ in $P_{max} = ?$

The axial stress is: $\sigma_{xx} = \dfrac{P_{max}}{(0.25)(2)} = 2P_{max}$

From Eq.(10-5a) we have: $k_{crit} = \sigma_{xx}\sqrt{\pi a_{crit}}$ or $20 = 2P_{max}\sqrt{\pi a_{crit}}$ or $\boxed{P_{max} = 32.6\text{kips}}$

10.64 A thin tube with a seam shown in Fig. P10.64 has a center-line radius of 100 mm and a thickness of 10 mm. The tube material has a critical stress intensity factor of 33 $\text{MPa}\sqrt{\text{m}}$.The applied torque on the tube is T = 100 kN-m, what is the critical length of a crack at which the tube should be taken out of service.

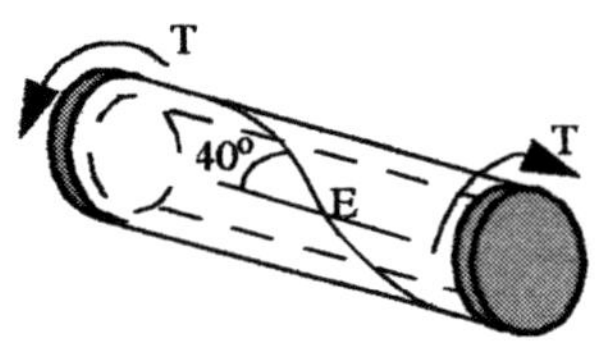

Fig. P10.64

Solution r= 100 mm t = 10 mm $k_{crit} = 33\ \text{MPa}\sqrt{\text{m}}$ T = 100 kN-m $2a_{crit} = ?$

The polar moment of inertia is: $J = \frac{\pi}{2}[(0.105)^4 - (0.095)^4] = 62.99(10^{-6})\ \text{m}^4$

The magnitude of maximum shear stress is

$$\tau = \frac{T(0.105)}{62.99(10^{-6})} = 1667T\ \frac{\text{N}}{\text{m}^2}$$ 1

For $T = 100(10^3)$ N-m we obtain $\tau = 166.7$ MPa. As all other stress components are zero, the principal stress one is $\sigma_1 = \tau = 166.7\ \text{MPa}$

From Eq.(10-5a) we obtain: $33 = 166.7\sqrt{\pi a_{crit}}$ or $a_{crit} = 0.01247$ m or $\boxed{2a_{crit} = 24.94\ \text{mm}}$

10.65 A thin tube with a seam shown in Fig. P10.65 has a center-line radius of 150 mm and a thickness of 10 mm. The tube material has a critical stress intensity factor of 33 $MPa\sqrt{m}$. Before the torque of T = 150 kN-m was applied, a crack was observed along the seam that is 40° to the axis of the shaft as shown in Fig. P10.64. (a) What is the critical length of a crack length along the seam? (b) What is the critical length of a crack at which the tube should be taken out of service?

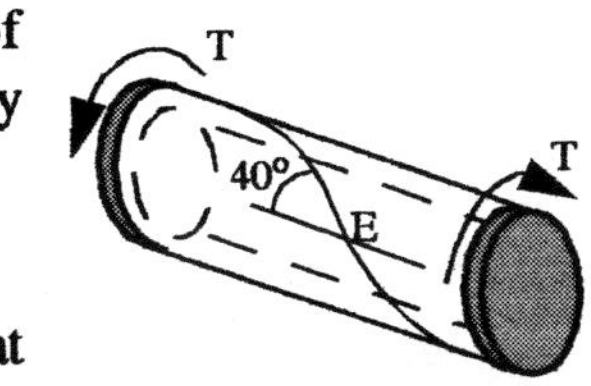

Fig. P10.65

Solution $r = 100$ mm $t = 10$ mm $k_{crit} = 33\ MPa\sqrt{m}$ $T = 150$ kN-m $2a_{seam} = ?$ $2a_{crit} = ?$

Substituting $T = 150(10^3)$ N-m in Eq.(1) in problem 10-64, we obtain $\tau = 250MPa$. Fig.(a) shows the direction of shear stress and Fig.(b) shows the stress cube.

(a) 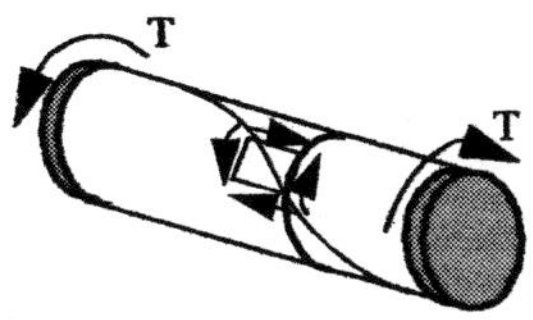

(b) 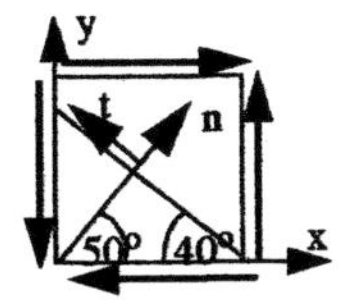

From Fig.(b) we have $\tau_{xy} = 250$ MPa and $\theta = 50°$. From Eqs.(8.1)and (8.2) we obtain

$\sigma_{nn} = 2(250)\cos 60 \sin 60 ksi = 246.2\ MPa$ and $\tau_{nt} = 250(\cos 50^2 - \sin 50^2) = -43.41\ MPa$

From Eqs.(10-5a) and (10-5b) we have :

$k_I = 246.2\sqrt{\pi a_{seam}} = 436.28\sqrt{a_{crit}}$ and $k_{II} = 43.41\sqrt{\pi a_{seam}} = 76.94\sqrt{a_{seam}}$

The equivalent stress intensity factor is

$k_{eq} = \sqrt{k_I^2 + k_{II}^2} = \sqrt{(436.28)^2 + (76.94)^2}(\sqrt{a_{seam}}) = 33$ or $a_{seam} = 0.00555\ m$

The critical crack length along the seam is: $\boxed{2a_{seam} = 11.1\ mm}$

The principal stress one is $\sigma_1 = 250MPa$. From Eq.(10-5a) we can write

$33 = 250\sqrt{\pi a_{crit}}$ or $a_{crit} = 0.00555\ m$ or $\boxed{2a_{crit} = 11.1\ mm}$

10.66 A thin tube with a seam shown in Fig. P10.66 has a center-line radius of 150 mm and a thickness of 10 mm. The tube material has a critical stress intensity factor of 33 $MPa\sqrt{m}$. The smallest cracks length of 5 mm represents the limit of the flaw detection system. What is the maximum torque that should be permitted to act on the tube, assuming there is no pre-existing crack?

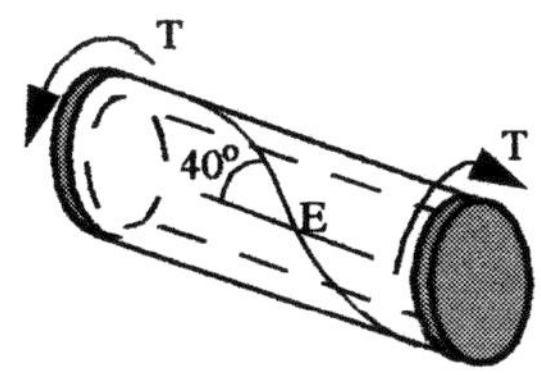

Fig. P10.66

Solution $r = 100$ mm $t = 10$ mm $k_{crit} = 33\ MPa\sqrt{m}$ $2a_{crit} = 5mm$ $T_{max} = ?$

From Eq.(1) in problem 10-64, we obtain $\tau = 1667T_{max}$. The principal stress one is $\sigma_1 = 1667T_{max}$

From Eq.(10-5a) we obtain: $(1667T_{max})\sqrt{\pi(0.0025)} = 33(10^6)$ or $\boxed{T_{max} = 223 kN - m}$

10.67 The beam shown in Fig. P10.67 is made from a material with a stress intensity factor of 33 $MPa\sqrt{m}$. The applied load on the beam is P = 150 N, what is the critical length of a crack at which the beam should be taken out of service?

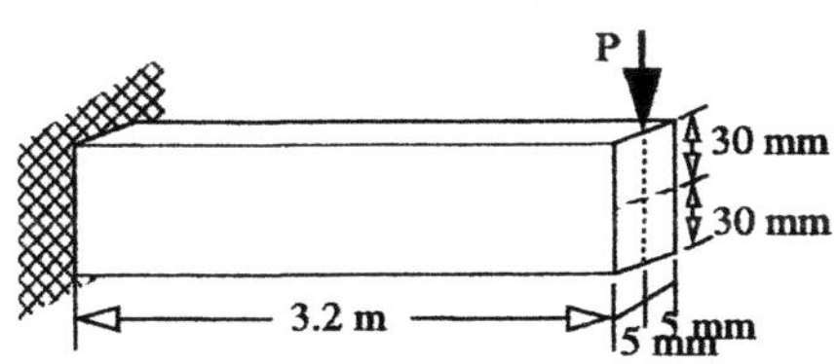

Fig. P10.67

Solution $k_{crit} = 33 MPa\sqrt{m}$ $P_{max} = 150$ N $2a_{crit} = ?$

The area moment of inertia is: $I = \frac{1}{12}(0.01)(0.06)^3 = 0.180(10^{-6})\ m^4$

The maximum moment is $M_{max} = 3.2$ P.

The maximum bending normal stress is: $\sigma_{xx} = \frac{M_{max}y_{max}}{I} = \frac{(3.2P)(0.03)}{0.18(10^{-6})} = 0.533P(10^6)\frac{N}{m^2}$

As the maximum bending shear stress is significantly smaller than the maximum bending normal stress, the maximum principal stress one is

$$\sigma_1 = 0.533P\ \ MPa \qquad 1$$

Substituting P = 150 N, we obtain $\sigma_1 = 80$ MPa

From Eqs.(10-5a) we have: $80\sqrt{\pi a_{crit}} = 33$ or $a_{crit} = 0.05416$ m or $\boxed{2a_{crit} = 108.3\ \ mm}$

10.68 The beam shown in Fig. P10.68 is made from a material with a stress intensity factor of 33 $MPa\sqrt{m}$. The smallest cracks length of 2 mm represents the limit of the flaw detection system. What is the maximum load P that should be permitted to act on the beam?

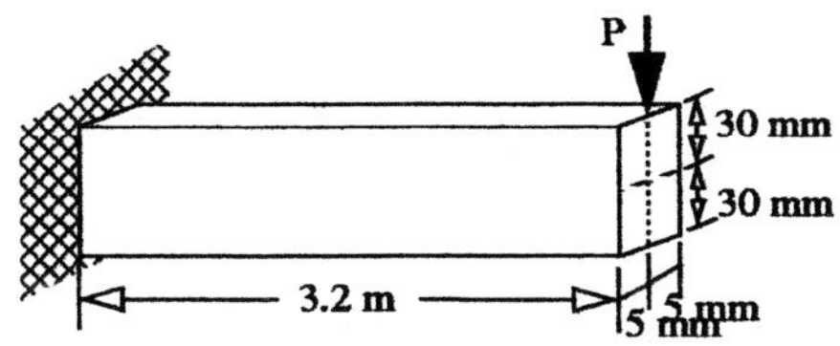

Fig. P10.68

Solution $k_{crit} = 33 MPa\sqrt{m}$ $2a_{crit} = 2$ mm $P_{max} = ?$

From Eq.(1) in problem 10-67, we have: $\sigma_1 = 0.533P_{max}MPa$

From Eq.(10-5a)we have: $0.533P_{max}\sqrt{\pi(0.001)} = 33$ or $\boxed{P_{max} = 1.1\ \ kN}$

10.69 The cylindrical tank shown in Fig. P10.69 has a mean diameter of 4 m has a wall thickness of 5 mm. The critical stress intensity factor of the tank material is 45 $MPa\sqrt{m}$. The pressure in the tank is 500 kPa, what is the critical length of a crack at which the cylinder should be taken out of service, assuming any leak can be ignored?

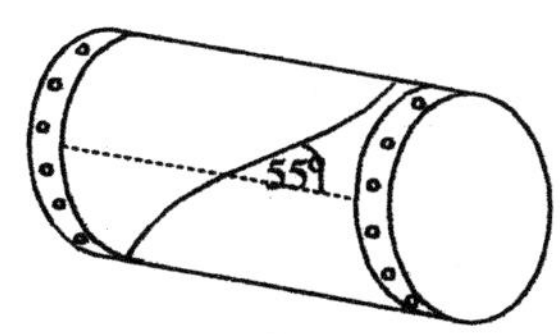

Fig. P10.69

Solution r= 4 m t = 5mm $k_{crit} = 45\ \ MPa\sqrt{m}$ p = 500 kPa$2a_{seam} = ?$ $2a_{crit} = ?$

The axial stress is: $\sigma_{xx} = \frac{p(2)}{(2)(0.005)} = 200p$

The hoop stress is: $\sigma_{\theta\theta} = \frac{p(2)}{0.005} = 400p$

We designate the tangential (hoop) direction as the y-direction.The state of stress is

$$\sigma_{xx} = 200p \qquad \sigma_{yy} = 400p \qquad \tau_{xy} = 0 \qquad 1$$

The stress σ_{yy} is principal stress one. Substituting $P = 500(10^3)\ N/m^2$, we obtain

$\sigma_1 = (400)(500)(10^3) = 200\ \ MPa$

From Eq.(10-5a) we have: $200\sqrt{\pi a_{crit}} = 45$ or $a_{crit} = 0.0161$ m or $\boxed{2a_{crit} = 32.2\ \ mm}$

10.70 The cylindrical tank shown in Fig. P10.70 has a mean diameter of 4 m has a wall thickness of 5 mm. The critical stress intensity factor of the tank material is 45 $MPa\sqrt{m}$. For a gas pressure of 750 kPa what would be the critical crack length along the weld line?

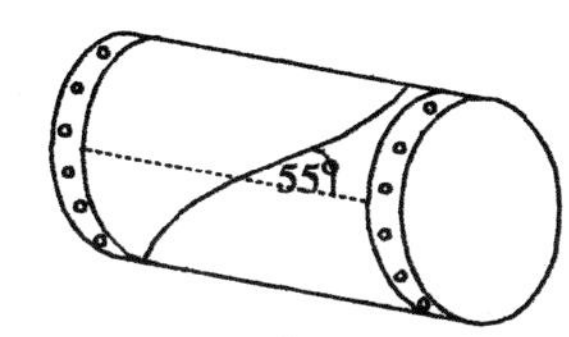

Fig. P10.70

Solution $d = 4m$ $t = 5\ mm$ $k_{crit} = 45\ MPa\sqrt{m}$ $p = 750kPa$ $2a_{crit} = ?$

Substituting $p = 750(10^3)\ N/m^2$ in Eq.(1) of problem 10-69, we obtain

$$\sigma_{xx} = 150MPa \qquad \sigma_{yy} = 300MPa \qquad \tau_{xy} = 0 \qquad 1$$

The angle of the normal to the weld line from the x-axis is $\theta = 55^o + 90^o = 145^o$. From Eqs.(8.1) and (8.2) we obtain the normal and shear stress on the weld surface as

$\sigma_{nn} = 150\cos 145^2 + 300\sin 145^2 = 199.35\ \ MPa$ and $_{nt} = -(150-300)\cos 145\sin 145 = -70.48\ \ MP$:

From Eqs.(10-5a) and (10-5b) we have :

$k_I = 199.35\sqrt{\pi a_{crit}} = 353.34\sqrt{a_{crit}}$ and $k_{II} = 70.48\sqrt{\pi a_{crit}} = 124.92\sqrt{a_{crit}}$

The equivalent stress intensity factor is: $k_{eq} = \sqrt{k_I^2 + k_{II}^2} = \sqrt{(353.34)^2 + (124.94)^2}(\sqrt{a_{crit}}) = 45$ or

$a_{crit} = 0.0144\ \ m$ or $\boxed{2a_{crit} = 28.8\ \ mm}$

10.71 The cylindrical tank shown in Fig. P10.71 has a mean diameter of 4 m has a wall thickness of 5 mm. The critical stress intensity factor of the tank material is 45 $MPa\sqrt{m}$. A critical crack length of 4 mm is used for estimating the maximum pressure the tank can sustain? What is the maximum permissible pressure in the tank?

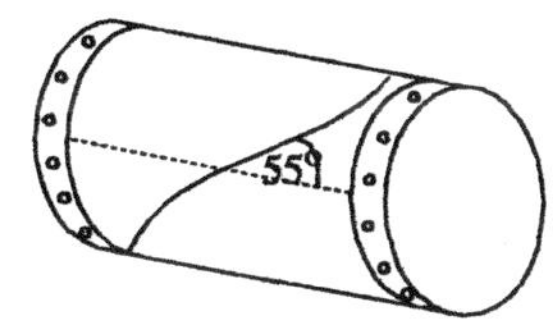

Fig. P10.71

Solution $d = 4\ m$ $t = 5\ mm$ $k_{crit} = 45MPa\sqrt{m}$ $2a_{crit} = 4\ mm$ $p_{max} = ?$

From Eq.(1) in problem 10-67, we have: $\sigma_{xx} = 200p_{max}$ $\sigma_{yy} = 400p_{max}$ $\tau_{xy} = 0$

The stress σ_{yy} is principal stress one. From Eq.(10-5a) we have :

$400p_{max}\sqrt{\pi(0.002)} = 45(10^6)$ or $\boxed{p_{max} = 1419\ \ kPa}$

10.72 An aluminum alloy (G = 4,000 ksi) hollow shaft has a critical stress intensity factor of 22 $ksi\sqrt{in}$. The outside diameter of the shaft is 2 inches and has a wall thickness of 1/4 inch. At what critical length of a crack should the shaft be taken out of service, if T = 60 in-kips?

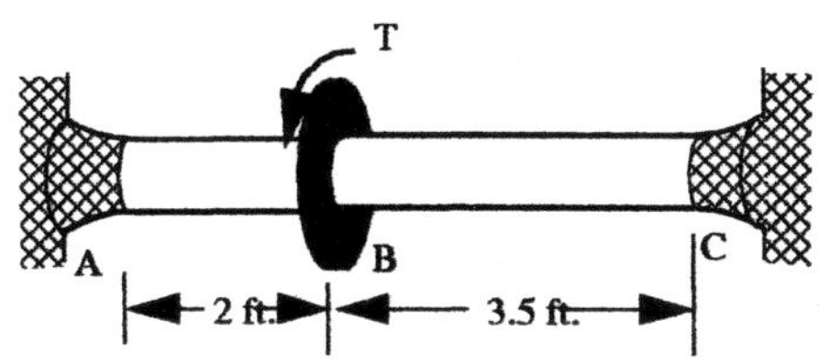

Fig. P10.72

Solution $G = 4000\ ksi$ $k_{crit} = 22ksi\sqrt{in}$ $d_o = 2\ in$ $t = 1/4\ in$ $T = 60$ in-kips $2a_{crit} = ?$

The polar moment of inertia is

$$J = \frac{\pi}{32}[2^4 - 1.5^4] = 1.074\ \ in^4 \qquad 1$$

Fig.(a) and (b) show free body diagrams for calculation of internal torque

By equilibruim of moment we obtain

$$T_{AB} = T_A \quad \text{and} \quad T_{BC} = T_A - T \qquad 2$$

(a) 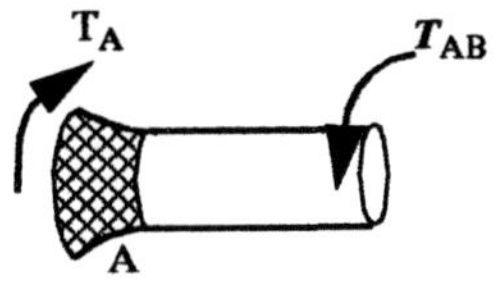

(b) 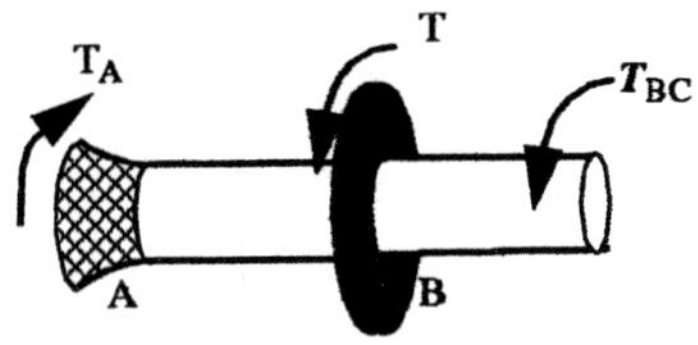

The realtive rotations of the ends of each segments can be written as

$$\phi_B - \phi_A = \frac{T_{AB}(x_B - x_A)}{GJ} = \frac{T_A(24)}{(4000)(1.074)} = 5.587T_A(10^{-3}) \qquad 3$$

$$\phi_C - \phi_B = \frac{T_{BC}(x_C - x_B)}{GJ} = \frac{(T_A - T)(42)}{(4000)(1.074)} = 9.778(T_A - T)(10^{-3}) \qquad 4$$

Adding Eqs.(3) and (4) and noting that $\phi_c = 0$ and $\phi_c = 0$

$$\phi_C - \phi_B = [5.587T_A + 9.778(T_A - T)](10^{-3}) = 0 \quad \text{or} \quad T_A = 0.636T$$

From Eqs.(2) we obtain: $T_{AB} = 0.636T$ and $T_{BC} = -0.364T$

The maximum shear stress will be in segment AB and its value is

$$\tau = \frac{T_{AB}\rho_{AB}}{J} = \frac{(0.636T)(1)}{1.074} = 0.593T \qquad 5$$

Substituting T = 60 in -kips , weobtain $\tau = 35.5\text{ksi}$ .The principal stress one is $\sigma_1 = 35.5$ ksi.

From Eq.(10-5a), we obtain: $35.5\sqrt{\pi a_{crit}} = 22$ or $a_{crit} = 0.122\text{in}$ or $\boxed{2a_{crit} = 0.244 \text{ in}}$

10.73 An aluminum alloy (G = 4,000 ksi) hollow shaft has a critical stress intensity factor of 22 $\text{ksi}\sqrt{\text{in}}$. The outside diameter of the shaft is 2 inches and has a wall thickness of 1/4 inch. If the smallest crack length that can be detected is 0.04 inch, determine the maximum torque that can be applied to the shaft.

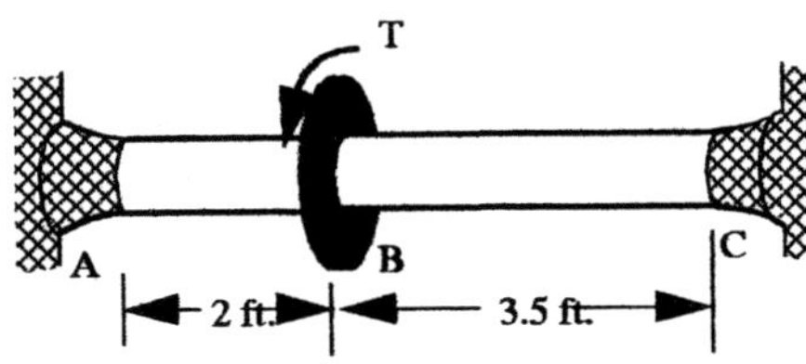

Fig. P10.73

Solution G = 4000 ksi $k_{crit} = 22\text{ksi}\sqrt{\text{in}}$ $d_o = 2$ in t = 1/4 in $2a_{crit} = 0.04$ $T_{max} = ?$

From Eq.(5) in problem 10-72, we have $\tau = 0.593T_{max}$

From Eq.(10-5a), we obtain: $0.593T_{max}\sqrt{\pi a_{crit}} = 22$ or $\boxed{T_{max} = 148 \text{ in} - \text{kips}}$

10.74 A thin cylindrical tube with an outer diameter of 5 inches is fabricated by butt-welding 1/16 inch thick plate along a spiral seam as shown. A through crack of 0.07 inch was observed in the seam. If the critical stress intensity factor for the material is 22 $\text{ksi}\sqrt{\text{in}}$, determine the maximum torque T that the tube can transmit.

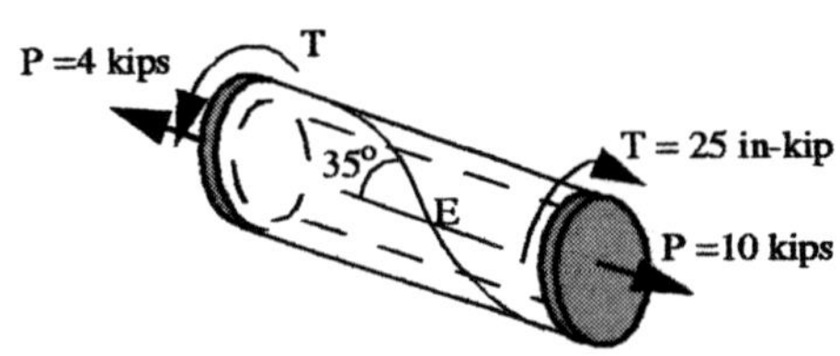

Fig. P10.74

Solution $d_o = 5$ in t = 1/16 in $k_{crit} = 22\text{ksi}\sqrt{\text{in}}$ ' $2a_{crit} = ?$

The cross-sectional area and polar moment of inertia are:

$A = \frac{\pi}{4}[5^2 - 4.875^2] = 0.9695\ \text{in}^2 \qquad J = \frac{\pi}{32}[5^4 - 4.875^4] = 5.9096\ \text{in}^4$

The axial stress is: $\sigma_{xx} = \frac{N}{A} = \frac{10}{0.9695} = 10.315\ \text{ksi}$

The torsional shear stress is: $\tau = \frac{T\rho}{J} = \frac{(25)(2.5)}{5.9096} = 10.576\ \text{ksi}$

Fig. (a) shows the direction of torsional shear stress and Fig. (b) shows the stress cube.

(a)

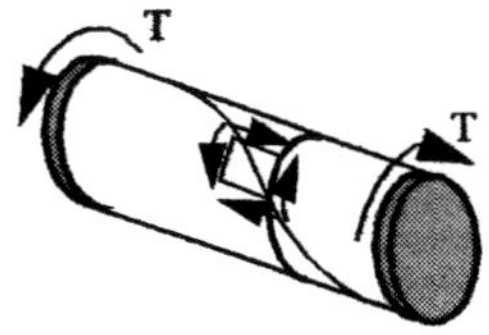

(b)

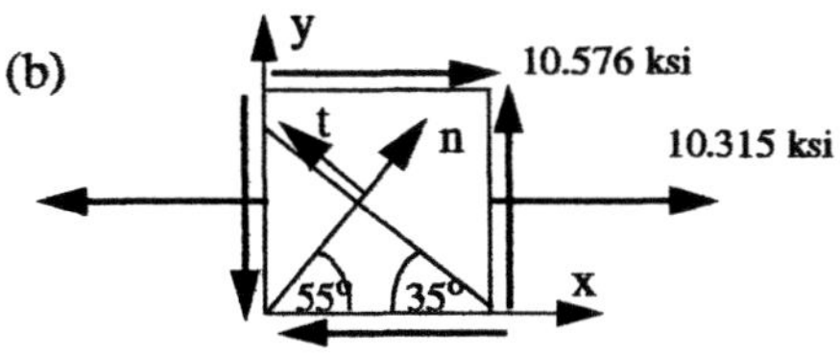

From Fig. (b) we obtain: $\sigma_{xx} = 10.315\ \text{ksi} \qquad \sigma_{yy} = 0 \qquad \tau_{xy} = 10.576\ \text{ksi}$

The outward normal to the seam surface makes an angle of 55° to the x-direction. The normal and shear stress on the seam can be found as:

$\sigma_{nn} = 10.315\cos^2 55 + (2)(10.576)\cos 55 \sin 55 = 13.33\ \text{ksi}$ and

$\tau_{nt} = -(10.315)\cos 55 \sin 55 + (10.576)(\cos^2 55 - \sin^2 55) = -8.46\ \text{ksi}$

From Eqs.(10-5a) and (10-5b) we obtain:

$k_I = 13.33\sqrt{\pi a_{crit}} = 23.63\sqrt{a_{crit}}$ and $k_{II} = 8.46\sqrt{\pi a_{crit}} = 14.99\sqrt{a_{crit}}$

The equivalent stress intensity factor is: $k_{eq} = \sqrt{k_I^2 + k_{II}^2} = \sqrt{(23.63)^2 + (14.99)^2}(\sqrt{a_{crit}}) = 22$ or

$a_{crit} = 0.618\ \text{in}$ or $\boxed{2a_{crit} = 1.236\ \text{in}}$

10.75 Due to a force P, the stress components at a critical point that is in plane stress were found to be $\sigma_{xx} = 10P\ \text{MPa(T)} \qquad \sigma_{yy} = 20P\ \text{MPa(C)} \qquad \tau_{xy} = 5P\ \text{MPa}$. The material has a yield stress of 160 MPa as determined in a tension test. If yielding must be avoided, predict the maximum value of the force P using: (a) Maximum Shear Stress theory. (b) Maximum Distortion Energy theory.

Solution: $\sigma_{xx} = 10$ MPa $\quad \sigma_{yy} = -20$ MPa $\quad \tau_{xy} = 5$ P MP $\quad \sigma_{yield} = 160$MPa

$P_{max} = ?$ Max. Shear $\quad P_{max} = ?$ Max. Distortion. $\quad$ Plane stress

---

The principal stresses can be found as:

$$\sigma_{1,2} = \frac{(10P - 20P)}{2} \pm \sqrt{\left(\frac{10P + 20P}{2}\right)^2 + (5P)^2} = -5P \pm 15.81P \text{ or } \sigma_1 = 10.81P \text{ and } \sigma_2 = -20.81P$$

The third principal stress $\sigma_3 = 0$. The maximum shear stress is:

$$\tau_{max} = \left|\frac{\sigma_1 - \sigma_2}{2}\right| = 15.81P \le \frac{\sigma_{yield}}{2} \text{ or } P \le 5.06 \qquad \boxed{P_{max} = 5\text{kN}}$$

The von-mises stresses is

$$\sigma_{von} = \frac{1}{\sqrt{2}}\sqrt{(\sigma_1 - \sigma_2)^2 + (\sigma_2 - \sigma_3)^2 + (\sigma_3 - \sigma_1)^2} \text{ or}$$

$$\sigma_{von} = \frac{P}{\sqrt{2}}\sqrt{(31.62)^2 + (20.81)^2 + (10.81)^2} = 27.836P \le 160 \text{ or } P \le 5.75 \text{ or } \qquad \boxed{P_{max} = 5.75\text{kN}}$$

10.76 Due to a force P, the stress components at a critical point that is in plane stress were found to be $\sigma_{xx} = 4P \text{ ksi(C)}$ $\sigma_{yy} = 3P \text{ ksi(T)}$ $\tau_{xy} = -5P \text{ ksi}$. The material has a tensile rupture strength of 18 ksi and a compressive rupture strength of 32 ksi. Determine the maximum value of force P using the modified Mohr's theory.

Solution: $\sigma_{xx} = -4P$ $\sigma_{yy} = 3P$ $\tau_{xy} = -5P$ $\sigma_T < 18 \text{ ksi}$ $\sigma_c < 32 \text{ ksi}$ $P_{max} = ?$

The principal stresses can be found as: $\sigma_{1,2} = \frac{(-4P+3P)}{2} \pm \sqrt{\left(\frac{-4P-3P}{2}\right)^2 + (-5P)^2} = -0.5P \pm 6.1P$ or

$\sigma_1 = 5.6P \text{ ksi (T)}$ and $\sigma_2 = -6.6P = 6.6P \text{ ksi (T)}$

From Mohr's modified theory: $\frac{\sigma_1}{\sigma_T} - \frac{\sigma_1}{\sigma_C} \le 1$ or $\frac{5.6P}{18} - \frac{6.6P}{32} \le 1$ or $\boxed{P_{max} = 9.5\text{kips}}$

10.77 A material has a tensile rupture strength of 18 ksi and a compressive rupture strength of 32 ksi. During usage a component made from this showed the following stresses on a free surface at a critical point: $\sigma_{xx} = 9 \text{ ksi(T)}$ $\sigma_{yy} = 6 \text{ ksi(C)}$ $\tau_{xy} = -4 \text{ ksi}$. Determine the factor of safety using the modified Mohr's theory.

Solution: $\sigma_{xx} = 9 \text{ ksi}$ $\sigma_{yy} = -6 \text{ ksi}$ $\tau_{xy} = -4 \text{ ksi}$ $\sigma_T = 18 \text{ ksi}$ $\sigma_c = 32 \text{ ksi}$ $k = ?$

The principal stresses can be found as

$\sigma_{1,2} = \frac{(9-6)}{2} \pm \sqrt{\left(\frac{9+6}{2}\right)^2 + (-4)^2} = 1.5 \pm 8.5$ or $\sigma_1 = 10\text{ksi(T)}$ and $\sigma_2 = -7\text{ksi} = 7\text{ksi(C)}$

From Mohr's modified theory: $F = \frac{\sigma_1}{\sigma_T} - \frac{\sigma_1}{\sigma_C} = \frac{10}{18} - \frac{7}{32} = 0.3368$

The factor of safety is $k = \frac{1}{F} = \frac{1}{0.336P}$ or $\boxed{k = 2.969}$

10.78 On a free surface of aluminum (E=10,000 ksi, $\nu = 0.25$, $\sigma_{yield} = 24$ ksi). the strains recorded by the three strain gages shown in Fig. P10.78 are $\varepsilon_a = -600 \ \mu \text{ in/in}$ $\varepsilon_b = 500 \ \mu \text{ in/in}$ $\varepsilon_c = 400 \ \mu \text{ in/in}$

By how much can the loads be scaled without exceeding the yield stress of aluminum at the point. Use Maximum shear stress theory.

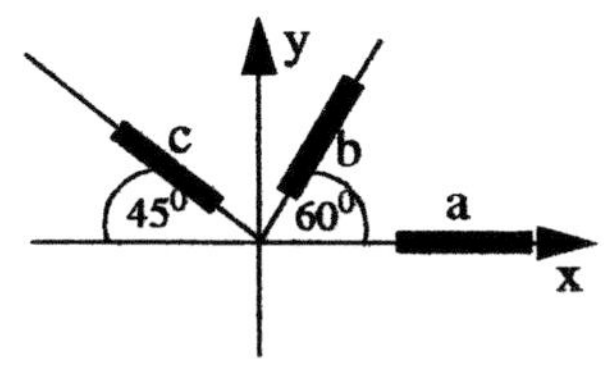

Fig. P10.78

Solution: $E = 10{,}000 \text{ ksi}$, $\nu = 0.25$, $\sigma_{yeild} = 24 \text{ ksi}$, $k = ?$

From Fig. P10.78 we have $\varepsilon_{xx} = -600\mu$ in/in. Substituting $\theta_b = 60°$ and $\theta_c = 135°$ in strain transformation equations we obtain.

$\varepsilon_b = -600(\cos 60)^2 + \varepsilon_{yy}(\sin 60)^2 + \sigma_{xy}(\cos 60)(\sin 60) = 500$ or

$$0.75\varepsilon_{yy} + 0.433\gamma_{xy} = 650 \quad \mathbf{1}$$

$\varepsilon_c = -600(\cos 135)^2 + \varepsilon_{yy}(\sin 135)^2 + \gamma_{xy}(\cos 135)(\sin 135) = 400$ or

$$0.5\varepsilon_{yy} - 0.5\gamma_{xy} = 700 \quad \mathbf{2}$$

Solving Eqs. (1) and (2) we obtain

$$\varepsilon_{yy} = 1061.8\mu \qquad \text{and} \qquad \gamma_{xy} = -338.1\mu \qquad 3$$

The shear modulus of elasticity is: $G = E/[2(1+\nu)] = 10000/[2(1.25)] = 4000\text{ksi}$

The shear stress is $\tau_{xy} = G\gamma_{xy} = -1.352\text{ksi}$. From generalised Hookes law we can unite,

$$\sigma_{xx} - \nu\sigma_{yy} = E\varepsilon_{xx} = (10000)(-600)(10^{-6}) = -6\text{ksi} \qquad 4$$

$$\sigma_{yy} - \nu\sigma_{xx} = E\varepsilon_{xy} = (10000)(1061.8)(10^{-6}) = 10.62\text{ksi} \qquad 5$$

Solving equations (4) and (5) with $\nu = 0.25$ we obtain

$$\sigma_{xx} = -3.568\text{ksi} \qquad \text{and} \qquad \sigma_{yy} = 9.728\text{ksi} \qquad 6$$

The principal stresses are: $\sigma_{1,2} = \dfrac{(-3.568+9.728)}{2} \pm \sqrt{\left(\dfrac{-3.568-9.728}{2}\right)^2 + (-1.352)^2} = 3.080 \pm 6.784$ or

$\sigma_1 = 9.864\text{ksi}$ and $\sigma_2 = -3.704\text{ksi}$. The third principal stress is zero $\sigma^3 = 0$. The maximum shear stress is $\tau_{max} = |(\sigma_1 - \sigma_2)/2| = 6.784\text{ksi}$. The failure shear stress is $= \sigma_{yield}/2 = 12\text{ksi}$

The scaling factor is $k = 12/6.784$ or **$k = 1.76$**

10.79 On a free surface of steel ($E = 200$ GPa, $\nu = 0.28$, $\sigma_{yield} = 210$ MPa) the strains recorded by the three strain gages shown in Fig. P10.79 are

$\varepsilon_a = -800\ \mu\ \text{m/m} \qquad \varepsilon_b = -300\ \mu\ \text{m/m} \qquad \varepsilon_c = -700\ \mu\ \text{m/m}$

By how much can the loads be scaled without exceeding the yield stress of steel at the point. Use Maximum Distortions Energy theory.

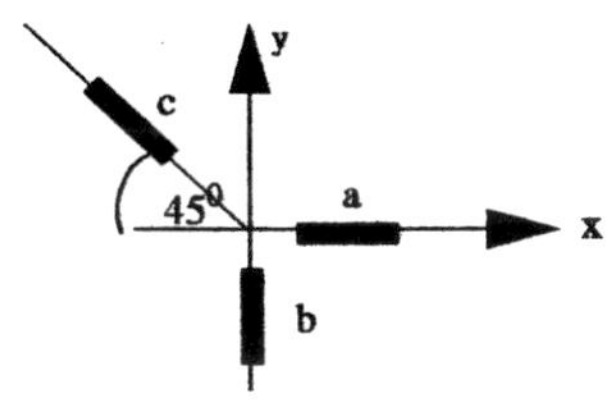

Fig. P10.79

Solution: $E = 200$ GPa, $\nu = 0.28$, $\sigma_{yield} = 210$ MPa, $k = ?$

From Fig. P10.80 we have $\varepsilon_{xx} = -800\mu$ m/m $\varepsilon_{yy} = -300\mu$ m/m. Substituting $\theta_c = 135°$ in strain transformation equation we obtain.

$\varepsilon_c = -800(\cos 135)^2 + (-300)(\sin 135)^2 + \gamma_{xy}(\cos 135)(\sin 135) = -700$ or $\gamma_{xy} = 300\mu$

The shear modulus of elasticity is: $G = 200/[2(1+0.28)] = 78.125\text{GPa}$

The shear stress is $\tau_{xy} = 300(10^{-6})(78.125)(10^9) = 23.44(10^6)\text{N/m}^2 = 23.44\text{MPa}$. From generalised Hooke's law we have,

$$\sigma_{xx} - \nu\sigma_{yy} = E\varepsilon_{xx} = (200)(10^9)(-800)(10^{-6}) = -160(10^6)\text{N/m}^2 = -160\text{MPa} \qquad 1$$

$$\sigma_{yy} - \nu\sigma_{xx} = E\varepsilon_{yy} = (200)(10^9)(-300)(10^{-6}) = -60(10^6)\text{N/m}^2 = -60\text{MPa} \qquad 2$$

Solving Eqs. (1) and (2) with $\nu = 0.28$ we obtain: $\sigma_{xx} = -191.8\text{MPa}$ and $\sigma_{yy} = -113.7\text{MPa}$

The principal stresses are: $\sigma_{1,2} = \dfrac{(-191.8-113.7)}{2} \pm \sqrt{\left(\dfrac{-191.8+113.7}{2}\right)^2 + (23.44)^2} = -152.75 \pm 45.55$ or

$\sigma_1 = -107.2\text{MPa}$ and $\sigma_2 = -198.3\text{MPa}$. The third principal stress is zero $\sigma_3 = 0$. The von-mises stresses is

$$\sigma_{von} = \frac{1}{\sqrt{2}}\sqrt{(\sigma_1-\sigma_2)^2 + (\sigma_2-\sigma_3)^2 + (\sigma_3-\sigma_1)^2} = \frac{1}{\sqrt{2}}\sqrt{(-107.2+198.3)^2 + (107.2)^2 + (198.3)^2} = 171.9\text{MPa}$$

The scaling factor is $k = \sigma_{yield}/\sigma_{von} = 210/171.9$ or **$k = 1.22$**

10.80 A thin-walled cylindrical gas vessel has a mean radius of 3 feet and a wall thickness of 1/2 inch. The yield stress of the material is 30 ksi. Using Von-Mises failure criterion determine the maximum pressure of the gas inside the cylinder if yielding is to be avoided.

Solution: $r = 3$ ft., $t = 1/2$ in., $\sigma_{yield} = 30$ ksi, $p_{max} = ?$

The axial stress is $\sigma_{xx} = \frac{p(36)}{2(0.5)} = 36p$

The hoop stress is $\sigma_{\theta\theta} = \frac{p(36)}{(0.5)} = 72p$

As there are no shear stresses the principal stresses are: $\sigma_1 = 72p$; $\sigma_2 = 36p$; $\sigma_3 = 0$

The von mises stresses can be written as:

$$\sigma_{von} = \frac{1}{\sqrt{2}}\sqrt{(\sigma_1-\sigma_2)^2+(\sigma_2-\sigma_3)^2+(\sigma_3-\sigma_1)^2} = \frac{P}{\sqrt{2}}\sqrt{(36p)^2+(36p)^2+(72p)^2} = 62.35p \le 30 \text{ or}$$

$p \le 0.481\,\text{ksi}$ $\boxed{p = 481\,\text{psi}}$

10.81 A thin-cylindrical boiler can have a minimum mean radius of 18 inches and a maximum mean radius of 36 inches. The boiler will be subjected to a pressure of 750 psi. A sheet metal with a yield stress of 60 ksi is to be used with a factor of safety of 1.5. Construct a failure envelope with the mean radius R and sheet metal thickness t as axis using Maximum Energy Distortion theory.

Solution: $18\text{in} < R < 36\text{in}$ $p = 750$ psi $\sigma_{yield} = 60$ ksi $k = 1.5$ Failure envelope: R and t

---

The allowable stress is: $\sigma_{max} = 60/1.5 = 40\text{psi}$

The axial stress is: $\sigma_{xx} = pR/(2t) = 375R/t$

The hoop stress is: $\sigma_{\theta\theta} = pR/t = 750R/t$

The principal stresses are: $\sigma_1 = 750R/t$; $\sigma_2 = 375R/t$; $\sigma_3 = 0$

The von mises stress is: $\sigma_{von} = \frac{1}{\sqrt{2}}\sqrt{(750-375)^2+(750)^2+(375)^2}\left(\frac{R}{t}\right) = 649.5\left(\frac{R}{t}\right) \le 40(10^3)$ or

$$(R/t) \le 61.58 \qquad (1)$$

The value of R and t that satisfy Eq. 1 and are in the range for r that is given can be found using a spread sheet as shown in table below. The failure envelope is shown in the figure below.

R (in)	t (in)
18	0.292
20	0.325
22	0.357
24	0.390
26	0.422
28	0.455
30	0.487
32	0.520
34	0.552
36	0.585

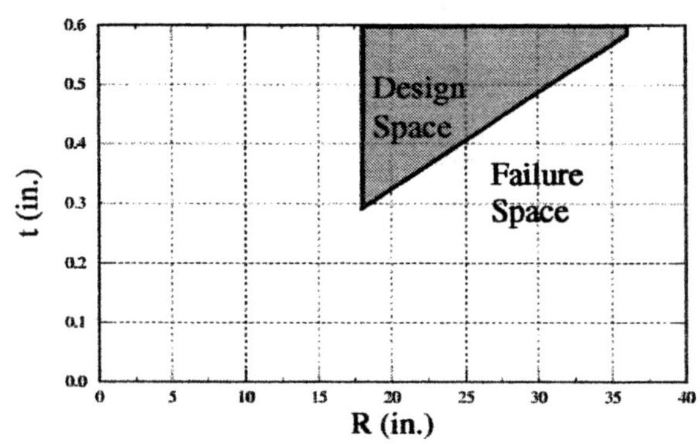

10.82 For *plane stress* show that the von-Mises stress representing the left-hand side of Equation 10.12

can be written as

$$\sigma_{von} = \sqrt{\sigma_{xx}^2 + \sigma_{yy}^2 - \sigma_{xx}\sigma_{yy} + 3\tau_{xy}^2} \qquad \textbf{10.17}$$

Solution

For plane stresses $\sigma_3 = 0$ and the other principal stresses are

$$\sigma_1 = \frac{(\sigma_{xx} - \sigma_{yy})}{2} + \sqrt{\left(\frac{\sigma_{xx} + \sigma_{yy}}{2}\right)^2 + (\tau_{xy})^2} \qquad \sigma_2 = \frac{(\sigma_{xx} - \sigma_{yy})}{2} - \sqrt{\left(\frac{\sigma_{xx} + \sigma_{yy}}{2}\right)^2 + (\tau_{xy})^2} \qquad 1$$

The von mises stresses with $\sigma_3 = 0$ is

$$\sigma_{von} = \frac{1}{\sqrt{2}}\sqrt{(\sigma_1 - \sigma_2)^2 + (\sigma_1)^2 + (\sigma_2)^2} \qquad 2$$

We can write the following:

$$(\sigma_1 - \sigma_2)^2 = 4\left[\frac{(\sigma_{xx} - \sigma_{yy})^2}{2} + (\tau_{xy})^2\right] = (\sigma_{xx})^2 + (\sigma_{yy})^2 - 2(\sigma_{xx})(\sigma_{yy}) + 4(\tau_{xy})^2 \qquad 3$$

$$(\sigma_1)^2 + (\sigma_2)^2 = 2\left[\frac{(\sigma_{xx} + \sigma_{yy})^2}{2} + \frac{(\sigma_{xx} - \sigma_{yy})^2}{2} + (\tau_{xy})^2\right] = 2\left[\frac{2(\sigma_{xx})^2}{4} + \frac{2(\sigma_{yy})^2}{4} + (\tau_{xy})^2\right] \text{ or}$$

$$(\sigma_1)^2 + (\sigma_2)^2 = (\sigma_{xx})^2 + (\sigma_{yy})^2 + 2(\tau_{xy})^2 \qquad 4$$

Substituting Eqs. (4) and (5) into Eq. (3) we obtain

$$\sigma_{von} = \frac{1}{\sqrt{2}}\sqrt{2(\sigma_{xx})^2 + 2(\sigma_{yy})^2 - 2(\sigma_{xx})(\sigma_{yy}) + 6(\tau_{xy})^2} = \sqrt{(\sigma_{xx})^2 + (\sigma_{yy})^2 - (\sigma_{xx})(\sigma_{yy}) + 3(\tau_{xy})^2} \qquad 5$$

Eq. (6) is same as Eq. 10.17

10.83 In cartesian coordinates the von-Mises stress in three dimension is given by:

$$\sigma_{von} = \sqrt{\sigma_{xx}^2 + \sigma_{yy}^2 + \sigma_{zz}^2 - \sigma_{xx}\sigma_{yy} - \sigma_{yy}\sigma_{zz} - \sigma_{zz}\sigma_{xx} + 3\tau_{xy}^2 + 3\tau_{yz}^2 + 3\tau_{zx}^2} \qquad \textbf{10.18}$$

Show that for plane strain the above equation reduces to the following:

$$\sigma_{von} = \sqrt{(\sigma_{xx}^2 + \sigma_{yy}^2)(1 + \nu^2 - \nu) - \sigma_{xx}\sigma_{yy}(1 + 2\nu - 2\nu^2) + 3\tau_{xy}^2} \qquad \textbf{10.19}$$

where, ν is the Poisson's ratio of the material.

Solution:

For plain strain $\varepsilon_{zz} = 0 = [\sigma_{zz} - \upsilon(\sigma_{xx} + \sigma_{yy})]/E$ or $\sigma_{zz} = \upsilon(\sigma_{xx} + \sigma_{yy})$

The shear strains $\gamma_{zx} = 0$ and $\gamma_{zy} = 0$ which implies $\tau_{zx} = 0$ and $\tau_{zy} = 0$. Substituting the stress values in Eq. 10.18 we obtain

$$\sigma_{von} = \sqrt{\sigma_{xx}^2 + \sigma_{yy}^2 + \sigma_{zz}^2 - \sigma_{xx}\sigma_{yy} - \sigma_{zz}(\sigma_{xx} + \sigma_{yy}) + 3\tau_{xy}^2} \text{ or}$$

$$\sigma_{von} = \sqrt{\sigma_{xx}^2 + \sigma_{yy}^2 - \sigma_{xx}\sigma_{yy} + \upsilon^2(\sigma_{xx} + \sigma_{yy})^2 - \upsilon(\sigma_{xx} + \sigma_{yy}) + 3\tau_{xy}^2} \text{ or}$$

$$\sigma_{von} = \sqrt{\sigma_{xx}^2 + \sigma_{yy}^2 - \sigma_{xx}\sigma_{yy} + (\upsilon^2 - \upsilon)[\sigma_{xx}{}^2 + \sigma_{yy}{}^2 + 2\sigma_{xx}\sigma_{yy}] + 3\tau_{xy}^2} \text{ or}$$

$$\sigma_{von} = \sqrt{(\sigma_{xx}^2 + \sigma_{yy}^2)(1 + \nu^2 - \nu) - \sigma_{xx}\sigma_{yy}(1 + 2\nu - 2\nu^2) + 3\tau_{xy}^2} \qquad 1$$

Eq. (1) is same as Eq. 10.18

10.84 Fracture Mechanics shows that the stresses in the vicinity of the crack tip for Mode I, are given by

Equations 10.20a,10.20b, and 10.20c. Notice at θ = π, i.e., at crack surface all stresses are zero. In terms of K_I and r, obtain the von-Mises stress at θ = 0 and θ = π/2.

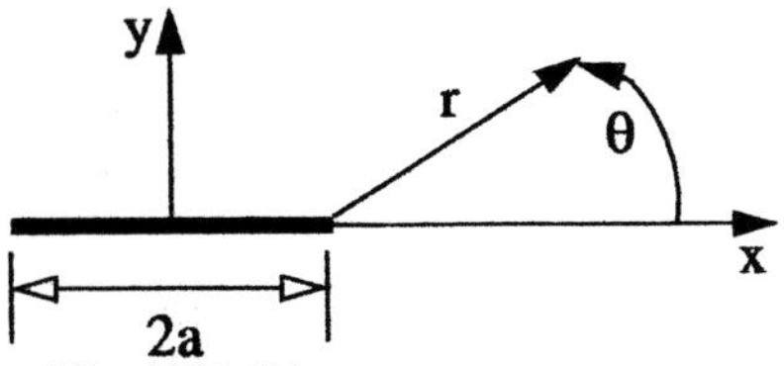

Fig. P10.84

$$\sigma_{xx} = \frac{K_I}{\sqrt{2\pi r}}\cos\frac{\theta}{2}\left[1 - \sin\frac{\theta}{2}\sin\frac{3\theta}{2}\right] \quad \textbf{10.20a}$$

$$\sigma_{yy} = \frac{K_I}{\sqrt{2\pi r}}\cos\frac{\theta}{2}\left[1 + \sin\frac{\theta}{2}\sin\frac{3\theta}{2}\right] \quad \textbf{10.20b}$$

$$\tau_{xy} = \frac{K_I}{\sqrt{2\pi r}}\sin\frac{\theta}{2}\cos\frac{\theta}{2}\cos\frac{3\theta}{2} \quad \textbf{10.20c}$$

Solution: $\sigma_{von} = ?$ at θ = 0 and θ = π / 2

Substituting θ = 0 in Eqs. (10.20a), (10.20b) and (10.20c) we obtain:

$$\sigma_{xx} = \frac{K_I}{\sqrt{2\pi r}}; \ \sigma_{yy} = \frac{K_I}{\sqrt{2\pi r}}; \ \tau_{xy} = 0$$

For plane stress the principal stresses are: $\sigma_1 = \sigma_2 = \frac{K_I}{\sqrt{2\pi r}}$ and $\sigma_3 = 0$.

The von-mises stress is: $\sigma_{von} = \frac{1}{\sqrt{2}}\sqrt{(\sigma_1 - \sigma_2)^2 + (\sigma_1)^2 + (\sigma_2)^2} = \frac{1}{\sqrt{2}}\frac{K_I}{\sqrt{2\pi r}}\sqrt{2}$ or $\boxed{\sigma_{von} = \frac{0.7071 K_I}{\sqrt{\pi r}}}$

Substituting θ = π / 2 in Eqs. (10.20a), (10.20b) and (10.20c) we obtain:

$$\sigma_{xx} = \frac{K_I}{4\sqrt{\pi r}}; \ \sigma_{yy} = \frac{3K_I}{4\sqrt{\pi r}}; \ \tau_{xy} = \frac{-K_I}{4\sqrt{\pi r}}$$

For plane stress the principal stresses are: $\sigma_3 = 0$. and

$$\sigma_{1,2} = \frac{K_I}{2\sqrt{\pi r}} \pm \sqrt{\left(\frac{K_I}{4\sqrt{\pi r}}\right)^2 + \left(\frac{-K_I}{4\sqrt{\pi r}}\right)^2} = \frac{K_I}{\sqrt{\pi r}}\left(\frac{1}{2} \pm \frac{\sqrt{2}}{4}\right) \quad \text{or } \sigma_1 = \frac{0.85355 K_I}{\sqrt{\pi r}} \text{ and } \sigma_2 = \frac{0.14645 K_I}{\sqrt{\pi r}}$$

The von-mises stress is:

$$\sigma_{von} = \frac{1}{\sqrt{2}}\sqrt{(\sigma_1 - \sigma_2)^2 + (\sigma_1)^2 + (\sigma_2)^2} = \frac{K_I}{\sqrt{2}\sqrt{\pi r}}\sqrt{(0.7071)^2 + (0.8535)^2 + (0.1464)^2} \text{ or } \boxed{\sigma_{von} = \frac{0.7906 K_I}{\sqrt{\pi r}}}$$

Problem Set 11.1

11.1 In terms of the linear spring constant k and the length of the rigid bar L, determine P_{cr}, the critical load value.The springs can be in tension or compression.

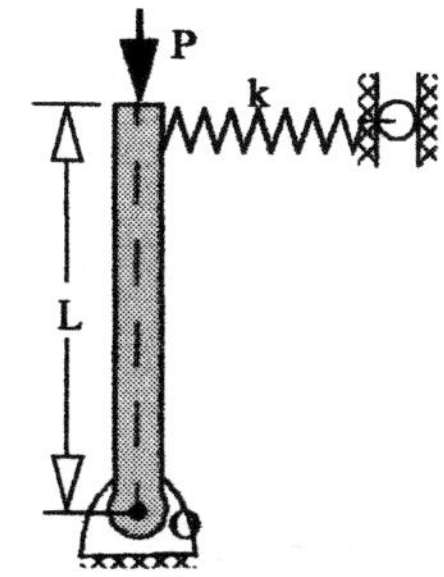

Fig. P11.1

Solution: $P_{cr} = f(k,L) = ?$

Fig. (a) shows the free body diagram at any angle θ.

(a)

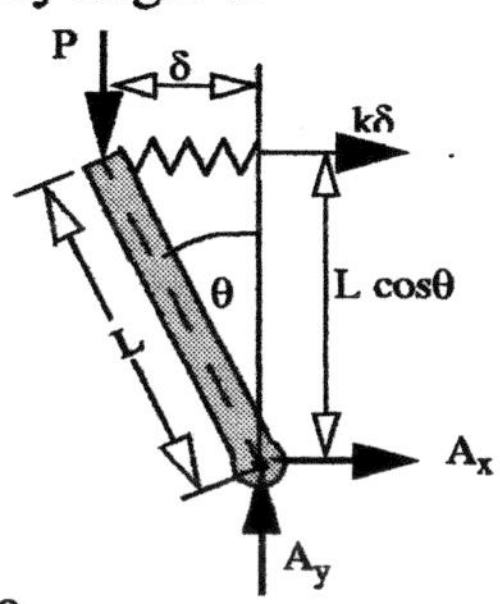

The deformation of the spring is $\delta = L \sin\theta$

The moment at point A is: $M_A = PL\sin\theta - (kL\sin\theta)L\cos\theta = L\sin\theta(P - kL\cos\theta)$

When P = kL then equilibrium condition of $M_A = 0$ cannot be met at any θ except θ = 0. The critical load is thus $\boxed{P_{cr} = kL}$

11.2 In terms of the linear spring constant k and the length of the rigid bar L, determine P_{cr}, the critical load value.The springs can be in tension or compression.

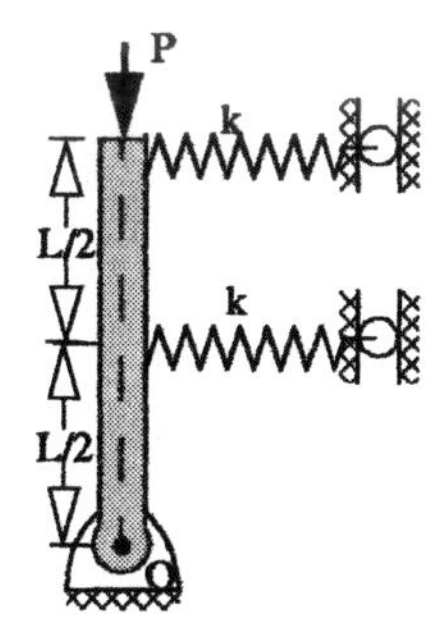

Fig. P11.2

Solution: $P_{cr} = f(k,L) = ?$

Fig. (a) shows the free body diagram at any angle θ

(a)

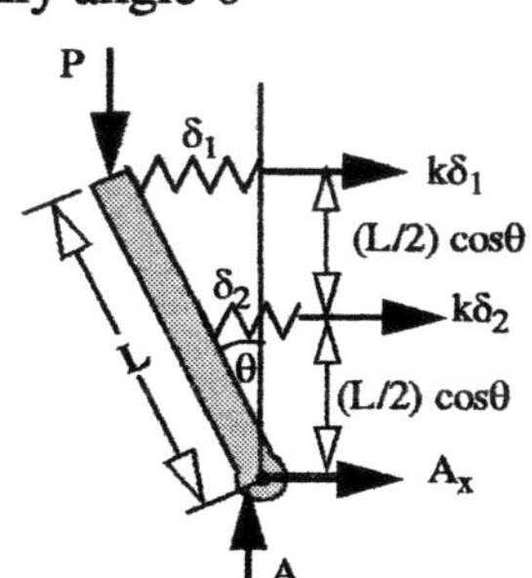

The deformation of the springs are $\delta_1 = L\sin\theta$ and $\delta_2 = (L/2)\sin\theta$

The moment at point A is

$$M_A = PL\sin\theta - (kL\sin\theta)L\cos\theta - \left(\frac{kL}{2}\sin\theta\right)\frac{L}{2}\cos\theta \text{ or } M_A = L\sin\theta\left(P - \frac{5kL}{4}\cos\theta\right)$$

When P=5kL/4 the equilibrium condition of $M_A = 0$ cannot be met at any θ except θ = 0. The critical load is thus $\boxed{P_{cr} = \frac{5kL}{4}}$

11.3 In terms of the linear spring constant k and the length of the rigid bar L, determine P_{cr}, the critical load value. The springs can be in tension or compression.

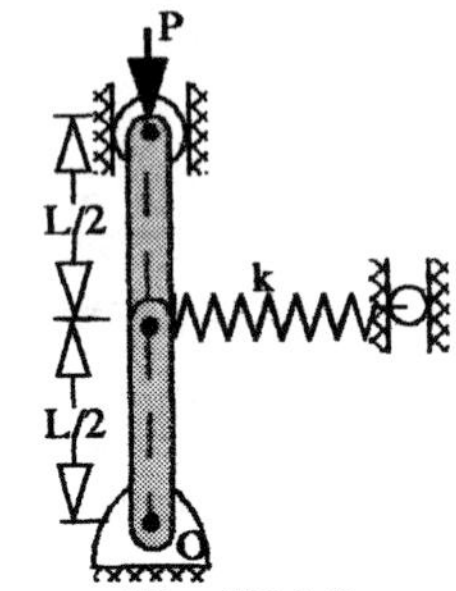

Fig. P11.3

Solution: $P_{cr} = f(k,L) = ?$

Fig. (a) and (b) show the free body diagram at any angle θ

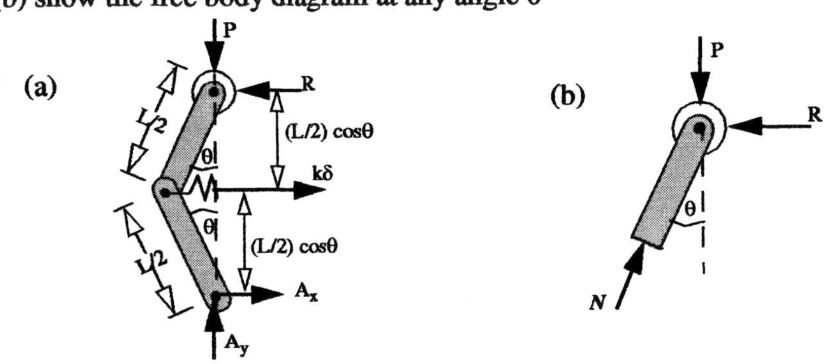

The deformation of the spring is $\delta = (L/2)\sin\theta$

By equilibrium of forces in Fig(b) we obtain $N\cos\theta = P$ and $R = N\sin\theta = P\tan\theta$

The moment at point A in Fig.(a) is

$$M_A = R\left(\frac{L}{2}\cos\theta + \frac{L}{2}\cos\theta\right) - \left(\frac{kL}{2}\sin\theta\right)\frac{L}{2}\cos\theta = P\tan\theta(L\cos\theta) - \frac{kL^2}{4}\sin\theta\cos\theta = L\sin\theta\left(P - \frac{kL}{4}\cos\theta\right)$$

When P= kL/4 the equilibrium condition of $M_A = 0$ cannot be met at any θ except θ = 0. The critical load is thus $\boxed{P_{cr} = kL/4}$

11.4 Linear deflection springs and torsional springs are attached to rigid bars as shown. The springs can act in tension or compression and resist rotation in either direction. Determine P_{cr}, the critical load value.

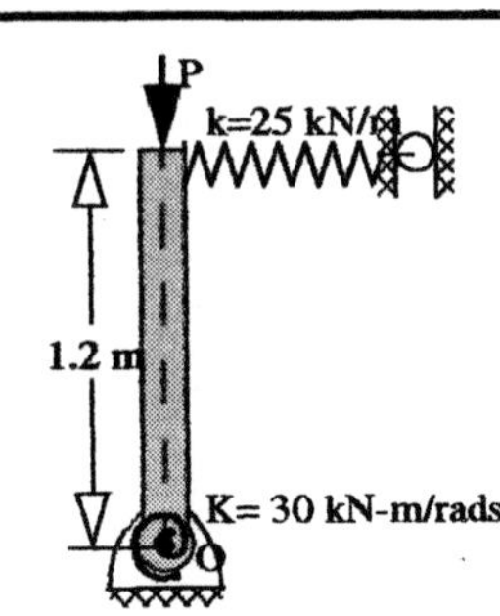

Fig. P11.4

Solution: $P_{cr} = ?$

Fig. (a) shows the free body diagram at any angle θ.

(a)

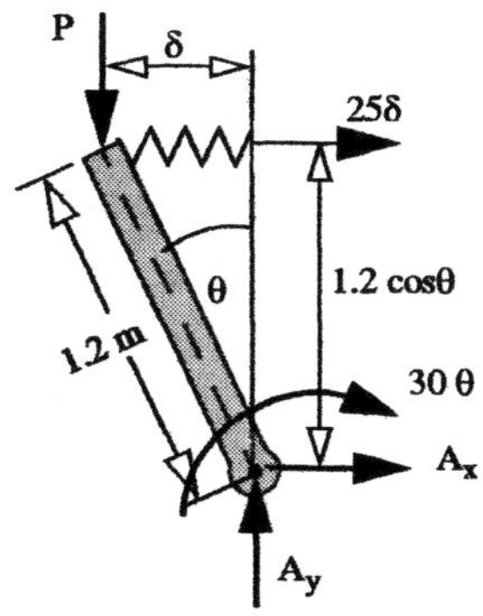

The deformation of the linear spring is δ=1.2 sinθ.

The moment at point A is: $M_A = P(1.2)\sin\theta - 25(1.2\sin\theta)(1.2\cos\theta) - 30\theta$

At θ = 0, the equilibrium condition $M_A = 0$ is met. For small perturbations from this equilibrium positions we can approximate sinθ = θ and cosθ = 1 to obtain

$$M_A = 1.2P\theta - 25(1.2)^2\theta - 30 = (1.2P - 36 - 30)\theta = (1.2P - 66)\theta$$

For P= 66/1.2= 55kN, the equilibrium condition of $M_A = 0$ is met at every θ, implying the system is in neutral equilibrium. The critical load is $\boxed{P_{cr} = 55\text{kN}}$

11.5 Linear deflection springs and torsional springs are attached to rigid bars as shown. The springs can act in tension or compression and resist rotation in either direction. Determine P_{cr}, the critical load value.

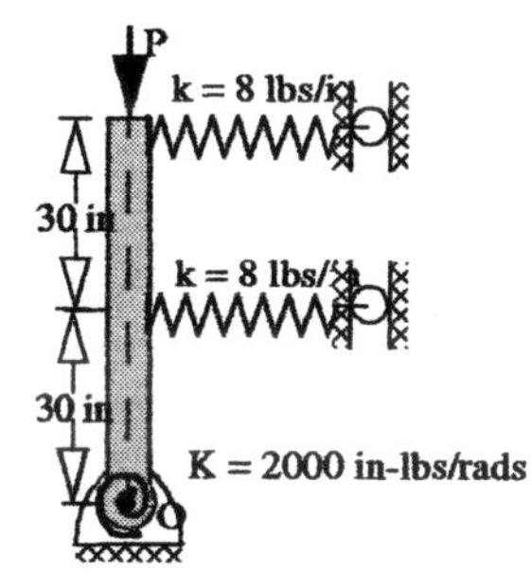

Fig. P11.5

Solution: $P_{cr} = ?$

Fig. (a) shows the free body diagram at any angle θ.

(a)

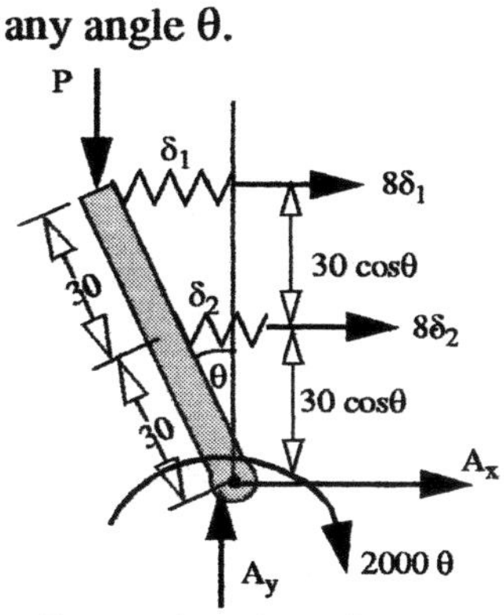

The deformation of the linear springs are $\delta_1 = 60\sin\theta$ and $\delta_2 = 30\sin\theta$

The moment at point A can be written as:

$$M_A = P(60)\sin\theta - 8(60\sin\theta)(60\cos\theta) - 8(30\sin\theta)(30\cos\theta) - 2000\theta$$

At $\theta = 0$, the equilibrium condition $M_A = 0$ is met. For small perturbations from this equilibrium positions we can approximate $\sin\theta = \theta$ and $\cos\theta = 1$ to obtain

$M_A \cong 60P\theta - 28800\theta - 7200\theta - 2000\theta = (60P - 38000)\theta$

For P= 38000/60 = 633.3 lb., the equilibrium condition of $M_A = 0$ is met at every θ, implying the system is in neutral equilibrium. The critical load is $\boxed{P_{cr} = 633.3 lb}$

11.6 Linear deflection springs and torsional springs are attached to rigid bars as shown. The springs can act in tension or compression and resist rotation in either direction. Determine P_{cr}, the critical load value

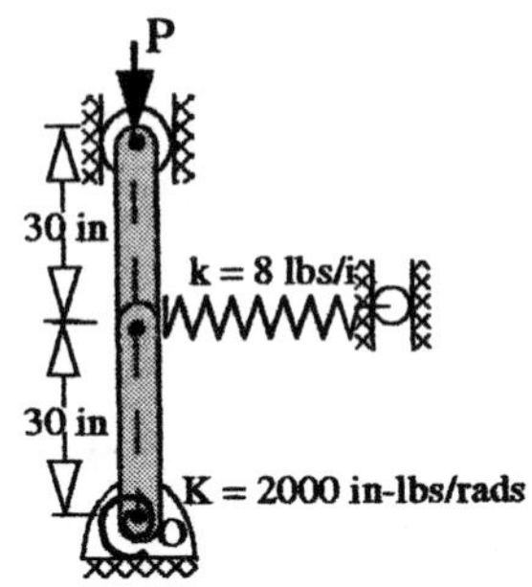

Fig. P11.6

Solution: $P_{cr} = ?$

Fig. (a) and (b) show the free body diagram at any angle θ

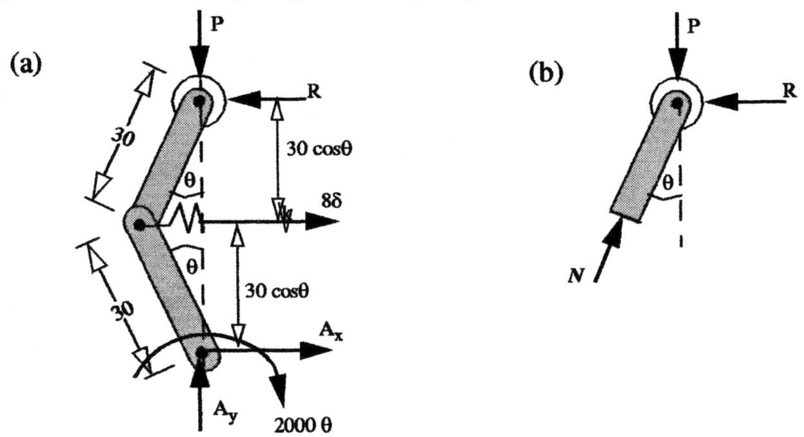

The deformation of the linear spring is $\delta = 30\sin\theta$

By equilibrium of forces in Fig.(b) we obtain $N = P\cos\theta$ and $R = N\sin\theta = P\tan\theta$

The moment at point A in Fig.(a) is

$M_A = R(30\cos\theta + 30\cos\theta) - 8(30\sin\theta)(30\cos\theta) - 2000\theta = 60P\sin\theta - 7200\sin\theta\cos\theta - 2000\theta$

At $\theta = 0$, the equilibrium condition $M_A = 0$ is met. For small perturbations from this equilibrium positions we can approximate $\sin\theta = \theta$ and $\cos\theta = 1$ to obtain $M_A = 60P\theta - 7200\theta - 2000\theta = (60P - 9200)\theta$

For P= 9200/60 = 153.3 lb., the equilibrium condition of $M_A = 0$ is met at every θ, implying the system is in neutral equilibrium. The critical load is $\boxed{P_{cr} = 153.3 lb}$

11.7 A hollow circular steel (E = 200 GPa) column is simply supported over a length of 5 m. The inner and outer diameters of the cross-section are 75 mm and 100 mm. Determine (a) The slenderness ratio. (b) The critical buckling load. (c) The axial stress at the critical buckling load. (d) If roller supports are added at the mid-point, what would be new critical buckling load.

Solution: E = 200 GPa L=5m d_i=75 mm d_o=100 mm

L/r = ? $P_{cr} = ?$ $\sigma_{cr} = ?$ $P_{cr2} = ?$

The cross-sectional area and area moment of inertia are:

$A = \frac{\pi}{4}[0.1^2 - 0.075^2] = 3.436(10^{-3})m^2$ and $I = \frac{\pi}{64}[0.1^4 - 0.075^4] = 3.356(10^{-6})m^4$

The radius of gyration is $r = \sqrt{\frac{I}{A}} = 31.25(10^{-3})m$

The slenderness ratio is $\frac{L}{r} = \frac{5}{31.25(10^{-3})}$ or $\boxed{\frac{L}{r} = 160}$

The critical buckling load is $(P_{cr}) = \frac{\pi^2 EI}{L^2} = \frac{\pi^2(200)(10^9)(3.356)(10^{-6})}{5^2} = 264.9(10^3)N$ or $\boxed{P_{cr} = 265kN}$

The axial stress at critical buckling load is: $\sigma_{cr} = \frac{P_{cr}}{A} = \frac{264.9(10^3)}{(3.436)(10^{-3})} = 77.1(10^6)N/m^2$ or

$\boxed{\sigma_{cr} = 77.1MPa(T)}$

The middle support will cause the column to fail at the next buckling mode

$P_{cr2} = 2^2\left(\frac{\pi^2 EI}{L^2}\right) = 4P_{cr}$ or $\boxed{P_{cr2} = 1060kN}$

11.8 A 30 ft. long hollow square steel (E = 30,000 ksi) column is built into the wall at either end. The column is constructed from 1/2 inch thick sheet and has an outer dimension of 4 in x 4 in. Determine (a) The slenderness ratio. (b) The critical buckling load. (c) The axial stress at the critical buckling load.

Solution: L = 30 ft E = 30000 ksi t= 0.5 in 4in x 4in built in ends

$L_{eff}/r = ?$ $P_{cr} = ?$ $\sigma_{cr} = ?$

The cross-sectional area and area moment of inertia are:

$A = (4)(4) - (3)(3) = 7in^2$ and $I = \frac{1}{12}[(4)(4)^3 - (3)(3)^3] = 14.583in^4$

The radius of gyration is $r = \sqrt{\frac{I}{A}} = 1.443in$

The effective length is $L_{eff} = 0.5L = (0.5)(30)(12) = 180in$

The slenderness ratio is $\frac{L_{eff}}{r} = \frac{180}{1.443}$ or $\boxed{\frac{L_{eff}}{r} = 124.7}$

The critical buckling load is $(P_{cr}) = \frac{4\pi^2 EI}{L^2} = \frac{4\pi^2(30000)(14.583)}{(360)^2}$ or $\boxed{P_{cr} = 133.3kip}$

The axial stress at critical buckling load is: $\sigma_{cr} = \frac{P_{cr}}{A} = \frac{133.3}{7}$ or $\boxed{\sigma_{cr} = 19.0ksi(C)}$

11.9 A 20 ft. long wooden (E = 1,800 ksi) column has a cross-section dimension of 8 in x 8 in. The column is built in at one end and simply supported at the other end. Determine (a) The slenderness ratio. (b) The critical buckling load. (c) The axial stress at the critical buckling load.

Solution: L = 20 ft E = 1800 ksi 8in x 8in built and simple support

$L_{eff}/r = ?$ $P_{cr} = ?$ $\sigma_{cr} = ?$

The cross-sectional area and area moment of inertia are:

$A = (8)(8) = 64in^2$ and $I = \frac{1}{12}(8)(8)^3 = 341.33in^4$

The radius of gyration is $r = \sqrt{\frac{I}{A}} = 2.309\text{in}$

The effective length is $L_{eff} = 0.7L = (0.7)(240) = 168\text{in}$

The slenderness ratio is $\frac{L_{eff}}{r} = \frac{168}{2.309}$ or $\boxed{\frac{L_{eff}}{r} = 72.7}$

The critical buckling load is $(P_{cr}) = \frac{(20.19)EI}{L^2} = \frac{(20.19)(1800)(341.33)}{(240)^2}$ or $\boxed{P_{cr} = 215.4\text{kip}}$

The axial stress at critical buckling load is: $\sigma_{cr} = \frac{P_{cr}}{A} = \frac{\mathbf{215.4}}{64}$ or $\boxed{\sigma_{cr} = 3.36\text{ksi(C)}}$

11.10 A W12 x 35 steel section (see Appendix E) is used for a 21 ft. column that is simply supported at each end. Use E = 30,000 ksi and determine: Determine (a) The slenderness ratio. (b) The critical buckling load. (c) The axial stress at the critical buckling load. (d) If roller supports are added at intervals of 7 feet, what would be the critical buckling load.

Solution:	W12x35	L = 21 ft	E = 30000 ksi	simple support
	$L/r = ?$	$P_{cr} = ?$	$\sigma_{cr} = ?$	$P_{cr3} = ?$

From Table E.1 we have the following:

The cross-sectional area A = 10.3 in^2.

Minimum area moment of inertia I = 24.5 in^4

The minimum radius of gyration is = 1.54 in

The slenderness ratio is $\frac{L}{r} = \frac{(21)(12)}{1.54}$ or $\boxed{\frac{L}{r} = 163.6}$

The critical buckling load is $(P_{cr}) = \frac{\pi^2 EI}{L^2} = \frac{\pi^2(30000)(24.5)}{(252)^2}$ or $\boxed{P_{cr} = 114.2\text{kip}}$

The axial stress at critical buckling load is: $\sigma_{cr} = \frac{P_{cr}}{A} = \frac{114.2}{10.3}$ or $\boxed{\sigma_{cr} = 11.1\text{ksi(c)}}$

11.11 A S200 x 34 steel section (see Appendix E) is used as a 6 m column that is built in at each end. E = 200 GPa and determine: Determine (a) The slenderness ratio. (b) The critical buckling load. (c) The axial stress at the critical buckling load.

Solution:	S200 x 34	L = 6 m	E = 200 GPa	built in ends
	$L_{eff}/r = ?$	$P_{cr} = ?$	$\sigma_{cr} = ?$	

From Table E.1 we have the following:

The cross-sectional area A = 4368 mm^2

Minimum area moment of inertia I = 1.794(10^6) m

The minimum radius of gyration is = 20.3 mm

The effective length is $L_{eff} = 0.5L = 3\text{m}$

The slenderness ratio is $\frac{L_{eff}}{r} = \frac{3}{(20.3)(10^{-3})}$ or $\boxed{\frac{L_{eff}}{r} = 147.8}$

The critical buckling load is $(P_{cr}) = \frac{4\pi^2 EI}{L^2} = \frac{4\pi^2(200)(10^9)(1.794)(10^{-6})}{6^2}$ or $\boxed{P_{cr} = 393.5(10^3)\text{N}}$

11.12 Columns made from an alloy will be used in a construction of a frame. The cross-section of the columns is a hollow-square of thickness 0.125 inch and outer dimension of 'a' inches. The Modulus of Elasticity is E = 9000 ksi and the yield stress is σ_{yield} = 90 ksi. Table 11.12 shows a list of the length 'L' and outer square dimension 'a'. Identify the long and the short columns. Assume the ends will be simply supported.

Table P11.12 Column Geometric properties.

L feet	a inches
1	1.125
1.5	1.5
2.0	1.75
2.5	2.75
3.0	3.0
3.5	3.0
4.0	3.0

Solution t = 0.125 in E = 90 ksi σ_{yield} = 90 ksi Long and Short Columns = ?

The cross-sectional area is: $A = a^2 - (a - 2(0.125))^2$

The area moment of inertia is: $I = [a^4 - (a - 2(0.125))^4]/12$

Equating the axial stress at critical buckling load we can obtain the critical slenderness ratio as shown below.

$$\sigma_{cr} = \frac{P_{cr}}{A} = \frac{\pi^2 E}{(L/r)_{cr}^2} = 90 \quad \text{or} \quad (L/r)_{cr} = \sqrt{\frac{\pi^2(9000)}{90}} = 31.41 \qquad 1$$

From the given values of L and a, the area moment of inertia I, the cross-sectional area A, the radius of gyration r, and the slenderness ratio can be found using spreadsheet as shown in the table below. Columns with slenderness ratio less than the critical slenderness ratio in Eq. 1 are short columns and those greater than the critical slenderness ratio are long columns as shown in the table below.

L (ft)	a (in)	I (in^4)	A (in^2)	r (in)	L/r	Column Type
1.0	1.125	0.085	0.500	0.411	29.167	Short
1.5	1.500	0.218	0.688	0.564	31.934	Long
2.0	1.750	0.360	0.813	0.665	36.071	Long
2.5	2.750	1.511	1.313	1.073	27.962	Short
3.0	3.000	1.984	1.438	1.175	30.643	Short
3.5	3.000	1.984	1.438	1.175	35.750	Long
4.0	3.000	1.984	1.438	1.175	40.857	Long

11.13 Columns made from an alloy will be used in a construction of a frame. The cross-section of the columns is a hollow-cylinder of thickness 10 mm and outer diameter of 'd' mm. The Modulus of Elasticity is E = 200 GPa and the yield stress is σ_{yield} = 300 MPa. Table 11.13 shows a list of the length 'L' and outer diameter 'd'. Identify the long and the short columns. Assume the ends of the column are built in.

Table P11.13 Column Geometric properties.

L m	d mm
1	60
2	80
3	100
4	150
5	200
6	225
7	250

Solution t = 10 mm E = 200 GPa σ_{yield} =300 MPa Long and Short Columns = ?

The cross-sectional area is: $A = \pi[d^2 - (d - 20)^2]/4 \text{ mm}^2$

The area moment of inertia is: $I = \pi([d^4 - (d - 20)^4]/64)$

Equating the axial stress at critical buckling load we can obtain the critical slenderness ratio as shown below.

$$\sigma_{cr} = \frac{P_{cr}}{A} = \frac{4\pi^2 E}{(L/r)_{cr}^2} = 300(10^6) \quad \text{or} \quad (L/r)_{cr} = \sqrt{\frac{4\pi^2(100)(10^9)}{600(10^6)}} = 81.11 \qquad 1$$

From the given values of L and d, the area moment of inertia I, the cross-sectional area A, the radius of gyration r, and the slenderness ratio can be found using as shown in the table below. Columns with slenderness ratio less than the critical slenderness ratio in Eq. 1 are short columns and those greater than the critical slenderness ratio are long columns as shown in the table below.

L (m)	d (mm)	I (mm^4)	A (mm^2)	r (mm)	L/r	Column Type
1	60	5.11E+05	1.57E+03	18.0	55.5	Short
2	80	1.37E+06	2.20E+03	25.0	80.0	Short
3	100	2.90E+06	2.83E+03	32.0	93.7	Long
4	150	1.08E+07	4.40E+03	49.6	80.6	Short
5	200	2.70E+07	5.97E+03	67.3	74.3	Short
6	225	3.91E+07	6.75E+03	76.1	78.8	Short
7	250	5.44E+07	7.54E+03	84.9	82.4	Long

11.14 Three column cross-sections are shown. The area of each of the three cross-section is equal to 'A'. Determine the ratios of critical loads P_{cr1}:P_{c2}:P_{cr3} assuming (a) the ends are simply supported. (b) the ends are built in. (c) How do you expect the ratios to change if the end conditions were as in Case 2 and 3 of Table 11.1?

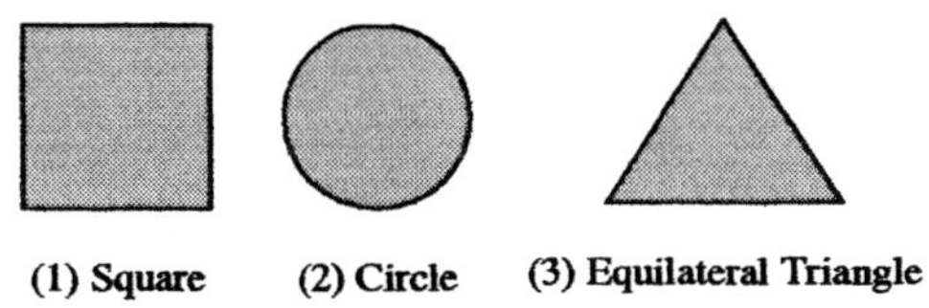

Fig. P11.14

Solution: P_{cr1}:P_{c2}:P_{cr3} = ?

Let a_s, a_c and a_t represent the side of a square, radius of the circle, and side of an equilateral triangle respectively. These dimensions in terms of cross-sectional area are:

For square: $a_s^2 = A$ or $a_s = \sqrt{A}$ The area moment of inertia is: $I_s = \frac{1}{12}a_s^4 = \frac{A^2}{12}$

For circle: $\pi a_c^2 = A$ or $a_c = \sqrt{\frac{A}{\pi}}$ The area moment of inertia is: $I_c = \frac{\pi}{12}a_c^4 = \frac{A^2}{4\pi}$

For triangle: $(1/2)a_t^2 \sin 60 = A$ or $a_t = \sqrt{\frac{4A}{\sqrt{3}}}$. The area moment of inertia is:

$$I_t = \frac{1}{36}a_t(a_t \sin 60)^3 = \frac{3\sqrt{3}}{288}a_t^4 = \frac{A^2}{6\sqrt{3}}$$

Noting that $(P_{cr}) = \frac{\pi^2 EI}{L^2}$ we obtain P_{cr1}:P_{c2}:P_{cr3} = $I_s : I_c : I_t$ =0.0833A^2: 0.0796A^2: 0.0962A^2 or

$$\boxed{P_{cr1};P_{cr2};P_{cr3} = 1;0.9549;1.1547}$$

The ratio of critical loads will not change with end conditions.

11.15 A force F= 750 lb. Both bars have a diameter of d = 1/4 inch, Modulus of Elasticity E = 30,000 ksi, and yield stress σ_{yield} = 30 ksi. Bar AP and BP have lengths of L_{AP}= 8 inches and L_{BP}= 10 inches, respectively. Determine the factor of safety for the two-bar structure.

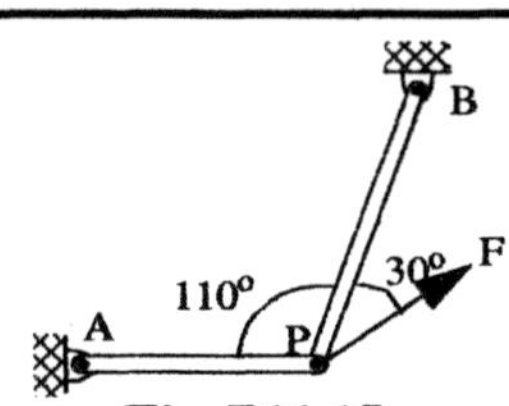

Fig. P11.15

Solution F=750 lb E = 30000 ksi d = 1/4 in σ_{yield}=30 ksi
L_{AP}= 8 in L_{BP} = 10 in K=?

The cross-sectional area and area moment of inertia for the two members are:

$$A = \frac{\pi}{4}\left(\frac{1}{4}\right)^2 = 0.0491\text{in}^2 \qquad I = \frac{\pi}{64}\left(\frac{1}{4}\right)^4 = 191.7(10^{-6})\text{in}^4$$

Fig(a) shows the free body diagram

(a)

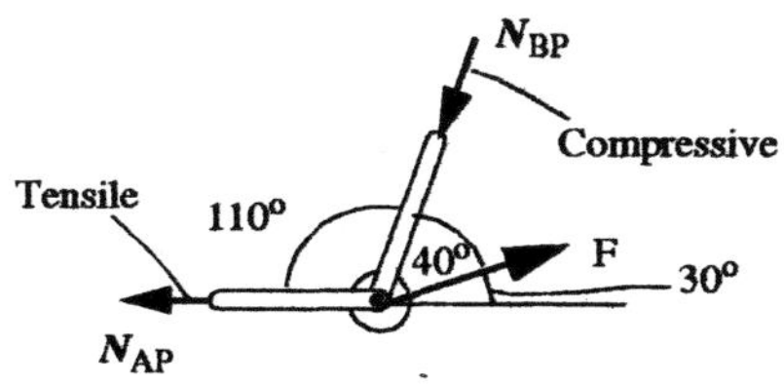

By equilibrium of forces we obtain

$N_{BP}\sin 70 = 750\sin 40$ or $N_{BP} = 513.031\text{b}$ and $N_{AP} = 750\cos 40 - N_{BP}\cos 70$ or $N_{AP} = 399.071\text{b}$

The axial stress in AP is: $\sigma_{AP} = \dfrac{399.07}{0.0491} = 8126.5\text{psi}$

The critical buckling load for member BP is: $(P_{cr})_{BP} = \dfrac{\pi^2 EI}{L_{BP}^2} = \dfrac{\pi^2(30)(10^6)(191.7)(10^{-6})}{10^2} = 567.61\text{b}$

The factor of safety with respect to failure due to yield stress K_y and with respect to failure due to buckling K_{cr} are: $K_y = \dfrac{\sigma_{yield}}{\sigma_{AP}} = 3.69 \qquad K_{cr} = \dfrac{(P_{cr})_{BP}}{N_{BP}} = 1.106$

The factor of safety for the structure is: $\boxed{K = 1.106}$

11.16 A force F= 750 lb. Both bars have a diameter of d = 1/4 inch, Modulus of Elasticity E = 30,000 ksi, and yield stress σ_{yield} = 30 ksi. Bar AP and BP have lengths of L_{AP}= 8 inches and L_{BP}= 10 inches, respectively. Determine the factor of safety for the two-bar structures.

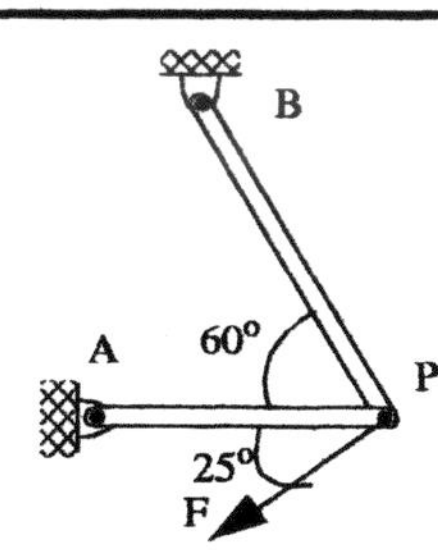

Fig. P11.16

Solution F=750 lb E = 30000 ksi d = 1/4 in σ_{yield}=30 ksi
L_{AP}= 8 in L_{BP} = 10 in K=?

Fig(a) shows the free body diagram

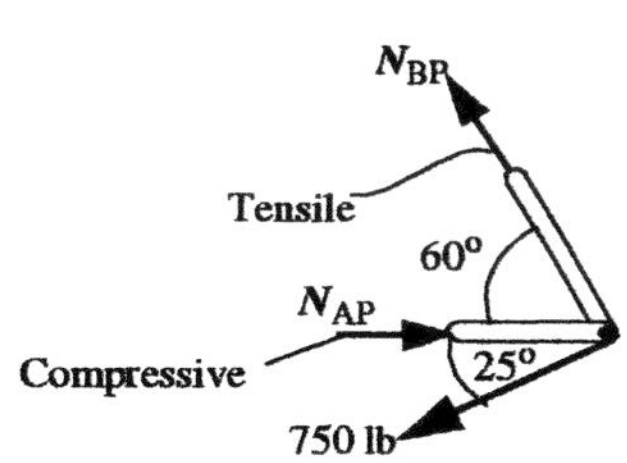

By equilibrium of forces we obtain:

$N_{BP}\sin 60 = 750\sin 25$ or $N_{BP} = 3661\text{b}$ and $N_{AP} = 750\cos 25 + N_{BP}\cos 60$ or $N_{AP} = 862.71\text{b}$

The axial stress in BP can be found using the cross-sectional area calculated in Eq.(1) in problem 11-15:

$$\sigma_{AP} = \frac{N_{AP}}{A} = \frac{862.7}{0.0491} = 17570\text{psi(C)}$$

The critical buckling load for member AP can be found using area moment of inertia calculated in Eq.(1) in problem 11.15:

$$(P_{cr})_{AP} = \frac{\pi^2 EI}{L_{AP}^2} = \frac{\pi^2(30)(10^6)(191.7)(10^{-6})}{8^2} = 887.88\text{lb}$$

The factor of safety with respect to failure due to yield K_y and with respect to failure due to buckling K_{cr} are:

$$K_y = \frac{\sigma_{yield}}{\sigma_{AP}} = 1.702 \qquad K_{cr} = \frac{(P_{cr})_{AP}}{N_{AP}} = 1.029$$

The factor of safety for the structure is: $\boxed{K = 1.02}$

11.17 A force F= 750 lb. Both bars have a diameter of d = 1/4 inch, Modulus of Elasticity E = 30,000 ksi, and yield stress σ_{yield} = 30 ksi. Bar AP and BP have lengths of L_{AP}= 8 inches and L_{BP}= 10 inches, respectively. Determine the factor of safety for the two-bar structures.

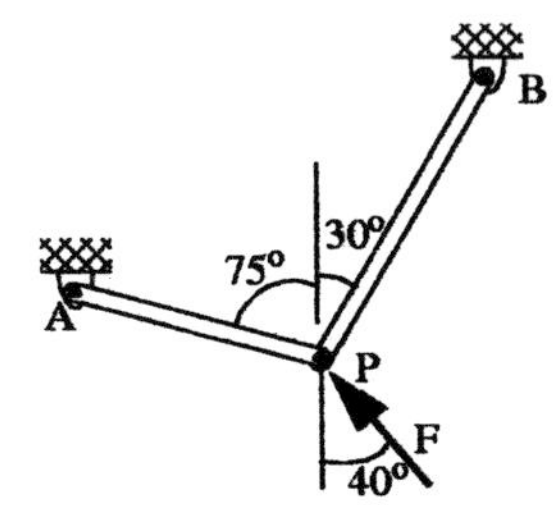

Fig. P11.17

Solution F=750 lb E = 30000 ksi d = 1/4 in σ_{yield}=30 ksi

L_{AP}= 8 in L_{BP} = 10 in K=?

Fig(a) shows the free body diagram

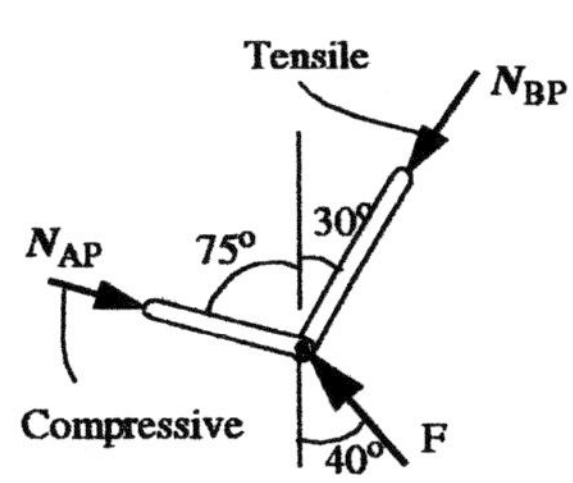

By equilibrium of forces we obtain

$$N_{AP}\cos 75 + N_{BP}\cos 30 = 750\cos 40 = 574.53 \quad 1$$

$$N_{AP}\sin 75 - N_{BP}\sin 30 = 750\sin 40 = 482.09 \quad 2$$

Solving Eqs.(1) and (2) we obtain

$$N_{AP} = 729.63\text{lb} \quad \text{and} \quad N_{BP} = 445.36\text{lb} \quad 3$$

Using the cross-sectional area calculated in Eq.(1) in problem 11-15 we obtain the axial stress in member AP as:

$$\sigma_{AP} = \frac{N_{AP}}{A} = \frac{729.63}{0.0491} = 14860\text{psi(C)}$$

Using the critical buckling load for member AP from Eq.(1) in problem 11.16 and for member BP in Eq.(1) in problem 11.15 we obtain the following Factor of safety:

$$K_y = \frac{\sigma_{yield}}{\sigma_{AP}} = 2.019 \qquad K_{cr1} = \frac{(P_{cr})_{AP}}{N_{AP}} = 1.217 \text{ an } K_{cr2} = \frac{(P_{cr})_{BP}}{N_{BP}} = 1.274$$

The factor of safety for the structure is: $\boxed{K = 1.2}$

11.18 A force F= 10 kN is applied to the roller that slides inside a slot. Both bars have a diameter of d = 10 mm, Modulus of Elasticity E = 200 GPa, and yield stress σ_{yield} = 360 MPa. Bar AP and BP have lengths of L_{AP} = 200 mm and L_{BP} = 300 mm respectively. Determine the factor of safety for the two-bar structures.

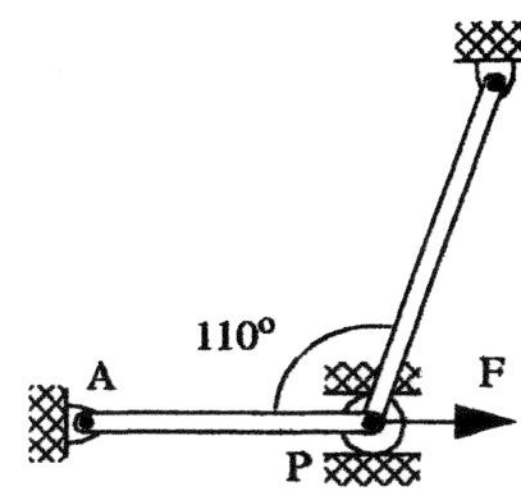

Fig. P11.18

Solution F=10 kN E = 200 GPa d = 10 m σ_{yield} = 360 MPa

L_{AP}= 200 mm L_{BP}= 300 mm K = ?

The cross-sectional area and area moment of inertia are:

$$A = \frac{\pi}{4}0.01^2 = 78.54(10^{-6})\text{m}^2 \qquad I = \frac{\pi}{64}0.01^4 = 0.4909(10^{-9})\text{m}^4$$ 1

We draw the free body diagram of the rigid bars as shown in Fig. (a)

(a)

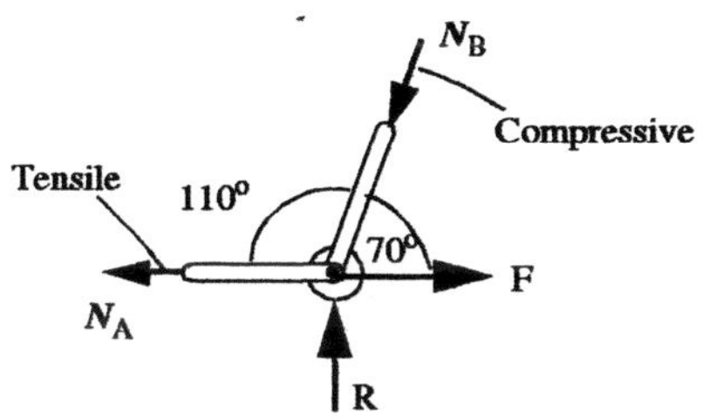

(b)

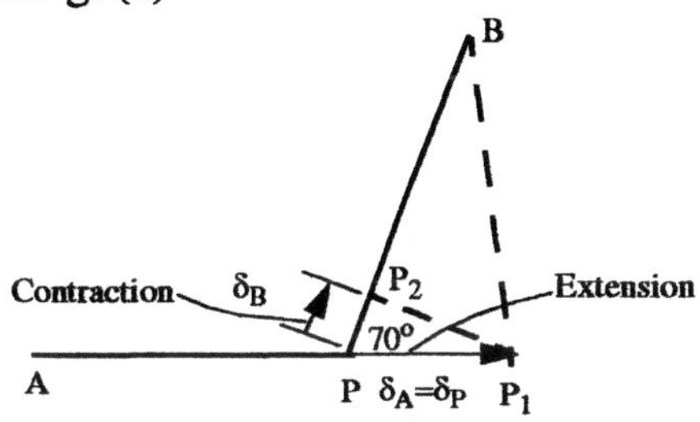

By force equilibrium in the x-direction in Fig.(a), we obtain the following.

$$N_A + N_B\cos 70 = \text{F} \qquad \text{or} \qquad N_A + 0.342N_B = 10(10^3)$$ 2

We draw the exaggerated deformed shape as shown in Fig. (b). As per small strain approximation, we need component of PP_1 in the original direction of AP and BP, i.e., PP_1 represents the deformation of bar AP and PP_2 represents the deformation of bar BP and are calculated as shown below.

$$\delta_A = \delta_P$$ 3

$$\delta_B = \delta_P\cos 70 = 0.342\delta_P$$ 4

From Eq 4.27, we have the following:

$$\delta_A = \frac{N_A L_A}{EA} = \frac{N_A(0.2)}{(200)(10^9)(78.54)(10^{-6})} = \delta_P \qquad \text{or} \qquad N_A = 78.54\delta_P(10^6)$$ 5

$$\delta_B = \frac{N_B L_B}{EA} = \frac{N_B(0.3)}{(200)(10^9)(78.54)(10^{-6})} = 0.342\delta_P \qquad \text{or} \qquad N_B = 17.91\delta_P(10^6)$$ 6

Substituting Eqs. (5) and (6) into Eq. (2), we obtain the displacement δ_P as shown below.

$$[(78.54\delta_P) + 0.342(17.91\delta_P)](10^6) = 10(10^3) \qquad \text{or} \qquad \delta_P = 0.1181(10^{-3})\text{m}$$ 7

Substituting Eq.(7) into Eqs. (5) and (6) we obtain:

$$N_A = (78.54)(0.1181)(10^3) = 9.277(10^3)\ \text{N} \qquad N_B = (17.91)(0.1181)(10^3) = 2.115(10^3)\ \text{N}$$ 8

The axial stress in member A is: $\sigma_A = \frac{N_A}{A} = \frac{9.277(10^3)}{(78.54)(10^{-6})} = 118.1(10^6)\ \text{N/m}^2 = 118.1\ \text{MPa(T)}$

The critical buckling load for member BP is:

$$(P_{cr})_B = \frac{\pi^2 EI}{L_B^2} = \frac{\pi^2(200)(10^9)(0.4909)(10^{-9})}{0.3^2} = 10.76(10^3)\ \text{N}$$ 9

The factor of safety with respect to failure due to yield K_y and with respect to failure due to buckling K_{cr} are:

$$K_y = \frac{\sigma_{yield}}{\sigma_A} = 3.048 \qquad K_{cr} = \frac{(P_{cr})_B}{N_B} = 5.091$$

The factor of safety for the structure is: $\boxed{K = 3.048}$

11.19 A force F= 10 kN is applied to the roller that slides inside a slot. Both bars have a diameter of d = 10 mm, Modulus of Elasticity E = 200 GPa, and yield stress σ_{yield} = 360 MPa. Bar AP and BP have lengths of L_{AP}= 200 mm and L_{BP}= 300 mm respectively. Determine the factor of safety for the two-bar structures.

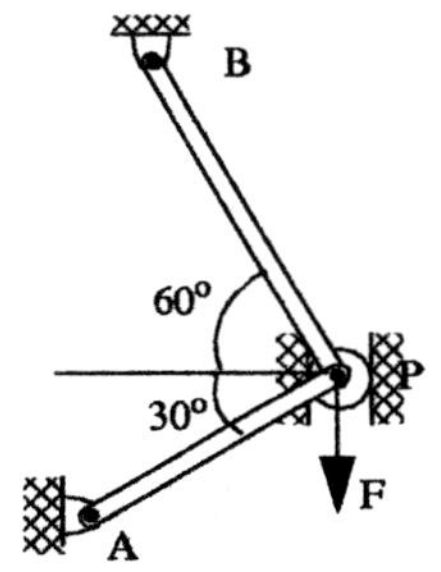

Fig. P11.19

Solution F=10 kN E = 200 GPa d = 10 m σ_{yield} = 360 MPa
L_{AP}= 200 mm L_{BP}= 300 mm K = ?

The cross-sectional area and area moment of inertia are:

$$A = \frac{\pi}{4}0.01^2 = 78.54(10^{-6})\text{m}^2 \qquad I = \frac{\pi}{64}0.01^4 = 0.4909(10^{-9})\text{m}^4$$ 1

We draw the free body diagram of the rigid bar as shown in Fig. (a)

(a)

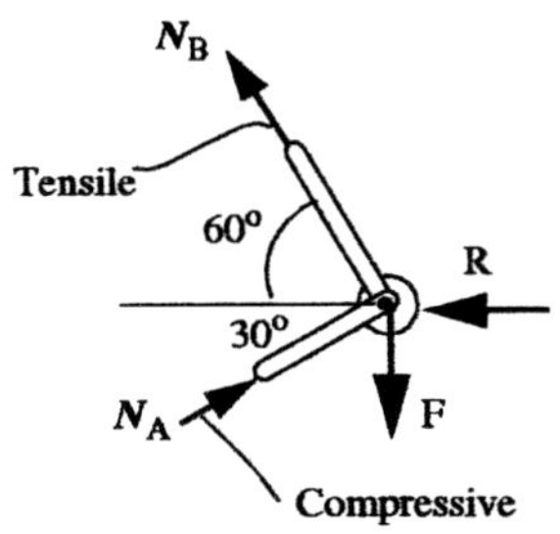

(b)

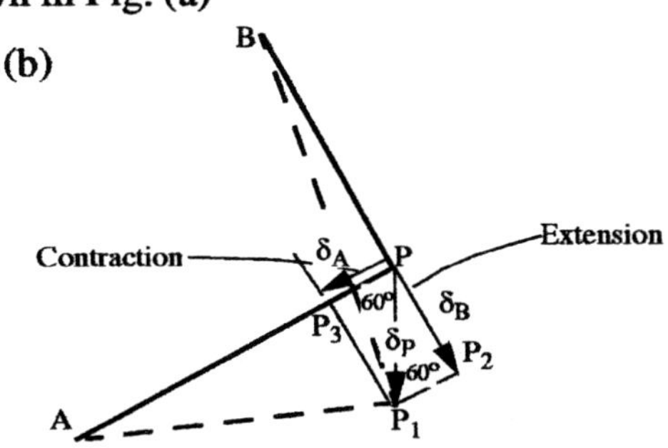

By force equilibrium in the y-direction in Fig.(a), we obtain the following.

$$N_A \sin 30 + N_B \sin 60 = F \qquad \text{or} \qquad 0.5N_A + 0.866N_B = 10(10^3)$$ 2

We draw the exaggerated deformed shape as shown in Fig. (b) As per small strain approximation, we need component of PP_1 in the original direction of AP and BP, i.e., PP_3 represents the deformation of bar AP and PP_2 represents the deformation of bar BP and are calculated as shown below.

$$\delta_A = \delta_P \cos 60 = 0.5\delta_P$$ 3

$$\delta_B = \delta_P \sin 60 = 0.866\delta_P$$ 4

From Eq 4.27, we have the following:

$$\delta_A = \frac{N_A L_A}{EA} = \frac{N_A(0.2)}{(200)(10^9)(78.54)(10^{-6})} = 0.5\delta_P \qquad \text{or} \qquad N_A = 39.27\delta_P(10^6)$$ 5

$$\delta_B = \frac{N_B L_B}{EA} = \frac{N_B(0.3)}{(200)(10^9)(78.54)(10^{-6})} = 0.866\delta_P \qquad \text{or} \qquad N_B = 45.34\delta_P(10^6)$$ 6

Substituting Eqs. (4) and (5) into Eq. (1), we obtain the displacement δ_P as shown below.

$$[0.5(39.27\delta_P) + 0.866(45.34\delta_P)](10^6) = 10(10^3) \quad \text{or} \quad \delta_P = 0.1698(10^{-3})\text{m}$$ 7

Substituting Eq.(6) into Eqs. (4) and (5) we obtain:

$$N_A = (39.27)(0.1698)(10^3) = 6.667(10^3)\ \text{N} \qquad N_B = (45.34)(0.1698)(10^3) = 7.697(10^3)\ \text{N}$$ 8

The axial stress in member B is: $\sigma_B = \dfrac{N_B}{A} = \dfrac{7.697(10^3)}{(78.54)(10^{-6})} = 98.01(10^6)\text{N/m}^2 = 98.01\ \text{MPa(T)}$

The critical buckling load for member AP is:

$$(P_{cr})_A = \frac{\pi^2 EI}{L_A^2} = \frac{\pi^2(200)(10^9)(0.4909)(10^{-9})}{0.2^2} = 24.22(10^3)\ \text{N}$$ 9

The factor of safety with respect to failure due to yield K_y and with respect to failure due to buckling K_{cr} are:

$$K_y = \frac{\sigma_{yield}}{\sigma_A} = 3.673 \qquad K_{cr} = \frac{(P_{cr})_A}{N_A} = 3.633$$

The factor of safety for the structure is: $\boxed{K = 3.633}$

11.20 A force F= 10 kN is applied to the roller that slides inside a slot. Both bars have a diameter of d = 10 mm, Modulus of Elasticity E = 200 GPa, and yield stress σ_{yield} = 360 MPa. Bar AP and BP have lengths of L_{AP}= 200 mm and L_{BP}= 300 mm respectively. Determine the factor of safety for the two-bar structures.

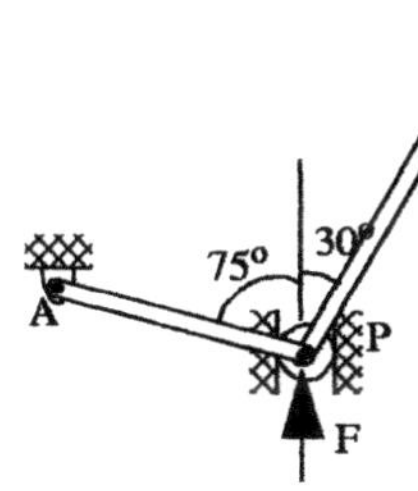

Fig. P11.20

Solution	F=10 kN	E = 200 GPa	d = 10 m	σ_{yield} = 360 MPa
	L_{AP}= 200 mm	L_{BP}= 300 mm	K = ?	

The cross-sectional area and area moment of inertia are:

$$A = \frac{\pi}{4}0.01^2 = 78.54(10^{-6})\text{m}^2 \qquad I = \frac{\pi}{64}0.01^4 = 0.4909(10^{-9})\text{m}^4$$ 1

We draw the free body diagram of the rigid bar as shown in Fig. (a)

(a)

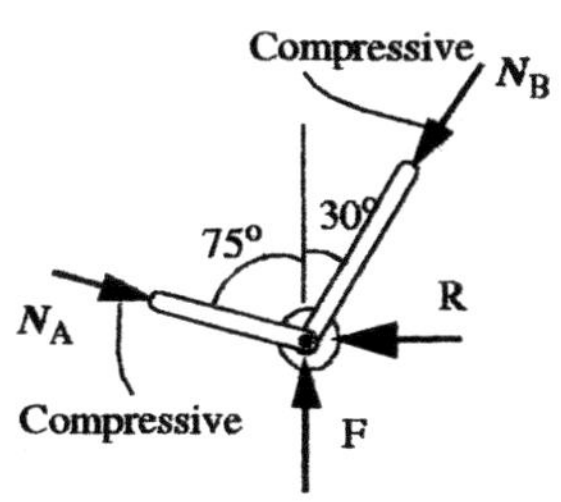

(b)

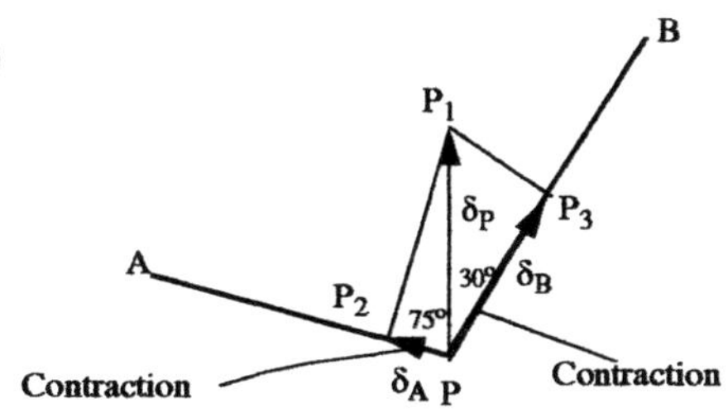

By force equilibrium in the y-direction in Fig.(a), we obtain the following.

$$N_A \cos 75 + N_B \cos 30 = F \qquad \text{or} \qquad 0.2588 N_A + 0.8660 N_B = 20(10^3)$$ 1

We draw the exaggerated deformed shape as shown in Fig. (b). As per small strain approximation, we need component of PP_1 in the original direction of AP and BP, i.e., PP_1 represents the deformation of bar AP and PP_2 represents the deformation of bar BP and are calculated as shown below.

$$\delta_A = \delta_P \cos 75 = 0.2588\delta_P$$ 2

$$\delta_B = \delta_P \cos 30 = 0.8660\delta_P$$ 3

From Eq 4.27, we have the following:

$$\delta_A = \frac{N_A L_A}{EA} = \frac{N_A(0.2)}{(200)(10^9)(78.54)(10^{-6})} = 0.2588\delta_P \quad \text{or} \quad N_A = 20.326\delta_P(10^6)$$ 4

$$\delta_B = \frac{N_B L_B}{EA} = \frac{N_B(0.3)}{(200)(10^9)(78.54)(10^{-6})} = 0.866\delta_P \quad \text{or} \quad N_B = 45.343\delta_P(10^6)$$ 5

Substituting Eqs. (4) and (5) into Eq. (1), we obtain the displacement δ_P as shown below.

$$[0.2588(20.326\delta_P) + 0.866(45.343\delta_P)](10^6) = 10(10^3) \text{ or } \quad \delta_P = 0.2246(10^{-3})\text{m}$$ 6

Substituting Eq.(6) into Eq. (4) we obtain:

$$N_A = (20.326)(0.2246)(10^3) = 4.564(10^3) \text{ N} \qquad N_B = (45.343)(0.2246)(10^3) = 10.183(10^3) \text{ N}$$

The axial stress in member B is: $\sigma_B = \frac{N_B}{A} = \frac{10.183(10^3)}{(78.54)(10^{-6})} = 129.65(10^6)\text{N/m}^2 = 129.65 \text{ MPa(C)}$

From Eq. 9 in problems 11.18 and 11.19 we have: $(P_{cr})_A = 24.22(10^3)$ N $\qquad (P_{cr})_B = 10.76(10^3)$ N

The factor of safety with respect to failure due to yield K_y and with respect to failure due to buckling K_{cr} are:

$$K_y = \frac{\sigma_{yield}}{\sigma_B} = 2.776 \qquad K_{cr_1} = \frac{(P_{cr})_A}{N_A} = 5.306 \qquad K_{cr_2} = \frac{(P_{cr})_B}{N_B} = 1.057$$

The factor of safety for the structure is: $\boxed{K = 1.057}$

11.21 (a) Solve the Boundary Value Problem for Case 2 in Table 11.1 and obtain the critical load value P_{cr} that is given in the table. (b) If buckling in mode one is prevented, then what would be the P_{cr} value.

Solution $\qquad P_{cr1} = ? \qquad P_{cr2} = ?$ for Case 2 in Table 11.1

The boundary value problem for Case 2 in Table 11.1 is:

$$EI\frac{d^2v}{dx^2} + Pv = Pv(L) \quad \text{or} \quad \frac{d^2v}{dx^2} + \lambda^2 v = \lambda^2 v(L)$$ 1

$$v(0) = 0$$ 2 $$\frac{dv}{dx}(0) = 0$$ 3

The homogenous solution to Eq. 1 is: $v_h(x) = A\cos\lambda x + B\sin\lambda x$

The particular solution to Eq. 1 is: $v_p(x) = v(L)$

The total solution to Eq. 1 is: $v(x) = v_h(x) + v_p(x)$ or

$$v(x) = A\cos\lambda x + B\sin\lambda x + v(L)$$ 4

Substituting x = 0 into Eq. 4 and using Eq. 2 we obtain:

$$v(0) = A + v(L) = 0 \quad \text{or} \quad A = -v(L)$$ 5

Taking the first derivative of Eq. 4 we obtain:

$$\frac{dv}{dx}(x) = -\lambda A\sin\lambda x + \lambda B\cos\lambda x$$ 6

Substituting x = 0 into Eq. 6 and using Eq. 3 we obtain:

$$\frac{dv}{dx}(0) = \lambda B = 0 \quad \text{or} \quad B = 0$$ 7

Substituting x = L into Eq. 4 and using Eq. 5 and Eq. 6 we obtain:

$$v(L) = A\cos\lambda L + B\sin\lambda L + v(L) = v(L)[-\cos\lambda L + 1] \text{ or}$$

$$\cos\lambda L = 0 \qquad \mathbf{8}$$

The roots of the characteristics equation above are

$$\lambda L = (2n+1)\frac{\pi}{2} \qquad \text{where } n = 0, 1, 2\ldots\ldots \qquad \mathbf{9}$$

The critical value corresponds to n = 0 is $\sqrt{\frac{P_{cr1}}{EI}} = \frac{\pi}{2L}$ or $\boxed{P_{cr1} = \frac{\pi^2 EI}{4L^2}}$

The critical value corresponding (next mode) to n = 1 is: $\sqrt{\frac{P_{cr2}}{EI}} = \frac{3\pi}{2L}$ or $\boxed{P_{cr2} = \frac{9\pi^2 EI}{4L^2}}$

11.22 Solve the Boundary Value Problem for Case 3 in Table 11.1 and obtain the critical load value P_{cr} that is given in the table. (b) If buckling in mode one is prevented, then what would be the P_{cr} value.

Solution $P_{cr1} = ?$ $P_{cr2} = ?$ for Case 3 in Table 11.1

The boundary value problem for Case 3 in Table 11.1 is:

$$EI\frac{d^2v}{dx^2} + Pv = R_B(L-x) \qquad \text{or} \qquad \frac{d^2v}{dx^2} + \lambda^2 v = \frac{R_B}{EI}(L-x) \qquad \mathbf{1}$$

$$v(0) = 0 \quad \mathbf{2} \qquad \frac{dv}{dx}(0) = 0 \quad \mathbf{3} \qquad v(L) = 0 \quad \mathbf{4}$$

The homogenous solution to Eq. 1 is: $v_h(x) = A\cos\lambda x + B\sin\lambda x$

The particular solution to Eq. 1 is: $v_p(x) = \frac{R_B}{\lambda^2 EI}(L-x)$

The total solution to Eq. 1 is: $v(x) = v_h(x) + v_p(x)$ or

$$v(x) = A\cos\lambda x + B\sin\lambda x + \frac{R_B}{\lambda^2 EI}(L-x) \qquad \mathbf{5}$$

Substituting x = 0 into Eq. 5 and using Eq. 2 we obtain:

$$v(0) = A + \frac{R_B L}{\lambda^2 EI} = 0 \qquad \text{or} \qquad A = -\frac{R_B L}{\lambda^2 EI} \qquad \mathbf{6}$$

Taking the first derivative of Eq. 5 we obtain:

$$\frac{dv}{dx}(x) = -\lambda A\sin\lambda x + \lambda B\cos\lambda x - \frac{R_B}{\lambda^2 EI} \qquad \mathbf{7}$$

Substituting x = 0 into Eq. 7 and using Eq. 3 we obtain:

$$\frac{dv}{dx}(0) = \lambda B - \frac{R_B}{\lambda^2 EI} = 0 \qquad \text{or} \qquad B = \frac{R_B}{\lambda^3 EI} \qquad \mathbf{8}$$

Substituting x = L into Eq. 5 and using Eqs. 6, 8, and 4 we obtain:

$$v(L) = A\cos\lambda L + B\sin\lambda L = \left[-\frac{R_B L}{\lambda^2 EI}\cos\lambda L + \frac{R_B}{\lambda^3 EI}\sin\lambda L\right] = \frac{R_B L}{\lambda^2 EI}(-\lambda L\cos\lambda L + \sin\lambda L) = 0 \text{ or}$$

$$\tan\lambda L = \lambda L \qquad \mathbf{9}$$

The two lowest roots of the above characteristics equation are

$$\lambda L = 4.4934 \qquad \text{and} \qquad \lambda L = 7.7253 \qquad \mathbf{10}$$

The critical value corresponds to smaller root $\sqrt{\frac{P_{cr1}}{EI}} = \frac{4.4934}{L}$ or $\boxed{P_{cr1} = \frac{20.19EI}{L^2}}$

The critical value corresponding (next mode) to the next root: $\sqrt{\frac{P_{cr2}}{EI}} = \frac{7.7253}{L}$ or $\boxed{P_{cr2} = \frac{59.68\pi^2 EI}{L^2}}$

11.23 Solve the Boundary Value Problem for Case 4 in Table 11.1 and obtain the critical load value P_{cr} that is given in the table.(b) If buckling in mode one is prevented, then what would be the P_{cr} value.

Solution $P_{cr1} = ?$ $P_{cr2} = ?$ for Case 4 in Table 11.1

The boundary value problem for Case 4 in Table 11.1 is:

$$EI\frac{d^2v}{dx^2} + Pv = R_B(L-x) + M_B \quad \text{or} \quad \frac{d^2v}{dx^2} + \lambda^2 v = \frac{R_B}{EI}(L-x) + \frac{M_B}{EI} \qquad 1$$

$$v(0) = 0 \quad 2 \qquad \frac{dv}{dx}(0) = 0 \quad 3 \qquad v(L) = 0 \quad 4 \qquad \frac{dv}{dx}(L) = 0 \quad 5$$

The homogenous solution to Eq. 1 is: $v_h(x) = A\cos\lambda x + B\sin\lambda x$

The particular solution to Eq. 1 is: $v_p(x) = \frac{R_B}{\lambda^2 EI}(L-x) + \frac{M_B}{\lambda^2 EI}$

The total solution to Eq. 1 is: $v(x) = v_h(x) + v_p(x)$ or

$$v(x) = A\cos\lambda x + B\sin\lambda x + \frac{R_B}{\lambda^2 EI}(L-x) + \frac{M_B}{\lambda^2 EI} \qquad 6$$

Substituting x = 0 into Eq. 6 and using Eq. 2 we obtain:

$$v(0) = A + \frac{R_B L}{\lambda^2 EI} + \frac{M_B}{\lambda^2 EI} = 0 \qquad 7$$

Taking the first derivative of Eq. 6 we obtain:

$$\frac{dv}{dx}(x) = -\lambda A\sin\lambda x + \lambda B\cos\lambda x - \frac{R_B}{\lambda^2 EI} \qquad 8$$

Substituting x = 0 into Eq. 8 and using Eq. 3 we obtain:

$$\frac{dv}{dx}(0) = \lambda B - \frac{R_B}{\lambda^2 EI} = 0 \quad \text{or} \quad \frac{R_B}{\lambda^2 EI} = \lambda B \qquad 9$$

Substituting Eq. 9 into Eq. 7 we obtain:

$$A + \lambda LB + \frac{M_B}{\lambda^2 EI} = 0 \quad \text{or} \quad \frac{M_B}{\lambda^2 EI} = -(A + \lambda LB) \qquad 10$$

Substituting x = L into Eq. 6 and using Eqs. 9, 10, and 4 we obtain:

$$v(L) = A\cos\lambda L + B\sin\lambda L + \frac{M_B}{\lambda^2 EI} = 0 \quad \text{or} \quad A(\cos\lambda L - 1) + B(\sin\lambda L - \lambda L) = 0 \qquad 11$$

Substituting x = L into Eq. 8 and using Eqs. 9, 10, and 5 we obtain:

$$\frac{dv}{dx}(L) = -\lambda A\sin\lambda L + \lambda B\cos\lambda L - \frac{R_B}{\lambda^2 EI} = 0 \quad \text{or} \quad -A\sin\lambda L + B(\cos\lambda L - 1) = 0 \qquad 12$$

Multiplying Eq. 11 by $\sin\lambda L$, Eq. 12 by $(\cos\lambda L-1)$ and adding we obtain:

$$B[(\cos\lambda L - 1)^2 + \sin\lambda L(\sin\lambda L - \lambda L)] = 0 \quad \text{or} \quad B[\cos^2\lambda L + \sin^2\lambda L + 1 - 2\cos\lambda L - \lambda L\sin\lambda L] = 0$$

or $B[2(1-\cos\lambda L)-\lambda L\sin\lambda L] = 0$. For non-trivial solution ($B \neq 0$) we obtain:

$$2(1-\cos\lambda L)-\lambda L\sin\lambda L = 0 \qquad \mathbf{13}$$

The above characteristic equation can be written as:

$$2[2\sin^2(\lambda L/2)]-\lambda L[2\sin(\lambda L/2)\cos(\lambda L/2)] = 0 \quad \text{or} \quad \sin\left(\frac{\lambda L}{2}\right)\left[\sin\left(\frac{\lambda L}{2}\right)-\frac{\lambda L}{2}\cos\left(\frac{\lambda L}{2}\right)\right] = 0 \qquad \mathbf{14}$$

Eq. 14 is satisfied if: (i) $\sin\left(\frac{\lambda L}{2}\right) = 0$ or if (ii) $\tan\left(\frac{\lambda L}{2}\right) = \frac{\lambda L}{2}$.

From the first possibility we obtain: $\frac{\lambda L}{2} = n\pi \qquad n = 1, 2\ldots\ldots$ The first two roots correspond to $\frac{\lambda L}{2} = \pi$ and $\frac{\lambda L}{2} = 2\pi$.

From the second possibility we obtain: $\frac{\lambda L}{2} = 4.4934$ and $\frac{\lambda L}{2} = 7.7253$

Thus, the two lowest roots of the equations are: $\frac{\lambda L}{2} = \pi$ and $\frac{\lambda L}{2} = 4.4934$

The critical value corresponds to smaller root $\sqrt{\frac{P_{cr1}}{EI}} = \frac{2\pi}{L}$ or $\boxed{P_{cr1} = \frac{4\pi^2 EI}{L^2}}$

The critical value corresponding (next mode) to the next root: $\sqrt{\frac{P_{cr2}}{EI}} = \frac{2(4.4934)}{L}$ or $\boxed{P_{cr2} = \frac{80.76EI}{L^2}}$

11.24 A torsional spring with a spring constant K is attached at one end of a column as shown. Assume that bending about y-axis is prevented. (a) Determine the Characteristic Equation for this buckling problem. (b) Show that for $K = 0$ and $K = \infty$ the critical load P_{cr} is as given in Table 11.1 for Cases 1 and 3, respectively.

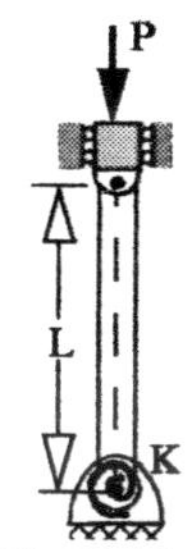

Fig. P11.24

Solution Characteristic Equation = ? $K = 0$ is Case 1 and $K = \infty$ is Case 3

Fig. (a) shows the free body diagram of the entire column. Fig.(b)shows the free body diagram after an imaginary cut through the deformed column.

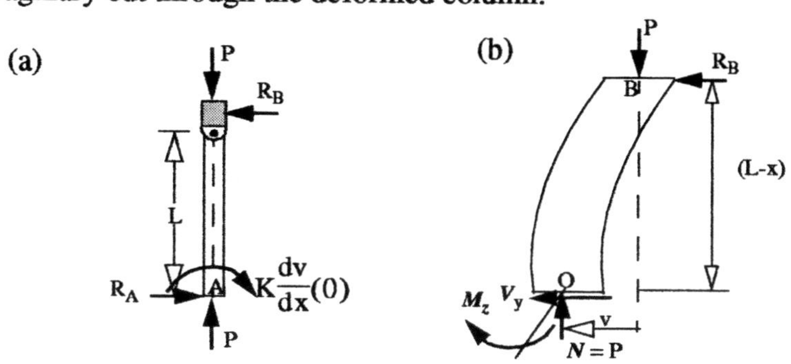

By equilibrium of moment about point A in Fig. (a) we obtain. $R_B L = K\frac{dv}{dx}(0)$ or $R_B = \frac{K dv}{L dx}(0)$.

By equilibrium of moment about point O we obtain: $M_z + Pv = R_B(L-x) = \frac{K dv}{L dx}(0)(L-x)$

The boundary value problem can be written as:

$$EI\frac{d^2v}{dx^2} + Pv = \frac{K}{L}\frac{dv}{dx}(0)(L-x) \quad \text{or} \quad \frac{d^2v}{dx^2} + \lambda^2 v = \frac{K}{EIL}\frac{dv}{dx}(0)(L-x) \qquad 1$$

$$v(0) = 0 \qquad 2 \qquad\qquad v(L) = 0 \qquad 3$$

The homogenous solution to Eq. 1 is: $v_h(x) = A\cos\lambda x + B\sin\lambda x$

The particular solution to Eq. 1 is: $v_p(x) = \frac{K}{\lambda^2 EIL}\frac{dv}{dx}(0)(L-x)$

The total solution to Eq. 1 is: $v(x) = v_h(x) + v_p(x)$ or

$$v(x) = A\cos\lambda x + B\sin\lambda x + \frac{K}{\lambda^2 EIL}\frac{dv}{dx}(0)(L-x) \qquad 4$$

Substituting x = 0 into Eq. 4 and using Eq. 2 we obtain:

$$v(0) = A + \frac{K}{\lambda^2 EI}\frac{dv}{dx}(0) = 0 \quad \text{or} \quad A = -\frac{K}{\lambda^2 EI}\frac{dv}{dx}(0) \qquad 5$$

Substituting x = L into Eq. 4 and using Eq. 3 we obtain:

$$v(L) = A\cos\lambda L + B\sin\lambda L = 0 \quad \text{or} \quad B\tan\lambda L = \frac{K}{\lambda^2 EI}\frac{dv}{dx}(0) \qquad 6$$

Taking the first derivative of Eq. 4 we obtain:

$$\frac{dv}{dx}(x) = -\lambda A\sin\lambda x + \lambda B\cos\lambda x - \frac{K}{\lambda^2 EIL}\frac{dv}{dx}(0) \qquad 7$$

Substituting x = 0 into Eq. 7 we obtain:

$$\frac{dv}{dx}(0) = \lambda B - \frac{K}{\lambda^2 EIL}\frac{dv}{dx}(0) \quad \text{or} \quad B = \frac{1}{\lambda}\frac{dv}{dx}(0)\left(1 + \frac{K}{\lambda^2 EIL}\right) \qquad 8$$

Substituting Eq. 8 into Eq. 6 we obtain: $\frac{1}{\lambda}\frac{dv}{dx}(0)\left(1 + \frac{K}{\lambda^2 EIL}\right)\tan\lambda L = \frac{K}{\lambda^2 EI}\frac{dv}{dx}(0)$ or $\boxed{\tan\lambda L = \frac{K\lambda L}{K + \lambda^2 EIL}}$

As $K \to 0$ we obtain $\tan\lambda L = 0$ or $\sin\lambda L = 0$, which is the characteristic equation of case 1 in Table 11.1.

The Characteristic equation can be written as: $\tan\lambda L = \frac{\lambda L}{1 + (\lambda^2 EIL)/K}$. As $K \to \infty$ we obtain $\tan\lambda L = \lambda L$ which is the characteristic equation of case 3 in Table 11.1

11.25 A torsional spring with a spring constant K is attached at one end of a column as shown. Assume that bending about y-axis is prevented. (a) Determine the Characteristic Equation for this buckling problem. (b) Show that for $K = 0$ and the critical load P_{cr} is as given for Case 2 in Table 11.1.(c) For $K = \infty$ obtain the critical load P_{cr}.

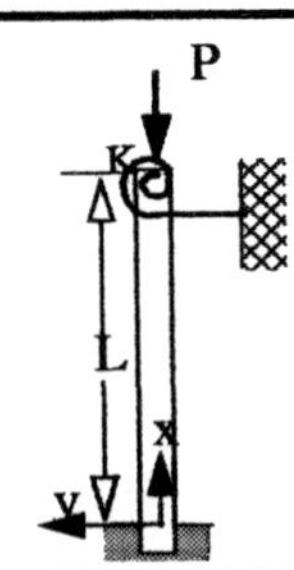

Fig. P11.25

Solution Characteristic Equation = ? $K = 0$ is Case 2 and $K = \infty$ is?

Fig.(a)shows the free body diagram after an imaginary cut through the deformed column.

(a)

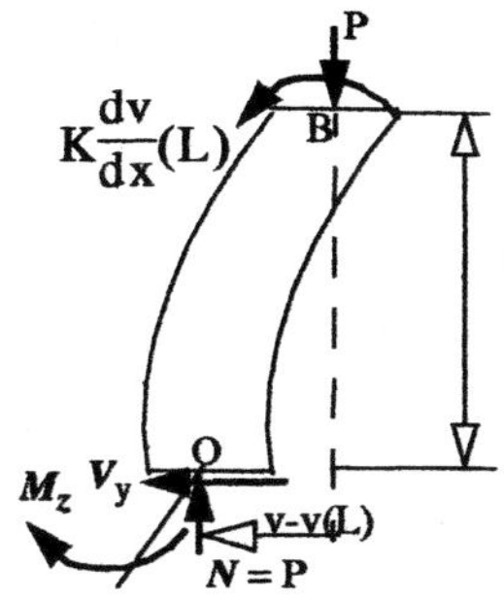

By equilibrium of moment about point O we obtain: $M_z + P(v - v(L)) - K\frac{dv}{dx}(L) = 0$

The boundary value problem can be written as:

$$EI\frac{d^2v}{dx^2} + Pv = Pv(L) + K\frac{dv}{dx}(L) \qquad \text{or} \qquad \frac{d^2v}{dx^2} + \lambda^2 v = \lambda^2 v(L) + \frac{K}{EI}\frac{dv}{dx}(L) \qquad 1$$

$$v(0) = 0 \qquad 2 \qquad\qquad \frac{dv}{dx}(0) = 0 \qquad 3$$

The homogenous solution to Eq. 1 is: $v_h(x) = A\cos\lambda x + B\sin\lambda x$

The particular solution to Eq. 1 is: $v_p(x) = v(L) + \frac{K}{\lambda^2 EI}\frac{dv}{dx}(L)$

The total solution to Eq. 1 is: $v(x) = v_h(x) + v_p(x)$ or

$$v(x) = A\cos\lambda x + B\sin\lambda x + v(L) + \frac{K}{\lambda^2 EI}\frac{dv}{dx}(L) \qquad 4$$

Substituting x = 0 into Eq. 4 and using Eq. 2 we obtain:

$$v(0) = A + v(L) + \frac{K}{\lambda^2 EI}\frac{dv}{dx}(L) = 0 \qquad \text{or} \qquad A = -\left[v(L) + \frac{K}{\lambda^2 EI}\frac{dv}{dx}(L)\right] \qquad 5$$

Taking the first derivative of Eq. 4 we obtain:

$$\frac{dv}{dx}(x) = -\lambda A\sin\lambda x + \lambda B\cos\lambda x \qquad 6$$

Substituting x = 0 into Eq. 6 and using Eq. 3 we obtain:

$$\frac{dv}{dx}(0) = \lambda B = 0 \qquad \text{or} \qquad B = 0 \qquad 7$$

Substituting x = L into Eq. 4 and using Eq. 5 and Eq. 6 we obtain:

$$v(L) = A\cos\lambda L + v(L) + \frac{K}{\lambda^2 EI}\frac{dv}{dx}(L) = -\left(v(L) + \frac{K}{\lambda^2 EI}\frac{dv}{dx}(L)\right)\cos\lambda L + v(L) + \frac{K}{\lambda^2 EI}\frac{dv}{dx}(L) \text{ or}$$

$$\frac{K}{\lambda^2 EI}\frac{dv}{dx}(L)(1 - \cos\lambda L) = v(L)\cos\lambda L \qquad \text{or} \qquad \frac{dv}{dx}(L) = \left(\frac{\lambda^2 EI}{K}\right)\frac{\cos\lambda L}{1 - \cos\lambda L}v(L) \qquad 8$$

Substituting x = 0 into Eq. 6 and using Eq. 8 we obtain

$$\frac{dv}{dx}(L) = \lambda\left[v(L) + \frac{K}{\lambda^2 EI}\frac{dv}{dx}(L)\right]\sin\lambda L \qquad \text{or} \qquad \left(\frac{\lambda^2 EI}{K}\right)\frac{\cos\lambda L}{1 - \cos\lambda L}v(L) = \lambda v(L)\left[1 + \frac{\cos\lambda L}{1 - \cos\lambda L}\right]\sin\lambda L \text{ or}$$

$$\boxed{\tan\lambda L = \lambda EI/K}$$

We can write the characteristic equation as $K\sin\lambda L = \lambda EI\cos\lambda L$. As $K \to 0$ we obtain $\cos\lambda L = 0$ which is the characteristic equation of case 2 in Table 11.1.

The characteristic equation can also be written as: $\sin\lambda L = \dfrac{\lambda EI\cos\lambda L}{K}$. As $K \to \infty$ we obtain $\sin\lambda L = 0$ which is the characteristic equation of case 1 in Table 11.1.

11.26 (a) Determine the critical buckling in terms of E, I, L, and α. (b) Show that when α = 0.5 the critical load corresponds to mode 2 as shown in Figure 11.7.

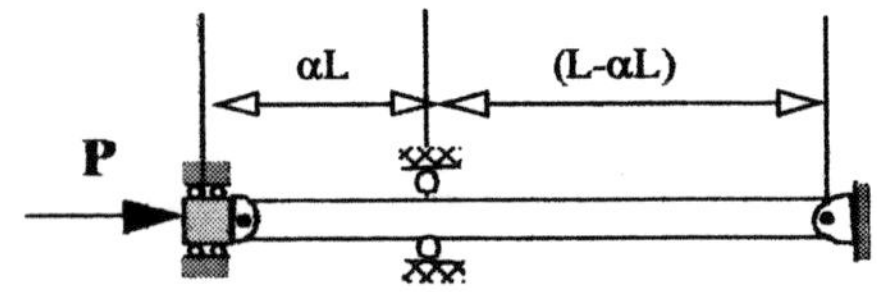

Fig. P11.26

Solution $P_{cr} = f(E, I, L, \alpha) = ?$ For $\alpha = 0.5$ $P_{cr} = ?$

In Fig. (a) the support in middle is replaced by a reaction force. Fig. (b) shows the free body diagram after an imaginary cut.

(a)

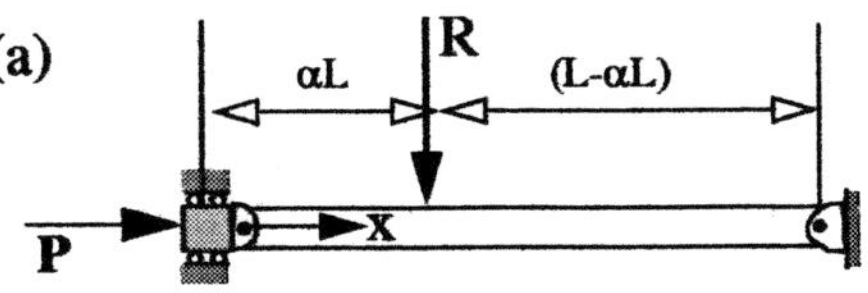

(b)

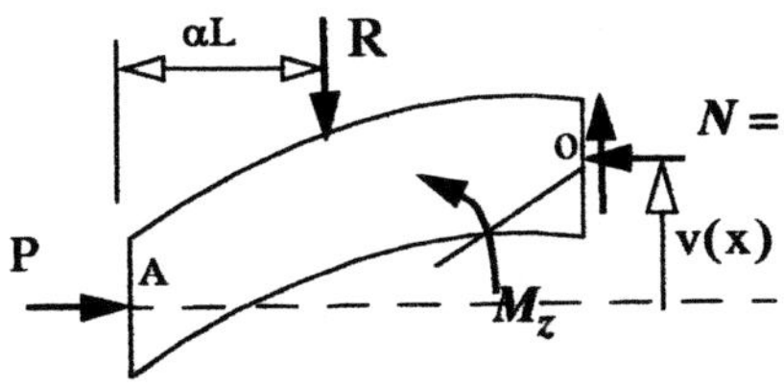

Using singularity functions we can write the moment equilibrium equation as: $M_z + Pv + R\langle x - \alpha L\rangle^1 = 0$

The boundary value problem can be written as:

$$EI\frac{d^2v}{dx^2} + Pv = -R\langle x - \alpha L\rangle^1 \quad \text{or} \quad \frac{d^2v}{dx^2} + \lambda^2 v = -\frac{R}{EI}\langle x - \alpha L\rangle^1 \qquad 1$$

$$v(0) = 0 \quad 2 \qquad v(L) = 0 \quad 3 \qquad v(\alpha L) = 0 \quad 4$$

The homogenous solution to Eq. 1 is: $v_h(x) = A\cos\lambda x + B\sin\lambda x$

The particular solution to Eq. 1 is: $v_p(x) = -\dfrac{R}{\lambda^2 EI}\langle x - \alpha L\rangle$

The total solution to Eq. 1 is: $v(x) = v_h(x) + v_p(x)$ or

$$v(x) = A\cos\lambda x + B\sin\lambda x - \frac{R}{\lambda^2 EI}\langle x - \alpha L\rangle^1 \qquad 5$$

Substituting x = 0 into Eq. 5 and using Eq. 2 we obtain:

$$v(0) = A - 0 = 0 \quad \text{or} \quad A = 0 \qquad 6$$

Substituting x = L into Eq. 5 and using Eq. 3 we obtain:

$$v(L) = B\sin\lambda L - \frac{R}{\lambda^2 EI}\langle L - \alpha L\rangle^1 = 0 \quad \text{or} \quad B\sin\lambda L = \frac{RL}{\lambda^2 EI}(1-\alpha) \qquad 7$$

Substituting x = αL into Eq. 5 and using Eq. 3 we obtain

$$v(L) = B\sin\lambda\alpha L = 0 \quad \text{or} \quad \sin\lambda\alpha L = 0 \quad \text{or} \quad \lambda\alpha L = n\pi \qquad n = 1, 2\ldots\ldots \qquad 8$$

The critical load value corresponds to n = 1 in Eq. 8. We obtain:

$$\lambda_{cr} = \sqrt{\frac{P_{cr}}{EI}} = \frac{\pi}{\alpha L} \quad \text{or} \quad \boxed{P_{cr} = \frac{\pi^2 EI}{\alpha^2 L^2}}$$

For $\alpha = 0.5$ we obtain $P_{cr} = \dfrac{4\pi^2 EI}{L^2}$, which corresponds critical value at mode 2.

11.27 In terms of the Modulus of Elasticity E, the length of the column L, the area moment of inertia I, the axial force P, and the parameter α, determine (a) the deflection at x = L. (b) the critical load P_{cr}.

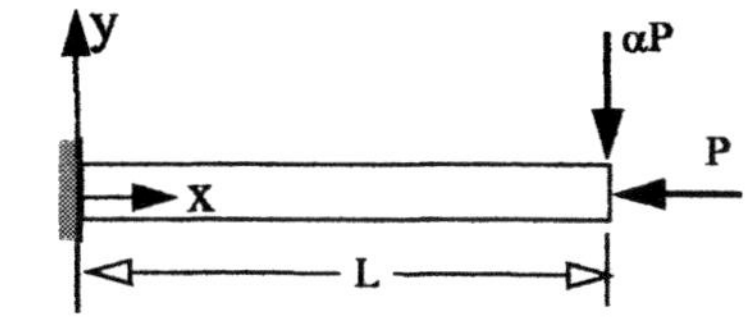

Fig. P11.27

Solution $v(L) = f_1(E, L, I, \alpha) = ?$ $P_{cr} = f_2(E, L, I, \alpha) = ?$

Fig. (a) shows the free body diagram of the column after an imaginary cut.

(a)

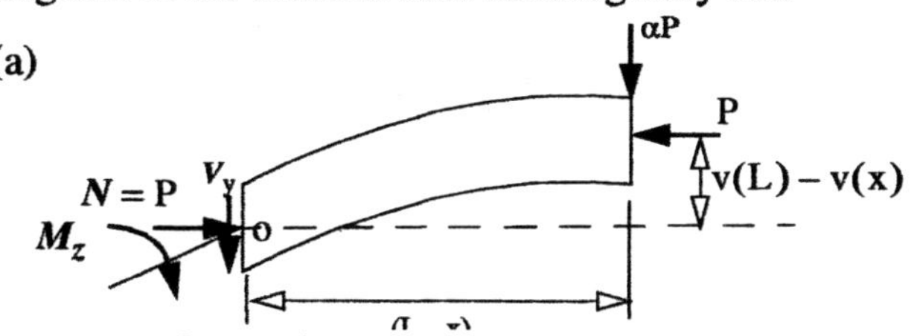

By equilibrium of moment about point O we obtain: $M_z - P(v(L) - v) + \alpha P(L - x) = 0$

The boundary value problem can be written as:

$$EI\frac{d^2v}{dx^2} + Pv = Pv(L) - \alpha P(L - x) \quad \text{or} \quad \frac{d^2v}{dx^2} + \lambda^2 v = \lambda^2 v(L) - \lambda^2 \alpha(L - x) \qquad 1$$

$$v(0) = 0 \qquad 2 \qquad\qquad \frac{dv}{dx}(0) = 0 \qquad 3$$

The homogenous solution to Eq. 1 is: $v_h(x) = A\cos\lambda x + B\sin\lambda x$

The particular solution to Eq. 1 is: $v_p(x) = v(L) - \alpha(L - x)$

The total solution to Eq. 1 is: $v(x) = v_h(x) + v_p(x)$ or

$$v(x) = A\cos\lambda x + B\sin\lambda x + v(L) - \alpha(L - x) \qquad 4$$

Substituting x = 0 into Eq. 4 and using Eq. 2 we obtain:

$$v(0) = A + v(L) - \alpha L = 0 \quad \text{or} \quad A = \alpha L - v(L) \qquad 5$$

Taking the first derivative of Eq. 4 we obtain:

$$\frac{dv}{dx}(x) = -\lambda A\sin\lambda x + \lambda B\cos\lambda x - \alpha \qquad 6$$

Substituting x = 0 into Eq. 6 and using Eq. 3 we obtain:

$$\frac{dv}{dx}(0) = \lambda B - \alpha = 0 \quad \text{or} \quad B = \frac{\alpha}{\lambda} \qquad 7$$

Substituting x = L into Eq. 4 and using Eq. 5 and Eq. 7 we obtain:

$$v(L) = A\cos\lambda L + B\sin\lambda L + v(L) = (\alpha L - v(L))\cos\lambda L + \frac{\alpha}{\lambda}\sin\lambda L + v(L) \text{ or } \boxed{v(L) = \alpha(\lambda L + \tan\lambda L)}$$

The tangent function, hence displacement v(L), becomes unbounded at $\lambda L = \pi/2$ thus, $\boxed{P_{cr} = \frac{\pi^2 EI}{4L^2}}$

11.28 In terms of the Modulus of Elasticity E, the length of the column L, the area moment of inertia I, the axial force P, and the parameter α, determine (a) the deflection at x = L. (b) the critical load P_{cr}

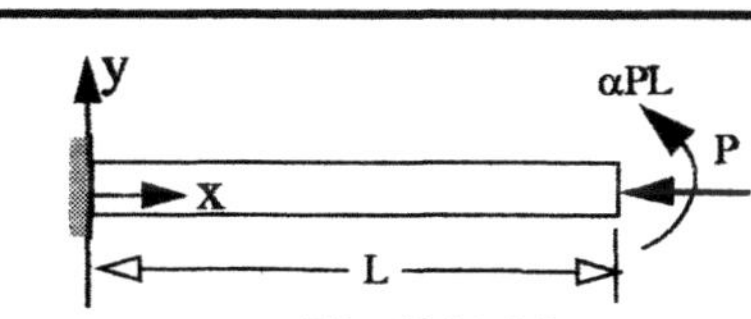

Fig. P11.28

Solution $v(L) = f_1(E, L, I, \alpha) = ?$ $P_{cr} = f_2(E, L, I, \alpha) = ?$

Fig. (a) shows the free body diagram of the column after an imaginary cut.

(a)

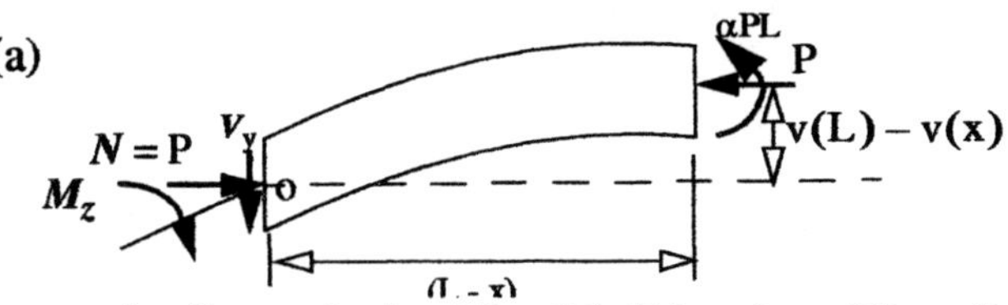

By equilibrium of moment about point O we obtain: $M_z - P(v(L) - v) - \alpha PL = 0$

The boundary value problem can be written as:

$$EI\frac{d^2v}{dx^2} + Pv = Pv(L) + \alpha PL \quad \text{or} \quad \frac{d^2v}{dx^2} + \lambda^2 v = \lambda^2 v(L) + \lambda^2 \alpha L \qquad \textbf{1}$$

$$v(0) = 0 \qquad \textbf{2} \qquad\qquad \frac{dv}{dx}(0) = 0 \qquad \textbf{3}$$

The homogenous solution to Eq. 1 is: $v_h(x) = A\cos\lambda x + B\sin\lambda x$

The particular solution to Eq. 1 is: $v_p(x) = v(L) + \alpha L$

The total solution to Eq. 1 is: $v(x) = v_h(x) + v_p(x)$ or

$$v(x) = A\cos\lambda x + B\sin\lambda x + v(L) + \alpha L \qquad \textbf{4}$$

Substituting $x = 0$ into Eq. 4 and using Eq. 2 we obtain:

$$v(0) = A + v(L) + \alpha L = 0 \quad \text{or} \quad A = -\alpha L - v(L) \qquad \textbf{5}$$

Taking the first derivative of Eq. 4 we obtain:

$$\frac{dv}{dx}(x) = -\lambda A\sin\lambda x + \lambda B\cos\lambda x \qquad \textbf{6}$$

Substituting $x = 0$ into Eq. 6 and using Eq. 3 we obtain:

$$\frac{dv}{dx}(0) = \lambda B = 0 \quad \text{or} \quad B = 0 \qquad \textbf{7}$$

Substituting $x = L$ into Eq. 4 and using Eq. 5 and Eq. 7 we obtain:

$$v(L) = A\cos\lambda L + v(L) + \alpha L = (-\alpha L - v(L))\cos\lambda L + v(L) + \alpha L \quad \text{or} \quad \boxed{v(L) = \frac{\alpha L(1 - \cos\lambda L)}{\cos\lambda L}}$$

The denominator is zero when $\lambda L = \pi/2$ thus,

$$\boxed{P_{cr} = \frac{\pi^2 EI}{4L^2}}$$

11.29 In terms of the Modulus of Elasticity E, the length of the column L, the area moment of inertia I, the axial force P, and the parameter α, determine (a) the deflection at $x = L$. (b) the critical load P_{cr}

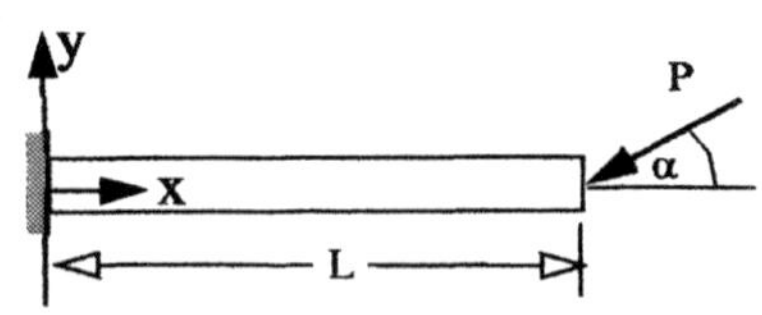

Fig. P11.29

Solution $v(L) = f_1(E, L, I, \alpha) = ?$ $P_{cr} = f_2(E, L, I, \alpha) = ?$

Fig. (a) shows the column of problem 11.27.

(a)

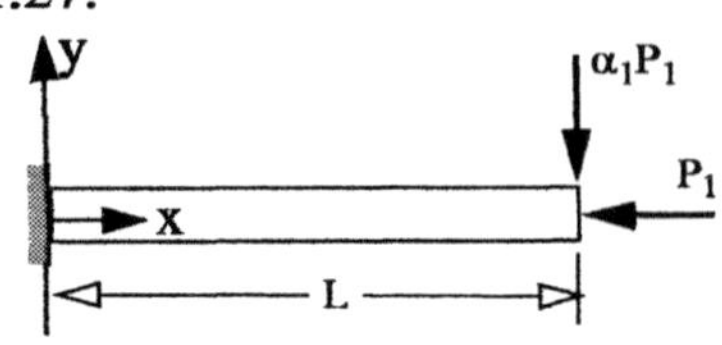

From the results of problem 11.27, the deflection at x = L is:

$$v(L) = \alpha_1(\lambda_1 L + \tan\lambda_1 L) \qquad \lambda_1 = \sqrt{P_1/EI} \qquad \textbf{1}$$

Resolving the force P in Fig. 11.29 into components we obtain:

$$P_1 = P\cos\alpha \qquad \textbf{2} \qquad\qquad \alpha_1 P_1 = P\sin\alpha \qquad \textbf{3}$$

From Eqs. 2 and 3, we obtain $\alpha_1 = \tan\alpha$. From Eq. 1 and 2 we obtain: $\lambda_1 = \sqrt{P\cos\alpha/EI} = \lambda\sqrt{\cos\alpha}$

The deflection at x = L can be written as: $\boxed{v(L) = \tan\alpha[\lambda L\sqrt{\cos\alpha} + \tan(\lambda L\sqrt{\cos\alpha})]}$

The tangent function, hence displacement v(L), becomes unbounded at $(\lambda L\sqrt{\cos\alpha}) = \pi/2$ thus,

$$\boxed{P_{cr} = \frac{\pi^2 EI}{4L^2\cos\alpha}}$$

11.30 A column built in on one end and free at the other end has a load that is eccentrically applied at a distance e from the centroid. Show that the deflection curve is given by the following equation.

$$v(x) = \frac{e(1-\cos\lambda x)}{\cos\lambda L}$$

where, λ is as given by Equation 11.3.

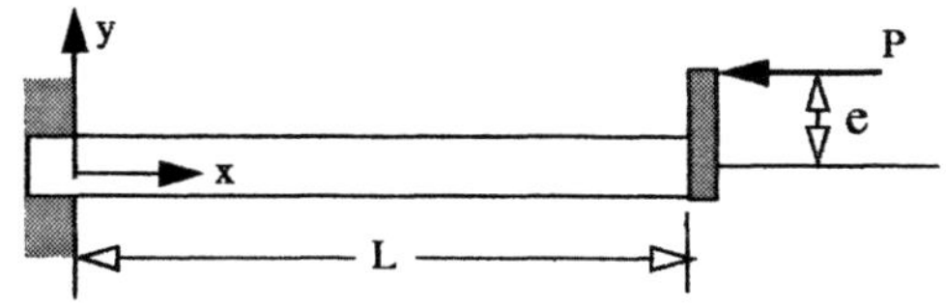

Fig. P11.30

Solution v(x) = ?

Fig. (a) shows an equivalent diagram to Fig. P11.30.

(a)

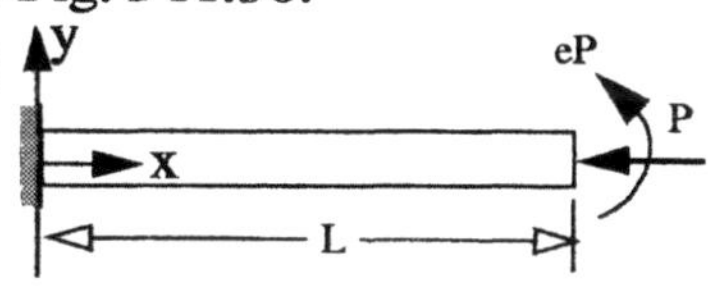

Comparing Fig. (a) to Fig. P11.28 we obtain: $\alpha L = e$. Substituting Eq. 5 and 7 of problem 11.28 into Eq. 4 of problem 11.28 and $\alpha L = e$ we obtain

$$v(x) = (-\alpha L - v(L))\cos\lambda x + v(L) + \alpha L = (v(L)+\alpha L)(1-\cos\lambda x) = (v(L)+e)(1-\cos\lambda x) \qquad \textbf{1}$$

Substituting x = L into Eq. 1 we obtain:

$$v(L) = (v(L)+e)(1-\cos\lambda L) \quad \text{or} \quad v(L) = \frac{e(1-\cos\lambda L)}{\cos\lambda L} \qquad \textbf{2}$$

Substituting Eq. 2 into Eq. 1 we obtain: $v(x) = \left(\frac{e(1-\cos\lambda L)}{\cos\lambda L} + e\right)(1-\cos\lambda x)$ or $\boxed{v(x) = \frac{e(1-\cos\lambda x)}{\cos\lambda L}}$

11.31 On the cylinder shown in Fig. P11.31 the applied load P = 3 kips and the length L = 5 feet. The Modulus of Elasticity is E = 30,000 ksi. What is the maximum stress and the maximum deflection?

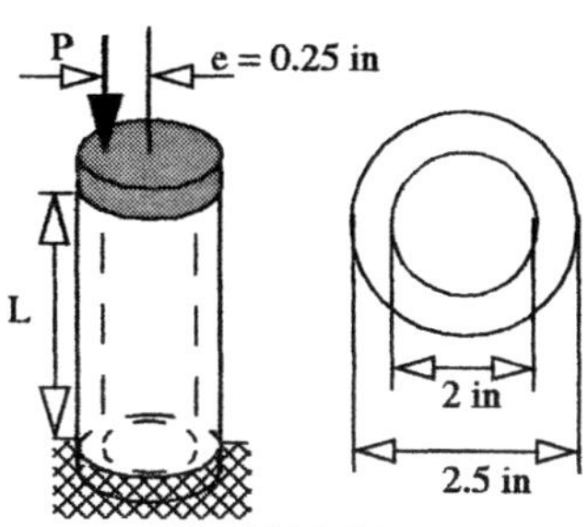

Fig. P11.31

Solution: P = 3 kips E = 30000 psi L= 5 ft ε = 0.25 σ_{max} = ?v_{max} = ?

The cross-sectional area and area moment of inertia are:

$A = \pi[2.5^2 - 2^2]/4 = 1.767\text{in}^2$ and $I = \pi[2.5^4 - 2^4]/64 = 1.132\text{in}^4$

The radius of gyration is: $r = \sqrt{\frac{I}{A}} = 0.8004\text{in}$

The eccentricity ratio is: $\frac{ec}{r^2} = \frac{(0.25)(1.25)}{(0.8004)^2} = 0.4878$

The slenderness ratio is $\frac{L_{eff}}{r} = \frac{(2)(5)(12)}{(0.8004)} = 149.92$

The critical buckling load is $P_{cr} = \frac{\pi^2 EI}{4L^2} = \frac{\pi^2(30000)(1.132)}{4(60)^2} = 23.37\text{kip}$

The maximum deflection is: $v_{max} = e\left[\sec\left(\frac{\pi}{2}\sqrt{\frac{P}{P_{cr}}}\right) - 1\right] = 0.25\left[\sec\left(\frac{\pi}{2}\sqrt{\frac{3}{23.37}}\right) - 1\right]$ or $\boxed{v_{max} = 0.0458\text{in}}$

The maximum normal stress is:

$$\sigma_{max} = \frac{P}{A}\left[1 + \frac{ec}{r^2}\sec\left(\frac{L_{eff}}{2r}\sqrt{\frac{P}{EA}}\right)\right] = \frac{3}{1.767}\left[1 + 0.4878\sec\left(\frac{149.92}{2}\sqrt{\frac{3}{(30000)(1.767)}}\right)\right] \text{ or } \boxed{\sigma_{max} = 2.68\text{ksi(C)}}$$

11.32 On the cylinder shown in Fig. P11.32 the applied load is P = 3 kips. The Modulus of Elasticity is E = 30,000 ksi. If the allowable normal stress is 8 ksi, what is the maximum permissible length L of the cylinder?

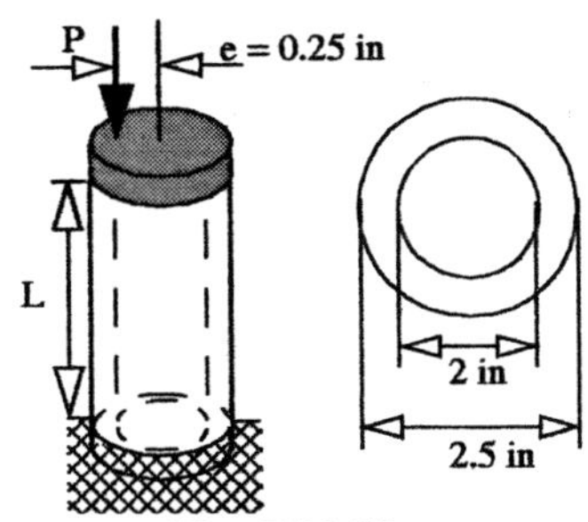

Fig. P11.32

Solution: $P = 3$ kips $\quad E = 30000$ ksi $\quad \sigma_{max} = 8$ ksi $\quad L_{max} = ?$

From equations (1), (2) and (3) of problem 11.31 we have $A = 1.767\ \text{in}^2$, $r = 0.8004$, $ec/r^2 = 0.4878$
The maximum normal stress is

$$\sigma_{max} = \frac{P}{A}\left[1 + \frac{ec}{r^2}\sec\left(\frac{L_{eff}}{2r}\sqrt{\frac{P}{EA}}\right)\right] = \frac{3}{1.767}\left[1 + 0.4878\sec\left(\frac{L_{eff}}{2(0.8004)}\sqrt{\frac{3}{(30000)(1.767)}}\right)\right] \text{ or}$$

$$\sigma_{max} = 1.6978[1 + 0.4878\sec(4.699L_{eff}(10^{-3}))] = 8 \text{ or}$$

$$\sec(4.699L_{eff}(10^{-3})) = 7.6097 \quad \text{or} \quad 4.699L_{eff}(10^{-3}) = 1.439 \quad \text{or} \quad L_{eff} = \frac{1.439}{4.699(10^{-3})} = 2L_{max} \text{ or}$$

$$\boxed{L_{max} = 154.1\text{in}}$$

11.33 The length of the cylinder shown in Fig. P11.33 is L = 5 feet. The yield stress of steel used in the cylinder is 30 ksi, and the Modulus of Elasticity is E = 30,000 ksi. Determine the maximum load P that can be applied. Use plot for steel in Figure 11.17.

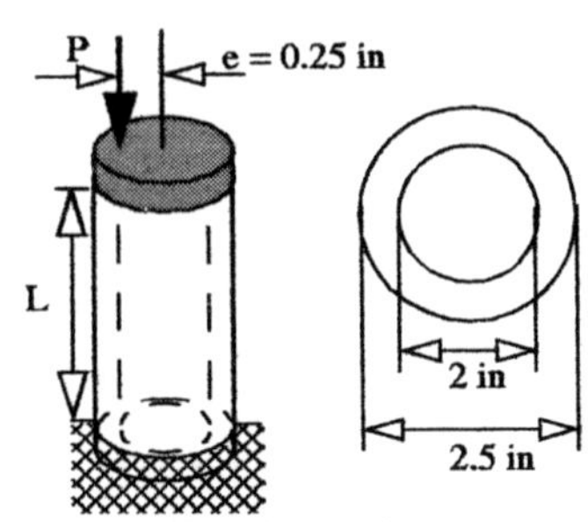

Fig. P11.33

Solution: $L = 5$ ft $\quad \sigma_{yield} = 30$ ksi $\quad E = 30000$ ksi $\quad e = 0.25 \quad P_{max} = ?$

From Equation (1), (3) and (4) in problem 11.31 we have: A= 1.767 in^2. $ec/r^2 = 0.4878$, $L_{eff}/r = 149.92$
We note that $\sigma_{yield}/E = 0.001$. From Figure 11.17 for steel we obtain the following estimates
$(P/A)/\sigma_{yield} = 0.32$
Thus $P_{max} = (0.32)(30)(1.767)$ or $\boxed{P_{max} = 16.96\text{kip}}$

11.34 On the column shown in Fig. P11.34 the applied load P = 100 kN and the length L = 2 m. The Modulus of Elasticity is E = 70 GPa. What is the maximum stress and the maximum deflection?

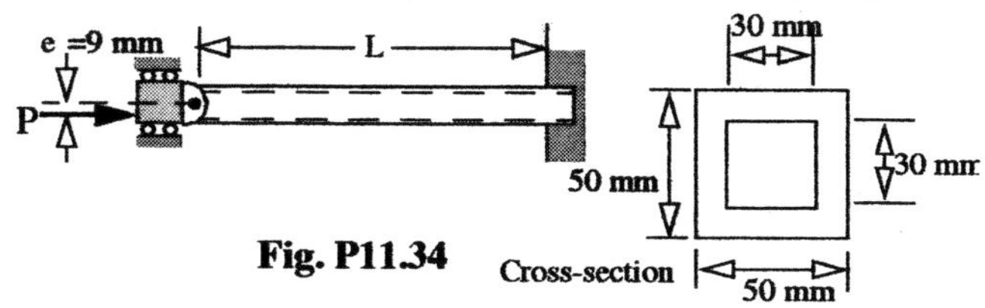

Fig. P11.34

Solution: $P = 100$ kN $\quad E = 70$ GPa $\quad L = 2$ m $\quad e = 0.009 \quad \sigma_{max} = ? \quad v_{max} = ?$

The cross-sectional area and area moment of inertia are:

$$A = (0.05)^2 - (0.03)^2 = 1.6(10^{-3})\text{m}^2 \text{ and } I = \frac{1}{12}[0.05^4 - 0.03^4] = 0.453(10^{-6})\text{m}^4$$

The radius of gyration is: $r = \sqrt{\frac{I}{A}} = 0.01683\text{m}$

The eccentricity ratio is: $\frac{ec}{r^2} = \frac{(0.009)(0.025)}{(0.01683)^2} = 0.7942$

The critical buckling load is: $P_{cr} = \frac{20.13EI}{L^2} = \frac{(20.13)(70)(10^9)(0.453)(10^{-6})}{2^2} = 159.69\text{kN}$

The maximum deflection is: $v_{max} = e\left[\sec\left(\frac{\pi}{2}\sqrt{\frac{P}{P_{cr}}}\right) - 1\right] = 0.009\left[\sec\left(\frac{\pi}{2}\sqrt{\frac{100}{159.69}}\right) - 1\right]$ or $\boxed{v_{max} = 18.96\text{mm}}$

The slenderness ratio is $\frac{L_{eff}}{r} = \frac{(0.7)(2)}{(0.01683)} = 83.172$

The maximum normal stress is:

$$\sigma_{max} = \frac{P}{A}\left[1 + \frac{ec}{r^2}\sec\left(\frac{L_{eff}}{2r}\sqrt{\frac{P}{EA}}\right)\right] = \frac{100(10^3)}{1.6(10^{-3})}\left[1 + 0.7942\sec\left(\frac{83.172}{2}\sqrt{\frac{100(10^3)}{(70)(10^9)(1.6)(10^{-3})}}\right)\right] \text{ or}$$

$\boxed{\sigma_{max} = 216.5\text{MPa(c)}}$

11.35 On the column shown in Fig. P11.35 the applied load is P = 100 kN kips. The Modulus of Elasticity is E = 70 GPa. If the allowable normal stress is 250 MPa, what is the maximum permissible length L of the column?

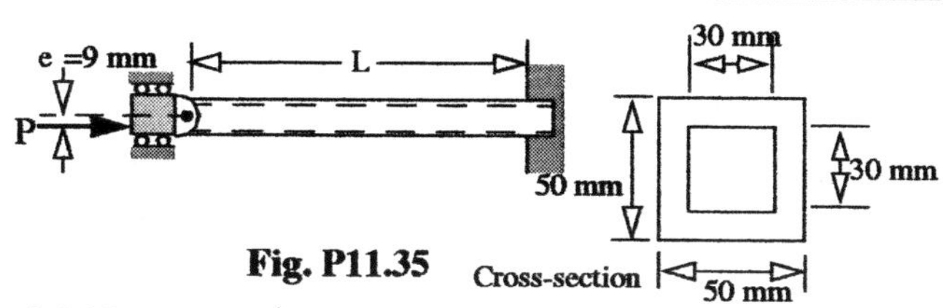

Fig. P11.35

Solution: $P = 100$ kN $\quad E = 70$ GPa $\quad \varepsilon = 0.009 \quad L_{max} = ? \quad \sigma_{max} = 250$ MPa

---

From Eqs. (1), (2) and (3) of problem 11.34 we have: $A = 1.6(10^{-3})$ m^2, $r = 0.01683$, $ec/r^2 = 0.7942$
The maximum normal stress is

$$\sigma_{max} = \frac{P}{A}\left[1 + \frac{ec}{r^2}\sec\left(\frac{L_{eff}}{2r}\sqrt{\frac{P}{EA}}\right)\right] = \frac{100(10^3)}{1.6(10^{-3})}\left[1 + 0.7942\sec\left(\frac{L_{eff}}{2(0.01683)}\sqrt{\frac{100(10^3)}{(70)(10^9)(1.6)(10^{-3})}}\right)\right] \text{ or}$$

$\sigma_{max} = 62.5(10^6)[1 + 0.7942\sec(0.8877L_{eff})] = 250(10^6)$ or $\sec(0.8877L_{eff}) = 3.777$ or

$0.8877L_{eff} = 1.3028$ or $L_{eff} = 1.4676 = 0.7L_{max}$ or $\boxed{L_{max} = 2.09\text{m}}$

11.36 The length of the column shown in Fig. P11.36 is L = 2 m. The yield stress of aluminum used in the column is 280 MPa, and the Modulus of Elasticity is E = 70 GPa. Determine the maximum load P that can be applied. Use the plot for aluminum in Figure 11.17.

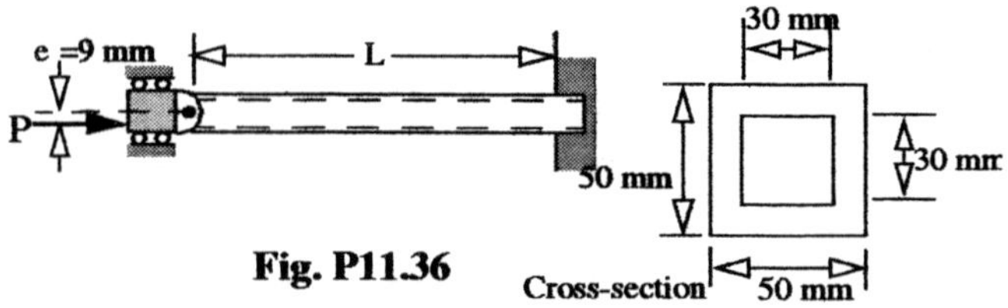

Fig. P11.36

Solution: $L = 2$ m $\quad \sigma_{yield} = 280$ MPa $\quad E = 70$ GPa $\quad e = 0.009$m $\quad P_{max} = ?$

From Eqs. (1), (3), and (4) problem 11.35 we have: $A = 1.6(10^{-3})\text{m}^2 \quad \frac{ec}{r^2} = 0.7942 \quad \frac{L_{eff}}{r} = 83.172$

We note that $\sigma_{yield/E} = 0.004$. From Figure 11.17 for Aluminum we obtain the following estimates $(P/A)/\sigma_{yield} = 0.23$

Thus $P_{max} = (0.23)(280)(10^6)(1.6)(10^{-3})$ or $\boxed{P_{max} = 103\text{kN}}$

11.37 A wide flange W 8 x18 member is used as a column as shown in Fig. P11.37. The applied load P = 20 kips and the length L = 9 ft. . The Modulus of Elasticity is E = 30,000 ksi. What is the maximum stress and the maximum deflection?

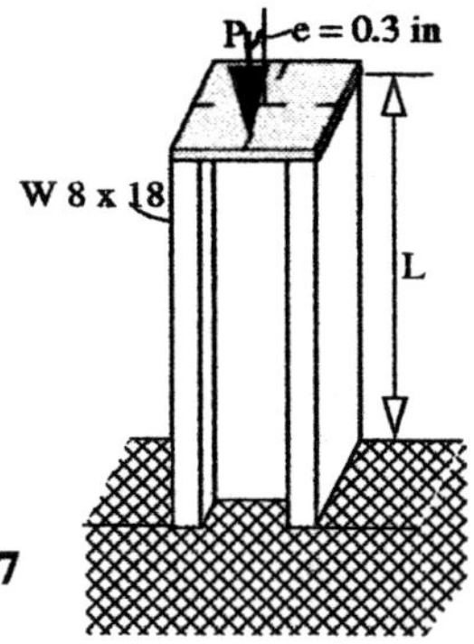

Fig. P11.37

Solution: W8 X 18 P = 20kips $\quad$ E = 30000 psi $\quad$ L= 9ft e = 0.3 in $\quad \sigma_{max} = ?\ v_{max} = ?$

From Table E-1 we have:

$A = 5.26\text{ in}^2, I = 7.97\text{ in}^4, r = 1.23\text{ in}, c = 5.25/2 = 2.625$

The eccentricity ratio is: $\frac{ec}{r^2} = \frac{(0.3)(2.625)}{(1.23)^2} = 0.5205$

The slenderness ratio is: $\frac{L_{eff}}{r} = \frac{(2)(9)(12)}{(1.23)} = 175.61$

The critical buckling load is $P_{cr} = \frac{\pi^2 EI}{4L^2} = \frac{\pi^2(30000)(7.97)}{4(108)^2} = 50.579\text{kip}$

The maximum deflection is: $v_{max} = e\left[\sec\left(\frac{\pi}{2}\sqrt{\frac{P}{P_{cr}}}\right) - 1\right] = 0.3\left[\sec\left(\frac{\pi}{2}\sqrt{\frac{20}{50.579}}\right) - 1\right]$ or $\boxed{v_{max} = 0.245\text{in}}$

The maximum normal stress is:

$\sigma_{max} = \frac{P}{A}\left[1 + \frac{ec}{r^2}\sec\left(\frac{L_{eff}}{2r}\sqrt{\frac{P}{EA}}\right)\right] = \frac{20}{5.26}\left[1 + 0.5305\sec\left(\frac{175.61}{2}\sqrt{\frac{20}{(30000)(5.26)}}\right)\right]$ or $\boxed{\sigma_{max} = 7.4\text{ksi(c)}}$

11.38 On the column shown in Fig. P11.38 the applied load is P = 20 kips. The Modulus of Elasticity is E = 30,000 ksi. If the allowable normal stress is 24 ksi, what is the maximum permissible length L of the column?

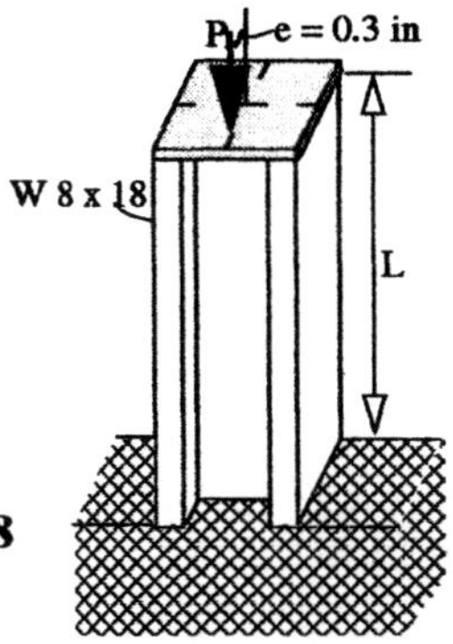

Fig. P11.38

Solution: W8 x 18 P = 20kips E = 30000 psi e = 0.3 in $\sigma_{max} = 24$ ksi L_{max} =?

From Table E-1 and Eq.(1) of problem 11-37 we have: $A = 5.26\ in^2$, $r = 1.23$ in, $ec/r^2 = 0.5205$
The maximum normal stress is:

$$\sigma_{max} = \frac{P}{A}\left[1 + \frac{ec}{r^2}\sec\left(\frac{L_{eff}}{2r}\sqrt{\frac{P}{EA}}\right)\right] = \frac{20}{5.26}\left[1 + 0.5205\sec\left(\frac{L_{eff}}{2(1.23)}\sqrt{\frac{20}{(30000)(5.26)}}\right)\right] \text{ or}$$

$$\sigma_{max} = 3.802[1 + 0.5205\sec(4.576L_{eff}(10^{-3}))] = 24 \text{ or } \sec(4.576L_{eff}(10^{-3})) = 10.205 \text{ or}$$

$$4.576L_{eff}(10^{-3}) = 1.473 \text{ or } L_{eff} = 321.8 = 2L_{max} \text{ or } \boxed{L_{max} = 160.9\text{in}}$$

11.39 The length of the column shown in Fig. P11.39 is L = 9 ft. The yield stress of steel is 30 ksi, and the Modulus of Elasticity is 30,000 ksi. Determine the maximum load P that can be applied. Use the plot for steel in Figure 11.17.?

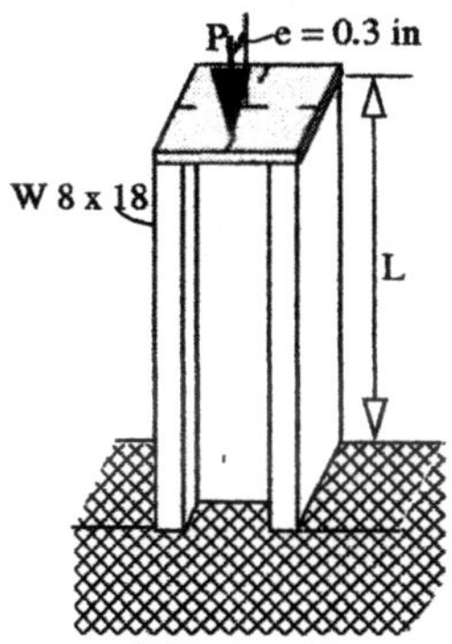

Fig. P11.39

Solution: W 8x 18 L = 9 ft E = 30000 psi e = 0.3 in $\sigma_{yield} = 30$ksi P_{max} = ?

From Table E-1 and Eqs.(1) and (2) of problem 11-37 we have:

$$A = 5.26\text{in}^2 \qquad \frac{ec}{r^2} = 0.5205 \qquad \frac{L_{eff}}{r} = 175.61$$ We note that $\sigma_{yield}/E = 0.001$.

From Figure 11.17 for steel we obtain the following estimate $(P/A)/\sigma_{yield} = 0.25$
Thus $P_{max} = (0.3)(30)(5.26)$ or $\boxed{P_{max} = 39.45\text{ksi(C)}}$

11.40 A hoist is constructed using two wooden bars to lift a weight of 5 kips. The Modulus of Elasticity for wood is E = 1,800 ksi and the allowable normal stress 3.0 ksi. Determine the maximum value of L to the nearest inch that can be used in constructing the hoist.

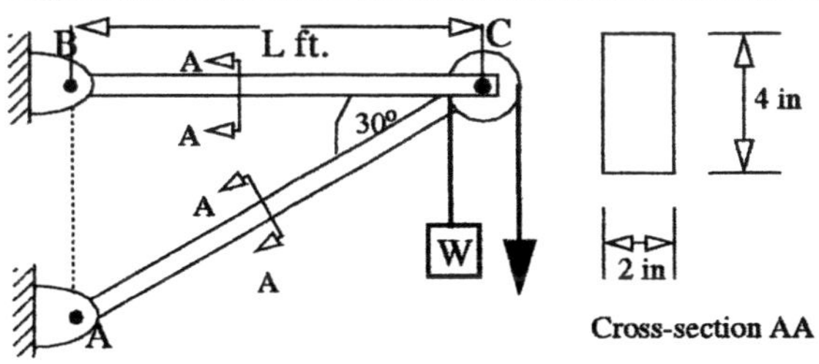

Fig. P11.40

Solution: $W = 5$ kip , $E = 1880$ ksi , $\sigma_{max} = 3$ ksi , $L_{max} = ?$ to nearest inch

---

Fig(a) shows the free body diagram of the pulley.

(a)

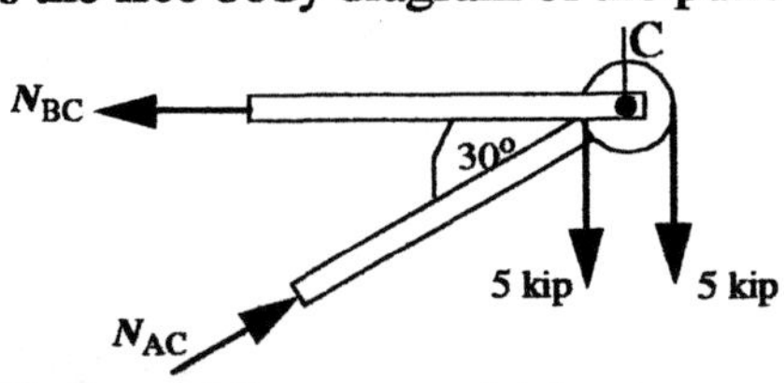

(b)

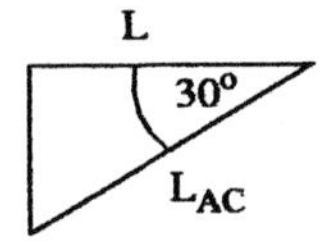

By the equilibrium of forces we obtain

$N_{AC}\sin 30 = 10$ or $N_{AC} = 20\text{kip}$ and $N_{BC} = N_{AC}\cos 30$ or $N_{BC} = 17.32\text{kip}$

The cross-sectional area and area moment of inertia for the two members are:

$$A = (2)(4) = 8\text{in}^2 \qquad I_{min} = \frac{1}{12}(4)(2)^3 = 2.667\text{in}^4$$

The maximum normal stress in AB is: $\sigma_{max} = 20/8 = 2.5\text{ksi}$ irrespective of the length.

The critical buckling load for member AC is

$$P_{cr} = \frac{\pi^2 EI_{min}}{L_{AC}^2} = \frac{\pi^2(1800)(2.667)}{L_{AC}^2} \geq N_{AC} \text{ or } L_{AC}^2 \leq \frac{\pi^2(1800)(2.667)}{20} \text{ or } L_{AC}^2 \leq 48.67\text{in}$$

From Fig(b) we have $L_{AC} = \frac{L}{\cos 30}$. Thus we obtain $\frac{L}{\cos 30} \leq 48.67\text{in}$ or $L \leq 42.15\text{in}$

The maximum value of L to the nearest inch is: $\boxed{L = 42\text{in}}$

11.41 Two steel cylinders ($E = 30{,}000$ ksi and $\sigma_{yield} = 30$ ksi) AB and CD are loaded as shown in Fig. P11.41. Determine the maximum load P, if a factor of safety of 2 is desired. Model the ends of the column AB as built in.

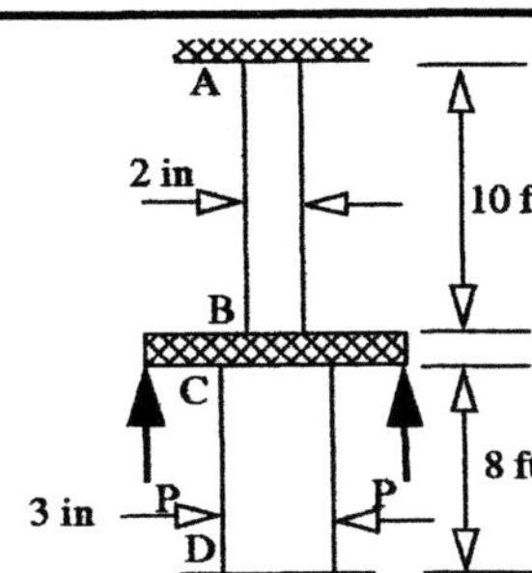

Fig. P11.41

Solution: $E = 30000$ psi $\sigma_{yield} = 30\text{ksi}$ $k = 2$ $P_{max} = ?$

Figs (a) and (b) show free body diagrams

(a)

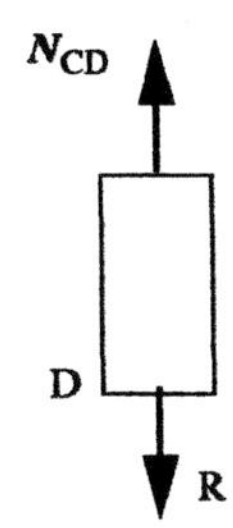

(b)

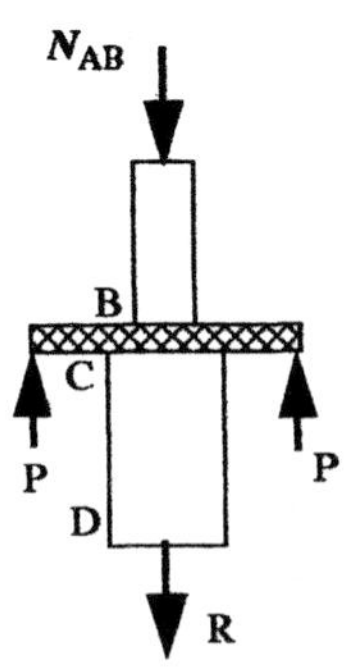

By equilibrium of forces we obtain

$$N_{CD} = R \text{ and } N_{AB} = 2P - R \qquad (1)$$

The deformation of the two bars is the same

$$\delta_{AB} = \delta_{CD} \qquad (2)$$

The cross-sectional area of the two bars is

$$A_{AB} = \pi(2)^2/4 = 3.141\text{in}^2 \qquad A_{CD} = \pi(3)^2/4 = 7.068\text{in}^2 \qquad 3$$

Eq. (2) can be written as

$$\frac{N_{AB}L_{AB}}{EA_{AB}} = \frac{N_{CD}L_{CD}}{EA_{CD}} \text{ or } \frac{R(120)}{3.141} = \frac{(2P-R)(96)}{7.068} \text{ or } 38.197R = 13.582(2P-R) \text{ or } R = 0.5246P \qquad 4$$

From Eq.(1) we obtain

$$N_{CD} = 0.5246P \qquad \text{or} \qquad N_{AB} = 1.4754P \qquad 5$$

The axial stress in both bars should be less than $\sigma_{yield}/k = 15\text{ksi}$

$$\sigma_{CD} = \frac{N_{CD}}{A_{CD}} = \frac{0.5246P}{7.068} \le 15 \qquad \text{or} \qquad P \le 202.1\text{kip} \qquad 6$$

$$\sigma_{AB} = \frac{N_{AB}}{A_{AB}} = \frac{(1.4754P)}{3.141} \le 15 \qquad \text{or} \qquad P \le 31.94\text{kips} \qquad 7$$

The area moment of inertia is $I_{AB} = \frac{\pi}{64}(2)^4 = 0.7854\text{in}^4$

The critical buckling load is

$$P_{cr} = \frac{20.13EI_{AB}}{L_{AB}^2} = \frac{20.13(30000)(0.7854)}{(120)^2} = 32.94 \qquad 8$$

with a factor of safety of 2

$$N_{AB} \le \frac{P_{cr}}{2} \text{ or } 1.4754P \le \frac{32.94}{2} \text{ or } P \le 11.162\text{kip} \qquad 9$$

The maximum value of P that satisfies Eqs. (6), (7), (9) is: $\boxed{P_{max} = 11162\text{lb}}$

11.42 A spreader is to be made from an aluminum pipe (E = 10,000 ksi) of thickness 1/8 inch and outer diameter of 2 inches. The pipe lengths available for design are from 4 ft. in six inch steps up to 8 ft. The allowable normal stress is 40 ksi. Develop a table for the length of pipe and the maximum force F the spreader can support.

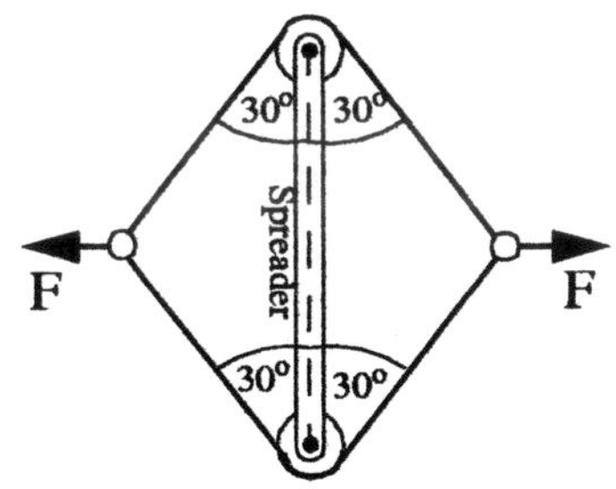

Fig. P11.42

Solution: E = 10000 psi t = 1/8in d_0= 2in $\sigma_{max} \le 40\text{ksi}$ L = ?

Figs (a) and (b) show free body diagrams of the roller and the ring

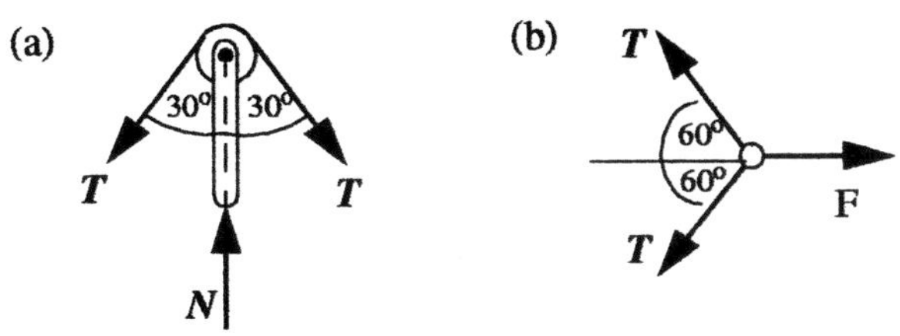

Figs (a) and (b) show free body diagrams of the roller and the ring
By equilibrium of forces we obtain

$$N = 2T\cos 30 \quad \text{and } F = 2T\cos 60 \text{ or } N = \frac{F\cos 30}{\cos 60} = 1.732F \qquad 1$$

The cross-sectional area and the area moment of inertia of the spreader are

$A = \pi(2^2 - 1.75^2)/4 = 0.7363 in^2 \qquad I = \pi(2^4 - 1.75^4)/64 = 0.325 in^4$

The axial stress in the spreader is:

$$\sigma = \frac{N}{A} = \frac{1.732F}{0.7363} = 2.352F \le 40 \qquad \text{or} \qquad F \le 17 kips \tag{2}$$

The critical buckling load can be written as

$$P_{cr} = \frac{\pi^2 EI}{L^2} = \frac{\pi^2(10000)(0.325)}{L^2} = \frac{32076.2}{L^2} \quad \text{or} \quad N \le P_{cr} \quad \text{or} \quad 1.732F \le \frac{32076.2}{L^2} \quad \text{or} \quad F \le \frac{18519}{L^2} \tag{3}$$

The table below shows the values of F calculated from Eq. 3 for values of L varying between 4ft to 8ft in increments of 6 inches. As the value of F in the table is always less than 17kips, Eq.(2) does not affect the solution.

L (ft)	F (kip)
4.00	8.04
4.50	6.35
5.00	5.14
5.50	4.25
6.00	3.57
6.50	3.04
7.00	2.62
7.50	2.29
8.00	2.01

11.43 Two 200 mm x 50 mm lumber (E = 12.6 GPa) form a part of a deck that is modeled as shown. The allowable stress for the lumber is 18 MPa. (a) Determine the maximum intensity of the distributed load w. (b) What is the factor of safety for bar BD corresponding to the answer in part (a).

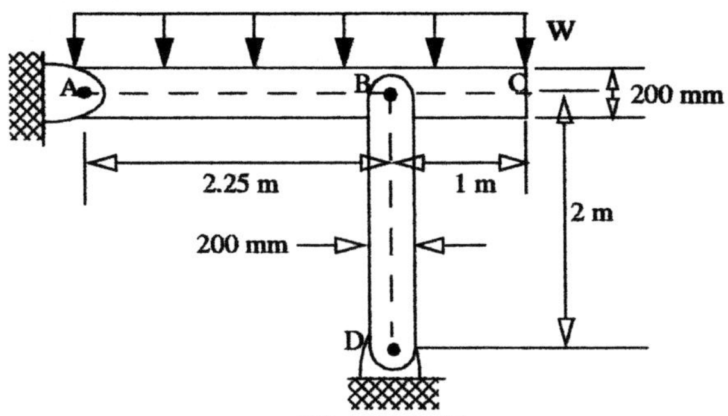

Fig. P11.43

Solution: E = 12.6GPa $\qquad \sigma_{max}$ = 18MPa $\qquad$ 200mm x 50mm $\qquad w_{max}$ = ? $\qquad k_{BD}$ =?

Figs (a) shows free body diagram of member ABC

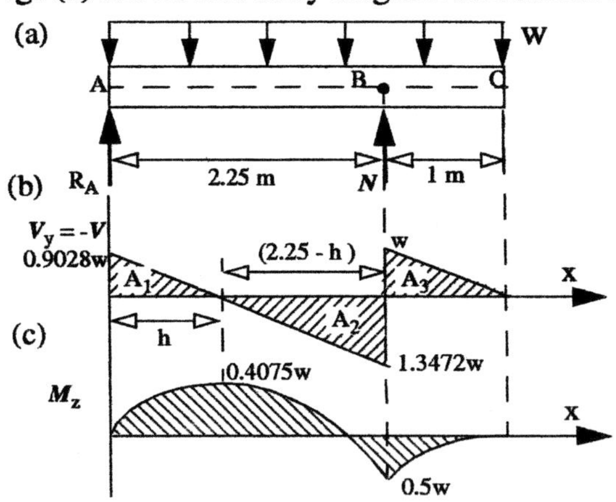

$$\frac{h}{0.9028w} = \frac{(2.25 - h)}{1.3472w}$$

$$h = 0.9028$$

$$A_1 = \frac{1}{2}(0.9028w)(0.9028) = 0.4075w$$

$$A_2 = \frac{1}{2}(1.3472w)(1.3472) = 0.9075w$$

$$A_3 = \frac{1}{2}(1)(w) = 0.5w$$

By equilibrium of moment about point A we obtain

$$N(2.25) - (3.25w)(1.625w) = 0 \quad N = 2.3472w \qquad \mathbf{1}$$

By equilibrium of forces we obtain: $R_A = 3.25w - N = 0.9028w$

Figs. (b) and (c) show the shear and moment diagrams.
The maximum moment is $M_{max} = 0.5w$

The area moment of inertia is: $I_{zz} = \frac{1}{12}(0.05)(0.2)^3 m^4 = 33.33(10^{-6})m^4$

The maximum bending normal stress will also be the maximum normal stress as bending shear stress is an order of magnitude less.

$$\sigma_{max} = \left|\frac{M_{max} y_{max}}{I_{zz}}\right| = \frac{(0.5w)(0.1)}{33.33(10^{-6})} = 1500w \le 18 \quad \text{or} \quad w \le 12(10^3)\frac{N}{m} \qquad \mathbf{2}$$

The minimum area moment of inertia to member BD is: $I_{min} = \frac{1}{12}(0.2)(0.05)^3 = 2.083(10^{-6})m^4$

The critical buckling load for member BD is

$$P_{cr} = \frac{\pi^2 EI_{min}}{L_{BD}^2} = \frac{\pi^2(12.6)(10^9)(2.083)(10^{-6})}{2^2} = 64.77 \qquad \mathbf{3}$$

From Eqs.(1) and (3) we obtain:

$$N \le P_{cr} \quad \text{or} \quad 2.3472w \le 64.76(10^3) \quad \text{or} \quad w \le 27.59(10^3)\frac{N}{m} \qquad \mathbf{4}$$

The maximum axial stress in BD is

$$\sigma_{BD} = \frac{N}{A} = \frac{2.3472w}{(0.05)(0.2)} = 234.7w \le 18(10^6) \quad \text{or} \quad w \le 76.7(10^3)\frac{N}{m} \qquad \mathbf{5}$$

The maximum value of w that can satisfy Eqs. (2), (4), (5) is $\boxed{w_{max} = 12\frac{kN}{m}}$

From Eq.(1) we obtain $N = (2.3472)(12)(10^3) = 28.17(10^3)$

The factor of safety of member BD is $k_{BD} = \frac{P_{cr}}{N} = \frac{64.76(10^3)}{28.17(10^3)}$ or $\boxed{k_{BD} = 2.3}$

11.44 Two 200 mm x 50 mm lumber (E = 12.6 GPa) form a part of a deck that is modeled as shown. The allowable stress for the lumber is 18 MPa. (a) Determine the maximum intensity of the distributed load w. (b) What is the factor of safety for bar BC corresponding to the answer in part (a).

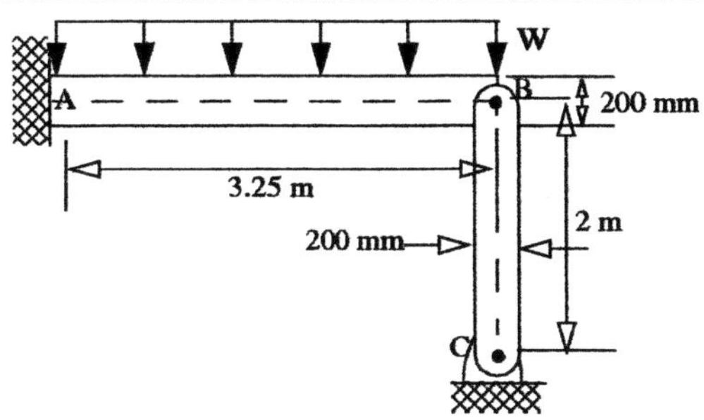

Fig. P11.44

Solution: 200mm x 50mm E = 12.6GPa $\sigma_{max} \le 18MPa$ w_{max} = ? k_{BD} =?

Beam AB can be considered as the sum of two beams below

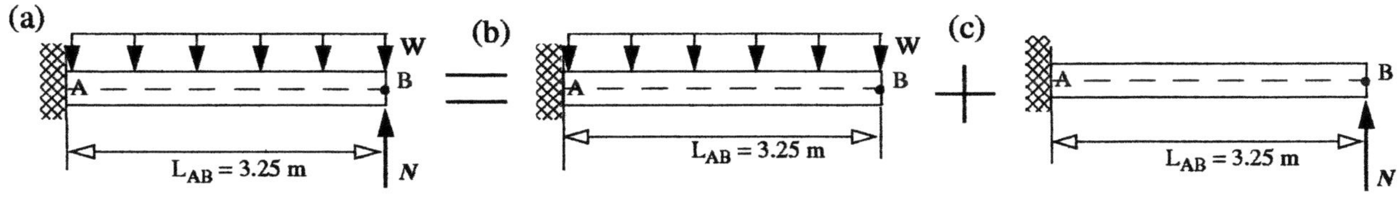

The deflection of point B can be written as superposition of deflection in case 1 and 3 of table 7.1 with a =L and b = 0

$$v_B = \frac{wL_{AB}^4}{8EI_{AB}} - \frac{NL_{AB}^3}{3EI_{AB}} \qquad 1$$

The axial deformation of bar BC can be written as

$$\delta_{BC} = \frac{NL_{BC}}{EA_{BC}} \qquad 2$$

The area moment of inertia of beam AB is $I_{AB} = \frac{1}{12}(0.05)(0.2)^3 = 33.33(10^{-6})m^4$

The cross-sectional area of bar BC is $A_{BC} = (0.05)(0.2) = 0.01m^2$

Equating Eqs.(1) and (2) we obtain

$$\frac{w(3.25)^4}{8(12.6)(10^9)(33.33)(10^{-6})} - \frac{N(3.25)^3}{3(12.6)(10^9)(33.33)(10^{-6})} = \frac{N(2)}{(12.6)(10^9)(0.01)} \quad \text{or}$$

$$33.2w(10^{-6}) - 27.24N(10^{-6}) = 15.87N(10^{-9}) \quad \text{or } N = 1.218W \qquad 3$$

Fig.(d) shows the free body diagram of the beam AB. By force equilibrium we obtain

$$R_A = 3.25w - 1.218w = 2.032w \qquad 4$$

By moment equilibrium we obtain

$$M_A - 3.25w\left(\frac{3.25}{2}\right) + 1.218w(3.25) = 0 \quad \text{or } M_A = 1.3228w \qquad 5$$

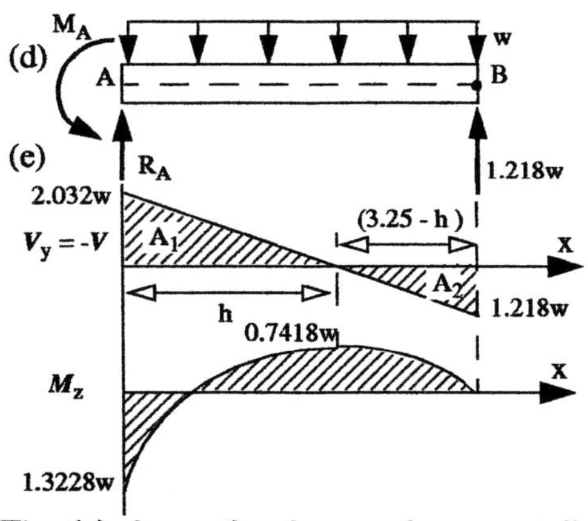

$$\frac{h}{2.032w} = \frac{(3.25 - h)}{1.218w}$$

$$h = 2.032$$

$$A_1 = \frac{1}{2}(2.032w)(2.032) = 2.0645w$$

$$A_2 = \frac{1}{2}(1.218w)(1.218) = 0.7418w$$

Fig. (e) shows the shear and moment diagram. The maximum moment is $M_{max} = 1.3228w$. The maximum bending normal stress will also be the maximum normal stress as bending shear stress is an order of magnitude less.

$$\sigma_{max} = \left|\frac{M_{max}y_{max}}{I_{AB}}\right| = \frac{(1.3228w)(0.1)}{33.33(10^{-6})} = 3.9684w(10^3) \le 18(10^6) \quad \text{or } w \le 4.538(10^3) \qquad 6$$

The critical buckling load for member BD is given by Eq.(3) in problem 11.44 as: $P_{cr} = 64.77(10^3)$

From Eqs.(3) and (5) we obtain

$$1.218w \le 64.77(10^3) \qquad \text{or} \qquad w \le 53.18(10^3) \qquad 7$$

The maximum value of w that can satisfy Eqs. (4) and (7) is $\boxed{w_{max} = 4.5kN}$

From Eq.(3) we have $N = (1.218)(4.5)(10^3) = 5.481(10^3)$

The factor of safety for bar BD is $k_{BD} = \frac{64.77(10^3)}{5.481(10^3)}$ or $\boxed{k_{BD} = 11.8}$

11.45 A rigid bar hinged at point O has a force P applied to it as shown. Bars A and B are made of steel with a Modulus of Elasticity E = 30,000 ksi and a allowable stress of 25 ksi. Bars A and B have a circular cross-section with areas of $A_A = 1$ in ^{2} and $A_B = 2$ in ^{2} respectively. Determine the maximum force P that can be applied.

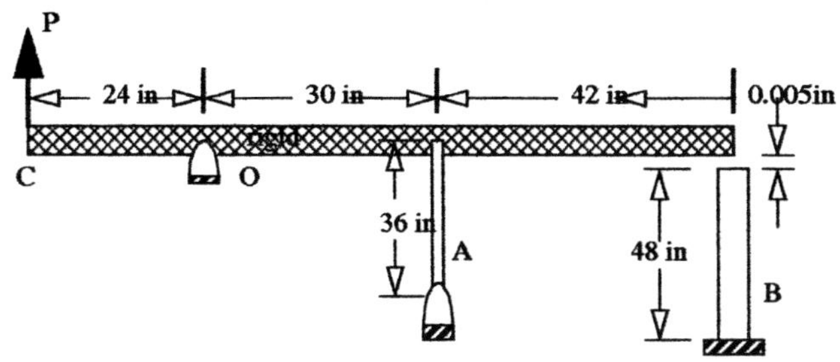

Fig. P11.45

Solution: $E = 30000\text{ksi}$ $\sigma_{max} \le 25\text{ksi}$ $A_A = 1\text{in}^2$ $A_B = 2\text{in}^2$ $P_{max} = ?$

From problem 4.80 Eq.(10) we have compressive axial force in members A and B as

$$N_A = (0.085P + 1.556)\text{kip} \qquad (1)$$

$$N_B = (0.2988P - 0.6483)\text{kip} \qquad (2)$$

The axial stress in A and B should be less than 25 ksi. we obtain

$$\sigma_A = \frac{N_A}{A_A} = \frac{(0.083P + 1.556)}{1} \le 25 \quad \text{or} \quad P \le 282.4\text{kips} \qquad (3)$$

$$\sigma_B = \frac{N_B}{A_B} = \frac{(0.2988P - 0.6483)}{2} \le 25 \quad \text{or} \quad P \le 169.5\text{kips} \qquad (4)$$

The diameters of bars A and B are $\pi d_A^4/4 = 1$ or $d_A = 1.128$ and $\pi d_B^4/4 = 2$ or $d_B = 1.595$

The area moment of inertia of bars A and B are

$$I_A = \pi {d_A}^4/64 = \pi(1.128)^4/64 = 0.0796\text{in}^4 \qquad (5)$$

$$I_B = \pi {d_B}^4/64 = \pi(1.595)^4/64 = 0.3183\text{in}^4 \qquad (6)$$

Member A can be modeled as pin connected and member B as built in on one end

The critical buckling loads are

$$(P_{cr})_A = \frac{\pi^2 E I_A}{L_A^2} = \frac{\pi^2(30000)(0.0796)}{(36)^2} = 18.18\text{kip} \qquad (7)$$

$$(P_{cr})_B = \frac{\pi^2 E I_B}{4L_B^2} = \frac{\pi^2(30000)(0.3183)}{4(48)^2} = 10.22\text{kip} \qquad (8)$$

Now $N_A \le (P_{cr})_A$ and $N_B \le (P_{cr})_B$ we obtain

$$0.083P + 1.556 \le 18.18 \quad \text{or} \quad P \le 200.3\text{kips} \qquad (9)$$

$$0.2988P - 0.6483 \le 10.22 \quad \text{or} \quad P \le 36.37\text{kips} \qquad (10)$$

The maximum value of P that satisfies Eqs. (3), (4), (9) and (10) is $\boxed{P_{max} = 36.3\text{kip}}$

11.46 Show that for a beam with a constant bending rigidity EI, the fourth order differential equation for solving Buckling problems is given by the equation below.

$$EI\frac{d^4v}{dx^4} + P\frac{d^2v}{dx^2} = p_y \qquad \textbf{11.20}$$

where, P is a compressive axial force and p_y is the distributed force in the y-direction.

Solution

Fig. (a) shows a differential element.

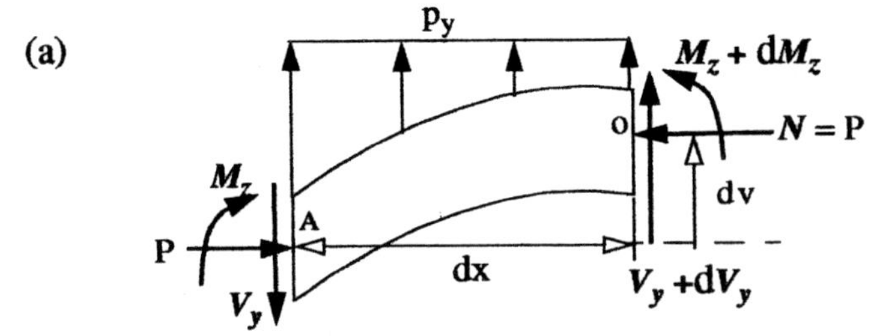

By equilibrium of forces we obtain: $V_y + dV_y - V_y + p_y dx = 0$ or

$$\frac{dV_y}{dx} = -p_y \quad (1)$$

By equilibrium of moment about point O we obtain: $(V_y + dV_y)dx + M_z + dM_z - M_z - (p_y dx)\frac{dx}{2} + PdV = 0$

or $V_y dx + dM_z + PdV = 0$ or

$$\frac{dM_z}{dx} + V_y + P\frac{dv}{dx} = 0 \text{ or } V_y = -\left(\frac{dM_z}{dx} + P\frac{dv}{dx}\right) \quad (2)$$

Substituting Eq. 2 into Eq. 1, we obtain:

$$-\frac{d^2M_z}{dx^2} - P\frac{d^2v}{dx^2} = -p_y \quad (3)$$

Substituting $M_z = EI\frac{d^2v}{dx^2}$ into Eq. (3) we obtain

$$EI\frac{d^4v}{dx^4} + P\frac{d^2v}{dx^2} = p_y \quad (4)$$

Eq. 4 is same as Eq. 11.20.

11.47 Using Equation 11.20 solve Example 11.4.

Solution: $v_{max} = ?$

In Example 11.4 $p_y = -w$. From Eq.11.20 and the simple support conditions, we obtain the following boundary value problem

$$EI\frac{d^4v}{dx^4} + P\frac{d^2v}{dx^2} = -w \quad \text{or} \quad \frac{d^4v}{dx^4} + \lambda^2\frac{d^2v}{dx^2} = -\frac{w}{EI} \quad (1)$$

$$v(0) = 0 \quad (2)$$

$$M_z(0) = EI\frac{d^2v}{dx^2}(0) = 0 \quad (3)$$

$$v(L) = 0 \quad (4)$$

$$M_z(L) = EI\frac{d^2v}{dx^2}(L) = 0 \quad (5)$$

Integrating Eq.(1) twice we obtain

$$\frac{d^2v}{dx^2} + \lambda^2 v = -\left(\frac{w}{EI}\right)\frac{x^2}{2} + C_1 x + C_2 \qquad 6$$

Substituting x = 0 in Eq.(6) and using Eq.(2) and (3) we obtain $C_2 = 0$

Substituting x = L in Eq.(6) and using Eq.(4) and (5) we obtain

$$-\frac{w}{EI}\left(\frac{L^2}{2}\right) + C_1 L = 0 \text{ or } C_1 = \frac{wL}{2EI}$$

Substituting C_1 and C_2 in Eq.(6) we obtain

$$\frac{d^2v}{dx^2} + \lambda^2 v = \frac{wLx}{2EI} - \frac{wx^2}{2EI} \qquad 7$$

Eqs (7), (2) and (4) are identical to the boundary value problem given by Eqs (E3), (E4), and (E5)
The maximum deflection is as given in Example 11.4

$$v_{max} = -\frac{wEI}{P^2}\left[\sec\left(\frac{L}{2}\sqrt{\frac{P}{EI}}\right) - 1\right] + \frac{wL}{8P}$$

11.48 Show that the critical change of temperature at which the beam shown will buckle is given by

$$\Delta T_{crit} = \frac{\pi^2}{\alpha(L/r)^2}$$

where, α is the thermal coefficient of expansion, and r is the radius of gyration.

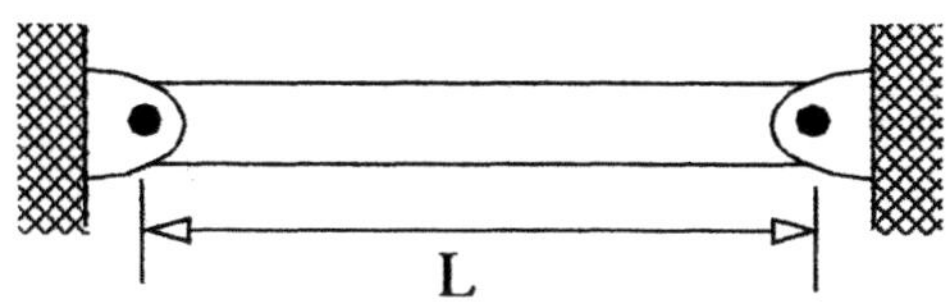

Fig. P11.48

Solution: $\Delta T_{crit} =$

The compressive axial reaction force P can be found by noting that the total axial strain is zero

$$\varepsilon = \frac{\sigma}{E} + \alpha\Delta T = 0 \quad \text{or} \quad \sigma = -E\alpha\Delta T = -\frac{P}{A} \quad \text{or} \quad P = E\alpha\Delta TA \qquad 1$$

At critical Temperature P would equal to the critical buckling load. Thus

$$P_{cr} = E\alpha\Delta T_{crit}A = \frac{\pi^2 EI}{L^2} \quad \text{or} \quad \Delta T_{crit} = \frac{\pi^2(I/A)}{\alpha L^2} = \frac{\pi^2 r^2}{\alpha L^2} \quad \text{or}$$

$$\Delta T_{crit} = \frac{\pi^2}{\alpha(L/r)^2}$$

11.49 A simply supported 6 ft. pipe has an outside diameter of 3 inches and a thickness of 1/8 inch. The pipe material has the stress-strain curve shown. Using Equation 11.21 determine the critical buckling load.

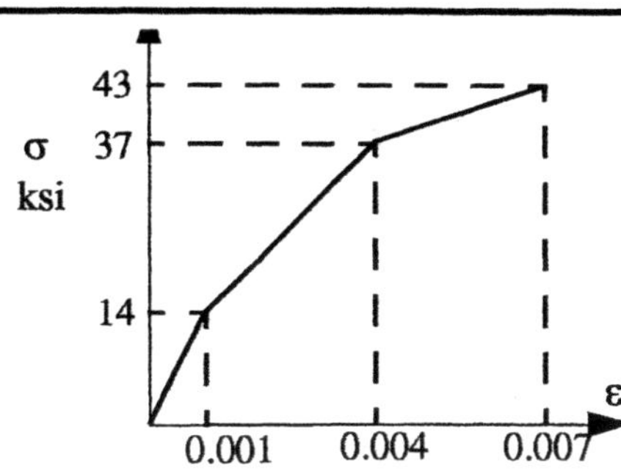

Fig. P11.49

Solution: L = 6 ft $\quad d_0 = 3$in $\quad t = 1/8$in $\quad P_{cr} = ?$

The cross-sectional area and the area moment of inertia are

$$A = \frac{\pi}{4}(3^2 - 2.75^2) = 1.129\text{in}^2 \qquad I = \frac{\pi}{64}(3^4 - 2.75^4) = 1.168\text{in}^4$$

The tangent modulus of elasticity for the three regions of stress- strain curve are

$E_1 = \frac{14}{0.001} = 14000\text{ksi}$, $E_2 = \frac{37-14}{0.004-0.001} = 7666.7\text{ksi}$, and $E_3 = \frac{43-37}{0.007-0.004} = 2000\text{ksi}$

The critical buckling load can be written as

$$P_{cr} = \frac{\pi^2 E_t(1.168)}{72^2} = 2.224E_t(10^{-3}) \qquad 1$$

The axial stress at critical buckling load is

$$\sigma_{cr} = \frac{P_{cr}}{A} = \frac{2.224E_t(10^{-3})}{1.129} = 1.9696E_t \qquad 2$$

Substituting $E_t = E_1$ we obtain: $\sigma_{cr} = (1.9696)(14000)(10^{-3}) = 27.57\text{ksi}$

As the above stress value is greater than 14ksi we next try $E_t = E_2$ and obtain

$\sigma_{cr} = (1.9696)(7666.7)(10^{-3}) = 15.1\text{ksi}$

The above stress value is in the second linear region. Substituting $E_t = E_2$ in Eqs.(2) we obtain

$P_{cr} = (2.224)(7666.7)(10^{-3})$ or $\boxed{P_{cr} = 17\text{kip}}$

11.50 A square box column is constructed from an aluminum sheet of thickness 10 mm. The outside dimension of the square is 75 mm x 75 mm and the column has a length of 0.75 m. The mainlining stress-strain curve is approximated as shown. Using Equation 11.21 determine the critical buckling load.

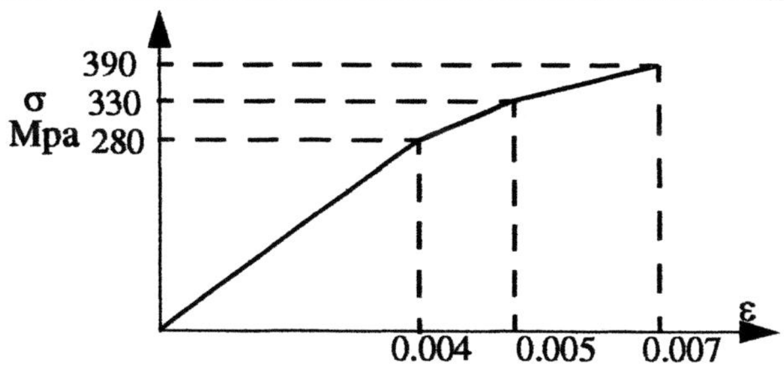

Fig. P11.50

Solution: t = 10mm 75mm x 75mm L = 0.75m P_{cr} = ?

The cross-sectional area and the area moment of inertia are

$A = (0.075^2 - 0.055^2) = 2.6(10^{-3})\text{m}^2$ $\quad I = \frac{1}{12}(0.075^4 - 0.055^4) = 1.874(10^{-6})\text{m}^4$

The tangent modulus of elasticity for the three regions of stress- strain curve are

$E_1 = \frac{280(10^6)}{0.004} = 70\text{GPa}$, $E_2 = \frac{(330-280)(10^6)}{0.005-0.004} = 50\text{GPa}$, and $E_3 = \frac{(390-330)(10^6)}{0.007-0.005} = 30\text{GPa}$

The critical buckling load can be written as

$$P_{cr} = \frac{\pi^2 E_t(1.874)(10^{-6})}{0.75^2} = 32.88E_t(10^{-6}) \qquad 1$$

The axial stress at critical buckling load is

$$\sigma_{cr} = \frac{P_{cr}}{A} = \frac{32.88E_t}{2.6(10^{-3})} = 12.65E_t(10^{-3}) \qquad 2$$

Substituting $E_t = E_1$ we obtain: $\sigma_{cr} = (12.65)(70)(10^9)(10^{-3}) = 885\text{MPa}$

As this stress is greater than 280MPa, we try $E_t = E_2$ and obtain:

$\sigma_{cr} = (12.65)(50)(10^9)(10^{-3}) = 632.5\text{MPa}$

Once more the stress level is greater than the upper limit of region two i.e greater than 330MPa.

We next try $E_t = 30\text{MPa}$ and obtain: $\sigma_{cr} = (12.65)(30)(10^9)(10^{-3}) = 379.5\text{MPa}$

The above stress is in the third linear region. Substituting $E_t = E_3$ in Eq.(1) we obtain

$P_{cr} = (32.88)(30)(10^9)(10^{-6})$ or $\boxed{P_{cr} = 986\text{kN}}$

11.51 A column that is pin held at its ends has a small initial curvature that is approximate by the Sine function shown. Show that the elastic curve of the column is given by the equation below.

$$v(x) = \left(\frac{v_o}{1-P/P_{cr}}\right)\sin\frac{\pi x}{L}$$

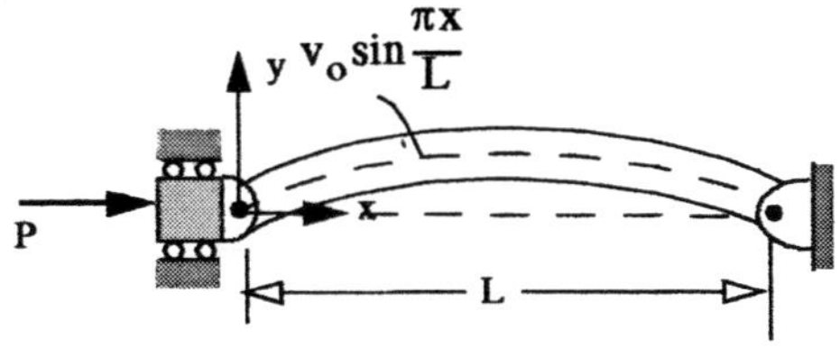

Fig. P11.51

Solution

Let v(x) represent the deflection from y = 0 in Figure 11.51. Fig. (a) shows the free body diagram after an imaginary cut.

(a)

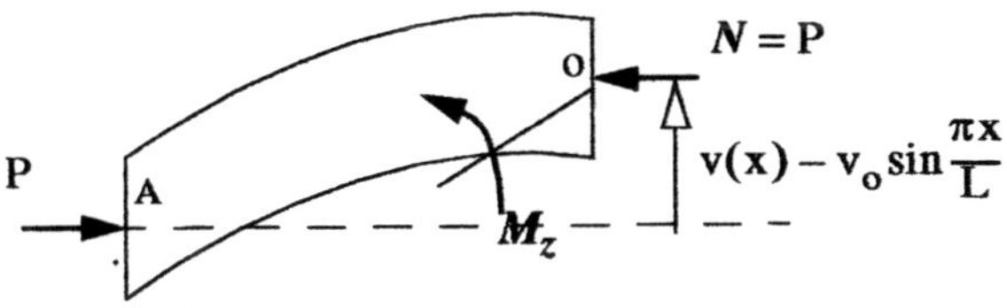

By equilibrium of moment about point O we obtain:

$$M_z + P\left(v - v_o \sin\frac{\pi x}{L}\right) = 0 \quad \text{or} \quad EI\frac{d^2 v}{dx^2} + Pv(x) = Pv_o \sin\frac{\pi x}{L} \quad \text{or}$$

$$\frac{d^2 v}{dx^2} + \lambda^2 v = \lambda^2 v_o \sin\frac{\pi x}{L} \qquad 1$$

The boundary conditions are:

$$v(0) = 0 \qquad 2 \qquad\qquad v(L) = 0 \qquad 3$$

The homogenous solution to Eq. 1 is: $v_h(x) = A\cos\lambda x + B\sin\lambda x$

The particular solution to Eq. 1 is: $v_p(x) = C\sin\frac{\pi x}{L}$. Substituting v_p in Eq. 1 we obtain:

$$\left[-C\left(\frac{\pi}{L}\right)^2 + \lambda^2 C\right]\sin\frac{\pi x}{L} = \lambda^2 v_o \sin\frac{\pi x}{L} \quad \text{or} \quad C = \frac{\lambda^2 v_o}{\lambda^2 - (\pi/L)^2} = \frac{v_o}{1-P/P_{cr}} \qquad 4$$

The total solution to Eq. 1 is: $v(x) = v_h(x) + v_p(x)$ or

$$v(x) = A\cos\lambda x + B\sin\lambda x + \frac{v_o}{1-P/P_{cr}}\sin\frac{\pi x}{L} \qquad 5$$

Substituting x = 0 into Eq. 4 and using Eq. 2 we obtain:

$$v(0) = A = 0 \quad \text{or} \quad A = 0 \qquad 6$$

Substituting x = L into Eq. 4 and using Eq. 3 we obtain:

$$v(L) = B\sin\lambda L = 0 \quad \text{or} \quad B = 0 \qquad 7$$

Substituting Eqs. 6 and 7 into 5 we obtain: $\boxed{v(x) = \left(\frac{v_o}{1-P/P_{cr}}\right)\sin\frac{\pi x}{L}}$

11.52 A column with a constant bending rigidity EI rests on an elastic foundation. The foundation modulus is k that exerts a spring force per unit length of kv. Show that the governing differential equation is

$$EI\frac{d^4v}{dx^4} + P\frac{d^2v}{dx^2} + kv = 0 \qquad \textbf{11.21}$$

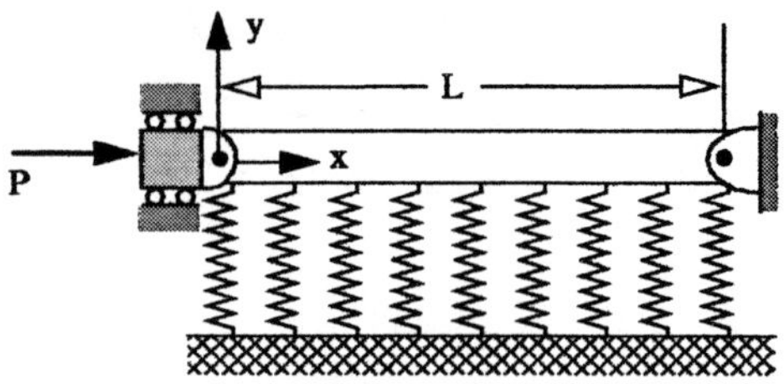

Fig. P11.52

Solution

Fig. (a) shows a differential element of the column.

(a)

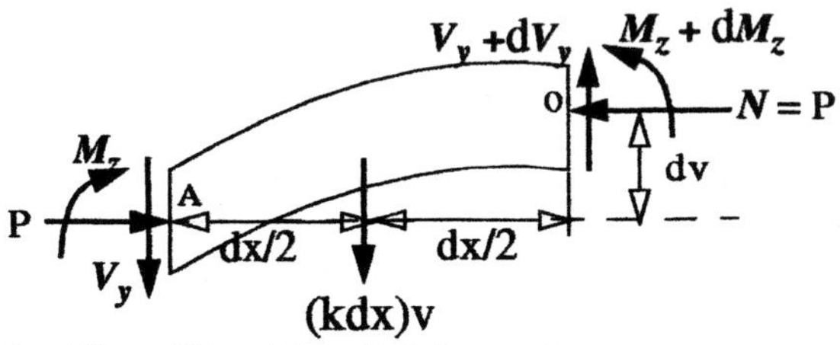

By equilibrium of forces we obtain: $V_y + dV_y - V_y - (kdx)v = 0$ or

$$\frac{dV_y}{dx} - kv = 0 \qquad 1$$

By equilibrium of moment about point O we obtain: $M_z + dM_z - M_z + V_y dx - (kdx)v\left(\frac{dx}{2}\right) + Pdv = 0$ or

$$\frac{dM_z}{dx} + P\frac{dv}{dx} = -V_y \qquad 2$$

Substituting Eq. 2 into Eq. 1, we obtain:

$$-\left(\frac{d^2M_z}{dx^2} + P\frac{d^2v}{dx^2}\right) - kv = 0 \quad \text{or} \quad EI\frac{d^4v}{dx^4} + P\frac{d^2v}{dx^2} + kv = 0 \qquad 3$$

Eq. 3 is same as Eq. 11.21.

11.53 Show that the Buckling Load (eigenvalues) for the column on elastic foundation described in problem 11.52 is given by the equation below.

$$P_n = \frac{\pi^2 EI}{L^2}\left[n^2 + \frac{1}{n^2}\left(\frac{kL^4}{\pi^4 EI}\right)\right] \qquad n = 1, 2, 3\cdot\cdot\cdot \qquad \textbf{11.22}$$

Note: for n = 1 and k = 0 the above equation gives the Euler Buckling load.

Solution

Consider the following solution for Eq. 11.21:

$$v(x) = A\cos\Lambda x + B\sin\Lambda x \qquad 1$$

The value of the second and fourth derivative is as shown below.

$$\frac{dv}{dx} = -A\Lambda\sin\Lambda x + B\Lambda\cos\Lambda x \qquad 2$$

$$\frac{d^2v}{dx^2} = -A\Lambda^2\cos\Lambda x - B\Lambda^2\sin\Lambda x = -\Lambda^2(A\cos\Lambda x + B\sin\Lambda x) = -\Lambda^2 v \qquad 3$$

$$\frac{d^4 v}{dx^4} = -\Lambda^2 \frac{d^2 v}{dx^2} = \Lambda^4 v \qquad 4$$

Substituting Eqs. 1, 3, and 4 into Eq. 11.21 we obtain:

$$(EI\Lambda^4 - P\Lambda^2 + k)v = 0 \quad \text{or} \quad EI\Lambda^4 - P\Lambda^2 + k = 0 \qquad 5$$

For simply supported column v(0) = 0 hence from Eq. 1 we obtain A = 0. We also have v(L) = 0. From Eq. 1 we obtain:

$$v(L) = B\sin\Lambda L = 0 \quad \text{or} \quad \sin\Lambda L = 0 \quad \text{or} \quad \Lambda L = n\pi \quad \text{or} \quad \Lambda = \frac{n\pi}{L} \qquad 6$$

Substituting Eq. 6 into Eq. 5 and letting $P = P_n$ we obtain:

$$EI\left(\frac{n\pi}{L}\right)^4 - P_n\left(\frac{n\pi}{L}\right)^2 + k = 0 \quad \text{or} \quad P_n\left(\frac{n^2\pi^2}{L^2}\right) = EI\left(\frac{n^4\pi^4}{L^4}\right) + k \quad \text{or} \quad P_n = EI\left(\frac{n^2\pi^2}{L^2}\right) + k\frac{L^2}{n^2\pi^2} \quad \text{or}$$

$$P_n = \frac{\pi^2 EI}{L^2}\left[n^2 + \frac{1}{n^2}\left(\frac{kL^4}{\pi^4 EI}\right)\right] \qquad 7$$

Eq. 7 is the same as Eq. 11.22.

11.54 For a simply supported column with a symmetric composite cross-section, show that the critical load P_{cr} is as given below:

$$P_{cr} = \left[\pi^2 \sum_{i=1}^{n} E_i I_i\right] / L_{eff}^2 \qquad \textbf{11.23}$$

where, L_{eff} = effective length of the column, E_i is the Modulus of Elasticity for the ith material, I_i is the area moment of inertia about the buckling axis, n is the number of materials in the cross-section.

Solution

Fig. (a) shows a simply supported column and Fig. (b) shows the free body diagram.

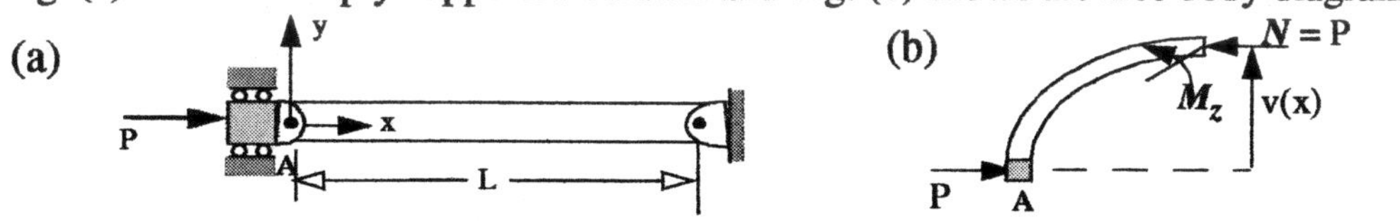

By equilibrium of moment about point A we obtain: $M_z + Pv = 0$. Substituting the moment-curvature relationship for composite given by Eq. 6.30, we obtain the differential equation below.

$$\left(\sum_{j=1}^{n} E_j (I_{zz})_j\right)\frac{d^2 v}{dx^2} + Pv = 0 \qquad 1$$

The above differential Eq. differs from the case 1 of Table 11.1 in the bending rigidity which is a constant. We may thus replace EI the bending rigidity in the Euler buckling load formula with the bending rigidity for the composite to obtain Eq. 11.23.

11.55 In *Double Modulus theory*, also known as *Reduced Modulus theory* for Intermediate columns it was recognized that the bending action during buckling increases the compressive axial stress on the concave side of the beam but decreases the compressive stress on the convex side of the beam.Thus, the use of Tangent Modulus of Elasticity E_t is appropriate on the concave side, but on the convex side of the beam it may be better to use the original Modulus of Elasticity. Modeling the cross-section material with two Modulus E_t and E, and using Equation 11.23 show:

$$P_{cr} = \frac{\pi^2 E_r I}{L_{eff}^2} \qquad E_r = E_t\frac{I_1}{I} + E\frac{I_2}{I} \qquad \textbf{11.24}$$

where, E_r is called the *Reduced Modulus of Elasticity*, I_1 and I_2 are the moment of inertia of the area on the concave and convex side of the axis passing through the centroid, and I is the moment of inertia of the entire cross-section.

Solution

We write the bending rigidity for composite beam as:

$$\sum_{j=1}^{2} E_j(I_{zz})_j = E_t I_1 + E I_2 = E_r I \quad \text{or} \quad E_r = E_t\frac{I_1}{I} + E\frac{I_2}{I} \qquad 1$$

Substituting Eq. 1 into Eq. 11.23 we obtain Eq. 11.24.

11.56 A composite column has a cross-section shown. The Modulus of Elasticity of the outside material is twice that of the inside material. In terms of E, d, and L, determine the critical buckling load.

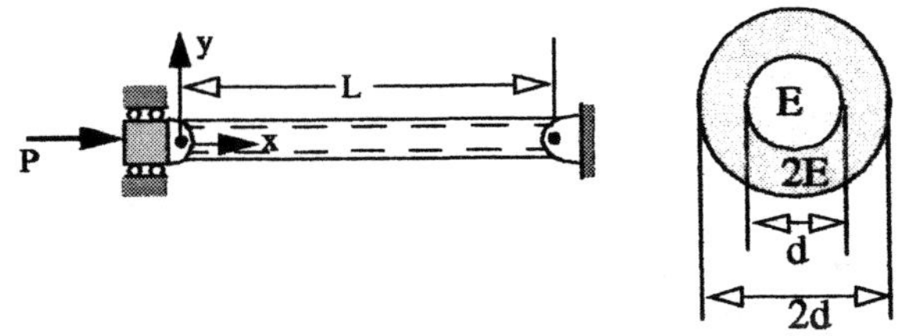

Fig. P11.56

Solution $P_{cr} = f(E, d, L)$

The area moment of inertias of inner and outer cylinder are:

$$I_i = \pi d^4/64 \qquad I_o = \pi((2d)^4 - d^4)/64 = (15\pi)d^4/64 \qquad 1$$

The bending rigidity of the cross-section is:

$$\sum_{j=1}^{n} E_j(I_{zz})_j = EI_i + 2EI_o = E\left(\frac{\pi}{64}d^4\right) + 2E\left(\frac{15\pi}{64}d^4\right) = \frac{31\pi}{64}d^4 \qquad 2$$

Substituting Eq. 2 into Eq. 11.23 and $L_{eff} = L$ we obtain:

$$P_{cr} = \left[\pi^2\left(\frac{31\pi}{64}d^4\right)\right]/L^2 \quad \text{or} \quad \boxed{P_{cr} = \frac{31\pi^3 d^4}{64L^2}}$$

11.57 Two strips of material of Modulus of Elasticity of 2E are attached to a material with a Modulus of Elasticity E to form a composite cross-section of the column as shown. In terms of E, a, and L, determine the critical buckling load. The column is free to buckle in any direction.

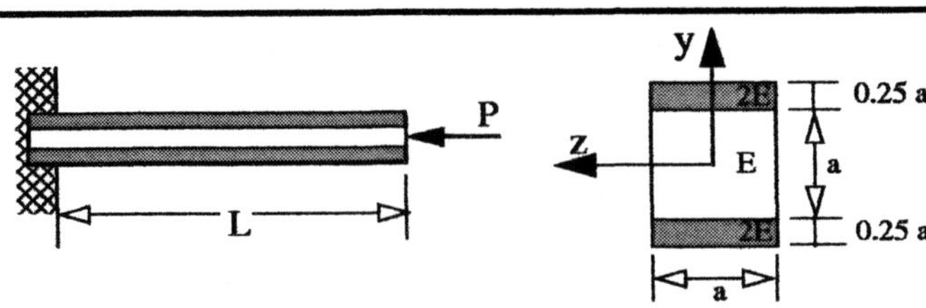

Fig. P11.57

Solution $P_{cr} = f(E, d, L)$

The minimum area moment of inertias will be about the y axis and the values of the inertias are

$$I_1 = \frac{1}{12}a^4 \qquad I_2 = \frac{1}{12}(0.25a)(a^3) = \frac{1}{48}a^4 \qquad 1$$

The bending rigidity of the cross-section is:

$$\sum_{j=1} E_j (I_{zz})_j = EI_1 + 2[2EI_o] = E\left(\frac{1}{12}a^4\right) + 4E\left(\frac{1}{48}a^4\right) = \frac{E}{6}a^4 \qquad 2$$

Substituting Eq. 2 into Eq. 11.23 and $L_{eff} = 2L$ we obtain:

$$P_{cr} = \left[\pi^2\left(\frac{E}{6}a^4\right)\right]/(2L)^2 \quad \text{or} \qquad \boxed{P_{cr} = \pi^2 a^4 E/(24L^2)}$$

11.58 A circular marble column of diameter 2 feet and length 20 feet has a load P applied to it at a distance of 2 inches from the center. The Modulus of Elasticity is 8,000 ksi and the allowable stress is 20 ksi. Determine the maximum load P the column can support assuming that both ends are (a) pinned (b) built in.

Solution: d = 2ft E = 8000 ksi L= 20ft e = 2in $\sigma_{max} \le 20\text{ksi}$

P_{max} = ? (a) pinned (b) built in

The cross-sectional area and area moment of inertia are:

$$A = \pi(12)^2 = 452.4\text{in}^2 \text{ and } I = \frac{\pi}{4}(12)^4 = 16286\text{in}^4$$

The radius of gyration is: $r = \sqrt{\frac{I}{A}} = 6\text{in}$. The eccentricity ratio is: $\frac{ec}{r^2} = \frac{(2)(12)}{6^2} = 0.6667$

(a) For pinned column L_{eff} = L = 240in. Let σ = P/A

The maximum normal stress can be written as:

$$\sigma_{max} = \sigma\left[1 + \frac{ec}{r^2}\sec\left(\frac{L_{eff}}{2r}\sqrt{\frac{\sigma}{A}}\right)\right] = \sigma\left[1 + 0.6667\sec\left(20\sqrt{\frac{\sigma}{8000}}\right)\right] = 20\text{ksi} \quad \text{or}$$

$$f_1(\sigma) = \sigma[1 + 0.6667\sec(0.2236\sqrt{\sigma})] - 20 = 0 \qquad 1$$

Table below shows the calculation of the root σ in Eq. 1. From the Table we obtain the root as σ = 7.12ksi

Thus $P_{max} = (10.56)(452.4)$ or $\boxed{P_{max} = 4777\text{kip}}$

(b) For built in ends L_{eff} = 0.5L = 120in.

The maximum normal stress can be written as: $\sigma_{max} = \sigma\left[1 + 0.6667\sec\left(10\sqrt{\frac{\sigma}{8000}}\right)\right] = 20\text{ksi}$ or

$$f_2(\sigma) = \sigma[1 + 0.6667\sec(0.1118\sqrt{\sigma})] - 20 = 0 \qquad 2$$

Table below shows the calculation of the root in Eq.2. From the table we obtain the root as σ = 10.57 ksi

Thus $P_{max} = (11.64)(452.4)$ or $\boxed{P_{max} = 5265\text{kip}}$

σ	$f_1(\sigma)$	σ	$f_1(\sigma)$	σ	$f_1(\sigma)$
1.00	-18.32	10.00	-1.23	10.50	-0.15
2.00	-16.60	10.10	-1.02	10.51	-0.13
3.00	-14.84	10.20	-0.80	10.52	-0.11
4.00	-13.04	10.30	-0.58	10.53	-0.09
5.00	-11.20	10.40	-0.37	10.54	-0.06
6.00	-9.31	10.50	-0.15	10.55	-0.04
7.00	-7.38	10.60	0.07	10.56	-0.02
8.00	-5.39	10.70	0.29	10.57	0.00
9.00	-3.34	10.80	0.50	10.58	0.02
10.00	-1.23	10.90	0.72	10.59	0.05
11.00	0.95	11.00	0.95	10.60	0.07

σ	$f_2(\sigma)$	σ	$f_2(\sigma)$	σ	$f_2(\sigma)$
4.00	-13.27	11.00	-1.13	11.60	-0.07
5.00	-11.56	11.10	-0.95	11.61	-0.05
6.00	-9.84	11.20	-0.78	11.62	-0.03
7.00	-8.12	11.30	-0.60	11.63	-0.02
8.00	-6.39	11.40	-0.42	11.64	0.00
9.00	-4.65	11.50	-0.25	11.65	0.02
10.00	-2.89	11.60	-0.07	11.66	0.04
11.00	-1.13	11.70	0.11	11.67	0.05
12.00	0.64	11.80	0.29	11.68	0.07
13.00	2.42	11.90	0.46	11.69	0.09
14.00	4.21	12.00	0.64	11.70	0.11

11.59 Determine the maximum load P to the nearest newton in problem11.36.

Solution E = 70 GPa $P_{max} = ?$

From problem 11.36 we have: $A = 1.6(10^{-3})\ m^2 \qquad \frac{ec}{r^2} = 0.7942 \qquad \frac{L_{eff}}{r} = 83.172$

The maximum normal stress can be written as:

$$\sigma_{max} = \frac{P}{A}\left[1 + \frac{ec}{r^2}\sec\left(\frac{L_{eff}}{2r}\sqrt{\frac{P}{EA}}\right)\right] = \sigma(10^6)\left[1 + 0.7942\sec\left(\frac{83.172}{2}\sqrt{\frac{\sigma(10^6)}{70(10^9)}}\right)\right] = 280(10^6)$$

where σ = P/A and is in MPa. The above equation can be re-written as:

$$f(\sigma) = \sigma\left[1 + 0.7942\sec\left(\frac{83.172}{2}\sqrt{\frac{\sigma}{70000}}\right)\right] - 280 = 0$$ 1

The roots of the above equation can be solved on a spreadsheet as shown in Appendix B. The results are reported in the table below.

σ=P/A (MPa)	f(σ)	σ=P/A (MPa)	f(σ)	σ=P/A (MPa)	f(σ)	σ=P/A (MPa)	f(σ)
50	-140.10	65	-41.65	69	-1.10	69.09	-0.08
55	-113.75	65.5	-37.06	69.01	-0.99	69.091	-0.06
60	-81.75	66	-32.35	69.02	-0.87	69.092	-0.05
65	-41.65	66.5	-27.50	69.03	-0.76	69.093	-0.04
70	10.61	67	-22.52	69.04	-0.65	69.094	-0.03
75	82.39	67.5	-17.40	69.05	-0.53	69.095	-0.02
80	188.44	68	-12.12	69.06	-0.42	69.096	-0.01
85	363.33	68.5	-6.69	69.07	-0.30	**69.097**	0.00
90	711.71	69	-1.10	69.08	-0.19	69.098	0.02
95	1767.79	69.5	4.66	69.09	-0.08	69.099	0.03
		70	10.61	69.1	0.04	69.1	0.04

From the above table we obtain:

$$\sigma = \frac{P}{A} = 69.1\ \text{MPa} \quad \text{or} \quad P = (69.097)(10^6)(1.6)(10^{-3}) = 110555.2\ \text{N or } \boxed{P = 110555\ \text{N}}$$

11.60 Determine the maximum load P to the nearest lb. in problem 11.39.

Solution E = 30000 ksi $P_{max} = ?$

From problem 11.39 we have: $A = 5.26\ in^2 \qquad \frac{ec}{r^2} = 0.5205 \qquad \frac{L_{eff}}{r} = 175.61$

The maximum normal stress can be written as:

$$\sigma_{max} = \frac{P}{A}\left[1 + \frac{ec}{r^2}\sec\left(\frac{L_{eff}}{2r}\sqrt{\frac{P}{EA}}\right)\right] = \sigma\left[1 + 0.5205\sec\left(\frac{175.61}{2}\sqrt{\frac{\sigma}{30000}}\right)\right] = 30$$

where σ = P/A and is in ksi. The above equation can be re-written as:

$$f(\sigma) = \sigma\left[1 + 0.5205\sec\left(\frac{175.61}{2}\sqrt{\frac{\sigma}{30000}}\right)\right] - 30 = 0$$ 1

The roots of the above equation can be solved on a spreadsheet as shown in Appendix B. The results are reported in the table below.

σ=P/A (ksi)	f(σ)	σ=P/A (ksi)	f(σ)	σ=P/A (ksi)	f(σ)	σ=P/A (ksi)	f(σ)
1	-28.40	7.000	-6.99	7.500	-1.00	7.560	-0.09
2	-26.62	7.100	-5.97	7.510	-0.85	7.561	-0.08
3	-24.56	7.200	-4.87	7.520	-0.70	7.562	-0.06
4	-22.06	7.300	-3.69	7.530	-0.55	7.563	-0.05
5	-18.85	7.400	-2.40	7.540	-0.40	7.564	-0.03
6	-14.34	7.500	-1.00	7.550	-0.25	7.565	-0.02
7	-6.99	7.600	0.54	7.560	-0.09	7.566	-0.00
8	8.48	7.700	2.22	7.570	0.06	7.567	0.01
9	72.62	7.800	4.09	7.580	0.22	7.568	0.03
10	-181.65	7.900	6.16	7.590	0.38	7.569	0.05
11	-70.95	8.000	8.48	7.600	0.54	7.570	0.06

From the above table we obtain:

$\sigma = \frac{P}{A} = 7.567 \text{ ksi}$ or $P = (7567)(5.26) = 39802 \text{ lb}$ or $\boxed{P = 39802 \text{ lb}}$

Solution to Static Review Exam 2

1. Determine

(a) the coordinates (y_c, z_c) of the centroid of the cross section.

(b) the area moment of inertia about an axis passing through the centroid of the cross section and parallel to the z-axis.

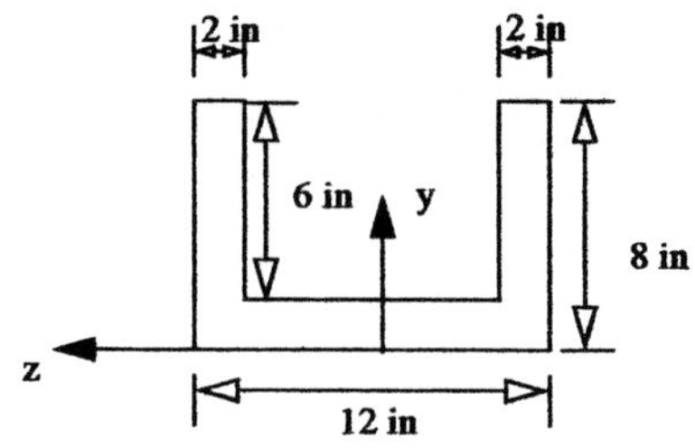

1. As the y-axis is the axis of symmetry, the centroid will lie on the y-axis. Thus $z_c = 0$ *1 point*.

Eqns A.5 and A.9 can be used for finding the y-coordinate of the centroid and area moment of inertia.

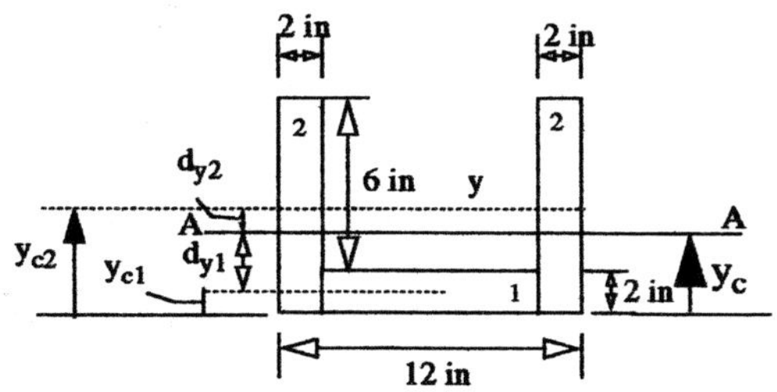

Calculation of the centroid.

section	y_{ci} in	A_i in^2	$y_{ci}A_i$ in^3
1	1	(8)(2)=16	16
2	4	2(8)(2)=32	128
Total		48	144

For each correct entry 1 point for a total of 8 points.

From Eq. A.5 we obtain: $y_c = 144/48$ or $\boxed{y_c = 3\text{in}}$ *1 point for the correct answer and units.*

section	$d_{zi} = y_c - y_{ci}$ in	$I_{zizi} = \frac{1}{12}a_i b_i^3$ in^4	$I_{zizi} + A_i d_{zi}^2$ mm^4
1	2 in	$(8)(2^3)/12=5.333$	69.333
2	1 in	$(2+2)(8^3)/12=170.66$	202.667
	1 point for each correct entry	*2 points for each correct entry*	*1 point for each correct entry*

From Eq. A.9 we obtain $\boxed{I_{AA} = 272\ \text{in}^4}$ *1 point for the correct answer*
1 point for the correct units.

---

2. A distributed load acts on a symmetric C section as shown. Determine the force F and its location (x_F, y_F coordinates) that is statically equivalent to the distributed load.

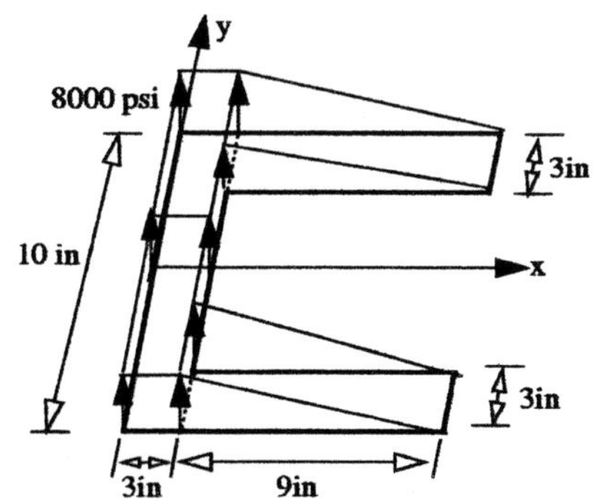

We can replace uniform and linear loading by an equivalent forces as shown in Figure A.8, then replace it by a single force as shown in Fig. (a). The magnitude of the forces are:

$$F_1 = (8000)(10)(3) = 240000\text{lbs} = 240\text{kip} \qquad F_2 = \left[\frac{1}{2}(8000)(9)(3)\right] = 108000\text{lb} = 108\text{kip}$$

For correct location of force F_1 and F_2, 3 points/force. For correct magnitudes of F_1 and F_2, 3 points/force.--Total 12 points

(a)

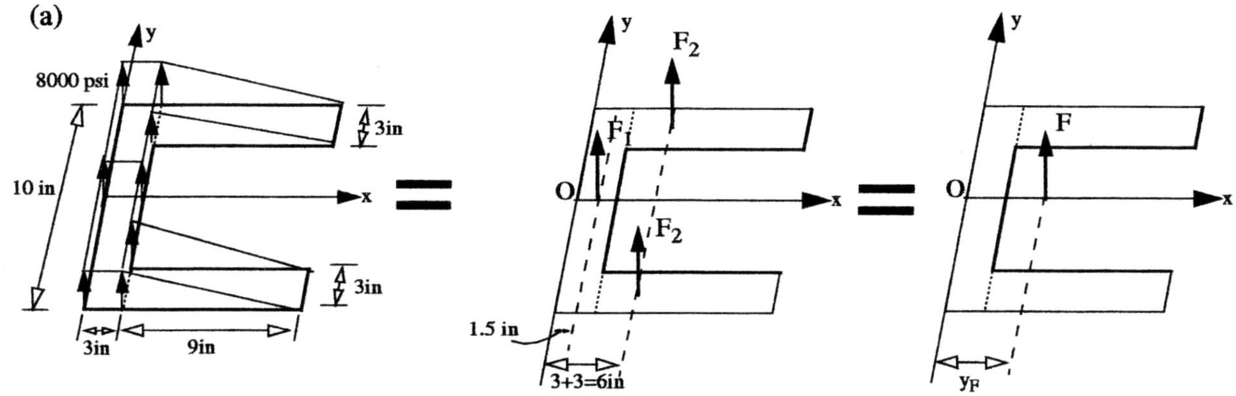

The resultant force for the two systems on the right must be the same. We obtain:

$F = F_1 + 2F_2 = 240 + 2(108)$ or $\boxed{F = 456 \text{ kips}}$ *2 points for the correct answer* *1 point for the correct units.*

The resultant moment about any point (point O) for the two systems on the right must be the same. We obtain: $1.5(F_1) + (2)(6)(F_2) = y_F(F)$ or $y_F = (1656)/456$ or $\boxed{y_F = 3.63 \text{ in}}$ *1 point for the correct answer* *1 point for the correct units.*

1 point for each correct entry in this equation.

---

3. Find the internal axial force (Indicate tension or compression) and internal torque (magnitude and direction) acting on an imaginary cut through point E

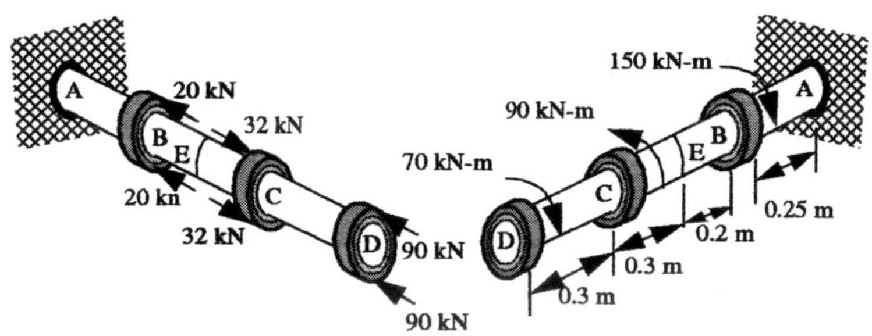

We make imaginary cut at E and draw free body diagrams as shown below

(a) Internal axial force calculations

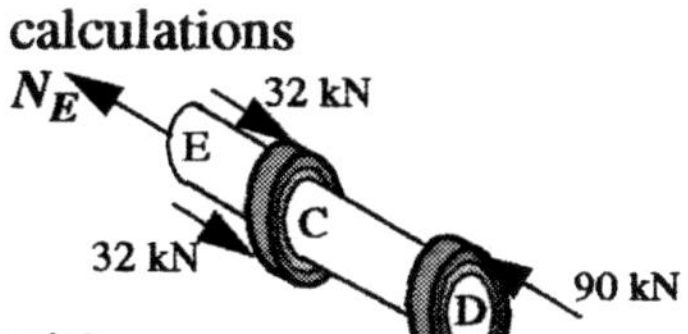

For drawing: 180 kN at D--2 points
64 kN at C ----2 points
Normal force at E ((either direction)---2 points

(b) Internal torque calculations

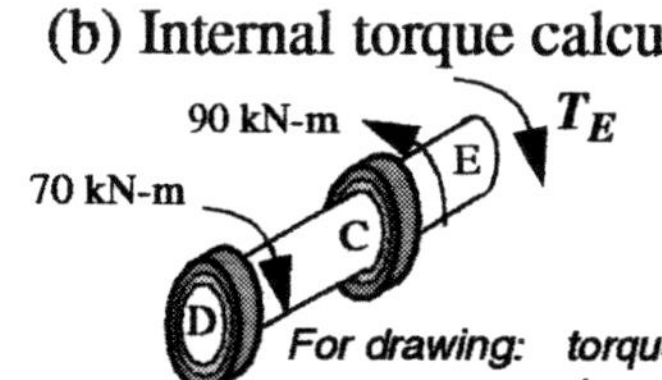

For drawing: torque at D--2 points
torque at C ----2 points
torque at E (either direction) ---2 pc

From Fig. (a) we obtain:

$N_E - 64 + 180 = 0$ *1 point for the correct equation.* $\boxed{N_E = 116\text{kN(C)}}$ *1 point for correct answer.* *1 point for correct units.* *1 point for reporting compression*

From Fig. (b) we obtain:

$T_E - 90 + 70 = 0$ *2 points for the correct equation.* $\boxed{T_E = 20 \text{ kN-m}}$ *1 point for correct answer.* *1 point for correct units.*

---

4. A simply supported beam is loaded by a uniform distributed force of intensity 0.1 kips/in applied at 60°as shown. Also applied is a force F at the centroid of the beam. Neglecting the effect of beam thickness determine at section C the internal axial force, the internal shear force, and the internal bending moment.

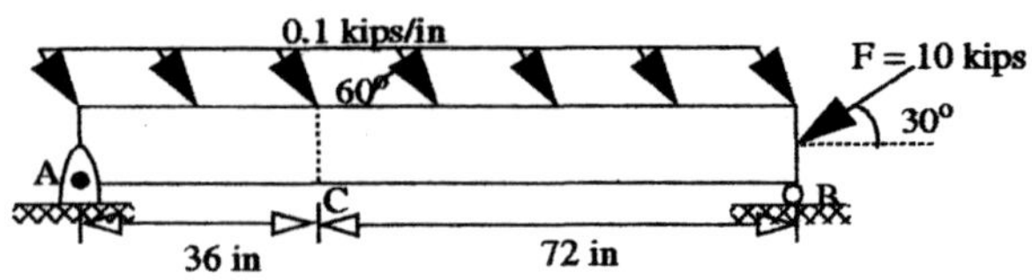

Fig. (a) shows the free body diagram of the entire beam.

(a)

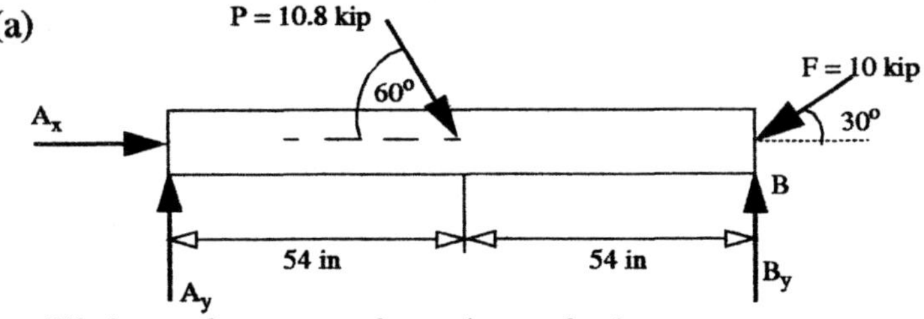

0.5 point for each force besides P
2 points for force magnitude and direction of P
1 point for correct location of P
Total 5 points.

By equilibrium of moment about A we obtain:

$B_y(108) - (10\sin 30)(108) - (10.8\sin 60)(54) = 0$ or $B_y = 9.676\text{kip}$ *0.5 point for each term in the equation for a total of 1.5 points.*

By equilibrium of forces in the y direction we obtain:

$A_y + B_y - 10.8\sin 60 - 10\sin 30 = 0$ or $A_y = 4.676\text{kip}$ *0.5 point for each term in the equation for a total of 1.5 points.*

By equilibrium of forces in the x-direction we obtain:

$A_x - 10\cos 30 + 10.8\cos 60 = 0$ or $A_x = 3.26\text{kip}$ *0.5 point for each term in the equation for a total of 1.5 points.*

We make an imaginary cut at C and draw the free body diagram as shown in Fig. (b) and replace the distributed load to get Fig. (c)

(b) (c)

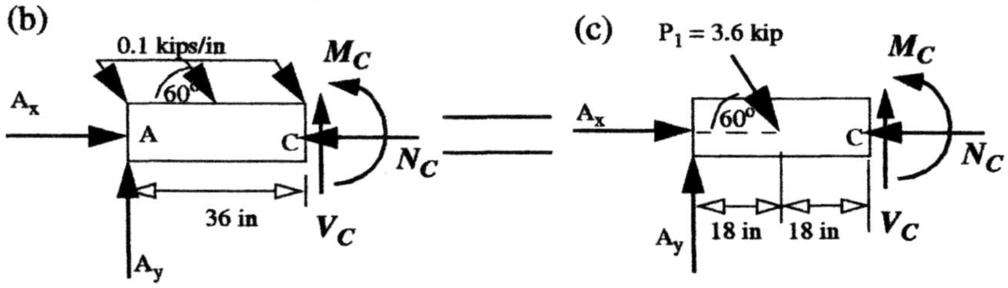

0.5 point for each force besides P1
0.5 point for moment
1.5points for force magnitude of P1
0.5 point for correct location of P1
Total 4.5 point.

By equilibrium of forces in the x-direction we obtain:

$N_C - 3.6\cos 60 - A_x = 0$ or $\boxed{N_C = 5.06\text{kip(C)}}$ *0.5 point for each term in the equation for a total of 1.5 points. 0.5 point for compression*

By equilibrium of forces in the y direction we obtain:

$A_y + V_C - 3.6\sin 60 = 0$ or $\boxed{V_C = 1.558\text{kip downward}}$ *0.5 point for each term in the equation for a total of 1.5 points. 0.5 point for downward*

By equilibrium of moment about C we obtain:

$M_C + (3.6\sin 60)(18) - A_y(36) = 0$ or $\boxed{M_C = 112.2\text{in} - \text{kip CCW}}$ *0.5 point for each term in the equation for a total of 1.5 points.0.5 point for CCW.*

5. A system of pipes is subjected to a force P as shown. By inspection (or by drawing FBD) identify the zero and non-zero internal forces and moments. Also indicate the coordinate direction in which the internal shear forces and internal bending moments act.

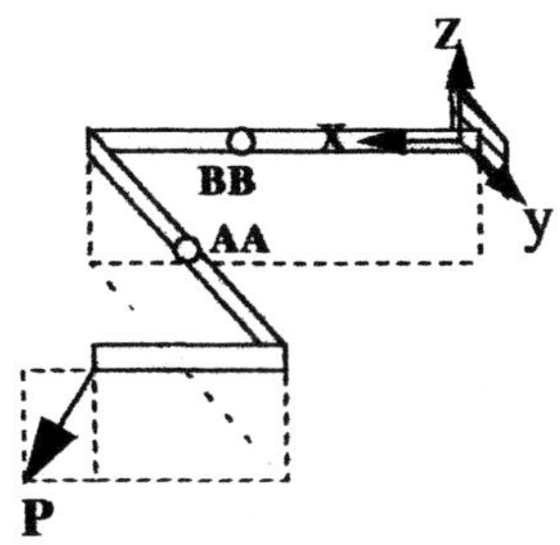

By inspection we can write the following answers.

Internal Force/Moment	Section AA zero/non-zero	Section BB zero/non-zero
Axial Force	zero	non-zero
Shear force	non-zero in xdirection.	non-zero in z direction.
Shear force	zero in y direction.	zero in y direction.
Torque	zero	non-zero
Bending Moment	non-zero in y direction.	non-zero in y direction.
Bending Moment	non-zero in z direction.	non-zero in z direction.

1 point for each correct zero/non-zero entry.

1 point for each correct direction. Total 20 points.